ESTUARINE AND COASTAL MODELING

PROCEEDINGS OF THE EIGHTH
INTERNATIONAL CONFERENCE

November 3–5, 2003
Monterey, California

EDITED BY
Malcolm L. Spaulding, Ph.D., P.E.

Published by the American Society of Civil Engineers

Library of Congress Cataloging-in-Publication Data

Estuarine and coastal modeling : proceedings of the eighth international conference, November 3-5, 2003, Monterey, California / edited by Malcolm L. Spaulding.
p. cm.
Includes bibliographical references and indexes.
ISBN 0-7844-0734-7
1. Estuaries--Mathematical models--Congresses. 2. Coast changes--Mathematical models-Congresses. 3. Hydrodynamics--Mathematical models--Congresses. 4. Water quality--Mathematical models--Congresses. 5. Sediment transport--Mathematical models--Congresses.
I. Spaulding, Malcolm L.

GC96.5.E775 2004
551.46'18'015118--dc22 2004017718

American Society of Civil Engineers
1801 Alexander Bell Drive
Reston, Virginia, 20191-4400

www.pubs.asce.org

ISBN 0-7844-0734-7
Manufactured in the United States of America.

Foreword

This conference represents the 8th in a biennial series to explore the development, testing, application, calibration, validation, and visualization of predictions from estuarine and coastal models. Application of models to problems in hydrodynamics, water quality, and sediment transport were presented. Attendance at the meeting was 145 and included representatives from both the US and many foreign countries. Participants were predominantly government and academic engineers and scientists, but also included a significant number of industry professionals.

As with the earlier conferences in the series, the goal of the present conference was to bring together a diverse group of model developers, users, and evaluators to exchange information on new directions in the field and the current state-of-the-art and practice in marine environmental modeling. The primary focus was on development of new models and the application of models to bays, sounds, lagoons, estuaries, embayments, bights, and coastal seas. The models were addressed at solving engineering and environmental impact assessment problems and also at better understanding circulation and pollutant transport in near shore waters. Model applications to address regulatory requirements for facility sighting and operation were also presented.

The conference included 12 oral sessions and 1 poster session, held over the two and one half day meeting period. Papers from both poster and oral sessions are included in the conference proceedings. Each paper in the proceedings was presented at the meeting, subjected to at least two external peer reviews, and accepted, if appropriate and after revision, by the proceedings editor.

The enthusiastic support and assistance of the organizing and advisory committees, whose names are listed below, is acknowledged. The conference was managed by Joseph Pittle and Connie Kenyon of the University of Rhode Island Conference and Special Programs Development Office who contributed greatly to the success of the meeting. The Wisdom Group Ltd provided the on-line registration and abstract submission and posting system. Thanks are also extended to the many other individuals who generously served as session chairs and moderators.

ORGANIZING COMMITTEE:

Professor Keith W. Bedford, Ohio State University, Columbus, OH.
Dr. Alan F. Blumberg, Stevens Institute of Technology, Hoboken, NJ.
Dr. Mark Dortch, US Army Corp of Engineers, Vicksburg, MS.
Dr. Ralph T. Cheng, U.S. Geological Survey, Menlo Park, CA.
Professor Malcolm L. Spaulding, University of Rhode Island, Narragansett, RI.

ADVISORY COMMITTEE:

Special thanks are extended to Dolores Provost, University of Rhode Island, Ocean Engineering Department who generously provided her time to perform the many administrative tasks necessary to plan the meeting, to support the paper peer review process, and to produce the conference proceedings. Thanks are also due to Peter Cook and Larry Simoneau who designed and maintained the conference web site.

Planning is currently in progress for the 9th International Estuarine and Coastal Modeling Conference (ECM 9) to be held in the Fall 2005.

Malcolm L. Spaulding
Narragansett, RI
Conference Chairman

Contents

Session 1: General Joint Session

No papers were presented for this session.

Session 2A: Coastal Ocean Observing and Forecasting Systems Narragansett Bay

Session 2B: Circulation Models

Session 2C: Bio-Geochemical Models

Session 3A: Coastal Ocean Observing and Forecasting Systems
Regional Examples

Session 3B: Regional Process Models

Session 3C: Bio-Geochemical Models

Session 4A: Coastal Ocean Observing and Forecasting Systems
Regional Examples

COASTMAP: A Globally Re-Locatable, Real Time, Marine Environmental Monitoring and Modeling System, with Application to Narragansett Bay and Southern New England Coastal Waters

Malcolm L. Spaulding[1], E. Howlett [2], M. Ward[2], and C. Galagan[2].

Abstract

Researchers at the University of RI, Ocean Engineering; Applied Science Associates, Inc (ASA); Drexel University, Civil and Architectural Engineering; the National Oceanic and Atmospheric Administration (NOAA), National Ocean Service; and Brown University have partnered in a three year (2000 to 2003), National Ocean Partnership Program (NOPP) sponsored study to develop a globally re-locatable, integrated system for real time observation, modeling, and data distribution for shelf, coastal sea, and estuarine waters. The core of the system, called **COASTMAP**, has been applied to Narragansett Bay and RI coastal waters to illustrate the practical use of the system. This paper provides an overview of the various components of the project and then highlights **COASTMAP**'s globally, re-locatable design architecture including: data server, model client, professional client and web-based client. The paper presents examples of **COASTMAP's** embedded geographic information system, shows its ability to collect, manage, and analyze data from a wide variety of sources, summarizes linkages to environmental models (hydrodynamic, meteorological, oil/chemical spill transport and fate, search and rescue, atmospheric plumes from chemical releases) to provide now-casting and forecasting capabilities. It also illustrates the extension to support crisis management and incident command systems.

Introduction

A Global Ocean Data Assimilation Experiment (GODAE) has been proposed by the international oceanographic community to demonstrate the practicality and feasibility of routine, real time global ocean data assimilation and prediction. In response to this initiative the US National Ocean Partnership Program (NOPP) has funded several partnerships to use the output of these global and basin scale now-casts and forecast models; in conjunction with high resolution models and observations (e.g. *in situ,* real time monitoring stations, remotely sensed data) to solve practical problems inshore of the shelf break. It is intended that the information products generated by this initiative should significantly improve our ability to address the circulation, productivity, and pollution transport and fate, and support marine transportation and research and rescue operations in the coastal ocean. Partnerships are also expected to explore and

[1] Professor, Ocean Engineering, University of RI, Narragansett, RI 02882
[2] Senior Scientists, Applied Science Associates, Inc., Narragansett, RI 02882

develop the market for these GODAE derived data products and to demonstrate by example what products, services, and data distribution mechanisms lead to the most effective use of GODAE products.

In a related effort, the National Ocean Partnership Program (NOPP) established Ocean.US in 2000 to prepare plans for the implementation of an integrated and sustained ocean observing system (IOOS) for the US. As envisioned, IOOS would include both global and coastal components. The immediate need for the system is to provide a unified, comprehensive, cost-effective approach for providing data and information required to: (1) improve the safety and efficiency of marine operations, (2) more effectively mitigate the effects of natural hazards, (3) improve predictions of climate change and its effects on coastal populations, (4) improve national security, (5) reduce public health risks, (6) more effectively protect and restore healthy coastal marine ecosystems, and (7) enable sustained use of marine resources.

Ocean.US held a workshop in March 2002 (Ocean.US, 2002a) to provide the information and guidance needed to formulate a phased implementation plan for the federal contribution to such a system (Ocean.US, 2002b). Based on input received at the workshop Ocean.US staff developed a phased, implementation plan (Ocean.US, 2002c). The plan calls for the coastal component of the Integrated Sustained Ocean Observing System (IOOS) to be structured as a national federation in which regional systems are nested in a national backbone. The backbone will measure, collection, process, and distribute core variables required in the regional systems and will provide national scale capabilities in now-casting and forecasting. The coastal backbone will also (1) establish a relatively sparse network of reference or sentinel stations; (2) provide economies of scale that will improve the cost-effectiveness of national and regional observing systems by investing in a system that minimizes redundancy and optimizes data and information exchange; (3) formulate and institute national standards and protocols for measurements, data dissemination, and management; (4) link the coastal ocean observing system to the global component of IOOS; (5) provide the means for a comparative analyses of system performance among the regions to learn of successes and failures; and (6) facilitate capacity building in the regions. The regional systems will increase the resolution at which core or common variables are measured, measure other variables of local interest, and provide data, information products, and predictions tailored to meet the needs of the user community in the region. Regional systems are tentatively planned to cover all the continental shelves of the US (Northeast, Mid Atlantic, Southeast, Gulf of Mexico, Southwest, Northwest, Alaska, and Hawaii/Pacific Islands) and the waters of the Great Lakes.

The ideal regional system will provide end-to-end capability (Oceans US, 2002b). The system will include: a *monitoring subsystem* (platforms, sensors, measurement techniques) to measure key variables on the space and time scales appropriate to the region and issues of concern, *a data communications and management subsystem* to collect, quality control, disseminate, and archive/store data, and *model products and a data analysis, modeling, and assimilation subsystem* to now-cast and forecast variables of principal concern to the regional/local user community.

The data and communications subsystem is the life-blood of the coastal IOOS, given that it links measurements to the applications and hence users. The design of national backbone of this subsystem (Ocean.US, 2003b) features HTTP as the network data transport protocol for data discovery, employs oceanographic middleware protocols for data acquisition, and federal standards for semantic metadata descriptions of the data.

Regional users will be able to obtain data (historical, real time), data products, and model predictions (with data assimilation), including hindcasts, now-casts, and forecasts. This information maybe obtained from the national backbone or the regional system or both. The regional systems will likely have local modeling capability that will allow predictions of currents, waves, oil and chemical spill transport and fate, and the drift of objects lost at sea (search and rescue).

In response to GODAE and the more recent Ocean.US coastal IOOS initiative, researchers at the University of RI, Ocean Engineering; Applied Science Associates, Inc (ASA); Drexel University, Civil and Architectural Engineering; the National Oceanic and Atmospheric Administration (NOAA), National Ocean Service; and Brown University have partnered in a three year (2000 to 2003), National Ocean Partnership Program (NOPP) sponsored study to develop a globally re-locatable, integrated system for real time observation, modeling, and data distribution for shelf, coastal sea, and estuarine waters. The system has been applied to Narragansett Bay and RI coastal waters as a demonstration of the practical use of the system to support environmental monitoring, marine pollutant transport and fate, marine transportation, and search and rescue operations, and to provide a foundation to advance our understanding and predictive capabilities for the bay. The final goal of the effort is to transition this system to the commercial marketplace and broaden the sources of support for the continued development and wide spread application of the system.

Results of the project are presented in five companion papers in this volume. The present paper gives an overview of **COASTMAP** (www.coastmap.com), the core of the globally re-locatable, real time, marine environmental monitoring and modeling system, and discusses its application to Narragansett Bay and southern New England coastal waters. Details of **COASTMAP**'s internet-based data acquisition and management server systems are presented in Opishinski (2004). Kelley et al (2004) present an evaluation of a high-resolution atmospheric analyses and forecast system for the Narragansett Bay region. In the concluding paper, Piasecki et al (2004) describe a web-based dissemination system for now and forecast data products generated during the project. Investigators at Brown University, led by Dr. John Mustard, provided high resolution LANDSAT based thermal imagery of Narragansett Bay and adjacent coastal waters. The results of that investigation are not presented here.

COASTMAP Architecture

Investigators at URI Ocean Engineering have been working on the development of **COASTMAP** for the past decade. Prior work on the system is summarized in Opishinski and Spaulding (1995, 2002), Opishinski et al (1996), Spaulding and Opishinski (1995) and Spaulding et al (1997, 2001). An extension of the system to land-based transportation applications, **TRANSMAP**, is provided in Spaulding et al (2000, 2001). The present paper provides an update of the current structure and operation of the system.

Figure 1 presents an overview of the basic **COASTMAP** architecture. The system is designed so that it can be readily applied to any area in the world. This is achieved by having the system interface and underlying programs written so that they all operate on a user supplied location data set. The location data set includes a base map of the area, either in vector or raster form, and all data sets and model systems that are available for the area. The system can be configured for as many locations as desired by the user. The size of the areas and the number are up to the user and simply depend on system storage capacity. The core of **COASTMAP** is the professional client. The professional client operates on a personal computer (Pentium IV, Windows 2000), that is linked to other system components via the internet. The professional client is envisioned as a high-end, professionally trained user, who has day-to-day responsibility for using the system in support of their operations. The professional client can access a fully integrated geographic information system, a range of vector and raster based charts, data management and analysis tools, a resource management system, and a suite of environmental models. Professional clients might include environmental regulators and decision-makers, wastewater treatment plant operators, power plant operators, emergency response planners and managers, search and rescue personnel, homeland security response teams, and marine vessel operators. The system could be used for routine monitoring, training, or to assist in response to given incidents. Users can also access the system via a web client. The development of the web client is presented in Piasecki et al (2004). An additional web client interface is discussed later in this paper. The web-based system has been designed to serve a larger audience with modest data and information needs. Users in this category might include the general public, the recreational community, fishermen, tourists, and marina operators. The web-based system could also be used, by the professional client, to post information for wider dissemination. This might include forecasts of wind and wave conditions, predictions of beach and shellfish closure areas, forecasts of areas that might be impacted by natural hazards (e.g. hurricanes, storm surges, tsunamis) or accidental spills or releases.

COASTMAP is configured to collect either real time or archived data from its own data server or directly from a variety of internet-based sources. Opishinski (2004) describes the **COASTMAP** data server in depth. The server application system is designed to acquire environmental data from Internet sites and subsequently process, archive, and distribute the data to support independent system clients. System processes are configured through an integrated data management system with acquisition and export processes monitored through an integrated status display. The system's asynchronous architecture supports multiple, simultaneous Internet connections to permit timely acquisition, processing, and distribution of data. The data server can also be configured to access data from instruments and sensors that are operated for the exclusive use of the system. These typically are private data collection systems, where access to the data is strictly limited. **COASTMAP** can also directly access sources via the internet, independent of its own data server. The later is useful for applications where data is readily available and the needs for setting up a data server are not warranted for the anticipated application. To obtain hindcasts, nowcasts or forecast of winds, waves, currents, or other parameters the data server can access a **COASTMAP** model server or model products available via the internet. In the former, the model server collects data from the data server on bathymetry, topography, boundary conditions, and forcing fields. The model server then executes the desired forecasts and provides the model products to the data server. The model server can be set for one time or routine operation.

Figure 2 shows the user interface for the **COASTMAP** professional client. The geographic based operational window of the system is shown on the right. Access to the geographic information system, and associated layers of data, are provided in the upper left and to the data management system in the lower left. Hydrodynamic model predicted tidal currents for the lower portion of the East Passage of Narragansett Bay are overlaid on the base map for January 25, 2003 in geographic window. Model predictions are provided from **COASTMAP**'s own hydrodynamic model, which operates on the model server and posts results to the data server. The insert in the upper right gives the current scale and time of the prediction. The color contours show the current speed in the area. Mapping tools and the GIS are accessed by the icons in the upper level corner of the screen. They allow zooming, panning, distance measuring, and adding or removing layers and or information from the various GIS layers. The GIS layers for streams, rivers, and roads are displayed. These GIS layers were acquired via the internet from the RI Geographic Information System web site. Access to data is organized via a tree structure starting from source, to location, to sensor system, and finally to individual sensor. The window in the figure shows the tree structure for wind and water level sensors for the Newport, RI, NOAA Physical Oceanographic Real Time System (PORTS) data. Pointing the cursor on an individual station provides the most recent data collected for that station including a time stamp. Double clicking allows one to access the web site where the data is available and to download the data. This could be either the NOAA PORTS web site or **COASTMAP**'s data server. The downloaded data is then appended to any prior data collected at that site. The automated download of data can also be implemented so that the data record in **COASTMAP** is continuously updated.

As configured for Narragansett Bay, **COASTMAP** routinely downloads NOAA PORTS data (water level, wind speed/direction, and where available Acoustic Doppler Current Profiler data) for Narragansett Bay, US Geological Survey stream flow data for the major rivers in the bay's watershed, forecasts of low frequency water levels from the NOAA National Weather Service, Extra-Tropical Storm Surge (ETSS) model, and forecasts of hydrodynamics and meteorology (Kelley et al, 2004) for the Narragansett Bay study area. The hydrodynamics model predictions are available from **COASTMAP**'s data server, while the meteorological forecasts are obtained via FTP from the NOAA/ National Ocean Survey.

As shown in Figure 3, **COASTMAP** can also be used to download and display data being collected by high frequency coastal radar systems. The figure shows observed surface currents in western Block Island Sound and the adjacent offshore region, collected by the University of Connecticut- University of Rhode Island CODAR system (Ullman et al, 2003) at 3:00 on April 24, 2003. Figure 4 shows the ability of the system to access and display satellite imagery and animations of weather systems passing through the area. In this case the data is from NOAA/National Weather Service GOES 8 satellite infrared radar image of the weather over the continental US (mosaic) on April 21, 2003. Figure 5 shows observations and harmonic based forecasts of the water levels at Providence, Conimicut Light, and Fall River for February 23, 2004. The data and plots are taken directly from the NOAA web site.

COASTMAP, as configured for Narragansett Bay, has its own hydrodynamic forecast model and can access the NOAA/NOS meteorological model predictions. Predictions of the currents (Figure 1) and winds (not shown) can be displayed, animated, and used as input to environmental models (oil or chemical spills, search and rescue planning, atmospheric releases). The system features an ability to allow model predictions to be rapidly and directly compared to observations. As an example, Figure 6 shows **COASTMAP** predicted water elevation at Newport, RI for March 5 and 6, 2003, compared with observations collected from the NOAA/NOS national water level monitoring network station through about 9:00 on March 6, 2003. Predictions are visually assessed to be in very good agreement with the observations. By clicking on the file icon in the upper level corner of the window a detailed statistical and correlation analysis can be very quickly performed to quantitatively assess model performance. These comparisons can be made between observations and model predictions for any user selected parameter and any location.

Once data is collected it is often desired to perform time series analysis to better understand the data. To this end **COASTMAP** features a suite of tools (written as MATLAB routines) to process the data. Data processing includes quality control (spike identification and removal, flagging suspect data, data gap filling), statistical analysis (means, medians, standard deviations, and kurtosis), time series analysis (filtering (low, high, and band pass), demeaning, power spectra, correlation), and harmonic analysis (decomposition and composition). These tools provide the user a powerful ability to analyze the data (or model products) immediately after they are generated. As an example, Figure 7 shows the observed water level versus time (upper window) at Newport, RI for May 2003. The lower window shows the power spectra for the data collected at this station. The power spectrum clearly shows significant energy at the principal semi-diurnal and diurnal tidal frequencies. Harmonics of the principal lunar semi-diurnal tide are clearly shown.

Web Client

As currently configured, **COASTMAP** has two web based applications, the IM2 (http://nopp.cae.drexel.edu/~web/beta/NOPP.html) application described in Piasecki et al (2004) and **COASTMAP** web (www.coastmap.com). Figure 8 shows the **COASTMAP** web-based user interface. The system has been designed using ESRI ArcIMS to allow the user a GIS based system with the ability to present information on user selectable base maps. This strategy is consistent with that employed in the professional client. The base maps can be in either vector or raster form. Map tools (zooming in/out, panning, interrogation, etc) are accessed in the upper left, available GIS layers in the lower left, and the most recent data for wind speed and direction from the NOAA PORTS system for Narragansett Bay on the right. In the present case, the web client can select GIS layers including bathymetry, locations of rivers and streams, drainage basin boundaries, shellfish closure areas, and a LANDSAT image of the study area. The layers displayed can be added or removed from the display and the map refreshed as desired. The center screen shows hydrodynamic model predictions of the tidal currents in lower Narragansett Bay on October 15, 2003 at 14:00. Access to archived NOAA PORTS data is also available in the lower right portion of the screen.

Tidal current predictions have been organized in the form of an atlas-based approach. In this strategy, hydrodynamic model predictions were performed for a variety of tidal ranges, varying from neap to spring tides, and stored at hourly intervals; referenced to the time of high tide at Newport, RI. Comparison of model predicted current speeds for the mean tidal range case were then compared to every other range case for each hour after high tide. This analysis showed that tidal current speeds throughout the study area, for all other tidal range cases, could be approximated by range scaling of the predictions for the mean tidal range case. Once the user selects the time for which he wants tidal predictions the system determines high tide for the day of interest (using harmonic forecasting techniques), determines the tidal range for that day, and calculates the ratio of the tidal range to the mean tidal range (range scaling factor). The tidal currents for the mean tidal range case and for the desired hour are then multiplied by the range-scaling factor, and the results presented to the web client. This simplified procedure allows rapid estimates of the tidal currents to be made for any time, with accuracy levels appropriate for the majority of web-based users. This strategy eliminates the need for the operation of a routine forecasting model and has proven very popular with the recreational and boating communities.

Demonstration and Feedback

To demonstrate, train, and illustrate practical applications to potential users, a Users Demonstration and Training Workshop was held. The protocol for the demonstration project followed that given in Szilagyi et al (2000). The invited participants included representatives from all major potential user groups including: US Coast Guard and Navy personnel, environmental regulators, government and academic researchers, water quality managers, recreational boaters, state legislature representatives, education and outreach specialists, private industry, harbor masters, environmental policy specialists, and non profit organizations. The morning session of the workshop featured presentations of the NOPP project and its application to Narragansett Bay. The afternoon session was devoted to hands on training in **COASTMAP** and the web-based system developed by Piasecki et al (2004). The workshop was followed by two formal evaluation periods, consisting of two, two week demonstration periods. The break between the first and second evaluation periods was used to modify the system output to address comments made during the first evaluation period. Users were invited to provide comments to the system developers, either directly or via evaluation forms provided on the web sites, for IM2 and **COASTMAP**.

The lessons learned from the demonstration and training session are that:

(1) User demand for and interest in the system is highly variable. In some cases, the users saw an immediate need that the system could fill in support of their day-to-day activities. For other users, the system was interesting but the need for it, in the context of their current operations, was not compelling.
(2) Allowing users to access either via PC or web based approaches met with general acceptance for the range of potential system users who participated in the evaluation.
(3) Delivery of data and information products must be carefully targeted to meet a given user groups needs. Information provided in the wrong format or that is hard to understand will not be used.
(4) As system developers, we overestimated the number of individuals whose needs could best be filled by the professional client and underestimated those whose needs could be met by web-based access. It is clear that much of the functionality in the professional client will, in time, be available to the web client.
(5) The users were universally impressed with the ability of the system to provide useful information; its architecture, flexibility, and scalability, and the low cost of the computational platform required to run the system.
(6) Users were enthusiastic about the geographic information system based user interfaces for the both **COASTMAP** and both of its web based systems.
(7) System developers placed significant value on the sophisticated forecasting capability of the system and improving its level of accuracy. Many system users (recreational users, marina operators, and those involved in training) were satisfied with simpler forecasting techniques (atlas based methods, empirical approximations) and were willing to accept the degradation in forecast accuracy for simplicity in operation.

Extension to Support Crisis Management Systems

In the last several years, there has been a rapid evolution of integrated modeling and resource response management systems to assist in training, planning for, and responding to accidents or incidents in coastal marine waters. These systems are normally focused on major ports. These Crisis Management Systems (CMS) typically provide access to real time environmental data or forecasts; to models of oil and chemical spill, search and rescue, atmospheric releases of contaminants, nuclear fallout plumes, and general marine accidents; and incident command systems (ICS). ICS is a highly developed and tested methodology used by many organizations for managing large, complex teams of personnel, from many different government and commercial agencies, and equipment in an integrated response effort. Howlett et al (2002) illustrate the use of **COASTMAP** to provide environmental data in support of the application of a CMS for the Port of Singapore. In the present study, **COASTMAP** and CMS have been applied to Narragansett Bay and adjacent coastal waters to provide a system for training emergency response personnel to respond to accidents and terrorist incidents. As a demonstration, Figure 9 shows CMS's chemical model predictions of the atmospheric plume for a methane release associated with an accident occurring as a tanker enters the lower East Passage of Narragansett Bay. The predictions are overlaid on a map of lower Narragansett Bay. The tanker path is shown by the dotted line and its current location by the cross. The atmospheric methane plume, with associated methane concentration contours, is shown emanating from the vessel. Wind data, used as input to the model, was obtained from a meteorological forecast for the area. The locations of all schools in nearby Newport, RI taken from the RI Geographic Information System (RIGIS), are shown to give the user a sense of which schools are at potential risk from the atmospheric plume. Also shown is the census data which will be helpful for identifying the potential number of residents that might be impacted by the methane release. The evolution of the event, from its inception until the release is stopped, can obviously be animated. The above example shows the ability of **COASTMAP** to be extended to provide the underlying data necessary to support crisis management training and response in coastal waters.

Conclusions

COASTMAP is maturing as globally re-locatable, integrated system for real time observations, modeling, and data distribution for marine waters. Application to Narragansett Bay has demonstrated the system's capacity to meet the original objectives of the NOPP project and those established for the emerging coastal component of the integrated ocean observing system. The global re-locatability has been demonstrated by its application to other areas including the ports of Hong Kong and Singapore (Howlett et al, 2002). Extension of **COASTMAP** to support a crisis management system has also demonstrated the flexibility of the approach and its ability to support rapidly evolving needs in port and maritime security training and response.

To date, the system has seen application as a real time water level and forecasting system for the Mexican naval and coastal defense forces, as the data collection and analysis module for real time monitoring of water quality conditions in upper Narragansett Bay, as a tool for monitoring water quality in the Belizean coral reefs, and as an environmental data collection and analysis to support Navy operational forecasting/modeling systems for homeland security activities within US waters and coastal warfare activities outside of the US. The system architecture is currently being extended to allow it to serve environmental data for the next generation of search and rescue planning tools for the US Coast Guard, for US coastal waters.

Acknowledgements

This research was made possible through a National Ocean Partnership Program (NOPP) grant, administered by NASA under grant number NAG13-00040. **COASTMAP** is a copyright of Applied Science Associates, Inc. Narragansett, RI.

References

Kelley, J. G. W., M. Tsidulko, and M. Ward, 2004. Evaluation of high-resolution atmospheric analyses and forecasts for the Narragansett Bay region, Proceedings of the 8 th International Conference on Estuarine and Coastal Modeling, M. L. Spaulding

Howlett, E, E. Anderson, and C. Galagan, 2002. Real time emergency response tools, 2 nd International Conference on Port and Maritime Research Development and Technology, Singapore.

Ocean.US, 2002a. Building Consensus; Towards an integrated and sustained ocean observing system, Ocean.US Workshop Proceedings, Airlie House, Warrenton, VA, March 10-15, 2002.

Ocean.US, 2002b. A multi-year phased implementation plan for an integrated ocean observing system for the US, prepared by Ocean.US, Arlington, VA.

Ocean.US, 2002c. An integrated and sustained ocean observing system (IOOS) for the US: design and implementation, prepared by Ocean.US, Arlington, VA, 21 pp.

Ocean.US, 2003a. Implementation of the initial US integrated ocean observing system, Part I Governance, prepared by Ocean.US, Arlington, VA. p 26.

Ocean.US, 2003b. The U.S. Integrated Ocean Observing System (IOOS) plan for data management and communications (DMAC), prepared by Ocean.US, Arlington, VA.

Opishinski, T. and M. L. Spaulding, 2002. Application of an integrated monitoring and modeling system to Narragansett Bay and adjacent waters incorporating internet based technology, Proceedings of the 7 th International Conference on Estuarine and Coastal Modeling, M. L. Spaulding (editor), New Orleans, LA.

Opishinski, T.B. and M.L. Spaulding, 1995. **COASTMAP**: An integrated system for environmental monitoring, modeling and management, 7th Annual New England Environmental Exposition, May 9-11, 1995, Boston, Massachusetts.

Opishinski, T.B., M.L. Spaulding, and C. Swanson, 1996. **COASTMAP**: An integrated system for environmental monitoring, modeling and management. Presented at the ASCE North American Water and Environmental Congress 1996 (NAWEC 1996), Anaheim, CA, June 22-28, 1996.

Piasecki M, L. Bermudez, S. Islam, and F. Sellerhoff, 2004. IM2: A Web-based Dissemination System for Now and Forecast Data Products, Proceedings of the 8 th International Conference on Estuarine and Coastal Modeling, M. L. Spaulding (editor). (this volume).

Spaulding, M. L., K. Korotenko, T.Opishinski, and C. Galagan, 2000. **TRANSMAP**: An integrated, real time environmental monitoring and forecasting system for highways and waterways in RI- Phase I, prepared for University of Rhode Island Transportation Center, Kingston, RI, URI-TC Project 536139.

Spaulding, M. L., T. Opishinski, and C. Galagan, 2001. **TRANSMAP**: An integrated, real time environmental monitoring and forecasting system for highways and waterways in RI- Phase II, prepared for University of Rhode Island Transportation Center, Kingston, RI, URI-TC Project 536139, p 30.

Spaulding, M. L. and T. Opishinski, 1995. COASTMAP: Environmental monitoring and modeling, Maritimes, Vol. 38, No. 1, Spring 1995, p. 17-21.

Spaulding, M. L., S. Sankaranarayanan, L. Erikson, T, Fake, and T. Opishinski, 1997. **COASTMAP**, An integrated system for monitoring and modeling of coastal waters: application to Greenwich Bay, Proceedings of the 5th International Conference on Estuarine and Coastal Modeling, American Society of Civil Engineers, Alexandria, VA, October 22-24, 1997.

Szilagyi, G. J., F. Aikman, and L. Breaker, 2000. Evaluation of the coastal marine demonstration, NOAA, National Ocean Survey, Silver Spring, MD, p. 40.

Ullman, D., J. O'Donnell, C. Edwards, T. Fake, D. Morschauser, M. Sprague, A. Allen and B. Krenzien, 2003. Use of coastal ocean dynamics application radar (CODAR) technology in the US Coast Guard search and rescue planning operations, Department of Homeland Security, US Coast Guard Research and Development Center, Groton, CT, Report Number: CG-D-09-03, June 2003.

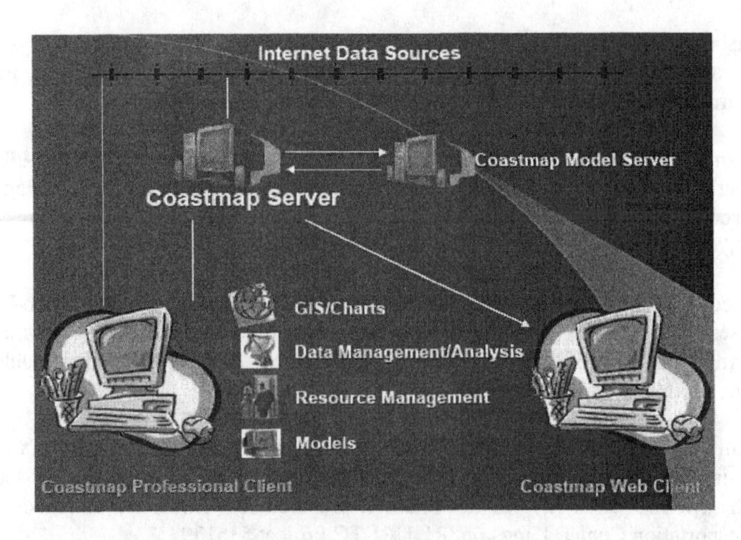

Figure 1: Schematic of the COASTMAP system

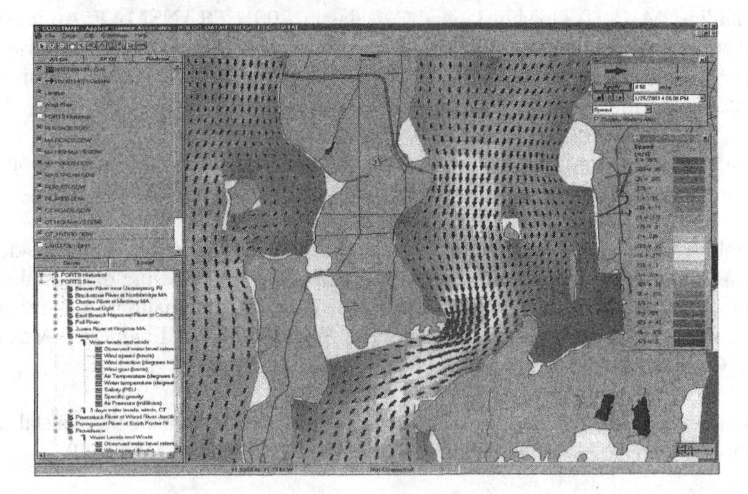

Figure 2: COASTMAP Professional Client user interface. A map of the current operational window for the study area is shown at the right. Access to the GIS is given in the upper left corner and to the data management system (tree structure) in the lower left corner.

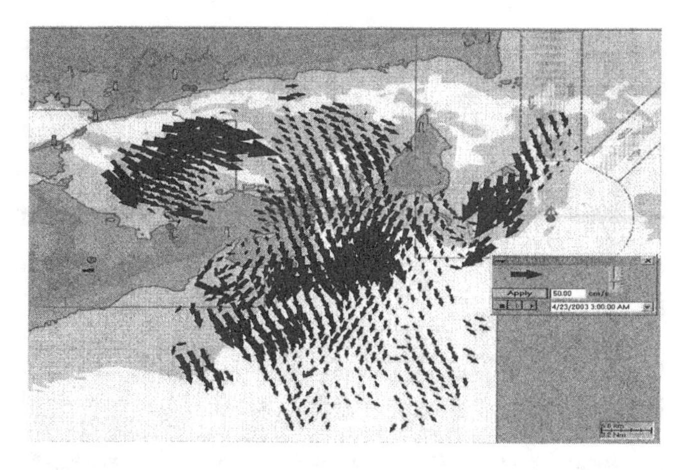

Figure 3: Current vectors from a coastal HF radar data collection system operating in Block Island Sound. The radar is operated by the University of Rhode Island and University of Connecticut (Ullman et al, 2002) with three base stations.

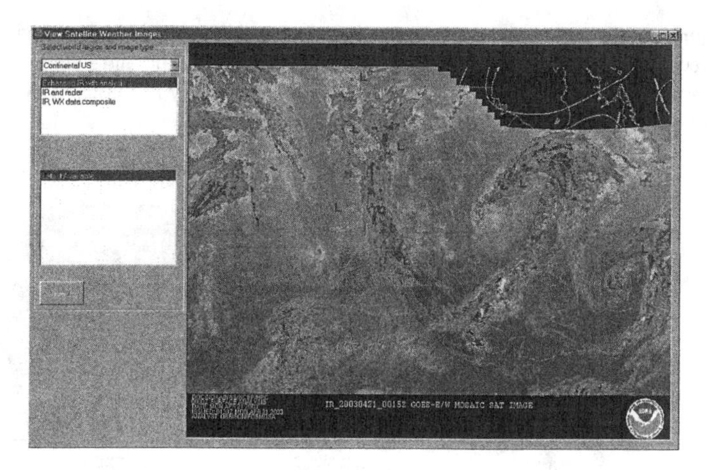

Figure 4: NOAA/National Weather Service GOES 8 satellite infrared radar image of the weather over the continental US (mosaic) on April 21, 2003.

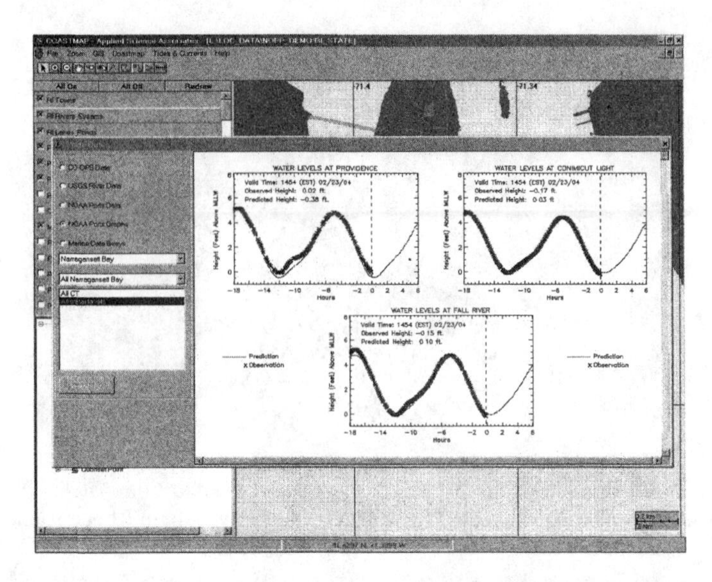

Figure 5: The window in the center shows observations and harmonic based forecasts of the water levels at Providence, Conimicut Light, and Fall River for February 23, 2004. The data and plots were taken directly from the NOAA web site: http://www.nos.noaa.gov/programs/coops/welcome.html

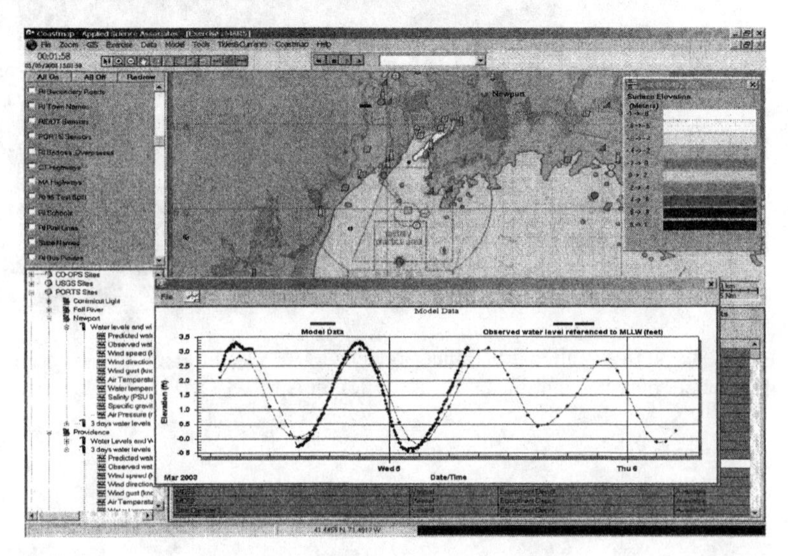

Figure 6: Output of **COASTMAP** showing a comparison of hydrodynamic model predicted to observed water elevation data at Newport RI for March 5 and 6, 2003. Observations are presented through 9:00 March 6 and predictions through the whole period.

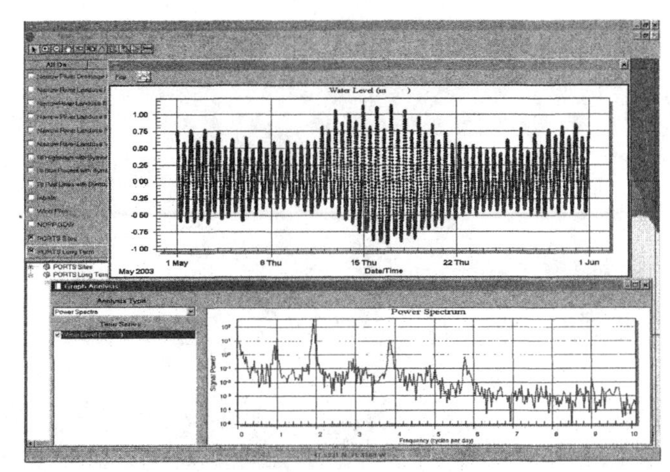

Figure 7: Display of water level data versus time (upper window) for May 2003 from Newport, RI. The power spectral energy versus frequency (lower window) is also provided for this data set.

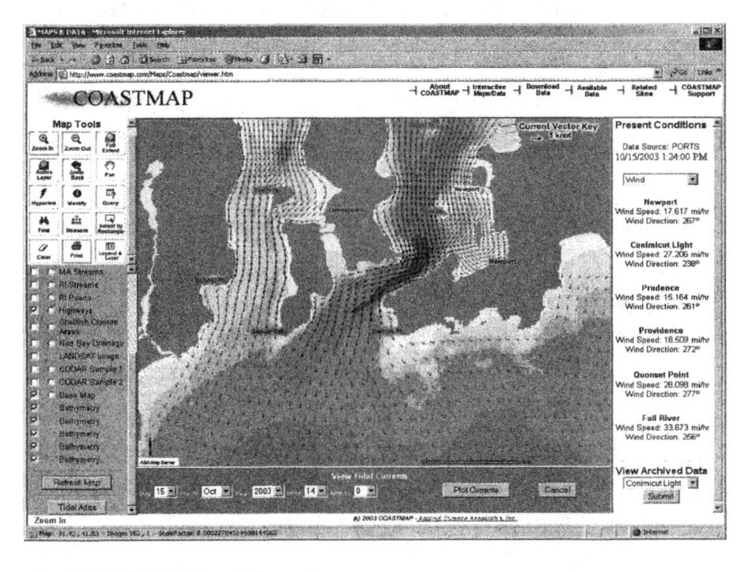

Figure 8: COASTMAP web interface.

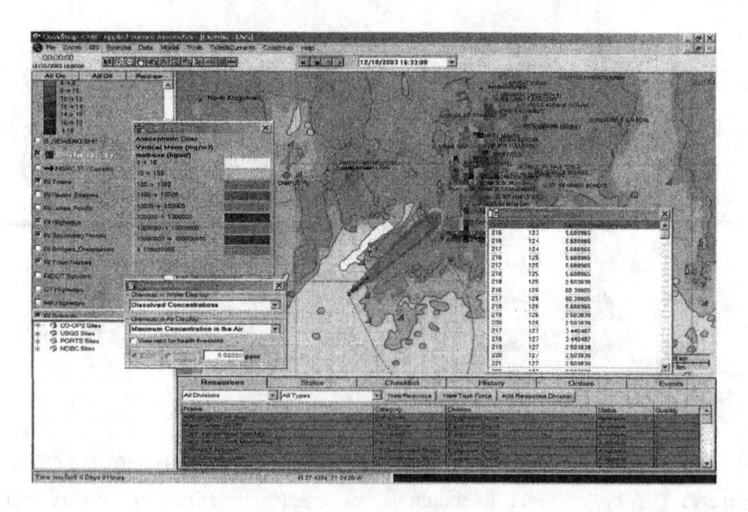

Figure 9: Predictions of the atmospheric concentrations of methane released from a tanker entering the lower East Passage of Narragansett Bay. The tanker is noted by the cross and its path by the dotted line. NOAA S57 electronic charts serve as the base map for the area.

An Internet-Based Data Acquisition and Management Server

By Thomas B. Opishinski[1]

Abstract

An Internet-based Data Acquisition and Management Server has been developed under a three year National Ocean Partnership Program (NOPP). The server application system is designed to acquire environmental data from Internet sites and subsequently process, archive and distribute the data to support independent client components developed under the project. System processes are configured through an integrated management system with acquisition and export processes monitored through an integrated status display. The system's asynchronous architecture supports multiple simultaneous Internet connections to permit timely acquisition, processing and distribution of data. To meet NOPP project objectives, the system has been configured to acquire environmental data available for Narragansett Bay from various Internet-based sources (e.g., NOAA PORTS, NOS and NWS nowcast/forecast models, Landsat 7 satellite images from Brown University) for analysis and modeling applications and for dissemination to the public. The multi-institution project includes researchers from Drexel University, Civil and Architectural Engineering; the University of Rhode Island, the National Oceanic and Atmospheric Administration (NOAA), the National Ocean Service (NOS), Brown University and Applied Science Associates. This paper provides a brief overview of the project and describes the functionality provided by the data server.

Introduction

The University of Rhode Island and Drexel University are leading a multi-institution partnership to develop an integrated system for real time observation, modeling, and data distribution for shelf, coastal sea, and estuarine waters. The project, funded by the United States National Ocean Partnership Program (NOPP), seeks to utilize Global Ocean Data Assimilation Experiment (GODAE) data and other global and ocean basin scale data products to produce model estimates of environmental conditions in the near shore and coastal regions. It is a necessary requirement then that consistent and reliable access to GODAE type data be

----

[1]Interactive Oceanographics, 81 Shippee Road, East Greenwich, RI 02818-1028; iocean@cox.net

established so that numerical models, utilizing this information, may produce useful information in the near shore environment. The Internet-Based Data Acquisition system provides access to such data permitting independent client applications to access the data to assist in hindcast, nowcast and forecast exercises. As a means to validate the models, the data server also provides access to data from near real time systems (e.g., NOAA/National Ocean Survey (NOS) Physical Oceanographic Real Time System (PORTS)) and historical archived data (e.g., satellite imagery of sea surface temperature) available on the Internet for the application area. In addition, near real time data collected by the Data Acquisition and Management Server may be distributed for public access via standard web browser applications

Internet-based monitoring and forecasting systems often post data for a limited time period so, in order to have continuous records from these systems, it is necessary that the data be extracted from the systems as it becomes available and archived in a local database for present or future use. These tasks can be embedded within model or web client applications but this would require each operational client application to maintain a continuous link to numerous Internet systems which leads to an increase in internet traffic and bandwidth requirements. By separating these tasks from the client components of the system it is only necessary to operate a single Data Acquisition and Management Server. Clients applications that are responsible for modeling activities may then access data from the Data Acquisition and Management Server on an as needed basis. The Data Acquisition and Management Server can also feed near real time data to a web portal through a single Internet communication path to provide public access to multiple sources of consolidated data representing present environmental conditions.

System Architecture

The Data Acquisition and Management Server is designed as the central component responsible for accessing and storing data from external Internet-based systems and then maintaining the information to provide a single point of distribution to related system components. The system operates on Microsoft Windows 2000® or XP® server in order to support incoming connections from client components. A high bandwidth Internet connection is required to support multiple simultaneous Internet connections and to retrieve continuous sets of data from systems that post data for a short time interval before overwriting the data with a new reading (e.g., access to data in PORTS flat file format).

Figure 1 details the architectural concept of the system components with data communication paths indicated by the arrows. Data and model forecasts are collected from the Internet-based data sources by the Data Acquisition and Management Server. The communication path between the Data Acquisition and Management Server and the Internet-based data sources is indicated by the upper arrow. Several examples of environmental data and forecasts available from Internet-based data sources include GODAE type model forecast products such as NOAA's Coastal Ocean Forecasting System (Kelley et al, 1997; Parker, 1997), local high resolution predictive models such as the NOAA/NOS Narragansett Bay Local Analysis and Prediction System (Kelley, 2003) and near real time data available from NOAA's Physical Oceanographic Real Time System (PORTS) for Narragansett Bay. Once data is acquired it is either processed (as in the case of the PORTS data) or is stored in its native format. An auxiliary database is maintained for each data set that details the content (e.g., parameter types stored for each data source) and availability (i.e., the time period).

The Data Acquisition and Management Server provides and maintains data for two system components. These components are represented by the two lower boxes in Figure 1. Environmental data is processed and reformatted immediately after it is acquired by the Data Acquisition and Management Server to create a metadata descriptor file and associated data file in a standardized XML format. Once created, the files are automatically sent by the Data Acquisition and Management Server over a Secure File Transfer Protocol (SFTP) connection to IM^2, a Web-based dissemination system developed and operated by Drexel University (Piasecki, 2003) for the present project to provide public access to the data products. After transferring the XML files to IM^2 they are deleted from the local database to optimize storage space. Immediate transfer of environmental data to IM^2 provides a timely access to the present environmental conditions for the operational area. IM^2 is designed as an operating system independent application and is accessed using any standard web browser. IM^2 provides a map-based interface allowing the various data sets to be georeferenced according to their collection site and includes the ability to view time series data and to animate model forecasted data such as wind vectors directly on the map.

The second component utilizing data gathered by the Data Acquisition and Management Server is COASTMAP (Spaulding, 2003). COASTMAP accesses the environmental databases maintained by the Data Acquisition and Management Server using a File Transfer Protocol (FTP) connection and locates the appropriate set of data by reading the auxiliary databases (e.g., header files) maintained on the server and subsequently downloading the file containing the desired data parameters. COASTMAP includes utilities to analyze and display environmental data as time series displays or to animate vector data within the framework of a geographic information system. It also utilizes the data to define forcing conditions for hindcast, nowcast and forecast modeling exercises for the operational area in Narragansett Bay or to validate model predictions.

Data Organization

A hierarchal tree structure was created to organize and manage the information available from the various Internet-based monitoring and forecasting systems. Figure 2 details the tree structure with the left panel showing the conceptual breakdown of the various levels. Each node in the tree structure represents a metadata object and is used to store information specific to that node. The highest level of the tree includes nodes that represent the organization supporting the system from which the data is collected. The next level includes nodes that represent the geographic region for which data is available. A geographic region is typically represented by a number of individual sites or is formed from a gridded domain and is depicted in the next lower level. The next level includes metadata objects that symbolize the instruments or sensors located at that particular site. Finally the lowest level includes nodes for each parameter that is measured and reported by the sensor. The data tree structure for the NOAA/National Ocean Survey (NOS) Physical Oceanographic Real Time System (PORTS) for Narragansett Bay is shown in the right hand panel and depicts the breakdown of the PORTS data source into levels, each consisting of one or more nodes. In the example given, a single coverage (i.e., Narragansett Bay) is defined under the NOAA/NOS PORTS data source. Individual monitoring sites (e.g., Conimicut Light, Newport, Providence, etc.) contained within the Narragansett Bay coverage occupy nodes of the next lower level. Nodes are inserted under each site for the sensors in operation at the site such as water level, CTD, meteorological, and Acoustic Doppler Current Profilers (ADCP). At the lowest level of the structure a discrete node defines each data stream or parameter.

Implementation of this basic metadata structure, as displayed by the Data Acquisition and Management Server, is presented in Figure 3 for the sites located within the Narragansett Bay PORTS system. The left panel shows the data structure configured in the Data Acquisition and Management Server, beginning with the local data server and branching to the data sources, then to data coverages, and finally to the level of the monitoring sites. In this example, the meteorological sensor for Conimicut Light, RI has been selected with the parameters measured by the meteorological sensor shown in the lower right panel along with additional attribute information. Attribute information such as the parameter name, classification, measurement units, minimum and maximum permitted values are stored in the metadata object for the Conimicut Light meteorological sensor. The data format was designed to be flexible and scalable, to allow each node to contain property structures specific to that node, to improve data handling and transfer, and to minimize storage requirements.

Storing information for a data source in individual nodes contained in the metadata structure facilitates management of attribute information and the communication interface used to access the data source. The information is entered into the system through a series of property dialog forms. Authenticated users may modify attributes in the properties dialog window; otherwise the properties are presented to the user but cannot be altered. Examples of the property dialog forms are shown in Figures 4, 5 and 6. Figure 4 details the property dialog form for the Internet-based data source (highest level in the hierarchal tree structure) which, in this case shown for the Narragansett Bay PORTS system. The attribute information that may be entered includes the name and description of the Internet-based system. An image may also be associated with the node and is displayed in the interface of the Data Acquisition and Management Server whenever the node for the Narragansett Bay PORTS system is selected. Skipping to the site level of the hierarchal structure

Figure 5 illustrates the property dialog form for the Conimicut Light PORTS site. Various types of attribute information may be defined for each site and are separated by category according to the tabs shown along the upper portion of the form. This includes "General Site Properties" (name, description and associated image of the site), "Geographic Location" (longitude, latitude and elevation), "Permissions" (monitoring on/off) and "Export Attributes". In Figure 5 the "Geographic Location" tab has been selected to display the data entry fields for the longitude, latitude and elevation attributes.

Figure 6 is the property dialog form for instruments which also includes a several tabs in order to separate the attributes into separate categories. The "Acquisition Properties" tab has been selected for this figure as it contains the information necessary to access and download the environmental data (e.g., communication protocol, server IP address, user name and password, etc.) from the Internet-based data source. Acquisition of data can be synchronized with the Internet-based data source to minimize delays in acquiring data. For example, NOAA PORTS posts new data every six minutes so the polling time used to initiate the download sequence may be set so that the Data Acquisition and Management Server is synchronized to poll for new data as soon as possible after the data is posted. If the time is not set accordingly a potential 6 minute delay can occur. Figure 7 shows the "Schedule Options" tab of the property dialog form for the Conimicut Light meteorological instruments and the settings for the polling schedule.

System Status

An integrated status display is provided within the graphic user interface for both import and export operations. The status displays details the status of individual connections made to the Internet-based monitoring sites during operations to download data files. Feedback is provided that shows the status of the attempted connection and once the file download executes the status details the progress of the download. If the connection or download fails a standardized Internet socket message is displayed that describes the failure. When a file is successfully retrieved the status will then monitor the processing of the file. In the case that there is a failure processing the file the status display will detail the error. The status display will also provide a notification if new data was not found on the server.

Figure 8 shows the integrated File Import Status Display that is accessible by clicking on the "Import Status" icon contained in the tree structure in the left panel of the application interface. The display provides several column listings with the first column detailing the names of the site and instrument for which the data file download has been configured. Reading across any row the successive columns provide the date/time of the next scheduled download, the date/time of the last successful download, the present status of the file transfer process, the status of the connection, the name of the file that was last transferred and the transfer interval (not shown). The status information is initially important to make sure that the configuration parameters have been set correctly. For instance, if the user name or password required to access the remote server were entered incorrectly a message would be returned stating "Invalid User Name or Password" allowing one to review the entries and make corrections. Once the configuration parameters have been set correctly other messages will be displayed in the status area to detail the success or failure of an attempted download.

Figure 9 shows the File Export Status Display and details the export processes for transmitting XML files to the IM^2 system at Drexel University. The File Export Status Display is accessible by clicking on the "Export Status" icon contained in the tree structure in the left panel of the application interface. Similar to the Import Status Display, the integrated File Export Status Display provides column listings with the first column detailing the names of parameter to be exported, the date/time of the next scheduled export, the date/time of the last successful export, the present status of the file transfer process and connection state. Columns that are not shown in the figure detail the remote server name and path, the communication protocol used to transfer the file and the file format. Again the information in the File Export Status Display allows the system administrator to monitor system activities related to exporting data and to make any corrections for invalid configurations. In the event that the export operation fails for a reason unrelated to the local configuration, the status information will provide a message detailing the error. For example, if the connection times out or the remote server in down for any reason an appropriate message is displayed according to standard Internet socket protocol.

Operational Considerations

During the course of the NOPP project, the various system components were tested in an application to Narragansett Bay and nearby coastal waters of Rhode Island to evaluate the practical use and effectiveness of the system. In addition, two formal demonstration periods were offered to local environmental managers, scientists and researchers not directly involved with the project to illicit feedback on the system. During the evaluation period the system performed well although limits of using the Internet as a communication medium for real time data acquisition were identified. The primary issue that detracts from the present system is that many remote systems from which environmental data is acquired do not maintain a temporary history of recent data measurements. In these cases, the remote system posts real time data which is overwritten when new measurements become available. In cases where the Data Acquisition and Management Server cannot successfully connect to the remote system to access data before it is overwritten there will be gaps in the data set. During the course of the system evaluation, heavy Internet traffic (resulting in connection timeouts) or complete Internet outages have been recognized as the common causes leading to missed data. Power outages present a greater problem since data collection will not resume until the computer has been powered up after electricity is restored and resulting in gaps of extended duration. With the exception of power outages, these errors are identified by an appropriate message on the status display. Unfortunately an appropriate action cannot be taken to fill in missed data once a connection is successfully made to the remote server following a failure because the data has been replaced with the most recent measurements.

Data gaps were present in the databases created from the near real time data acquired in Flat File Format from the PORTS system. Other systems from which environmental data was acquired during this project maintained a short history (usually one or two weeks) of recent data. Gaps in the data sets acquired from these systems were avoided since the Data Acquisition and Management System is designed to retrieve all available data since the last successful download in order to fill in the gaps and maintain a continuous record.

Summary

To support the data requirements of several groups participating in a NOPP funded project, the development of an Internet-based Data Acquisition and Management Server was undertaken. The Data Server was designed to acquire data from several Internet-based systems providing near real time environmental data, GODAE type ocean-basin scale data or high resolution model forecasted data for the coastal region. By applying an efficient, systematic and dedicated routine of data acquisition, the system is able to minimize bandwidth requirements of client and web-based components, ease labor and handle the time-consuming tasks associated with data searching and acquisition of data from multiple Internet sites. The system is easily configured to acquire different types of data from the Internet and automatically processes the data and archives it in local databases. The system can be expanded to include functionality to automate processing for other types of data as well.

The main disadvantage identified during the evaluation of the system is that continuity of the data sets, acquired from remote systems, requires a temporary history of recent environmental measurements to be maintained by the remote system. Unfortunately this issue can only be addressed with the assistance of the organization operating the remote data source. A secondary disadvantage is that access to near real time measurements is affected by power and Internet outages. This is identified as the major issue in critical applications where access to near real time data is essential.

References

Kelley, J. 2003. Evaluation of High-Resolution Atmospheric Analyses and Forecasts for the Narragansett Bay Region, Proceedings of 8th International Conference on Estuarine and Coastal Modeling (ECM 8), November 3 to 5, 2003, Monterey, CA.

Kelley, J., D. Behringer, H. Thiebaux, D. Chalikov, and B. Balasubramaniyan 1997. Development of data assimilation for the coastal ocean forecast system (COFS) (abstract), 5th International Conference on Estuarine and Coastal Modeling, Alexandria, VA., October 1997.

Parker, B.B., 1998. Development of model-based regional now-casting and forecasting, Proceedings of the 5th International Conference on Estuarine and Coastal Modeling, M. Spaulding and A. F. Blumberg (editors), p.355-373.

Piasecki, M. 2003. IM2: A Web-based Dissemination System for Now and Forecast Data Products, Proceedings of 8th International Conference on Estuarine and Coastal Modeling (ECM 8), November 3 to 5, 2003, Monterey, CA.

Spaulding, M. L. 2003. COASTMAP: A globally re-locatable, real time, marine environmental monitoring and modeling system, with application to Narragansett Bay and southern New England coastal waters, Proceedings of 8th International Conference on Estuarine and Coastal Modeling (ECM 8), November 3 to 5, 2003, Monterey, CA.

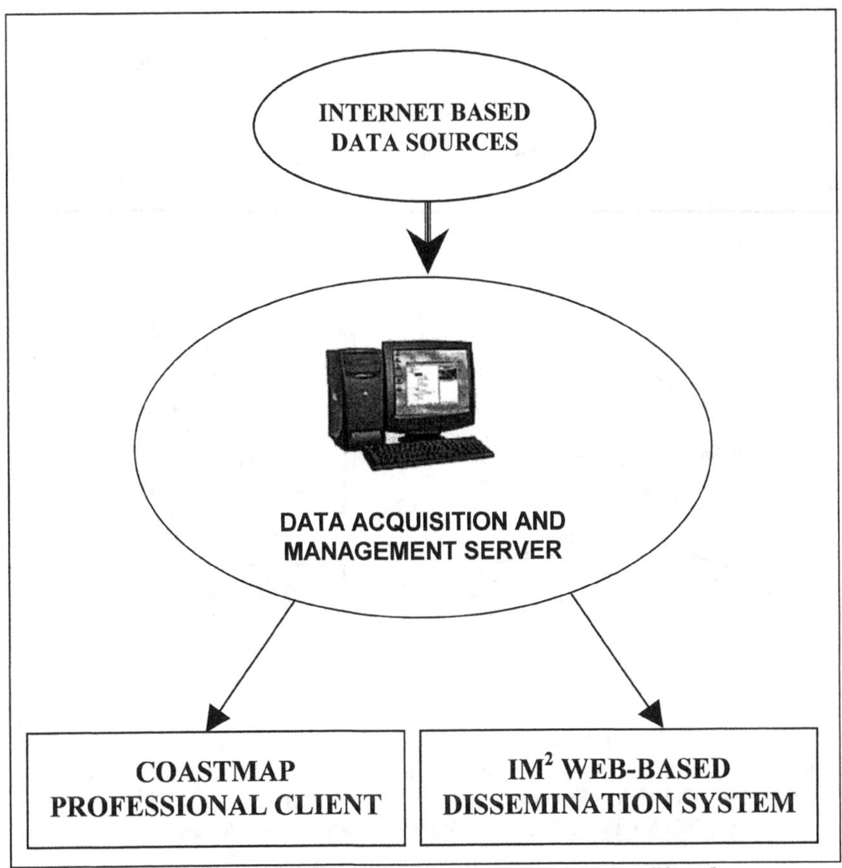

Figure 1: Architectural structure of the system detailing the communication paths between various system components.

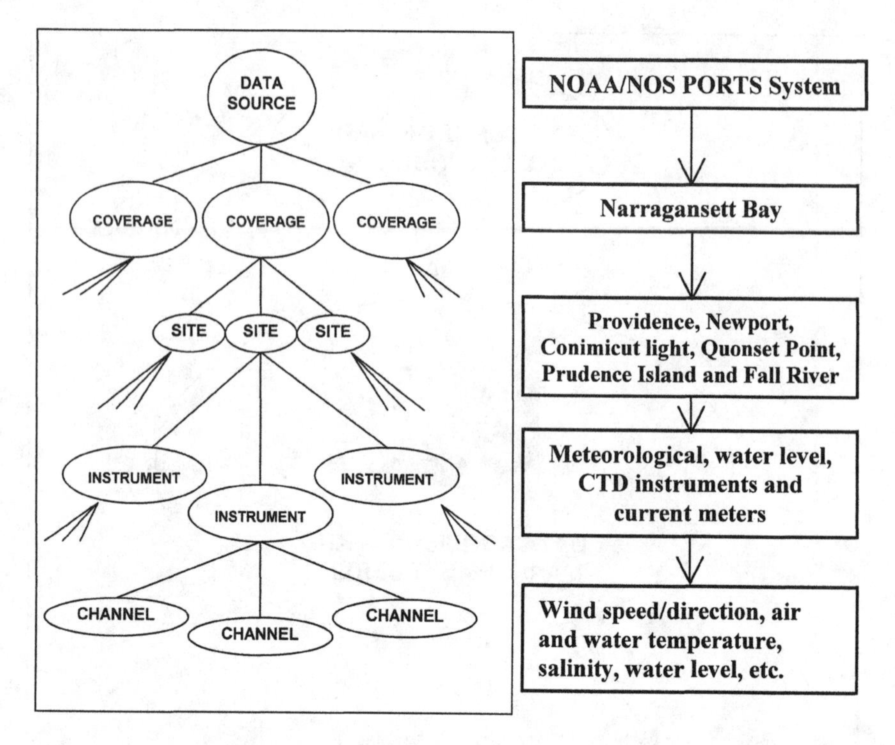

Figure 2: The tree structure, depicting the hierarchy of the metadata objects implemented in the Data Acquisition and Management Server, is shown in the left panel. The right panel of figure illustrates, as an example, a breakdown of the NOAA PORTS system for Narragansett Bay for each level.

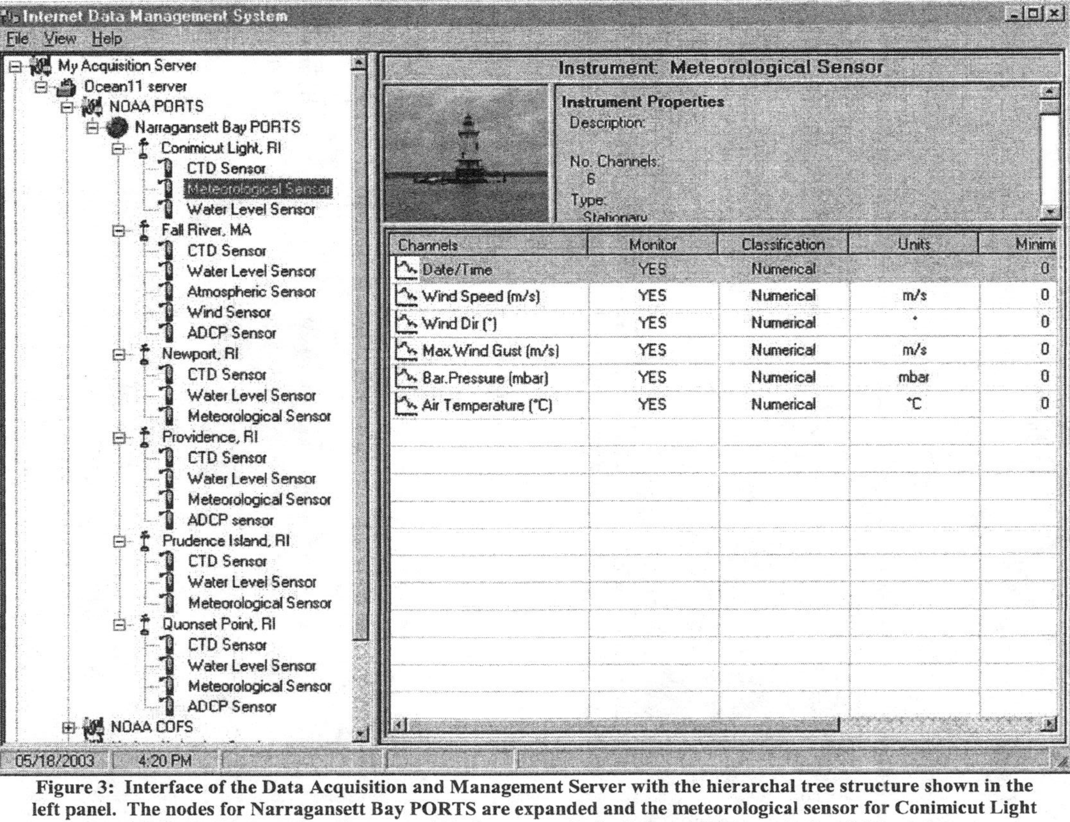

Figure 3: Interface of the Data Acquisition and Management Server with the hierarchal tree structure shown in the left panel. The nodes for Narragansett Bay PORTS are expanded and the meteorological sensor for Conimicut Light selected. The parameters measured by the meteorological sensor are shown in the right panel.

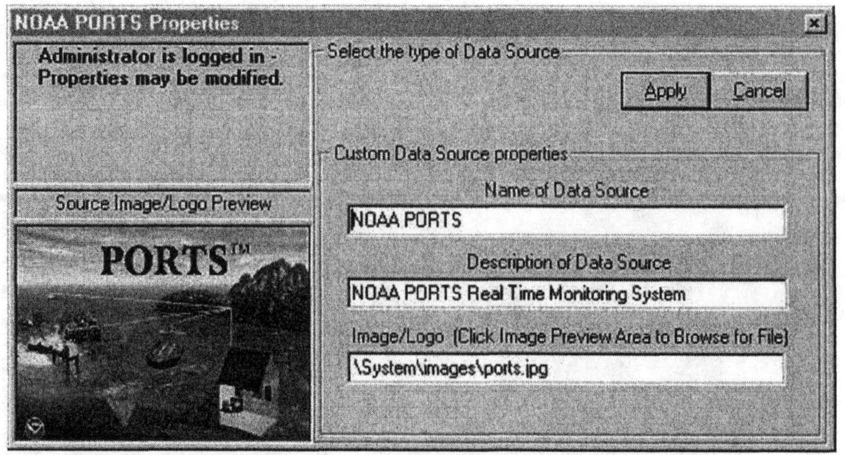

Figure 4: Property dialog form used to define attributes of the data source or organization from the data originates.

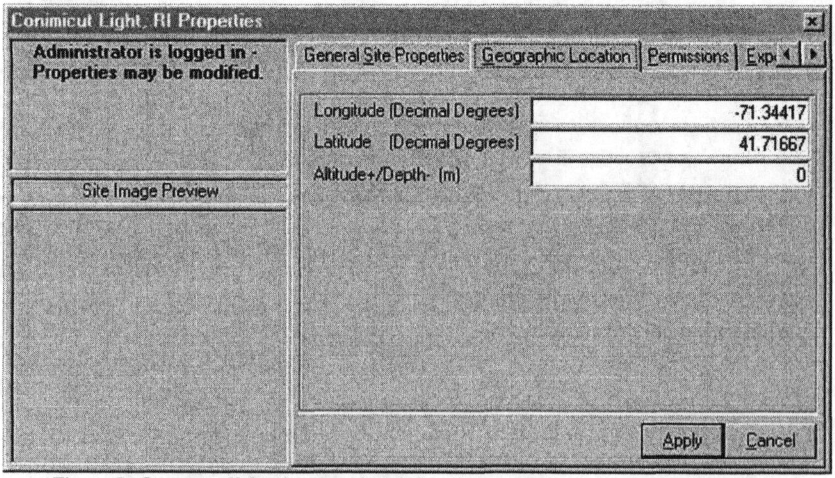

Figure 5: Property dialog form used to define the properties for a monitoring site. In this case the attributes for the PORTS monitoring site at Conimicut Light in Rhode Island have been entered.

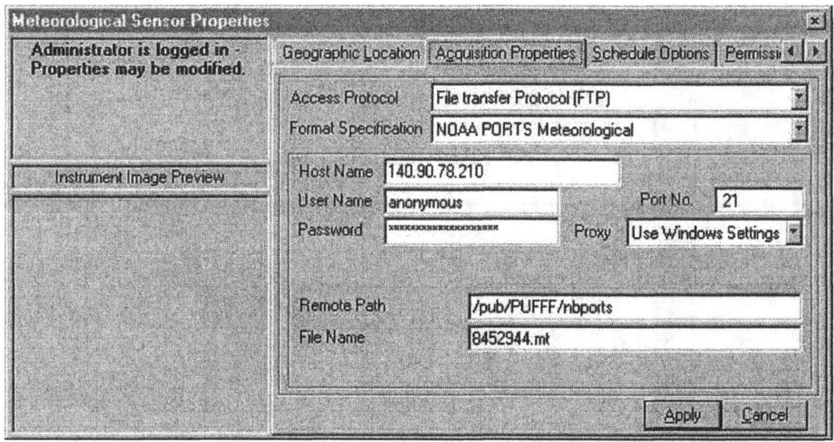

Figure 6: Property dialog form for an instrument. In this figure the "Acquisition Properties" tab has been selected and the parameters used by the system to retrieve the data for the meteorological instruments at the PORTS Conimicut Light site are shown.

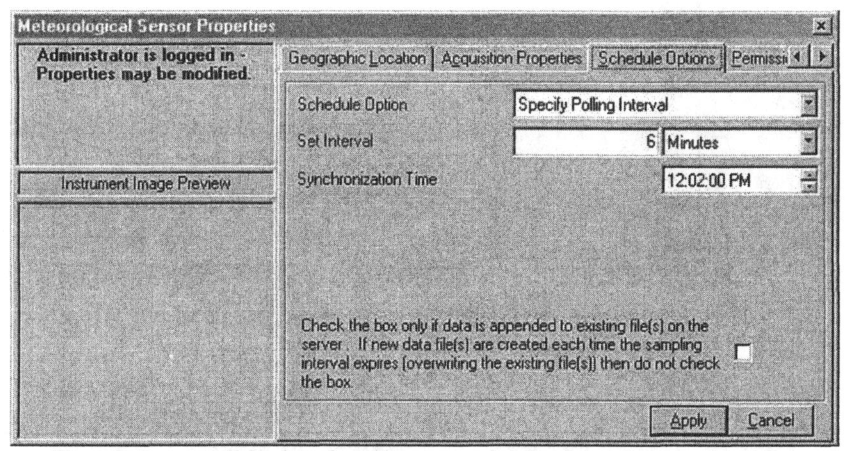

Figure 7: Property dialog form for an instrument showing the attributes used to define the "Schedule Options" used to synchronize the retrieval of the meteorological data available for the PORTS Conimicut Light site.

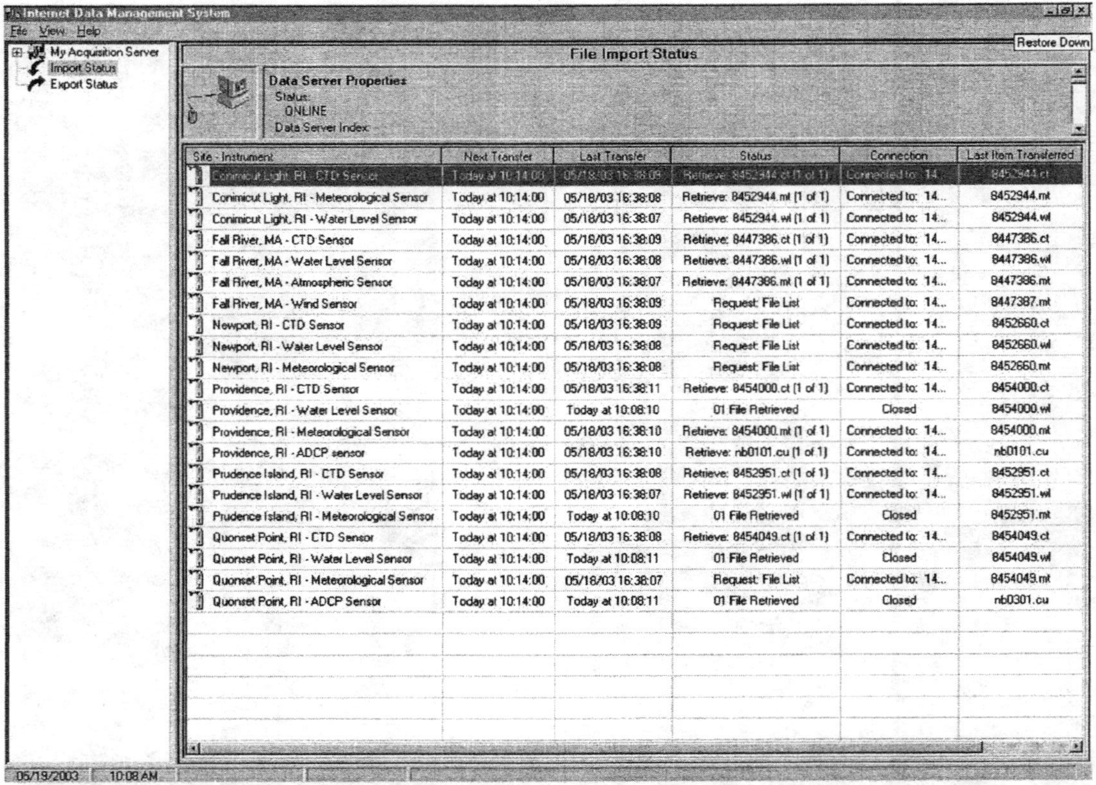

Internet Data Management System

File View Help

My Acquisition Server
Import Status
Export Status

File Import Status

Data Server Properties
Status:
ONLINE
Data Server Index:

Site - Instrument	Next Transfer	Last Transfer	Status	Connection	Last Item Transferred
Conimicut Light, RI - CTD Sensor	Today at 10:14:00	05/18/03 16:38:09	Retrieve: 8452944.ct (1 of 1)	Connected to: 14...	8452944.ct
Conimicut Light, RI - Meteorological Sensor	Today at 10:14:00	05/18/03 16:38:08	Retrieve: 8452944.mt (1 of 1)	Connected to: 14...	8452944.mt
Conimicut Light, RI - Water Level Sensor	Today at 10:14:00	05/18/03 16:38:07	Retrieve: 8452944.wl (1 of 1)	Connected to: 14...	8452944.wl
Fall River, MA - CTD Sensor	Today at 10:14:00	05/18/03 16:38:09	Retrieve: 8447386.ct (1 of 1)	Connected to: 14...	8447386.ct
Fall River, MA - Water Level Sensor	Today at 10:14:00	05/18/03 16:38:08	Retrieve: 8447386.wl (1 of 1)	Connected to: 14...	8447386.wl
Fall River, MA - Atmospheric Sensor	Today at 10:14:00	05/18/03 16:38:07	Retrieve: 8447386.mt (1 of 1)	Connected to: 14...	8447386.mt
Fall River, MA - Wind Sensor	Today at 10:14:00	05/18/03 16:38:09	Request: File List	Connected to: 14...	8447387.mt
Newport, RI - CTD Sensor	Today at 10:14:00	05/18/03 16:38:09	Request: File List	Connected to: 14...	8452660.ct
Newport, RI - Water Level Sensor	Today at 10:14:00	05/18/03 16:38:08	Request: File List	Connected to: 14...	8452660.wl
Newport, RI - Meteorological Sensor	Today at 10:14:00	05/18/03 16:38:08	Request: File List	Connected to: 14...	8452660.mt
Providence, RI - CTD Sensor	Today at 10:14:00	05/18/03 16:38:11	Retrieve: 8454000.ct (1 of 1)	Connected to: 14...	8454000.ct
Providence, RI - Water Level Sensor	Today at 10:14:00	Today at 10:08:10	01 File Retrieved	Closed	8454000.wl
Providence, RI - Meteorological Sensor	Today at 10:14:00	05/18/03 16:38:10	Retrieve: 8454000.mt (1 of 1)	Connected to: 14...	8454000.mt
Providence, RI - ADCP sensor	Today at 10:14:00	05/18/03 16:38:10	Retrieve: nb0101.cu (1 of 1)	Connected to: 14...	nb0101.cu
Prudence Island, RI - CTD Sensor	Today at 10:14:00	05/18/03 16:38:08	Retrieve: 8452951.ct (1 of 1)	Connected to: 14...	8452951.ct
Prudence Island, RI - Water Level Sensor	Today at 10:14:00	05/18/03 16:38:07	Retrieve: 8452951.wl (1 of 1)	Connected to: 14...	8452951.wl
Prudence Island, RI - Meteorological Sensor	Today at 10:14:00	Today at 10:08:10	01 File Retrieved	Closed	8452951.mt
Quonset Point, RI - CTD Sensor	Today at 10:14:00	05/18/03 16:38:08	Retrieve: 8454049.ct (1 of 1)	Connected to: 14...	8454049.ct
Quonset Point, RI - Water Level Sensor	Today at 10:14:00	Today at 10:08:11	01 File Retrieved	Closed	8454049.wl
Quonset Point, RI - Meteorological Sensor	Today at 10:14:00	05/18/03 16:38:07	Request: File List	Connected to: 14...	8454049.mt
Quonset Point, RI - ADCP Sensor	Today at 10:14:00	Today at 10:08:11	01 File Retrieved	Closed	nb0301.cu

05/19/2003 10:08 AM

Figure 8: The Integrated File Import Status display for the Data Acquisition and Management Server is shown above. The display details the state of the Internet connections, file download status and success or failure of the data processing.

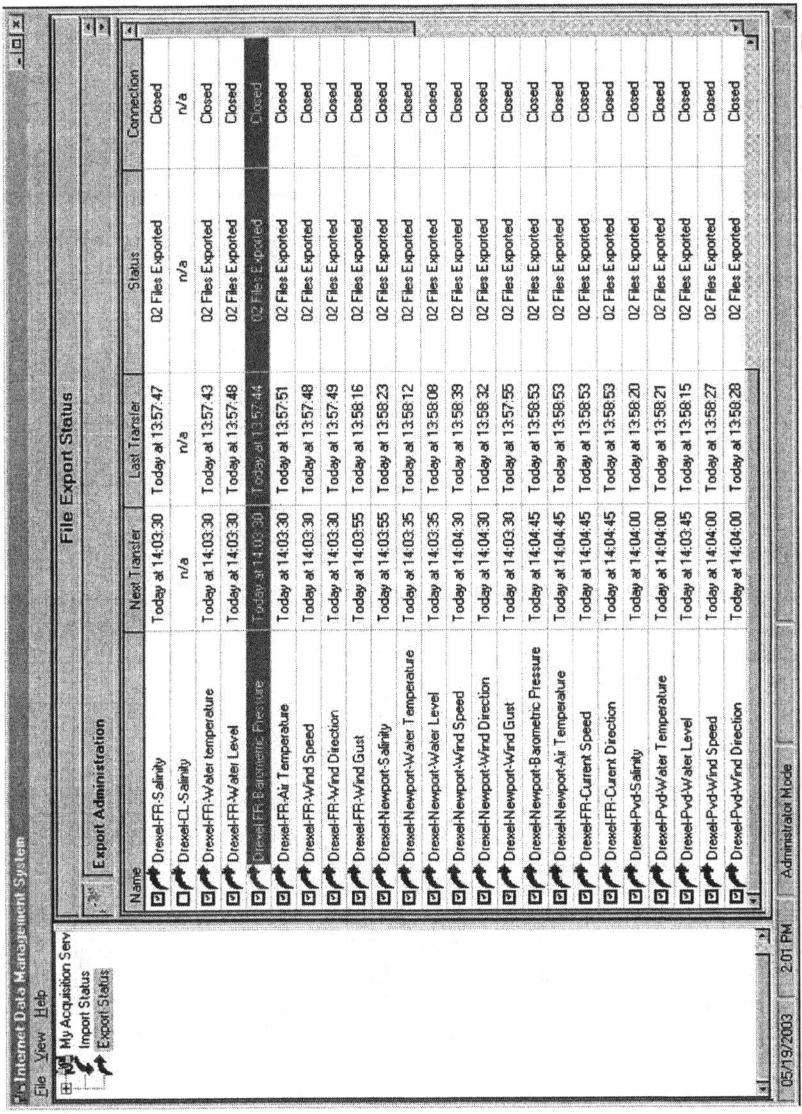

Internet Data Management System

File View Help

My Acquisition Serv
Import Status
Export Status

File Export Status

Export Administration

Name	Next Transfer	Last Transfer	Status	Connection
Drexel-FR-Salinity	Today at 14:03:30	Today at 13:57:47	02 Files Exported	Closed
Drexel-CL-Salinity	n/a	n/a	n/a	n/a
Drexel-FR-Water temperature	Today at 14:03:30	Today at 13:57:43	02 Files Exported	Closed
Drexel-FR-Water Level	Today at 14:03:30	Today at 13:57:48	02 Files Exported	Closed
Drexel-FR-Barometric Pressure	Today at 14:03:30	Today at 13:57:44	02 Files Exported	Closed
Drexel-FR-Air Temperature	Today at 14:03:30	Today at 13:57:51	02 Files Exported	Closed
Drexel-FR-Wind Speed	Today at 14:03:30	Today at 13:57:48	02 Files Exported	Closed
Drexel-FR-Wind Direction	Today at 14:03:30	Today at 13:57:49	02 Files Exported	Closed
Drexel-FR-Wind Gust	Today at 14:03:55	Today at 13:58:16	02 Files Exported	Closed
Drexel-Newport-Salinity	Today at 14:03:55	Today at 13:58:23	02 Files Exported	Closed
Drexel-Newport-Water Temperature	Today at 14:03:35	Today at 13:58:12	02 Files Exported	Closed
Drexel-Newport-Water Level	Today at 14:03:35	Today at 13:58:08	02 Files Exported	Closed
Drexel-Newport-Wind Speed	Today at 14:04:30	Today at 13:58:39	02 Files Exported	Closed
Drexel-Newport-Wind Direction	Today at 14:04:30	Today at 13:58:32	02 Files Exported	Closed
Drexel-Newport-Wind Gust	Today at 14:03:30	Today at 13:57:55	02 Files Exported	Closed
Drexel-Newport-Barometric Pressure	Today at 14:04:45	Today at 13:58:53	02 Files Exported	Closed
Drexel-Newport-Air Temperature	Today at 14:04:45	Today at 13:58:53	02 Files Exported	Closed
Drexel-FR-Current Speed	Today at 14:04:45	Today at 13:58:53	02 Files Exported	Closed
Drexel-FR-Curent Direction	Today at 14:04:45	Today at 13:58:53	02 Files Exported	Closed
Drexel-Pvd-Salinity	Today at 14:04:00	Today at 13:58:20	02 Files Exported	Closed
Drexel-Pvd-Water Temperature	Today at 14:04:00	Today at 13:58:21	02 Files Exported	Closed
Drexel-Pvd-Water Level	Today at 14:03:45	Today at 13:58:15	02 Files Exported	Closed
Drexel-Pvd-Wind Speed	Today at 14:04:00	Today at 13:58:27	02 Files Exported	Closed
Drexel-Pvd-Wind Direction	Today at 14:04:00	Today at 13:58:28	02 Files Exported	Closed

05/19/2003 2:01 PM Administrator Mode

Figure 9: The Integrated File Export Status display for the Data Acquisition and Management Server is shown above. The display details the state of the Internet connections and file export progress.

Evaluation of High-Resolution Atmospheric Analyses and Forecasts for the Narragansett Bay Region

John G. W. Kelley[1], Marina Tsidulko[2], and Matthew Ward[3]

Abstract

NOAA's Local Analysis and Prediction System (LAPS) was implemented for the Narragansett Bay region to provide surface forcing for an oceanographic forecast model for the bay and adjacent coastal waters. The forecast model used by LAPS was the Pennsylvania State University/National Center for Atmospheric Research (NCAR) mesoscale model (MM5). The Narragansett Bay, with its complex shoreline and numerous islands, provided a difficult challenge for an atmospheric model in properly forecasting mesoscale phenomena such as sea and land breezes. MM5 forecasts (primarily its surface wind predictions) will be evaluated using observations from a variety of different federal, state, university, and cooperative networks. When considered together, these networks provide a dense coverage of observations essential for evaluating mesoscale forecasts in the coastal zone.

Introduction

The National Ocean Service's (NOS') Coast Survey Development Laboratory implemented the Local Analysis and Prediction System (LAPS) for Narragansett Bay and adjacent coastal waters. The LAPS implementation was part of a 3-year project funded by the National Ocean Partnership Program and led by the University of Rhode Island and Drexel University. The LAPS analyses and forecasts were used as surface forcing by a real-time oceanographic prediction system being developed for Narragansett Bay and adjacent waters by project partner ASA, Inc. The forecast model used in LAPS was MM5 (Grell et al. 1995). This paper describes the LAPS implementation for Southern New England and examines one sea-breeze case in May 2003.

[1] Coast Survey Development Laboratory, NOAA-UNH Joint Hydrographic Center, NOAA/National Ocean Service, Durham, NH 03824; John.Kelley@noaa.gov
[2] Mesoscale Modeling Branch, National Centers for Environmental Prediction, NOAA/National Weather Service, Camp Springs, MD 20746
[3] Applied Science Associates, Inc., Narragansett, RI 02882

Overview of LAPS

LAPS was developed by the National Oceanic and Atmospheric Administration's (NOAA's) Forecast System Laboratory (FSL). LAPS produces high-resolution, real-time analyses and short forecasts of local surface and upper-air atmospheric conditions (Snook et al., 1998). LAPS consists of an analysis scheme and a mesoscale forecast model.

Analysis Scheme

The analysis portion of the Narragansett Bay LAPS (NBLAPS) is based on LAPS version 0-17-14 distributed by FSL. The NBLAPS analysis scheme used a 125 x 125 point mesh centered over Southern New England with a horizontal grid increment of 4 km. The analysis scheme was run on a dual processor workstation. NBLAPS used data from traditional surface and upper-air observing stations in the generation of its 3-D analyses. The surface observations were obtained from surface airway stations, buoys, Coastal-Marine Automated Network stations, NOS Physical Oceanographic Real-Time System (PORTS), ships, and automated NWS cooperative climate stations. The upper-air observations included data from radiosondes, boundary layer wind profilers, and automated aircraft weather reports.

Forecast Model

The National Centers for Environmental Prediction's (NCEP's) workstation version of its non-hydrostatic Eta atmospheric prediction model was first used as the forecast model of NBLAPS. However, highly unrealistic nighttime air temperatures were predicted over land by the workstation Eta model during the cold season. Thus, the workstation Eta model was replaced near the end of the project with MM5. MM5 was configured using a single grid domain consisting of 82 x 102 grid points with a horizontal resolution of 4 km with 37 vertical sigma levels. An explicit cloud water and ice parameterization scheme is used to calculate grid-scale precipitation and the Grell cumulus parameterization was used for convective precipitation. The boundary layer model selected was the one used by NCEP's Medium-Range Forecast Model (MRF). The MRF scheme includes a five-layer soil model. A 30 arc-second topography and land use dataset was used.

Lateral boundary and initial conditions were obtained from the forecasts of NCEP's operational Eta-12 km model. Sea surface temperatures (SSTs) for the coastal ocean and estuaries were based on NCEP's daily, real-time global SST analysis which uses both in-situ SST observations and multi-channel SST retrievals from satellites. The model was run on a workstation using up to 6 processors to generate 24 hours forecasts once a day.

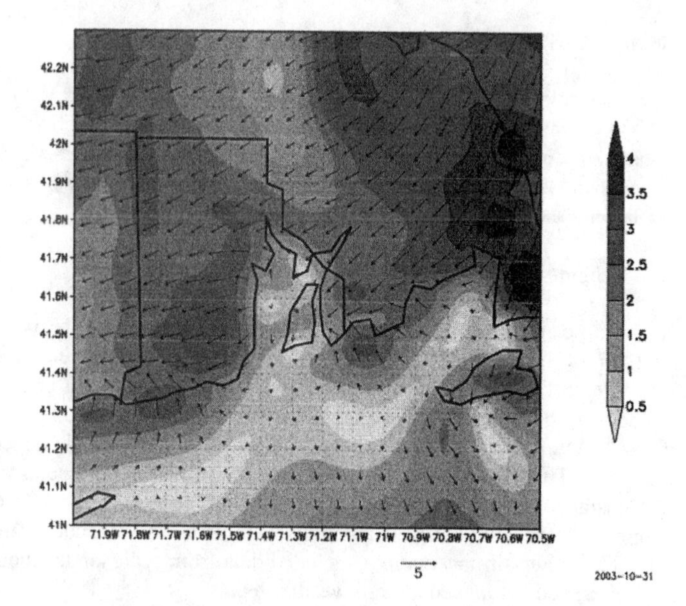

Figure 1 Surface wind velocity (m/s) forecast from MM5 valid at 1200 local time on 19 May 2003.

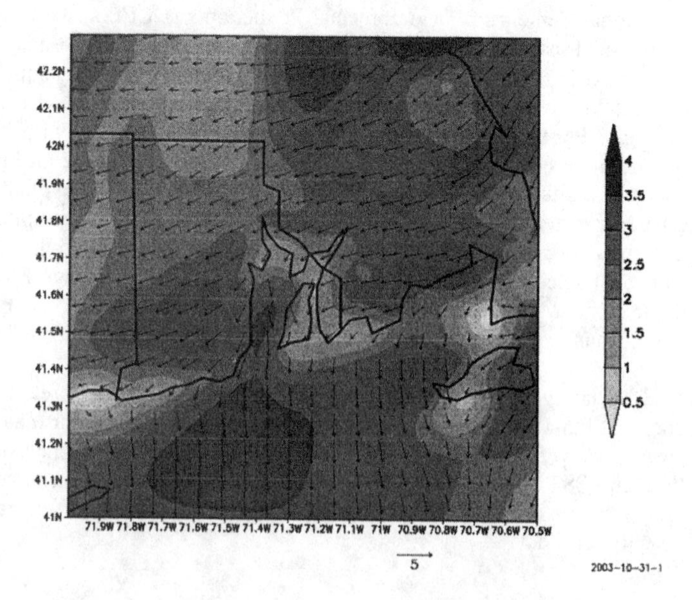

Figure 2 Surface wind velocity (m/s) forecast from MM5 valid at 1400 local time on 19 May 2003.

Evaluation

Since MM5 replaced the workstation Eta model in NBLAPS late in the project, an extensive evaluation of MM5 surface forecasts for the region has not yet been conducted. A non-quantitative comparison was performed for the sea breeze event that occurred on 19 May 2003. This case demonstrated many of the same characteristics of the Narragansett Bay sea breeze described by U.S. Navy (1968). MM5 forecasted the development of a sea breeze at 1200 local time (LT) on the 19th in the Bay and along the southern coasts of southern Connecticut, Rhode Island, and Massachusetts (Fig. 1). However, MM5 did not maintain the sea breeze circulation. By 1400 LT, the surface wind direction was offshore (Fig. 2). Surface observations throughout the bay (not shown) showed that the sea breeze circulation began around 12LT and lasted till at least 17LT. A comparison of MM5's hourly forecasts vs. observations at the NOS PORTS Port of Providence station is given in Fig. 3. MM5 failed to properly forecast the sea breeze at Providence in terms of wind direction, speed, and air temperature.

A quantitative evaluation of MM5 forecasts will be conducted for the sea breeze event of 19 May. In addition, an evaluation will be performed for wind events that caused high water levels in the bay in the spring and summer of 2003. MM5 will be evaluated using observations from traditional observing systems as well as mesonets operated by state and federal agencies and local universities.

Acknowledgements

The implementation of LAPS and MM5 for Narragansett Bay and adjacent waters was funded by a grant from National Ocean Partnership Program.

References

Grell, G. A., J. Dudhia, and D. R. Stauffer, 1995: A description of the fifth-generation Penn State/NCAR mesoscale model (MM5). NCAR Tech. Note TN-398 + STR, 122 pp. [Available from UCAR Communications, P. O. Box 3000, Boulder, CO 80307.]

Snook, J. S., P. A. Stamus, J. Edwards, Z.Christidis, and J. A. McGinley, 1998: Local-domain mesoscale analysis and forecast model support for the 1996 Centennial Olympic Games. *Wea. Forecasting*, **13**, 138-150.

U.S. Navy, 1968: Local - Area Forecaster's Handbook. Fleet Weather Facility, Naval Air Station, Quonset Pt, RI.

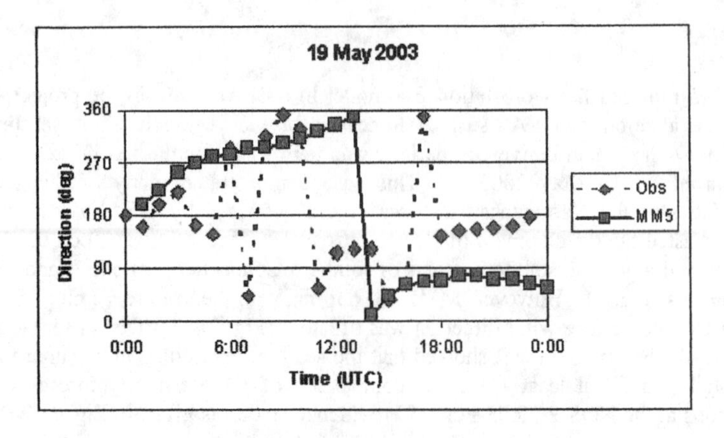

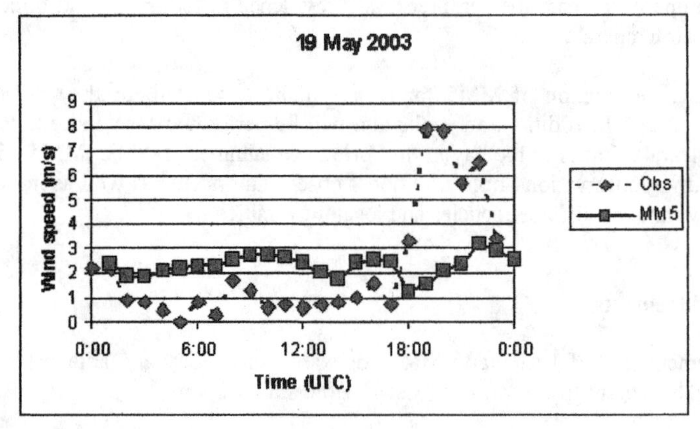

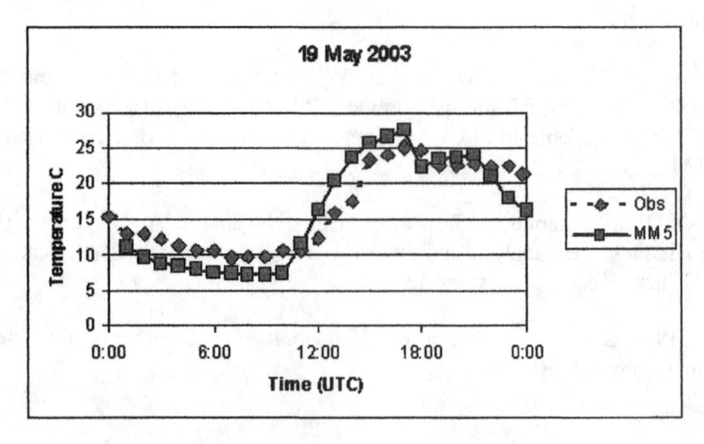

Figure 3 MM5 vs. surface observations at the NOS Port of Providence station for 19 May 2003.

IM2: A Web-based Dissemination System for Now and Forecast Data Products

Michael Piasecki[1], Luis Bermudez[2], Saiful Islam[3], Frank Sellerhoff[4]

Abstract

This paper describes the various steps to develop a web-based information system for now- and forecast oceanographic and meteorological data products. An initial user survey delineates the user needs and requirements and determines to what extent and level the web-system would need to be developed. In order to have a globally re-locatable system it needs to be independent of the operating system (achieved through the use of JAVA) and to have an underlying data structure that separates data content from data description (achieved through the use of metadata concepts). At the same time, it must permit data display of various kinds, for example, time series for stations, contour plots for spatially continuous data variables, and animations over time. Also, the system needs to be flexible to accept various raw data sets in "pull" and "push" mode, be easily adaptable to different data products, and at the same time permit speedy access and response times for the user. The latter involved the use of data compression techniques, and also the use of "thin" client configurations to accommodate less powerful client-end machines.

Introduction

Web-based information systems have become more and more state-of-the-art in all fields of science and engineering. Funded through the National Oceanographic Partnership Program (NOPP, 2003), and under the leadership of Drexel/URI, a group of researchers have developed a globally relocatable information system including a Web-based dissemination sub-system (IM2) for now- and forecast data products. The test bed for this is the Narragansett Bay and adjacent RI coastal waters, which serve as a demonstration of the practical use of the system to support environmental monitoring, marine pollutant transport and fate, marine transportation, and search and rescue operations.

The development objectives for the IM2 system are twofold. First, it is supposed to cater to a wide user audience with varying demands on data analysis, hardware set-up, and a variety of desirable data products. Secondly, it needs to be a system that is fast (small wait times for client and server communications), is operating system independent and flexible in its setup requirements (therefore globally relocatable), and that is user-friendly by providing an intuitive user interface that does not require a steep learning curve. Also, the system was to be designed that no additional software requirements would arise through licensing fees of proprietary components.

[1] Assist. Prof., Dept. of Civil, Architectural & Environ. Engr., Drexel University, Philadelphia, PA 19104

[2] Research Assist., Dept. of Civil, Architectural & Environ. Engr., Drexel University, Philadelphia, PA 19104

[3] Research Assist., Dept. of Civil, Architectural & Environ. Engr., Drexel University, Philadelphia, PA 19104

[4] CEO smile consult GmbH, Vahrenwalder Str. 7, D-30165 Hannover, Germany

This paper describes the development of IM2 using a number of approaches to address the above objectives, more specifically i) the use of user surveys to identify the needs and priorities of the community, ii) the use of state-of-the-art technology like JAVA and XML to develop an OS-independent information system, iii) the use of object oriented paradigms to facilitate the development of the software, iv) the use of a Metadata standard to develop the system specific subset of data descriptors, v) the use of a flat file system for storage, compression techniques, and a multi-server configuration to increase the performance of client-server communications, and vi) the utility of a workshop and demonstration phase to receive feedback from the user community to improve the system functionality.

User Survey

Due to the large and different amount of data that is either collected or produced, the decision was made to investigate, via a survey, what exactly the data needs in the Narragansett Bay user community are. After collection and compilation of the survey returns, the results of this survey were used to aid the Drexel research team to focus on specific issues that relate to the design, content, and most of all functionality of the proposed data dissemination web-site. To this end, a small survey was developed that sought:

- to identify the user group at large and in detail
- to identify the data variable that is being sought
- to identify the time mode, i.e. hind- now- forecasting
- to identify the display frequency
- to identify the preferred display mode, i.e. time-series plot, contours, etc.

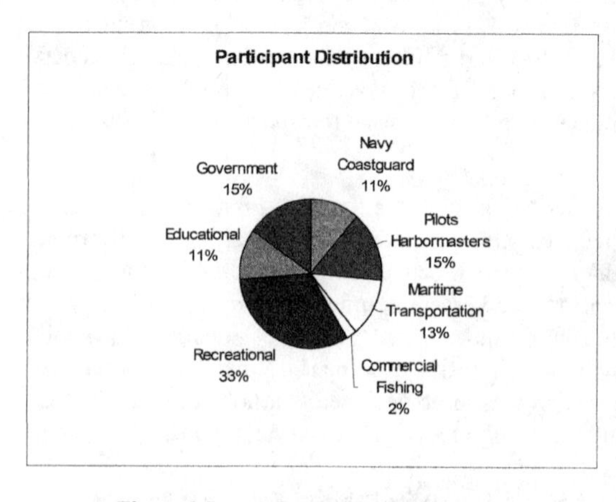

Figure 1 Survey Participant Distribution

The survey was sent out to a number of institutions, groups, individuals, and governmental departments in order to capture a large and most encompassing cross-section of potential users in the Narragansett Bay area. Because of the relatively fine-grained initial identification of user community groups, participant numbers from group to group varied substantially (numbers from the recreational community were quite high, while participants from Navy and Pilot organizations remained limited). However, it was felt that as many identified groups, even when low in numbers should be polled separately in order to have a more detailed description as far as the need for specific data products are concerned.

The survey results are based on a response rate of about 50% resulting in 47 surveys that were returned. While this number may not seem to be very high and therefore representative, it was felt that the number of returned surveys provided a sufficient basis for a good assessment of the survey effort, particularly in light of the large diversity among the participants, i.e., the institutional affiliation, which is shown in Figure 1.

In order to understand better what state variables the user community prefers to see disseminated or displayed, the survey asked to identify and rank the available variable information on a scale from 1 (least) to 5 (most). The ranks were added up and then divided by the number of participants to compute an average ranking. From Figure 2 it becomes clear that wind, waves, currents, visibility, water, and air temperature are the most desired states (in that order) by the community.

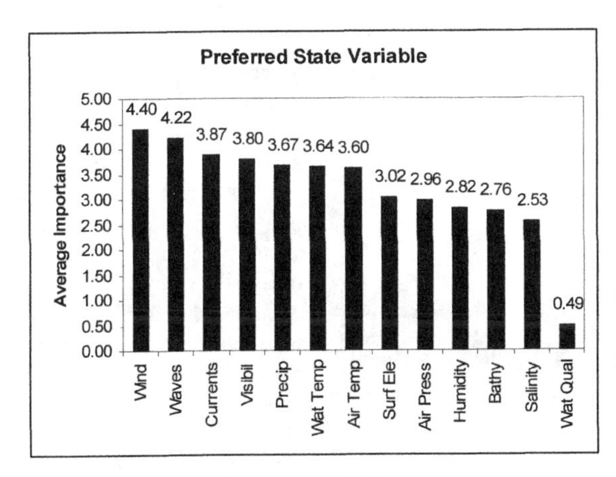

Figure 2 Data Preference

While the overall preference is an important indicator for what needs to be made available, it is also important to notice that preferences change from user group to user group. For instance, surveys submitted by members of the Naval Station (Submarine Warfare Center) indicated that bathymetry, salinity, and water temperature are of significant importance, which can be attributed to the mission of the Warfare Center. Additionally, while water quality received only a small score, members of the scientific community (both University and governmental) placed high marks on the importance of knowing about the water quality situation even though the list of potential pollutants is long and would certainly need to be more specific. Another important target variable was the reporting of wave conditions, coming in second, which was not included in the initial list of available products.

The survey also tried to identify what casting mode is preferred by the user community. The choice was fore-casting, real-time observations, and access to historical records. This part was particularly important for identifying what data products needed to be generated for the user community, and also whether a history record was needed to mandate the provision of a large data storage facility. The results showed a clear overall average preference for real-time and forecasting information. As for the data products (variables) preferences in the previous section, this result is dominated by the structure of the user community that, in general, preferences real-time and fore-casting data products, while historical records are only important for the scientific and regulatory bodies of the Narragansett Bay community. These results clearly indicated that the proposed development of fore-casting models and the resulting fore-casting data products was prudent and reflected a data product line that met the need of potential users. As a result, the development of the WEB-portal, IM2, needed to include options for both real-time and fore-casting data display.

The last item addressed in the survey was aimed at finding out how frequent the data should be made available. The strong desire for real-time data (see above) clearly indicated that continuous data display should be a high priority. Figure 4 shows the results in percent of participants that ranked a certain frequency as highest priority, where the 33% vote for continuous display reflects the desire for real-time observations.

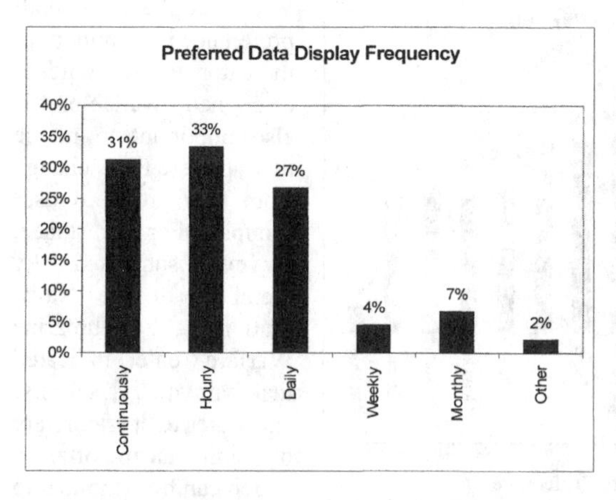

Figure 3 Preferred Data Display Frequency

Clearly, most of the participants thought that data dissemination on an hourly basis would satisfy their needs best, followed by daily updates. Since the daily frequency is a subset of an hourly display, it was concluded that data should be made available on an hourly basis. This was an important decision because some of the data is collected or computed with time-steps that are much shorter hence creating an overhead that would not be used. The only exception was the real-time mode where data was displayed as often as possible, pending the frequency with which it was sent to the data server and with which it could be processed prior to display.

Metadata and Data Storage

The concept of metadata is to describe data in a standardized form, hence, "data to describe data" (Gilliland-Swetland, 2000). Instead of searching through piles of collected data (often insufficiently described), data availability is searched based on the metadata description of the raw data set. The objective for this project was to identify a minimum metadata set that would permit the exchange of data sets among the collectors/creators and the web-based system and at the same time permit the end-user an easy and intuitive access to the data holdings of the IM2 system. In addition, the metadata set had to be simple and short to reduce capacity demands on the data transfer bandwidth when pushing or pulling the data sets through the internet. On the other side, the set had to be extensive enough to sufficiently describe the given data sets and flexible enough to be expanded or altered when used for a different task elsewhere.

Currently, there are a number of metadata standards in circulation, all of which are in a state of continued expansion and development. The Federal Geospatial Data Committee (FGDC, 2002) has been leading the effort to develop metadata standards for the US-based scientific and public community. The FGDC coordinates the development of the National Spatial Data Infrastructure (NSDI, 2002), and hosts about a dozen sub-committees and workgroups that attempt to develop metadata standards for a variety of disciplines that span the entire spectrum of the scientific community, e.g., Ground Transportation, Marine and Coastal Spatial Data, Soils, Cadastral, Geologic, and Remote Sensing, to name just a few. Perhaps the most intensive and globally reaching metadata initiative is carried out by the International Standards Organization (ISO) through its Technical Committee 211 (TC211). The ISO TC211 combines over 50 participating and observing member states and their respective Standard Organizations. The scope of the TC211 is the general "Standardization in the field of digital geographic information". Findings and recommendations of the TC211 are disseminated through the publication of ISO norms, of which the 19115 norm deals with metadata. This metadata standard defines a much larger set of descriptive elements that are grouped in mandatory, optional, and voluntary sets, and which has also become the Draft International Standard (DIS) (ISO, 2002).

While the above standards suggest a comprehensive coverage and therefore appear to be an obvious choice, these standards are designed to be quite generic and to cover a large spectrum of geo-science communities. As a result, their strengths are also their weaknesses, as the given metadata element set is too extensive to work well for a system framework that requires only a bare minimum set. In fact, the objectives of the system were to merely identify a metadata set such the data products could be easily found in the IM2 system. A more suitable standard for this task is the Dublin Core Metadata Standard (Dublin Core Metadata Initiative DCMI, 2002) that emerged in 1995 and has since seen a wide acceptance in the library community. Its advantage is the small number of required elements and the ease with which it can be extended to include descriptions that are not part of the standard set. A list of the DCMI elements used for the development of the IM2 system in addition to the extensions is given in Table 1. All data sets use this boiler plate template to fully describe the received datasets. All internal querying, parsing, retrieval, and subsequent task execution is based on the metadata information provided in the table.

Table 1 Dublin Core Metadata Elements used in IM2

Metadata Element	Purpose	Type
Title	General Title for Data-set	character string
Subject	Variable identifier	character string
Description	What data variable is collected?	character string
Publisher	Who posts it on the WEB?	character string
Contributor	Who collected/prepared it?	character string
Date	When the data collected?	real
Type	What type of data?	character string
Format	Is the data text/ASCII?	character string
Identifier	Pointer to the data file	URL
Source	Where did the data originate from?	character string
Language	What language is being used?	character string
Coverage	Temporal/spatial coverage of the data set?	
Box	What is the Geo-spatial extent?	
Northlimit	Northernmost limit	real
Southlimit	Southernmost limit	real
Westlimit	Westernmost limit	real
Eastlimit	Easternmost limit	real
Period	What is the temporal extent?	
Start	Time for first data digit [millisecs]	integer
End	Time for last data digit [millisecs]	integer
Frequency	Δt between one data item and the next	Real
Rights	Copyright identifier	character string

The white cells show standard DCMI elements, while the cells highlighted in grey are DCMI qualifiers i.e., extensions to the given set generated by following the extension rules of DCMI.

Data Storage

In view of the objective to develop a globally-relocatable web-based information system, much of the initial planning centered on the need how to facilitate this objective. Important aspects to be considered were: costs (license fees), maintenance, performance (fast access, querying & retrieval), and the need to accommodate a variety of platforms, like WINDOWS, MAC, LINUX, and several other UNIX derivatives (Lehfeldt and Piasecki, 2002; Bermudez and Piasecki, 2002). While several public domain Rational Database Management Systems (RDBMS) are available to the user community (MySQL, 2002; PostgrSQL, 2003) and as such would not demand a license fee (unlike systems as ORACLE, Dbase, ACCESS), the decision to install and implement a RDBMS must take into account all other aspects as well. To aid in the decision making process, we set up a decision matrix to weigh the different components against each other, as shown in Table 2.

Table 2 Decision Matrix for Choice of Storage Model

Database Type	Pros	Cons
Proprietary Database	Support, functionality	Costly, a lot of overhead
Pub Domain Database	Free, less overhead (lean)	No support, vulnerable
File System Database	Very lean, no costs, easy	Limitation on structure

Since the complexity of the data structure is quite manageable (that is the number of different data files (data variables) is limited and there is little to no cross-referencing necessary among the different data items), the choice was an easy one to make. Because of its leanness, ease of installation, no additional training requirements, and no-cost allure, the File System DataBase (FSDB) was selected and implemented in the server system, as shown schematically in Figure 4.

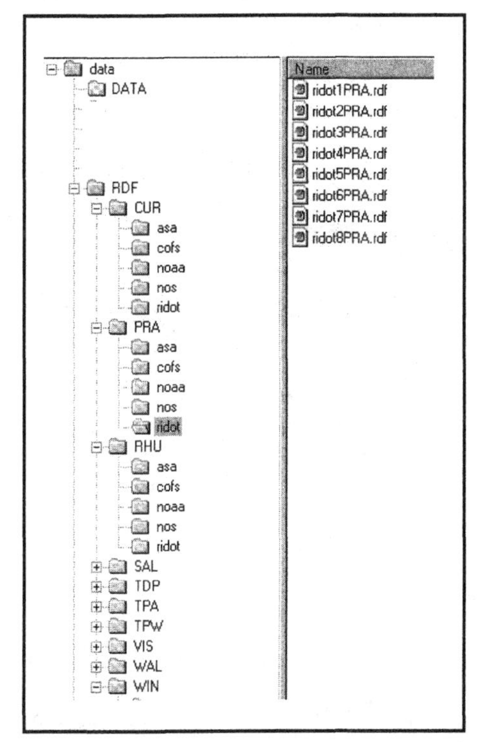

Figure 4 Schematic of the File System Database

The data-root (**data**) is essentially divided up into two branches: one for the actual data files called **DATA**, and one for the data file descriptions called **RDF**. These two directories are mirror images of each other as far as the subdirectory structure in each of these branches is concerned. They also contain the exact same number of files, i.e. each file containing "real" data has a complementary one located in the RDF branch.

Each branch is then subdivided into the 11 data variables (Wind, Salinity, Air Pressure, Relative Humidity, Water Surface Elevation, Bathymetry, Wind Gusts, Visibility, Water Temperature, Wave Height, and Dew Point Temperature) accounted for, which are then divided into each of the data providers. Some of these directories are empty since a certain variable may not be supplied by a specific data provider, but it was reasoned that it is better to maintain a symmetrical structure both for maintenance reasons and future expansion potential. At any given time during the operation of the system, several thousand files are stored in the directory tree, of which the RDF part is parsed based on the queries sent by the client machine (then processed by SERVLETS residing on the TOMCAT server). Despite this large number of files, the parsing process is very fast, taking only several thousand milliseconds to parse through 10,000 RDF files.

Encoding

While the elements of the DCMI metadata set define the semantic requirements of description framework, the syntactic base, or encoding approach, remains to be defined so the metadata descriptions can be machine read or "parsed". The syntactic framework can be provided by using the Extensible Markup Language (XML, 2003). It is considered a meta-language (similar to HTML but much more powerful) that can be used to describe any kind of information and that is platform independent. It is purely syntactic and therefore constitutes a "grammar" framework that a user can utilize to embed his/her specific semantic (e.g., a metadata catalogue) requirements to form a new markup language suitable for a specific purpose. The language is easy to read and ideally suited for structured documents, i.e., repetitive occurrences of certain information pieces, as can be found in data sets (e.g., seismic array data).

Its flexible structure makes it an ideal vehicle for the metadata descriptions. Once the metadata elements have been identified, they can be coded in XML, which in turn allows these data descriptions to be disseminated across the intra- or internet. Dissemination or querying is facilitated by placing the XML-encoded in a separate file, which in turn is then stored on the server's local directory structure for querying purposes. An example **RDF**-file for the salinity field of the ASA model is shown in Figure 5.

The use of the XML language to encode metadata descriptions is favorable for a number of reasons. First, the data exchange between the various modules can be accomplished using XML documents, i.e., the modules communicate through these documents. Second, XML allows sending data files (or descriptions of these files) across the internet allowing for a standardized data exchange among a variety of users through the internet. Third, XML can easily be extended in case future applications demand an extension of the functionality of the IM2 system.

System Design and Functionality

Performance

The fact that much of the data is computed and measured at time levels that are neither synchronized, nor collected exactly at the top of the hour, mandated the establishment of a common clock or time reference. In order to standardize the time levels at which data will be displayed, we developed an automated and server-based interpolation routine that maps any data delivered at any time to a time-frame that uses the full hour mark as reference point. Hence, data is displayed at 1.00am, 2.00am, 3.00am, ..., 10.00pm, 11.00pm, 12.00pm, for a total of 24 data points per day


```
<?xml version="1.0" encoding="UTF-8" ?>
 <rdf:RDF xmlns:rdf="http://www.w3.org/1999/02/22-rdf-syntax-ns#"
 xmlns:rdfs="http://www.w3.org/2000/01/rdf-schema#"
 xmlns:dc="http://purl.org/dc/elements/1.1/"
 xmlns:eor="http://dublincore.org/2000/03/13/eor#"
 xmlns:dcq="http://purl.org/dc/terms/extent/">
 <rdf:Description rdf:about="DATA/SAL/asa/01010600noaa1SAL.dat">
 <dc:title>Salinity values of ASA - Model from January 1, 2001 at 03:03-
        05:00 to January 5, 2001 at 04:03-05:00</dc:title>
 <dc:subject>Salinity</dc:subject>
 <dc:description>Salinity values, in PSU, collected every 1hr </dc:description>
  <dc:publisher>Drexel University</dc:publisher>
  <dc:contributor>University of Rhode Island, NOAA</dc:contributor>
  <dc:date>2001-01-06T00:00-05:00</dc:date>
  <dc:type>Dataset</dc:type>
  <dc:format>text/ascii</dc:format>
  <dc:identifier>DATA/SAL/asa/01010600asa1SAL.dat</dc:identifier>
  <dc:source>http://www.appsci.com/</dc:source>
  <dc:language>en</dc:language>
  <dc:coverage>
  <dcq:box>
   <northlimit units="degrees">42.0</northlimit>
   <eastlimit units="degrees">-72.0</eastlimit>
   <southlimit units="degrees">41.0</southlimit>
   <westlimit units="degrees">-70.0</westlimit>
  </dcq:box>
  <dcq:period>
    <start>20010101030305 00</start>
    <end>"2001010504030500</end>
    <frequency units="hr">1</frequency>
  </dcq:period>
 </dc:coverage>
 <dc:rights>Coyright 2001, ASA</dc:rights>
 </rdf:Description>
 </rdf:RDF>
```


Figure 5 Example XML-encoded RDF file for Salinity Array

for any grid or measurement point in the domain. An exception to this rule were the time series data received from the University of Rhode Island's COASTMAP server which collected and passed data from the Narragansett Bay PORTS program on to the collection bin on the Drexel server whenever it became available, typically in 6 minute intervals. Because the data storage requirements were relatively small it was decided that an interpolation onto full and half hour marks would not serve any higher purpose and the time intervals as well as the delivery times were kept as is. The IM2-

server was configured such that it would routinely check in the server's delivery bin for the new incoming data, move it into a temporary storage area, execute the recognition and interpolation routines, and then pass on the processed data to the respective directories in the FSDB.

In order to further improve the performance of the IM2 system, it was decided that the density of the results of the hydrodynamic simulations, executed by ASA on a curvilinear gird comprised of approximately 15,000 nodes, could be reduced without loosing a significant amount of information content. This was accomplished by "thinning" out the ASA grid, mostly in the regions of high resolution, i.e., inside the bay. This reduction resulted in a "Drexel-display-grid" with only approximately 5000 grid points, i.e., a reduction to one-third of the original grid. The server routines (servelets executed by the TOMCAT server engine) would take the ASA files and interpolate the results based on the 15K node grid onto the 5K node reduced grid (shown in Figure 6) using a bi-variate interpolation. The interpolation of a complete 24-hour file (for one variable) just took a few seconds, hence a complete 24-hour forecast dataset would be ready for display in less than 5 minutes after it had been delivered from the ASA server. The grid used for the atmospheric simulations (by NOAA) is comprised of a regular 4*4 kilometer mesh size, which did not lend itself for further reduction. Hence, the results of all atmospheric simulations were retained for that grid.

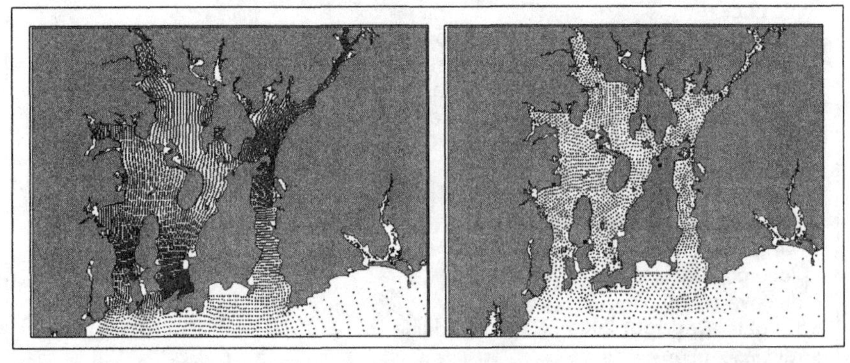

Figure 6 Original ASA Grid (left panel) and reduced Drexel Grid (right panel)

Two additional aspects needed attention during the development phase of the IM2 portal. First, a decision had to be made regarding the uncertainty of what type of browser the end user might be using. Since the targeted user community was very diverse and a prediction therefore quite difficult, the development team had to assume a mixture of the most commonly used browsers (NetScape, Windows Explorer, and Mozilla). This posed a problem, because of the inconsistency of JAVA versions used and supported by any of these browsers. While the Explorer supported JAVA version 1, JAVA2 was not supported but supplanted by a slightly different Microsoft version of JAVA. This made the use of JAVA2 API like SWING difficult, because Explorer users would have to disable the embedded Java Virtual Machine, and download the

JVM from SUN Microsystems instead. While this is a fairly simple action to execute, it provides a considerable level of inconvenience to the end user. Nevertheless, the advantages of using JAVA2 were apparent, so development of the portal proceeded using the JAVA2 system. Fortunately in early 2003, a court ruling appeared that prohibited Microsoft to continue using its own JAVA derivative and as a result cleared the way for all browsers now using JAVA2 and a compatible JVM to run the applet applications.

Second, while much of the data transfer occurred over the internet and then inside high performance networks typically employed at university campuses and governmental institutions, it remained unclear how the end user might be connected to the WWW, i.e. how the "last mile" would be covered. Considering that many of the potential end users may be recreational boaters, the likelihood of 56KB modem connections was high, which prompted the development team to consider seriously the choice of a light- or thin- versus a heavy-weight solution. More specifically, the design of the portal had to balance the advantage of a system that would receive the raw data and then execute all calculations on the client end (heavy-weight) versus the advantage of having all calculations executed on the (powerful) server side, leaving the portal rather lean, but then requiring a considerable amount of client-server communications. In view of a commonly accepted "patience threshold", i.e., the time that should at most elapse between a click of the mouse button and the results of this action showing on the screen, of 6 seconds, it was decided to adopt a middle-of-the-road approach. This approach focused on leaving the most computationally intensive computations on the server, but giving the portal (client machine) some computations to carry out, which would reduce the frequency of communication. Key to this approach was to try to reduce the size of the data objects sent between the client and the server. The adoption of a short integer notation, a binary format, and the use of JAVA's JAR classes (a compression system like ZIP) accomplished this goal in a very satisfactory manner. Some of the data files sent during a session were reduced by over 90%, as shown in Table 3.

Table 3 Example Data Files and their Reduction

Data File	Original Size	Compressed Size
Coastline Polygon	292KB (float, ascii)	114KB (sint, bin, jar)
ASA single WSE data set	91KB (float, ascii)	4KB (sint, bin, jar)
ASA grid	1.2MB (float, ascii)	148KB (sint, bin, jar)
Drexel grid	794KB (float, ascii)	60KB (sint, bin, jar)

The portal was designed to permit background uploading, i.e., while some of the graphics appear in the viewer pane, other data continues to be transferred. Also, once the initial data was uploaded it stayed active throughout the duration of the client session, reducing the data upload needs to just the result files, which were small (for example 4KB).

IM2-System Configuration

The initial setup of the system included a single server running the LINUX operating system on Pentium 3 processor and having 512MB of RAM. While this configuration was sufficient during the initial development phase, it became clear in the course of the system expansion that the LINUX machine reached its limits handling the raw data streams, processing of those, hosting the web-site, delivering the portal and maintaining sufficient speed during a client-server session. An upgrade to 1 GB solved the problem only partially as the main performance problem was encountered during client sessions. In order to mitigate the problem the development team decided to pursue upgrade options for the existing hardware and turned to SUN Microsystems by applying for an Academic Equipment Grant (AEG). This grant, totaling $128K, was installed in phases. In phase 1 the development team received a 280R server with 2 SPARCIII processors, which greatly improved the computational speed available to the IM2 system. The second phase saw the installation of 0.5 terraBytes of hard-disk and tape-backup in addition to 3 SUNBLADE 2000 workstations that were used for continued development and testing. Also, it permitted the IM2 system to incorporate 2 servers, one server for collecting the data, hosting the website (APACHE Web server, v. 2.0.40), and delivering the IM2 portal to the user, while the other server would handle the number crunching intensive part, i.e. the interpolations, reformatting, and running the servlet engine (Jarkarta TOMCAT server, v. 5.0.1), as shown in Figure 7.

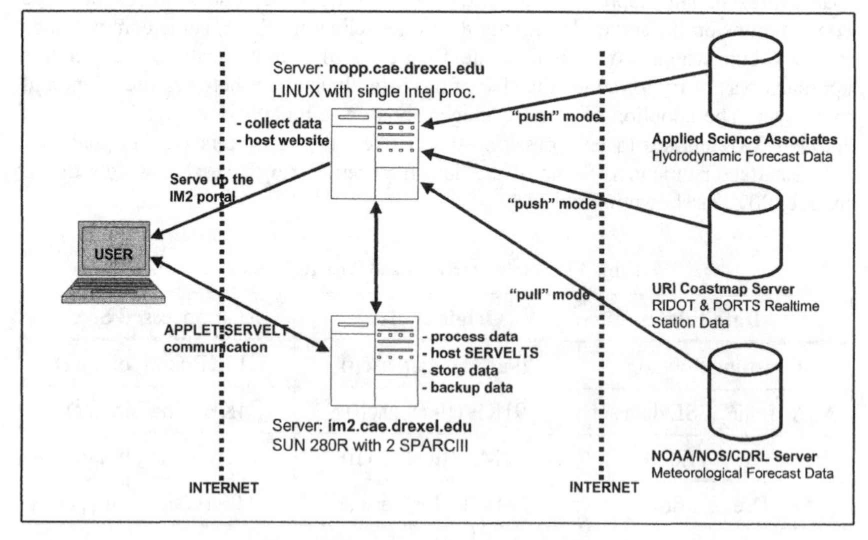

Figure 7 Components of the IM2 System

The IM2 system is configured such that it receives both *.DAT and *.RDF files from Applied Science Associates, ASA, (hydrodynamic forecast data) as well as from the

COASTMAP server at the University of Rhode Island, URI, (real-time station data collected from the PORTS and RIDOT systems). The frequency for the ASA data sets is once per 24 hours received during the early morning hours, while the station data from URI arrives in the collecting bin whenever it becomes available (i.e., about every 6 minutes). In contrast, the meteorological data sets are "pulled" from the NOAA server and delivered in netCDF encoding. These data sets become available around 5.00 a.m. every day and contain 24 hours of forecast information presented at the full hour. The IM2 system is fully automated (i.e., all data collection procedures, recognition, interpolation, and subsequent storage is carried out either using SOLARIS scripts or executing servlets on the TOMCAT server). Despite occasional interventions and subsequent necessary restarts of the system, the configuration has proven to be quite stable with only a few disruptions during the course of its operation.

Portal Design and Functionality

One of the main objectives of the IM2 development was to make sure that the system is user friendly and could be used with only a small period of learning. This mandated the need to keep the portal well organized and structured and also the need to minimize the number of available options and buttons that could be used to alter, modify, and select display settings as well as the variable selection, as shown in Figure 8.

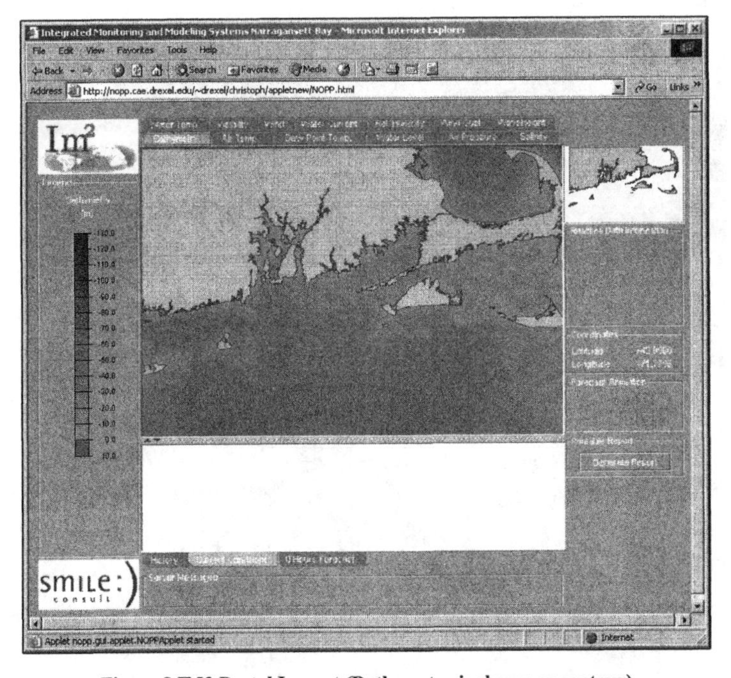

Figure 8 IM2 Portal Layout (Bathymetry is shown as contour)

In addition, the system was designed such that the appearance, as well as the functionality, would be controlled through configuration files. In other words, each change or expansion of the IM2 portal is controlled in a central configuration module that makes sure that relevant changes trickle through and modify the system accordingly. Also, much of the IM2 portal design was influenced by the outcome of the user survey, particularly the selection of desired state variables, display frequency, and display options. It was clear that both real-time and forecast data were foremost on the mind of the user group, displayed every hour, and presented as either time-series, contours, or vector plots (for current and wind velocities). The latter two were preferably animated (i.e., cycling through the forecast hours).

The view-pane dimensions (about 480 by 300 pixels) was aligned with a physical view box bounded with limits of 70 to 72 degrees West and 40.75 to 42 degrees North. All geographical references with respect to grid points and data stations are given in LON/LAT coordinates and are transformed to the horizontal view plane using a Universal Transverse Mercator (UTM) transformation (region 17). During zoom operations, the screen coordinates are transferred back to LON/LAT coordinates applying a reverse UTM transformation. There is also a small viewer pane tracking the mouse movements, so the FSDB can be parsed using LON/LAT coordinates.

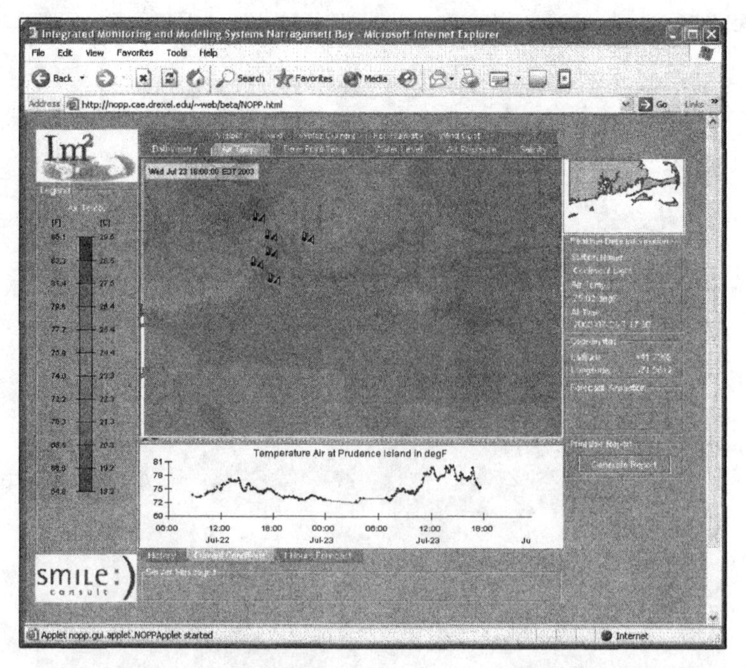

Figure 9 Area Contour and Station History Data for current Air Temperature Distribution

The IM2 portal furthermore contains a pane for the legend (left side), a view pane for time histories (below main pane), and an action pane (on the right) where the user can control animations by stepping through the time levels one by one, or automatically animating them. There is also a button that can be used for controlling the display speed. The variables can be selected from a set of tabs on the top pane, while real-time, forecast or historic data can be selected from tabs at the bottom of the main view pane. Also there is an option to select a specific station (real time data) and obtain a data summary of all data that is currently being selected (the latest available data bit will be displayed). Also, the summary report plots a history of the last 60 hours of data that has been selected at that station.

The system was designed such, that it would automatically display icons for the measurement stations if the selected variable was present both in the forecast simulations and real-time collections (for example, air temperature), as shown in Figure 9. In addition, a sub-pane in the right action pane automatically displays what the current value is (for the selected variable) on a mouse-over of one of the displayed station icons. It should be noted that real-time (or current) data measures are not necessarily synchronized between the various sources. Since forecast data is available only at the hour and station data becomes available about every 6 minutes, the system uses the convention that "real-time" is defined as either the last available data set (station data) or the contour time-step closest to the actual local time. In other words, the real-time display originating from the forecast calculations may be off by 30 minutes while the one from the stations may just encounter a 6-minute delay.

Demonstration Periods and Updates

In order to test and improve the IM2 system (also the COASTMAP system), it was decided by the research group that 2 demonstration periods should be identified interspersed by a workshop to which the expected user-community would be invited. IM2 went online for the first demonstration in March of 2003, which was concluded with a training and user workshop on April 22, 2003 held at the University of Rhode Island's Narragansett Bay Campus. The purpose of this workshop was to give the potential user community an opportunity to sit down and use the system hands on, while providing feedback to the development team on how the system could be improved to better suit the needs of the various user groups.

The workshop proved to be a success as the users took to the IM2 with ease and positively commended on the ease-of-use, layout of the portal, and information content of the data displayed. However, some suggestions were made to improve the system, which were largely related to the fact that different users or user communities have different expectations as far as the data display conventions are concerned. Not surprisingly, the comment made most concerned the unit system available in the legend pane as well as the vector display of wind speed and direction. The former addressed the fact that, depending on what home institution the individual would come from, SI or English units were preferred while others (Navy, Pilots, USCG) preferred current and wind speeds in knots rather than feet/s or m/s. Also, it was noted that the magnitude of the original vector display (vectors scaled to a reference vector) was very hard to discern, which prompted the team to change the vector plot to include a color fill that would be matched in a color coded legend, as shown in Figure 10.

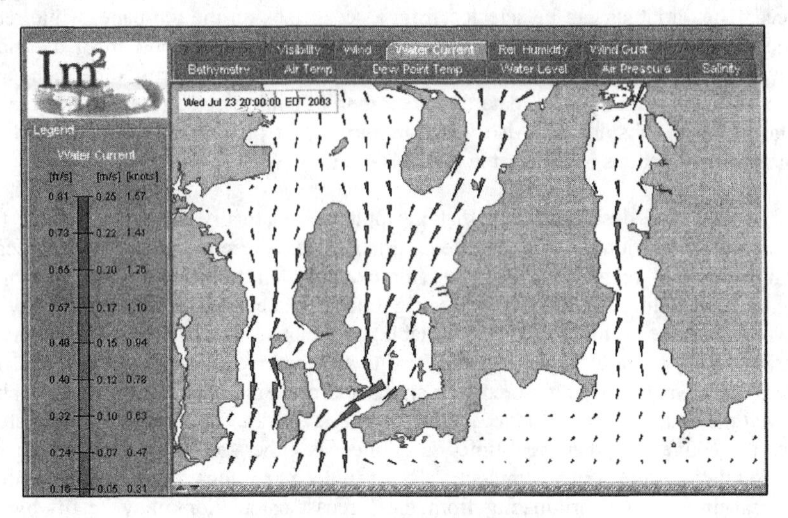

Figure 10 Vector Plot for Currents in the NB Area

Also notice that as a result of the suggestions made the legend was changed to display all unit systems. It should be noted that the vector density (i.e., vectors plotted per screen segment) is fixed to one vector per 20*20 pixel box. This feature was implemented, because of the dense grid distribution inside the bay. If a vector were to be plotted for each node, no information would be discernable because i) the vector heads would have to be extremely small, or ii) the vectors would overlap to a degree that they are indistinguishable. The result is a fixed 20*20 pixel grid that is associated with information from the underlying grid points, irrespective of what the zoom level is. This effectively means that low zoom levels data is being interpolated from a fine grid (model grid with high density of nodes per 20*20 square) to a coarse pixel grid. On the other hand, during high zoom levels the pixel grid might become denser than the underlying model grid and consequently data is interpolated from a coarse to a fine grid. If the fine-to-coarse and coarse-to-fine grid ratios are too high (or too small), interpolation tends to act as a filter and information detail is either lost or a detail is suggested that does not really exist. Yet, the development team felt that it was, and is, more important to discern data trends that, even though slightly inaccurate at times, will fulfill the data needs of the user.

One other concern voiced during the workshop was how the wind field is displayed. The meteorological community prefers a convention that indicates wind direction with a straight line and associates magnitude with a set of perpendicular line segments attached at the end of the base (direction) line segment. The number of these perpendicular line segments then depicts the wind speed bracket. An example of the wind field plot is shown in Figure 11, together with a legend showing all unit systems.

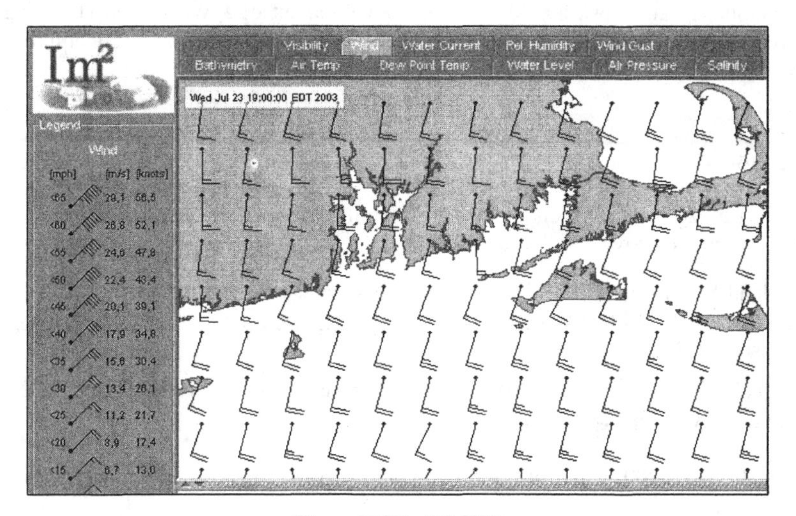

Figure 11 Wind Field Plot

The system is fully operational at and continues to receive data from the ASA, NOAA, and URI servers. Some of the data items, however, are not consistently being delivered, like Air Pressure and Visibility, as well as Currents and Wave Height. These data products depend on the continued execution of the forecast models which, without funding can only be maintained for a limited period of time. Efforts are underway to make this system an integral part of other modeling domains (i.e., to transfer the basic concepts and apply them, for example, to environmental information systems).

The IM2 portal can be accessed at http://nopp.cae.drexel.edu/~web/beta/NOPP.html using "beta" and "beta" as loginID and passwd when prompted. The site also contains an online comment-submission feature so users can continue to submit their comments for further improvement of the system.

Acknowledgements

This research was made possible through a National Ocean Partnership Program (NOPP) grant, administered by NASA under grant number NAG13-00040. The research team would also like to acknowledge the generous Academic Equipment Grant (AEG) received from SUN Microsystems, who provided an essential building blocks of the server system (280R, tape backup system, additional hard-disk storage, free-of-charge software licenses, 3 SunBlade 2000 workstations for system development).

References

Bermudez, L. and Piasecki, M. (2002), "Development of a Web-based Hydrologic Information System", American Water Research Association Annual Conference, Philadelphia, Nov. 3-7, 2002.

DCMI (2002), Dublin Core Metadata Initiative, Retrieved Sep. 2002 from http://dublincore.org/.

FGDC (2002), Federal Geographic Data Committee. Retrieved Nov. 2002 from http://www.fgdc.gov/

Gilliland-Swetland, A.J.(2002), "Introduction to Metadata: Setting the Stage", Pathways to Digital Information, http://www.getty.edu/research/institute/standards/, accessed 2002.

ISO, (2002). International Organization for Standardization. Retrieved Dec. , 2002 from http.//www.iso.ch/

MySQL v.4.0, (2002) http://www.mysql.com/, accessed Sep. 2002.

National Spatial Data Infrastructure (NSDI) (2002); http://www.fgdc.gov/nsdi/nsdi.html, accessed 2002.

National Ocean Partnership Program, NOPP (2000), accessed Sep. 2003 http://www.nopp.org/

PostgrSQL v.7.3 (2003) http://www.postgresql.org/, accessed Sep. 2003

World Wide Web Consortium (W3C), (2003), "Extensible Markup Language", http://www.w3.org/XML/, accessed Sep. 2003.

R. Lehfeldt, and M. Piasecki, (2002), "Components of Web Portals in Coastal Engineering", 5th International Conference on Hydroinfomatics, July 1 - 5, 2002, Cardiff and Bristol, Wales.

Modeling of Tidal Circulation and Pollutant Transport in Port Angeles Harbor and Strait of Juan de Fuca

Zhaoqing Yang,[1] Tarang Khangaonkar,[2] and Steve Breithaupt[1]

Abstract

Strait of Juan de Fuca is a high tidal energy pathway, which connects the Northeast Pacific Ocean continental shelf to the Strait of Georgia and Puget Sound. Complex circulation patterns, such as strong tidal fronts and eddies are observed in this high tidal energy regime, especially near the complex coastlines. These tidal fronts and eddies have significant effect on the dilution, trapping and transport of effluent discharges. A three-dimensional hydrodynamic model was developed to simulate the tidal circulation and pollutant transport in the eastern region of Strait of Juan de Fuca using the Environmental Fluid Dynamic Code (EFDC). In the present study, the model was driven by major semi-diurnal and diurnal tidal constituents, as well as surface winds. The model domain consists of multiple open boundaries, which connect to the ocean water and a number of estuarine systems. Historical water surface elevation, ADCP data, and surface drogue trajectories were used to calibrate the model over a period covering the spring and neap tidal cycle. To study the tidal eddy and transport processes near the Port Angeles Harbor located on the southeast shores of Strait of Juan de Fuca, a localized high-resolution boundary-fitted model grid was generated. The model successfully reproduced the tidal dynamics in the study area and good agreements between model results and observed data were obtained. The calibrated model was applied to simulate the transport and deposition of total suspended solids (TSS) discharged within the Port Angeles Harbor using weighted average of multiple particle sizes corresponding to typical secondary pulp mill effluent. The TSS deposition rates corresponding to different outfall characteristics were calculated and sediment impact zones in the Port Angeles Harbor area were analyzed.

Introduction

Rayonier Inc. (Rayonier) operated an ammonia-based, acid-sulphite process, dissolving grade pulp mill at Port Angeles, Washington from 1937 through 1997.

[1] Principal Research Scientist - Water Resources Modeling. Battelle Memorial Institute. 4500 Sand Point Way NE, Seattle, WA

[2] Practice Leader - Water Resources Modeling. Battelle Memorial Institute. 4500 Sand Point Way NE, Seattle, WA

The pulp mill was located on the northern coast of Washington's Olympic Peninsula, near the Strait of Juan de Fuca. The mill operations produced approximately 36 million gallons per day of process water that was discharged to Port Angeles Harbor. Port Angeles Harbor is formed by Ediz Hook, which is a depositional spit that extends from the mainland to the sea in a west to east direction (Figure 1). From 1937 to 1972, the mill discharged raw effluent from five separate outfalls near the dock. During the period from 1972 to 1978, following the installation of primary treatment and construction of the deepwater outfall, the mill discharged primary treated effluent from the single consolidated outfall. Secondary treatment upgrade was completed in 1978 and the secondary effluent was discharged from 1978 through 1997.

Following an investigation by U.S. EPA, the former Rayonier Pulp Mill site was prioritized for cleanup and, consequently, a remedial investigation/feasibility study (RI/FS) was initiated under the provisions of an agreed-upon order with Washington State Department of Ecology. The focus of the RI/FS activity is on collecting, developing, and evaluating existing information to select a cleanup action. A sampling and analysis plan for the marine environment was developed to fill specific data gaps because considerable information had been collected through previous investigations at the site (Foster Wheeler, 2001; EPA, 1998). These investigations characterized marine sediment quality and extent of woody debris at the harbor bottom. The combined data were used to characterize chemicals of potential concern and their area of potential concern with respect to sediment imparting. The information was used to identify the data gaps and to design the sampling and analysis plan to provide the required information. The design of the sediment-sampling portion of the sampling and analysis plan was based upon existing data, professional judgment, and preliminary steady-state modeling. However, there was a concern that the sampling stations mostly focused on the nearfield and perhaps did not cover far field areas potentially subjected to effluent sedimentation. The concern was that plumes from both the nearshore and deepwater outfalls could have been transported by the tidal circulation well beyond the immediate mixing zone region. The agencies also questioned the validity of the reference stations at the Dungeness Spit which are located approximately 15 miles northeast of the former mill site, since it was not clear whether the sites could have been affected by the Rayonier effluent plume.

This study presents a detailed, high-resolution, dynamic modeling evaluation of effluent fate and transport to provide information for addressing the issues discussed above using a combination of plume dilution, 3-D hydrodynamic circulation, and sediment transport and deposition models. The analysis includes dilution and dispersion of the effluent, and transport and deposition of total suspended solids (TSS) discharged by the former Rayonier Pulp Mill from both the nearshore and deepwater outfalls. Initial dilution of the effluent was conducted using EPA-approved plume model UDKHDEN, and the hydrodynamic and transport analysis was conducted using three-dimensional model Environmental Fluid Dynamic Code (EFDC). The model results provide information about deposition rates of sediment and particulate matter discharged from the Rayonier outfalls in the Port Angeles Harbor, the Dungeness Spit region, and Strait of Juan de Fuca. The model results are expected to provide information that helps refine the design of the sediment sampling plan and address concerns related to the adequacy of the sampling plan.

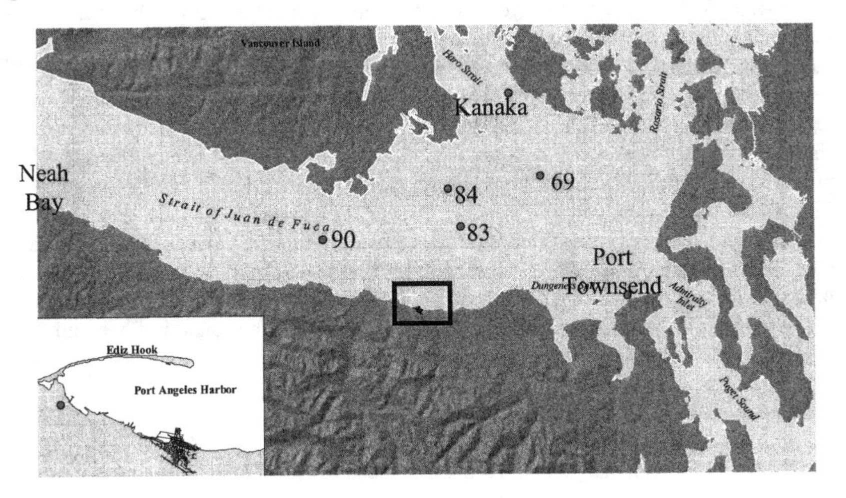

Figure 1: Study Area - Port Angeles Harbor and Strait of Juan de Fuca

Model Setup

Model Domain and Model Description

The circulation in the Strait of Juan de Fuca is dominated by tides. Tidal waves propagate into the Strait of Juan de Fuca from the Pacific Ocean and are altered by the complicated multiple-channel estuarine system formed by the Puget Sound in the southeast and the Strait of Georgia in the northeast. The dominant astronomical tidal components in the Puget Sound area are the semi-diurnal tide M_2, and the diurnal tide K_1. The type of tide in the Strait of Juan de Fuca is a mixed, semi-diurnal tide with large inequalities in range and time between the two high tides and two low tides each day. The tidal waves in the Strait of Juan de Fuca also experience strong spring and neap tidal cycle variations on a 14-day period. The tidal range in the study area is about 3.0 m in spring tide and 1.0 m in neap tide. Tidal currents in narrow waterways generally have good correlation with tidal heights. However, in shallow water, bottom friction usually causes phase lag in the slack and tidal currents relative to tidal heights. Tidal currents are relatively strong in the middle deep channel of Strait of Juan de Fuca, and maximum velocity magnitudes during spring tides could exceed 2.0 m/s. Density stratification caused by freshwater runoff from the Fraser River in British Columbia, Canada, and Puget Sound may affect the surface circulation during high runoff seasons (Labrecque et al., 1994).

However, because the Strait of Juan de Fuca is at the outer boundaries of the Puget Sound estuarine system, the effect of freshwater runoff on the general circulation pattern may not be significant. In addition, wind stresses also affect the area's circulation.

The three-dimensional hydrodynamic and transport model EFDC is designed for the simulation of hydrodynamic and transport processes in rivers, reservoirs, lakes, and estuaries and coastal waters (Hamrick, 1992). EFDC simultaneously solves the 3-D equations of motion for velocities and transport equations for temperature, salinity, effluent tracer and suspended sediment concentrations in a finite difference framework. A second-moment turbulence closure model is solved to provide the vertical turbulent eddy viscosity in the model (Mellor and Yamada 1982). The model uses the boundary-fitted curvilinear-orthogonal coordinate system in the horizontal plane to represent the horizontal geometry accurately. A sigma-stretched coordinate system in the vertical direction is used to best fit the model bottom boundary to the topography. Applications of EFDC in modeling of hydrodynamics and transport processes in estuaries and coastal waters can be found in Shen and Kou (1999), Ji et al. (2001), and Yang and Hamrick (2003).

Model Grid and Boundary Conditions

A model grid was generated to cover the study area, from Crescent Bay to the entrance of Rosario Strait in the east-west direction, and from the entrance of Admiralty Inlet to the entrance of Haro Strait in the south-north direction (see Figure 2). The horizontal model grid is curvilinear-orthogonal and boundary-fitted to the coastline. The grid size varies from 50 m by 50 m to 1700 m by 1700 m, with high grid resolution in the Port Angeles Harbor. The grid resolution gradually reduces away from the Port Angeles Harbor to the open boundaries in order to maintain the computational efficiency of the model. The total number of horizontal cells is 11,982. In order to simulate the vertical structure of currents and TSS concentrations, five vertical layers were considered. The model domain covers a region of about 80 km along the channel and about 50 km across the channel of Strait of Juan de Fuca. The model grid consists of four open boundaries and four land boundaries, respectively. The locations of the open boundaries were specified sufficiently far away from the Rayonier Pulp Mill outfall locations so that boundary effects on the area of interest and interior region were minimized.

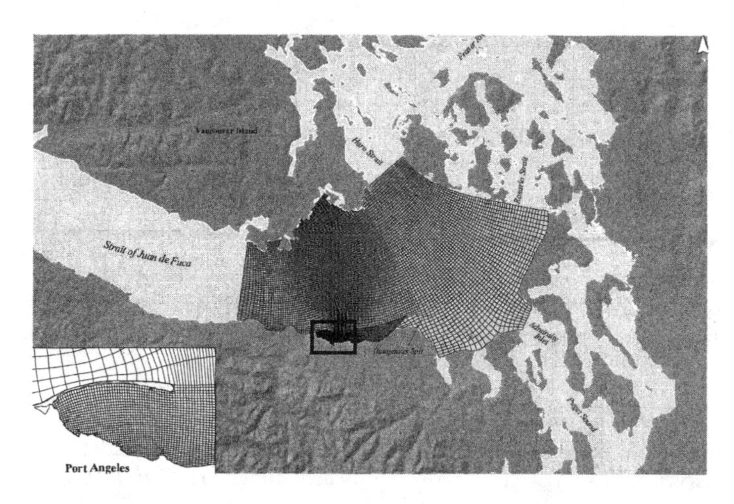

Figure 2: EFDC Model Grid of Port Angeles Harbor and Strait of Juan de Fuca

One of the major challenges in this study is the specification of the open boundary conditions. For the hydrodynamic model, tidal consituents must be specified on all the model open boundaries. However, no field data observations were conducted specifically for this modeling effort. Therefore, the modeling study completely relied on the limited existing data from difference sources. The model was set up for the period of September 1975 to October 1975, corresponding to the period of available data collected by NOAA for model calibration purpose. However, there are only two NOAA tide gage stations near the open boundaries providing good data records: Kanaka near the northern open boundary and Port Townsend near the southern open boundary. NOAA also maintains a real-time station at the Neah Bay at the entrance of the Strait Juan de Fuca and a station in the Port Angeles Harbor. Harmonic analysis was conducted to extract the tidal constituents along the northern and southern open boundaries based on Kanaka and Port Townsend tidal data. Tidal constituents on the eastern and western open boundaries were estimated by extensive analysis and interpolation of Kanaka, Port Townsend, and Neah Bay tidal data. The tide data at Port Angeles were used for model calibration.

At the water surface, wind stress was specified. Wind data from the NOAA meteorological station at Smith Island were used. Wind stress was applied uniformly to the entire model domain. Because the atmospheric heating and air-sea interaction are not the focus of this study, heat exchange between the air-sea interface was not considered. Review of existing temperature and salinity data in the study area indicates that the density stratification in the study area is generally not significant. Therefore, the density effect was not considered in the present modeling study.

Model Calibration

Once the completion of model setup, model simulation was conducted to calibrate the model. The period selected for model calibration is October 10, 1975 to November 9, 1975 (Julian day 282 to 312), covering the period of available NOAA observed data (Parker and Bruce, 1980). Calibration of the model was conducted primarily by adjusting the tidal phases on the open boundaries to provide the best match between the model-predicted tidal elevations and currents to the observed data. The magnitude of the tidal elevation at the eastern open boundary was also adjusted to account for the elevation change as tidal waves propagated from Neah Bay at the entrance of the Strait Juan de Fuca to Crescent Bay at the model eastern open boundary.

The comparison of model-predicted tidal elevations and the observed data at Port Angeles Harbor is shown in Figure 3. It can be seen that the model results match the observed data very well. The model-predicted high and low tidal phases are in strong agreement with observed data. The spring-neap tidal cycle and the diurnal inequality are well reproduced. Comparisons of model-predicted velocities and observed data at selected stations are presented in Figure 4. The water depths at Stations 69, 83, 84, and 90 are 157m, 135m, 98m, and 159 m, respectively. The current meter depths 4.5 meter (15 feet) below mean lower low water. In general, the model-predicted velocities match the observed velocities reasonably well, especially at the high flood and ebb tides. The spring and neap tidal cycle and the diurnal inequality were also reproduced successfully in the velocity predictions. The model tended to over-predict the low peak in velocity. The mismatch between model results and field data is most likely due to lack of exact tidal elevations data for the specification of the boundary conditions. The model calibration can be further improved by obtaining synoptic tide data at the open boundaries, along with currents data within the study area of interests. However, for the purpose of simulating effluent dilution and transport, the level of accuracy in the currents simulated is considered sufficient.

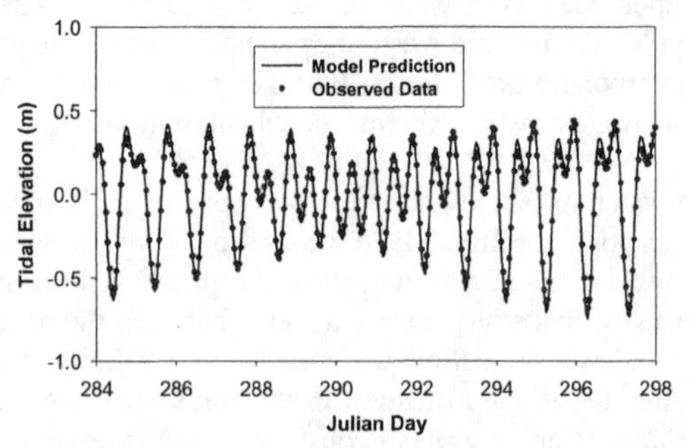

Figure 3: Comparison of Predicted and Observed Tidal Elevation at Port Angeles Harbor

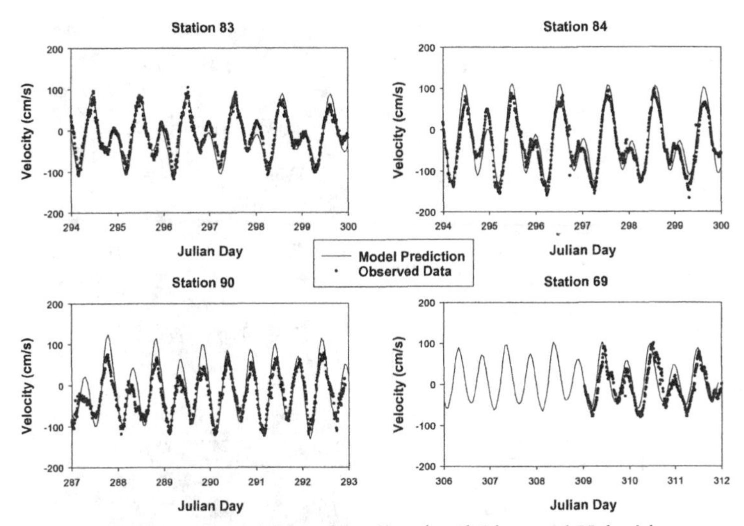

Figure 4: Comparisons of Predicted and Observed Velocities in Strait of Juan de Fuca

The hydrodynamic model was also validated against observed drogue data and physical model data in previous studies. Drogue trajectories were simulated in the hydrodynamic model and compared to the observed drogue study data near the Port Angeles Harbor. The Drogue study data were obtained from the Outfall Location Study (ITT Rayonier, 1971). Numerical drogues were released at the same location and time corresponding to the observed data. Figure 5 shows the comparisons of drogue trajectories between model results and field observed data for different drogue release time in a tidal cycle (high tide, mid-ebb, low tide and mid-flood). Good agreement between model-simulated and field-observed drogue trajectories was obtained. One particularly interesting feature is that the drogues released near the mouth of Port Angeles Harbor and close to the shore moved westward into Port Angeles Harbor, even at flood tide (Figure 5d). This is because of the effect of a strong tidal eddy generated at the north end of the mouth of Port Angeles Harbor due to the geometry effect of Ediz Hook.

Model-predicted flow streamlines were also compared to results of a physical model study that was also conducted as part of the study on the dispersion of pulp mill effluent in Port Angeles Harbor (Ebbesmeyer, 1979). Figure 6 shows that the model-predicted flow streamlines match the distribution pattern from the physical model result. The fact that model correctly simulates drogue motion and particle trajectories, is a further confirmation that the model calibration was satisfactory.

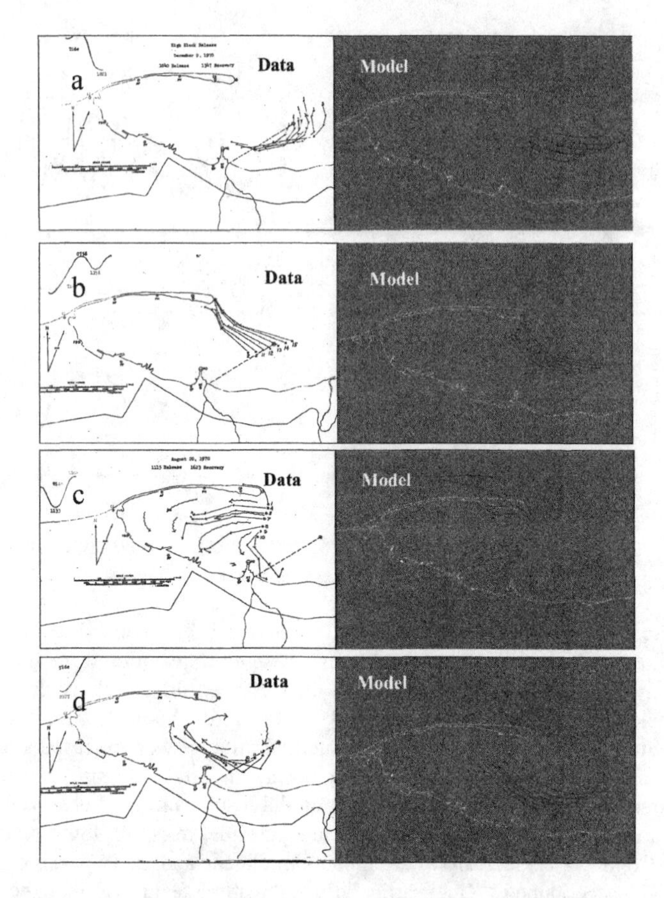

Figure 5: Comparisons of Predicted and Observed Drogue Trajectories in Strait of Juan de Fuca: (a) High Slack; (b) Ebb; (c) Low Slack; (d) Flood

To show the general velocity distribution patterns in the Port Angeles Harbor area, instantaneous velocity vector distributions at flood and ebb are given in Figure 7. Velocities in the harbor and the former Rayonier Pulp Mill site are relatively small. Tidal phases in the Port Angeles Harbor area also shifted and due to the influence of Ediz Hook. A strong clockwise tidal eddy is present southeast of the tip of Ediz Hook and the duration of the eddy typically lasts 6 hours during flood period. This eddy was observed in both the drogue data (ITT Rayonier, 1971) and the physical model (Ebbesmeyer et al., 1979). The maximum velocity magnitude distribution over a one-day simulation is presented in Figure 8. It shows clearly that tidal currents within the Port Angeles Harbor are very small.

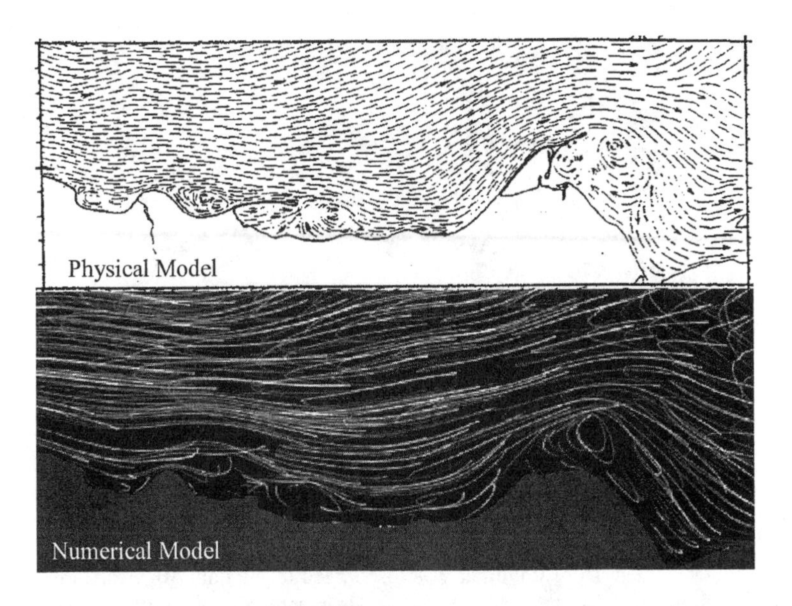

Figure 6: Comparisons of Physical Model and Numerical Model Streamlines in Strait of Juan de Fuca

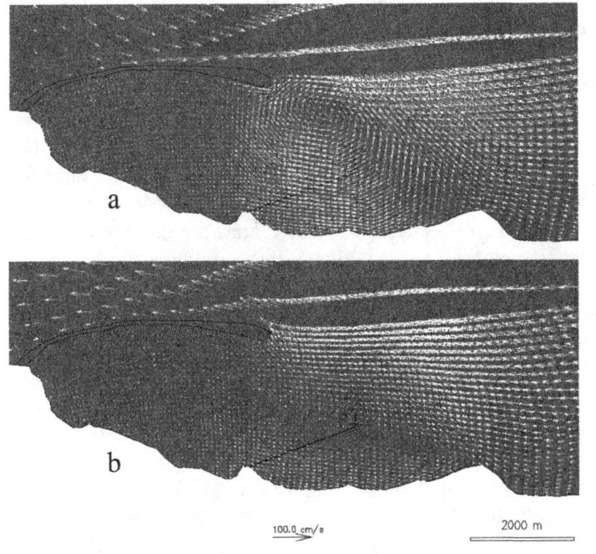

Figure 7: Predicted Velocity Distribution in Port Angeles Harbor: (a) Flood; (b) Ebb

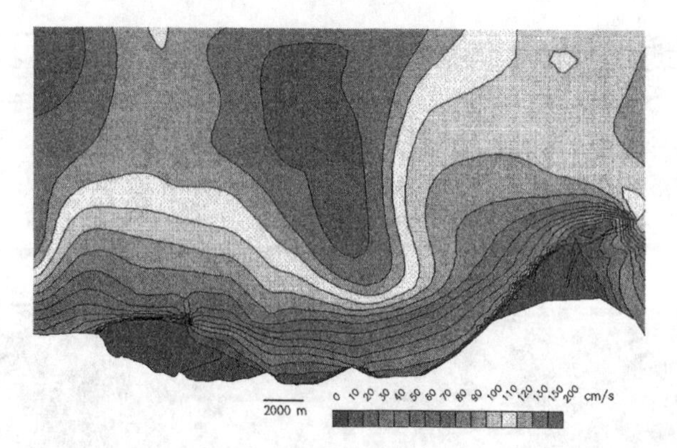

Figure 8: Predicted Maximum Velocity Distribution in Port Angeles Harbor and Strait of Juan de Fuca

Effluent near-field Dilution and Far-field Transport

Raw effluent from Rayonier was discharged to the Port Angeles Harbor from five nearshore outfalls until 1972, when a deepwater outfall with a diffuser was installed and used to discharge consolidated treated effluent. The effluent contained varying levels of TSS depending on the nature of treatment during these periods. One major objective of this study is to characterize the sediment deposition rates and sedimentation patterns associated with these discharges. The deposition rates are directly related to effluent flow and solids loads, the dispersion/dilution of the effluent, and its dynamic transport, and sediment settling characteristics.

When the effluent enters the receiving water, its dilution, transport, and dispersion occurs in two distinct phases, an initial dilution phase followed by a far-field dispersion phase. During the initial dilution phase, the effluent is diluted to a large degree. As the effluent moves away from the ports, the jet momentum and the velocity reduce, and the waste field stabilizes, marking the end of the initial dilution phase. In the presence of tidal currents, the waste field is swept in the upstream and downstream directions over a tidal cycle. In the second phase of plume mixing, the transport and dispersion occurs through advection and diffusion processes only, and is referred to as far-field dilution. The process of initial dilution is dominant for deepwater submerged outfalls such as Rayonier's 980-feet deepwater diffuser. In contrast, the effluent discharged at the surface through open-ended outfalls does not benefit from mixing and dilution of the buoyancy or jet momentum. The initial dilution component of the overall mixing is small for these surface discharges. The strategy adopted, therefore, was to assume that the initial dilution provided by the nearshore outfalls at Rayonier was small and that dispersion of the effluent through far-field transport was the only mechanism of dilution for the nearshore outfalls. Initial dilution modeling was therefore conducted only for the deepwater outfall.

Effluent Initial Dilution and Transport Synchronization

The initial dilution model used in this study is UDKHDEN, which belongs to the family of EPA-approved plume models. In this study, three different types of discharges were simulated in this study, based on different locations of outfalls and the different types of effluents discharged:

- Deepwater outfall – secondary effluent (1979 to 1997)
- Deepwater outfall – primary effluent (1972 to 1978)
- Nearshore outfall – raw effluent (1956 to 1972)

An average flow rate corresponding to each outfall for the specified period was selected as representative of the operating conditions (Figure 9). Water density and flow rate are the effluent parameters with the strongest influence on the initial dilution. Effluent salinity was assumed to be zero representing freshwater. An effluent temperature 28.2° C was specified to represent critical summer conditions. The tidal elevation and current profiles were specified at each time step (15-minute intervals) to generate a dilution ratio time history for a 14-day period. The currents and tidal elevation were provided by the hydrodynamic model EFDC.

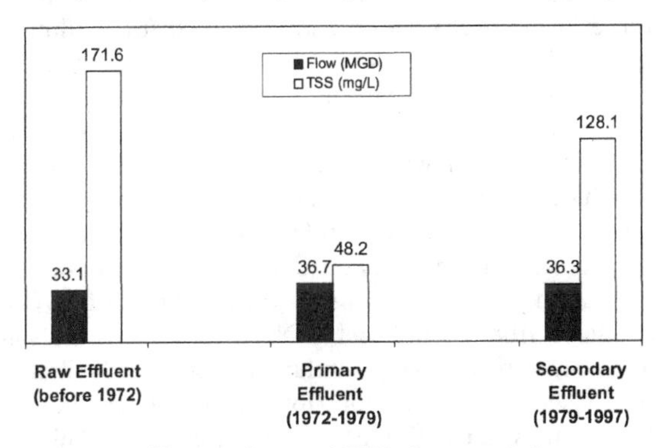

Figure 9: Effluent Flow and TSS characteristics

Figure 10 shows the predicted initial dilution as a function of ambient currents. The results show that with increasing currents the initial dilution increases linearly until approximately 10 cm/s and then the rate of increase drops off due to factors such as merging of plumes, and reduced trapping depth. Initial dilution can also be affected by variation in water depth and current velocity over a neap-spring cycle. Note that the dilution will increase further as ambient currents carry the plume past the acute and chronic mixing zone boundaries 35.5 feet and 355 feet away from the diffuser ports, respectively.

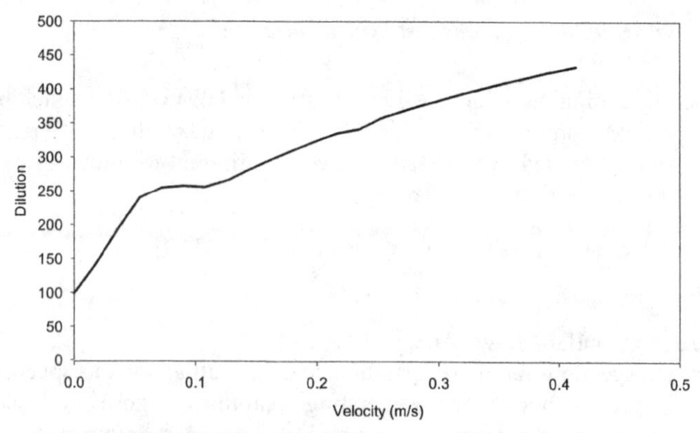

Figure 10: Predicted Initial Dilution as a Function of Velocity

To simulate the far-field effluent transport correctly, effluent concentration at the grid cell where the outfall is located must be specified properly. In EFDC, effluent concentration at the outfall cell is calculated by the mixing of the discharged effluent concentration with the water in the cell. The grid cell size and the vertical distribution of effluent discharge were adjusted so that the dilution in the outfall cell matched the results obtained from the initial dilution model (UDKHDEN). Simulation for the secondary deepwater outfall was conducted and compared to the initial dilution model results. The averaged effluent flow rate (36.3 MGD) and concentration (128.1 mg/L) were specified and assumed constant throughout the simulation. Effluent tracer was assumed to be neutrally buoyant, and settling velocity was set to zero. A comparison of a 14-day time history (spring and neap tidal cycle) of dilutions calculated from the EFDC model at the outfall cell and from the UDKHDEN initial dilution model is presented in Figure 11. Figure 11 shows that dilutions calculated from EFDC and UDKHDEN are in the same range, and the trends of the time histories are very similar. The predicted initial dilution results over the same period show variation from low values of approximately 1:100 to high values of 1:400. It is noted that EFDC produces slightly higher peaks and lower troughs in the dilution time series. This is due to in EFDC the initial dilution occurs over a fixed volume, whereas in UDKHDEN the initial dilution volume varies as a function of ambient velocity.

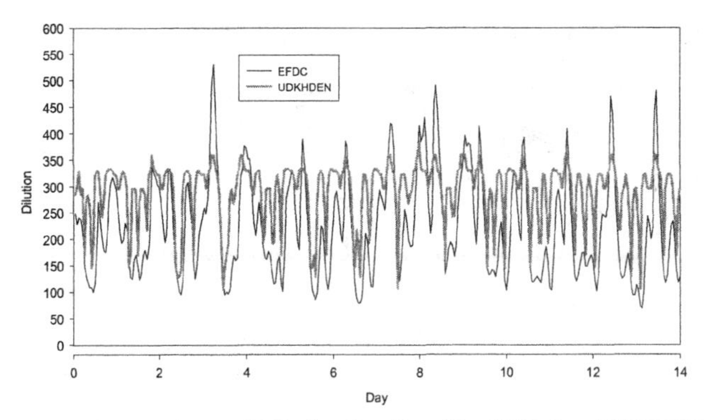

Figure 11: Comparison of Dilution Predicted by EFDC and UDKHDEN

Far-field Transport

Similar to the hydrodynamic model, an effluent transport model was set up to simulate the effluent TSS plume transport for the Rayonier Pulp Mill outfalls. Following the completion of the initial dilution calculation, the far-field effluent dispersion simulations were performed for the effluent discharged from different outfalls. The effluent transport model used in this study is the transport module within EFDC. Effluent concentrations and flow rates, and initial and boundary conditions are required as inputs for the model to conduct the effluent plume transport and deposition simulations. On the land boundaries, zero-flux boundary conditions were specified. On the open boundaries, zero inflow concentrations of TSS were assumed because the open boundaries are far enough away from the effluent outfalls.

To simulate TSS effluent transport and deposition, a solids particulate settling velocity is required as an input parameter to the model. However, effluent settling characteristics data were not available for the former Rayonier Port Angeles Pulp Mill site. As part of a special study focused on effluent solids deposition, settling velocity distribution of effluent solids was measured by a similar pulp mill: the Ketchikan Pulp Company (KPC) (ENSR, 1996). Particle settling velocities and particle size distributions were estimated from sequences of video images of particles settling through quiescent salt water. This relationship between particle size and settling velocity is summarized in Table 1. Because only one single class of effluent particle was simulated in this study, a mean settling velocity was calculated based on the KPC data in Table 1. The weighted averaged settling velocity of 4.5×10^{-5} m/s was calculated from the sum of all settling velocities weighted by the percentage of mass for each class. It is comparable to the settling velocity of silt or clay particles.

Table 1. Effluent Solids Settling Characteristics from Ketchikan Pulp Mill

Settling Velocity (cm/s)	Nominal Diameter (mm)						Total	
	0.0316	0.0502	0.0796	0.1262	0.2	0.1585		
	N%	N%	N%	N%	N%	N%	N%	M(N%)
0.0003	0.5	0.5	0.0	0.0	0.0	0.0	1	0.1
0.0004	1.2	0.9	0.0	0.0	0.0	0.0	2.1	0.1
0.0006	2.3	1.9	0.0	0.0	0.0	0.0	4.2	0.3
0.0010	2.8	2.6	0.2	0.0	0.0	0.0	5.6	0.5
0.0016	3.5	5.8	4.0	1.4	0.5	0.2	15.4	9.1
0.0025	4.7	9.8	9.6	5.4	1.6	0.2	31.3	26.7
0.0040	3.3	6.8	7.2	5.8	2.3	0.0	25.5	30.5
0.0063	0.9	1.6	2.1	3.3	2.1	0.2	10.2	22.2
0.0100	0.2	0.5	0.9	1.4	0.9	0.2	4.3	10.5
0.0158	0.0	0.2	0.2	0.0	0.0	0.0	0.4	0.1
Total	**19.4**	**30.6**	**24.3**	**17.3**	**7.5**	**0.9**	**100.0**	**100.0**

N% = percentage number of particles
M(N%) = percentage of mass of particles

Before the far-field transport model is applied to simulate the TSS effluent transport, it must be calibrated, using observed data or a well-established analytical solution. Since no observed TSS or dye concentrations data near the study area were available, the far-field transport model was calibrated to the Brook's Law dilution model for this study. The dye dispersion was simulated using an effluent tracer at the secondary deepwater outfall. Dilution ratios downstream along the centerline of the plume were calculated. Figure 12 shows the comparison of dilution ratios as a function of distance calculated from EFDC and Brook's Law. The figure shows that the far-field dilution ratios from EFDC are comparable to Brook's Law dilution ratios. Brook's Law conservatively assumes that effluent is fully mixed vertically and dilution is through lateral spreading only. EFDC considers vertical spreading plus lateral dilution and, therefore, the ranges of EFDC dilution values are expected to be higher than Brook's Law.

For comparison, a study (Carpenter et al., 1986) on sediment accumulation rates in the Puget Sound region was examined. The calculated sediment accumulation rates vary from 460 to 12000 g/m^2/year, and the accumulation rates at two stations closest to the study area are 460 and 760 g/m^2/year.

Once the effluent transport model was set up and calibrated, it was used to simulate the TSS transport and deposition for effluents discharged from Rayonier's outfalls.

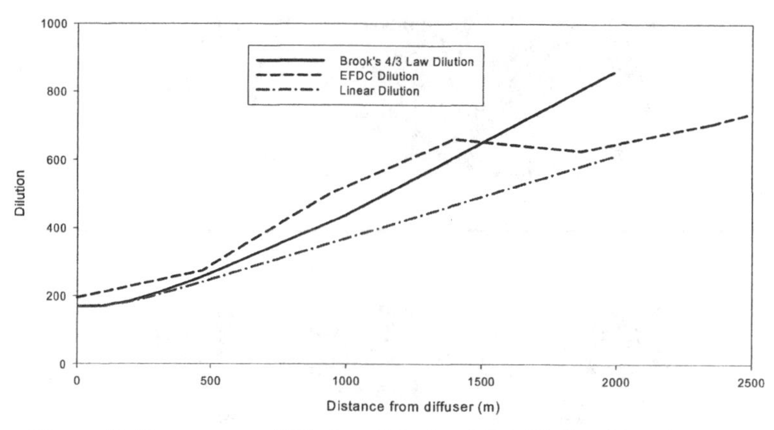

Figure 12: Comparison of Dilution Calculated by EFDC and Brook's Law

Deepwater Outfall – Secondary Effluent

For the secondary effluent discharge from the deepwater outfall, the flow rate and TSS concentration used were 36.3 MGD and 128.1 mg/L, respectively. One of the main objectives in this study was to investigate deposition rates due to the effluent discharges from the Rayonier outfalls. There are two important parameters in the sediment transport model that control the deposition and erosion processes. The first is the critical shear stress for deposition, which allows deposition to occur when the bottom shear stress is less than the critical shear stress. The other parameter is the critical shear stress for erosion, which allows erosion to occur when the bottom shear stress is greater than the critical shear stress. The critical shear stress for sediment deposition was set to a large value to allow the deposition process to occur continuously, while the critical shear stress for erosion was set to a large value so that the erosion process could never occur. The effluent transport model was run for 14 days to cover a spring and neap tidal cycle and to ensure that the effluent plume was developed fully. The surface TSS concentrations at flood and ebb tides are plotted in Figure 13. The deposition rate of TSS effluent was also calculated in the model. A contour plot of the TSS deposition is given in Figure 14. It shows that the maximum deposition rate is about 180 g/m^2/year with the secondary deepwater outfall. The deposition zone is in a narrow band shape and aligned to the tidal excursion direction.

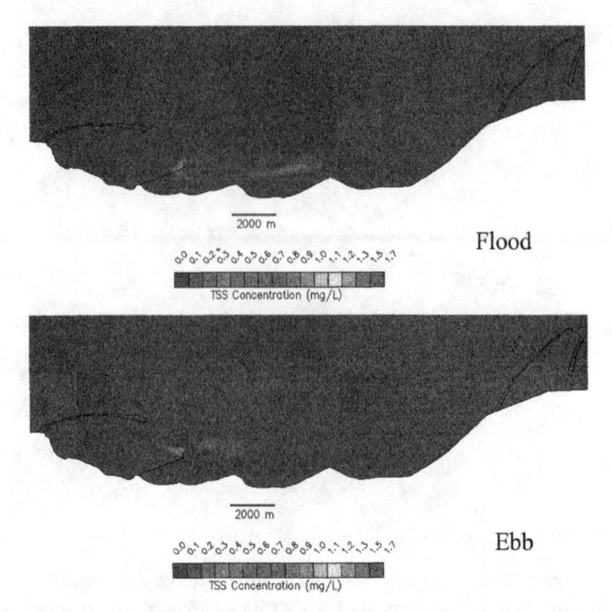

Figure 13: Model Predicted Surface TSS Concentrations for Secondary Effluent

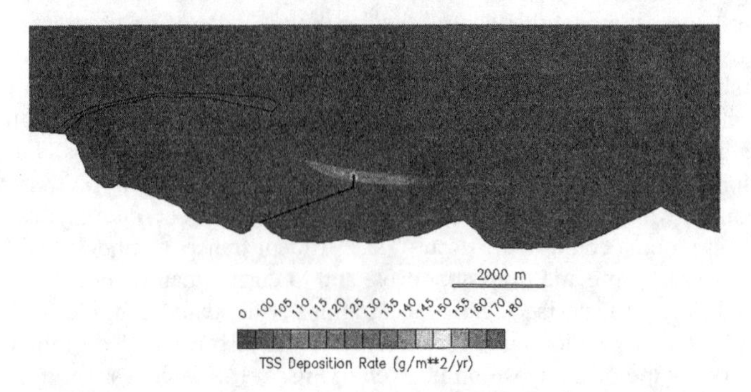

Figure 14: Model Predicted TSS Deposition Rate for Secondary Effluent

Deepwater Outfall – Primary Effluent

The primary deepwater outfall was simulated using the average effluent flow rate of 36.7 MGD and averaged TSS concentration of 48.2 mg/L. It is noted that the discharged TSS concentration in the primary outfall is considerably lower than that in the secondary outfall. Because the discharged TSS concentration is lower, the size of the TSS effluent plume is also smaller in comparison to that in the secondary outfall. Figure 15 shows the TSS deposition rate for primary deepwater outfall. The distribution pattern is very similar to the secondary deepwater outfall except the deposition rate is lower. The maximum deposition rate is about 65 g/m^2/year and much smaller than the background sediment deposition rate.

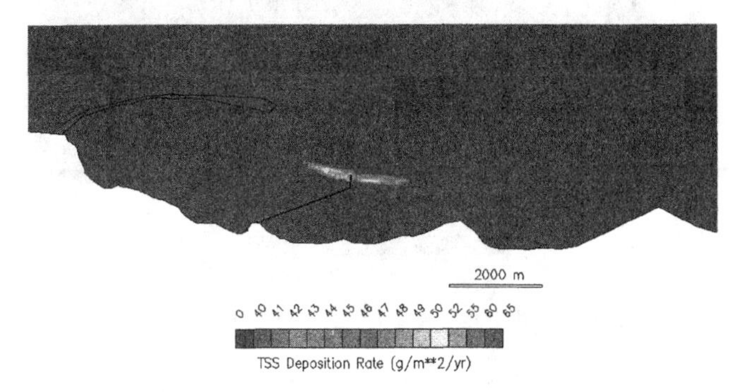

Figure 15: Model Predicted TSS Deposition Rate for Primary Effluent

Nearshore Outfalls – Raw Effluent

To simulate the nearshore TSS effluent transport, flow rates and TSS concentrations at the five original nearshore outfalls were specified. The flow rates for nearshore outfall A, B, C, D, and E are: 1.0, 8.7, 20.4, 2.9, and 2.6 MGD, respectively. The TSS concentration for outfall A, B, C, D, and E are 1,096.5, 353.2, 43.0, 158.0, and 83.9 mg/L, respectively. The main effluent discharge is from outfall A, B, and C because of their high flow rates or concentrations. Even though the model grid resolution is fine enough near the nearshore outfalls to specify the five effluent outfalls in different grid cells, they are so close to each other that effluents discharged from all the outfalls are seen to quickly merge together into a single plume. The plume mainly distributes along the coastline east of the outfalls. The predicted TSS deposition rate is plotted in Figure 16. The maximum deposition rate is about 17,300 g/m^2/year, which is considerably higher than those predicted in the secondary and primary deepwater outfalls. This is mainly because the velocities in the ambient water near the nearshore outfalls are much smaller than those in the deepwater outfalls, nearshore outfalls do not receive the benefit of initial dilution, and the total combined discharged TSS concentration, is also higher (Figure 9). The lower bound cutoff value was set to 500 g/m^2/year in the contour plot, which is within the range of background values (460 to 760 g/m^2/year). Figure 16 indicates that the deposition zone around the nearshore outfalls is limited within an area about 4,500 m in the longshore direction and 500 m in the offshore direction.

Summary and Conclusion

A numerical modeling tool was developed for the former Rayonier Pulp Mill site at Port Angeles, Washington, to predict the tidal circulation, the effluent plume transport, and sediment deposition. The modeling system consists of near-field initial dilution model (UDKHDEN) and a 3-D hydrodynamic and far-field transport model

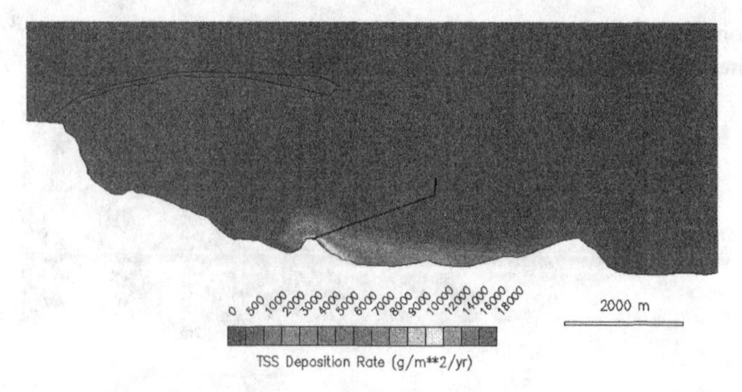

Figure 16: Model Predicted TSS Deposition Rate for Raw Effluent

(EFDC). The transport of the TSS plume was simulated with two steps: a) the initial dilution using the near-field dilution model UDKHDEN, and b) the far-field transport using EFDC. TSS effluent concentrations and deposition rates were calculated for the nearshore and deepwater outfalls, corresponding to different effluent discharge periods and characteristics. The major conclusions of this modeling study are as follows:

- The 3-D hydrodynamic model was calibrated using the existing historical current data and validated by the drogue study and physical model data. The model predictions match reasonably well with the observed data.
- The far-field effluent transport model was successfully calibrated to the Brook's Law dilution model results. The model was applied to simulate the effluent plumes transport and TSS deposition for the three different periods of Rayonier's operational history, consisting of raw, primary, and secondary effluent discharges.
- Model results indicated that the effluent plume moved up and down along the coast with tides. The effluent plume was diluted over a 1,000-fold, well before the plume exited Port Angeles Harbor.
- The predicted TSS deposition rates were generally very small, compared to the background sediment accumulation rate of 460 to 760 g/m^2/year. The predicted TSS deposition rate in the deepwater outfall was negligible when the settling velocity 4.5×10^{-5} m/s was used. The deposition area in the nearshore outfall is within an area of 4,500 m by 500 m. The deposition zone in the deepwater outfall is smaller than that in the nearshore outfall and in a narrow band shape oriented to the tidal excursion direction

Overall, the hydrodynamic and effluent transport model developed for the former Rayonier Port Angeles site, based on available data, provided very promising preliminary results. The model can be further improved when more data for model boundary condition specifications and TSS settling characteristics are obtained.

Nonetheless, these model results provide useful information that can be used to design the sediment sampling plan near the Rayonier effluent outfalls.

References

Carpenter, R, M. L. Peterson and J. T. Bennett, 1986. ^{210}Pb-Derived Sediment Accumulation and Mixing Rates for the Greater Puget Sound Region. Marine Geology, 64, 291-312.

Ebbesmeyer, C. C., J. M. Cox, J. M. Helseth, L. R. Hinchey, and D. W. Thomson, 1979. Dispersion of pulp Mill Effluent in Port Angeles Harbor and Vicinity. In: History and Effect of Pulp Mill Effluent Discharges, Port Angeles, Washington, G. B. Shea, ed., Draft Report. Northwest Environmental Consultants, Inc., Seattle, Washington.

ENSR, 1996. Study of Solids Deposition in Ketchikan Pulp Company, Ketchikan, Alaska. ENSR Consulting and Engineering, Redmond, Washington.

EPA, 1998. *Rayonier Pulp Mill Expanded Site Inspection*. TDD: 97-06-0010. Prepared for EPA Region 10, Superfund Technical Assessment and Response Team, Seattle, Washington. by Ecology and Environment. October 1998.

Foster Wheeler Environmental, April 2001. Volume III: Sampling and Analysis Plan – Marine Environment for Conducting a Remedial Investigation at Former Rayonier Pulp Mill Site, Port Angeles, Washington. Prepared for Rayonier Inc., Jacksonville, FL.

Hamrick, J. M., 1992. A Three-Dimensional Environmental Fluid Dynamics Computer Code: Theoretical and Computational Aspects. Virginia Institute of Marine Science,VA. Special Report 317. 63pp.

ITT Rayonier Inc., 1971. Outfall Location Studies, Port Angeles, Washington.

Labrecque, A. J. M., R. E. Thomson, M. W. Stacey, and J. R. Buckley, 1994. Residual Currents in Juan de Fuca Strait. *Atmosphere-Ocean*, 32 (2), 375-394.

Mellor, G. L. and T. Yamada, 1982. Development of a turbulence closure model for geophysical fluid problems. *Rev. Geophys. Space Phys*. 20: 851-875.

Parker, B. B. and J. T. Bruce, 1980. Puget Sound Approaches Circulation Survey, NOAA, NOS oceanographic Circulatory Survey Report No 3. Rockville, MD.

Shen, J. and A.Y. Kuo, 1999. Numerical investigation of an estuarine front and its associated topographic eddy. Journal of Waterway, Port, Coastal, and Ocean Engineering, 125, pp. 127-135.

Ji, Z.-G., M. R. Morton, and J. M. Hamrick, 2001. Wetting and drying simulation of estuarine processes, Estuarine Coastal Shelf Sci., 53, 683-700.

Yang, Z., and J.M. Hamrick, 2003. Variational Inverse Parameter Estimation in a Cohesive Sediment Transport Model: an Adjoint Approach. *J. Geophys. Res*, 108(C2), 3055, doi:10.1029/2002JC001423, 2003

Development of a hierarchy of nested models to study the California Current System.

I. Shulman[1], J. C. Kindle[1], S. deRada[2], S. C. Anderson[2], B. Penta[1] and P. J. Martin[1]
[1]Naval Research Laboratory, Stennis Space Center, MS 39529
[2]JE/Sverdrup, Stennis Space Center, MS 39529

Abstract

The Naval Research Laboratory has developed a hierarchy of differing resolution data assimilating models in the Pacific Ocean, which includes global models, regional U.S. West Coast models, and high resolution coastal models such as for the Monterey Bay area. The three regional U. S. West Coast models (from 30°N to 49°N), designed to study the California Current System, are based on the Princeton Ocean Model (POM), the Navy Coastal Ocean Model (NCOM), called NCOM-CCS, and the Hybrid Coordinate Ocean Model (HYCOM), respectively. The NCOM-CCS formulation is a parallel version model capable of running reliably on many computer platforms. The model has nesting capabilities and offers the choice of using the sigma or a hybrid (sigma-z) vertical coordinate. The NCOM-CCS model also includes a coupled ecosystem model based on Chai et al., 2002.

A variety of scientific issues related to model initialization, forcing, open boundary conditions, and model spin up are discussed. The focus of this paper is on: the sensitivity of the horizontal resolution of atmospheric forcing on the NCOM-CCS model predictive skills; the impact of open boundary conditions and coupling with global models on reproducing major hydrographic conditions in the California Current System; and the analysis of the model mixed layer predictions and data assimilation issues. Qualitative and quantitative comparisons are made between observations and model predictions for October 2000 - December of 2001 period.

1. Introduction

To accommodate the wide range of horizontal scales of physical and biological oceanic processes, the Naval Research Laboratory (NRL) uses a nested modeling approach in which global models provide boundary data to regional scale models which, in turn, feed coastal models and so on (Figure1). Three large-scale models are shown in Figure 1: the global Navy Layered Ocean Model (NLOM, Wallcraft et al, 2002), the global model based on Navy Coastal Ocean Model (NCOM, Rhodes et al, 2002), and Pacific Ocean model based on Hybrid Coordinate Ocean Model (HYCOM, Bleck, 2002). The global NLOM is an isopycnal model with seven vertical levels (six active dynamical levels and mixed layer model) and horizontal grid resolution of 1/16° to better resolve fronts and eddies (Hurlburt et al, 1996). The NLOM assimilates satellite SSH and SST and is run as an operational model for the US Navy, some results of which may be viewed online (www7320.nrlssc.navy.mil/global_nlom). The NCOM based global model uses a hybrid vertical coordinate system (sigma levels up to 150m depth, z levels at the bottom) with 1/8° horizontal resolution and is run in real-time (www7320.nrlssc.navy.mil/global_ncom); the model is presently in the final stages of validation and evaluation for operational use (Rhodes et al, 2002). Global NCOM has been spun up from a climatological state to the present, using a combination of NOGAPS (Navy Operational Global Atmospheric Prediction System) forcing and the assimilation of 3-dimensional temperature and salinity observations derived from the Modular Ocean Data Assimilation System (MODAS; Fox et. al, 2002). The Pacific Ocean HYCOM (Hybrid Coordinate Ocean Model) has 1/12° horizontal resolution and 20 vertical levels; this research model is a proxy for the future development of a real-time global HYCOM with 1/12° resolution in 2006. HYCOM uses the hybrid vertical isopycnal-sigma-z level coordinate system (Bleck, 2002; Chassignet, 2003; oceanmodeling.rsmas.miami.edu/hycom).

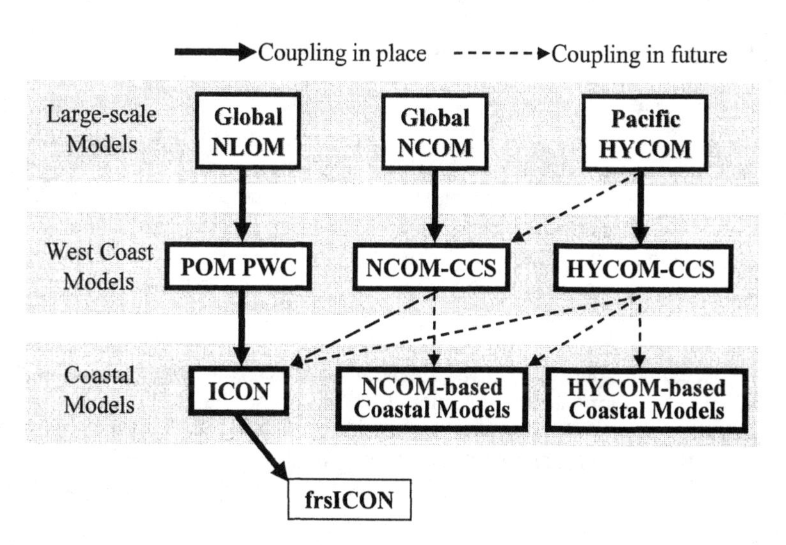

Figure 1. Diagram of the hierarchy of NRL models.

The second row in Fig. 1 presents three basin scale models of the US Pacific West Coast: POM-PWC (Haidvogel et al, 2000) is the model based on Princeton Ocean Model (Blumberg and Mellor, 1987), NCOM-CCS (California Current System) is the model based on the NCOM model (see below for more information), and the third is the HYCOM-based model (HYCOM-CCS). Both POM-PWC and NCOM-CCS have the same horizontal 9km resolution while HYCOM-CCS uses a grid resolution of ~7-8 km; all three models have the same domain which extends from 30°N to 49°N of latitude and from the coast to 135°W of longitude. POM-PWC and NCOM-CCS have 30 sigma vertical levels, and HYCOM-CCS has 20 vertical levels.

Figure 1 also presents two fine resolution models of the Monterey Bay. One of the models was initially developed under National Oceanic Partnership Program (NOPP) "Innovative Coastal-Ocean Observing Network" (ICON) project (Shulman et al, 2002). This model is based on the Princeton Ocean Model, and has a variable horizontal resolution ranging from 1-4 km and 30 vertical sigma levels. Another fine resolution Monterey Bay area model (frsICON) is nested inside of the ICON model, and has variable horizontal resolution ranging from 0.5 km to 1.5 km (this model was used for fine-resolution modeling of bioluminescence distributions in the Monterey Bay (Shulman et al., 2003)).

One of the objectives for the development of this hierarchy of different resolution models is the testing and evaluation of POM, NCOM and HYCOM based models in basin-scale and coastal environments. Another objective is determining the best combination of these nested models for the development of coupled bio-optical and physical models aimed at studying the California Current System. Additionally, the use of nested models permits one to use atmospheric forcing products that range from global to regional/local in scale. The impact of higher resolution atmospheric products is an important open question in ocean modeling. There are many different issues that arise from the development of these nested models (Fig. 1) such as those associated with coupling models with different physics, vertical coordinate systems and numerics, the sensitivity to the model's initialization and data assimilation schemes, and the sensitivity to the resolution of the atmospheric forcing. The goals of this paper are to briefly describe the nested system at NRL for the US west coast and to relate experiences in examining the impacts of our choices in the development of such nested models.

Given space limitations, the primary focus of the paper is on modeling issues using the NCOM-CCS formulation. In section 2, additional details of NCOM-CCS model are provided. Section 3 illustrates the sensitivity of the model predictions to the different resolution of atmospheric forcing. Coupling between NCOM-CCS and global NCOM are discussed in the Section 4, and some results from the California Current Ecosystem modeling are presented in Section 5. Finally, Section 6 discusses conclusions and future plans.

2. NCOM-CCS

The NCOM-CCS has a free surface and is based on the primitive equations and the hydrostatic, Boussinesq, and incompressible approximations (Martin, 2000; Rochford and Martin, 2001). The model uses an Arakawa C grid and is leapfrog in time with an Asselin filter to suppress time splitting. The propagation of surface waves and vertical diffusion are treated implicitly. A choice of the Mellor-Yamada Level 2 or Level 2.5 turbulence models is provided for the parameterization of vertical mixing. The horizontal resolution of the model is around 9km, and the model has 30 vertical sigma levels. The model assimilates ~14 km daily Multi-Channel Sea Surface Temperature (MCSST). The model is one-way coupled to the 1/8° global NCOM (details of coupling scheme are presented in section 4).

The NCOM-CCS also includes a 9-component ecosystem model; the biological model was implemented into the NCOM in collaboration with Dr. Fei Chai (Chai et al, 2002). The biological model, a 9-component ecosystem formulation originally developed for the equatorial Pacific upwelling system, includes three nutrients (silicate, nitrate and ammonia), two phytoplankton groups, two zooplankton grazers, and two detritus pools.

3. Sensitivity of NCOM-CCS model predictions to the resolution of atmospheric forcing.

A variety of atmospheric products were used to assess the sensitivity of the NCOM-CCS model to the resolution of the wind forcing. Wind products derived from the European Center for Medium Range Forecasting (ECMWF; 1.125° resolution), NOGAPS (1° resolution atmospheric model) and the Navy's Coupled Ocean Atmosphere Mesoscale Prediction System (COAMPS™; Hodur, 2002) atmospheric models were used in sensitivity studies with NCOM-CCS. For the COAMPS™ product, winds from the COAMPS™ Reanalysis for the Eastern Pacific (EP) were used (Kindle et al, 2002); these atmospheric winds exist on a triply nested 81/27/9 km grid beginning in November of 1998. The COAMPS™ EP Reanalysis fields on the native Lambert Conformal grid were interpolated to the latitude-longitude coordinate ocean model grid with ~ 9km resolution. A weighted-average bilinear interpolator (WABI) with ocean/land-discrimination was developed and used to interpolate (project) atmospheric forcings from 9/21/81 grids of the COAMPS™ to the grid of the NCOM CCS.

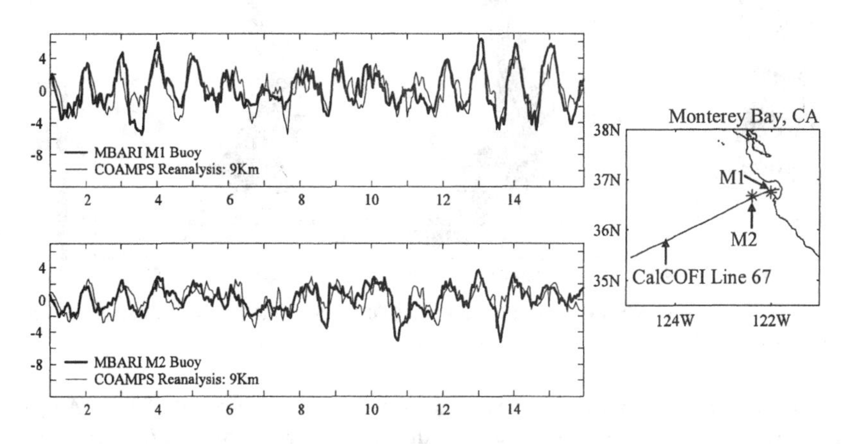

Figure 2. Observed and model-predicted wind velocities (projected on principal axis) at moorings M1 and M2.

Figure 2 demonstrates the quality of winds from COAMPS™ 9km predictions. Filtered observed and model predicted wind velocities (projected on principal axis) are shown at two mooring locations (M1 and M2) in the Monterey Bay area. COAMPS™ model reproduced nicely observed diurnal variability, and observed shoreward intensification of the diurnal amplitude (see also, Kindle et al., 2002).

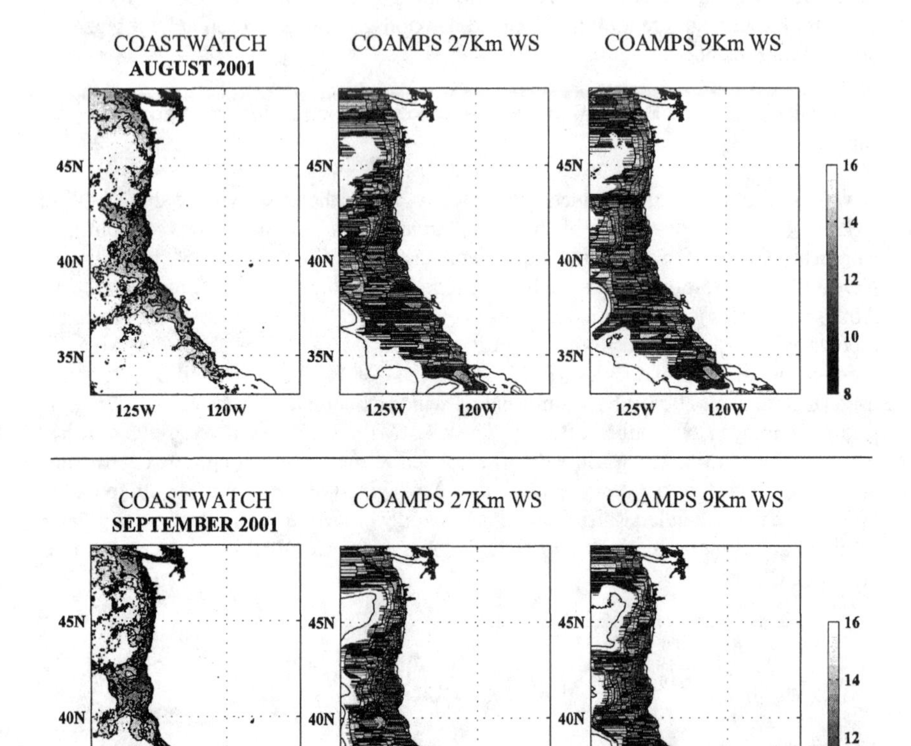

Figure 3. Comparisons between monthly mean NCOMCCS model SST and CoastWatch SST monthly composites.

Figure 3 shows comparisons of mean model-predicted and observed SST for August and September of 2001. One can see that model forced with 27km and 9km COAMPS™ winds nicely reproduce the meandering front between the warmer water of California Current and the upwelled water along the coast. At the same time, this front and observed features have a sharper representation in model simulations with finer 9km resolution wind. Quantitatively, the sensitivity of the NCOM-CCS model predictions to different resolution wind forcing was studied with a comparison of model predicted and observed velocities at mooring M1 (location of which is shown on Fig. 2). It should be noted that it is a very challenging task for any resolution physical model without data assimilation to show a good predictive skill of the velocity field along the central California coast, particularly near the Monterey Bay region. Mooring M1 is located in the transition zone between the California Current and upwelled water, and the front between these water masses--sometimes in the shape of the anticyclonic eddy-- migrates onshore-offshore depending on relaxation or upwelling favorable wind conditions.

The magnitudes of complex correlation coefficients and angular displacements between the observed and model-predicted currents were used for comparisons. The magnitude ρ of the complex correlation coefficient between the ADCP and NCOM-CCS model currents and the angular displacement θ (phase angle, average veering) for a particular depth were estimated by using the approach outlined in Kundu (1976). The magnitude of ρ is estimated using the following formula: $\rho = \sqrt{Re^2 + Im^2}$, where

$$Re = \frac{\sum_t \left(u_t^o u_t^m + v_t^o v_t^m\right)}{\sqrt{\sum_t \left(\left(u_t^o\right)^2 + \left(v_t^o\right)^2\right) \sum_t \left(\left(u_t^m\right)^2 + \left(v_t^m\right)^2\right)}}, \text{ and}$$

$$Im = \frac{\sum_t \left(u_t^o v_t^m + v_t^o u_t^m\right)}{\sqrt{\sum_t \left(\left(u_t^o\right)^2 + \left(v_t^o\right)^2\right) \sum_t \left(\left(u_t^m\right)^2 + \left(v_t^m\right)^2\right)}}.$$

The corresponding angular displacement θ (phase angle, average veering) for particular depth is estimated according to:

$$\theta = \tan^{-1} \frac{\sum_t \left(u_t^o v_t^m - v_t^o u_t^m\right)}{\sum_t \left(u_t^o u_t^m + v_t^o v_t^m\right)},$$

where, u_t^m, v_t^m and u_t^o, v_t^o are demeaned model and observed east-west and north-south components of velocity. The angular displacement θ gives the average counterclockwise angle of the NCOM-CCS currents with respect to the ADCP currents. The value of θ is only meaningful if ρ is significant. Standard statistical techniques (see, for example, Emery and Thomson, 1998) were used for estimating 95% significance levels of the correlation. For the period of 15 June – 15 September of 2001, the significant level is 0.144. Figure 4 shows the magnitude of complex correlation and angular displacements between observed and model predicted currents up to 120m depth. In Figure 4a, the comparisons are shown for model runs when direct (atmospheric model predicted) wind stresses were used. On the bottom (Figure 4b), wind stresses were estimated from 10m wind velocity by using the Large and Pond (1981) formulation. Comparisons are shown for three model runs forced with COAMPS™ (81km/27km/9km) resolution winds, respectively. The magnitude of the correlation between model-predicted and observed currents increases as wind resolution increases. Also, the average angle between model and observations is improved with increase in wind resolution. Use of Large and Pond derived stresses degraded the predictions with coarser resolution winds (81 and 27km). However, it improved model predictions at the surface when the model is forced with 9km wind. Overall, though, the model shows low correlation with observed currents. In previous research (Shulman et al, 2001, Paduan and Shulman, 2003) it was demonstrated that assimilation of HF radar-derived surface currents improves the coastal Monterey Bay (ICON) model currents predictions. Hence, it is expected that the assimilation of HF radar-derived surface currents from installations along the coast will improve the NCOM-CCS currents predictions, as well. HF radar-derived surface currents assimilation into the regional scale NCOM-CCS model is a topic for future research.

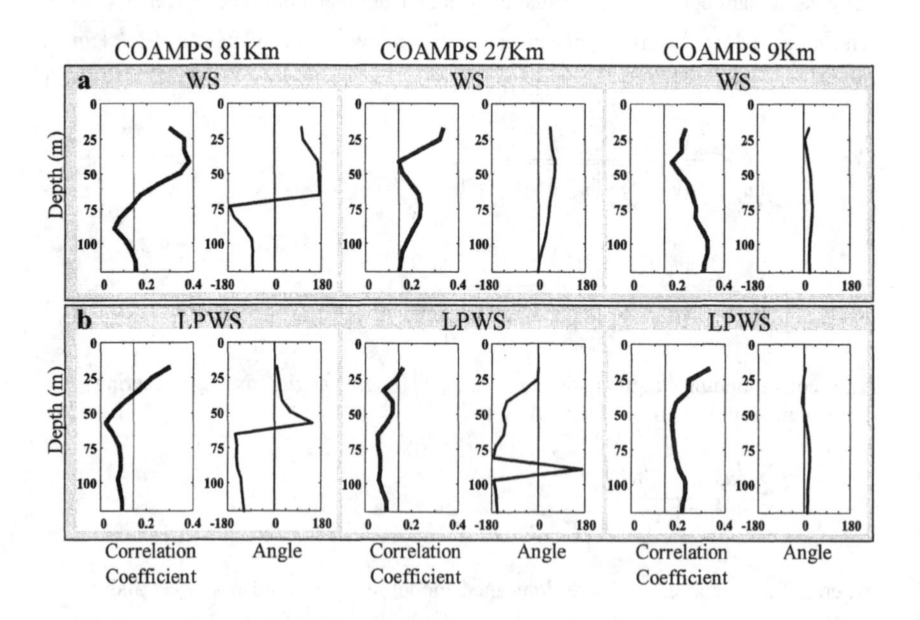

Figure 4. Magnitude of complex correlation and angular displacements between observed and model predicted currents for three model runs with different resolution wind forcing. a - Direct Stresses (WS), b - Large & Pond Stresses (LPWS)

4. Coupling NCOM-CCS with NCOM based global model.

Table 1 presents boundary conditions which are used for one way coupling between NCOM-CCS and global NCOM. Sea surface elevations and vertically averaged velocities (barotropic information) from the global NCOM and NCOM-CCS are coupled through the Flather (1976) boundary condition for the NCOM-CCS model (this open boundary condition represents a radiation condition on differences between NCOM-CCS and global model sea level elevations and transports). The baroclinic coupling consists of using the vertical structure of velocity, temperature, and salinity from the global model in specification of the vertical distributions of NCOM-CCS fields (see Table 1).

The coupling between NCOM-CCS and the global model was evaluated by comparing model-predicted and observed sea level elevations at tidal stations along the coast. Figure 5a shows comparisons at the San Diego station which is the closest station to the southern open boundary of the NCOM-CCS model domain. On Figure 5a, a 3-day moving average of observed and model-predicted sea level elevations are shown when barotropic, as well as baroclinic couplings were used according to Table 1. The model demonstrated a reasonably good skill in prediction of sea level elevation; the correlation is around 0.82, and skill score (estimated according to Murphy and Epstein, 1989) is around 0.62. Also, on Figure 5a, the comparisons are shown for the model run, when the radiation condition was used for normal baroclinic velocity instead of the advectional open boundary condition (Table 1).

Table 1. Coupling NCOMCCS with NCOM global.

Variables	Open Boundary condition
Elevation and normal barotropic velocity	Flather radiation condition. Radiation condition for differences between NCOMCCS and NCOM global elevations and normal barotropic velocities.
Normal baroclinic velocity	Advectional boundary condition Overflow case – velocity from first interior grid is advected to the NCOMCCS open boundary. Inflow case – NCOM global velocity is advected to the NCOMCCS open boundary
Tangential barotropic and baroclinic velocity	Zero Gradient condition NCOMCCS tangential velocity is set equal to the values at the first interior grid next to the open boundary.
Tracers (T, S and biology constituents)	Advectional boundary condition Outflow case – values from first interior grid are advected to the NCOMCCS open boundary Inflow case – NCOM global values (T and S) and derived from external source biology data are advected to the NCOM-CCS open boundary

In this case, no external baroclinic information about structure of the baroclinic velocity was used in specification of open boundary conditions for the baroclinic velocity of the NCOM-CCS. It is evident that the model predictive skill at San Diego station was significantly degraded without the use of baroclinic velocity from the global NCOM model. At the same time, the importance of baroclinic coupling diminishes as one moves from the southern to the northern boundary of the domain. In Figs 5b and 5c, comparisons are shown for stations at Crescent City (which is somewhat in the middle of the model domain) and for Neah Bay station, which is closest to the northern open boundary of the model domain.

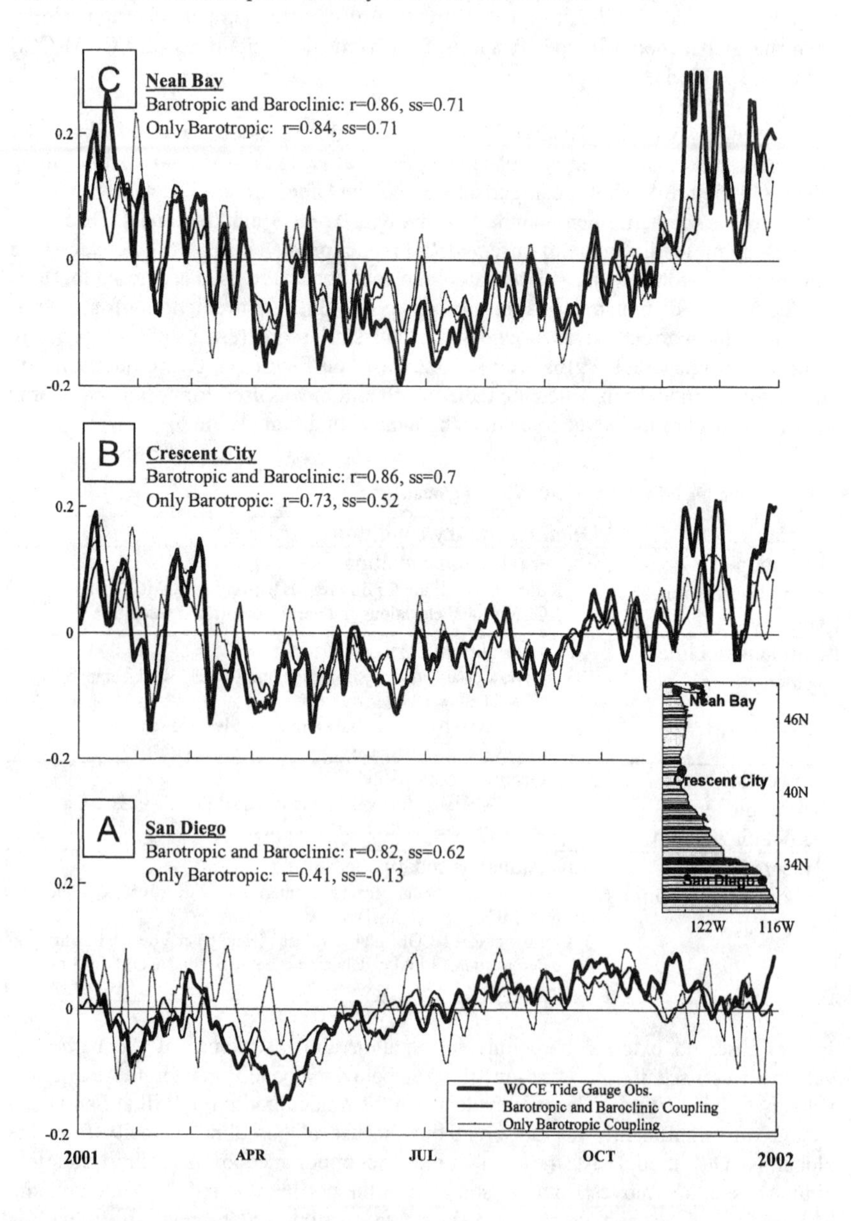

Figure 5. Sea surface height anomaly comparison with WOCE tide gauge stations. Data shown are 3-day average of daily data.

At the Neah Bay station (Fig. 5c), the model skills for runs with and without baroclinic coupling are very similar. This can be explained by the fact that remote baroclinic signal (supplied by global model to NCOM-CCS) propagates from south to north and diminishes along the coastal area of the model domain.

Fig. 6a shows comparisons of observed and model-predicted mixed layer depths (MLD) at the mooring M1. MLD was defined as depth at which the water temperature becomes 0.1°C less than SST (Martin, 1985). The plotted observed MLD (Figure 6a) has very low variability during summer time because of poor vertical resolution of available observations in upper 20 meters. The model reproduces the shallowing of mixed layer during the summer time, and deepening of MLD during transition to the winter time. However, without the baroclinic coupling, the model misses some observed events of shallowing MLD during summer time and deepening during the transition to the winter.

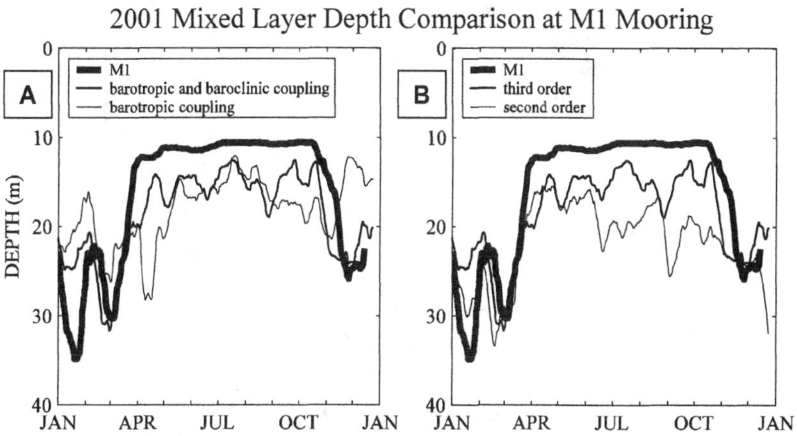

Figure 6. Observed and model-predicted mixed layer depth (MLD) comparison at the M1 Mooring.

Comparisons of MLD for two model runs, in which third and second order advection schemes were used, are displayed in Figure 6b. The MLD for the run with second order advection has significantly less variability during summer and during the transition to the winter. This supports the well-known fact that frontal structures are more diffused by second order advective schemes than third order schemes. Overall, despite the MCSST assimilation into the NCOM-CCS model, model-predicted MLD is deeper than the observed. The assimilation of MCSST does not adequately compensate for the over-mixing in the model. This points to the necessity of including of heat fluxes, especially evaporative heat flux, into the NCOM-CCS surface boundary conditions to produce thinner mixed layers during the summertime.

5. Modeling of California Current Ecosystem with NCOM-CCS.

The NCOM-CCS model also includes coupling to the 9-component ecosystem model implemented in collaboration with Dr. Fei Chai (Chai et al, 2002). The biological model, a 9-component ecosystem formulation originally developed for the equatorial Pacific upwelling system, includes three nutrients (silicate, nitrate and ammonia), two phytoplankton groups, two zooplankton grazers, and two detritus pools. Figure 7 shows a comparison of model-predicted and SeaWIFS ocean color satellite derived chlorophyll fields for July, 2001. The model is able to represent the horizontal scale of the phytoplankton bloom due to summertime upwelling favorable winds. The model is also able to reproduce the approximate magnitude and offshore extension of the coastal blooms near Cape Blanco (~ 42.5°N) and Heceta Bank (~ 44.5°N).

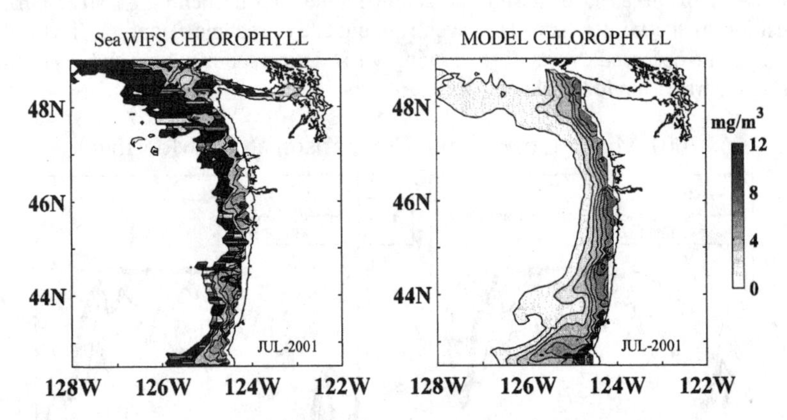

Figure 7. Surface chlorophyll comparison of observed SeaWifs (July 2001) composite and mean monthly chlorophyll from NCOMCCS model embedded with a 9-compartment biological model.

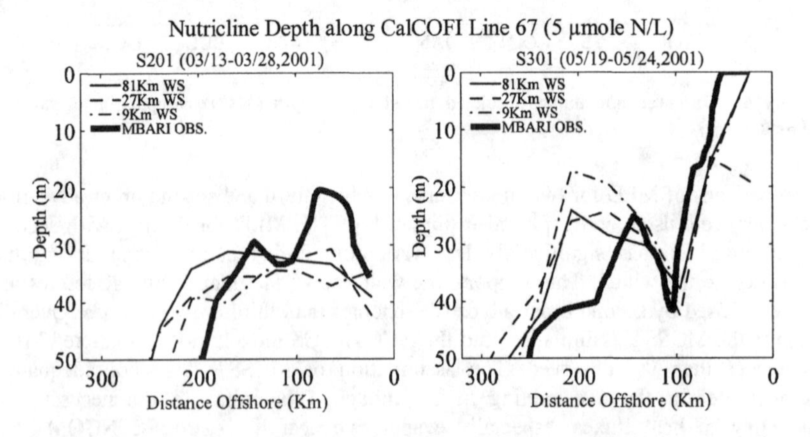

Figure 8. Comparison of observed and model predicted nutricline depths along CalCOFI Line 67. Three model runs forced with different resolution wind stresses are shown.

Figure 8 demonstrates comparisons of model predicted and observed nutricline depths. The nutricline depth is defined as the depth where the concentration exceeds 5 μmol N m^{-3}. Comparisons are shown along California Cooperative Oceanic Fisheries Investigations (CalCOFI) line 67 (see map on figure 2) for CalCOFI surveys in March and May of 2001. Without any assimilation of biological observations we should not expect one to one correspondence between model and observations. However, the model is able to reproduce the observed variability and depth of the nutricline well. There are three model runs forced with different resolution wind stresses are shown on Fig. 8. All three runs show similar results and therefore weak dependence of the nutricline depth on resolution of the wind forcing.

6. Discussions and future plans

The NCOM-CCS model simulations demonstrated a good predictive skill in reproducing features and observed variability in SST and SSH fields. The correlation between the model-predicted and observed ADCP currents is higher than the 95% significance level when the model is forced with COAMPS™ 9km resolution wind. The model reproduces the seasonal variations in mixed layer depth: shallowing of mixed layer during the summer time, and deepening of MLD during transition to the winter time. The NCOM-CCS model predictions demonstrated the importance of the baroclinic coupling with the NCOM global model for accurate predictions of SSH variability along the US West Coast (especially, in prediction of SSH at southern portion of the model domain). NCOM model has flexible options for choosing different order advective schemes. Seasonal variability of the model-predicted depth of the mixed layer agreed better with observations when the third order advective scheme was used. The NCOM-CCS model also includes coupling to the 9-component ecosystem model implemented in collaboration with Dr. Fei Chai (Chai et al, 2002). The model was able to represent the horizontal scale of the phytoplankton bloom due to summertime upwelling favorable winds, and the model was able to reproduce the observed variability of the nutricline depth.

At the same time, comparisons between observations and model results show that even with high resolution atmospheric forcing, the model shows low correlation with observed currents. The model predicted mixed layer depth is deeper than what was observed during the summer time. The above and other discrepancies between model predictions and observations suggest future research items aimed at improvement of the NCOM-CCS model predictions:

- Inclusion of heat fluxes and air-sea coupling with COAMPS™ predictions
  Coupling with atmospheric model is needed for the improvement of model predictions of mixed layer depth, especially during the summer time.
- Assimilation of MODAS synthetic temperature and salinity profiles.
  The global model (which assimilates MODAS fields) provides boundary conditions for NCOM CCS. Without assimilation of MODAS fields into the NCOM CCS, the model dynamics inside of

the model domain might drift significantly from open boundary values, which will lead to the development of artificial waves along the open boundaries. This is why assimilation of MODAS fields in NCOM CCS is probably needed for long duration runs.

- Assimilation HF radar derived surface currents from installation along the US West Coast.

  This is needed for the improvement of model-predicted currents close to the coast and for a better correlation with observations.
- Doubling of horizontal resolution up to 4.5 km.

  Transition of the NCOM CCS model to 4 -4.5 km of horizontal resolution will be a significant improvement in physical as well as ecosystem modeling.
- Assimilation of bio-optical observations.

  With a high level of uncertainty in values of the ecosystem model parameters and ecosystem modeling, assimilation of observations, for example, ocean color measurements is needed.

Acknowledgments

This research was funded through the Naval Research Laboratory (NRL) 6.1 project, "Coupled bio-optical and physical processes in the coastal zone" under program element 61153N sponsored by the Office of Naval Research and the NOPP "Simulations of Coastal Ocean Physics and Ecosystems" (through NASA grant NAG13-00042) project. Appreciation is extended to F. Chavez of MBARI for providing mooring and CalCOFI line 67 data. We thank R. Hodur and J. Doyle of NRL for providing COAMPS™ fields and Drs. Fei Chai and Lei Shi of U. Maine for providing assistance with the implementation of the ecosystem model. Computer time for the numerical simulations was provided through a grant from the Department of Defense High Performance Computing Initiative. This manuscript is NRL contribution # PP/7330-03-20.

References

Bleck, R., 2002: An oceanic general circulation model framed in hybrid isopycnic-cartesian coordinates, *Ocean Modeling*, **4**, 55-88.

Blumberg, A., and G. L. Mellor, 1987: A description of a three-dimensional coastal ocean circulation model. In *Three Dimensional Coastal Models*, N.S. Heaps, Ed., Coastal and Estuarine Sciences, 4, *Am. Geophys. Un.*, 1-16.

Chai F., R. C. Dugdale, T.-H. Peng, F. P. Wilkerson, and R. T. Barber, 2002: One Dimensional Model of the Equatorial Pacific Upwelling System, Part I: Model Development and Silicon and Nitrogen Cycle, *Deep Sea Res., Part II.,* 49 (2002) 2713-2745.

Chassignet, E.P., L.T. Smith, G.R. Halliwell, and R. Bleck, 2003: North Atlantic simulation with the HYbrid Coordinate Ocean Model (HYCOM): Impact of the vertical coordinate choice, reference density, and thermobaricity. *J. Phys. Oceanogr.*, **33**, 2504-2526.

Emery, W. J and R. E. Thomson, 1998: Data Analysis Methods in Physical Oceanography, Pergamon Press, 634 pp.

Flather, R. A., 1976: A tidal model of the northwest European continental shelf. Mem. Soc. Roy. Sci. Liege 6 (10) , 141-164.

Fox D. N., C. N. Barron, M. R. Carnes, M. Booda, G. Peggion and J. Van Gurley, 2002: The Modular Ocean Data Assimilation System. *Oceanography*, 15, 1, 22-28.

Haidvogel, D.B., J. Blanton, J.C. Kindle and D.R. Lynch, 2000, Coastal Ocean Modeling: Processes, and Real-time Systems. *Oceanography*, 13, 1, 35-46.

Hodur, R. M., J. Pullen, J. Cummings, X. Hong, J.D. Doyle, P. J. Martin, M.A. Rennick, 2002: The Coupled Ocean/Atmospheric Mesoscale Prediction System (COAMPS) *Oceanography*, Vol 15, No. 1, 88-98.

Hurlburt, H. E., A. J. Wallcraft, W. J. Schmitz, P. J. Hogan, E. J. Metzger, 1996: Dynamics of the Kuroshio/Oyashio current system using eddy-resolving models of the North Pacific Ocean. *J. Geophys. Res.*, 101, 941-976.

Large, W.G and S. Pond, 1981: Open Ocean Momentum Flux Measurements in Moderate to Strong Winds. *Journal of Physical Oceanography*, 11, 324-336.

Kindle, J.C., R. Hodur, S. deRada, J. Paduan, L.K. Rosenfeld, and F. P. Chavez, 2002, A COAMPS^{tm} reanalysis for the eastern Pacific: Properties of the diurnal sea breeze along the central California coast, *Geophys. Res. Lett.*, 29(24), 2203, doi:10.1029/2002GL015566.

Kundu, P. K., 1976. Ekman veering observed near the ocean bottom. *Journal of Physical Oceanography*, 6, 238-242.

Martin P.J., 1985. Simulations of the mixed layer at OWS November and papa with several models. *Journal of Geophysical Research*, 90, 903-916.

Martin, P.J., 2000: Description of the Navy Coastal Ocean Model Version 1.0, NRL/FR/732--00—9962, Naval Research Laboratory, Stennis Space Center, Mississippi.

Murphy, A.H., and E.S. Epstein, 1989: Skill scores and correlation coefficients in model verification. *Mon. Wea. Rev.*, 117, 572-581.

Paduan J. D. and I. Shulman, 2003: HF Radar Data Assimilation in the Monterey Bay Area. *Journal of Geophysical Research*, to appear.

Rochford, P. A., and P. J. Martin, 2001: Boundary Conditions in the Navy Coastal Ocean Model, NRL/FR/7330--01-9992, 33 pp., Naval Research Laboratory, Stennis Space Center, Mississippi.

Rhodes, R.C., H.E. Hurlburt, A.J. Wallcraft, C.N. Barron, P.J. Martin, O.M. Smedstad, S. Cross, J.E. Metzger, J. Shriver, A. Kara, and D.S. Ko, 2002: Navy Real-Time Global Modeling Systems, *Oceanography*, Vol 15, No. 1, 29-43.

Shulman I., S. H. D. Haddock, D.J. McGillicuddy, J. D. Paduan, W. P. Bissett, 2003: Numerical Modeling of Bioluminescence Distributions in the Coastal Ocean, *Journal of Atmospheric and Oceanic Technology*, 20, 1060- 1068.

Shulman, I., C.R. Wu, J.K. Lewis, J.D. Paduan, L.K. Rosenfeld, J.C. Kindle, S.R. Ramp and C.A. Collins, 2002: High resolution Modeling and Data Assimilation in the Monterey Bay Area. *Continental Shelf Research*, 22, 1129–1151.

Shulman, I., C. R. Wu, J. D. Paduan, J. K. Lewis, L. K. Rosenfeld, S. R. Ramp, 2001: High Frequency Radar Data Assimilation in the Monterey Bay. In *Estuarine and Coastal Modeling*, M. L. Spaulding, Ed., 434-446.

Wallcraft, A.J., A.B. Kara, H.E. Hurlburt, and P.A. Rochford, 2001: The NRL Layered Global Ocean Model (NLOM) with an embedded mixed layer sub-model: Formulations and model simulations. *J. Atmos.Oceanic Technol.,* v.20, N. 11, pp. 1601-1615.

Modeling the effects of bed drag coefficient variability under wind waves in South San Francisco Bay

By Jeremy D. Bricker[1], Satoshi Inagaki[2], and Stephen G. Monismith[3] (Associate Member, ASCE)

Abstract

In this paper, we report the results of a study of the variation of shear stress and the bottom drag coefficient C_D with sea state and currents at a shallow site in San Francisco Bay. Via field experiments, we found that the model of Styles & Glenn (2000), though formulated to predict C_D under ocean swell on the continental shelf, accurately predicted enhanced drag under wind waves in an estuary (Bricker et al, 2004).

Knowing this, we apply the enhanced drag coefficient determined by Styles & Glenn to the estuarine circulation model TRIM-3D of Casulli & Cattani (1994) and Gross et al (1999), and use it to examine the effects of variable roughness on contaminant and sediment transport in South San Francisco Bay. We also investigate the importance of the wave model used in TRIM-3D, by comparing the results of Inagaki et al's (2001) fetch and wave model with those of SWAN (Booj et al, 1999). The different wave models we use generate noticeably different trends in sediment transport and roughness variation.

Given a wave model, we find that tidal stage in South San Francisco Bay is quite insensitive to variability in roughness, but tidally-averaged RMS and residual currents are quite sensitive (modified by about 10%), and channel-shoal asymmetry is enhanced. Along with this, the deep channel experiences more erosion of sediment, and the shoals more deposition. Furthermore, variable roughness extends the hydraulic residence time (inhibits flushing) south of the Dumbarton Bridge from 18 days to 19 days.

Introduction

An understanding of the bottom drag coefficient is essential for accurate modeling of hydrodynamics and sediment transport in estuaries, rivers, lakes, and on the continental shelf. The bottom drag coefficient C_D is typically defined by

$$\tau_c = \rho C_D |U| U \qquad (1)$$

where τ_c is the shear stress at the bed, ρ is water density, and U is the velocity of the mean current at height z_r. C_D is nearly equivalent to the physical bed roughness

----

[1]Environmental Fluid Mechanics Laboratory, Department of Civil and Environmental Engineering, Stanford University, Stanford, CA 94305-4020, USA; now Research Associate in the Department of Civil Engineering, Kobe University, 1-1 Rokkodai-cho, Nada-ku, Kobe 657-8501, JAPAN; phone +81-78-803-6444; email bricker@stanfordalumni.org

[2]Senior Research Engineer, Environmental Engineering Department, Kajima Technical Research Institute, 2-19-1, Tobitakyu, Chofu-shi, Tokyo 182-0036, JAPAN

[3]Professor, Environmental Fluid Mechanics and Hydrology, Department of Civil and Environmental Engineering, Stanford University, Stanford, CA 94305-4020, USA

length scale k_b, or the zero-velocity roughness height z_0 (related by $z_0=k_b/30$), that appears in the log law

$$U = \frac{u_*}{\kappa}\ln\left(\frac{30z_r}{k_b}\right) \tag{2}$$

where u_* is the shear velocity and $\kappa \approx 0.41$ is Von Karman's constant. The log law assumes that flow is taking place in the region of the steady bottom boundary layer (the inner or overlap layers) in which shear stress is constant ($\overline{u'w'} = u_*^2$) and eddy viscosity is increasing linearly away from the bed (or, alternatively, that shear stress is linearly varying and eddy viscosity is parabolic, as is often the model used for open channel flow). Combining Equation 1 with Equation 2 results in

$$C_D = \left[\frac{\kappa}{\ln(30z_r/k_b)}\right]^2 \tag{3}$$

as discussed in Gross et al (1999).

In lieu of stratification, the value of C_D depends upon bed sediment grain size or bedforms. Drag is further enhanced when surface waves are long enough to reach the bed ($\lambda_{wave}/2$>depth). In this case, the thin, oscillatory wave bottom boundary layer experiences greater turbulence than the thicker, steady current bottom boundary layer (bbl). More flow momentum is leaked to the bed, resulting in a very high "apparent roughness" outside the wave bbl (Grant & Madsen, 1979).

Physical bed roughness has been studied well in the field (Cheng et al, 1999), but apparent roughness has only been measured on the continental shelf (Cacchione et al, 1994; Green & McCave, 1995), not in estuaries. Grant & Madsen (1979) and Styles & Glenn (hereafter called SG2000) provided the standard model for determining apparent roughness via theoretical means. In this model, apparent roughness is the manifestation of the wave bbl's enhanced turbulence on the region above the wave bbl. Enhanced turbulence within the wave bbl causes more momentum from the mean flow (above the wave bbl) to be transported to the bed, resulting in greater drag on the mean flow. Since circulation models used in coastal and environmental engineering do not resolve the wave bbl itself, correct prediction of apparent roughness is necessary for these models to accurately determine mean flow over shallow regions (Davies & Lawrence, 1995; Kagan et al, 2003). The theory presented in SG2000 was formulated for ocean swell propagating over the continental shelf, however, and not estuaries, where the wind-driven steady surface boundary layer overlaps with the frictional steady bottom boundary layer. The questions we set out to answer were thus:

1. "how well does SG2000, while formulated for ocean swell over the continental shelf, predict shear stress and the bed drag coefficient under wind waves in an estuary?"
2. "what are the effects of variable roughness on hydrodynamics and contaminant and sediment transport in South San Francisco Bay?"

The first question was addressed in Bricker et al (2004), and the results will be briefly summarized here. The second question is the topic of the remainder of this paper.

Experiments

To investigate variation of water column shear stresses and the bed drag coefficient under wind waves on the shoals of a tidal estuary, we ran experiments at Coyote Point in South San Francisco Bay (Figure 1) during June of 2000 and June-July of 2002. Each experiment was run for 2 weeks to capture a full spring-neap tidal cycle. Coyote Point experiences mixed semidiurnal/diurnal tides with water depth of 1 m at low spring tide, 4 m at high spring tide, and currents up to 30 cm/s during peak flood. During the summertime, a northwesterly diurnal sea breeze of 10-15 m/s blows almost every afternoon, resulting in wind waves reaching the Point with about 2-second period, 50 cm height, and frequent spilling whitecaps. Using a variety of instruments, including pointwise Acoustic Doppler Velocimeters (ADVs), an Acoustic Doppler Current Profiler (ADCP), pressure sensors, and a capacitance wave gauge, we recorded tidal and wind-driven currents, sea state, and turbulence parameters under a variety of conditions.

Methods for shear stress determination

Using velocity time series obtained from each of the ADVs (at varying heights above the bed), we calculated Reynolds stresses $\overline{u'w'}$ directly from wave-turbulence decomposed data. This wave-turbulence decomposition was carried out by the methods of Benilov & Filyushkin (1970) and Shaw & Trowbridge (2001), as well as the Phase Lag method described in Bricker et al (2004). Reynolds stresses were also calculated from spatial profiles of velocity recorded by the ADCP via nonlinear least-squares fitting to a logarithmic velocity profile (Bricker et al, 2004). Knowing these Reynolds stresses, the bed drag coefficient under various conditions was then calculated from the ADCP and the ADVs nearest the bed as

$$C_D = \frac{u_*^2}{U^2} \quad (4)$$

Theoretical shear stresses at the bed, as well as the bed drag coefficient, were also calculated under all conditions via SG2000.

Effect of waves on stresses

Shear stresses computed in the ways described above are plotted in Figure 2 for all conditions seen during the June-July 2002 experiment (results from the June 2000 experiment were similar). Figure 3 indicates that the drag coefficient, calculated via Equation 4, using stresses from the ADV nearest the bed, was significantly affected by waves, as Coyote Point is shallow enough for waves to "feel" the seabed. Shear stresses obtained through the ADVs, ADCP, and the model of SG2000, all agreed in trend. All drag coefficients converged to a value on the order of the canonical drag coefficient at 1 meter of 0.0025 (Dronkers, 1964) in the limit where mean current velocity was much greater than the maximum near-bed wave-induced orbital velocity. All methods also revealed an increase in C_D of up to two orders of magnitude over the canonical value when mean current velocity was 1/100 the near-bed orbital velocity. SG2000, though formulated to predict enhanced drag under ocean swell on the continental shelf, was thus shown to accurately predict bottom drag on the mean current under wind waves in an estuary as well, despite the influence of the surface wind-driven boundary layer. Stresses higher up in the water column, however, are affected by the wind stress as well, due to the overlap of the steady boundary layers at the bed and surface. Bricker et al (2004) discusses this in detail.

One item to note is that our analysis did not consider the effects of suspended-sediment-induced stratification near the bed. Under summertime conditions in San Francisco Bay, sediment concentrations are not great enough to cause such stratification. However, at times when such stratification can occur, such as during wintertime storms, this stratification may have a significant effect on the bed shear stress and drag coefficient (Green & McCave, 1995).

Effects of enhanced drag on circulation in South San Francisco Bay

We incorporated the bottom boundary layer model SG2000 into the tidal circulation model TRIM-3D (Gross et al, 1999; Casulli & Cattani, 1994) to investigate the effect of wave-induced enhanced drag on simulation of tidal flow in South San Francisco Bay. The inputs to SG2000 are a current field and a wave field (near-bed wave-induced orbital velocity and excursion). TRIM supplied the former, while a wave model had to be used to supply the latter. We tried two wave models for this purpose.

Wave models

One wave model we used was that of the US Army Corps of Engineers (1994) and Inagaki et al (2001). Here, significant wave height and period were determined by the equations

$$\frac{gH}{U_A^2} = 0.283\tanh\left[0.530\left(\frac{gd}{U_A^2}\right)^{3/4}\right]\tanh\left\{\frac{0.00565\left(\frac{gF}{U_A^2}\right)^{1/2}}{\tanh\left[0.530\left(\frac{gh}{U_A^2}\right)^{3/4}\right]}\right\} \quad (5)$$

$$\frac{gT}{U_A^2} = 7.54\tanh\left[0.833\left(\frac{gd}{U_A^2}\right)^{3/8}\right]\tanh\left\{\frac{0.0379\left(\frac{gF}{U_A^2}\right)^{1/3}}{\tanh\left[0.833\left(\frac{gh}{U_A^2}\right)^{3/8}\right]}\right\} \quad (6)$$

In these equations, H is significant wave height (m), T is wave period (s), U_A is wind speed (m/s), F is the fetch at the location in question (m), and h is water column depth (m).

This wave model has shortcomings, however. One flaw is that it predicts larger and longer waves with increasing fetch, but this should be limited by wave breaking. Equations 5 and 6, which predict significant wave height and period (Figures 4 and 5), do not account for wave breaking, however. Whitecapping is pervasive on the shoals of San Francisco Bay on windy days, because waves are depth-limited, but Equations 5 and 6 predict only fetch-limited waves. Therefore, the waves predicted on the eastern shoals during a westerly wind have a very long fetch, and are larger and longer than reasonable, because the model does not account for whitecapping. Depth-induced steepening and breaking (shoaling) is not accounted for in this model either, but this process is less important than whitecapping (wind-induced steepening and breaking over a relatively flat bed) in South San Francisco Bay.

Conversely, the wave prediction model of the Corps of Engineers (COE) underestimates wave height and period in some regions because it does not account for wave refraction. At Coyote Point, many of the the waves incident on the Point appeared to have refracted in from the main channel of San Francisco Bay. This enhanced the wave height above what was predicted by the Army Corps of Engineers' equations alone. Refraction also affects the direction of propagation of waves, which, along with the direction of mean currents, is essential in determining enhanced roughness in SG2000. However, the Corps of Engineers' simplified equations assume all waves travel in the direction of the wind; ie the Shore Protection Manual's wave model doesn't account for refraction.

The alternative wave model we used was SWAN (Booij et al, 1999). This model calculates the spectral components of waves at each grid cell in a bathymetric domain based on the conservation of wave action density, and it accounts for the processes of wave breaking and refraction. Figures 6 and 7 show the wave field predicted by SWAN under the same wind conditions used with the Corps of Engineers' equations to produce Figures 4 and 5. Figures 8 and 9 show the difference in the average wave field predicted by each model over a 2 week period. In short-fetch regions (the Bay's western shore), SWAN predicted waves with greater height and longer period than the Corps of Engineers' equations did, because SWAN accounts for refraction of waves from the main channel onto the shoals. In long-fetch regions (the Bay's eastern shore), SWAN predicted waves with smaller height and shorter period than the Corps of Engineers' equations did, because it accounts for energy loss due to breaking of waves as they travel across the Bay.

Figure 10 shows comparisons of model results to observations of sea state at Coyote Point during the experiments we ran there in June of 2000. Wave height and period predicted by SWAN closely match that predicted by the Corps of Engineers' model. Notice in Figures 8 and 9 that the north side of Coyote Point (where the experiments were carried out) is a location at which the sea states predicted by SWAN and by the Corps' of Engineers' equations do not differ significantly. At other locations, the difference is greater, and we expect SWAN to give more accurate results in these areas, because SWAN incorporates physics that the Corps of Engineers' model does not. However, using SWAN is much more computationally expensive than using the Corps of Engineers' model.

The reason for jumps in significant wave period in SWAN in Figure 7 is not an error. It is due to the fact that SWAN calculates its wave field on a discrete spectral domain, using discrete frequencies. For the sake of computational efficiency, we chose a relatively coarse spectral resolution (34 lograthmically-distributed discrete frequencies between 0.1 Hz and 5 Hz). The significant wave period SWAN calculates is actually the period at which the spectral peak in wave energy exists. The significant wave period predicted by SWAN can therefore only take values corresponding to any one of these discrete frequencies.

Enhanced drag under wind waves

After investigating the differences between these wave models, we investigated the effect of enhanced bed drag under waves on circulation in the estuary. After each time step of a TRIM run (Table 1), SG2000 determined the near-bed steady current shear stress when waves were present. The drag coefficient experienced by the steady current at the following time step was then determined by Equation 4, and this enhanced drag coefficicent was applied to the bottom boundary condition of TRIM-3D (Equation 1), affecting the steady current itself.

Since SG2000 predicts the increase of the drag coefficient over its calm-seas value, we still had to specify the physical (base) roughness z_0 of the bed. We used a base z_0 of 1.34 mm for the calm-seas roughness value throughout South Bay. This is a representative value for roughness in the channel of South Bay found by Cheng et al (1999) via fitting of ADCP-derived velocity profiles to Equation 2. We forced the model with winds recorded at San Francisco International Airport (courtesy Francis Ludwig, Doug Sinton, and the National Weather Service). Tides at the open boundary were taken from Dean Pentcheff's (2003) web interface for David Flater's Xtide program. Modeled currents within the domain agreed well with data from the experiments at Coyote Point and from US Geological Survey monitoring stations throughout the Bay (Tables 2-4, data supplied by Gartner (2003), described in detail in Bricker (2003)).

Since each wave model predicted different wave fields, the enhanced roughness determined by SG2000 was dependent on the wave model used. The inset of Figure 1 shows the bathymetry of South San Francsico Bay, differentiating shoals (1-2m depth) from the main channel (10-20m depth). Figure 11 shows how the drag coefficient calculated by SG2000 with the SWAN wave model was enhanced above the base value when heavy seas were present. Figure 12 shows the average difference in C_D predicted by SG2000 using SWAN vs. that predicted by SG2000 using the COE wave model over a 2 week period.

Figure 12 shows that, even on a day with rough seas, SG2000 with the COE wave model predicted a high drag coefficient only on shoals which have a large fetch, as it is here that wave height and period, and therefore also maximum wave-induced near-bed oribital velocity, were largest. Using the SWAN wave model, on the other hand, generates a large drag coefficient over shoals with shorter fetch as well, because it accounts for the fact that waves refract from the channel out onto these shoals. Another difference is that, on shoals with the longest fetch (on the eastern shore of the Bay), C_D predicted by the model using SWAN was smaller than C_D predicted by the model using the Corps of Engineers' equations. This occurred because SWAN accounts for wave breaking, while the Corps of Engineers' equations do not. The Corps of Engineers' equations thus allowed waves in this region to grow too large and too long, causing apparent roughness to grow too large as well. The other difference between the two wave models is the wave propagation direction predicted. Since apparent roughness depends in part on the angle between waves and currents, this affects apparent roughness as well, though not as drastically as refraction and wave-breaking effects do.

Enhanced drag varies in time as well as space. During calm mornings, the drag coefficient remains at its base value throughout the entire Bay. During times of strong winds, however, it grows by more than an order of magnitude over the shoals (see Figures 11 and 13). These results are similar to those observed during the Coyote Point experiments (Figure 3).

Despite the large modulation of the drag coefficient by waves that we observed, accounting for this physics does not have an immediately noticeable effect on tidal elevation (Figure 10, top) or tidal currents (Tables 2-4) predicted by the model. Effects of enhanced drag become apparent, however, when we look at tidally-averaged RMS currents, Eulerian residuals, and contaminant and sediment transport trends.

Root Mean Square (RMS) velocities

With base roughness only (SG2000 not used), TRIM predicted depth-averaged RMS velocities in the channel of about 50 cm/s, and over the shoals of about 20 cm/s (see Figure 14). Using the enhanced roughness model, RMS velocities were about 2 cm/s faster in the channel and 2 cm/s slower over the shoals than those predicted by the model using base roughness only (Figure 15).

The reason for this is the enhancement of the drag coefficient over its base value that occurred over shoals during times of heavy seas (as in Figure 11). The bay's larger average roughness caused a larger longitudinal pressure gradient to develop between the bay's mouth (where the tidal free surface boundary condition is applied) and head. Over the shoals, momentum lost due to enhanced roughness more than offset this larger pressure gradient, so the flow rate there was reduced from what it was in the case of uniform roughness. Since the channel is too deep for waves to "feel" the seabed, however, roughness there remained at its base value at all times. Because of this, the enhanced pressure gradient resulted in a greater flow rate through the channel than occurred in the uniform roughness case.

Since summertime winds over the Bay are diurnal, the shoals experience enhanced roughness a large portion of each day. This roughness caused the observed decrease in RMS velocity over the shoals, and the increase in RMS velocity in the channel. This could lead to larger cross-channel shear and thus enhanced mixing between the channel and shoals, as well as enhanced longitudinal shear-flow dispersion.

Eulerian residual velocities

TRIM predicted that tidally-averaged Eulerian residual depth-averaged velocities were generally downwind over the shoals and upwind in the channel (Figure 16). This is what we expect when tidal flows are averaged out, and a strong wind-driven flow dominates the residual signal.

In a simulation with variable roughness, residuals at all locations were weakened by approximately 10% (Figure 17). By weakening residual velocities, enhanced roughness could reduce the flushing rate of this estuary.

Passive scalar transport and flushing

To study the net effect of variable roughness on flushing, we replicated the experiment of Gross et al (1999). We released a constant 100 kg/s flow of passive scalar at the San Jose sewage treatment plant at the southern end of the Bay. As stated by Gross, "this mass was added to a cell without any volume of water, as if at each time step a constant powdered tracer mass were mixed uniformly into the... water column." The model was then run until a "dynamic steady state" was reached, in which the tidally averaged scalar field was nearly unchanging and decayed monotonically away from the source. Hydraulic residence time was calculated via the release of a passive tracer as

$$\tau_{hydraulic} = \frac{M}{\dot{M}} \tag{7}$$

where M is the total mass in the domain (see Figure 18), and $\dot{M}$ is the steady-state mass flow rate through the domain (Gross et al, 1999). For the case of base roughness only, the model predicted a hydraulic residence time south of the Dumbarton Bridge of 18 days. With roughness from SG2000, the hydraulic residence time increased to 19 days. Signell & List (1997) observed that variable roughness caused a similar decrease in flushing rates in their study of Massachusetts Bay. Glorioso & Davies (1995) also observed this phenomenon in their numerical studies, strengthening the case that accounting for enhanced roughness is important for the prediction of residence times (and for the formulation of policies dependent on residence time prediction) in a variety of locales.

Effects of variable roughness on sediment transport

Variable roughness had a significant effect on the integrated deposition minus erosion (D-E) of sediment predicted by the model. Sediment was assumed to be cohesive, as clays and silts dominate the suspended load in this region of San Francisco Bay. The model of McDonald & Cheng (1997) specified settling velocity

$$\begin{aligned} w_s &= w_{s,base}, C \le C_{cr} \\ w_s &= w_{s,base}\left(\frac{C}{C_{cr}}\right)^{2/3}, C > C_{cr} \end{aligned} \tag{8,9}$$

as well as deposition and erosion rates,

$$\begin{aligned} Deposition &= w_s C\left(1 - \frac{\tau_{b,cw}}{\tau_d}\right), \tau_{b,cw} < \tau_d \\ &= 0, \tau_{b,cw} > \tau_d \\ Erosion &= P\left(\frac{\tau_{b,cw}}{\tau_e} - 1\right), \tau_{b,cw} > \tau_e \\ &= 0, \tau_{b,cw} < \tau_e \end{aligned} \tag{10,11}$$

which are functions of the combined wave-current bed shear stress $\tau_{b,cw}$ within the wave bottom boundary layer. In these equations, w_s is settling velocity, C is sediment concentration, C_{cr} is the critical concentration for flocculation, P is an erosion rate constant, τ_e is the critical bed shear stress for erosion, and τ_d is the critical stress for deposition.

Van Rijn (1993) showed that the combined wave-current bed stress can be broken down into a wave-dominated (oscillatory) component $\tau_{b,w}$ and a current-dominated (steady) component $\tau_{b,c}$

$$\tau_{b,cw} = \tau_{b,c} + \tau_{b,w} \tag{12}$$

To investigate the effect of enhanced roughness, we held the unsteady component of the bed shear stress constant, while the steady component (Equation 1) varied with the enhanced C_D of SG2000. This comparison revealed that, in the case with variable roughness, the Bay experienced more erosion (or less deposition) in the channel and more deposition (or less erosion) over the shoals than the case with uniform roughness (see Bricker (2003) for color figures, as grayscale are not available for this paper). The magnitude of the difference averaged 10% of the D-E seen by either model.

Given the effect of variable roughness on RMS currents in South Bay, this difference is expected. Stronger currents in the channel lead to greater bed shear stress (Equation 1) there, which enhanced erosion and inhibited deposition. Deposition in the channel was further reduced by the reduced influx of sediments from the shoals, due to the retarded currents there. Over the shoals, however, reduced transport due to reduced currents acted to retain sediments close to the location at which they were first scoured from the bed, inhibiting erosion at these locations, and preventing transport of sediments away to the channel during ebb. Signell & List (1997) also observed reduced transport of sediments during times of peak wave activity, when the suspended load was greatest.

Keeping roughness and the steady component of bed shear stress in Equation 12 constant, it is interesting to note that the choice of wave model has a significant effect on the oscillatory component of bed shear stress and thus also on D-E. Bricker (2003) investigated the difference in D-E predicted with each wave model due to variation in $\tau_{b,w}$ only (again, grayscale figures not available for this paper), resulting from the differences in wave height and period determined by each wave model. The difference in predicted D-E is substantial, being on the order of the D-E predicted by each model itself.

Conclusions

"What are the effects of variable roughness on hydrodynamics and contaminant and sediment transport in South San Francisco Bay?"

Hydrodynamic modeling results showed that the variable bed drag coefficient predicted by SG2000 had an effect on flushing and sediment transport in South San Francisco Bay, especially because it is a channel-shoal system. Since the drag coefficient under heavy seas took a larger value over shoals only, tidally averaged RMS velocities were enhanced in the channel and reduced over shoals. This resulted in more scouring (~10%) of sediments from the channel. Since enhanced drag retarded currents over the shoals when wave-induced erosion was largest, transport of sediments from the shoals to the channel was also reduced, inhibiting erosion over the shoals, and further reducing deposition in the channel.

By releasing a passive tracer from the San Jose POTW at a constant rate, we determined that variable roughness has the net effect of reducing the flushing rate in South San Francisco Bay. The hydraulic residence time south of the Dumbarton Bridge was increased from 18 days to 19 days when variable roughness was accounted for. Flow modulation of this sort has important ramifications which policy makers must account for when setting maximum mass loadings and maximum contaminant levels in a shallow urban estuary subject to wind waves.

Tables and figures

Table 1. Parameters used in TRIM-3D simulations. Derived from Table 6-1 of Inagaki (2000).

Parameters (general)	Domain and Bathymetry	See figure 1
	Number of grid points	116(x) • •268(y)
	Mesh interval	200 m*1
	Calculation Time Step	30 sec
	Horizontal Viscosity	0 m^2/s
	Maximum number of vertical cells	30
	Vertical resolution	1 m
	Simulation periods	
	For comparisons with USGS data	10/1 to 11/8/1994
	For comparisons with year 2000 Coyote Point Experiment	6/1 to 6/15/2000
	For comparisons with year 2002 Coyote Point Experiment	6/19 to 7/9/2002
Sediment transport	Base settling velocity $w_{s,base}$	0.0004 m/s
	Critical shear stress for erosion τ_e	0.3 N/m^2
	Critical shear stress for deposition τ_d	0.15 N/m^2
	Erosion rate constant P	0.0025 g/m^2
Initial conditions	Deflection from mean sea level	0.0 m
	Velocity u,v	0
	Suspended sediment concentration	0
Boundary conditions	Suspended sediment concentration	0 (constant)
	Tidal condition (at open boundary)	From harmonic constants at Bay Bridge
	Wind condition	Measured Wind at SFO

Table 2. Normalized mean square error (ε^2) of TRIM-generated depth-averaged currents, as compared to ADCP data, during October 1994 over San Bruno Shoal.

Model	ε^2 current speed	ε^2 current direction
Base (uniform) C_D only	0.06	0.21
SG2000 w/ COE waves	0.06	0.22
SG2000 w/ SWAN	0.06	0.21

Table 3. Normalized mean square error (ε^2) of TRIM-generated depth-averaged currents, as compared to ADCP data, during October 1994 at the San Mateo Bridge.

Model	ε^2 current speed	ε^2 current direction
Base (uniform) C_D only	0.06	0.10
SG2000 w/ COE waves	0.07	0.10
SG2000 w/ SWAN	0.07	0.10

Table 4. Normalized mean square error (ε^2) of TRIM-generated depth-averaged currents, as compared to ADCP data, during October 1994 at the Dumbarton Bridge.

Model	ε^2 current speed	ε^2 current direction
Base (uniform) C_D only	0.49	0.22
SG2000 w/ COE waves	0.49	0.22
SG2000 w/ SWAN	0.49	0.22

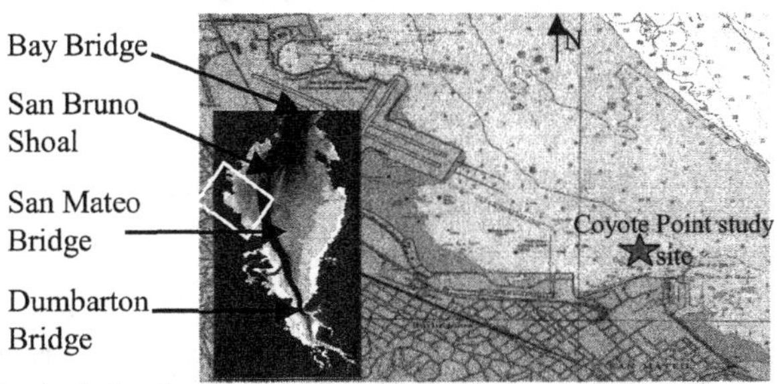

Figure 1. South San Francisco Bay and Coyote Point. Chart courtesy of NOAA Coast Survey. Chart is not to be used for navigational purposes. Inset indicates deep channel (dark shade) and shoals (light shade).

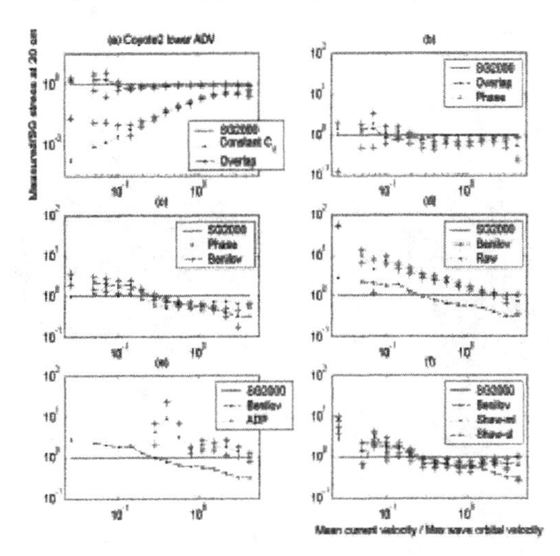

Figure 2. Ratio of shear stress obtained via various methods to shear stress obtained via SG2000 at 20 cm above the bed vs. the ratio of mean current velocity to near-bottom wave-induced orbital velocity during June-July 2002. Pluses are error bars representing 95% confidence intervals on the mean value of the stress ratio in each evenly spaced bin on the logarithmic x-axis. Error bars represent variability in stress due to variation in the independent parameters of orbital excursion and water column depth, as well as instrument error.

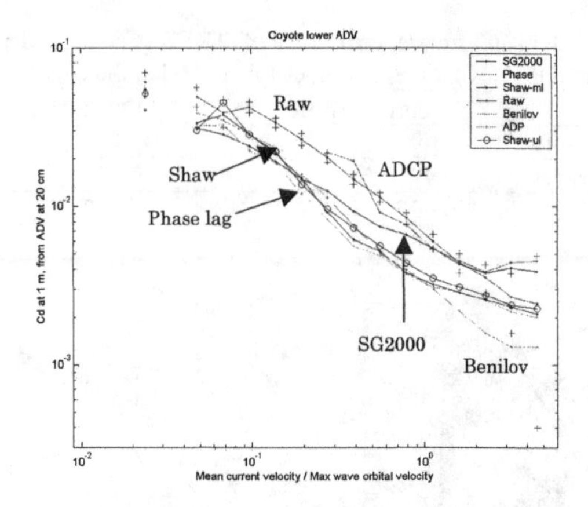

Figure 3. Drag coefficient at 1 m, as derived from the ADV 20 cm above the bed, vs. the ratio of mean current velocity to near-bottom wave-induced orbital velocity during June-July 2002.

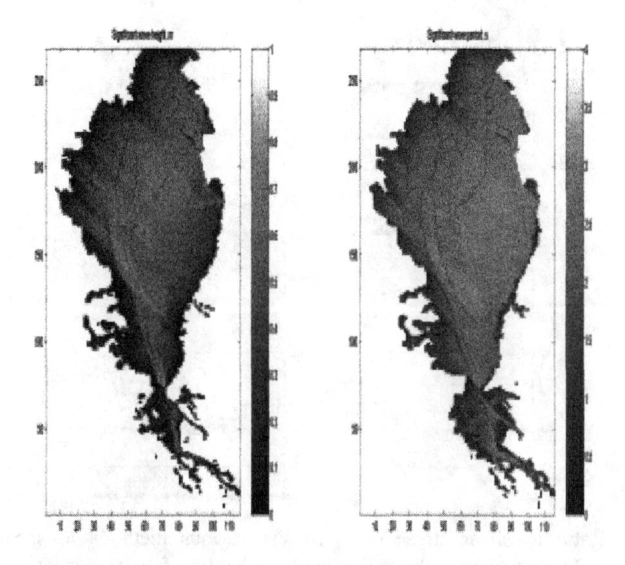

Figure 4 (left). Significant wave height (m) from the COE wave model during 14 m/s westerly winds.
Figure 5 (right). Significant wave period (s) from the COE wave model during 14 m/s westerly winds.

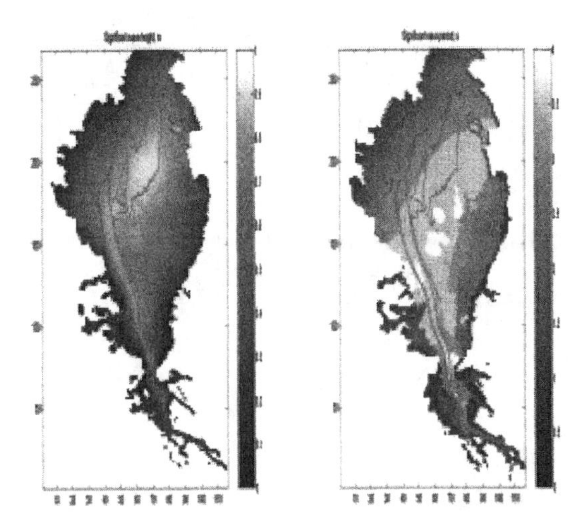

Figure 6 (left). Significant wave height (m) from the SWAN wave model during 14 m/s westerly winds.
Figure 7 (right). Significant wave period (s) from the SWAN wave model during 14 m/s westerly winds.

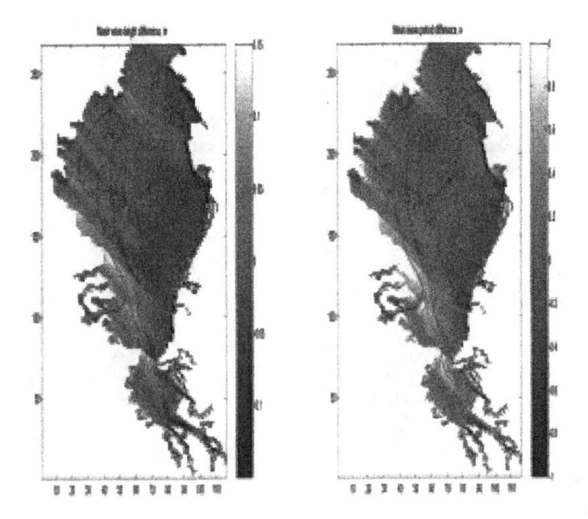

Figure 8 (left). Time-averaged wave height difference (m) between SWAN and the COE wave model over 2 weeks.
Figure 9 (right). Time-averaged wave period difference (s) between SWAN and the COE wave model over 2 weeks.

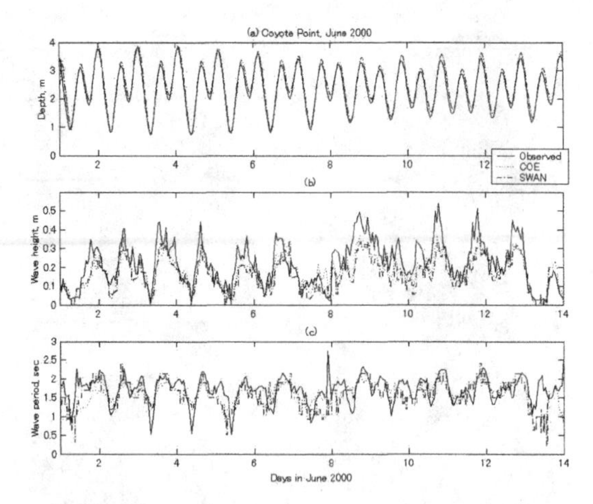

Figure 10. Total depth, significant wave height, and significant wave period at Coyote Point vs. days in June 2000. Solid line is observed data from SeaBird SBE 26 Sea Gauge. Dotted line is simulation result with the COE wave model used in Inagaki et al (2001). Dash-dot line is simulation result with the SWAN wave model.

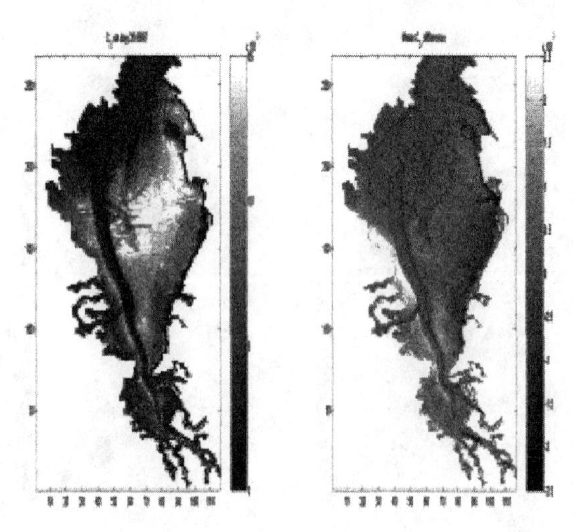

Figure 11 (left). C_D at 1 meter above the bed (mab) determined by SG2000 using the SWAN wave model under heavy seas (westerly wind ≈ 14 m/s).
Figure 12 (right). Time-averaged drag coefficient (at 1 mab) difference between runs using SWAN and runs using the COE wave model over 2 weeks.

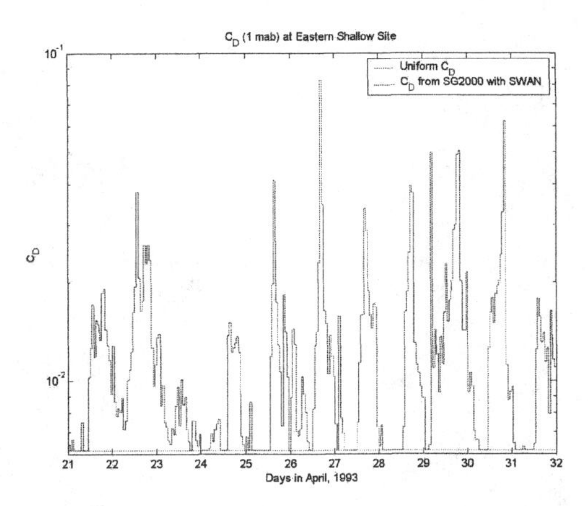

Figure 13. Enhanced C_D from SG2000 at 1 mab as a function of time over shoals. During calm mornings, C_D reduces to the base C_D.

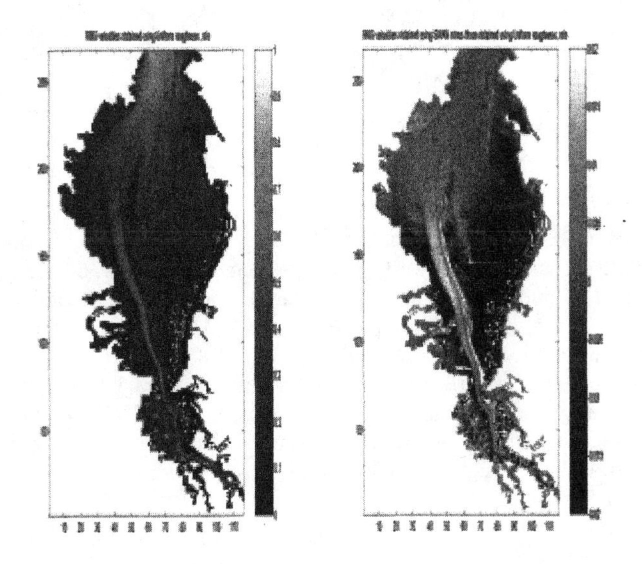

Figure 14 (left). Depth-averaged RMS velocity field (m/s) obtained by using base roughness only. Velocities are RMS averaged over 14 days of simulation.
Figure 15 (right). Difference between the depth-averaged RMS velocity field (m/s) obtained using roughness from SG2000 and the SWAN wave model, and the depth-averaged RMS velocity field obtained using base roughness only.

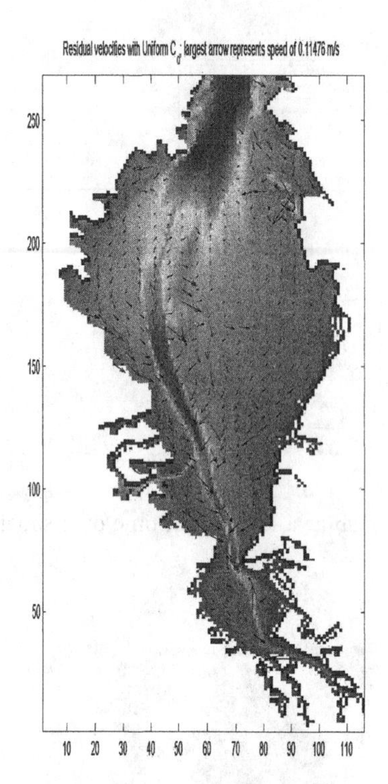

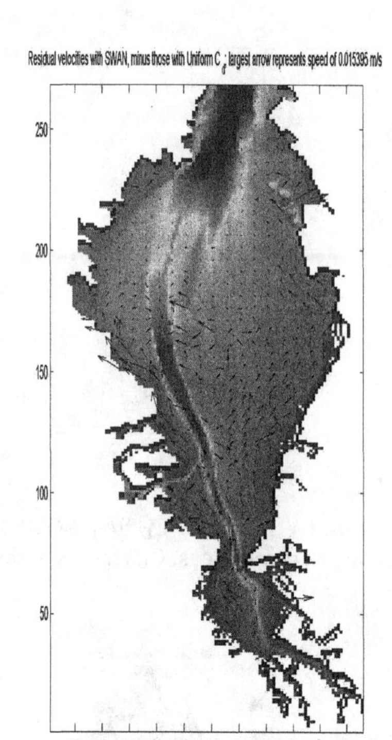

Figure 16 (left). Tidally-averaged Eulerian residual depth-averaged velocities obtained using base roughness only. Velocities from 14 days of simulation are low-pass filtered with a Butterworth filter (n=7, cutoff period = 36 hours, as in Lacy, 2000) and then averaged in time. The largest arrow on this plot represents a velocity of 11.5 cm/s. Shading represents depth. Darkest and lightest regions are the channel, while medium gray are shoals.

Figure 17 (right). Difference between tidally-averaged Eulerian residual depth-averaged velocities obtained using roughness from SG2000 with the SWAN wave model, and tidally-averaged Eulerian residual depth-averaged velocities obtained using base roughness only. The largest arrow on this plot represents a velocity of 1.5 cm/s.

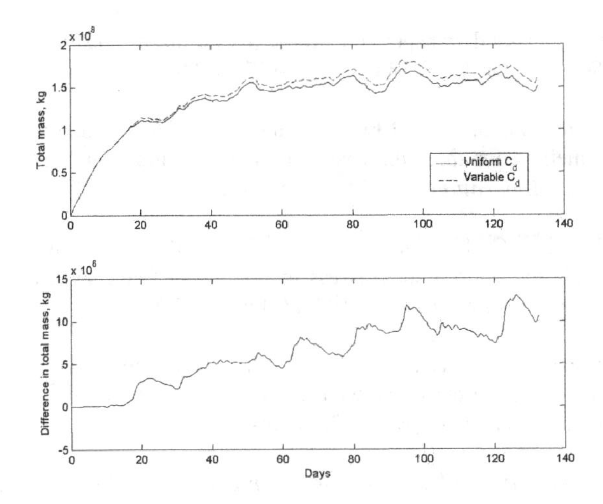

Figure 18. Top: plot of tidally averaged total scalar mass south of the Dumbarton Bridge vs. days of simulation. Bottom: plot of the difference in total scalar mass south of the Dumbarton Bridge predicted by a simulation using variable roughness from SG2000 and the COE wave model, and that predicted by a simulation using base (uniform) roughness only.

Acknowledgements: Funding for this work was provided by the UPS foundation and by the Office of Naval Research (ONR grant N00014-99-1-0292-P00002 monitored by Dr. Louis Goodman).

References

Benilov, A. Y. and Filyushkin, B. N. (1970). "Application of methods of linear filtration to an analysis of fluctuations in the surface layer of the sea." *Izvestiya Atmospheric and Oceanic Physics.* v6 n8 pp810-819.

Booij, N., Ris, R. C., and Holthuijsen, L. H. (1999). "A third-generation wave model for coastal regions 1. Model description and validation." *Journal of Geophysical Research.* v104 nC4 pp7649-7666.

Bricker, J.D. (2003). *Bed drag coefficient variability under wind waves in a tidal estuary: field measurements and numerical modeling.* Stanford University Ph.D. thesis.

Bricker, J.D., Inagaki, S., Monismith, S.G.. (2004). "Bed drag coefficient variability under wind waves in a tidal estuary: field measurements." Under review for publication in *ASCE Hydraulics*.

Cacchione, D. A., Drake, D. E., Ferreira, J. T., and Tate, G. B. (1994). "Bottom stress estimates and sand transport on northern California inner continental shelf." *Continental Shelf Research.* V14 n10/11 pp1273-1289.

Casulli, V., and Cattani, E. (1994). "Stability, accuracy, and efficiency of a semi-implicit method for three-dimensional shallow water flow." *Computers and Mathematics with Applications.* v 27 n 4 pp 99-112.

Cheng, R. T., Ling C. H., and Gartner J. G. (1999). "Estimates of bottom roughness length and bottom shear stress in South San Francisco Bay, California." *Journal of Geophysical Research.* v 104 n C4 pp 7715-7728.

Davies, A.M. and Lawrence, J. (1995). "Modeling the effect of wave-current interaction on the three-dimensional wind-driven circulation of the eastern Irish sea." *Journal of Physical Oceanography.* v25 pp29-45.

Dronkers, J. J., (1964). *Tidal Computations in Rivers and Coastal Waters.* North Holland, New York.

Glorioso, P.D., and Davies, A.M. (1995). "The influence of eddy viscosity formulation, bottom topography, and wind wave effects upon the circulation of a shallow bay." *Journal of Physical Oceanography.* v25 pp1243-1264.

Gartner, J. (2003). Written communication. USGS, Menlo Park, CA.

Grant, W. D., and Madsen, O. S. (1979). "Combined Wave and Current Interaction with a Rough Bottom." *Journal of Geophysical Research.* v84 nC4 pp1797-1808.

Green, M., and McCave. (1995). "Seabed drag coefficient under tidal currents in the eastern Irish Sea." *Journal of Geophysical Research.* v100 nC8 pp16057-16069.

Gross, E. S., Koseff, J. R., and Monismith, S. G. (1999). "Three-dimensional salinity simulations of South San Francisco Bay." *ASCE Journal of Hydraulic Engineering.* v 125 n 11 pp 1199-1209.

Inagaki, S. (2000). *Effects of a proposed San Francisco airport runway extension on hydrodynamics and sediment transport in South San Francisco Bay.* Engineer's Degree Thesis, Stanford University.

Inagaki, S., Monismith, S.G., Koseff, J.R., and Bricker, J.D. (2001). "Sediment transport simulation in South San Francisco Bay." *Proceedings of Coastal Engineering, JSCE,* v48, pp641-645 (in Japanese).

Kagan, B. A., Alvarez, O., Izquierdo, A., Mananes, R., Tejedor, B., and Tejedor, L. (2003). "Weak wind-wave/tide interaction over a moveable bottom: results of numerical experiments in Cadiz Bay." *Continental Shelf Research.* v23 pp435-456.

Lacy, Jessica R. (2000). *Circulation and Transport in a Semi-enclosed Estuarine Subembayment.* Ph.D. Thesis, Stanford University.

McDonald, E. T., and Cheng, R. T. (1997). "A numerical model of sediment transport applied to San Francisco Bay, California." *J. Marine Lm. Eng.* v 4 pp 1-41.

Pentcheff, Dean. (2003). *WWW tide and current predictor.* http://tbone.biol.sc.edu/tide.

Shaw, W. J., and Trowbridge, J. H. (2001). "The Direct Estimation of Near-bottom Turbulent Fluxes in the Presence of Energetic Wave Motions." *Journal of Atmospheric and Oceanic Technology.* v 18 pp 1540-1557.

Signell, R. P., and List, J. H. (1997). "Effect of wave-enhanced bottom friction on storm-driven circulation in Massachusetts Bay." *Journal of Waterway, Port, Coastal, and Ocean Engineering.* v123 n5 pp233-239.

Styles, R., and Glenn, S. M. (2000). "Modeling stratified wave and current bottom boundary layers on the continental shelf." *Journal of Geophysical Research.* v 105 n C10 pp 24119-24139.

United States Army Corps of Engineers. (1994). *Shore Protection Manual.*

van Rijn, L. C. (1993). *Principles of Sediment Transport in Rivers, Estuaries, and Coastal Seas.* Aqua Publications. The Netherlands.

Development and Application of Combined 1-D and 3-D Modeling System for TMDL Studies

Venkat S. Kolluru[1] and Mike Fichera[2]

Abstract
A combined 3-D and 1-D modeling system was developed to include many streams, ponds and lakes that enter multiple receiving water bodies. The combined model will be used as a management tool for TMDL (Total Maximum Daily Load) studies in the Delaware's Inland Bays. The Inland Bays TMDL study region consist of 19 river networks with two to ten tributaries in each network, five ponds and three water bodies that were listed as impaired waters under section 303(d) of the Clean Water Act for nutrients, dissolved oxygen or bacteria. Both the hydrodynamic / transport and water quality models of the combined system were used for TMDL studies. The combined modeling system also uses HSPF (Hydrological Simulation Program-FORTRAN) watershed model outputs as time varying non-point source loadings in the rivers. The combined 3D and 1D modeling system was calibrated for the years 1998, 1999 and verified for the year 2000 using the available and synthesized input data. Delaware's Inland Bays have been sampled and measured frequently over the past decade. For the purposes of this model, data retrieval was concentrated upon the years 1998 to the present. Recalibration is being performed using the HSPF model outputs for the Inland Bays watershed regions. The calibrated and verified model with HSPF results will be applied to perform TMDL scenario studies in near future.

Introduction
A proper selection of modeling tools is necessary for successful TMDL studies. Often the modeling tools have to be either obtained from public domain or commercially available software. These tools frequently have to be updated with additional provisions such as site specific algorithms for various water quality processes, new outputs or modeling methods as part of TMDL studies. J. E. Edinger Associates, Inc. (JEEAI) of Wayne, Pennsylvania have enhanced the existing hydrodynamic and water quality model called GEMSS (Generalized Environmental Modeling System for Surface waters) as a modeling tool for TMDL application on the Delaware Inland Bays (Figure 1). The enhancement includes modeling of streams and ponds, multiple

[1] J. E. Edinger Associates, Inc., Suite 609, 983 Old Eagle School Road, Wayne, PA 19087-1711
[2] ENTRIX, Inc, 10 Corporate Circle, Suite 100, New Castle, DE 19720

receiving waterbodies with different vertical datum, use of HSPF watershed model outputs as loadings for the streams and special output processing tools for TMDL scenario studies. GEMSS was previously applied to evaluate different design alternatives to improve the flushing of impaired waters in the Indian River and Rehoboth Bays and Little Assawoman Bays (ENTRIX, 2001).

The interlocked Delaware Inland Bay System includes two main water bodies: Indian River Bay and Rehoboth Bay. Both water bodies are shown in Figure 1. The Indian River Bay is connected to the Atlantic Ocean on the east via the Indian River Inlet and to Little Assawoman Bay to the south via the Little Assawoman Canal. Rehoboth Bay is connected to Delaware Bay to the north via the Lewes-Rehoboth Canal and to Indian River Bay to the south. The western portion of Indian River Bay, referred to as the Indian River, terminates at Millsboro Dam. The streams included in the current project are from the State's impaired waters listed under section 303(d) of the Clean Water Act for nutrients, dissolved oxygen or bacteria within the Inland Bays watershed and are shown in Figure 1. There are 19 impaired stream networks as listed under section 303(d) of the Clean Water Act. It also includes five ponds. The geological and hydrographical information of the Bay were obtained from the United States Army Corps of Engineers report (Cerco, 1994).

Description of GEMSS

The models used in this study are the hydrodynamic and transport module, GEMSS-HDM, the water quality module GEMSS-WQM and the recently developed 1-D module, GEMSS-1DM (Kolluru, 2002). These models are embedded in a geographic information and environmental data system (GIS), grid generator and editor, control file generator, 2-D and 3-D post processing viewers and additional tools that include meteorological and time varying data generators to support 3-D modeling. Customization of the suite of hydrodynamic, transport and water quality models is achievable through the use of a modular design reflecting the needs of each user's application. The GEMSS software main window is shown in Figure 2.

GEMSS-HDM

The GEMSS-HDM module uses GLLVHT (Generalized, Longitudinal-Lateral-Vertical Hydrodynamic and Transport), which is a state-of-the-art three-dimensional numerical model that computes time-varying velocities, water surface elevations, and water quality constituent concentrations in rivers, lakes, reservoirs, estuaries, and coastal water bodies. The hydrodynamic and transport relationships used in GLLVHT are developed from the horizontal momentum balance, continuity, constituent transport and the equation of state. The basic relationships are given in Edinger and Buchak (1980 and 1995). The hydrodynamic equations are semi-implicit in time. The semi-implicit integration procedure has the advantage that computational stability is not limited by the Courant condition $\Delta x/\Delta t$, $\Delta y/\Delta t < (gh_m)^{1/2}$ where h_m is the maximum water depth that can lead to inefficiently small time steps of integration. Since the solutions are semi-implicit (for example, explicit in the constituent transport and the time lagged momentum terms) the stability is controlled by the Torrence condition ($U\Delta t/\Delta x$, $V\Delta t/\Delta y < 1$; Δx and Δy are grid sizes in x and y directions, respectively). Hence, the integration time step can be chosen to realistically represent the details of the boundary data, typically 15 minutes for tides and up to one hour for meteorological data.

Transport computation is explicit in time. It is developed in such a way that transport coefficients can be computed once and used for all constituents during that time step at a given "n", "k" location ("n" represents the cell location in the horizontal direction; "k" represents the cell location in the vertical direction). The solution time is not sensitive to the number of constituents being examined. Constituent computations are performed using a higher order transport QUICKEST-ULTIMATE (Leonard, 1988) scheme in all three directions. The model is built to accept a large number of transport constituents and constituent relationships depending on the water quality model, sediment transport model, oil and chemicals model and toxic model being used.

GEMSS-WQM

A series of water quality models have been developed for coupling with the computationally efficient HDM module of GEMSS with different level of usage.

The most useful model to determine relative changes in the water quality constituents due to any individual source is the linear coefficient model called WQLCM. This model, an extension of approach proposed by Thomann and Muellar (1987), cycles nitrogen components organic nitrogen, ammonia, and nitrate (ON, NH4, NO3) through algae, and includes dissolved oxygen depression due oxidation of NH4 to NO3, and of CBOD. The algae take up NH4 and NO3, and then die at a specified rate to release ON and OC (as CBOD) to complete the cycle. It requires only specifying the constant transfer rates between the various compartments. In this model all of the components are computed relative to the sources incorporated, so that it gives the incremental change in each constituent due to the sources.

The simplest WQM is the EPA supported EUTRO5 (Ambrose et al., 1993) model that was assembled from earlier models in 1983. This model is called WQE5M in GEMSS. This model simulates the interaction of eight water quality state variables, i.e., the dissolved forms of ammonia nitrogen, nitrate nitrogen, inorganic phosphorus, carbonaceous BOD, organic nitrogen, organic phosphorus, and; phytoplankton and dissolved oxygen. They can be considered as four interacting systems: phytoplankton kinetics, the phosphorus cycle, the nitrogen cycle, and the dissolved oxygen balance. This requires solar radiation, etc and does try to compute actual dissolved oxygen levels. Actually, we have found this model to be relatively unstable in real time simulations partially because it assumes that particulate constituents are a constant multiplier of the dissolved constituents.

The next most complex model is the WQDPM model, which is an extension of EUTRO5. WQPDM includes both the dissolved and particulate forms of nitrogen, phosphorus and CBOD. As seen on the process diagram shown in Figure 3, it also allows for properly including zooplankton excretion contributions through the particulate forms. Settling is applied only to the particulate components of the constituents. This model level of modeling is near state of the art and is being used on many estuaries including the Delaware River Estuary. This model is a preferred starting place, but we do not have a breakdown between dissolved and particulate concentrations for the constituents in the waste load sources much less in the river.

The carbon based water quality model WQCBM (Wu et al., 1998) is also incorporated in GEMSS. It includes separate organic carbon, organic nitrogen and organic phosphorous pathways through algae and both fast and slow rates for the initial transformations, multiple algal compartments with different growth and decay properties, diurnally vertically migrating dynoflagellates, zooplankton, etc. This model can also be linked to an available sediment diagensis routine. It is the framework for ultimate level of WQ modeling and the model is designed in such a way that it can be expanded to incorporate other components.

A number of auxiliary functions have been developed from field studies, such as light attenuation as a function of surface solar radiation algae concentration and depth; reaeration rate as a function of wind speed and temperature; zooplankton grazing; and, relationships for sediment exchange for ammonia, nitrate nitrogen, inorganic phosphorus, and sediment-oxygen demand (SOD). These relationships are used as guidelines for applications of the water quality models to other water bodies.

GEMSS-1DM

The GEMSS-1DM module was developed to incorporate multiple long and narrow streams, ponds and lakes and structures such as weirs. It uses generalized longitudinal hydrodynamic and transport (GLHT) computation derived from the three-dimensional equations of fluid motion and continuity as given in Edinger and Buchak (1980). It is the sectionally homogenous form of the hydrodynamic and transport relationships and applies to flows and constituents that vary in time and in the longitudinal direction. It is applicable to shallow and relatively narrow rivers. Both wind and bottom shear stress terms are included in the model. Specific applications of GEMSS-1DM have been accepted by both state and federal agencies in various regulatory proceedings. The water quality models for GEMSS-1DM were derived from the three dimensional counter parts (GEMSS-WQM). Similar approach was used to other modules. A 1-D grid generation module is developed to set up stream segments, cross-sections and slopes. This module allows the user to add and delete stream branches into the water body. Additional tools were included to set up time varying loadings for different streams and structures such as weirs. Water quality rates and constants can be applied to each river network using a spreadsheet based user interface. All 1D hydraulic models were tested to determine if they give correct hydraulic properties along different stream reaches through comparisons to available tributary field data.

Modeling Framework in GEMSS

The WQDPM of GEMSS-WQM is linked and run simultaneously with the HDM and 1DM modules of GEMSS on the same grid (Figure 3). This has been found to be an important step since many short-term processes in tidal situations, such as vertical water velocities, influence short-term particle settling and light dependent processes. Using a reasonably detailed three-dimensional grid, the model can simulate the combined hydrodynamic and water quality conditions for one year in an over night run on a 3.0 GHz PC machine. The combined framework modeling approach eliminates the need to save hydrodynamic outputs and tedious steps involved in the comparison of volumetric transport across different sections in the water quality grid and the hydrodynamic grid for accuracy and correctness. The conceptual diagram showing the link between 1DM and HDM is shown in Figure 2.

Mulitple Coastal Water Bodies
The Indian River – Rehoboth Bay and Little Assawoman Bay are at different elevations above MSL. So, the algorithms used to simulate drying and wetting of grid cells due to tidal actions, have been updated so that both the waterbodies are linked hydraulically.

Linkage of GEMSS with HSPF Model
GEMSS was also updated to use loadings (flows and water quality constituent concentrations) generated from HSPF model. A graphical user interface was designed to generate loadings input data from HSPF model outputs to different reaches of the streams and to the different shoreline extents of the Inland Bays. The linkage of GEMSS with HSPF Model is shown in Figure 4. The HSPF text output files are converted into Microsoft Access database files using field data import format available in GEMSS. Microsoft Access database files are viewed through a tool called DBViewer. Time varying loading files are created using this tool along with another tool called TVDGen.

Review and Assessment of Input Data
Delaware's Inland Bays have been sampled and measured frequently over the past decade. For the purposes of this model, data retrieval has concentrated upon the years 1998 to the present. During these years, the have been the highest concentration of data collections covering covered the broadest range of relevant model input requirements. Some model inputs have already been gathered and incorporated into the GEMSS field data format during the Inland Bays Flushing Study. The following lists the databases that have been gathered for this modeling exercise including those previously gathered for the existing GEMSS model.

Bathymetry and Stream Depths: DNREC (2002), USGS, ACOE, ENTRIX, Martin and Miller (1993)
Meteorological Data: NOAA NCDC (1965-2001)
Stream Flow: USGS
Point Source Discharges: DNREC
Tide Gage Data: NOAA, Nautical Software, Inc. (1996)
Sediment: DNREC, Seitzinger and DeKorsey, 1994, Owens and Cornwell, 2002
Water Quality
DNREC Water Quality Monitoring: DNREC

University of Delaware: CISNet Inland Bays TMDL Database
Pfiesteria Monitoring Water Quality Data: DNREC (1999)
Citizen Monitoring Data: University of Delaware Sea Grant
Conectiv 316(a): Demonstration Indian River Survey: ENTRIX (2001)
STORET: EPA (1999)

The amount of data available was so enormous that it needed a better way of organizing the data. HTML documentation was one such approach used to compile and display the data. All the ASCII data was converted into Microsoft ACCESS databases for data analysis using GEMSS. After checking for goodness of data spatially and temporally, GEMSS-WQDPM model was selected to perform Delaware Inland Bays TMDL studies.

GEMSS Model Grid for Inland Bays

The 3-D model grids for the Indian River (IR), Rehoboth Bay (RB) and Little Assawoman (LAW) and 1-D model grids for all the impaired streams were developed using the automated 3-D grid generation tool available in GEMSS. The bathymetric and shoreline data for the waterbodies and the streams, described in the data inventory section of this paper was used to set up the grids. The water bodies are connected via existing IR-LAW canal. The combined grid system is shown in Figure 5. There are 45 grids in the east-west direction and 37 grids in the north-south direction. Instead of creating new grid cells for LAW waterbody, empty grid cells available in the 45 x 37 rectangular region were used. This approach provided a significant reduction in the model computational time. There are about 335 and 79 active cells at mean higher high water (MHHW) for the IR-RB and LAW, respectively. There are 19 z-level layers with variable thicknesses in the vertical direction. The vertical layer thickness was set at a constant value of 0.5 m for the tidal range and to a depth of 2 m below mean lower low water (MLLW), 1.0 m for the depth in the range 2 m and 9 m below MLLW, 2.0 m for the depth in the range 9 m and 13 m and 3.0 m for the rest of the bathymetry.

1-D stream segments were generated using the newly developed 1-D grid generator in GEMSS. There are about 19 stream networks with a total of 98 branches and 5 ponds. Along with the stream segments, cross-section and slope properties were also included in the 1-D grid. The slope information was supplied as elevations which were measured for different sections of the stream using USGS quad maps available for the study region. Most of the streams in the Millsboro Pond watershed and west of the Rehoboth Bay have large slopes compared to the region south of the IR-RB region. The number of 1-D segments varies in the range of 10 to 30. The number of branches varies in the range of 1 to 15.

Statistical Error Analysis

Error analysis was carried out to evaluate the performance of the enhanced 1-D and 3-D hydrodynamic and water quality model. The quality of the time varying boundary conditions data considerably influences the model results. Many of the assumptions have been made to generate the model time varying input file for setting up 1-D and 3-D boundary conditions. First level of checks normally done during calibration is to graphically view the model predicted results and field data. Visual error analysis is sufficient if model is able to reproduce field data values. Otherwise, statistical error analysis is needed to quantify the confidence level in the calibration. A new statistical error analysis tool was developed in GEMSS to speed up the calibration effort.

The statistical error analysis conducted in this study was focused on the difference of the model results and field data. The computed statistical properties were number of observations, mean error, standard deviation error, relative mean error, absolute mean error, average error, standard deviation error and sums of model predicted and field data values for the simulation time period and relative error based on the averaged values for the simulation time period. These statistical properties were performed for each freshwater creek and each water body region wherever field data were available.

Model Calibration & Verification

As per the TMDL requirement, Years 1998 and 1999 were set as calibration time periods and Year 2000 was set as verification time period. The Scenario using HSPF output data as loadings to the model for the year 1999 was also considered as verification time period. The initial conditions for elevation, salinity and temperature and water quality constituents were obtained from the time varying data available for the fresh water inflows and tidal conditions at the IR Inlet and at the Fenwick landing in LAW. Most of the input data start from March, 1998 and has at the most 4 records for the year 1998 from STORET. For the preliminary calibration, the required input data for the first three months of the year 1998 were assumed same as the March 1998 data record. Also, the data requirement at the entrance of the stream had to be derived from available field stations downstream of the stream.

The model was calibrated by making minor adjustments to the model bathymetry, developing methodologies to extrapolating the boundary profile data to model layers above mean low water, adjusting the Chezy friction coefficient and tuning up the water quality kinetics rates and constants. In the present modeling system, the water quality rates and constants can be changed for each 1-D stream network as well as for different regions of the receiving water bodies. Variations in bathymetry changed the model velocity amplitudes and variations in the Chezy coefficient changed the model to data phase comparisons. Chezy cofficient was varied between 40 $m^{1/2}/s$ and 60 $m^{1/2}/s$ for the streams depending on the stream bed roughness and a fixed value of 40 $m^{1/2}/s$ for both the IR and LAW. The response temperature computed by bringing the water bodies in equilibrium with the exchange of heat from the surface to the atmosphere was used for point sources.

In the preliminary calibration stage, major streams' main branches were included in the 1D model. Once the preliminary calibration was established to a certain degree of accuracy, all tributaries were included by proportioning the flow and hence the loadings using their respective drainage areas. The sediment exchange of nutrients and dissolved oxygen were not available for all the streams. Sediment exchange values were obtained using the measured data available for IR, RB and LAW receiving waterbodies. These values were assumed the same for the entire stream. Again these values were also adjusted for proper calibration.

Hydrodynamic Model Calibration

Tidal Elevations

The tidal elevations available at Rosedale Beach were compared with model predictions for the year 1998. The field and model predicted elevations overlap each other when plotted for the entire one year. So, the model predicted elevations were plotted and compared with field data for two weeks arbitrarily selected from each half-year in Figure 6. These results clearly show that the model is able to predict the tidal elevations with reasonable accuracy. Similar analysis was performed for the other years.

Volumetric Flow Rate

The water quality model requires an accurate prediction of the volumetric transport that drives the mixing and exchange process within the bay and between the bay and the ocean. The exchange of flow between the Atlantic Ocean and the Indian River Bay is through the Indian River Inlet. Model grid widths were made larger than the actual widths of the inlet to allow reasonably larger time steps for the water quality model simulations. Such an approach is reasonable since the volumetric transport through the inlet is more important than absolute velocities for long term predictions of water quality constituents' concentrations. Therefore the calibration was restricted to volumetric flow rate, and the results are shown in Figure 7. Volumetric flow rate is velocity multiplied by area, so the increase in area is offset by the decrease in velocity. Calibration of inlet temperatures and salinities did not affect by this approach. The exact dates for these plots are not shown because the volumetric flow rates taken from Raney et al. (1990) report are for the end of July 1988, while the model results are for the year 1998.

Water Temperature

Model predicted time varying water temperatures at 60 field stations were obtained from each yearly simulations. Comparison was made between the available field data and model results for the years 1998, 1999 and 2000. Surface water temperatures in streams show more diurnal variations as compared to water bodies. Calibration was performed at places where there was discrepancy by selecting the appropriate shading function and wind sheltering coefficient in the heat budget calculations. Since many streams use STORET data which has only four records for every year, model predicted water temperatures tend to show larger values during the winter and fall quarterly time periods. Figures 8 shows model-field data comparisons of temperatures at different stations along IR for the year 1999. This figure graphically illustrates that the model is able to predict water temperatures in the Inland Bays with minimum errors.

The errors were quantified by performing statistical analysis for the years 1998, 1999 and 2000. A summary of the statistical properties for all 3D waterbodies and 1D streams is shown in Table 1. This table clearly shows the model is able to predict water temperatures both in the waterbodies and the streams with reasonable accuracy. The average error for water temperature in the waterbodies is 0.65, 1.37 and 0.66 °C for the years 1998, 1999 and 2000, respectively. The average error in water temperature for all the streams is 0.30, 0.11 and 0.42, for the years 1998, 1999 and 2000, respectively.

Salinity

Correct transport of salinity in different regions of the water body is very important for the water quality modeling. Most of the time, salinity calibration can be achieved to greater degree of accuracy by capturing all the bathymetric features within the water body. During the calibration simulations, bottom bathymetry in the Indian River was adjusted to improve the intrusion of saline waters into it. Figure 9 shows model-field data comparisons of salinities at different stations along IR for the year 1999 and the comparison is good. As seen in Table 1, the model reproduced the salinity in the waterbodies reasonably well. The average error in salinity for all the waterbodies is 1.57, 2.12 and 1.25 ppt for the years 1998, 1999 and 2000, respectively.

Water Quality Model Calibration

Typically, the water quality model calibration involves checking all loadings that enter into the water body. The loadings have to be checked for any unrealistic values. In addition the seasonal changes in the loadings have to be analyzed for any missing data and find a method to fill in missing data using the seasonal trends. If the water body under study involves tidal exchange with open ocean, then it is very important to check all the input water quality concentrations at the tidal boundary for correctness and time varying trend. Once the loadings are analyzed and corrected properly, then a series of calibration runs will be performed using the loadings and by adjusting the rates and constants.

Model parameter selection is an iterative process, where parameters are adjusted within a range of feasible values to obtain the best agreement between predicted and observed values of the state variables. The feasible range is ideally determined from site-specific field measurements or experiments. When site-specific data are unavailable, the feasible range is determined from other similar field measurements, experiments, models, or by the best professional judgment of the modeler.

Effects of Inflow and Boundary Data

The initial water quality simulations showed how sensitive the water quality model is to the boundary conditions and inflow source data. Preliminary simulations showed that the constituent concentrations in the up estuary portions of the Indian River are generally proportional to the loadings of the numerous surrounding non-point sources from the major tributary basins. In Indian River their influence extended from below the mouth of Pepper Creek to the head of Indian River. This region of the estuary also had a long flushing time indicating that most of the non-point source loadings are retained there. The water quality constituents brought in by the tides dominate the parts of Indian River and Rehoboth Bays that experience short flushing times. An initial simulation with available data showed anomalous results in nutrient concentrations over one period of time at the Indian River Bay station closest to the mouth. An analysis of the tidal boundary input data showed that nutrient concentrations at the boundary during this period of time were generally high on visual inspection, and resulted in a very high mean and standard deviation of their values over the whole simulation record. Replacement of these high values with the mean of the data without them brought the record back in line, and it had a statistical variance similar to the other boundary water quality data.

Calibration of Non-Tidal Systems

Simulation of 98 streams along with 5 ponds and 2 water bodies and needed outputs take approximately 40 hrs on a 3 GHz personal computer with 1 GB RAM. Because of the long computation time, efforts were made to find an efficient method for water quality modeling of Inland Bays. The Longwood Pond, Burton Pond and Millsboro Pond are non-tidal systems and the rest of the stream networks are tidal systems. So, calibration was performed on these three ponds individually by switching off the other streams in the model. Fine tuning was performed by adjusting the kinetics coefficients and increasing the loadings. Increasing the loadings was performed when the time varying input data at the stream entrance was obtained from the downstream field stations. Graphical and statistical comparisons were made for all the three ponds for the years 1998, 1999 and 2000. A graphical comparison plot is shown in Figure 10 for the year 1999 at the exit of the Millsboro Pond. This figure clearly shows that the model predicted results match reasonably well with the field data. The time varying data at the exit of the Millsboro Pond, Longwood Pond and Burton Pond were saved at every 15 minutes for temperature. Salinity and all the water quality constituents were obtained from these simulations for use in the calibration of the remaining tidally influenced streams of the Inland Bays.

Tidal Systems

Calibration was performed on the remaining streams that are tidally connected to the Indian River and Rehoboth Bay and the Little Assawoman Bay. The model predicted water temperatures, salinities and 14 water quality constituents (all modeled water quality constituents plus derived constituents) at various field locations selected from the available field databases were obtained for the years 1998, 1999 and 2000. The derived constituent, total nitrogen was computed by summing all the nitrogen components of the water quality constituents used in the model (NH3-N, NO3-N, ON_D, ON_P). The derived constituent, total phosphorous was obtained by summing all the phosphorous components of the water quality constituents used in the model (PO4, OP_D and OP_P). The model-field data comparisons for all the sixteen constituents were first compared graphically for all the 60 stations for all the years. Figure 11 shows a comparison plot at Station WQ6 for the year 1999. Model is able to predict the average trend for most of the constituents.

Statistical error analysis was performed on the differences between the model results and all the available field data. According to the Technical Guidance Manual for Performing Waste Load Allocations (USEPA, 1990), an acceptable relative error is 15% for dissolved oxygen and 45% for nutrient parameters (nitrogen, phosphorous and carbon). The overall error statistics for the Delaware Inland Bays were 12 % for dissolved oxygen, 20 % for total nitrogen and 25% for total phosphorous. The relative error statistics for the Delaware Inland Bays meets the general guidance criteria published in USEPA(1990).

As part of TMDL studies, model predicted output data for the sixteen variables (salinity, temperature, 12 water quality constituents and two derived constituents) were saved at every hour for all the streams and ponds used in the model. The saved data was then processed to obtain minimum, average and maximum dissolved oxygen, total nitrogen and total phosphorous for all the streams. All these results were plotted for each stream along with the available field data. One such plot (Figure 12) shows dissolved oxygen variation in along various streams in the Love Creek network. The dissolved oxygen range does not change much along the streams because of small residence time in the creeks due to the large stream slopes.

The variation of maximum, minimum and average dissolved oxygen, chlorophyll-a, total nitrogen and phosphorous along the Indian River and Indian River Bay for the years 1998, 1999 and 2000 are analyzed and is shown in Figure 13 for the year 1999. Also plotted in the figure are the available field data from STORET, Pfiesteria, UDEL and Conectiv 316a databases. It is observed that the model predicted results are well within the observed values.

Conclusion

A three dimensional hydrodynamics and transport model called GEMSS which was previously used in the Indian River flushing studies, has been enhanced with additional models and graphical user interfaces to simulate multiple streams and ponds in the Delaware Inland Bays. The model was also updated with many additional post processing tools needed for TMDL analysis. Hydrodynamic and water quality calibration was performed on the combined 3-D and 1-D modeling system for the years 1998 and 1999 and verified for the year 2000. Model predicted water quality concentrations at selected 60 stations in the Inland Bays for all the years show reasonable comparison with available limited forcing data (time varying loads) for the model. The calibrated model will be rerun with HSPF loadings for TMDL studies in early 2004.

References

Ambrose, R.B., T.A. Wool, and J.L.Martin. 1993. The Water Quality Analysis Simulation Program. WASP5: Part A: Model Documentation: Part B: The WASP5 Model Input Dataset. EPA, Athens, GA. September.

Cerco, C.F., B. Bunch, M.A. Cialone, H. Wang. 1994. Hydrodynamics and Eutrophication Model Study of Indian River and Rehoboth Bay, DE. Technical

Report EL-94-5, U.S. Army Engineer Waterways Experiment Station, Vicksburg, MS.

Delaware Department of Natural Resources and Environmental Control (DNREC), Envrionmental Consultants, Inc., and Environmental Resources, Inc. 2002. The Assawoman Canal Dredging Project Assessment Report. Prepared for Delaware DNREC, Division of Parks and Recreation. July 2002. Dover, DE.

Delaware Department of Natural Resources and Environmental Control (DNREC). 1999. Inland Bays Pfiesteria Study.

Edinger, J. E. and E. M. Buchak. 1980. Numerical Hydrodynamics of Estuaries in Estuarine and Wetland Processes with Emphasis on Modeling. Plenum Press, New York, New York, pp. 115-146.

Edinger, J. E. and E. M. Buchak. 1995. Numerical Intermediate and Far Field Dilution Modelling. Journal Water Air and Soil Pollution 83: 147-160,1995. Kluwer Academic Publishers, The Netherlands.

Enivronmental Protection Agency (EPA). 1999. Storet (Storage and Retrieval) Environmental Data System Database. February 26, 2001. <http://www.epa.gov/storet>.

ENTRIX, Inc. Jan. 2001. An Ecological Risk-Based 316(a) Demonstration for the Indian River Power Plant. Prepared for Conectiv, Wilmington, DE

ENTRIX, 2001. Hydrodynamic and Water Quality Modeling and Feasibility Analysis of Indian River, Rehoboth Bay and Little Assawoman Bay. Prepared for Delaware Department of Natural Resources and Environmental Control as part of Delaware's Inland Bays Flushing Study Project.

Kolluru, V. S., 2002. GEMSS – Technical Document on Generalized Environmental Modeling System for Surface waters, J. E. Edinger Associates, Wayne, Pennsylvania. Available at http://www.jeeai.com/Software-GEMSS.htm

Leonard, B. P. 1988. Universal Limiter for Transient Interpolation Modeling of the Advective Transport Equations: The Ultimate Conservative Difference Scheme, NASA Technical Memorandum 100916, ICOMP-88-11.

Martin, C.C, Miller, R.W. 1993. Delaware's Public Ponds. Division of Fish and Wildlife Dover, DE. Document No. 40-05-93-04-03. pgs. 77.

Mellor, G. L. and T. Yamada. 1982. Development of a Turbulence Closure Model for Geophysical Fluid Problems. Rev. Geophysical Space Physics. 20:851-875.

National Climatic Data Center (NCDC). 1965-2001. Record of Climatological Observations. Online. Available: http://www.ncdc.noaa.gov.

Nautical Software, Inc. 1996. Tides & Currents for Windows™. Third edition CD ROM, February 1996.

Okubo, A. 1971. Oceanic Diffusion Diagrams. Deep-Sea Res. 18:789.

Owens, M. and J.C. Cornwell. February 2002. Delaware Coastal Bays Sediment - Water Exchange Study: Data Summary and Interpretation. Report Submitted to DNREC February 2002 by Chesapeake Biogeochemical Associates.

Raney, D.C., J.O. Doughty, and J. Livings. 1990. Tidal prism numerical model investigation of the Indian River Inlet scour study. Bureau of Engineering Research Report 500-13, University of Alabama.

Seitzinger, S. P., R. DeKorsey, 1994. Sediment-Water Nutrient Interactions in Rehoboth and Indian River Bays," Academy of Natural Sciences Division of Environmental Research, Philadelphia, PA, submitted to the Inland Bays Estuary Program, Dover DE, June 1994.

Thomann, R.V. and Mueller, J.A. 1987. Principles of Surface Water Quality Modeling and Control.. Harper & Row. New York.

USEPA. 1990. Technical Guidance Manual for Performing Waste Load Allocations. Book III Estuaries, Part 2, Application of Estuarine Waste Load Allocation Models. U. S. Environmental Protection Agency, Office of Water. EPA823-R-92-003. May 1990.

Wu, J., V.S. Kolluru, J.E. Edinger. 1998. Combined Hydrodynamic and Water Quality Modeling for Waste Water Impact Studies. Proceedings of the Mid-Atlantic Industrial Wastes Conference. Lee Christensen, Editor. Villanova University, Villanova, PA. July 13, 1998.

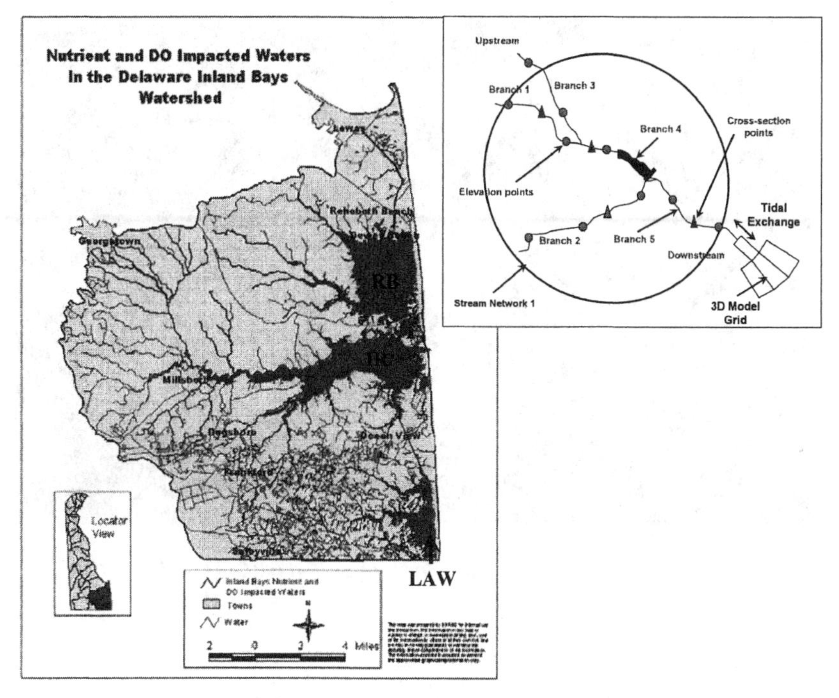

Figure 1 Delaware Inland Bays showing impaired waters. Also shown is the conceptual diagram for interfacing 3-D and 1-D models

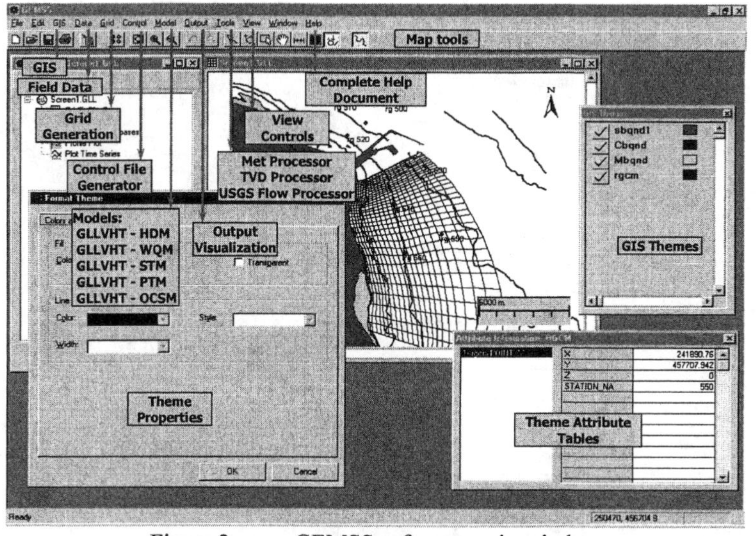

Figure 2 GEMSS software main window

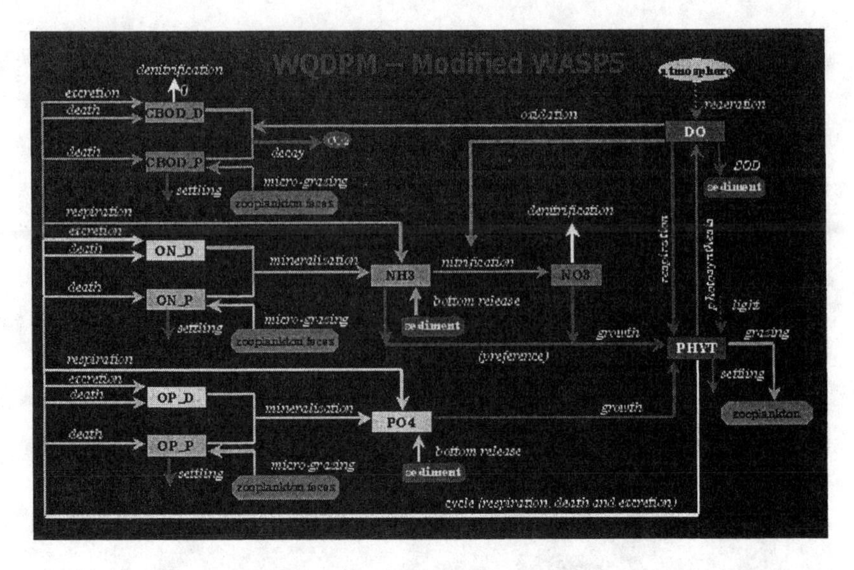

Figure 3 Processes diagram for WQDPM model of GEMSS-WQM

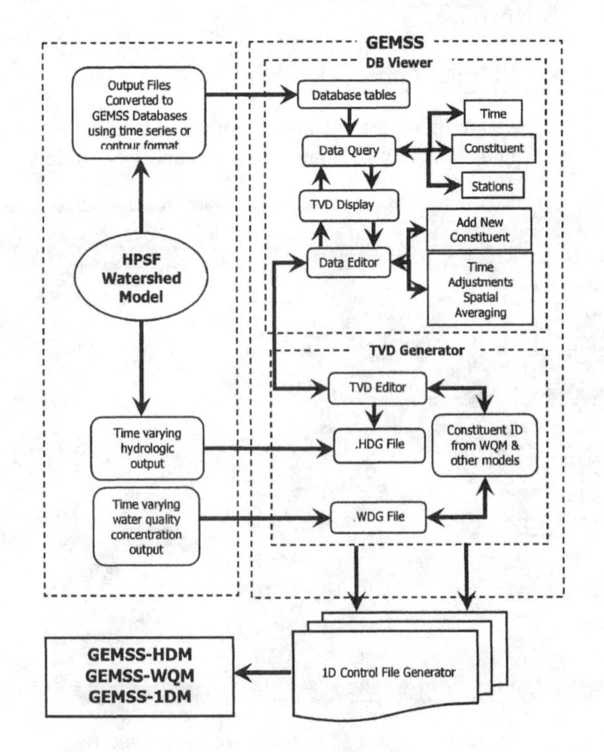

Figure 4 Flowchart showing GEMS-HSPF interface

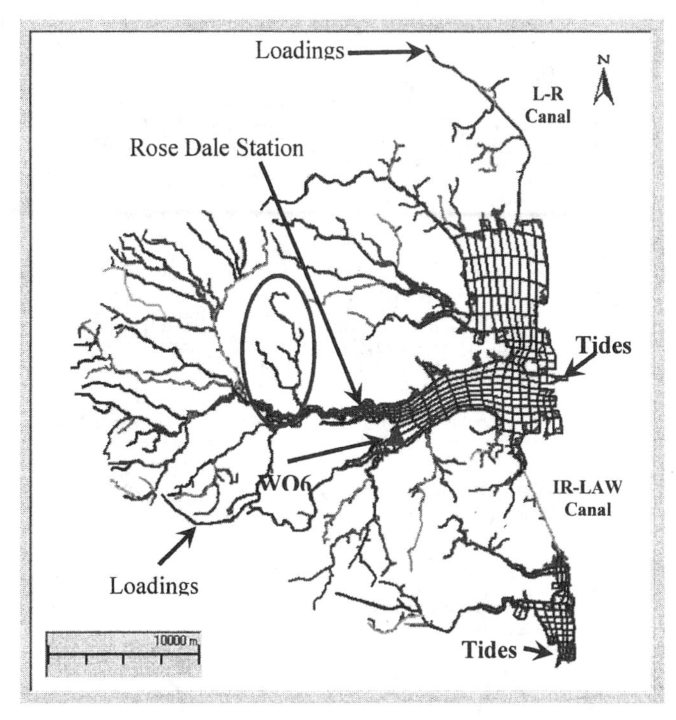

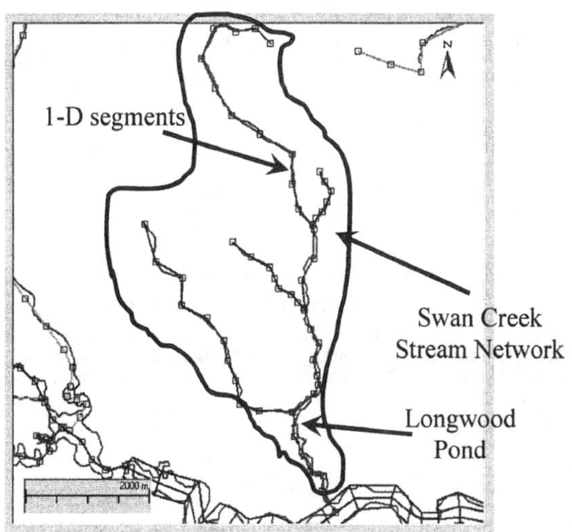

Figure 5 3-D and 1-D model grids for the Delaware Inland Bays. Streams are shown in different colors for easy identification in a stream network

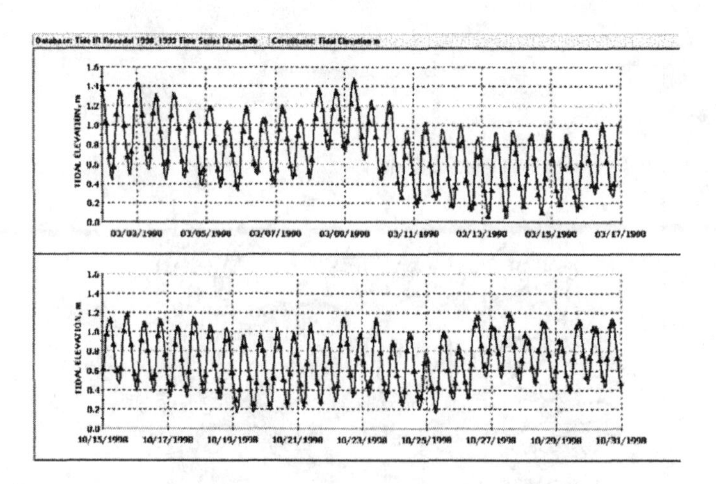

Figure 6 Comparison of model predicted tides with observed tides at Rose Dale

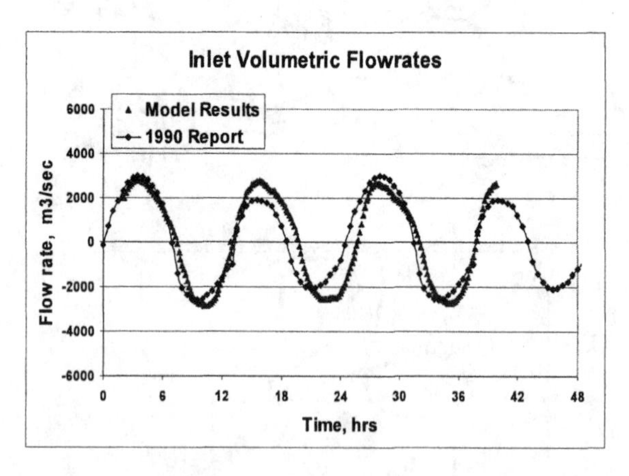

Figure 7 Comparison of inlet volumetric flow rates with the data published in the Raney et al. (1990) report

Table 1 Statistical comparison for temperature and salinity

Year	Constituent	Region	AVE	STD	REme	REabs
1998	Temperature	All 3D	-0.65	1.99	-0.03	0.07
1999	Temperature	All 3D	-1.37	1.90	-0.06	0.08
2000	Temperature	All 3D	0.66	2.30	0.03	0.06
1998	Temperature	All streams	0.30	2.94	0.02	0.12
1999	Temperature	All streams	0.11	2.49	-0.01	0.13
2000	Temperature	All streams	0.42	4.85	0.06	0.31
1998	Salinity	All 3D	-1.57	4.28	-0.07	0.15
1999	Salinity	All 3D	-2.12	5.69	-0.11	0.24
2000	Salinity	All 3D	1.21	2.82	0.05	0.10

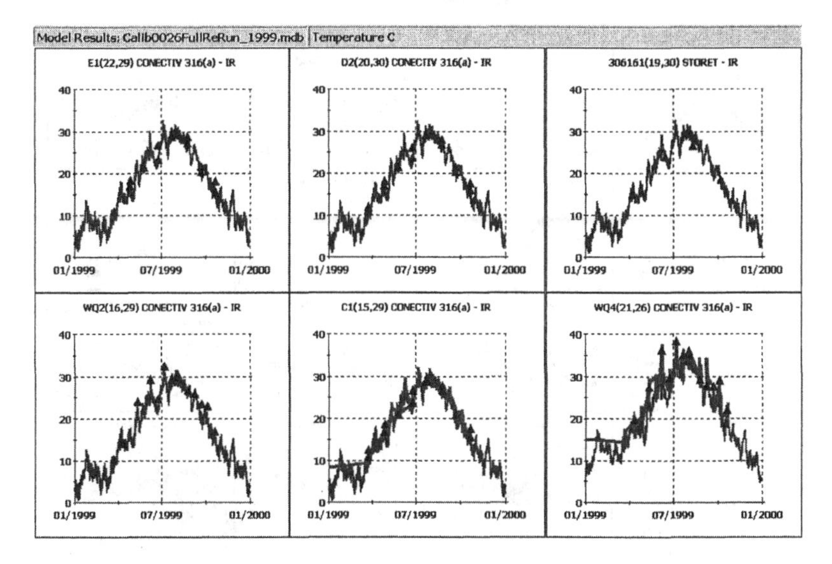

Figure 8 Model versus field data comparison of temperatures in the middle region of Indian River for the year 1999

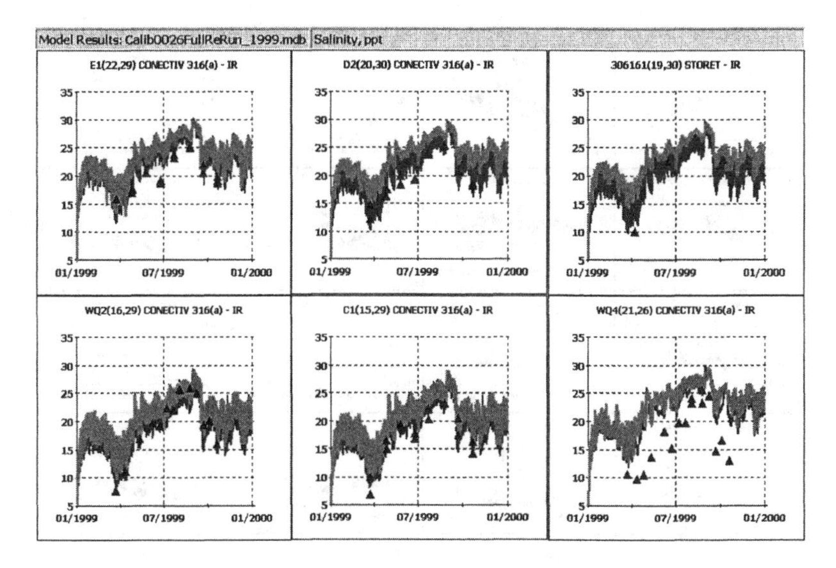

Figure 9 Model versus field data comparison of salinities in the middle region of Indian River for the year 1999

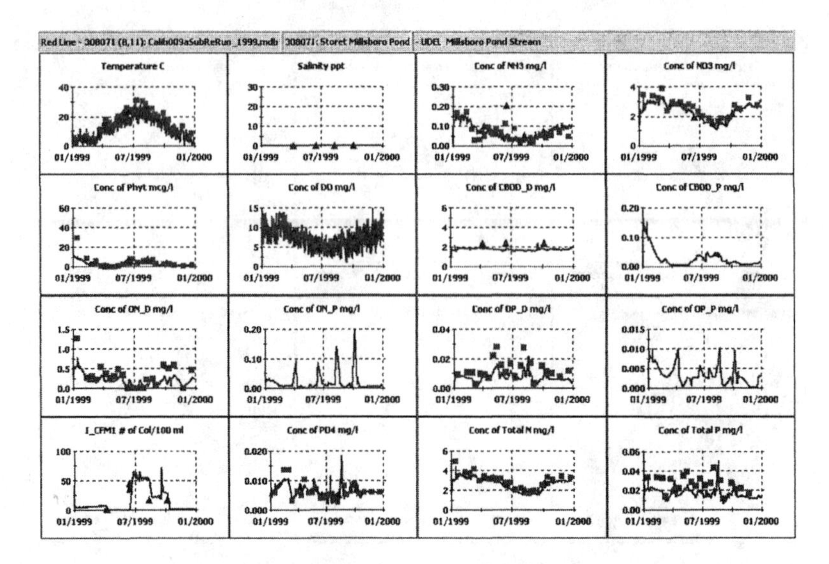

Figure 10 Model versus field data comparison of water quality constituents at the exit of Millsboro Pond for the year 1999

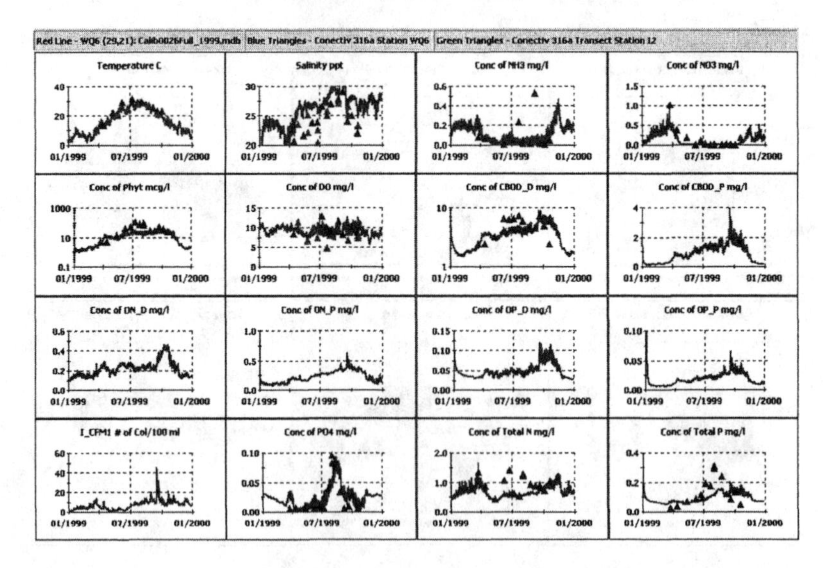

Figure 11 Model versus field data comparison of water quality constituent in the middle region of Indian River for the year 1999

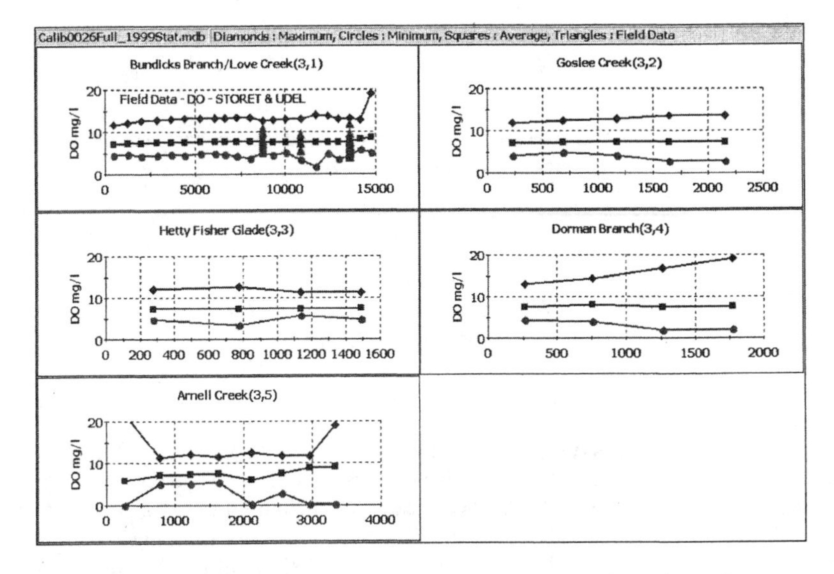

Figure 12 Maximum, minimum and average DO along the various streams in Love Creek network for the year 1999

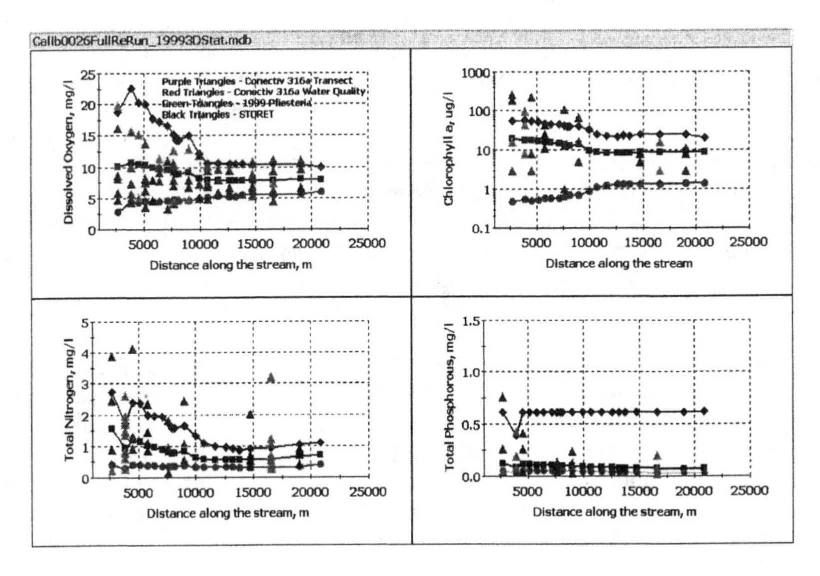

Figure 13 Maximum, minimum and average dissolved oxygen along the Indian River for the year 1999

SST Assimilation and Assessment of the Predicted Temperature from the GoMOOS Nowcast/Forecast System

Huijie Xue*
Lei Shi
Stephen Cousins
School of Marine Sciences, University of Maine

Abstract

A circulation nowcast/forecast system was developed for the Gulf of Maine as an integral component of the Gulf of Maine Ocean Observing System (GoMOOS) technical program. The system has been used to generate nowcasts and short-term forecasts of the circulation and physical properties in the Gulf of Maine. One of the expectations is that the model can provide consistent SST to fill in AVHRR gaps and eventually produce reliable 3D temperature fields for fishery applications.

This paper first presents the framework of the nowcast/forecast system, which includes an algorithm to assimilate satellite derived SST. It then discusses the performance of the system be comparing the predicted and the observed temperatures (both *in situ* and satellite derived). In general, the assimilation algorithm is stable and produces SST patterns mimicking the AVHRR. However, the modeled temperature appears to be colder than the *in situ* temperature at buoy locations. Elimination of the exceptionally cold satellite SST at cloud edges before it is assimilated corrects for a large portion of the cold bias in the model. Furthermore, seasonal variation in temperature is well reproduced and the predicted synoptic temperature variation is significantly correlated with its counterpart from the mooring measured temperature.

1. Introduction

The Gulf of Maine is a productive, marginal sea (Figure 1). Its exchange with the open ocean is largely controlled by the geometry of banks and channels that characterize its open boundary. Inside the Gulf there are three basins separated at the 200 m depth, namely the Jordan Basin and the Georges Basin in the Eastern Gulf of Maine and the Wilkinson Basin in the Western Gulf of Maine. A circulation schematic based on the satellite-tracked drifters and hydrographic observations delineates two distinct cyclonic gyres centered over the two basins in the eastern Gulf

* *Corresponding author address*: School of Marine Sciences, University of Maine, Orono, ME 04469-5741, USA
hxue@maine.edu, (207)581-4318

of Maine and a complex and well-developed cyclonic coastal current system encompassing the gyre pair. It has been documented repeatedly that the gulf-wide circulation is the strongest and most coherent in the summer, but lacks a recognizable pattern in the winter (*e.g.*, Bumpus and Lauzier, 1965, Vermersch et al., 1979, Brown and Irish, 1992), most likely related to the evolving density structure as suggested by Brooks and Townsend (1989) and Brown and Irish (1992).

Processes that influence the density distribution inside the Gulf of Maine include surface heat flux, tidal mixing, river runoff, and the inflow of the Scotian Shelf Water (SSW) and the slope water. For example, Pettigrew et al. (1998) found that the thermohaline structure in the eastern Gulf of Maine was substantially modified by the cold and less saline SSW. Brooks (1990) observed that the slope water spread over Lindenkohl sill and moved from Georges Basin towards northwest. Xue et al (2000) suggested that the surface heat flux plays an important role in regulating the annual cycle of the circulation in the Gulf of Maine by eroding the stratification in the upper water column due to winter cooling and reestablishing the stratification due to summer warming. The Gulf of Maine and Bay of Fundy system is well known for its nearly resonant semi-diurnal tidal responses and vigorous tidal stirring keeps the water vertically well mixed over Georges Bank, the western shelf of Nova Scotia (Loder and Greenberg 1986), and the eastern Maine coast (Pettigrew et al. 1998).

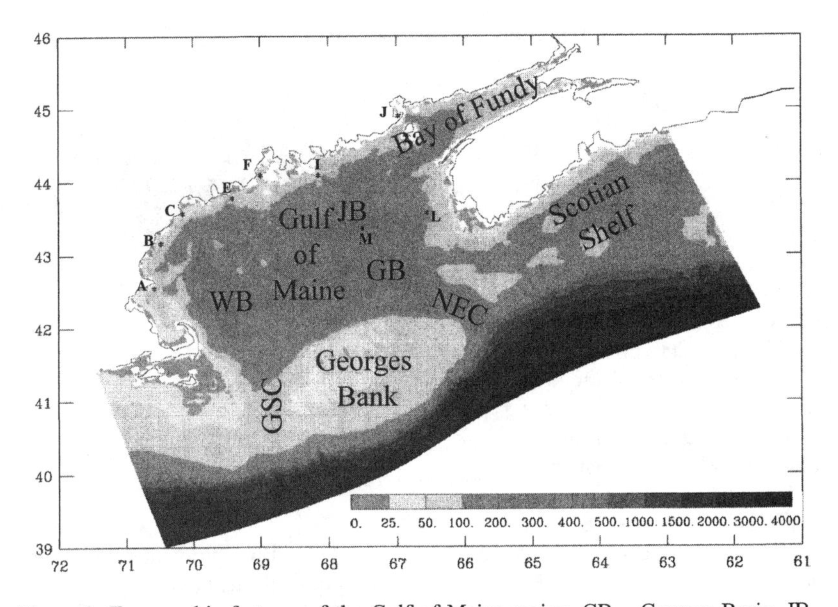

Figure 1. Topographic features of the Gulf of Maine region. GB – Georges Basin, JB – Jordan Basin, WB – Wilkinson Basin, NEC – the Northeast Channel, GSC – the Great South Channel. Also marked in asterisks are the approximate locations of GoMOOS buoys.

The GoMOOS circulation nowcast/forecast system is developed from the comprehensive Gulf of Maine circulation model of Xue et al. (2000). The goal is to establish an operational numerical prediction system for the Gulf of Maine, which produces forecasts of the ocean condition by coupling with available meteorological forecasts and makes them available via the web in real time. One of the expectations is that the model, in a long run, can provide short-term and long-term, three-dimensional temperature variations, which are known to affect the fisheries in the Gulf of Maine. For example, a correlation between warm temperature anomalies and lobster larval settlement has been found in the Penobscot Bay region.

Skillful forecasts are made often by using data to constrain the model in the assimilative manner. The Gulf of Maine circulation nowcast/forecast system now includes assimilation of the satellite SST, and it will also include in the future assimilation of the CODAR measured sea surface velocity, which is now under extensive tests and evaluation. In this paper, only the SST assimilation is discussed. The following section describes the current version of the GoMOOS circulation nowcast/forecast system. Section 3 discusses a SST assimilation scheme and comparisons of predicted and observed temperatures in the upper ocean of the GOM. Section 4 includes a summary of the operational system and a brief outline of the ongoing and planed activities.

2. The Gulf of Maine circulation nowcast/forecast system

The Gulf of Maine circulation nowcast/forecast system is based on the three-dimensional Princeton Ocean Model in a curvilinear grid (Figure 2). It is driven at the surface by heat, moisture, and momentum fluxes from the National Center for Environmental Prediction (NCEP)'s Eta mesoscale atmospheric forecast model. Boundary forcing includes daily river outflows from St. John, Penobscot, Kennebec, Androscoggin, Saco, and Merrimack, tidal (M_2, S_2, N_2, K_1, O_1, and P_1) and subtidal forcing from the open ocean, which is interpolated from the daily nowcast of the NCEP Regional Ocean Forecast System (ROFS).

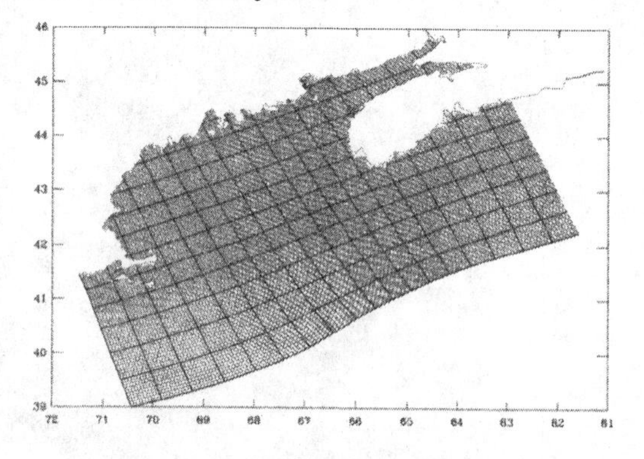

Figure 2. Grid of the Gulf of Maine circulation nowcast/forecast system.

The daily procedure, which starts at 6:00 am everyday, includes three consecutive jobs: preprocessing, model integration, and post processing. Preprocessing consists of a series of automated scripts, which download the river discharge, AVHRR, Eta and ROFS forecasts, and interpolate them to the Gulf of Maine grid. Handling missing data is a critical step. For short-term disruption of river discharge data, the last valid number is carried forward. Climatological monthly mean is used during extensive ice period during winter. For AVHRR, a composite of the last eight days is formed to minimize the cloud cover and, if there are still clouds, an 8-day climatology from the same Julian days is used to fill in the cloudy spots. Missing ROFS data is usually substituted using the ROFS output of the last available date with tidal correction, which work well for interruptions up to a week. Similarly, the last available Eta forecast is used to substitute for missing Eta data. This, however, can result in considerable errors.

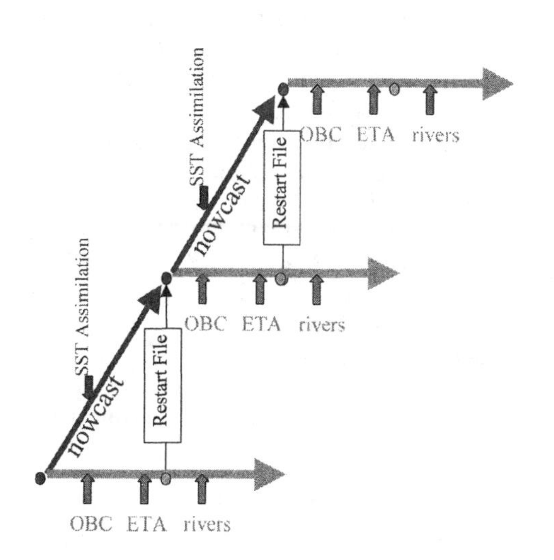

Figure 3. Schematic of the daily operation procedure.

Upon the completion of preprocessing, which supplies necessary boundary conditions, the automated daily procedure calls for the model integration session (figure 3). A 24-hour nowcast cycle assimilates the satellite SST (see details in section 3) and prepares the initial condition for the next 48-hour forecast.

Post-processing includes data storage, web interface, and error assessment (see details below). Model output of every 3 hours has been archived since 1 January 2001, and daily restart files have been saved incrementally for hindcasts of specific events. Temperature, salinity and velocity at three levels and surface elevation are shown graphically on the web at a 3-hour interval (http://gomoos.org). The GoMOOS circulation forecast system is currently running on both the SGI Origin 3200 and a dual processor PC. The daily procedure takes about 20 minutes on the SGI and about 2 hours on the PC. The system is robust, seldom needs human intervention over the last three years other than handling missing forcing data.

Understanding forecast errors is an integral component of the nowcast/forecast system. The 1st level of examination includes the comparison of the modeled and *in situ* temperature and velocity at GoMOOS buoy sites (see Figure 1 for buoy locations). In particular, an 8-day running mean is applied to the temperature time series to separate the synoptic and seasonal variations (see section 3b for details), while harmonic analysis is applied to the velocity time series to separate the tidal current from the residual flows. The quality of predicted tidal current varies from buoy to buoy. In general, the model performs better at three shelf buoys (B, E, and I) but rather poorly at buoy J, which is located in Cobscook Bay where the model does not have adequate resolution. Figure 4 compares the modeled and the observed surface velocity at buoy E and I. The modeled velocity has similar magnitude and variance as the observed velocity. Furthermore, the model appears to catch the timing of seasonal transition of the coastal current. The 2nd level of Quality Assessment (QA) subsystem is currently in the development, which includes the analysis of bias, variances, and spectral properties.

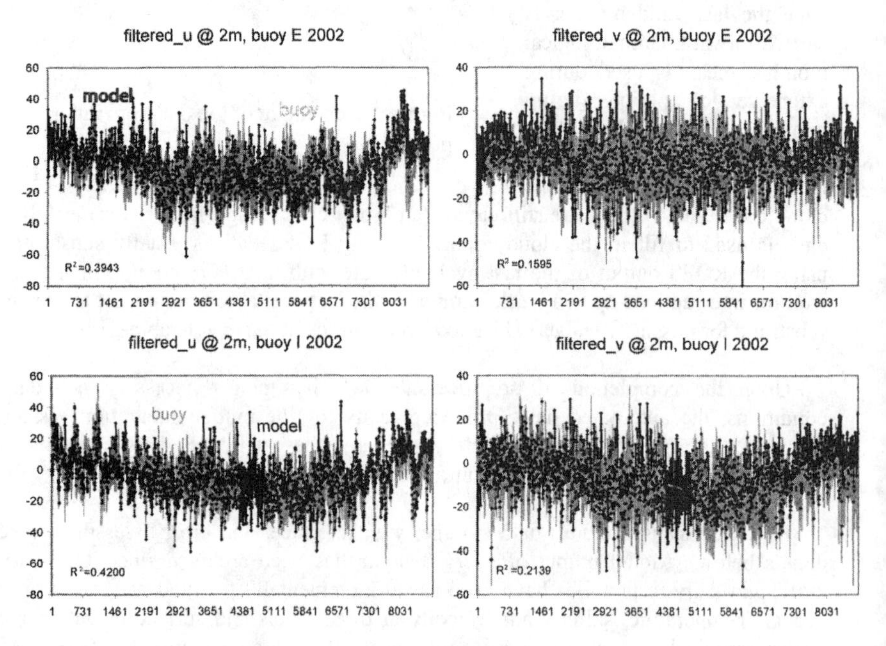

Figure 4. Comparisons of tidal residual velocity at buoy E and I.

3. SST assimilation

In order for a model to produce SST correctly, it is essential to have exact heat balance at the surface. The Eta heat fluxes appear to contain significant errors as suggested by the discrepancy between the observed and the predicted air temperature as seen in Figure 5. As a result, the modeled SST is considerably lower in the summer months and does not show the typical east-west contrast often seen in the AVHRR (Figure 6a and 6b). Furthermore, heat fluxes affect not only the SST in the model, but also the subsurface temperature and the circulation, especially when and where strong mixing occurs. Secondly, since there are periods of extensive cloud covers especially during winter, satellite AVHRR data contains wide gaps. It is expected that the model can be used as a dynamically consistent interpreter to fill in the data gaps.

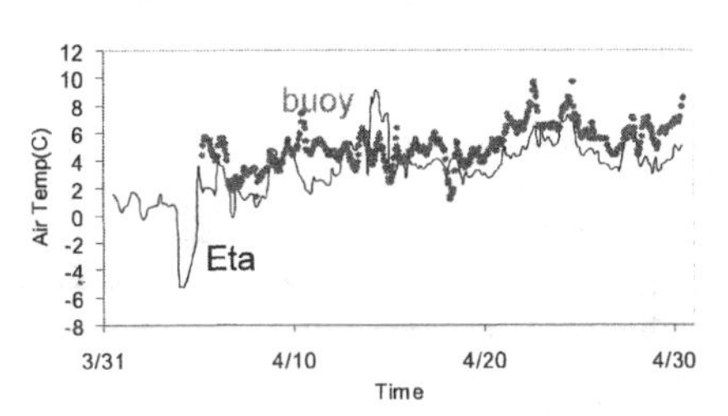

Figure 5. Comparison between the observed and Eta forecasted air temperature at NDBC buoy 44005 in April 2002.

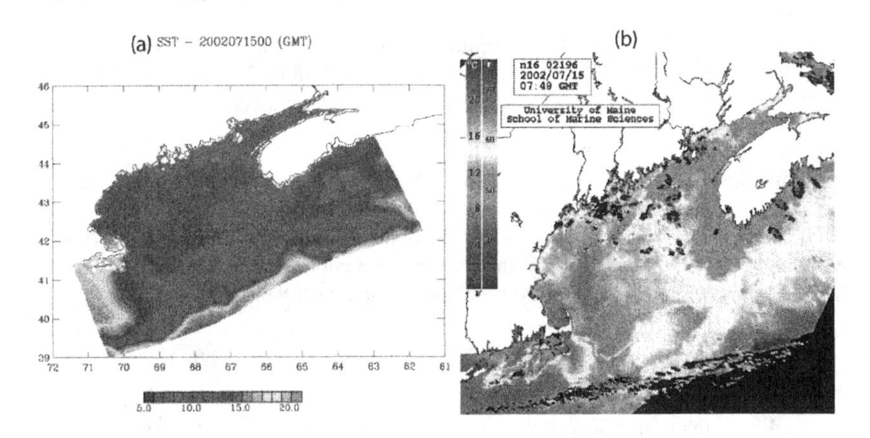

Figure 6. Without SST assimilation, the modeled sea surface temperature (a) appears to be colder and fails in reproducing the typical summertime SST pattern, which has higher temperature in the western Gulf of Maine as seen in the AVHRR (b).

To assimilate satellite AVHRR data into the model, an algorithm similar to that of Kelley et al. (1999) has been incorporated into the GoMOOS nowcast/forecast system since June 2001. The algorithm, chosen mainly because it is very efficient computationally, consists of three elements: an optimal interpolation scheme (Derber and Rosati, 1989; Behringer *et al.*, 1998); a mixed-layer adjustment (Chalikov *et al.*, 1996); and a Newtonian nudging.

3a. Assimilation scheme

The first task is to produce observed daily temperature field. There are 4 passes per day including both ascending and descending tracks of n12 and n14. Data from all 4 passes in the past 8 days are combined to form a composite. It is not unusual that there is still considerable amount of cloud cover in the composite. Two different approaches have been used to fill the sea surface temperatures over the cloud-covered areas as seen in figure 7. One is to use the climatological 8-day composite, another is to use the modeled SST. The latter is effectively without data assimilation.

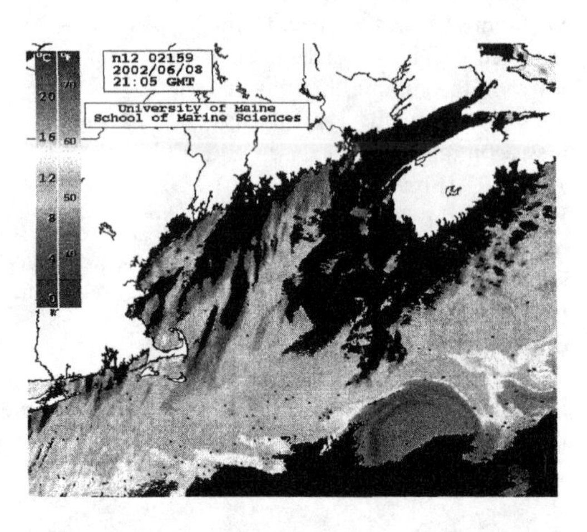

Figure 7. Sea surface temperature used to form the composite on 08 June 2002. Areas covered by clouds are shown in black.

Upon the deriving of the observation field (T_o), the optimal interpolation procedure is to determine a correction field for the model's top layer temperature by simultaneously minimizing two difference fields. One is the difference between the corrected temperature field (T_c) and the model temperature (T_m), and the other is between the corrected temperature field and the observed temperature (T_o).

$$I = (T_c - T_m)^t \Re (T_c - T_m) + (\aleph T_c - T_o)^t \Im (\aleph T_c - T_o) \tag{1}$$

where the superscript, t, denotes the transpose matrix. $\Re$ and $\Im$ are the error covariance matrices for the model data and the observations, respectively. Following Kelley et al (1999), $\Im$ is a diagonal matrix (assuming the observational error is uncorrelated). The value of the diagonal elements of $\Im$ is the observational error variance. A time factor is incorporated into $\Im$ such that more weight is placed on the observations with time closer to the assimilation date. The model error covariance is approximated, for any two grid-points, as $\Re_{ij} = a\ exp[-(r_{ij}/b)^2]$, where r_{ij} is the horizontal distance between two model grid points. a is the first guess error variance and b is the estimate of the correlation spatial scale of the model error, and they are set to 0.50 $^{o}C^2$ and 60 km, respectively. $\aleph$ is the transformation matrix that converts values at the model grid points to the observation locations. Minimization is achieved by using a preconditioned conjugate gradient algorithm (Gill *et al.*, 1981, Golub and Van Loan, 1989), which finds the solution iteratively. Figure 8a is the resultant correction field corresponding to the composite shown in Figure 7.

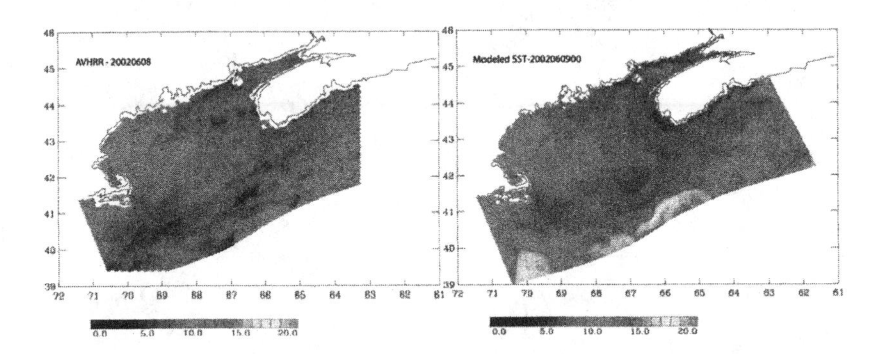

Figure 8. The correction field (a) that is constructed with the optimal interpolation approach, and the SST produced after 1-day integration in the nowcast cycle (b).

SST, however, is not a strong dynamic constraint. Temperature below the surface needs to be adjusted accordingly. A simple procedure is currently employed in the GoMOOS nowcast/forecast system such that when the corrected surface temperature is warmer than the model surface temperature, the correction is distributed throughout the mixed layer, while when the corrected temperature is colder, the corrected temperature replaces the model temperatures down to the depth where they become equal. Newtonian nudging is used to slowly apply the correction field to the model temperature over the nowcast cycle. Figure 8b is the SST resulted from the nowcast cycle using the correction field shown in Figure 8a, which serves as the initial SST in the 9 June 2002 integration.

3b. The modeled temperature

The algorithm is steady and has produced robust sea surface temperature patterns since it was implemented. As seen from Figure 8b, the typical summertime east-west contrast in the sea surface temperature is well simulated, as well the cold temperature on Georges Bank. However, comparisons with *in situ* temperature show that the modeled temperature is colder (Figure 9). This is probably due to the artifact of the exceptionally cold temperature at cloud edges from the satellite SST (see Figure 7) and the assimilation scheme especially the mixed layer adjustment. Note also in this experiment the cloud-covered areas was filled in using the climatological 8-day composite derived from 15 years of AVHRR, which was found later to have a cold bias of ~2°C in the winter. Furthermore, vertical adjustment of the temperature was limited to the top 200m. Several experiments were carried out to test various aspects

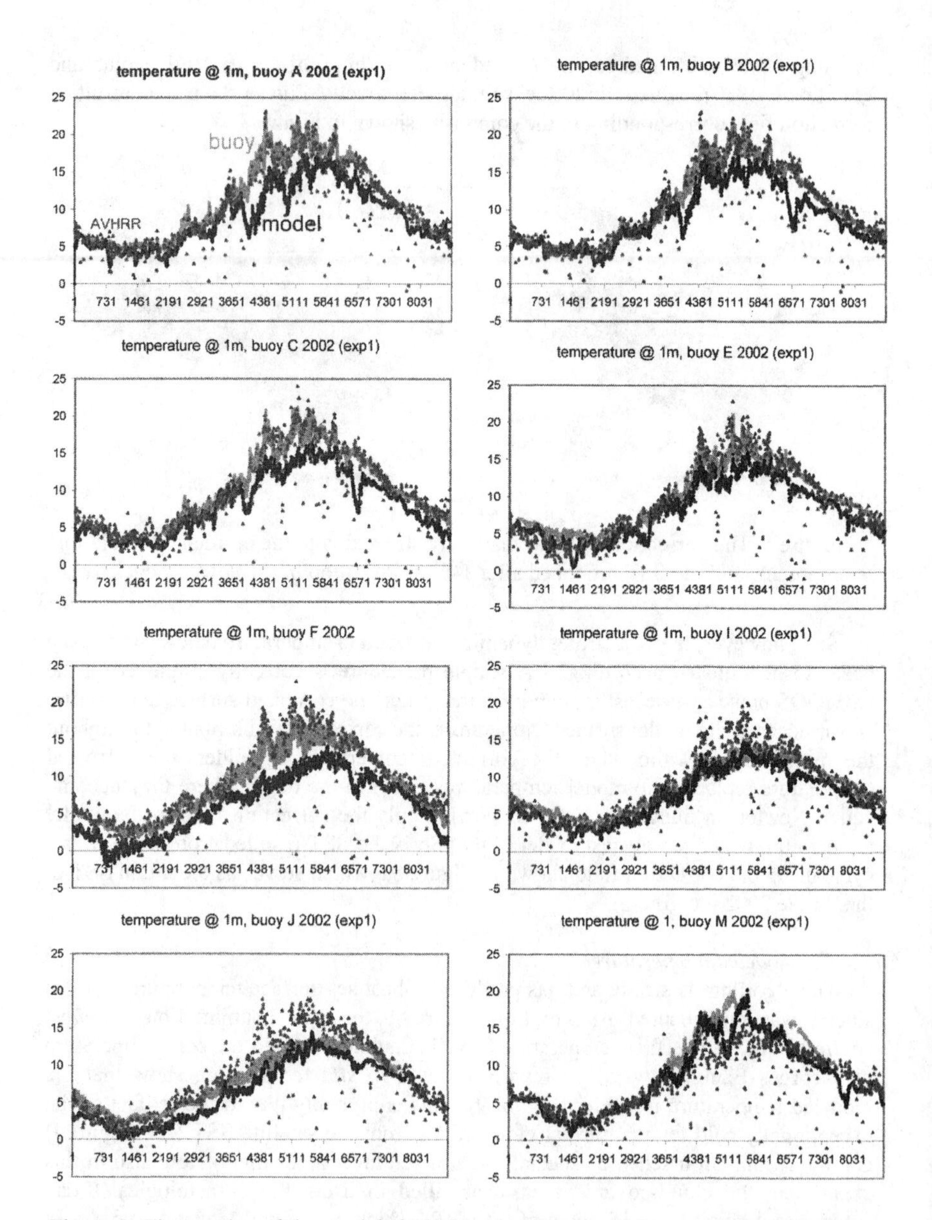

Figure 9. Time series of the modeled (dark curve), in situ (light curve), and satellite (Scattered dots) sea surface temperature at the GoMOOS buoys A, B, C, E, F, I, J, and M.

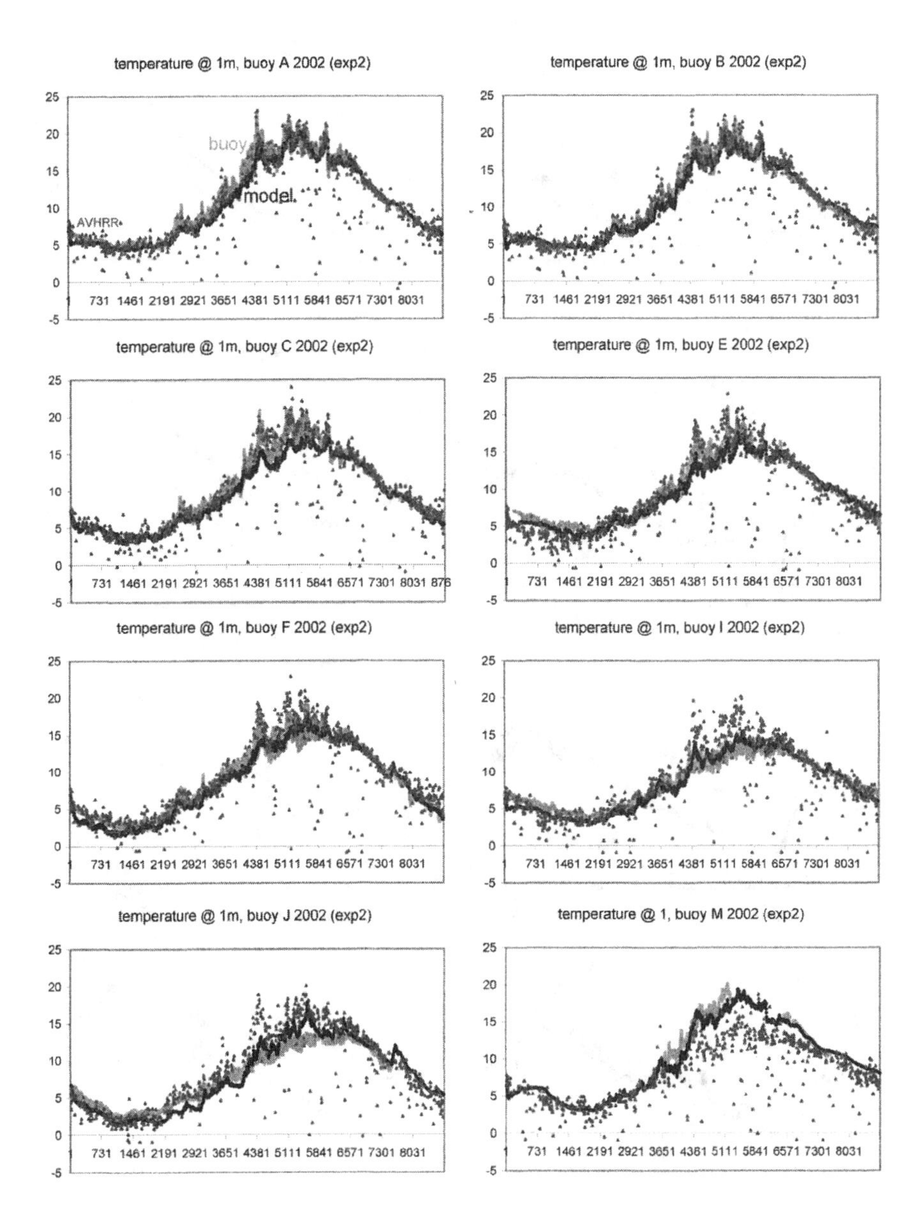

Figure 10. Similar to Figure 9 but for experiment 2. See text for differences between experiment 1 and 2.

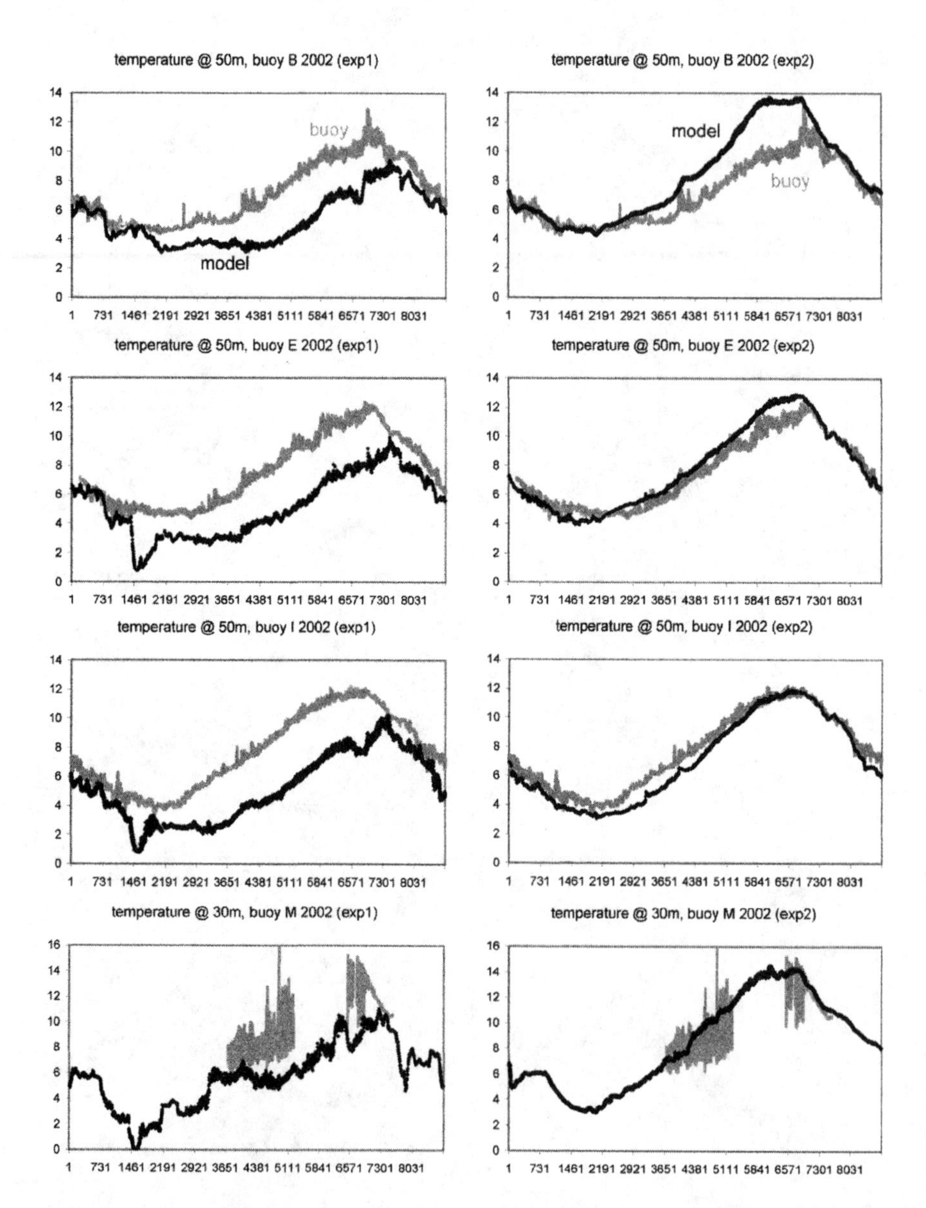

Figure 11. Time series of the modeled and in situ temperature at 50m at the GoMOOS buoys B, E, I, and M. Left panels are for experiment 1, and right panels experiment 2. See text for differences between these two experiments.

of the assimilation scheme. Figure 10 shows similar comparisons of the sea surface temperature for the second experiment in which, 1) the SST correction was limited to 2°C so that the exceptionally cold temperature (artifact of cloud shadows) was excluded; 2) the modeled SST instead of the climatological satellite 8-day composite was used for cloud-covered area in the daily composite, and 3) the vertical adjustment is limited to the top 50m. It is clear that the modeled temperature matches much better with the observed temperature at all buoys. Figure 11 shows the comparisons of the temperature at 50m for both experiments at the buoys with at least 3 months of data in 2002. Although the 50m temperature is too warm at buoy B during summer months, the modeled temperature appears to agree better with the observed temperature at 50m in the second experiment even for buoy B, suggesting that this experiment produces better-fit vertical temperature profiles as well.

Obviously, the largest variability in temperature is the annual cycle, which is then separated from the time series using a simple 8-day running mean. The low-pass filtered data (time series – its 8-day running mean) consists of higher frequency variations including those in the weather band and those with longer periods of several weeks. An example of this remnant time series is shown in Figure 12. Most of the events appear to be well reproduced in the model.

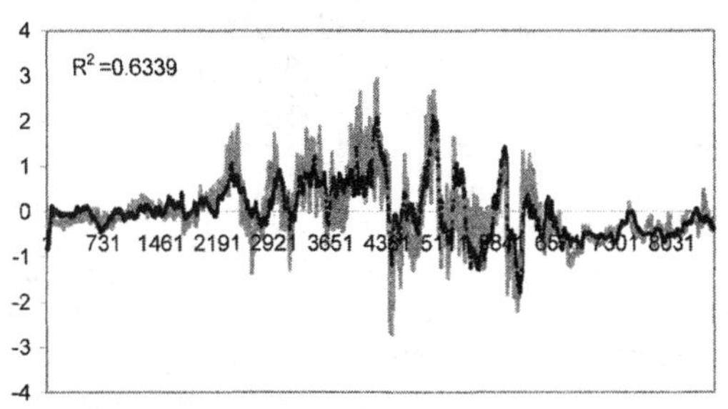

Figure 12. Comparison of the temperature variability (model – dark curve, buoy – light curve) in the weather band.

Another way to measure the ability of the model in predicting the seasonal and synoptic variations in the temperature is the correlation coefficient (Table 1). For the annual cycle, R^2 is greater than 0.90 between the model temperature and the *in situ* temperature for all buoys except M. Note that the observed record at buoy M is significantly shorter. R^2 in the synoptic band ranges from 0.33 to 0.70, again with the exception of buoy M at 30m where R^2 was 0.0606. This low in correlation is due to large amplitude high frequency variations in the observed temperature at this location (see lower right panel in Figure 11).

Spatially, the correlation is higher at the 4 buoys located in the western Gulf of Maine (namely, A, B, C, and E), having to do with the eta model assimilating more information from the land-based weather stations, which is also probably why correlation is higher again at buoy I since it is near the C-MAN on Mount Desert Rock. Correlation is lower in the bays (F and J), which is also consistent with the generally lower correlations for eta predictions in the bays. Finally, the correlation between the modeled temperature and the observed temperature decreases from the surface downward.

Table 1. Correlation coefficients between the modeled and the observed temperature.

	R^2-seasonal	R^2-synoptic band
Buoy A – 1m	0.9824	0.6339
Buoy B –1m	0.9847	0.6984
Buoy C – 1m	0.9698	0.6444
Buoy E – 1m	0.9710	0.5168
Buoy F – 1m	0.9859	0.3780
Buoy I – 1m	0.9856	0.4278
Buoy J – 1m	0.979	0.2951
Buoy M – 1m	0.8593	0.4638
Buoy B –50m	0.9039	0.3274
Buoy E – 50m	0.9615	0.3722
Buoy I – 50m	0.9885	0.4794
Buoy M – 30m	0.8054	0.0606

In general, the annual cycle in temperature is well reproduced with R^2 well above the 95% significance level. The correlation in the synoptic band is much lower for several reasons. First of all, the correlation between the buoy and the satellite temperature in the same frequency band is in the range of 0.3 to 0.4. Secondly, a daily composite of the satellite SST is formed and then assimilated in the model. As a result, the diurnal variation is greatly suppressed in the model. The objective correlation should thus be between the 24hour lowpass filtered series. Thirdly, R^2 between the Eta forecasted and buoy measured wind and air temperature varies from 0.45 to 0.8 with the higher correlation attained in winter. This is probably due to the fact that northwesterly and northeasterly winds dominate in winter and more dense coverage of weather stations on the land supply enough observations to the Eta model and it hence produces more reliable forecasts. In spring and summer, however, a large percentage of the weather systems that affect the Gulf of Maine come from south and the small number of weather buoys in the sea provides only limited information to the Eta model, which results in less satisfactory meteorological forecast. The disassociation between the AVHRR and the in situ SST and the disassociation between the Eta predicted wind and the observed wind both affect the quality of model predicted sea surface temperature, much so in the synoptic frequency band. Nevertheless, the correlation coefficients are above the 95% significance level (except the 30m depth at buoy M) as the time series contain about 2900 data points with degree of freedom over 100, suggesting that the model has certain ability in predicting temperature variability in the synoptic frequency band.

4. Summary

The GoMOOS nowcast/forecast system based on the 3-dimensional Princeton Ocean Model generates daily and short-term forecasts of the circulation and physical properties in the Gulf of Maine. With the realtime meteorological forcing and river discharge, as well as the model (*i.e.*, ROFS) predicted open ocean boundary condition, the nowcast/forecast system is able to produce realtime, 3-dimension distributions of the circulation and water properties for the Gulf of Maine, Bay of Fundy and the adjoining Georges Bank region. Preliminary analyses of the results in 2002 indicate that the model produces realistic seasonal variations of the surface temperature and the coastal current. In the synoptic frequency band, the correlation coefficients between the modeled temperature and the *in situ* temperature are higher in the western Gulf of Maine than in the eastern Gulf of Maine, higher in the shelf than in the bays. More importantly, they are above the 95% significance level and generally comparable or higher than the correlation coefficients between the satellite SST and the *in situ* SST at the same location. As such, it is concluded that the model generated surface temperature can be used to fill in the data gaps of the satellite derived sea surface temperature.

We are continuing to examine the assimilation scheme and the model produced subsurface temperature and stratification. For example, the subsurface temperature in the western Gulf of Maine in particular at buoy B tends to be too warm in summer months. It is noted that the strength of winter convection is stronger in the western Gulf of Maine and the deep convection in winter could have lingering effects on the subsurface temperature in summer. Moreover, buoy M was the only mooring with CTD at greater depths (80, 130, 180 and 240m), and there were about 2 month data in June and July and about 1 month data in October 2002. It appears that the model has a larger seasonal variation at depths. The model results will be compared with NMFS hydrographic data (Taylor et al., 2003). A spatially and temporally varying adjustment depth might be needed and a viable approach may be the feature model of Gangopadhyay et al. (2002). We hope to reach a similar benchmark for the subsurface temperature.

Another on going task is to establish the benchmark for velocity comparisons in different frequency bands, i.e., the tidal, synoptic, and seasonal variability. It is important to understand the source of errors and their attributions to the errors in the open boundary condition, surface wind forcing, and the modeled hydrographic structure and mixing processes. It is hoped that these analyses can lead to new insights in terms of predictability of the model and guide the assimilation of the sea surface velocity measured by CODARs.

CODARs are the high frequency radar units that can be used to map ocean surface currents. To assimilate the CODAR data into numerical ocean models, Lewis et al. (1998) used a shear stress approach, in which the modeled velocity is nudged towards the CODAR measured velocity by imposing an additional shear stress. It was noted that errors in the CODAR data could cause unrealistic horizontal divergence and sea level in the model. More recently, Lipphardt et al. (2000) found that the unrealistic divergence could be limited by first filtering the CODAR velocity field. A CODAR data assimilation scheme similar to that of Oke *et al.* (2000) has been developed. It is a simplified Kalman filter data assimilation system that assimilates low-pass filtered CODAR velocity. The non-homogeneous and non-isotropic forecast error covariance was empirically derived from an ensemble of typical model simulations.

Coastal ocean forecasting is one of the major challenges that the oceanography community faces. Using data to constrain models has been recognized as a key element for successful forecasts. On the other hand, benchmarks for successful predictions need to be established, although they may vary from region to region as limited by the available open boundary condition from larger scale models and by the meteorological forecasts of different resolutions. It is hoped that overall quality of coastal ocean forecasts would be improved as the GODAE products become available and more standardized and as we accumulate the experiences with coastal ocean forecasts.

Acknowledgments. The authors would like to thank Prof. Andrew Thomas and Mr. Ryan Weatherbee at the University of Maine for providing the AVHRR data and Prof. Neal Pettigrew and his group at the University of Maine for providing the buoy data. This study is supported by the Gulf of Maine Ocean Observing System and by the NASA grant (NAG5-10620) to the University of Maine.

4. References

Behringer, D. W., J. Ming, and A. Leetma, (1998): An improved coupled model for El Nino prediction and implications for ocean initialization. Part I: The ocean data assimilation system. *Mon. Wea. Rev*., 126, 1013-1021.

Bloom, S. C., L. L. Takacs, A. M. Da Silva, and D. Ledvina, (1996): Data assimilation using incremental analysis updates. *Mon. Wea. Rev*., 124, 1256-1271.

Brooks, D. A. and D. W. Townsend, 1989: Variability of the coastal current and nutrient pathways in the GOM. *J. Geophys. Res.*, **47**, 303-321.

Brown, W. S., and J. D. Irish, (1992): The annual evolution of geostrophic flow in the Gulf of Maine. *J. Phys. Oceanogr.*, **22**, 445-473.

Bumpus, D. F., and L. M. Lauzier, (1965): Surface circulation on the continental shelf of eastern North America between Newfoundland and Florida. *Amer. Geogra. Sco. Serial Atlas of the Marine Environmet*, Folio 7, Amer. Geogra. Sco., 4pp.

Chalikov, D., L. Breaker and L. Lobocki, (1996): Parameterizatio of mixing in the upper ocean. NWS/NCEP, 40 pp.

Derber, J., and A. Rosati, (1989): A global oceanic data assimilation system. *J. Phys. Oceanogr.*, 19, 1333-1347.

Gangopadhyay, A., A.R. Robinson, P.J. Haley, W.G. Leslie, C.J. Lozano and James J. Bisagni, (2002): Feature Oriented Regional Modeling and Simulation (FORMS) for the Gulf of Maine and Georges Bank. Submitted to *Continental Shelf Research*.

Gill, P. E., W. Murray, and M. H. Wright, (1981): *Practical Optimization*. Academic Press Inc., 401 pp.

Golub, G. H. and C. F. Van Loan, (1989): *Matrix Computations*. The John Hopkins University Press, 642 pp.

Kelley, J. G. W., D. W. Behringer, and H. J. Thiebaux, (1999): Description of the SST data assimilation system used in the NOAA coastal Ocean Forecast System (COFS) for the U.S. east coast. NOAA/NWS/NCEP, OMB Contribution No. 174. 49pp.

Lewis, J. K., I. Shulman, and A. E. Blumberg, (1998): Assimilation of Doppler radar current data into numerical ocean models. *Cont. Shelf Res.*, 18, 541-559.

Lipphardt, B. L., Jr., A. D. Kirwan, Jr., C. E. Grosch, J. K. Lewis, and J. D. Paduan, (2000): Blending HF radar and mdoel velocities in Monterey Bay through normal mode analysis. *J. Geophys. Res.*, 105, 3425-3450.

Oke, P. R., J. S. Allen, R. N. Miller, G. D. Egbert, and P. M. Kosro, (2000): Assimilation of surface velocity data into a primitive equation coastal ocean model. *Submitted to J. Geophys. Res.*

Pettigrew, N. R, D. W. Townsend, H. Xue, J. P. Wallinga, P. J. Brickley, and R. D. Hetland, (1998): Observations of the Eastern Maine Coastal Current and its Offshore Extensions. *J. Geophys. Res.*, **103**, 30623-30640.

Taylor, M. H., C. Bascuñán, and J. P. Manning, (2003): Description of the 2002 Oceanographic Conditions on the Northeast Continental Shelf, Northeast Fisheries Science Center Reference Document 03-05. http://www.nefsc.noaa.gov/nefsc/publications/crd/crd0305/

Vermersch, J. A., R. C. Beardsley, and W. S. Brown, (1979): Winter circulation in the Western Gulf of Maine: Part 2. Current and pressure observations. *J. Phys. Oceanogr.*, **9**, 768-784.

Xue, H., F. Chai, and N. R. Pettigrew, (2000): A model study of seasonal circulation in the Gulf of Maine. *J. Phys. Oceanogr.*, 30, 1111-1135.

A Forecast Circulation Model of the St. Johns River, Florida

Edward P. Myers[1], Frank Aikman[1] and Aijun Zhang[1]

Abstract

We present a nowcast/forecast hydrodynamic model of the St. John's River, Florida. The model is one of the tools being developed for the pilot project of the Coastal Storms Intiative (CSI), an interdisciplinary program NOAA is implementing to enhance the tools and resources available to coastal communities during storm events. CSI provides a one-stop shopping approach for coastal managers to access information they need to make timely informed decisions. While CSI is developing new tools and collecting new data, it is also critical that it build upon existing resources. Such is the case with the circulation modeling of the river, as the St. Johns River Water Management District (SJRWMD) had previously developed a well calibrated application of the Environmental Fluid Dynamics Code (EFDC) in the St. Johns River. To build upon this work, we ported their application to NOAA and integrated it into a real-time system for producing nowcasts and forecasts. Realtime data for winds, water levels, salinity and temperature are used to force the model nowcasts, and forecasts of wind and subtidal coastal water levels are used to drive the forecasts. The output is likewise made available in realtime on a website that will be accessible through the CSI home page. We are also developing an application of ELCIRC (Eularian-Lagrangian Circulation model) in the St. Johns River to examine the potential for simulating storm inundation in flood-prone areas of the watershed.

Introduction

NOAA's Coastal Storms Inititiave (CSI) was developed to help coastal communities better prepare for and respond to the hazards that can occur during storm events. With over half of the U.S. population living in coastal areas, coastal managers need access to a variety of resources to make decisions affecting the safety of their communities. Coastal areas also serve as centers of commerce for tourism, transportation, recreation, fishing and other industries. Coastal storms can therefore cause substantial costs to this infrastructure, which is inextricably linked to the U.S. economy. CSI is designed to provide communities with a portal to critical tools and data that will help them protect lives and minimize damage.

----

[1]National Ocean Service, NOAA, Office of Coast Survey, SSMC-3, N/CS13, 1315 East-West Highway, Silver Spring, MD 20910-3282; phone (301) 713-2809 x103; Edward.Myers@noaa.gov

The pilot project for CSI is in the St. Johns River, Florida. The St. Johns River watershed encompasses sixteen counties and is the longest river within the state. Florida' smost populous city, Jacksonville, lies next to the river, as does the Mayport Naval Station and the Jacksonville Naval Air Station. During a storm event, realtime and forecasted information regarding water levels and currents in the river will therefore affect decisions regarding navigational safety for commercial, military, fishing, and recreational users of the river. Similarly, estimates of flooding potential will influence all homes, businesses and resources in flood-prone areas. Release of hazardous materials into the river will be dispersed, and forecasts of the currents would help to estimate the spatial extent of the spill. Considering all of these factors, an operational nowcast/forecast system of river conditions was determined to be an integral part of the CSI for the St. Johns River.

In meetings with local agencies and users in northern Florida, the St. Johns River Water Management District discussed with NOAA their application of EFDC (Hamrick, 1992a; 1992b) to the St. Johns River. The model application had been well calibrated for a simulation of the river for 1995-1998 (Sucsy and Morris, 2001). To build upon this existing resource, the SJRWMD shared this model application with the NOAA Coast Survey Development Laboratory for development of an operational nowcast/forecast system of the river. The three-dimensional model application computes water levels, salinity and temperature on a curvilinear, orthogonal grid with six vertical sigma layers.

The grid for the EFDC application does not extend onto saltmarshes and other low-lying areas that experience flooding and drying. The lack of resolution in these regions was sacrificed so as to maintain a reasonable runtime for the model hindcast simulation. However, in order to facilitate modeling of the wetting and drying, the ELCIRC model (Myers and Baptista, 2000) is being calibrated for the same domain. ELCIRC is a finite volume/finite difference baroclinic model for unstructured grids. It offers benefits that include: 1) an unstructured grid that can easily be modified to add/coarsen refinement where needed; 2) an advection algorithm that permits large time steps and thus affords more grid refinement if necessary; and 3) a natural treatment of wetting and drying. The finite volume treatement of the continuity equation results in an algorithm that is mass-conservative. The flexibility in adding additional refinement enables grid extension not only into the saltmarshes that experience daily wetting and drying, but also to flood-prone areas along the river that would be vulnerable to storm surge inundation.

Development of a Nowcast/Forecast System with EFDC

The EFDC application was developed by the SJRWMD as part of a program to establish water quality goals in the lower St. Johns River. The model was calibrated for the period 1/1/1995-11/30/1998 and linked to a water quality model to simulate transformation of nutrients and eutrophication in the river. Calibration parameters for the hydrodynamic model included bathymetry, friction, open ocean tides, ocean salinity, the number of vertical sigma layers, and specification of a non-reflective upstream boundary condition.

The primary difference between the nowcast/forecast simulations and the original hindcast simulation lies in the sources of data used as boundary forcings. The data used to force the SJRWMD hindcast simulation were not necessarily all available in realtime. For the nowcasts and forecasts this information needs to be garnered from either realtime data or from forecasts of appropriate variables, either of which need be readily available in a public medium (ftp, website, etc.). We will first look at the model inputs, including the grid and boundary forcings, and will delineate any differences between the nowcast/forecast input specifications and those used in the SJRWMD hindcast simulation. The nowcasts simulate conditions in the river over the past 24 hours and are made every hour in the experimental system. Forecasts of river conditions extending 36 hours into the future are made four times a day.

Grid

The SJRWMD model uses a curvilinear, orthogonal, boundary-fitted grid that extends from the Atlantic Ocean to the upstream boundary at Buffalo Bluff, where the United States Geological Survey (USGS) maintains a river gauge measuring discharge. There are 2,210 water cells embedded within a transformed 188 x 105 rectangular computational grid (Figures 1 and 2). Cell length sizes range from 81m to 2,040m. Six sigma vertical layers were determined to be necessary to produce acceptable stratification.

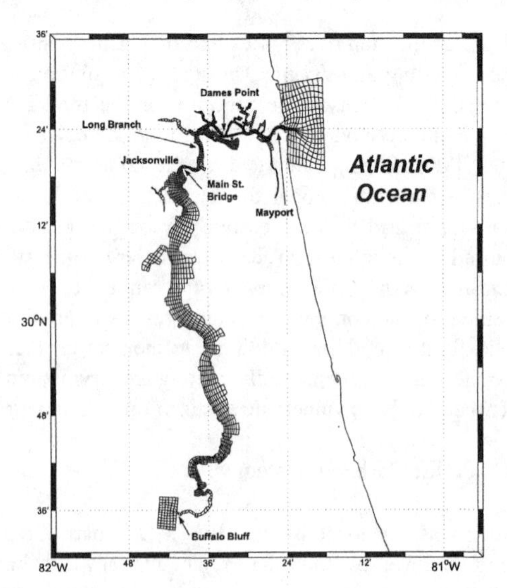

Figure 1. Orthogonal grid used by EFDC.

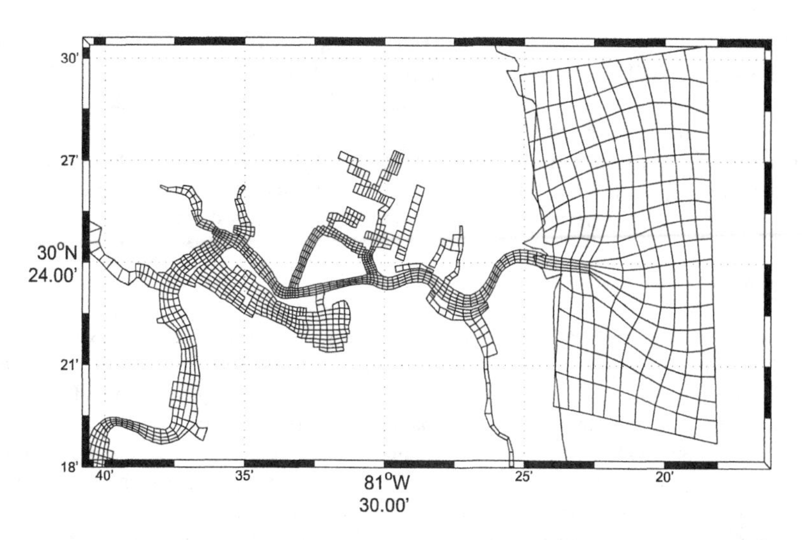

Figure 2. Zoom of EFDC grid in the lower St. Johns River.

Open Ocean Boundary Forcings

For the hindcast simulation of 1/1/1995-11/30/1998, the open ocean boundary of the grid is forced using observations made at NOAA observational instruments in Mayport, Florida. The SJRWMD used time series measurements of total water levels here and extracted the subtidal signal using a FFT low-pass filter for periods less than 30 hours. The tidal signal was taken as the harmonic analysis of the observational time series. Since the Mayport instrument location does not lie in close proximity to the open boundary, the SJRWMD calibrated the tidal harmonics at the boundary by amplifying the amplitudes from Mayport by approximately 5%. This provided the optimal fit between modeled and observed harmonic constituents inside the river. Tidal phases were not adjusted due to the close proximity of the open boundary to Mayport. Optimal salinity boundary forcings were also determined by the SJRWMD to be a linearly varying function of the depth, with a surface value of 35 PSU and a bottom value of 36 PSU.

A similar approach was used for assigning open boundary conditions in the nowcasts and forecasts. For the nowcasts, the total water levels measured in realtime at the NOAA CO-OPS (Center for Operational Oceanographic Products and Services) gauge in Mayport are amplified by a factor of 1.05 for the open boundary condition. This is in slight contrast to the hindcast simulation, where only the tidal amplitudes were increased. Sensitivity tests indicated that both approaches show similar levels of accuracy. Forecasts for the open boundary condition add the tidal water levels from harmonic constituents at Mayport to forecasted subtidal water levels at Fernandina Beach, Florida made by the

National Weather Service Extratropical Storm Surge Model (ETSS). ETSS is an application of the SLOSH model (Jelesnianski et al., 1992) in the Western North Atlantic Ocean. Salinity forcing along the ocean boundary for both the nowcasts and forecasts is the same as that used in the hindcast simulation. Also, while water temperature is held constant throughout the domain in the hindcast simulation, the nowcasts use realtime temperatures from the Mayport data as an open boundary condition. Persistence of the last recorded water temperature at Mayport is used in the forecast simulations.

Upstream River Boundary Conditions

At the upstream boundary of Buffalo Bluff, there is a 28.4 cm tidal range. Without information regarding water levels this far upstream, an upstream boundary condition that effectively dissipates the tidal energy needs to be implemented to avoid artificial reflection of the waves. The SJRWMD applied a sponge boundary condition in the small box of grid cells located at the upstream boundary. Within the sponge, bottom roughness coefficients were set to a large value to damp any incoming long waves. The depth of each of these upstream grid cells was set equal to one meter. The sponge boundary condition was found to be efficient given that river flowrates were also being specified at the upstream boundary.

The USGS maintains a gauge at Buffalo Bluff that monitors river discharge, salinity and temperature. For the hindcast, river discharge and salinity were specified along the western edge of the upstream box of grid cells. While the salinity here is not influenced by the ocean, measurements indicate above-normal levels of chlorine than is typically found in freshwater runoff. Since the long wave tides and oceanic subtidal water levels do propagate as far as Buffalo Bluff, there are actually periods of time when the net discharge in the river at the upstream boundary is flowing out of the grid domain (south, or upstream). The hydrodynamic models therefore need to be able to handle such situations where there is a negative discharge (i.e. flowing out of the grid).

The nowcast simulations use the realtime information (river flowrates, salinity and temperature) from Buffalo Bluff that is posted on the USGS website. The only difference in the boundary condition from the hindcast is that temperatures are also forced. The forecasts presently persist the river discharge, salinity and temperature from the last measurements available. In order to derive a better approach than persistance at the upstream boundary, the CSI framework enabled a unique collaborative project between the National Ocean Service (NOS) and NWS that will be further discussed later in the section titled "NOS-NWS Collaboration on River Forecasts."

Tributary Input to the Model Domain

The main-stem river discharge measured at Buffalo Bluff accounts for about 62% of the total flow entering the lower St. Johns River. The remainder comes from tributaries carrying freshwater drainage from the 424 sub-basins of the lower St. Johns River watershed (Adamus et al., 1997). Roughly 85% of this tributary flow into the lower river is monitored by USGS river gauges. These gauges measure inflow from Dunns Creek, Black Creek, Rice Creek, Sixmile Creek, Deep Creek, Ortega River, and Big Davis Creek. For the hindcast simulation, the SJRWMD estimated discharge from the remaining ungauged areas using a GIS-based hydrologic model (Adamus and Bergman, 1993). The hydrologic model uses land-use and soil types to determine coefficients for translating rainfall into seasonal runoff in each of the tributary basins.

The nowcasts used the realtime discharge data from the USGS river gauges available on the USGS website. A limitation to the nowcasts (and forecasts) is the inability to incorporate a similar approach for estimating the discharge from ungauged basins due to realtime data access limitations. The forecasts currently use persistence of the last available USGS data, but the NOS-NWS collaborative efforts to be discussed later are aiming to ameliorate this approach with new forecast points in the lower St. Johns River.

Atmospheric Forcings

Wind data from the Jacksonville Naval Air Station just south of Jacksonville were used by the SJRWMD as input of the wind forcing to the hydrodynamic model. The hourly winds were specified as spatially-constant throughout the model domain. The site provides reliable unobstructed observations of the wind in the central portion of the domain. However, this data could not be accessed in realtime for input to the nowcast model runs. Therefore, realtime wind observations from the CO-OPS Mayport station are used as spatially-constant, time-varying input to the nowcast simulations. Likewise, while rainfall and evaporation were incorporated into the SJRWMD hindcast simulation, such data were not readily accessible in realtime for the nowcasts.

The forecast simulations interpolate winds from the Eta-12 atmospheric model (Ferrier et al., 2002; Black, 1994) forecasts. Thus, the forecasts differ from the hindcast and nowcasts in using a spatially-variable wind forcing. As part of the CSI suite of improved tools for the region, the NWS office in Jacksonville is developing a high-resolution WRF (Weather Research and Forecasting) model for northeast Florida. This model is implemented on a Linux cluster and will soon supplant the Eta-12 model in the hydrodynamic model forecast simulations.

NOS-NWS Collaboration on River Forecasts

The NWS Southeast River Forecast Center (SERFC) produces forecasts of river stages at select river forecast points. These forecast points are determined by local needs for river forecasts. Presently there are no river forecast points in the model domain, from the mouth of the river to Buffalo Bluff. The USGS realtime gauges downstream of Buffalo Bluff account for an additional 23% of the total riverflow into the lower St. Johns River.

Ideally for the hydrodynamic model forecasts, one would like to have forecasts of the river flows at the same locations where realtime USGS data is provided. While this need does not necessarily match the local demand for new river forecast points, the CSI framework provided a unique opportunity to collaborate efforts between the hydrodynamic modeling NOS is performing and the NWS river forecast modeling.

Therefore, collaboration between NOS and NWS has led to the development of six new river forecast points in the lower St. Johns River. NWS uses a hydrologic model to predict the flow entering the river from the tributaries. This information is then fed into a dynamic routing model, FLDWAV (Fread and Lewis, 1988), to predict a profile of water surface elevations along the river. As part of this collaborative project, NWS will calibrate the FLDWAV model for the six new river forecast points and will develop a flood mapping capability with FLDVIEW (Cajina et al., 2002) to eventually integrate the NOS circulation model and storm surge model output. Thus, through the CSI, tools being developed in different line offices are being effectively linked synergistically to improve the ensemble product available to the coastal communities.

Model Results

Hindcast

As documented by Sucsy and Morris (2001), the SJRWMD hindcast results agreed well with the observations of water levels, currents, salinity, and river discharge. One of the first tests performed by the SJRWMD was to examine the accuracy of the M2 tidal propagation in the river. Their percent errors in the M2 amplitude at 26 active and historic tide stations ranged from 0.6-14% (average error was 5.6%), while the maximum M2 phase error was 6 degrees (average phase error of 2.8 degrees). Thus, the model correctly accounted for frictional effects on the tidal propagation. Tidal velocities were similarly well reproduced in their model to within 10% of the observed velocities at four ADCP sites deployed by NOAA in the lower estuary. An example comparison of surface and bottom velocities near Dames Point is displayed in Figure 3 for one of these NOAA stations where ADCP measurements were available. The SJRWMD also analyzed modeled and observed subtidal water levels. These rms errors ranged from 2.8-5.6 cm and correlation coefficients with the data were all above 0.94.

The salinity can vary significantly in the lower St. Johns River, yet the SJRWMD model was able to accurately capture tidally-averaged values (rms errors of 0.02-2.4 psu), monthly averaged values (rms errors of 0.1-1.7 psu), and long-term averaged values (generally within 0.2 psu). An example comparison of modeled and observed salinity at four locations is shown in Figure 4. The model was able to reproduce well the changes in salinity at tidal periods and the daily salinity range for downstream locations that are more influenced by oceanic salinity.

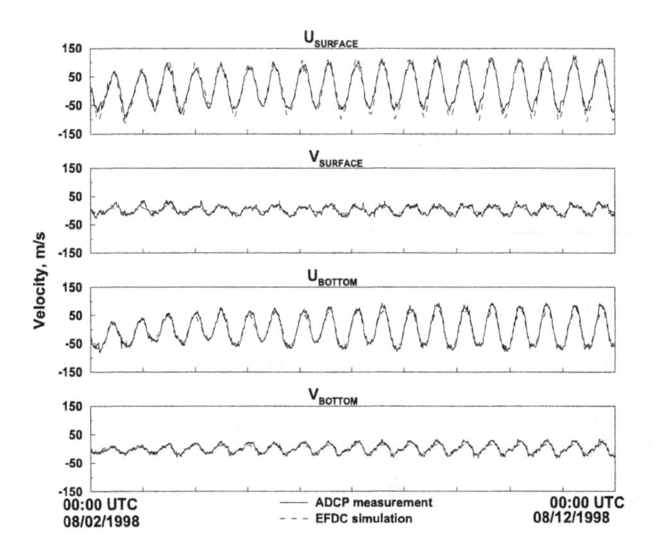

Figure 3. Hindcast model velocities compared with observations at the surface and bottom near Dames Point.

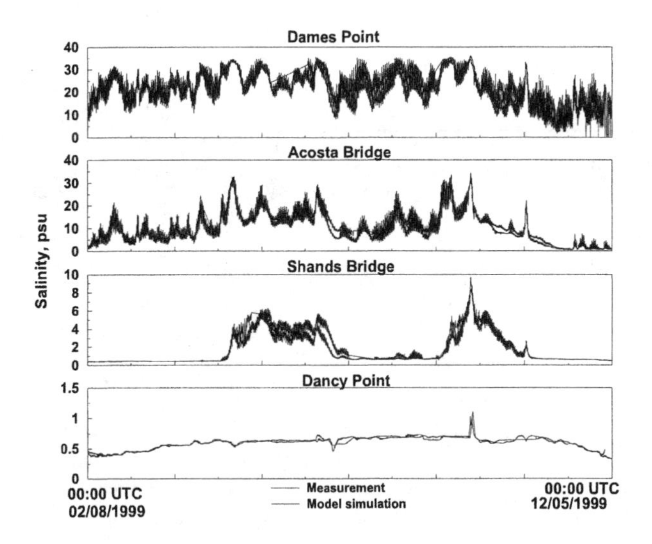

Figure 4. Comparison of hindcast model results and observed salinity at four locations.

Nowcast/Forecast Model Results

The hourly nowcasts and the four forecasts made daily are now run experimentally within the NOAA Coast Survey Development Laboratory (CSDL). Figures 5a-b show example plots of output from the experimental St. Johns River nowcast/forecast system. Results were compared with available realtime data and/or tidal predictions. During this phase of the development, the models were standardized in terms of input and output. They will also be validated according to NOS skill assessment standards (NOS, 1999). Upon successful standardization in both these respects, the model will then be ported to CO-OPS for operational implementation.

The St. Johns River nowcast/forecast system has been adjusted to use a standard template for accessing input data for the model and for disseminating results to output files. This is accomplished using the NGOFS (Next Generation Operational Forecast System) approach under development within CSDL. For input to all the CSDL models, NGOFS is designed to provide a standard mechanism of retrieving input data for the models. Likewise, all CSDL models will output files formatted using uniform netcdf standards.

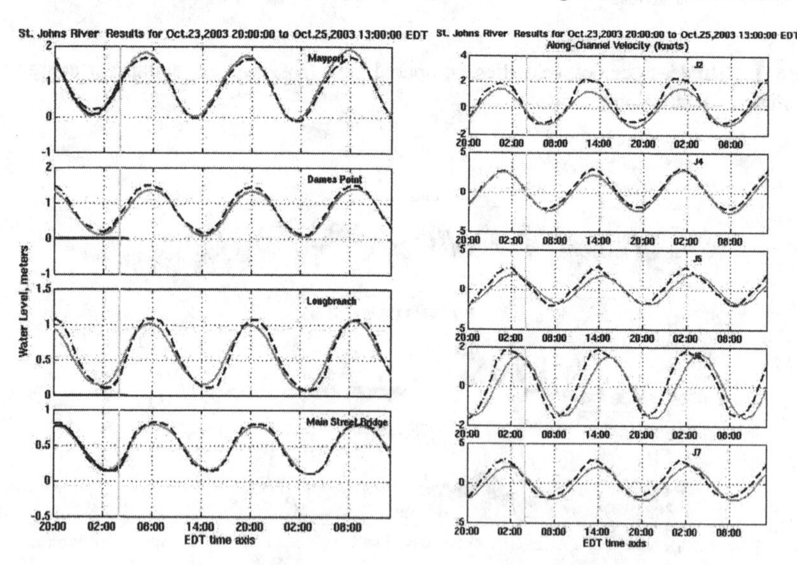

Figures 5a-b. (a, left column) Web page graphic comparing nowcast/forecast water levels with observations and tide predictions. (b, right column) Comparison of along-channel velocities between the nowcasts/forecasts and tidal predictions. In both graphs, the dark solid lines are observations, the solid gray line curve is the tide prediction, the vertical gray line is the ending time of the nowcast, and the dashed lines are the nowcasts and forecasts. The along-channel velocities currently do not have observations available.

To properly validate the models in a systematic manner, CSDL and CO-OPS have developed standards for evaluating the skill assessment of NOS hydrodynamic models (NOS, 1999). These standards are currently being updated within NOS, and the St. Johns River model is presently being evaluated under these new standards (Hess et al., 2003). The general approach of these skill assessment standards is to measure the performance of the model in simulating astronomical tidal variability, simulating total water level variability in both the model development stage and the operational environment, and in providing a more accurate forecast than the tide tables and/or persistance. The primary variables assessed in the skill assessment include the magnitude of the water level, times and amplitudes of high and low water, speed and direction of the currents, times and amplitudes of the maximum flood and ebb currents, start/end times of slack water, and water density effects on buoyancy.

Future Directions

Data

The CSI is also improving the available data in the St. Johns River watershed. The NWS National Data Buoy Center, CO-OPS, and the State of Florida Department of Environmental Protection are working together to standardize and integrate their efforts. This enhanced network of data in northeast Florida will provide improved wind, barometric pressure, air temperature, water level, water current, water temperature and conductivity measurements. These data can serve to improve both the inputs to the hydrodynamic model as well as to provide more validation of the model performance. Assimilating much of this data into the realtime model simulations will also be investigated. CSI has enabled the NOS Office of Coast Survey to collect new bathymetric data that will supplant the existing bathymetry in the model grid.

Modeling Wetting/Drying of Flood-Prone Areas

In order to mitigate the risks of coastal storms, predictions of potential overland flooding are an invaluable tool to the local community. To facilitate the modeling of flooding during storm events, the hydrodynamic model ELCIRC is being examined for use in the nowcast/forecast system. Using NOS skill assessment standards, ELCIRC and EFDC will be evaluated as to their performance accuracy in St. Johns River applications. ELCIRC offers the flexibility to easily modify the numerical grid composed of triangular elements. The grid can also be extended onto land, so that wetting and drying of elements can occur during the model simulation. A finite volume treatment of the continuity equation permits a natural treatment of wetting and drying in the algorithm. In addition, ELCIRC uses a Lagrangian treatment of advection that allows larger time steps to be used, thus providing flexibility in the computation time to add more (horizontal or vertical) refinement to the grid. An example unstructured grid that may be used by ELCIRC for the St. Johns River is displayed in Figure 6. In preliminary calibrations of ELCIRC to the St. Johns River, it was found that larger time steps allowed significantly more horizontal and vertical refinement (10389 nodes and 34 vertical levels) to be added without compromising computational time.

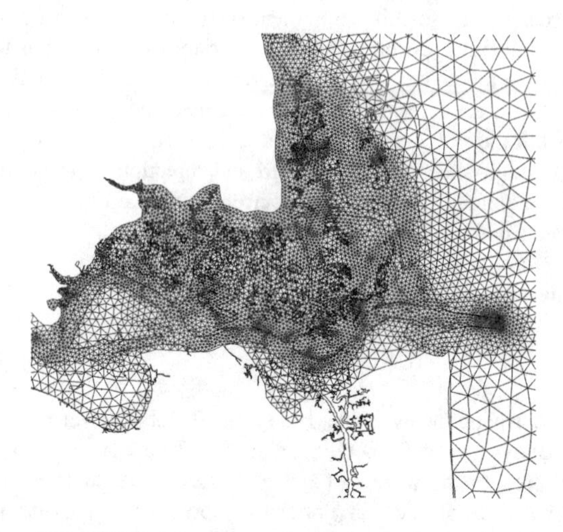

Figure 6. Portion of ELCIRC' s unstructured grid for the lower St. Johns River.

CSI Products, Protecting Communities

The experimental nowcast/forecast hydrodynamic model for the St. Johns River is available online at http://nauticalcharts.noaa.gov/csdl/op/stjohns.htm. The experimental model is still under development within CSDL and will eventually be ported to CO-OPS, once skill assessment and standardization procedures have been completed. The model results will soon be available on the CSI website, http://www.csc.noaa.gov/csi/. Other CSI projects include collecting new bathymetry, improving observations, ecological assessment, improving weather prediction, a risk and vulnerability tool, an inland flood evacuation tool, outreach and extension, and data access and standards. All of these resources and tools will be available from the CSI website, thus providing a one-stop shopping approach for coastal managers to access information that can help them mitigate the impact of incoming storms.he river that would be vulnerable to storm surge inundation.

Acknowledgements

We are very grateful to Peter Sucsy, Frederick Morris and Tim Cera of the St. Johns River Water Management District for their help in transferring their St. Johns River application of EFDC to the NOAA Coast Survey Development Laboratory. This work has benefited from close interaction with the NOAA Southeast River Forecast Center and the Jacksonville Weather Forecast Office and has been supported through the NOAA Coastal Storms Initiative.

References

Adamus, C., D. Clapp and S. Brown, 1997. Surface Water Drainage Boundaries, St. Johns River Water Management District: A Reference Guide. *Tech. Publ. SJ97-1*, St. Johns River Water Management District, Palatka, Florida.

Adamus, C. and M. Bergman, 1993. Development of a Nonpoint Source Pollution Load Screening Model, *Tech. Mem. No. 1,* Department of Surface Water Programs, St. Johns River Water Management District, Palatka, Florida.

Black, T., 1994: The new NMC mesoscale Eta model: Description and forecast examples. *Wea. Forecasting*, 9, 265-278.

Cajina, N., J. Sylvestre, E. Henderson, M. Logan and M. Richardson, 2002. FLDVIEW: The NWS flood forecast mapping application, *18th International Conference on IIPS, Interactive Symposium on AWIPS*, Orlando, Florida.

Ferrier, B. S., Y. Jin, Y. Lin, T. Black, E. Rogers, and G. DiMego, 2002: Implementation of a new grid-scale cloud and precipitation scheme in the NCEP Eta Model. *Preprints, 19th Conference on Weather Analysis and Forecasting/15th Conference on Numerical Weather Prediction*, San Antonio, TX, Amer. Meteor. Soc., 280-283.

Fread, D. L. and J. M. Lewis, 1988. FLDWAV: A Generalized Flood Routing Model. *Hydraulic Engineering, Proceedings of 1988 Conference*, HY Div, American Society of Civil Engineers, Colorado Springs, CO, pp. 668-673.

Hamrick, J.M., 1992a. A Three-Dimensional Environmental Fluid Dynamics Computer Code: Theoretical and Computational Aspects. *Special Rept. 317*. The College of William and Mary, Virginia Inst. of Marine Sciences, Virginia.

Hamrick, J.M., 1992b. Estuarine environmental impact assessment using a three-dimensional circulation and transport model. In *Estuarine and Coastal Modeling, Proceedings of the 2nd International Conference*, ASCE, New York.

Hess, K.W., T.F. Gross, R.A. Schmalz, J.G.W. Kelley, E. Wei and M. Vincent, 2003. NOS Standards for Evaluating Operational Nowcast and Forecast Hydrodynamic Model Systems, *NOAA Technical Report NOS CS 17*, 48 pp.

Jelesnianski, C.P., J. Chen and W.A. Shaffer, 1992. SLOSH: Sea, Lake, and Overland

Surges from Hurricanes, *NOAA Technical Report NWS 48*, 71 pp.

Myers, E.P. and A.M. Baptista, 2000. Baroclinic Modeling of River-Dominated Tidal Inlets and Plumes. *XIII International Conference on Computational Methods in Water Resources*, Calgary, Alberta, Canada, June 2000.

NOS, 1999. NOS Procedures for Developing and Implementing Operational Nowcast and Forecast Systems for PORTS. *NOAA Technical Report NOS CO-OPS 30*, 33 pp.

Sucsy, P.V. and F.W. Morris, 2001. Calibration of a Three-dimensional Circulation and Mixing Model of the Lower St. Johns River, *St. Johns River Water Management District*, Palatka, Florida.

Towards An All Weather Nowcast/Forecast System for Galveston Bay

Richard A. Schmalz, Jr.[1], M. ASCE

Abstract

The National Oceanic and Atmospheric Administration has developed an experimental nowcast/forecast system over Galveston Bay using a modified version of the Blumberg-Mellor (1987) three-dimensional hydrodynamic model as discussed by Schmalz (1998a,b,c; 2000a,b,c). In addition, a one-way coupled fine resolution Houston Ship Channel model (Schmalz, 1998a; 2001) has also been incorporated into the system. The nowcast component works directly from the PORTS universal flat file format (PUFFF) files, while during the forecast the NWS Aviation and Extratropical Storm Surge Models are used to provide the meteorological and Gulf of Mexico subtidal water level residual forcings, respectively. Nowcast and forecast results have been assessed (Schmalz and Richardson 2002) over the one-year period April 2000 through March 2001 based on the NOS (1999) formal acceptance statistical criteria. Event analysis was also performed for both water levels and principal component direction currents (Richardson and Schmalz, 2002a,b).For water levels, a majority of the formal acceptance targets were met and considerable skill was achieved in forecasting low water level events associated with winter cold frontal passages. Currents were more problematic especially at Morgans Point at the head of the Bay, where rainfall/runoff flows exhibit considerable impact on currents within the Port of Houston.

To improve the current response and to move towards an all weather capability, the following processes have been considered: 1) rainfall/runoff inflows from four major basins within the City of Houston, 2) overland flooding, 3) tropical storm and hurricane wind and pressure fields, and 4) surface waves. Herein, the algorithms used to describe each of the above processes are first developed. Next, the design of the all weather nowcast/forecast system is presented. Initial test plans for Tropical Storm Allison in 2001 and for more extreme storms are then discussed. Finally, future enhancements of a generalized vertical coordinate and incorporation of wave/current interactions are outlined.

Introduction

The National Ocean Service (NOS) installed a Physical Oceanographic Real Time System (PORTS), patterned after the system described by Appell et al. (1994), in June 1996 to

1 Oceanographer, National Oceanic and Atmospheric Administration, National Ocean Service, Coast Survey Development Laboratory, Marine Modeling and Analysis Programs, 1315 East-West Highway, Rm 7824, Silver Spring, Maryland 20910. (301)-713-2809 x104; e-mail: Richard.Schmalz@noaa.gov.

provide water surface elevation, currents at prediction depth, which is considered 4.7m below MLLW, as well as near-surface and near-bottom temperature, near-surface salinity, and meteorological information at the locations shown in Figure 1. To complement the PORTS, a nowcast/forecast system has been developed based on the NOS Galveston Bay three-dimensional hydrodynamic model and the National Weather Service (NWS) Aviation atmospheric, Western Gulf River Forecast Center (WGRFC) river, and Extratropical Storm Surge (ETSS) models.

A three-dimensional sigma coordinate Galveston Bay and near shelf (GBM) model has been developed (Schmalz, 1996) as shown in Figure 2 based on a version of the Blumberg and Mellor (1987) model extended to orthogonal curvilinear coordinates. The GBM computational grid consists of 181x101 horizontal cells (dx = 254-2482m, dy= 580-3502m) with 5 levels in the vertical. In addition, a high resolution Houston Ship Channel model (HSCM) has been incorporated as shown in Figure 3 to provide finer spatial resolution (Schmalz, 1998a; 2000a). The_refined channel grid was developed in three sections. Each grid section was linked in order to develop the final composite channel grid consisting of 71 x 211 horizontal cells (dx=63-1007m, dy=133-1268m) with the same 5 sigma levels as in the GBM. The two models were then nested in a one-way coupling scheme, wherein GBM water surface elevation, salinity, temperature, turbulent kinetic energy, and turbulent length scale time histories were saved at 6-minute intervals to provide boundary conditions to drive the HSCM. For salinity, temperature, turbulent kinetic energy, and turbulent length scale a one-dimensional (normal to the boundary) advection equation is used. On inflow GBM values are advected into the HSCM domain, while on outflow HSCM internal values are advected through the boundary. The experimental nowcast/forecast system has been further described and demonstrated by Schmalz (1998a,b,c; 2000a) and consists of a daily 24 hour nowcast followed by a 36 hour forecast based on the 00 UTC NWS cycle. The experimental system has been evaluated over the one year period April 2000 through March 2001 using the NOS (1999) formal procedures. Event analysis was also performed for both water levels and principal component direction currents (Richardson and Schmalz, 2002a,b). In general, the water surface elevation nowcast and forecast results meet or exceed the acceptance criteria, except for the timings of high and low waters. For principal component direction currents, the acceptance criteria are generally met, except for the timings of the zero crossings (slack before ebb and slack before flood), peak ebb, and peak flood currents.

To further address navigation safety concerns, improvements in the current response were sought by including additional inflows in the Port of Houston. The extension of the nowcast/forecast system to handle hurricanes, tropical storms, rainfall/runoff events is not only desirable from a navigation and public safety perspective, but also to determine the Bay salinity stratification response to freshwater pulses which effect the Bay residence times. To move towards these objectives, the following processes have been investigated: 1) rainfall/runoff inflows from four major basins within the City of Houston, 2) overland flooding, 3) tropical storm and hurricane wind and pressure fields, and 4) surface waves. Herein, the algorithms used to describe each of the above processes are first developed. Next, the design of the all weather nowcast/forecast system is presented, followed by test plans for Tropical Storm Allison in 2001, Hurricane Carla in 1961, and Hurricane Alicia in 1983. Finally, future enhancements of a generalized vertical coordinate and incorporation of wave/current interaction are outlined.

Rainfall/Runoff

In conjunction with the Houston Urban Runoff Program, the USGS has obtained streamflow and rainfall data for major drainage basins throughout the City of Houston from 1964-1989. In an effort to characterize the influence of development on drainage characteristics, sets of regression equations for basins north (based on 408 storms) and south (based on 331 storms) of Buffalo Bayou have been developed by Liscum (2001) for the major descriptors of the runoff as given in Table 1. To apply these relations to the City of Houston, four major basins are considered with basin characteristics given in Table 2.

Table 1. Major Runoff Descriptors for use in USGS rainfall/runoff regression relations

Stormwater Runoff Characteristics

QPEAK = Peak flow (cfs)
RODUR= Runoff duration (hrs)
TRISE=Time of rise from base flow to QPEAK (hrs)
Q75DUR=Duration of flow that equals or exceeds 75 percent of QPEAK (hrs)
Q50DUR=Duration of flow that equals or exceeds 50 percent of QPEAK (hrs)
TRECES=Time of recession from QPEAK to base flow (hrs)
BLAG=Duration from centroid of storm rainfall to centroid of direct runoff (hrs)

Rainfall Event Characteristics

RTOT=Total rainfall (in)
R60MAX=Maximum 60-minute rainfall (in)
RDUR=Rainfall duration (hrs)
R85DUR=Shortest 85 percent RDUR (hrs)
RI=Antecendent rainfall index (in)

Table 2. Basin Characteristics for Major Runoff in City of Houston. Note Region 1 is located North and Region 2 South of Buffalo Bayou.

Characteristic	Greens Bayou	Brays Bayou	Sims Bayou	Hunting Bayou
USGS Gage No.	08076700	08074760	08075500	08075770
Region	1	2	2	1
Total Basin Drainage Area (DA) (mi^2)	182.0	94.9	20.2	16.1
Gage Drainage Area (DA) (mi^2)	182.0	94.9	64.0	16.1
Basin Development Factor (BDF) (0-12)	7	8	8	8
Basin Slope (SL) $(ft/(ft/mi)^{0.5}$	4.5	3.9	3.9	3.9
Base flow (cfs)	0.0	0.0	0.0	0.0

Daily hyetographs are separated into *nevent*, rainfall events by specifying a no rainfall minimum duration of 12 hours between events. Then for each event, the rainfall event characteristics in Table 1 are computed for use in the regression equations given in Table 3.

The antecedent rainfall index, R^i_t , is a key parameter and is determined at the start of each event based on the hourly rainfall of the previous 5 days using the following relation:

$$RI_t = RI_{t-1}k^{\Delta t} + r_{\Delta t} \qquad t = 1,\dots n \tag{1}$$

where

RI_0=Initial value (inches)
RI_t=Reduced value of the index t hours later (inches)
k=Recession factor set to 0.9 after Linsley et al. (1982)
$\Delta t = T_t - T_{t-1}$ (hrs)
$r_{\Delta t}$=Rainfall during Δt

Table 3. USGS Runoff Characteristic Regression Equations from Liscum (2001)

Region	Equation	R^2 (percent)
1	$QPEAK=312\ DA^{0.728}(13\text{-}BDF)^{-1.04}RTOT^{0.984}RI^{0.135}$	87
2	$QPEAK=312\ DA^{0.735}(13\text{-}BDF)^{-1.07}RTOT^{0.837}RI^{0.093}$	87
1	$RODUR=13.7\ DA^{0.199}(13\text{-}BDF)^{0.227}RTOT^{0.298}R85DUR^{0.154}$	67
2	$RODUR=10.7\ SL^{0.410}(13\text{-}BDF)^{0.439}RTOT^{0.274}R85DUR^{0.117}$	66
1	$TRISE=1.93\ DA^{0.199}RTOT^{0.741}R60MAX^{-0.519}R85DUR^{0.255}$	55
2	$TRISE=1.63\ DA^{0.278}RTOT^{0.501}R60MAX^{-0.357}R85DUR^{0.266}$	50
1	$Q75DUR=0.374\ DA^{0.294}(13\text{-}BDF)^{0.678}RTOT^{0.678}R60MAX^{-0.425}$	77
2	$Q75DUR=0.367\ DA^{0.274}(13\text{-}BDF)^{0.646}RTOT^{0.682}R60MAX^{-0.431}$	55
1	$Q50DUR=0.665\ DA^{0.287}(13\text{-}BDF)^{0.723}RTOT^{0.625}R60MAX^{-0.454}$	80
2	$Q50DUR=0.624\ DA^{0.271}(13\text{-}BDF)^{0.711}RTOT^{0.665}R60MAX^{-0.471}$	57
1	$TRECES=9.50\ DA^{0.223}(13\text{-}BDF)^{0.337}RTOT^{0.282}R85DUR^{0.084}$	62
2	$TRECES=9.17\ SL^{0.435}(13\text{-}BDF)^{0.473}RTOT^{0.314}RI^{0.091}$	58
1	$BLAG=0.720\ DA^{0.333}(13\text{-}BDF)^{0.781}RTOT^{0.126}R85DUR^{0.106}$	77
2	$BLAG=0.693\ SL^{0.481}(13\text{-}BDF)^{0.969}RTOT^{0.296}RI^{0.110}$	62

For each $t=1,\dots n$, the event flows Q^i_t for $i=1,nevents$ are summed to obtain the basin hydrograph.

To test the scheme, basin rainfall records were obtained from 3-12 June 2001 during Tropical Storm Allison from the Harris County Flood Control District. Computed average daily runoff flows using the above method were compared with the USGS records as shown in Table 4.

Table 4. Average Daily Flow (cfs) Comparison June 3-12, 2001 during Tropical Storm Allison. Note Pred=Prediction, Obs=Observation, and e following observation denotes an estimated value.

Date 2001	Greens Bayou		Brays Bayou		Sims Bayou		Hunting Bayou	
	Pred	Obs	Pred	Obs	Pred	Obs	Pred	Obs
June 3	0	-	0	107	0	8.8	0	4.8
June 4	0	-	0	106	0	7.3	0	5.5
June 5	0	1110	166	3100	32	862	0	378
June 6	6438	13500	4927	960	876	809	866	1290
June 7	5520	2540	2013	5560	476	3210	787	332
June 8	2938	2380	1001	3000	200	939	257	601
June 9	37609	59300e	12377	14000e	1021	4650	4172	2950
June 10	17886	41100e	4839	530e	358	109	405	1320
June 11	1240	4240e	210	210	110	27	17	103
June 12	54	702	9	137	5	17	1	45
Total Rainfall(in)	34.07		21.07		12.90		24.11	
Total Pred Runoff (in)	13.82		9.49		5.39		14.06	
Ratio 1	0.406		0.450		0.418		0.583	
Ratio 2	0.352		0.655		0.479		-	

Ratio 1== Total Predicted Runoff/Total Rainfall June 3-12, 2001 ; Ratio 2==Average Yearly Runoff/Average Yearly Rainfall with averages computed over 1965-1989.

Results given in Table 4 were computed by applying the reduction factors based on event rainfall totals to TRISE, TRECES, BLAG, Q50DUR, and Q75DUR given in Table 5. This was necessary to improve the timing of the runoff flows, since the regression equations were developed for rainfall events of less than 5 inches.

Table 5. Empirical Reductions Factors for Large Rainfall Events

Rainfall total (in)	5	10	15	20	25	50
Reduction factor	1.	0.5	0.25	0.33	0.29	0.2

Overland Flooding Scheme

The scheme was developed to supplement the drying/wetting scheme previously developed (Schmalz, 1998a; 2001). The original scheme allowed water areas to dry and subsequently wet (tidal flat problem) but did not allow land areas to flood. To accommodate this feature, a digital elevation model or test topography is used in which land elevations are set to a maximum of –1 m. Along the grid border, land elevations are set to –100m. Since in the scheme, it is necessary to perform computations over dry grid cells, a minimum water depth, *dflood*, is specified. Subsequent cell and cell face flags properly mask any undesired results. For surface temperature specification, a value tflood=29.5°C is used for the overland flood cells. The flood scheme is actuated each internal mode time step and consists of the following check for a typical u-velocity face to see if it should be activated. Activation is based on the water level in the wet cell exceeding the dry cell bed elevation by a critical depth, *dcrit*, equal to *10dflood* plus the depth needed to account for the water initially placed on the dry cell as given in the relation below. Following Mellor's (2003) notation:

$$\begin{aligned}
&\text{If } dum_{i,j} = 0,\ fsm_{i,j} = 0,\ \text{and } fsm_{i-1,j} = 1, \quad \text{then we compute}\\
&dil = dflood{*}art_{i,j}/art_{i-1,j}.\\
&\text{If } el^{m}_{i-1,j} + h_{i,j} > dcrit + dil \text{ and } d^{m}_{i-1,j} > 2dcrit \text{ then}\\
&fsm_{i,j} = iover_{i,j} = dum_{i,j} = 1 \text{ and the cell face is activated}\\
&\text{and } F^{m}_{i,j,k} = F^{m}_{i-1,j,k}, \quad F^{m-1}_{i,j,k} = F^{m-1}_{i-1,j,k}, \quad k = 1,...kbm1, \text{ where } F = (T,S,Q^2,Q^2l)
\end{aligned} \tag{2}$$

A similar procedure is used for the case, $dum_{i,j}=0$, $fsm_{i,j}=1$, and $fsm_{i-1,j}=0$ and for the v-cell face. Once both u and v faces have been activated, the potential exists on a u-face for both $fsm_{i,j}=fsm_{i-1,j}=1$ with $dum_{i,j}=0$. If this occurs, the average u-cell face water depth, *dbar*, is computed as $dbar = 0.5(el^{m}_{i,j} + el^{m}_{i-1,j} + h_{i,j} + h_{i-1,j})$. If $dbar > dcrit + dflood$ then $dum_{i,j}=1$. An analogous procedure is used for the v-face. To monitor the computations and determine the areal extent of overland flooding, the following procedures were implemented. First an additional cell mask, $imask_{i,j}$, was created and set to 1. If the cell bed elevation becomes negative, $imask_{i,j}$=-1. If overland flooding occurs, $iover_{i,j}$=1 and $imask_{i,j}$=0. This allows water depths over the complete water area, or over just the overland flooded portions of the grid, to be printed/plotted.

Two test applications were employed in which *dflood*=25mm and *dcrit*=25cm. In the first application a test topography was specified based on the cell's distance from mid-Bay as given in Table 6. In the second application, the USGS 3-arc second DEM for Houston-West and Houston-East was used to specify the overland topography. In both applications, the 8 Sept 1999 JD 251 24 hour nowcast cycle (249.75-250.75) was simulated with the test surge given in Table 7 imposed. Model mean water surface elevations at three locations are given in Table 8 for each case. While the model mean and maximum water surface elevations were very close at the stations in Table 8, the areal extent of the flooding was vastly different. For the test topography, only order 10 grid cells in the lower eastern portion of the Bay mean sea level boundary were inundated with flood depth levels of order 0.5m. For the DEM topography, order 1000 grid cells over major portions of the Bay east and west mean sea level boundaries were inundated with flood depth levels of order 2.2m. The river input flow cells were surrounded by floodwater.

Table 6. Test Topography based on cell centered distance from mid-Bay (29.4°N, 95.0°W).

Distance (km)	0	5	10	15	20	25	30	35	40	45	50
Bed elevation (m)	-2.	-2.5	-3.0	-4.0	-5.0	-6.0	-7.0	-8.0	-10.0	-12.0	-15.0

Table 7. Test Storm Surge Specification based upon Hurricane Carla and Alicia surge levels.

Elapsed Time (Hours)	0	6	12	48	72
Surge Level (m)	0.	2.5	3.5	2.0	1.0

Table 8. Simulated Mean Water Surface Elevation (m)- JD 249.75-250.75 Nowcast Cycle with Test Storm Surge.

Case	Galveston Pleasure Pier	Port Bolivar	Galveston Pier 21
Test Topography	3.07	2.23	2.60
DEM Topography	3.05	2.51	2.66

Hurricane Windfield and Atmospheric Pressure Algorithms

Initially, the work of Schmalz (1986a,b,c) was reviewed to consider the Standard Project Hurricane (SPH) and the Tetra Tech (1979) models. A more recent parametric model developed by Holland (1980) and further modified by Sinha and Mandal (1999) was also considered. Each is outlined below and was studied using a hypothetical storm track and parameter set (R= radius to maximum winds, P_o=far field atmospheric pressure, and ΔP= atmospheric pressure deficit). Each approach uses an inflow angle, α, as given by Graham and Nunn (1959) in the following relationship:

$$\alpha = \begin{cases} 10d/R & d \le R \\ 10 + 15(d-R)/1.2R & R < d < 1.2R \\ 25 & d \ge 1.2R \end{cases} \quad (3)$$

where d=Distance from the storm center

Each approach also uses the Schwerdt et al. (1979) asymmetry factor to account for storm forward speed as follows:

$$\begin{aligned} \theta_{max} &= \theta_h + 90 + \alpha \\ \theta_a &= \theta_s - \theta_{max} \\ V_a &= 1.5 V_f^{1.63} \cos\theta_a \end{aligned} \quad (4)$$

where

θ_h=Storm direction bearing
α=Inflow angle
θ_s=Storm center bearing
V_f=Storm forward speed
θ_{max}=Bearing of maximum wind
θ_a=Asymmetry angle
V_a=Asymmetry speed adjustment

For the Standard Project Hurricane (SPH) , the maximum gradient windspeed is determined from the following relation:

$$V_{gx} = 67\Delta P^{1/2} - 1800\Omega R \tag{5}$$

where
ΔP=Central pressure deficit
Ω=Earth rotation

The maximum sustained windspeed at 10m, $V_{max}=0.9V_{gx}$.
Next a reduction factor, f_r is determined by fits to observed radial wind profiles in Schwerdt et al. (1979) and is given by:

$$f_r = \begin{Bmatrix} (d/R)^3 & d \le R \\ e^{-(d/7R+1/7)} & d > R \end{Bmatrix} \tag{6}$$

where d and R are as previously defined. The complete windspeed, V, including asymmetry effects and Schloemer (1954) pressure profile, P, are given by:

$$\begin{aligned} V &= f_r V_{\max} + V_a \\ P &= P_0 + \Delta P(e^{-R/d} - 1) \end{aligned} \tag{7}$$

where quantities on the right hand sides have been previously defined.

For the Tetra Tech approach, the algorithm is the same as the SPH algorithm with the exception of the relation for f_r, which is replaced by the following relation developed by Collins and Viehman (1971).

$$f_r = \begin{Bmatrix} 0 & 0 < d \le R/3 \\ 3/2(d/R - 1/3) & R/3 < d \le R \\ 1/c_1 d^k \log(R/c_2 d^m) & d > R \end{Bmatrix} \tag{8}$$

where $c_1 = 3.354,\ c_2 = 1.265\text{x}10^{-3}, k = -0.15128,\ m = 1.607.$
For the Holland approach, the storm category, k_{cat}, is first determined based on the central pressure deficit using the following relation:

$$k_{cat} = \begin{Bmatrix} 1 & \Delta P \le 30 \\ 2 & 30 < \Delta P \le 50 \\ 3 & 50 < \Delta P \le 70 \\ 4 & \Delta P > 70 \end{Bmatrix} \tag{9}$$

$$\begin{aligned} &R_1^m = 28. \quad R_2^m = 26. \quad R_3^m = 22. \quad R_4^m = 18. \\ &b_1 = 1.0 \quad b_2 = 1.1 \quad b_3 = 1.2 \quad b_4 = 1.3 \\ &pf_1 = 1.6 \quad pf_2 = 3.1 \quad pf_3 = 1.3 \end{aligned} \tag{10}$$

Next the Coriolis parameter, f, is determined based on the hurricane latitude, λ_h, by:

$$\begin{aligned} f &= 2\Omega \sin \lambda_h \\ a_{kcat} &= Rb_{kcat} \end{aligned} \tag{11}$$

Note a_{kcat} is determined from b_{kcat} rather than independently specified.
The complete windfield and pressure description is then determined as given below:

$$V_{gx} = (a_{kcat} b_{kcat} \Delta P e^{-a_{kcat}/d^{b_{kcat}}} / \rho_{air} d^{b_{kcat}} - d^2 f^2 / 4)^{1/2} - df / 2$$
$$f_r = 0.65 p f_1 + e^{-(p f_2 R / R^m_{kcat} + p f_3 R^m_{kcat} / R)} \tag{12}$$
$$P = P_0 + \Delta P (e^{-a_{kcat}/d^{b_{kcat}}} - 1)$$

The three approaches are contrasted in Table 9 for a storm moving due North over (29.4°N, 95°W) with a constant radius to maximum winds of 25 nm and central pressure deficit of 25 mb.

Table 9. Hurricane and Pressure Fields For an Arbitrary Storm based on SPH, Tetra Tech, and Holland Methods. Note longitude of the storm track is constant at 95°W, radius to maximum winds is 25nm, and central pressure deficit is 25mb with a far field pressure of 1013mb. Note f=all weather nowcast/forecast system, 1=SPH, 2=Tetra Tech, and 3=Holland.

Track Index	Location (lat °N)	Distance (nm)	Forward Speed (kts)	Windspeed (kts) Method				Wind Direction (oT)	Sea Level Pressure (mb)
				f	1	2	3		
1	28.0	84	3.5	32.5	30.1	30.7	21.7	-115	1007.7
2	28.35	63	3.5	36.9	35.2	34.9	27.7	-115	1006.2
3	28.7	42	3.5	42.4	41.2	40.2	36.1	-115	1003.5
4	29.05	21	3.5	50.8	48.8	48.2	49.8	-101	997.6
5	29.4	0	3.5	3.3	3.3	3.3	3.3	0	988.0
6	29.65	15	3.25	33.2	21.4	31.5	55.1	83	994.6
7	29.9	30	3.0	49.0	47.6	46.6	45.3	65	1000.8
8	30.15	45	2.75	45.1	42.8	41.9	37.0	65	1004.0
9	30.4	60	2.5	40.1	38.5	38.1	30.8	65	1005.9

A slightly modified approach was selected for incorporation within the all weather nowcast/forecast system in which for the Holland method, V_{gx} is evaluated for $d=R$ and f_r is based on the Collins and Viehman (1971) method used by Tetra Tech (1979) to determine the reduction factor. The Holland (1980) pressure profile is used.

Wave Algorithms

Schmalz (2003) has recently compared two parametric models with the finite difference based Donelan (1977) wave model for wind events over Galveston Bay. The two parametric models considered were developed by the USACE and are outlined as follows.

CETN-I-6 (1981): The following two equations are considered for significant wave height, H_s (ft) and period, T_s (sec):

$$\frac{gH_s}{U_A^2} = 0.283 \tanh[0.530[gd / U_A^2]^{0.75}] \tanh[\frac{0.00565[gF / U_A^2]^{0.5}}{\tanh[0.530[gd / U_A^2]^{0.75}]}] \tag{13}$$

$$\frac{gT_s}{U_A} = 7.54\tanh[0.833[gd/U_A^2]^{0.375}]\tanh[\frac{0.0379[gF/U_A^2]^{0.333}}{\tanh[0.833[gd/U_A^2]^{0.375}]}] \quad (14)$$

using the following notation:

g = Gravitational acceleration (32.2 ft/sec^2)
U_A = Windspeed (kts)
F = Fetch (nm)
d = Depth (ft)

CW-167 (Project CW-167, 1955): A second approach utilizes curves developed during the comprehensive Project CW-167 on Lake Okeechobee, Florida, which is a shallow water system similar to Galveston Bay in size and depth. Relations for the significant wave height (ft), H_s, and for significant wave period, T_s (sec) are as follows:

$$H_s = \frac{U_A^2}{g}\min[0.00356(\frac{gF}{U_A^2})^{0.448}, 0.1844(\frac{gd}{U_A^2})^{0.690}]$$

$$T_S = 6.262\frac{U_A}{g}(\frac{gd}{U_A^2})^{0.4057} \quad (15)$$

where the previous notation is used. Further details on the Donelan (1977) wave model may be found in Schwab et al. (1984). Governing equations as well as details of the initial testing may be found in Schmalz (2003).

The January 25-30, 1997 time period was selected for further study because of the availability of USACE wave measurements off Eagle Point. Application of the models to this period revealed that despite the incorporation of shallow water in the finite difference based Donelan model effects by a linear reduction in transfer of wind to wave momemtum as described in Schmalz (2003), best results were achieved by using a combination of the two simpler parametric wave models. CETN-I-6 relations were used for significant wave height, while CW-167 relations were used for significant wave period. Total significant wave height was computed as the sum of the CETN-I-6 results plus the swell, which was determined at the open boundary from the NDBC Buoy 42035 measurements. The total significant wave period was taken as equal to the CW-167 result alone. Corresponding results are given in Figure 4 at Eagle Point and in Figure 5 at NDBC Buoy 42035, respectively. Note, the peak at Eagle Point is no longer delayed relative to the observations as experienced using the Donelan model with the shallow water adjustment and the peak at NDBC Buoy 42035 is well reproduced.

A robust and computationally efficient scheme, which will provide wave conditions representative of mid-Bay at order 0.5 km resolution, is required. The wave model should require no more than order 0.5 hours of CPU on modest workstations to effect a 24 hour nowcast followed by a 36 hour forecast cycle. The wave model will employ the same curvilinear grid as used by the Galveston Bay circulation model and the same windfield, thereby allowing it to be incorporated as a separate subroutine within the circulation model as was done by Schmalz (1986) for Lake Okeechobee. For the swell open boundary condition separate date resources are required on nowcast and forecast.

For the nowcast, data from NDBC Buoy 42035 may be obtained from the Texas Automated Buoy System (TABS) Project sponsored by the Texas General Land Office (TGLO) at: http://tabs.gerg.tamu.edu /Tglo/DailyData/Data/42035_met.shtml.

For the forecast, the NWS Wavewatch III model forecast results at 42035 can be utilized at:
ftp://polar.wwb.noaa.gov/pub/waves/latest_run/wna.42035.bull.

Design of the All Weather Nowcast/Forecast System

The NOS experimental all weather nowcast/forecast system uses the separate nowcast/forecast set-up program to establish hydrodynamic model nowcast and forecast inputs with minor modifications (indicated by * and italics in the list below). The modified set-up program utilizes the following twelve step procedure, where the steps 1-10 constitute the original procedure:

*1) Setup 24 hour nowcast and 36 hour forecast time periods and grid parameters,
2) Predict astronomical tide,
3) Predict astronomical currents,
4) Read PUFFF files and develop station time series,
5) Develop GBM subtidal water level signal,
6) Assimilate PORTS salinity and temperature data into GBM and HSCM initial conditions,
*7) Establish GBM and HSCM salinity and temperature boundary conditions,
8) Establish GBM and HSCM SST forcings,
*9) Establish USGS observed and NWS/WGRFC forecast freshwater inflows,
*10) Establish PORTS based and Aviation Model wind and pressure fields,
11) Establish rainfall/runoff for City of Houston inflows, and
12) Establish wave swell characteristics

Step 1 was modified to include the grid modifications to incorporate the four additional City of Houston inflows. Step 7 was modified to include salinity and temperature specification for these inflows. Step 9 was modified to set the City of Houston inflows to zero for no rainfall/runoff. Note if rainfall/runoff occurs (storm track file specified) these flows are determined in Step 11. In Step 10, hurricane wind and pressure fields are developed if the storm track file exists. If a storm track file does not exist, the set-up program skips to Step 12. Otherwise, the wind and pressure fields are then blended into the the Aviation Model far fields over a distance from the storm center of five times the radius to maximum winds. In Step 11, the rainfall/runoff is developed if a storm track file is present. A 2-5 day antecedent rainfall description is used to determine the rainfall moisture index. A Quantitative Precipitation Forecast (QPF) will be used to provide the rainfall input over the forecast period. In Step 12, the open boundary wave swell height, direction, and period are specified based on measurements at Buoy 42035 and the NCEP Wavewatch forecast at Buoy 42035.

The GBM is modified to incorporate the overland flooding algorithm and includes the four additional freshwater inflows. The combined parametric wave model is included as a subroutine and uses the same wind fields. An open boundary condition swell boundary is applied to incorporate Gulf of Mexico wave swell conditions. The fetch along the HSCM boundary is written to a transfer file for input to the HSCM. The HSCM modifications are similar. The overland flooding algorithm is included and the wave algorithm appears as a separate subroutine. The fetch from the GBM boundary and additional fetch specification within the Port of Houston is interpolated over the grid for input to the wave subroutine. The City of Houston inflows are incorporated as in the GBM.

Initial Test Plans

The storms shown in Table 10 will be used to test the all-weather nowcast/forecast system on a nowcast/forecast cycle basis. Initially, Tropical Storm Allison will be considered since the rainfall/runoff has already been developed as given in Table 4. A storm track file will be constructed and the modified set-up program tested. Hydrodynamic simulations using both the GBM and HSCM will be performed with and without the storm track file. Water level and current predictions will be compared to assess the impact of including rainfall/runoff flows from the City

Table 10. Storm Characteristics over Galveston Bay, Texas.

Storm/Source	Central Pressure (mb)	Windspeed (kts)	Radius to Max Winds (nm)	Rainfall (in)	Storm Surge (ft)
Allison 9-11 June 2000/ NWS(2001); Stewart (2002)	990	20-30	30	9.8-35.1	1.8-2.1
Carla 3-15 Sept 1961/ Dunn and staff (1962)	970-975	75-80	30	5.0-10.0	8.8-9.3
Alicia 15-21 Aug 1983/ Case and Gerrish (1984)	963-965	80-100	30	7.8-10.7	8.9

of Houston. Wave growth and degree of flooding will be assessed. Next, Hurricanes Carla and Alicia will be simulated to consider the effects of more severe storms. In this context, the older bathymetry file will be used based on measurements prior to 1988. The focus will be on storm surge prediction ability for the more severe storm events.

Future Enhancements

To seek improvement in the ability of the HSCM to assess the impacts of storm induced freshwater pulses on salinity stratification, a generalized vertical coordinate was investigated. Mellor et al. (2002) report on the development of a generalized s vertical coordinate within the Princeton Ocean Model (POM), in which the Cartesian coordinate system (x,y,z,t) is transformed to the s coordinate system, (x^*,y^*,k,t^*), according to the following relation:

$$\begin{aligned} x &= x^* \\ y &= y^* \\ t &= t^* \\ z &= \eta(x^*,y^*,t^*) + s(x^*,y^*,k,t^*) \end{aligned} \tag{16}$$

Here k is a continuous variable in the range. Note kb corresponds to the lowest level, which follows the bottom depth as in the standard POM σ coordinate, in which, $k=1$ represents the top level and $1<k<kb$, with levels fixed in proportion to each other independent of water surface elevation or water depth. Here when the differential equations are discretized, k will also be discrete and will again label the vertical level. η is the free surface, so $s=0$ at $k=1$. The transformed s coordinate equations are given in Mellor et al. (2002) with further details.

Within this framework a mixed σ-z coordinate system has been constructed. A 5m depth cutoff was used for the upper 5 sigma levels, which occupy the bathymetry in the range (0.5-5.0m). Note the majority of the Bay outside the Houston Ship Channel is order 2m in depth. Below 5m, 1m depth increments are used. The lowest vertical level depth increment is increased to reach the bottom. A maximum of 15 levels are used with the upper 5 levels in sigma space.

The JD253 1999 hindcast was used as an ideal test case. Bay model (GBM) to sigma coordinate channel model (HSCM) and to generalized s coordinate channel model (HSCS) linkages for the test case were revised to set boundary water surface elevations to zero and maintain boundary conditions for (S,T,Q^2,Q^2l) at their initial values. Initial conditions were zero velocity and water surface elevation and a horizontally uniform depth dependent temperature and salinity stratification. River inflows and wind and atmospheric pressure anomaly were set to zero. SCLIM, TCLIM, and RMEAN, refer to Mellor (2003), were computed and used in both channel models in the horizontal diffusion of (S,T) and in the baroclinic pressure gradient terms. A turbulent Prandtl number of unity was used. A 24 hour simulation was performed with each model and the maximum changes in S and T were noted in a grid patch located in the Bay Entrance as given in Table 11. Based on these initial results, the generalized s coordinate is better able to maintain the stratification and will be further pursued.

Table 11. Vertical Coordinate System Model Results for Ideal Test Case After 24hrs of Simulation. Note all entries in the table should ideally remain zero. Results shown are the maximum values over the grid patch (I,J)=(36-44,26-34).

Vertical Coordinate	T Change (°C)	S Change (PSU)	Maximum WSE (mm)	Maximum U (cm/s)	Maximum V (cm/s)
σ	0.829	2.016	-0.505	10.95	8.20
s	0.558	0.283	0.220	2.66	2.99

The incorporation of the wave algorithm as a subroutine allows for further experimentation with wave-current interaction. The effect of waves on surface wind stress and on vertical mixing can be incorporated along with generation of radiation stresses. Their complete specification is considered in a recent paper by Mellor (2002) and awaits further research. Note with a three-dimensional all-weather modeling approach, the effect of return flows can be simulated and the potential exists for more accurate surge prediction with the wave effects included as well. These issues are being examined in conjunction with the NOAA Coastal Storms Initiative (NOAA, 2002).

Acknowledgments

Dr. Bruce B. Parker, Chief of the Coast Survey Development Laboratory, conceived of this project and provided leadership and critical resources. Dr. Frank Aikman provided the day to day direction and further information on the Coastal Storms Initiative. Philip H. Richardson assisted in the application of the wave algorithms to the January 25-30, 1997 period.

References

Appell, G. F., T. N. Mero, T. D. Bethem, and G. W. French, 1994: The development of a real-time port information system, *IEEE, Journal of Oceanic Engineering*, 19,149-157.

Blumberg, A. F., and G. L. Mellor, 1987: A description of a three-dimensional coastal ocean circulation model. *Three-Dimensional Coastal Ocean Models*, (ed. Heaps), American Geophysical Union, Washington, DC., 1 - 16.

Case, R.A. and H.P. Gerrish, 1984: Atlantic hurricane season of 1983, *Monthly Weather Review*, Vol. 112, 1083-1092.

Coastal Engineering Technical Note, CETN-I-6, 1981: Revised method for wave forecasting in shallow water, Coastal Engineering Research Center, Vicksburg, MS.

Collins, J.I. and Viehman, M.J., 1971: A simplified empirical model for hurricane wind fields, Offshore Technology Conference, Vol 1, Preprints, Houston, TX.

Donelan, M.A., 1977: A simple numerical model for wave and wind stress prediction, *Report, National Water Resources Institute*, Burlington, Ontario.

Dunn, G.E. and staff, 1962: The hurricane season of 1961, *Monthly Weather Review*, Vol. 134, 107-119.

Graham, H.E. and Nunn, D.E., 1959: Meteorological considerations pertinent to the Standard Project Hurricane, Atlantic and Gulf Coasts of the United States, *NHRP No. 33*, National Weather Service, Silver Spring, MD.

Holland, G.J., 1980: An analytic model of the wind and pressure profiles in hurricanes, *Monthly Weather Review*, Vol 108, 1212-1218.

Linsley, R.K, Kohler, M.A., and Paulhus, J.L.H., 1982: *Hydrology for engineers*, New York, McGraw Hill.

Liscum, F., 2001: Effects of urban development on stormwater runoff characteristics for the Houston, Texas metropolitan area, *USGS, Water-Resources Investigations 01-4071*, Austin, TX.

Mellor, G.L., 2002: The three-dimensional, current and surface wave equation, (submitted). (Available at www.aos.princeton.edu/WWWPUBLIC/htdocs.pom/publications.htm)

Mellor, G.L., S. Hakkinen, T. Ezer, and R. Patchen, 2002: A generalization of a sigma coordinate ocean model and an intercomparison of model vertical grids, In: *Ocean Forecasting: Conceptual Basis and Applications*, N. Pinardi and J.D. Woods (eds.), Springer, Berlin, 55-72.
(Available at www.princeton.edu/WWWPUBLIC/htdocs.pom/publications.htm)

Mellor, G.L, 2003: Users guide for a three-dimensional, primitive equation, numerical ocean model, Progr Atmos and Ocean Sci, Princeton University. (Available at www.aos.princeton.edu/WWWPUBLIC/htdocs.pom/publications.htm)

NOS, 1999: NOS procedures for developing and implementing operational nowcast and forecast systems for PORTS, *NOAA Technical Report NOS CO-OPS 0020*, Silver Spring, MD.

NOAA, 2002: Coastal Storms Initiative, Coastal Science Center, Charleston, SC. (Available at www.csc.noaa.gov/csi)

NWS, 2001: Service Assessment Tropical Storm Allison Heavy Rains and Floods Texas and Louisiana June 2001, National Weather Service, Silver Spring, MD.

Project CW-167, 1955: Waves and wind tides in shallow lakes and reservoirs: summary report, U.S. Army Corps of Engineers, Jacksonville, FL.

Richardson, P. H. and R. A. Schmalz, 2002a: Water Level Event Analysis: Program Documentation, *CSDL Informal Technical Note 1*, Silver Spring, MD.

Richardson, P. H. and R. A. Schmalz, 2002b: Principal Component Direction Event Analysis: Program Documentation, *CSDL Informal Technical Note* 2, Silver Spring, MD.

Schloemer, R.W., 1954:Analysis and synthesis of hurricane wind patterns over Lake Okeechobee, Florida, *Hydromet Report No. 31*, Washington, DC.

Schmalz, R. A., 1986a: A numerical investigation of hurricane induced water level fluctuations in Lake Okeechobee, Report 1: Forecasting and Design, *Miscellaneous Paper CERC-86-12*, Coastal Engineering Research Center, Vicksburg, MS.

Schmalz, R. A., 1986b: A numerical investigation of hurricane induced water level fluctuations in Lake Okeechobee, Report 1: Forecasting and Design Appendicies, *Miscellaneous Paper CERC-86-12*, Coastal Engineering Research Center, Vicksburg, MS.

Schmalz, R. A., 1986c: A numerical investigation of hurricane induced water level fluctuations in Lake Okeechobee, Report 2: Joint Probability Theoretics and Application, *Miscellaneous Paper CERC-86-12*, Coastal Engineering Research Center, Vicksburg, MS.

Schmalz, R. A., 1996: National Ocean Service Partnership: DGPS-supported hydrosurvey, water level measurement, and modeling of Galveston Bay: development and application of the numerical circulation model. NOAA, National Ocean Service, Office of Ocean and Earth Sciences, *NOAA Technical Report NOS OES 012*, Silver Spring, MD.

Schmalz, R. A., 1998a: Development of a Nowcast/Forecast System for Galveston Bay, *Proceedings of the 5th Estuarine and Coastal Modeling Conference*, ASCE, Alexandria, VA, 441-455.

Schmalz, R. A., 1998b: Design of a Nowcast/Forecast System for Galveston Bay, *Proceedings of the 2nd Conference on Coastal Atmospheric and Oceanic Prediction and Process*, AMS, Phoenix, AZ, 15-22.

Schmalz, R. A., 1998c: Initial Evaluation of a Nowcast/Forecast System for Galveston Bay during September 1997, *Proceedings of the Ocean Community Conference*, MTS, Baltimore, MD, 426-430.

Schmalz, R. A., 2000a: A Nowcast/Forecast System for Galveston Bay, *Proceedings of the 2000 Joint Conference on Water Resources Engineering and Water Resources Planning and Management*, ASCE, General Conference, Flood Hazard Modeling Applications, CDROM.

Schmalz, R. A., 2000b: Demonstration of a Nowcast/Forecast System for Galveston Bay, *Proceedings of the 6th Estuarine and Coastal Modeling Conference*, ASCE, New Orleans, LA, 868-883.

Schmalz, R. A., 2000c: High-Resolution Houston Ship Channel ADCP and CTD Survey: Model-Data Intercomparisons, *NOAA Technical Report NOS CS 6*, Silver Spring, MD.

Schmalz, R. A., 2001: Three-Dimensional Hydrodynamic Model Developments for a Galveston Bay Nowcast/Forecast System, *NOAA Technical Report NOS CS 9*, Silver Spring, MD.

Schmalz, R. A., 2003: Development of a Real Time Wave Model for Galveston Bay, *Proceedings of the World Water & Environmental Resources Congress*, ASCE, Philadephia, PA, CDROM.

Schmalz, R. A. and P. H. Richardson, 2002: Evaluation of the NOS Experimental Nowcast/Forecast System for a Galveston Bay, *NOAA Technical Report NOS CS 14*, Silver Spring, MD.

Schwab, D. R., J.R. Bennett, P.C. Liu, and M.A. Donelan, 1984: Application of a Simple Numerical Wave Prediction Model to Lake Erie, *Journal of Geophysical Research*, Vol. 89, C3, 3586-3592.

Schwerdt, R. et al., 1979: Criteria for Standard Project Hurricane and probable maximum hurricane wind field, Gulf and East Coast United States, *NOAA Technical Report NWS-23*, Silver Spring, MD.

Sinha, J. and G.S. Mandal, 1999: An analytical model of over-land surface windfield in cyclone for the Indian coastal region, National Seminar on Wind Engineering-01, IIT, Kharagpur.

Stewart, S.R., 2002: Tropical Storm Allison, 5-17 June 2001, National Hurricane Center Tropical Storm Report, (http://www.nhc.noaa.gov/2001/allison.html), Miami, FL.

Tetra Tech, Inc., 1979: Coastal Flooding Storm Surge Model Part I: Methodology, Part II: Codes and User's Guide, Federal Emergency Management Agency, Washington, DC.

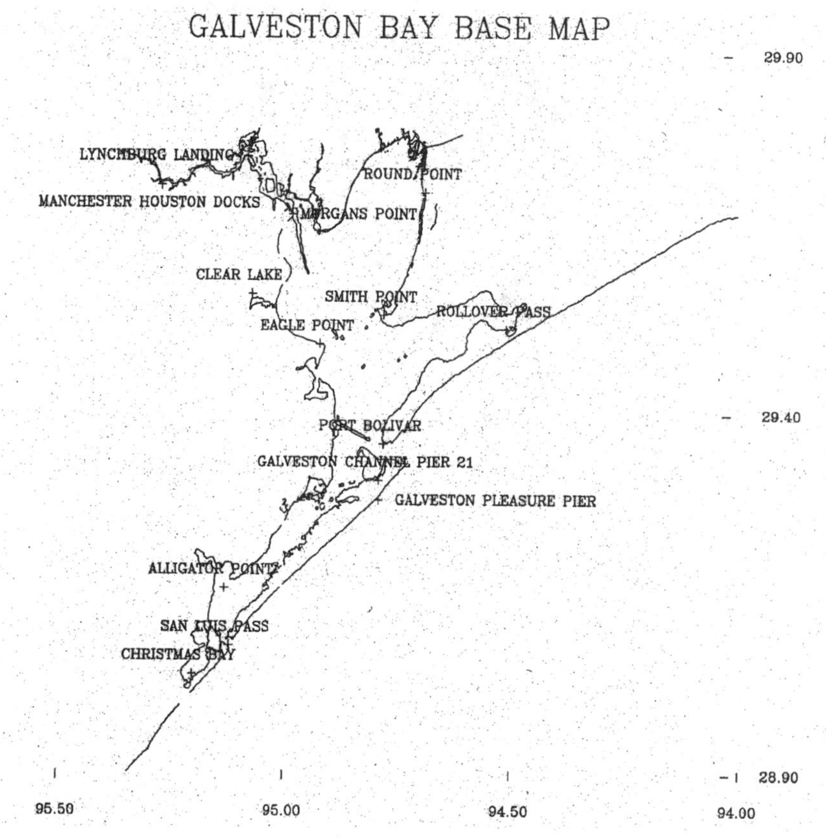

Figure 1. Galveston Bay PORTS Station Locations. Note NDBC Buoy 42035 is located at 29.25 N 94.41 W approximately 22 nm SE of Galveston

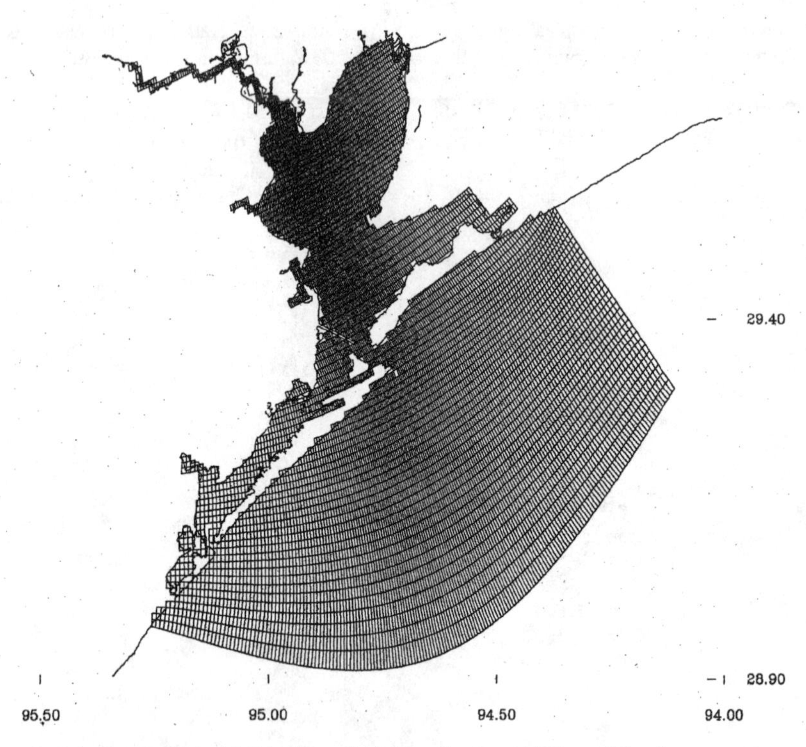

Figure 2. Galveston Bay Model Curvilinear Grid. Note the same grid is used for the parametric wave model.

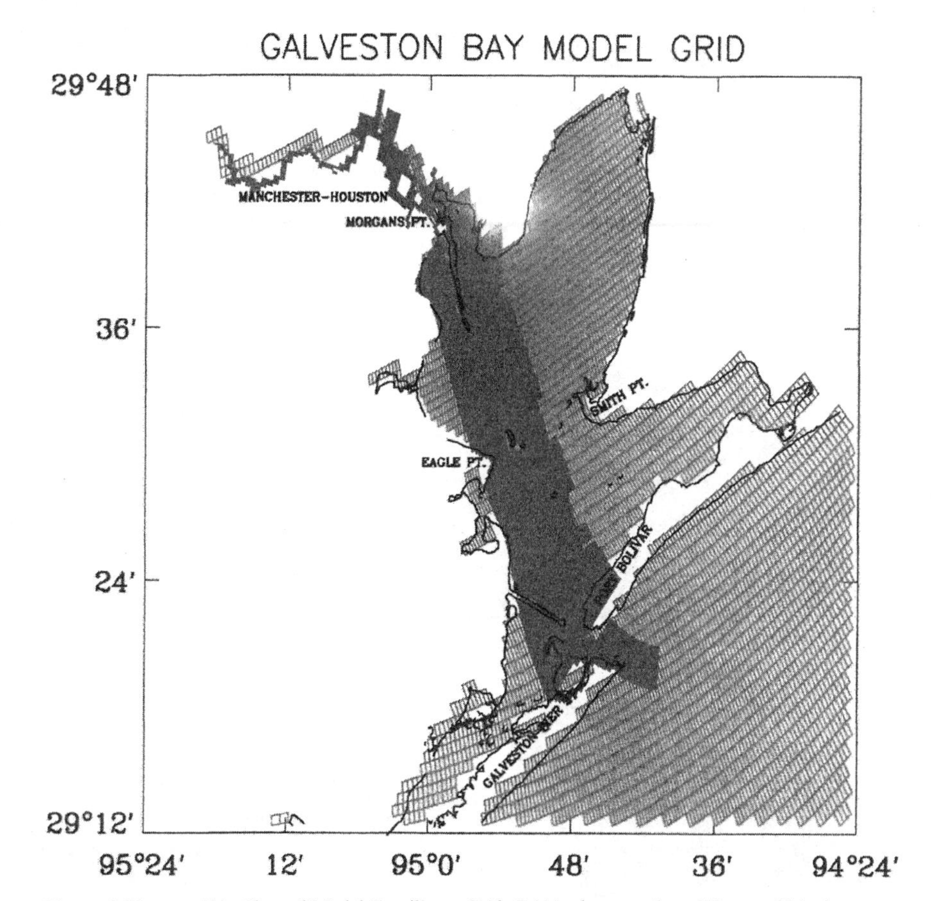

Figure 3. Houston Ship Channel Model Curvilinear Grid. Grid is shown as the solid area within the Galveston Bay Grid.

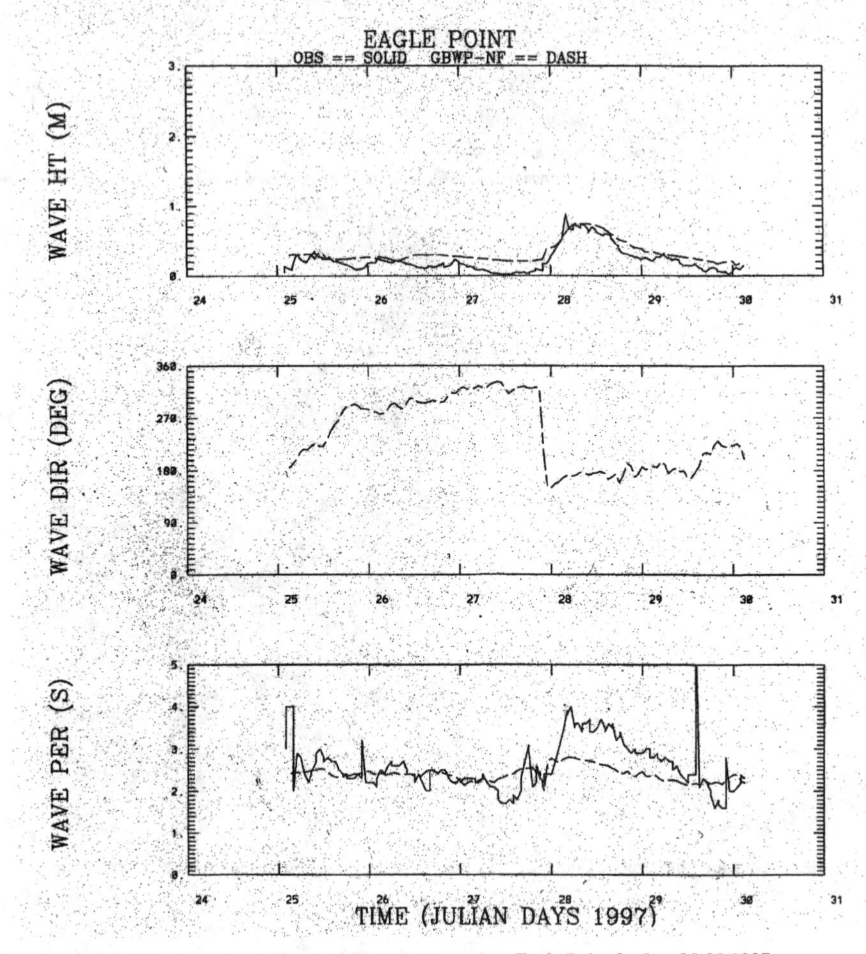

Figure 4. Parametric Model vs Observed Wave Parameters at Eagle Point for Jan. 25-30 1997.

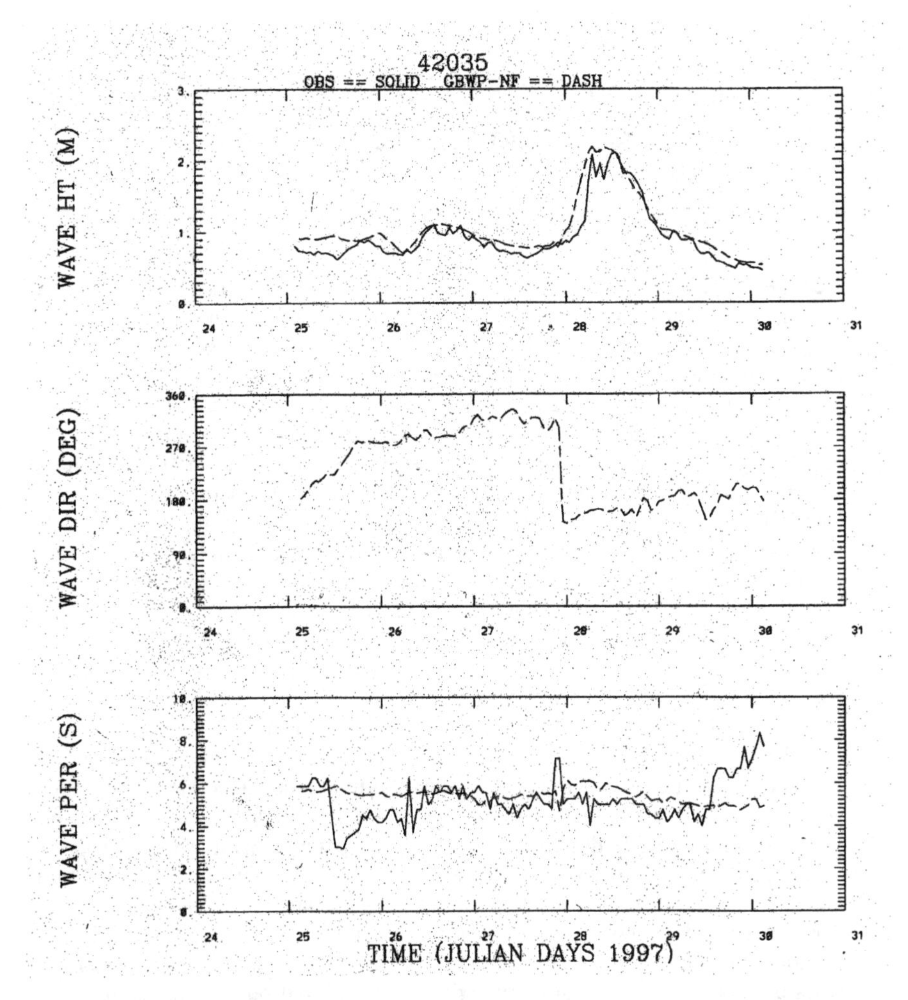

Figure 5. Parametric Model vs Observed Wave Parameters at 42035 for Jan. 25-30 1997.

COMMUNITY MODEL CONCEPT AND IMPLEMENTAION IN THE SURFACE WATER MODELING SYSTEM

Alan K. Zundel[1], Russell D. Jones[2], and Nicholas C. Kraus[3]

ABSTRACT: The U.S. Army Corps of Engineers has begun an initiative called regional sediment management that promotes consideration of multi-project engineering and environmental benefits within sediment-sharing water bodies. Often, multiple navigation, environmental enhancement, and shore-protection projects reside within the same physical system, such as a watershed with its coastal segments between inlets, rivers, bays, and estuaries. Consideration of regional processes streamlines the study, design, and execution of projects. Further, it minimizes unintended consequences at locations beyond the formal limits of a directly authorized (local) project. In support of this initiative, the Corps' Coastal Inlets Research Program and Regional Sediment Management Program are developing informatics infrastructure within the Surface-water Modeling System (SMS) to input, manage, archive, visualize, and output model-related data of various types, including input geometry, hydraulic data, model specific parameters, and metadata or documentation. The "Community Model" concept refers to the capability of archiving all data pertinent to a region in an accessible, sharable and self documenting way. This data includes measured quantities such as raw geometric survey data, decimated data, and observed hydraulic conditions which can be used both as model forcing and model verification data. It also includes model grids and meshes for representing various processes and multiple design scenarios, model parameters, and model solution data. Finally it also includes relational information defining how various local models interact such as circulation, surface waves, and sediment transport. The Community Model concept provides much greater flexibility, portability and economy. The files are portable across operating systems and languages. Various types of data are contained in a single file, and that file is binary with compression capabilities to minimize its size. Further, any of the data can be accessed quickly with a variety of data mining techniques for fast utilization and management. The Community Model concept is being implemented in the SMS, with a first revision

1) Environmental Modeling Research Laboratory, Brigham Young University, 242 Clyde Building, P.O. Box 24080, Provo, UT, 84602-4080. Tel: (801) 378-2812; zundel@byu.edu.
2) Environmental Modeling Research Laboratory, Brigham Young University, 242 Clyde Building, P.O. Box 24080, Provo, UT, 84602-4080. Tel: (801) 378-2812; jonesr@et.byu.edu.
3) U.S. Army Engineer Research and Development Center, Coastal and Hydraulics Laboratory, 3909 Halls Ferry Road, Vicksburg, MS, 39180. Tel: (601) 634-2139; Nicholas.C.Kraus@erdc.usace.army.mil.

to be released in the fourth quarter of 2003. This version will consist of a documented format and API for geometric data storage including scattered data sets, structured grids, and unstructured grids. A single file will be able to store multiple data objects along with functional data sets associated with those objects. The data sets may be scalar or vector, and may include multiple time steps. In addition, metadata comments will be included for each object type. In the future, checks will be available to assure that all required documentation to ensure standardized data is specified.

INTRODUCTION

Coastal and estuarine models often require the management of many types of data. This is especially true when studies involve several interrelated processes such as circulation and waves. A complete project requires input data, verification data, and model solution data. Input data consists of any or all of the following: bathymetry, tidal information, offshore wave data, boundary conditions, spatially varied parameters such as Manning's n values, and global model parameters. This data may be at multiple resolutions, low resolution for large scale regional portions of a project and high resolution for small scale or local portions. Verification data includes known values within the computational domain to compare with model solutions. This may be single data points as a specified time, or transient data for a range of times. Model solutions include computed values at specified locations or across the grid.

Much of the data that must be managed is spatially and possibly temporally variable. For example, model bathymetry by its nature must be defined at every cell or node in the model domain. Similarly, model solutions often include sets of values for each node or cell. This paper refers to these types of data simply as data sets. A data set is a scalar or vector data at every computational location (node, vertex, or cell) for the domain. If the data set represents a measurement that changes over time, it may include multiple time steps to show this variation.

Simply put, a massive amount of data is involved. A typical project could include several numerical simulations in addition to the input data. Each simulation and the input data include thousands of data points and dozens of data sets. Each data set could potentially have hundreds of time steps. Add to this that multiple scenarios may be included in a project and each model involved requires multiple input files and creates multiple output files and the data management problem becomes significant. Further add that some of the files that make up a project are specific to a computing platform, and others may include many gigabytes of uncompressed data and the problems can become unmanageable. A first approach to organize this situation is the use of metadata or data that describes the data. Unfortunately, this increases the number of files that must be dealt with, and provides another piece of the puzzle that can be lost.

This paper describes an approach to mitigate the data management nightmare. Management is provided through what we refer to as a community grid file. The community grid approach allows the user to determine the number of data files the data should be partitioned into, as well as controlling what data is parceled to individual files. Further, the metadata information for data in the project is also included in the same file and is easily extracted for evaluation.

The following sections describe the data storage and management techniques that make a community grid possible, the methodology for interacting with a community grid file, a specific discussion on the incorporation of metadata into the community grid, and a sample application.

Data Storage and Management

A community grid file is built and used through a set of functions or programming library. Applications that wish to use the data directly from a community grid file must be modified to call these functions to retrieve input data, and store output data. Alternatively, access applications such as SMS may be used as an intermediary between an application and the community grid. SMS can retrieve the data from a community grid, save the appropriate files for an analysis code, then read the solution data and store it in the community grid file. Ideally, all applications will eventually access the community grid files directly.

When designing this library, several goals were kept in mind. These include:

- The library must be accessible from multiple programming languages. Specifically, both the C and the FORTRAN languages must be supported.
- The library must be accessible from a variety of platforms including PC-windows based, UNIX based, and high performance computing machines.
- The library must be able to store data in compressed and uncompressed modes.
- The library must be able to retrieve data in random order at high speeds. This includes chunks or slices of data.

To accomplish the first goal, two versions of the library were developed. One is written in C and the other in FORTRAN. To accomplish the other three goals, the library was built upon the Hierarchal Data Format (HDF5) developed at the National Center for Supercomputing Applications at the University of Illinois at Urbana-Champaign (NCSA). This means that community grid files are a type of HDF5 files.

The HDF5 library creates binary files which are portable across platforms. By storing data in binary format, resulting files require less disk space than the traditionally used ASCII format (National Center for Supercomputing Applications, 2003). The HDF5 library overcomes the platform specific problems related to binary files such as the bit order problem by writing files in a tagged format. The data in the file includes tags (information) which specify the type of data, bit order, precision, and any additional information necessary to enable the files to be written and read in a platform independent manner. This is handled without specification from the application that is using the HDF5 library. The HDF5 library also adds capabilities for additional data compression, and data extraction options that are accessed through the library created for community grids.

The principal components in a community grid file are spatial data objects. Types of spatial data objects include: unstructured grids (also referred to as finite element meshes), finite difference or structured grids (both Cartesian and curvilinear), triangulated irregular networks or TINs (2D scattered data sets), and layers of geographical information system objects (known as coverages).

Each spatial data object includes a geometric definition. This consists of nodes and element definitions for an unstructured grid, it consists of grid extents, cell dimensions and depth values at each cell for a structured grid, and it consists of data points or vertices for a TIN. In addition to geometry, spatial data objects may contain attributes and data sets. Attributes may be assigned on a spatial data object level or to components of the spatial object. Figure 1 shows an example of this for an unstructured finite element grid. The grid contains attributes defining model parameters. Each component of the grid includes attributes for that component of the object. The attributes may be single values, curves, or data sets.

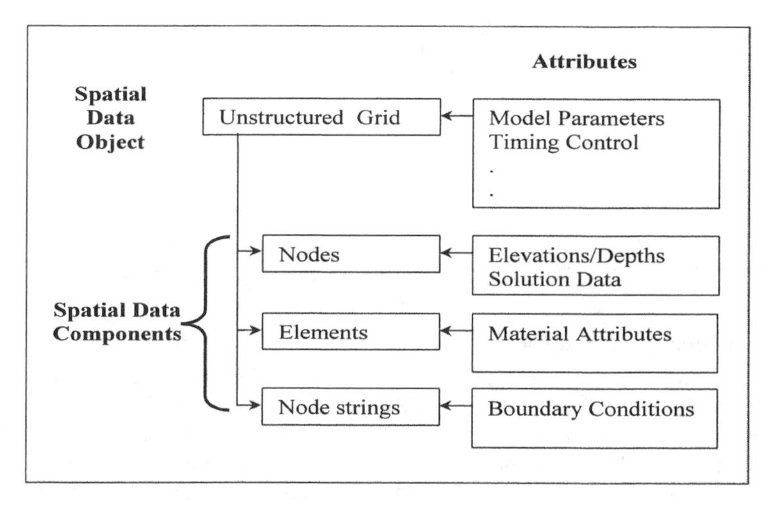

Figure 1 Spatial Data Object and attributes

Spatial data objects with associated attributes and data sets are the building blocks of community grid files. All input data, model parameters, boundary conditions, geometry, solutions and verification/calibration data can be stored using these building blocks. A community grid file is organized much like files on a disk. A spatial data object is stored in a folder utilizing sub-folders for parts of a spatial object. The files spatial objects are organized by type of data. The data contained in an example community grid file is shown as it appears in an HDF5 viewer is shown in Figure 2. The community grid file name is "TexasCommunityGrid.h5". It includes four spatial data objects. One is an unstructured grid or mesh. The other three include scattered data points or survey data. Each of these objects includes components (elements or triangles and nodes) and datasets. Each object also has a set of defined properties (attributes).

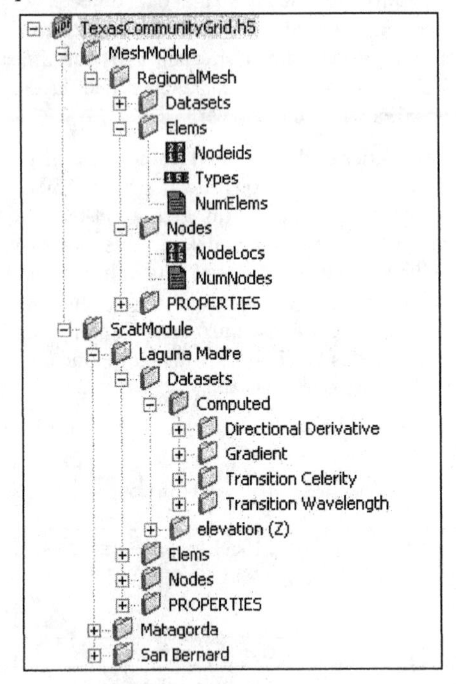

Figure 2 Community grid file structure (viewed in HDF5 browser)

Community Grid File Tools

The tools to visualize, apply, and manage the various types of data included in a community grid file exist inside the Surface Water Modeling System (SMS). The data is visualized inside of SMS in two ways. First, the organization of the community grid file is displayed in a tree structure as shown in Figure 3. The first level of the data tree identifies the type of spatial data objects included in this file. Each object is then listed with its accompanying data sets, and attributes.

The second method of visualizing the data in SMS is to view the data in a graphical display. Entities can be flagged to be displayed, or ignored in the tree (the toggles shown in Figure 3). Those flagged for display are plotted to a display window and can be visualized as shown in Figure 4 for a project covering the Gulf of Mexico.

In addition to visualization, SMS provides tools to create, edit, and format data related to a project or specific numerical simulation in a project (Environmental Modeling Research Laboratory, 2002).

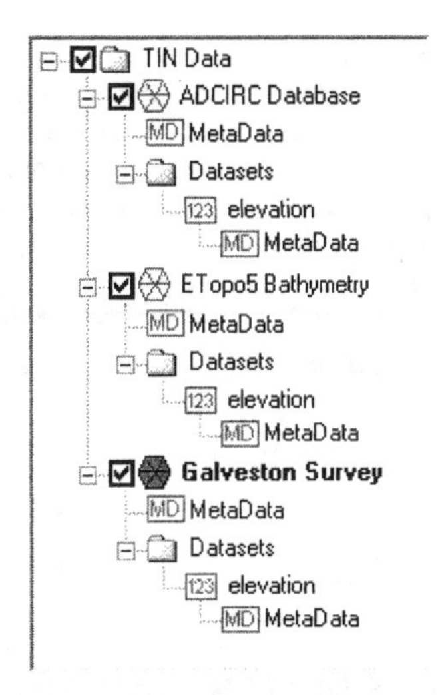

Figure 3 – Sample Data tree

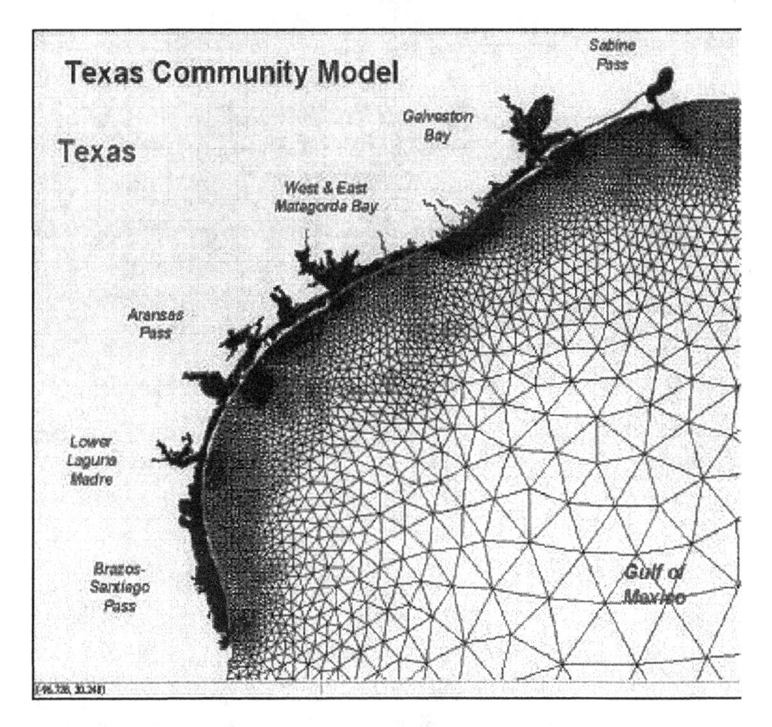

Figure 4- Sample content of graphics

Data Documentation (Metadata)

Community grid files contain a variety of data that may come from a variety of sources and cover various portions of the study area. To increase reliability, the data should be accompanied by historical information defining how it was gathered, how it is used and comments on the quality of information. This information is the metadata associated with the project data, and should reside with the data to prevent it from being lost. It should also be available to anyone who uses the data. As data is added to a project, accompanying metadata should be added as well. To assist in the archival and sharing of this information, community grid files may contain metadata. This includes:

- general information about the data
- spatial information
- contact information and data sources
- historical comments on the data

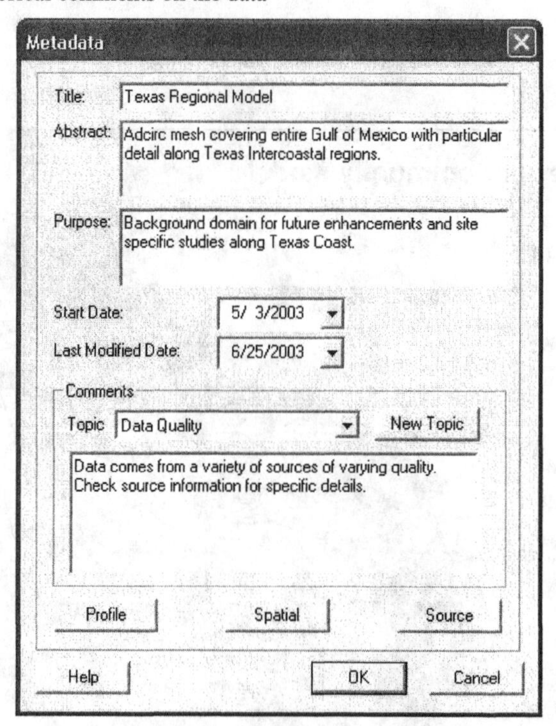

Figure 5 – General Metadata dialog.

Full metadata is written for each spatial data object in a community grid file. Metadata objects can be accessed through the SMS data tree (see Figure 3) and can be queried and added to throughout the life of the project.

General information includes an object title, abstract, beginning and ending editing dates, and user defined fields. User defined fields allow the modeler to include comments and other information not included by default. This should include information regarding use and quality of data. The general metadata record for the previous example is shown in Figure 5.

Three buttons near the bottom of the general information dialog show links to other metadata included in a metadata entry. This includes the profile of the individual responsible for adding this information, the spatial information, and comments of the source of the data.

The profile includes contact information (name, address, phone number, and email) for the person responsible for the spatial data. Like all other metadata, this is specified for each specific spatial object, which allows for multiple contact points for different data inside of a single community grid file. When SMS is used to manage community grid files, SMS will store the profile for the person using the program. When new spatial data objects are added to a project, the profile information for this object is added automatically.

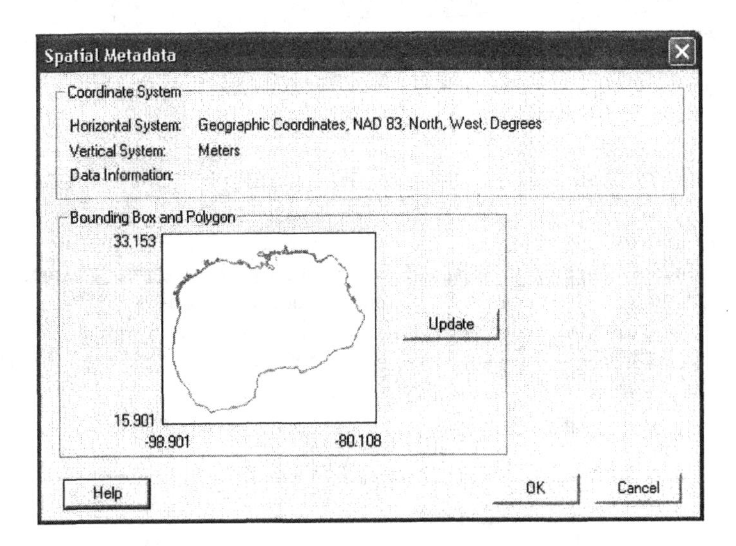

Figure 6 – Spatial metadata dialog.

The spatial information (Figure 6) defines the coordinate system and boundary information for the spatial object. The coordinate system includes the horizontal and vertical units. The boundary information includes the bounding box and also a bounding polygon. This information can be used to quickly determine if ones area of interest is included in a particular study. A database of community grid files could use this information to find previous studies that overlap the area of interest for a study in progress.

The final type of metadata included for a spatial object is the source data. An object usually has more than one source. The metadata for all parent sources are copied into the metadata record of the spatial object. For example, a finite element mesh is generally created from a conceptual model for meshing parameters and a scattered data set for bathymetry. The conceptual model may have originated from a GIS database. The scattered data set may have been created from merging data from several sources. The metadata for the finite element mesh will include a copy of the metadata records for all the parent sources including the original scattered data (Figure 7). This may be useful to verify source and quality of bathymetry data even if the original data is no longer available.

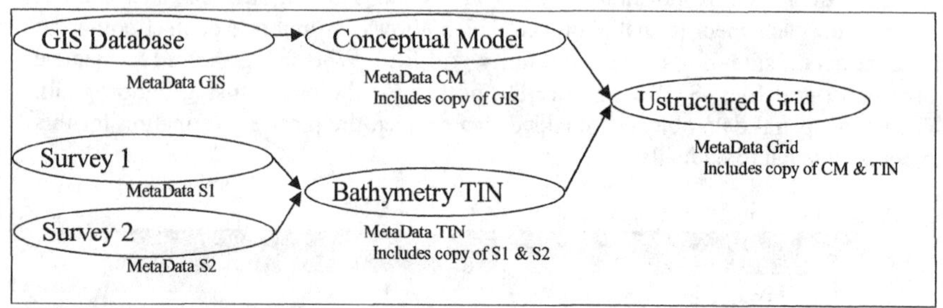

Figure 7 Source Metadata

COMMUNITY GRID FORMAT APPLICATION IN TEXAS REGIONAL MODEL

The Waterways Experiment Station is involved in ongoing studies of sites along the Texas Coast and throughout the Gulf of Mexico. Numerical models have been used to model circulation, waves, and sediment transport processes. The input, model, and verification data associated with these studies is extensive. Data from these studies has been passed between different individuals working on portions of the study. This data (or a portion received by the Environmental Modeling Research Lab) includes:

- 6 files of coastline data.
- 20 files containing surveyed bathymetric data.
- 8 files of conceptual models.
- 13 files containing verification data from gages.
- Several images.
- 20 ADCIRC simulations (multiple files each).
- 1 STWAVE simulation (multiple files).

Each of these files, while useful in itself, comprises only a small fraction of the information available for the region. How the files fit together, what coordinate system the data is in, and how they came to be is not explained at all when data is passed in this manner.

The Community grid methodology combines all of these various files into a single, self documenting file. Since the data was not managed using the community grid methodology, much of the metadata information is missing and must be replaced. If the data had been managed using the community grid methodology, the data and associated metadata would be have been compiled together.

To create the community grid file, the data is imported into SMS and then stored into a community grid file. Inside SMS, the different types of data are separated into different modules based upon the types of data included. The data for the Texas regional community grid data resides primarily in following three modules: the Map Module, the 2D Scattered Data Module, and the Mesh Module.

The Map Module data includes GIS like data stored in coverages (or layers), images, and DXF data. The map module for Texas regional model includes coverage data which contain the coastline information, conceptual models (domain extents, features and parameters), and calibration/verification data. The map module also includes several images. The tree representing the map module data is shown in Figure 8.

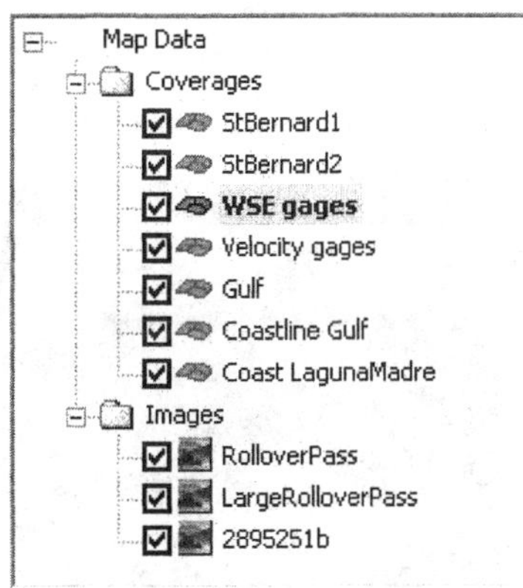

Figure 8 Map module data for Texas Regional Model

The 2D Scattered Data Module includes triangulated irregular networks with datasets. The Texas regional community grid data uses 2D scattered data primarily for bathymetry. It is also likely that 2D scattered data sets were used to guide the mesh creation processes but such datasets are no longer part of the data. Several bathymetry surveys included information about how the data was gathered, dates, and coordinate system information. Unfortunately because the community grid methodology was not used from the beginning of the project, we do not have metadata for much of the bathymetry data. Several of the ADCIRC simulations included in Texas community files cover the entire Gulf of Mexico. Certainly the bathymetry data for these simulations came from multiple sources. Likely the deep water data and data far from the areas of interest would come from a large database while shallow areas near the area of interest would be from local area surveys. If the community grid methodology had been used, each ADCIRC simulation would include a history of where the data came from. This would provide a record of which surveys were used and all metadata associated with each of these surveys. The SMS tree of scattered data sets included in the Texas community grid files is shown in Figure 9.

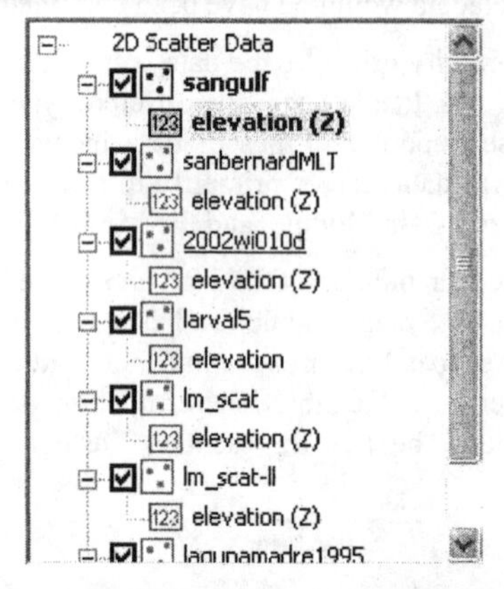

Figure 9 SMS tree for scattered data used in Texas studies

Figure 10 Input data used for numerical models

Finite element meshes are generally constructed from data in the map and scattered dataset module. The map module uses coastline and image data to create model domains and scattered data sets are used for bathymetry. Figure 10 shows some of the data used to construct a finite element mesh for the Texas studies. For clarity, only one image, coastline, and set of bathymetry points is displayed, although any number may be contained inside of a community grid file.

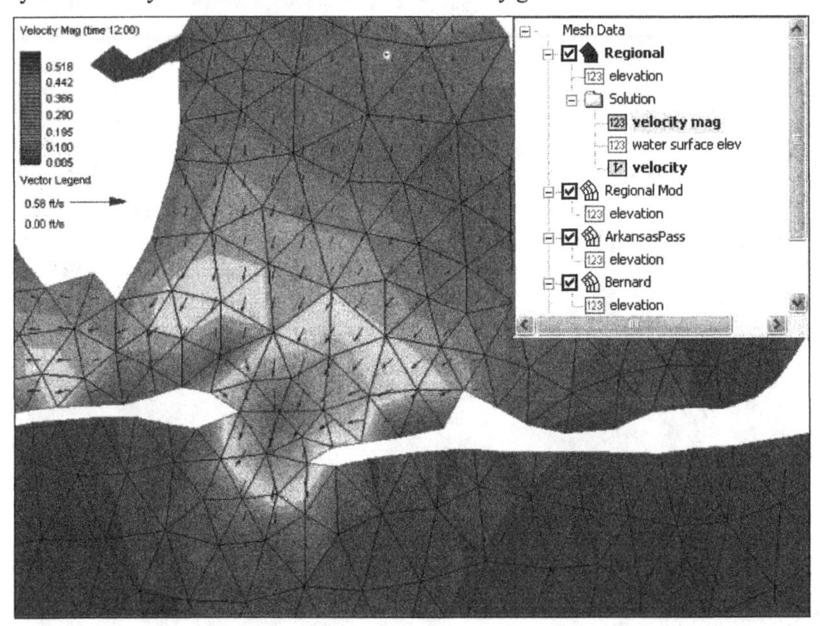

Figure 11 Finite element mesh, contours, vectors, and tree representation

Finite element meshes are edited and visualized inside of SMS in the mesh module. The mesh module includes the interface to the ADCIRC circulation model which is the principal model used in the Texas studies. ADCIRC simulations often encompass a large area with refinements in local areas of interest. Along with the finite element mesh for geometry the mesh module includes the model parameters, boundary conditions, and solutions for the ADCIRC simulations. The ADCRIC solutions included in the Texas community grid files include water surface elevations and velocities at every node in the mesh (Luettich, 2000). Using community grid files for the solution datasets makes the data storage more efficient and provides faster retrieval for visualization. Figure 11 shows a finite element mesh, a scalar dataset represented by contours, a vector dataset represented with arrows based upon direction and magnitude, and the tree with data from the Texas community grid files.

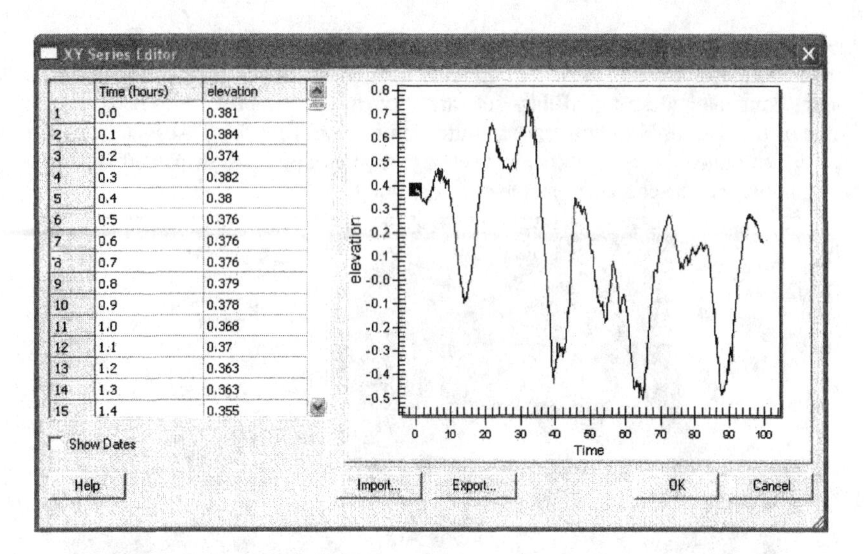

Figure 12 Water surface elevation data for calibration.

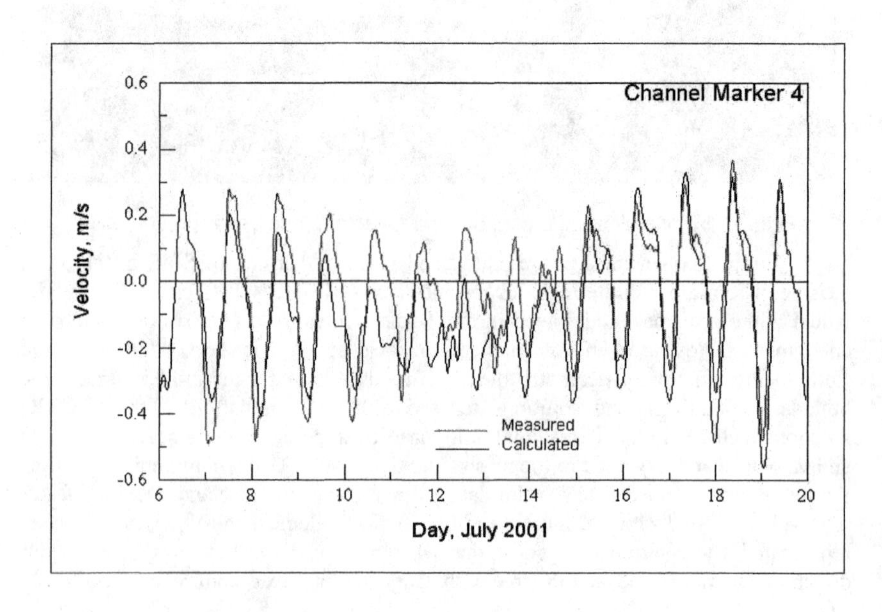

Figure 13 Measured vs. calculated plot used for verifying a model.

Once computed, numerical model solutions are compared with data gathered at specific locations for calibration and verification. This calibration data is stored in the map module inside of SMS. Figure 12 shows water-surface calibration data from the Texas studies at a specific location. SMS can interpolate solution data to a point and generate plots comparing the solution data with calibration data. An example of this type of plot is shown in Figure 13.

Using the community grid methodology, metadata for each object would be stored in the file. This metadata includes source, spatial, general, and contact information for each portion of the data. This data was gathered for the Texas Community grid file where available. However, much of this information is now unavailable. Were the community grid methodology in place for the entire study, we would have metadata for all of our spatial data objects.

CONCLUSIONS

The amount of data that goes into a series of numerical studies can easily become overwhelming. As time passes, the body of data continues to grow. Managing, storing, sharing and editing the information in a useful manner requires improved methods over what has been available. The Community Grid approach provides a solution to this situation. Not only is the data catalogued and organized, it can be easily shared and distributed. The tools described in this paper are currently implemented in the Surface-water Modeling System (SMS), and the community grid data files including the metadata are stored using this system. The API used to store the data will be publicly available through the USACE-ERDC web page.

ACKNOWLEDGEMENTS

Preparation of this paper was supported by the Inlet Modeling System and Technology Transfer Work Units, Coastal Inlets Research Program of the U.S. Army Corps of Engineers (USACE). Permission was granted by Headquarters, USACE, to publish this information.

REFERENCES

Environmental Modeling Research Laboratory. 2002. *Surfacewater Modeling System v8.0 Reference Manual.* Brigham Young University, Provo, Utah.

Luettich, R. A., and Westerink, J. J. (2000). ADCIRC Users Manual. Retrieved October 26, 2003 from http://www.marine.unc.edu/C_CATS/adcirc/.

Luettich, R. A, Westerink, J. J., and Scheffner, N. W. 1992. ADCIRC: An Advanced Three-dimensional Circulation Model for Shelves, Coasts, and Estuaries; Report 1, Theory and Methodology of ADCIRC-2DDI and ADCIRC-3DL. Technical Report DRP-92-6, U.S. Army Engineer Waterways Experiment Station, Vicksburg, Mississippi.

National Center for Supercomputing Applications. 2003. HDF5: API Specification Reference Manual. http://hdf.ncsa.uiuc.edu/HDF5/doc/RM_H5Front.html

REGIONAL CIRCULATION MODELING OF THE TEXAS COAST AND INLAND COASTAL WATERS

Mitchell E. Brown[1], Nicholas C. Kraus[2], and James M. Kieslich[3]

Abstract: In support of regional sediment management initiatives by the U.S. Army Corps of Engineers, a regional circulation model has been developed for the coast of Texas and its inland coastal waters and waterways. This large and complex hydrodynamic and sediment-sharing system encompasses 8 bays, 4 lagoons, 7 larger rivers, and the 560-km long Gulf Intracoastal Waterway. Mesh development rests on numerous bathymetry sources, and model validation benefits from availability of more than 20 water level stations and numerous coastal wind gauges along the coast and inland waters. Practicalities of regional model development are described, together with preliminary validation. Availability of the regional model has benefited several local projects.

INTRODUCTION

The Gulf of Mexico coast and adjacent inland coastal waters of Texas are comprised of 8 bays, 4 lagoons, 7 larger rivers, and the 560-km long Gulf Intracoastal Waterway (GIWW), all of which communicate with the Gulf of Mexico through 8 federal inlets and 8 non-federal inlets or river mouths. The Texas coast is a challenging study site for establishment of regional circulation and transport processes models because of the numerous water bodies, multiple-inlet bays, strong wind, flooding and drying of large tidal flats and marshes, and large spatial extent. The coast is also frequented by tropical storms that rearrange sediments and create breaches across the barrier islands.

In this complex flow-sharing and sediment-sharing environment, the U.S. Army Corps of Engineers, Galveston District conducts numerous navigation, environmental preservation, and flood-protection projects. Engineering flexibility and environmental sustainability are enhanced if multiple-project uses can be promoted

[1] Civil Engineer Technician, U.S. Army Engineer Research and Development Center (ERDC), Coastal and Hydraulics Laboratory (CHL), 3909 Halls Ferry Road, Vicksburg, MS 39180, Mitchell.E.Brown@erdc.usace.army.mil.

[2] Senior Scientist, U.S. Army ERDC, CHL, 3909 Halls Ferry Road, Vicksburg, MS 39180, Nicholas.C.Kraus@erdc.usace.army.mil

[3] Chief, Operations Division, U.S. Army Corps of Engineers, Galveston District, 2000 Fort Point Road, Galveston, TX 77550. James.M.Kieslich@swg02.usace.army.mil

and executed within a rational and defensible framework. In addition, modifications or actions taken at one project site may hold consequences for other project sites and for the coastal system in general. For these reasons, the Galveston District has initiated a Regional Sediment Management initiative to develop numerical modeling technology to address wide–area and short- to long-term processes. This initiative is compatible with the Corps' national programs for regional sediment management (http://www.wes.army.mil/rsm/). The regional Texas mesh is one of the prototypes for the Community Model concept introduced in the Corps' Coastal Inlets Research Program for documenting mesh and model metadata for ease in sharing information among organizations and projects (Zundel et al. 2003; Militello and Kraus 2003). Local and wider area wave and sediment transport models (e.g., Lin et al. 2003) are being combined with the regional circulation model described here.

Development of a regional circulation model for the Texas coast greatly benefits from the products of a water level and wind data-collection program called the Texas Coastal Ocean Observation Network (TCOON), supported by the Texas General Land Office and the Galveston District. The TCOON collects high-quality measurements of simultaneous tidal and historical water-surface elevation and wind, which can be supplemented by data on the current and bathymetry available from site-specific projects. Since the early 1990s, the TCOON has operated more than 20 water-level gauges (Fig. 1) and from 5 to 10 coastal near- or over-water wind gauges along the inland and coastal waterways during any given year. This paper describes experience with regional modeling and selected local project applications with regard to the Texas Regional Model. Central to the development is regional validation by means of a wide-area network of tidal and wind measurements.

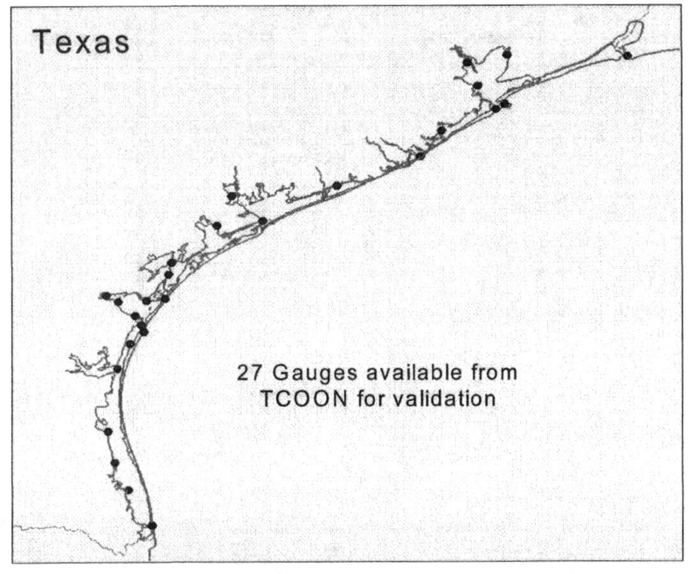

Fig.1. Locations of 27 gauges where TCOON data are available.

HYDRODYNAMIC MODEL AND MEASUREMENTS

For the regional circulation model, an ADCIRC (Luettich et al. 1991) finite-element mesh covering the Texas coast, inlets, and inland coastal waters was established for the Texas coast. It was derived in part from the mesh of the East Coast and Gulf of Mexico developed for the East Coast 2001 Database of Tidal Constituents (Mukai et al. 2002). The model mesh encompasses the entire Gulf of Mexico and includes 8 bays, 4 lagoons and 7 rivers, in addition to the GIWW.

A coastline was downloaded from the National Oceanic and Atmospheric Administration (NOAA) Coastline Extractor website, and the NOAA National Ocean Service (NOS) Medium Resolution database was incorporated to provide the new Texas coastline with adequate resolution (Fig.2). Many relatively small areas were refined based on recent aerial photograph coverage obtained by the Galveston District and the Corps' Coastal Inlets Research Program. The regional model presently contains approximately 95,000 nodes and high resolution in areas of ongoing engineering projects, such as the Colorado River, Aransas Pass, and East and West Matagorda Bays (Fig. 3), and will be refined in Community Model applications.

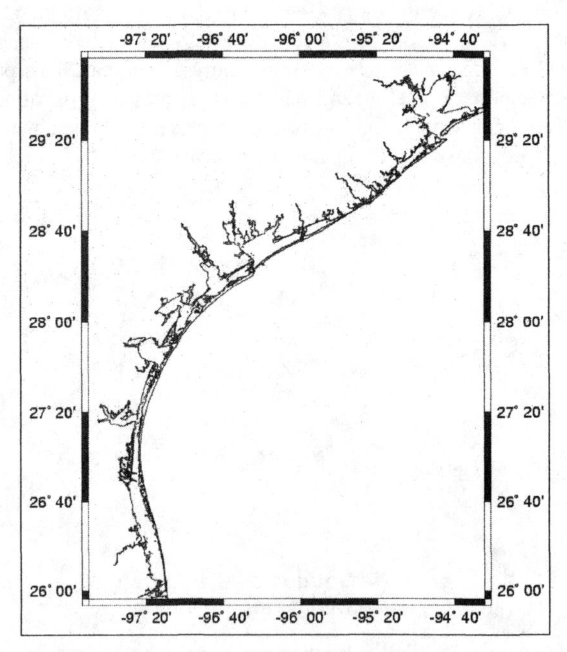

Fig. 2. Coastline extracted from NOAA/NOS medium-resolution database.

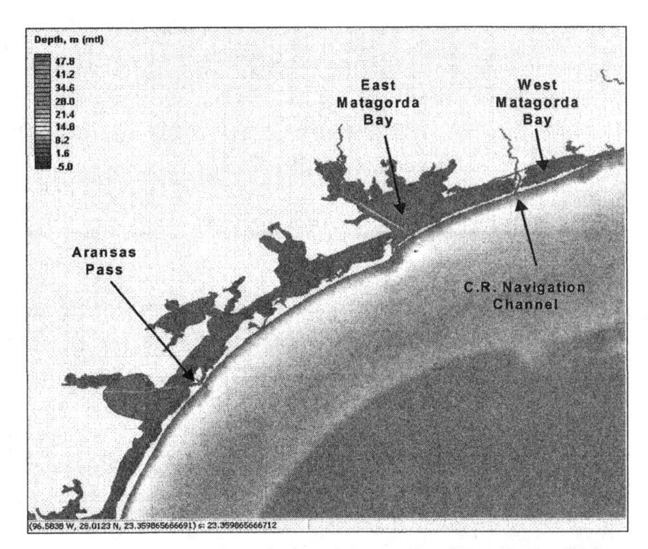

Fig 3. Portion of regional mesh containing areas of interest for this paper.

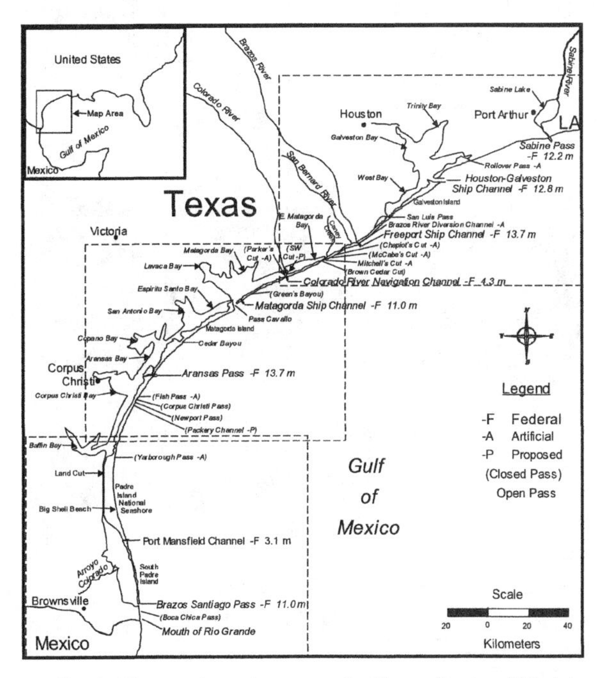

Fig. 4. Three sub-system areas for Texas Regional Model.

Once most of the structure of the mesh was generated, the basic bathymetry for the entire domain was extracted from the East Coast 2001 tidal database mesh (Mukai et al. 2002) (Fig. 5). Subsequently, old and new bathymetry surveys for individual areas were added. The modeling effort includes representation of the surface wind field, as well as tidal constituent forcing specified by the East Coast 2001 database. Regional modeling validation is being undertaken through comparison with simultaneous tidal and historical water-surface elevation measurements and wind measurements acquired from the TCOON.

Wind plays a major role in controlling the water level and circulation in the shallow coastal waters of Texas (cf. review in Militello and Kraus 2001). The prevalent southeasterly summer winds, sea breeze, and intermittent winter northerly wind fronts also force the circulation, set up, and set down in Texas bays, as well as the water flow through its inlets and river mouths. Wind information for this study was acquired from the Gulf of Mexico wind field database (1/4-deg resolution) for 1990-1999 developed recently by the U.S. Army Engineer Research and Development Center, Coastal and Hydraulics Laboratory.

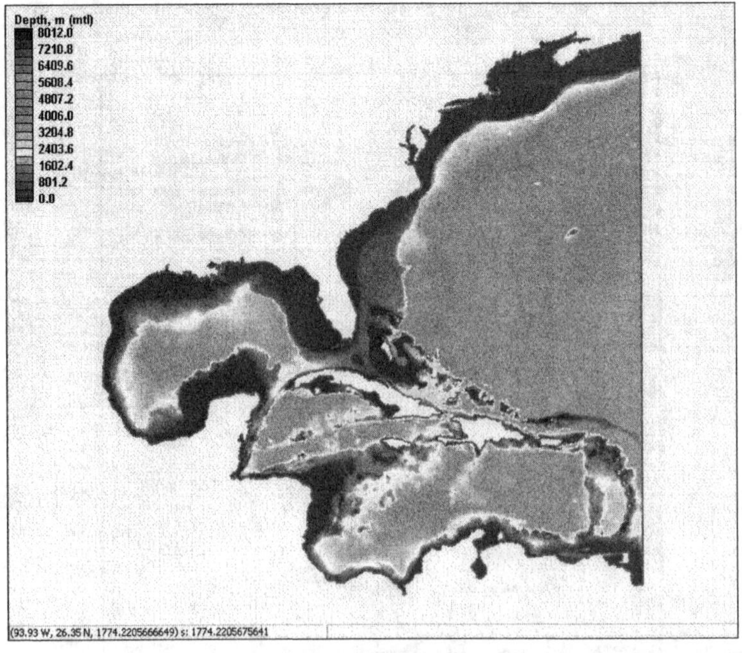

Fig. 5. EastCoast 2001 tidal database domain.

The present unstructured Texas Regional Model ADCIRC mesh is shown in Fig. 6 and represents the bathymetry of Sabine Bay, Galveston Bay, East Matagorda Bay, (west) Matagorda Bay, Aransas Bay, Corpus Christi Bay, and the Lower Laguna

Madre, as well as the GIWW throughout the Texas coast (Fig. 4). The smaller bays are encompassed within the larger bays, and the regional mesh is being treated as three subsystems for bathymetry refinement and detailed validation. Model set up used a 0.75-sec time step and default control parameters (τ_o = 0.01 and bottom friction coefficient = 0.0025). Two tidal forcing boundaries were applied, one located between the Yucatan Peninsula, Mexico and Cuba and the other between Cuba and the tip of Florida. Because the forcing boundaries were remotely located, all connected inlets could be treated as a regional interacting system. At the present time, lack of reliable bathymetry data in many areas precludes detailed validation.

As examples of local model detail, the mesh at the Mouth of the Colorado River (MCR) and Colorado River Navigation Channel area is shown in Fig. 7, including the junction of the Colorado River and GIWW, the diversion channel, and the two GIWW boat locks bracketing the Colorado River. Figure 8 depicts element resolution of the Matagorda Ship Channel and GIWW within Matagorda Bay. This regional mesh has elements with edge lengths ranging from 70 km in the middle of the Gulf of Mexico, down to 7 m in the Colorado River region of Texas.

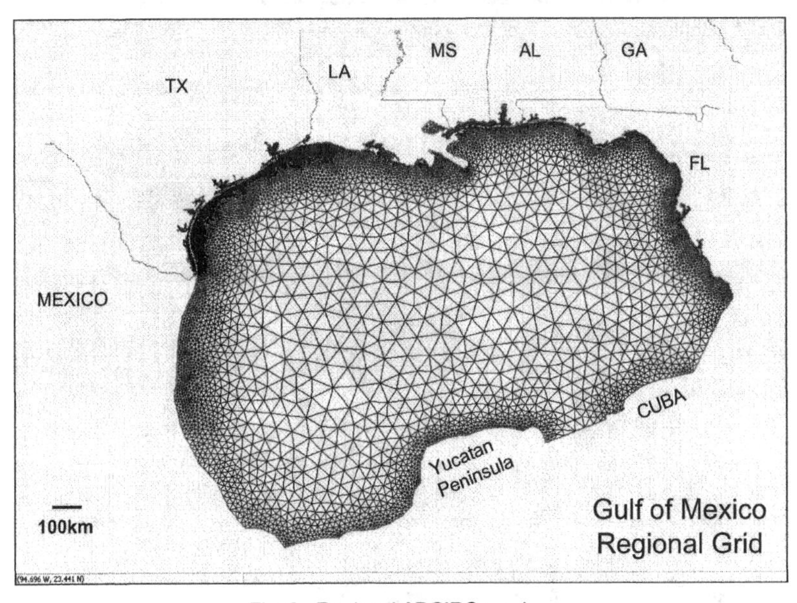

Fig. 6. Regional ADCIRC mesh.

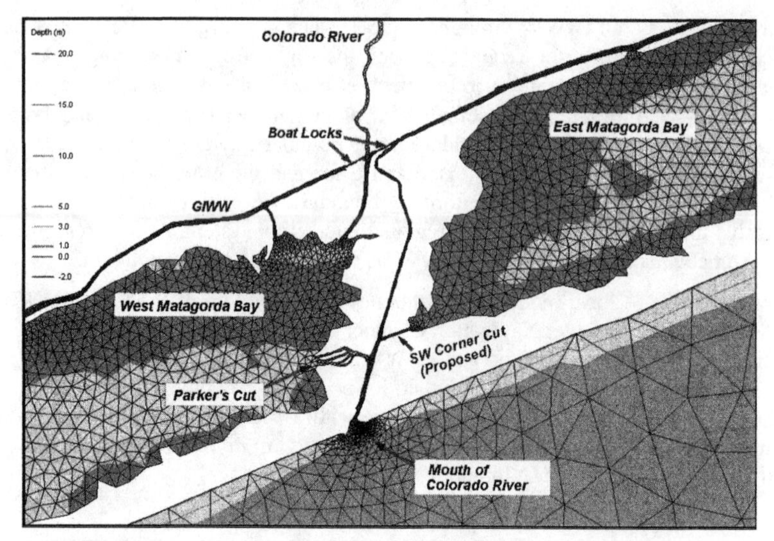

Fig. 7. Mouth of Colorado River area of regional mesh, illustrating project-level detail.

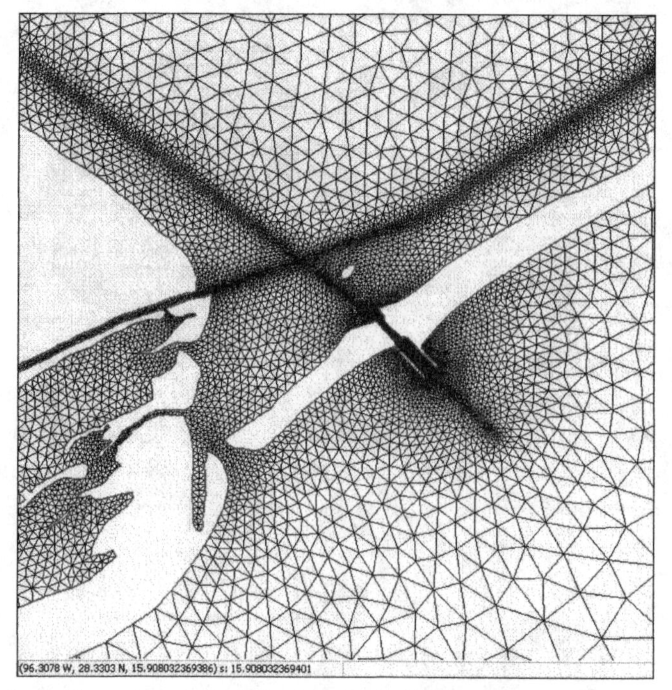

Fig. 8. Element resolution of the Matagorda Ship Channel, Matagorda Bay.

REGIONAL COMPARISONS

The time period of March 6-20, 2003 was selected for preliminary model validation because of compatibility with recent bathymetric surveys at several inlets. The purpose of the preliminary validation is to identify limitations in bathymetry data and mesh resolution. From the 27 available TCOON gauges, 20 gauges were operational during the specified time frame. As representative results, water-surface elevation calculations and measurements from March 7-14, 2003 are presented here for five locations – Rockport, Freeport, Port Aransas, Port Isabel, and Pleasure Pier (Figs. 9-13). Correlation coefficients were calculated to quantify the correspondence between the measurements and simulation. The root-mean square (RMS) difference

$$\mathrm{RMS}(\%) = \sqrt{\frac{1}{N}\sum_{i}^{N}\left(\frac{(x_i - y_i)}{y_i}\right)^2} \times 100\%$$

was also calculated as an absolute measure of difference between measurement and simulation, where N =number of data points compared; x_i = calculated water surface elevation; and y_i = measured water surface elevation. Table 1 lists the tide range in meters, RMS difference, and correlation coefficient for 10 TCOON gauges. These gauges were selected as the northern-most (Sabine) and southernmost (Port Isabel) locations of measurements, with other listed gauges representing water bodies and locations in-between.

The large RMS difference at Sabine Pass is attributed to the coarseness of the regional mesh in the surrounding area and to the lack of accurate bathymetry. Identification of such "weak" areas of the regional model was one of the purposes of the preliminary validation effort. The RMS difference and correlation coefficient are judged to be reasonable for this regional mesh with default values of ADCIRC parameters and will be improved as bathymetry measurements are made and resolution is increased in future projects.

The preliminary validation uncovered general issues that will need to be resolved at the next stage of refinement within the three sub-regions (Fig. 4). Although water level range and phase are overall well represented at times when the wind was weak, many strong wind events are not represented well in the model, indicating that local wind coverage and, perhaps, frequency of wind input need to be examined. In addition, longer-term simulations (not shown here) do not capture the two inter-annual regional increases and decreases in water elevation. This low-frequency variation has implications for navigation in shallow-draft channels, as well as on sediment transport.

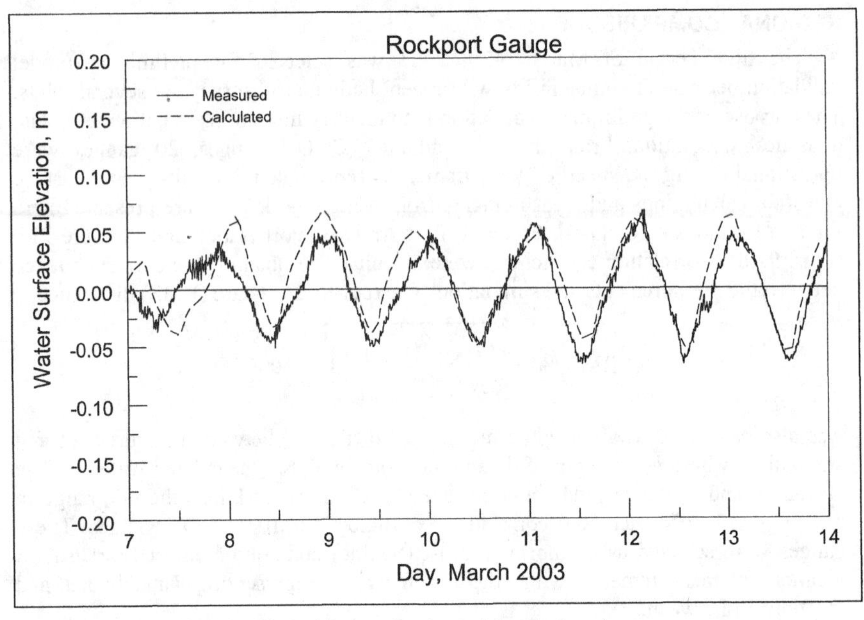

Fig. 9. Water-level comparison at the Rockport gauge.

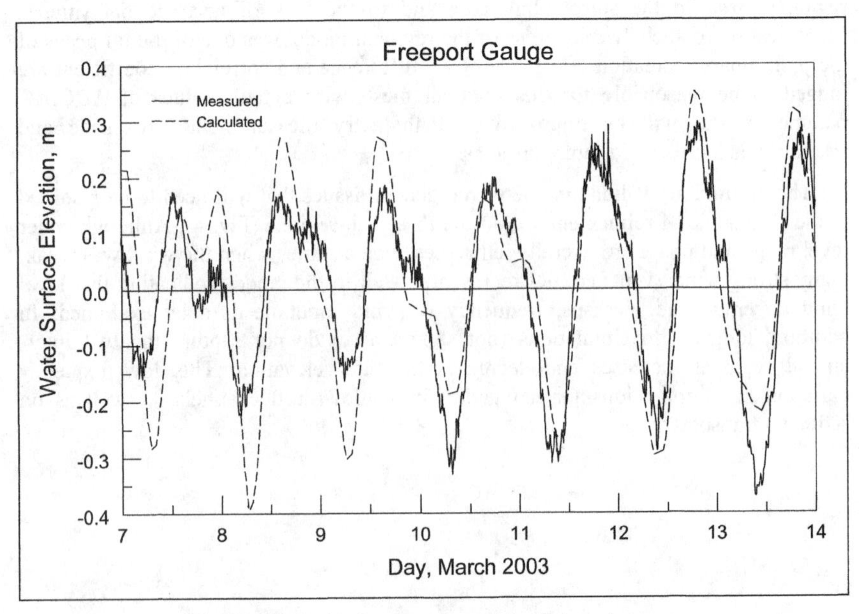

Fig. 10. Water-level comparison at the Freeport gauge.

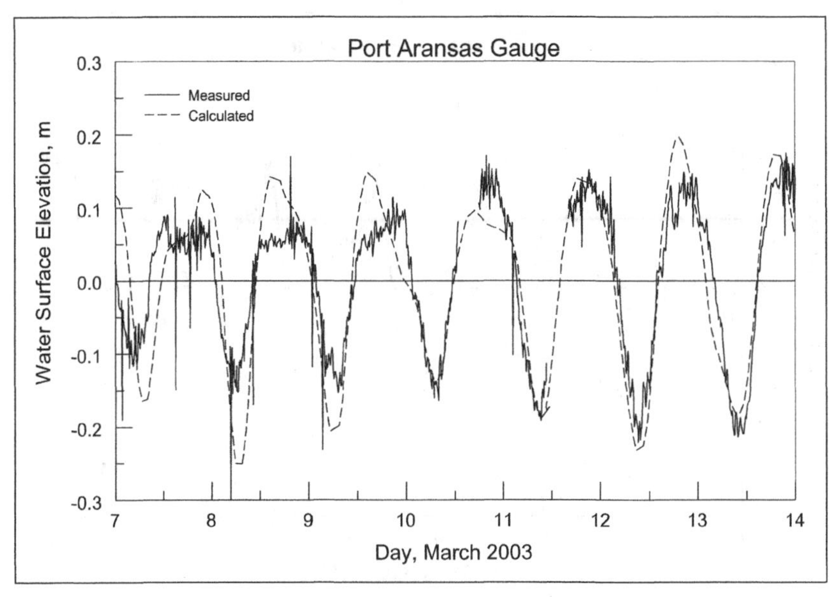

Fig. 11. Water -level comparison at the Port Aransas gauge.

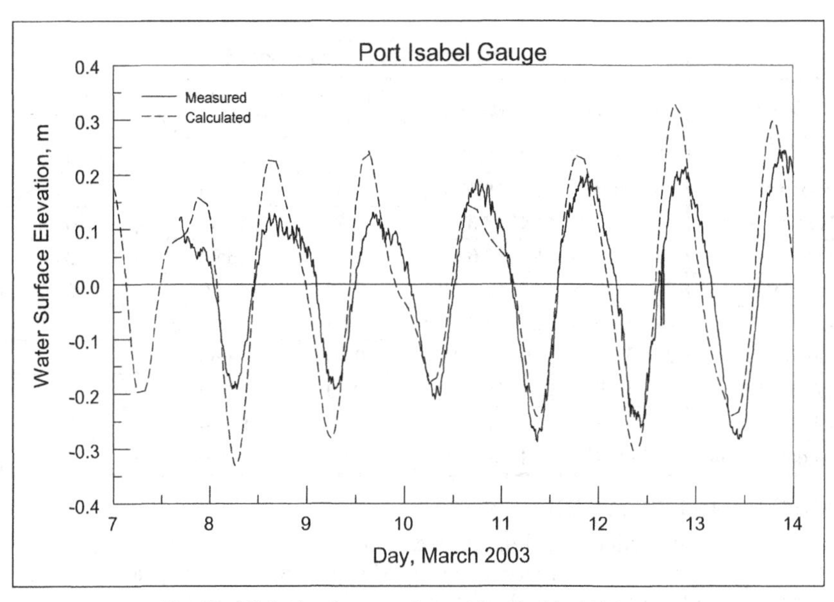

Fig. 12. Water level comparison at the Port Isabel gauge.

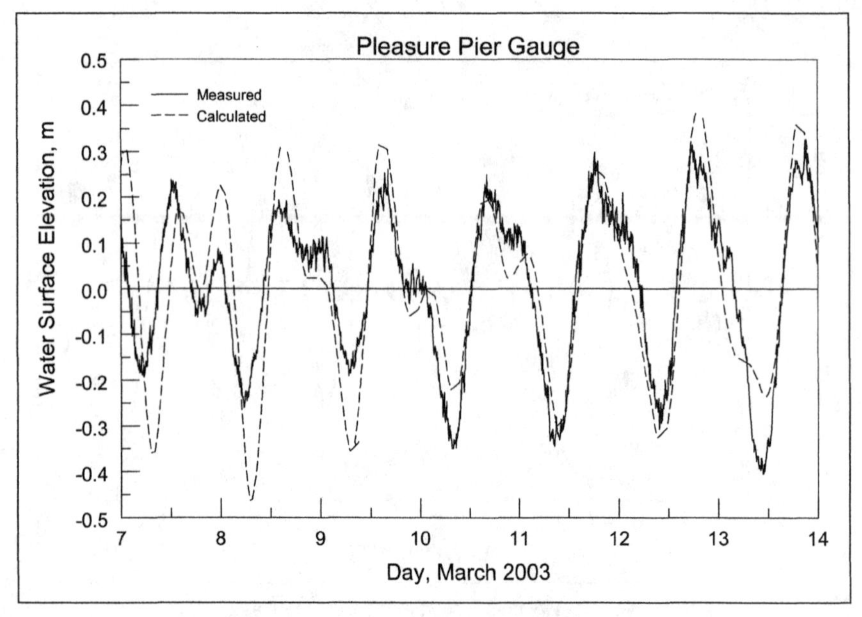

Fig. 13. Water level comparison at the Pleasure Pier gauge.

Table 1. Water Surface Elevation, RMS Difference, and Percent Error

Station	Latitude	Longitude	Tide Range, m	RMS Difference, (%)	Correlation Coef.
Rockport	28° 1.3'N	97° 2.8' W	0.13	9.38	0.93
Freeport	28° 56.9'N	95° 18.5' W	0.31	11.35	0.84
Pleasure Pier	29° 17.1'N	94° 47.3' W	0.92	11.92	0.83
Port Aransas	27° 50.3'N	97° 3.6' W	0.50	6.94	0.85
Port Isabel	26° 3.6'N	97° 12.9' W	0.53	12.40	0.80
Bob Hall Pier	27° 34.9'N	97° 13.0' W	0.83	13.83	0.89
Port Lavaca	28° 39.0'N	96° 35.7' W	0.40	4.66	0.90
Rollover Pass	29° 30.9'N	94° 30.8' W	0.60	10.36	0.73
Sabine Pass	29° 43.8'N	93° 52.2' W	0.74	51.54	0.73
Morgan's Point	29° 40.9'N	94° 59.1' W	0.60	7.89	0.88

CONCLUDING DISCUSSION

A regional circulation model has been established for the coast of Texas and its inland coastal waters. The Texas Regional Model is already being applied and proving its utility in Galveston District projects. As a focus of effort according to resources and priorities, the regional model is being refined in three subsystems. Each subsystem will encompass the same domain as this regional model, however each will have finer resolution in areas of interest for that particular sub-region and coarser resolution outside. Each new subsystem mesh will be updated with more accurate coastlines and bathymetry in specific areas.

This Texas Regional Model functions as a framework for the Inlet Modeling System (IMS) developed by the U.S. Army Corps of Engineers' Coastal Inlets Research Program. The IMS consists of coupled calculations of waves, circulation, sediment transport, and morphology change. Modeling undertaken within the IMS, such as for the Mouth of the Colorado River (Lin et al. 2003) benefit from availability of calibrated regional model for accurate local forcing.

Three benefits accrue by establishing circulation models at a regional scale. One is the reliable boundary condition for a local model that can be obtained from the entire regional domain. In operating a regional model, the forcing boundaries are at great distance from the project area, allowing the model to generate non-linearities, such as the overtides, as well as phase differences close to the coast. Second, by availability of regional models, regionally validated forcing information and preliminary model meshes are available to address new projects or to support emergency, fast-track actions that become necessary. A third benefit is that regional models allow calculation of water and sediment-sharing systems with their interactions to evaluate the consequences of actions taken at one site at locations within the region of influence of that site. A priori, the range of influence may not be known.

The preliminary validation described here identified the region-wide needs of better representation of wind forcing and inter-annual fluctuations in water level, as well as limitations in the calculation mesh developed to date.

ACKNOWLEGEMENTS

This study was supported by the U.S. Army Engineer District, Galveston, as part of its Regional Sediment Management initiative, and by the Coastal Morphology Modeling Work Unit, Regional Sediment Management Program. We appreciate assistance provided by Mr. Michael Tubman, and Dr. Lihwa Lin in developing the bathymetry input and model setup, and by Mr. Ronnie Barcak in reviewing the regional mesh. Permission was granted by Headquarters, U.S. Army Corps of Engineers, to publish this information.

REFERENCES

Cialone, M. A., and Kraus, N. C. (2001). Engineering Study of Inlet Entrance Hydrodynamics: Grays Harbor, Washington, USA, *Proc. 4th Coastal Dynamics Conf.*, ASCE, 413-422.

Cialone, M. A. and Kraus, N. C. (2002). Coupling of Wave and Circulation Numerical Models at Grays Harbor Entrance, Washington, USA, *Proc. 28th Coastal Engineering Conf.*, World Scientific, 1279-1291.

Kraus, N. C., and Militello, A. (1999). Hydraulic Study of Multiple Inlet System: East Matagorda Bay, Texas. *J. Hydraulic Eng.*, 125(3), 224-232.

Lin, L., Kraus, N. C., and Barcak, R. G. (2003). Modeling Coastal Sediment Transport at the Mouth of the Colorado River, Texas. *Proc. 8th Estuarine and Coastal Modeling Conf.*, ASCE, submitted.

Luettich, R.A., Westerink, J.J., and Scheffner, N.W. (1991). ADCIRC: An Advanced Three-dimensional Circulation Model for Shelves, Coasts and Estuaries; Report 1: Theory and methodology of ADCIRC-2DDI and ADCIRC-3DL, Technical Report DRP-92-6, U.S. Army Engineer Waterways Experiment Station, Vicksburg, MS.

Luettich, R. A., and Westerink, J. J. (2000). ADCIRC Users Manual. Retrieved October 26, 2003 from http://www.marine.unc.edu/C_CATS/adcirc/.

Mukai, A. Y., Westerink, J. J., Luettich, R. A., and Mark, D. (2002). East Coast 2001, A tidal constituent database for Western North Atlantic, Gulf of Mexico, and Caribbean Sea, Technical Report ERDC/CHL TR-02-24, U.S. Army Engineer Research and Development Center, Vicksburg, MS.

Militello, A., and Kraus, N. C. (2001). Generation of Harmonics by Sea Breeze in Nontidal Water Bodies, *J. Physical Oceanog.*, 31(6), 1,639-1,647.

Militello, A., and Kraus, N. C. (2003). Regional Circulation Model for Coast of Long Island, New York, *Proc. 8th Estuarine and Coastal Modeling Conf.*, ASCE, submitted.

Zundel, A. K., Cialone, M. A., and Moreland, T. J. (2002a). The SMS Steering Module for Coupling Waves and Currents, 1: ADCIRC and STWAVE, Coastal & Hydraulics Eng. Tech. Note IV-41, U.S. Army Eng. Research and Development Center, Vicksburg, MS.

Zundel, A. K., Jones, R. D., and Kraus, N. C. (2003). Community Model Concept and Implementation within the Surfacewater Modeling System, *Proc. 8th Estuarine and Coastal Modeling Conf.*, ASCE, submitted.

REGIONAL CIRCULATION MODEL FOR COAST OF LONG ISLAND, NEW YORK

Adele Militello[1] and Nicholas C. Kraus[2]

ABSTRACT: During the past 5 years, the U.S. Army Corps of Engineers' Coastal Inlets Research Program has been taking a systematic regional modeling approach to calculating combined scales of nearshore circulation and providing established models, called "Community Models," for new projects that can involve modeling by different organizations. Here, we describe a regional circulation model for the coast of Long Island, NY. Domain coverage includes the New York Bight and Long Island Sound. The two-dimensional finite-element circulation model ADCIRC was applied because of its stability and flexibility in element sizes. Placement of tidal boundaries far from the area of interest allows free propagation of the tides throughout the domain, preserving the natural amplitude and phase variation. Since development of the original Long Island regional model, it has been applied to numerous studies around the north and south shores of Long Island. The community model benefited from bathymetry updates and increases in resolution for each project. Specific study applications are: flood shoal mining at Shinnecock Inlet; erosion control at Jones Inlet; storm surge calculations from Fire Island to Montauk Point; proposed relocation of Fire Island Inlet; and stability of two small inlets on the north shore of Long Island. Because each project could obtain, adapt, and apply the existing regional model and rigorous forcing conditions, considerable time and effort were saved.

---

1) Coastal Analysis LLC, 4886 Herron Road, Eureka, CA, 95503. CoastalAnalysis@cox.net.

2) U.S. Army Engineer Research and Development Center, Coastal and Hydraulics Laboratory, 3909 Halls Ferry Road, Vicksburg, MS, 39180-6199. Nicholas.C.Kraus@erdc.usace.army.mil.

INTRODUCTION

During the past 5 years, the U.S. Army Corps of Engineers' Coastal Inlets Research Program (CIRP) has been taking a systematic regional modeling approach to calculating combined scales of motion and providing established models for new projects that can involve modeling by different organizations. Six regional circulation models have been developed by the CIRP to date. Areas of coverage are: coast of Long Island (Militello and Kraus 2001a), coast of Texas (Brown et al. 2004), east coast of Florida (Militello and Zarillo 2000), central U.S. east coast extending from Virginia to Delaware, California coast, and the Washington coast (Militello et al. 2000a; Cialone and Kraus 2001). The regional models are at the center of a community model concept (Zundel et al. 2004), in which the mesh can be adopted, refined, documented, and applied in new local projects, with each application improving the mesh and benefiting from regional calibration.

The two-dimensional version of the finite-element model ADCIRC (Luettich et al. 1992; Luettich and Westerink 2000) is being applied for these regional modeling efforts because of its flexibility in mesh resolution, robust stability including under wide-area flooding and drying, efficiency of calculation, and accuracy over a wide range of temporal and spatial scales (Militello et al. 2000b). ADCIRC is versatile in capability of forcing input, including coupling with the steady spectral wave model STWAVE (Smith et al. 1999) for calculation wave-current interaction and wave-driven current (Zundel et al. 2000). A regional wave model with nested grids has also been developed for the south shore of Long Island and verified with measurements (Grosskopf et al. 2001). In this paper, we focus on the regional circulation model for the coast of Long Island.

Multiple scales of motion can be represented in a regional model. Tides are forced at the seaward boundaries, and storm wind and pressure fields can be applied over the domain to compute surge along the entire coast of Long Island. Wave forcing has also been calculated over selected regions of the domain, such as the inlets, to compute wave-driven currents. All of these forcings can take place in a single simulation, yielding hydrodynamic response to motions ranging from seconds to days. An example showing inlet ebb jet (local scale) interacting with the coastal tidal current (large scale) is presented to illustrate multiple-scale interactions.

Regional models give an established foundation for calculating local-scale hydrodynamics with either fine resolution specified within the regional model or with a local nested model. The CIRP has developed an integrated package called the Inlet Modeling System (IMS) in which regional models can be linked to STWAVE and to the hydrodynamic model M2D (Militello et al. 2004) through boundary condition specification (Militello and Zundel 2002). Both local- and regional-scale models can be coupled with STWAVE, providing flexibility. The IMS operates within the Surfacewater Modeling System (SMS), and connectivity between models is specified there, including development of input files and specification of boundary conditions.

In this paper, we describe the regional Long Island model, engineering projects that have benefited from its availability, and calculation of multiple scales of motion.

COAST OF LONG ISLAND REGIONAL MODEL

The regional Long Island mesh covers the New York Bight, Long Island, and Long Island Sound (Fig. 1). Development of the Long Island regional model was initially focused around Shinnecock Inlet, the easternmost of six permanent inlets on the south shore of Long Island. The extent of the regional domain was selected so that tidal boundaries were far from the area of interest, which in the community model concept, could be anywhere on Long Island. Mesh resolution varied greatly over the domain, with largest elements in the ocean and smallest elements at Shinnecock Inlet. The ratio of largest to smallest element area is 8.2×10^6.

Fine resolution of the inlet, bay, flood shoal, ebb shoal, and channels was specified on the model grid (Fig. 2). The Shinnecock Canal, a tide-controlled waterway connecting Shinnecock Bay and Great Peconic Bay that allows 1-way flow into Shinnecock Bay, was also implemented. Tidal gate opening and closing is triggered by differences in water-surface elevation on each end of the canal. Water-level measurements collected at the ends of the canal, together with records of times of gate opening and closing, gave representative water-level differences of 30 and -5 cm for opening and closing, respectively. Other fine-resolution features included are the Ponquogue Bridge, and the Quogue Canal that connects Shinnecock Bay and Moriches Bay. Details of the bathymetry and model resolution at the inlet are shown in Fig. 3. The model was calibrated and verified for tidal forcing (LeProvost et al. 1994; Militello and Zundel 1999) with measurements at four water-level stations and two velocity stations within Shinnecock Inlet and Bay (Militello and Kraus 2001a).

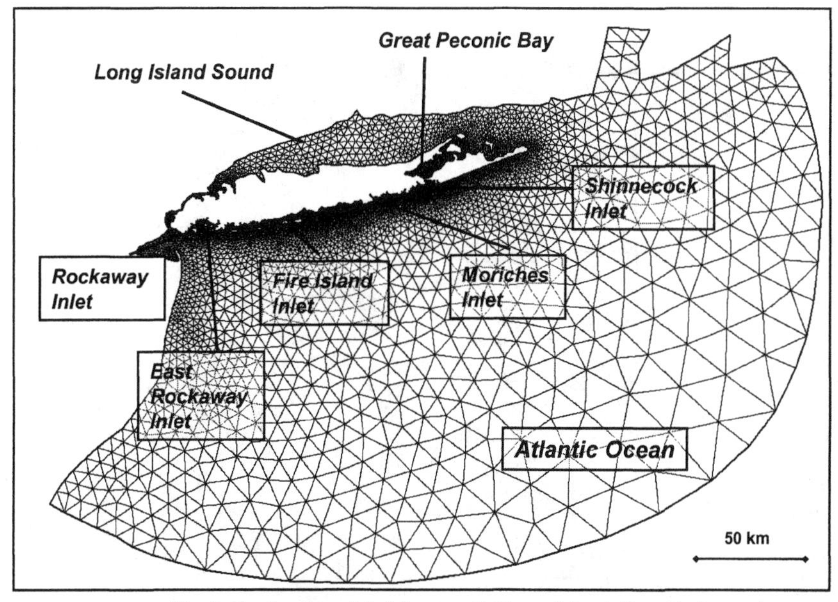

Fig. 1. Regional model of the coast of Long Island, New York

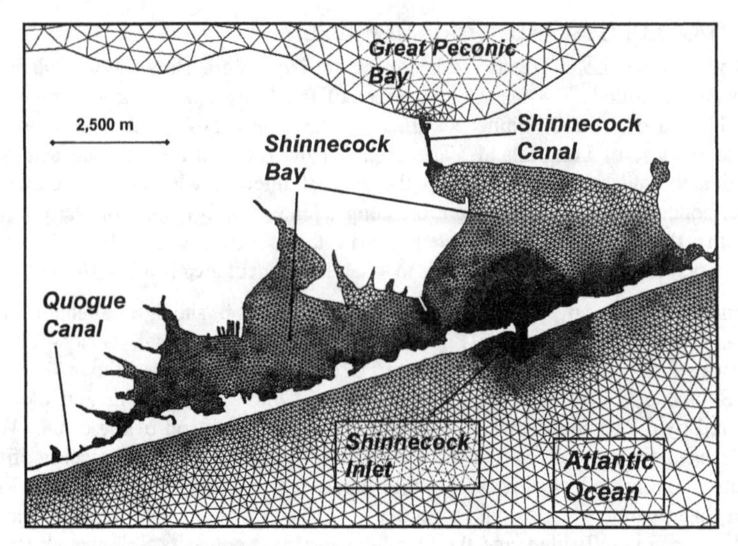

Fig. 2. Model detail at Shinnecock Inlet and Bay

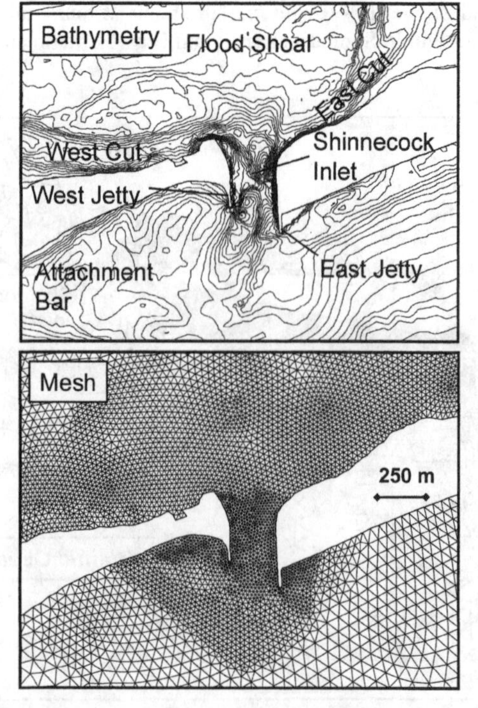

Fig. 3. Fine-scale resolution of Shinnecock Inlet and bathymetry

Five sets of bathymetric data were integrated into the model. In areas of overlapping coverage, the most recent data set was applied. Table 1 lists sources of bathymetric data and their regions of application.

Table 1. Sources of Bathymetric Data and Regions of Application

Data Source	Region of Application
NOAA[1]	Open ocean
GEODAS[2]	Long Island Sound, all bays on South Shore of Long Island with the exception of Shinnecock Bay
Traditional Survey[3]	Shinnecock Bay, except flood shoal
SHOALS Survey[4]	Shinnecock Inlet, Shinnecock Bay flood shoal, nearshore area, Moriches Inlet
USACE Traditional Survey[5]	Shinnecock Inlet, Jones Inlet

1. National Oceanic and Atmospheric Administration.
2. GEOphysical Data System developed by the National Geophysical Data Center, http//www.ngdc.noaa.gov/mgg/geodas/geodas.html.
3. Traditional boat survey conducted in 1998 by SUNY Stony Brook.
4. Shinnecock survey conducted in 1998; Moriches survey conducted in 1996.
5. Traditional boat survey data at Shinnecock Inlet supplemented SHOALS data.

Although initial development was aimed at resolving hydrodynamic details in the vicinity of Shinnecock Inlet, the remaining five Federal inlets and their associated back bays were represented in the model. Efforts in several subsequent projects have improved representation of these five inlets and bay areas. Examples are:

1. Fire Island Inlet proposed relocation: Mesh resolution increased at Fire Island Inlet and in Great South Bay near the inlet, and existing inlet closed with opening of the relocated inlet for evaluation of the proposed inlet relocation (Kraus et al. 2003).

2. Jones Inlet erosion control study: Mesh resolution increased at Jones Inlet, bathymetry updated by 2000 traditional survey conducted by the U.S. Army Corps of Engineers, New York District.

3. Fire Island to Montauk Point Reformulation Study: All inlet and bay areas where newer bathymetric data were available were updated. Mesh resolution was modified over most of the inlet and bay areas on the south shore to be compatible with storm surge calculations (Irish et al. 2003). Specific areas of further refinement and bathymetric improvements were Fire Island Inlet, Great South Bay, Moriches Inlet, Moriches Bay, and Quogue Canal.

4. Jamaica Bay wetlands restoration study. Bathymetry data for Jamaica Bay were updated to better represent existing wetland representation.
5. Mattituck Inlet (Federal) and Goldsmith Inlet (Town of Southold) improvement studies. Mesh resolution was increased to represent these inlets, and local bathymetry was updated with new soundings.

Each of these projects benefited from the previously developed regional model, such that time and cost were saved because project needs required model updates rather than full mesh development. In turn, the regional model was improved by these projects and can be accessed in future projects, furthering time and cost savings.

The regional model has also served as a source of information for local nested models. As an example, a local-scale research modeling effort is presently being undertaken at Shinnecock Inlet, in which information has been implemented from the Long Island regional model, including bathymetry and shorelines. The local M2D grid has been developed within the IMS system and includes the nearshore, inlet, and back bay. The model is being applied to calculate morphology change in response to tide and wave forcing (Buonaiuto and Militello 2004). Boundary conditions for the local model have been obtained from a regional solution and preserve tidal phase variation across the ocean boundary, which increases accuracy in the hydrodynamic and sediment transport calculations, providing calculations of morphology change that closely follow those measured by bathymetric surveys.

REGIONAL MODELING: PROCESSES AND SCALES OF MOTION

Calculation of coastal hydrodynamics requires representation of processes ranging in scale from days (tide, storm surge, large-basin seiching) to seconds (wind waves), and corresponding spatial scales of 10s to 100s of kilometers to meters. Representative physical processes and engineering actions are characterized in Fig. 4 by their compatible or feasibly calculated temporal and spatial scales. Requirements of any modeling effort provide constraints on model development and forcing, depending on the particular application. For example, calculation of tide-forced water levels along the coast might allow spatial resolution on the order of 100s of meters, whereas details of the velocity field in a tidal inlet may require mesh resolution in the range of 10 to 20 m to replicate bathymetric and structural control on the current field, to resolve development of eddies, and to accurately calculate flow through channels (Militello 1998).

Multiple temporal and spatial scales of motion can be represented within a regional model. Spatial mesh resolution must be compatible with motion on scales of interest. Boundary input, wave stress, and wind stress must also be prescribed at sufficient resolution to represent the forcing processes of interest for a particular application. By selective specification of mesh resolution and forcing input, motion on combined scales can be calculated. For example, a simulation may include tidal propagation (macro scale) and wave forcing (micro scale). In the nearshore area, these combined forcings would provide a longshore current that is modulated by tidal elevation and temporal change in wave properties. At an inlet, the wave-driven current would interact with the tidal flow, possibly enhancing flood velocity and retarding, deforming, or deflecting the ebb jet.

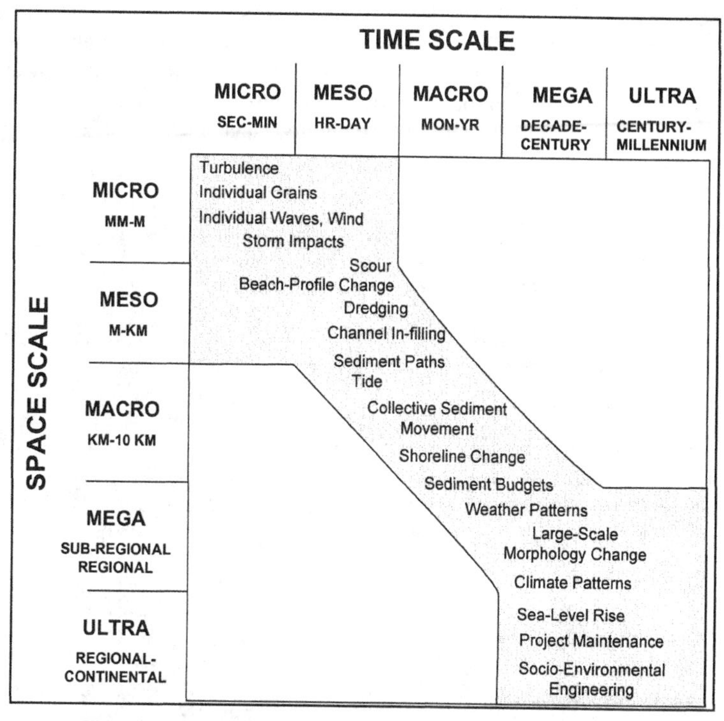

Fig. 4. Compatible scales of spatial and temporal motion (modified after Larson and Kraus 1994)

Calculation of multiple scales of motion is most effectively conducted by a regional model or by providing forcing information from the regional model and prescribing it as boundary input to a local model. The reason is that long-wave propagation, primarily tide and storm surge, can be successfully calculated with a regional model because the waves can freely move throughout the model domain. Thus, spatial variation in amplitude, phase, and velocity are retained. The value in this approach is minimization of error by retaining of these properties, while capturing phase-dependent phenomena that may be especially significant in multiple-inlet bays and in the vicinity of tidal inlets. Without a regional model, local modeling efforts typically apply predicted tides lacking phase and amplitude variation across the boundary, or a single measured time series of water level (possibly modified with longer-period variation superimposed) over the seaward boundary. By raising and lowering the ocean boundary uniformly, coastal currents induced by gradients in long-wave amplitude and in phase are omitted, and substantial error can be introduced.

Circulation patterns at Shinnecock Inlet are modified by the coastal tidal current. During ebb, the westward-moving tidal current advects the inlet ebb jet with it at an average rate of 10 cm/sec, enhancing jet migration toward the west. Calculation of this process by the regional Long Island model is shown in Fig. 5. In this figure, current patterns are shown at hourly intervals and plotted over bathymetric contours. Jet migration distance is largest at the seaward tip of the jet and reduced toward the jetties, which act as a pivot point. Measurements taken over an ebb cycle at Shinnecock Inlet verified the pattern of jet migration calculated by the regional model (Militello and Kraus 2001a,b).

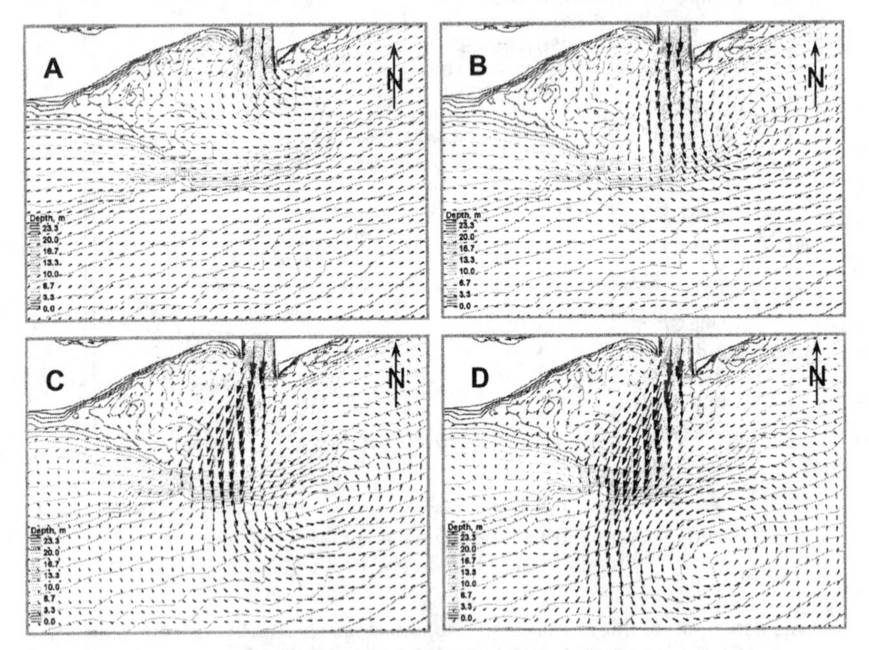

Fig. 5. Calculated migration of the ebb jet.

Consecutive bathymetric surveys at Shinnecock Inlet have revealed a systematic pattern of channel thalweg migration toward the west (Fig. 6). In 1989, the inlet thalweg trended southwest from the jetty tips, then was dredged in 1990 to extend directly seaward from the inlet. By 1991, the thalweg had migrated toward the southwest, and further migration had taken place by 1992. This migration is controlled, at least in part, by the jet migration (waves also contribute), which occurs twice daily and exerts strong stresses on the seabed (Militello and Kraus 2001b). Recent circulation and sediment transport calculations with the local model M2D in which boundary conditions were obtained from the regional ADCIRC model (Buonaiuto and Militello 2004) reproduce the westward thalweg migration.

If phase variation near the inlet is not included in calculation of circulation patterns at the inlet, the jet migration will not be as strong as it occurs in nature, and its influence on the westward propagation of the thalweg would be underestimated. Thus, preservation of the phase variation in nearshore calculations plays a significant role in accurate calculation of circulation patterns there. This situation illustrates one advantage of reproducing multiple scales of motion in a single simulation.

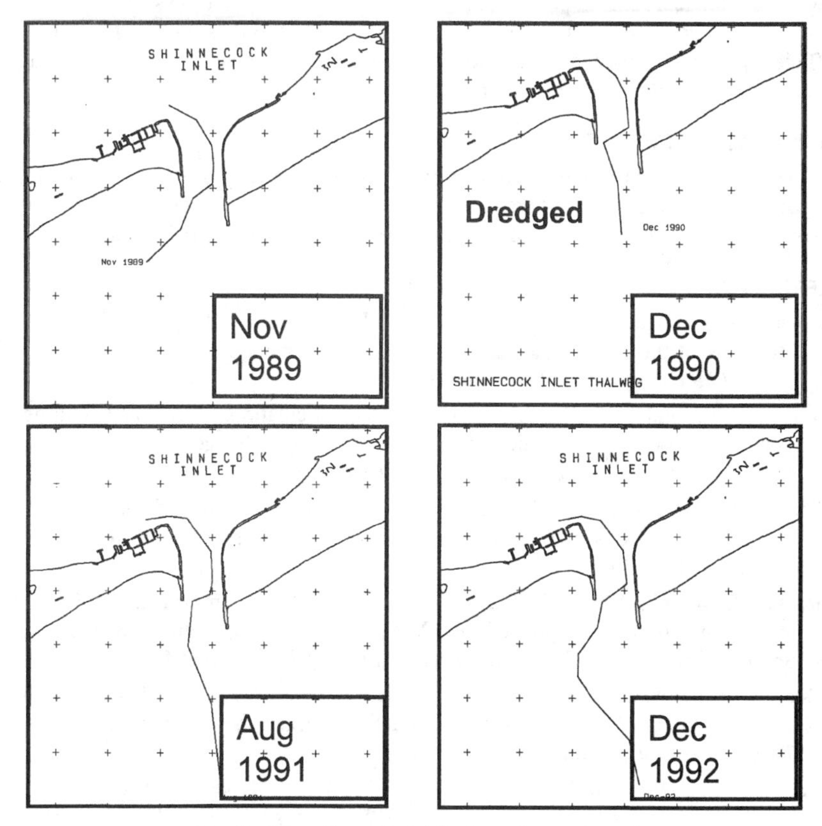

Fig. 6. Migration of inlet thalweg toward west

CONCLUSIONS

A regional circulation model has been developed by the CIRP that covers the entire coast of Long Island, and is has finest detail at Shinnecock Inlet and Bay. After development and tidal calibration at Shinnecock Inlet and Bay, the model was applied in several other projects along the south shore of Long Island in exercising of a community model approach. These projects benefited from the regional model because it provided a foundation from which local mesh updates were made, thus saving time and cost. Also, regional calibration was also performed, allowing the new modeling projects to progress rapidly. These updates provide an improved model for future efforts.

The regional approach to circulation modeling holds three advantages for accuracy of calculations. Boundary conditions are located far from project areas, which allows long-wave motion, such as tide and storm surge, to freely propagate through the domain. Phase and amplitude variation are preserved, which can be significant for reproducing processes at local sites. Second, multiple scales of motion and their interactions can be represented within a single simulation. Finally, efficiency and accuracy of new modeling projects are greatly increased through availability of a calibrated regional model.

ACKNOWLEDGEMENTS

This paper was prepared as an activity of the Inlet Modeling System Work Unit, Coastal Inlets Research Program, U.S. Army Corps of Engineers (USACE). Permission was granted by Headquarters, USACE, to publish this information.

REFERENCES

Brown, M. E., Kraus, N. C., and Kieslich, J. M. 2004. Regional Circulation Modeling of the Texas Coast and Inland Coastal Waters, *Proc. 8th Estuarine and Coastal Modeling Conf.* (this volume).

Buonaiuto, F., and Militello, A. 2004 Coupled Circulation, Wave, and Sediment Transport Modeling for Calculation of Morphology Change, Shinnecock Inlet, New York, *Proc. 8th Estuarine and Coastal Modeling Conf.* (this volume).

Cialone, M. A. and Kraus, N. C. 2001. Engineering Study of Inlet Entrance Hydrodynamics: Grays Harbor, Washington, USA," *Proc. 4th Coastal Dynamics Conf.,* ASCE, 413-422.

Grosskopf, W. G., Kraus, N. C., Militello, A., and Bocamazo, L. M. 2001. Implementation of a Regional Wave Measurement and Modeling System, South Shore of Long Island, New York. *Proc. Waves 2001 Conf.*, ASCE, 610-619.

Irish, J., Williams, B., Militello, A., and Mark, D. Abstract accepted September 2003. Regional-Scale Storm-Surge Modeling of Long Island, New York, USA. *Twenty-ninth International Conf. on Coastal Engineering.*

Kraus, N. C.. Zarillo, G. A., and Tavolaro, J. F. 2003. Hypothetical Relocation of Fire Island Inlet, New York, *Proc. Coastal Sediments 03*, CD-ROM Published by World Scientific Publishing Corp. and East Meets West Productions, Corpus Christi, TX, USA. ISBN 981-238-422-7.

Larson, M., and Kraus, N. C. 1994. Cross-Shore Sand Transport Under Random Waves at SUPERTANK Examined at Mesoscale. *Proc. Coastal Dynamics '94*, ASCE, 204-219.

Le Provost, C., Genco, M. L., Lyard, F., Vincent, P., and Cenceill, P. 1994. Spectroscopy of the World Ocean Tides from a Hydrodynamic Finite Element Model. *J. Geophysical Res.*, 99(C12), 24,777-24,797.

Luettich, R. A, Westerink, J. J., and Scheffner, N. W. 1992. ADCIRC: An Advanced Three-dimensional Circulation Model for Shelves, Coasts, and Estuaries; Report 1, Theory and Methodology of ADCIRC-2DDI and ADCIRC-3DL. Technical Report DRP-92-6, U.S. Army Engineer Waterways Experiment Station, Vicksburg, MS.

Luettich, R. A., and Westerink, J. J. 2000. ADCIRC: A (Parallel) Advanced Circulation Model for Oceanic, Coastal and Estuarine Waters. http://www.unc.edu/depts/marine/C_CATS/adcirc/.

Militello, A. 1998. *Grid Development for Modeling Two-Dimensional Inlet Circulation.* Coastal Engineering Technical Note CETN-IV-14, U. S. Army Engineer Waterways Experiment Station, Vicksburg, MS.

Militello, A., and Kraus, N. C. 2001a. Shinnecock Inlet, New York, Site Investigation, Report 4, Evaluation of Flood and Ebb Shoal Sediment Source Alternatives for the West of Shinnecock Interim Project, New York. Coastal Inlets Research Program Technical Report ERDC-CHL-TR-98-32. U. S. Army Engineer Research and Development Center, Vicksburg, MS.

Militello, A., and Kraus, N. C. 2001b. Re-Alignment of an Inlet Entrance Channel by Ebb-Tidal Eddies. *Proc. Coastal Dynamics 01*, ASCE, 423-432.

Militello, A., Kraus, N. C., and Brown, M. E. 2000b. Regional Coastal and Inlet Circulation Modeling with Application to Long Island, New York, *Proc. 13th Annual Conf. on Beach Preservation Technology*, Florida Shore & Beach Preservation Association, Tallahassee, FL, 191-201.

Militello, A., Reed, C. W., and Zundel, A. K. 2004. Two-Dimensional Circulation Model M2D: Version 2.0, Report 1, Technical Documentation and User's Guide. Coastal Inlets Research Program Contractor's Report ERDC-CHL-CR-__-_, U.S. Army Engineer Research and Development Center, Vicksburg, MS (in press).

Militello, A., Scheffner, N. W., Bratos, S., Brown, M. E., and Fenical, S. 2000a. Circulation and Transport Modeling, in *Study of Navigation Channel Feasibility, Willapa Bay.* Kraus, N. C., ed., Technical Report ERDC-CHL-TR-00-6, U.S. Army Engineer Research and Development Center, Vicksburg, MS.

Militello, A., and Zarillo, G. A. 2000. Tidal Motion in a Complex Inlet and Bay System, Ponce de Leon Inlet, Florida. *J. Coastal Research*, 16(3), 848-860.

Militello, A., and Zundel, A. K. 1999. *Surfacewater Modeling System Tidal Constituents Toolbox for ADCIRC.* Coastal Engineering Technical Note CETN-IV-21, U. S. Army Engineer Research and Development Center, Vicksburg, MS.

Militello, A., and Zundel, A. K. 2002. *Coupling of Regional and Local Circulation Models ADCIRC and M2D.* Coastal and Hydraulic Engineering Technical Note ERDC/CHL CHETN-IV-42, U. S. Army Engineer Research and Development Center, Vicksburg, MS.

Smith, J.M., Resio, D.T., and Zundel, A. K. 1999. STWAVE: Steady-State Spectral Wave Model, Report 1: User's Manual for STWAVE Version 2.0. Instructional Report CHL-99-1, U.S. Army Engineer Waterways Experiment Station, Vicksburg, MS.

Zundel, A. K., Jones, R. D, and Kraus, N. C. 2004. Community Model Concept and Implementation in the Surface Water Modeling System, *Proc. 8th Estuarine and Coastal Modeling Conf.* (this volume).

Zundel, A. K., Militello, A., and Gonzales, D. S. 2000. Numerical Model Steering Module – Let the Communication Begin. *Proc. 4th International Conf. on Hydroinformatics*, International Association of Hydraulic Research, Cedar Rapids, IA, Abstract Volume, CD-ROM -IT-3227-1099.pdf.

Developing a Tidal Constituent Database for the Eastern North Pacific Ocean

Emily Spargo[1], Joannes Westerink[2], Rick Luettich[3], Dave Mark[4]

Abstract

This paper describes the development of the Eastern North Pacific (ENPAC) 2003 tidal database of the elevations and velocity components of eight major tidal constituents. This database was developed using the two-dimensional, depth integrated form of the coastal hydrodynamic model, ADCIRC, which consists of the shallow water equations in the generalized wave continuity equation form. The final ENPAC 2003 model incorporates the most accurate bathymetric data available. The resolution of the unstructured finite element mesh was designed to provide high levels of resolution along the continental slope and throughout the coastal waters to resolve the non-linear hydrodynamics that dominate the area. The domain of this model extends beyond the continental shelf, into deep ocean waters but does not include any amphidromes located in the Pacific Ocean. At the open boundary, the forcing conditions are extracted from global ocean models, in the area where these models are most accurate. A brief discussion comparing the results from the global ocean models with the results from the ADCIRC model is included in this paper. The results of the ENPAC 2003 model are approaching the error levels found in the station data that is used for model validation.

Introduction

Providing accurate prediction of the ocean tides is necessary for many coastal engineering applications including defining water depths for navigation, calculating the energy potential from tides, and determining pollutant and sediment movement. The shallow water equations that govern tidal processes cannot be solved analytically without making many unphysical assumptions and therefore, these equations must be solved numerically. Over the past twenty-five years, finite element methods have

----

[1]Coast Survey Development Laboratory, NOS, NOAA; [2] Corresponding Author: University of Notre Dame, 156 Fitzpatrick Hall, Notre Dame, IN 46556, jjw@photius.ce.nd.edu; [3]University of North Carolina at Chapel Hill; [4]Coastal and Hydraulics Laboratory, ERDC, Army Corps of Engineers

been successfully employed to obtain highly accurate solutions to these equations (Lynch 1983; Westerink and Gray, 1991; Kolar and Westerink, 2000; Mukai, et al., 2001).

Engineers and scientists have the greatest need for tidal information in coastal waters, but creating small, regional models to describe these near-shore areas must be done with caution. The boundary conditions for these models are often selected from global ocean models. These models are highly accurate in the deep ocean, but typically do not provide the necessary resolution over the continental shelf and through coastal waters to correctly resolve both astronomical and the associated non-linear tides that exist in this area. If the regional models extract their boundary conditions from these global models, they may be forcing the smaller, regional model with inaccurate information. To provide accurate boundary conditions for regional models, a larger, transitional scale model can be employed. This model will be able to place the open boundary in the deep ocean where boundary conditions can be extracted from the global ocean model where they have the highest level of accuracy. Using unstructured grids, a transitional scale model can provide a high level of resolution over the shelf break and along the coasts in order to capture the complicated physics that dominate these regions. This technique has been successfully employed to provide accurate information for many coastal regions (Lynch and Gray, 1979; Foreman, 1986; Blain et al., 1994; Westerink et al., 1994, 1996; Luettich et al., 1992; Kolar et al., 1996; Mukai et al., 2001).

The objective of this particular study was to produce a tidal model of the coastal waters along the United States west coast and lower Alaska. A so-called transitional scale domain was employed for this model. The resulting domain extends into the more stable deep ocean waters of the Eastern North Pacific Ocean, and this model is aptly named the Eastern North Pacific (ENPAC) 2003 tidal model. This model was developed using the two-dimensional depth integrated finite element code, ADCIRC (Luettich et al., 1992, Luettich and Westerink, 2003).

Model Features

The two-dimensional, depth-integrated (2DDI) form of the ADCIRC (Advanced Circulation) model was used to perform the hydrodynamic computations (Luettich, et al, 1992). This model uses the depth-integrated mass and momentum equations with Boussinesq and hydrostatic pressure approximations. For this application, a hybrid form of the standard quadratic parameterization for bottom stress was used that provides a friction factor that increases as the depth decreases in shallow water similar to a Manning relationship. Baroclinic processes were neglected, including any expansion and contraction due to radiational heating. The primitive continuity and momentum equations were expressed in a spherical coordinate system (Flather 1988; Kolar, et al. 1994b).

Solving the shallow water equations in the primitive form, results in spatial oscillations in the solution on the order of $2\Delta x$ (Lynch and Gray, 1979; Kolar et al., 1994, Kolar and Westerink, 2000). These oscillations can be eliminated by the addition of non-physical damping by viscous terms. To avoid the addition of non-physical characteristics to the solution, a reformulation of these equations was proposed by Lynch and Gray (1979) and modified by Kinmark (1986). This reformulation is the Generalized Wave Continuity Equation (GWCE). The spatially differentiated form of the conservative momentum equation is substituted into the time derivative of the primitive continuity equation. Then, the primitive continuity equation multiplied by the numerical weighting parameter, G is added to create the GWCE. The numerical parameter, G, sets the balance between a pure wave equation, where G equals zero, and the primitive continuity equation, where G is much greater than zero. The GWCE was solved in conjunction with the primitive momentum equations.

The finite element method was then applied to discretize these equations in space. Triangular elements were used for elevation, velocity and depth. The temporal discretization of the GWCE was solved using a finite difference, weighted three-time level implicit scheme except for the nonlinear, Coriolis, atmospheric pressure forcing and tidal potential terms. These latter terms were treated explicitly. In the momentum equations, a two-time level Crank-Nicolson scheme was applied to all terms except a portion of the bottom friction and convective terms which were treated explicitly. This time stepping scheme does require the time steps to be Courant limited for the nonlinear terms to remain stable. The decoupling of the time and space discrete forms of the equations, the time independent and/or tri-diagonal system matrices, and the full vectorization of all major loops leads to a highly efficient code.

The parallel version of this code applies domain decomposition, a conjugate gradient solver and MPI (message passing interface). Benchmark tests have been performed to show linear and even super-linear performance rates on up to 128 processors for grids of approximately 300,000 nodes.

Initial Model Development

Initial Domain, Bathymetry and Grid. The initial model domain is shown in Figure 1. It extends from Seal Cape on Unimak Island, Alaska in the north to Acapulco, Mexico in the south. The eastern land boundary includes the Pacific coastlines of Alaska, Canada, United States west coast, and Mexico. The area of the entire domain is 1.79 x 107 square kilometers. Dense NOS sounding data was incorporated into the grid using a cluster averaging approach. In this method, the cluster area includes one node and all of the elements surrounding it. All of the bathymetric sounding points in this larger cluster area are averaged together to define the bathymetry at the node. This procedure automatically filters the available bathymetric data to the existing grid scale. Where the NOS sounding data was unavailable, bathymetry from the ETOPO-2 database was used (NGDC, 1998). Since the NOS sounding points were found only in the near-shore U.S. waters, most of the bathymetric data in the grid was from the ETOPO-2 database.

The flexibility inherent in the finite element method makes it possible to provide increased levels of grid resolution where necessary. To ensure that all key hydrodynamic features were resolved, it was important to consider several mesh development techniques. The first criterion examined was the wavelength to grid size ratio which is based on linear, frictionless constant-depth wave theory (Kashiyama and Okada, 1992). For a constant acceleration of gravity, g, and a given wave period, T, to maintain a constant wavelength to grid size, $\lambda/\Delta x$, ratio, Equation 4 shows that as the water depth, h, decreases, the grid size must also decrease.

$$\frac{\lambda}{\Delta x} = \frac{\sqrt{gh}}{\Delta x} T \quad (4)$$

Based on this criterion, the resulting grid will need increasing resolution over decreasing depths. Studies have demonstrated that a wavelength to grid size ratio of at least 25 should be maintained for accurate results. However, this criterion alone is not sufficient to achieve the desired accuracy. In fact, areas of sharp bathymetric gradients, such as the continental shelf break, are under-resolved if grid development is based exclusively on the wavelength to grid size criterion (Westerink et al., 1994; Hannah and Wright, 1995; Luettich and Westerink, 1995, Hagen, 2001a, Hagen et al., 2001b).

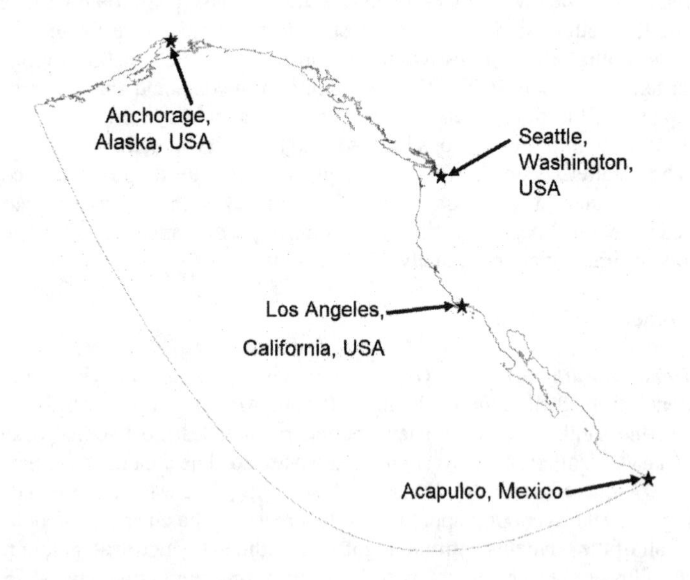

Figure 1. ENPAC 2002 Domain Boundary with Reference Locations

To address this problem the introduction of a topographic length scale (TLS) criteria, as in Equation 5, is necessary.

$$\Delta x \leq \frac{\alpha h}{h_{,x}} \quad (5)$$

The grid size, Δx, is now determined by the water depth, h, in addition to the bathymetric gradient, $h_{,x}$. The mesh generation criterion, α is designed so that α does not exceed a maximum value within the domain. The problem with the TLS criteria is that where there is little or no bathymetric change (i.e. when $h_{,x}$, --> 0) the grid size is allowed to increase to infinity. To create a grid on which convergent results will be produced, both the wavelength to grid size ratio and the TLS criterion need to be considered. Grids that incorporate both of these criterions are similar to grids that are designed per the more complex local truncation error analysis (LTEA). LTEA examines the truncation error in the governing equations. For this grid design process, a limit is set for the maximum allowable localized truncation error (Hagen, 2001a; Hagen et al., 2001b). The ENPAC 2002 model did not involve LTEA, but this level of resolution was simulated by incorporating both the wavelength to grid size and the TLS criterion.

Figure 2 shows a section of the finite element grid. The grid consists of 290,715 nodes and 567,145 elements. Approximately 85 percent of the total number of elements lies between the toe of the continental slope and the coastal shoreline boundary demonstrating the high level of resolution needed to adequately resolve these areas of steep bathymetric gradients and shallow waters. The element size in kilometers ranges from 25 to 40 kilometers in the deep ocean to less than one kilometer in select coastal areas. Figure 2 clearly shows this gradation in element size. For the M2 tide (with a tidal period of 12.42 hours), in waters deeper than 1.0 meters, the wavelength to grid size ratio ranges from 27 to 1939; above the lowest recommended wavelength to grid size ratio of 25. For the TLS criterion, based on the given depth, slope and element size, throughout most of the domain, the α-values are very low. Especially in deep ocean waters, the values are well below 0.5. A band of higher α-values (from 1.0 to 1.5), exists along the continental slope. This indicates that more refinement would be needed over the continental slope to produce the same α-values associated with areas of small bathymetric gradient, namely the deep ocean and coastal waters. Another way to maintain a constant α-value would be to decrease the resolution in the deep ocean and coastal waters. Although it may be possible to decrease the resolution in the deep ocean and still achieve an accurate solution, decreasing the resolution in the coastal waters was not appropriate. As discussed above, the wavelength to grid size criterion calls for a high level of resolution in these shallow waters in order to capture the physics that are unique to this area due to non-linearities trapped in the coastal waters by the continental shelf break. This demonstrates both the impossibility of maintaining a constant α-value over the entire domain, and the need to incorporate both the TLS and wavelength to grid size criterion when developing a finite element mesh.

Model Input Parameters. Sixty-day tidal simulations were run with the time history computed at every node in the domain. For the last 30 days of the simulation, data was recorded every 12 minutes and used for the harmonic analysis. Thirty-seven frequencies and steady state were used in this process, including the main diurnal, , K_1, O_1, P_1, Q_1, and semi-diurnal, M_2, S_2, N_2, K_2. To account for non-linear interactions between these constituents, four overtides and 25 compound tides were included in the harmonic decomposition. This was necessary to ensure the accuracy of the least squares based analysis.

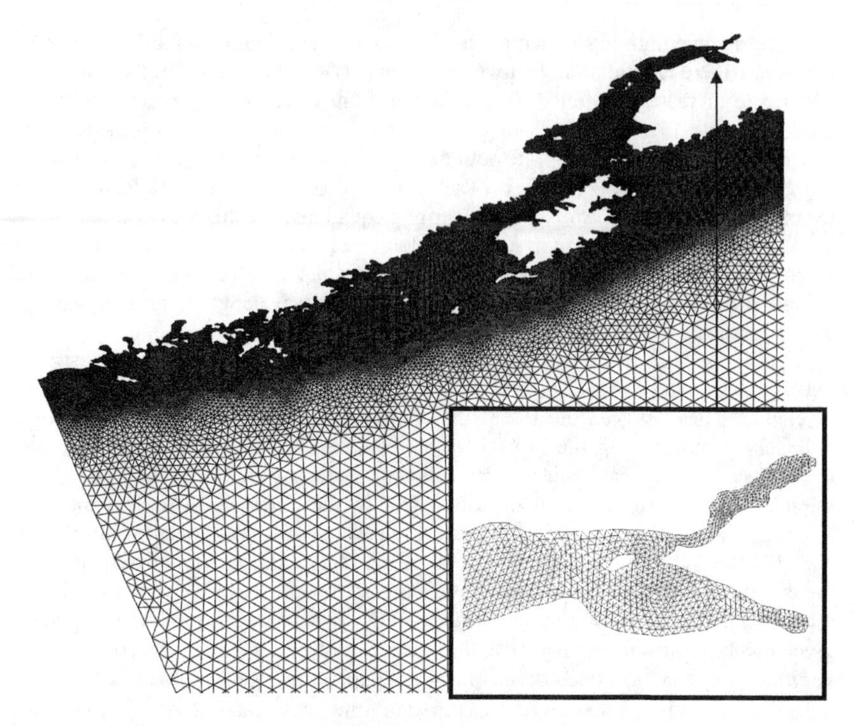

Figure 2. Section of ENPAC 2002 Finite Element Grid Along Aluetian Islands Showing Gradation in Element Size from 40 km in the Deep Ocean to 2 km along the Coastline and Inset of Grid in the Cook Inlet, Alaska Area.

In the final runs, a six-second time step was used to insure stability by keeping the Courant number below 1.0 for the entire domain. Due to the large area of the domain, a spatially variable Coriolis parameter was used. A fully non-linear hybrid bottom friction law was incorporated. This formulation of bottom friction is similar to a Manning-type friction law where friction increases as the depth decreases. The lateral eddy viscosity coefficient was set at 20 m/s. Since tests have shown that the ideal *G* value (the parameter that sets the sets the balance between a pure wave equation and the primitive continuity equation) is two to ten times t_*, the bottom friction parameter, a variable *G* was chosen that reflects the changes in the importance of the bottom friction over the domain (Kolar, 1994a). When H was greater than 10 meters, $G = 0.005$, and when H was less than 10 meters, $G = 0.02$. The non-linear finite amplitude terms were enabled and wetting and drying was allowed to occur. Since the domain did not include many strongly converging or diverging flow regimes, such as through an inlet, it was reasonable to linearize the advective terms. The elevation specified boundary conditions were applied using a hyperbolic tangent ramp function for the first 15 days of the model run. This ramp function increased the boundary forcings from zero to almost their full value at day

15. The hyperbolic tangent ramp function was used to alleviate the problems that occur when a model was shocked from the zero η elevation and the u and v velocity response initial conditions.

The model was forced with the amplitude and phase of four diurnal, K_1, O_1, P_1, Q_1, and four semi-diurnal, M_2, S_2, N_2 and K_2, constituents. The amplitude and phase data for these constituents was extracted from Oregon State University's TPXO.2, TPXO.5 and TPXO.6 global ocean tidal models (Egbert et al., 1994; Egbert, 1997; Egbert and Erofeeva, 2002). The TPXO models use a generalized inversion (GI) method which is basically a large linear least-squares analysis to assimilate TOPEX/Poseidon satellite data. This approach tries to minimize a quadratic penalty function that is developed from trying to find the best fit between the hydrodynamic equations and the satellite (or other) data. Along the open boundary, in general, the TPXO.5 and TPXO.6 models match each other closely while the TPXO.2 model shows a difference in many of the tidal constituent amplitudes. The phase output from all three models was very similar. The major differences between all three of the TPXO models were apparent in the coastal waters, continental shelf break, and continental slope leading into the Aleutian trench. Some discrepancies were also seen on the narrow slope and coastal waters off the coast of Mexico. Not only do the models disagree with each other, but the semi-diurnal constituents can appear physically unrealistic in shallow coastal waters. At the open boundary in the Alaskan cross shelf area, unrealistic circulation cells developed in the initial ADCIRC model runs. As the model run progressed, these cells would advect into the domain and cause the model to become unstable. To compensate for the unrealistic open boundary forcings and to eliminate the model instability, the ADCIRC model had to be run with the advective terms turned off.

The interior domain was forced with four diurnal, K1, O1, P1, Q1, and four semi-diurnal, M2, S2, N2, K2, constituents. The Newtonian tidal potential and Earth Tide reduction were included, but the self attraction/load tide forcings were not. The Newtonian tidal potential amplitudes (in centimeters) are listed from largest to smallest for the eight constituents: $M_2 = 24.2334$, $K_1 = 14.1565$, $S_2 = 11.2841$, $O_1 = 10.0514$, $P_1 = 4.6843$, $N_2 = 4.6398$, $K_2 = 3.0704$, and $Q_1 = 1.9256$.

Error Analysis

Field Data Description and Analysis. The ENPAC 2003 model was validated using 91 pelagic pressure and coastal tide gauges located throughout the domain. The model results, at these 91 locations, were compared to the amplitude and phase results available from the station data records. The station data was available from several sources, specifically the International Hydrography Organization (IHO), National Ocean Service (NOS) and the Pacific Marine Environmental Laboratory (PMEL) (Eble and Gonzalez, 1991). The locations of all 91 stations are shown in Figure 3. To examine the errors on a regional basis, the location of the gauges was divided into 5 regions: (1) Mexico, (2) U.S. west coast, (3) Canada, (4) Alaska, and (5) Deep ocean.

Figure 3. Location of 91 Pelagic Pressure and Coastal Tide Gauge Stations Used for Model Validation.

Fourteen stations throughout the entire domain had station data from two different sources. Two stations had data from all three sources. At the stations with duplicate (but often differing) records, the amplitude and phase values for different tidal constituents were used to perform an estimation of the station data error. For the amplitude values, a standard proportional deviation was calculated according to Equation 6.

$$E_{j-amp}^{obs1-obs2} = \left[\frac{\sum_{l=1}^{L} \left[\eta_j^{obs1}(x_l, y_l) - \eta_j^{obs2}(x_l, y_l) \right]^2}{\sum_{l=1}^{L} \left[\eta_j^{obs2}(x_l, y_l) \right]^2} \right]^{1/2} \quad (6)$$

In Equation 6, η_j^{obs1}= the observed amplitude at a given (x_l, y_l) location for constituent j from the first data source, η_j^{obs2}= the observed amplitude at a given (x_l, y_l) location for constituent j from the second data source, and L = number of elevation recording stations in the domain (or sub-region) with more than one data set. When IHO and NOS data were available, IHO data was considered the first data source (*obs1*) and the NOS data was considered the second data source (*obs2*). When IHO and PMEL data were available, IHO data was considered the first data source and PMEL data was considered the second data source. When NOS and PMEL data area available, NOS data was considered the first data source and PMEL data was considered the second data source. When all three data sources were available, IHO data was considered the first data source and PMEL data was considered the second data source.

For the phase data, an average absolute difference was calculated according to Equation 7.

$$E_{j-pha}^{obs1-obs2} = \frac{1}{L}\Sigma_{l-1}^{L}\left|\varphi_j^{obs1}(x_l, y_l) - \varphi_j^{obs2}(x_l, y_l)\right| \tag{7}$$

In Equation 7, φ_j^{obs1} = the observed phase at a given (x_l, y_l) location for constituent j from the first data source, φ_j^{obs2} = the observed phase at a given (x_l, y_l) location for constituent j from the second data source, and L = number of elevation recording stations in the domain (or sub-region) with more than one data set.

The standard proportional deviation amplitude errors and the average absolute difference phase errors for eight tidal constituents are reported in Table 1 for the entire domain (thirty-four stations), as well as the U.S. West Coast (twenty stations), Canada (three stations), Alaska (nine stations), and Deep ocean sub-domains (three stations). Since no duplicate station records were found in the Mexico Stations sub-domain, no errors are reported.

In general, this analysis shows lower error levels in the diurnal than semi-diurnal constituents. In particular, the highest error levels were seen in the S_2 and N_2 constituents for the U.S. west coast and Alaska stations. We will later note that similar high error levels were found in the computed values of these constituents in these locations. This may correspond to shifting of the semi-diurnal amphidromes off the U.S. west coast by such phenomenon as shifting currents or changes in the Pacific Ocean environment due to the El Nino Southern Oscillation.

Model Validation by Field Data. For the ENPAC model, the amplitude error is defined as the proportional standard deviation error between the model output and the station data. The phase error is defined as an absolute average between the model output and the station data. This follows the same form as Equations 6 and 7, except that the "obs1" data is replaced with the ENPAC model results and the "obs2" data is replaced with the most current station data available at that location. The proportional standard deviation error used to define the amplitude error in the station data and the computed model results can be interpreted as an estimate of the mean percent deviation from the measured error. The advantage of the proportional standard deviation error measure over a standard percentage error is that the proportional standard deviation incorporates the uncertainty in the data. The phase error is similar, but does not normalize the results.

Table 1. Standard Proportional Deviation Amplitude Errors (Dimensionless) and Average Absolute Difference Phase Errors (Degrees) for Eight Tidal Constituents Over the Entire Domain and the U.S. West Coast, Canada, Alaska, and Deep Ocean Sub-Domains for the Station Data Records

Amp.	**Overall (34 stations)**	**U.S. west coast (20 stations)**	**Canada (3 stations)**	**Alaska (9 stations)**	**Deep ocean (3 stations)**
K_1	0.0277	0.0246	0.0081	0.0395	0.0038
O_1	0.0241	0.0291	0.0064	0.0193	0.0061
P_1	0.0542	0.0295	0.0076	0.0972	0.0000
Q_1	0.0231	0.0409	0.0082	0.0156	0.0017
M_2	0.0346	0.0627	0.0029	0.0290	0.0088
S_2	0.0810	0.1641	0.0196	0.0172	0.0028
N_2	0.1205	0.1020	0.0054	0.1372	0.0398
K_2	0.0519	0.0597	0.0330	0.0485	0.0110
Phase					
K_1	2.41	0.97	0.30	6.63	0.03
O_1	2.27	1.12	0.37	5.77	0.13
P_1	3.19	1.92	1.63	7.13	0.10
Q_1	4.60	2.19	0.37	12.06	0.03
M_2	4.41	1.66	0.63	12.33	0.17
S_2	5.02	2.33	0.60	12.61	2.03
N_2	6.40	4.95	1.27	12.62	0.43
K_2	2.95	2.02	1.10	6.27	0.00

ENPAC 2002 Results

Table 2 lists the standard proportional deviation amplitude error and the absolute average phase error for each tidal constituent as calculated for the overall domain (i.e. at all 91 stations). The station data error, as determined at the stations with at least two data sets available, is also shown. Overall, there is a much higher error level in the semi-diurnal constituents. Examining the errors broken down by regions, it can be seen that there are consistently high levels of error along the U.S. West coast. It is likely that these errors correspond to the semi-diurnal amphidromic cluster to the west of this region.

Comparing the ENPAC 2002 (with the TPXO.5 open boundary forcing) amplitude and phase results with OSU's TPXO.5 model amplitude and phase results over the entire domain, few differences were seen between the diurnal constituents, but major differences existed between the semi-diurnal constituents. The largest differences between the diurnal constituents were found along the continental slope and the coastal waters, which would be expected since the ENPAC 2002 model has a much higher level of resolution and was designed to better capture the non-linear physics which exist in that area. Unexpected, though, were the large differences seen between the two models in the area of the semi-diurnal amphidrome locations, off the U.S. west coast. Although the exact source of this difference is not known it should be noted that the largest error in the station data records is also in the semi-diurnal constituents. This suggests that the amphidromes may be shifting position due to large scale dynamic forces such as current systems or even the El Nino Southern Oscillation which causes many changes in the Pacific Ocean over seasons, years and decades. Although the OSU global ocean models assimilate data, which can compensate for missing physics not included in the model, this does not eliminate all sources of error. There are several reasons why both models may lack the physics to correctly model this ocean system. Neither the OSU models nor the ADCIRC model incorporate baroclinic effects, or other 3-dimensional effects. Additionally, the ADCIRC model does not include the load and self attraction tide. None of the models, but particularly the global ocean models, have the high level of resolution needed to include the vast number of bays, estuaries, and rivers (especially in the southern Alaskan and Canadian waters) which are a major dissipative mechanism for this domain given the narrow shelf. Finally, none of the processes believed to be causing the shift in the amphidromes' locations, are included in any of the models.

Table 2. Proportional Standard Deviation Amplitude Errors and Average Absolute Difference Phase Errors for 91 Stations for the ENPAC 2002 Model With TPXO.2, TPXO.5, and TPXO.6 Open Boundary Forcing and Station Data

Amp.	**ENPAC 2002 TPXO.2**	**ENPAC 2002 TPXO.5**	**ENPAC 2002 TPXO.6**	**Station Data**
K_1	0.0791	0.0650	0.0659	0.0277
O_1	0.0571	0.0573	0.0575	0.0241
P_1	0.0935	0.0641	0.0623	0.0542
Q_1	0.0813	0.0830	0.0896	0.0231
M_2	0.1221	0.0635	0.0643	0.0346
S_2	0.1529	0.1807	0.2120	0.0810
N_2	0.1311	0.0747	0.0730	0.1205
K_2	0.1478	0.2625	0.2370	0.0519
Phase				
K_1	4.21	4.06	4.09	2.41
O_1	3.57	3.75	3.73	2.27
P_1	4.49	4.11	4.20	3.19
Q_1	7.00	4.39	4.42	4.60
M_2	11.86	12.54	12.78	4.41
S_2	23.38	17.57	20.31	5.02
N_2	15.49	13.86	13.07	6.40
K_2	18.72	14.83	16.78	2.95

To see if the placement (or possibly misplacement) of the semi-diurnal amphidromic points in the ENPAC 2002 domain were the major source of the remaining tidal constituent database errors, additional restrictions in the domain were made.

ENPAC 2003 Model Development

ENPAC 2003 Domain Definition. The ADCIRC model physics with open boundary conditions extracted from the OSU TPXO global ocean models was not placing the semi-diurnal amphidromes correctly. To eliminate this problem, changes were made in the ENPAC 2002 domain to order to avoid this cluster of amphidromes off the west coast of the United States. This new domain will henceforth be called the ENPAC 2003 domain. This smaller domain extends from Seal Cape on Unimak Island, Alaska in the north to Chamela, Mexico in the south. The eastern land boundary includes the Pacific coastlines of Alaska, Canada, United States west coast, and some of Mexico. The area of the entire domain is 7.57 x 106 square kilometers. Although 42 percent of the ENPAC 2002 domain was removed, the U.S. west coast and Alaska coastlines remained intact. The approximate location of the amphidromes off the U.S. west coast along with the original ENPAC 2002 and the revised ENPAC 2003 domain are shown in Figure 4.

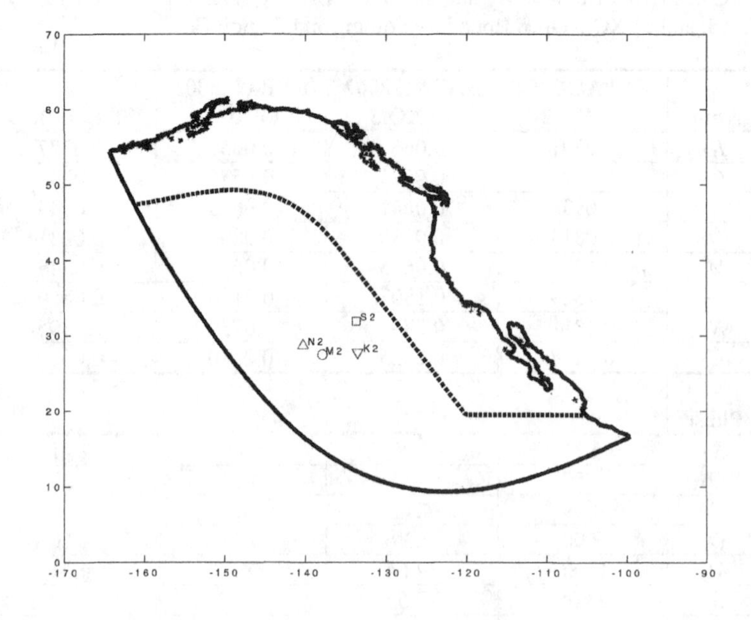

Figure 4. Approximate Location of Amphidromes Within the ENPAC 2002 Domain (solid line) and the ENPAC 2003 Domain (dashed line)

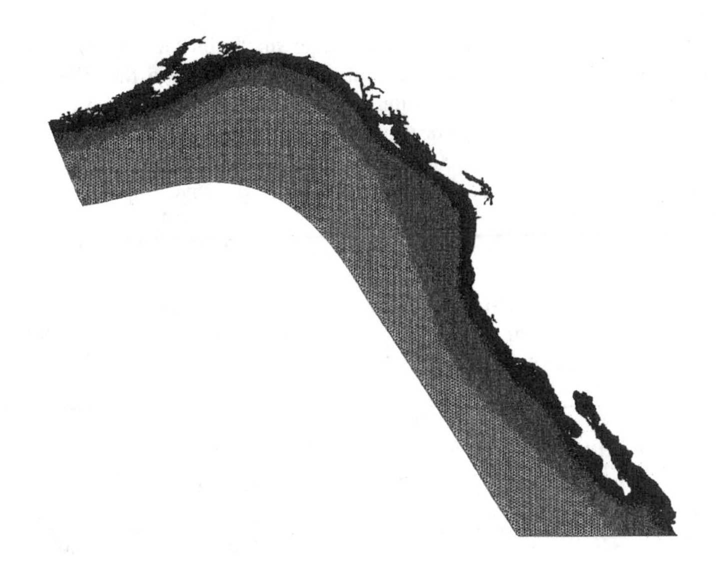

Figure 5. ENPAC 2003 Finite Element Grid

Although a sizeable portion of the ENPAC 2002 model area was removed to create the ENPAC 2003 model domain, nearly 94 percent of the elements were retained in the finite element grid. This was due to the fact that the highest concentration of elements exists along the coastal areas - the areas that were retained in the finite element grid of the ENPAC 2003 model domain. Figure 5 shows the finite element grid representing the ENPAC 2003 model domain. This finite element mesh contains 531,680 elements and 272,913 nodes. The bathymetry remained the same as that of the ENPAC 2002 model domain, as a combination of NOS and ETOPO-2 bathymetric data. Due to the narrow continental shelf, the open boundary of the new domain remains in the deep ocean where accurate boundary conditions can be extracted from global ocean models.

ENPAC 2003 Model Results. Table 3 lists the standard proportional deviation amplitude error and the absolute average phase error for each tidal constituent as calculated for the overall domain (i.e. at all 91 stations).

The largest improvements over the ENPAC 2002 model were in the U.S. West coast region with in the semi-diurnal constituents. This clearly demonstrates the advantage of eliminating the semi-diurnal amphidromes from the ENPAC 2003 domain. Still, the exact physics behind this observation are obscure. It is interesting to note that the ADCIRC model of the Western North Atlantic Tidal (WNAT) did not have as much trouble correctly placing the amphidromes as did the ADCIRC model for the ENPAC 2002 domain (Mukai et al., 2001). One of the differences between these domains is the width of the continental shelf. The Eastern North Pacific physical domain has a very narrow shelf and many complex bays, estuaries and rivers, particularly in the Canadian and lower Alaskan waters, while the Western

North Atlantic physical domain has a much wider shelf with fewer bays and estuaries. The wide continental shelf, which was sufficiently resolved in the WNAT models, provides much more natural dissipation than the narrow shelf in the ENPAC 2002 model. The bays and estuaries, where much of the natural dissipation will occur, are under-resolved in the ENAPC 2002 model. But whatever the physical reason behind the problems in the ENPAC domain, eliminating the sensitive amphidromes from the model domain and forcing the open boundary with reasonably accurate deep water values, has eliminated much of the error seen in the ENPAC 2002 results.

Table 3. Proportional Standard Deviation Amplitude Errors and Average Absolute Difference Phase Errors for 91 Stations for the ENPAC 2003 Model With TPXO.2, TPXO.5, and TPXO.6 Open Boundary Forcing and Station Data

Amp.	ENPAC 2003 TPXO.2	ENPAC 2003 TPXO.5	ENPAC 2003 TPXO.6	Station Data
K_1	0.0643	0.0658	0.0648	0.0277
O_1	0.0562	0.0547	0.0557	0.0241
P_1	0.0736	0.0660	0.0641	0.0542
Q_1	0.1061	0.0902	0.0870	0.0231
M_2	0.0723	0.0615	0.0615	0.0346
S_2	0.1304	0.1090	0.1088	0.0810
N_2	0.0757	0.0695	0.0694	0.1205
K_2	0.1198	0.1098	0.1236	0.0519
Phase				
K_1	4.12	3.52	3.82	2.41
O_1	3.52	3.47	3.50	2.27
P_1	4.15	4.35	4.27	3.19
Q_1	6.96	4.36	3.72	4.60
M_2	5.99	5.94	5.95	4.41
S_2	6.35	6.21	6.34	5.02
N_2	6.34	6.23	6.11	6.40
K_2	10.01	8.26	10.67	2.95

To determine which OSU global ocean model open boundary forcing produced the best ENPAC 2003 model, a weighted average, based on the Newtonian equilibrium tidal potential amplitude, was computed for the overall amplitude and phase errors. The results from this analysis are shown in Table 4.

Table 4. Amplitude and Phase Weighted Average Error for the ENPAC 2003 model with TPXO.2, TPXO.5, and TPXO.6 Open Boundary Forcing

	TPXO.2 Forcing	TPXO.5 Forcing	TPXO.6 Forcing
Amplitude Error	0.0806	0.0722	0.0725
Phase Error	5.4518	5.1562	5.3089

Table 4 shows that the lowest error is found with the TPXO.5 open boundary forcing. Due to the dominance of the M2 equilibrium tidal potential amplitude, the weighted average error was dominated by the error in that constituent. A straight average, where each constituent was considered equally, was also calculated and the results were shown in Table 5.

Table 5. Amplitude and Phase Straight Average Error for the ENPAC 2003 model with TPXO.2, TPXO.5, and TPXO.6 Open Boundary Forcing

	TPXO.2 Forcing	TPXO.5 Forcing	TPXO.6 Forcing
Amplitude Error	0.0873	0.0783	0.0794
Phase Error	5.9312	5.2938	5.5465

The results from the straight average were very similar, again producing not only comparable values to the weighted average analysis, but also showing the TPXO.5 open boundary forcing to produce the best results.

Discussion and Conclusion

The ENPAC 2003 model with the TPXO.5 open boundary forcing provides accurate tidal amplitude and phase data for the ENPAC 2003 domain. The overall proportional standard deviation amplitude errors for the tidal constituents, as compared to pelagic pressure and coastal tide gauge data, ranges from 0.0546 for the O_1 constituent (or approximately 5 percent) to 0.1098 for the K_2 constituent (or approximately 10 percent). The overall average absolute difference phase error, as compared to the station data, ranges from 3.19 degrees for the K1 constituent to 7.60 degrees for the K_2 constituent. In comparison, the overall proportional standard deviation amplitude error for the recorded station data ranges from 0.0231 for the O_1 constituent (or approximately 2 percent) to 0.1205 for the N_2 constituent (or approximately 12 percent). The overall average absolute difference phase error for the recorded station data ranges from 2.27 degrees for the O_1 constituent to 6.40 degrees for the N_2 constituent. In Cook Inlet, AK, an area that is frictionally dominated and sees a very large tidal variation, the model compares favorably with measured data. Along the U.S. west coast where there is a much smaller tidal variation, the model results were not as accurate, probably due to the influence of nearby semi-diurnal amphidromes.

The diurnal constituents produce a field of smooth gradation of increasing amplitude from the south-west to north-east with extreme values in the Vancouver Island and Cook Inlet areas. The semi-diurnal constituents demonstrate similar amplitude properties, with increasing amplitude along the same direction. Semi-diurnal constituents show the nearness of amphidromic points along the open boundary off the west coast of the United States with areas of extremely low amplitude and converging phase lines. Additionally, a degenerate amphidrome appears in the northern portion of the Gulf of California. The diurnal constituents' phase values increase from the south-east to the north-west. Semi-diurnal phase results are similar and increase with a concentrated area of phase lines converging at an assumed amphidromic point near the boundary of the domain off the U.S. west coast. This is particularly evident in the K_2 and S_2 constituents, which show some unphysical anomalies near the amphidrome locations.

Although the overall errors in the ENPAC 2003 model results are low, they are still, in some cases, greater than the data error, thus indicating that additional improvements may be possible. Future work on this project could incorporate improvements in three categories: bathymetry, grid resolution and physics. The comparison between the NOS bathymetric sounding data and the ETOPO-2 bathymetric data showed a significant difference between these two sources of data, especially in the shallow coastal waters. Since the ETOPO-2 data is not as dense as the NOS sounding data, it is not possible for the ETOPO-2 to capture the small-scale bathymetric differences in these shallow waters. Future work should include high density bathymetric data, like the NOS data, in more areas of the domain. The final ENPAC 2003 domain particularly lacks this level of bathymetric resolution particularly along the Mexican and Canadian coasts.

Another important issue is the grid resolution. Little dissipation is inherent in the deep water areas, and due to the narrow continental shelf in this domain, little dissipation is inherent on shelf. Instead, much of the dissipation may occur in the bays, estuaries and rivers that exist all along the coast, but in a particularly high density in the Canadian and lower Alaskan waters. Adding the shallow channels and estuaries into the domain and continuing to further refine those that exist, is key to providing the appropriate physical dissipation. This, in turn, should lead to even more accurate results throughout the domain, but particularly in the Canada and Alaska sub-regions.

Better understanding the physics that govern this system is important to further improvement in this model. As seen with the ENPAC 2002 runs, the ADCIRC model had many difficulties placing the amphidromes in the appropriate place. Part of this problem could be from inaccurate open boundary forcing conditions. Although global models that assimilate satellite altimetry data are, in some ways, compensating for the missing physics, this does not produce an error-free solution. No model can incorporate all of the physics which govern ocean hydrodynamics. Specifically, neither the ADCIRC nor the global models incorporate major current systems, baroclinic effects or any three-dimensional effects. The global ocean models have a relatively coarse resolution along the coasts (especially when compared to the ENPAC 2003 model) and do not include the bays, estuaries and rivers that are important in providing dissipation for this domain. The propagation of errors into the ENPAC 2002 domain from the boundary (when this model was run with the advective terms turned on) is a testimony to this. While the boundary conditions may be propagating inaccurate information into the domain, the ADCIRC model itself, is lacking the physics mentioned above as well as the load and self attraction tide. It was shown that the ADCIRC model did not project the amphidromes into the same location as the OSU TPXO.5 model, but lack of the load and self attraction tide may not be the only reason behind this error. While it was assumed that the global ocean models placed the amphidromes in their correct locations, the complications of changing current patterns or events as the El Nino Southern Oscillation, which causes many changes in many Pacific Ocean parameters such as surface water level and water temperature, allow the possibility that the global ocean models may also be wrong. The possibility of shifting amphidromes can be supported by the examination of the station data errors. The station data showed particularly high errors in the semi-diurnal constituents off the U.S. West coast indicating possible changes in the location of these troublesome amphidromes. The questions of domain definition, amphidrome locations, and the physics incorporated into the ADCIRC and global ocean models is a complicated issue which will require more in-depth study.

Although improvements can be made in the ENPAC 2003 model, the results produced were in the desired output range of less than 10 percent amplitude error and less than 10 degrees phase error. The error levels in the field data were highest in the U.S. west coast and Alaska regions, approaching (and in the case of the N2 and S2 amplitude errors off the U.S. west coast, exceeding) the error of the ENPAC 2003 model. This high level of field data error seems to indicate the advent of shifting amphidrome locations and demonstrates the need for up-to-date station records to be used in the validation process. While the field data error was much lower than the ENPAC 2003 model error in the Canadian and deep ocean stations, overall, the field data errors were similar to the error levels in the ENPAC 2003 model. This indicates that this model is approaching a point, and in some constituents is at the point, where validating the accuracy of the model is constrained by the error in the field data. While further improvements could be achieved in the ENPAC 2003 model by incorporating more accurate bathymetry, increasing grid resolution and incorporating the complicated physics that govern such complex processes as shifting amphidromes, the ENPAC 2003 model has produced results with an acceptable error level. The final results of this model will be provided as the ENPAC 2003 tidal constituent database for users to extract elevation and velocity information for the eight major tidal constituents (K_1, O_1, P_1, Q_1, M_2, S_2, N_2, and K_2) at any location within the domain.

References

Eble, M.C. and F.I. Gonzalez. (1991). "Deep-ocean bottom pressure measurements in the northeast Pacific Ocean," *J. Atmos. Oceanic. Technol.*, 8(2), 221-33.

Egbert, G. A., Bennett, M. Foreman. (1994). "TOPEX/Poseidon tides estimated using a global inverse model," *J. of Geophys. Res.* 99(C12), 24,821-52.

Egbert, G.A. (1997). "Tidal data inversion: interpolation and inference," *Prog. Oceanog.* 40, 53-80.

Egbert, G.A., and S.Y. Erofeeva. (2002). "Efficient inverse modeling of barotropic ocean tides," *J. Atoms. Ocean. Tech.* 19, 183-204.

Flather, R.A. (1988). "A numerical model investigation of tides and diurnal-period continental shelf waves along Vancouver Island," *J. Phys. Oceanog.* 18, 115-139.

Hagen, S.C. (2001a). "Estimation of the Truncation Error for the Linearized, Shallow Water Momentum Equations," *Engineering with Computers* 17, 354-362,

Hagen, S.C., J.J. Westerink, R.L. Kolar, and O. Horstmann. (2001b). "Two-dimensional, unstructured mesh generation for tidal models," *Int. J. Numer. Meth. Fluids* 35, 669-86.

Hannah, C.G. and D.G. Wright. (1995). "Depth dependent analytical and numerical solutions for wind-driven flow in the coastal ocean," *Quantitative Skill Assessment for Coastal Ocean Models*. D.R. Lynch and A.M. Davies, eds., A.G.U., 47, 125-152.

Kolar, R.L., J.J. Westerink, M.E. Cantekin and C.A. Blain. (1994a). "Aspects of nonlinear simulations using shallow water models based on the wave continuity equation," *Computers and Fluids* 23(3), 523-38.

Kolar, R.L., W.G. Gray, J.J. Westerink, and R.A. Luettich. (1994b). "Shallow water modeling in spherical coordinates: equations formulation, numerical implementation and application," *J. of Hydraulic Res.* 32(1), 3-24.

Kolar, R.L., W.G. Gray, and J.J. Westerink, "Boundary conditions in shallow water models – an alternative implementation for finite element codes," *Int. J. Numer. Methods. Fluids* 22, 603-18.

Kolar, R.L., J.J. Westerink. (2000). "A look back at 20 years of GWC-based shallow water models," *Comput. Meth. In Water Res. XIII.* Bently et al., eds., Balkema, Rotterdam, 899-906.

Luettich, R.A., J.J. Westerink and N.W. Scheffner. (1992). "ADCIRC: an advanced three-dimensional circulation model of shelves, coasts, and estuaries, Report 1: theory and methodology of ADCIRC-2DDI and ADCIRC-3DL", Technical Report DRP-92-6, Department of the Army, Vicksburg, MS.

Luettich, R.A., J.J. Westerink. (1995). "Continental Shelf Scale Convergence Studies with a Barotropic Tidal Model," *Coastal and Estuarine Studies* 47, 349-71.

Luettich, R.A., J.J. Westerink. "Theory Report: Formulation and Numerical Implementation of the 2D/3D ADCIRC finite element model version 43.XX," http://www.marine.unc.edu/C_CATS/adcirc/adcirc.htm (3 Mar 2003).

Lynch, D.R. and W.G. Gray. (1979) "A wave equation model for finite element tidal computations," *Computers and Fluids* 7, 207-28.

Lynch, D.R. (1983). "Progress in hydrodynamic modeling, review of U.S. contributions," *Reviews of Geophysics and Space Physics* 21(3), 741-54.

Mukai, A.M., J.J. Westerink, R.A. Luettich. (2001). "Guidelines for using the Eastcoast 2001 database of tidal constituents within the Western North Atlantic Ocean, Gulf of Mexico and Caribbean Sea," Technical Note IV-XX, Department of the Army, Vicksburg, MS.

National Geophysical Data Center, (1998). "GEODAS CD-ROM hydrographic survey data, *Data Announc.* 98-MGG-03, Natl. Oceanic and Atmos. Admin., U.S. Dep. Commer., Boulder, CO.

Kashiyama, K. and T. Okada. (1992). "Automatic mesh generation method for shallow water flow analysis," *Int. J. Numer. Methods Fluids* 15, 1037-57.

Kinnmark, I., (1986). *The shallow water wave equations: formulation, analysis and application.* Springer-Verlag Berlin, Heidelberg.

Westerink, J.J. and W.G. Gray. (1991). "Progress in surface water modeling," *Reviews of Geophys. Supplement* 210-217.

Westerink, J.J., R.A. Luettich, J.C. Muccino. Modeling tides in the western North Atlantic using unstructured graded grids. *Tellus*, 46A:178,199, 1994.

Westerink, J.J., R.A. Luettich, and R.L. Kolar, "Advances in finite element modeling of coastal ocean hydrodynamics," *Comp. Methods in Water Res. XI*, A.A. Aldama et al., eds., 2, 313-22.

Model Assessment of Dissolved Oxygen and Flow Dynamics in the San Joaquin River near Stockton, California

Haridarshan Rajbhandari[1]

DSM2 water quality model characterizes the spatial and temporal distributions of important water quality variables in an estuarine system. The model is capable of simulating the dynamics of dissolved oxygen including primary production and temperature. Using a dynamic flow field obtained from the companion hydrodynamics model, the model performs advective and dispersive steps of mass transport including net transfer of energy at the air-water interface. Changes in mass of constituents due to decay, growth and biochemical transformations are simulated utilizing relationships derived from the literature.

Calibration and validation of the model were performed using field observations of water quality parameters. The model results matched well with dissolved oxygen and temperature observed in the San Joaquin River (SJR) near Stockton, California where dissolved oxygen levels frequently fall below 5 mg/l during warm, dry months. Low dissolved oxygen levels are of concern because they may adversely affect resident fish and other aquatic life. The current work is one of several projects established through the Total Maximum Daily Load (TMDL) stakeholder process aimed at exploring the ways of improving water quality of SJR. The TMDL stakeholder process was created for this portion of the SJR to meet the water quality standards established by the Federal Clean Water Act. Through evaluations of different scenarios, the DSM2 model can aid in developing potential management strategies to address low DO issues in the estuary. One such scenario is presented here. Historically, rock barriers are installed each year in the South Delta to protect San Joaquin salmon migrating through the Delta and provide adequate agricultural water supply in terms of quantity, quality, and increased channel water levels to meet the local needs. In this study, low head pumps are utilized to transfer water across the barriers and increase the flow in the SJR near Stockton.

The impact of the above operation on the dissolved oxygen levels of the SJR Deep Water Ship Channel (DWSC) was modeled using DSM2. The magnitude of pumping was determined such that a desired target flow rate is maintained in the SJR. No

[1] Senior Engineer, Bay-Delta Office, California Department of Water Resources. 1416 Ninth Street, Sacramento, CA 95814; phone 916-657-5171; hari@water.ca.gov

pumping occurs at the times these flow targets are already met. Scenarios consisting of a base case with no pumping and two alternatives with 1500 cfs and 2500 cfs target flows were examined. Simulations for 1996 through 2000 showed that the oxygen levels in the SJR DWSC improved significantly with the higher target flow rate.

Introduction

Low dissolved oxygen (DO) levels are of concern in the San Joaquin River (SJR) Deep Water Ship Channel (DWSC) in the vicinity of Stockton, California (Figure 1). The DO levels frequently fall below the U.S. Environmental Protection Agency (EPA) standard of 5 mg/l for aquatic health and the Regional Water Quality Control Board standard of 6 mg/l for upstream migration of fall-run Chinook salmon. The 6 mg/l limit is applicable from September 1 through November 30 each year. The Total Maximum Daily Load (TMDL) Stakeholder process was created for this portion of the SJR to meet the water quality standards established by the Federal Clean Water Act. In another study, a link-node model was used to evaluate effectiveness of various hypothetical management scenarios devised to raise the DO above the 5 mg/l standard (Chen and Tsai, 2001). The present study evaluated additional options in greater detail. This paper presents an overview of modeling based on Delta Simulation Model 2 (DSM2) and examines the potential DO water quality benefits of adding auxiliary flow pumps across the Grant Line Canal Barrier (shown in Figure 2). A brief description of the model, followed by the concept of auxiliary flows is described below.

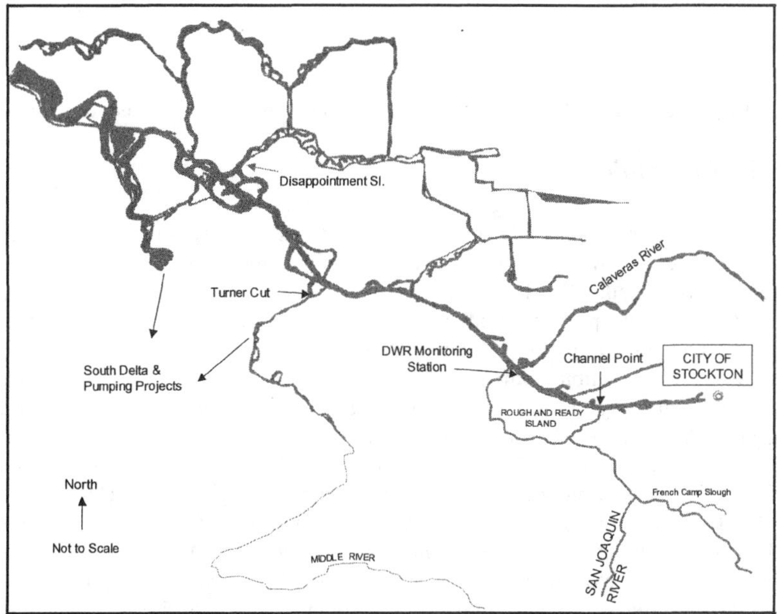

Figure 1. Deep Water Ship Channel near Stockton (Reproduced from Gowdy, 2002).

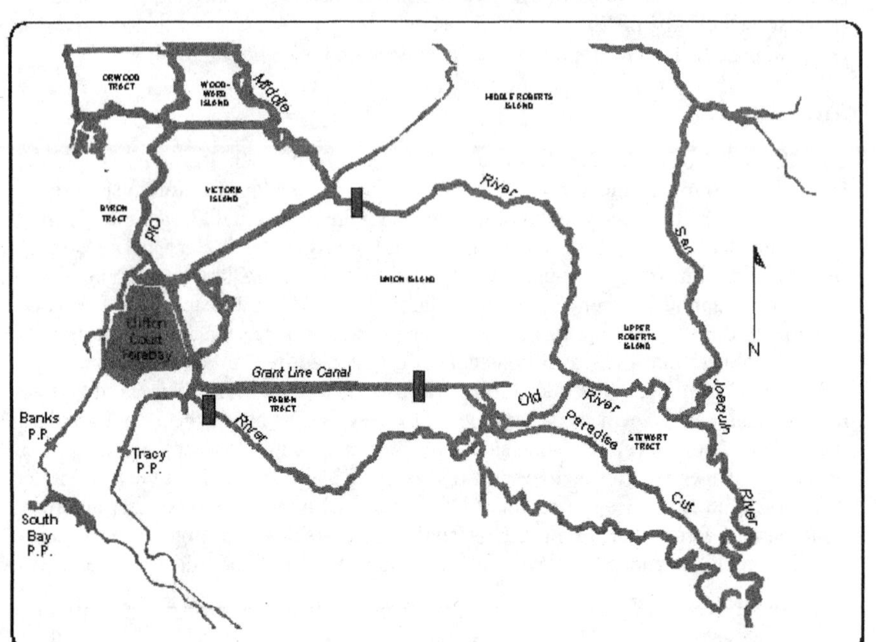

Figure 2. South Delta Agricultural Barriers (Indicated by Rectangles)

Model Description

DSM2 is a one-dimensional mathematical model for dynamic simulation of hydrodynamics and water quality in a network of riverine or estuarine channels. The water quality module QUAL is Lagrangian based and is capable of simulating DO dynamics including primary production and temperature. The major advantage of using the Lagrangian method is the reduction in numerical oscillation and instability when steep concentration gradients occur. The advantages of the Lagrangian schemes are documented by Jobson and Schoellhamer (1987) whose work provided computational framework for QUAL as described in Rajbhandari (1995, 1998). QUAL uses flow information provided by HYDRO module, a one-dimensional, implicit, four-point finite-difference model developed originally by DeLong et al. (1997). Typically both HYDRO and QUAL use a time step of 15 minutes. The Delta is divided into about 500 channel segments. Simulation of DO by QUAL requires information on water temperature, carbonaceous biochemical oxygen demand (CBOD), chlorophyll, organic nitrogen, ammonia nitrogen, nitrite nitrogen, nitrate nitrogen, organic phosphorus, and dissolved phosphorus (ortho-phosphate) in the Delta. Interaction among water quality variables in DSM2 is shown in Figure 3. The rates of mass transfer (shown by the arrows) are functions of temperature. It is important that temperature simulation be included in the DO simulation. The sources and sinks of DO are indicated in Figure 3.

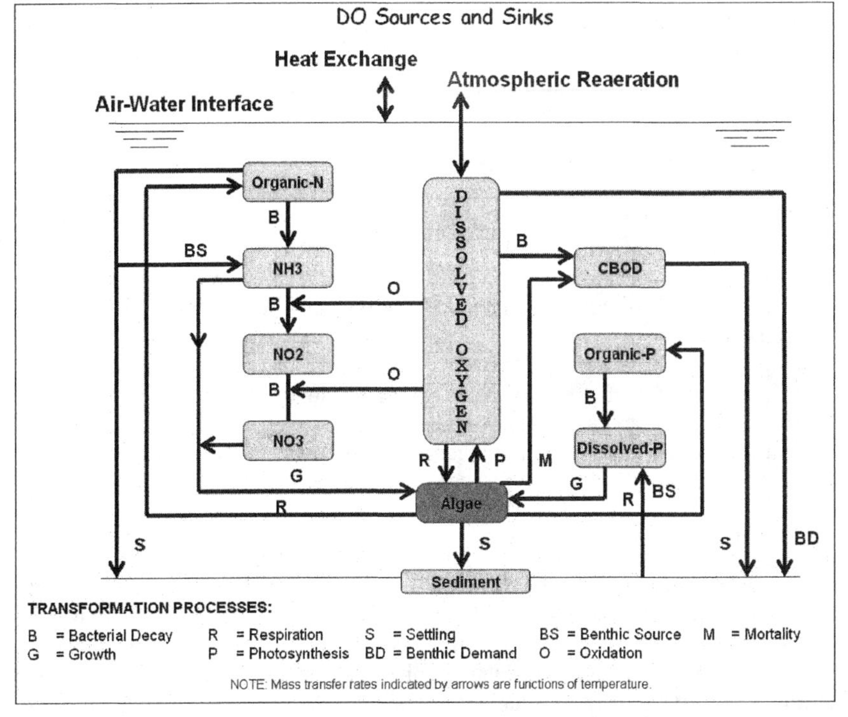

Figure 3: Interaction between DO and Related Parameters.

The following mass balance equation is used in the model for DO mass balance.

$$\frac{?[O]}{?t} = \frac{\partial}{\partial \xi}\left[E_x \frac{\partial [o]}{\partial \xi}\right] - (k_1 + k3)\,L + k_2\,(O_s - [O]) - \alpha_5\,k_n\,[NH3]$$

Dispersion CBOD Reaeration NH3 oxidation

$$- \alpha_6\,k_{ni}\,[NO_2] + \alpha_3\,\mu\,[A] - \alpha_4\,\rho[A] - K4/d$$

NO2 oxidation Photosynthesis Respiration SOD

where [O] = dissolved oxygen concentration, mg/L or, g/m^3; k_1 = CBOD decay rate, day^{-1}; K_3 = rate of loss of CBOD due to settling day^{-1}; L = CBOD conc., mg/L; k_2 = reaeration coefficient, day^{-1}; O_s = DO at saturation, mg/L; k_n = ammonia decay rate, day^{-1}; $[NH_3]$ = ammonia conc. as N, mg/L; $\alpha 5$ = amount of oxygen consumed in conversion of ammonia to nitrite; $\alpha 6$ = amount of oxygen consumed in conversion of nitrite to nitrate; k_{ni} = nitrite decay rate, day^{-1}; $[NO_2\}$ = nitrite conc. as N, mg/L; $\alpha 3$

= amount of oxygen produced per unit of algal photosynthesis; μ = phytoplankton growth rate, day^{-1}; $\alpha 4$ = amount of oxygen consumed per unit of algae respired; ρ = phytoplankton respiration rate, day^{-1}; [A] = phytoplankton conc., mg/L; k_4 = benthic or sediment oxygen demand (SOD), $g/m^2\ day^{-1}$; d = mean channel depth, m; E_x = longitudinal dispersion coefficient; and ξ = distance from the parcel, the Lagrangian distance coordinate.

$$\xi = x - x_o - \int_{to}^{t} u\, d\tau$$

where x_o = location of the fluid parcel at time t_o; x = Eulerian distance coordinate; t = time; u = velocity and t_o = initial time.

Equations similar to the above are included in the model for all eleven constituents. Rajbhandari (1998) discusses the DO kinetics used in QUAL in greater detail. This report can be found at:
http://modeling.water.ca.gov/delta/reports/annrpt/1998/1998Ch3.pdf. A more complete description of DO kinetics and model development is available in Rajbhandari (1995) and Rajbhandari et al. (1996). The equations are solved using the finite difference method. Source sink terms are updated using Modified Euler method.

Auxiliary Flows

The objective for the auxiliary pumping is to meet the flow target in San Joaquin River downstream from the head of Old River. Historically, rock barriers are installed each year at Grant Line Canal (GLC), Middle River, Old River, and head of Old River to protect San Joaquin salmon migrating through the Delta and provide adequate agricultural water supply in terms of quantity, quality, and increased channel water levels. The barriers are installed, as needed, at certain times each year. In the scenarios based on future management plans, the first three barriers were installed each year from April through November[2].

In the pump scenarios, low head pumps are utilized to transfer water across the GLC barrier and increase the flow in the SJR near Stockton. Two flow targets at the San Joaquin River below the head of Old River were considered: a) 1500 cfs [43 m^3/s] and b) 2500 cfs [71 m^3/s]. The simulation period covered July 1996 through December 2000. This period was chosen because the data needed for simulating DO was available. Unfortunately, this period does not include extreme dry periods, which are typically associated with extreme low DO levels in DWSC.

Figure 4 shows the magnitude of the required auxiliary pumping on the GLC barrier to maintain 1500 cfs and 2500 cfs flow targets. Since the selected hydrologic period

---

[2] These runs do not include the installation of the barrier at the head of Old River. Normally a barrier at the head of Old River is installed every spring and fall. It is expected that with the inclusion of the three barriers and the auxiliary pumping at Grant Line Canal, there would be no need for the Head of Old River Barrier.

happens to be one with somewhat higher flows than often occur in the SJR, the target flows are met except for the four summer-fall periods shown. The year 1998 especially had significantly higher flows, so flow targets were already met for the entire period without auxiliary flows.

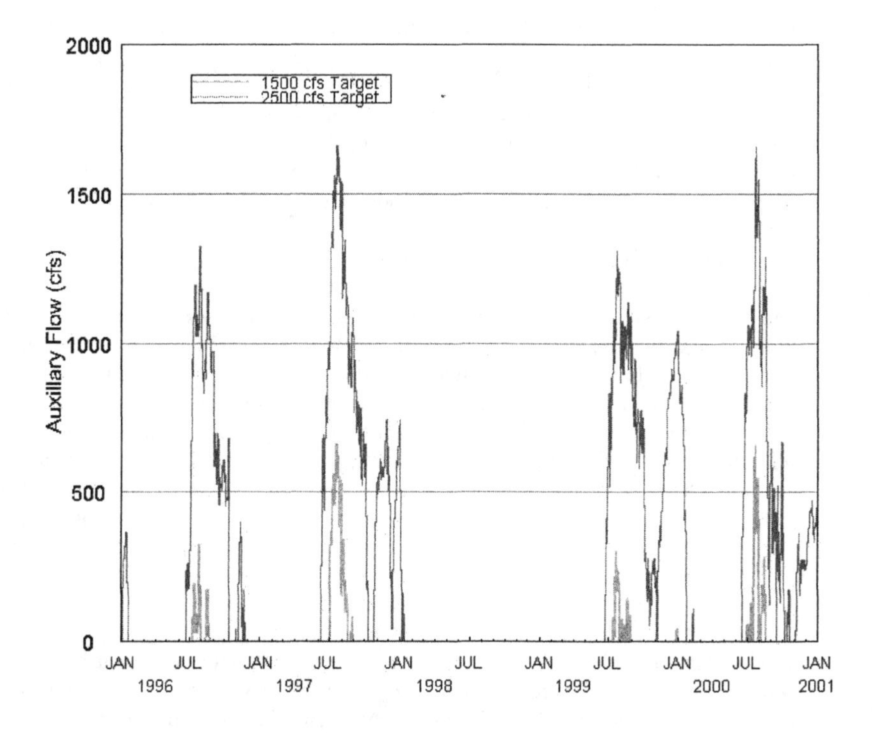

Figure 4. Auxiliary Flows over the Grant Line Canal Barrier

Model Scenarios and Input

Four sets of hydrodynamic runs were conducted representing the following scenarios:

a) Historical: replicates historic conditions to verify model against field data. Barriers operate from a few weeks to a few months, mostly during spring and summer[3].
b) Base: uses hydrology identical to historical, but barriers are in place April – November; no auxiliary pumping.
c) Alternative 1: auxiliary pumping based on the target flow of 1500 cfs [43 m^3/s]; barriers same as in Base.
d) Alternative 2: auxiliary pumping based on the target flow of 2500 cfs [71 m^3/s]; barriers same as in Base.

[3] Except in 1998 when the barriers were not constructed due to high flow conditions in the San Joaquin River.

In the base and alternative one and two scenarios, the three barriers are assumed to be operating April through November. Auxiliary pumps operate only during the months when the three barriers are present. Output from DSM2-HYDRO was used to run DSM2-QUAL.

DO and temperature data collected at hourly intervals provide boundary information needed by QUAL. A combination of hourly temperature and DO data in Sacramento River at Freeport and Rio Vista were provided for the Sacramento River model boundary. The historical record of DO and temperature at Martinez was used for the downstream model boundary. Because continuous data were not available at Vernalis, hourly values of DO and temperature from the nearby station at Mossdale were used to approximate these quantities for the boundary inflow at Vernalis. Because the flows at Vernalis are primarily unidirectional and the hydraulic residence time is relatively short, this assumption is reasonable.

Data on effluent flows from the City of Stockton's Regional Wastewater Control Facility were obtained from the City of Stockton Municipal Utilities Department (Huber, 2001). Flow, BOD, and temperature data were available on a daily basis. The data for ammonia nitrogen were available on approximately a two-day interval. These data were interpolated to obtain daily estimates. EC, organic nitrogen, nitrite nitrogen, and nitrate nitrogen data were available on weekly intervals, and interpolated to daily intervals. For most of these constituents, the values were sometimes given as "less than" a detection limit. Approximations were made based on the preceding and the subsequent known values and sensitivity analysis indicated a negligible impact on the simulations.

Nutrient data at Vernalis were approximated from the San Joaquin River TMDL measurements sampled at weekly intervals in 1999. The nutrient data at Freeport on the Sacramento River were approximated from the U.S. Geological Survey report (USGS, 1997) and chlorophyll data were approximated from CDWR (1999). Estimates of flow and water quality of agricultural drainage returns at internal Delta locations were based on earlier CDWR studies.

Hourly or 3-hour interval air temperature, wet bulb temperature, wind speed, cloud cover, and atmospheric pressure data was provided by the National Climatic Data Center starting in July 1996 and was used as QUAL input to simulate water temperature. However, for most of 1996, only the minimum and maximum values for temperature and wind speed were available. For this period, hourly values for temperature and wind speed were approximated based on the daily maximum and minimum values.

Calibration and Validation

QUAL was previously calibrated and validated for simulating DO; however, it was based on data from 1998 to 1999 (Rajbhandari, 2001). HYDRO calibration is described in Nader-Tehrani (2001) and at the site http://iep.water.ca.gov/dsm2pwt/calibrate/. At the time of this previous DO calibration and validation, hourly time series data were available only at Rough and Ready Island (RRI) (Figure 5); thus calibration and validation were limited to DO and temperature comparisons at that location. Under normal flow conditions, the DO levels in the SJR at RRI depend mainly on SJR flow and quality. However, for the scenarios that may involve flows from the South Delta region, it is necessary to extend the validation of DO to include the South Delta region.

This extended calibration and validation was achieved by comparing model DO against field data available for the year 2000 at the three South Delta locations, two in Old River and one in Middle River (Figure 5). It was necessary to calibrate QUAL primarily in the South Delta region. During DO calibration, the rate coefficients for algae (growth and mortality rates) and sediment oxygen demand were adjusted. Calibrated coefficients (Table 1) are within the range suggested in the literature (Bowie et al., 1985; Brown and Barnwell, 1987; Thomann and Mueller, 1987). Table 2 summarizes the continuous monitoring stations used in the calibration and validation of QUAL.

Table 1: Calibrated Coefficients in the South Delta

Maximum algal growth rate	2.5 per day
Algae mortality rate	0.7 per day
SOD rate	3 g/m^2 per day

Table 2: Summary of Continuous DO and Temperature Monitoring Stations.

Map Location	Field Station Name	Calibration/ Validation/Boundary	Monitoring Start Date	Figure
1	Middle River at Howard Road	Calibration	2000	Figure 6
2	Old River at Tracy Wildlife Association	Calibration	2000	Figure 7
3	Old River near DMC	Calibration	2000	Figure 8
4	SJR at Rough & Ready Island	Validation	1983	Figure 9 Figure 10
5	SJR at Mossdale[1]	Boundary	1983	-
6	Sacramento River at Freeport[2]	Boundary	1999	-
7	Sacramento River at Rio Vista[2]	Boundary	1983	-
8	Sacramento River at Martinez	Boundary	1983	-
9	Sacramento River at Mallard Slough	Validation	1983	Figure 11
10	SJR at Antioch	Validation	1983	Figure 12

(1) Mossdale data was used to fill in missing values for the Vernalis boundary condition.
(2) Rio Vista data was used to fill in missing values for the Freeport boundary condition.

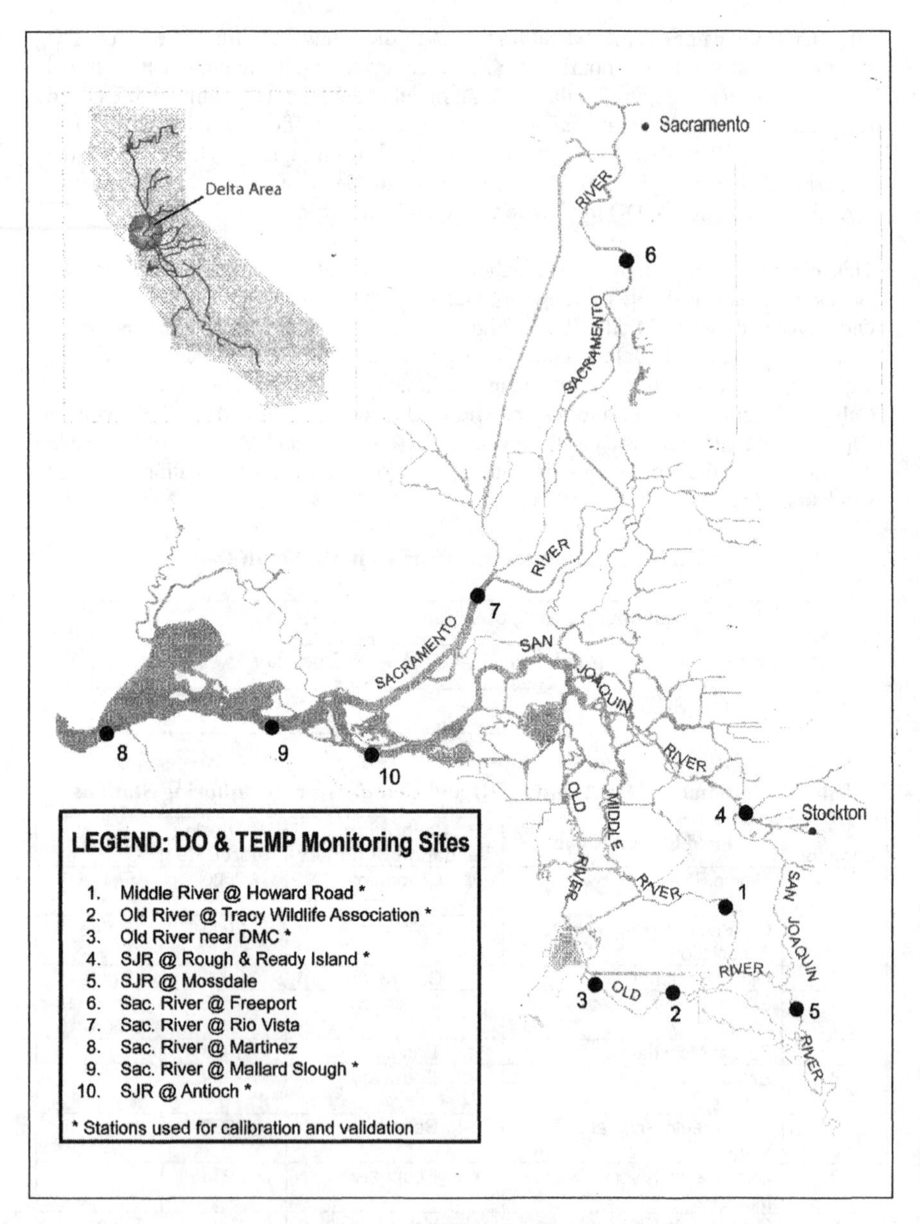

Figure 5: DO and Temperature Water Quality Stations in the Delta.

Model Calibration

Figure 6 compares modeled results with measured hourly DO values in Middle River at Howard Road. For the spring and summer months the model diurnal range tends to be much shorter than the measured values, but the general trend and the low DO values appear to be in fairly good agreement. Supersaturated levels of DO observed in the field data in early June and at certain times in August and September were not reproduced in the model results.

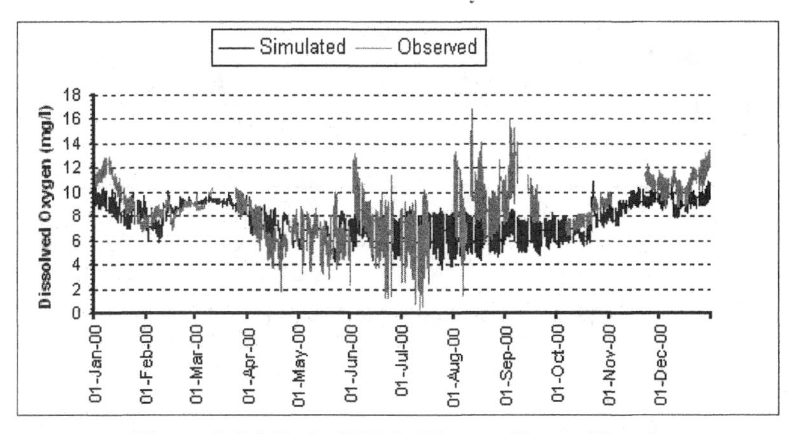

Figure 6: DO in the Middle River at Howard Road.

A comparison of model DO with field DO in Old River at Tracy Wildlife Association shows a fair agreement during most months (Figure 7). QUAL tends to under-predict the diurnal range. Highly supersaturated DO levels that occurred in early June and November were under-predicted by QUAL.

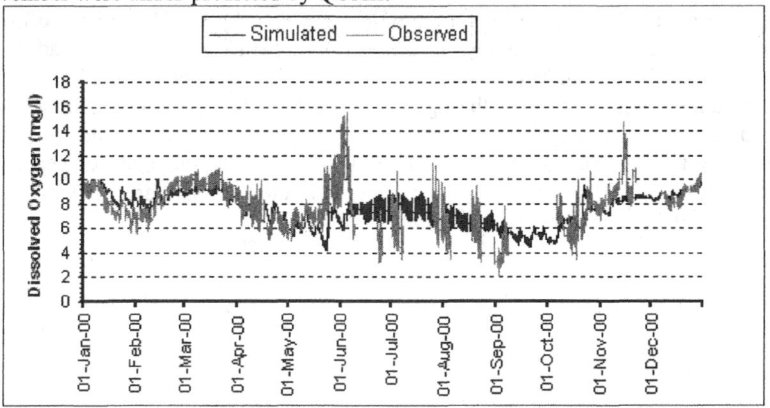

Figure 7: DO in the Old River at the Tracy Wildlife Association.

Field data for Old River at Delta Mendota Canal (DMC) is available only for May through November 2000 and is shown with the model results in Figure 8. The model tends to capture the monthly trend with better agreement in the lower range.

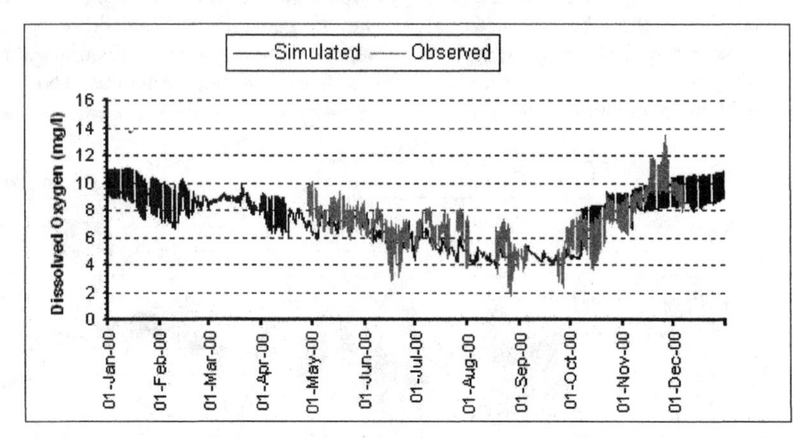

Figure 8: DO in the Old River at DMC.

Model Validation

Figure 9 presents the comparison of model DO and field observations in the San Joaquin River near Rough and Ready Island (RRI). The model seems to do well in representing the DO levels that fall below the required standard of 5 mg/l. In general, the differences between model and field DO were within 1 mg/l at the lower end of diurnal range and for the summer months. Seasonal highs and lows appear to be in phase. DSM2 was not able to reproduce the supersaturated values of DO observed during summer and fall 2000. These conditions indicate algal bloom. Model has difficulty in simulating algal bloom episodes primarily because there is insufficient amount of algal data for model boundary input. Much of the data had to be interpolated, and may have introduced errors. The EPA requires that the DO must be greater than or equal to 5 mg/l throughout the year, while the Water Quality Control Board requires that the DO levels remain at 6 mg/l or above for the months of September through November. As a result, it is desirable that the model be capable of predicting the low DO levels more accurately than the supersaturated DO values, so the under-prediction of supersaturated values of DO is not critical.

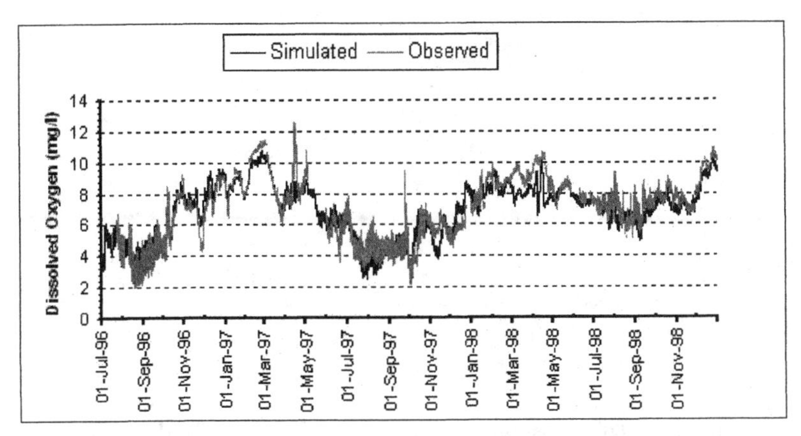

Figure 9a: DO in the San Joaquin River at Rough and Ready Island, 1996-1998.

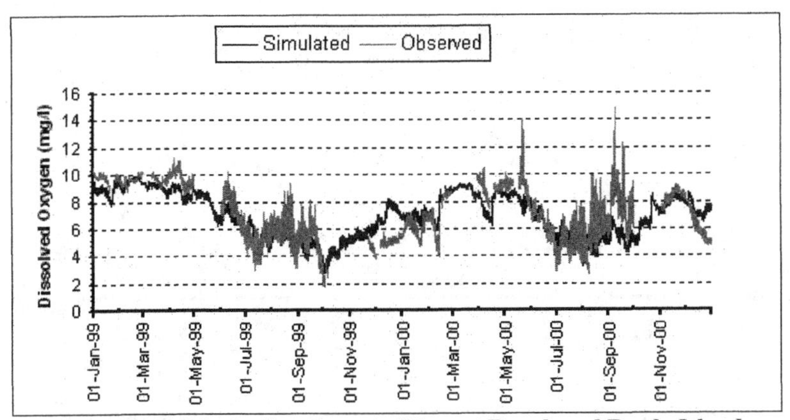

Figure 9b: DO in the San Joaquin River at Rough and Ready Island, 1999-2000.

Figure 10 compares simulated water temperature with field data at the continuous monitoring station at the San Joaquin River near RRI. In general, DSM2 seems to underestimate the observed data but the differences are generally less than 1° Celsius. The diurnal range in temperature simulation results is generally smaller than the range for the field data, especially in the summer months; however, tests showed low DO sensitivity to small variations in temperature.

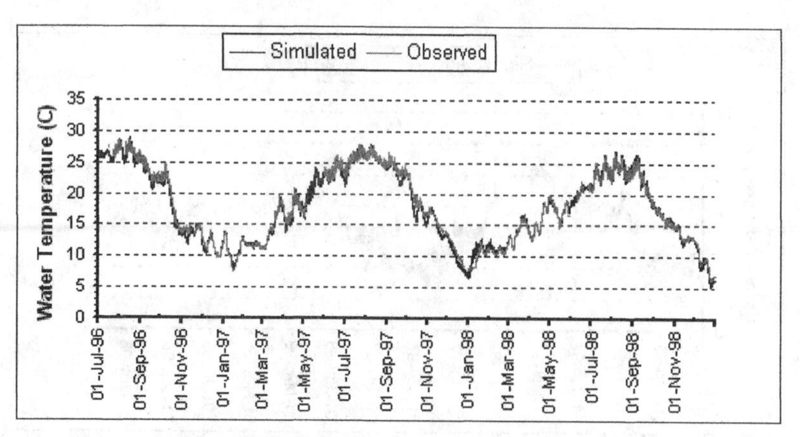

Figure 10a: Water Temperature in the San Joaquin River at Rough and Ready Island, 1996-1998.

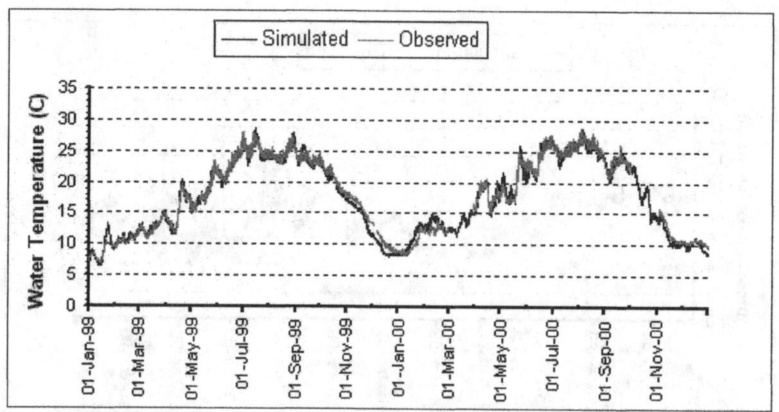

Figure 10b: Water Temperature in the San Joaquin River at Rough and Ready Island, 1999-2000.

Modeled DO at Mallard Slough (Figure 11) captures the seasonal variation of DO in the measured data. Except for summer through fall 1999, modeled DO was within 0.5 mg/l of the data for the low DO periods of the summer months. Comparison of simulated and field DO in the San Joaquin River at Antioch is shown in Figure 12. The agreement was good and generally within 0.5 mg/l except for fall 1997 and winter 1998, when the model overestimated DO by up to 3 mg/l.

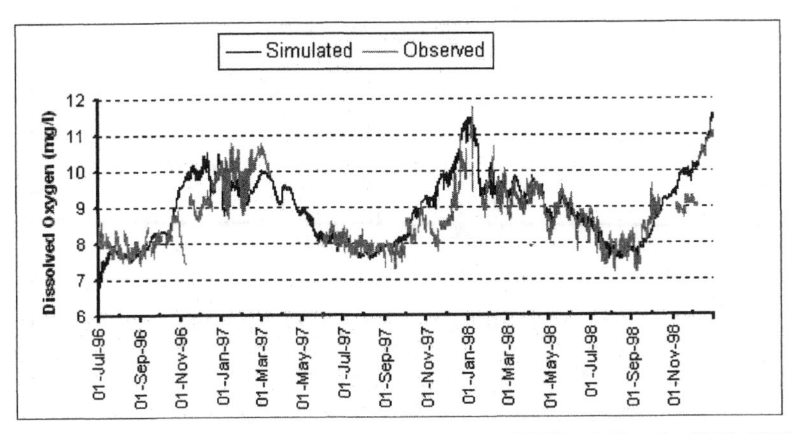

Figure 11a: DO in the Sacramento River at Mallard Slough, 1996-1998.

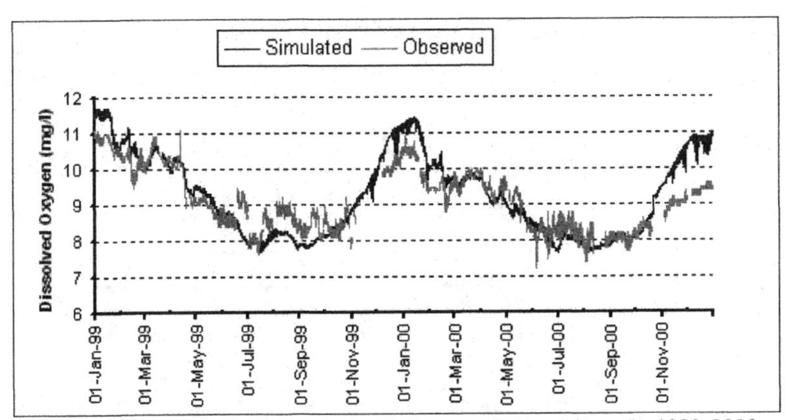

Figure 11b: DO in the Sacramento River at Mallard Slough, 1999-2000.

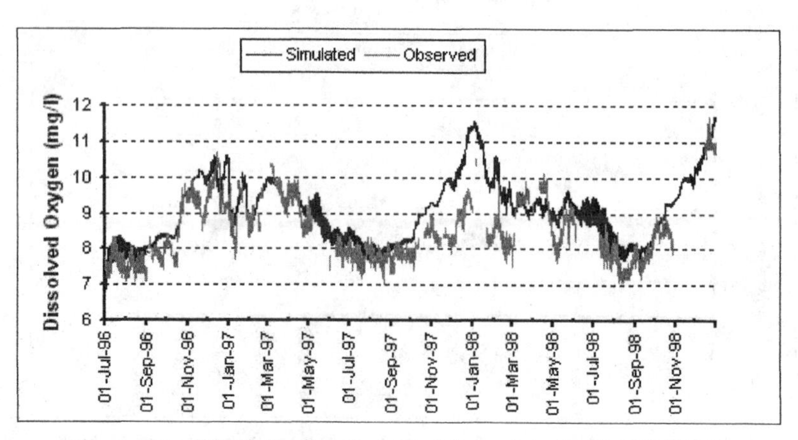

Figure 12a: DO in the San Joaquin River at Antioch, 1996-1998.

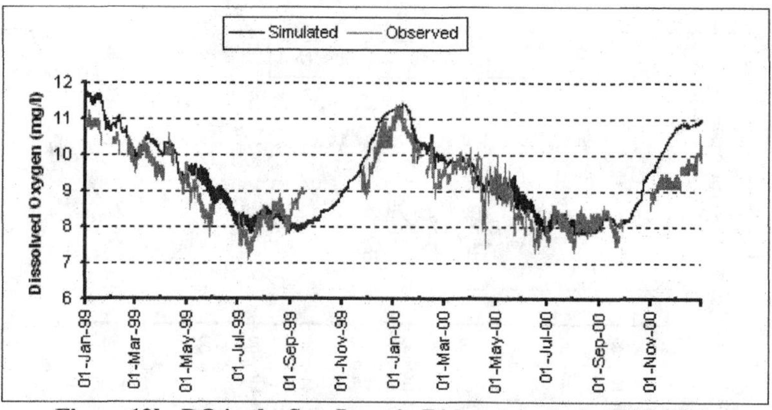

Figure 12b: DO in the San Joaquin River at Antioch, 1999-2000.

DO Simulations for Base and Auxiliary Flow Scenarios

DSM2 was used to simulate DO levels at Rough and Ready island (RRI) for all four scenarios: historical, base, and target flows of 1500 cfs and 2500 cfs. Compared to the historical scenario, the base scenario produced higher dissolved oxygen levels for all low DO periods that occurred during the summers from 1996 through 2000 (Figure 13). Even without the auxiliary pump flows DO levels were improved because the three barriers operating for eight months (April-November) held flows in the SJR. While the scenario with auxiliary flows designed for a target flow of 1500 cfs showed higher DO levels at RRI, the DO benefits compared to the base scenario were noticeable only for the summers of 1997 and 2000 (Figure 14). This is because for the rest of the period flows in SJR were already at or above 1500 cfs range. It is expected that in dry years, when SJR typically has flows at about 1000 cfs or lower, the impact will be more evident. As expected, the scenario with auxiliary flows designed for a target flow of 2500 cfs resulted the highest DO benefits. The DO improvements are influenced primarily by water quality of the auxiliary flow and the lower hydraulic residence time at the DWSC.

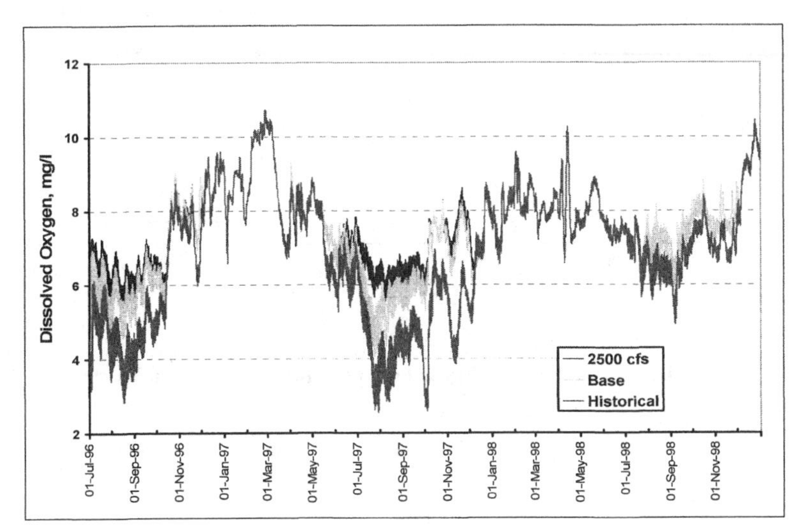

Figure 13a. Impact of 2500 cfs Flows on DO in the San Joaquin River at RRI, 1996-1998.

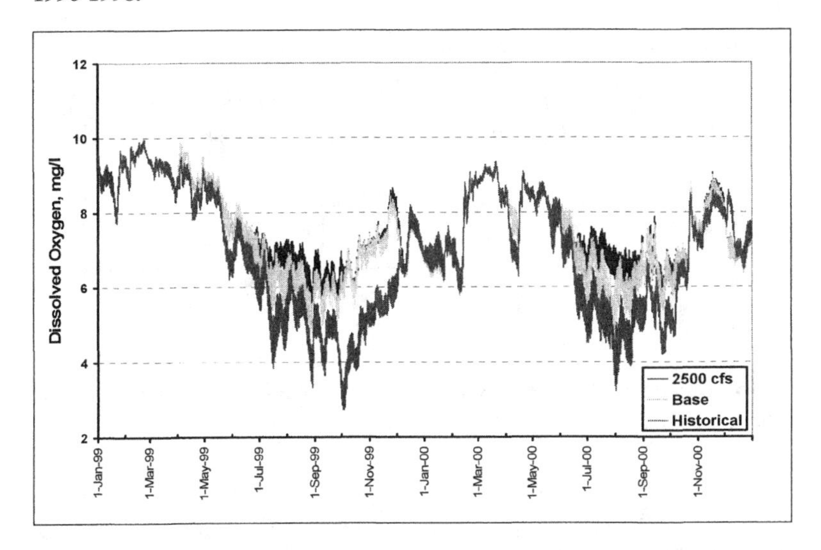

Figure 13b: Impact of 2500 cfs Flows on DO in the San Joaquin River at RRI, 1999-2000.

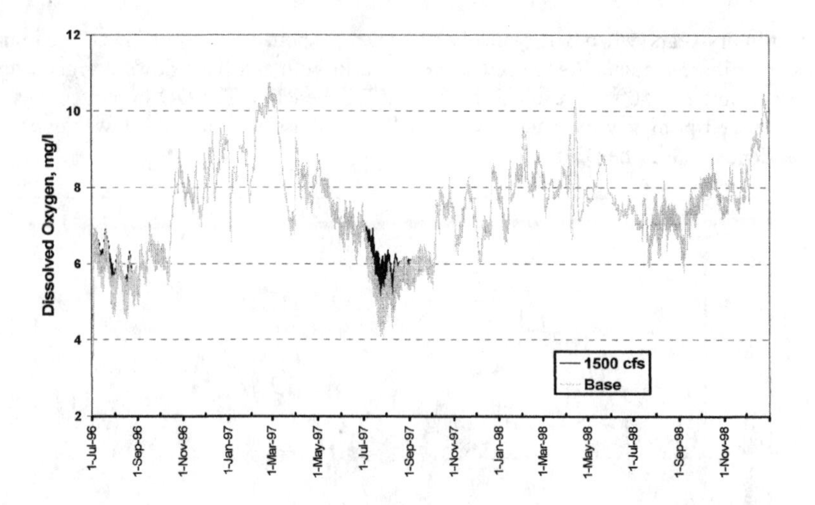

Figure 14a: Impact of 1500 cfs Flows on DO in the San Joaquin River at RRI, 1996-1998.

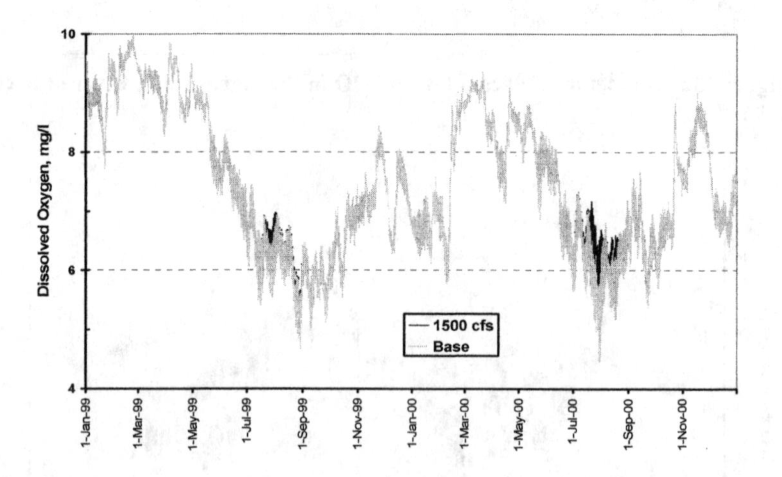

Figure 14b: Impact of 1500 cfs Flows on DO in the San Joaquin River at RRI, 1999-2000.

Summary

- DSM2 was used to evaluate changes in DO levels in Stockton DWSC by adding auxiliary pumping across GLC barrier. The net effect of auxiliary pumping is to increase flows in the San Joaquin River.
- DSM2 was calibrated based on the DO data at three South Delta locations (Figures 6-8). Model validation was based on the stations in San Joaquin River at RRI and Antioch, and Sacramento River at Mallard Slough (Figures 9-12).
- DSM2 was used to simulate DO levels (July 1996 - Dec 2000) for three alternatives: I) Base case, II) Base case + 1500 cfs target flow in main stem of San Joaquin River, III) Base case + 2500 cfs. Model showed some improvement with 1500 cfs and significant improvement with 2500 cfs (see Figures 13 and 14).

Future Directions

A program to supplement the existing monitoring program and develop the needed database for future calibration is desirable. A minimum data collection program should include both spatial and temporal characterizations of the primary quality constituents, e.g., dissolved oxygen, temperature, chlorophyll-a, and nitrogen species, under at least two distinct hydrologic and hydrodynamic conditions.

Surveys should be conducted of sediment deposits along the DWSC to determine spatial variations in benthic/sediment oxygen demand, and the nitrogen and phosphorus content in the sediments to improve calibration of the model. Subject to a consistent expansion of the database, future extensions in the model to add additional variables, such as zooplankton and benthic algae, are likely to result in improvement in model performance. Extension of model should include a dynamic interaction of sediments with simulated constituents.

Other potential applications of DSM2 may include the following projects:

- SJR Modeling upstream of Vernalis:

  A proposal to develop and calibrate the DSM2-SJR model for DO and the related parameters, as a part of the Proposal for Upstream Monitoring 2004-2006 has been approved by California Bay Delta Authority (formerly known as CALFED). The DSM2-SJR model, a multi-agency effort, is expected to provide an essential link to understanding the SJR algae growth processes that create a substantial load of organic material that may contribute to DO decline episodes in the DWSC.

- Detailed/Multi-dimensional model analysis:

  Special studies may require a more refined analysis that would be best served by two-dimensional (2D) or three-dimensional (3D) models. DSM2 can be utilized in a way that would exchange information with the 2D/3D models that already exist, or are being developed by the other agencies.

Acknowledgement

This study was conducted at the California Department of Water Resources with support from CALFED. The author expresses appreciation for Dr. Parviz Nader for insightful suggestions during this project and Dr. Bijaya Shrestha for conducting hydrodynamic simulations.

References

Bowie, G.L., W.B. Mills, D.B. Porcella, C.L. Campbell, J.R. Pagenkopt, G.L. Rupp, K.M. Johnson, P.W.H. Chan, and S.A. Gherini. (1985). *Rates, Constants and Kinetics Formulations in Surface Water Quality Modeling, 2nd Ed.* US EPA. Athens, Georgia. EPA 600/3-85/040.

Brown, L.C. and T.O. Barnwell. (1987). *The Enhanced Stream Water Quality Models QUAL2E and QUAL2E-UNCAS; Documentation and Users Manual.* US EPA. Athens, Georgia. EPA 600/3-87/007.

California Department of Water Resources. (1999). *Water Quality Conditions in the Sacramento-San Joaquin Delta during 1995.* Division of Local Assistance.

Chen, C.W. and W. Tsai (2001). *Improvements and Calibrations of Lower San Joaquin River DO Model.* Report from Systech Engineering, Inc for CALFED 2000 Directed Action Grant.

Delong, L.L., D.B. Thompson, and J.K. Lee (1997). The computer program FOURPT (Version 95.01) - A Model for Simulating One-Dimensional, Unsteady, Open-Channel Flow. U.S. Geological Survey Water-resources Investigations Report 97-4016, 69 pp.

Gowdy, M. (2002). The San Joaquin River Dissolved Oxygen TMDL – A Stakeholder Process. In *Helping Irrigated Agriculture Adjust to TMDLs.* Proceedings of the *USCID Water Management Conference,* Sacramento, Calif.

Huber, L. (2001). City of Stockton Municipal Utilities Department. Personal Communication.

Jobson, H. E., and D.H. Schoellhamer (1987). "Users Manual for a Branched Lagrangian Transport Model." US Geo. Survey, *Water Resour. Investig. Rep.,* 87-4163.

Nader-Tehrani, P. (2001). "Chapter 2: DSM2 Calibration and Validation." *Methodology for Flow and Salinity Estimates in the Sacramento-San Joaquin Delta and Suisun Marsh. 22nd Annual Progress Report to the State Water Resources Control Board.* California Department of Water Resources. Sacramento, Calif.

Rajbhandari, H.L. (2001). "Chapter 6: Dissolved Oxygen and Temperature Modeling using DSM2." *Methodology for Flow and Salinity Estimates in the Sacramento-San Joaquin Delta and Suisun Marsh. 22nd Annual Progress Report to the State Water Resources Control Board.* California Department of Water Resources. Sacramento, Calif.

Rajbhandari, H.L. (1998). "Chapter 3: DSM2 Quality." *Methodology for Flow and Salinity Estimates in the Sacramento-San Joaquin Delta and Suisun Marsh. 19th Annual Progress Report to the State Water Resources Control Board.* California Department of Water Resources. Sacramento, Calif.

Rajbhandari, H. L., G. T. Orlob and F. I. Chung, 1996. *Simulation of Dissolved Oxygen in Sacramento - San Joaquin Delta.* Proceedings of North American Water and Environment Congress, American Society of Civil Engineers, Anaheim, Calif.

Rajbhandari, H.L. (1995). *Dynamic simulation of Water Quality in Surface Water Systems Utilizing a Lagrangian Reference Frame.* Ph.D. Dissertation. University of California, Davis.

Thomann, R.V., and J.A. Mueller. (1987). *Principles of Surface Water Quality Modeling and Control.* Harper and Row, New York.

US Geological Survey. (1997). "Water Resources Data, California, Water Year 1997." Vol. 4.

Acronyms and Abbreviations

1D – 1-dimensional
2D – 2-dimensional
3D – 3-dimensional
BOD – Biochemical Oxygen Demand
CALFED – State (CAL) and federal (FED) agencies participating in the Bay-Delta Accord
CBOD – Carbonaceous Biochemical Oxygen Demand
CDWR – California Department of Water Resources
cfs – cubic feet per second
DO – Dissolved Oxygen
DSM2 – Delta Simulation Model 2
DWSC – Deep Water Ship Channel
EPA – U.S. Environmental Protection Agency
GLC – Grant Line Canal
HYDRO – DSM2 Hydrodynamics Model
m3/s – cubic meter per second
QUAL – DSM2 Water Quality Model
RRI – Rough and Ready Island
SJR – San Joaquin River
TMDL – Total Maximum Daily Load
USGS – U.S. Geological Survey

Physical Versus Anthropogenic Control of Nutrient Concentrations in the Irish Sea

E.F. Young[1*], J.T. Holt[1], G.L. Evans[2], P. Williams[2]

Abstract

A three-dimensional, density resolving model (the Proudman Oceanography Laboratory Coastal Ocean Modelling System; POLCOMS) has been applied to the study of long term variability in salinity and nutrient concentrations in the Irish Sea. Analyses of time series of observational data collected at two sites in the Irish Sea considered the variability to reflect either changes in anthropogenic loading or climatic variability. In this modelling study, the latter supposition is investigated in more detail by considering the response of salinity and tracers in the Irish Sea to variability in climatic forcing.

A series of theoretical simulations investigated the roles of tide, wind, baroclinic and far field forcing in the transport and dispersion of nutrients in the Irish Sea. Far field forcing acted to transport and retain the tracer close to the main sampling site in the central Irish Sea (Cypris station). The baroclinic component of flow restricted westward transport of nutrients, with the dominant flow predicted along the coast northwards and out of the region. Direct wind forcing caused high frequency movement of near coastal water masses with high tracer concentrations, thus inducing high frequency variability in predicted concentrations at the Cypris station. Increases in freshwater inflows also caused a shift in these high concentration waters, but at a lower frequency.

Observational analyses have shown an inverse relationship between salinity and winter nitrate concentration in the Irish Sea. Thus, by simulating and understanding long term variability in salinity, it is possible to gain insight into the dominant influences on nutrient concentrations. The results of a 40 year hindcast simulation demonstrated the ability of the model to reproduce temperatures throughout the model domain for the 40 year period, and more specifically the observed variability in temperature at the Cypris station. Trends in salinity variability were also generally reproduced, although absolute values of salinity were generally too high. Considered in conjunction with the inverse relationship between salinity and nitrate concentrations, these results suggest that the observed nutrient variability at the

[1] Proudman Oceanographic Laboratory, Bidston Observatory, Bidston Hill, Prenton, Merseyside, CH43 7RA, UK
[2] University of Wales, Bangor, School of Ocean Sciences, Menai Bridge, Anglesey, LL59 5AB, UK
[*] e-mail: eyoung@pol.ac.uk tel. : +44 (0) 151 653 1532 fax : +44 (0) 151 653 6269

Cypris station is significantly influenced by climatic, as opposed to anthropogenic, influences. In particular, the effect of a moving freshwater/seawater dilution zone on observations at fixed sampling points suggests that a degree of caution should be exercised when analysing such data.

Introduction

The Irish Sea is a semi-enclosed shelf sea bounded by the UK and Ireland (Fig. 1). There is a residual northward flow through the region, with Atlantic water entering via St. George's Channel in the south, and leaving via the North Channel between Scotland and Northern Ireland. These Atlantic waters annually transport over 100 million tonnes of nitrogen and 28 million tonnes of phosphorus into the Irish Sea. The Irish Sea itself covers 47,000 km^2 and receives freshwater from a 43,000 km^2 catchment area (Bowden, 1980), which includes several dense industrial centres and conurbations. Industrial, sewage and river sources contributed an estimated 86,800 tonnes of nitrate and 9100 tonnes of phosphate (OSPAR Commission, 2001) to the Irish Sea in 1999, which appear relatively small amounts compared to the Atlantic contribution. However, the higher concentrations found in river and effluent waters can have a major localised impact on coastal water quality. Short term studies have shown waters in Liverpool Bay in the eastern Irish Sea to be enriched with nitrogen and phosphate (e.g. Foster, 1984), and this region has also been shown to support enhanced phytoplankton production and standing stock (Gowen *et al.*, 2000).

European and national pressures to improve the quality of coastal waters are growing. For example, the EU Urban Waste Water Treatment Directive (EC 91/271) requires member states to identify marine areas according to their sensitivity to eutrophication, and calls for higher standards of wastewater treatment where coastal waters are identified as 'sensitive areas'. Liverpool Bay in the eastern Irish Sea is one such region. However, whilst the need for remediation is clear, the dynamics of nutrients and their impact on phytoplankton communities in the region are poorly understood, thus it is difficult to judge the likely effect of changes in anthropogenic nutrient sources on water quality. The work described in this paper is aimed at further developing our understanding of the key processes affecting nutrient concentrations in the Irish Sea.

1) Observational Background

Two temporally extensive datasets have been collected in the Irish Sea, representing a highly valuable resource for understanding long term changes in this region. Since 1954, data have been collected by Port Erin Marine Laboratories (PEML) at a fixed sampling point (Cypris station) approximately 5 km off the southwest coast of the Isle of Man (Fig. 1). These data comprise samples at 5 m intervals through the water column, taken at fortnightly intervals, and analysed for physical parameters, nutrients, and chlorophyll. The second dataset was collected by the School of Ocean Sciences, Bangor as part of commissioned research, postgraduate and project studies, and only recently collated into a database that spans the period 1963-2002 (Evans *et al.*, 2003).

These measurements were taken from the Menai Straits, the body of water separating Anglesey from mainland North Wales (Fig. 1), and comprise physical, chemical and biological observations. Although the data are discontinuous and irregularly spaced in time, they provide a valuable additional resource for understanding variability in the Irish Sea.

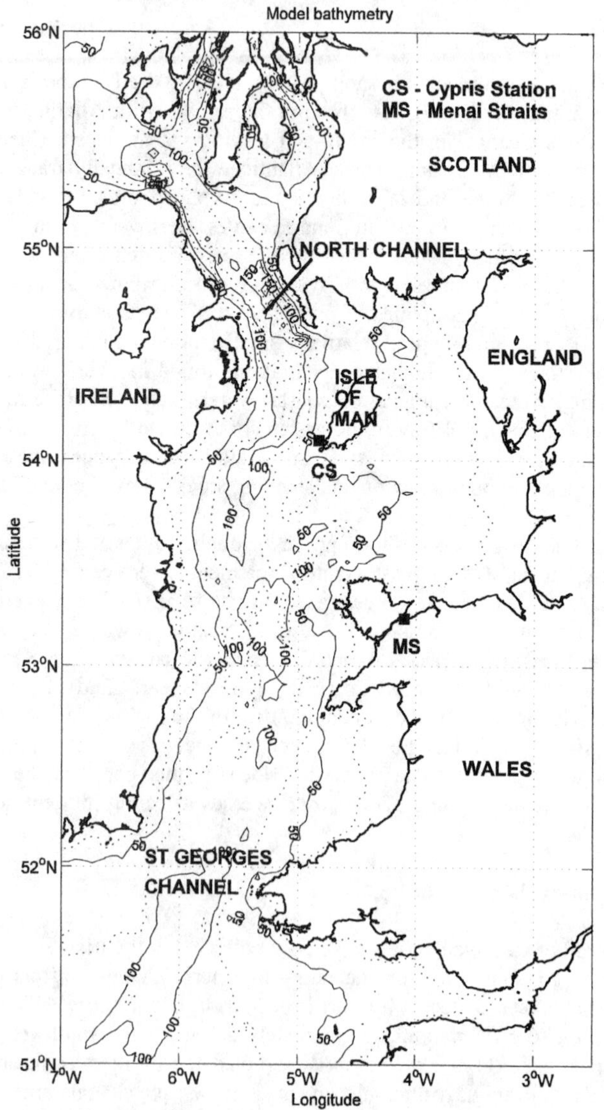

Figure 1: Map illustrating model bathymetry (depths in metres) and locations referred to in the text.

The trends in winter nitrate concentrations measured at the Cypris station and the Menai Straits are shown in Fig. 2. Quadratic models have been fitted to the data, as discussed in Evans *et al.*, 2003. Nitrate data at the Menai Straits describe a pattern of low concentration in the 1960s, a peak in the late 1970s and 1980s followed by a decline in the 1990s. The nitrate time series for the Cypris station show a rise from 1955 to 1991 with a slight decline in the 1990s. Analyses of the Cypris Station data attributed observed trends in nutrient concentrations in the Irish Sea to changes in anthropogenic loading (Allen *et al.*, 1998; Gowen *et al.*, 2002). Thus an examination of the nutrient loading in the freshwater end-member should show a rise from the 1960s to the mid 1980s and then a fall in the 1990s. However, analysis of the four largest riverine nutrient sources into the eastern Irish Sea found that concurrent changes in freshwater nitrate loading did not appear to follow the in situ marine concentrations (Evans *et al.*, 2003). In particular, nitrate levels in the Mersey are likely to have risen in the 1990s as improvements in sewage treatment have resulted in a reduction in denitrification in the river due to the lessening organic loads.

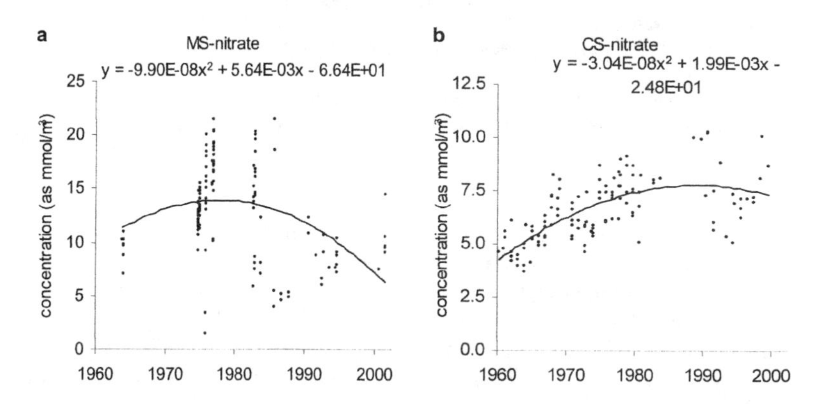

Figure 2: Winter nitrate measurements from the Menai Strait (MS) and the Cypris station (CS). Figure courtesy of Evans *et al.*, 2003.

A more recent analysis considering both the Cypris Station and Menai Straits data series has suggested that climate also has a significant influence (Evans *et al.*, 2003). The analyses have found a strong negative correlation between salinity and nutrient concentration over a decadal timescale. Salinity variability is clearly not derived from anthropogenic activity. Thus it has been proposed that observed variability may be due to the interaction of the Eulerian nature of the sampling (fixed geographical locations) with climatically induced shifts in the water mass boundaries between regions of Atlantic and freshwater influence. Movement back and forth of the freshwater/seawater dilution zone causes salinity and nutrient concentrations to vary inversely. This effect would appear as a local change in salinity and nutrient concentration. This is conceptualised in Fig. 3, which depicts the effect of a shift in the position of the dilution zone on the salinity and nutrient concentrations at a fixed sampling site.

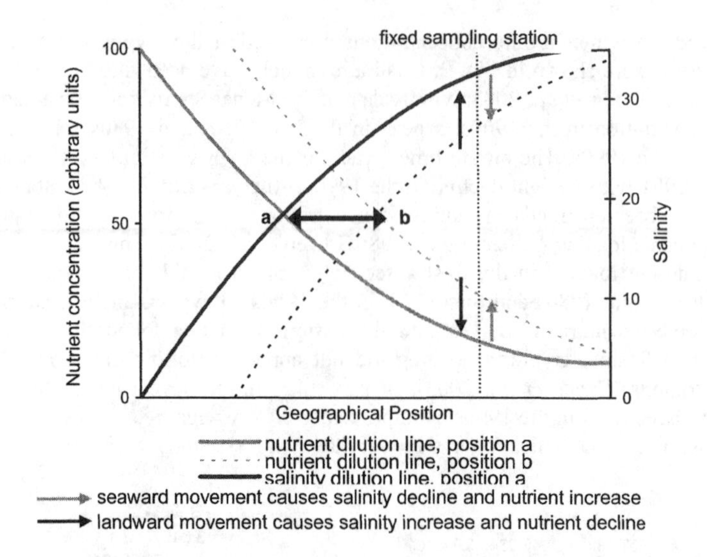

Figure 3: Conceptualised spatial changes in nutrient concentration and salinity from position a to position b. The horizontal arrow indicates a shift in the position of the seawater/freshwater gradient, the vertical arrows the consequent shift in salinity and nutrient concentration. Figure courtesy of Evans *et al.*, 2003.

The geographical position of the mixing zone is governed by the balance between Atlantic inflow and riverine fluxes of fresh water. Shifts in this balance are climatic in nature, occurring through either variability in river discharge (and thus through variability in precipitation), or changes in Atlantic inflow. These environmental factors are linked to variability in the North Atlantic Oscillation (NAO) and the Gulf Stream so water mass shifts in the Irish Sea are likely to be associated with these larger scale influences.

A three-dimensional baroclinic model, the Proudman Oceanographic Laboratory Coastal Ocean Modelling System (POLCOMS), has been applied to the Irish Sea region to further investigate the role of climate in the observed variability of salinity and nutrient concentrations. In section (3) the relative influences of tides, winds, baroclinic flows and far field forcing on the transport and dispersion of a passive tracer will be discussed. Section (4) will present the results of a forty-year hindcast simulation. Comparisons with temperature and salinity measured at the Cypris station will be presented and the implications of these results for the simulation of nitrate concentration variability in the Irish Sea will be discussed.

2) Model description

POLCOMS is a three-dimensional finite difference primitive equation model formulated in spherical coordinates on an Arakawa (1972) B-grid with a terrain following sigma coordinate system employed in the vertical. Both temperature and salinity are treated as prognostic variables. Key aspects of the numerical scheme include: calculation of vertical diffusivities for momentum and scalars using a Mellor-Yamada-Galperin level 2.5 turbulence closure scheme; calculation of horizontal pressure gradients by interpolation onto horizontal planes, which significantly reduces the numerical accuracy commonly found in sigma-coordinates over steep topography; and use of the 'Piecewise Parabolic Method' (PPM), (James, 1996) for horizontal advection of physical variables, which is a conservative scheme that minimizes numerical diffusion and has excellent front preserving properties. The ability of this model to reproduce the observed barotropic and baroclinic variability on the northwest European shelf has been demonstrated in a range of applications, including Proctor and James (1996), Holt and James (2001) and Holt *et al.* (2001).

In this study, the model grid has a horizontal resolution of 1/20° longitude by 1/30° latitude, and has 25 layers in the vertical distributed evenly throughout the water column. The domain extends from 7° W to 2.65° W, and from 51° N to 56° N, and thus extends into the Celtic Sea in the south and beyond the North Channel of the Irish Sea in the north (Fig. 1). This domain thus limits the influence of the open boundary on predictions in the region of interest, the Irish Sea. Model bathymetry was obtained by smoothing a fine resolution (approximately 1 km) bathymetry digitized from Admiralty Fair Sheets (where available) and Admiralty Charts supplied by the U.K. Hydrographic Office (Brown et al., 1999).

The initial temperature field was extracted from a winter climatological temperature field of the northwest European shelf. The initial salinity field was derived by constructing a climatological, winter salinity distribution from ICES data. At the open boundary, imposed tidal elevations and depth mean currents were obtained from an analysis of the POL storm surge model (Flather et al., 1991) and interpolated to provide forcing terms at each open boundary point. Time series of residual elevations and currents, which represent the far field forcing, were extracted from hindcast simulations of this model for the 1960-1999 period. The storm surge model uses atmospheric pressures and surface winds from the Norwegian Meteorological office (DNMI). Therefore, to ensure consistency with the residual elevations and currents imposed at the open boundary, atmospheric pressure and wind fields for the forty year hindcast simulation of this Irish Sea model were also spatially and temporally interpolated from the DNMI data. Surface fluxes of heat were calculated using the bulk formulae discussed in Gill (1982). Short wave solar radiation was calculated from equations describing the dependence of solar radiation on the position of the sun, adjusted for cloud cover (as described in COHERENS user documentation, Luyten *et al.*, 1999). Spatially and temporally varying air temperature, relative humidity and cloud cover were interpolated from data provided by the National Centres for Environmental Prediction (NCEP, http://www.cdc.noaa.gov/)

Predictions of salinity were of particular importance in this study, thus the representation of open boundary values, precipitation and riverine fluxes of freshwater were of concern. Salinities and temperatures on the open boundaries of the model were interpolated from an average annual cycle of the POL northeast Atlantic S12 model, obtained by averaging the 1998-2002 simulations. Although the use of a model-predicted average annual cycle is not ideal, there were insufficient observational data available with which to derive spatially and temporally varying boundary forcing data. The potential impact of the representation of the open boundary salinity on model predictions is discussed in section (4). There were insufficient long term rainfall data to derive spatially varying precipitation fields for the forty year hindcast simulation. However, it was possible to construct daily precipitation rates by averaging the available data from 10 sites around the Irish Sea coast. Evaporation rates were calculated using the evaporative loss term of the bulk heat flux formulae.

Previous applications of POLCOMS to the northwest European shelf region have used a combination of daily discharge data and average annual fluxes to provide the freshwater riverine inputs (e.g. Holt and James, 2001). However, whilst Britain maintains a dense network of gauging stations (currently around 1400), the density varies considerably from region to region with relatively few stations in remote (e.g. western Scotland) or technically more challenging areas (e.g. in the flat terrain of East Anglia). Thus direct measurements account for only around 65% of the outflow from Great Britain. To obtain a more accurate representation of freshwater flows, some estimate of the contribution from the ungauged areas is required, including consideration of major rivers in ungauged basins, minor streams draining directly to the sea, groundwater outflows, and in a few areas (e.g. the Thames estuary) sewage effluent. Researchers at the Centre for Ecology and Hydrology (CEH), Wallingford, have developed an adjustment technique based on the observed association between river flow and catchment size. A detailed description of the technique may be found in Marsh and Sanderson (2003). In summary, the method requires the identification of basic areal units for which outflows need to be aggregated, known as Hydrometric Areas (HAs). The 97 HAs identified for mainland Britain are shown in Figure 4. For each of these, representative index catchments are identified and weighting factors are derived to allow for those outflows that are not measured. These take the initial first-order estimates of runoff based on catchment size and further refine them by weighting according to the relation between mean runoff from the gauged and ungauged areas. The resulting weighting factors were used as multipliers for the gauged river data. Daily data for the full 40 year period were not available for all the gauged rivers. For these rivers, a mean annual cycle was calculated from the available data and used to fill the gaps.

3) Process Studies

A series of model simulations were performed to investigate the roles of tidal, wind, density, and far field forcing on the transport patterns of nutrients in the Irish Sea. For these process orientated simulations, the nutrients were treated as passive tracers from sources with constant fluxes of 1 per day (arbitrary units) at the mouths of two significant riverine sources of nitrate in the eastern Irish Sea, the Mersey and the Ribble. The model was started from a state of rest and run for a period of two years, with output commencing after the first year, thus allowing a year for background concentration fields to develop. The simulation years were arbitrarily chosen as 1994 to 1995, and the river

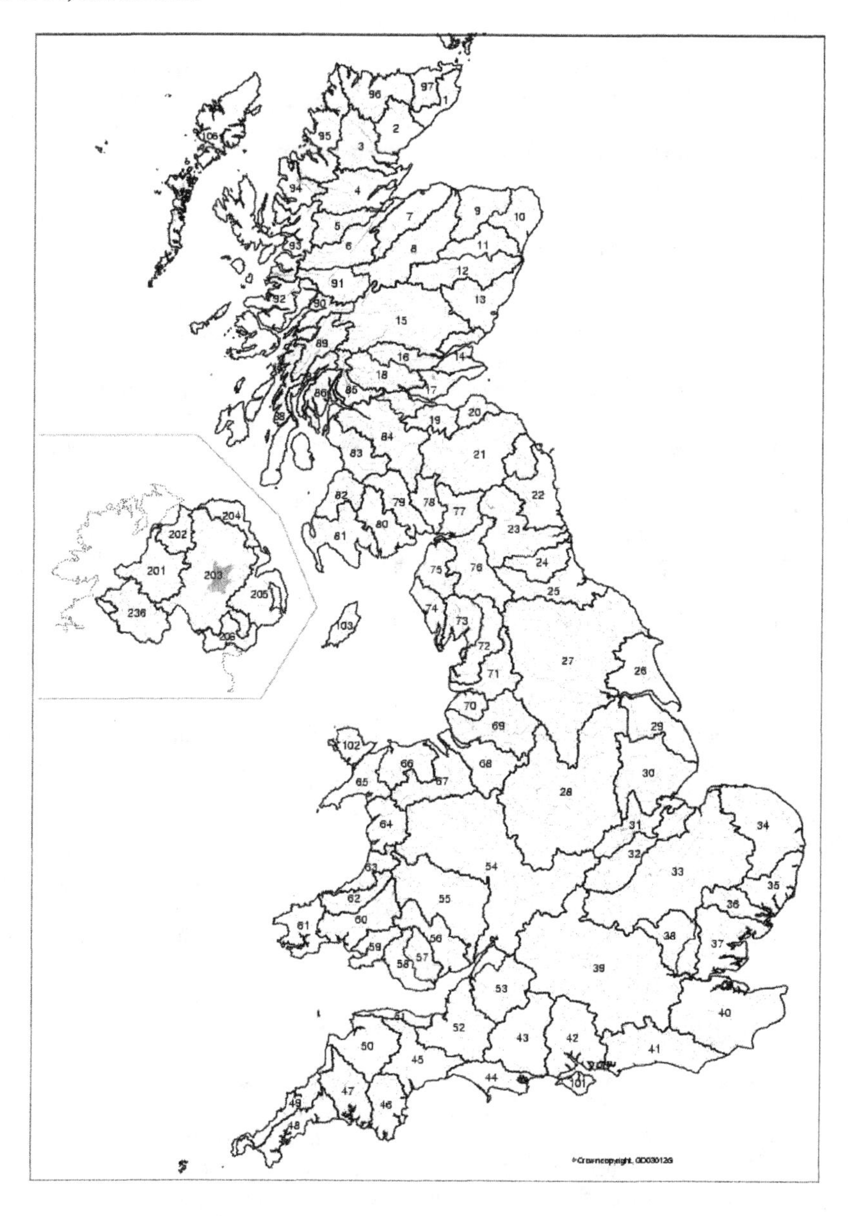

Figure 4: UK hydrometric areas. Each area is labelled with a unique number identifier (shown). Map courtesy of Marsh and Sanderson (2003).

flows, tide, atmospheric, and open boundary forcing were appropriate for these years. To simplify the interpretation of the results, evaporation and precipitation were neglected.

For the tidal only run, atmospheric and open boundary forcing, and river flows were set to zero, and the baroclinic components of the model were switched off. The resultant tracer distribution patterns indicated very little transport away from the source locations; virtually all the tracer stayed within the confines of the eastern Irish Sea, with just a weak plume transported anti-cyclonically towards Anglesey and then northwards to the east of the Isle of Man. These results indicate that tides have a negligible role in the observed nutrient concentration variability described in section 1.

The second simulation incorporated tide, wind and far field forcing (represented as a residual elevation and depth mean current on the open boundary), but with the baroclinic component of the model remaining switched off. The temporal evolution of the tracer concentration field displayed bursts of transport of material both westwards and (more often) northwards from the source locations. Tracer was dispersed throughout the Irish Sea, with a significant amount retained within the western Irish Sea. The residual transport pattern was northwards in the eastern Irish Sea, then westwards to the north of the Isle of Man, followed by either a transport northwards through the North Channel, and hence out of the modelled region, or southwards to the west of the Isle of Man, the latter eventually joining tracer in the eastern Irish Sea. Thus tracer was predicted to reach the locality of the Cypris station through two main pathways. Material that reached the western Irish Sea was more diffuse and delivered a lower, more constant concentration of tracer to the site. The occasional pulses of higher concentration material driven westwards from the source region were a more intermittent source. These pulses were primarily produced by direct wind forcing. Far field effects are unlikely to have had a large influence on these high frequency events as simulation 3, with no far field forcing, predicted similar pulses.

The third simulation neglected the far field forcing on the open boundary but included the baroclinic component of the flow, and freshwater inputs. The predicted retention within the Irish Sea was considerably reduced relative to the previous simulation, with far less tracer transported into the western Irish Sea. Thus the contribution of tracers in the western area contributed little to the concentrations near the Cypris Station site. The baroclinic flows acted to transport tracer northwards to the east of the Isle of Man, concentrated closer to the coast than the previous simulation. There was evidence of a distinct front in the southern half of the eastern Irish Sea between high concentration near-coastal waters and Atlantic water transported north through St George's Channel with negligible tracer concentrations. This front moved back and forth in a northeast-southwest direction, driven by variability in the wind forcing. Thus the front intermittently impacted the Isle of Man region, and pulses of higher tracer concentration at the Cypris station were observed. The variability in the daily river flows appeared to have little impact on these intermittent pulses, but gross increases in river flow (e.g. by doubling the flow rates) generated a general westward shift of the tracer concentration front.

The final simulation incorporated the full suite of tide, wind, density and far field forcing. The results showed an interesting combination of simulations 2 and 3; the front generated by the baroclinic flows in the eastern Irish Sea was still evident, but the inclusion of the far field forcing allowed more material to be retained in the western Irish Sea. Overall, more material was therefore retained in the Irish Sea than for simulation 3, but the inclusion of the baroclinic flows caused more material to be transported northwards through the North Channel than for simulation 2. The pulses of high concentration were again predicted at the Cypris station site, driven primarily by direct wind forcing, whilst the retention of material in the western Irish Sea due to far field effects generated a weak background concentration at the site.

These simulations have demonstrated that winds and far field forcing could have a significant influence on the high frequency variability of nutrient concentrations at fixed sampling sites in the Irish Sea, and on the retention of material. Baroclinic flows, in part due to freshwater inputs, generate a front in the eastern Irish Sea with a strong gradient in nutrient concentration. This will contribute to the appearance of pulses of higher concentration material at fixed sampling sites, as the front is transported back and forth. Whilst high frequency (daily) variability in river flows did not appear to have a large influence on the predicted tracer distributions, gross increases in river flows caused a westward shift in the tracer concentration front, and would therefore cause an apparent increase in observed concentrations at the Cypris station site.

4) Hindcast simulations

The observational analyses by Evans *et al.* described in section (1) found an inverse relationship between salinity and winter nitrate concentration in the Irish Sea. Thus by simulating and understanding processes affecting long term salinity variability, it is possible to gain insight into the dominant physical (climatic) influences on nutrient concentrations.

A fortyyear hindcast simulation has been completed for the period 1960-1999. All observational (CTD) data collected in the modelled region over these years were obtained from the British Oceanographic Data Centre (BODC) to perform a statistical evaluation of the model accuracy. The resulting statistical comparison of observed and modelled temperatures is shown in Table 1 (5566 CTD profiles). A comparison of modelled and observed near-surface temperatures at the Cypris station is illustrated in Fig. 5, where the observations include sea surface temperatures (SST) derived from AVHRR satellite images. With overall mean and RMS errors of 0.002 °C and 0.773 °C respectively, it can be seen that the model does an excellent job of reproducing observed temperatures throughout the model's domain over the 40 year period. The model also succeeds in simulating the observed variability at Cypris station, with just a limited tendency to underpredict the observed winter temperatures.

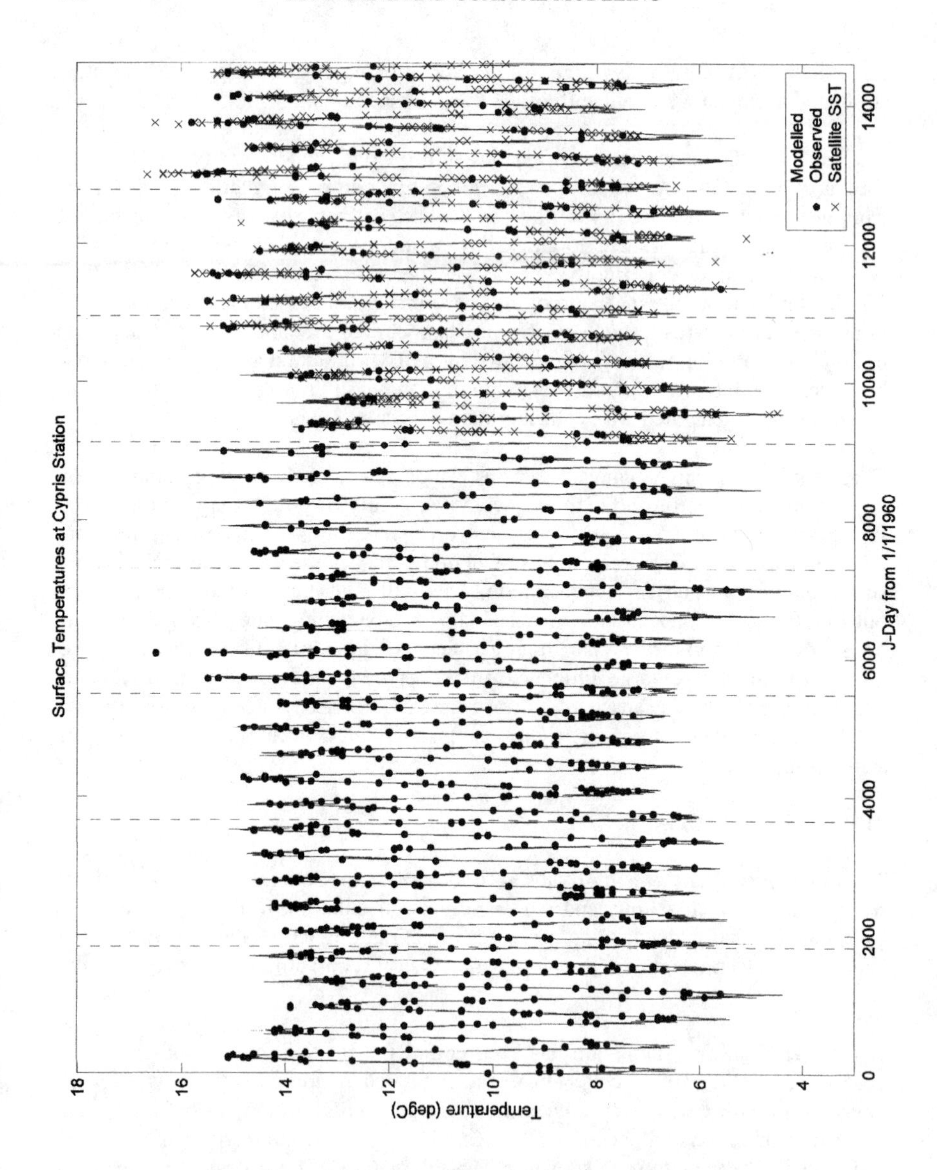

Figure 5: The results of a 40 year hindcast simulation: comparison of predicted and observed near-surface temperatures at the Cypris station.

The statistical comparison of observed and modelled salinities is shown in Table 1. With overall mean and RMS errors of 0.672 and 1.033 respectively, it can be seen that the model does not reproduce the observed salinities as accurately as the temperatures and is generally too salty. A comparison of modelled and observed near-bed salinities at the Cypris station is illustrated in Fig. 6. Here, the monthly mean salinities are presented, with the annual mean salinities overlaid, as the high frequency variability in the modelled daily data made comparison and interpretation difficult. Again it can be seen that the model tends to be too salty for the majority of the 40 year period. This is perhaps not surprising; this version of POLCOMS does not include a wetting and drying algorithm, thus a minimum water depth of 10 m is prescribed. In the near coastal regions of the eastern Irish Sea where the tidal range exceeds 8 m and there are extensive tidal flats, this prescribed depth is deeper than the observed bathymetry. Thus, the saltwater volume with which the river water is mixed in the grid cells at the river mouths will be too high. Even assuming the imposed freshwater fluxes are accurate, this will result in a higher predicted salinity than observed. The incorporation of a wetting and drying algorithm into POLCOMS has recently been completed and it is intended to test this hypothesis by re-running the 40 year hindcast with more accurate near-coastal bathymetry.

	TEMPERATURES	
	Mean	*RMS*
All points	0.002	0.773
Near-surface	-0.049	0.718
Near-bed	0.007	0.850
	SALINITY	
	Mean	*RMS*
All points	0.672	1.033
Near-surface	0.588	1.073
Near-bed	0.689	1.015

Table 1 : Statistical comparison of observed temperatures and the corresponding predicted values for the simulated period, 1960-1999 (5566 CTD profiles). Results presented as mean error and root mean square (RMS) error.

However, although the model is generally too salty, many of the observed trends in interannual variability are reproduced. For example, the sections marked on Fig. 6 between Jdays 500 and 6000 (**A**), and between Jdays 12500 and 14500 (**D**) show very similar trends in the modelled and observed data. Concurrent peaks at Jdays 7000 (**B**) and 9300 (**C**) are also marked. This suggests that for much of the 40 year period, whilst the model may not reproduce the absolute values of observed salinity, it contains the dominant physical processes for the simulation of salinity variability, and thus of physical influences on nutrient variability.

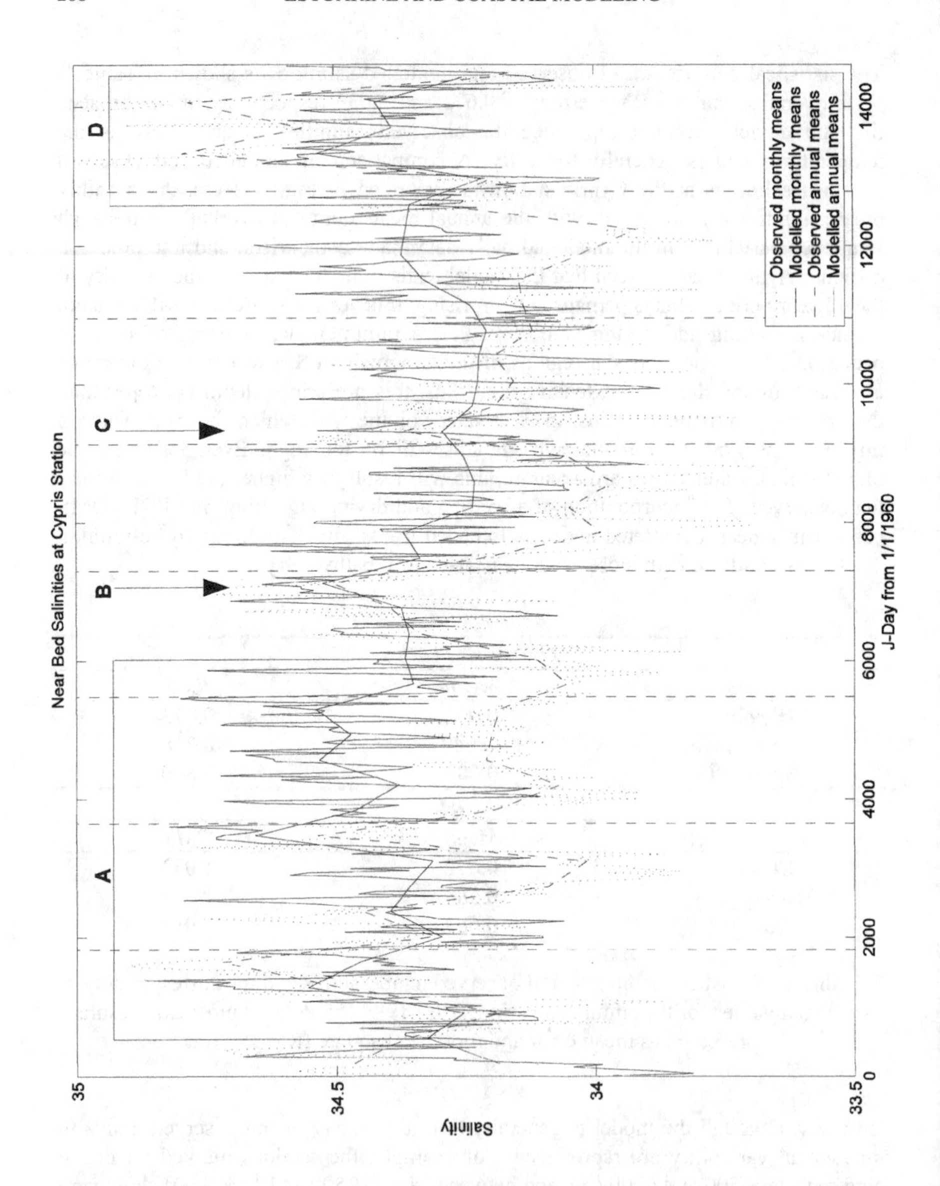

Figure 6: Results from a 40 year hindcast simulation: comparison of predicted and observed mean monthly, and mean annual near-bed salinities. A-D indicate areas of good agreement, referred to in the text.

Although many of the observed trends in salinity variability are reproduced, the model predictions between Jdays 6000 and 12500 are not generally a good representation of the variability in the observed salinity. It is therefore important to consider the possible sources of error in the predictions of long term salinity variability. There are three main influences on long-term salinity predictions at the Cypris station, namely:

1) Freshwater inputs.
2) Evaporation-precipitation (E-P) balance.
3) Far field variability in salinity.

The method used to prescribe accurate freshwater inputs in the model was described in section (2). Although this provides excellent estimates of the freshwater flux for the temporal periods when gauged river data exists, the substitution of an annual average cycle when data are unavailable is not ideal, as it neglects interannual variability. Simulations using only average annual cycles found the observed salinity at the Cypris station was very poorly reproduced so the inclusion of the interannual variability clearly has a very significant influence on model predictions. However, the approach used, whilst it has limitations, is the best available at present.

As discussed in section (2), precipitation in the model is temporally varying, but spatially constant. The inclusion of E-P had a significant influence on reducing the general over-prediction of salinity in the model, so it is clearly important to include this forcing term. However, it is not clear what effect neglecting the spatial variability of the precipitation has on model predictions. Unfortunately, there is insufficient data available with which to derive the spatial variability so this cannot be assessed.

The residual transport through the Irish Sea region is northwards, thus water impacting the Irish Sea originates further south and is transported through St George's Channel. Variability in the salinity of the waters to the south could therefore influence observed salinity at the Cypris station. A time series of observations exists for the period 1960-1987 from a location in the Celtic Sea, just off the southwest tip of England. A comparison of the observed salinity variability at this site with that observed at the Cypris station was performed, using time series analysis and including consideration of the transport time between the two sites. For the range of temporal lags considered (up to 100 days), the maximum correlation was 0.169. This suggests that observed salinity at the Cypris station is not correlated with the observed salinity in the Celtic Sea. It is possible that it is correlated with the salt flux, rather than the salinity. However, there are no long term measurements of volume fluxes through St. George's Channel, so this hypothesis cannot be tested.

Volume fluxes through St George's Channel could also be influencing observed salinity at the Cypris station in a more indirect manner. The Eulerian nature of the observations discussed in section (1) suggests that inaccurate predictions in the position of the saltwater/freshwater boundary could influence model/observed data comparisons. The long term position of this boundary is governed by the balance between the volume fluxes of saltwater through St George's Channel, and of freshwater from coastal sources. Thus inaccuracies in the representation of the far field forcing in the model could affect the model predictions of salinity variability at the Cypris station. In section (3) it was concluded that the high frequency variability was predominantly caused by direct wind forcing. Long term shifts in predominant weather patterns, and in wind direction, could therefore also exhibit as interannual variability in observed salinity.

Summary

The modelling study described in this paper has investigated the physical influences on salinity and nutrient concentrations in the Irish Sea, including the role of climate variability. Previous observational work had noted an inverse relationship between salinity and nitrate concentrations in the Irish Sea, and highlighted the effect of moving water mass boundaries on observations at a fixed sampling site. The process studies described here have demonstrated how high frequency pulses of lower salinity, high nutrient concentration waters could be observed at a fixed sampling site near the Isle of Man (the Cypris station), mainly through the effects of direct wind forcing. Far field forcing for the two arbitrary years used in the study acted to retain material in the western Irish Sea, thus providing a weak background concentration of nutrients to the sampling site. Although daily variability in river flows did not appear to have a significant influence on the high frequency variability at the Cypris station, an arbitrary doubling of river inflows caused a shift to the southwest of the water mass containing higher tracer concentrations. Thus longer term changes in freshwater inflows, for example through increases in precipitation in positive NAO years, could have a significant effect on the decadal scale salinity variability observed at the Cypris station by causing a westwards movement of the boundary between fresh and saltwater masses.

To further investigate influences on the long term variability of salinity at the Cypris station, a 40 year hindcast simulation was completed for the period 1960-1999. A comparison of modelled temperatures with observed values from 5566 CTD profiles throughout the 40 years, demonstrated the ability of the model to reproduce the observed seasonal and longer term cycles, with mean and RMS errors of 0.002 °C and 0.773 °C. The long term temperature variability at the Cypris station was also reproduced. The predicted salinities were generally too high, with mean and RMS errors of 0.672 and 1.033, probably due to the artificially deep grid cells in the near coastal regions of the eastern Irish Sea. However, many of the trends in the observed salinity variability were reproduced, suggesting that the dominant physical processes influencing salinity and nutrient concentration variability in the Irish Sea are generally well represented in the model. With the observed inverse correlation between salinity and nitrate, the fact that the model resolves many of the trends in salinity variability suggests that it will also resolve those in the nutrient concentrations. This provides further evidence for the premise of Evans *et al.* (2003) who suggest that climate variability may be a significant influence on the nutrient concentrations observed at the Cypris station, although they do not disregard the effect of changes in anthropogenic loading.

These results suggest that observations taken at fixed sampling sites in active environments such as the Irish Sea should be treated with a degree of caution. Although anthropogenic loading undoubtedly has some influence on nutrient concentrations, much of the observed variability may be due to climatic variability, in particular the combined effects of freshwater inflows (dependent on precipitation) and far field forcing of saltwater masses on the geographical position of the water mass boundary.

Acknowledgements

The bathymetry used for this work was funded by the Department of Transport and the Regions and the Ministry of Agriculture Fisheries and Food (Brown et al., 1999). AVHRR data were obtained from the NASA Physical Oceanography Distributed Active Archive Center at the Jet Propulsion Laboratory, California Institute of Technology. NCEP Reanalysis data was provided by the NOAA-CIRES Climate Diagnostics Center, Boulder, Colorado, USA, from their Web site at http://www.cdc.noaa.gov/ .

References

Allen, J.R., Slinn, D.J., Shammon, T.M., Hartnoll, R.G. (1998). Evidence for eutrophication of the Irish Sea over four decades. *Limnology and Oceanography* 43, 1970-1974.

Arakawa, A. (1972). Design of the UCLA general circulation model. *Tech. Rep.* 7, Univ. of Calif., Los Angeles.

Bowden, K.F. (1980). Physical and dynamical oceanography of the Irish Sea. In F.T. Banner, W.B.Collins, & K.S. Massie (Eds.), *The North West European Shelf Seas: The sea bed and the sea in motion, Physical and chemical oceanography and physical resources* Vol. II (pp. 391-413). Amsterdam: Elsevier.

Brown, J., Joyce, A.E., Aldridge, J.N., Young, E.F., Fernand, L., Gurbutt, P.A., (1999). Further identification and acquisition of bathymetric data for Irish Sea modelling. DETR research contract CW0753.

Evans, G.L., Williams, P.J. le B., Mitchelson-Jacob, G. (2003). Physical and anthropogenic effects on observed long-term nutrient changes in the Irish Sea. *Estuarine, Coastal and Shelf Science* 57, 1159-1168.

Flather, R.A., R. Proctor, J. Wolf (1991). Oceanographic forecast models. In: *Computer Modelling in the Environmental Sciences*, D.G. Farmer and M.J. Rycroft, editors, IMA Conference Series 28, Clarendon Press, Oxford, pp.15-30.

Foster, C. (1984). Nutrient distributions in the winter regime of the northern Irish Sea. *Marine Environmental Research* 13, 81-95.

Gill, A.E. (1982). Atmosphere-Ocean Dynamics. Academic Press, New York, 662pp.

Gowen, R.J., Mills, D.K., Trimmer, M., Nedwell, D.B. (2000). Phytoplankton production and its fate in two coastal regions of the Irish Sea: the influence of anthropogenic nutrients. *Marine Ecology Progress Series* 208, 51-64.

Gowen, R.J., Hydes, D.J., Mills, D.K., Stewart, B.M., Brown, J., Gibson, C.E., Shammon, T.M., Allen, M., Malcolm, S.J. (2002). Assessing trends in nutrient concentrations in coastal shelf seas: a case study in the Irish Sea. *Estuarine, Coastal and Shelf Science* 54, 927-939.

Holt, J.T., James, I.D. (2001). An *s* coordinate density evolving model of the northwest European continental shelf 1, Model description and density structure. *Journal of Physical Oceanography* 106 (C7), 14015-14034.

Holt, J.T., James, I.D., Jones, J.E. (2001). An *s* coordinate density evolving model of the northwest European continental shelf 2, Seasonal currents and tides. *Journal of Physical Oceanography* 106 (C7), 14035-14053.

James, I.D. (1996). Advection schemes for shelf sea models. *Journal of Marine Systems* 8, 237-254.

Luyten, P.J., Jones, J.E., Proctor, R., Tabor, A., Tett, P., Wild-Allen. K. (1999). COHERENS - A Coupled Hydrodynamical-Ecological Model for Regional and Shelf Seas : User Documentation. MUMM report, Management Unit of the Mathematical Models of the North Sea, 914pp.

Marsh, T.J., Sanderson, F.J. (2003). Derivation of daily outflows from Hydrometric Areas. Report prepared for the Proudman Oceanographic Laboratory, National River Flow Archive, CEH Wallingford, July 2003, 14 pp.

OSPAR Commission (2001). *Data report on the comprehensive study of riverine inputs and direct discharges (RID) in 1999*. Assessment and monitoring (21 pp.). Oslo and Paris Commissions, ISBN 0946956618.

Proctor, R., James, I.D. (1996). A fine-resolution 3D model of the southern North Sea. *Journal of Marine Systems* 8, 285-295.

THE POL COASTAL OBSERVATORY – Methodology and some first results

Roger Proctor, John Howarth, Philip J. Knight[1] and David K. Mills[2]

ABSTRACT

We describe progress with a pilot Coastal Observatory (2002-2006) in Liverpool Bay (Eastern Irish Sea, UK) that integrates (near) real-time measurements with coupled models in a pre-operational coastal prediction system. The aim is to understand a coastal sea's response to natural forcing and the consequences of human activity. The foci are the impacts of storms, seasonality, and variations in river discharge (freshwater and nutrients) on the functioning of Liverpool Bay. The present measurement suite includes: moorings for high frequency in-situ surface waves, temperature, salinity, turbidity, nutrients and chlorophyll; vertical profiles of current; 4-6 weekly regional surveys with CTD, suspended particulate matter and nutrient measurements; an instrumented ferry measuring surface properties (temperature, salinity, turbidity and chlorophyll; coastal tide gauges; satellite data – infra-red (for sea surface temperature) and visible (for chlorophyll and suspended sediment). A suite of nested 3-dimensional models (the Proudman Oceanographic Laboratory Coastal Ocean Modeling System - POLCOMS) is run daily, focusing on the Observatory area by covering the ocean/shelf of northwest Europe (at 12 km resolution) and the Irish Sea (at 1.8 km). All measurements and model outputs are displayed on the Coastal Observatory web-site (coastobs.pol.ac.uk). Initial ways in which this system is being used to support an ecosystem-based approach to marine management are described.

INTRODUCTION

Coastal ocean observing systems are now technically feasible (e.g. Glenn et al., 2000 on the east coast of the USA; Blaha et al., 2000 in the Gulf of Mexico; Buch and Dahlin, 2000, in the Baltic Sea; IEEE J. Oceanic Eng, vol 27 for a summary of recent developments in the US, Edson et al., 2002). No single organisation in the UK (or Europe) alone has sufficient capability to design or support such a system to address the full range of marine issues (Prandle and Lane, 2000). POL, however, is uniquely placed in the UK, through its measurement and modeling capabilities and existing interactions with key agencies, to act as the focus for the development of a pilot coastal zone observing and monitoring system. This represents a major UK innovation in the approach to testing process understanding in shelf seas.

Through collaborations with the UK Met Office (operational numerical weather predictions (NWP) and ocean/shelf circulation models), the Environment Agency

---

[1] Proudman Oceanographic Laboratory, Bidston Observatory, Wirral CH43 7RA, UK

[2] Centre for Environment, Fisheries and Aquaculture Science, Lowestoft, NR33 0HT, UK

(routine monitoring of river discharge, nutrients and contaminants) and Natural Environment Research Council (NERC) facilities (operational production of regional remote sensing data) the necessary inputs required by a pre-operational coastal ocean modeling system can be secured. Regulatory bodies, namely Centre for Environment, Fisheries and Aquaculture Science (CEFAS) and the Environment Agency (EA), are partners in the project through deliberate alignment of their own measurement programmes with the Coastal Observatory.

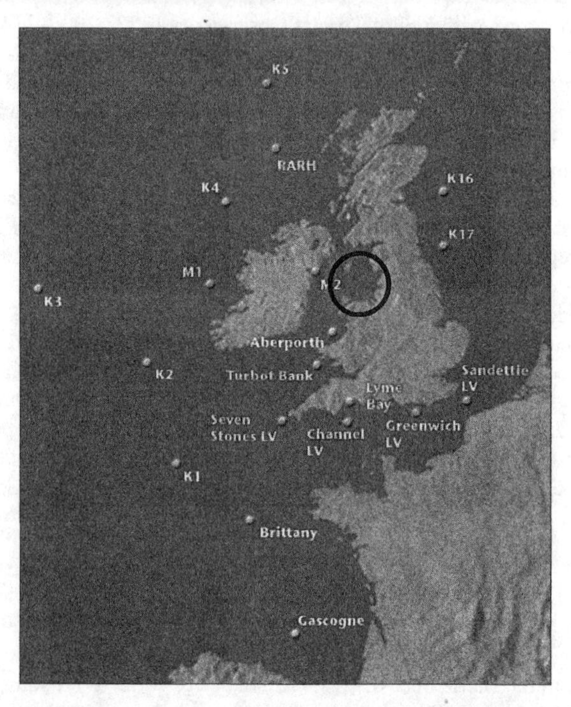

Figure 1: North West European continental shelf showing Coastal Observatory domain in Liverpool Bay, Irish Sea (circled). Also shown are the Met Office Moored Automatic Weather Stations (MAWS) used to validate the operational ocean/shelf model.

THE PILOT COASTAL OBSERVATORY

The Observatory is located in Liverpool Bay in the Irish Sea (Figure 1). This region is an archetypal coastal sea system with strong tides, occasional large storm surges and waves, freshwater input (the rivers Mersey, Dee and Ribble), stable and unstable stratification, intertidal regions with exposed banks, high suspended sediment concentration and biogeochemical interaction. The Bay is also stressed (near eutrophication from river-borne nutrients and subject to river-borne pollutants) and so of concern to regulatory agencies. The Irish Sea is connected to the North West European Shelf through the North Channel and St. Georges Channel in the south and so experiences interaction and exchanges with the North East Atlantic Ocean.

The Observatory integrates (near) real-time measurements with coupled models into a pre-operational coastal prediction system whose results are displayed on a website (http://coastobs.pol.ac.uk). The concept is founded on obtaining data in (near) real-time, using acoustic transmission to relay the underwater data to the sea surface then telemetry (via ORBCOMM) to relay the data to POL and finally to a website ('armchair oceanography'). This, the aspiration of every oceanographer, is now feasible, and there are an increasing number of prototype observing systems, employing a range of technological solutions, around the world.

The emphasis is on a modest, pragmatic approach in the initial establishment of the Observatory (drawing on existing technology and partnerships). The Observatory will evolve as the concept and its effectiveness becomes established in the UK context, all the while building up long time series of measurements. The focus of the Observatory will be on the impacts of storms, variations in river discharge (especially the Mersey), seasonality, and blooms in Liverpool Bay.

MEASUREMENTS

An extensive array of different measurement techniques are planned to complement each other and provide information on the four-dimensional variability in Liverpool Bay (Figure 2). The first measurements were started in August 2002.

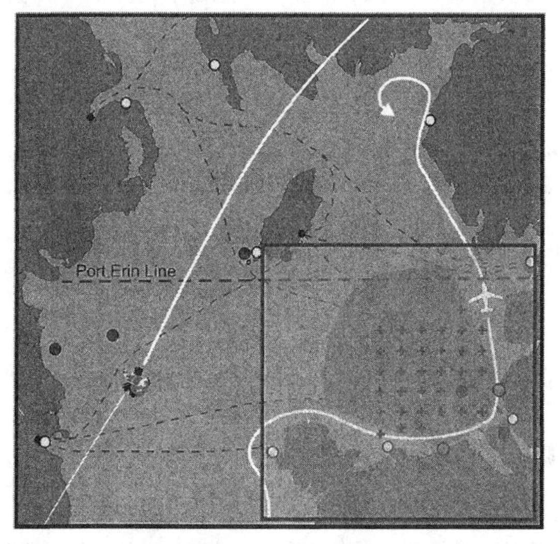

Figure 2. Planned Irish Sea monitoring system. Yellow dots = tide gauges; red dots = moorings; red dotted lines = ferry routes; blue dotted lines = university monitoring routes; shaded area = HF radar coverage; black crosses = CTD, SPM, nutrients survey points; red square = meteorological station; white lines = airborne (satellite, lidar) monitoring. Red box highlights the focus of the Coastal Observatory in Liverpool Bay.

The present measurement array includes:

- A central mooring at 53° 32' N, 3° 21.8' W, installed in August 2002, providing *in situ* time series of current, temperature and salinity profiles. An expanded measurement set (including turbidity and chlorophyll) is eventually planned. At present data is not available in real-time, being stored and then downloaded on the routine maintenance cruises carried out by the RV Prince Madog. Sub-surface acoustic transmission has been successfully tested and installation is in progress so that the data can be transmitted in real-time.

- A CEFAS SmartBuoy (www.cefasdirect.co.uk/monitoring), installed in November 2002. This records surface properties including salinity, temperature, turbidity, nutrients and chlorophyll and transmits the data in real-time.

- A WaveNet directional wavebuoy was also installed in November 2002. This transmits spectral wave components in real-time (www.cefas.co.uk/wavenet). Both the SmartBuoy and the WaveNet are located adjacent to the central mooring.

- Mooring maintenance is carried out by the RV Prince Madog at approximately six week intervals (four weeks in the summer to overcome biofouling of the SmartBuoy sensors). Spatial surveys of Liverpool Bay are carried out each cruise (black crosses in Fig. 2, comprising 34 vertical profiles of CTD, SPM, some bed sediment sampling, and, recently, nutrients).

- The Liverpool - Isle of Man ferry ('The Lady of Mann' operated by the Isle of Man Steam Packet Company) has been equipped with instruments for near surface (5m depth) temperature, salinity, turbidity, chlorophyll. The measurements began in January 2003. This ferry is one of the nine ferry routes under study in the EU FERRYBOX project (www.ferrybox.com). This ferry boat only runs from September to March each year, so a second ferry route that is operational all year (Liverpool – Belfast, Northern Ireland, operated by Norse Merchant) is presently being equipped with a measurement capability.

- The UK Tide Gauge Network has been upgraded to allow real-time transmission of data from all tide gauges around the UK. The tide gauges in the Irish Sea, with additional sensors for met, waves, temperature and salinity where appropriate, are being incorporated into the Observatory.

- Bidston Observatory has been a meteorological recording station since 1867. This provides local real-time weather information (atmospheric pressure, wind speed and direction, cloud cover, rainfall). Unfortunately, the impending move of POL away from the Observatory to the Liverpool University campus (in Liverpool) will result in this series ending. A new station on Hilbre Island in the Dee Estuary, 10km to the north west of the Observatory, is planned.

- Satellite data - infra-red (for sea surface temperature) and visible (for chlorophyll and suspended sediment) is provided by the Remote Sensing Data Acquisition Service (RSDAS) of NERC. Daily and weekly composite images are obtained via ftp but the extensive cloud cover over the Irish Sea means the daily images are often of little use. Weekly composite images are usually 90-100% complete.

INITIAL OBSERVATIONS

The seasonal cycle of temperature and salinity at the mooring site can be seen in Fig. 3.

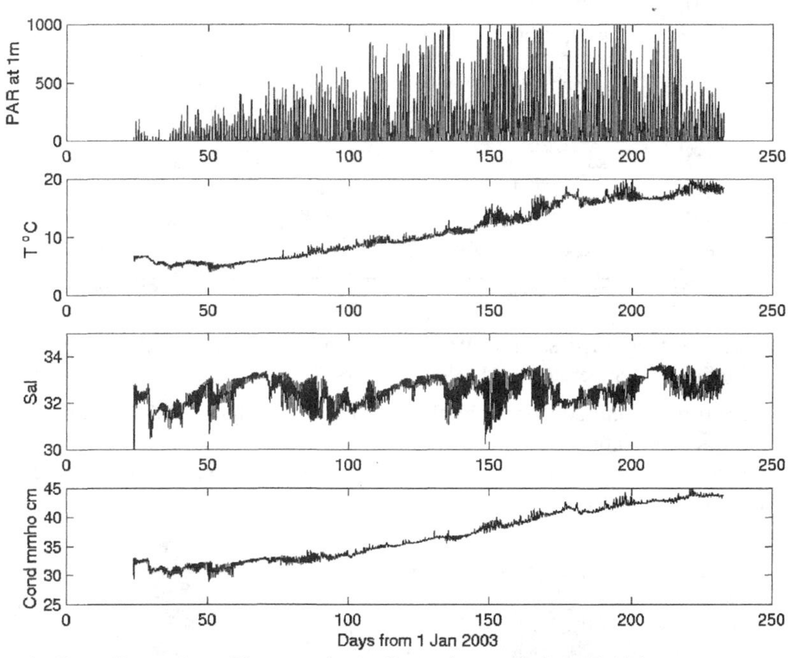

Figure 3: SmartBuoy data: Photosynthetically active radiation (PAR), temperature (T), salinity (Sal), Conductivity (Cond)) from 1 January – 20 August 2003.

This figure indicates the seasonal cycle of surface temperature (minimum in February, maximum in August) with a range of ~15°C. The tidal currents are strong at the mooring location with speeds of 1m/s during spring tides, resulting in a well-mixed water column for most of the year, although vertical stratification can occur on neap tides. The salinity record shows fluctuations around 32 within a +/- 2 psu band, but with no clear seasonal cycle. Both salinity and temperature are important for controlling the water density. Close examination of the period 25 July – 20 August 2003 (Fig. 4) reveals strong tidal advection in both temperature and salinity signals, with variations within a tidal cycle of up to +/- 1 °C and +/- 0.5 psu. These oscillations indicate the influence that the Dee / Mersey River outflows have upon the water column structure in this area.

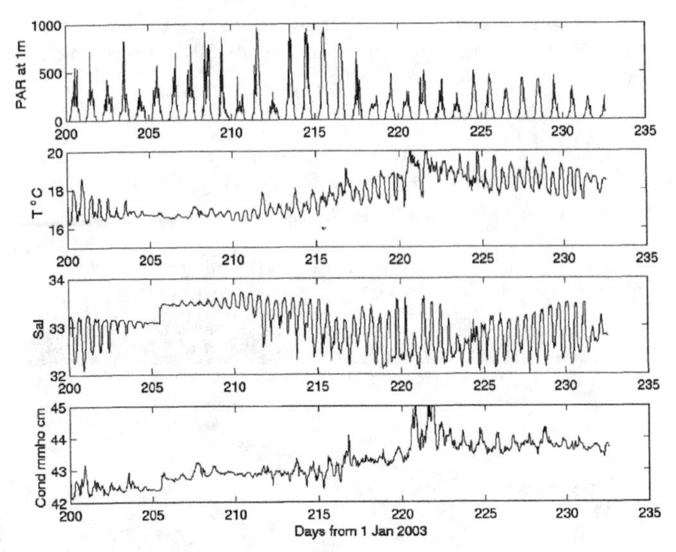

Figure 4: Magnified section of fig. 3 (25 July – 20 August 2003) showing tidal oscillations in temperature and salinity.

The current structure recorded by the ADCP during August 2003 (fig. 5) indicates an estuarine circulation – surface currents offshore (mean value 3 cms^{-1}), bottom currents onshore (mean value 3.5 cms^{-1}), as expected close to the Mersey estuary at this time of the year.

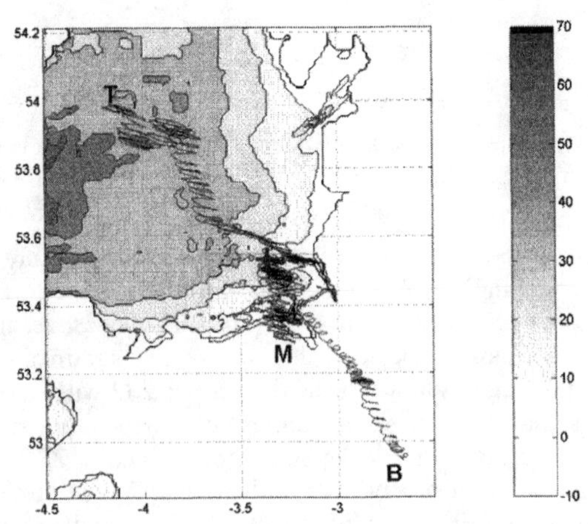

Figure 5: Progressive vector diagram, from 25 July to 20 August 2003, of ADCP current structure throughout the water column, showing outflow near the surface (T) and inflow near the bed (B). (M=mid-water vector). Shaded areas indicate bathymetry in metres. Horizontal axis is longitude (°E), Vertical axis is latitude (°N).

Fine scale surface temperature, salinity, turbidity and chlorophyll can be obtained from the ferry measurements. The Liverpool – Isle of Man ferry records data every 10 seconds, which equates to a spatial resolution of less than 100 metres. An example of the fine scale structure obtainable is shown in Fig.6, where the Mersey river plume is clearly identifiable as salinities drop below 33 psu.

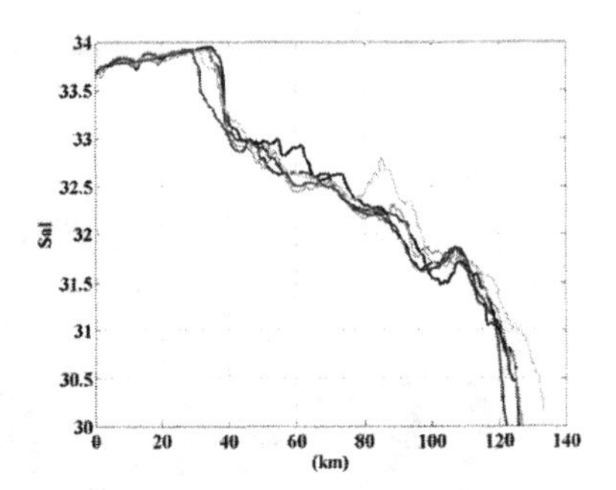

Figure 6: Salinity measured on 5 crossings between Douglas, Isle of Man (0 km), on the left, and Liverpool (120 km), on the right, on 10 and 11 January, 2003.

MODELING

The Coastal Observatory uses POLCOMS (the **P**roudman **O**ceanographic **L**aboratory **C**oastal **O**cean **M**odeling **S**ystem, see www.pol.ac.uk/home/research/polcoms), a 3-dimensional modeling system whose main elements, schematised in Figure 7, are a 3-dimensional baroclinic hydrodynamic model (Holt and James, 2001) linked to a surface wave model, WAM (Wolf *et al.*, 2002), a sediment resuspension and transport model (Holt and James, 1999) and an ecosystem model (ERSEM - European Regional Seas Ecosystem Model (Baretta et al., 1995)) with benthic and pelagic components.

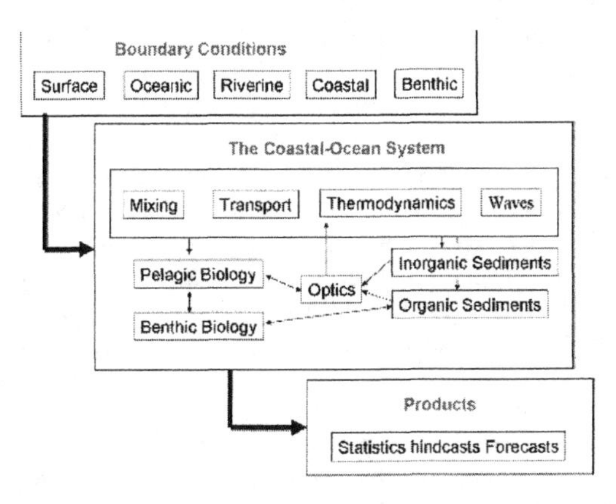

Figure 7: Schematic of POLCOMS system.

This modeling system has been developed over the last 4 years primarily to investigate physical-biogeochemical interactions in shelf seas, see for example Allen *et al.* (2001), Proctor *et al.* (2002, 2003) for a fuller description of the system.

In collaboration with the UK Met Office nested 3-dimensional models will cover the ocean/shelf of the northwest Europe (12km resolution), Irish Sea (1.8km) and Liverpool Bay (at 100-300m resolution), Fig. 8.

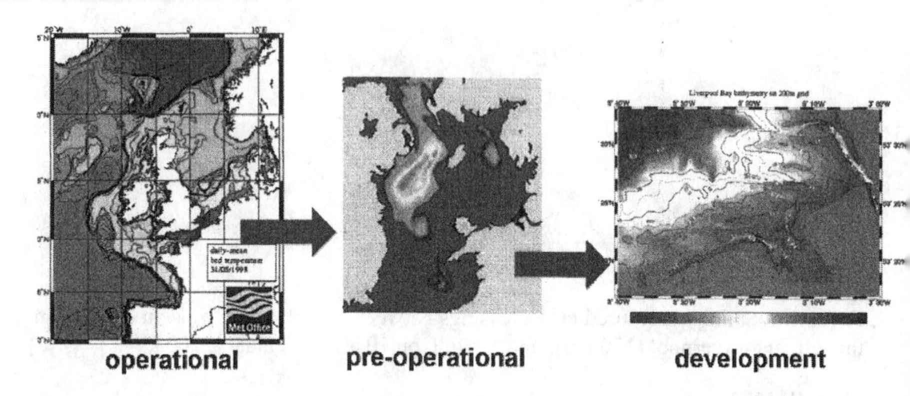

Figure 8: Nested set of POLCOMS models. Left = Met Office ocean-shelf (12km) operational model, centre = Irish Sea (1.8km) pre-operational model, right = Liverpool Bay (200m) model under development.

At the Met Office POLCOMS on an ocean/shelf domain (20W-15E, 40-65N), driven by surface fluxes from the Met Office mesoscale (12km) unified meteorological model and lateral ocean fluxes from the North Atlantic 1/3rd degree FOAM (Forecast Ocean Assimilation Model, Cattle *et al.*, 1998), has been operational since December 2002. This model produces daily hindcasts (24-hour) and forecasts (48-hour) and provides the boundary conditions (temperature, salinity, elevations, currents) for the Irish Sea model. At present the Irish Sea model, brought on-line in July 2003, runs in pre-operational mode, with a planned move to operational status on the POL cluster in 2004. Models run daily in near-real time, either at the Met Office or at POL with the necessary forcing information transferred by ftp between the two computers. The Irish Sea model in turn provides boundary conditions to the Liverpool Bay model which is currently under development and includes drying and wetting processes to accommodate the significant intertidal regions which arise from the 12m spring tidal range. Local river discharges are planned for inclusion in real-time through a link-up to the Environment Agency river-flow network; at present monthly climatological inflows are prescribed.

Present implementation of the 3-dimensional baroclinic models for the ocean/shelf and Irish Sea domains involves standard sea level and current responses driven by tides, winds, surface heat and salt fluxes and climatological river discharges (e.g. Holt and Proctor, 2003).

An example of the comparison between observations and models is given in Fig. 9. This shows timeseries of pressure, surface temperature and bottom temperature for July/August 2003.

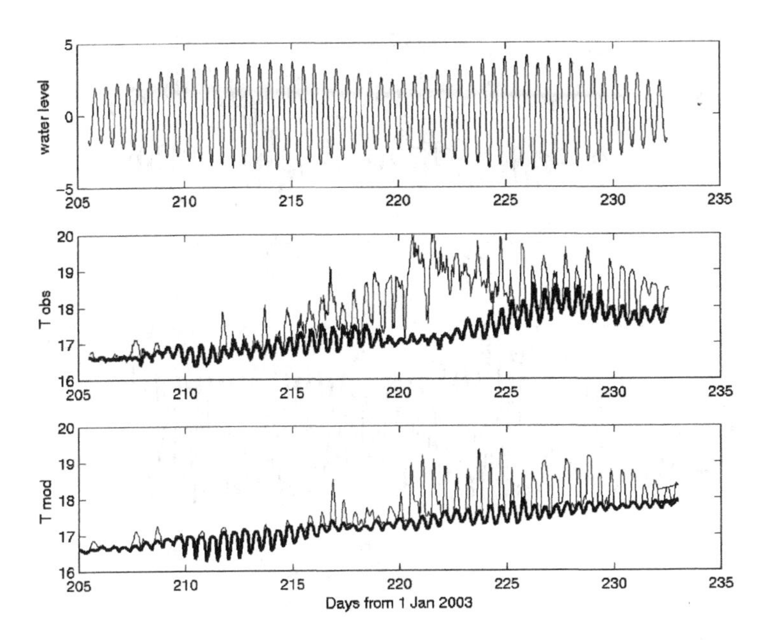

Figure 9: Observed surface pressure (shown as sea surface level in metres relative to mean sea level), observed (mid panel) and modelled (lower panel) surface (thin line) and bottom (thick line) temperatures (°C).

The surface elevation indicates the tidal range varying between ~8m at spring tides and ~6m at neap tides. The observed temperatures (surface temperature from the SmartBuoy, bottom temperature from the Microcat sensor on the ADCP mooring) indicate stable stratification at neap tide and well mixed conditions at spring tides. The model (the Irish Sea model, 1.8km resolution with 18 sigma levels in the vertical) only shows weak stratification at neap tides although the absolute temperature variations over the month agree quite well. It is interesting to see that model and observation are in good agreement for the first 10 days or so, then diverge. The salinity plot of the same variables (not shown) gives a similar picture, with model salinities developing larger surface tidal oscillations than seen in the observations. As yet we are unsure of the reason for this deviation but there are two likely contenders. One possibility is the fact that at present the model only has climatological river discharges, and that, as August was an unusually dry month the climatological input is too strong, leading to excess freshwater stratification. Clearly, real-time freshwater inputs are preferable. The second possibility arises because the model surface currents are stronger than those observed (fig. 10).

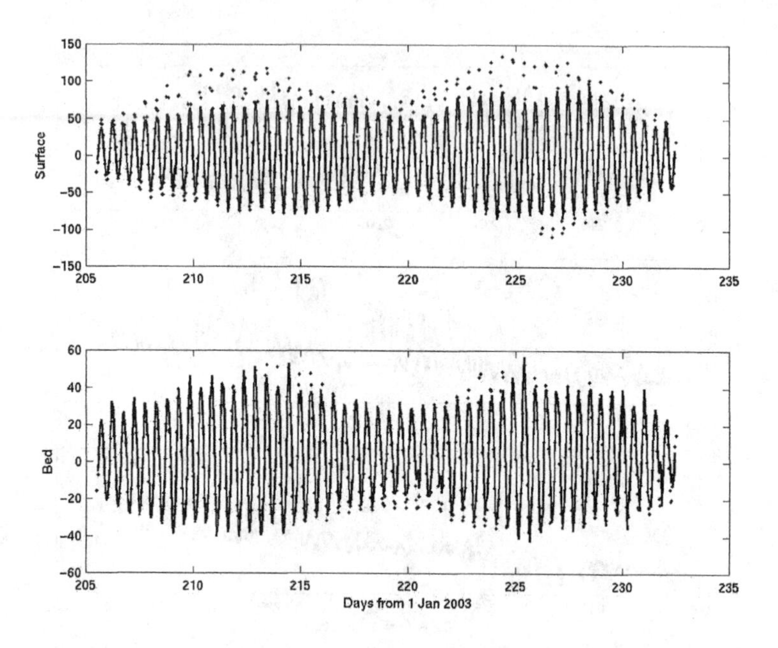

Figure 10: Surface and bottom east-directed currents during 25 July to 20 August 2003. Observed is solid line, model is dots. (Surface = 2m below surface, Bed = 2.65m above bed). Units = cms^{-1}.

Although bottom currents are in good agreement, modelled surface currents, especially flood tide (positive) currents, are significantly stronger than those observed. In fact, modelled surface currents at neap tide are as strong as those observed at spring tides, which is likely to be a contributory factor to the model not showing the same sustained stratification observed at neaps (fig. 9). Investigations into both of these factors continues. Understanding the surface current structure will be aided by the WERA HF radar system (16 MHz) purchased in mid-2003 which is currently being installed; this will have a 100km radius from Liverpool providing surface currents in 2km x 2km bins at 20 minute intervals and surface waves in 5km x 5km bins at 60 minute intervals, see Fig. 11. An X-band radar (9 GHz) with a range of 2km recording surface waves at 84 second intervals is also being installed as part of a near shore coastal sedimentation experiment in the Dee Estuary off Hilbre Island, Wirral.

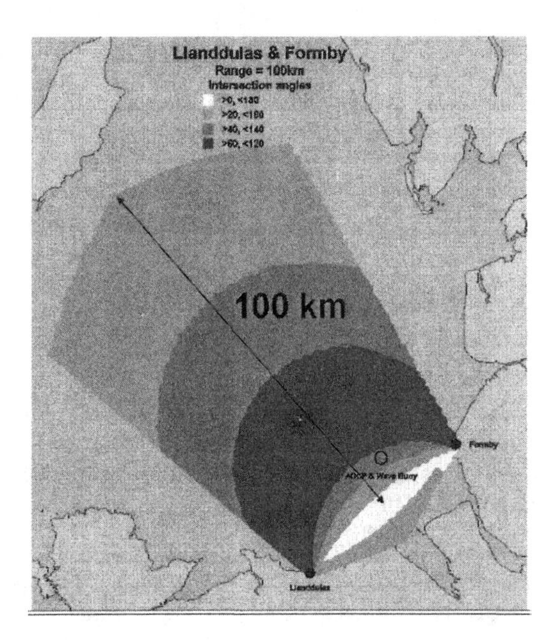

Figure 11: Planned HF radar coverage using sites at Llandulas and Formby.

WEB SITE

A major component of the Observatory is the dissemination of the results through a web site (http://coastobs.pol.ac.uk), both as figures and as data. The web site aims to target several different audiences – scientists, coastal zone managers and the general public by producing outputs tailored to their requirements. The different model nowcasts and predictions (up to 48 hours ahead for most variables, further ahead for tidal sea levels) are displayed (e.g. meteorological data, waves and sea surface heights, daily mean sea surface and sea bed temperatures, currents) and compared where possible with measurements. A schematic of the website is shown in Fig. 12.

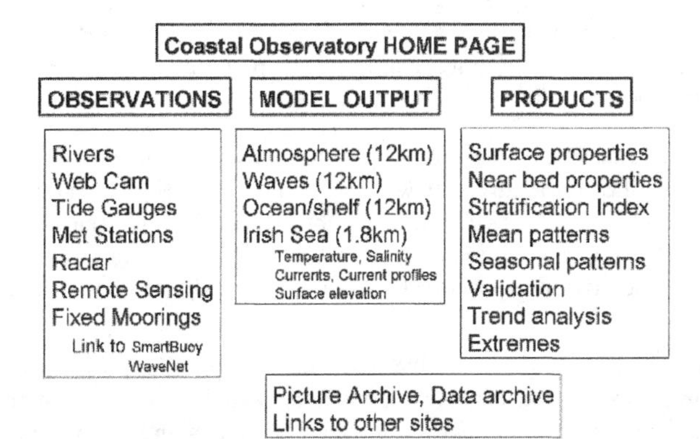

Figure 12: Schematic of Coastal Observatory website (http://coastobs.pol.ac.uk)

ECOSYSTEM BASED APPROACH TO MARINE MANAGEMENT

At about the same time as the Coastal Observatory began collecting measurements, the Department of Environment, Fisheries and Rural Affairs (DEFRA) published a strategy document ("Safeguarding our seas" (2002)) which highlighted the need for an ecosystem approach to marine stewardship. A report of the Review of Marine Nature Conservation (RMNC) recommended the setting up of a pilot study in the Irish Sea to test the potential for an ecosystem approach to managing the marine environment at a regional sea scale. The Irish Sea is considered to be one of the most ecologically-distinct regional seas around the UK. The pilot aims to examine the potential for regional sea management over the whole Irish Sea.

The Irish Sea Pilot project was set up by the Joint Nature Conservation Committee (JNCC), a part of DEFRA, with the following objectives:

- test ways of integrating nature conservation into key sectors in order to make an effective contribution to sustainable development on a regional basis
- test the framework proposed by the paper "An implementation framework for the conservation, protection and management of nationally important marine wildlife in the UK" (Laffoley *et al.*, 2000)
- determine the potential of existing regulatory and other systems for delivering effective marine nature conservation, and identify any gaps
- recommend measures to fill the gaps identified.

The pilot is charged with trialing a proposed new marine nature conservation framework in the Irish Sea, as prepared by the conservation agencies of the UK, namely English Nature, Scottish Natural Heritage, Countryside Council for Wales, and the Environment and Heritage Service (Northern Ireland). One of the key outputs from the Study will be a refined 'marine landscapes' classification of ecological units, along with a suite of habitats and species for conservation action within the pilot area. It will examine the degree to which this framework can contribute to wider sustainable development in the marine environment by setting objectives for nationally important conservation features to help guide the future actions of national, regional and local regulatory bodies and users.

In a review and testing of legislation, governance and enforcement in the UK marine environment, the pilot investigates the manner in which nature conservation objectives could be integrated into the objectives of other marine interest sectors (fisheries, oil and gas, shipping, etc) in practice. This requires discussions with those sectors, the identification of practical solutions and the reporting on difficulties encountered.

POL was invited to contribute to the pilot study by providing hydrodynamic outputs, derived from the Coastal Observatory and earlier research, which would help categorise the 'marine landscapes'. As such, POL has already provided information in the form of maps and statistics detailing hydrodynamical parameters such as bed stress distribution (maximum bed stress shown in Fig. 13), tidal front positions and their variability, water column stratification potential and its interannual variability, and we are in discussion on what other parameters could be usefully integrated. This input has helped to define the set of ecological units (also called seabed types or physiographic units) of the Irish Sea, giving the equivalent of the terrestrial 'countryside map' of the seabed. It shows the seabed equivalent of mountains and moorland, forests and grasslands. The set contains 15 ecological units, including 6 sediment categories and 9 physiographic features.

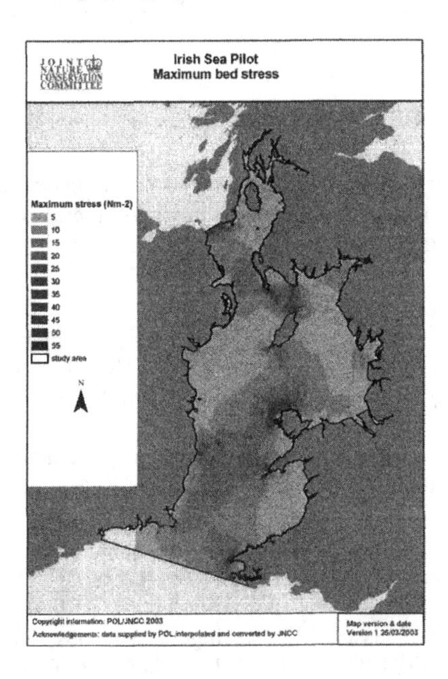

Figure 13: JNCC map of maximum bed stress produced from POL model output

The pilot study will report on its findings in 2004. For further information see http://www.jncc.gov.uk/marine/irishsea_pilot.

NEXT STEPS FOR THE COASTAL OBSERVATORY

Planned for early 2004 the Irish Sea and Liverpool Bay models will include wave-current interaction provided by two-way linking between wave (WAM) and the 3-dimensional currents (e.g. Wolf *et al.*, 2002) with performance checked against the in-situ temperature, salinity, current and wave measurements and coastal sea-level measurements received in real-time. Methods of data assimilation are being explored both to enhance the value of the data and to improve the accuracy of the model forecasts. Initially, for example, AVHRR sea surface temperature will be assimilated (e.g. Annan and Hargreaves, 1999); assimilation of sea surface currents and waves derived from the HF radar will be considered later.

In the latter part of 2004 nutrients and plankton dynamics will be simulated by including the coupling to ERSEM and the sediment transport and resuspension module. The sediment module is important for estimating concentrations of suspended particulate matter, an important influence on light attenuation and hence biological processes in the shallow Case 2 waters of the eastern Irish Sea. The SmartBuoy, the ferries and the SeaWifs / MODIS satellite ocean colour sensors will provide validation data.

ACKNOWLEDGEMENT

Thanks to Martin Holt at the UK Met Office for contributing operational products to the project, and to Chris Lumb of the Joint Nature Conservation Committee for permission to show Figure 13.

REFERENCES

Allen, J.I., Blackford, J., Ashworth, M.I., Proctor, R., Holt, J.T. and J. Siddorn, 2001. A highly spatially resolved ecosystem model for the northwest European continental shelf. *Sarsia,* **86**: 423-440.

Annan, J. D., and J. C. Hargreaves, 1999. Sea Surface temperature assimilation for a three-dimensional baroclinic model of shelf areas. *Continental Shelf Research*, **19**:1507-15020.

Baretta, J.W., Ebenhoeh, W. and P. Ruardij, 1995. The European Regional Seas Ecosystem Model, a complex marine ecosystem model. *Netherlands Journal of Sea Research*, **33**, (3-4): 233-246. ISSN: 0077-7579

Blaha, J.P., Born, G.H., Guinasso, N.L., Herring, H.J., Jacobs, G.A., Kelly, F.J., Leben, R.R., Martin, R.D., Mellor, G.L., Niiler, P.P., Parke, M.E., Patcheon, R.C., Schaudt, K., Scheffner, N.W., Shum, C.K., Ohlmann, C., Sturges III, W., Weatherly, G.L., Webb, D. and H.J. White, 2000. Gulf of Mexico Ocean Monitoring System. *Oceanography*, **13**, 2: 10-17.

Buch, E., and H. Dahlin, Eds., 2000. The BOOS Plan: Baltic Operational Oceanographic System, 1999-2003. *EuroGOOS Publication No. 14.* Southampton Oceanography Centre, Southampton. ISBN 0-904175-41-3

Cattle, H., Bell, M.J. and M.W. Holt, 1998. Operational analysis and forecasting of the global ocean - the UK Met Office FOAM system. *Oceanology International 98: The global ocean*, held 10-13 March 1998, Brighton, U.K. Volume 1: 161-169. ISBN: 0900254203.

DEFRA, 2002. Safeguarding our seas: A strategy for the conservation and sustainable development of our marine environment. HMSO, 82pp.

Edson, J,B., Chave, A.D., Dhanak M. and F.K. Duennebier, 2002. Guest Editorial. *IEEE J. Oceanic Egineering*, **27**, 145.

Glenn, S.M., Biocourt, W., Parker, B. and T.D. Dickey, 2000. Operational observatoion networks for ports, a large estuary and on an open shelf. *Oceanography*, **13**, 1: 12-23.

Holt, J.T., and I.D. James, 1999. A Simulation of the Southern North Sea in comparison with measurements from the North Sea Project. Part 2: Suspended Particulate Matter, *Continental Shelf Research*, *19*: 1617-1642.

Holt, J.T. and I.D. James, 2001. An s-coordinate density evolving model of the north west European continental shelf. Part 1: Model description and density structure. *Journal of Geophysical Research,* **106**, C7: 14015-14034.

Holt, J.T. and R. Proctor, 2003. The role of advection in determining the temperature structure of the Irish Sea. *Journal of Physical Oceanography*, **33**(11): 2288-2306.

Laffoley, D. D'a., Baxter, J., Bines, T., Bradley, M., Connor, D.W., Hill, M., Tasker, M. & M. Vincent, 2000. An implementation framework for conservation, protection and management of nationally important marine wildlife in the UK. Prepared by the statutory nature conservation agencies, Environment Heritage Services (Northern Ireland) and JNCC for the DETR Working Group on the Review of Marine Nature Conservation. Peterborough: English Nature Research Reports, No. 394. 29 pp.

Prandle, D. and A. Lane, Eds., 2000. Special issue: Operational oceanography in coastal waters. *Coastal Engineering,* **41**(1-3): 1-359.

Proctor, R., Holt, J.T., Anderson, T.A., Kelly-Gerreyn, B.A. Blackford, J. and F. Gilbert, 2002. Towards 3-D ecosystem modelling of the Irish Sea. *Proc. 7th International Conference on Estuarine and Coastal Modelling*, 3-5 November 2001, St. Petersburg, Florida. ASCE publication.

Proctor, R., J. T. Holt, J. I. Allen and J. Blackford, 2003. Nutrient fluxes and budgets for the North West European Shelf from a three-dimensional model. *Science of the Total Environment*, **314-315**: 769-785.

Wolf, J., Wakelin, S.L. & J.T. Holt, 2002. A coupled model of waves and currents in the Irish Sea. *Proceedings of the Twelfth (2002) International Offshore and Polar Engineering Conference*, Kitakyushu, Japan, May 26–31, 2002. **Vol. 3**:108-114.

Recent applications of 3D wind-driven circulation modeling to a semi-enclosed sea: Hauraki Gulf, New Zealand

Robert G. Bell[1], John W. Oldman[1], Scott A. Stephens[1]

Abstract

Two- and three-dimensional hydrodynamic models of the Hauraki Gulf (Auckland, New Zealand) have been developed over several years, verified against several observational datasets for currents, temperature and tides. The modeling has reached a stage where it is now being used in a wide range of applications to support resource management and maritime operations. The underpinning modeling relates to both tidal and wind-generated currents and circulation in the Gulf over a variety of timescales from hours to multiple decades. Applications of the model described in this paper relate to sustainability of fisheries and aquaculture, surface wind drift (search & rescue, outfall plumes, water quality) and yacht racing (e.g., Americas Cup). The next stage in utilizing the models is the development of a nowcasting and forecasting hydrodynamic system.

Introduction

Knowledge of coastal tide- and wind-generated currents and their spatial and temporal variability is fundamental to managing an urbanized semi-enclosed sea like the Hauraki Gulf off Auckland. The Gulf is heavily utilized for a range of competing commercial and recreational activities that require knowledge of both tidal and wind-generated residual currents. Resource uses for the Hauraki Gulf include a snapper fishery, aquaculture farms, yacht racing (e.g., America's Cup, round-the-world events), sewage and dredge-material disposal, sand extraction for the building industry, and the associated wider issue of marine ecosystem sustainability. There is a also a growing need for hindcasting and forecasting of surface currents and waves to support search & rescue, shipping operations (e.g., supertanker operations at the Marsden Point refinery), oil spill clean-up operations and recreational water-use safety.

Three-dimensional modeling of coastal/shelf hydrodynamics is now commonplace to assist resource managers and maritime operations. Such models include

[1] National Institute of Water & Atmospheric Research (NIWA), PO Box 11-115, Hamilton, New Zealand. Email: r.bell@niwa.co.nz Web: www.niwa.co.nz

wind-generated currents and associated circulation patterns. However, there continue to be challenging modeling aspects of shelf circulation such as oceanic-shelf interactions, how density differences drive circulation, simulating density fronts, baroclinic (internal) tides and coastally-trapped waves (Csanady, 1997), all of which effect the Hauraki Gulf.

Tidal currents dominate the hydrodynamics in the more constricted inner Hauraki Gulf (Figure 1), due to numerous islands, headlands and harbor entrances. Wind-generated currents, waves and density currents are more dominant in open, deeper waters. Tidal currents range from around 1.0 m/s in the vicinity of headlands, 0.4 to 0.5 m/s in more constricted channels, but mostly are less than 0.2 m/s in the inner Gulf and less than 0.1 m/s in the outer Gulf.

The hydrodynamics of the Gulf have been modeled extensively over two decades—initially with depth-averaged (2D) models, and recently with three-dimensional models (e.g., Bowman & Chiswell, 1982; Greig & Proctor, 1988; Proctor & Greig, 1989; Black et al., 2000). The interaction of oceanic waters with the shelf waters, upwelling/downwelling, oceanic intrusions, and internal tides have been the subjects of several studies (e.g., Sharples, 1997; Sharples & Greig, 1998; Black et al., 2000; Sharples et al., 2001). Tides around the New Zealand shelf including the Hauraki Gulf have also been modeled with an unstructured finite-element model (TIDE2D) by Walters et al. (2001). Recent applications to the Hauraki Gulf revolve around three-dimensional models on a regular 750-m grid (Figure 2), mainly using either a) an explicit finite-difference solver; 3DD (Black, 1995), or b) an implicit finite-difference solver; MIKE3 (DHI Water & Environment). Both models incorporate forcing by tides, winds, and buoyancy (density from either or both temperature and salinity gradients). Details of the 3DD model and its application to the Gulf are covered by Black et al. (2000). Surface waves have also been modeled on the same 750 m grid using the shallow-water wave model SWAN, that complements the larger ocean spectral wave model (WAM) used to develop a 20-year hindcast wave climate for New Zealand including the outer Gulf (Gorman et al., 2003). Other model grids of higher resolution have been utilized for specific and localized applications e.g., harbor water quality studies, mussel farm impacts and ocean outfall discharges.

Given this background of hydrodynamic modeling for the Hauraki Gulf, this paper focuses on the utilization of these calibrated and verified hydrodynamic models to support a diverse range of resource management and maritime operational requirements. The paper also describes ongoing efforts to improve the calibration and verification of the three-dimensional models of the Gulf, particularly in situations when the water column is partially stratified. This work is complemented by specific field measurements to support both the model verification and the adaptation of the modeling results to the resource management applications.

Data and Models

Sea-level data and tidal harmonic constituents were obtained from 16 tide-gauge sites ("C" series in Figure 1). Current velocities were measured at other sites throughout the

Gulf, shown in Figure 1. Wind velocity is measured at two automatic weather stations (stars in Figure 1) located in the inner Gulf (Whangaparaoa) and the outer Gulf (Mokohinau Islands), and then converted to an equivalent 10-m height.

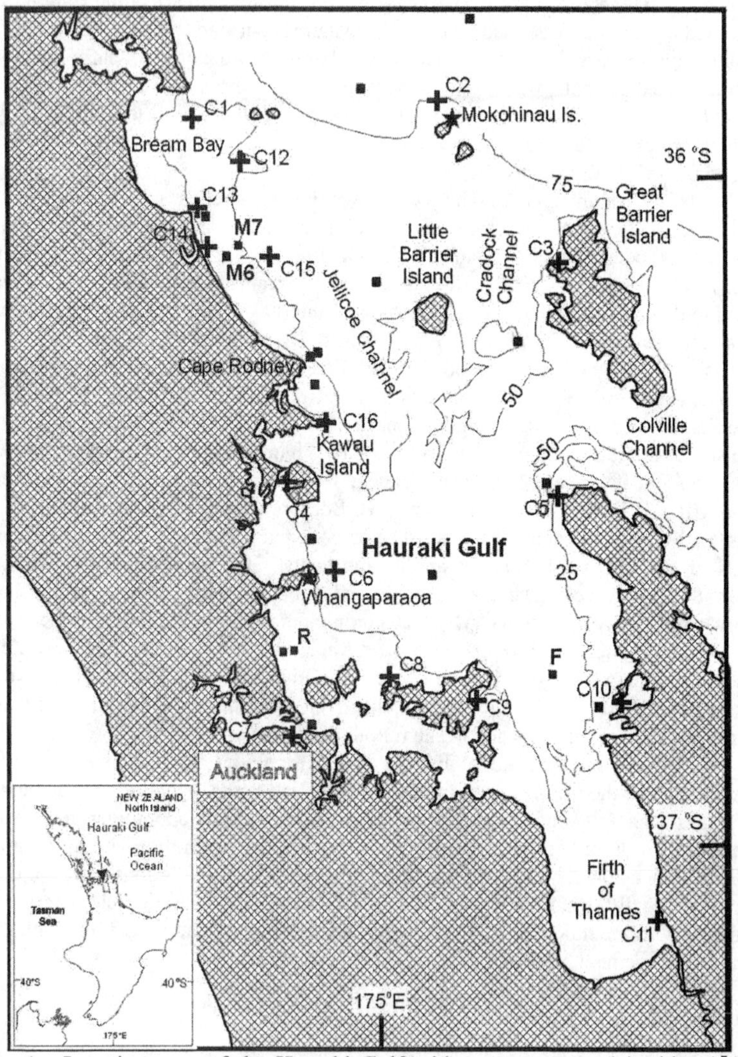

Figure 1. Location map of the Hauraki Gulf with current meter (■), tides (✚) and wind station (★) sites and depth contours (m)

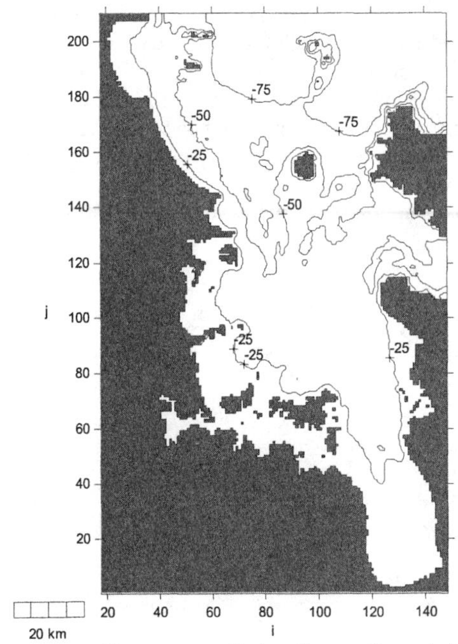

Figure 2. 750 m regular-grid coverage (i, j cell coordinates) for the Hauraki Gulf (Auckland) oriented to True North, with shading for depth (m).

Eight major tidal constituents were obtained from a tidal model of New Zealand's Exclusive Economic Zone (Walters et al. 2001) to provide tidal forcing along the north and east offshore boundaries of the 750-m grid models (Figure 2).

The calibration of the hydrodynamic model 3DD for tides and its use in investigating wind-induced upwelling and stratification in the Gulf are fully discussed by Black et al. (2000). Following this study, further calibration of the 3DD model for wind-generated currents was undertaken following the deployment of two acoustic Doppler current profiler (ADCP) instruments in 22 and 35 m water depth along a cross-shore transect off Mangawhai in the outer Gulf (M6 & M7 sites shown in Figure 1). Winds from Mokohinau Island were used to force the 3DD model (excluding tides) covering the 1-month period of the ADCP deployment. The vertical eddy diffusivity within 3DD is based on a mixing length and Richardson number formulation (Black et al, 2000) and was set to 1×10^{-3} m^2/s based on thermistor-string data obtained from the outer Gulf (Black et al., 2000).

Simulated wind-generated currents from three representative layers (near-surface, mid-depth and near-bottom) were compared with the non-tidal residual current from the relevant ADCP bins (Figure 3). The ADCP velocities were first processed by removing the six main tidal harmonic constituents, then subjected to low-pass digital filtering using a tapered tanh filter with a cut-off frequency of 0.07 cycles per hr or a period of 15 h (shorter than normal to partially preserve a short 12-hour wind impulse event). (Despite the extraction of tides and filtering, some 12 to 24 hour non-tidal oscillations remain in the field data that were not covered by the local wind forcing.) The model predictions improved throughout the water column (Figure 3) as well-mixed water column conditions set in during the latter half of the deployment period. A short 12-hour wind impulse of up to 20 m/s winds from the northwest occurred on 19-April-1997.

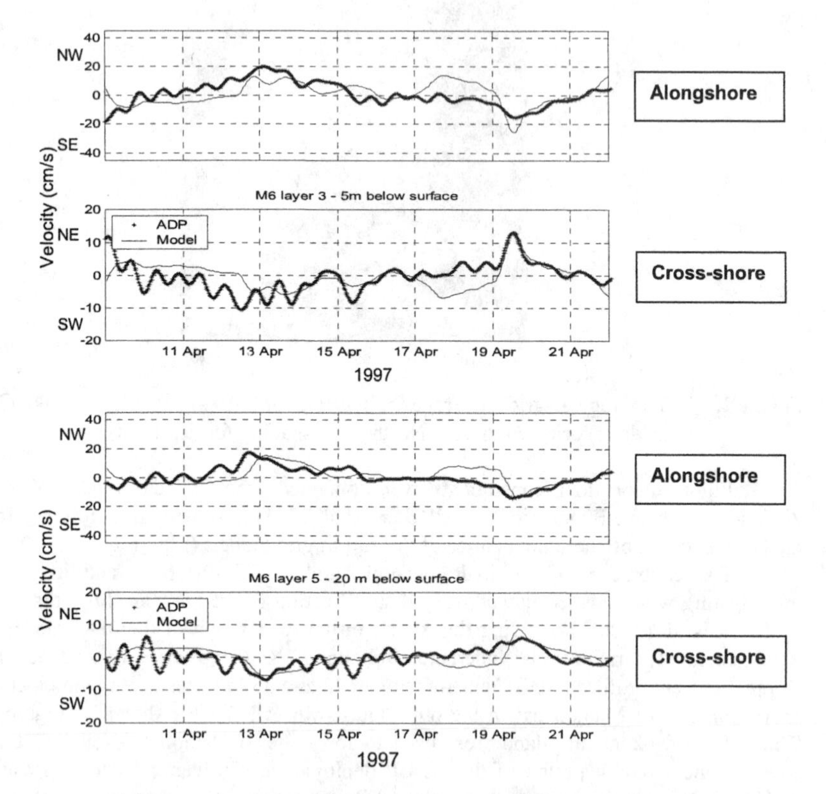

Figure 3. Comparison of alongshore (NW positive) and cross-shore (NE positive) velocities at site M6 (outer Gulf) from the ADCP filtered data (heavy line) and the 3DD model (thin line) for: (upper), layer 3 at 5 m below the surface, and (lower), layer 5 at 20 m below the surface (15 m above bed).

The wind impulse caused a strong downwind flow of 0.20 m/s measured at 5 m below the surface (Figure 3; top panel). This is not entirely reflected in the plotted observed results, perhaps due to the partial damping from the low-pass filter. The offshore Ekman near-surface drift is evident (to the left in southern latitudes) in the second panel of Figure 3. The results presented in Figure 3 indicate that in the outer Gulf, realistic model predictions of low-frequency wind-generated currents may be achieved solely by the use of offshore wind measurements from the Mokohinau Islands weather station. This paves the way for the use of this lengthy 30-year wind record for modeling long-term residual circulation, and simplifies the path towards developing a nowcast/forecast system.

The DHI MIKE3 model, based on a semi-implicit ADI solver, has recently been applied on the same 750-m grid for the Gulf (Figure 2), achieving a similar close match to the calibrated tidal results from the 3DD model, shown in Table 1.

Table 1. Calibration of the Hauraki Gulf 3DD (Black et al., 2000) and MIKE3 models relative to observations of the mean tide (M_2), listed as differences between modeled and observed M_2 tidal height amplitudes (cm) and phases (°, NZST) recorded at "C-series" site positions shown in Figure 1.

Site	Observed		Observed–model (3DD)		Observed–model (MIKE3)	
	Amplitude (cm)	Phase (°)	Amplitude (cm)	Phase (°)	Amplitude (cm)	Phase (°)
C1	89	209	4	4	6	8
C2	85	203	0	1	4	4
C3	86	199	-3	-2	0	1
C4	91	203	-11	-5	-10	1
C5	91	202	-3	0	0	10
C6	103	208	1	2	2	9
C7	116	204	4	-8	4	0
C8	102	198	-7	-10	-8	-3
C9	116	202	5	-5	7	3
C10	101	206	-11	2	-12	4
C11	127	204	3	-2	4	4
C12	88	205	2	1	4	3
C13	88	206	2	2	5	5
C14	89	205	6	2	6	3
C15	90	204	5	2	6	3
C16	97	204	3	1	11	6

Further, a realistic calibration of the MIKE3 model for sea temperatures through the water column was obtained during a 20-day summer simulation of Hauraki Gulf. The simulation incorporated both temperature and salinity stratification on vertical layers of 2 m thickness; and included forcing by tide, wind stress, air temperature and river inputs (Figure 4). The modeled thermocline level was close to that measured, but about 2 m shallower. The observed heat pulse in the surface water on days 7 to 14 and subsequent cooling was reproduced with a lag of about 1-day in the model (Figure 4). This temperature calibration is encouraging, considering the complexities of thermal heat transport, and bodes well for future three-dimensional circulation modelling that accounts for water-column stratification.

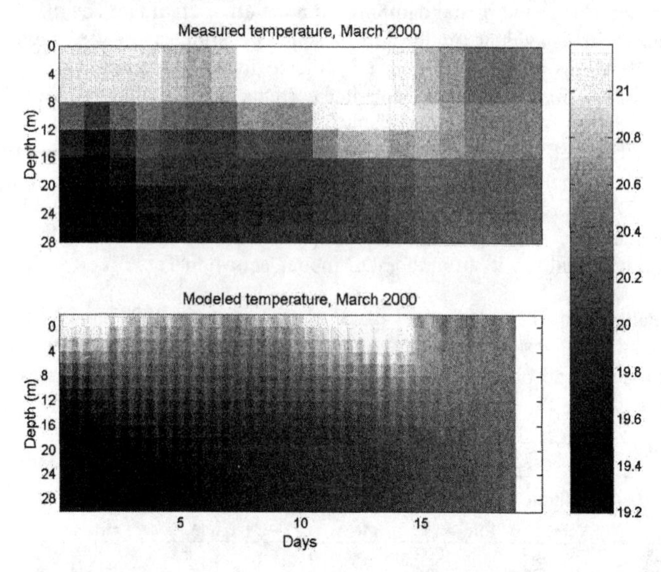

Figure 4. Comparison between measured (upper) and modeled (lower) sea temperatures (°C) at the long-term mooring site F (Figure 1) during March 2000. Measurements were interpolated to 4 m depth intervals, but those above 8 m were set equal to the uppermost available thermistor measurement, where the water column was mostly well mixed.

Recent model applications

Long-term wind circulation. Knowledge of the long-term wind circulation in a semi-enclosed sea such as the Hauraki Gulf is helpful in understanding long-term ecosystem functioning including nutrient budgets and preferred flushing mechanisms. However, modelling such long timescales is difficult to achieve, hampered by short-term current meter deployments and the fact that three-dimensional model simulations for periods of even a few years remains constrained by computer resources. Consequently, a simplified approach was devised to utilize long-term wind records, in the form of compass rose-probability distributions for different wind cases. Essentially, the region's wind rose (e.g., Figure 5) was categorised into 56 bins with specific wind speed ranges (0 to 24 m/s; in 4 m/s increments) and sectors (eight main points of the compass; North, Northeast, East etc), and an overall probability of occurrence assigned to each bin. Then 56 separate 3DD model simulations were undertaken, one for each case, running through to a steady-state equilibrium condition in the Gulf. To obtain the long-term residual wind-driven circulation, the predicted current velocity for each model grid cell from each of the 56 simulations was weighted by the wind-rose probability of that wind case occurring. Finally, the 56 weighted velocities at each grid node were vector-averaged to produce a net residual velocity. This approach does have some limiting assumptions, namely all time history and hysteresis is removed and time to reach equilibrium is not accounted for, although for the stronger winds, it is just a few hours and considerably shorter than for light breezes. Full details are outlined in Oldman et al. (submitted). The "wind-rose probability" approach was verified against a full 1-month time-domain simulation using the 3DD model.

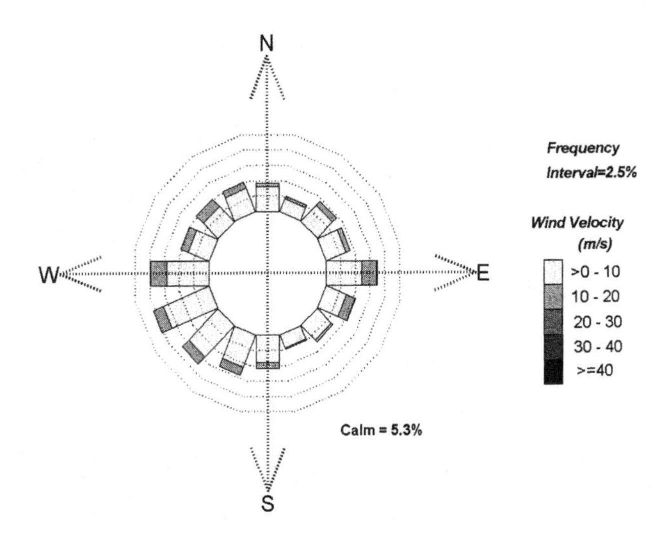

Figure 5. Wind rose (Mokohinau Islands) for periods 1961–1984, 1993–2000.

The "rose-probability" approach results in small residuals of around 0.02 m/s in open Gulf waters. The low net wind-residual velocities arise from a) predominant, slow winds cancel each other out around the wind rose (Figure 5), and b) moderate to high winds partially cancel each other out from the northeast and southwest wind bi-modal distribution. The procedure was repeated for wind velocity threshold of 8 m/s (mean wind speed), 12 m/s (75th percentile wind speed) and 15 m/s (90th percentile wind speed). For the lower wind velocity thresholds, the predicted long-term pattern of wind-residual velocities is very similar to those for the complete wind record. However at the higher wind velocity threshold of 15 m/s, a more distinct residual circulation pattern emerges over the 30-year period that helps explain the long-term mode of flushing into and out of the Gulf during high winds, as summarised in Figure 6. The long-term net wind-driven residual pattern integrated for all strong winds above 15 m/s reveals the more dominant role northeast winds have on strong-wind flushing events over long timescales of years to decades. As shown in Figure 6, the integrated strong-wind near-surface flow is to the southwest in near-surface waters, counterbalanced by an outflow at depth through the Jellicoe Channel to the north, and secondary outflows through the Cradock and Colville Channels to the northeast and east (Oldman et al., submitted). However, for much of the time at shorter timescales of days to weeks, the prevailing southwest winds of slow to moderate wind speeds preferentially induce slow to moderate near-surface flow to the northeast in the Gulf, that is added to the tidal flows.

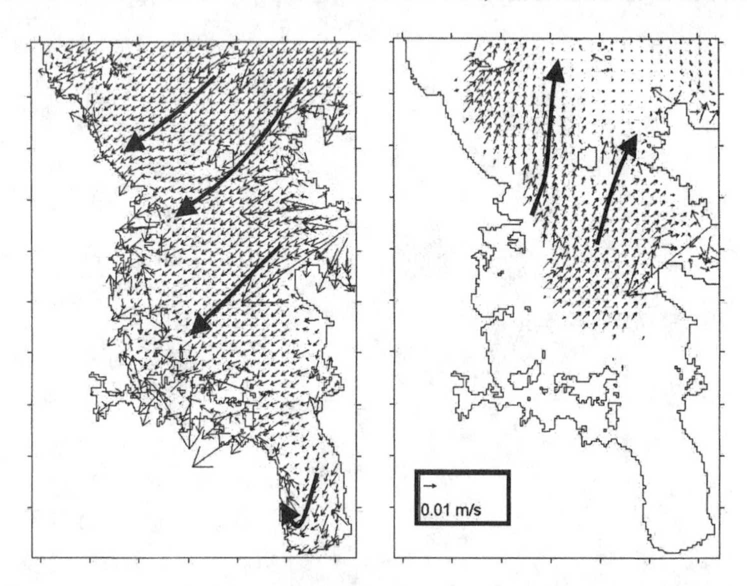

Figure 6. Integrated long-term wind-generated residual pattern for all strong winds above 15 m/s in the Hauraki Gulf for: (left) the near-surface layer, and, (right) the layer between 40–50 m below the surface. (Based on a 30-year wind record from Mokohinau Islands via the rose-probability approach.)

Seasonal variability in wind-driven circulation and possible implications for ENSO. Variability at seasonal timescales is an important influence on marine ecosystem response and nutrient cycling in semi-enclosed seas. A major component of variability in the Gulf is the differences in wind-generated residual currents during different seasons and the variability induced by the 2 to 4 year El Niño–Southern Oscillation (ENSO) system. Winds blowing from the west and southwest are more prevalent during El Niño years and winds blowing from the east and northeast are more prevalent during La Niña years. On a seasonal basis, more southwest winds tend to occur in winter, and more northeast, onshore winds are experienced during summer. Accurate three-dimensional model simulations over several ENSO timescales are not practical. So to help explain to resource managers what the effects of both seasonal and ENSO would be on proposed developments of mussel farms within the Firth of Thames (location; Figure 1) representative wind time series in summer and winter were used to mimic typical El Niño and La Niña winds. Then these results were extended by inference to describe the combination of seasonal circulation and the weaker interannual ENSO effects.

MIKE3 simulations of northeast winds during stratified summer (March 2000) conditions showed residual surface currents directed southwest with a monthly-averaged clockwise near-surface circulation in the lower Firth of Thames, and deep currents returning toward the north (Figure 7a). During southwest winds (favoured more in strong El Niño episodes) in the same summer period, surface currents were directed downwind (northeast) with a time-averaged anticlockwise circulation in the lower Firth, and deep currents returning toward the southwest (Figure 7b). In winter, stratification was weak and wind-driven currents experienced less vertical shear but more horizontal variability.

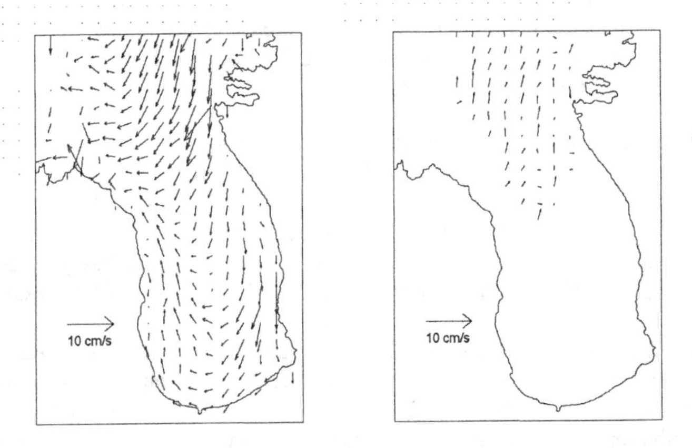

Figure 7(a). Monthly-averaged currents forced by tides, northeast-dominated winds, river discharges and temperature. Temperature inputs were those of late summer (March 2000). Plots are for horizontal layers, 0–2 m (left) and 18–20 m (right) below mean sea level.

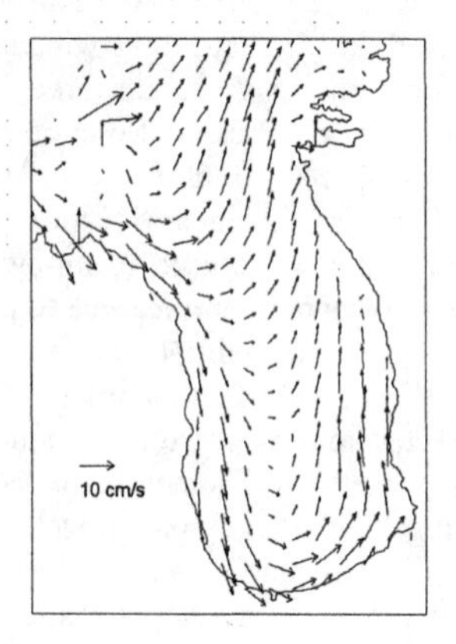

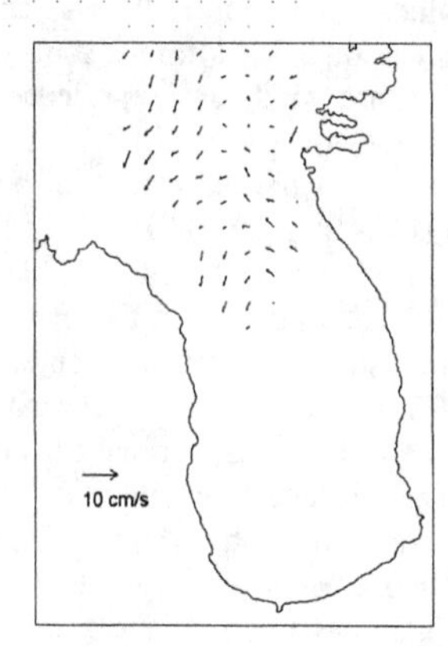

Figure 7(b). Monthly-averaged currents forced by tide, southwest-dominated winds, river discharges and temperature. Temperature inputs were those of late summer (March 2000). Plots are for horizontal layers, 0–2 m (left) and 18–20 m (right) below mean sea level.

Wind variability between ENSO events is subtler than seasonal wind trends; therefore for rigorous long-term modelling of wind circulation, short-term (e.g., 1-month) simulations cannot adequately represent ENSO variability. However, we observed that La Niña/El Niño wind patterns tend to weakly mirror the summer/winter seasonal wind trends, so as a guide to resource managers, the likely weaker ENSO effects can be inferred from the "typical" summer and winter Gulf model results.

Seasonal effects on snapper fishery. The calibrated 3DD model was also used recently to examine the role that surface and shelf-derived bottom waters have on the pelagic ecosystem of the Gulf (Zeldis et al., in press). Phytoplankton, zooplankton and fish larval distribution were measured within the Gulf over three consecutive summers (Nov-Jan 1985/86, 1986/87 and 1987/88). The highest abundance of plankton and fish larvae was found in the summer of 1987/88, so modeling was undertaken to help explain this puzzling anomaly beyond the fact that the 1987/88 summer coincided with the early onset of a La Niña episode. Integrated results from each of the three sets of "summer" hydrodynamic simulations showed that there was a greater incursion of near-bed continental shelf waters (layers 3 & 4) from the north into Hauraki Gulf during the 1985/86 and 1986/87 El Niño summers when westerly winds prevailed (Figure 8). The increased exchange of cooler shelf waters into the Gulf (and the cooler sea-surface temperature (SST) anomalies associated with the El Niño summers) appears to hinder snapper larval production. During the summer of 1987/88, winds were more evenly balanced between east and west and, combined with rising SST from the onset of a La Niña episode, enhanced snapper larval production.

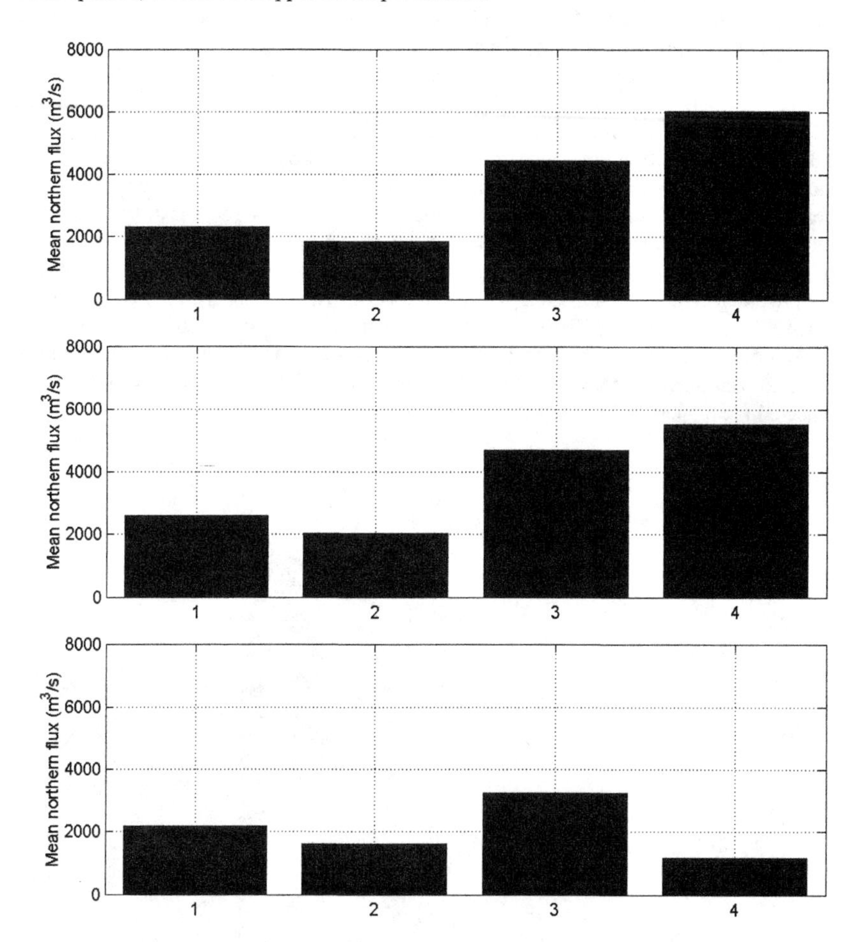

Figure 8. Average flux into the Gulf from the north through Jellicoe and Cradock Channels distributed through four 4 lumped layers of the 3DD model (layer 1 = 0–2 m, layer 2 = 2–5 m, layer 3 = 5–15 m, layer 4 = 15–50 m) for the: (upper) 1985/1986; (middle) 1986/1987 (bottom) 1987/1988 simulations.

Near-surface wind and tide currents (Americas Cup). Americas Cup yacht races were held in the inner Hauraki Gulf off Auckland's North Shore in the summers of 1999/2000 and 2002/03. The Volvo Ocean Race also visited Auckland in 2002. Through consultancy contracts with Cup syndicates, these events provided further opportunities to improve the verification of wind-generated currents within the top 1-m layer of the 3DD model. The model was used to predict tidal and wind-driven near-surface currents centred 1 m below the water surface, where they act on a yacht's hull and keel. The model nodes selected throughout the Cup course are shown in Figure 9. Combined wind and tidal currents were simulated for a combination of spring and neap tide and two wind speed categories from four compass quarters that covered the expected summer racing wind conditions. For each node, velocity results were summarised for syndicates into nautical "tidal-diamond" look-up tables (i.e. speed and direction on the hour relative to high water time (t = 0) at a standard port–in this case Auckland). The tables comprised current speed and direction at 1-m depth, with one set for tidal currents only, one set for wind-generated currents only, and a final table for the combined wind and tide conditions for the cases simulated. This provided some flexibility in interpolating tide and wind-drift currents on race day based on forecast tides and winds for the expected racing period.

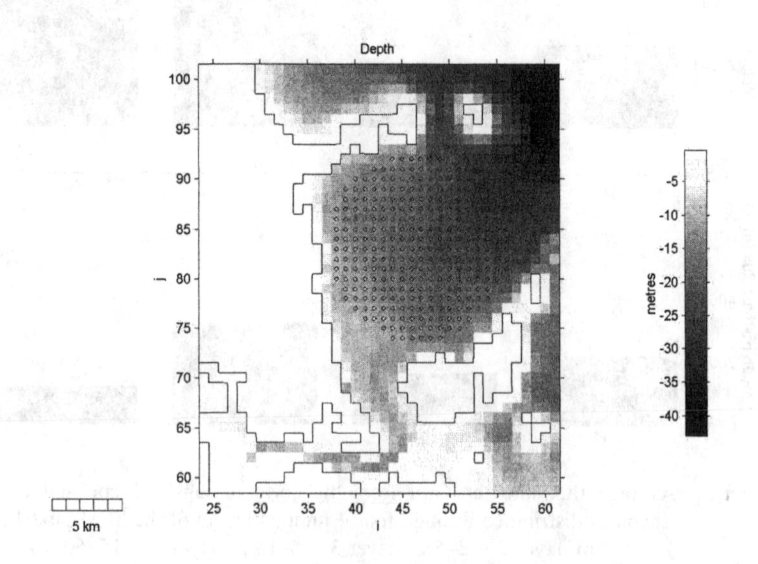

Figure 9. Sub-section of 750-m Hauraki Gulf model grid off the North Shore's East Coast Bays, showing the "tidal-diamond" output nodes marked by 'o'.

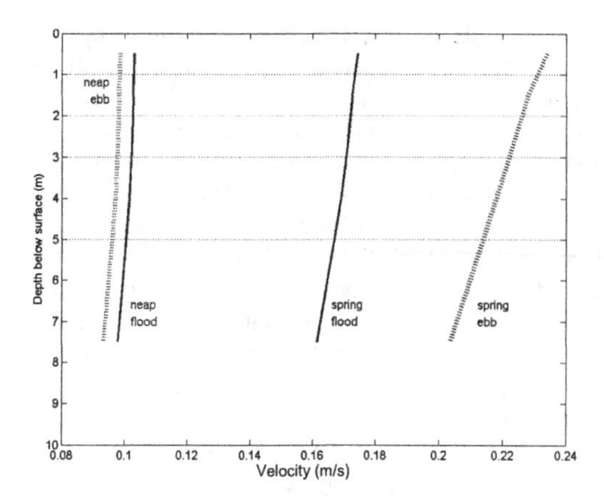

Figure 10(a). Depth variations in peak spring- and neap-tidal current magnitudes (no wind) in the center of the Americas Cup zone [grid cell i= 46, j=83]

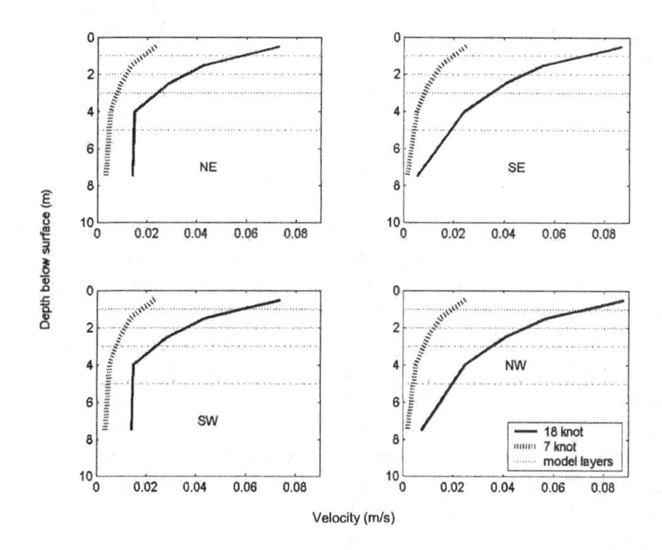

Figure 10(b). Depth variation in wind-driven current velocity magnitudes (no tide) for a 7 and 18 knot (3.6 and 9.3 m/s) wind at the same cell as Fig. 10(a). Elapsed simulation time for these steady-state runs was 6 hours.

Peak model-predicted, near-surface ebb tidal currents in the central East Coast Bays embayment of the inner Gulf reached approximately 0.2 m/s and 0.1 m/s during spring and neap tides, respectively (Figure 10(a)). Tidal currents were strongest to the south of the Cup course, where the offshore island constricts the main shipping channel. Wind-driven near-surface currents were approximately 0.1 m/s for a 9.3 m/s wind (Figure 10(b)), and can markedly change near-surface flow patterns compared to tide-only or quiescent wind conditions. Predicted wind-driven currents varied between 0.7% to 1.1% of the 10-m wind speed when averaged over the 1-m deep surface layer of the water column and depended as expected, on the wind fetch lengths and local topographically-steered three-dimensional circulation within the embayment.

The modeling of wind-generated currents in the near-surface layer was complemented by two sets of field experiments (SURFEX-I & II) in 2001 and recently in 2003, to quantify surface currents in the sheltered inner Gulf waters. The aim was to provide datasets that can be used to improve model predictions of near-surface currents and further increase our understanding of near-surface currents in a mixed tidal-wind forced system. The key field techniques used were a vessel-mounted twin set of RDI ADCPs (one looking down, and the other horizontally mounted) along with surface float mats and drogues to measure an ensemble of near-surface drift currents 30-40 m out from the vessel.

The development of the method for using a horizontally-directed ADCP to measure an ensemble of near-surface velocities and ship wakes is described in Marmarino & Trump (1999). Preliminary results from the SURFEX-I study show near-surface currents that are consistent with the 5-8 m/s south-westerly wind blowing at the time of a slack tide (Figure 11). Around slack tide, drogues set at 0.7 m depth moved downwind at 0.6–0.9% of the local wind speed, while the concurrent results from the ADCP for the upper 1-m layer (Figure 11) produced wind-driven currents of 0.8–1.1% of the local wind speed. This range for wind-induced currents closely matches the model results presented above (Figure 10(b)). These local %wind-speed factors will vary markedly throughout the Gulf, depending on a variety of factors including wind fetch/exposure, bathymetry (which slopes downwards to the eastern side), coastal and island planform, and oceanographic conditions. That is why a three-dimensional model is essential to accurately predict near-surface currents in a complex semi-enclosed sea like the Hauraki Gulf. For example, a usual rule of thumb for open waters is that the surface velocity generated is approximately 3% of the 10-m wind speed. This simplified expression would greatly overestimate wind-driven currents by a factor of 3 in this case. SURFEX-II data are yet to be analyzed but extend SURFEX-I instrumentation by including concurrent wave and surface-current measurements from a vessel-mounted microwave radar. Improvements in mounting stability and introduction of pitch and roll filtering information will enhance the data quality using the horizontal-ADCP technique in the future.

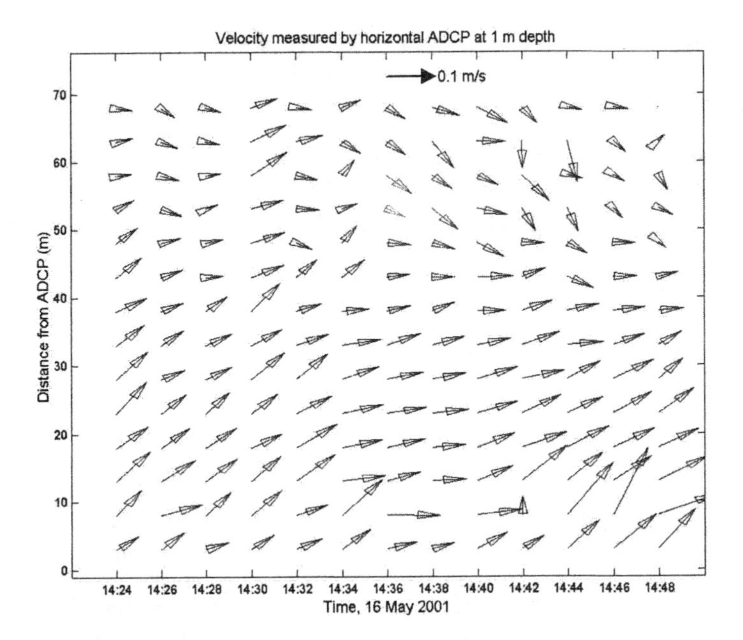

Figure 11. Near-surface velocities measured from a horizontally-mounted ADCP at 1 m depth during a 5-8 m/s wind from the southwest (16 May 2001; around slack tide). Positive direction on the plot corresponds to True North. Velocities have been averaged into 5 m horizontal bins and 60 s ensembles prior to plotting. Beam interference with the water surface occurred beyond 45–50 m from the vessel, with some noise due to boat movement.

Near-surface wind drift tracks. Given a calibrated three-dimensional model, surface drift tracks can be produced for maritime and resource management applications e.g., search and rescue (SAR) or oil-spill clean-up operations. Such products include look-up maps of realistic wind-drift track plots over a few days for various wind directions and speed categories. A Lagrangian particle tracking technique, based on the formulation of Black and Gay (1988), was used to develop wind-drift tracks over a 4-day window for each depth layer in the three-dimensional model (3DD). Release sites were chosen that were representative of the different sub-environments within the Gulf (Figure 1):

- Firth of Thames (southern section)—shallow estuarine circulation due to the Waihou and Piako rivers;
- Firth of Thames (open northern section)—estuarine with oceanic intrusion under certain wind conditions;

- Western coastal embayments—sheltered, shallow, inner Gulf waters, e.g., East Coast Bays where the Americas Cup racing was staged;
- Outer Gulf—deep sites in outer Gulf with more oceanic/shelf characteristics.

Some examples of surface wind-drift tracks are shown in Figure 12. Generally, wind drift steers tracks downwind in open outer Gulf waters, until they interact with nearshore waters. Bathymetric steering of surface wind drift is evident in the middle section of the Gulf, while an anti-clockwise (south or west winds) or clockwise (north or east winds) surface circulation is set up in the shallow confined southern section (Firth of Thames). Predicted tidal currents from the tidal model can be added to these drift plots to produce a product that can assist search and rescue operations in this popular recreational boating area served by the Auckland Coastguard.

Conclusions

Over a period of two decades of numerical model developments and associated field programmes, a stage has now been reached where the tide and wind-forced behaviour of the Hauraki Gulf can be predicted, particularly under well-mixed conditions. This paper has covered diverse range of applications of the 750-m regular grid three-dimensional models that simulate the hydrodynamic behaviour of the Gulf. Supporting field experiments to quantify near-surface wind drift are producing valuable datasets that are enhancing our confidence in the ability of the models to predict tidal and wind-driven surface currents for operational maritime and resource management applications.

Some of the critical outcomes or remaining issues that are pertinent to the Hauraki Gulf modeling are:

- Given the small financial resources (NZ$500K spread over 8 years), we adopted the advice of "start simple and don't apologize". However, we could have progressed earlier to three-dimensional models as substantial differences in wind-generated current circulation between two-dimensional depth-averaged and three-dimensional layered modeling were later demonstrated for the Gulf (Black et al. 2000). Initially, it was believed that the Gulf is reasonably well-mixed most of the time. However, subsequent modeling and field measurements have shown that thermal stratification in summer can induce marked variations in the wind-generated currents of the outer Gulf, particularly in the deeper layers. Baroclinic processes are now being simulated to improve the predictability of wind-generated currents;
- Maintenance of the offshore wind station is critical. In this case it was used as a basis for comparison to modelled wind-generated currents through the water column in the outer Gulf. For the inner Gulf, there should be further extension of the work to include orographic surface wind modeling using CALMET (Gimson, 2000).
- Fortunately, there was an extensive field dataset to assist in calibrating and verifying the models, along with detailed tidal constituents from the New Zealand tidal model (Walters et al., 2001) to provide stable continuously varying boundary conditions.

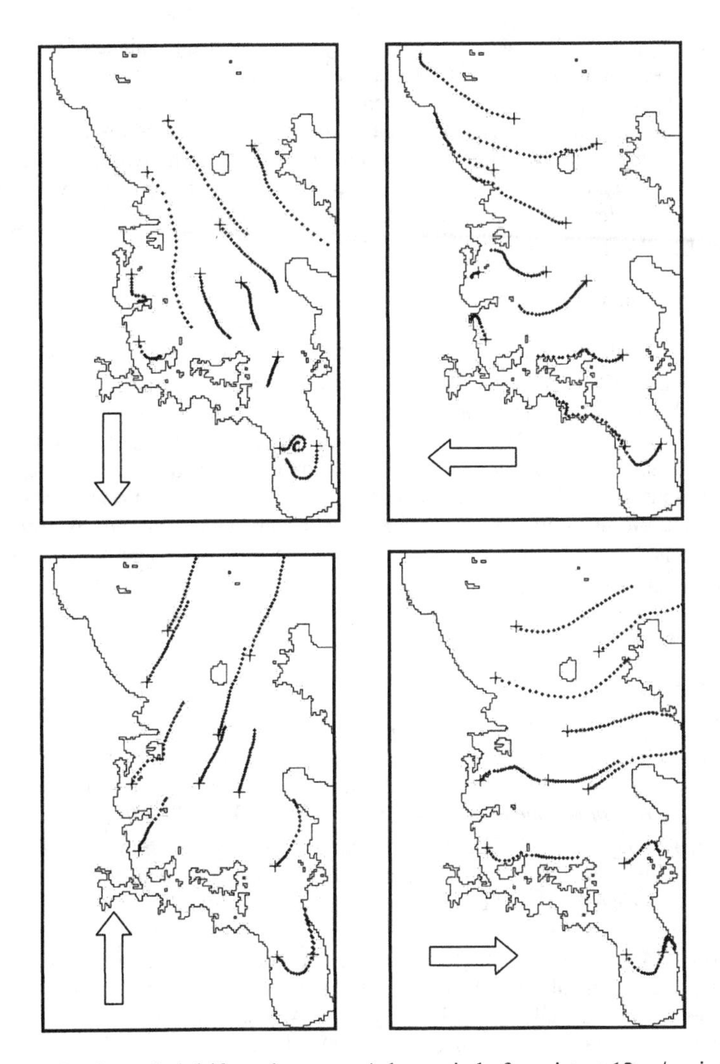

Figure 12. Surface wind-drift tracks over a 4-day period of persistent 12 m/s wind from each of the four main compass points (large arrow) for eleven release sites within the Hauraki Gulf.

Given the degree of calibration, field verification and knowledge gained of the tidal and wind processes within Hauraki Gulf, the stage has been reached where modeling could form the basis of an operational system to nowcast and forecast hydrodynamic conditions in the Gulf. This also includes coupling with the SWAN wave model. However, assimilation of remotely-sensed data (e.g., SST from Advanced Very High-Resolution Radiometer (AVHRR) satellite sensors) and long-term mooring sites will need to be considered to support accurate forecast of currents within the Hauraki Gulf, particularly during stratified summer conditions.

Acknowledgements

The authors thank Dr. Kerry Black (now ASR Ltd.) for his earlier development of the Hauraki Gulf 3DD model, Dr. John Zeldis (NIWA) who lead the snapper fishery work and supplied Firth of Thames temperature data, and Rick Liefting and Rod Budd for their assistance in the SURFEX-I field experiments. This work was largely funded by Contract C01X0218 from the NZ Foundation for Research Science and Technology, with contributions from various consultancy projects including the Western-Firth Mussel Consortium, North Shore City Council, Alinghi and Stars & Stripes Americas Cup syndicates and the Illbruck Challenge Volvo Ocean Race syndicate, whose assistance enabled further tuning of the hydrodynamic models.

References

Black, K.P. (1995). The hydrodynamic model 3DD and support software. Occasional Report 19, Dept. of Earth Sciences, University of Waikato, New Zealand. 53 p.

Black, K.P.; Bell, R.G.; Oldman, J.W.; Carter, G.S.; Hume, T.M. (2000). Features of 3-dimensional barotropic and baroclinic circulation in the Hauraki Gulf, New Zealand. *NZ Journal of Marine & Freshwater Research*, 34(1), 1–28.

Black, K.P.; Gay, S.L. (1988). A numerical scheme for determining trajectories in particle models. *In*: Bradbury, R.H. (ed.) *Acanthaster and the coral reef: A theoretical perspective*. Lecture Notes in Biomathematics, 88, 151–156, Springer-Verlag.

Bowman, M.J.; Chiswell, S.M. (1982). Numerical tidal simulations within the Hauraki Gulf, New Zealand. *In*: Nihoul, J.C.J. (ed.), *Hydrodynamics of semi-enclosed seas*. Elsevier Oceanography Series No. 34, 349–384, Elsevier.

Csanady, G.T. (1997). On the theories that underlie our understanding of continental shelf circulation. *Journal of Oceanography*, 53, 207–229.

Gimson, N.R. (2000). Model simulation of the air quality of Auckland. *Clean Air and Environment*, 34, 37–42.

Gorman, R.M.; Bryan, K.R.; Laing, A.K. (2003). Wave hindcast for the New Zealand region: nearshore validation and coastal wave climate. *NZ Journal of Marine & Freshwater Research*, 37, 567–588.

Greig, M.J.; Proctor, R. (1988). A numerical model of the Hauraki Gulf, New Zealand. *NZ Journal of Marine and Freshwater Research*, 22, 379–390.

Marmorino, G.O.; Trump, C.L. (1999). Near-surface current measurements using a ship-deployed "horizontal" ADCP. *Journal of Atmospheric and Oceanic Technology*, 16,1456-1463.

Oldman, J.W.; Bell, R.G.; Black, K.P. (submitted). Hindcasting residual currents from long-term wind records in the Hauraki Gulf, New Zealand. Submitted to the *NZ Journal of Marine & Freshwater Research.*

Proctor, R.; Greig, M.J.N. (1989). A numerical model investigation of the residual circulation in Hauraki Gulf. *NZ Journal of Marine and Freshwater Research*, 23, 421–442.

Sharples, J. (1997). Cross-shelf intrusion of subtropical water into the coastal zone of northeast New Zealand. *Continental Shelf Research*, 17(7), 835–857.

Sharples, J.; Greig, M.J.N. (1998). Tidal currents, mean flows, and upwelling on the north-east shelf of New Zealand. *NZ Journal of Marine and Freshwater Research*, 32, 215–231.

Sharples, J.; Moore, C.M.; Abraham, E.R. (2001). Internal tide dissipation, mixing, and vertical nitrate flux at the shelf edge of NE New Zealand. *Journal of Geophysical Research*, 106 (C7), 14,069–14,081.

Walters, R.A.; Goring, D.G.; Bell, R.G. (2001). Ocean tides around New Zealand. *NZ Journal of Marine and Freshwater Research 35*: 567–579.

Zeldis, J.; Oldman, J.W.; Ballarra, S.L; Richards L. (in press). Cross-shelf exchange, pelagic ecosystem structure, and larval fish survival in Hauraki Gulf, New Zealand. *Marine Ecology Progress Series.*

Adjoint data assimilation into a 3D unstructured mesh coastal finite element model

F. Fang * (1), C.C. Pain (1), M.D. Piggott (1), G.J. Gorman (1), G.M. Copeland (2), C.R.E. de Oliveira (1), A.J.H. Goddard (1), I. Gejadze (2)

(1) Department of Earth Science and Engineering, Imperial College, Prince Consort Road, London, SW7 2AZ, U.K.
* Corresponding email: f.fang@imperial.ac.uk
(2) Department of Civil Engineering, University of Strathclyde, Glasgow, U.K.

Abstract

An adaptive mesh adjoint model has been developed to invert for uncertain parameters in the simulation of coastal flows. These may include, for example, open boundary conditions and initial conditions. Our objective in this paper is to obtain a better representation of water flows around and over coastal topography. In this paper a first attempt has been made at using mesh adaptivity in both transient forward and adjoint models. Although the resulting method has inconsistent discrete schemes for the forward and adjoint models, the inversion results shown here are nevertheless good. The derivation of the continuous adjoint model is presented. A description of the adaptive mesh adjoint technique is also given. As a study case, the method is applied to the inversion of boundary conditions for flow past a headland.

1. Introduction

There are often many uncertain factors in numerical ocean models, e.g. initial and boundary conditions, and turbulent and other sub-grid scale modelling parameters. With the continued introduction of advanced observational techniques, more attention has been giving to data assimilation techniques to improve the predictive capabilities of ocean models, see for example (Bennett et al., 2000; Navon et al., 1992; Giles and Pierce, 2000; Moore, 1991; Bosseur et al., 2002; Lakshmivarahan et al., 2003; Martine et al., 2002). The adjoint method is a technique used to calculate the gradient of a cost function (gauging the difference between observed and modelled results) while taking into account the dynamics of a simulated ocean flow. The cost function describes the fit of the numerical results to observations, and is optimized by adjusting the uncertain control variables in an ocean model over both space and time (Sandu et al. 2003; Alekseev and Navon, 2003). In this investigation, an adjoint method is used with a 3D unstructured mesh ocean model. This model is non-hydrostatic, uses finite elements to discretise in space on 3-D anisotropic adapting meshes. Optimal domain decomposition methods are used for parallel computing.

Here a continuous adjoint model is derived from the continuous forward model, this is discretised to obtain our discrete adjoint model. In this case, the discrete adjoint model cannot be guaranteed to be consistent with the discrete forward model. Other approaches to obtain discrete adjoint models, formed directly from the discrete forward model, would be consisitent. The inconsistency however allows us here to use different transient evolving meshes for both the forward and adjoint models which may have widely varying characteristics and requirements in terms of resolution and mesh anisotropy. For example, for the forward flow higher resolution may be desirable near a headland or coastline shedding an eddy, whereas the adjoint model may require higher resolution near the detectors where observational data is available. Mesh adaptivity algorithms may then be employed to yield meshes which allow the efficient and accurate simulation of both the forward and adjoint solutions. This seems a sensible approach whose advantages may outweigh the fact that consistency has been sacrificed. This paper represents the first investigation of the viability of such an approach.

The rest of this paper is organized as follows. In Section 2, the inversion problem is presented and the optimization techniques introduced. Sections 3-5 describe the forward model and the development of the adjoint model. In Section 6, we present the implementation of the adaptive mesh adjoint technique in practice. A study case is presented in Section 7, followed by a discussion and some conclusions.

2. The inverse problem

1) The objective function

The objective of the optimization, which is achieved by adjusting the uncertainties in the model, is to minimize an objective function. This function is defined as the misfit between the numerical solution and available observations. A term in the objective function is also introduced to penalize unreasonably large values of the control variables (Gunzburger, 2003; Xie et al., 2002; Zhu and Navon, 1999). The objective cost function used here for the inversion of boundary conditions can be defined as follows:

$$\Im(\mathbf{u},\mathbf{m}) = \frac{1}{2}\int_t\int_\Omega\sum_{k=1}^{Nos}(\mathbf{u}-\mathbf{u}_{o,k})^T \cdot W_k \cdot (\mathbf{u}-\mathbf{u}_{o,k})d\Omega dt + \frac{1}{2}\lambda\int_t\int_{\partial\Omega}\mathbf{u}_b{}^T \cdot \mathbf{u}_b d(\partial\Omega)dt. \quad (1)$$

Where, k represents an index of detectors which yield observations and Nos is the total number of detectors; $\mathbf{m}$ represents control variables, here these are taken to be boundary conditions to be inverted for, i.e. $\mathbf{m} = \mathbf{u}_b$; $\mathbf{u} = (u, v, w)$ is the flow velocity vector in the forward model, which is a function of (x, y, z, t); x, y, z are the spatial coordinate values and t represents time; $\mathbf{u}_{o,k} = (u_{o,k}, v_{o,k}, w_{o,k})$ is the observed velocity vector at detector k; subscript 'o' represents an observation; W_k is a scalar weight associated with the misfit between the numerical solution and observations at detector k; $\mathbf{u}_b$ is the numerical flow velocity vector at the boundary, which is unknown and will be inverted for using the adjoint method; λ is a suitable scalar for the penalty term; Ω is the computational domain and $\partial\Omega$ is its boundary which contains the unknown boundary conditions (control variables). In this work the adjoint method is used to calculate the gradient of the objective function (1), this is then used to help minimize (1).

2) The optimization technique

The minimum of the function $\Im(\mathbf{u},\mathbf{m})$ is achieved on the basis of a succession of steps,

$$\mathbf{m}_{i+1} = \mathbf{m}_i + \Delta\mathbf{m}_i, \quad (2)$$

where i is an iteration index. In this work, we use the nonlinear conjugate gradient algorithm (Bishop, 1995). Giving the conjugate direction $\mathbf{d}_i$ at the iteration i and the step length α, the increment of the control vector $\Delta\mathbf{m}_i$ will be $\alpha\mathbf{d}_i$. Eq. (3) can then be written

$$\mathbf{m}_{i+1} = \mathbf{m}_i + \alpha\mathbf{d}_i. \quad (3)$$

The search is initialized with the direction $\mathbf{d}_1 = -\mathbf{g}_1$, where $\mathbf{g}_1$ is the gradient of the functional (1) at the first conjugate gradient iteration. The gradient is calculated directly from the adjoint solution. We find the optimal step length α using a line search method (Bishop, 1995). The line search procedure is then applied along a set of conjugate directions

$$\mathbf{d}_{i+1} = -\mathbf{g}_i + \beta_i\mathbf{d}_i, \quad (4)$$

the coefficient β_i can be written in the Polak-Ribiere form (Bishop, 1995)

$$\beta_i = \frac{\mathbf{g}_i^T(\mathbf{g}_{i+1} - \mathbf{g}_i)}{\mathbf{g}_i^T\mathbf{g}_i}. \quad (5)$$

The whole nonlinear conjugate gradient procedure is terminated when the change in the control variables is below a specified tolerance.

3. Forward model

The forward model here is described by the Navier-Stokes equations, for which the momentum equations are

$$\frac{\partial u}{\partial t} + u\frac{\partial u}{\partial x} + v\frac{\partial u}{\partial y} + w\frac{\partial u}{\partial z} - fv + \frac{1}{\rho}\frac{\partial p}{\partial x} - D_x = 0, \quad (6)$$

$$\frac{\partial v}{\partial t} + u\frac{\partial v}{\partial x} + v\frac{\partial v}{\partial y} + w\frac{\partial v}{\partial z} + fu + \frac{1}{\rho}\frac{\partial p}{\partial y} - D_y = 0, \quad (7)$$

$$\frac{\partial w}{\partial t} + u\frac{\partial w}{\partial x} + v\frac{\partial w}{\partial y} + w\frac{\partial w}{\partial z} + g + \frac{1}{\rho}\frac{\partial p}{\partial z} - D_z = 0, \quad (8)$$

where D_x, D_y, D_z are the viscous terms.

As well as the continuity equation

$$\frac{\partial u}{\partial x}+\frac{\partial v}{\partial y}+\frac{\partial w}{\partial z}=0. \tag{9}$$

The boundary conditions for the forward model can be written

$$u|_{\partial\Omega}=u(x,y,z,t)|_{\partial\Omega}, \qquad v|_{\partial\Omega}=v(x,y,z,t)|_{\partial\Omega}, \qquad w|_{\partial\Omega}=w(x,y,z,t)|_{\partial\Omega}. \tag{10}$$

We also employ zero normal flow velocity boundary conditions at coastal boundaries, at outflow the natural boundary conditions of zero normal and shear stress are applied, and at inflow boundaries Dirichlet velocity boundary conditions are imposed. These Dirichlet conditions are inverted for in this work by adjoint data assimilation.

The initial conditions are given as

$$u=u(x,y,z,0), \qquad v=v(x,y,z,0), \qquad w=w(x,y,z,0). \tag{11}$$

In the applications presented here the fluid is assumed to be initially at rest.

4. Tangent linear model

By differentiating (6)-(9) with respect to perturbations of the control variables **m**, the tangent linear model is obtained:

$$\frac{\partial \bar{u}}{\partial t}+u\frac{\partial \bar{u}}{\partial x}+v\frac{\partial \bar{u}}{\partial y}+w\frac{\partial \bar{u}}{\partial z}-f\bar{v}+(\bar{u}\frac{\partial u}{\partial x}+\bar{v}\frac{\partial u}{\partial y}+\bar{w}\frac{\partial u}{\partial z})+\frac{1}{\rho}\frac{\partial \bar{p}}{\partial x}-\bar{D}_x=0, \tag{12}$$

$$\frac{\partial \bar{v}}{\partial t}+u\frac{\partial \bar{v}}{\partial x}+v\frac{\partial \bar{v}}{\partial y}+w\frac{\partial \bar{v}}{\partial z}+f\bar{u}+(\bar{u}\frac{\partial v}{\partial x}+\bar{v}\frac{\partial v}{\partial y}+\bar{w}\frac{\partial v}{\partial z})+\frac{1}{\rho}\frac{\partial \bar{p}}{\partial y}-\bar{D}_y=0, \tag{13}$$

$$\frac{\partial \bar{w}}{\partial t}+u\frac{\partial \bar{w}}{\partial x}+v\frac{\partial \bar{w}}{\partial y}+w\frac{\partial \bar{w}}{\partial z}+(\bar{u}\frac{\partial w}{\partial x}+\bar{v}\frac{\partial w}{\partial y}+\bar{w}\frac{\partial w}{\partial z})+\frac{1}{\rho}\frac{\partial \bar{p}}{\partial z}-\bar{D}_z=0, \tag{14}$$

$$\frac{\partial \bar{u}}{\partial x}+\frac{\partial \bar{v}}{\partial y}+\frac{\partial \bar{w}}{\partial z}=0, \tag{15}$$

where $\bar{u}$, $\bar{v}$, $\bar{w}$ are the variations of u, v, w with respect to perturbations of **m**, i.e.

$$\bar{u} = \frac{\partial u}{\partial m}, \quad \bar{v} = \frac{\partial v}{\partial m}, \quad \bar{w} = \frac{\partial w}{\partial m}.$$

5. Adjoint model

We adopt a standard approach to derive the continuous adjoint equations, by first introducing the adjoint variables u^*, v^*, w^*, p^*, and defining the Lagrangian function subject to the strong dynamical constraints implied by the flow (6)–(9),

$$L(\mathbf{u},\mathbf{m},\mathbf{u}^*) = \Im(\mathbf{u},\mathbf{m}) + \int_t\int_\Omega (E_1 u^* + E_2 v^* + E_3 w^* + E_4 p^*) d\Omega dt, \tag{16}$$

where E_1, E_2, E_3 and E_4 are the left hand sides of (6) – (9) respectively. By differentiating (16) and considering the tangent linear model (12) – (15), which we assume is satisfied exactly (i.e. the application of strong dynamical constraints), we obtain

$$\frac{\partial L(\mathbf{u},\mathbf{m},\mathbf{u}^*)}{\partial m} = \frac{\partial \Im(\mathbf{u},\mathbf{m})}{\partial m}$$

$$+\int_t\int_x\int_y\int_z [\frac{\partial \bar{u}}{\partial t} + u\frac{\partial \bar{u}}{\partial x} + v\frac{\partial \bar{u}}{\partial y} + w\frac{\partial \bar{u}}{\partial z} - f\bar{v} + (\bar{u}\frac{\partial u}{\partial x} + \bar{v}\frac{\partial u}{\partial y} + \bar{w}\frac{\partial u}{\partial z}) + \frac{1}{\rho}\frac{\partial \bar{p}}{\partial x} - \bar{D}_x] \cdot u^* dzdydxdt$$

$$+\int_t\int_x\int_y\int_z [\frac{\partial \bar{v}}{\partial t} + u\frac{\partial \bar{v}}{\partial x} + v\frac{\partial \bar{v}}{\partial y} + w\frac{\partial \bar{v}}{\partial z} + f\bar{u} + (\bar{u}\frac{\partial v}{\partial x} + \bar{v}\frac{\partial v}{\partial y} + \bar{w}\frac{\partial v}{\partial z}) + \frac{1}{\rho}\frac{\partial \bar{p}}{\partial y} - \bar{D}_y] \cdot v^* dzdydxdt$$

$$+\int_t\int_x\int_y\int_z [\frac{\partial \bar{w}}{\partial t} + u\frac{\partial \bar{w}}{\partial x} + v\frac{\partial \bar{w}}{\partial y} + w\frac{\partial \bar{w}}{\partial z} + (\bar{u}\frac{\partial w}{\partial x} + \bar{v}\frac{\partial w}{\partial y} + \bar{w}\frac{\partial w}{\partial z}) + \frac{1}{\rho}\frac{\partial \bar{p}}{\partial z} - \bar{D}_z] \cdot w^* dzdydxdt$$

$$+\int_t\int_x\int_y\int_z [\frac{\partial \bar{u}}{\partial x} + \frac{\partial \bar{v}}{\partial y} + \frac{\partial \bar{w}}{\partial z}] \cdot p^* dzdydxdt. \tag{17}$$

However, in the discrete sense the tangent linear equations are not satisfied exactly (last form terms in this equation are not zero). We however still assume the form (17) (and assume the last four terms are zero) for the construction of the discrete adjoint model. This is the route cause of the inconsistency between the forward and adjoint models.

By taking an assimilation time window of [0, T] and integrating (17) by parts, we obtain an expression involving $\bar{u}$ given by

$$\int_x\int_y\int_z [u^*\bar{u}]\Big|_0^T dzdydx$$

$$+\int_t\int_y\int_z [2uu^* + vv^* + ww^* + p^*]\bar{u}\Big|_{\Gamma_x} dzdydt + \int_t\int_x\int_z [vu^*\bar{u}]\Big|_{\Gamma_y} dzdxdt + \int_t\int_x\int_y [wu^*\bar{u}]\Big|_{\Gamma_z} dydxdt$$

$$+\int_t\int_y\int_z [N_h \frac{\partial \bar{u}}{\partial x} u^*]_{\Gamma_x} dzdydt - \int_t\int_y\int_z [N_h \bar{u} \frac{\partial u^*}{\partial x}]_{\Gamma_x} dzdydt$$

$$+\int_t\int_x\int_z [N_h \frac{\partial \bar{u}}{\partial y} u^*]_{\Gamma_y} dzdxdt - \int_t\int_x\int_z [N_h \bar{u} \frac{\partial u^*}{\partial y}]_{\Gamma_y} dzdxdt$$

$$+\int_t\int_x\int_y [N_z \frac{\partial \bar{u}}{\partial z} u^*]\Big|_{\Gamma_z} dydxdt - \int_t\int_x\int_y [N_z \bar{u} \frac{\partial u^*}{\partial z}]\Big|_{\Gamma_z} dydxdt$$

$$+\int_t\int_x\int_y\int_z [-\frac{\partial u^*}{\partial t} - (\frac{\partial uu^*}{\partial x} + \frac{\partial vu^*}{\partial y} + \frac{\partial wu^*}{\partial z}) + fv^* - \frac{\partial p^*}{\partial x} - D_x^* - (u\frac{\partial u^*}{\partial x} + v\frac{\partial v^*}{\partial x} + w\frac{\partial w^*}{\partial x})] \cdot \bar{u} dxdzdydt. \quad (18)$$

analogously we obtain an expression involving $\bar{v}$ given by

$$\int_x\int_y\int_z [v^*\bar{v}]\Big|_0^T dzdydx$$

$$+\int_t\int_x\int_z [uu^* + 2vv^* + ww^* + p^*]\bar{v}\Big|_{\Gamma_y} dzdxdt + \int_t\int_y\int_z [uv^*\bar{v}]\Big|_{\Gamma_x} dzdydt + \int_t\int_x\int_y [wv^*\bar{v}]\Big|_{\Gamma_z} dydxdt$$

$$+\int_t\int_y\int_z [N_h \frac{\partial \bar{v}}{\partial x} v^*]_{\Gamma_x} dzdydt - \int_t\int_y\int_z [N_h \bar{v} \frac{\partial v^*}{\partial x}]_{\Gamma_x} dzdydt$$

$$+\int_t\int_x\int_z [N_h \frac{\partial \bar{v}}{\partial y} v^*]_{\Gamma_y} dzdxdt - \int_t\int_x\int_z [N_h \bar{v} \frac{\partial v^*}{\partial y}]_{\Gamma_y} dzdxdt$$

$$+\int_t\int_x\int_y [N_z \frac{\partial \bar{v}}{\partial z} v^*]\Big|_{\Gamma_z} dydxdt - \int_t\int_x\int_y [N_z \bar{v} \frac{\partial v^*}{\partial z}]\Big|_{\Gamma_z} dydxdt$$

$$+\int_t\int_x\int_y\int_z [-\frac{\partial v^*}{\partial t} - (\frac{\partial uv^*}{\partial x} + \frac{\partial vv^*}{\partial y} + \frac{\partial wv^*}{\partial z}) - fu^* - \frac{\partial p^*}{\partial y} - D_y^* - (u\frac{\partial u^*}{\partial y} + v\frac{\partial v^*}{\partial y} + w\frac{\partial w^*}{\partial y})] \cdot \bar{v} dxdzdydt, \quad (19)$$

and for $\bar{w}$ given by

$$\int_x\int_y\int_z [w^*\bar{w}]\Big|_0^T dzdydx$$

$$+\int_t\int_x\int_y [uu^* + vv^* + 2ww^* + p^*]\bar{w}\Big|_{\Gamma_z} dxdydt + \int_t\int_y\int_z [uw^*\bar{w}]\Big|_{\Gamma_x} dzdydt + \int_t\int_x\int_z [vw^*\bar{w}]\Big|_{\Gamma_y} dzdxdt$$

$$+\int_t\int_y\int_z [N_h \frac{\partial \bar{w}}{\partial x} w^*]_{\Gamma_x} dzdydt - \int_t\int_y\int_z [N_h \bar{w} \frac{\partial w^*}{\partial x}]_{\Gamma_x} dzdydt$$

$$+\int_t\int_x\int_z [N_h \frac{\partial \bar{w}}{\partial y} w^*]_{\Gamma_y} dzdxdt - \int_t\int_x\int_z [N_h \bar{w} \frac{\partial w^*}{\partial y}]_{\Gamma_y} dzdxdt$$

$$+\int_t\int_x\int_y [N_z \frac{\partial \bar{w}}{\partial z} w^*]\Big|_{\Gamma_z} dydxdt - \int_t\int_x\int_y [N_z \bar{w} \frac{\partial w^*}{\partial z}]\Big|_{\Gamma_z} dydxdt$$

$$+\int_t\int_x\int_y\int_z [-\frac{\partial w^*}{\partial t} - (\frac{\partial uw^*}{\partial x} + \frac{\partial vw^*}{\partial y} + \frac{\partial ww^*}{\partial z}) - \frac{\partial p^*}{\partial z} - D_z^* - (u\frac{\partial u^*}{\partial z} + v\frac{\partial v^*}{\partial x} + w\frac{\partial w^*}{\partial z})] \cdot \bar{w} dxdzdydt. \quad (20)$$

Where D_x^*, D_y^*, D_z^* are simply the self-adjoint terms D_x, D_y, D_z respectively and finally the terms involving $\bar{p}$ are given by

$$\int_t\int_y\int_z (\frac{1}{\rho} u^* \bar{p})\Big|_{\Gamma_x} dzdydt + \int_t\int_x\int_z (\frac{1}{\rho} v^* \bar{p})\Big|_{\Gamma_y} dzdxdt + \int_t\int_x\int_y (\frac{1}{\rho} w^* \bar{p})\Big|_{\Gamma_z} dydxdt +$$

$$\int_t \int_x \int_y \int_z -[\frac{\partial(\frac{1}{\rho}u^*)}{\partial x} + \frac{\partial(\frac{1}{\rho}v^*)}{\partial y} + \frac{\partial(\frac{1}{\rho}w^*)}{\partial z}] \cdot \bar{p} dz dy dx dt \, .$$

(21)

Therefore, the adjoint model equations are defined as follows:

$$-\frac{\partial u^*}{\partial t} - (\frac{\partial uu^*}{\partial x} + \frac{\partial vu^*}{\partial y} + \frac{\partial wu^*}{\partial z}) + fv^* - \frac{\partial p^*}{\partial x} - D_x^* - (u\frac{\partial u^*}{\partial x} + v\frac{\partial v^*}{\partial x} + w\frac{\partial w^*}{\partial x}) = (u - u_o)\delta(x - x_o)\delta(y - y_o)\delta(z - z_o),$$

(22)

$$-\frac{\partial v^*}{\partial t} - (\frac{\partial uv^*}{\partial x} + \frac{\partial vv^*}{\partial y} + \frac{\partial wv^*}{\partial z}) - fu^* - \frac{\partial p^*}{\partial y} - D_y^* - (u\frac{\partial u^*}{\partial y} + v\frac{\partial v^*}{\partial y} + w\frac{\partial w^*}{\partial y}) = (v - v_o)\delta(x - x_o)\delta(y - y_o)\delta(z - z_o),$$

(23)

$$-\frac{\partial w^*}{\partial t} - (\frac{\partial uw^*}{\partial x} + \frac{\partial vw^*}{\partial y} + \frac{\partial ww^*}{\partial z}) - \frac{\partial p^*}{\partial z} - D_z^* - (u\frac{\partial u^*}{\partial z} + v\frac{\partial v^*}{\partial x} + w\frac{\partial w^*}{\partial z}) = (w - w_o)\delta(x - x_o)\delta(y - y_o)\delta(z - z_o),$$

(24)

$$\frac{\partial u^*}{\partial x} + \frac{\partial v^*}{\partial y} + \frac{\partial w^*}{\partial z} = 0. \quad (25)$$

where, u_o, v_o, w_o are the observed velocities at position (x_o, y_o, z_o). The adjoint model starts from $u^* = v^* = w^* = 0$ at the end of the time domain and is marched backwards in time from $t = T$ to $t = 0$. The boundary terms in (18)-(20) are used to construct the adjoint system with appropriate boundary conditions, which are dependent on the inversion problem being solved. The solution of these equations allows us to obtain the gradient of the objective function, examples of which are given below. Details of the construction of the boundary conditions for the adjoint model have been described by other authors (Gunzburger, 2003; Giles and Pierce, 2000; Navon, 2003). For example suppose we wish to invert for the boundary conditions at an inlet, i.e. here the control variables **m** consist of the velocities (boundary conditions) at the inlet. By giving natural boundary conditions at the outflow in the forward model, $u^* = 0$ for the adjoint model at the outflow, and substituting (18) into (17), and satisfying the gradient of the Langrangian function (16) is zero, i.e., ***DL/Dm = 0***, the gradient at the inflow boundary will be

$$\left(\frac{\partial \Im}{\partial \mathbf{u}(x,y,z)}\right)_x\Big|_{x=0} = -2uu^*\Big|_{x=0} \, . \quad (26)$$

If we include a penalty term in the formulation of the cost function, then the gradient is

$$\left.\left(\frac{\partial \Im}{\partial \mathbf{u}(x,y,z)}\right)_X\right|_{x=0} = (-2uu^* - \lambda u)\big|_{x=0}. \tag{27}$$

Here λ is set to the value 0.1, although results are reasonably insensitive to this choice.

6. Implementation of the adaptive mesh adjoint method

(1) The forward model

In this work the forward model is given by a finite element discretisation of the non-hydrostatic Boussinesq equations in 3D. The model uses piecewise-linear basis functions for the components of velocity and pressure. The time-stepping is typically given by the second-order Crank-Nicolson scheme, with a projection method for pressure ensuring the flow stays divergence-free whilst decoupling the computations for the momentum and continuity equations. A filter is employed to remove spurious modes in the pressure field, and the option for Petrov-Galerkin weighting is used to remove instabilities due to the convectively dominated nature of the flow. For more details see (Ford et al 2003, Pain et al. 2001). Other approaches to forward modelling of coastal flow using finite elements may be found in for example (Lynch and Gray 1979; Lynch et al. 1996), often spurious free surface modes are removed using a wave equation reformulation of the continuity equation. The inclusion of free surface dynamics in the current model is the subject of current ongoing research.

(2) Numerical error measurement and mesh adaptivity

To control the quality and resolution of simulations we impose certain parameter values, in particular the maximum and minimum element sizes (here these are taken to be 2000m and 200m respectively) in the mesh, as well as an allowable tolerance in the error (here taken to be 0.05m) implied by the use of linear interpolants to approximate smooth fields. The interpolation errors are computed directly via the Hessian of the solution fields which are then superimposed to obtain an error metric. This metric then controls both the resolution and anisotropy of the mesh. In the simulations presented the metric is actually normalized with respect to the solution fields so as not to overly bias mesh resolution towards large values of the fields. In the simulations to be presented the mesh is initially structured for ease of generation, although more sophisticated mesh generation algorithms may be employed (Gorman et al. 2004). The mesh adaptivity algorithm then alters the mesh through a sequence of local mesh connectivity changes and node movement, as guided by the constructed metric. For more details see (Pain et al. 2001). The adapted meshes obtained using this approach generally have a smooth transition in element sizes, important for avoiding spurious wave reflections. However a hard constraint on element size ratios may also be built in and is the subject of continued research.

(3) Interface for the forward and adjoint models

An important aspect of the implementation of the adaptive mesh inversion code is to establish an interface module - interfacing the forward and adjoint models. As stated above, after the mesh is adapted in space and time, the optimal mesh in the forward model can be different from the optimal mesh used in the adjoint model, although they may start from the same mesh. The interface has been designed to undertake the following tasks:

(a) The treatment of control variables

In the discretized system, the control vector is $\mathbf{m} = (\mathbf{m}^1, ..., \mathbf{m}^k, ..., \mathbf{m}^{Ntime})$, where $\mathbf{m}^k$ are the controls at time level *k*, in which $k \in \{1, 2, ..., Ntime\}$, *and Ntime* is the number of time levels. For each time level $\mathbf{m}^k = (m_1^k, ..., m_j^k, ..., m_{Nb_k}^k)$, in which Nb_k is the number of boundary nodes, which may well be changing from one time step to the next. The total number of controls **m** is $\sum_{k=1}^{Ntime} Nb_k$ in this case. The interface performs the following sequence for control variables:

- The forward mesh *(from time level 1 to Ntime)* at the beginning of the line search is chosen for the calculation of $\mathbf{m}_{i+1}$ in (3) at each nonlinear iteration, here called the Control Variable Mesh (CVM). This mesh dynamically changes with time;
- At each nonlinear conjugate gradient iteration, the CVM will be updated, i.e. the new mesh for the controls is that from the forward model mesh of the previous nonlinear conjugate gradient iteration. The search direction in (4) will be calculated on the new CVM. Note that the previous gradient $\mathbf{g}_i$ is interpolated from the previous CVM to the new CVM;
- The gradients $\mathbf{g}_i$ and $\mathbf{g}_{i+1}$ are calculated from the adjoint solutions, which are interpolated from the current adjoint mesh (*also dynamically changing with time)* onto the CVM;
- When running the forward model, the updated control vector $\mathbf{m}_{i+1}$ will be re-interpolated from the CVM onto the current forward mesh.

(b) *The treatment of u, v, w in the adjoint model*

To run the adjoint model we require the discrete solution variables from the forward model to be located at the same locations as the adjoint model on the adjoint mesh. Because mesh adaptivity is being used in both the forward and adjoint models the two meshes are in general different. Therefore an interpolation step is required to map forward solution variables such as ***u***, ***v*** and ***w*** onto the same mesh being used by the adjoint model. For this work we use a grid-to-grid linear interpolation technique which combines advancing front vicinity, local vicinity and brute force searches to provide a very fast and robust search method to establish the elements in the source mesh contain the vertices in the destination mesh. Values are then interpolated using the finite element bases function, further details may be found in (Löhner, 1995). The search is performed once for each new forward-adjoint mesh pair and the search result then stored for interpolation until a new mesh pair is generated. The use of linear interpolation does introduce some smoothing into the numerical algorithm. However mesh adaptivity is not invoked at each time step, only when the solution fields have evolved a significant amount, thus the spurious smoothing is deemed negligible. In addition, since linear interpolation is also used between forward and adjoint meshes repeatedly, again some smoothing is introduced. This can again be minimized through the use of more sophisticated higher-order shape-preserving interpolation algorithms. However these shall be the subject of future investigation and the matter is not considered further here.

A computational flow-chart is shown in Fig. 1.

7. Inversion results for flow past a headland

To test the feasibility of the inconsistent discrete system, we apply the adaptive mesh adjoint model to invert boundary conditions for flow past a headland (Fig. 2). The computational domain is 30,000m by 15,000m, with a simple triangular headland measuring 4000m by 3750m, see Fig. 2. Based upon the length scale of the headland and a typical maximum velocity of 1m/s the Reynolds number of the flow is chosen to be 800. Ten detector positions are set at $x = 17500 + 500(i+1)\,\mathrm{m}$, $y = 7500\,\mathrm{m}$. Velocity observations are assumed which yield expected boundary velocities at the inlet of $u = A\sin(t/T), v = 0, w = 0$. In the simulation to be presented below the amplitude A is set to 0.5m and the period T is 1 hour. The pseudo-observations at the detectors are obtained here by running the forward model with mesh adaptivity using the same options and mesh parameters used in the forward/adjoint data assimilation procedure. This single forward run is forced by the exact expected boundary condition, and numerical results at the ten detector positions stored at each time level. Since the pseudo-observations are obtained on an adapting mesh which may not coincide exactly with the observation locations, the data is spread throughout the domain via convolution with a Gaussian function.
Fig. 3 (a) and (b) show the adaptive meshes for the forward and adjoint models respectively, which are adapted from the original regular mesh (Fig. 2) according to their own flow characteristics, as implied by the curvature, or Hessian, of the flow fields, see section 6. We can see that both meshes conform well to local flow features. It is worth mentioning that Fig. 3 (a) shows the adaptive mesh only at a particular time level. At a later time the location of higher resolution in the adapted mesh has moved to the other side of the headland due to the specified periodic forcing.
The corresponding velocity and adjoint fields are also displayed in Figs. 4 (a) and (b). The flow field (Fig. 4(a)) is forced by the inverted inflow at the inlet boundary. We run the models with and without the mesh adaptivity. It is found that local eddies near the headland can be simulated only when the mesh adaptivity technique is used.
The inversion of the boundary conditions can be seen in Fig. 5. The inversion was started with an initial guess of the boundary condition: $x|_{x=0} = -4.0(t/T - 0.5)^2 + 1.0$. The final inverted boundary conditions at the inlet are close to the expected boundary conditions. The method is particularly good in the first quarter of simulated time. After the third nonlinear iteration, the maximum misfit of velocities between the inverted and expected boundary conditions is reduced from ~ 1.0 m/s to 0.2 m/s. The adjoint model was started from the end of simulation time ($t = T$) when the initial values of the adjoint variables are set to zero. This means that the adjoint solution, as well as the gradient of the cost function (25) and (26), is close to zero around the simulation time ($t = T$). This explains why the control variables (i.e. the inflow boundary conditions) in (2) – (4) cannot be inverted for near the simulation time $t = T$, and that there is a lag in phase between the

inverted and expected boundary conditions during the simulation period $t = [0.5T, T]$, see Fig. (5).

8. Conclusions

In this investigation, a new adaptive mesh adjoint model, designed for costal flow, is presented. Its adjoint discretised equations are not consistent with the discretised forward equations. This means that the adjoint discrete system is very flexible and can be computed on an adaptive mesh well suited to the unique requirements of the adjoint solution e.g. with mesh refinement in the vicinity of the observations used in the inversion procedure. The inversion can thus be performed with a dynamically adaptive mesh. This also enables a hierarchy of increasingly fine inversion models to be easily used to accelerate the inversion. An interface module for the communication between the forward and adjoint model has been established. The interface is used to store and interpolate the information from the forward model to the adjoint model variables, and is also used to pass information regarding the controls to the forward model. These controls are calculated from the adjoint solution.

The adjoint model was tested on the problem of boundary condition inversion for flow past a headland. The results show that the meshes for the forward and the adjoint models can adapt well to resolve their own flow features. After the first three nonlinear CG iterations, the inverted boundary conditions are quite close to those used to generate the synthetic data (used in the inversion). This is encouraging and demonstrates the feasibility of using an adjoint model that is not necessarily consistent with the forward model. However much further work is required to investigate fully the advantages, and increases in efficiency of the approach outlined in this paper over, for example, a consistent adjoint fixed mesh method.

Acknowledgments.
The authors would like to thank Professor I.M. Navon for advice and many useful discussions related to this work. This research was carried out under funding from EPSRC under grant GR/60898. We appreciate all useful comments from the three reviewers.

References:

Alekseev, A. K. and Navon, I. M. (2003). "Calculation of uncertainty propagation using adjoint equations." *International Journal of Computational Fluid Dynamics,* accepted.

Bennett, A. F., Chua, B. S., Harrison, D. Ed. and McPhaden, M. J. (2000). "Generalized Inversion of Tropical Atmosphere-Ocean Data and a Coupled Model of the Tropical Pacific. II. The 1995-96 La Nina and 1997-1998 El Nino." *Journal of Climate,* 11, 2770-2785.

Bishop, C. M. (1995). Neural Networks for Pattern Recognition. Oxford University Press.

Bosseur, F., Flori, F., and Orenga, P. (2002). "Identification of boundary conditions in a nonlinear shallow water flow." *Computer and Mathematics with application,* 43, 1559-1573.

Ford, R., Pain, C .C., Piggott M. D., Goddard A. J. H., de Oliveira C. R. E. and Umpleby A. P., (2003). "A non-hydrostatic finite element model for three-dimensional stratified oceanic flows." *Monthly Weather Review,* submitted.

Giles, M. B. and Pierce, N. A. (2000). "An introduction to the adjoint approach to design." *Flow, Turbulence and Combustion,* 65(3-4), 393-415.

Gorman, G.J, Piggott, M.D., Pain, C.C., de Oliveira, C.R.E., Umpleby, A.P., and Goddard, A.J.H. "Optimal bathymetric representation through constrained unstructured mesh adaptivity." Ocean Modell. Submitted.

Gunzburger M. D.,(2003), Perspectives in flow control and optimisation, *Siam, Advance in design and control,* Editor, H. Thomas.

Lakshmivarahan, S., Honda, Y. and Lewis, J. M. (2003). "Second-order approximation to the 3DVAR cost function: application to analysis/forecast." *Tellus,* 55A, 371-384.

Löhner, R.(1995) "Rubust, vectorized search algorithms for interpolation on unstructured grids." *Journal of computational physics,* 118, 380-387.

Lynch, D.R. and Gray, W.R. (1979), "A wave equation model for finite element tidal computations." *Comput. Fluids,* 7, 207-228.

Lynch, D.R., Ip, J.T.C., Naimie, C.E. and Werner, F.E. (1996). "Comprehensive coastal circulation model with application to the Gulf of Maine." Contin. Shelf Res. 16, 875-906.

Moore, A. M. (1991) "Data assimilation in a quasi-geostrophic open-ocean model of the gulf stream region using the adjoint method." *Journal of Physical Oceanography,* 21(3), 398-427.

Martin, M. J., Bell, M. J. and Nichols, N. K. (2002) "Estimation of system error in an equatorial ocean model using data assimilation." *International Journal for Numerical Methods in Fluids,* 40, *435-444.*

Navon, I. M. (2003). Adjoint equations and inverse CFD Problems, Internal communication.

Navon, I. M., Zou, X., Derber, J. and Sela, J. (1992), "Variational data assimilation with an adiabatic version of the NMC spectral Model." *Monthly Weather Review,* 120(7), 1433-1446.

Pain, C. C., Umpleby, A. P., de Oliveira, C. R. E. and Goddard, A. J. H. (2001). "Tetrahedral mesh optimisation and adaptivity for steady-state and transient finite element calculations." *Comput. Methods Appl. Mech. Engrg.,* 190, 3771-3796.

Sandu, A., Daescu D. N., and Carmichael G. R. (2003). "Direct and adjoint sensitivity analysis of chemical kinemical kinetic systems with KPP: Part I----theory and software tools." accepted by *Atmospheric Environment.*

Xie, Y., Lu, C. and Browning, G. L. (2002). "Impact of formulation of cost function and constraints on three-Dimensional variational data assimilation." *Monthly Weather Review,* 130, 2433-2447.

Zhu, Y. and Navon, I. M. (1999). "Impact of parameter estimation on the performace of the FSU global spectral model using its full-physics adjoint." *Monthly Weather Review,* 127, 1497-1517.

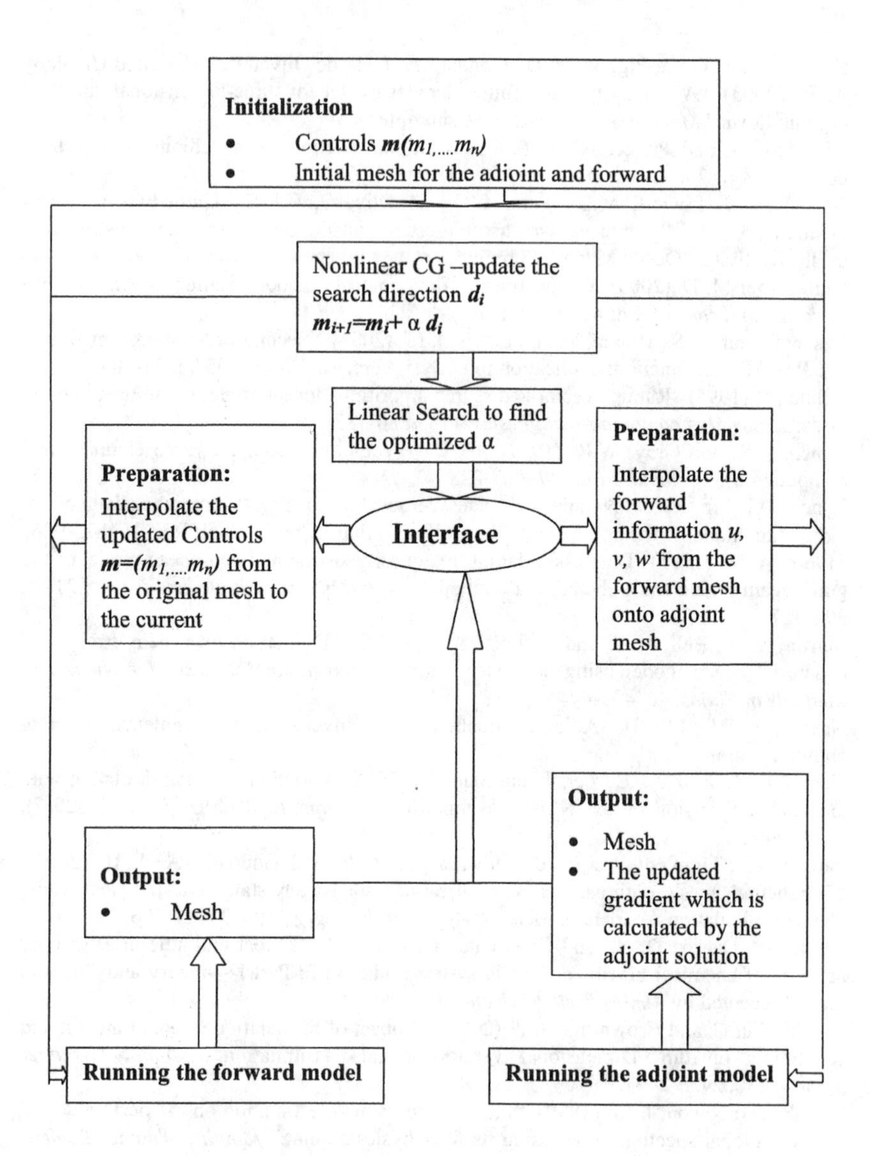

Figure 1. Flow chart of the mesh adaptive inversion procedure

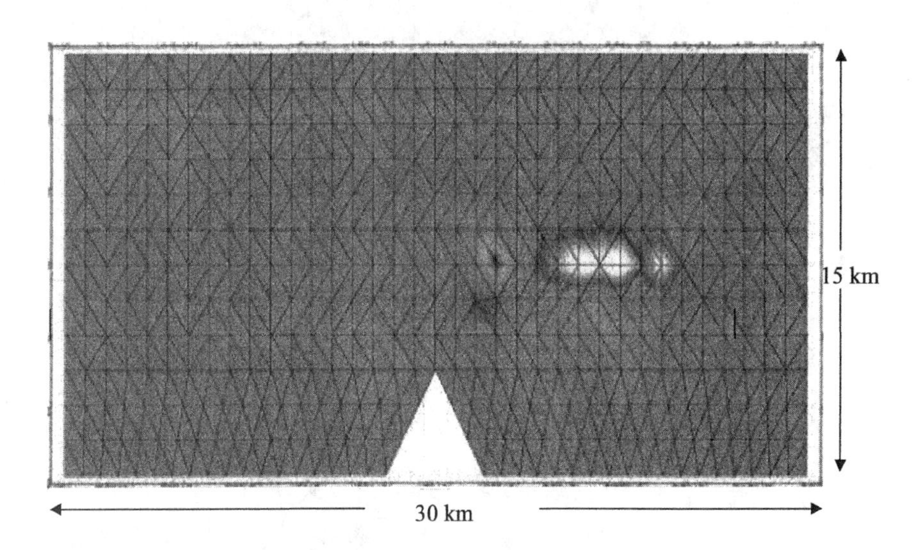

Figure 2. 2D flow past a headland - the original mesh with stream wise adjoint variable shown at an instance in time half way through the time domain.

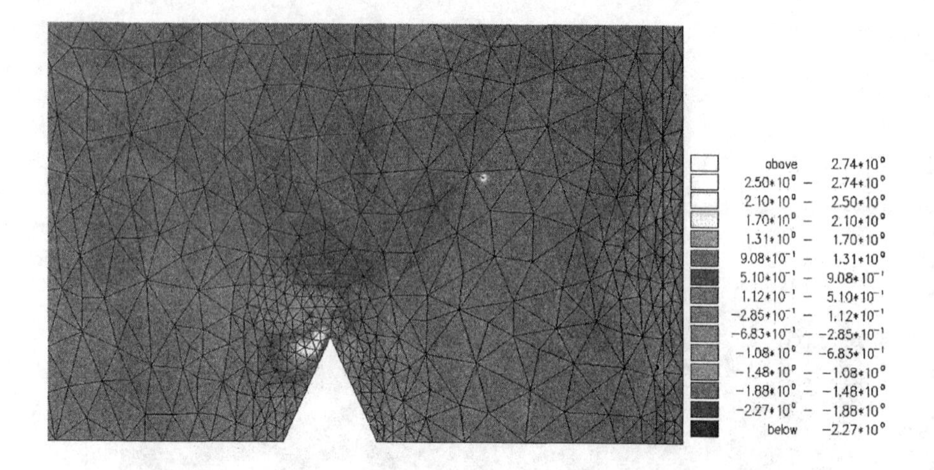

Figure 3(a). The stream-wise forward model velocity component u (at a time half way through the assimilation window) superimposed onto its forward model mesh

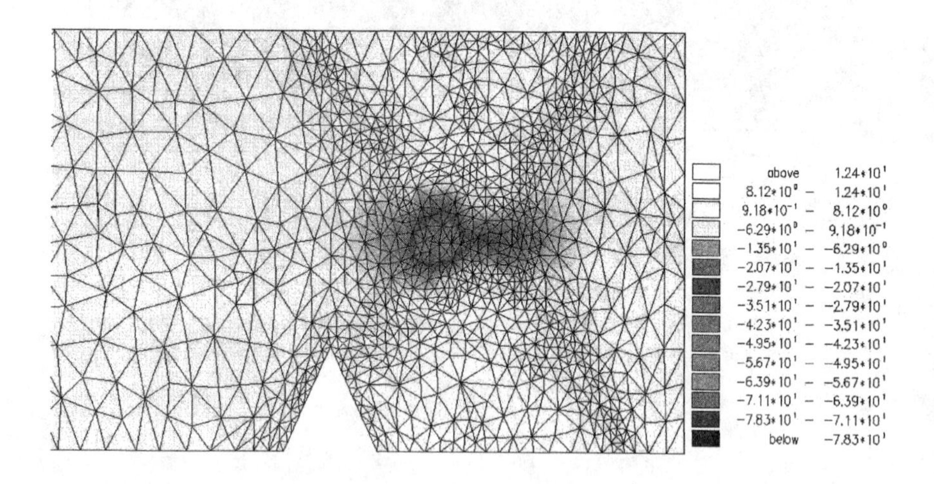

Figure 3(b). The stream-wise adjoint model velocity component u^* (at a time half way through the assimilation window) superimposed onto its adjoint model mesh

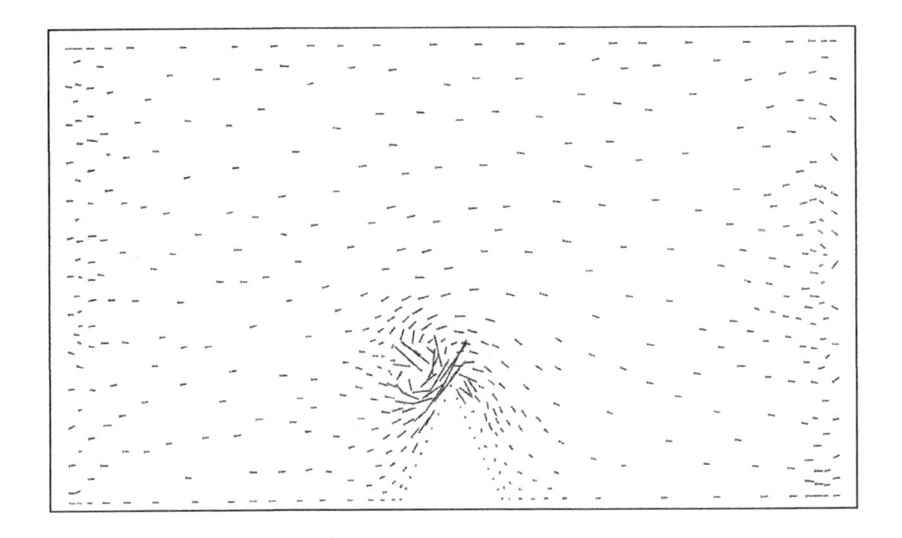

Figure 4(a). The velocity vector field for the forward model at an instant half way through time domain

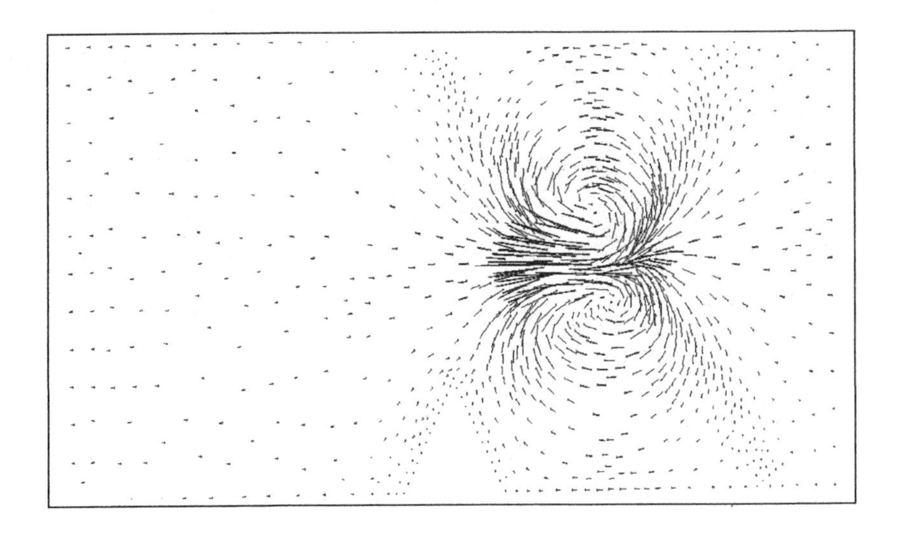

Figure 4.(b). The adjoint velocity vector field for the adjoint model at an instant half way through time domain

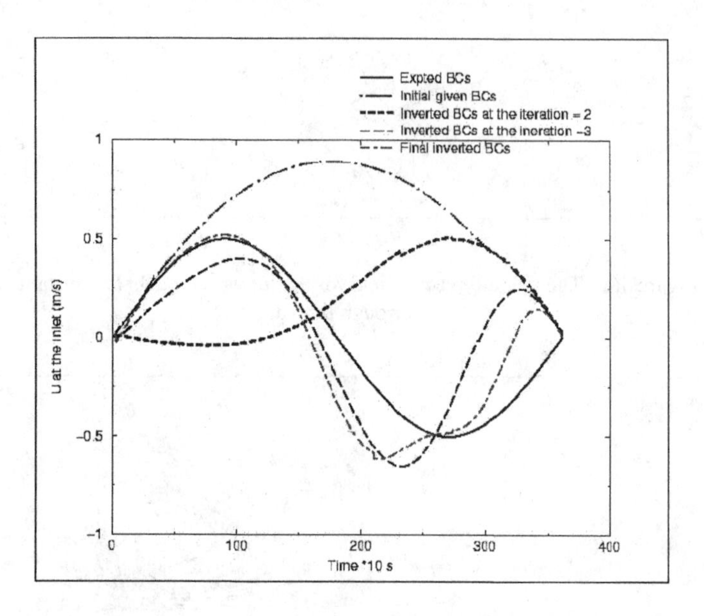

Figure 5. The mean inverted boundary conditions at the inlet for the first three conjugate gradient iterations compared with the exact boundary result.

Characteristics of Coastal Circulation in the Semi-Enclosed Seto Inland Sea Using A Regional Atmosphere-Ocean Model

Hidemi Mutsuda[1], Yoshitaka Ichii[1], Yoshiaki Akiyama[2] and Yasuaki Doi[1]

ABSTRACT

Accurate coastal weather and ocean forecasting are essential to ensure economic well-being and public safety in coastal regions. In this study, a three-dimensional mesoscale atmosphere-ocean model, was developed by coupling the Advanced Regional Prediction System (ARPS) with the Princeton Ocean Model (POM) to provide atmosphere-ocean prediction. The model was used to study coastal land-air-sea interaction in the Seto Inland Region, western part of Japan. This paper describes characteristic features of the coastal circulation in the Seto Inland Sea. The coastal circulation in the Seto Inland Sea is explained by the evolution and migration of the orographic wind system.

INTRODUCTION

The exchange of heat, moisture, momentum and particles across the air-sea interface is fundamental to an understanding of atmosphere-ocean systems, especially in a semi-enclosed sea. The semi-enclosed sea has a number of distinctive characteristics, including complex coastline and terrain variations, strong surface temperature gradients due to upwelling, and intense air-sea-land interaction. Wind-induced currents play a major role in circulation in coastal regions around the world (Smith 1995). To better understand these complicated flows, a coastal environmental system requires high-resolution observations and a coupled atmosphere-ocean modeling.

1 Dept. of Social and Environmental Engineering, Hiroshima University,
1-4-1, Kagamiyama Higashi-Hiroshima City, Hiroshima 739-8527, JAPAN
Tel : +81-824-24-7778, Fax : +81-824-22-7194, E-mail :mutsuda@naoe.hiroshima-u.ac.jp
2 Japan Meteorological Agency

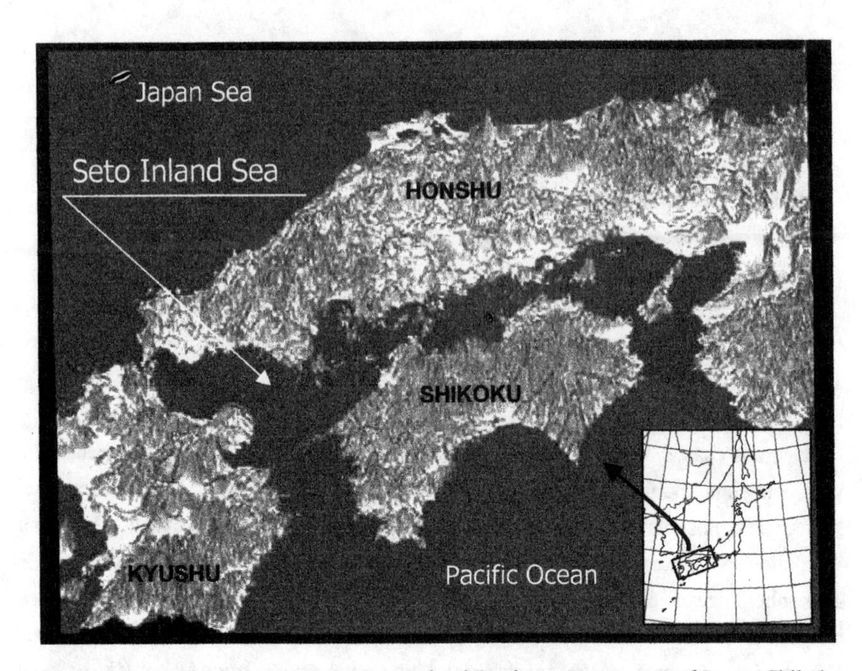

Fig.1. Geographical map over the Seto Inland Region, western part of Japan. Shikoku and Honshu are elongated in east-to-west direction. Mt. Shikoku runs from east to west.

The Seto Inland Sea is the largest semi-enclosed coastal sea in Japan, covering an area of 23,203 km^2 surrounded by Honshu, Shikoku and Kyushu, as shown in Fig.1. The Seto Inland Sea is 15-55 km wide and 450 km long and consists of deep basins with steep slopes, shallow straits, medium-deep semi-enclosed basins and approximately 1,015 islands. Nutrients from land are well mixed vertically and horizontally and efficiently recycled in the Seto Inland Sea because of the existence of many straits (Takeoka, H., 2002).

Type of atmospheric forcing in the Seto Inland Sea is characterized by the developed orographic wind all over the Seto Inland Region in relation to the well-known severe winds, called 'Hijikawa-Arashi', 'Rokko-Oroshi', 'Hirodo-Kaze' and 'Yamaji-Kaze'. Especially, terrain shape of Shikoku has a gentle slope on the southern side and a steep on the northern side, and appears to be suitable for an occurrence of a downslope surface wind when a southerly wind is present. The surface wind plays a key role in the water circulation and coastal upwelling and downwelling in the Seto Inland Sea. When it blows hard, wind-induced mixing occurs in the sea and dense nutrient salts and cysts mix up from bottom layer of the eutrophic sea. Diffusion of these materials can plays an important role triggering an occurrence of red-tide and shellfish toxicity. Coastal environments in the Seto Inland Sea are under an influence of the atmospheric forcing as wind stress on a variety of space and time scales from mesoscale to synoptic scale. However, relation between the developed orographic wind and the wind-induced current in the Seto Inland Sea is not clear. Therefore, detailed studies of the physical and biological environmental processes in the Seto Inland Sea is important for accurate coastal weather and ocean forecasting (e.g. Hayashi *et al.*, 2001 ; Minato, 1996).

The goal of this study is to develop the coupled regional atmosphere-ocean model in the Seto Inland Region, including the Seto Inland Sea, and to understand the physical estuarine process related to the air-sea interaction and the coupling of the ocean and atmosphere on the regional scale. In this study, the major focus is on the wind-driven mesoscale (2-5 km horizontal space scale, a few day time scale) coastal oceanographic processes (e.g., alongshore coastal jets, upwelling and downwelling fronts, eddies) as influenced by temporal and spatial variability of the atmospheric forcing. The coupled atmosphere-ocean model in the Seto Inland Region is applied to investigate the physical mechanism of generation of mesoscale wind-induced current influenced by the three-dimensional orographic wind.

REGIONAL ATMOSPHERE-OCEAN MODEL

The three-dimensional mesoscale atmosphere-ocean model is developed by coupling the Advance Regional Prediction System (ARPS) with the Princeton Ocean Model (POM). Details of two model configurations used are presented below.

Atmosphere model

The ARPS model (Ming Xue *et al.*, 1995) at the Center for Analysis and Prediction of Storms, University of Oklahoma, is used to estimate the three-dimensional atmospheric forcing field. The ARPS model is a state of the prognostic non-hydrostatic numerical model with advanced parameterizations of turbulence, radiation, microphysics, and other physical processes. The ARPS uses a terrain-following vertical coordinate system. The governing equations of the atmospheric model include the compressible Navier-Stokes equations, heat, mass, turbulent kinetic energy and the equation of state. These equations are represented in an orthogonal, curvilinear coordinate system in the Seto Inland Region. These equations are also expressed in a fully conservative form and solved in a rectangular computational space. The model variables are staggered on an Arakawa C-grid. The spatial difference terms are second-order accurate. The advection can be either second- or forth-order accurate. The time step uses a mode-splitting time integration system. This technique divides a big integration time step into a number of computationally inexpensive small time steps and updates the acoustically active terms every time step while computing all the other terms only once every big time step.

Ocean model

In this study, the Princeton Ocean Model (POM), developed by Blumberg and Mellor (1987), is applied to the Seto Inland Sea using rectangular cartesian coordinates. We use a time dependent, primitive equation circulation model on a three dimensional grid to investigate the estuarine circulation in response to the orographic wind on the Seto Inland Sea (See Fig.1). The POM is a three-dimensional, nonlinear, primitive equation finite difference ocean model. The governing equations are written in the sigma vertical coordinate system and include a turbulence closure parameterization, with an implicit time scheme for vertical mixing. An important simplification is the Boussinesq approximation with respect to density variations.

The ocean model is specifically designed to accommodate mesoscale phenomenon commonly found in estuarine and coastal oceanography. Nonlinear advection affects the formation of mesoscale eddies, especially, three-dimensional eddy structures and the resulting vertical mixing. In this study, the POM model is expanded with a higher-order advection scheme using the Constrained Interpolation Profile (CIP) method (Yabe *et al.*, 2001). The CIP method is a less-diffusive and more stable algorithm in comparison with any other high-order, finite difference, convective schemes. The most important point in the CIP method is a way to determine a velocity profile, which is given by a Hermite cubic spline function including a spatial derivative. The spatial derivative of a physical value is also determined to predict a velocity profile at the next time step. Therefore, the CIP method can track sharply, even at a discontinuous interface such as associated with shear flow and with strong temperature and salinity gradients.

Numerical conditions

The mesoscale atmospheric forcing is provided by the ARPS with 4km horizontal resolution and 14 vertical levels. The model configuration consists of a regional 4 km resolution to resolve the small-scale wind flow structures associated with many mountains over the Seto Inland Region. Vertical grid stretching by a hyperbolic tangent function is used to define the spacing of the vertical layers, from 10 m near the land surface to 4000m in the upper atmosphere. The calculation domain has 116×116 grid points. Since the atmosphere model described by the governing equations is compressible, the time step limited by the acoustic wave speed is 1 second.

Fig.2. Bottom topography of the Seto Inland Sea, Japan

The mesoscale ocean model is expected to treat the steep topography and the complex coastline in the Seto Inland Sea as shown in Fig.2. The model bathymetry includes the realistic topography of the Seto Inland Sea and is interpolated to the POM grid with 1-2 km horizontal resolution and 15 sigma levels conforming to the realistic bottom topography. The time step of the external mode is 5 seconds and that of the internal mode is 60 seconds. In this study, the wind-driven estuarine circulation is focused to make clear the influence of atmospheric forcing as wind stress although the tidal current is very important.

Numerical experiments are performed for cases without a general wind. However, three dimensional wind fields in the Seto Inland Region are estimated from 16 model runs where the model is forced with uniform logarithmic winds from 16 different but typical wind directions at initial conditions because the wind varies seasonally. The model is integrated for one day from an initial at rest state with a three-dimensional wind stress. It is sufficient to be a quasi steady-state wind field (Cronin *et al.*, 2000).

The steady-state wind field for each wind direction is modeled with the coupled ocean model, taking into account wind-induced currents and surface elevations. The surface fluxes of momentum across the air-sea interface happen at every ocean time step. The wind field at the lower layer of the atmosphere model is imposed as the driving force for the surface layer of the ocean model. To simplify and understand the characteristics of the wind-driven current influenced by the orographic wind in the Seto Inland Sea, heat and mass flux adjustments are not used.

RESULTS AND DISCUSSIONS

Validation

In order to validate the predictions of the present ocean model, comparisons between the calculated results of the model and that of other ocean models are presented in this section.

Figure 3 shows comparison of the sea surface elevation between the calculated results marked by solid line and that of other ocean models (POM results marked by dashed line ; Delft3D (1999) results marked by dotted) at each location. The present results are good agreement with two ocean models. This suggests that the model performed sufficiently well to provide useful estimates of the sea surface elevation in the Seto Inland Sea.

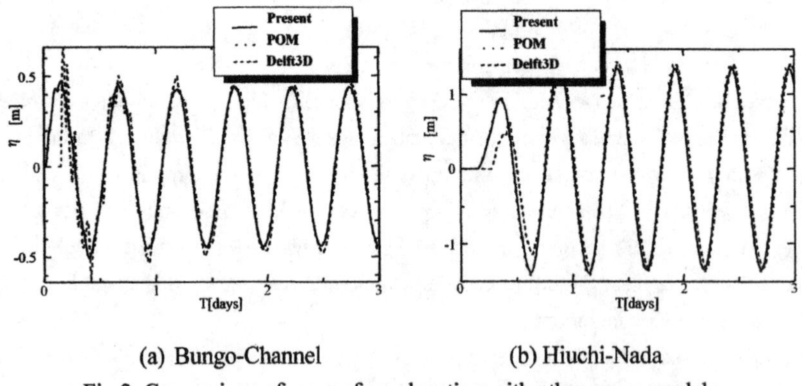

(a) Bungo-Channel (b) Hiuchi-Nada

Fig.3. Comparison of sea surface elevation with other ocean models

Another validation of the calculated results of the model is by comparing the observed data and the predicted horizontal fields of tidal current at the surface layer. Figure 4 shows comparison of the calculated and the observed (Higo *et al,* 1980) tidal current ellipses for M_2 tide at Aki-Nada, Kurushima-Strait, Hiuchi-Nada and

Bisan-Strait in the Seto Inland Sea. The calculated tidal current ellipses generally agree with observations.

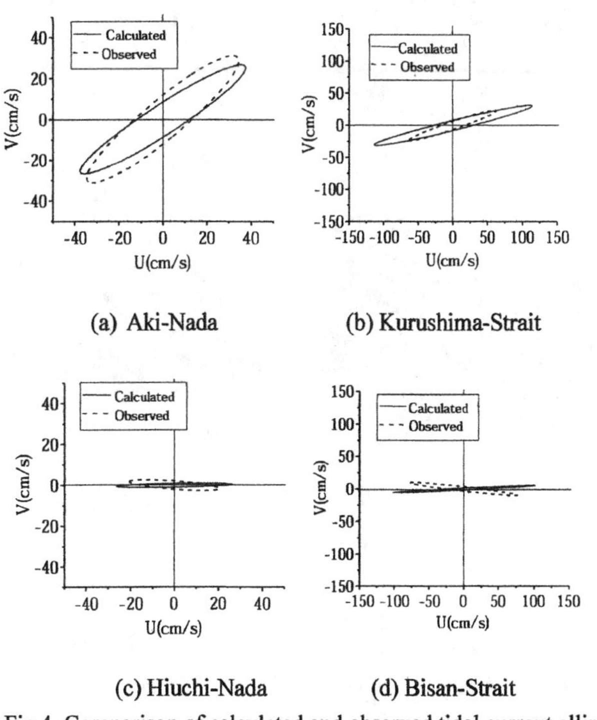

(a) Aki-Nada (b) Kurushima-Strait

(c) Hiuchi-Nada (d) Bisan-Strait

Fig.4. Comparison of calculated and observed tidal current ellipses

Figure 5 shows the instantaneous tidal current at the surface layer in Kurushima-Strait. This is one the most narrowest straits including complex coastline and terrain variations. Therefore, in the narrow strait, the maximum tidal speeds approach 5 m/s in the Seto Inland Sea (Yanagi and Higuchi, 1981). The basic features of the current in Kurushima-Strait are in reasonably good agreement with what is known from observations and this is in sharp contrast to other lower resolution models.

It is not appropriate to compare directly the calculated wind fields with observed wind ones because the numerical experiments are performed for cases without a general wind field. However, radar charts of the wind velocity will show reasonable qualitative agreement with what is known from observations as mentioned below. In the following sections, although we neglect the effect of tidal currents to simplify, characteristics of the wind-induced current influenced by the atmospheric forcing over the Seto Inland Sea will be made clear.

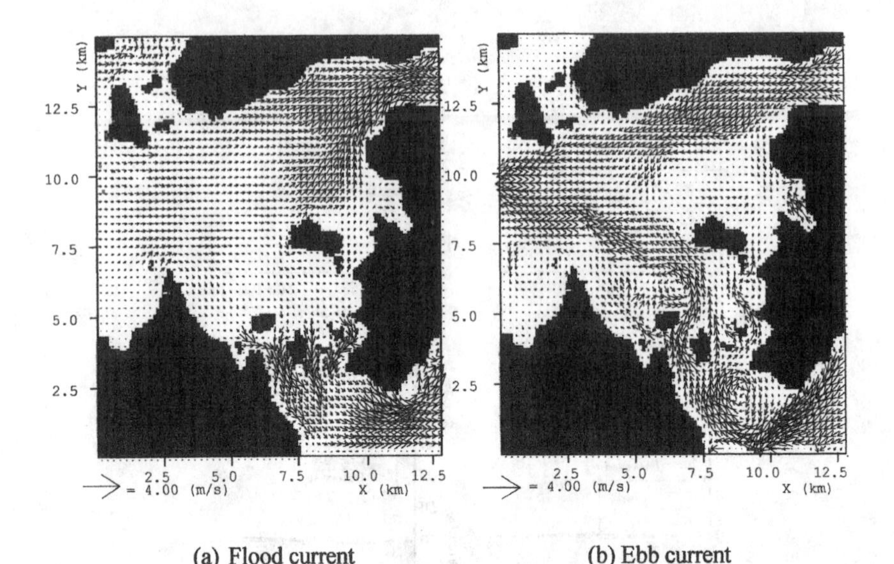

(a) Flood current (b) Ebb current

Fig.5. Calculated tidal current in Kurushima-Strait

Atmospheric forcing field and wind-induced current

We perform several numerical experiments for the atmospheric flow over the Seto Inland Sea to examine the response of water currents to surface wind stress.

Figure 6 shows the atmospheric flow at the lower layer of the atmosphere model after reaching the quasi-steady state. Figure 7 shows the instantaneous wind-induced current at the surface layer of the ocean model in the case of northwesterly wind during typical winter. This case is one of the calculated results induced by winds from 16 different but typical wind directions. The southeastward wind through Bungo- and Kii-Channel becomes stronger and the eastward wind over the center of the Seto Inland Sea becomes weaker. In spite of these changes, the eastward wind-induced current through Bisan-Strait becomes stronger than the currents through Bungo- and Kii-Channel because the currents through Bisan- and Kurushima-Strait are largely compensated by the changes of mass transport through Bungo- and Kii-Channel.

The wind fields are broadly similar in main features for each of the 16 different but typical wind directions although significant quantitative differences are evident.

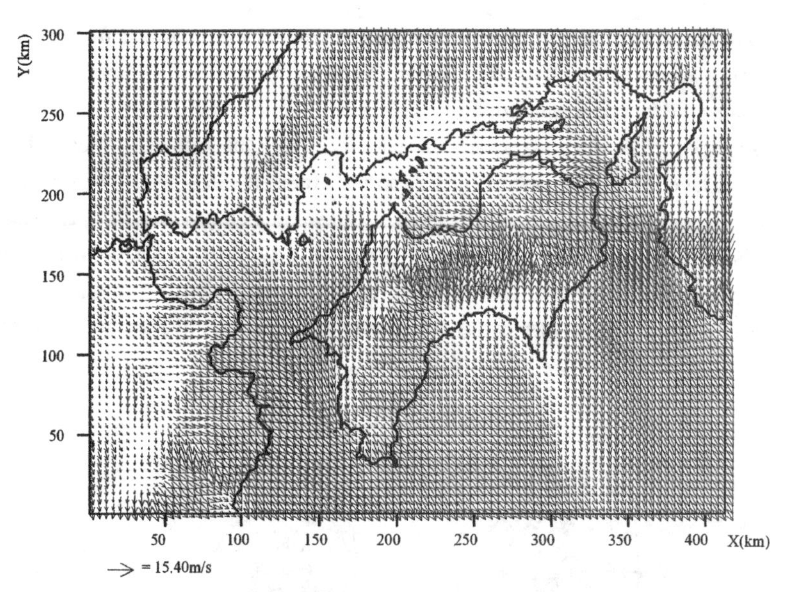

Fig.6. Surface wind distributions in the case of northwesterly wind

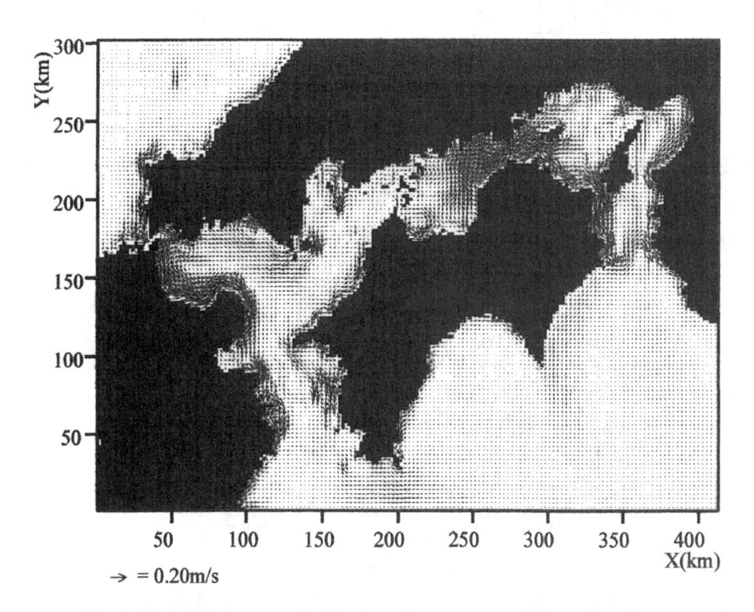

Fig.7. Instantaneous wind-induced current at the surface layer

Effect of the terrain shape of the Seto Inland Region on the mountain flow

In this subsection, the effect of the terrain shape of Honshu, Shikoku and Kyushu on the mountain flow is investigated.

Figures 8-10 show the radar charts of relation between the averaged surface wind at the lower layer of the atmosphere model and the averaged wind-induced current at the surface layer of the ocean model for each wind direction after reaching the quasi-steady state. Solid circle denotes the east-to-west components and triangle symbol denotes the north-to-south components. The radar chart of the surface wind velocity shows reasonable qualitative agreement and typically stronger than the observation-based fields.

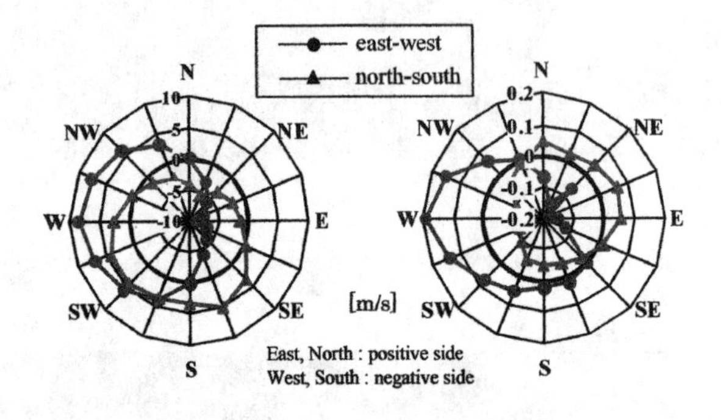

(a) Wind (b) Wind-induced current

Fig.8. Radar chart of relation between wind velocity and wind-induced current for each wind direction in Kurushima-Strait

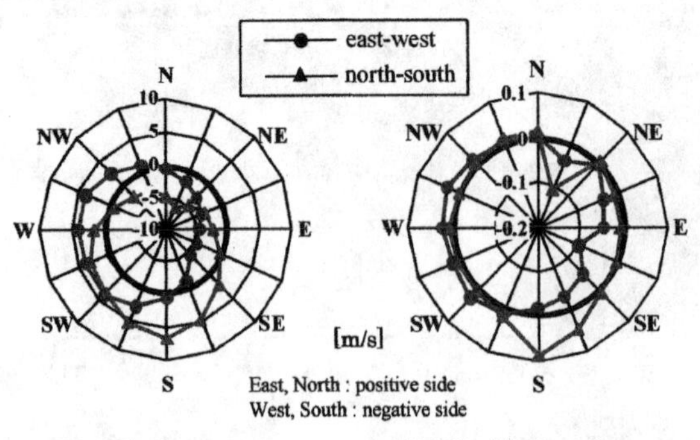

(b) Wind (b) Wind-induced current

Fig.9. Radar chart of relation between wind velocity and wind-induced current for each wind direction in Hiroshima Bay

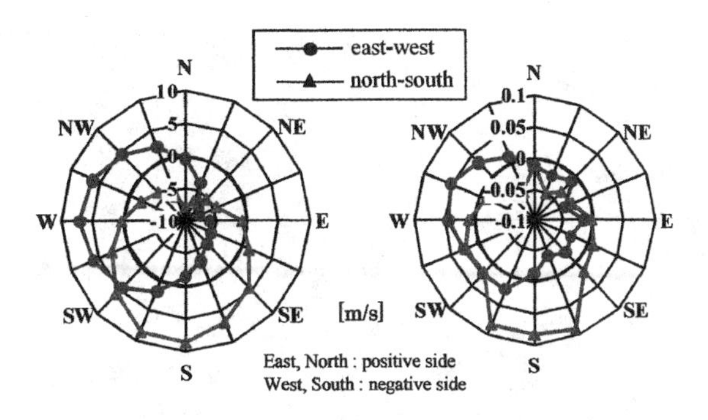

(a) Wind (b) Wind-induced current

Fig.10. Radar chart of relation between wind velocity and wind-induced current for each wind direction in Suo-Nada

This suggests the evidence of the coupled mesoscale air-sea interaction processes associated with the complex ocean bottom topography and the terrain shape of the Seto Inland Region if the radar chart of the wind-induced current is different from that of the wind. In Figs.8-10, the results suggest the importance of the three-dimensional valley structures. The surface wind is influenced by the complicated three-dimensional effect of the actual terrain shape. The orographic wind over the Seto Inland Sea is related to the effect of the terrain shape of mountains in Honshu, Shikoku and Kyushu.

Characteristics of coastal circulation induced by the local downslope wind

In this section, characteristics of coastal circulation induced by the local downslope wind are presented to investigate the relation between the orographic wind and the wind-induced current in the Seto Inland Sea.

Figure 11 shows the instantaneous atmospheric flow at the lower layer in the case of southerly wind during summer after reaching the quasi-steady state. The most outstanding is the maximum wind speed over the top of the mountains and the narrower valley generates the stronger wind velocity in the valley. The wind velocity increases on the lee side near the top of the mountains, while it remains weak due to the blocking effect on the upstream side of the mountains in Shikoku and Honshu. Furthermore, the wind influenced by the terrain shape of the mountains drives mesoscale eddies especially at the north of Shikoku and over Hiuchi-Nada (the center part of the Seto Inland Sea).

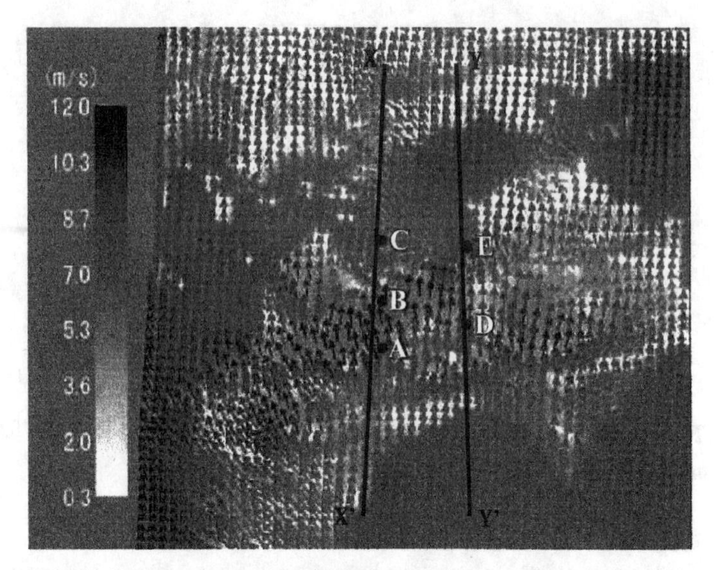

Fig.11. Instantaneous atmospheric flow in the case of southerly wind

To investigate the effect of the terrain shape of the mountains in Shikoku, the vertical cross sections of the deviation of potential temperature along X-X' and Y-Y' lines on the five locations of A-E are shown in Fig.12. The locations of A, B and D are the top of the mountain in Shikoku and that of C and E are on the sea surface near the Shikoku as shown in Fig.11. The inversion of the potential temperature is found on the lee side of the mountain. In general, inversion layers are usually found when a strong downslope wind is observed over a mountain. According to Lilly and Klemp (1979), a strong downslope wind tends to occur on the steep of an asymmetric mountain such as Shikoku. The entire sea surface in the leeward side of the mountains in Shikoku is covered by the orographic wind influenced by the asymmetric mountain because the downslope wind blows over the northern coastal plain of Shikoku and moves to Hiuchi-Nada. This pattern is quite similar to the case of the 'Yamaji-Kaze' influenced by the complicated three-dimensional effect of the terrain shape (Akiyama, T., 1956).

Figure 13 shows the instantaneous wind-induced current at the surface and bottom layers in Hiuchi-Nada. At the northern part of Hiuchi-Nada, the northeastward current induced by the downslope wind at the surface and bottom layers becomes stronger. The flow at the surface layer is dominated by the regional atmospheric forcing associated with the terrain shape of the mountain although its physical validation in a realistic wind forcing is required. The southeastward inflow from Kurushima-Strait at the surface layer becomes stronger in the southern part of Hiuchi-Nada, while it does not become at the bottom layer.

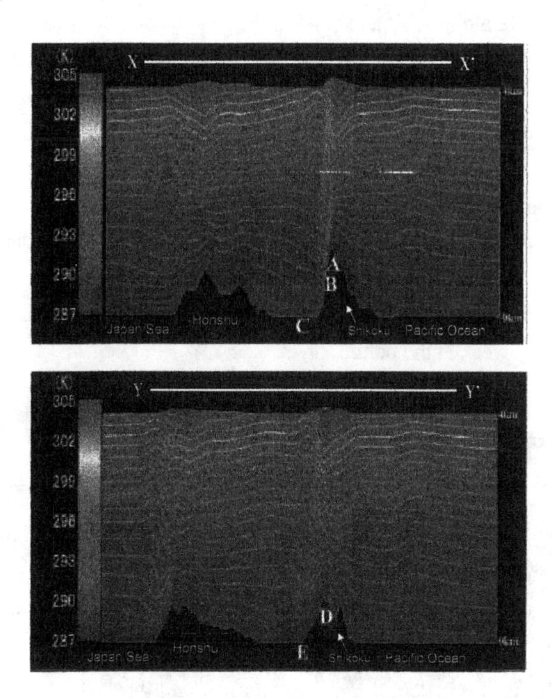

Fig.12. Vertical cross sections of the deviation of potential temperature along X-X' and Y-Y' lines on the five locations of A-E as shown in Fig.11

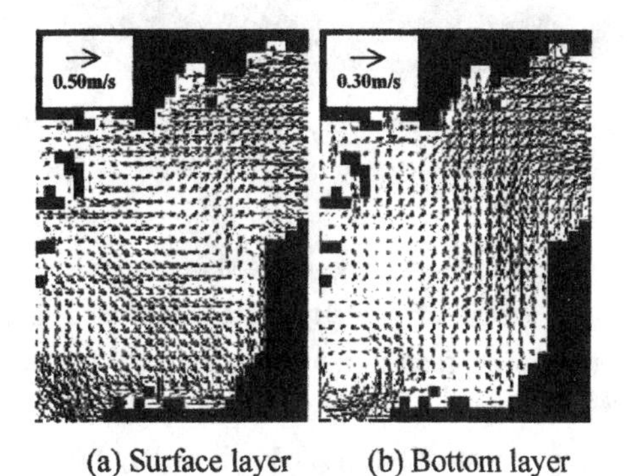

(a) Surface layer (b) Bottom layer

Fig.13. Instantaneous wind-induced current at the surface and bottom layers in Hiuchi-Nada

Figure 14 shows the instantaneous atmospheric flow at the lower layer in the case of northwesterly wind. Figure 15 shows the vertical cross section of the deviation of potential temperature along Z-Z' as shown in Fig.14. Strong downslope wind is found on the lee side of the mountain and the inversion layer appears to rise around Hiroshima city. The strong downslope wind velocity is greater than 10m/s. On the other hand, the deceleration of the wind velocity over Hiroshima Bay becomes clear. The almost perpendicular wind direction to the initial wind appears and the area of the strong surface wind no longer spreads leeward.

Figure 16 shows the instantaneous wind-induced current at the surface and bottom layers in Hiroshima Bay. At the northern part of Hiroshima Bay, the eastward current influenced by orographic wind can be seen at the surface layer, while its current is somehow relatively weak at the bottom layer.

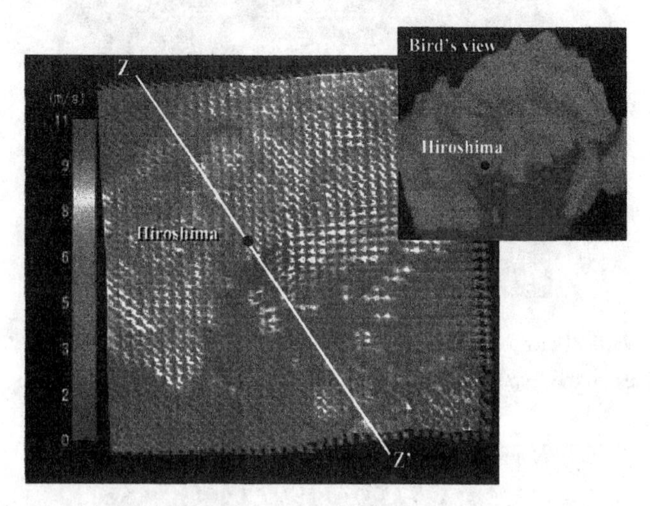

Fig.14. Instantaneous atmospheric flow at the lower layer over the Hiroshima Bay in the case of northwesterly wind

Figure 17 shows the instantaneous wind-induced current at the surface and bottom layers in Suo-Nada in the case of westerly wind. This wind direction is a typical condition of the average wind for the month of January. At the surface layer, the wind-induced current is eastward and it responds to the easterly wind field. However, at the bottom layer, the large counterclockwise eddy is mainly located and it is relatively weak. Therefore, the waters of Suo-Nada are vertically well mixed. The estuarine circulation in shallower coastal regions in Suo-Nada is visible in this wind situation.

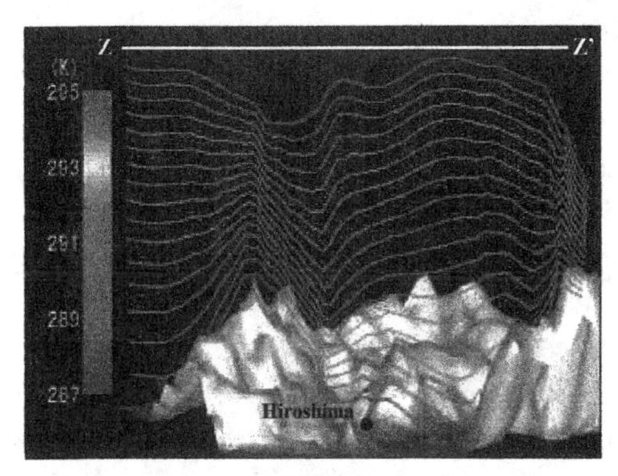

Fig.15. Vertical cross section of the deviation of potential temperature along Z-Z' as shown in Fig.14

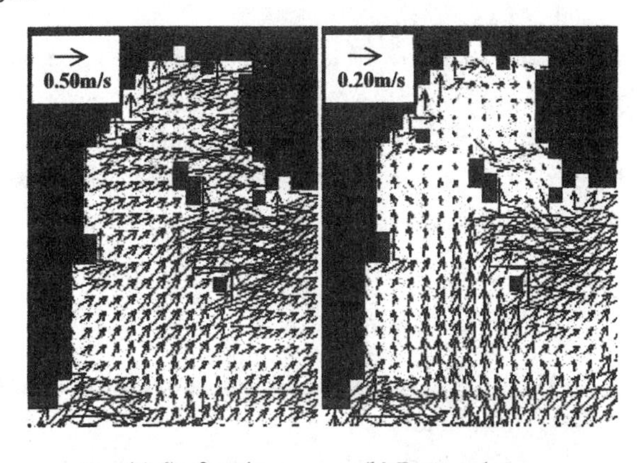

(a) Surface layer (b) Bottom layer

Fig.16. Instantaneous wind-induced current at the surface and bottom layers in Hiroshima Bay

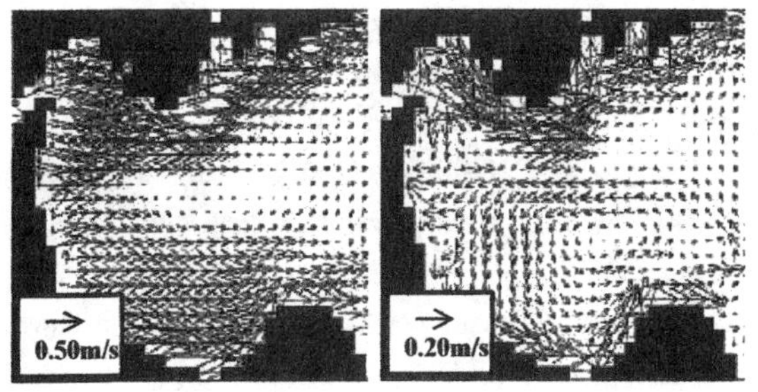

(a) Surface layer (b) Bottom layer

Fig.17. Instantaneous wind-induced current at the surface and bottom layers in Suo-Nada

CONCLUSIONS

The three-dimensional mesoscale atmosphere-ocean model was developed to examine the simulation of the coastal circulation in the Seto Inland Sea and to investigate the relation between the orographic wind and the wind-induced current.

The characteristic features of the coastal circulation in the Seto Inland Sea can be explained by the evolution and migration of the local downslope wind. The strength of the orographic wind system seems to be a very important factor in the generation of the coastal circulation in the Seto Inland Sea.

Although the numerical model is still in the preliminary stage and the validation of the three-dimensional results of the wind-induced current is quantitatively required, the interactions between the orographic wind field and the wind-driven current were qualitatively simulated.

Because the simplified conditions made, the model needs to be addressed with complex models with the exchange of heat, moisture and mass between air and sea. In the near future, the next stage will be planned by using a high-resolution simulation in the Seto Inland Region to compare directly the simulated surface wind with the observed wind. Moreover, the data assimilation technique will be needed to improve the accuracy of the model.

REFERENCES

[1] Akiyama, T. (1956) : On the occurrence of the local severe wind 'Yamaji', *Part 1, J. Meteor. Res.*, **8**, 627-641 (in Japanese).

[2] Blumberg, A. F., and G. L. Mellor (1987) : Diagnostic and prognostic numerical circulation studies of the South Atlantic Bight, *J. Geophys. Res.*, **88**, 4579-4592.

[3] Cronin, M. F., and M. J. McPhaden (2000) : Wind-forced reversing jets in the western equatorial pacific, *Journal of Physical Oceanography*, **Vo.30**, 657-675.

[4] Delft Hydraulics (1999) : Delft3D-FLOW, A simulation program for hydrodynamic flows and transport in 2 and 3 dimensions, release 3.05.

[5] Hayashi, M., T. Yanagi and T. Hashimoto (2001) : Analysis of phosphorus cycling in the inner part of Osaka Bay using a numerical ecosystem model, *Oceanogr. Japan*, **10**, 203-217 (in Japanese).

[6] Higo T., Y.Takasugi and H.Tanabe (1980) : Tidal current in the Seto Inland Sea, *Reports of the Government Industrial Research Institute, Chugoku*, **No.12**, 81-120.

[7] Lilly, D. K. and J. B. Klemp (1979) : The effects of terrain shape on non-linear hydrostatic mountain waves, *J. Fluid Mech.*, **95**, 241-261.

[8] Minato, S. (1996) : Numerical simulation of tide and storm surges in the Seto Inland Sea using the Princeton Ocean Model, *Papers in Meteor. and Geophys.*, **47**, 103-114.
[9] Ming Xue, Kelvin K. Droegemeier, Vince Wong, Alan Shapiro and Keith Brewster (1995) : ARPS Version 4.0 User's Guide, *Center for Analysis and Prediction of Storms (CAPS)*.
[10] Smith, R. L. (1995) : The physical processes of coastal upwelling systems. Upwelling in the Ocean : *Modern processes and ancient records*, 39-64.
[11] Takeoka, H. (2002) : Progress in Seto Inland Research, *Jour. of Oceanography*, **Vol. 58**, 93-107.
[12] Yabe, T., Feng Xiao and T.Utsumi (2001) : The Constrained Interpolation Profile Method for Multiphase Analysis, *Journal of Computational Physics*, **169**, 556-593.
[13] Yanagi, T. and H. Higuchi (1981) : Tide and tidal current in the Seto Inland Sea, *Proc. 28th Conf. Coast. Eng., JSCE*, 555-558 (in Japanese).

A New Two-Way Nested-Grid Ocean Modelling Technique Applied to the Scotian Shelf and Slope

Xiaoming Zhai[1], Jinyu Sheng[1], and Richard Greatbatch[1]

Abstract

A nested-grid ocean circulation model is developed for the Scotian Shelf and adjacent slope. The nested model consists of a fine-resolution inner model embedded inside a coarse-resolution outer model. The inner model covers the Scotian Shelf and slope, with a resolution of approximate 7 km. The outer model is the northwest Atlantic Ocean model developed by Sheng et al. (2001), with a horizontal resolution of approximate 25 km. The unique feature of the nested-grid model is its use of the semi-prognostic method to exchange information between the two different grids. The nested-grid ocean model is forced by the monthly mean COADS (Comprehensive Ocean-Atmosphere Data Set) wind stress and surface heat flux. The model sea surface salinity is restored to the monthly mean climatology. The nested-grid model is integrated for two years and the model results in the second year are presented in this paper. The inner model produces not only large-scale circulation features which are consistent with observations and those produced by the outer model, but also more meso-scale features than those in the outer model.

Introduction

The Scotian Shelf (SS) and adjacent Slope Water (SW) are bounded by the Laurentian Channel to the east, the Northeast Channel to the west, and deep waters to the south (Figure 1). The Scotian Shelf is about 700 km long and 200 km wide, with an average depth of approximate 90 m. The bottom topography is characterized by deep basins and marginal troughs of greater than 150 m depth over the inner shelf and long chains of shallow banks over the outer shelf. The offshore banks on the SS form a natural barrier to exchange between oceanic and coastal waters. Water

[1]Department of Oceanography, Dalhousie University, Halifax, Nova Scotia, Canada, B3H 4J1. E-mail: zhai@phys.ocean.dal.ca

mass distributions on the SS are affected primarily by seasonally varying air-sea heat and freshwater fluxes, the Cabot Strait outflow from the Gulf of St. Lawrence, and the shelf-break current that is the continuation of the Labrador Current from the Newfoundland Shelf. The Gulf Stream also influences the Scotian Shelf indirectly via Slope Water and transient rings. In addition, the SS and SW are frequently affected by storms, especially during the winter months. It follows that the ocean circulation and temperature/salinity distributions on the SS and SW have significant spatial and temporal variability at various time and length scales.

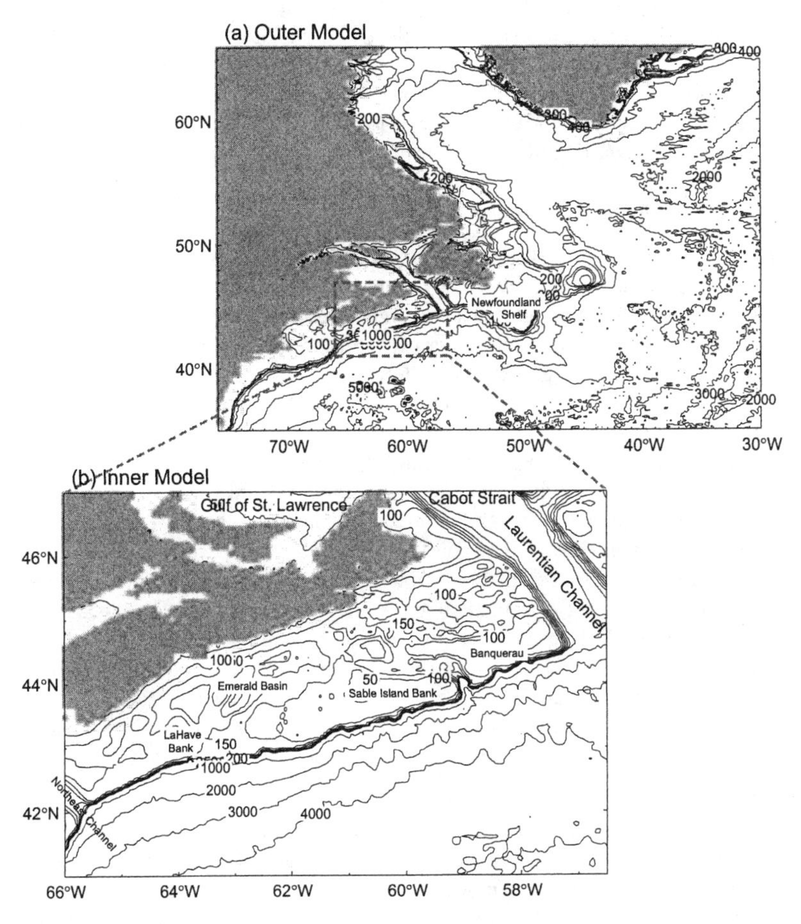

Figure 1: Bathymetric features within (a) the outer model domain of the northwest Atlantic Ocean, and (b) the inner model domain of the Scotian Shelf and Slope Water.

Ocean circulation models are useful tools for understanding and eventually predicting circulation over the shelf seas. Significant effort has been made over the last twenty years in developing numerical shelf circulation models for the SS and SW, ranging from two-dimensional barotropic models to sophisticated three-dimensional models (Greenberg, 1983; Wright et al., 1986; Hannah et al., 1996; Han et al., 1997; Thompson and Sheng, 1997; Sheng et al., 2001). Most of these models have a regional shelf domain with a specification of the transport and water mass distribution at the model open boundaries. Since the physical processes affecting circulation on the SS and SW operate over a wide range of temporal and spatial scales, to fully resolve all these different scales using a single-grid model is a formidable task. For any single-grid numerical model, one is usually faced with a choice between a large-scale simulation at a relatively coarse resolution, or a limited-area simulation with a very high resolution in which a prior knowledge is required for physical processes at work along the model open boundaries (Ginis et al., 1998). Furthermore, in limited-area simulations, the large-scale circulation features must be specified, and interactions between the large and small scales are prohibited.

The main advantage of a nested-grid model is that it allows the large-scale flow over a large domain to be simulated by a coarse resolution model, while circulation over a small domain is simulated at a higher resolution by a submodel that has essentially the same physics and numerical structure as the coarse-resolution model. The two-way interaction between two different model grids can be achieved in many ways. A fairly common nesting technique is to exchange information between the two grids in a narrow zone (dynamic interface) near the grid interface (e.g., Kurihara et al., 1979). The coarse grid model variables, such as velocity, temperature, salinity and associated fluxes, are interpolated at the dynamic interface onto the fine grid to provides the time-dependent boundary conditions for the fine-grid model. The fine-grid model variables are interpolated back onto the coarse grid to update the coincident coarse-grid values in the narrow zone. The main difficulty of this nesting scheme is a compatibility problem (e.g., mass conservation) at the grid interface (Ginis et al., 1998). Furthermore, there is undesirable numerical noise resulting from the change of the grid resolution at the grid interface and additional damping is required (Fox and Maskell, 1995). It should be noted that the dynamic interface of this nesting scheme can be considered as an internal boundary of the coarse-grid model, and the coarse-grid integration is not necessary over the subregion covered by the fine-grid domain.

An alternative nesting technique developed by Oey and Chen (1992) is to embed a fine grid (inner model) inside a coarse grid (outer model) and use the inner model variables to replace the outer model variables over the subregion where the two grids overlap. In comparison with other schemes, Oey and Chen's nesting scheme has the advantage of allowing a two-way interaction not only at the grid interface but also directly inside the common subregion where the two grids overlap. In this study, we follow Oey and Chen (1992) and embed a fine-grid model inside a coarse-grid model. Different from Oey and Chen's nesting technique, we use the semi-prognostic method (Sheng et al., 2001; Eden et al., 2004; Greatbatch et al., 2004) to exchange information between the two grids over the common subregion. (A detailed discussion of the new nesting technique can be found in Sheng et al. (2004).) The main objective of this study is to demonstrate the advantage of the new nesting technique in simulating the mean and seasonal circulation and temperature/salinity distributions on the Scotian Shelf and adjacent Slope Water.

The arrangement of this paper is as follows. The next section briefly reviews the ocean circulation model. This is followed by a presentation of the new two-way nesting technique based on the semi-prognostic method. We discuss the nested grid model results and compare them with results produced by other nesting techniques. The final section is a summary and conclusion.

Ocean Circulation Model

The ocean circulation model used in this study is the three-dimensional primitive equation z-level model known as CANDIE (Sheng et al., 1998). This model has been successfully used to study various physical processes on the shelf, including wind-driven circulation over an idealized coastal canyon (Sheng et al., 1998), nonlinear dynamics of a density-driven coastal current (Sheng, 2001), tidal circulation in the Gulf of St. Lawrence (Lu et al., 2001) and wind-driven circulation over a stratified coastal embayment (Davidson et al., 2001). Most recently, CANDIE has been applied to the western Caribbean Sea by Sheng and Tang (2003). The governing equations of the model can be written in spherical coordinates as

$$\frac{\partial u}{\partial t} + \mathcal{L}u - \left(f + \frac{u \tan\phi}{R}\right)v = -\frac{1}{\rho_o R\cos\phi}\frac{\partial p}{\partial \lambda} + \mathcal{D}_m u + \frac{\partial}{\partial z}\left(K_m \frac{\partial u}{\partial z}\right), \quad (1)$$

$$\frac{\partial v}{\partial t} + \mathcal{L}v + \left(f + \frac{u \tan\phi}{R}\right)u = -\frac{1}{\rho_o R}\frac{\partial p}{\partial \phi} + \mathcal{D}_m v + \frac{\partial}{\partial z}\left(K_m \frac{\partial v}{\partial z}\right), \quad (2)$$

$$\frac{1}{R\cos\phi}\left(\frac{\partial u}{\partial \lambda} + \frac{\partial (v\cos\phi)}{\partial \phi}\right) = -\frac{\partial w}{\partial z}, \quad (3)$$

$$\frac{\partial p}{\partial z} = -\rho g, \quad (4)$$

$$\frac{\partial T}{\partial t} + \mathcal{L}T = \mathcal{D}_h T + \frac{\partial}{\partial z}\left(K_h \frac{\partial T}{\partial z}\right), \quad (5)$$

$$\frac{\partial S}{\partial t} + \mathcal{L}S = \mathcal{D}_h S + \frac{\partial}{\partial z}\left(K_h \frac{\partial S}{\partial z}\right) \quad (6)$$

where u, v and w are the east (λ), north (ϕ) and vertical (z) components of the velocity vector $\vec{u}$, p is pressure (see below), and ρ is the density. Here K_m and K_h

are vertical eddy viscosity and diffusivity coefficients, f is the Coriolis parameter, ρ_o is a reference density, R and g are the Earth's radius and gravitational acceleration, $\mathcal{L}$ is an advection operator defined as

$$\mathcal{L}q = \frac{1}{R\cos\phi}\frac{\partial(uq)}{\partial\lambda} + \frac{1}{R\cos\phi}\frac{\partial(vq\cos\phi)}{\partial\phi} + \frac{\partial(wq)}{\partial z} \tag{7}$$

and $\mathcal{D}_m$ and $\mathcal{D}_h$ are diffusion operators defined as

$$\mathcal{D}_{(m,h)}q = \frac{1}{R^2}\left[\frac{1}{\cos^2\phi}\frac{\partial}{\partial\lambda}\left(A_{(m,h)}\frac{\partial q}{\partial\lambda}\right) + \frac{\partial q}{\partial\phi}\left(\cos\phi A_{(m,h)}\frac{\partial q}{\partial\phi}\right)\right] \tag{8}$$

where A_m and A_h are horizontal eddy viscosity and diffusivity coefficients, respectively. The model uses the subgrid-scale mixing parameterization of Smagorinsky (1963) for the horizontal eddy viscosity (A_m) and diffusivity (A_h), and the schemes proposed by Large et al. (1994) for the vertical mixing coefficients K_m and K_h.

As mentioned in the introduction, the nested-grid modelling system has a fine-resolution inner model embedded inside a coarse-resolution outer model (Figure 1). The fine-resolution inner model covers the Scotian Shelf and slope between 54°W and 66°W and between 39°N and 47°N, with a horizontal resolution of one eleventh degree in longitude (about 7 km). The coarse-resolution outer model is the northwest Atlantic Ocean model developed by Sheng et al. (2001) which covers the areas between 30°W and 76°W and between 35°N and 66°N, with a horizontal resolution of one third degree in longitude (about 25 km). There are 31 unevenly spaced z levels in both the outer and inner models with the centers of each level located at 5, 16, 29, 44, 61, 80, 102, 128, 157, 191, 229, 273, 324, 383, 450, 527, 615, 717, 833, 967, 1121, 1297, 1500, 1733, 2000, 2307, 2659, 3063, 3528, 4061, and 4673 m, respectively. Both the outer and inner models are initialized with the January mean temperature and salinity and forced by the monthly mean COADS (Comprehensive Ocean-Atmosphere Data Set) wind stress and surface heat flux. The model sea surface salinity is restored to the monthly mean climatology at every time step. In addition, at the inner model open boundaries, the temperature and salinity fields are first radiated from the inner model using an explicit Orlanski radiation condition (Orlanski, 1976) and then restored to the outer model values at each z-level with the time scale of 2 days.

A New Nesting Technique Based on the Semi-Prognostic Method

The semi-prognostic method was introduced originally by Sheng et al. (2001), for the purpose of adjusting model momentum equations to correct for model systematic errors (see Greatbatch et al., 2004 for an overview). The adjustment is accomplished by replacing the density variable ρ in the hydrostatic equation (4) by a linear combination of the model-computed density ρ_m and an input density ρ_c:

$$\rho = \alpha\rho_m + (1-\alpha)\rho_c \equiv \rho_m + (1-\alpha)(\rho_c - \rho_m) \tag{9}$$

where $\rho_m = \rho(T, S, p_{ref})$ is the density calculated from the model potential temperature T and salinity S described in (5) and (6), and p_{ref} is the reference pressure at the center of each z level, and α is the linear combination coefficient with a value between 0 and 1. In Sheng et al. (2001), the input density ρ_c is computed from a climatology of hydrographic data, but it might also be density from another model, as in the nesting technique to be described. Using (9), the hydrostatic equation can be rewritten as:

$$\frac{\partial p}{\partial z} = -g\rho_m - g(1-\alpha)(\rho_c - \rho_m), \qquad (10)$$

where the second term on the RHS of (10) is the correction term used to correct for model systematic error and unresolved processes. As stated in Sheng et al. (2001), the above procedure is equivalent to adding a forcing term to the horizontal momentum equations. To demonstrate this, we decompose the model pressure variable p into two terms:

$$p = p^* + \hat{p} \qquad (11)$$

where p^* is the traditional pressure variable satisfying

$$\frac{\partial p^*}{\partial z} = -g\rho_m \qquad (12)$$

with $p^* = g\rho_0\eta$ at the sea surface, and $\hat{p}$ is a correction term satisfying

$$\frac{\partial \hat{p}}{\partial z} = -g(1-\alpha)(\rho_c - \rho_m) \qquad (13)$$

with $\hat{p} = 0$ at the sea surface. Using (11), the horizontal momentum equations can be rewritten as:

$$\frac{\partial \vec{u}}{\partial t} = -\frac{1}{\rho_o}\nabla_h p^* - \frac{1}{\rho_o}\nabla_h \hat{p} + \ldots \qquad (14)$$

where $\vec{u}$ is the horizontal velocity vector and ∇_h is the horizontal Laplacian operator. Therefore, the semi-prognostic method is equivalent to adding a forcing term $(-\frac{1}{\rho_o}\nabla_h \hat{p}\ )$ to the model horizontal momentum equations. It is important to note that the semi-prognostic method is adiabatic, leaving the temperature and salinity equations unchanged, i.e. as in (5) and (6) (Greatbatch et al., 2004).

The original semi-prognostic method (OSP hereinafter) introduced by Sheng et al. has drawbacks of damping the mesoscale eddy field and reducing wave propagation speeds. Eden et al. (2004) recently introduced the smoothed semi-prognostic method (SSP hereinafter) by applying the correction term only on large spatial scales:

$$\frac{\partial p}{\partial z} = -g\rho_m - g(1-\alpha) < \rho_c - \rho_m > \qquad (15)$$

where <> represents spatial averaging. Eden et al. (2004) demonstrated that the SSP method effectively reduces the damping effect of the OSP method on the mesoscale eddy field.

In this study, we use a new two-way nesting technique based on the smoothed semi-prognostic method (referred to as the SSP nesting technique; Sheng et al., 2004). First, the outer model variables are interpolated onto the fine grid. The interpolated fields are then used to provide boundary conditions for the inner model. Second, the outer model density is used to adjust the inner model in the region where the domains of the two models overlap based on

$$\frac{\partial p}{\partial z} = -g\rho_{inner} - g(1-\beta_i) < \hat{\rho}_{outer} - \rho_{inner} > \quad \text{(for the inner model)} \qquad (16)$$

where p is the pressure variable carried by the inner model, ρ_{inner} is the inner model density, $\hat{\rho}_{outer}$ is the density computed from the outer model T and S fields after interpolation to the fine grid, and β_i is a linear combination coefficient with a value between 0 and 1, and $<>$ is the smoothing operator (for the results shown here, the correction term is smoothed over 16 grid points; that is, 112 km).

In the same way, the inner model density is used to adjust the outer model in the overlapping subregion based on

$$\frac{\partial p}{\partial z} = -g\rho_{outer} - g(1-\beta_o) < \hat{\rho}_{inner} - \rho_{outer} > \quad \text{(for the outer model)} \qquad (17)$$

where p is the pressure variable of the outer model, $\hat{\rho}_{inner}$ is the density calculated from the inner model T and S fields after interpolation to the coarse grid, and β_o is a linear combination coefficient with a value between 0 and 1, and $<>$ is a smoothing operator (usually different from that used in (16)). For the present application, this second smoothing operation is not applied. (A complication in the present application is that the outer model is also being corrected using climatological data. Details of how to apply these two corrections simultaneously are given in Sheng et al. (2004).)

The above two-way interaction can be applied at each outer model time step. However, since the present nested-grid model is driven by monthly forcings, it is sufficient for the correction term to be updated once per day. Also, for simplicity, β_i and β_o are set to be 0.5.

It should be noted that Sheng and Tang (2004) used the original semi-prognostic method in the development of a two-way nested-grid model for the Meso-American Barrier Reef System of the Caribbean Sea. Their nesting technique (referred as to the OSP nesting technique) differs from the above SSP technique only in that the correction terms used to exchange information between the two grids are not spatially smoothed. We show a comparison between the SSP and OSP methods below. We also compare the SSP method with what we call "conventional one-way" nesting (COW for short). In the COW method, the boundary conditions for the inner model are taken from the outer model exactly as in the SSP method. The difference is that the inner model only knows about the outer model through the boundary conditions. In particular, there is no transfer of information into the interior of the inner model domain, and there is also no transfer of information back from the inner model to the outer model.

Model Results

The ocean circulation models are integrated for two years. The model results in the second year are used to calculate the annual mean volume transport streamfunctions over the outer and inner model domains, respectively, for the SSP and COW methods (Figure 2). Since there is no influence from the inner to the outer model in the COW case, the field shown for the outer model in Figure 2d is the same as that in Sheng et al. (2001). In the SSP case, the annual mean streamfunctions produced by inner and outer models are very similar over their common domain, and also very similar to the outer model when run on its own, without feedback from the inner model (see Figure 2d). The annual mean volume transport produced by the outer model is about 50 Sverdrup (Sv) for the subpolar circulation of the North Atlantic, 90 Sv for the Gulf Stream, and 10 Sv for the recirculation in the Slope Water region between the Scotian Shelf and the Gulf Stream, comparable to estimates found in diagnostic modelling studies and in observations (see Mellor et al., 1982; Greatbatch et al., 1991). Over the Slope Water region, offshore from the Scotian Shelf, the two-way nested-grid outer model produces a slightly stronger recirculation in comparison with Sheng et al. (2001) (Figures 2b,d), indicating the effect of the feedback from the inner model to the outer model in the two-way nested modelling system.

In the COW method, the inner model is constrained by the outer model only along the boundary of its domain, and the interior of the inner model domain is not constrained directly by the outer model. The result is that the inner model can drift away from the outer model, as is evident here. The one-way nested inner model generates an unrealistically large and broad eastward transport over the SW and adjacent deep water regions and fails to produce the widely recognized recirculation in the SW region (Figure 2c).

The model results in the second year are also used to calculate the annual mean near-surface currents at 16 m depth in the SSP, OSP and COW cases (Figure 3). Figure 3a, for the outer model, shows a southwestward coastal flow known as the Scotian Current over the coastal region, a narrow current known as the shelf break jet over the outer SS, and a narrow jet over the SW that first flows northeastward over the SW region off the southwestern SS and then turns eastward to the deep water off the southeastern SS. Further south of the SW, there is a strong and broad eastward flow as part of the Gulf stream. The nested-grid inner model using the SSP nesting technique reproduces the large-scale features of the near-surface currents produced by the outer model in the overlapping region (Figures 3a,b), but with much better resolution.

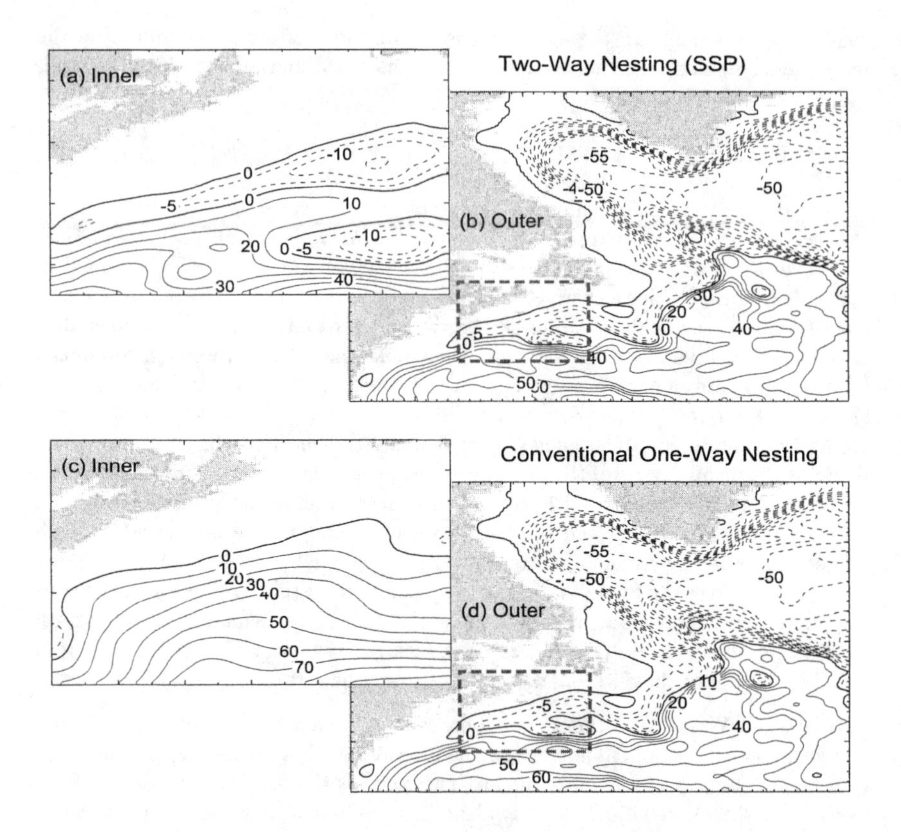

Figure 2: Annual mean volume transport streamfunctions computed from the second year model results produced by (a) the inner and (b) outer model of the two-way nested grid modelling system using the smoothed semi-prognostic nesting technique and by (c) the inner and (d) outer model using the conventional one-way nesting technique. The streamfunction shown in (d) is the same as that produced by the single-domain Northwest Atlantic Ocean model of Sheng et al. (2001). Contours are labeled in units of Sv ($= 10^6$ m^3 s^{-1}).

To assess the performance of the nested-grid model, we compare the model calculated annual mean near-surface currents with the time mean currents inferred from the observed trajectories of near-surface drifters by Fratantoni (2001) over the SS and SW region. Overall, the model calculated currents in the SSP case agree with the observed currents (Figures 3a,b). To further quantify the misfit between the observed and model-calculated near-surface currents, we follow Sheng and Tang (2003)

Table 1: The γ^2-values that measure the misfit between the observed and the model simulated near-surface currents at 15 m in terms of different nesting techniques.

Technique	γ^2
Outer (SSP)	0.73
SSP	0.62
OSP	0.65
COW	1.01

and define γ^2 as

$$\gamma^2 = \frac{\sum_k{}^N[(u_k{}^o - u_k{}^s)^2 + (v_k{}^o - v_k{}^s)^2]}{\sum_k{}^N[(u_k{}^o)^2 + (v_k{}^o)^2]} \tag{18}$$

where $(u_k{}^o, v_k{}^o)$ are the horizontal components of the observed near-surface currents at the kth location estimated by Fratantoni (2001), $(u_k{}^s, v_k{}^s)$ are the horizontal components of the simulated near-surface currents at the same location as the observations, and N is the total number of locations where observed estimates are available. Clearly, the smaller γ^2, the better the model results fit the observations. For the two-way nested-grid modelling system using the SSP nesting technique, the value of γ^2 is about 0.73 for the outer model and 0.62 for the inner model (Table 1), indicating that the inner model performs better than the outer model in reproducing Fratantoni's observed currents.

The overall distributions of the annual mean near-surface currents produced by the inner model using the OSP technique agree with those produced by the inner model using the SSP technique, with some differences in the small-scale features (Figure 3c). The inner model using the OSP technique generates relatively weaker southwestward currents over the western Scotian Shelf and relatively weaker eastward currents in the deep water off the SW region, in comparison with those using the SSP technique (Figures 3b,c). The inner model using the OSP nesting technique also reproduces Fratantoni's time-mean observed currents in the SS and SW region reasonably well, with the γ^2 value of about 0.65 (Table 1), which is slightly larger than the value for the SSP technique.

The annual mean near-surface currents at 16 m produced by the inner model using the COW nesting technique differ significantly from the inner model results using the OPS and SSP nesting techniques (Figure 3d). In addition, the inner model results using the COW technique agree poorly with Fratantoni's time-mean observed near-surface currents, with the γ^2 value of about 1.01, which is about 60% larger than that of the inner model results using the SSP or OSP nesting techniques.

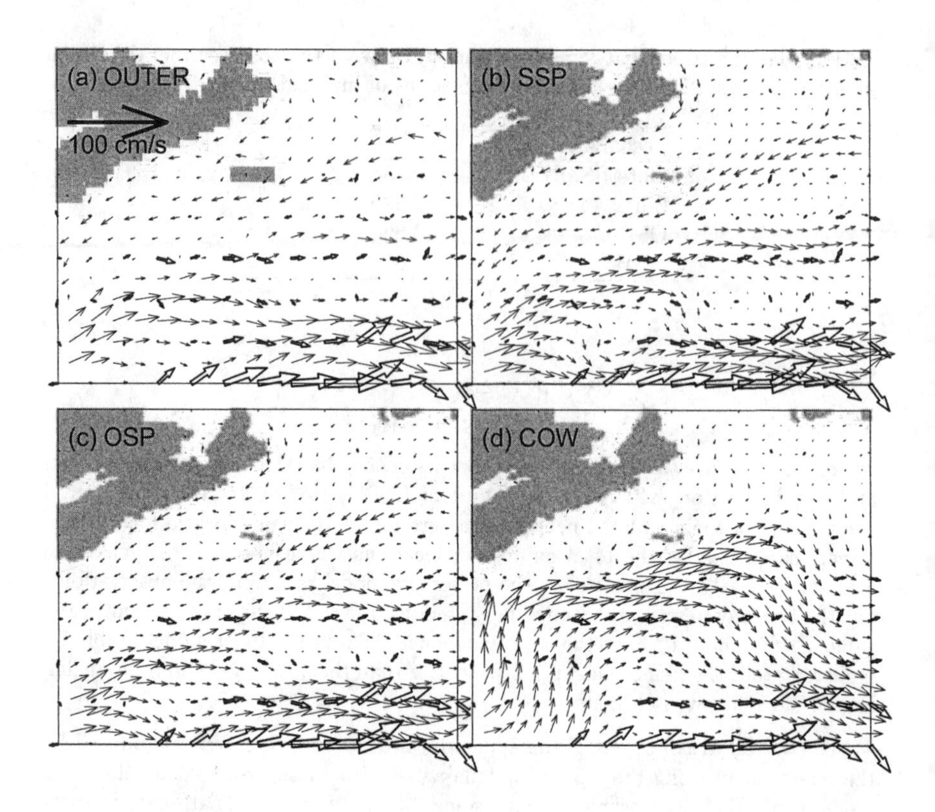

Figure 3: Comparison of modeled (solid arrows) and observed (open arrows) near-surface currents over the SS and SW region. The observed currents are the gridded time-mean near-surface currents during the 1990s inferred from trajectories of 15 m-drogued satellite-tracked drifters by Fratantoni (2001) on a 1° grid. The modeled currents are the annual mean currents at 16 m computed from the second year model results generated by (a) the outer model and (b) inner model of the two-way nested-grid modelling system using the smoothed semi-prognostic (SSP) nesting technique; (c) the two-way nested-grid inner model using the original semi-prognostic (OSP) nesting technique; and (d) the conventional one-way (COW) nested-grid inner model.

We next compare the instantaneous sub-surface currents and temperatures at 50 m at day 690 (end of November of the second model year) produced by the outer and inner models using the SSP, OSP and COW nesting techniques (Figure 4). In the SSP and OSP cases, the large-scale features of the sub-surface currents and temperature produced by the inner model compare well with those produced by the outer model, except that the inner model produces more meso-scale eddies over the SW region, especially in the SSP case (Figures 4a,b). Both the circulation and the temperature field in the COW case are quite unrealistic. In particular, the COW nested inner model fails to generate the well-known Scotian Current near the coast and the shelf-break jet over the outer SS. Instead, the model generates unrealistically large and broad northeastward flow over the SW region. The inner model using the COW nesting technique also overestimates the sub-surface SW temperature significantly (Figure 4d).

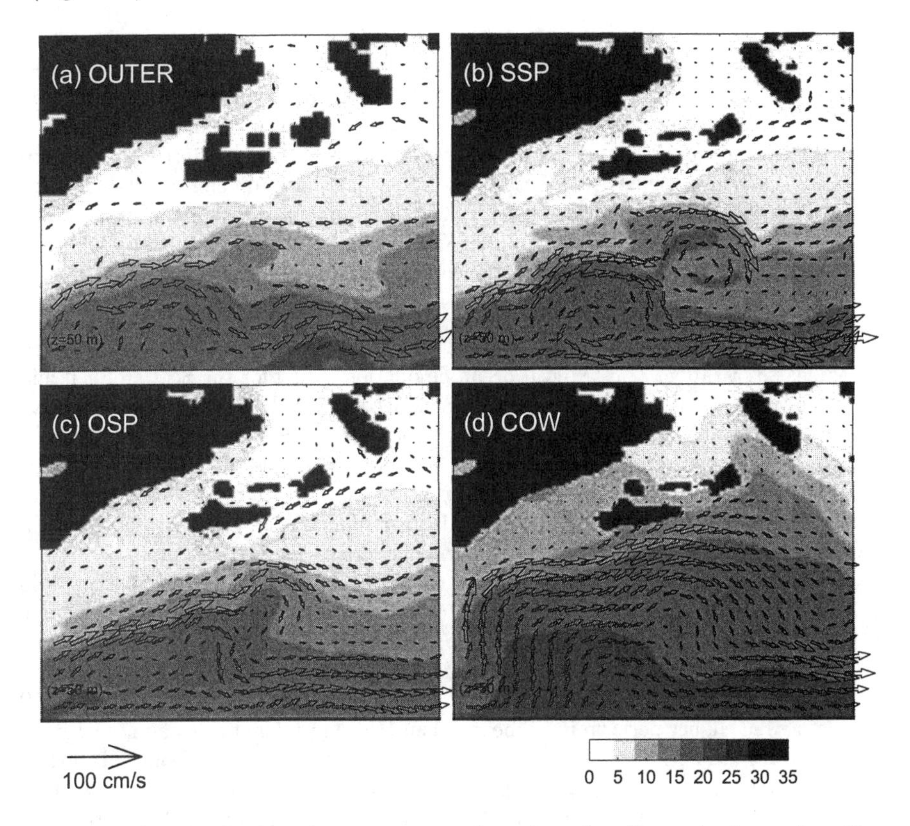

Figure 4: Simulated sub-surface temperature (gray image) and currents (arrows) at 50 m over the SS and SW region at day 690 produced by (a) the outer model and (b) the inner model of the nested system using the SSP nesting technique; (c) the nested-grid inner model using the OSP nesting technique (d) the conventional one-way nested inner model. Velocity vectors are plotted at every four model grid points for the inner model and every two model grid points for the outer model.

As mentioned above, one of the most important circulation features over the SS and SW region is the spreading of the Labrador Current from the Grand Banks to the outer Scotian Shelf. The inner model using the SSP nesting technique successfully reproduces the cold tongue associated with the Labrador Current along the shelf-break (Figure 4b), indicating that advection is important for the presence of the cold intermediate layer on the shelf (Hachey, 1938).

Summary and conclusion

We have described a new two-way interactive nesting technique based on the smoothed semi-prognostic (SSP) method suggested by Eden et al. (2004) (see also Greatbatch et al., 2004). In this paper we demonstrated that both the SSP and OSP methods can be used to exchange information between the sub-components of a nested-grid modelling system by adding a two-way interaction term to the horizontal momentum equations of each sub-model. The main difference between these two nesting techniques is that the SSP nesting technique uses a spatially smoothed (large-scale) interaction term, while the OSP nesting technique uses an unsmoothed interaction term. Therefore, the SSP nesting technique reduces the damping effect of the OSP nesting technique on the mesoscale eddy field.

The two-way nesting technique based on the SSP method was used in the development of a two-way nested-grid ocean circulation model for the Scotian Shelf and Slope Water, with a high-resolution inner model embedded inside a coarse-resolution outer model. The inner model covers the Scotian Shelf (SS) and Slope Water (SW), with a resolution of approximate 7 km. The outer model is the northwest Atlantic Ocean model developed earlier by Sheng et al. (2001), with a horizontal resolution of approximate 25 km. The nested-grid inner model using the SSP nesting technique reproduces many well-known large-scale circulation features over the SS and SW region and is fully compatible with the outer model. In addition, the fine-grid inner model using the SSP technique produces more small-scale features (e.g., meanderings, warm rings etc.) than those produced by the coarse-grid outer model. In comparison with the time mean near-surface currents inferred from the observed trajectories of near-surface drifters at 16 m by Fratantoni (2001), both the SSP and OSP nesting techniques perform much better than the conventional one-way nested inner model over the SS and SW region. Comparison of the instantaneous sub-surface circulation and temperature field produced by the inner models demonstrates that the SSP nesting technique performs better than the OPS and much better than the COW nesting techniques.

Acknowledgments

We wish to thank Carsten Eden, Liqun Tang, Jim Chuang and Jun Zhao for their useful suggestions and comments. We thank David Fratantoni for providing the near-surface currents determined from the 10 m-drogued satellite-tracked drifters in the North Atlantic. This work has been supported by funding from CFCAS and NSERC. R.J.G. and J.S. are also supported by NSERC, MARTEC (a Halifax based company), and the Meteorological Service of Canada (MSC) through the NSERC/MARTEC/MSC Industrial Research Chair in "Regional Ocean Modelling and Prediction".

References

Davidson, F., R. J. Greatbatch, and B. deYoung, Asymmetry in the response of a stratified coastal embayment to wind forcing, *J. Geophys. Res., 106*, 7001-7016, 2001.

Eden, C., R. J. Greatbatch, and C. W. Böning, Adiabatically correcting an eddy-permitting model using large-scale hydrographic data: Application to the Gulf Stream and the North Atlantic Current, *J. Phys. Oceanogr.*, in press, 2004.

Fox, A. D., and S. J. Maskell, Two-way interactive nesting of primitive equation ocean models with topography, *J. Phys. Oceanogr., 25*, 2977-2996, 1995.

Fratantoni, D. F., North Atlantic surface circulation during the 1990's observed with satellite-tracked drifters, *J. Geophys. Res., 106*, 22067-22093, 2001.

Ginis, I., A. Richardson, and L. Rothstein, Design of a multiply nested primitive equation ocean model, *Mon. Wea. Rev., 126*, 1054-1079, 1998.

Greatbatch, R. J., A. F. Fanning, A. D. Goulding, and S. Levitus, A diagnosis of interpentadal circulation changes in the North Atlantic, *J. Geophys. Res., 96*, 22009-22023, 1991.

Greatbatch, R. J., J. Sheng, C. Eden, L. Tang, X. Zhai, and J. Zhao, The semi-prognostic method, *Continental Shelf Res.*, in press, 2004.

Greenberg, D. A., Modelling the mean barotropic circulation in the Bay of Fundy and the Gulf of Maine, *J. Phys. Oceanogr., 13*, 886-904, 1983.

Hachey, H. B., The origin of the cold water layers of Scotian Shelf, *Trans. R. Soc. Can. Ser.,3*, 29-42, 1938.

Han, G., C. Hannah, J. W. Loder, and P. C. Smith, Seasonal variation of the three dimensional mean circulation over the Scotian Shelf, *J. Geophys. Res., 102*, 1011-1025, 1997.

Hannah, C. G., J. W. Loder, and D. G. Wright, Seasonal variation of the baroclinic circulation in the Scotia-Maine region, In *Buoyancy effects on coastal dynamics, Coastal Estuarine Stud., 53*, 1996.

Kurihara, Y., G. J. Tripoli, and M. A. Bender, Design of a movable nested-mesh primitive equation model, *Mon. Wea. Rev., 107*, 239-249, 1979.

Large, W. G., J. C. McWilliams, and S. C. Doney, Oceanic vertical mixing: A review and a model with a nonlocal boundary layer parameterization, *Reviews of Geophysics, 32*, 363-403, 1994.

Lu, Y., K. R. Thompson, and D. G. Wright, Tidal currents and mixing in the Gulf of St. Lawrence: an application of the incremental approach to data assimilation, *Can. J. Fish. Aquat. Sci., 58*, 723-735, 2001.

Mellor, G. L., C. R. Mechoso, and E. Keto, A diagnostic model of the general circulation of the Atlantic Ocean, *Deep Sea Res., Part A, 29*, 1171-1192, 1982.

Oey, L. Y., and P. Chen, A nested-grid ocean model: with application to the simulation of meanders and eddies in the Norwegian Coastal Current, *J. Geophys. Res., 97*, 20063-20086, 1992.

Orlanski, I., A simple boundary condition for unbounded hyperbolic flow, *J. Comput. Phys., 21*, 251-269, 1976.

Sheng, J., Dynamics of a buoyancy-driven coastal jet: The Gaspé Current, *J. Phys. Oceanogr., 31*, 3146-3163, 2001.

Sheng, J., D. G. Wright, R. J. Greatbatch, and D. Dietrich, CANDIE: a new version of the DieCAST ocean circulation Model, *J. Atm. and Oceanic Tech., 15*, 1414-1432, 1998.

Sheng, J., R. J. Greatbatch, C. Eden, X. Zhai and L. Tang, A new two-way interactive nesting technique based on the semi-prognostic method, to be submitted, 2004

Sheng, J., R. J. Greatbatch, and D. G. Wright, Improving the utility of ocean circulation models through adjustment of the momentum balance, *J. Geophys. Res., 106*, 16711-16728, 2001.

Sheng, J., and L. Tang, A numerical study of circulation in the western Caribbean Sea, *J. Phys. Oceanogr., 33*, 2049-2069, 2003.

Sheng, J., and L. Tang, A two-way nested-grid ocean circulation model for the Meso-American Barrier Reef System, *Ocean Dynamics*, in press, 2004.

Smagorinsky, J., General circulation experiments with the primitive equations. I. The basic experiment, *Mon. Wea. Rev., 21*, 99-165, 1963.

Thompson, K. R., and J. Sheng, Subtidal circulation on the Scotian Shelf: Assessing the hindcast skill of a linear, barotropic model, *J. Geophys. Res., 102*, 24987-25003, 1997.

Wright, D. G., D. A. Greenberg, J. W. Loder, and P. C. Smith, The steady-state response of the Gulf of Maine and adjacent regions to surface wind stress, *J. Phys. Oceanogr., 16*, 947-966, 1986.

Modeling the Rio de la Plata Circulation

ISMAEL PIEDRA-CUEVA[1] and HUGO RODRÍGUEZ[2].

Abstract

A two-dimensional (2-D) depth-integrated finite element numerical model was applied to study the circulation in the Rio de la Plata estuary in the vicinity of the Port of Montevideo. In this paper, the application of the model and its calibration is presented.

The hydrodynamic regime in the area of interest comes from the response of the entire estuary to different and interacting driving forces. These circumstances conduce to the need for a numerical model of the hydrodynamic regime of the whole Rio de la Plata and adjacent continental shelf.

The Rio de la Plata is located on the east coast of South America, between 34° and 36° South Latitude and 54° 50' and 58° 30' West Longitude, between Uruguay and Argentina. It flows from the northwest towards the southeast for approximately 280 km. Its surface area is about 35,000 km^2, its depth varies from 3 to 20 m and its width increases from about 2 km to approximately 220 km at its mouth.

The two major tributaries are the Parana and Uruguay Rivers, with annual average discharges of 16,000 m^3/s and 6,000 m^3/s, respectively. The model open boundary condition was hourly distributions of water surface elevations obtained from two coastal stations.

The finite element model was calibrated in two stages. First, bottom roughness and astronomical forcing were calibrated with astronomical tide components obtained from tide observation at seven stations along both coasts. Even though wind effects were found not to be the primary forcing for surges, the wind drag coefficient was calibrated with comparisons of surge water levels observed at Montevideo, and then verified with surge water levels measured at Torre Oyarvide.

Comparisons of calibrated model output with observational data for the Rio de la Plata have shown that the model is capable of simulating water surface elevation and current speed and direction with a relatively high degree of accuracy.

[1] Institute of Fluid Mechanics and Environmental Engineering, College of Engineering, J. Herrera y Reissig 565, Montevideo, Uruguay. Tel. (5982)-711-5277, email: ismaelp@fing.edu.uy - **Contact person.**

[2] Tetra Tech, Inc., 2110 Powers Ferry Road, Atlanta, Georgia 30339. Tel. 770-850-0949, email: hugo.Rodriguez@tetratech-ffx.com.

Introduction

The Rio de la Plata is located on the east coast of South America, between 34° and 36° South Latitude and 54° 50' and 58° 30' West Longitude, between Uruguay and Argentina. It flows from the northwest towards the southeast for approximately 280 km. Its surface area is about 35,000 km^2 and its width increases from about 4 km near Carmelo to approximately 220 km at its mouth.

The two major tributaries are the Parana and Uruguay Rivers, with annual average discharges of 16,000 m^3/s and 6,000 m^3/s, respectively, with minor tributary discharges being several orders of magnitude smaller.

Two main regions can be identified based on the morphology and dynamics of the Rio de la Plata (CARP 1989, Framinan and Brown1996). Barra del Indio, a shallow area located in the line Punta Piedras and Montevideo (Figure 1), separates the inner from the outer region. The inner region has a fluvial regime. In the outer region, the river increases in width and currents forced by fresh water discharges are weaker. This change in the geometry implies a new forcing balance that generates complex flow patterns.

The general circulation of atmospheric conditions in the region is controlled by the influence of the quasi-permanent South Pacific high pressure that generates southwestern and southern winds. This general circulation is modified by the action of a low-pressure system located at northern Uruguay that generates winds from the northwest and southeast. Other important events occurring in the Rio de la Plata region are frontal systems with southwest-northeast trajectories that interact with littoral low-pressure systems. This interaction can generate strong winds over 30 m/s (Framinan and Brown 1996) from the southeast and storms that affect the area for several days. This phenomenon is called sudestada (southeastern) and is responsible for disastrous flooding in the littoral of the Rio de la Plata area.

The astronomical tides in the Rio de la Plata have semi diurnal regime and are classified as micro tidal. The amplitude of the main semi diurnal component (M_2) inside the Rio de la Plata oscillates between 0.30 m and 0.10 m, and is larger over the Argentinean coast. Water levels can be strongly modified by meteorological events (Balay 1961). Storm surges of over 3-4 m have been registered during southeastern events. Tidal currents are on the order of 0.5 m/s with higher values at the southern coast. Currents are also strongly modified by meteorological events. Values ranging between 1.0 and 1.5 m/s have been observed in Montevideo during periods with southeasterly winds.

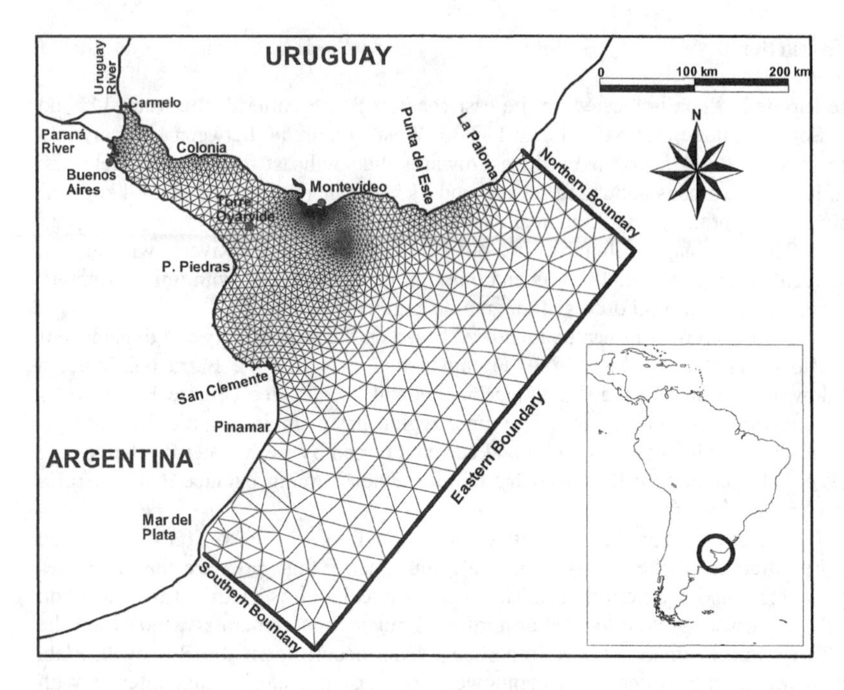

Figure 1. The Rio de la Plata study area. Bathymetry in m below Mean Sea Level.

Mathematical model and numerical solution.

Since the Rio de la Plata is a shallow, well-mixed system, the unsteady, two-dimensional (2D) vertically integrated, shallow water equations have been used to model tides and tide induced currents.

In this study, the two-dimensional finite element hydrodynamic numerical model RMA-2 has been implemented (King 1998). It computes water surface elevations and horizontal velocity components for free-surface two-dimensional flow fields. Friction can be calculated with the Manning or Chezy equations, and eddy viscosity coefficients are used to define turbulence characteristics. The equation of momentum and mass conservation are solved by the finite element method using the Galerkin Method of weighted residuals (Donnell et al. 2000).

Application to the Rio de la Plata

The modeled domain is the Rio de la Plata limited by the coasts of both shores and the continental shelf until the 200 m isobath. Fresh water is input at the cross section defined by the line that goes from Carmelo on the Uruguayan coast to the Parana River delta on the Argentinean coast. The open boundary extends into the continental shelf until approximately the 200 m isobath. The southern open boundary, located at Mar del Plata, has been selected so that the wave front of the semidiurnal M_2 arriving from the south enters the model domain parallel to it. The northern open boundary, located 80 km north of La Paloma, was selected parallel to the southern boundary. The eastern (outer) open boundary was selected perpendicular to the southern and northern open boundaries.

The extent of the modeled domain would produce a large amount of horizontal cells if a uniform grid were used. By using a non-uniform Finite Element grid, a reasonable number of horizontal cells can be obtained. The Finite Element grid depicted in Figure 1 includes 10,156 horizontal nodes and 4,870 triangular cells. The resolution ranges from 10 km on the continental shelf to 500 m near Montevideo.

Hydrodynamic model calibration

The 2-D model application requires the specification of the open boundary forcing, bottom friction coefficient and wind drag coefficient. For the purpose of calibration these parameters were adjusted to obtain the best fit between measured and simulated values.

First, the hydrodynamic model was forced using astronomical tide projections for January and February of 1983 at the ocean open boundary. Model results were compared to astronomical tide projections, obtained from the harmonic analysis, at interior points within the estuary. This calibration was used to assure that the model was accurately representing the progression of the tidal wave through the system. Amplitudes and phases of astronomical tides along the open boundary were obtained from the global surface model FES95.2 (Le Provost et al. 1998). Ocean tide models, although reliable, do not have sufficient accuracy in order to perform the calibration procedure and thus amplitude and phase were then fine-tuned as required.

The bottom friction, using Manning's coefficient, and the open boundary astronomical tidal forcing were adjusted by calibrating the tidal wave propagation through the estuary.

The following figures present comparisons between the simulated and measured conditions during the simulation period. These values represent the best overall results achieved under the calibration adjustments listed above. The final value of the bottom roughness coefficient was n=0.013.

The astronomical tide modeling was performed with the five principal tidal components M_2, N_2, S_2, O_1 and Q_1 that account for more than 95% of the tidal energy inside the estuary.

Figure 1 shows the locations of the gauge stations used for calibration, La Paloma, Punta del Este, Montevideo and Colonia on the Uruguayan coast and Pinamar, Torre Oyarvide and Buenos Aires on the Argentinean coast. Figures 2 and 3 present the comparisons between the projected and simulated water surface elevations. Examination of the plots shows that the model captures both

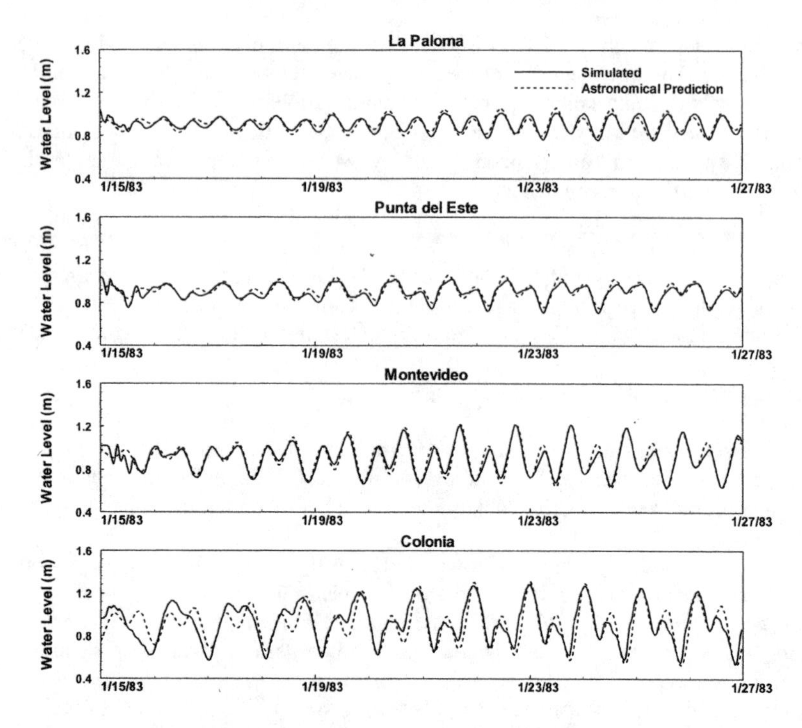

Figure 2. Simulated vs. Projected Water Surface Elevation at a La Paloma, Punta del Este, Montevideo and Colonia.

the phasing and amplitude of the tidal wave at each location. Table 1 presents the normalized RMS error, computed using a time interval of one hour (the same time step of registered water levels). The absolute errors are normalized to the local tidal range in order to provide intercomparison between the stations, and project an evaluation of the error as a percent of the overall signal strength.

Results show normalized RMS errors below 10 percent, except for La Paloma and Punta del Este (where tidal range is very small) and Colonia, where tidal signal is strongly distorted by a very shallow bank (Ortiz Bank) present near this station.

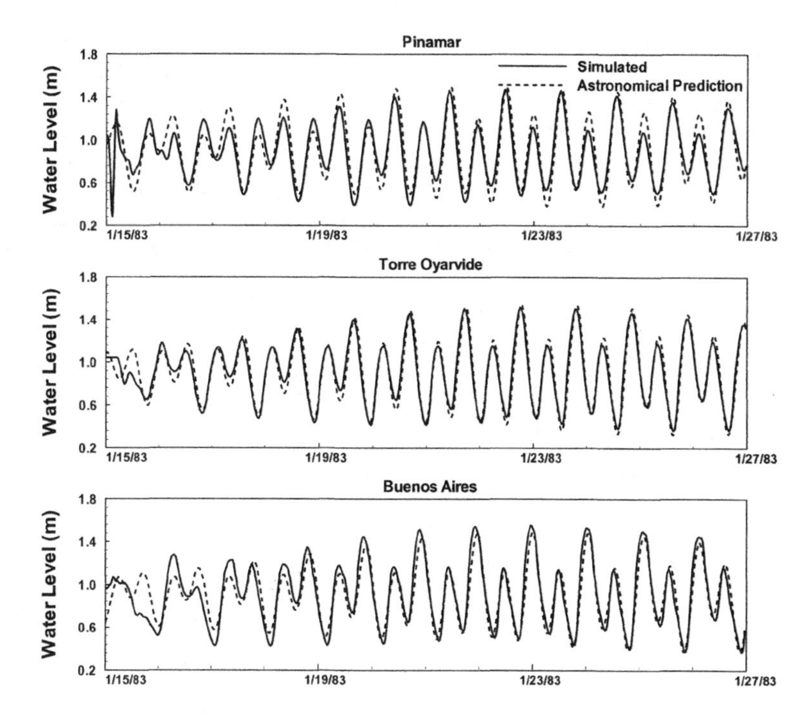

Figure 3. Simulated vs. Projected Water Surface Elevation at Buenos Aires, Torre Oyarvide and Pinamar.

Table 1: RMS Errors for Water Surface Elevation Stations

Station Location	RMS Error	Tidal Range	Normalized RMS Error
	(m)	(m)	(%)
La Paloma	0.05	0.28	19.53
Punta del Este	0.05	0.31	16.55
Montevideo	0.04	0.61	6.85
Colonia	0.10	0.78	12.32
Buenos Aires	0.10	1.12	9.22
Torre Oyarvide	0.07	1.22	5.55
Pinamar	0.11	1.12	9.84

In order to calibrate the wind drag coefficient, a storm that occurred between March 5 and 14, 1987 was selected. Wind effects alone over the modeled domain are not sufficient to properly reproduce storm surge levels in the interior of the Rio de la Plata. During the present study, it was found that meteorological effects are mainly forced by storm waves created in the continental shelf. In order to simulate storm water level with an acceptable degree of accuracy the storm wave that penetrates through the open boundary had to be considered. Registered water levels at Mar del Plata and La Paloma were linearly interpolated to generate the open boundary condition.

Hourly winds registered in the Carrasco Station (Montevideo) were used. Wind shear stresses were calculated by using van Dorn formulation given by the following expression:

$$\tau = \alpha^2 W^2 + \beta^2 (W - W_c)^2$$

with:

$$\alpha = 0.0012$$
$$\beta = 0.0016$$
$$W_c = 9.0\, m/s$$

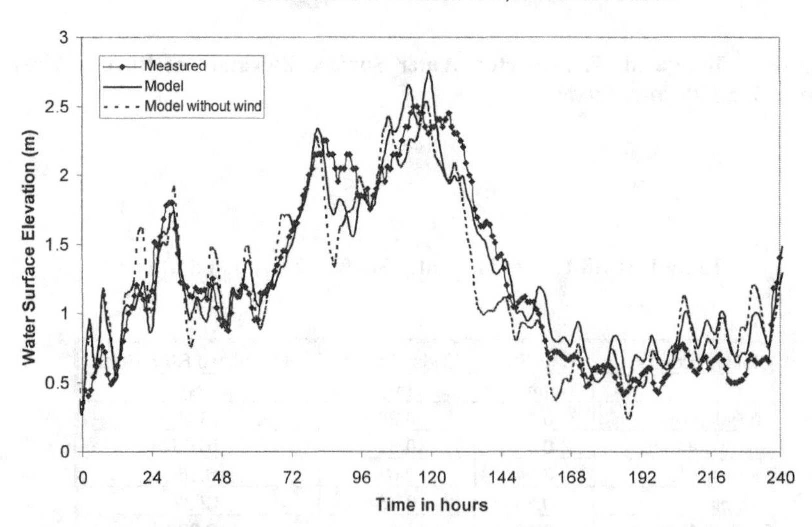

Figure 4. Simulated vs. Measured Water Surface Elevation at Montevideo Station.

Figure 4 shows the water levels registered and simulated that occurred in Montevideo during the selected storm. A good fit was obtained both in phase and peak of values of the wave. Figure 4 present also the water surface elevation obtained without considering wind effects. It is observed that wind effects are indeed not the primary forcing for surges and that remote forcing had to be accounted for.

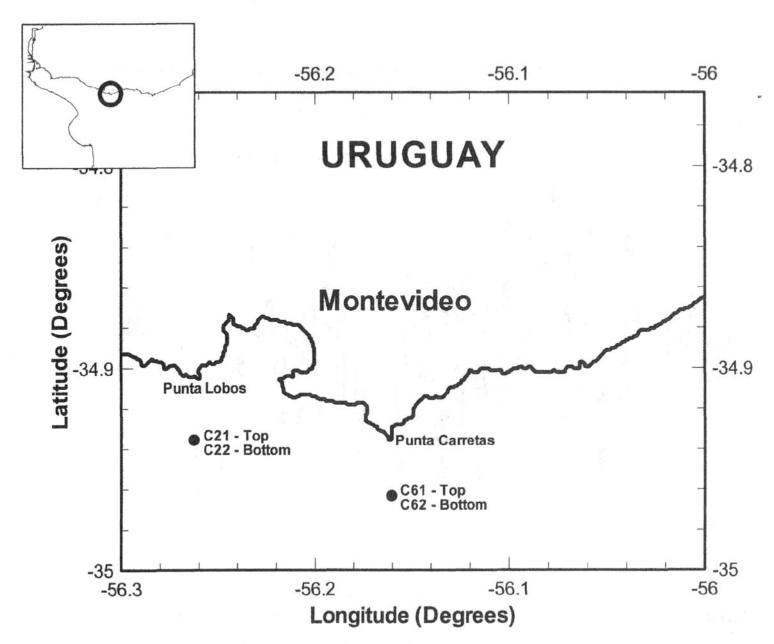

Figure 5. Locations of Velocity Sampling Stations

Once the calibration of the global model was carried out, a simulation was done to verify the predictive capacity of the model for velocities. For verification a period of time was selected during which velocity data near Montevideo was available. This period was from January 27 to February 13, 1983 and consists of recorded velocities at various stations at two depths every half hour. Figure 5 shows the locations of the velocity gauge stations. Station C21 (C22) at Punta Lobos is located approximately 3200 m from the coast, measurements of velocity at two different depths were carried out: surface (3 m of depth), and bottom (6 m of depth). The total depth at the location was 8.1 m. Station C61 (C62) at Punta Carretas is located 3350 m offshore. Two measurements at different depths were also taken at this location: surface (3 m of depth), and bottom (8.7 m of depth). The total depth was 10.7 m.

The station Punta Lobos is in a zone where the coast is aligned approximately in an East-West direction. It is expected that velocities in the East-West direction dominate over the North-South component. On the contrary, the station at Punta

Carretas is close to a very pronounced point in the coast, thus the velocity field is expected to have a more complex behavior.

Figures 6 to 9 present the comparison of the observed and simulated currents.

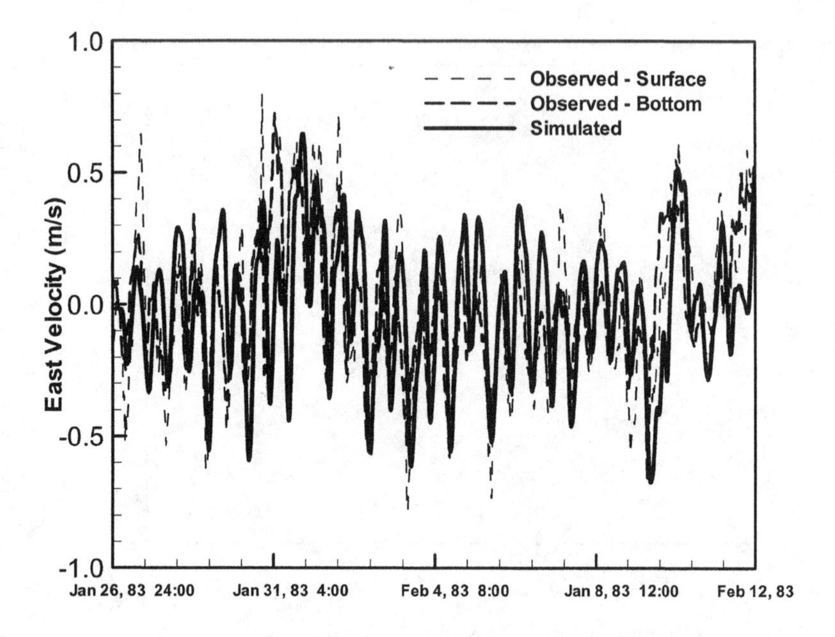

Figure 6. Simulated vs. Observed Velocities at Punta Lobos (component East).

Figure 6 shows that the surface and bottom measured velocities at Punta Lobos follow a similar pattern and that the modeling results present a good agreement with the observed values, both in phase and amplitude of the oscillation. Overall, the modeled results fall between the surface and bottom measured values, indicating that the model suitably predicts a depth average value. Figure 7 shows the comparison of the North-South components of the velocity. As the coast in this location is aligned in an East-West direction, the North-South component is sensibly smaller than the East-West component. A general agreement is observed, but the differences for this component are more significant.

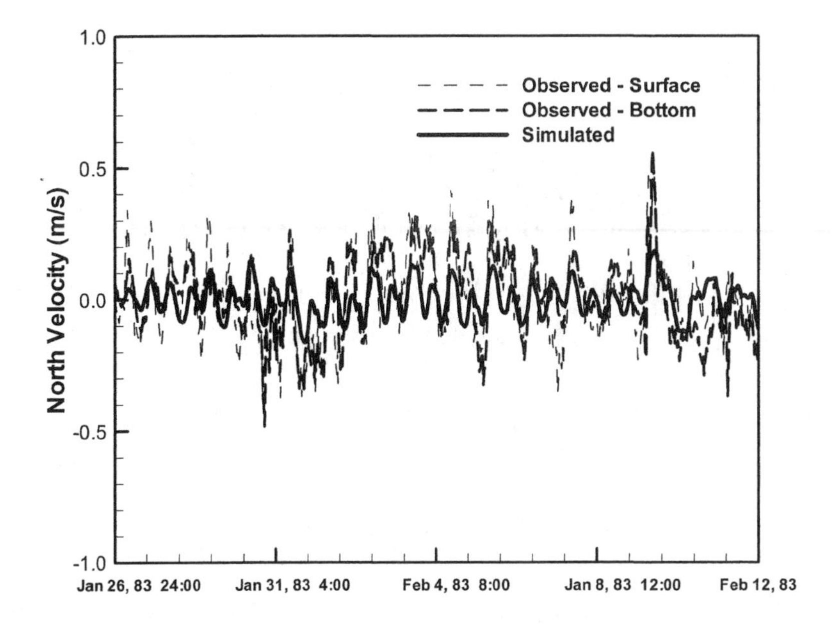

Figure 7. Simulated vs. Observed Velocities at Punta Lobos (component North).

Figure 8 shows a comparison of the measured and simulated East-West component of the velocity at Punta Carretas. A general good agreement of the modeled values with measured data is noted with modeled values falling between the measured values at the surface and bottom. Figure 9 shows the comparison of model results with measurements for the North-South component.

The agreement between simulation and measurements is also better for the East-West component. Considering the complexity of the hydrodynamics at the zone a good agreement in water surface elevations and in velocities exists for the data available.

The series of comparisons discussed previously, indicates that the implemented hydrodynamic model presents a good predictive capacity both in water levels and currents.

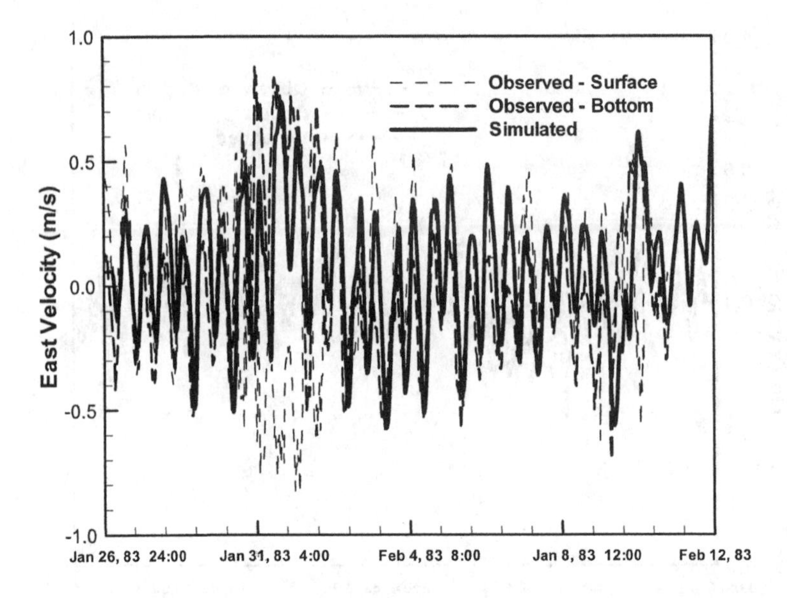

Figure 8. Simulated vs. Observed Velocities at Punta Carretas (component East).

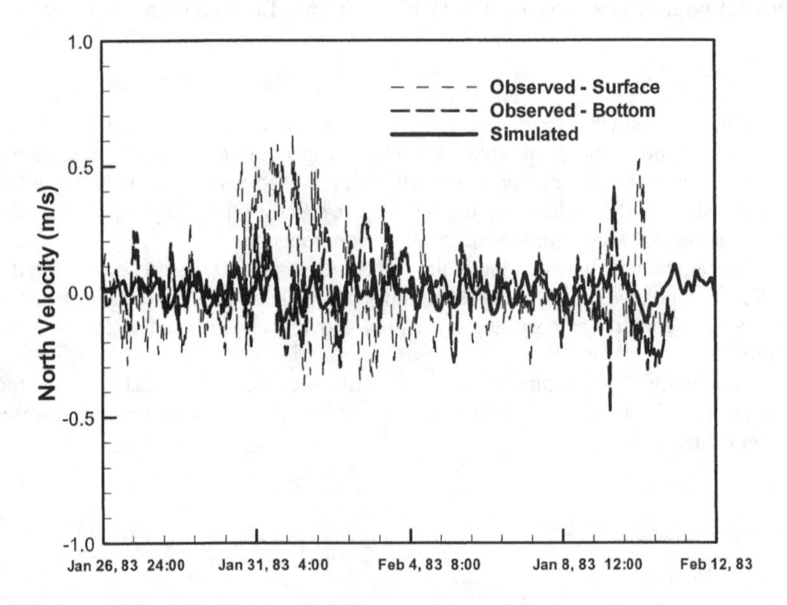

Figure 9. Simulated vs. Observed Velocities at Punta Carretas (component North).

Discussion

The RMA-2 model is a reasonable model for a shallow well-mixed system. However, the Rio de la Plata is not always a well-mixed system. In fact, an analysis of two months non-consecutive velocity data (1440 half-hourly data) show that 19.3% of the observed surface velocities and bottom velocities recorded in Punta Lobos Station are in opposite directions, while this value is 24.4% for Punta Carretas Station. Thus, it is expected that about twenty five per cent of time 3D effects are important, which can not be modeled wit the 2D approach.

Figure 10 a) present the water surface elevation at Mar del Plata and La Paloma Stations, used as boundary conditions. Figure 10 b) present water surface elevations recorded at Montevideo and Torre Oyarvide Stations, while figure 10 c) shows the wind velocity series. The inflow was set constant and equal to 40,000 m^3/s for all simulations.

It is noted that the water level series at Montevideo Station between 26-30 January shows a strong surge immediately followed by a strong drop, which could have triggered the 3D effect, which follow it.

Moreover water levels at Torre Oyarvide Station, which is only about 90 Km away, show a rather different behavior. While in Montevideo water levels present a strong drop, at the same time in Torre Oyarvide a strong surge is observed, reaching about 3.0 m high (in the region the mean water level is 0.91 m).

After January 30, winds are blowing mainly from the north, and a drop of water levels occurs as could be expected. However, before this date, wind is not clearly forcing the surge registered at Montevideo Station. Pressure gradients could be responsible of this fact.

At present, detailed velocity profiles are being measured by two ADCP, which will enable a better understanding of the stratifications effects.

Summary and Conclusions

A two-dimensional hydrodynamic model, using the RMA-2 code, was developed to simulate the Rio de la Plata estuary. The goal was to extend previous modeling efforts done at Instituto de Mecanica de los Fluidos e Ingenieria Ambiental (IMFIA) that include tide and storm surges dynamic simulations. The model extended from the continental shelf, up the Rio de la Plata.

Model inputs included tidal fluctuations at the ocean border, freshwater inflow at the headwaters of the Rio de la Plata, and winds over the simulated domain.
The model was calibrated for the summer of 1983 when water level and current measurements were available. The hydrodynamic model was calibrated for the water surface elevation and velocities generated by astronomical tide and storm surges.

The hydrodynamic calibrations show that the model is a valuable tool for the evaluation of the general circulation patterns of a complex system such as the Rio de la Plata.

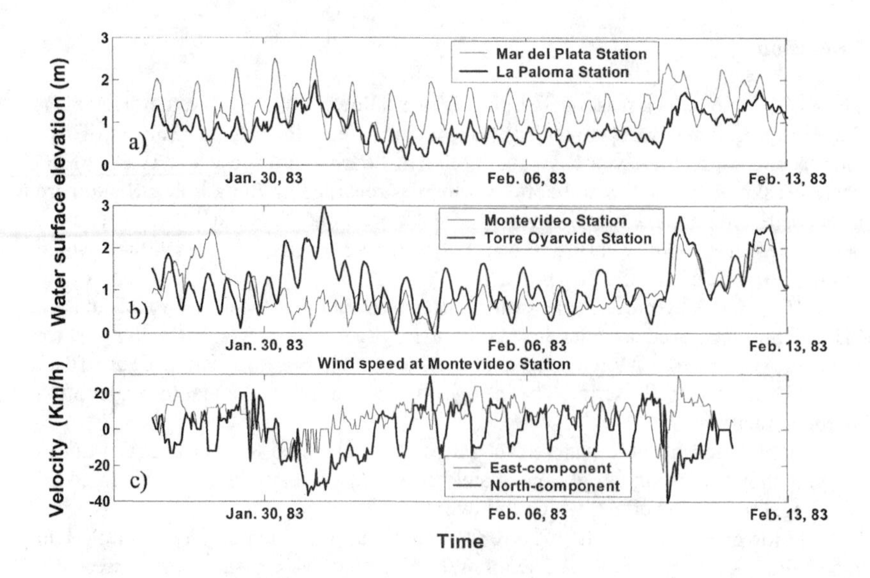

Figure 10. Observed water surface elevation at Mar del Plata and La Paloma Stations (a), Montevideo and Torre Oyarvide Stations (b), and wind velocity recorded at Montevideo (c).

Acknowledgments

This study was partially supported by the Municipality of Montevideo and the Project FREPLATA: Protección Ambiental del Río de la Plata y su Frente Marítimo.

Bibliography

Balay, M.A. (1961). El Rio de la Plata entre la atmósfera y el mar, Publicación H-621, Servicio de Hidrografía Naval, Armada Argentina, Buenos Aires, 153 pp.

C.A.R.P. (1989) Estudio para la evaluación de la contaminación en el Rio de la Plata. Comisión Administradora del Rio de la Plata. Montevideo, Buenos Aires, 422 pp.

Donnell, B. P.; Letter, J. V.; McAnally, W. H.; and others. (2000). "User's Guide for RMA2 Version 4.5," US Army, Engineer Research and Development Center Waterways Experiment Station, Coastal and Hydraulics Laboratory.

Framinan, M. N.; Brown, O. B. (1996). Study of the Rio de la Plata turbidity front, Part I: spatial and temporal distribution. Continental Shelf Research, vol. 16, Nº 10, pp 1259-1282.

King, I. P. (1998). RMA2 – A two dimensional finite element model for flow in estuaries and streams. Dept. of Civil Engineering. University of California. Davis, California.

Le Provost, C., F. Lyard, J.M. Molines, M.L. Genco, and F. Rabilloud, (1998). A hydrodynamic ocean tide model improved by assimilating a satellite altimeter-derived data set, *J. Geophys. Res.*, *103*, 5513-5529.

A High-Resolution Coastal Circulation Model for Lunenburg Bay, Nova Scotia

Jinyu Sheng[1] and Liang Wang[1]

Abstract

A high-resolution coastal circulation model is used to study the three-dimensional barotropic circulation in Lunenburg Bay and adjacent Upper and Lower South Coves, Nova Scotia. The model is driven by tidal forcing and shelf waves specified at the model open boundaries and wind stress applied at the sea surface. The tidal forcing at the model open boundaries is inferred from the tidal sea level prediction at Lunenburg Harbour. The remotely generated sub-inertial shelf waves that propagate into the model domain through the model open boundaries are calculated by a coarse-resolution storm surge model for the eastern Canadian seaboard (Bobanovic and Thompson, 2001). The high-resolution coastal circulation model is used to investigate the nonlinear tidal dynamics in the study region. The model results demonstrate that tidal circulation in the bay and the two coves is highly nonlinear with strong tidal asymmetry between flooding and ebbing, with an intense narrow jet flowing outward from Upper South Cove to Lunenburg Bay during the ebb. The coastal circulation model is also used to simulate the barotropic circulation in Lunenburg Bay during Hurricane Gustav in the second week of September, 2002. The model results demonstrate strong interactions between the local wind stress, tidal forcing, and remotely generated shelf waves during this period.

Introduction

Lunenburg Bay is a shallow coastal embayment situated on the south shore of Nova Scotia (Figure 1). It is about 8 km long, 5 km wide and at most 25 m deep.

[1]Department of Oceanography, Dalhousie University, Halifax, Nova Scotia, Canada, B3H 4J1. E-mail: jinyu.sheng@dal.ca, liangw@phys.ocean.dal.ca

Lunenburg Bay (LB) is connected to Upper South Cove (USC) and Lower South Cove (LSC) via a narrow channel known as Corkum's Channel (CC). Circulation in LB and the two coves is dominated by the semi-diurnal tide (Thompson et al., 1998), with spring and neap tidal ranges of about 2 m and 1 m, respectively. Due to the hydraulic control of the narrow mouth between the two coves (Figure 1), the tidal circulation in USC differs significantly from that in LB and LSC.

Various physical and biological studies have been made in LB and the two coves in the past (e.g., Grant et al., 1993; Dowd, 1997). In the early 1990s, a group of physical and biological oceanographers conducted an interdisciplinary research program to examine physical and biological conditions affecting scallop dispersal and mussel aquaculture in LB and USC (Grant et al., 1993). Most recently, Lunenburg Bay was chosen as a testbed for a marine environmental observation and prediction system (http://www.cmep.ca) for the coastal regions of Atlantic Canada, using data assimilative and coupled models guided directly by real-time observations. The high-resolution coastal ocean circulation model presented in this paper is part of the prediction system currently under development.

Sturley et al. (1993) were the first to use a three-dimensional tidal circulation model to study the tidal circulation in LB and the two coves. Their tidal model

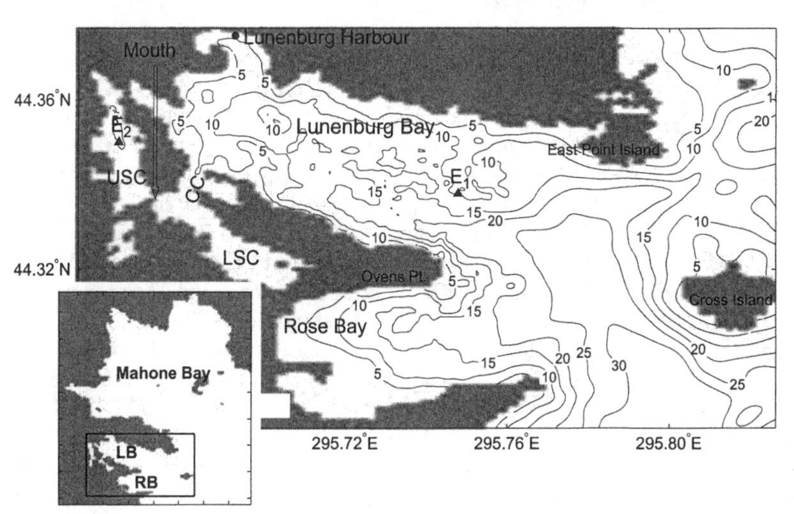

Figure 1. Selected bathymetric features within the domain of the Lunenburg Bay model. Contours are labeled in meters. Abbreviations are used for Upper South Cove (USC), Lower South Cove (LSC), Corkum's Channel (CC), Lunenburg Bay (LB) and Rose Bay (RB). The two locations marked by E_1 and E_2 are used in the presentation of model results. Inset shows the coastal sea of Mahone, Lunenburg and Rose Bays.

reproduced many well-known tidal circulation features in the region, including the intense tidal jet from USC to LB through Corkum's Channel and tidal asymmetry between flooding and ebbing. Using the same model, Sturley and Bowen (1996) demonstrated that the tidal circulation in LB and the two coves is highly nonlinear and the advective terms in the momentum equations and the nonlinear quadratic bottom friction are important in producing the hydraulic control and tidal distortion in LB. Thompson et al. (1998) numerically simulated the whorl-like circulation features observed by the synthetic aperture radar over the western LB. In this paper, we use the high-resolution coastal circulation model to study the nonlinear tidal dynamics and interactions of local wind, tides and remotely generated shelf waves that propagate into the model domain through the model open boundaries.

The arrangement of this paper is as follows. The next section briefly discusses the ocean circulation model. This is followed by a description of circulation features generated by the M_2 tide and associated overtides M_4 and M_6 in LB. We then present the barotropic circulation forced by wind, tides, and subtidal shelf waves during Hurricane Gustav in September, 2002. The final section is a summary and conclusion.

Ocean Circulation Model

The ocean circulation model used in this study is the three-dimensional primitive-equation z-level ocean circulation model known as CANDIE (Sheng et al., 1998, [http://www.phys.ocean.dal.ca/programs/CANDIE]). This model has been successfully applied to various modelling problems on the shelf, including the wind-driven circulation over an idealized coastal canyon (Sheng et al., 1998), nonlinear dynamics of a density-driven coastal current (Sheng, 2001), tidal circulation in the Gulf of St. Lawrence (Lu et al., 2001a), wind-driven circulation over a stratified coastal embayment (Davidson et al., 2001), and seasonal circulation in the northwestern Atlantic Ocean (Sheng et al., 2001). Most recently CANDIE has been applied to the western Caribbean Sea (Sheng and Tang, 2003, 2004). The standard version of CANDIE uses the rigid-lid approximation which excludes surface gravity waves. In studying the internal tide generation over topography, Lu et al. (2001b) developed a free-surface version of CANDIE by adding a linear free surface to the original CANDIE code based on the implicit free surface method suggested by Dukowicz and Smith (1994). In this study, we use a nonlinear free-surface version of CANDIE developed recently by Sheng and Wang (2004) which includes the nonlinear terms in the continuity equation and in the kinematic boundary condition at the sea surface.

The governing equations of the model with the nonlinear free surface can be written in spherical coordinates as

$$\frac{\partial u}{\partial t} + \mathcal{L}u - \left(f + \frac{u \tan\phi}{R}\right) v = -g \frac{\partial(\eta + \hat{p}/(\rho_o g))}{R\cos\phi\partial\lambda} + \mathcal{D}_m u + \frac{\partial}{\partial z}\left(K_m \frac{\partial u}{\partial z}\right) \quad (1)$$

$$\frac{\partial v}{\partial t} + \mathcal{L}v + \left(f + \frac{u \tan\phi}{R}\right) u = -g\frac{\partial(\eta + \hat{p}/(\rho_o g))}{R\partial\phi} + \mathcal{D}_m v + \frac{\partial}{\partial z}\left(K_m \frac{\partial v}{\partial z}\right) \tag{2}$$

$$\frac{\partial \eta}{\partial t} = -\frac{1}{R\cos\phi}\left(\frac{\partial \int_{-h}^{\eta} u dz}{\partial\lambda} + \frac{\partial(\int_{-h}^{\eta} v \cos\phi dz)}{\partial\phi}\right) \tag{3}$$

$$\frac{\partial w}{\partial z} = -\frac{1}{R\cos\phi}\left(\frac{\partial u}{\partial\lambda} + \frac{\partial(v\cos\phi)}{\partial\phi}\right) \tag{4}$$

$$\hat{p} = g\int_z^0 \rho dz' \tag{5}$$

where u, v, w are the east (λ), north (ϕ) and vertical (z) components of the velocity vector $\vec{u}$, η is the surface elevation, $z = -h(x, y)$ is the position of the sea bottom, $\hat{p}$ is the density-driven internal pressure, ρ is density, K_m is the vertical eddy viscosity coefficient, f is the Coriolis parameter, ρ_o is a reference density, R and g are the Earth's radius and gravitational acceleration. In the above equations, $\mathcal{L}$ is an advection operator defined as

$$\mathcal{L}q = \frac{1}{R\cos\phi}\frac{\partial(uq)}{\partial\lambda} + \frac{1}{R\cos\phi}\frac{\partial(vq\cos\phi)}{\partial\phi} + \frac{\partial(wq)}{\partial z}, \tag{6}$$

and $\mathcal{D}_m$ is a diffusion operator defined as

$$\mathcal{D}_m q = \frac{1}{R^2}\left[\frac{1}{\cos^2\phi}\frac{\partial}{\partial\lambda}\left(A_m\frac{\partial q}{\partial\lambda}\right) + \frac{\partial q}{\partial\phi}\left(\cos\phi A_m \frac{\partial q}{\partial\phi}\right)\right] \tag{7}$$

where A_m is the horizontal eddy viscosity coefficient. The model uses the fourth-order numerics (Dietrich, 1997) and Thuburn's (1996) flux limiter to discretize the nonlinear advection terms.

In this study we run the ocean circulation model in barotropic mode, with the model temperature and salinity (and therefore the model density) set to be invariant in time and space. The model horizontal resolution is 60 m and the vertical resolution is 2 m, except at the top z-level where it is 3 m. The horizontal eddy viscosity coefficient A_m is set to 3×10^4 cm^2 s^{-1}, and the drag coefficient of the quadratic bottom stress is set to C_d=2.5×10^{-3}. For the vertical eddy viscosity coefficient K_m, we follow Davies et al. (1997) and use

$$K_m = K_o|\vec{U}|\phi(z) \tag{8}$$

where $K_o = 2.5\times10^{-3}$ is a dimensionless coefficient, $|\vec{U}|$ is the amplitude of the depth-integrated horizontal velocity vector, and $\phi(z)$ is a prescribed vertical-structure function, which is set to be unity from surface to $z = -0.8h(x, y)$ and then decrease linearly to 0.01 at the sea bottom of $z = -h(x, y)$ (Davies et al., 1997).

At the model lateral closed boundaries, the no-slip condition is applied to the horizontal currents. Along the model open boundaries, the radiation condition suggested by Davies and Flather (1978) is used. At the eastern open boundary, for example, we have

$$u_e = u_t + u_s + \frac{c}{h}(\eta_e - \eta_t - \eta_s) \tag{9}$$

where (η_e, u_e) are the model-calculated surface elevation and normal flow at the eastern open boundary, (η_t, u_t) and (η_s, u_s) are respectively the tidal and storm surge surface elevation and normal currents at the eastern open boundary, h is the local water depth, and c is the phase speed. In this study we follow Lu et al. (2001b) and set $c = 0.16\sqrt{gh}$, where again g is the gravitational acceleration and h is local water depth.

Nonlinear Tidal Circulation in Lunenburg Bay

Previous studies in the region demonstrated that the semi-diurnal tide (M_2) is the principal tidal constituent in LB, accounting for about 80% of the total observed variance of the bottom pressures and 60% of the total observed variance of the currents in the region (Thompson, et al., 1998; Sheng and Wang, 2004). In this section, we examine the nonlinear tidal dynamics in LB and the two coves by forcing the high-resolution coastal circulation model with the semi-diurnal tidal surface elevation η_t specified at the model open boundaries. For simplicity, we assume η_t to be spatially uniform along the model open boundaries. Since there were no direct current measurements at the open boundaries that could be used to extract the M_2 current, we set u_t in Eq. (9) to zero. The local wind forcing and boundary forcing associated with the storm surge (i.e., η_s and u_s in Eq. (9)) are set to zero in this numerical experiment. We integrate the model for ten M_2 tidal cycles and discuss the model results of the last 8 tidal cycles in this section.

The model calculated surface elevations have negligible horizontal variations in each subregion. There are, however, large differences in phase and magnitude between the simulated surface elevations in LB and USC (Figure 2). The surface elevation of M_2 in LB peaks prior to that in USC by about one hour and forty minutes, which is comparable with, and slightly longer than, the observed phase lag of about one hour and thirty minutes. The model-calculated amplitudes of surface elevations are about 0.63 m and 0.44 m in LB and USC, respectively, which are comparable well with the observed amplitudes of 0.64 m in LB and 0.44 m in USC (Thompson et al., 1998). It should be noted that the simulated surface elevation in LB is highly symmetric. In comparison, the surface elevation in USC is less symmetric, with a fast rise to high water and slower fall to low water due to the nonlinear tidal dynamics in the area (Figure 2).

We calculate the near-surface current ellipses of the M_2 constituent using tidal harmonic analysis of the model results during the tenth M_2 cycle. The near-surface M_2 tidal current ellipses are clockwise and roughly alongshore (southeastward) in LB, with a typical semi-major axis of about 10 cm s^{-1} (Figure 3a). In Corkum's Channel (CC) and adjacent northwestern LB, the M_2 current ellipses are relatively large and roughly northeastward, with a maximum semi-major axis of about 50 cm s^{-1}. The near-surface current ellipses of M_2 are also relatively large in the southern part of USC and the northern part of LSC and along the narrow channel connecting the two coves. The typical semi-major axis of the M_2 tide is about 5 cm s^{-1} in the two coves.

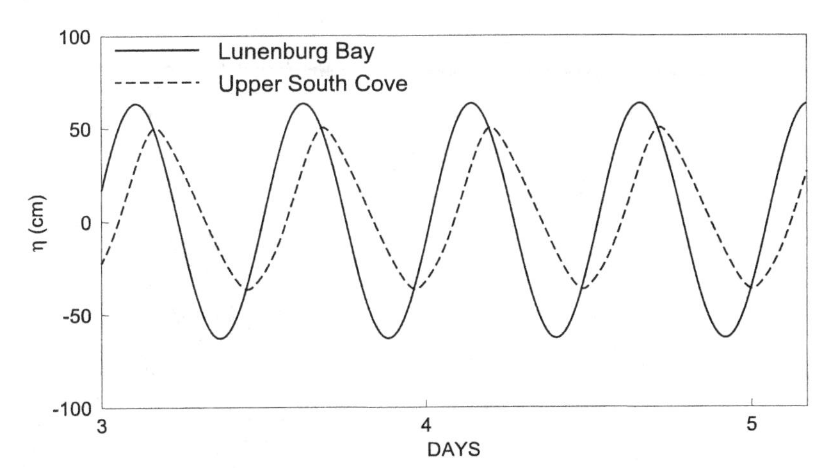

Figure 2: Model-calculated surface elevations at location E_1 in Lunenburg Bay and at location E_2 in Upper South Cove (Figure 1). The model is forced by the M_2 semi-diurnal tidal surface elevation specified at the model open boundaries.

Previous studies demonstrated that the overtide M_4 of the semi-diurnal constituent is produced mainly by the nonlinear terms in the momentum and continuity equations, whereas the overtide M_6 is produced mainly by the nonlinear bottom friction (Parker, 1991). Figures 3b and c show the near-surface current ellipses of the two overtides M_4 and M_6 calculated from the model results during the tenth M_2 tidal cycle. The near-surface tidal currents of the overtide M_4 are relatively large in the southern part of USC, northern part of LSC, Corkum's Channel and adjacent northwestern LB, with a typical semi-major axis of about 5 cm s^{-1}. In comparison, the near-surface M_4 tidal currents are relatively small in the outer and inner LB. Therefore, the nonlinear advection terms and nonlinear free surface terms play an important role mainly in USC, LSC, Corkum's Channel and adjacent western LB.

The near-surface currents of the overtide M_6 are relatively large in Corkum's Channel and in both sides of the narrow mouth connecting USC and LSC, with a typical semi-major axis of about 3 cm s^{-1}. The near-surface M_6 tidal currents are relatively small in the rest of the the study region. Therefore, the nonlinear bottom friction plays an important role in the narrow channel connecting LB to USC.

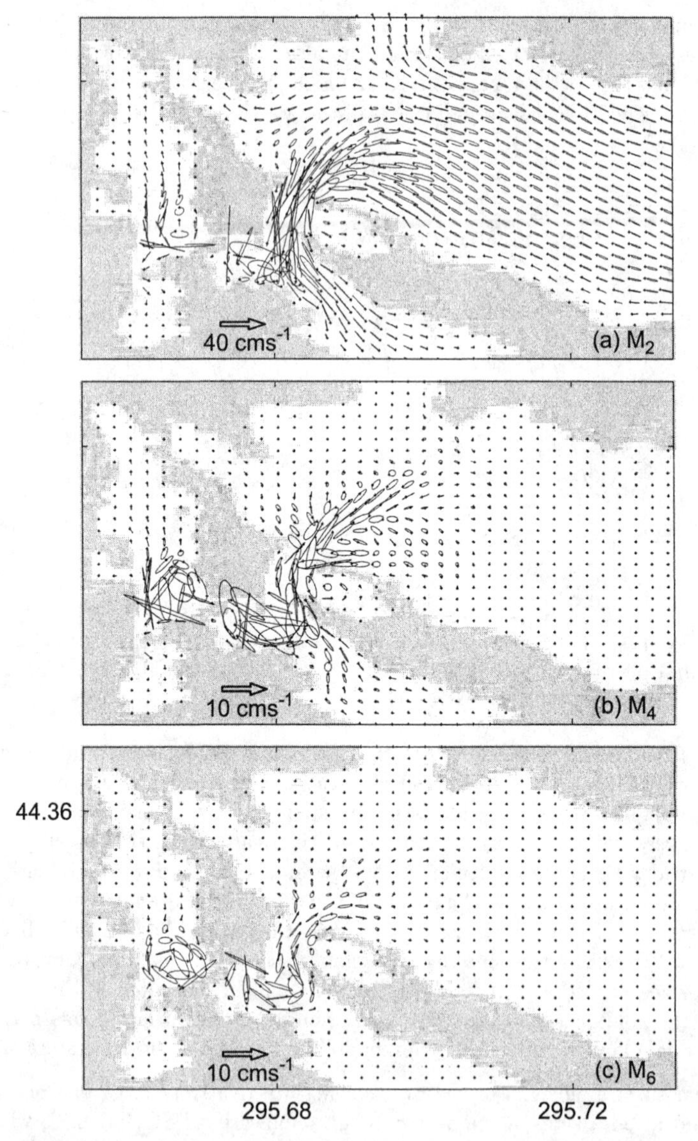

Figure 3. Near-surface current ellipses of (a) the semi-diurnal constituent M_2, (b) the overtide M_4 and (c) overtide M_6 of the M_2 constituent calculated from the model results during the tenth M_2 tidal cycle. Current ellipses are plotted at every third model grid point. The symbol + represents the starting time, which is the same for each ellipse.

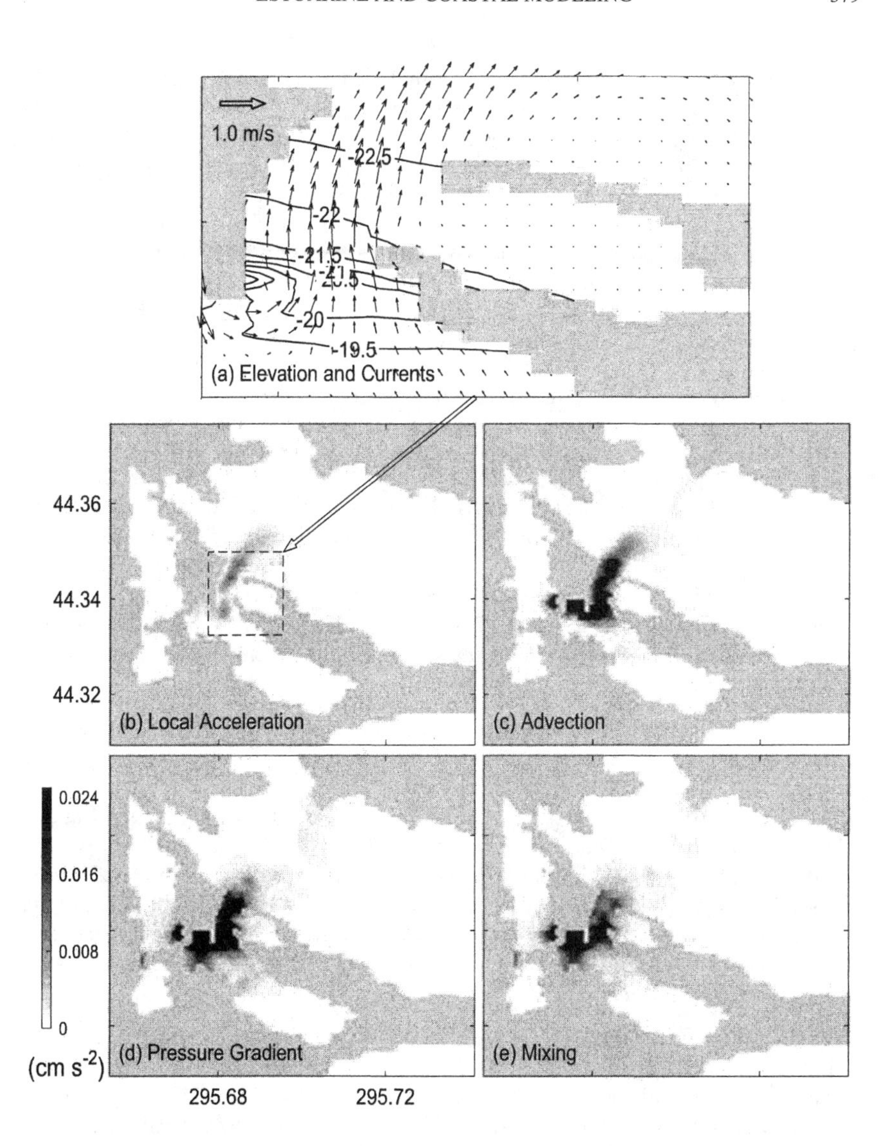

Figure 4. (a) Surface elevation and near-surface currents (at 1.5 m) during the ebb tide of the semi-diurnal constituent in LB, and magnitudes of (b) transient local acceleration; (c) transient advection of the oscillatory flow (i. e., the nonlinear interaction of tidal oscillatory currents with local bathymetry); (d) transient surface pressure gradient; and (e) lateral and bottom tidal dissipation in the top z-level at the same time. A 3-point moving average filter is used to eliminate small-scale features that are not well presented in (b-e).

We next examine the transient momentum balance during the ebb tide of the M_2 constituent in LB. At this time, an intense narrow jet flows northeastward through Corkum's Channel toward the northwestern LB, associated with a large surface elevation drop of about 3 cm over a distance of about 1.2 km (Figure 4a). Therefore, the pressure gradient plays a very important role in driving the outward jet along the channel (Figure 4d). The nonlinear advection term at this time is of the same order of the amplitude as the pressure gradient (Figure 4c). The tidal mixing term is large mainly over Corkum's Channel and the narrow mouth connecting the two coves. The local acceleration at this time is relatively small and the Coriolis term (not shown) is much smaller, as expected.

Circulation Forced by Wind, Tides and Storm Surge

In this section, we simulate the barotropic circulation in LB and the two coves during a 6-day period from September 10 (Julian Day 253) to September 15 (Julian Day 258), 2002. At beginning of this period, a tropical storm Gustav raced northeastward after buffeting the coasts of North Carolina and Virginia. Gustav intensified significantly and upgraded from a tropical storm to a hurricane on September 11, 2002, with maximum sustained winds of about 120 km h^{-1}. Hurricane Gustav approached the southwest Scotian Shelf at about 17:00 local time on September 11, and made landfall over the southern coast of Cape Breton, Nova Scotia at 1:30 am on September 12, 2002.

The model external forcing includes the local wind stress at the sea surface, and tidal forcing and remotely generated shelf waves specified at the model open boundaries. Figure 5a shows the time-series of wind stress in LB calculated from the 3 hourly forecast wind speeds provided by the Meteorological Service of Canada (MSC) based on the bulk formula of Large and Pond (1981). Since the MSC forecast winds have a horizontal resolution of 0.25 degree, which is about 17 km by 27 km in the study region, the wind stress used in this study has negligible spatial variability in LB and the two coves. The wind stress, however, has significant temporal variability, with a sustained southwestward stress of about 0.4 N m^{-2} on day 255 associated with Hurricane Gustav. Wind was relatively weak after day 256.

Hurricane Gustav also caused significant localized coastal flooding as it moved northeastward over the Canadian Maritime waters. There was a storm surge of more than 1 m reported along the coasts of Prince Edward Island, northern and eastern Nova Scotia and eastern New Brunswick. The storm surge surface elevation η_s at the model open boundaries shown in Figure 5b was calculated by a depth-averaged storm surge model for the eastern Canadian shelf developed by Bobanovic and Thompson (2001), with a horizontal resolution of about 1/12 degree, or about 7 km by 9 km in the LB region. This coarse-resolution storm surge model was forced by the same MSC forecast winds described above. The storm surge surface elevation at the open boundaries of the coastal circulation model increased from day 254 and reached a maximum surge elevation of about 50 cm on day 255 (Figure 5b). This large storm surge event is strongly correlated with the strong wind event shown in Figure 5a.

The tidal surface elevation η_t at the model open boundaries during this period (Figure 5b) was inferred from the tidal sea level prediction at Lunenburg Harbour based on the simplified incremental approach described by Sheng and Wang (2003). The tidal sea level prediction at Lunenburg Harbour was produced by the Canadian Hydrographic Service using analysis of more than 50 tidal constituents determined from the historical sea level observations at the harbour. Figure 5b confirms that the semi-diurnal constituent is the dominant component of η_s in the study region.

The near-surface currents driven by the combination of the local wind, tidal and storm surge forcing at day 255.003 (00:04:20, September 12) are strong and roughly southwestward over the western LB, outer LB, Rose Bay and areas around Cross

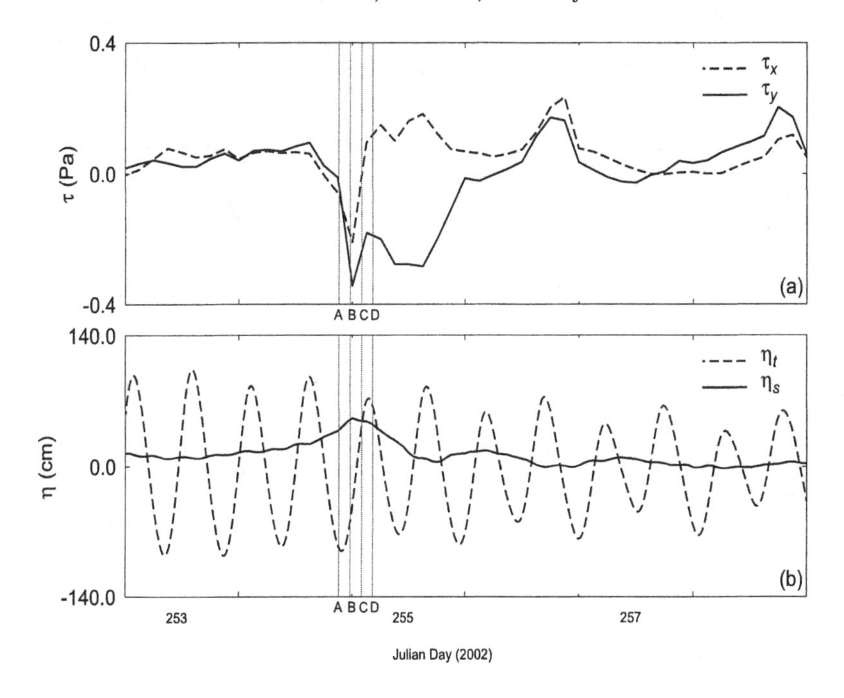

Figure 5. Time series of (a) eastward (τ_x) and northward (τ_y) components of surface wind stress in Lunenburg Bay, and (b) tidal (η_t) and storm surge (η_s) surface elevation at model open boundaries from September 10 (day 253) to 15 (day 258), 2002. Model results are presented at the times marked by four lines (A-D).

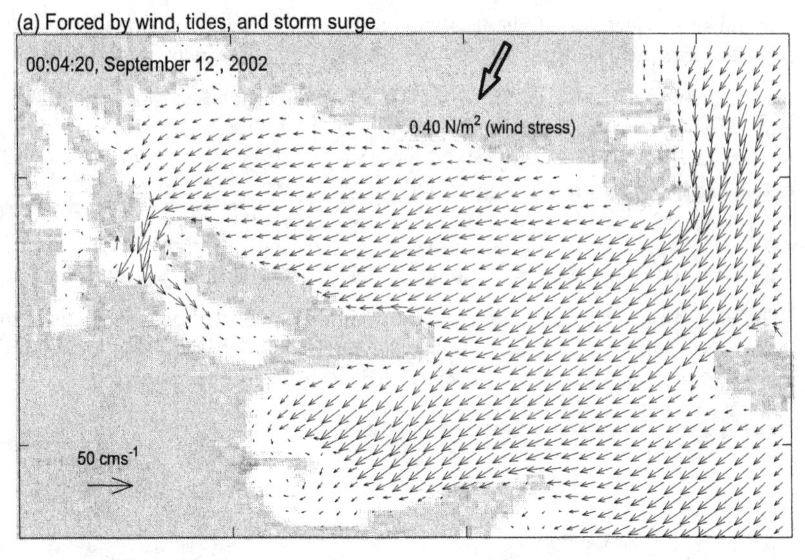

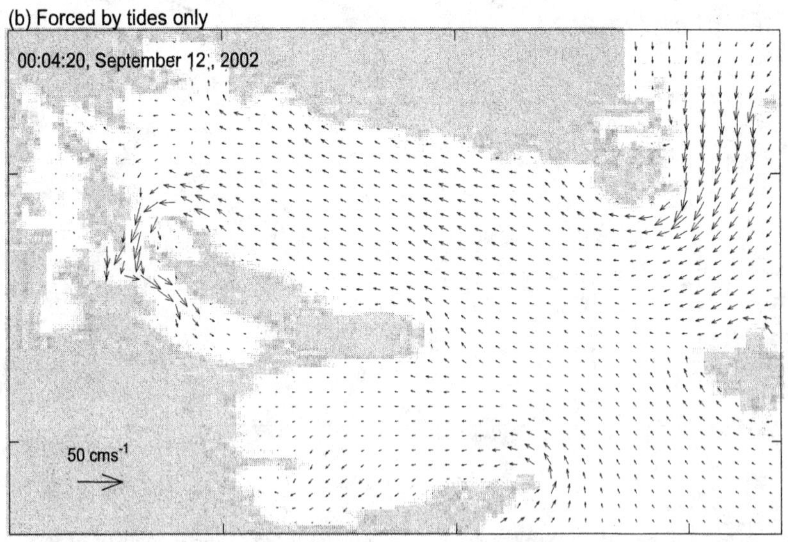

Figure 6. Model calculated near-surface currents at day 255.003 (00:04:20, September 12), 2002 forced by (a) the combination of the local wind stress at the sea surface, tidal and storm surge surface elevations at the model open boundaries and (b) the tidal forcing only at the model open boundaries. Velocity vectors are plotted at every fifth model grid point.

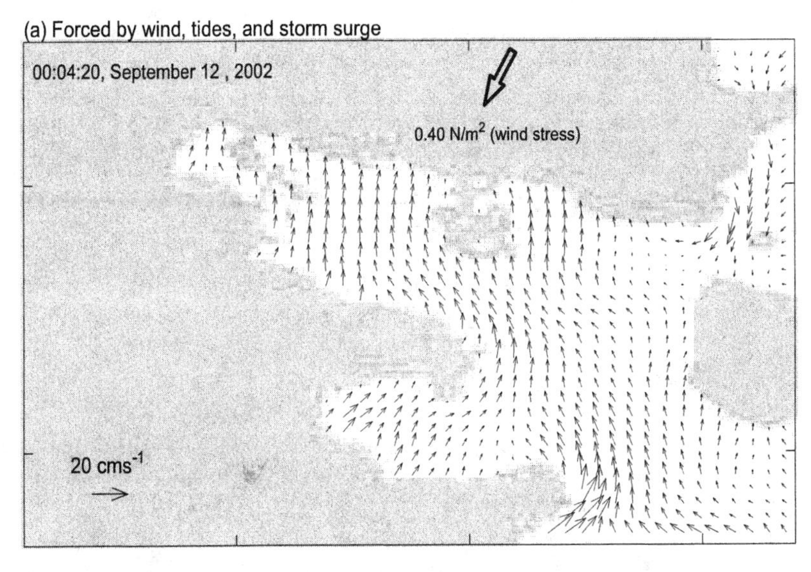

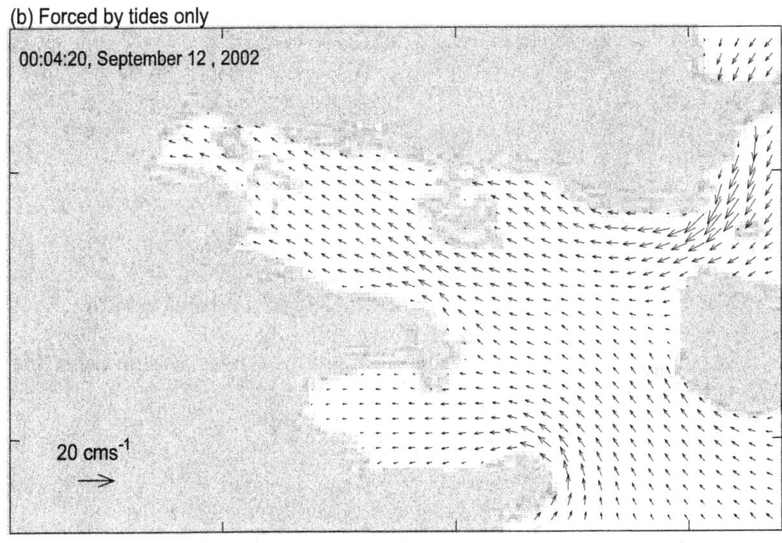

Figure 7. Model calculated sub-surface currents at a depth of 10 m at day 255.003 forced by (a) the combination of the local wind stress at the sea surface, tidal and storm surge surface elevations at the model open boundaries and (b) the tidal forcing only at the model open boundaries. Velocity vectors are plotted at every fifth model grid point.

Island (Figure 6a). Over the coastal areas of the western and northwestern LB, the near-surface currents are much weaker and roughly along-shore. There is a strong inward flow through Corkum's Channel at this time. This inward flow separates into two branches upon reaching the west end of the channel, with one branch flowing southeastward into LSC and the other branch flowing northwestward into USC.

In comparison, the near-surface currents driven only by the tidal forcing are much weaker and roughly northwestward in LB except for the northwest part of the bay (Figure 6b), indicating the dominant role of the surface wind stress in driving the near-surface circulation in LB at this time. However, the near-surface circulation in Corkum's Channel and the two coves forced by the combined forcing are highly comparable with those forced only by the tidal forcing, indicating the tidal forcing plays the dominant role in these areas.

The model calculated sub-surface currents at 10 m depth forced by the combined forcing are roughly northward over LB, with a typical speed of about 8 cm s^{-1} at day 255.003. In comparison, the sub-surface currents forced only by the tidal forcing at this time are roughly northwestward over LB with a typical speed of 5 cm s^{-1}. Therefore, both the local wind forcing and tidal forcing are equally important in driving the subsurface circulation in LB at this time.

To examine the nonlinear dynamics of the circulation forced by the local wind, tidal forcing and remotely generated shelf waves specified at the model open boundaries, we conduct an additional numerical experiment by forcing the linear version of CANDIE with the same model external forcings. We calculate the differences in the surface elevations and near-surface currents between the nonlinear model and linear model (Figure 8). At day 254.884 (time A in Figure 5), the near-surface current differences are relatively small in the study region, except for Corkum's Channel, the narrow mouth connecting the two coves and coastal areas. At this time, the local wind stress is relatively weak and the storm surge surface elevation is moderate. Therefore, the nonlinear dynamics at this time are mainly due to the nonlinear tidal dynamics. At day 254.983 (time B in Figure 5), the wind stress increases to 0.35 N m^{-2}, which is near the maximum wind stress caused by Hurricane Gustav, and the storm surge surface elevation reaches a maximum of 50 cm (Figure 5). The current differences at this time are moderate and relatively uniform in the study region except in the coastal areas. The surface elevation differences are about 4 cm in LB at this time. Therefore, the nonlinear dynamics at time B are associated mainly with the local wind and remotely generated storm surge surface elevations. At day 255.082 (time C in Figure 5), the differences in the surface elevations and near surface currents are largest over the study region, except for coastal areas, indicating strong nonlinear interaction between the local wind, tides, and remotely generated shelf waves that propagate into the study region through the model open boundaries.

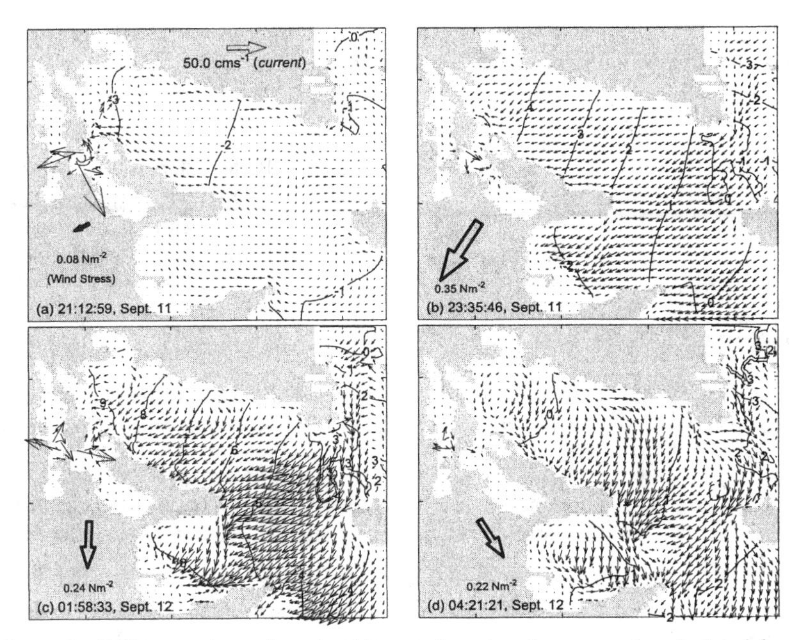

Figure 8. Differences in surface elevations and near-surface currents calculated by the fully non-linear model and those calculated by the linear model at (a) day 254.884, (b) day 254.983 (c) day 255.082, and (d) day 255.181. Velocity vectors are plotted at every fifth model grid point.

Summary and Discussion

We have developed a high-resolution barotropic coastal circulation model for Lunenburg Bay and adjacent two coves using the nonlinear free-surface version of CANDIE, as part of the development of a data-assimilative and coupled marine environmental observation and prediction system for the coastal regions of Atlantic Canada. In this study, we used this high-resolution coastal circulation model to study the nonlinear tidal dynamics in LB and adjacent Upper and Lower South Coves. The model results reproduce reasonably well the intensive jet during the ebb tide and phase and amplitude differences of surface elevations between Lunenburg Bay and Upper South Cove. The model results also demonstrate that the nonlinear terms in the momentum and continuity equations play an important role in the tidal circulation in the study region.

We also used the high-resolution coastal circulation model to simulate the barotropic circulation in Lunenburg Bay during Hurricane Gustav in September, 2002. We forced the model with the local wind stress at the sea surface, and the tidal and storm surge surface elevations specified at the model open boundaries. The model results demonstrate that all three forcings play an important role in driving circulation during this period with strong nonlinear interactions.

Acknowledgments

We wish to thank Tony Bowen, Serge Desjardins, Richard Greatbatch, Harold Ritchie, Keith Thompson, Alex Hay and two anonymous reviewers for their useful suggestions. We also thank Charles O'Really for providing the sea level tidal prediction at Lunenburg Harbour. This project is part of the CMEP project supported by CFI and CFCAS.

References

Bobanovic, J., and K. R. Thompson, The influence of local and remote winds on the synoptic sea level variability in the Gulf of Saint Lawrence, *Continental Shelf Res.*, **21**, 129-144, 2001.

Davidson, F., R. J. Greatbatch, and B. deYoung, Asymmetry in the response of a stratified coastal embayment to wind forcing, *J. Geophys. Res.*, **106**, 7001-7016, 2001.

Davies, A. M., and R. A. Flather, Computing extreme meteorologically induced currents, with application to the northwest European continental shelf, *Continental Shelf Res.*, **7**, 643-683, 1978.

Davies, A. M., S. C. M. Kwong, and R. A. Flather, A three-dimensional model of diurnal and semidiurnal tides on the European shelf, *J. Geophys. Res.*, **102**, 8625-8656, 1997.

Dietrich, D. E., Application of a modified Arakawa 'a' grid ocean model having reduced numerical dispersion to the Gulf of Mexico circulation, *Dyn. Atmos. and Oceans*, **27**, 201-217, 1997.

Dowd, M., On predicting the growth of cultured bivalves, *Ecological Modelling*, **104**, 113-131, 1997.

Dukowicz, J., K., and R. D. Smith, Implicit free-surface method for the Bryan-Cox-Semtner ocean model, *J. Geophys. Res.*, **99**, 7991-8014, 1994.

Grant, J., M. Dowd, K. Thompson, C. Emerson, and A. Hatcher, Perspectives on field studies and related biological models of bivalve growth and carrying capacity, In: *Bivalve Filter Feeders and Marine Ecosystem Processes*, Dame, R., Ed., Springer Verlag, New York, 371-420, 1993.

Large, W. G., and S. Pond, Open ocean momentum flux measurements in moderate to strong winds, *J. Phys. Oceanogr.*, **11**, 324-336, 1981.

Lu, Y., K. R. Thompson, and D. G. Wright, Tidal currents and mixing in the Gulf of St. Lawrence: an application of the incremental approach to data assimilation, *Can. J. Fish. Aquat. Sci.*, **58**, 723-735, 2001a.

Lu, Y., D. G. Wright, and D. Brickman, Internal tide generation over topography: experiments with a free-surface z-level ocean model, *J. Atmos. and Ocean. Tech.*, **18**, 1076-1091, 2001b.

Parker, B. B., The relative importance of the various nonlinear mechanisms in a wide range of tidal interactions (review), *Tidal Hydrodynamics*, 237-268, 1991.

Sheng, J., Dynamics of a buoyancy-driven coastal jet: The Gaspé Current, *J. Phys. Oceanogr.*, **31**, 3146-3163, 2001.

Sheng, J., R. J. Greatbatch, and D. G. Wright, Improving the utility of ocean circulation models through adjustment of the momentum balance. *J. Geophys. Res.*, **106**, 16711-19728, 2001.

Sheng, J., and L. Tang, A numerical study of circulation in the western Caribbean Sea, *J. Phys. Oceanogr.*, **33**, 2049-2069, 2003.

Sheng, J., and L. Tang, A two-way nested-grid ocean circulation model for the Meso-American Barrier Reef System, *Ocean Dynamics*, (in press), 2004.

Sheng, J., and L. Wang, Numerical study of tidal circulation in Lunenburg Bay, Nova Scotia, *J. Phys. Oceanogr.*, submitted, 2004.

Sheng, J., D. G. Wright, R. J. Greatbatch, and D. E. Dietrich, CANDIE: A new version of the DieCAST ocean circulation model. *J. Atmos. and Ocean. Tech.*, **15**, 1414-1432, 1998.

Sturley, D. R. M., and A. J. Bowen, A model for contaminant transport in Lunenburg Bay, Nova Scotia, *The Science of the Total Environment*, **179**, 161-172, 1996.

Sturley, D. R. M., K. R. Thompson, and A. J. Bowen, Fine-scale models of coastal inlets: a physical basis for water quality/biological models, *Proceedings of the 1993 Canadian Coastal Conference*, 387-404, 1993.

Thompson, K. R., D. E. Kelley, D. Sturley, B. Topliss, and R. Leal, Nearshore circulation and synthetic aperture radar: an exploratory study, *Int. J. Remote Sensing*, **19**, 1161-1178, 1998.

Thuburn, J., 1996: Multidimensional flux-limited advection schemes. *J. of Comput. Phys.*, **123**, 74-83, 1996.

Influence of Multi Algal Groups in the Calibration of a Water Quality Model

John Eric Edinger[1], Charles D. Boatman[2] and Venkat S. Kolluru[3]

Abstract
A management tool based on scientific inquiry rather than politics is needed for policy makers to make environmentally sound decisions regarding difficult issues, which will only occur more frequently as coastal populations continue to increase. Typically, a well-calibrated hydrodynamic and water quality model is used as the management tool. The development of a water quality and circulation model is based on field data that encompasses the factors affecting the water body, and the model is used to evaluate the potential effects of changing point source loadings under varying operational and environmental conditions. The level of predictability depends on the goodness of field data needed for model calibration and for setting up input data, assuming all the relevant water quality processes are simulated correctly. But there are times when good field data are available, yet the calibration is not good. In such a situation, it is conventional to revisit the algorithms used for the various water quality processes. This approach was attempted in a recent project where a three-dimensional hydrodynamic and water quality model called GEMSS was applied to predict the quality of water in the Budd Inlet, located in the Southern Puget Sound (Washington, USA). Within GEMSS, the use of the water quality model called WQDPM, which is a modified version of EPA's Eutro5 water quality model, did not predict the vertical structure of dissolved oxygen and phytoplankton at different locations in the Budd Inlet. The phytoplankton was modeled in Eutro5 as a single algal group. In order to improve the calibration, the alternate water quality model called WQCBM available in GEMSS was used. This model includes different forms of organic carbon that can be related to sediment exchange processes. WQCBM simulates five interacting sub-systems: net phytoplankton production, the phosphorus cycle, the nitrogen cycle, the dissolved oxygen balance, and the particulate organic carbon balance. The carbon based model was updated to include dinoflagellates and diatoms for the simulation of phytoplankton dynamics. The ability of dinoflagellates to undergo diel vertical

[1] Principal Scientist, J.E.Edinger Associates, Wayne, PA.
[2] Formerly Aura Nova, Bellview, WA. See acknowledgements.
[3] Senior Scientist, J.E.Ednger Associates, Wayne, PA.

migration and to actively take up nutrients at night greatly improved the prediction of dissolved oxygen and chlorophyll vertical structures. The concept of using spores and cysts as primary sources for algal blooming is discussed using numerical tank simulations.

Introduction

Phytoplankton processes in eutrophication models interact with and influence all nutrient cycling and dissolved oxygen concentrations computed over time and space within a water body (Krallis and Buchak, 2002). Many eutrophication models use multiple phytoplankton groups each with different growth and light limiting properties to produce observed phytoplankton and water quality distributions. Other processes controlling phytoplankton densities are being examined including zooplankton grazing rates (Edinger, et al., 2003), phytoplankton vertical migration, and production of phytoplankton from spores and cysts. The latter two processes are examined here.

Plankton Vertical Migration

One of the processes known to produce chlorophyll and DO maxima below the water surface is the vertical movement of phytoplankton. Cerco (2000) indicates that vertical migration of phytoplankton was one of the missing processes in Chesapeake Bay eutrophication modeling. Some phytoplankton position themselves in the water column where the penetrating light intensity is nearly equal to their optimal light saturation value. Some phytoplankton adjust their density to achieve and maintain this location. The density dependent movement of phytoplankton has not been fully formulated for application in water quality models.

Many different phytoplankton undergo diel vertical migration (DVM) in response to light. One formulation of DVM has been developed by Kamylkowski et al. (1988). This model is based on experiments of the effects of temperature and light on the dinoflagellate swimming speed. The model uses three equations that represent temperature dependence, light dependence, and sinking due to buoyancy using Stokes' law. These equations vectorally add up to the total swimming speed, which is applied in the vertical direction. The temperature and light dependent equations are given below.

$$V_T = V_{TMAX}\left(1 - \exp[-b\{T - T_L\}]\right)\left(1 - \exp[-c\{T_H - T\}]\right) - V_{off} \tag{1}$$

where:

V_T	=	temperature dependent velocity (m/s)
V_{TMAX}	=	maximum temperature dependent velocity (m/s)
b, c	=	empirical constants (b = 0.632, c = 2.0)
T	=	temperature (°C)
T_L	=	lower temperature constant (°C) (T_L = 11.68)
T_H	=	higher temperature constant (°C) (T_H = 33.0)

V_{off} = swimming speed enhancement due to light during experiments (m/s) (V_{off} = 0.35e-4)

and:

$$V_L = V_{LMAX} \tanh\left[\alpha I 4.15 / V_{LMAX} 10^6 \right] \tag{2}$$

where:

V_L = light dependent velocity (m/s)
V_{LMAX} = maximum light dependent velocity (m/s)
α = empirical constant (μm m^2/μEinst) ($\alpha = 10.0$)
I = light intensity (watts/m^2)

Equation 2 includes conversion of the ambient light intensity from μEinst m^{-2} s^{-1} to watts/m^2 which is used in many models. The empirical constants (b, c, α), temperature constants (T_H, T_L), and swimming speed enhancement value (V_{off}) given above are from Kamyłkowski and McCollum (1986) and Yamazaki and Kamylkowski (1991).

The final equation for the swimming function is:

$$V_{total} = V_L + V_T + V_{Sinking} \tag{3}$$

where the velocities are vertical vectors.

The application logic of the swimming function used to simulate the DVM in response to light is shown in Figure 1. The main logic split is between daytime and nighttime, where dinoflagellates swim upward during daytime with both the light and temperature velocities, and they swim downward at night with the temperature velocity. The sinking velocity is always acting downward. Figure 1 also shows the special case of the bottom model cell which does not allow swimming into the bottom sediments.

Laboratory and field studies have shown that dinoflagellates may form subsurface maximum concentrations where the irradiance is optimal for photosynthesis (Cullen and Horrigan, 1983; Harris et al., 1979; Eppley et al., 1968). In the case where nutrients are depleted throughout the water column, they will migrate to the surface (Cullen and Horrigan, 1983). Other possible vertical migration responses may be due to nutrient levels as well as irradiance.

Application of Vertical Migration Formulations

Chlorophyll and DO maxima below the water surface were often observed in the presence of dinoflagellates during intense water quality surveys in the Budd Inlet, an arm of Puget Sound. The hydrodynamic, transport and water quality modeling of the Budd Inlet has been described in Wu, et al. (1998). The process diagram of the WQCBM water quality model used in the Budd Inlet studies is shown in Figure 2. The oval shape at the bottom right quadrant of Figure 2 shows the use of two algal groups 1) diatoms and 2) dinoflagellates in the water quality kinetics. A detailed description of WQCBM is available in Kolluru (2001). Dinoflagellates with vertical migration were incorporated in the water quality and their significance was examined in two ways. First a comparison was made to time series of dissolved oxygen values with and without dinoflagellates incorporated in the model. Second, comparisons were made to chlorophyll vertical profile data with dinoflagellates incorporated in the model. The 3-D model grid used for water quality simulations in the Budd Inlet is shown in Figure 3. Also shown in the same figure are some field data locations where time series and vertical profile data were compared with observations. The time series of dissolved oxygen with and without dinoflagellates is given in Figure 4 along with observed data at Stations BI-6, BI-5 and BI-4. The model results with dinoflagellates reproduce the dissolved oxygen observations quite well. The latter are affected by water column depths, or tidal elevations that continually change the distance through which the dinoflagellates migrate, the depth of critical light intensity, and the temperature and salinity profiles.

Profiles of chlorophyll concentrations are given in Figure 5 and Figure 6 at Stations BC-2 and BI-5, respectively. They show that inclusion of dinoflagellate vertical migration leads to reproducing the chlorophyll maxima below the water surface. The larger maxima occur at times of higher water surface elevations, which confirms the tidal dependency of dissolved oxygen given in Figure 4. Dissolved oxygen vertical profiles are shown in Figure 7 for Stations BC-2. This figure clearly shows that the inclusion of dinoflagellate vertical migration predicts the availability of low dissolved oxygen at the bottom.

Algal sources from spores and cysts

Studies of Chesapeake Bay dinoflagellate data by Tyler and Seliger (1978) showed their seasonal distribution and blooms in the bay were related to the baroclinic bottom water up-bay transport in a vegetative form from the offshore waters and that blooms occurred in the same region of the bay year after year. These studies also showed that the off-shore "seed" forms were deposited from the outflowing surface layers in the fall and winter. Presently, quantitative eutrophication models assume that there is always a very small density of plankton present as "seed", and kinetics related to the drifting plankton such as temperature parameters and light saturation values are varied to get growth to occur at the proper times of year.

Greer and Amsler (2002) examined the generation of algae from their source as spores on aquatic vegetation, and Bravo and Ramilo (1998) examined the generation of algae from cysts in the sediment. Godhe (2002a, 2002b, 2003) is examining the dynamics between the planktonic and benthic stages of cyst forming species of dinoflagellates. These studies describe the factors triggering planktonic dinoflagellates to produce resting spores and cysts, their length of survival in the sediment, the external environmental conditions causing the cyst to germinate, and the role of benthic stages in inducing blooms of dinoflagellates. Investigation of sediment cores showed that dinoflagellate cysts can remain alive in the sediment for several decades.

If cysts of toxic species can trigger blooms of toxic species, this is of ecological importance when considering the effects of trawling and dredging, or even storm stirring of shallow sediments. Inclusion of spores and cysts as source terms along with properties for release including temperature, bottom scour, and bottom water mineral chemistry, would be a major step forward in describing the onset of algal formation, and possibly estimates of occurrence of algal blooms.

A time series of plankton densities typically available from historical data is given in Figure 8. These data result from sampling at 60- to 90-day intervals. The record shows at least three peaks in plankton densities that might be considered blooms in 7 years. Closer sampling intervals would identify more peaks as well as the approach to and decline from them. In Indian River Bay and adjoining Assawoman Bay, chlorophyll values over the survey years averaged 20 to 30 ug/l (1.4 mgC/l to 2.1 mgC/l). They reached maximum values of 300 to 1400 ug/l (21.0 to 100 mgC/l) in any one year. Thus, peak chlorophyll due to blooms was about 15 to 70 times the average values. Possible explanations for the peaks are generation from spore/cysts in the sediment, transport of algal cells and vegetative cells from another region, accelerated in-situ growth, and physical accumulation/mechanisms. Investigation of the transport and retention mechanisms require coupling the hydrodynamic and water quality model as presented in Kolluru and Fichera (2003).

The inclusion of the production of algae from spores and cysts in the eutrophication model requires a formulation of the processes that might be involved. McGillicuddy, et al. (2003) shows that coastal blooms of dinoflagellates can be related to dormant cysts located in offshore sediments. They hypothesize a model for vegetative cells that swim upward in the water column after being generated from resting cysts, a germination rate and an initiation time. McGillicuddy, et al. (2003) suggest that the cysts germinate under the right temperature and light conditions according to an internal clock, and once germinated, the cells grow in response to temperature, light, salinity, and nutrient conditions within the water. A similar model is presented here for initial investigations, but extended to include the transformation of the vegetative cells resulting from cysts/spores to algae and its effects on water quality parameters.

The proposed model includes the release of spores/cyst vegetative cells from the sediment at a designated time, their transport through the water column, and their transformation to algae. The vegetative cell transport and transformation is written as:

$$DCvc/Dt = Vvc\ \partial Cvc/\partial z - Kvc\ Cvc + Svcsed(z{=}d)/\Delta z \qquad (4)$$

where:

Cvc = Vegetative cell density, mgC/l
Vvc = Vertical vegetative cell velocity, m/s
Kvc = Rate of transfer of vegetative cells to phytoplankton, per sec.
Svcsed = rate of release of vegetative cells from sediment spore/cysts, mgC/sq.m/sec
Δz = Finite difference bottom layer thickness, m.

The algal density is increased by the vegetative cell transfer as:

$$DCphy/Dt = [Gp(N,P,I) - Dd - Dr]Cphy - Grz(t) + Kvc\ Cvc \quad (5)$$

where:

Cphy = algal density, mgC/l
Gp(N,P,I) = Nitrogen, phosphorous and light limited algal growth rate, per sec
Dd = algal death rate
Dr = algal respiration rate
Grz(t) = zooplankton grazing rate

The water quality nutrient cycling model presented in Edinger, et al. (2003) was extended to include Equations 4 and 5. A numerical demonstration of the algal densities produced by the spore release formulation was set up for a tank simulation initialized with dissolved nutrients and in which an instantaneous bottom release of spore is made during the simulation. The tank is set up with bottom SOD and releases of NH_3 and PO_4. Incident light is specified at the surface and surface oxygen exchange is included. The use of tank simulations to study the properties of water quality models is presented in (Edinger, 2002).

The algal and water quality data examined by Tyler and Seliger (1978) showed that there can be a sharp increase in dissolved oxygen and a decrease in total ammonia during a bloom followed by their returning to levels found before the bloom. The change in dissolved oxygen would be due to photosynthesis, and the change in ammonia would be due to its uptake by phytoplankton. The distribution of algal and water quality data examined by Tyler and Seliger (1978) extended over many weeks. The tank test simulation is designed to take place over a one-month period.

The time series of phytoplankton density with a spore release at 250 hours is given in Figure 9 and its variation with depth before and at bloom peak is shown in Figure 10. It shows that previous to the spore release, the phytoplankton density is a maximum near the depth where the incident light equals the phytoplankton light saturation value. The spore release from the tank bottom results in a high surface density shortly after release that disappears over time similar to the manner in which an algal bloom would appear and disappear. The equilibrium phytoplankton density returns to its starting levels after the spore release.

Figure 11 shows the change in dissolved oxygen through the bloom and its variation with depth before and at bloom peak is shown in Figure 12. At the surface the dissolved oxygen decreases following the bloom then increases again. At greater depths it increases from its initial equilibrium value to a peak following the bloom, and then decreases afterward.

Figure 13 shows the change in total ammonia through the bloom, and its variation with depth before and at bloom peak is shown in Figure 14. At the surface there is a very small decrease in total ammonia following the peak algal densities, however, the decrease is more pronounced at greater depths. The total ammonia concentration recovers to its initial equilibrium values following the bloom.

Conclusions

Water quality model calibration involves more than adjusting parameters in existing models. It requires a close examination of field data and an investigation of processes that are not included in the models. The most complex part of existing eutrophication models is the plankton component and the various processes required to describe their mobility and sources.

Two plankton processes examined here for inclusion in eutrophication models are their diel vertical migration and their production from spore/cysts. The formulation and significance of dinoflagellete diel vertical migration was demonstrated by application to the Budd Inlet. Also, a formulation for inclusion of spore/cysts as a source of phytoplankton was investigated in a tank model. Both have many potential uses for improving eutrophication models.

Acknowledgements

Chuck Boatman passed away shortly after developing the formulations of dinoflagellate vertical migration processes for inclusion in the Budd Inlet studies. He left us with the important modeling principle of searching for additional processes when the eutrophication modeling does not properly describe field data. This led us to investigate further the actual sources and "seeds" of plankton in a water body, and methods for including them in the modeling. The authors thank Charles Stock of the MIT/Woods Hole Oceanographic Joint Program in Oceanography/Applied Ocean Sciences and Engineering for his comments on this paper. The Budd Inlet project was managed by Christopher Cleveland of Brown and Caldwell in Olympia, Washington.

References

Bravo, I. and I. Ramilo, 1998. Distribution of microreticulate dinoflagellate cysts from the Galician and Portuguese coast, Scientia Marina, 63(1): 45-50.

Cerco, C.F. 2000. Phytoplankton Kinetics in the Chesapeake Bay Eutrophication Model. Water Quality and Ecosystem Modeling. 1, 5-49. 2000. Kluwer Academic Publishers.

Cullen, J.J. and S.G. Horrigan. 1983. Effects of Nitrate On The Diurnal Vertical Migration, Carbon to Nitrogen Ratio, and the Photosynthetic Capacity of The Dinoflagellate Gymnodinium Splendens. Marine Biology. 62:81-89.

Edinger, J.E. 2002. Waterbody Hydrodynamic and Water Quality Modeling: An Introductory Workbook to Numerical Three Dimensional Modeling. ASCE Press, Reston, VA. ISBN 0-7844-0550-6.

Edinger, J.E., S.Dierks and V.S.Kolluru. 2003. Density Dependent Grazing In Estuarine Water Quality Models. Water, Air and Soil Pollution 147: 163-182, 2003. Kluwer Academic Publishers, the Netherlands.

Eppley, R.W. O. Holm-Hansen and J.D.H. Strickland. 1968. Some Observations on the Vertical Migration of Dinoflagellates. J. Phycol. 4:333-340.

Godhe A., A-S Rehnstam-Holm, I. Karunasagar & I. Karunasagar (2002a). PCR Detection of Dinoflagellate Cysts in Field Sediment Samples From Tropic and Temperate Environments. Harmful Algae, Vol 1(4), 361-373.

Godhe A, S. Svensson & A-S Rehnstam-Holm (2002b) Oceanographic Settings Explain Fluctuations in Dinophysis Spp. and Concentrations of Diarrhetic Shellfish Toxin in the Plankton Community Within a Mussel Farm Area on the Swedish West Coast. Marine Ecology Progress Series, Vol. 240, 71-83.

Godhe A. & M.R. McQuoid (2003). Distribution of Dinoflagellate Cysts in Recent Marine Sediments in Relation to Benthic and Pelagic Environmental Factors Along the Swedish West Coast. Accepted for publication Aquatic Microbial Ecology.

Greer, S.P. and C.D. Amsler. 2002. Light boundaries and the Coupled Effects of Surface Hydrophobicity and Light on Spore Settlement in the Brown Alga *Hincksia irregularis* (Phaeophyceae). Journal of Phycology 38: 116-124.

Harris, G.P. S.I. Heaney and J.F. Talling. 1979. Physiological and Environmental Constraints in the Ecology of the Phytoplankton Dinoflagellate *Ciratium hirundinella*. Freshwater Biol. 9:413-428

Kamylkowski, D. and S.A. McCollum. 1986. The Temperature Acclimatized Swimming Speed of Selected Marine Dinoflagellates. J. Plankton Res. 8:275-287.

Kamylkowski, D., S.A. McCollum and G.J. Kirkpatrick. 1988. Observations and a Model Concerning the Translational Velocity of a Photosynthetic Marine Dinoflagellatesflagellate Under Variable Light Conditions. Limnol. Oceangr. 33:55-78.

Krallis, G.A. and E.M.Buchak. 2002. Systematic Calibration of a Water Quality Model. ASCE Engineering Mechanics Division Conference on Environmental Fluid Mechanics, Columbia University. Conference Proceedings, ASCE Publications, Reston, VA. June 2002.

Kolluru, V.S. and Fichera 2003. Development and Application of 1D and 3D Modeling System for TMDL Studies. ASCE Estuarine and Coastal Modeling Conference, ECM8, Monterey, CA November 2003.

Kolluru, V.S. 2001. Technical Documentation on GEMSS Software and available models. http://www.jeeai.com/Software-GEMSS.htm

McGillicuddy, D.J., Signell, R.P., Stock, C.A., Keafer, B.A., Keller, M.D., Hetland, R.D., and Anderson, D.M. 2003. A Mechanism for Offshore Initiation of Harmful Algal Blooms in the Coastal Gulf of Maine. Journal of Plankton Research. Vol. 95. Number 9. Pgs. 1131-1138. 2003.

Tyler, M.A. and H.H. Seliger. 1978. Annual Subsurface Transport of a Red Tide Dinoflagellate to its Bloom Area: Water Circulation Patterns and Organism Distributions in the Chesapeake Bay. Limnology and Oceanography. March 1978, Vol 23(2). Pp. 227-246.

Wu, J., V.S. Kolluru and J.E. Edinger. 1998. Combined Hydrodynamic and Water Quality Modeling for Waste Water Impact Studies. Proceedings of the Mid-Atlantic Industrial Wastes Conference. Lee Christensen, Editor. Villanova University, Villanova, PA. July 13, 1998.

Yamazaki, H. and D. Kamykowski. 1991. The Vertical Trajectories of Motile Phytoplankton in a Wind-Mixed Water Column. Deep-Sea Res. 38:219-241.

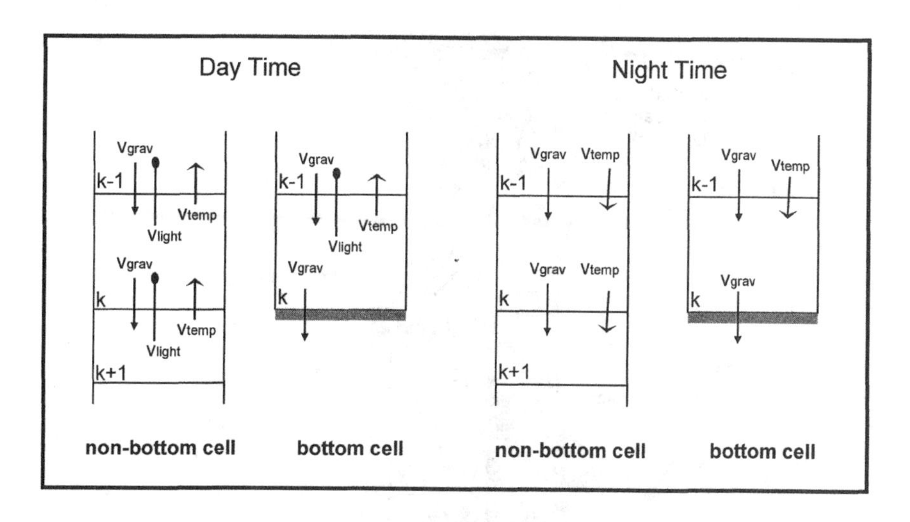

Figure 1 Dinoflagellates moving mechanism during day and night times

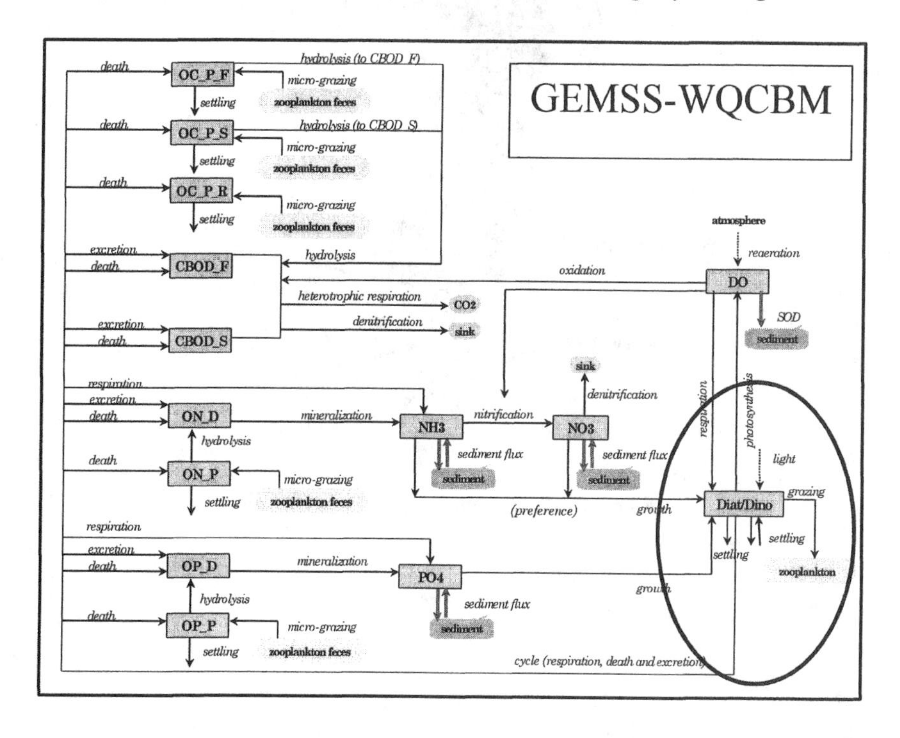

Figure 2 Water quality processes diagram for WQCBM

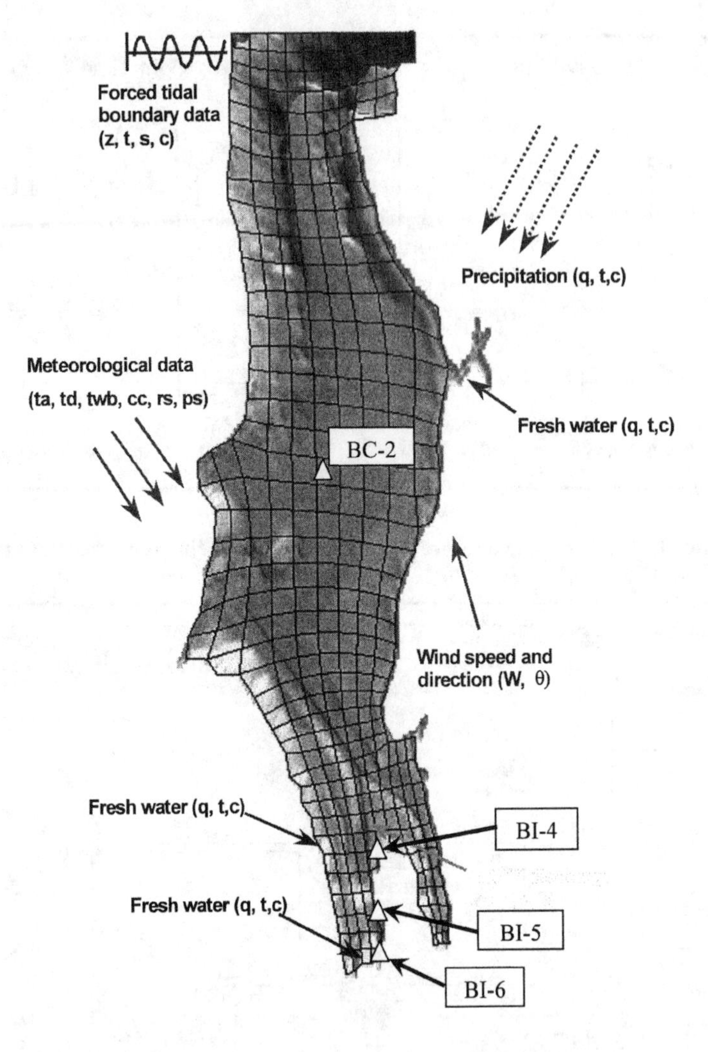

Figure 3 3-D model grid for Budd Inlet

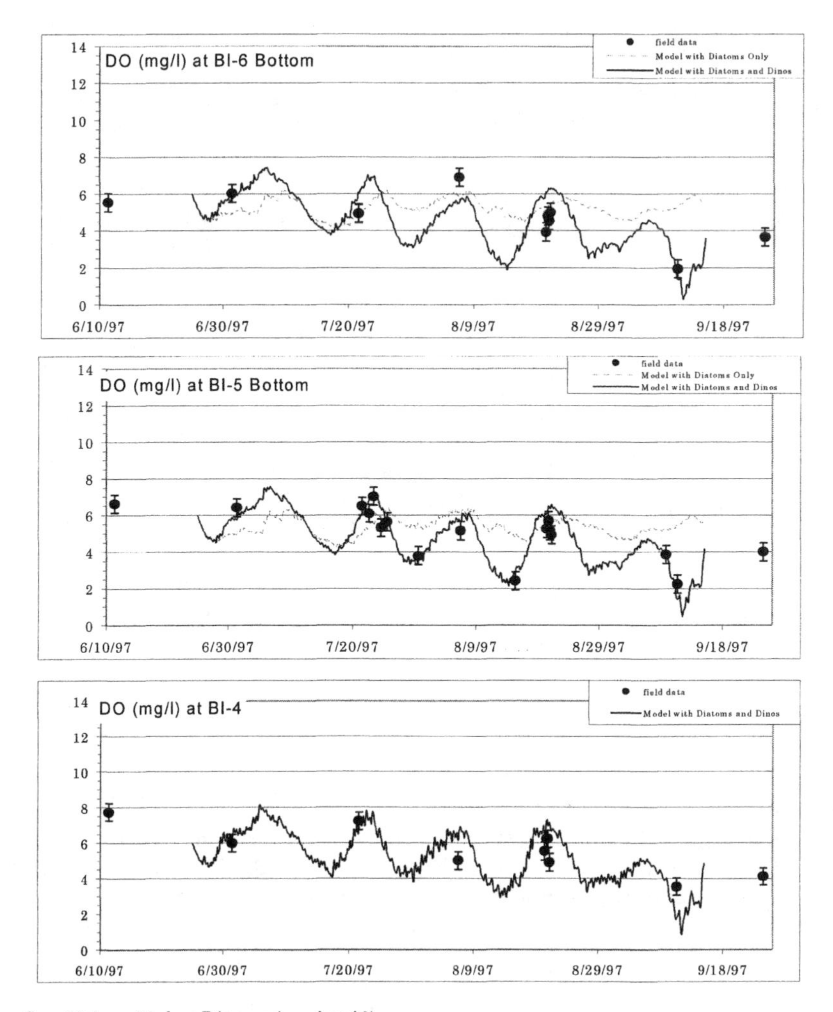

Figure 4 Time series of dissolved oxygen computed without and with dinoflagellate vertical migration along with observed values.

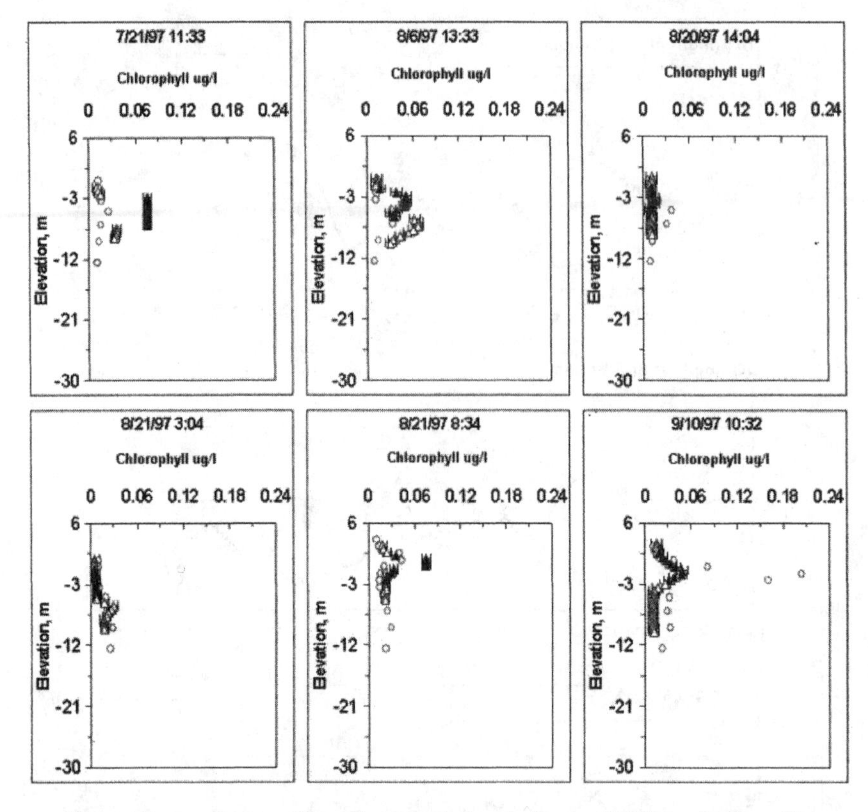

Figure 5 Comparison of computed and observed chlorophyll profiles through different daily cycles at Station BC-2 for case 59. The computed values are indicated by circles, and the observed values by triangles

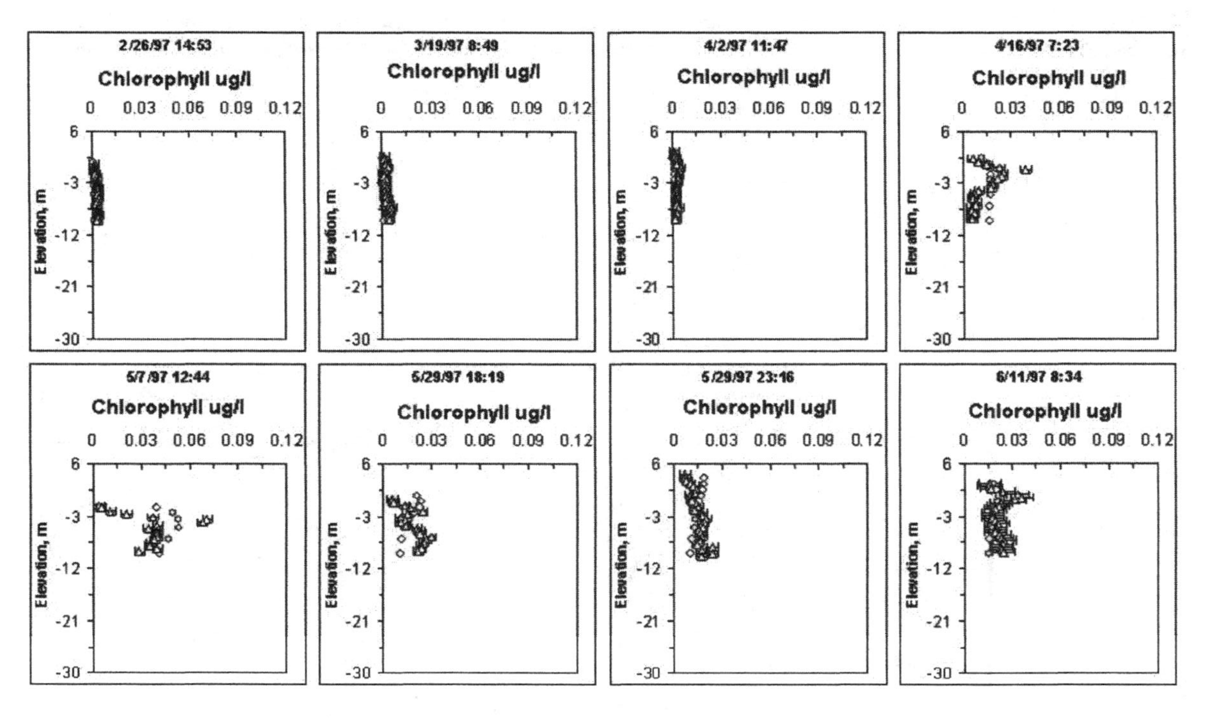

Figure 6 Comparison of computed and observed chlorophyll profiles through different daily cycles at Station BI-5. The computed values are indicated by circles, and the observed values by triangles

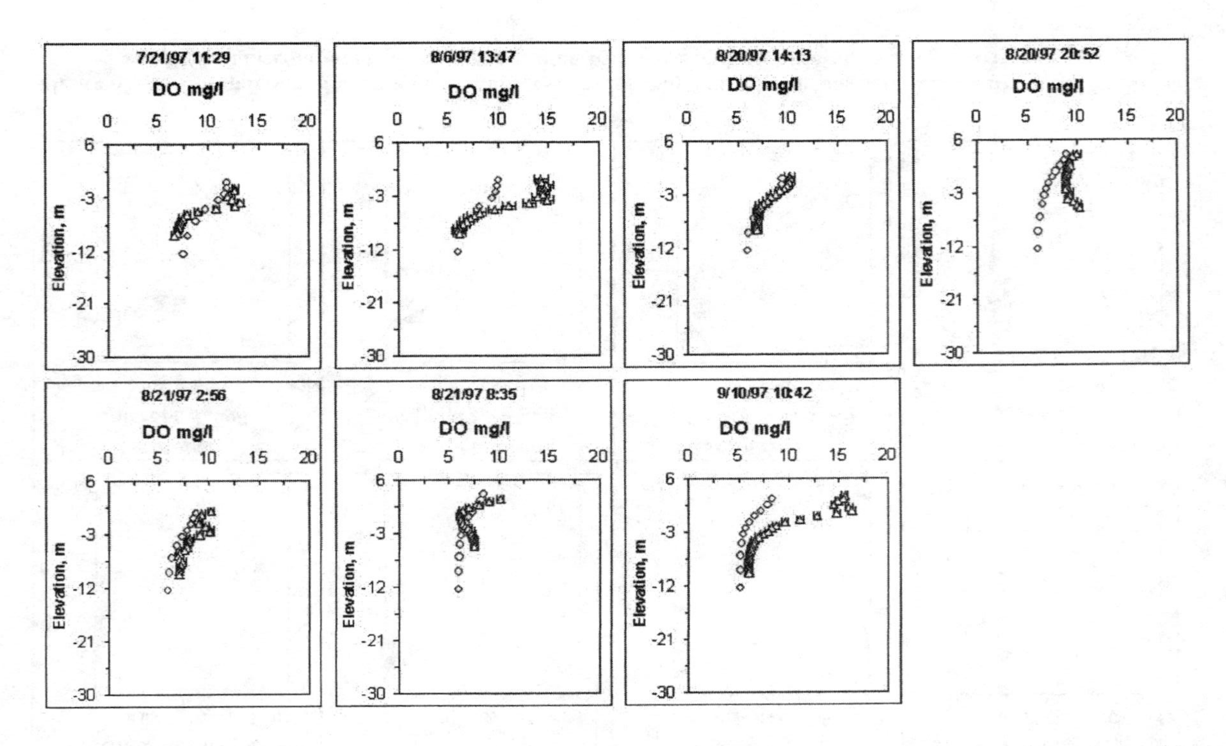

Figure 7 Comparison of computed and observed dissolved oxygen profiles through different daily cycles at Station BC-2. The computed values are indicated by circles, and the observed values by triangles

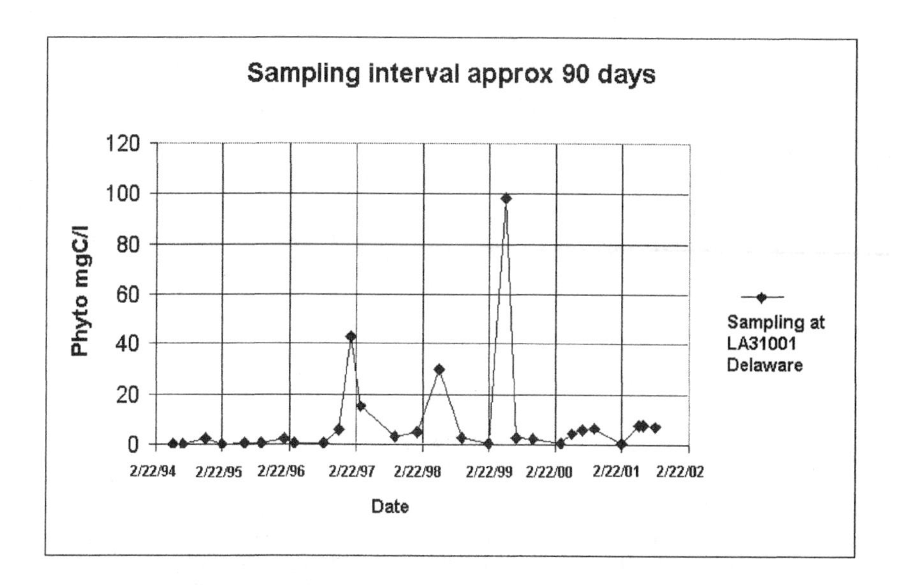

Figure 8 Time series of observed phytoplankton densities estimated from chlorophyll densities. Observations from EPA Storet data at Indian River Embayment Station LA31001, Delaware.

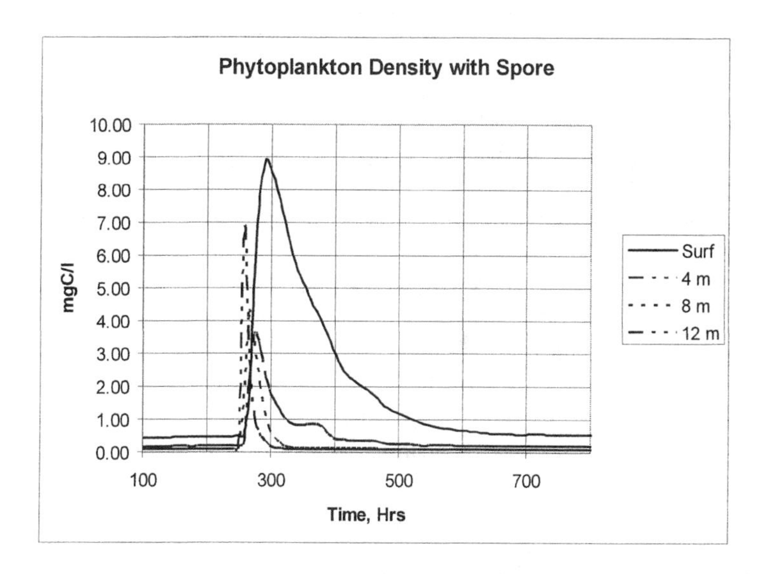

Figure 9 Tank simulation of phytoplankton densities at different depths with an instantaneous spore release.

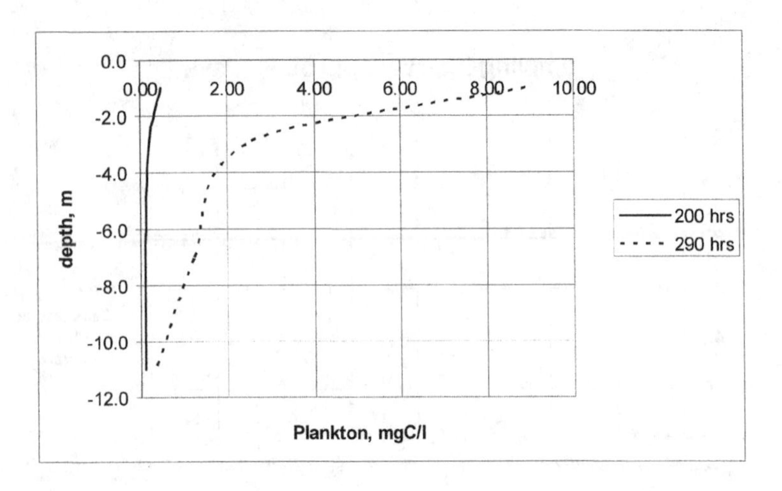

Figure 10 Tank simulation of phytoplankton density profiles before and at bloom peak

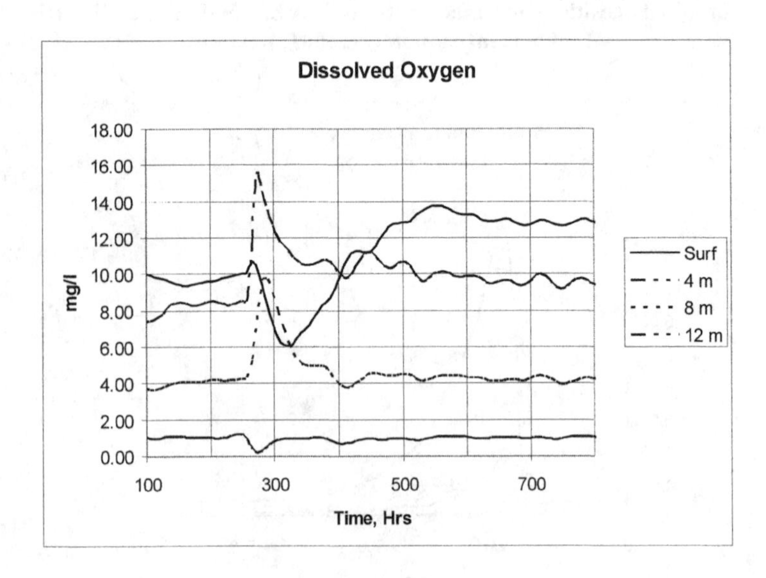

Figure 11 Tank simulation of dissolved oxygen at different depths with an instantaneous spore release

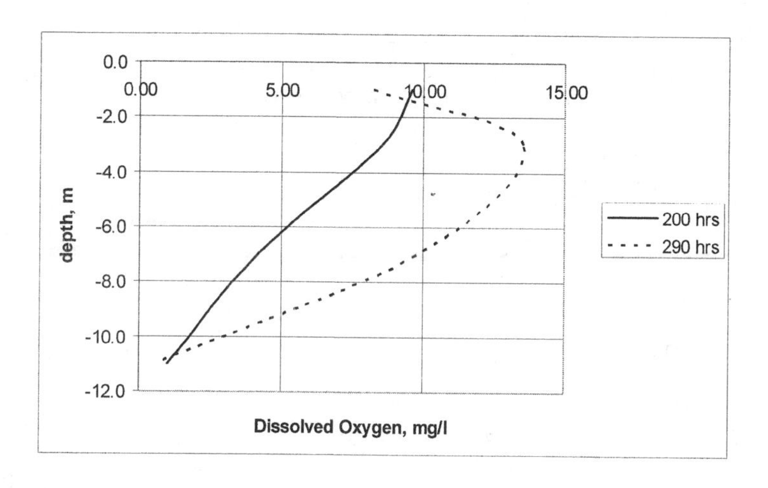

Figure 12 Tank simulation of dissolved oxygen profiles before and at bloom peak

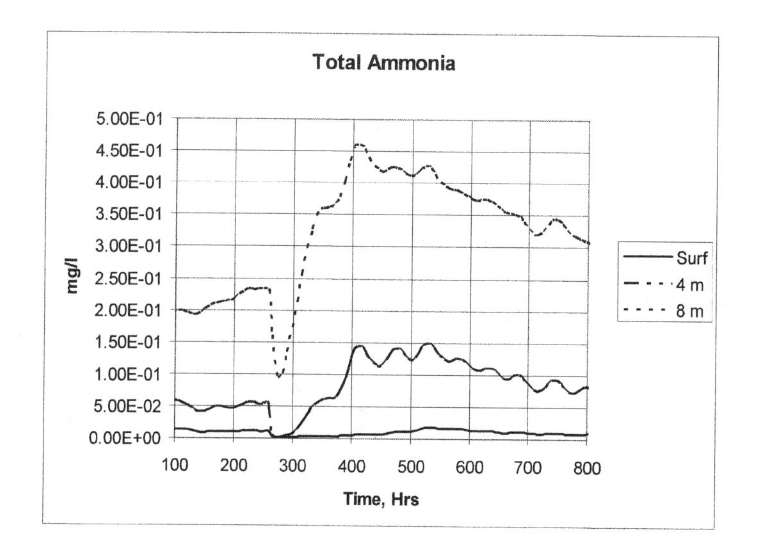

Figure 13 Tank simulation of total ammonia at different depths through a spore release

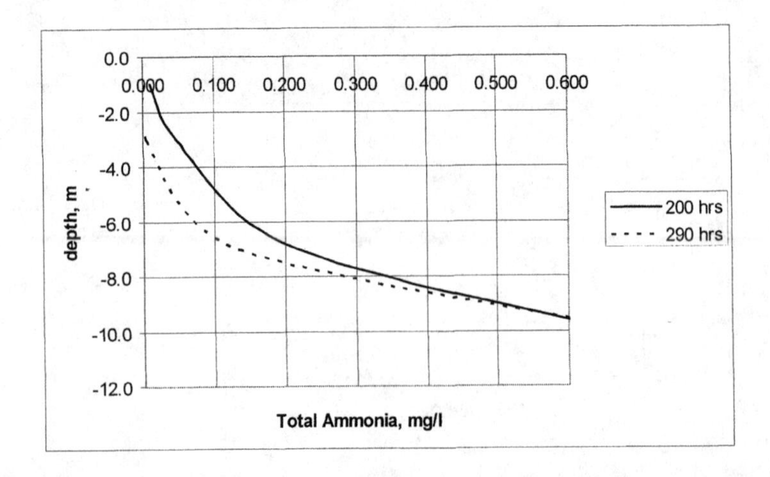

Figure 14 Tank simulation of total ammonia profiles before and at bloom peak

Integrated-Process and Integrated-Scale Modeling of Large Coastal and Estuarine Areas

Y. Peter Sheng[1], Justin R. Davis, Vladimir Paramygin, Kijin Park, Taeyun Kim, Vadim Alymov
Civil & Coastal Engineering Dept., University of Florida, Gainesville, FL 32611-6580

ABSTRACT

To predict the response of coastal and estuarine ecosystems to anthropogenic and natural changes, it is necessary to conduct integrated-process and integrated-scale modeling of large coastal and estuarine areas. This paper presents an integrated modeling system, CH3D-IMS (http://ch3d.coastal.ufl.edu/), which includes models of circulation, wave, particle trajectory, sediment transport, water quality dynamics, light attenuation, and seagrass dynamics. The CH3D-IMS has been and continues to be validated with data from various estuaries in Florida. A 3-D variable-density groundwater flow model and a fishery model are being coupled to the CH3D-IMS. We present example applications of the CH3D-IMS including: (a) simulation of the Indian River Lagoon and trajectory of Shuttle *Columbia* debris in North and Central Florida Atlantic Coastal water, (b) simulation of storm surge in Tampa Bay, Sarasota Bay and adjacent Gulf of Mexico, and (c) simulation of circulation in Charlotte Harbor and adjacent Gulf of Mexico water. As the integrated modeling system continues to be applied to ever more complex problems over increasingly larger coastal areas, it requires more computational resources and disciplinary expertises which are often unavailable in any single institution. To facilitate integrated-process and integrated-scale modeling by multiple institutions, the development of an infrastructure - a regional modeling "grid"- is proposed.

INTRODUCTION

More than 50% of the U.S. and world population live within 100 miles from the coastline. By 2025, it is expected that more than 75% of the population will live in the coastal zone. Changes in estuarine and coastal ecosystems have occurred due to anthropogenic and natural causes. On the one hand, increased pollutant loadings have led to deteriorated water quality, increased incidences of harmful algal bloom, and loss of fishery habitat. On the other hand, climatic change and episodic storms have affected ecosystems. To protect and restore ecosystems and to mitigate storm-induced damages, it is essential to be able to predict the response of coastal ecosystems to anthropogenic and natural climatic changes. To this end, integrated modeling systems that include a variety of physical and biogeochemical processes and cover a wide range of spatial (from boundary layers to basin/global scale) and temporal (from turbulence to decadal) scales are needed. A calibrated and validated integrated modeling system, e.g., the CH3D-IMS (Sheng et al., 2002) can be used by resource management agencies to set science-based loading limits, e.g., pollutant load reduction goal (PLRG), total maximum daily load (TMDL), and minimum flow and level (MFL) criteria, to a water body. Similarly, a validated integrated storm surge modeling system can be used to quantitatively delineate flood zones and guide hurricane evacuation.

---

[1]Member, ASCE. Email: pete@coastal.ufl.edu

Sheng (2000) presented the framework of an integrated modeling system, CH3D-IMS (Figure 1). Earlier versions of the CH3D-IMS were developed to study the effect of reduced nitrogen loading to Roberts Bay and upper Little Sarasota Bay in Florida (Sheng et al., 1996a), the effect of nitrogen load reduction on hypoxia and seagrass in Tampa Bay (Sheng and Yassuda, 1996b; Yassuda and Sheng, 1998), and the preliminary effect of nutrient load reduction on Indian River Lagoon (Sheng, 1997). Sheng et al. (2002) presented the validation CH3D-IMS using extensive data from the Indian River Lagoon collected during 1998. In this paper, we first present the continued development of CH3D-IMS, followed by some recent applications of CH3D-IMS to several Florida coastal and estuarine ecosystems of Florida (Figure 2). We then discuss two difficult issues related to the computational and interdisciplinary aspects of integrated modeling systems, followed by a proposal for the development of a regional "grid" (Foster et al., 2001) to facilitate integrated-process and integrated-scale modeling by multiple institutions.

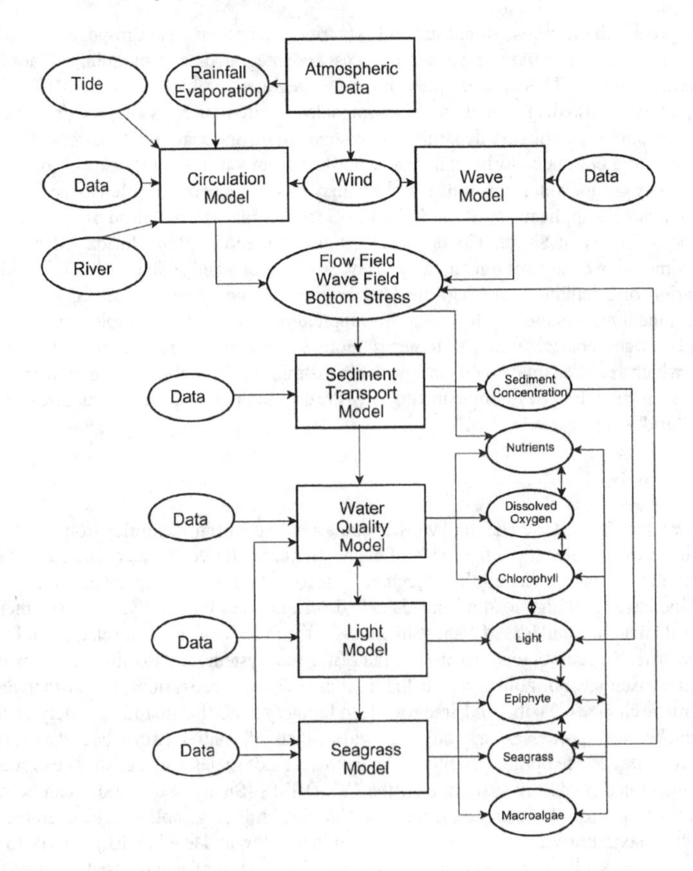

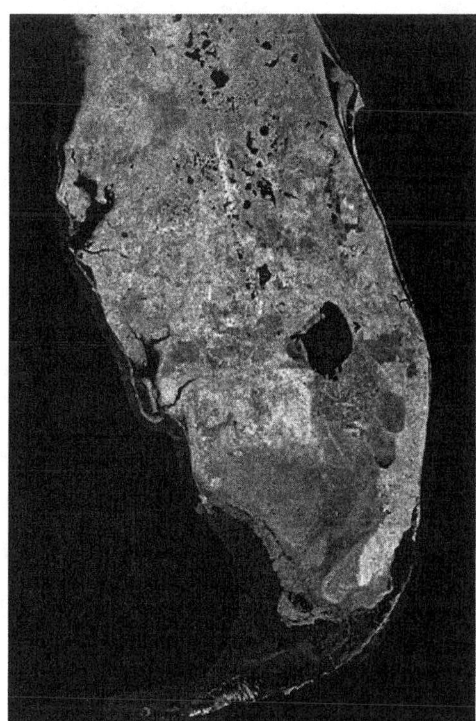

Figure 2. Central and South Florida with several estuarine and coastal ecosystems: Tampa Bay, Sarasota Bay, and Charlotte Harbor on the West coast, Indian River Lagoon on the East coast, and Florida Bay and Biscayne Bay on the South coast.

AN INTEGRATED MODELING SYSTEM: CH3D-IMS

The cornerstone of the CH3D-IMS is the 3-D Curvilinear-grid Hydrodynamic (CH3D) circulation model originally developed by Sheng (1987; 1990) and significantly enhanced during the past 17 years at the University of Florida. The CH3D model at the University of Florida now includes such features as wetting-and-drying, non-hydrostatic pressure distribution, current-wave interaction, vegetation canopies, and a GIS-based graphic user interface. In addition to CH3D (Sheng and Davis, 2002), the CH3D-IMS now includes a wave model, a 3-D sediment transport model CH3D-SED3D (Sheng et al., 2002a), a 3-D water quality model CH3D-WQ3D (Sheng et al., 2002b), a light attenuation model CH3D-LA (Sheng et al., 2002c, Christian and Sheng, 2003), and a seagrass model CH3D-SAV (Fong et al., 1997; Sheng et al., 2003).

The integrated modeling system is comprehensive and uses the same curvilinear grid and time step for all component models. The boundary-fitted grid does not have to be orthogonal, and hence can fit the complex coastal and estuarine boundaries more accurately than orthogonal grid models. The circulation model CH3D simulates the circulation due to wind, tide, wave, and density gradients. The sediment transport model simulates the transport, mixing, deposition, and resuspension of both fine and coarse grained sediments. The model includes resuspension of sediments by both slowly varying currents and wave-induced oscillatory currents. The water quality model includes a nitrogen model, a phosphorus model, a dissolved oxygen (DO) model, a phytoplankton model, and a zooplankton model. A carbon model, a silica model, and a 3-species phytoplankton model have recently been added to the water quality model. The water quality model includes a fully 3-D model for the water column, plus a vertical two-layer model for the sediment column: an aerobic layer on top of an anaerobic layer. The light model includes the absorption of light by water, chlorophyll, non-algal particulate matter (tripton), and dissolved organic matter, as well as scattering by tripton. The seagrass model includes the influence of light, nutrients, temperature, and salinity on the biomass of several relevant species in Florida (*Thallasia, Halodule*, and *Serongodium*), plus epiphyte and macroalgae.

The component models of CH3D-IMS have been completely integrated for operation on one computer system. Both one-way and two-way interactions exist among the various component models. For example, the sediment transport model includes computation of bottom shear stress generated by currents (computed by CH3D) and wave-orbital currents (computed by the wave model). Suspended sediments near the sediment-water interface may cause large density gradients which can significantly affect the bottom shear stresses. Sediment resuspension and deposition fluxes at the sediment-water interface, computed by the sediment transport model, are fed into the water quality model to allow calculation of the nutrient fluxes at the sediment-water interface. The light attenuation model calculates the light attenuation due to water, chlorophyll (calculated by the water quality model), non-algal particulate matter (calculated by the sediment transport model and water quality model), and dissolved organic matter (based on data). The light attenuation in turn affects the algal growth rate in the water quality model. The seagrass biomass is influenced by light (calculated by the light model), nutrients (calculated by the water quality model), temperature and salinity (calculated by the circulation model). The integrated modeling system uses the same curvilinear grid/spacing and time step for all component models, thus eliminating the need for any spatial and temporal averaging of fine-scale component model results to be used as input for other coarse component model(s), which tend to result in violation of mass conservation and loss of information from the fine scale component model.

To facilitate model improvement and model comparison, the CH3D-IMS has been made modular such that every module (i.e., component model) can be readily replaced by a newly improved version of the same module or a model developed by another research group. To improve the computational efficiency, the CH3D-IMS has been made flexible by using numerous *macros* and *cpp's* (C-preprocessors) to allow flexible execution of one or more component models that a user chooses (Davis and Sheng, 2000; 2002). To allow efficient simulations, the CH3D-IMS has been parallelized by using *OpenMP* for shared memory computers (Davis and Sheng, 2002) and *MPI* (Davis and Sheng, 2002; 200) for Beowulf clusters.

Sheng et al. (2002) presented the validation of the CH3D-IMS using 1998 data from the Indian River Lagoon. In the following, we present some further validations and applications of the CH3D-IMS to Indian River Lagoon (IRL), Tampa Bay (TB), and Charlotte Harbor (CH), Florida, and adjoining coastal ocean waters.

Indian River Lagoon and Atlantic Ocean

Validation of Hydrodynamic, Sediment, and Water Quality Models

Sheng et al. (2002) used the horizontal numerical grid with a total of 20988 (477x44) cells (Figure 3) and 4 to 8 vertical layers for the entire IRL. Validation of the circulation model, sediment transport model, and water quality model were presented. Basically, circulation in the IRL is driven by tides from Ponce, Sebastian, Ft. Pierce, and St. Lucie inlets, as well as wind and evaporation/precipitation. In the northern IRL and Banana River, where tides are negligible, circulation is driven by wind and evaporation/precipitation. Based on comparison with long-term data at 10 stations inside the IRL, simulated water levels are within 5% of the measured data, and simulated salinity values are within 10-25% of measured data. Based on measured currents during storms and within inlets, simulated currents are within 10-20% of observed values. During storm events and spring/fall/winter, sediments are readily resuspended by the combined action of slowly varying currents and high frequency wave-induced orbital currents. These resuspension events are often followed by release of nutrients from particulate phase into the water column. The sediment transport model simulates the sediment concentration during storm events and one year cycle with 30-50% error, due to lack of complete understanding and uncertainty in some model parameters and boundary/initial conditions. The water quality model simulates the annual variability of dissolved oxygen and dissolved nutrients with 10-35% error and particulate nutrients with 30-70% error, due to uncertainties associated with the sediment and nutrient processes.

Validation of Light and SAV Models and Ecological Forecasting

Since 2002, we have further validated the light attenuation model (Sheng et al., 2002c; Christian and Sheng, 2003) and the SAV model (Sheng et al., 2003), using the same numerical grid shown in Figure 3. Figure 4 shows the simulated light (Photosynthetically Available Radiance) at the bottom of water column, epiphyte biomass, and *Halodule* biomass in the IRL on June 30, 1998. The simulated Halodule biomass compares qualitatively well with the seagrass map produced from aerial photographs (SJRWMD, 2002). Using the validated CH3D-IMS for the IRL, we conducted ecological forecasting, i.e., predicting the response of the IRL ecosystem to various changes in pollutant (nutrients and total suspended solids, i.e., TSS) loading and hydrological structures (e.g., removal of causeways). For example, Figure 5 shows the *Halodule* distribution in the IRL on June 30, 1998 with and without loadings of nutrients and TSS. While the impact of load reduction on seagrass distribution appears to be rather insignificant for the most part of IRL, there is some appreciable changes in the vicinity of major tributaries. The impact of pollutant load reduction is expected to be more significant over periods longer than six months. Simulations of 5-20 years are being planned.

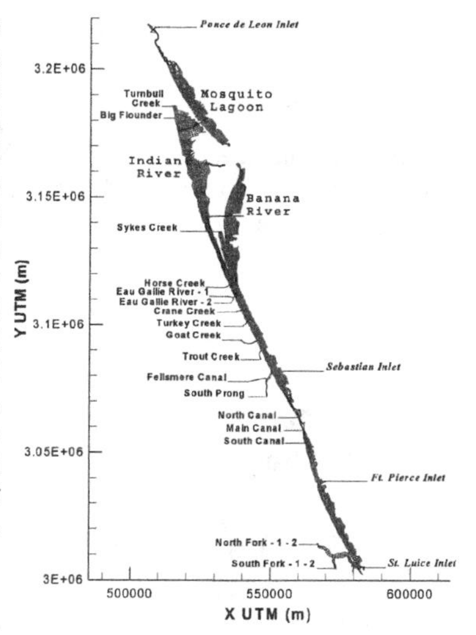

Figure 3. A numerical grid for IRL.

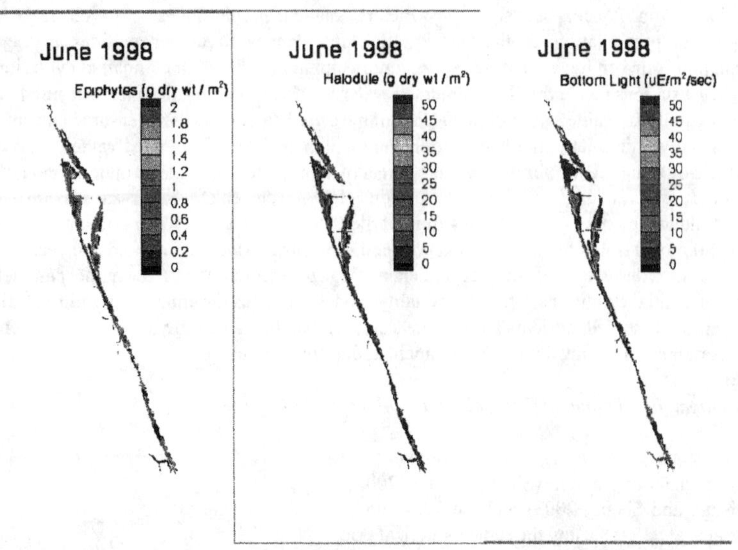

Figure 4. Simulated bottom PAR (photosynthetically available radiance) (right panel), epiphyte biomass (left panel), and Halodule boimass (middle panel) in the Indian River Lagoon on June 30, 1998.

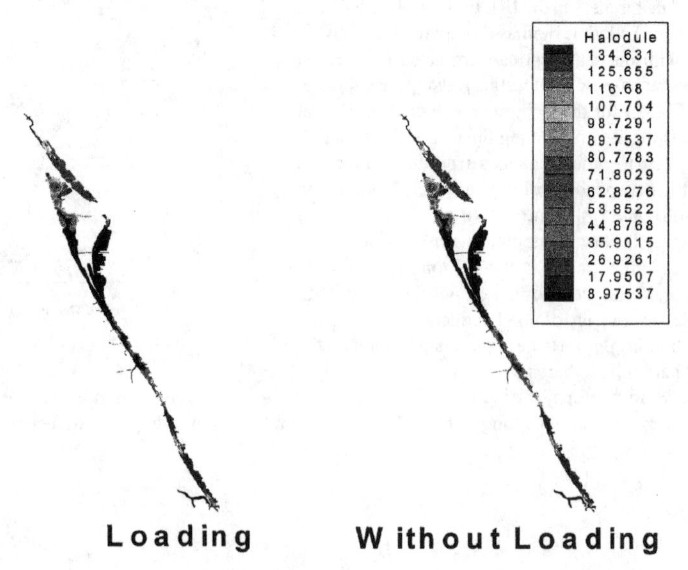

Figure 5. Simulated Halodule biomass with (left panel) and without loading (right panel) of nutrients and TSS in the Indian River Lagoon on December 31, 1998.

Simulation of Trajectory of Columbia Shuttle Debris in Atlantic Ocean

The particle trajectory model of the CH3D-IMS was recently used to simulate the trajectory of debris from space shuttle *Columbia* during its launch in January 2003. The flow field during the model simulation period is based on the output of the NOAA NCEP's Regional Ocean Forecast System (ROFS), which uses the horizontal numerical grid shown in Figure 6. Near the coast, the grid size is about 8 km. Shortly after the launch of shuttle *Columbia*, debris fell off the shuttle and impacted the Atlantic Ocean. Since the precise time of impact is unknown to us, the particle trajectory model of the CH3D-IMS was run for five particles which impacted the ocean at various time instants, ranging from 49 to 300 sec. after the launch. The results show the trajectories of the five particles for one week after their entries into water. The locations of the five particles are significantly different after one week. The results, however, could not be verified due to the classified nature of data.

Extension of the IRL Grid to Include St. Johns River and Continental Shelf

The IRL grid shown in Figure 3 only includes the estuary but not the coastal water. To enable investigation of estuary-shelf exchanges, a coastal grid must be added to the IRL grid. The coastal grid used by the debris trajectory calculation, however, is relatively coarse - 8 km. To allow investigation of estuarine and shelf processes with adequate resolution, we created a new high resolution estuary-shelf grid shown in Figure 8. The new grid has a total of 50,961 (866 x 138) horizontal grid points, vs. the 4,921 (477 x 43) grid points in the IRL grid shown in Figure 3. The new grid takes 0.92-2.2 sec per time step (60 sec), while the IRL grid takes 0.1-0.3 sec per time step, using 1-4 CPU's on our SGI Origin-300 system. This grid is sufficiently fine to resolve estuary-shelf exchange processes, by further coupling it to the ROFS grid, shown in Figure 6.

To demonstrate the feasibility of the extended grid, we simulated the tidal circulation in the entire domain during 1998. While there were water level data at the major tidal inlets and at 8-10km offshore of the inlets, there were no direct water level data at the open boundaries of the model grid shown in Figure 8. We conducted harmonic analysis of the water level data at the tidal inlets and the offshore stations, and extrapolated the water level data to the open boundaries by adjusting the amplitudes and phases of major tidal constituents such that the simulated water level at the tidal inlets agree with the measured data. Figure 9 shows the comparison between the simulated and measured water level at the Ponce, Sebastian, and Ft. Pierce inlets during Julian days 180-215.

Figure 6. Numerical grid used by the NOAA Regional Ocean Forecast System (ROFS).

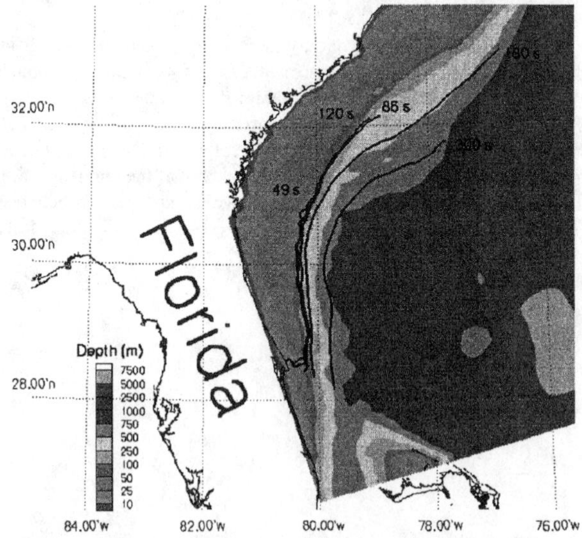

Figure 7. One-week trajectory of five hypothetical particles impacting the Atlantic ocean at various times (between 49 and 300 seconds after the shuttle launch).

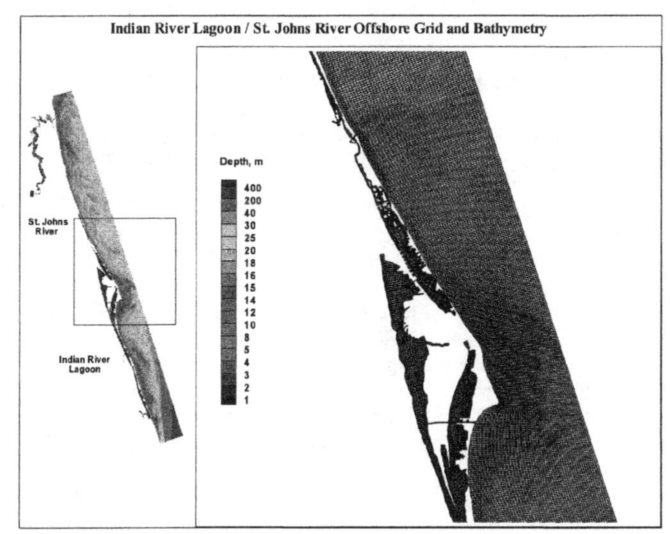

Figure 8. A horizontal numerical grid for Indian River Lagoon, St. Johns River, and a large coastal area.

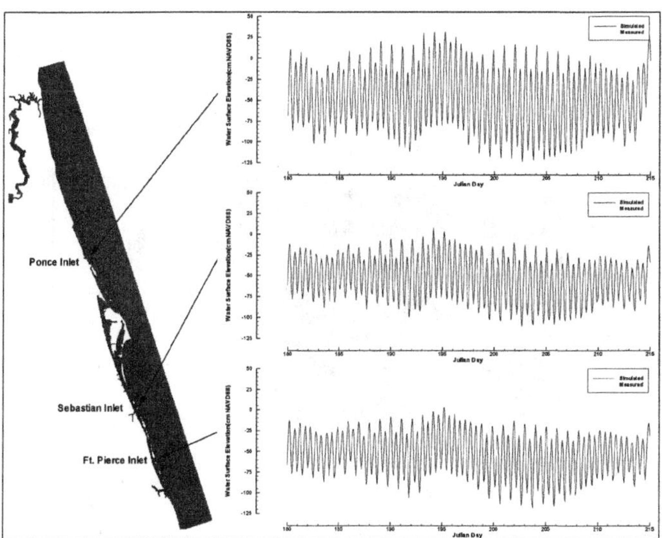

Figure 9. Simulated and measured water level at three inlets in the IRL+St.Johns+Offshore grid.

Tampa Bay,

Sarasota Bay, and Gulf of Mexico

Tampa Bay is a large estuary (second largest in Florida) situated along the Southwest coast of Florida, and connected to a coastal lagoon - Sarasota Bay. Pollutant loading has resulted in hypoxia and loss of seagrass in the past 50 years. Since the mid 1990's, however, nutrient load reduction has led to improved water quality and seagrass recovery (Tomasko, 2002). The area is subjected to many storms and hurricanes.

CH3D-IMS has been applied to simulate the circulation, water quality, light, and seagrass biomass in the Tampa Bay estuarine system (Sheng et al., 1996). The model successfully simulated the hypoxia event in Hillsborough Bay during 1991 (Yassuda and Sheng, 1998) and was applied to simulate the effect of nutrient load reduction on the Tampa Bay estuarine system (Sheng et al., 2001). In addition, the 2-D and 3-D versions of CH3D were coupled to the SWAN (Ris et al., 1996) and REF/DIF to simulate the storm surge and coastal flooding in the Pinellas County (Sheng et al., 2002). In the following, the results of storm surge simulation in the area, using a wetting-and-drying version of CH3D and a total of 54,476 horizontal grid cells and four vertical layers, are shown. Using one cpu, it takes 1.06 or 1.28 sec per time step (60 sec) for the model run. The results (Figure 11) show extensive flooding.

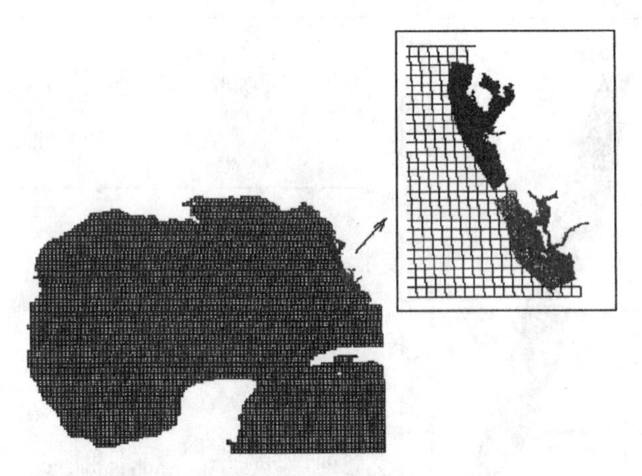

Figure 10. Numerical grids of Tampa Bay and Charlotte Harbor, which are coupled to a Gulf of Mexico grid.

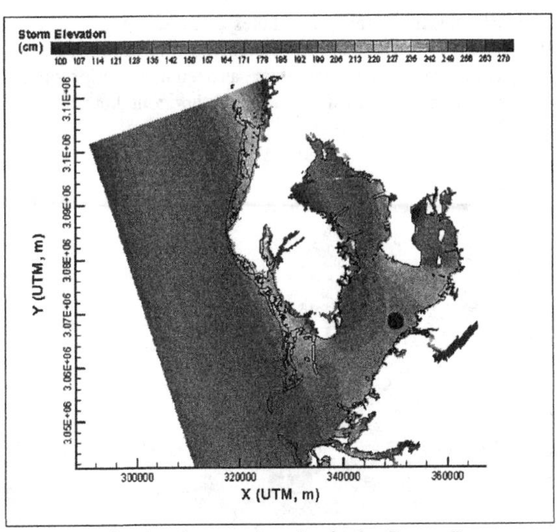

Figure 11. Simulated storm elevation in the Tampa Bay area during a hypothetical storm with uniform wind of 75 mph. The red dot indicates the center of the storm.

Charlotte Harbor and Gulf of Mexico

Charlotte Harbor is a large estuarine system along the Southwest Florida coast. One of the major issues concerning Charlotte Harbor is the determination of Minimum Flow and Level (MFL) criteria for the three major rivers (Peace, Myakka, and Caloosahatchee) due to concern of the impact of increasing demand for and withdrawal of freshwater from upstream of the estuary. The CH3D-IMS is being applied to simulate the flow and salinity distribution inside the entire estuarine system. The CH3D-IMS has been applied to simulate the circulation measured by USGS in 1986 and the water quality data measured by the State of Florida in 1996 and 2000 (Sheng and Park, 2002; 2003). Figure 12 shows the horizontal numerical grid (with 92 x 129 grid cells) used for the model simulation. Using eight vertical layers, there are a total of 94,944 grid points. The grid includes a large coastal region to allow simulation of dynamic exchanges between the estuary and the Gulf of Mexico. Using 2 CPU's on the SGI Origin-300, it takes 0.78 sec to run one time step (60 sec) of CH3D.

Recently, a real-time physical oceanographic observing system was installed near Station CH09B (Figure 13) by the University of Florida. The data include air temperature, wind velocity, relative humidity, vertical profile of horizontal currents (via ADCP), water temperature, and conductivity. The data are collected every 15 minutes and downloaded via cell phone onto a computer located at the University of Florida. Graphic forms of the data are presented on http://chharbor.coastal.ufl.edu/. The observing system has been in operation since April 2003. The following example shows the simulation of water level and horizontal currents during 6/03-8/03.

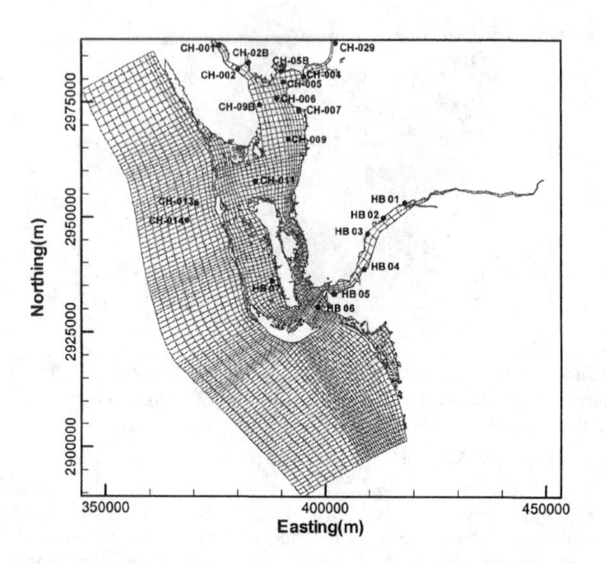

Figure 12. The horizontal numerical grid used by CH3D-IMS for Charlotte Harbor simulations, with water quality data stations labeled.

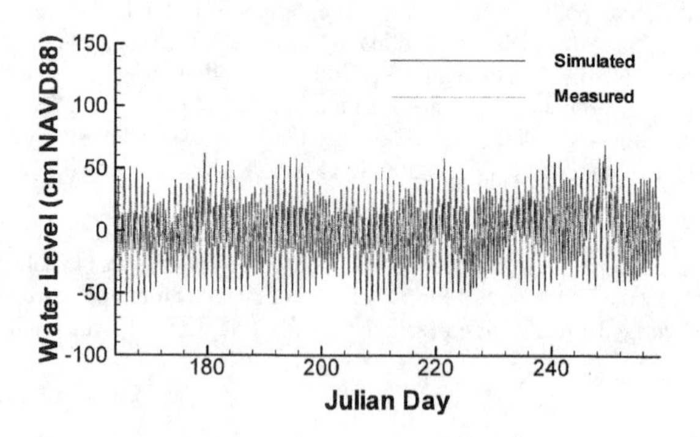

Figure 13. Simulated and measured water level at the UF station in 6/2003-8/2003.

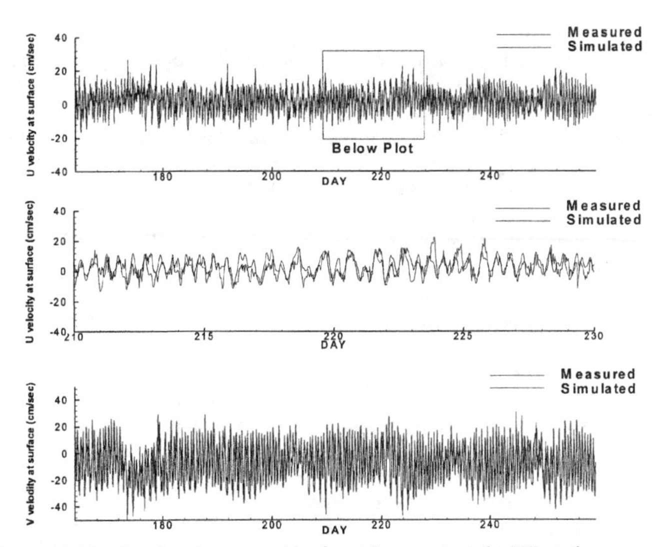

Figure 14. Simulated and measured horizontal currents at the UF station.

DISCUSSION

We have presented the development and recent applications of an integrated modeling system, CH3D-IMS, to several estuarine and coastal ecosystems in Florida. The integrated modeling system has been developed by the Advanced Coastal Environment Simulation (ACES) Laboratory at the University of Florida by integrating various component models and validating them with extensive field data from several estuarine and coastal ecosystems. The example applications showed the feasibility of using CH3D-IMS for predicting the response of estuarine and coastal environment to anthropogenic and natural changes. However, many aspects of the integrated modeling system need to be continually improved to make the modeling system more robust and more efficient, due to the wide range of scales and numerous processes involved in such simulations. We discuss several major issues in the following.

Computational Resource Requirement

As shown in the examples, integrated-scale and integrated-process modeling over large estuarine and coastal areas require high resolution and long-term simulations. While we are using 100-400 m grid spacing and a 60 sec time step, it is expected that resolution of 10-50 m with a smaller time step will be needed for more accurate simulations in coastal and nearshore areas. To allow efficient simulation, models should be parallelized, modular, and portable. To this end, we have created parallel versions of the CH3D-IMS for both shared memory computers and Beowulf clusters. All the example simulations of CH3D-IMS discussed required CPU time no more than 1/60 of the realtime. However, as we couple the estuarine and coastal models to the basin scale model (e.g., Gulf of Mexico model or Atlantic Ocean model) or global model, the required computational resources will be much greater and may exceed the capabilities of our computers (three SGI Origin systems and one Beowulf cluster). As the problem gets more complicated, it may become necessary to use an unstructured grid to reduce the total number of grid points for a large computational domain, and to develop fully parallel models using *MPI.* To meet the increasing demand for more computational resources, some institutions, e.g., University of Florida, are developing super Beowulf clusters with more than 1,000 CPU's. Others, which may be the majority, will continue to rely on the few supercomputing centers (with super clusters) in the nation. An attractive alternative is to link several mini-clusters into a regional "grid" (Foster et al., 2001).

Interdisciplinary Team Effort

The development of an integrated modeling system requires the integration of numerous disciplinary models, e.g., hydrodynamic, sedimentary, water quality, ecological, atmospheric, and groundwater models. This is a difficult task, since it is not only difficult to assemble an interdisciplinary modeling team, but also to achieve consensus among the team as to how to integrate the various disciplinary models. Different disciplinary models may contain very different spatial and temporal scales. Moreover, various disciplinary models are often developed with different programming languages on different computers with different operating systems. Although we have achieved the goal so far within a single research group, as the problem becomes more complicated, it will become increasingly difficult for us to integrate more disciplinary models and couple to larger scale models within the single research group.

An Infrastructure to Facilitate Collaboration Among Multiple Institutions

To deal with the two difficult issues pointed out in the last two paragraphs, it is believed that the modeling community can take advantage of the recent dramatic growth in networking and grid computing (Foster et al,. 2001). For example, within a year, the National Lambda Rails will bring 80 Gb/sec bandwidth to many universities and research laboratories in the nation. The rapid development and maturity of middleware, thanks to the NSF Middleware Initiative (NMI), will enable the development of computational grid for estuarine and coastal ocean modeling. An example coastal ocean modeling grid has been tested within the University of Florida (Davis et al., 2004). A prototype regional coastal ocean modeling grid is being developed by the University of Florida, Louisiana State University, and College of William and Mary, based on the approach proposed by Sheng and Davis (2003). Mini-clusters consisting of a combination of 1-processor, 2-processor, and 4-processor computers at each of the three institutions will be linked together into a regional "grid" via networking, middleware, and virtual machine software. This grid will allow virtual sharing of computer resources, models, data, and expertise among the three institutions.

Conclusions

Integrated-process and integrated-scale modeling of large estuarine and coastal areas is becoming increasingly important for coastal zone management and mitigation of coastal hazard. We have presented the continued development and recent applications of an integrated modeling system, CH3D-IMS. Due to increasing demand for higher resolution and longer-term simulations of complex interdisciplinary problems, the integrated modeling system needs to be continually improved to meet the challenges. To facilitate the rapid development and enhancement of integrated modeling systems, a regional coastal ocean modeling grid is proposed to provide regional virtual sharing of computer resources, models, data, and expertise. Integrated-process and integrated-scale modeling can be conducted by running different disciplinary models and different scale models at different institutions simultaneously on the modeling grid. This will revolutionize the coastal ocean modeling as we know it.

ACKNOWLEDGMENT

Recent development and applications of the integrated modeling system has been supported by various sponsors, including the St. Johns River Water Management District, the South Florida Water Management District, the Southwest Florida Water Management District, Pinellas County, and U.S. Environmental Protection Agency's Science To Achieve Results (STAR) Program.

REFERENCES

Christian, D. and Y.P. Sheng, 2003: "Relative Importance of Color, Chl_a, and Tripton in Affecting Light Attenuation in the Indian River Lagoon, FL, " in *Estuarine, Coastal, and Shelf Science*, Elsevier.

Davis, J.R. and Y.P. Sheng, 2000: "High Performance Estuarine and Coastal Environmental Modeling, Part I" in *Estuarine and Coastal Modeling , VI,* ASCE, pp. 470-484.

Davis, J.R. and Y. P. Sheng, 2002: "High-Performance Estuarine and Coastal Environmental Modeling, Part II" in *Estuarine and Coastal Modeling, VII,* American Society of Civil Engineers.

Davis, J.R. and Y.P. Sheng, 2003: "A Parallel Storm Surge Model," *Int'l Journal Numerical Mtds Fluids.*

Davis, J.R., Y.P. Sheng, and R. Figuereido, 2004: "Grid-Based Particle Tracking in Florida Bay," This Proceeding.

Fong, P., and M. E. Jacobson, M. C. Mescher, D. Lirman, and M. C. Harwell, 1997: "Investigating the management potential of a seagrass model through sensitivity analysis and experiments." *Ecological Applications*, 7(1), pp. 300-315.

Foster, I., Kesselman, C., and S. Tuecke, 2001: "The Anatomy of the Grid: Enabling Scalable Virtual Organizations," *Int'l Journal of Supercomputing Applications,* 15(3), pp. 200-222.

http://ch3d.coastal.ufl.edu/ Frequently asked questions about CH3D and CH3D-IMS.

Ris, R.C., N. Booij, L.H. Holthuijsen, and R. Padilla-Hernandez, 1996: *SWAN Cycle 2 User Manual,* Department of Civil Engineering, Delft University of Technology, Delft, the Netherlands.

Sheng, Y.P., 1987: "On Modeling Three-Dimensional Estuarine and Marine Hydrodynamics,'" *Three-Dimensional Models of Marine and Estuarine Dynamics* (J.C.J. Nihoul and B.M. Jamart, Eds.), Elsevier Oceanography Series, Elsevier, pp. 35-54.

Sheng, Y.P., 1990: "Evolution of a Three-Dimensional Curvilinear-Grid Hydrodynamic Model for Estuaries, Lakes and Coastal Waters: CH3D," *Estuarine and Coastal Modeling* (M.L. Spaulding, Ed.), ASCE, pp. 40-49.

Sheng, Y.P., 1997: "A Preliminary Hydrodynamic and Water Quality Model of the Indian River Lagoon," Coastal & Oceanographic Engineering Department, University of Florida, Gainesville, FL 32611.

Sheng, Y.P., 1998: "Pollutant Load Reduction Models for Estuaries," in *Estuarine and Coastal Modeling, V,* American Society of Civil Engineers, pp. 1-15.

Sheng, Y.P., 2000: "A Framework for Integrated Modeling of Coupled Hydrodynamics-Sedimentary-Ecological Processes," in *Estuarine and Coastal Modeling, VI,* ASCE, pp. 350-362.

Sheng, Y.P., E.A. Yassuda, and C. Yang, 1996a: "Modeling the Effect of Nutrient Load Reduction on Water Quality," in *Estuarine and Coastal Modeling, IV*, American Society of Civil Engineers, 644-658.

Sheng, Y.P., E.A. Yassuda, and X. Chen, 1996b: "On Hydrodynamics and Water Quality Dynamics in Tampa Bay," in *TAMPA Basis-3,* pp. 295-314.

Sheng, Y.P., J.R. Davis, D. Sun, C. Qiu, K. Park, T. Kim, Y. Zhang, 2002: Application of An Integrated Modeling System for Estuarine and Coastal Ecosystems to Indian River Lagoon, Florida, in *Estuarine and Coastal Modeling, VII,* ASCE.

Sheng, Y.P. and J.R. Davis, 2002: Indian River Lagoon Pollutant Load Reduction Model Development Project, Volume I: A 3-D Hydrodynamics and Salinity Model CH3D. University of Florida.

Sheng, Y.P. and J.R. Davis, 2003: *A Proposed Prototype Coastal Ocean Modeling Grid for SURA-SCOOP*, Presented at the SURA Workshop on Coastal Ocean Modeling Grid, University of Florida, July 31, 2003.

Sheng, Y.P., D. Sun, Y.F. Zhang, 2002a: Indian River Lagoon Pollutant Load Reduction Model Development Project, Volume II: A 3-D Sediment Transport Model CH3D-SED3D. University of Florida.

Sheng, Y.P., C. Qiu, K. Park, and T. Kim, 2002b: Indian River Lagoon Pollutant Load Reduction Model Development Project, Volume III: A 3-D Water Quality Model CH3D-WQ3D. University of Florida.

Sheng, Y.P., D. Christian, and T. Kim, 2002c: Indian River Lagoon Pollutant Load Reduction Model Development Project, Volume IV: A Light Attenuation Model CH3D-LA. University of Florida.

Sheng, Y.P., D. Christian, and T. Kim, 2003: Indian River Lagoon Pollutant Load Reduction Model Development Project, Volume V: A Submerged Aquatic Vegetation (SAV) Model CH3D-SAV. University of Florida.

U.S. Army Corps of Engineers, Coastal Engineering Research Center, 1984: *Shore Protection Manual*, Volume I and II, 4th Edition, Vicksburg, Mississippi.

Van Rijn, L.C., 1989: *Handbook, Sediment Transport by Currents and Waves.* Delft Hydraulics Report H461, Delft Hydraulics, Delft, the Netherlands.

Yassuda, E.A. and Y.P. Sheng, 1998: "Modeling Dissolved Oxygen Dynamics in Tampa Bay during the Summer of 1991," in *Estuarine and Coastal Modeling, V,* American Society of Civil Engineers, pp. 35-50.

Numerical Simulation of Tsunami Generation, Propagation, and Runup

Roy A. Walters[1]

Abstract

A numerical model is applied to calculate runup and inundation along the east coast of New Zealand arising from tsunami generated locally along the New Zealand coastal margin. In general, tsunami can be generated by a sudden movement of the ocean bed or by objects such as subaerial landslides and bolides falling into the ocean; however, this study is restricted to fault ruptures and submarine landslides. The model is based on the Reynolds-averaged Navier-Stokes (RANS) equation and uses a finite element spatial approximation, implicit time integration, a semi-Lagrangian advection approximation, and several different methods for treating pressure variations. These methods include the hydrostatic approximation, a simplified pressure interpolation scheme, and a full solution with pressure Poisson equation. The different methods of approximation are being evaluated against test problems for wave runup and submarine avalanches. Although these results are preliminary, the results with a simplified pressure model are encouraging in that they provide a realistic approximation to non-hydrostatic effects while remaining competitive with the efficiency of depth-averaged models.

Introduction

Tsunami have occurred relatively frequently along the coast of New Zealand in historic times (de Lange & Fraser, 1999). They are transient oscillations of sea level with periods much longer than wind waves but shorter than tides, and can be classified as long waves. They can cause coastal flooding, erosion, and loss of life in extreme events, and therefore can be significant coastal hazards. The particular problem being addressed is the runup and inundation due to both remote tsunami generated along the west coast of South America, and local tsunami generated along the New Zealand coastal margin. Historically, the two largest remote tsunami are the 1868 and the 1960 Chile tsunami with maximum runup heights of about 5 m at Lyttleton (de Lange and Healy, 1986). Locally generated tsunami have several sources such as fault movement, submarine landslides, and volcanic eruptions. Local

[1] National Institute of Water and Atmospheric Research, PO Box 8602, Christchurch, New Zealand. email: r.walters@niwa.co.nz

tsunami tend to be larger than remote tsunami and the maximum runup height identified from a paleotsunami deposit is 32 m above mean sea level (Nichol et al., 2003).

In a separate study, long wave resonances were evaluated along the coast of New Zealand using a frequency domain model (Walters, 2002a). Those results, coupled with historic and paleotsunami data, provide a consistent picture of the interaction between long waves and the coastal bathymetry (Walters and Goff, 2003). That study in turn has provided a general perspective from which we can now target smaller areas with local studies. In this paper, the focus is on runup from local sources, in particular fault ruptures and submarine landslides.

The objective is to develop an accurate and efficient numerical model that can be used to simulate generation, propagation, and runup from both remote and local tsunami. A finite element spatial approximation with unstructured grids allows a variable spatial resolution that can resolve the changing wavelength as the wave shoals. Depending on the specific problem, there are several wave theories that may be appropriate such as linear theory for ocean basin scale remote tsunami, hydrostatic (non-dispersive) theory for sufficiently large local fault ruptures, or non-hydrostatic (dispersive) wave theory for small scale events. It is difficult trying to maintain efficiency while including dispersive effects.

The next section contains a description of the model as currently developed. The section following that contains a comparison of model results with several runup and benchmark tsunami generation test cases. Following that is a case study of local tsunami near Kaikoura, on the east coast of the South Island, New Zealand.

Model Description

In developing a numerical model, the three most important decisions are choosing a spatial discretization that will resolve the bathymetry and important physical processes, choosing a time approximation that will resolve the important time scales, and choosing a numerical method that is accurate, robust and is sufficiently efficient to allow a large number of simulations and adequate sensitivity testing. In the current study, a finite element spatial approximation is chosen to resolve the intricate and irregular topography of rivers, estuaries, and coastal oceans. A semi-implicit time discretization is chosen for stability and flexibility of time-step size. The solution method was selected to provide the required robustness and efficiency (Walters and Casulli, 1998; Walters, 2002b).

The resulting model (RiCOM- River and Coastal Ocean Model) is based on the Reynolds-averaged Navier-Stokes equations (RANS) that are time-averaged over turbulent time scales. In addition, these equations are averaged over space to derive double-averaged equations (Finnigan, 2000; Nikora et al, 2001). In this manner, sub-grid spatial effects (vegetation, bottom roughness, etc) can be included in a rigorous manner with the inclusion of spatial averaged terms that are similar in concept to

Reynolds averages in time. Although the full capabilities of spatial averaging are not brought out in the examples presented later, it is a necessary ingredient in our work with bottom roughness and emergent vegetation, and important for runup through coastal vegetation.

The double-averaged equations are derived by spatially averaging the Reynolds-averaged Navier-Stokes equation and the continuity equation using

$$\langle F \rangle = \frac{1}{A_f} \iint_{A_f} F(x', y', z, t) dx'dy' \tag{1}$$

where A_f is the area parallel to the bed occupied by fluid out of a total area of A_0. For the flow region above bottom roughness crests, the equations can be written in matrix notation as

$$\begin{aligned} \frac{\partial \langle \overline{u}_i \rangle}{\partial t} + \langle \overline{u}_j \rangle \frac{\partial \langle \overline{u}_i \rangle}{\partial x_j} &= -\frac{1}{\rho} \frac{\partial \langle \overline{p} \rangle}{\partial x_i} + g\delta_{i3} \\ &\quad - \frac{\partial}{\partial x_j} \langle \overline{u_i' u_j'} \rangle - \frac{\partial}{\partial x_j} \langle \tilde{u}_i \tilde{u}_j \rangle + \nu \frac{\partial^2 \langle \overline{u}_i \rangle}{\partial x_j \partial x_j} \end{aligned} \tag{2}$$

$$\frac{\partial \langle \overline{u}_i \rangle}{\partial x_i} = 0 \tag{3}$$

where the overbar denotes a time average, t is time; $\overline{u}_i(x, y, z, t)$ are Reynolds-averaged velocity components in the x, y, and z-axis directions (east, north, and upward) as i=*1,2,* and *3*, respectively; $u_i'(x, y, z, t)$ is the velocity perturbation defined by $u_i' = u_i - \overline{u}_i$, p is pressure, g is the acceleration of gravity, δ_{ij} is the Kronecker delta which has the value 1 if $i = j$ and 0 otherwise, ν is kinematic viscosity, the brackets denote a space average, $\tilde{u}_i(x, y, z, t)$ is the spatial perturbation in velocity defined by $\tilde{u}_i = \overline{u}_i - \langle \overline{u}_i \rangle$, and the convention is used where repeated indices are summed. There is another equation for the flow below roughness crests that also includes wake drag terms (Nikora et al, 2001) but it is not reproduced here. That flows are coupled using a stress condition at the top of the roughness layer.

The final set of governing equations is derived with the use of the Boussinesq approximation, and by introducing a rotating frame of reference. In the following equations, the brackets and overbars are dropped because all quantities are time and space-averaged; *i.e.*, $u = \langle \overline{u} \rangle$. In differential form, the equations of momentum and mass conservation have the form

$$\frac{\partial u}{\partial t}+u\frac{\partial u}{\partial x}+v\frac{\partial u}{\partial y}+w\frac{\partial u}{\partial z}-fv=-\frac{1}{\rho}\frac{\partial p}{\partial x}+\frac{\partial}{\partial z}A_v\frac{\partial u}{\partial z}+\nabla_h\cdot(A_h\nabla_h u) \quad (4)$$

$$\frac{\partial v}{\partial t}+u\frac{\partial v}{\partial x}+v\frac{\partial v}{\partial y}+w\frac{\partial v}{\partial z}+fu=-\frac{1}{\rho}\frac{\partial p}{\partial y}+\frac{\partial}{\partial z}A_v\frac{\partial v}{\partial z}+\nabla_h\cdot(A_h\nabla_h v) \quad (5)$$

$$\frac{\partial w}{\partial t}+u\frac{\partial w}{\partial x}+v\frac{\partial w}{\partial y}+w\frac{\partial w}{\partial z}=-\frac{1}{\rho}\frac{\partial p}{\partial z}-g+\frac{\partial}{\partial z}A_v\frac{\partial w}{\partial z}+\nabla_h\cdot(A_h\nabla_h w) \quad (6)$$

$$\frac{\partial u}{\partial x}+\frac{\partial v}{\partial y}+\frac{\partial w}{\partial z}=0 \quad (7)$$

where w is vertical and u,v are horizontal velocity components; f is the Coriolis parameter; $\nabla_h=\vec{i}\partial/\partial x+\vec{j}\partial/\partial y$ is a vector operator in the horizontal x-y plane; and A_v and A_h are the vertical and horizontal viscosity coefficients, respectively. The viscosity coefficients A_v and A_h are defined by the stress terms in (2) so that the appropriate closures may be used. These closures are for the Reynolds stress terms (turbulence closure) and the spatial correlation terms. Alternatively, the spatial correlation terms can just be included as a body force in the equations, and the stress terms would then represent the conventional Reynolds stresses..

The equation for the free surface is derived by an integration of the continuity equation over depth and application of the kinematic free surface boundary condition:

$$\frac{\partial \eta}{\partial t}+\frac{\partial}{\partial x}\left[\int_h^\eta u dz\right]+\frac{\partial}{\partial y}\left[\int_h^\eta v dz\right]=0 \quad (8)$$

where $\eta(x,y,t)$ is the water-surface elevation measured from the vertical datum, $h(x,y)$ is the land elevation (top of roughness crests) measured from the vertical datum, and $H=\eta(x,y,t)-h(x,y)$ is the water depth. The vertical datum is arbitrary, but is usually a value approximately equal to the outflow water surface elevation (sea level). This choice tends to minimize truncation errors in the calculation of the water surface gradients.

For the hydrostatic version of the model, the vertical momentum equation (6) is approximated with the hydrostatic equation (the first 2 terms on the right-hand side of (6)). The continuity equation (7) is then used as a diagnostic equation for vertical velocity

$$w=-\frac{\partial}{\partial x}\left[\int_h^z u dz\right]-\frac{\partial}{\partial y}\left[\int_h^z v dz\right] \quad (9)$$

These equations are solved subject to conventional boundary conditions that include stress and no flow conditions at solid boundaries, stress and pressure conditions at the free surface, discharge or water-level conditions at river sources, and sea level and radiation conditions at open boundaries.

The finite element method was used because of its flexibility in spatial discretization. Raviart-Thomas elements of lowest order were used; these have a piecewise constant basis for sea level and linear bases for velocity (Raviart and Thomas, 1977; Arbogast, 1995). Sea level is defined at the element centroid and velocity is defined as a normal velocity at the center of the element edges. The advantage of using this element is that it has no spurious modes and can thus be used with the primitive equations. In addition, the continuity equation has a finite volume form so that mass is conserved locally as well as globally. For a regular grid, sea level is second-order accurate at the centroid which is similar to traditional finite difference methods (C grid). Implementation details can be found in Walters and Casulli (1998) and Walters (2002b).

The governing equations are discretized after defining a set of 2-dimensional unstructured triangular or quadrilateral elements in the horizontal plane. Vertical lines of nodes are created below the element vertices so that the three-dimensional elements are pie or brick-shaped. The vertical coordinate can be either a z coordinate or mapped to a sigma or terrain-following coordinate, depending on the specific application. Horizontal velocity is discretized with linear bases in the vertical and the nodal values are defined at the center of the element edges. Vertical velocity is defined at the centroid of the upper and lower faces of the element.

The time discretization is a semi-implicit scheme that removes stability constraints on gravity wave propagation (Casulli and Cattani, 1994). The advection scheme is semi-Lagrangian, which is robust, stable, and efficient (Staniforth and Côte, 1991). Wetting and drying occur in a straightforward manner with this formulation as a consequence of the finite volume form of the continuity equation and the method of calculating fluxes through the element faces. For wetting dry surfaces, the timestep is limited by a maximum Courant number of 1 on that side (the front cannot propagate faster than one element per timestep).

The continuity equation is combined with the horizontal momentum equation to form a discrete wave equation that is solved with a conjugate gradient iterative solver. A tridiagonal system of equations for velocity is solved first and used to eliminate velocity at the new time level from the continuity equation. The continuity equation then becomes a wave equation in the discrete sense. The resultant two-dimensional set of equations for sea level yields a symmetric, positive definite, and diagonally dominant matrix (see Casulli and Walters, 2000). The solution for sea level is then back-substituted into the momentum equations to solve for velocity. Details of the model may be found in Walters (2002b).

Finally, the pressure solution can be treated in three different ways. One of these is the standard hydrostatic model described in Walters (2002b). After solving for horizontal velocity as described above, vertical velocity is calculated from continuity through (9). This method is simple; however, it is not accurate for dispersive waves.

The second pressure solution technique follows an approximation method suggested by Stelling and van Kester (2000). All momentum equations are retained and a spline approximation is used for reduced pressure (total pressure minus hydrostatic pressure), where the pressure is defined at the centroid of the upper and lower faces of the element. This procedure can be used in a depth-averaged model and is an alternative to the Boussinesq equations (see Lynnett and Liu (2002) for a description of these equations). It has the advantage of better accuracy in the vertical than standard methods (next paragraph) so fewer vertical node points are required. Moreover, initial testing has shown that it is more efficient that assembling the fully non-linear Boussinesq equations.

The third pressure solution technique is a full non-hydrostatic implementation following Casulli (1999). Pressure is defined at the element centroids. In addition, the horizontal component of velocity is defined at the center of the vertical element faces (rather than edges), thereby leading to a finite difference approximation in the vertical coordinate. This leads to a matrix for the pressure solution that contains 7 diagonals, is symmetric, positive definite, and can be solved efficiently with conjugate gradient methods.

The tsunami generation process is modeled either by an initial condition for the water surface (appropriate for rapid fault movement), by incident waves generated through boundary integrals for sea level, or by a moving bed for landslides or slow fault displacement. Fault movement and incident waves are considered as input to the model and are not dealt with explicitly.

In the moving bed case, there are a variety of landslide models that can be used ranging from rigid sliding blocks (Watts et al, 2000), to two-layer fluids (Jiang and LeBlond, 1994) , to two-phase (Assier Rzadkiewicz et al, 1997) and particle models. Several landslide models have been tested starting with the simpler solid blocks. At this stage, measured laboratory-scale data from the benchmark tsunami generation test cases (Watts et al, 2000) have been compared with the model. In addition, a 2-layer fluid landslide model was compared to the solid block data using the same initial geometry, and is used in the Kaikoura case study described later. Currently, a solid/fluid continuum model for the landslide is being developed, and this model will be evaluated against laboratory experiments.

Testcases

The model has been tested against the classic solutions for wave runup on a sloping beach (Carrier and Greenspan, 1958), solitary wave runup (Synolakis, 1987), and with several runup benchmark test cases (Yeh et al, 1996).

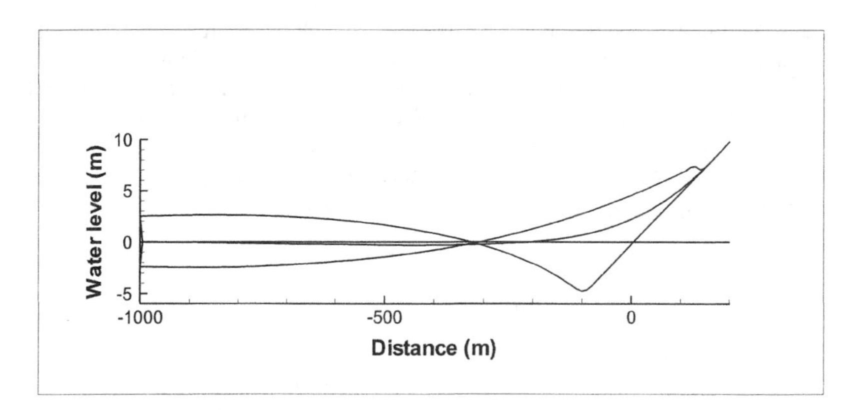

Figure 1. Wave runup on a beach with slope 0.05 for the case described by the periodic Carrier and Greenspan (1958) solution. The different lines represent the solution at different times during a cycle: minimum elevation, mean elevation and rising, and maximum elevation at the boundary.

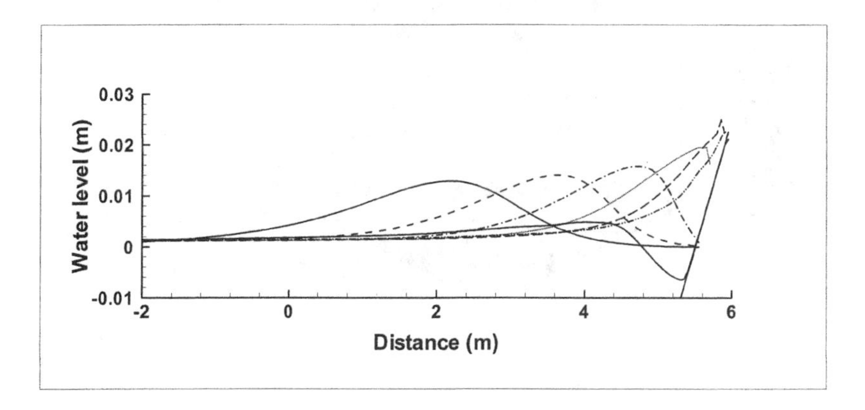

Figure 2. Solitary wave runup on a beach with slope 1:19.85 for the case $\eta/h = 0.04$ described in Synolakis (1987). The different lines represent water surface profiles at different times as the wave propagates shoreward.

For these examples, there is generally good agreement between the model (Fig. 1 and 2) and the published results. For the first two examples there is little difference between the hydrostatic and non-hydrostatic versions as has been observed in other studies; e.g., Bradford and Sanders (2002). For the case of solitary wave runup on a sloping beach (Fig. 2), the wave was initialized at a distance from the toe of the beach equal to the characteristic half-width of the wave (see Synolakis, 1987). However, there is considerable distortion of the solitary wave with the hydrostatic model if the wave is initialized farther from shore. Again, this is well known. Of interest is that the simplified (single layer) non-hydrostatic version propagates a solitary wave with minor distortion (a small trailing wave). For the third example, this behaviour is essential in order to correctly reproduce the laboratory results for benchmark 3, a solitary wave that is generated by a wavemaker and propagates to an inclined beach (Yeh et al, 1996).

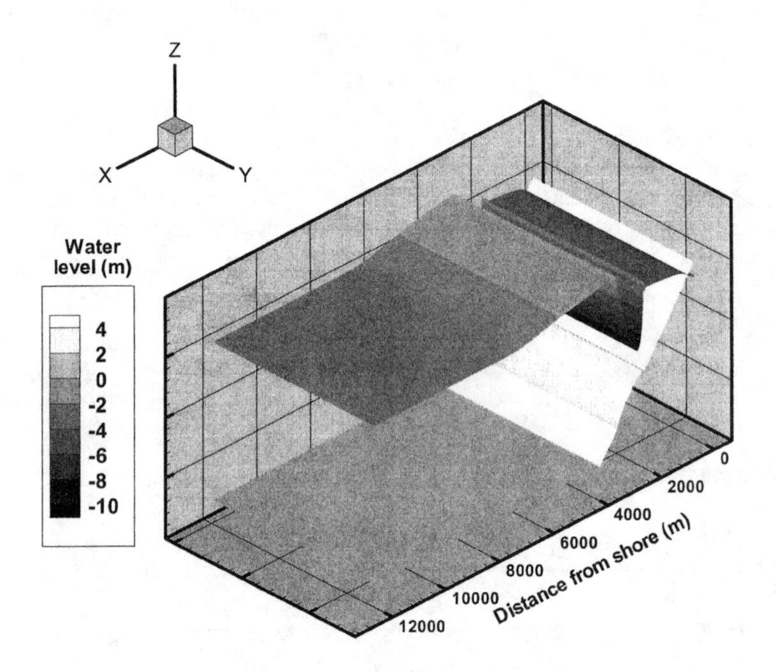

Figure 3. Numerical simulation of surface waves generated by an elliptical block sliding down an inclined plane. There is vertical exaggeration of the surface wave. The bottom of the slope is at a depth of 960 m.

Data from three benchmark tsunami generation problems (Watts et al, 2000) were also compared to the model. For these cases, a solid semi-elliptical body slides down a submerged slope with three different inclinations. All the solution techniques reproduce the shape of the surface wave: a bow wake leading the submerged body, a trough, a trailing wake, and a shoreline drawdown. Figure 3 shows the model results for one of the underwater landslide scenarios. A semi-elliptical block is shown in shades of grey as a disturbance in the planar geometry of the slope. The original shoreline is at a distance zero on the right. The sliding block creates a forced wave peak before it, and a trailing wave trough behind it. Eventually, the wave propagates as a free wave out to sea (toward the lower left), and a backward propagating wave runs up on the shore.

From a comparison of the different model versions, the hydrostatic version overestimates the depth of the wave trough, has too much drawdown along the shore, and the trough is slightly advanced toward the direction of movement. All of these effects can be attributed to the direct coupling of the bottom distortion to the water surface by hydrostatic pressure. Similar results are reported by Lynett and Liu (2002).

Kaikoura Study Results

The objective of this case study is an assessment of the potential impact of tsunami events on the Kaikoura District coastal margins (northeast coast, the South Island, New Zealand), and what hazards they pose to lifelines. The study area for this project extends from Oaro in the south, to Kekerengu in the north, including the populated area around Kaikoura Peninsula (Fig. 4). In this study, we have focused on locally generated tsunami and do not consider remote tsunami such as from across the Pacific Ocean. Although rather limited, historical data for the largest remote tsunami show that they do not have a significant impact in the study area (de Lange and Healy, 1986), so that locally generated tsunami are the major potential hazard.

The results show that the major threat is from a tsunami generated by a submarine landslide in Kaikoura Canyon and would affect the area from Oaro to South Bay.
Kaikoura Canyon is unusual (Lewis and Barnes, 1999). It has cut across the continental shelf to within 500 m of the present coast. Off Goose Bay south of Kaikoura, it comes to within 200 m of rocky projections from the shore platform and the canyon rim is only 18 m deep. Its head is well within the zone where modern sediments from big rivers are commonly moved northwards by southerly storms. Unlike most of the world's canyons, it has been highly active over the last few millennia.

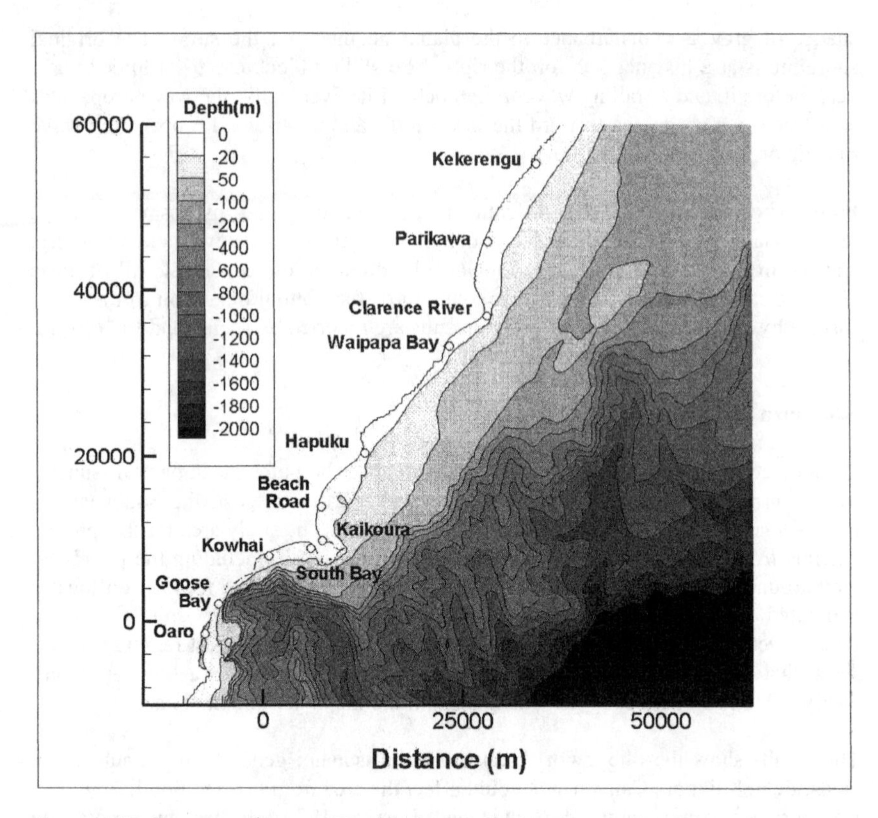

Figure 4. Kaikoura study area. The head of Kaikoura Canyon is offshore from Goose Bay and cuts across the continental shelf.

A discussion of submarine landslides in the Kaikoura Canyon and estimates of landslide volumes is found in Lewis and Barnes (1999). We have used those measurements and estimates to define a pre-landslide and a post-landslide topography. The pre-landslide topography is identical to the reference topography that is interpolated onto the model grid. We also have a secondary model grid that contains the post-landslide topography (slip surface). Both these sets of data are used within the model to define the initial bottom topography and the mass that fails and starts sliding down the head of the canyon (Figure 5). The landslide dynamics are modelled as a 2-layer fluid (Jiang and LeBlond, 1994).

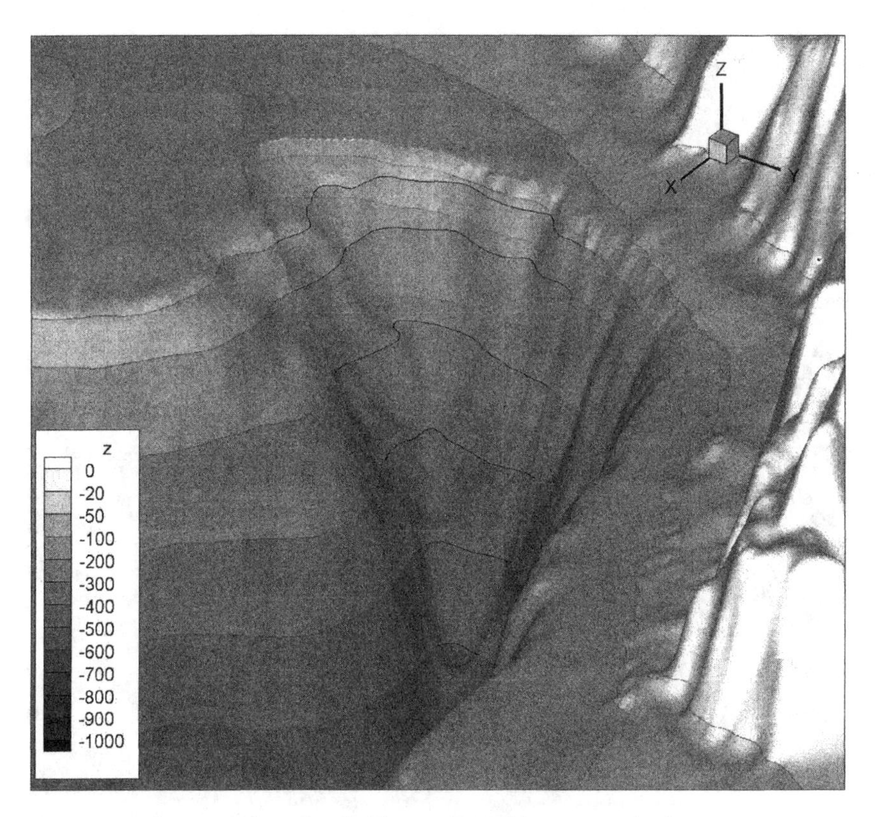

Figure 5. Kaikoura Canyon head. The pre-landslide topography is transparent over the post-landslide topography. The area between contour lines represents the landslide accumulation.

At the start, the landslide flows downslope and develops a bulbous head. This creates a surface depression behind the head and a bow-wake that is forced by the moving landslide. A snapshot of the surface wave after it has propagated away from the source area is shown in Figure 6. The wave arrival times are relatively short- a few minutes at Goose Bay with a wave crest height of about 10 m, to 7 minutes at South Bay with a wave crest of about 4 m. More than anything, this result underscores the need to educate the public and make them "tsunami aware" to the same degree as the coastal residents of Peru, Chile, and Japan.

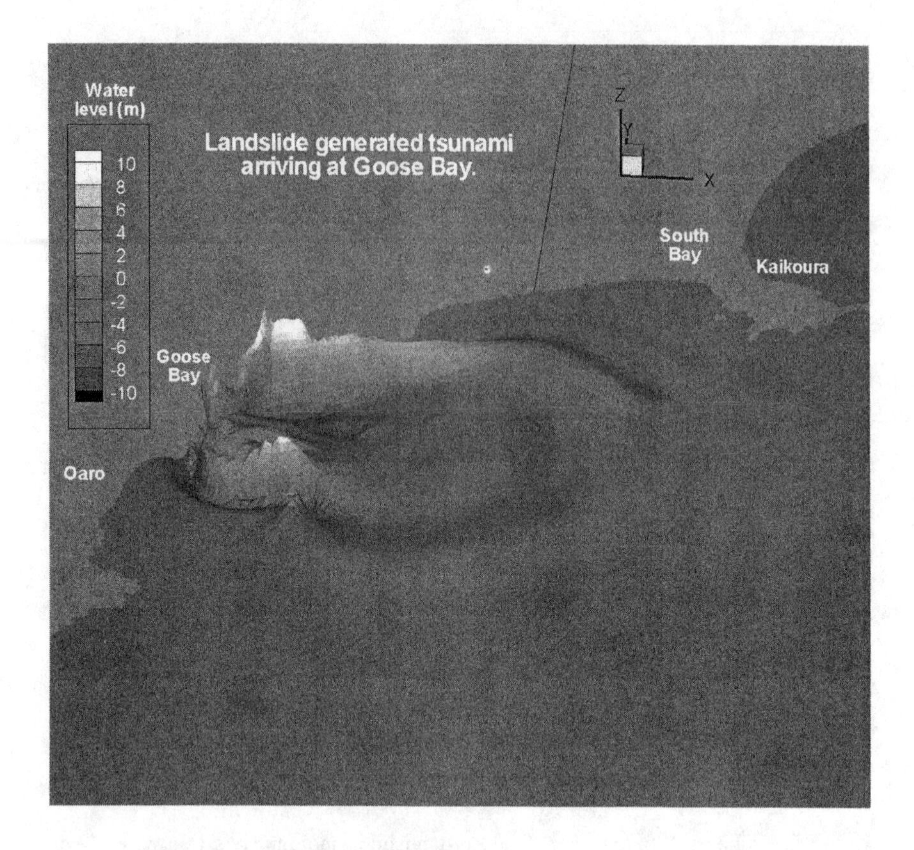

Figure 6. Kaikoura Canyon landslide results.

Another major threat is from a tsunami generated by a potential rupture of the Kekerengu Bank Fault (Fig 7). The northwest-dipping Kekerengu Bank thrust is located at a water depth of about 1000 to 1500 m beneath the Marlborough continental slope. This fault has a late Quaternary slip rate that can be estimated, on the basis of its bathymetric expression, to be of the order 0.5 to 1.5 mm/year.

Normally, fault movement is rapid when compared to time scales for wave propagation. Thus, due to the incompressible nature of water, the instantaneous initial conditions on the water surface are the same as the fault displacement. Essentially, the tsunami starts with zero velocity and a surface displacement given by the estimates for seabed rupture, and the wave evolves in time as a long gravity wave. At the open (sea) boundaries, a radiation condition is enforced so that the outgoing wave will not reflect back into the modelled area.

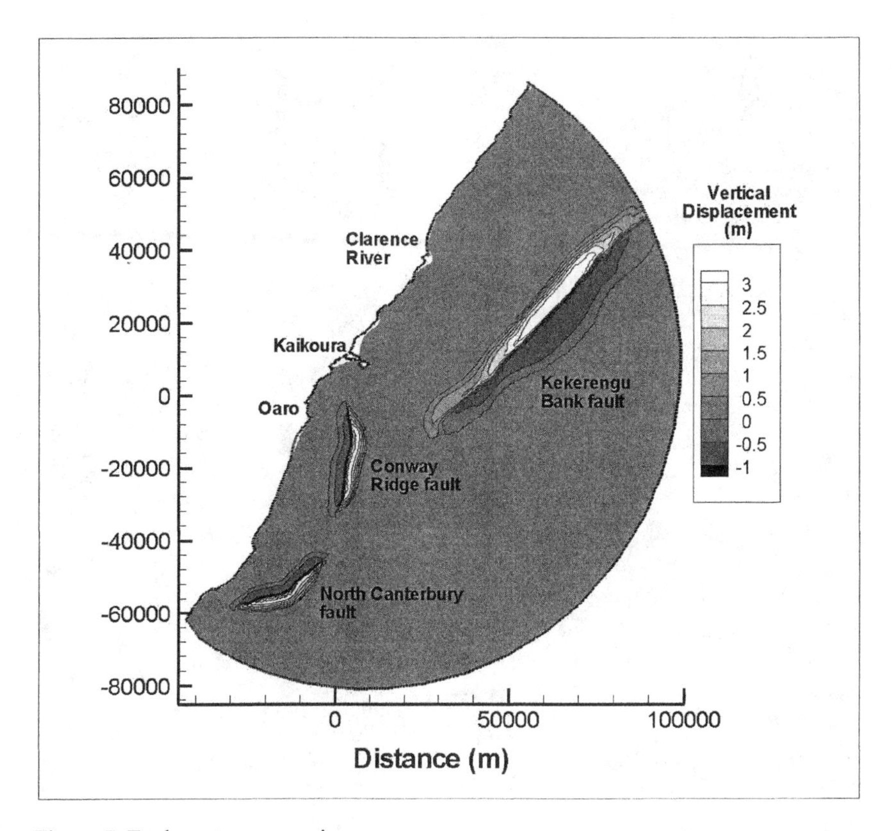

Figure 7. Fault rupture scenarios.

The initial bottom displacement is about 3 m in the landward overthrust area and -1 m in the seaward underthrust (Fig. 7). The resultant wave decomposes into 2 waves moving in opposite direction, one moving onshore that causes runup and one moving offshore that is radiated outward at the open boundaries. The landward propagating tsunami primarily affects the area from Clarence River northward, with maximum runup heights of about 7 m. Figure 8 shows the envelope of maximum water surface elevation during the simulation. In particular, the initial wave near the center of the figure and the shoaling effect on the wave are visible.

Most of the area is affected by the primary wave with a period of about 3 minutes, and there are minor waves for several hours afterward. However, when the primary wave reflects from the coast, it forms an edge wave which then travels down the coast and eventually excites a resonance near Kaikoura. The largest waves at Kaikoura (about 1.5 m) have a 10 minute period and appear about 1.5 hours after the tsunami was generated.

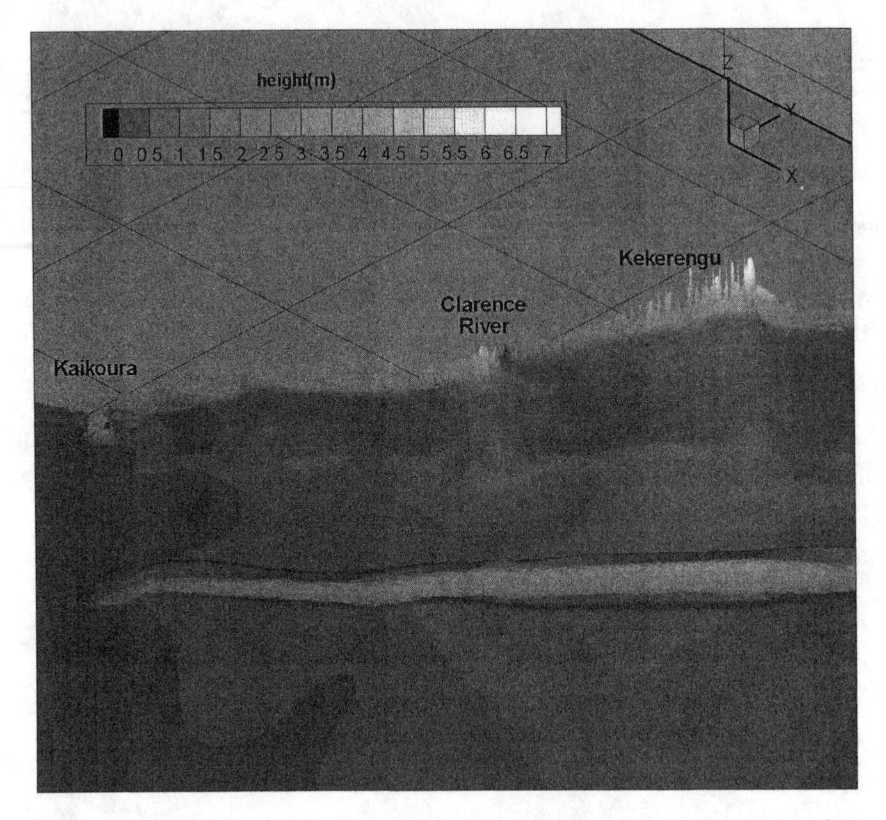

Figure 8. Kekerengu Bank Fault rupture results. Envelope of maximum water surface elevation.

Concluding Remarks

The numerical model presented here has many favourable attributes including stability, efficiency, and robustness. The particular finite elements used contain no spurious modes so that the primitive equations can be used directly. A benefit of this is that both local and global mass conservation is assured because the continuity equation takes on a finite volume form. Ongoing research is aimed towards evaluating the various methods of approximating non-hydrostatic pressure.

Although the results presented here are preliminary, results using the simplified model are encouraging in that they provide a realistic approximation to non-hydrostatic effects while retaining much of the efficiency of depth-averaged models. As the problems become more difficult, the model can be refined in the vertical in a straightforward manner to approximate the full 3-dimensional RANS equations.

Acknowledgements

This study is partially funded by the Marsden Fund from The Royal Society of New Zealand under contract NIW101 (Submarine Avalanches), and by the Foundation for Research Science and Technology (New Zealand) under contract C01X0218 (Mitigating Weather, Flooding and Coastal Hazards). The Kaikoura study was funded by Environment Canterbury.

References

Arbogast, T. (1995). Mixed methods for flow and transport problems on general geometry. pp. 275-286. In: Finite Element Modeling of Environmental Problems, G.F. Carey (Ed.). Wiley and Sons, New York.

Assier Rzadkiewicz, S., Mariotti, C. and Heinrich, P. (1997). Numerical simulation of submarine landslides and their hydraulic effects. *J. Waterways, Port, Coastal, and Ocean Eng.* 123: 149-157.

Bradford, S. and Sanders, B. (2002) Modeling flows with moving boundaries due to flooding, recession, and wave run-up. Proc. 7th Int. Conf. Est. and Coastal Modeling. ASCE.

Carrier, G.F. and Greenspan, H.P. (1958) Water waves of finite amplitude on a sloping beach. *J. Fluid Mech.* 17: 97-110.

Casulli, V. (1999). A semi-implicit finite difference method for non-hydrostatic, free-surface flows. *Int J. Numer. Meth. Fluids* 30: 425-440.

Casulli, V. and Cattani, E. (1994). Stability, accuracy, and efficiency of a semi-implicit method for three-dimensional shallow water flow. *Comput. Math. Apl.* 27: 99-112.

Casulli, V. and Walters, R.A. (2000). Unstructured grid. *Int J. Numer. Meth. Fluids* 32: 331-348.

de Lange, W.; Fraser, R.J. (1999). Overview of tsunami hazard in New Zealand. *Tephra 17*: 3–9.

de Lange, W.P.; Healy, T.R. (1986). New Zealand tsunamis 1840-1982. *New Zealand Journal of Geology and Geophysics 29*: 115-134.

Finnigan, J.J. (2000). Turbulence in plant canopies. *Annual Rev. Fluid Mech.* 32: 519-571.

Jiang, L. and LeBlond, P.H. (1994). The coupling of a submarine slide and the surface waves which it generates. *J. Geophys. Res* 97(C8): 12,713-12,744.

Lewis, K. and Barnes, P.M. (1999) Kaikoura Canyon, New Zealand: active conduit from nearshore sediment zones to trench-axis channel. *Mar. Geo.* 162: 39-69.

Lynett, P. and Liu, P. (2002) A numerical study of submarine-landslide-generated waves and run-up. *Proc. R. Soc. Lond. A* 458: 2885-2910.

Nichol, S., Goff, J, and Regnauld, H. (2003). Cobbles to diatoms: facies variability in a paleo-tsunami deposit. Proceedings of the Coastal Sediments 2003 "Crossing Disciplinary Boundaries" The Fifth International Symposium on Coastal Engineering and Science of Coastal sediment processes, Florida, USA. 1-11.

Nikora, V.; Goring, D.; McEwan, I.; Griffiths, G. 2001. Spatially-averaged open-channel flow over a rough bed. *J. Hydraul. Engrg*. 127(2): 123-133.

Raviart, P.A., and Thomas, J. M. (1977) A mixed finite element method for 2nd order elliptic problems. *Mathematical aspects of the finite element method, Lecture Notes in Math*, Springer-Verlag, Berlin.

Staniforth, A., and Côte J. 1991. Semi-Lagrangian integration schemes for atmospheric models—A review. *Mon. Weather Review* 119: 2206-2223.

Stelling, G. and van Kester, J. (2000). A new vertical approximation for the numerical simulation of non-hydrostatic free-surface flows. Delft Hydraulics, November 2000.

Synolakis, C.E. (1987). The runup of solitary waves. *J. Fluid Mech*. 185: 523-545.

Walters, R.A. (2002a). Long wave resonance on the New Zealand coast. NIWA Technical Report 109, 32 pp.

Walters, R.A. (2002b). From River to Ocean: A Unified Modeling Approach. Estuarine and Coastal Modeling: Proc. 7th Int. Conf., November 5-7, 2001, St. Petersburg, Florida, edited by M.L. Spaulding, ASCE: 683-694.

Walters, R.A., and Casulli, V. 1998. A robust, finite element model for hydrostatic surface water flows. *Communications in Numerical Methods in Engineering* 14: 931-940.

Walters, R. and Goff, J. (2003). Assessing tsunami hazard along the New Zealand coast. *Science of Tsunami Hazards* 21(3) p 147-153.

Watts, P., Imamura, F., and Grilli, S.T. (2000). Comparing model simulations of three benchmark tsunami generation cases. *J. Sci. Tsunami Hazards* 18(2) p 107-123.

Yeh, H. Liu, P. and Synolakis, C. (1996). Long wave runup models. World Scientific.

Modeling the transport of larval yellow perch in Lake Michigan

Dmitry Beletsky[1], David Schwab[2], Doran Mason[2], Edward Rutherford[3], Michael McCormick[2], Henry Vanderploeg[2], and John Janssen[4]

Abstract

The transport of larval yellow perch (*Perca flavescens*) in Lake Michigan is studied with a 3D particle trajectory model. The model uses 3D currents generated by the Great Lakes version of the Princeton Ocean Model driven by observed momentum and heat fluxes in June-August 1998, 1999 and 2000. Virtual larvae were released in the nearshore region with the most abundant preferred substrate for yellow perch spawning, rocks. We also investigated the potential for physical transport mechanisms to affect recruitment of Lake Michigan yellow perch by coupling hydrodynamic models with individual-based particle models of fish larvae to study variation in larval distributions, growth rates, and potential recruitment. Larval growth rates were simulated using a bioenergetics growth model with fixed consumption rates. Results indicate that lake circulation patterns are critical for understanding interannual variability in Great Lakes fish recruitment.

---

[1]Department of Naval Architecture and Marine Engineering, University of Michigan, Ann Arbor, MI. 48109

[2]NOAA Great Lakes Environmental Research Laboratory, Ann Arbor, MI. 48105

[3]Institute for Fisheries Research, School of Natural Resources and Environment, University of Michigan, Ann Arbor, MI. 48109

[4]Great Lakes Water Institute, University of Wisconsin-Milwaukee, Milwaukee, WI. 53204

Introduction

Yellow perch (*Perca flavescens*) is ecologically and economically important species in Lake Michigan that has suffered recruitment failures over the last decade (Francis et al., 1996). Causes for poor recruitment are not understood, but are believed to be a result of high mortality during the larval stage. Yellow perch have an early life history similar to many marine coastal species, which is a pelagic larval stage that is susceptive to physical transport. In inland lakes the pelagic stage, begins shortly after larvae hatch and lasts for approximately 40 days (Whiteside et al., 1985). In Lake Michigan the length of the pelagic period may be even longer, and at least some individuals can be captured in the pelagia after about 75 days. This extended pelagic period means the influence of lake physics on recruitment of yellow perch could be significant due to complex lake-wide circulation patterns.

The importance of meso-scale oceanographic features to fish population structure and recruitment variability has long been recognized for marine fish populations (e.g., Hjort, 1914). However, it has only been in more recent years that hydrodynamic models have been used to understand the impacts of marine physics on the recruitment. For example, Werner et al. (1996) used a 3-D hydrodynamics model, turbulence model, and models of foraging and growth to demonstrate the complementary interactions between circulation, trophodynamic processes and recruitment of cod *Gadua morhua* and haddock *Melanogrammus aeglefinus* on Georges Bank. Heath and Gallego (1998) coupled a particle-tracking model with an individual-based model of larval growth and survival to investigate the spatial and temporal patterns in recruitment processes of North Sea haddock *Melanogrammus aeglefinus*. In yet another study, hydrodynamic models were used to explain transport mechanisms for fish larvae recruiting from offshore to coastal estuaries in the South Atlantic Bight (Crowder and Werner, 1999).

Most studies quantifying the impact of physics on fish recruitment have been in estuarine and marine, with few in freshwater ecosystems. The Great Lakes, the largest freshwater system in the world, are frequently called inland seas because they experience similar physical processes as observed in marine systems. These physical processes occur over various temporal and spatial scales, and are very complex due principally to the wind driven forcing that generates these currents. Given the complexity and magnitude of the physical processes in Lake Michigan, it is very likely that these physical processes may play an important role in structuring the recruitment dynamics of Lake Michigan fishes. The goal of this paper is to begin exploring the effects of physical factors on recruitment variability of important Lake Michigan fishes in order to gain insight into the decline in the yellow perch populations and factors causing poor recruitment. The Great Lakes hold an important advantage over the coastal ocean because dispersal of fish larvae is relatively confined in the Great Lakes compared to dispersal in wide-open marine ecosystems like Georges Bank, the south Atlantic Bight or the North Sea.

In this paper, we apply a model-based Langrangian approach that utilizes 3-D circulation and thermal processes, physiology and ecology of fish larvae, and trophodynamics for understanding recruitment dynamics in Lake Michigan specifically, and the other Great Lakes in general. In the initial set of numerical experiments we focus on the transport of larval yellow perch hatched in the area in southern Lake Michigan known for high concentrations of rocky habitat preferred by yellow perch spawners (Figure 1). We couple models of circulation and temperature, with individual-based particle models of fish larvae to study variation in larval distributions, growth and survival rates, and potential recruitment.

Models

Hydrodynamic model

A 3-dimensional circulation model of Lake Michigan (Beletsky and Schwab, 2001) is used to calculate lake circulation. The model is based on the Princeton Ocean Model (Blumberg and Mellor, 1987) and is a nonlinear, hydrostatic, fully three-dimensional, primitive equation, finite difference model. The model uses time-dependent wind stress and heat flux forcing at the surface, free-slip lateral boundary conditions, and quadratic bottom friction. The drag coefficient in the bottom friction formulation is spatially variable. It is calculated based on the assumption of a logarithmic bottom boundary layer using constant bottom roughness of 0.1 cm. Horizontal diffusion is calculated with a Smagorinsky eddy parametrization (with a multiplier of 0.1) to give a greater mixing coefficient near strong horizontal gradients. The Princeton Ocean Model employs a terrain following vertical coordinate system (sigma-coordinate). The equations are written in flux form, and the finite differencing is done on an Arakawa-C grid using a control volume formalism. The finite differencing scheme is second order and centered in space and time (leapfrog). The model includes the Mellor and Yamada (1982) level 2.5 turbulence closure parameterization. The hydrodynamic model of Lake Michigan has 20 vertical levels with finer spacing near the surface and the bottom and a uniform horizontal grid size of 2 km (Figure 1).

Particle trajectory model

The 3-d particle trajectory code is a combination of the Princeton Ocean Model subroutine TRACE written by Jarle Berntsen (Institute of Marine Research, Bergen-Nordnes, Norway) and the second order accurate horizontal trajectory code described by Bennett and Clites (1987). The horizontal currents are first interpolated from velocity points to grid square corners on the Arakawa-C grid. By assuming bilinear variation of the horizontal currents within a grid square, the Taylor series expansion of velocities about the particle position in the trajectory equations yields a pair of simultaneous equations for the new particle position. The time step is chosen to limit the maximum excursion of a particle to 1/8 the distance between horizontal grid points. Vertical velocity is interpolated to z-coordinates from sigma coordinates based on bilinear interpolation of the depth to the particle position within a grid box. Particles are prevented from crossing the lake bottom or free surface, as well as horizontal boundaries. This method generally predicts more realistic trajectories than traditional first-order horizontal methods and does not allow particles to accumulate in 'stagnation' zones at grid square corners along the shoreline.

Biological model

We grow larval yellow perch in the model by using a bioenergetics growth model. The bioenergetics model is a species-specific, energy-balance approach that describes the flow of energy through an individual fish and how that energy is partitioned among consumption, growth (somatic and reproductive), and losses (respiration, egestion, excretion, and specific dynamic action). Energy per unit time is related to weight per unit time by a specific energy density for predator and prey (calories per unit weight). The basic form of the bioenergetics model in terms of weight specific growth rate ($g \cdot g \cdot d^{-1}$) is,

$$\frac{1}{W}\frac{dW}{dt} = \phi - (R_{resp} + SDA + F + U)$$

where W is weight of the individual predator, t is time in days, ϕ is feeding rate, R_{resp} is respiration, SDA is specific dynamic action, F is egestion, and U is excretion. Each of the terms in the equation is a function of water temperature, thus both feeding rate ϕ and temperature drive growth. Parameter values were adapted from a previously published bioenergetics model for yellow perch (Rose et al.1999). For this exercise, we assume that ϕ is constant and growth rate is driven primarily by temperature, which differs across space and time.

Results

Physics

Hydrodynamic model run begins on January 1, 1998 and ends on December 31, 2000. Initial currents are set to zero. In the beginning of model run the lake is thermally homogeneous vertically, but there are small horizontal temperature gradients with warmest temperatures located offshore and coldest temperatures nearshore. Based on available observations, the model was initialized with water temperatures in the 2-4.5^{o} C range. Hourly meteorological data (wind speed and direction, air temperature, dew point and cloud cover) for 1998-2000 were obtained from 18 National Weather Service stations around Lake Michigan and National Data Buoy Center (NDBC) buoys 45002 and 45007 (Figure 1). These observations form the basis for generating gridded meteorological fields. Details of heat and momentum flux calculations are presented in Beletsky and Schwab (2001), and details of a new spatial interpolation technique for meteorological data used in 1998-2000 simulations are presented in Beletsky et al. (2003).

Because temperature plays a crucial role in larval growth we present here monthly average surface temperature patterns for each summer in 1998-2000 (Figure 2). There is a general north-south temperature gradient seen in all months in all years. Another prominent feature of lake temperature patterns is a wind-driven upwelling at the west coast typical of summer conditions in Lake Michigan (Beletsky and Schwab, 2001). In southern Lake Michigan, surface temperature steadily increased from about 17 °C in June to 22 °C in July to 23 °C in August of both 1998 and 1999. In all summer months of 2000 lake surface temperature was about 1-2 °C lower which should have important implications on larval growth as will be shown in the next section.

Lake circulation patterns can also play an important role in larval transport and survival. The characteristic short-term circulation pattern in a lake, which is driven by a spatially uniform wind, consists of two counter-rotating gyres; a counterclockwise-rotating (cyclonic) gyre to the right of the wind and a clockwise-rotating (anticyclonic) gyre to the left (Bennett, 1974). The strongest currents are downwind in the coastal regions while weaker upwind return currents develop in the deeper waters. This simple pattern is sometimes seen in monthly mean circulation patterns in winter, but in summer buoyancy effects make circulation more complex. Thus in southern Lake Michigan for most months, the circulation pattern consisted of two gyres with the cyclonic gyre confined to the deep area and the anticyclonic gyre in the shallow southernmost area (Figure 3). This circulation pattern was seen in previous simulations (Beletsky and Schwab, 2001), however, it is difficult to detemine if this pattern is typical due to the scarcity of long-term current observations in Lake Michigan (Beletsky et al., 1999). If an anticyclonic gyre is persistent in summer in southern Lake Michigan, it can trap yellow perch larvae in that part of the lake. During some months (August 1998, June and August 2000), a strong northward current from southern Lake Michigan penetrated the northern basin with significant implications for larval transport. The speed of mean surface currents varied from 10 to 20 cm/s.

In the particle trajectory model, 246 particles were released north of Chicago (Figure 1) at bathymetric depths of less than 10 m. This reflects the fact that yellow perch prefer to spawn on the rocky habitat available at this location. Particles were distributed uniformly with depth: near the surface, at 1/3 and at 2/3 of a grid cell's depth. Particle model runs began on June first of each year (1998, 1999, and 2000) and ended in the end of August. Particle locations at the end of each month are shown in Figure 4. Because all particles were released in very shallow waters they have a tendency to stay relatively close to the surface (0-20 m). There was no significant difference in movement of particles released near the surface and ones that were released closer to the bottom. Thus, we focused on surface currents and not depth-averaged currents. Overall, particle movement matches the monthly mean surface current pattern rather well: particles initially move offshore and then continue to circulate in southern Lake Michigan in an anticyclonic fashion. Under certain conditions (like a case of a particularly strong northward coastal current in August 1998) a significant number of particles escape the southern basin and penetrate the northern basin of Lake Michigan. The proximity of particles to shore at the end of the model run in August can also be critical for larval survival. Larvae about this time begin to metamorphose into their adult characteristics and move into adult habitat, which is near bottom and near shore. As model results show, in 1998 the number of particles reaching nearshore waters in August was significantly higher than in 1999 and 2000 which may provide a significant advantage for surival.

Biology

Biological model runs also began in June of each year. The hydrodynamic model supplies information on the three-dimensional temperature field along the larval path predicted by the particle trajectory model. Another critical parameter for larval development is food availability (primarily zooplankton). Unfortunately, there is very little information available on the spatial distribution of zooplankton in south Lake Michigan in summer. Therefore, in all biological model runs we assumed that there is no spatial gradient in food (zooplankton) available for larval yellow perch.

All larvae were assumed to have initial length of 6 mm at hatching. Movement and growth of larvae in the model were followed from hatching to 30 mm, the length at which they settle and become demersal; larvae metamorphose into juveniles at 20 mm, and by 30 mm take on the characteristics of adult fish. Growth rates and time to settlement were predicted assuming two different food availability scenarios. This was done by multiplying maximum consumption by 1.0 (Scenario 1) and 0.5 (Scenario 2). Locations of larvae which reached 30 mm length under the maximum consumption conditions (Scenario 1) are shown in Figure 5. No larvae reached 30 mm before the end of the first month except a small number in 1999. On the contrary, all larvae reached 30 mm by the end of the second month and therefore July and August locations match exactly those of Figure 4. In case of the reduced consumption scenario 2 (more typical of realistic Lake Michigan conditions) the situation changes dramatically. As expected, no larvae reached 30 mm by the end of June but by the end of July only a few larvae reached 30 mm in 2000 while in 1998 and 1999 more than 60 % of larvae grew to their settlement length (Figure 6). This is undoubtedly the result of much cooler water temperatures in 2000 predicted by the hydrodynamic model and confirmed by surface temperature observations at the NDBC buoy 45007.

Discussion and conclusions

This is the first physical-biological model of larval fish developed for the Great Lakes. The model shows significant interannual variability in particle transport during three years of study with implications for larval yellow perch growth and settlement. Currently, we have not included mortality and zooplankton fields (larval fish food) in our model. Future efforts will focus on these areas. Yet another issue to resolve is the assumption that fish larvae are passive particles; larval fish swimming ability increases with size, and swimming competence along with other behavioral patterns that can influence their horizontal and vertical distributions but are not captured in our model. Despite these apparent shortcomings, our modeling approach is consistent with many recent observations in Lake Michigan.

The hydrodynamic dispersal of young yellow perch provides a mechanism for the genetic homogeneity of the Lake Michigan yellow perch population. Miller (2003) reported that collections from southern basin collection sites in Michigan, Wisconsin, and Indiana were homogenous and not much different from those from Lake Michigan's northern basin. They were distinct from populations from Green Bay and inland lakes. The distinctiveness relative to Green Bay is likely due to a spawning period that occurs approximately one month earlier in Green Bay, and due to limited water exchange between Green Bay and Lake Michigan.

The timing for yellow perch to become demersal may be delayed for yellow perch in Lake Michigan. Whiteside et al. (1985) found that yellow perch from Lake Itasca, Minnesota, became demersal at about 25 mm total length. A similar length, 30 mm, was reported for Lake St George, Ontario (Post and McQueen 1988) and Lake Erie (Wu and Culver 1992). In contrast, trawling and seining records for Michigan (D. Jude, University of Michigan, Pers. Comm.), Illinois (J. Dettmers, Illinois Natural History Survey, Pers. Comm.), and Wisconsin (P. Hirethota, Wisconsin Department of Natural Resources, Pers. Comm.) indicate that demersal yellow perch smaller than about 40 mm are rare and most fish are 50 mm or more. Moreover, preliminary pelagic trawling has collected yellow perch from about 20 to 35 mm in late July and as large as 70 mm in mid-September. More detailed sampling and estimates of larval age and growth obtained from otolith increment analysis will be needed to determine how long perch remain pelagic in Lake Michigan. Assuming that the pelagic stage of perch is extended, it remains to be determined whether this will result in increased mortality.

It appears that the hydrodynamic conditions may produce a "source and sink" recruitment dynamic. The rocky habitat, preferred for spawning (Dorr 1982, Robillard and Marsden 2001) and presumably for feeding (Wells 1977, 1980, Janssen et al. in press, Janssen and Luebke in press) is primarily on the western side of Lake Michigan, and is perhaps most extensive in Illinois (Powers and Robertson 1968,

Fucciolo 1993, Janssen et al. in press). The present modeling effort suggests that larvae originating from this preferred habitat would be mostly transported to the sandier and generally unconsolidated substrate along the eastern side of Lake Michigan. Much of the habitat along eastern Lake Michigan is now depauperate of potential forage for newly settled juvenile perch (Nalepa et al. 1998), which may impact survival in later life stages.

Our modeling effort represents a first step in integrating lake physics for understanding fish recruitment in the Great Lakes. Moreover, the modeling exercises presented here have shed light on how lake physics may modify and impact larval fish growth and survival in Lake Michigan. In addition, it has also exposed new questions in understanding the behavior (as driven movement) and ecology (size at which fish go demersal) of larval yellow perch, which may differ from other inland lakes.

References

Beletsky,D., D.J. Schwab, P.J. Roebber, M.J. McCormick, G. S. Miller, and J.H. Saylor. 2003. Modeling wind-driven circulation during the March 1998 sediment resuspension event in Lake Michigan. *J.Geophys. Res.*, 108(C2), 20-1 to 20-13.

Beletsky, D., and D.J. Schwab. 2001. Modeling circulation and thermal structure in Lake Michigan: Annual cycle and interannual variability. *J.Geophys. Res., 106, 19745-19771.*

Beletsky, D., J.H. Saylor, and D.J. Schwab. 1999. Mean circulation in the Great Lakes. *J. Great Lakes Res.,* 25, 78-93.

Bennett, J.R., 1974.On the dynamics of wind-driven lake currents. J. Phys. Oceanogr. 4, 400-414.

Bennett, J.R., and A.H. Clites, 1987 Accuracy of trajectory calculation in a finite-difference circulation model. J.Comp.Phys., 68, 272-282.

Blumberg, A.F., and G.L. Mellor. 1987. A description of a three-dimensional coastal ocean circulation model. In: Three dimensional ocean models, coastal and estuarine sciences, 5, N.S. Heaps (ed). American Geophysical Union, Washington, DC. pp. 1-16.

Crowder, L.B. and F.E. Werner (eds.). 1999. Fisheries oceanography of the estuarine-dependent fishes of the south Atlantic Bight. Fisheries Oceanography, 8(Suppl 2.):1-252.

Dorr, J.A. 1982. Substrate and other environmental factors in reproduction of the yellow perch (*Perca flavescens*). Ph.D. dissertation, University of Michigan, 292 pp.

Francis, J.T., Robillard, S.R., and Marsden, J.E. 1996. Yellow perch management in Lake Michigan: a multi-jurisdictional challenge. Fisheries (Bethesda). 21: 18-20.

Fucciolo, C.S. 1993. Littoral zone habitat classification and mapping of Illinois Lake Michigan coastal areas: bathymetry and distribution of bottom materials. Illinois Geological Survey OFS, 73 pp.

Heath, M.R., and Gallego, A. 1998. Biophysical modeling of the early life stages of haddock, Melanogrammus aeglefinus, in the North Sea. Fisheries Oceanography 7:110-125.

Hjort, J. 1914. Fluctations in the great fisheries of northern Europe. Rapports et Procés Verbaux des Reunion Conseil Internationale Exploration de la Mer, 20:1-228.

Janssen, J. and M. Luebke. In press. Preference for rocky habitat by young-of-the-year yellow perch and alewives. J. Great Lakes Res.

Janssen, J., M. Berg, and S. Lozano. in press. Submerged terra incognita: the abundant, but unknown rocky zones. In: (T. Edsall and M. Munawar, eds.) The Lake Michigan ecosystem: ecology, health, and management. Academic Publishing, Amsterdam.

Mellor, G.L., and T. Yamada. 1982. Development of a turbulence closure model for geophysical fluid problems. Reviews of Geophysical and Space Physics 20:851-875.

Miller, L. M. 2003. Microsatellite DNA loci reveal genetic structure of yellow perch in Lake Michigan. Trans. Amer. Fish. Soc. 132: 503-513.

Nalepa, T.F., D.J. Hartson, D.L. Fanslow, G.A. Lang, and S.J. Lozano. 1998. Declines in benthic macroinvertebrate populations in southern Lake Michigan, 1980-1993. Can. J. Fish. Aquatic Sci. 55: 2402-2413.

Post, J. R. and D. J. McQueen. 1988. Ontogenetic changes in the distribution of larval and juvenile yellow perch (*Perca flavescens*): a response to prey or predator? Can. J. Fish. Aquat. Sci. 45: 1820-1826.

Powers, C.F. and A. Robertson. 1968. Subdivisions of the benthic environment of the upper Great Lakes, with emphasis on Lake Michigan. J. Fish. Res. Board Can. 25: 1181-1197.

Robillard, S. R. and J. E. Marsden. 2001. Spawning substrate preferences of yellow perch along a sand-cobble shoreline in southwestern Lake Michigan. N. Am. J. Fish. Manage. 21:208-215.

Rose, K.A., E.S. Rutherford, J.L. Forney, E.L. Mills, and D. SinghDermott. 1999. Individual-based model of yellow perch and walleye populations in Oneida Lake, New York. Ecological Monographs 69:127-154.

Wells, L. 1977. Changes in yellow perch (*Perca flavescens*) populations in Lake Michigan, 1954-1975. J. Fish. Res. Board Can. 34: 1821-1829.

Wells, L. 1980. Food of alewives, yellow perch, spottail shiners, trout-perch, and slimy and fourhorn sculpins in southeastern Lake Michigan. Technical paper 98, U.S. Fish and Wildlife Service. Ann Arbor, MI 48107, 12 pp.

Werner, F.E. R.I. Peryy, R.G. Lough, and C.E. Naimie. 1996. Trophodynamic and advective influences on Georges Bank larval cod and haddock. Deep Sea Research II, 43:1793-1822.

Whiteside, M.C., C.M. Swindoll, and W.L. Doolittle. 1985. Factors affecting the early life history of yellow perch, *Perca flavescens*. Env. Biol. Fish. 12: 47-56.

Wu, L. and D. A. Culver. 1992. Ontogenetic diet shift in Lake Erie Age-0 yellow perch (*Perca flavescens*): a size-related response to zooplankton density. Can. J. Fish. Aquat. Sci. 49: 1932-1937.

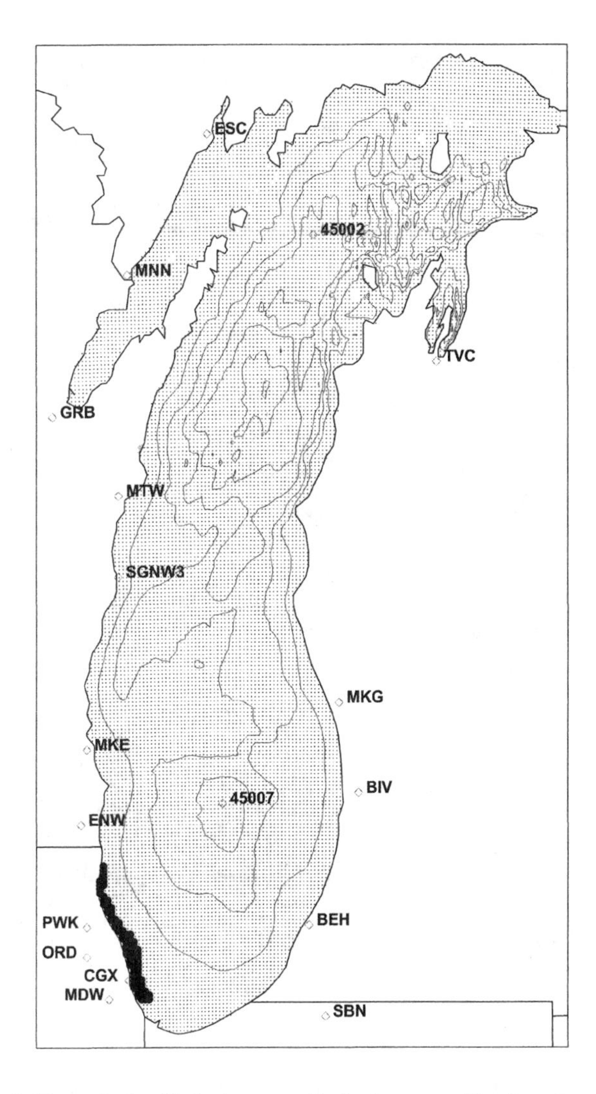

Figure 1. Numerical grid, bathymetry (isobaths every 50 m), and meteorological stations. Initial particles location in southern Lake Michigan is also shown.

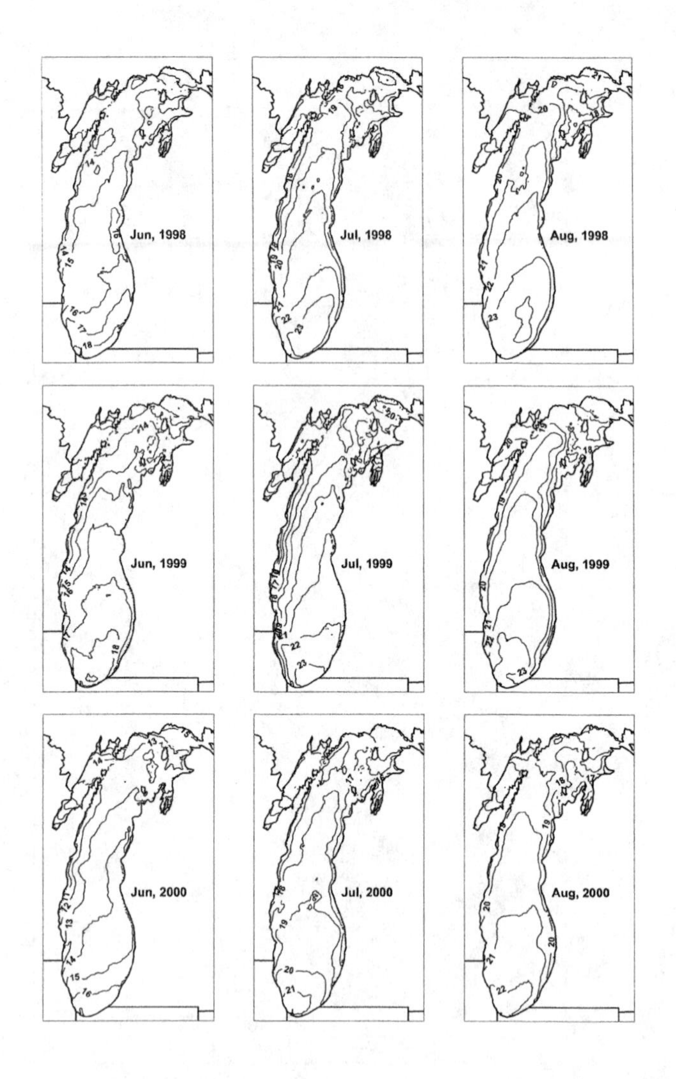

Figure 2. Lake surface temperature in 1998-2000

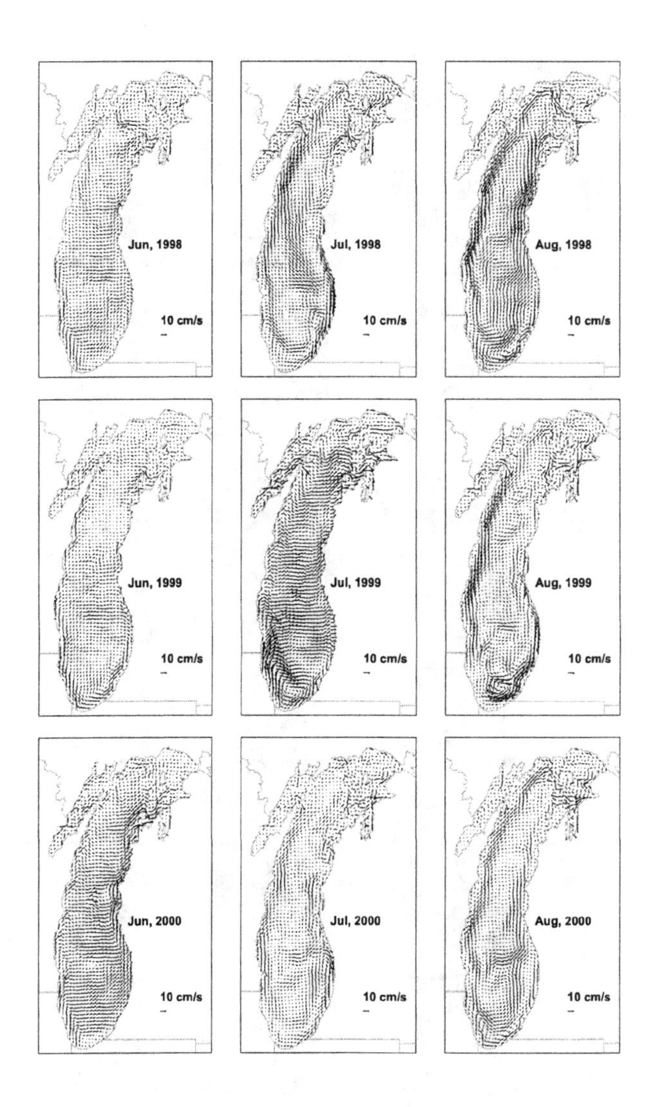

Figure 3. Surface currents in 1998-2000

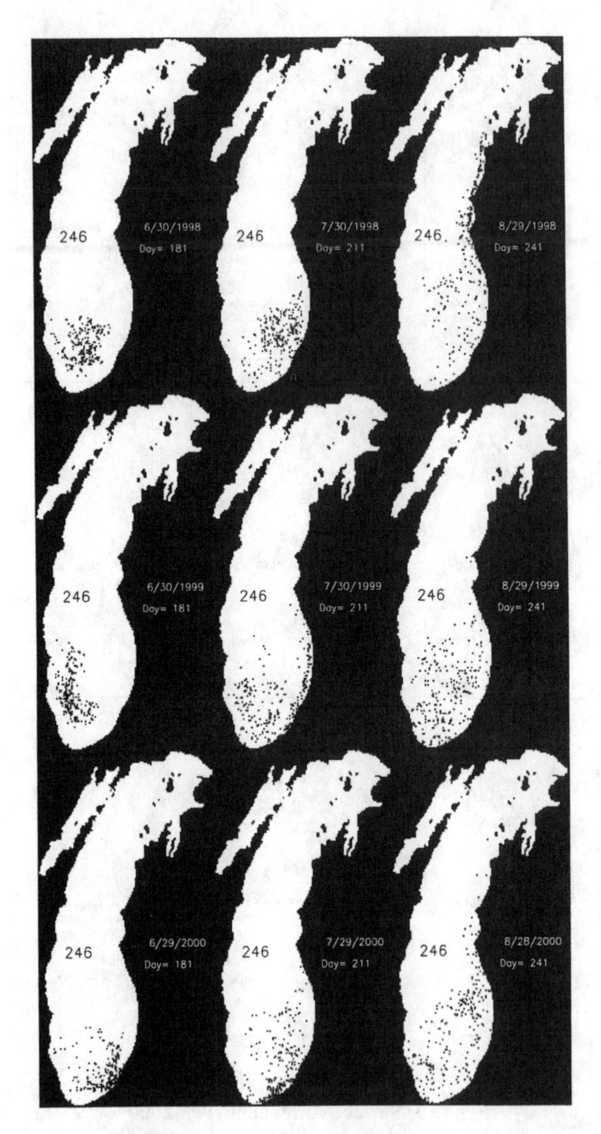

Figure 4. Particle transport in 1998-2000, total number of particles shown.

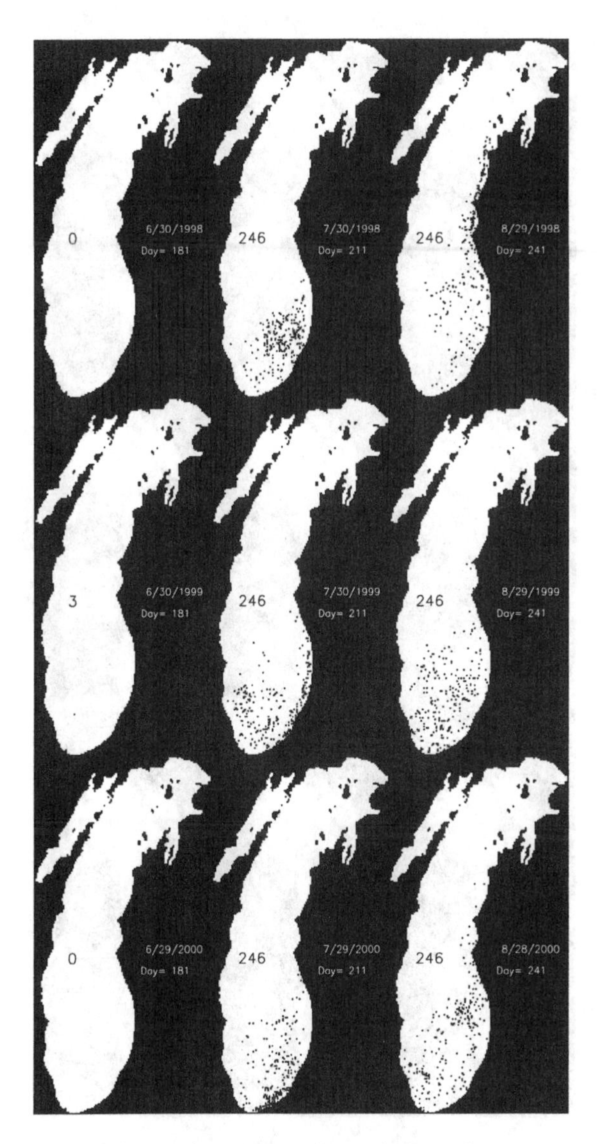

Figure 5. Larval transport and growth, Scenario 1 (see explanation in text). Total number of larvae reached 30 mm also shown.

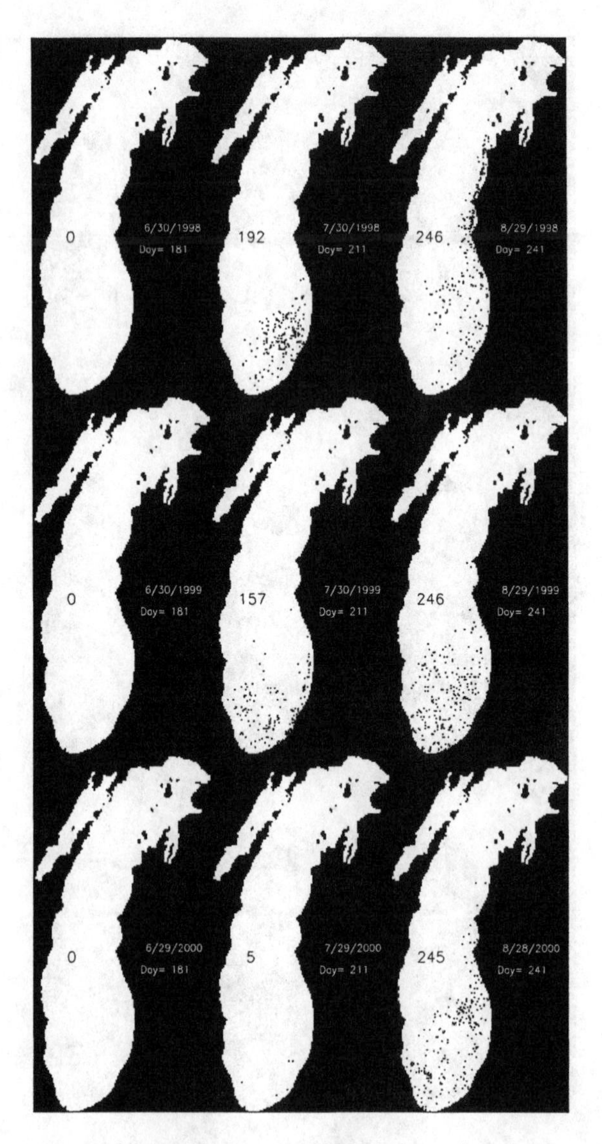

Figure 6. Larval transport and growth, Scenario 2 (see explanation in text). Total number of larvae reached 30 mm also shown.

Toxic Modeling in an Industrial Harbor - A Case Study for Baltimore Harbor

Jing Lin[1], Harry V. Wang[2], G. McAllister Sisson[2], and Jian Shen[2]

ABSTRACT

Baltimore Harbor, a tributary in the northwest portion of Chesapeake Bay, is historically a major industrial harbor, and has been known for its toxic contaminant pollution from adjacent industries. In order to investigate the fates and the processes of sediment-bound heavy metals in Baltimore Harbor, a three-dimensional hydrodynamic and sediment transport model (CH3D-SED), linked to a toxic model (ICM/TOXI), is applied to the Upper Chesapeake Bay and the sub-domain of Baltimore Harbor. The extended model domain of hydrodynamic and sediment transport, including the entire Upper Chesapeake Bay, provides a significant advantage for investigating interactions between the Harbor and the Upper Bay, while the limited domain for ICM/TOXI (including only Baltimore Harbor) allows more efficient model linkages and computation. A bed sediment sub-model, contained within ICM/TOXI and coupled with the water column model, computes the temporal variation of the sediment concentrations as the result of sediment resuspension, deposition, and physical mixing between different bed layers. Three types of heavy metals, Cr, Pb, and Zn, are simulated in Baltimore Harbor. The model results indicate that with current point source and nonpoint source loadings, the surface sediment contaminant concentration is decreasing gradually. This suggests that the sediment contaminant problem today in Baltimore Harbor is a legacy-related problem. The system is currently in a natural recovery process with regard to the aforementioned three metals. The rate of recovery is a function of current loadings of the concerned materials, the sediment dynamics, and the interaction between the Harbor and the Upper Bay mainstem.

[1]Dept. Marine Earth and Atmospheric Sciences, North Carolina State University, Raleigh, NC 27695

[2]Virginia institute of Marine Science, School of Marine Science, College of William and Mary, Gloucester Point, VA 23062

1. INTRODUCTION

Located along the western side of the Upper Chesapeake Bay about 250 km from the Bay mouth, Baltimore Harbor was a major industrial harbor in the East Coast of the United States for nearly two and a half centuries (Figure 1). The heavy industries built around the harbor discharge toxic materials into the harbor and have been the sources of different types of toxic pollution. Based on 1985-1987 Chesapeake.Bay monitoring data and a 1987 report by the Maryland Department of the Environment (formerly, the Office of Environmental Program), it was shown that high concentrations of contaminants such as PAH, chromium, copper, lead and zinc, existed in the Harbor and its adjacent areas in the Bay. The averaged bottom sediment concentrations of Zn, Chromium, Copper, and Lead measured in parts per million (ppm), as shown in Figure 2, were measured in eight different locations around the Upper Chesapeake Bay. Given that the highest concentrations for all trace metals occur in the Harbor, and that spatial gradients of the concentrations decrease with distance from the Harbor, it clearly indicate that the sediment concentrations of trace metals are associated with heavily industrialized areas in Baltimore Harbor.

A 1996 sediment mapping survey in the Harbor further confirmed that high concentration of contaminants, including heavy metals such as chromium, lead and zinc, insecticide such as chlordane, and organic compounds classed as PCBs (polychlorinated biphenyls), are the major toxic problems in the Harbor (Baker et al., 1997). Between 1996 – 2000, an extensive chemical sediment data collection dubbed CHARM (Comprehensive Harbor Assessment and Regional Modeling) was continued inside the Harbor (Baker et al., 2003). Purposes of the monitoring program were: (1) to use suitably sensitive and accurate sampling and analytical methods to measure the concentrations of the targeted analytes in industrial and municipal outfalls, (2) to monitor target analytes in the major storm water outfalls, and (3) to combine these chemical measurements with outfall flow record to estimate chemical loadings from each of the sampled outfalls into the Harbor.

Once toxic substances are discharged from the outfalls, they follow through transport pathway to have phase change, chemical reaction and exhibit its temporal and spatial distribution including those in the sediment bed. Thus, transport pathway is an important process, which cannot be neglected for addressing the fate of the toxic material. The Harbor's length is about 20 km and the width 2 to 5 km with bathymetry characterized by wide shallow shore areas and deep channels. The Harbor receives very limited fresh water from Gwynns Fall, Jones Fall, and Patapsco River. The water circulation inside the Harbor is generally regulated by tidal and local wind forces and in a less degree by river discharge (Garland, 1952). The hydrodynamics is usually greatly impacted by that of the main Chesapeake Bay. The Harbor often experiences a unique three-layer residual circulation pattern, which is different from most the other tributaries of Chesapeake Bay (Boicourt and Olson, 1982). Many tributaries of the Chesapeake Bay experiences two-layer circulation, in which buoyant freshwater runs seaward on the surface, and dense saline water is pushed land-ward at the bottom of the water column. In Baltimore Harbor, the

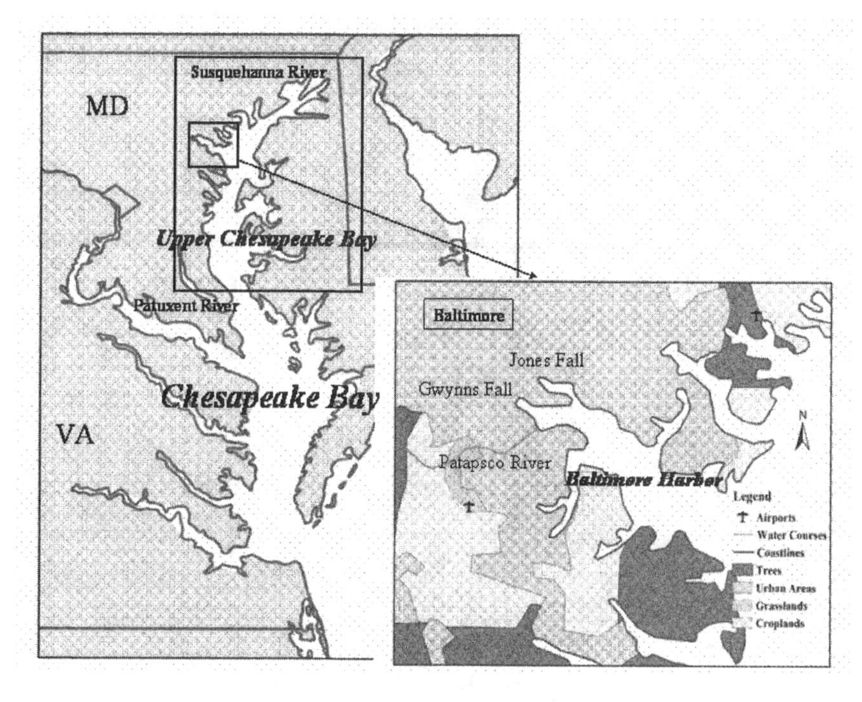

Figure 1. Study area.

freshwater input from the upland is very limited, hence the freshet from Susquehanna River (the biggest fresh water input in the Chesapeake Bay, Figure 1) often brings fresher water into the Harbor from the mouth of the Harbor. Thus, fresh and dense saline water flows into the Harbor at the surface and at the bottom respectively, while the intermediate density water flows out of the Harbor in the mid-depth of the water column. Due to the unique transport pattern, it is intriguing to simulate the three-layer circulation and its effect on the toxic substance pathway.

The present toxic modeling study of Baltimore Harbor focuses on three heavy metals components, namely, chromium (Cr), lead (Pb), and zinc (Zn). The water column concentrations of the three heavy metals described above are normally below their state-regulated limits due to the aggressive reductions of industrial release chemicals in the Harbor for the last 15 years. However, the contaminant concentration in the sediment bed is often exceeding their Effect Range Medium (ERM) values, especially near the Inner Harbor and Bear Creek regions. The highest concentrations for certain locations were observed in the deep sediment from 0.5 to 1 m depth (Sinex and Helz, 1982). This indicates that the toxic contaminants are accumulated in the sediment during the historic development of the City of Baltimore.

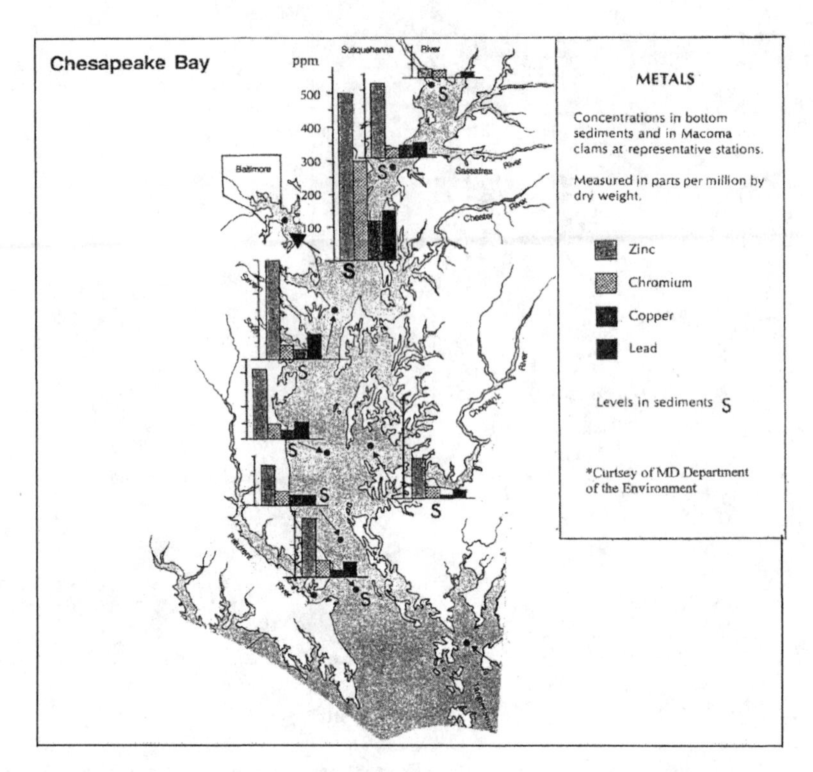

Figure 2. The bottom sediment concentration for Zn, Cu, and Pb measured in various locations in the Upper Bay during 1985-1987

The particular aims of this paper are to address questions as to: (1) whether the sediment contamination in the Harbor is purely a historic legacy problem or is still augmented by the current industrial loadings, and (2) if it is a legacy problem, what is the time frame for the natural recovery of the system below ERM. Section 2 describes the approach and section 3 illustrates the simulation results for 1996 (wet) and 2000 (dry) years, as well as the recovery rates from a 10 years run. Section 4 discusses the uncertainties and limitation of the model, and section 5 concludes the paper.

2. APPROACH

An integrated modeling system, which includes a coupled hydrodynamic, sediment transport model (CH3D-SED), and a toxic model (ICM/TOXI) was used in this study (Figure 3). Since most of the contaminants have a tendency to adsorb to the suspended sediments, a good representation of suspended sediments dynamics thus is important for simulating the fates of contaminants in the system.

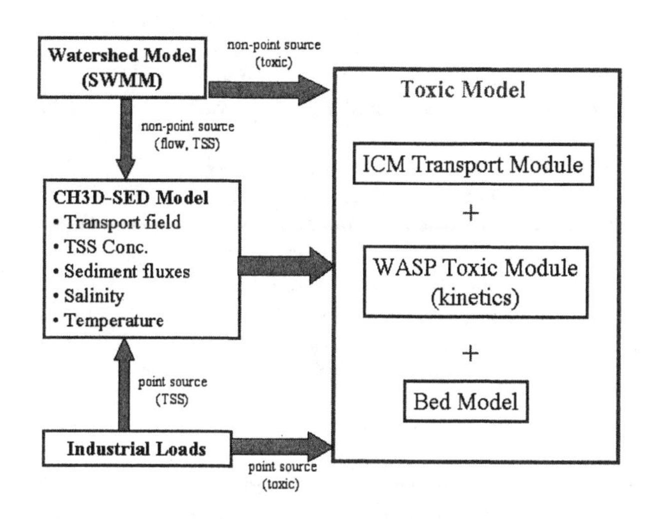

Figure 3. Flow chart of the integrated model system.

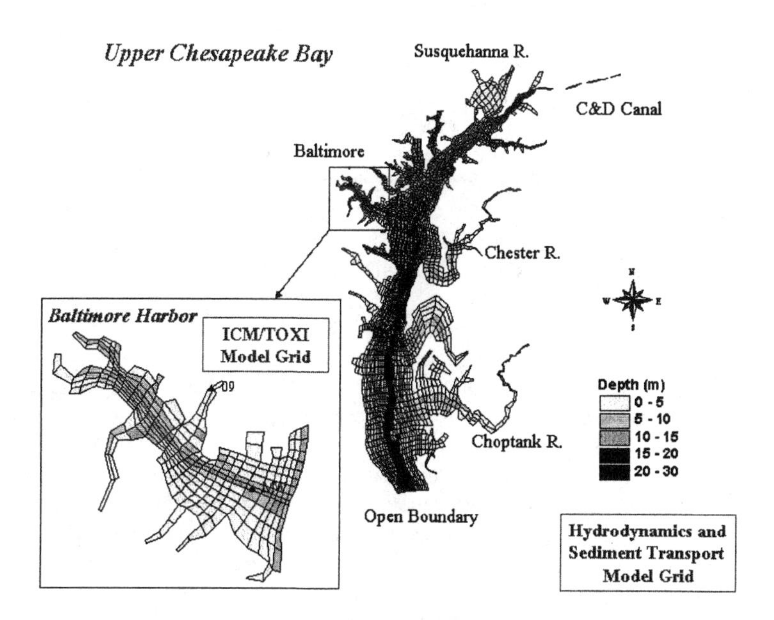

Figure 4. Model grid.

The major external sources of contaminants in Baltimore Harbor are from the bordering industries and the watershed while the major external sources of TSS are from the Chesapeake Bay. As a result, CH3D-SED was applied to the entire upper Chesapeake Bay to simulate the flow field and the suspended sediment transport, as well as the sediment fluxes between the sediment-water interfaces. The extended model domain of the upper Chesapeake Bay provides a significant advantage for investigating interaction between the Harbor and the Upper Bay. With the information of flow field, suspended sediment concentrations, as well as sediment deposition and erosion rates provided from CH3D-SED, the ICM/TOXI was applied to Baltimore Harbor to simulate the temporal and spatial variations of sediment contaminant concentrations. The limited model domain of Baltimore Harbor for the ICM/TOXI allows more efficient model linkages and computation (Figure 4).

Point Sources and Non-point Sources

The releases of contaminants loading (and TSS) from a confined, discrete source such as a pipe, ditch, tunnel, well, outfalls or floating craft in the Harbor are incorporated in the model as point sources. They include industrial, municipal and federal facilities. Based on the data provided by the Maryland Department of the Environment (MDE), there are 26 outfalls with respect to the afore mentioned three metals in the Harbor, of which 8 are major industrial outfalls including industries such as Bethlehem Steel, Millennium Inorganic-Hawkins, Cox Creek facility etc., whose discharges are reported either monthly or quarterly. The point source loadings from each outfall of the industries are distributed to their closest model cells and the loadings are equally divided among the cell's vertical layers. In the cases data are not available throughout several continuous months, the long-term (1998-2002) averaged value is used. In the cases data are not available for one or two months during the year, then the averaged value for that particular year is used to replace the missing value. The highest toxic discharge among the three metals is Zn with 11.6 tons/yr, followed by Cr with 4.5 tons/yr, and Pb with 1.2 tons/yr.

The release of contaminants (and TSS) loadings from the surrounding watershed, which cannot be defined as discrete points, are incorporated in the model as non-point source loadings. Included in the non-point sources are the urban storm water runoffs whose unit area of urban land contributes varying amounts of surface runoff and chemicals depending on the surface properties. The SWMM watershed model was used by MDE to provide the detailed information for flow, TSS and the contaminant material loading. Below the fall line, the loadings from each segment of the watershed model are equally divided into its adjacent surface model cells. Above the fall line, the loadings from the three tributaries of the Baltimore Harbor (i.e., Patapsco River, Gwynns Fall, and Jones Fall) are equally divided among vertical layers of their adjacent cells. The highest contributions from non-point source is Zn with 37.5 tons/yr, followed by Pb with 6.5 tons/yr, and Cr with 1.3 tons/yr. In terms of the distribution between point versus non-point source for each trace metals,

Figure 5 displays a relative distribution, showing that Pb and Zn are dominated by non-point sources while Cr is dominated by point sources.

Hydrodynamic and Sediment Model

The Three-dimensional Curvilinear Hydrodynamic model (CH3D) was used for this study. CH3D was originally developed by Sheng (1983) and Johnson et al. (1993). It has been successfully applied in many tidal rivers and coastal bays (Cerco et al., 1999; Hood et al., 1999; Wang and Johnson, 2002). CH3D simulates a three-dimensional free surface flow with a curvilinear boundary-fitted grid. Vertically, it uses z-coordinates with maximum 19 layers. The surface layer depth is initially set to 5 feet to overcome tidal variations and each of the rest sub-surface layers has a depth of 3 feet. In order to resolve the deep channel in Baltimore Harbor, the minimum horizontal grid size is about 300 m. A 3-minute time-step is used for this study.

The sediment sub-model (CH3D-SED) is directly coupled with CH3D using the same grid and time step. Three classes of suspended sediments are simulated in the sediment sub-model, emulating slowly settling background particles, silt-like intermediate particles, and more rapidly settling fine sand particles. The diameters of the three sediment classes are 3.5, 15, and 65 μm, respectively. The mass conservation equation is solved for each of the sediment classes separately with different settling velocities. The total suspended sediment concentration is the summation of those of the three classes.

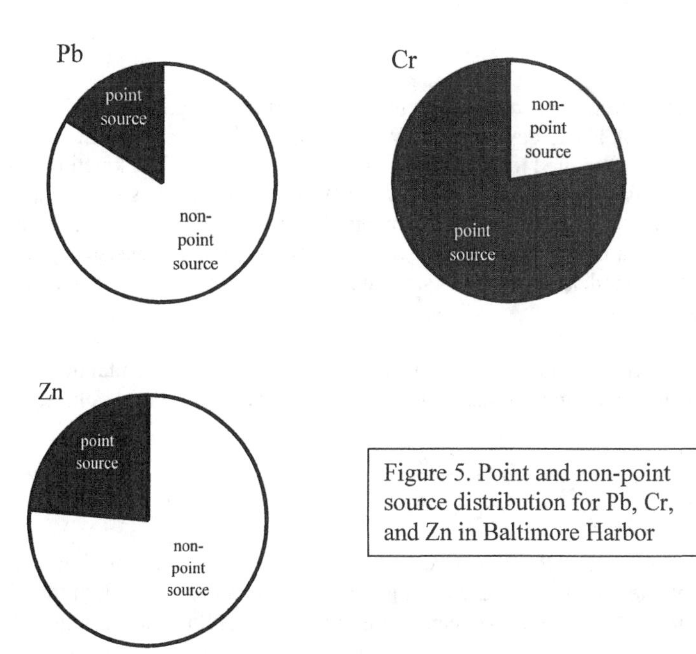

Figure 5. Point and non-point source distribution for Pb, Cr, and Zn in Baltimore Harbor

A good representation of the sediment fluxes between the water and sediment interfaces and hence the fate of the suspended sediments is essential in predicting the fate of contaminants inside the Harbor due to the close association between the contaminants and the sediments. The simulations of sediment deposition and erosion rates are examined carefully. At the interface between water and the sediment bed, the erosional and depositional sediment fluxes are calculated using two approaches. The first approach is based on Spasojevic and Holly (1994), which is applied to the entire model domain of the upper Chesapeake Bay, except for the areas inside the Baltimore Harbor. In-situ measurements of sediment erosional characteristics have been carried out in Baltimore Harbor (Sanford and Maa, 2001). Observations show that a fluffy layer of very soft sediment exists on the top of the sediment bed. Erosion occurs when bottom flow is relatively very weak, which suggests that the critical shear stress for a freshly deposited sediment bed is very low, much lower than the value used outside of Baltimore Harbor. In order to better simulate the suspended sediments distributions in Baltimore Harbor, a second approach is used to model the erosional and depositional fluxes in Baltimore Harbor. The empirical formulae proposed by Sanford and Maa (2001) is incorporated into the model, which permits the incorporation of the deposition and erosion history at the bed into the erosion process with an implicit bed model (Lin et al., 2003). A detailed description of the hydrodynamic and sediment model set up can be found in Wang et al. (2003). A wave prediction method suggested by the Shore Protection Manual (1984) is used in the model. The wave-induced bed shear stress is calculated using a quadratic drag formation (Nakagawa et al., 2000) and linear wave theory. The total bed shear stress is obtained by summing the wave-induced and the current-induced bed shear stresses.

Toxic Model

A three-dimensional contaminant model for surface water, CE-QUAL-ICM/TOXI or, ICM/TOXI, developed by Wang et al. (1996), is used to simulate the distributions of the impaired toxic materials in Baltimore Harbor. The kinetic portion of the model is based on the toxic sub-model of Water Quality Analysis Simulation Program (ToxiWasp) by the U.S. Environmental Protection Agency. The numerical scheme for physical transport is the same as CH3D-SED to ensure consistency. A sediment bed sub-model was added to simulate the fate and transport of sediment contaminants.

ICM/TOXI, as defined in its name, is an integrated compartmental model, which solves the three-dimensional mass balance equation for a control volume.

$$\frac{\delta(V_j C_j)}{\delta t} = \sum_{k=1}^{n} Q_{jk} C_k + \sum_{k=1}^{n} A_k D_k \frac{\delta C}{\delta x_k} + \sum S_j \tag{1}$$

where V_j is the volume of the jth control volume (m^3), C_j is the concentration in the jth control volume (gm/m^3), Q_{jk} is the volumetric flow across flow face k of the jth control volume (m^3/s), C_k is the concentration in flow across flow face k (g/m^3), A_k is the area of flow face k (m^2), D_k is the diffusion coefficient at flow face k (m^2/s), S_j is the external loads and kinetic sources and sinks in the jth control volume (gm/s), and n is the number of flow faces surrounding the jth control volume.

The transport fields including the volume, the volumetric flow across flow faces, the area of flow faces and the diffusion coefficients are imported from a binary file produced by CH3D-SED.

A variety of kinetic processes are simulated in ICM/TOXI, which are integrated in the model as the term $\sum S_j$ in Eq. (1) These include physical kinetic processes such as hydrophobic sorption, volatilization, and sedimentation; chemical kinetic processes such as ionization, hydrolysis, photolysis, and oxidation; and biological processes such as biodegradation. ICM/TOXI simulates most of the processes explicitly. Sorption-desorption is a very important process for simulating heavy metals. Equilibrium sorption-desorption is assumed in the model, and the fractions of dissolved and particulate phases of contaminants are calculated using partition coefficients.

ICM/TOXI simulates temperature, salinity, three classes of suspended solids, and three types of chemicals. Salinity is treated as a conservative material, and the simulated results are compared with those from CH3D-SED in calibration processes to ensure numerical mass conservation of the model. Temperature is simulated using the equilibrium temperature approach to describe surface heat exchange. Three size classes of solids as well as a sediment bed model are included to allow predictions of the transport of sorbed contaminants and the impact of sorption on other kinetic processes. The sediment transport is simulated in CH3D-SED and the results (including concentration, settling velocity and sediment fluxes between the sediment-water interfaces) are imported into ICM/TOXI as a binary file.

There are 3 sediment layers in the benthic sub-model. The surface sediment layer is 3 cm in depth, and both the 2 sub-surface layers are 6 cm thick. The sediment bed processes simulated in the benthic sub-model include sediment accretion due to deposition from the water column, resuspension, burial, pore water diffusion, and consolidation with pore water extrusion, partitioning among particulate and dissolved concentrations, and biochemical kinetic processes. Contaminant mass in the water column and bed are dynamically linked through deposition, resuspension, and dissolved phase diffusion and extrusion across the sediment-water interfaces. The sediment deposition and resuspension fluxes are imported from CH3D-SED, and the corresponding particulate contaminant fluxes between the water-sediment interface are calculated using partition coefficients. Processes such as sorting and armoring are implicitly considered in CH3D-SED. It is assumed that all benthic transport and mixing processes are in the vertical direction only. Therefore, horizontal advection and diffusion of solids and water are not represented in the benthic sub-model.

3. MODEL RESULTS

The hydrodynamic and sediment transport model was set up to run for the years of 1996 and 2000 to cover both very wet (1996) and very dry (2000) hydrological years. Since the dominant mechanisms of hydrodynamics and sediment transport in Baltimore Harbor may be very different in different hydraulic years, a good representation of water flow and sediment dynamics in both 1996 and 2000 indicates a robust hydrodynamic and sediment model.

The 3-layer residual flow in Baltimore Harbor is a well-known phenomenon to physical oceanographers and the model successfully reproduced this flow pattern (Figure 6). During spring seasons of both 1996 and 2000, the monthly averaged simulated flow velocities in Baltimore Harbor showed a distinct 3-layer circulation pattern. At the surface layer of the water column, the freshet from the Susquehanna River creates a barotropic force, which accelerates water from the Bay into the Harbor. At the middle layer, due to the shallowness at the landside of the Harbor, vertical mixing makes the water saltier and hence denser than that at the bayside; thus, a baroclinic force drives the net flow outside the Harbor into the main Bay. At the bottom layer of the deep channel, returning flow enters the Harbor from the Bay.

Temporal variations of simulated TSS concentrations are compared with long-term monitoring data in stations inside Baltimore Harbor as well as stations in the main Bay. The long-term monitoring data were collected on a bi-weekly basis and maintained by the EPA Chesapeake Bay Program. Both the field data and model results show that the TSS concentrations are generally around 10 to 20 mg/l with higher concentrations usually existing during spring. In addition, spatial distributions of simulated TSS concentrations were examined. A turbidity maximum usually occurs in the main channel of the Upper Chesapeake Bay, about 20 to 40 km downriver from the mouth of the Susquehanna River. Bottom TSS concentrations are usually above 50 mg/l in the region of the turbidity maximum.

For the purpose of further verifying the model within a short-term time scale, a one-month period in the Spring of 2000 were simulated. The model results were compared with both the mooring velocity and TSS data inside Baltimore Harbor. The monitoring station is located on the north side of the channel approximately 1.5 km from the Harbor mouth with about a 16-m depth. The current velocities were obtained from an S4 current meter about 1.5 m above the bottom. TSS concentrations were calibrated from two transmission meters deployed about 1 m above the bottom and 2.5 m below the surface (Baker et al., 2003).

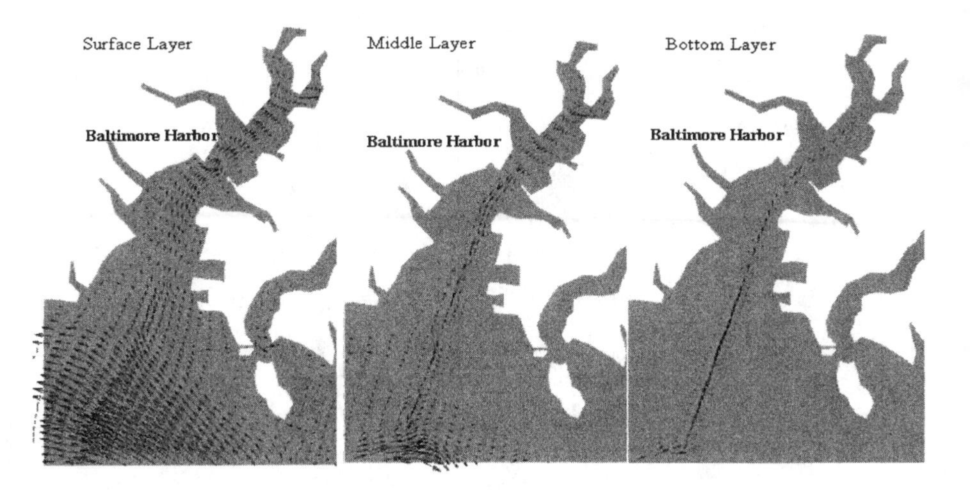

Figure 6. Model results of monthly averaged flow fields in Baltimore Harbor in April, 1996.

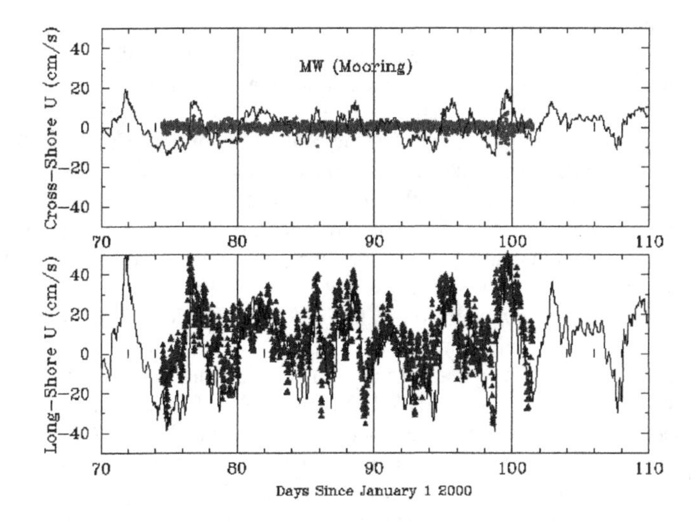

Figure 7. Model (lines) and data (symbols) comparisons of across- and along-channel flow velocities at a mooring station inside Baltimore Harbor. Data were obtained from Sanford (2003).

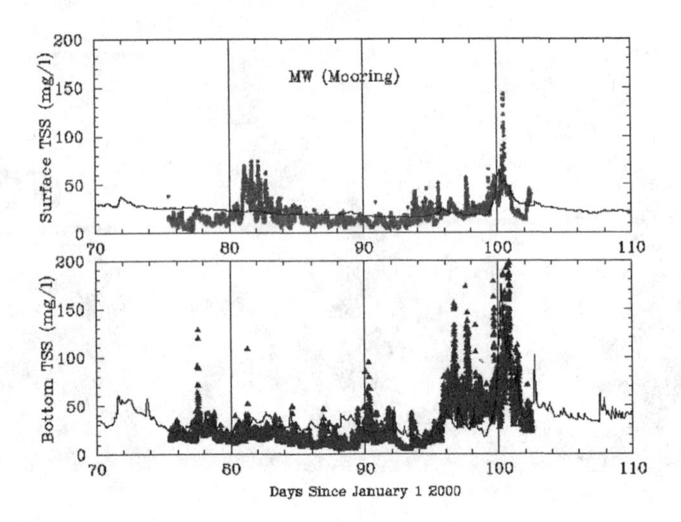

Figure 8. Model (lines) and data (symbols) comparisons of surface and bottom total suspended sediment (TSS) concentrations at a mooring station inside Baltimore Harbor. Data were obtained from Sanford (2003).

Figure 7 shows the comparison of bottom velocity in both cross-shore and long-shore directions. The simulated cross-shore bottom flow velocities tend to over-predict the observed values. The simulated long-shore flow velocities, however, agree very well with the observations. The positive values of long-shore velocities indicate flows from the Bay into the Harbor. The big peak of bottom flow intrusion from the Bay around Julian day 100 was well represented by the model. Concurrent with the high bottom velocities, it comes with strong peaks of TSS concentration. Figure 8 shows that the model captures the Julian day 100 high TSS concentration event with the right timing and the magnitude. A close analysis comparing the means of the observations and the model results show that the model seems to under-predict the bottom flow intrusions and over-predict the TSS concentrations. The means of the modeled and observed long-shore bottom velocity for the period are 0.6, and 9.1 cm/s, respectively. For the cross-shore bottom velocity, they are 1.9 and 0.1 cm/s, respectively. The means of the modeled and observed surface TSS are 27.2, and 19.5 mg/l, respectively. The means of the modeled and observed bottom TSS are 38.5, and 28.3 mg/l, respectively. The discrepancies partially are due to the fact that the data are collected at one point, but the model results represent the spatially averaged value over a certain model cell. The modeled bottom flow velocity and TSS concentration represent the averaged values in a 1.5 m high box on the bed (bottom layer of the water column). The observed bottom flow velocity indicates velocity at about 1.5 m above the bed and the observed bottom TSS concentration show sediment concentrations about 1 m above the bed. Away from the bed, flow velocities increase sharply and sediment concentrations decrease sharply. It is expected that the measured bottom flow velocity could be higher than the model simulated bottom layer flow velocity, while the measured bottom TSS concentration could be lower than the simulated values in the bottom layer of the model grid.

A critical factor in simulating the fate of the suspended sediment and the associated contaminants is the sedimentation rate, since it ultimately determines the retention of the contaminants inside the system. The simulated sedimentation rates were compared with those suggested by Pb210 data in different locations inside Baltimore Harbor (Figure 9). During the wet year of 1996, the simulated sedimentation rates are usually a little higher than those of Pb210 data, and the simulated sedimentation rates during the dry year of 2000 are usually a little lower than those suggested by Pb210 data. The difference may be caused by the much larger suspended sediments loads which were introduced from the Susquehanna River during a wet year. Part of the TSS are transported into Baltimore Harbor and deposited there. In contrast, much less TSS are introduced from the Susquehanna River and transported into the Harbor during a dry year. The sedimentation rate suggested by Pb210 data are the long-term results, which cover various hydraulic conditions and hence are averaged between wet and dry years.

The results of both hydrodynamic and sediment transport for the year 2000 were output by CH3D-SED and imported into the toxic model. The toxic model was set up to run for 10 years with a repeated flow and TSS settings of year 2000. The advantage of choosing year 2000 over 1996 is two-fold. First, the water column concentrations of the target contaminants were measured during 2000. In addition, the point source loadings of contaminants were available for 2000, but not for 1996. Since there is only a single time of the sediment mapping data for contaminants in 1996, and these were used as the initial condition of the sediment bed model, direct calibration of the temporal variations of sediment contaminant concentrations is not available. The calibration was conducted towards the major factors that affect the fate of the contaminants inside the system, which include the sedimentation rate (calibrated in the sediment model) and flushing rate. The flushing rate is a combined result of hydrodynamics (and hence calibrated in the hydrodynamic model) and the spatial distribution of water column contaminant concentrations. The temporal variations of the simulated water column contaminant concentrations were compared with the field data in different stations inside Baltimore Harbor. The field data all fall in the range of the model predicted values (Figures 10-12). The water column concentrations for dissolved Cr are usually within 2 μg/l with some higher values observed in Bear Creek and the Inner Harbor. The water column concentrations of dissolved Pb are generally within 4 μg/l, and the concentrations of dissolved Zn are generally within 10 μg/l, with higher values observed in some places at both the Outer Harbor and Inner Harbor.

The 10-year simulations of the toxic model suggest that, given the loadings and hydrodynamic settings of 2000, the contaminant concentrations at the sediment bed are generally decreasing for all three metals: Cr, Pb and Zn (Figure 13). Sediment concentrations of Cr and Zn are highest in Bear Creek throughout the 10-year simulation. At the end of the simulation, the spatially-averaged Cr and Zn concentrations in Bear Creek remain much higher than the state criteria (mean ERM quotient below 0.5). For other places, the contaminant concentrations are either close to or below the state criteria at the end of the 10-year simulation. In general, the model results indicate that the current high concentrations of Cr, Pb and Zn at the sediment bed are mainly a historical legacy problem, and the natural recovery rate is on a time scale of decades.

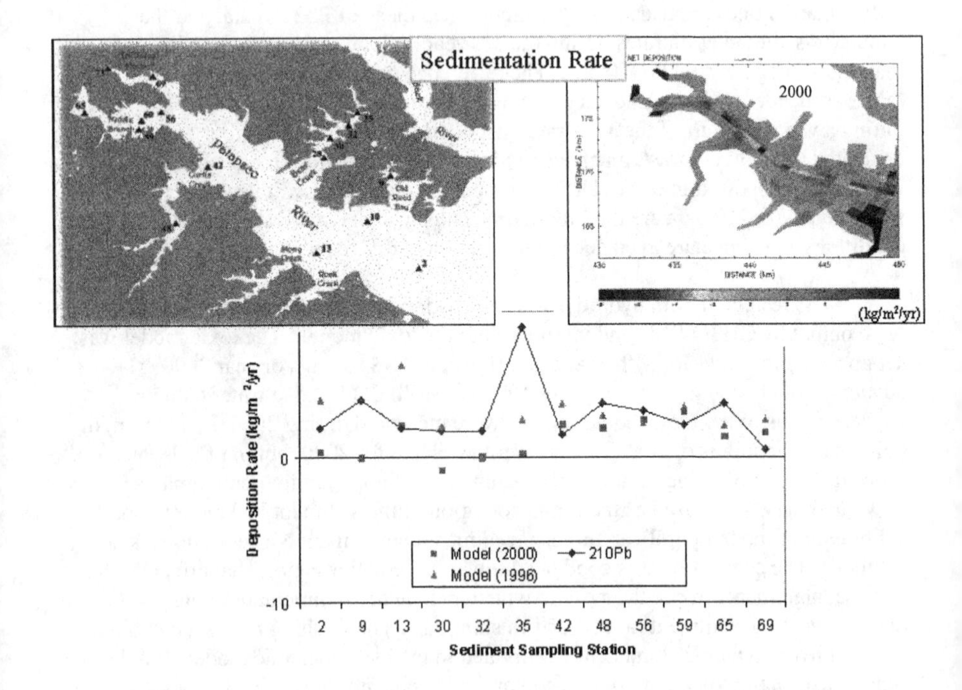

Figure 9. Comparison of sedimentation rates between model results in years 1996 and 2000, and 210Pb data. The upper left figure shows the station locations and the upper right figure shows the model results of spatial distributions of sedimentation rates in 2000. (The 210Pb data are from Cornwell unpublished data).

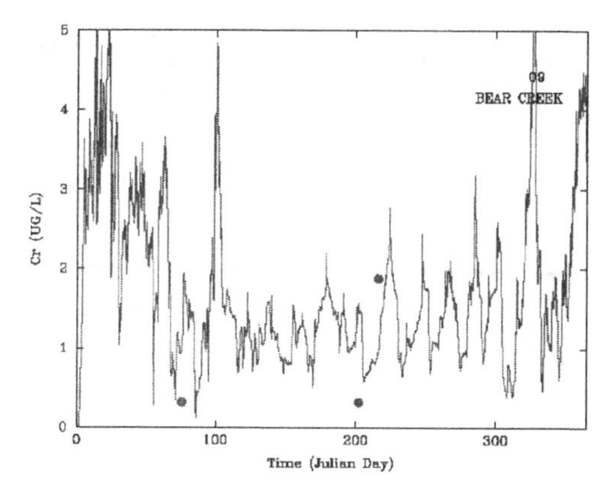

Figure 10. Comparisons between model results and field data of dissolved Cr concentrations (model: solid line; data: solid circles) at about 1 m below the surface of station 09 inside Bear Creek.

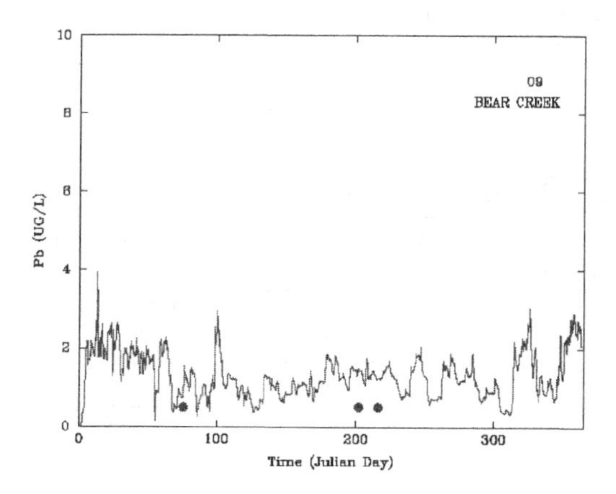

Figure 11. Comparisons between model results and field data of dissolved Pb concentrations (model: solid line; data: solid circles) at about 1 m below the surface of station 09 inside Bear Creek.

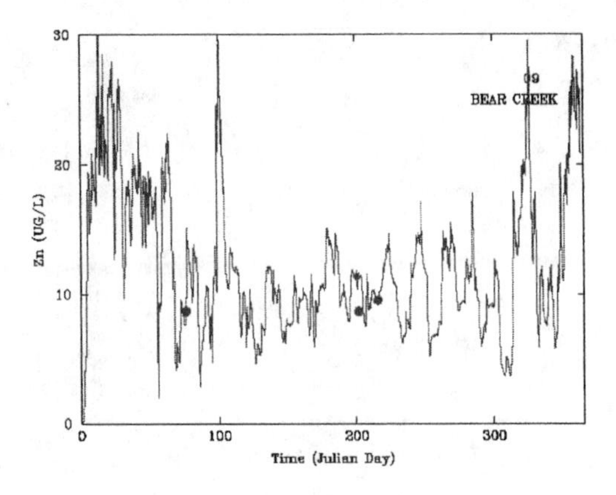

Figure 12. Comparisons between model results and field data of dissolved Zn concentrations (model: solid line; data: solid circles) at about 1 m below the surface of station 09 inside Bear Creek.

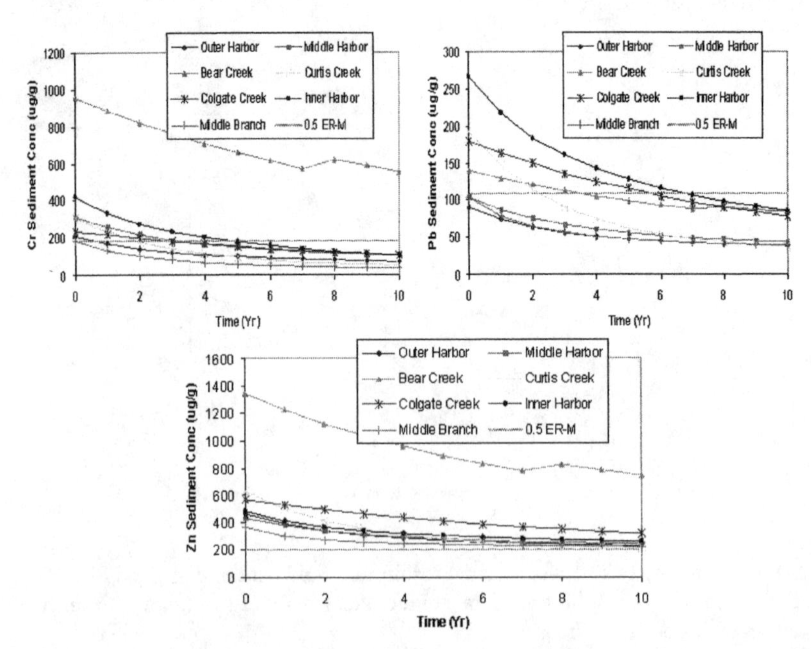

Figure 13. Model results of temporal variations of spatially-averaged Cr (upper left), Pb (upper right), and Zn (bottom) concentrations at the sediment bed in different sub-regions of Baltimore Harbor.

4. DISCUSSION

The uncertainties caused by different choices of parameters in the model are examined using sensitivity model tests. When water flows into the Harbor, constant values are used for contaminant concentrations at the open boundary. Although the values are chosen based on field observation, due to the scarcity of the data, temporal variations of the contaminant concentrations are not available. Model sensitivity tests were conducted to use 0 and 3.5 μg/l as the Zn contaminant concentrations at the open boundary. The results show that by assuming a completely clean ambient environment from the Bay (i.e. 0 μg/l), the sediment contaminant concentrations at the Outer Harbor could drop to below the state-regulated limits within 7 years. In contrast, by using the value 3.5 μg/l, as the field data suggested, the sediment contaminant concentrations at the Outer Harbor remain above the limits until the end of the model run (i.e. 10 years).

The model assumes equilibrium sorption-desorption and uses partition coefficients to calculate different phases of the contaminants. However, the error caused by the assumption is sometimes not negligible. Both a ten-fold increase and reduction of the original values (obtained through regression of field data) were chosen to assess the model responses to different partition coefficients. The rate of recovery (decreased sediment contaminant concentrations) is delayed when a higher partition coefficient is used in the model since, instead of being flushed out to the Chesapeake Bay as dissolved phases, more contaminants are adsorbed onto sediments and are retained inside Baltimore Harbor. By using a ten-fold increase of the original partition coefficient for Pb, the difference of contaminant concentrations at the end of the 10-year simulation are about 10 to 15% of the initial concentrations.

The benthic sub-model within the toxic model has 3 sediment layers, representing a total depth of 20 cm for the sediment bed. It is assumed that the active sediment bed is within the 20 cm depth. It is further assumed that the contaminants are well-mixed within individual sediment layers, and that mixing between different layers only occurs through molecular diffusion or sediment layer re-organization. The assumptions and also the representations of the depth-profile contaminant concentrations are to be examined when more field data are available.

The channel inside Baltimore Harbor is regularly dredged. The exposure of contaminants previously resident in deep sediment due to dredging is not considered in the model. By introducing the dredging effect and including deeper sediment beds with higher contaminant concentrations, it is possible to reduce the recovery rate suggested by the model results. Future work is needed to include the factors discussed here.

5. CONCLUSIONS

An integrated modeling system, which consists of a coupled hydrodynamic sediment transport model and a toxic model, was used to simulate the variations of Cr, Pb and Zn concentrations in the water column and in the sediment bed. The calibration includes the sedimentation rate at the sediment-water interfaces, and the comparison of temporal and spatial distributions between the model results and the measured data. Both long-term and short-term comparisons were conducted. Given the complicate interaction between TSS and contaminant concentration in the water column and in the sediment bed, which are functions of mixing rate, flow field, and sediment dynamics within the hydrodynamic and sediment models, the toxic model produces reasonable results.

Due to the fact that the major sediment source comes from outside the Harbor, especially that from the Susquehanna, application of the hydrodynamics and sediment transport models throughout an extended model domain (including the entire Upper Chesapeake Bay) is essential. It allows the sediment fluxes interchanging between the Harbor and the Bay to be properly simulated. The toxic model was then applied to the limited domain in the Baltimore Harbor using a radiation boundary condition for the toxic concentration; the flow and TSS were provided by the large domain model results. The calculations of ICM/TOXI on limited domain were quite reasonable as compared to the available observation data, with the advantage of fast computation and efficient model execution for long-term simulation.

The toxic model results indicate that, with current point source and nonpoint source loadings, the surface sediment contaminant concentration is decreasing gradually. It thus suggests that the sediment contaminant problem today in Baltimore Harbor is predominantly a legacy problem. The diagnostic run by the toxic model further indicates that the recovery rate is sensitive to the sediment dynamics; namely, the deposition will enhance the recovery rate through the burial process while erosion tends to expose the toxic from the deep sediment and thus slow the recovery rate. Since the sediment source in the Harbor derives mainly from the Upper Bay, it thus depends on the sediment interactions between the Harbor and the Upper Bay main-stem.

ACKNOWLEDGEMENTS

The authors are grateful to the support from the Maryland Department of the Environment: Dr. Miaoli Chang, Mr. Scott Macomber, and Mr. Leonard Schugam provided the watershed model results used by this study. The authors are thankful to Drs. Lawrence P. Sanford and Joel Baker for their helpful advice and valuable field observation data. Finally, the authors greatly appreciate Dr. Mark S. Dortch of Army Engineering Research and Development Center for his kindness and technical assistance in providing CE-QUAL-ICM/TOXI.

REFERENCES

Baker, J, Mason, R., Cornwell, J., Ashley, J., Halka, J., and Hill, J. (1997) "Spatial Mapping of sedimentary Contaminants in the Baltimore Harbor/Patapsco River/Back River System." Report submitted to Maryland Department of the Environment. University of Maryland, Center for Environmental Science, Chesapeake Bay Biological Laboratory, Reference no. UMCES [CBL] 97-142.

Baker, J, Chang, M. L., and Sanford, L. P. (2003) "Comprehensive Harbor Assessment and Regional Modeling study (CHARM)." University of Maryland, Center for Environmental Science, Chesapeake Bay Biological Laboratory, Technical Report Series no. TS-378-03-CBL.

Boicourt, C.W. and Olson, P.L. (1982) "A hydrodynamic study of the Baltimore harbor system I. Observations on the circulation and mixing in Baltimore Harbor." Baltimore, Maryland: Chesapeake Bay Institute, The Johns Hopkins University, *Reference 82-10, Bulletin 1*, 131 pp.

Cerco, C.F., Bunch, B., and Letter, J. (1999). Impact of flood-diversion tunnel on Newark Bay and adjacent waters. *Journal of Hydraulic Engineering*, 125(4), 328-338.

Garland, C.F. (1952). "A study of water quality in Baltimore Harbor." Publication No. 96, Chesapeake Biological Laboratory, Department of Research and Education, Solomon's Island, MD.

Hood, R.R., Wang, H.V., Purcell, J.E., Houde, E.D., Harding, L.W. Jr. (1999) "Modeling particles and pelagic organisms in Chesapeake Bay: Convergent features control plankton distributions." *Journal of Geophysical Research*, 104(C1), 1223-1243.

Johnson, B.H., Kim, K.W., Heath, R.E., Hsieh, B.B. and Butler, H.L. (1993) Validation of three-dimensional hydrodynamic model of Chesapeake Bay. *Journal of Hydraulic Engineering*, 119(1), 2-20.

Lin, J., Wang, H.V., Oh, J.-H., Park, K., Kim, S.-C., Shen, J., and Kuo, A. Y. (2003) "A new approach to model sediment resuspension in tidal estuaries." *Journal of Coastal Research*. 19(1): 76-88.

Nakagawa, Y., Sanford, L.P. and Halka, J.P. (2000) "Effect of wind waves on distribution of muddy bottom sediments in Baltimore Harbor, USA." *Proceedings of the 27th International Conference on Coastal Engineering,* (Sydney, Australia), pp. 3516-3524.

Sanford, L.P. and Maa, J.P.-Y., (2001) "A unified erosion formulation for fine sediments." *Marine Geology*, 179, 9-23.

Sheng, Y.P. (1983) "Mathematical modeling of three-dimensional coastal currents and sediment dispersion: model development and application." Vicksburg, Mississippi: US Army Engineer Waterways Experiment Station, *Technical Report CERC-83-2*, 288pp.

Shore Protection Manual (1984) Coastal Engineering Research Center. Vicksburg, Mississippi: Department of the Army, Waterways Experiment Station, Corps of Engineers.

Sinex S.A. and Helz, G.R. (1982) "Entrapment of zinc and other trace elements in rapidly flushed industrialized harbor." *Environ. Sci. Technol*, 16: 820-825

Spasojevic, M. and Holly, F.M. Jr. (1994) "Three-dimensional numerical simulation of mobile-bed hydrodynamics." New Orleans, Louisiana: U.S. Army Engineering District, *Contract Report HL-94-2*, 158pp.

Wang, H.V. and Johnson, B.H. (2002) "Validation and Application of the Second Generation Three Dimensional Hydrodynamic Model of Chesapeake Bay." *Water Quality and Ecosystem Modeling*, 1: 51-90.

Wang, H.V., Lin, J., and Sisson, G.M. (2003) "Toxic modeling in Baltimore Harbor." TASA, The Maryland Department of Environment, Draft report.

Wang, P. F., Martin, J. L., Wool, T. and Dortch, M. S. (1996) "CE-QUAL-ICM/TOXI, A three-dimensional contaminant model for surface water: Model theory and User guide." Environmental Laboratory, Army Corps of Engineers, Draft report, 170 pp.

Coastal Circulation and Effluent Transport Modeling at Cherry Point, Washington

Tarang Khangaonkar[1] and Zhaoqing Yang[2]

Abstract

A 3-D hydrodynamic circulation and effluent transport model was developed using a combination of the far field model (EFDC) and near-field plume models for the 9 miles of Cherry Point coastline from Point Whitehorne to Sandy Point in the Strait of Georgia, Washington. This study was initiated by the Cherry Point Industries to specifically evaluate the potential for cumulative effects from multiple sources of industrial effluent in the study area. The EFDC model was setup and calibrated using oceanographic data that was collected specifically for the model development. Sensitivity analysis was conducted to evaluate the importance of baroclinic forcing in comparison to barotropic motion. The near-field dilution and far-field effluent plume transport components were calibrated using historical dye study data from the site.

The study showed conclusively that accumulation of effluent does not occur. Effluent concentrations reach a dynamic steady state within a few days from the start of the discharge. The distribution patterns for all constituents consistently indicate that water quality standards will not be exceeded in the study area due to the Cherry Point discharges.

[1] Practice Leader - Water Resources Modeling. Battelle Memorial Institute. 4500 Sand Point Way NE, Seattle, WA 98105

[2] Principal Research Scientist - Water Resources Modeling. Battelle Memorial Institute. 4500 Sandpoint Way NE, Seattle, WA 98105

Introduction

The Cherry Point Industries, consisting of ARCO Products Company (ARCO), TOSCO Refining Company (Tosco) and INTALCO Aluminum Corporation (Intalco) are located near Ferndale, Washington. Treated wastewater, including process water and stormwater from the three facilities is discharged to the Strait of Georgia through four outfalls located near the piers at each facility (i.e., three process water outfalls, one from each facility, and one additional stormwater outfall at Intalco) (see Figure 1).

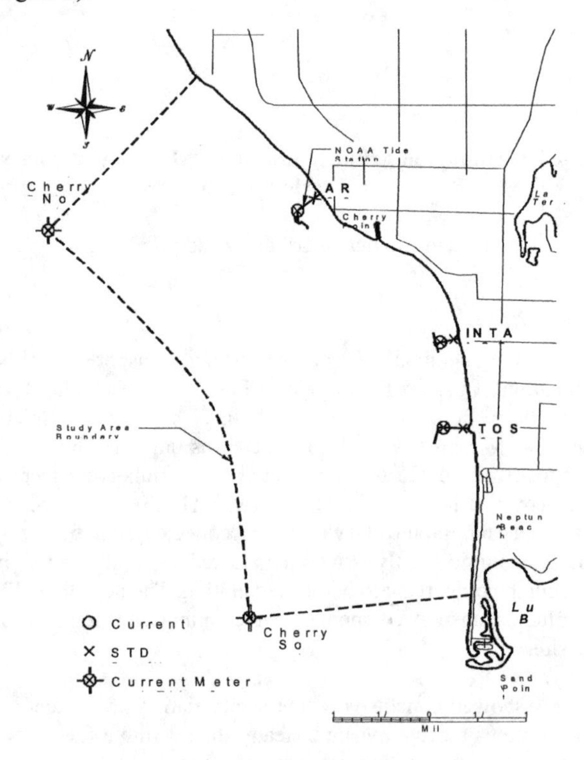

Figure 1: Study Area

The objective of this study was to evaluate if accumulation of effluent could lead to water quality exceedance. Effluent monitoring conducted by the industry and the mixing zone studies conducted at each facility showed clearly that water quality criteria were met within approximately 350 feet from the outfalls (ENSR 1991, 1992, and 1995). However, those studies were conducted as part of National Pollutant

Discharge Elimination System requirements, and evaluated each discharge independently using stand alone steady state models. The focus of this study was to assess cumulative effects of the multiple discharges using a 3-D hydrodynamic and transport model applied over multiple tidal cycles.

Hydrodynamic Model Setup

The circulation in the Strait of Georgia is dominated by tides. Tidal waves propagate into the Strait of Georgia from the Pacific Ocean and are altered by the complicated waterway systems. The dominant astronomical tidal components in the Vancouver, British Columbia and Puget Sound areas are the semi-diurnal tide M_2, and the diurnal tide K_1. The tide in the Strait of Georgia is a mixed, semi-diurnal tide with large inequalities in range and time between the two high tides and low tides each day. Tidal currents in narrow waterways have good correlation with tidal heights. However, in shallow water regions, bottom friction usually causes phase lag in the slack and tidal currents relative to tidal heights. In addition to tides, wind stresses and density stratification also affects the area's circulation. The hydrodynamic model selected for this study is the Environmental Fluid Dynamic Code (EFDC) (Hamrick, 1992). EFDC is well suited to simulate tidal circulation in coastal regions, and was specifically recommended by the Washington State Department of Ecology.

Model Description

The hydrodynamic model EFDC is designed for the simulation of hydrodynamics and transport processes in estuaries and coastal waters (Hamrick, 1992 and Hamrick and Wu, 1996). EFDC simultaneously solves the 3-D equations of motion for velocities and transport equations for temperature, salinity, dye tracer and suspended sediment concentrations in a finite difference framework. A second-moment turbulence closure model is solved to provide the vertical turbulent eddy viscosity in the model (Mellor and Yamada 1982). The model utilizes the boundary-fitted curvilinear-orthogonal coordinate in the horizontal plane. A sigma-stretched coordinate system in the vertical direction is used to best fit the model bottom boundary to the topography. Applications of EFDC in hydrodynamics, water quality, and sediment fate and transport can be seen in Shen and Kuo (1999), Yang et.al (2000); Ji et. al. (2001); and Yang and Hamrick (2002).

Model Grid

A curvilinear-orthogonal model grid was developed for this study and is shown in Figure 2. The model domain covers a region of about 7 miles along the shore and extends 4 miles offshore. The model domain consists of three open boundaries (south, west, and north) and a land boundary (east). The locations of the open boundaries were specified sufficiently far from the Cherry Point Industries' outfalls such that boundary effects on the outfalls and interior region were minimized. There were 120 cells in the longshore direction and 50 cells in the offshore direction. The minimum and maximum horizontal cell sizes were 30 m and 260 m, respectively. Five vertical layers were considered in the model in the sigma-stretched coordinate.

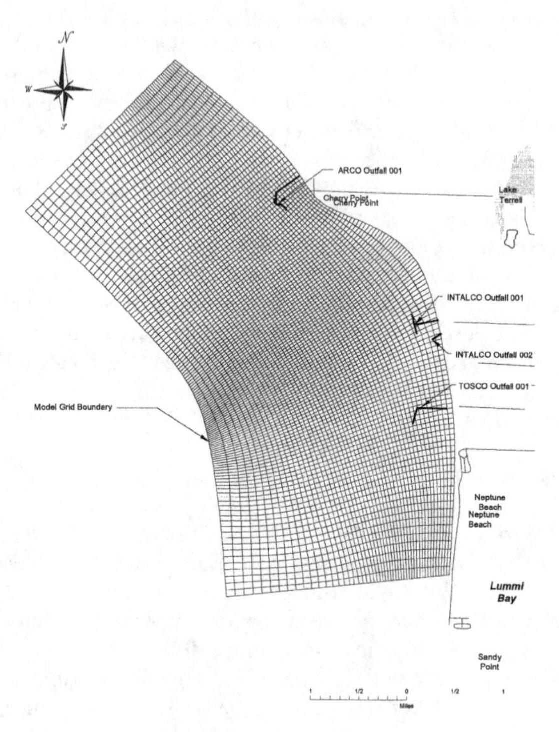

Figure 2: Model Grid

Model Boundary Conditions

Currents in the Strait of Georgia are subject primarily to three forcing mechanisms. The dominant forcing mechanism driving the circulation is astronomical tide. The Fraser River plume could be a significant factor in inducing the density-driven currents in the high river discharge season. The wind forcing in general is important in the surface circulation. Therefore, these mechanisms should be considered in the specification of model boundary conditions.

At the eastern land boundary, a no-slip boundary condition was specified. On the three open boundaries, water surface elevations were specified using observed water surface elevation data. The water surface elevation time series were read into the model at the data sample interval of 10 minutes and then interpolated at the model run time interval. To simulate the baroclinic motion induced by density-stratification, salinity and temperature boundary conditions are required along the open boundaries. In order to specify the salinity and temperature boundary conditions accurately, salinity and temperature time histories are required. However, only spot profile measurements for salinity and temperature were available in this study. Due to the limited amount of observed data, the salinity and temperature profiles (Figure 3) were used to represent the mean salinity and temperature stratification pattern for the period of May 1999. Surface wind stresses were specified at the water free surface. The wind data from Intalco pier for the period of May 1999 were used in the model.

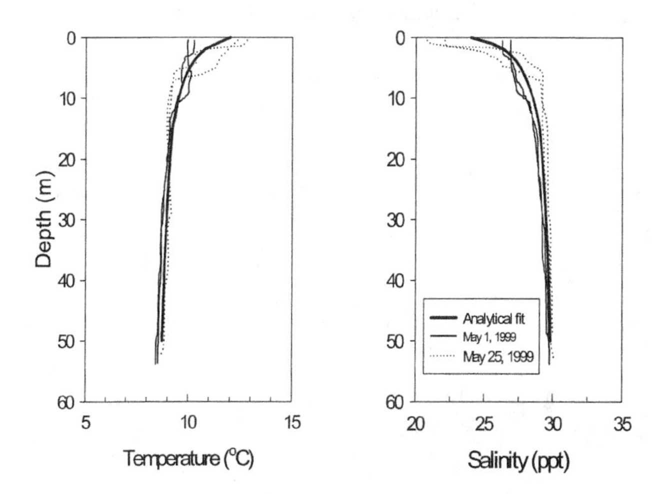

Figure 3: Mean Temperature and Salinity Analytical Fit

Model Calibration – Barotropic Mode

The hydrodynamic model was set up to simulate tidal circulation for the period of May 2, 1999 to May 25, 1999 and calibrated with the observed tide and current data. Due to the limited data for salinity and temperature along the open boundary conditions, the model was first calibrated under barotropic conditions without considering the density-driven circulation due to density stratification. The density effect was considered later as part of a sensitivity test using the representative salinity and temperature profiles for May (Figure 3). The background eddy viscosity and bottom roughness were adjusted by comparing the model water surface elevation to the NOAA tide data at the ARCO dock and model velocities were compared to the measured velocities at all observation stations. The water surface elevation comparison between model predicted results and observed data for a 15-day period at the ARCO station is shown in Figure 4. The model-predicted water surface elevation matches the observed water levels very well. The high and low tide phases from the model are in good agreement with observed data. The spring-neap tidal cycle and the diurnal inequality are well reproduced. Comparisons between model velocity predictions and field observations for a 23-day period in May 1999 are shown in Figures 5 to 7 for Tosco, ARCO and Intalco stations, respectively. In general, predicted velocity magnitude and phase match the observed values quite well. The spring-neap cycle and diurnal inequality are also well reproduced in the predicted velocities. The model also successfully predicted the current directions at all the observed stations.

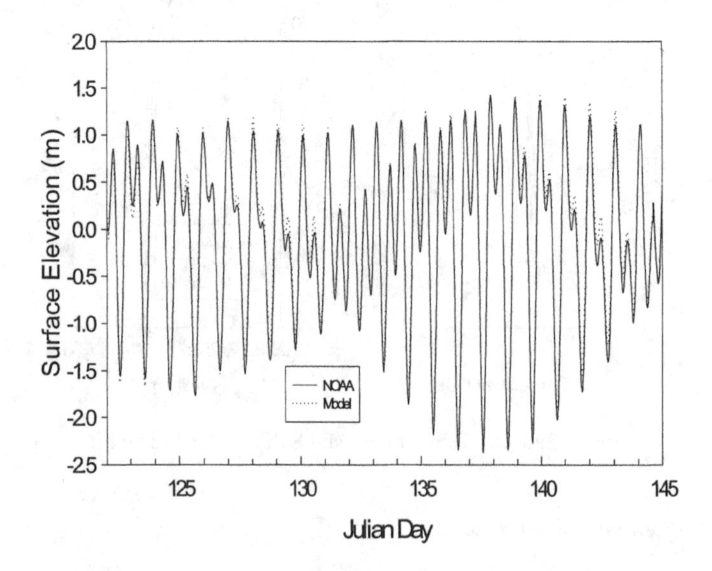

Figure 4: Comparison of Predicted and Observed Water Surface Elevation at ARCO Station

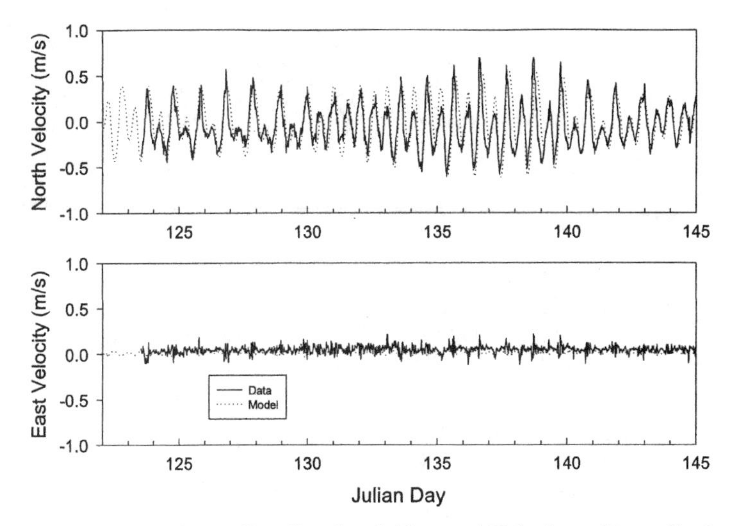

Figure 5: Comparison of Predicted and Observed Velocity at Tosco Station

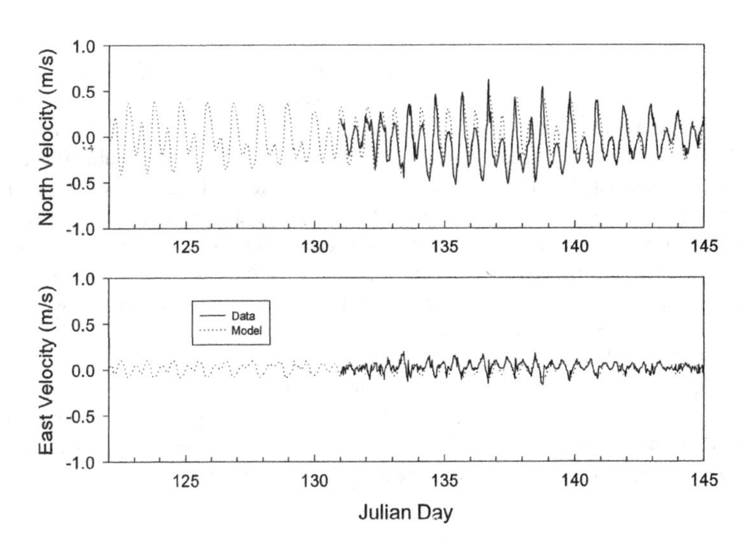

Figure 6: Comparison of Predicted and Observed Velocity at Intalco Station

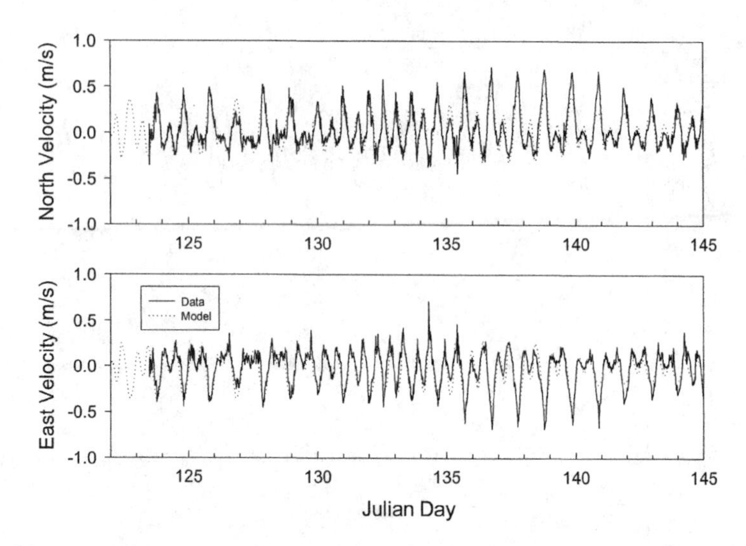

Figure 7: Comparison of Predicted and Observed Velocity at ARCO Station

Sensitivity Analysis – Baroclinic Mode

To investigate the effect of density stratification on circulation in the Cherry Point Industries area, salinity and temperature were simulated and coupled with the hydrodynamic model. As discussed in a previous section, time series information on salinity and temperature variations in the water column at the open boundaries are not available. As a consequence, a constant, analytically derived temperature and salinity profiles were used instead for the model run. This approximation, although not predictive of realistic baroclinic circulation conditions, does allow the model to provide general information about how density variations would affect velocity distributions. A comparison of velocities at the surface, middle, and bottom layers with and without density effects at the ARCO station on May 16 is shown in Figure 8.

The bottom velocities are quite similar for both cases. However, vertical velocity shear and surface velocities are stronger when density effects are considered. This sensitivity analysis indicates that velocity predictions might be improved if time dependent, stratification variations are properly incorporated into the model.

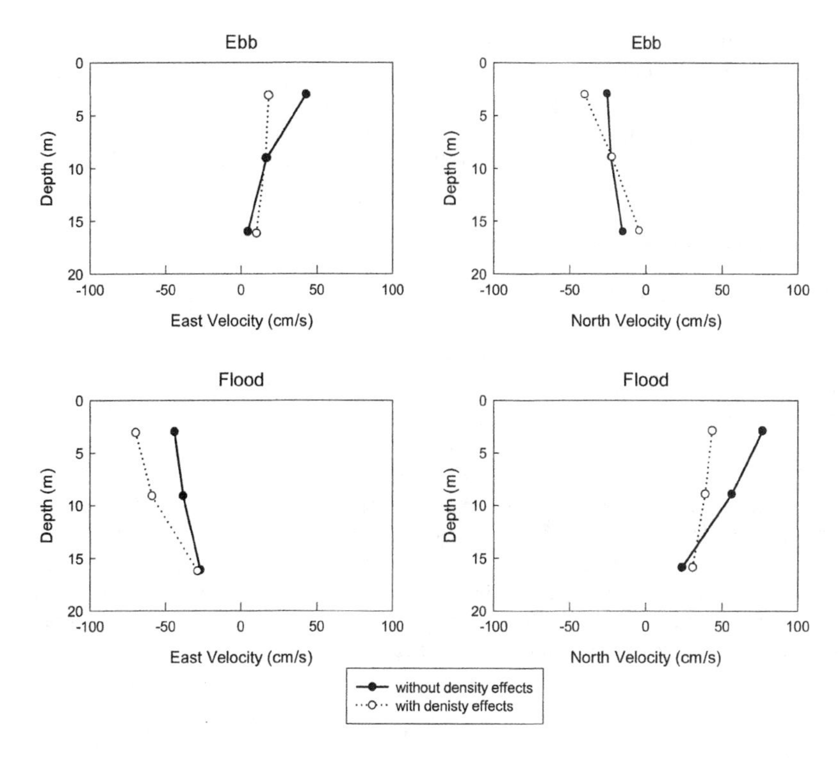

Figure 8: Comparisons of Barotropic and Baroclinic Velocities at ARCO Station

Effluent Plume Dilution and Transport Modeling

Effluent from the Cherry Point Industries enters the study area from four NPDES regulated discharges as shown in Figure 1. These include three process water outfalls (ARCO Outfall 001, Intalco Outfall 001, and Tosco Outfall 001) and one stormwater outfall (Intalco Outfall 002).

The effluent fate and transport calculations consisting of initial dilution, far-field transport, and long-term accumulation were conducted in the following steps.

1. Effluent loads (flows and temperature representing May conditions, and peak-measured concentrations representing the worst case scenarios) were introduced into the study area through the Cherry Point Industries outfalls.

2. Initial dilution was continuously computed for the four outfalls for the period of May 11 through May 26, 1999.
3. Resultant effluent concentrations were used by the far-field model (EFDC) for effluent transport and dispersion throughout the study area. This was conducted simultaneously for effluent from all outfalls. The model allows effluent plumes to merge and calculates cumulative concentrations.

Initial Dilution Modeling

Effluent from ARCO's treatment plant and stormwater is transported through a 20-inch pipeline to a diffuser outfall. The 13-port diffuser is 52 feet long and has a 20-inch diameter. It is located under the pier approximately 3 feet above the bottom (see Figure 9). Previous dilution study was conducted at ARCO was conducted using UDKHDEN model and was therefore retained for this study (Muellenhoff, W.P. et. al 1985).

Initial dilution ratio modeling for Tosco was conducted using the UM model (U.S. EPA 1994). Tosco's outfall consists of a 1.8-foot diameter steel outfall pipe that extends from the shoreline along the pier for approximately 1,000 feet, then bends downward to a submerged diffuser approximately 0.6 feet above the bottom. The plan and profile view of the Tosco outfall is shown in Figure 9 a. Tosco's four-port diffuser is modeled as a single port discharge with the assumption that a quarter of the effluent flow is discharged through each port.

Intalco's process wastewater outfall line 001 is suspended from the pier and extends approximately 1,100 feet from the edge of the shoreline. The 12-inch diameter diffuser is 120 feet long with 12 ports. The stormwater outfall 002 is approximately 800 feet south of the process outfall line and extends approximately 250 feet from the shoreline. The open-ended pipe is a 30-inch diameter fiberglass reinforced plastic pipe. Intalco's outfalls 001 and 002 are also shown in Figure 9. Dilution ratio calculations for Intalco's outfall 001 were conducted using UDHKDEN, and outfall 002 dilutions were simulated using CORMIX predictions, to maintain consistency with previous mixing zone studies at this site.

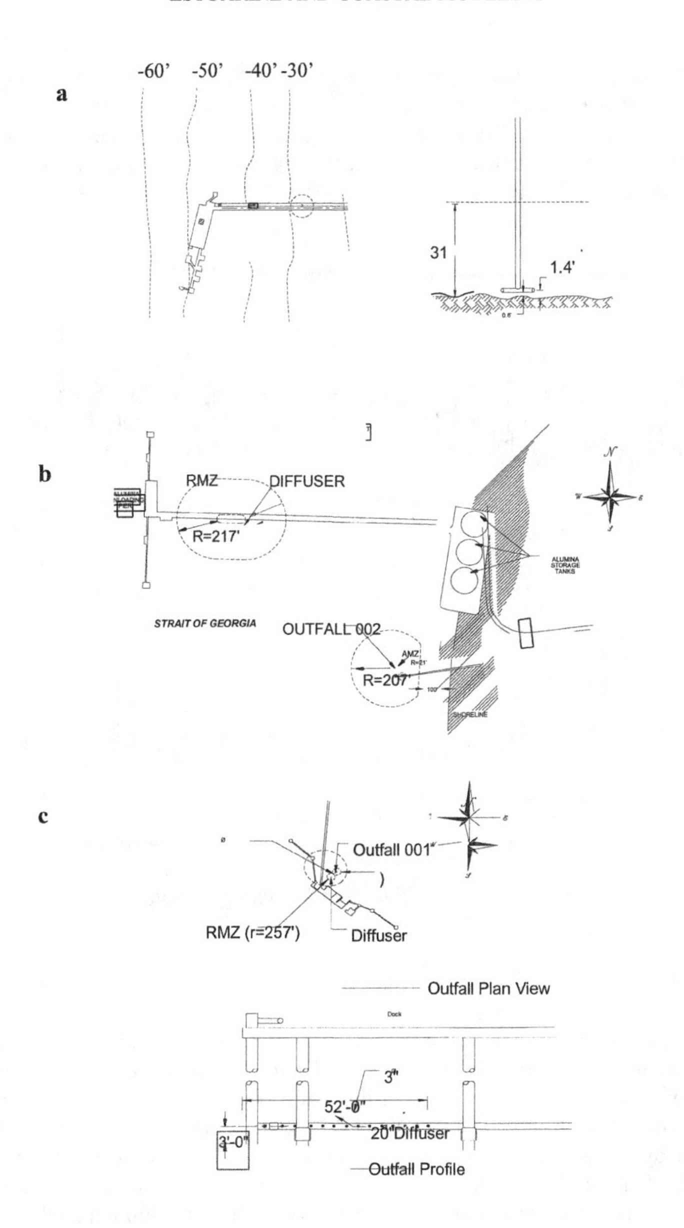

Figure 9: Out Fall Plan and Profile View: (a) Arco (b) Intalco; (c) Tosco

The predicted initial dilutions fluctuated with current velocity and water depth. With other factors held constant, it appears that current velocities had a strong effect on the fluctuation of dilution. ARCO had, by far, the highest mixing, with Tosco's and Intalco's shallower diffusers achieving less. The dilution time series are presented for four different outfalls in Figure 10.

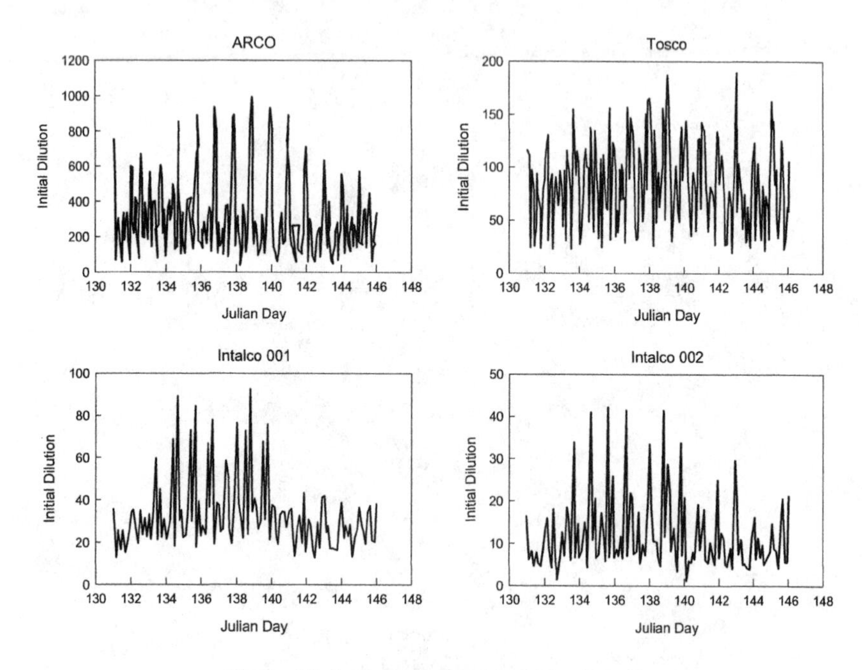

Figure 10: Initial Dilution Modeling Results

Far-field Transport Model Setup and Calibration

The transport module within EFDC was used to simulate the effluent plume transport in the Cherry Point Industries area. Therefore, the far-field plume transport model was set up using the same grid as the hydrodynamics model. Currents used in the transport model for the effluent far-field plume transport were simultaneously calculated by the hydrodynamics component of EFDC. The effluent is introduced into the far-field model as a mass flux, and the model grid was adjusted iteratively such that predicted concentration matched the near-filed initial dilution model results.

The far-field transport model was calibrated with the results of a previous dye study (Ebasco 1990). The process of model calibration involved adjusting the model dispersion coefficient until the best match between dye data and model results was obtained. In EFDC, the horizontal dispersion is calculated by the Smagorinsky formula. The background dispersion value was adjusted during the model calibration and the final model background dispersion coefficient was set as 1 m^2/s. Figure 11 shows the dilution ratio comparisons of dye data, Brook's Law dilution model results, and EFDC model results. The figure shows that the model results are in good agreement to the previous dye study results.

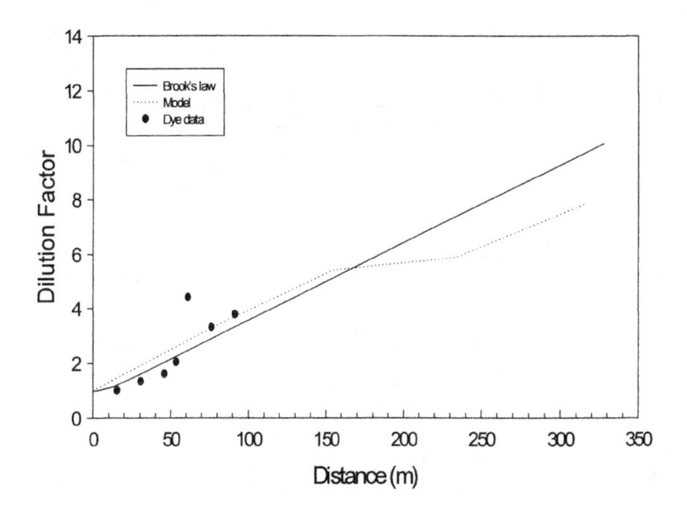

Figure 11: Calibration of Far Field Effluent Transport Model

Model Application

Following effluent transport model calibration, the model was applied to evaluate potential for accumulation of effluent constituents. The flow rate at each outfall in the Cherry Point Industries area was specified and considered as constant in time during the entire simulation period. All the effluent constituents were treated as neutrally buoyant and conservative in the EFDC transport module. Table 1 shows thelist of various constituents of interest. On the east land boundary, a zero-flux boundary condition was specified. For those effluent constituents with available data on background concentrations, the inflow open boundary values are set to the background values. Otherwise, the parameter's open boundary value was set to zero. The outflow effluent concentrations were calculated from the interior values immediately inside the open boundary using upwind method. When the flow at the open boundary changed direction from outflow to inflow, the inflow concentration was linearly interpolated from the last outflow value and the ultimate inflow value over a specific return interval.

Table 1 Effluent Characteristics and Compliance Evaluation

	Maximum Concentration (μg/L)				Water Quality	Dilution Required for Compliance			
Analyte	Tosco	Arco	Intalco 001	Intalco 002	Std (μg/L)	Tosco	Arco	Intalco 001	Intalco 002
Copper	29	30	16	24	3.1	9.3	9.7	5.1	7.7
Mercury	1.97	0.7	0.5	0.6	0.025	78	28	20	24
Nickel	7	208	9	10	8.2	0.8	25	1.1	1.2
Thallium	150	50	1	1	6.3	24	7.9	0.2	0.2
Cyanide	19	62	32	5	2.8	6.8	22	11	1.8
Benzo(a)-Pyrene	0.02	0.02	0.152	0.317	0.031	0.6	0.6	4.9	10.2
Benzo(b)-fluoranthene	0.02	0.02	0.192	0.535	0.031	0.6	0.6	6.2	17.3

The effluent transport model was run for a 14-day period from May 11, 1999 to May 25, 1999. Based on the length scale (approximately 10 km separation between the outfalls) and a residual current magnitude of 2 cm/s to the south, the tidal flushing time is less than 6 days. The 14-day simulation period is therefore considered sufficient for establishment of dynamic stable wastefield in the study area. Effluent constituents with high discharge concentration or high dilution requirement for compliance were selected for simulation (see Table 1). The background water concentration for each simulated constituent was used as the initial condition and incoming open boundary condition. Results for the predicted concentrations are plotted as time series for the entire 14-day simulation. A representative model result for mercury is shown in Figure 12. Model results showed that the maximum steady-state concentration for mercury is 0.0128 μg/L, which is well below the water quality standard of 0.025 μg/L (Table 1). Similar results were obtained for other constituents.

Conclusions

Effluents representative of May temperature and flow conditions containing peak, measured analyte concentrations were introduced into the study area through the various outfalls along the Cherry Point coastline. The initial dilution was continuously calculated for the four outfalls for the period May 11 through May 26, 1999. The far-field EFDC model used the resultant effluent concentrations to transport and disperse the plumes throughout the study area. The model allowed the effluent plumes to merge and calculated cumulative concentrations.

The conclusions of this study are as follows:

- The 3-D hydrodynamic model was successfully setup and calibrated using synoptic data collected during the month of May. The model predictions match reasonably well with the observed data.

- The effluent transport model was successfully calibrated to dye study data collected from the study area. The far-field model was linked to the near-field initial dilution models to continuously simulate the combined effluent field from ARCO, Intalco and Tosco.

- The cumulative effluent plume modeling results show the following:
  - Accumulation of effluent does not occur. The transport steady state is reached within four days, and concentrations reach a dynamic steady state.
  - Distribution patterns for all constituents consistently indicate that water quality standards are not exceeded in the study area due to Cherry Point discharges.

Acknowledgements

This study was funded by the Cherry Point Industries consisting of ARCO Products Company, TOSCO Refining Company, and INTALCO Aluminum Corporation. We would like to thank Ms. Karen Payne, Mr. Bill Yancey, and Ms. Elizabeth Daly of ARCO, Mr. John Webb and Ms. Rosanne Paris of Tosco, and Mr. Charlie Jurges and Ms. Michelle Evans of Intalco for providing available data and numerous technical reviews, and useful comments throughout the duration of the study. We would also like to than Ms. Kimberly Wigfield and Mr. Greg Pelletier of

Washington State Department of Ecology for their help in design and implementation of the study.

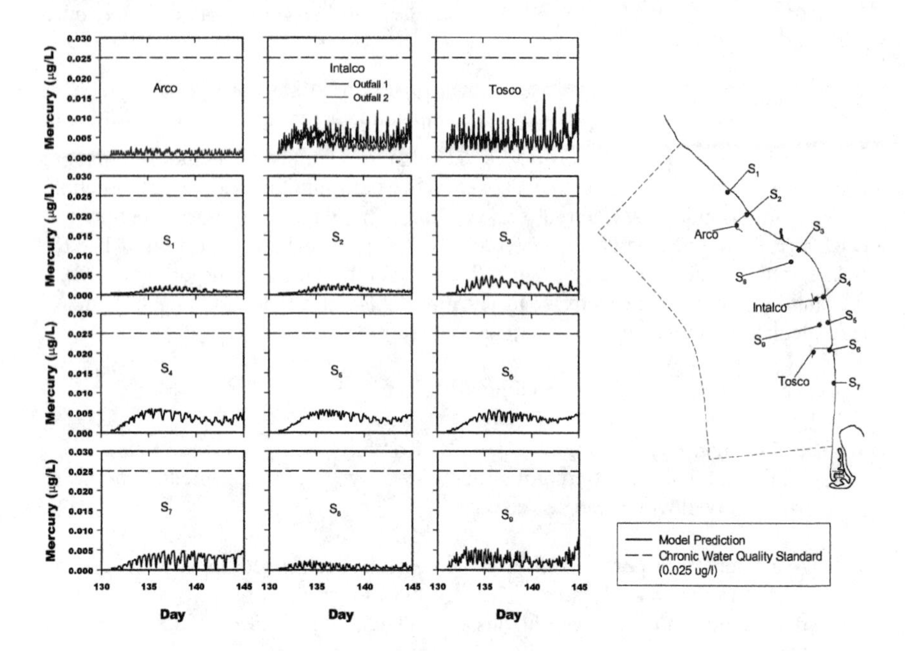

Figure 12: Mercury Concentration Time Histories

References

Ebasco Environmental. 1990. Dilution Ratio Study for BP Ferndale Refinery. Final Report. Prepared for BP Oil Company, Ferndale, Washington. NPDES Permit No. 000298-4.

ENSR Consulting and Engineering. 1991. *NPDES Effluent and Water Quality Monitoring Study Dilution Ratio Study, ARCO Petroleum Products Cherry Point Refinery*. February 1991. ENSR Document No. 0480-086-200.

ENSR Consulting and Engineering. 1992. *Dilution Ratio Study for Outalls 001 and 002*. Intalco Aluminum Corporation, Ferndale, Washington. October 1992. ENSR Document No. 3745-005-210.

ENSR Consulting and Engineering. 1995. *Dilution Ratio and Reasonable Potential Analysis*. Tosco Refining and Marketing, Ferndale, Washington. November 1995. ENSR Document No. 6752-010-400.

Hamrick, J.M. 1992: A Three-Dimensional Environmental Fluid Dynamics Computer Code: Theoretical and Computational Aspects. Virginia Institute of Marine Science, 317,VA. Special Report 317. 63pp.

Hamrick, J.M. and T.S. Wu. 1996. Computational Design and Optimization of the EFDC/HEM3D Surface Water Hydrodynamic and Eutrophication Models. In: Next Generation Environmental Models Computational Methods. G. Delic and M. F. Wheeler (eds). Proceedings in Applied Mathematics. 87.

Ji, Z.-G., M. R. Morton, and J. M. Hamrick: 2001: Wetting and drying simulation of estuarine processes, Estuarine Coastal Shelf Sci., 53, 683-700.

Mellor, G.L. and T. Yamada. 1982. Development of a Turbulence Closure Model for Geophysical Fluid Problems. *Rev. Geophys. Space Phys.* 20: 851-875.

Schwartz, M.L., R.C. Fackler, E.A. Hoerauf, C.E. Larsen, K.L. Lingbloom, and M.A. Short. 1971. Nearshore Currents in Southeastern Strait of Georgia. *Syesis*. 5:117-130.

Shen, J. and A.Y. Kuo. 1999: Numerical investigation of an estuarine front and its associated topographic eddy. Journal of Waterway, Port, Coastal, and Ocean Engineering, 125, pp. 127-135.

Smogarinsky, J. 1963. General Circulation Experiments with the Primitive Equations. I. The Basic Experiment. *Mon. Weather Rev.* 91: 99-164.

U.S. EPA. March 1994. Dilution Models for Effluent Discharges. EPA/600/R-94/086.

Yang, Z., T. Khangaonkar, C. DeGasperi and K. Marshall. 2000. Three-dimensional Modeling of Temperature Stratification and Density-driven Circulation in Lake Billy Chinook, Oregon, Estuarine and Coastal Modeling. In: Proceedings of the 6th International Conference, M. L. Spaulding, et al., (eds.). American Society of Civil Engineers, New Orleans, LA, pp. 422-425

Yang, Z., and J.M. Hamrick. 2002. Variational Inverse Parameter Estimation in a Long-term Tidal Transport Model. Water Resources Research, 38 (10), 1204, doi:10.1029/2001WR001121, 2002

A decision support system for optimising engineering and operational design of nearshore mining operations based on numerical simulations of nearshore waves, currents and water levels

Roy van Ballegooyen[1], Geoff Smith[1], Stephen Luger[1], Andre van der Westhuysen[2] and Sahil Patel[1].

Abstract

Designing economically viable mining operations in the normally hostile environment of the nearshore zone requires that a balance be struck between minimising capital costs of infrastructure (*e.g.* mining vehicle/jack-up rigs) and maximising operational time, but in a manner that ensures minimal safety risk (*e.g.* structural failure) and/or risk of disruption of the operations as planned.

To assist in determining specific design requirements for such nearshore mining operations a decision support system (DSS) has been developed whereby various design/operational thresholds (e.g. a wave height threshold, a current speed threshold, *etc*) can be specified and the resulting available operational time determined for the mining area of interest. By iteratively selecting design/operational thresholds and obtaining the associated operational time distribution across the mining area, the mining operator can ensure a rig/vehicle design and an associated design of mining operations that provides an optimal balance between minimising capital costs and maximising operational time. The economic viability of mining operations based on this optimised design can then be assessed.

The utility of the DSS in assessing the economic viability of a proposed operation is not limited to only mining operations and could be extended to other operations in the nearshore zone such as construction activities. Furthermore, a natural extension to the DSS is a forecast system to optimise safety and operational efficiency of operations in the nearshore zone.

This paper describes the DSS and associated numerical modelling effort used to assess a proposed nearshore mining operation on the west coast of southern Africa. The DSS is based on both simulated and measured time series of environmental conditions of a one year duration that are considered to be representative of the long-term environmental conditions at the location(s) of the proposed nearshore mining operations.

1 CSIR, P.O. Box 320, Stellenbosh, 7599, South Africa, Tel: +27 21 888 2574, Fax: +27 21 888 2693
Corresponding author email: rvballeg@csir.co.za .

2 Delft University of Technology, Julianalaan 134, postbus 5, 2600 AA, Delft, The Netherlands

Introduction

Operations in the nearshore zone (i.e. in waters shallower than 20 to 30 m) require that the structures and/or vehicles operating in this zone are optimally designed for the normally hostile environment within which they are to operate. Typically there is a requirement to determine appropriate engineering and environmental design thresholds that ensure a safe and viable operation where the risk of structural failure and/or disruption of mining operations is minimised.

Where existing vehicles/rigs are used in operations such as mining in the nearshore zone, the engineering design thresholds are pre-determined and specific to the vehicle/rig under consideration. Under these circumstances the viability of mining operations in the nearshore zone will be determined primarily by the cost of deployment of the specific mining vehicles/rigs and the available operational time for mining operations as determined by the prevailing environmental conditions. However, where the vehicles/rigs to be used in the mining operations are not yet in existence, the opportunity exists to design a more cost-effective operation by developing a design that is optimised according to the environmental conditions prevailing in the region of interest. Under these circumstances there is greater flexibility in developing a balance between minimising capital costs of infrastructure and maximising operational time in a manner that is both cost-effective and that ensures minimal risk of disruption of the operations as planned (see Figure 1).

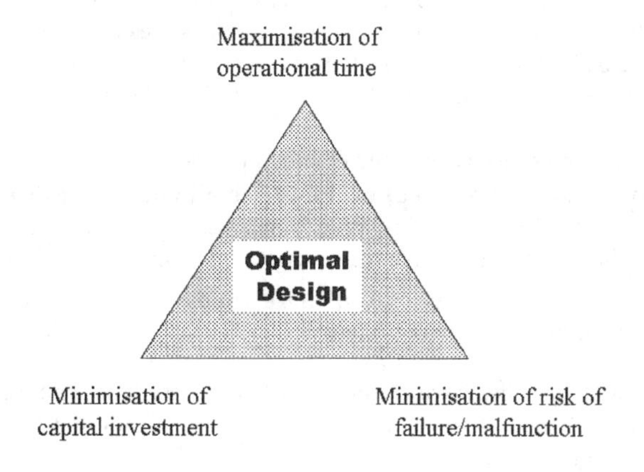

Figure 1: Design elements for an optimal design.

Specifically, the mining operator needs to provide design specifications (expressed in terms of environmental variables such as maximum wave height, currents, winds, water levels, *etc.*) to the designers of mining vehicles/rigs that ensure adequate operational time for the envisaged mining operation to remain economically viable. However, it is difficult to specify these design specifications if the impact of various specified environmental design limits on available operational time (or on access to various zones in the nearshore region) has not been clearly defined.

To assist in determining these precise design specifications, a decision support system (DSS) has been developed, whereby various design/operational thresholds (e.g. a wave height threshold, a current speed threshold, *etc*) can be specified and the operational time available and operational window persistences corresponding to these specified design/operational thresholds determined for the mining area of interest. By iteratively selecting design/operational thresholds and obtaining the associated operational time distribution across the mining area (as well as detailed information on the distribution of a range of operational "window" durations), the mining operator can then ensure a rig/vehicle design and an associated design of mining operations in a given mining area that is optimal in terms of economic viability and safety.

An important feature of the decision support system is that it provides *detailed spatial information* in addition to the temporal information typically used to develop design criteria. This resolution of spatial variability is necessary due to the rapid cross-shore change in environmental parameters experienced on moving shorewards through the surf-zone.

Approach and Methodology

The Decision Support System (DSS) developed calculates total operational window durations (in a year or season) for mining operations in a number of mining concessions for various combinations of environmental parameters, based on a one year duration of both simulated and measured time series of environmental conditions that are deemed representative of offshore conditions in the region of interest (i.e. in this case the nearshore region off the west coast of southern Africa).

The one-year time series used in the DSS comprises:

- **simulated (and derived) nearshore conditions** such as wave heights, wave period, total water level relative to mean sea level (MSL) or total water depth, combined wind and wave-driven currents, predicted tidal levels, and
- a combination of **hindcast offshore winds** and **measured coastal winds** in the vicinity of the area of interest.

The simulated nearshore wave time series are based on one year of deepwater directional wave measurements measured in a water depth of 175 m just offshore of the region of interest (see Figure 2).

Available bathymetry information in the region, comprising high resolution surveys of offshore bathymetry and beach profile surveys extending into shallow waters, have been used to set-up the Delft-3D suite of models used to transform waves from offshore to nearshore (incorporating the effects of wave refraction, shoaling and breaking) and to simulate the nearshore currents due to wind-, wave- and tidal forcing.

The above model-predicted time series, together with the measured data time series, provided the spatial time-series database, for each area of interest, from which the required total operational time and persistence information have been estimated. The relevant thresholds used in the DSS are based on a combination of the exact nature of the nearshore operations envisaged and the model predictions of nearshore conditions.

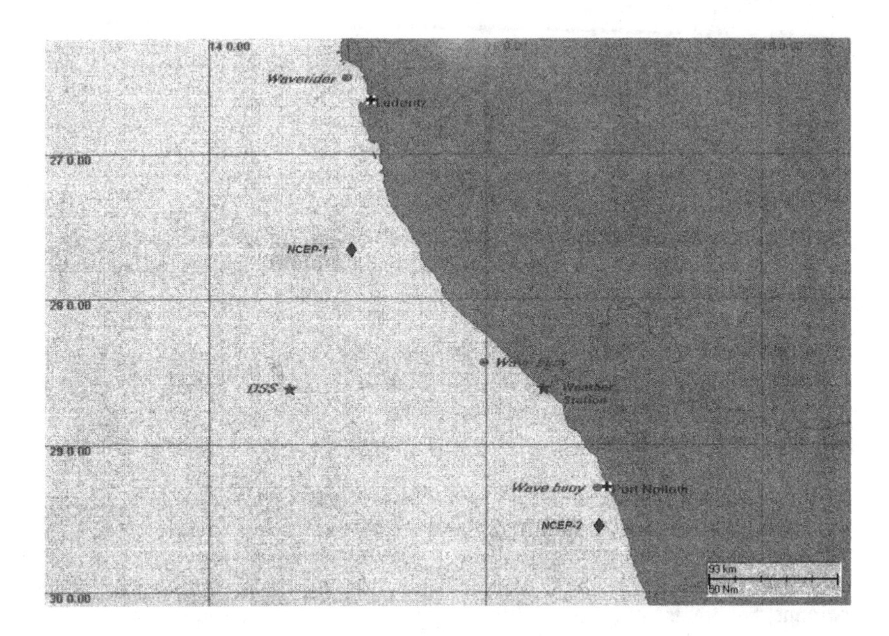

Figure 2: Map showing the location of the NCEP hindcast wind data locations (denoted NCEP-1 and NCEP-2), the measured wind data site (denoted Weather Station), the offshore wave data (denoted DSS) and the nearshore wave data (Wave Buoy site just north of the Orange River) used in the calibration of the wave model and as input to the DSS.

Model sensitivity tests have been conducted to investigate the sensitivity of the DSS outputs to the various input parameters and consequently the required accuracy of these input parameters, while analyses of longer-term wind and wave data sets have been undertaken to provide an indication of how representative the DSS results are expected to be of long-term environmental conditions.

Simulation of the one year time series used as input to the DSS.

Due to the complexity and interdependencies between the relevant processes, the model simulations have been performed using the DELFT3D suite of integrated environmental models from WL|Delft Hydraulics in the Netherlands.

Wave model description

The study area (Figure 2) is exposed to predominantly south-westerly ocean swells primarily generated in the *Roaring Forties* and is also often subject to local seas generated by strong, mainly southerly winds prevailing along the West Coast of southern Africa. On approaching the shore these waves are transformed due to refraction, shoaling and energy dissipation due to wave breaking. Obliquely incident waves generate significant wave-driven currents in the surfzone. To accurately simulate wave heights, wave periods, total water level and wave-driven currents in the nearshore zone, it is necessary to simulate accurately both the spatial and temporal variations in wave conditions. To transform a deepwater incident wave condition to the local conditions in the nearshore requires that the effects of refraction due to bathymetry, depth-induced breaking, depth-induced shoaling and bottom friction be taken into account.

The third-generation wave generation and refraction model SWAN (**S**imulating **Wa**ves **N**earshore) was applied (Booij, *et al.*, 1999) and was run within the DELFT3D-WAVE environment (WL|Delft Hydraulics, 2000), which provides a convenient interface for pre- and post-processing and for including wave-current interactions. The SWAN model computes the evolution of random, short-crested waves in coastal regions with deep, intermediate and shallow water and ambient currents. SWAN is based on the discrete spectral action balance equation and is fully spectral (in all directions and frequency), implying that short-crested random wave fields propagating simultaneously from widely different sources can be accommodated, e.g. a swell with superimposed wind sea. The SWAN model accounts for refractive propagation due to currents and depth and represents the processes of wave generation by wind, dissipation by whitecapping, bottom friction, depth-limited wave breaking and non-linear wave-wave interactions (quadruplets and triads) explicitly with state-of-the-art formulations. Wave blocking by currents is also explicitly represented in the model. Neither wave diffraction nor wave reflections are included in the version of SWAN utilised here, since both of these processes are considered to be of little relevance to this study.

Hydrodynamic model description

Currents in the nearshore zone of the study area are wave, wind, and tidally driven. For all but the lowest wave conditions, wave-driven currents predominate in the nearshore zone. However, in depths greater than approximately 10 to 15 m, and beyond the surf-zone, wind-driven currents are expected to predominate. Inertial currents are particularly strong in the study area (Simpson *et al.*, 2002). Tidally driven currents are generally weak except in the immediate vicinity of rapidly changing depths and complicated bathymetry. The influence of large-scale offshore, deep water currents is considered to be negligible in the region of interest (i.e. the nearshore zone at depths of less than approximately 15 m) as the large-scale offshore flows typically are only significant at and beyond the shelf break (i.e. at depth greater than 150 m or more). Consequently only the tidally and wind- and wave-driven flows over the continental shelf (i.e. inshore of approximately the 150 m isobath) have been included in the model simulations.

The currents have been simulated using a three dimensional hydrodynamic model, Delft3D-FLOW, running in a two-dimensional mode. The model as implemented includes formulations and equations that take into account tidal forcing, wind forcing, wave forcing in the surf zone, the effect of the earth's rotation (Coriolis force) and combined bed shear stresses due to currents and waves. The wave forcing in DELFT3D-FLOW is obtained from the wave energy dissipation rate computed by the SWAN wave refraction model. Enhanced bed stresses due to wave effects are incorporated in the model using the friction formulation of Fredsøe (1984).

For the sake of brevity neither the equations constituting the SWAN wave refraction model nor the DELFT3D-FLOW hydrodynamic model are repeated here. The equations for the SWAN model and their numerical implementation are described in detail in WL|Delft Hydraulics (2000) while those for DELFT3D-FLOW hydrodynamic model are described in detail in WL|Delft Hydraulics (1999).

Input data to the models

The data most critical to developing the one year simulated times series of waves and currents used as input to the model simulations are the *offshore directional wave data*. These data constitute a necessary input into the wave model, while nearshore wave data measured in the study area are required to calibrate and verify the wave model. The measurement locations and measurement durations of the model input and calibration wave data are indicated in Figure 2 and summarised in Table 1 below.

Table 1 Table of directional wave measurements offshore of Oranjemund.

Station	Coverage Period	Water Depth	Sampling Interval	Data Return	Actual data availability
Nearshore	1/10/96 to 26/11/98 (intermittent)	20 m	6 hourly	21 months of data	28/4/98 to 26/11/98
Offshore	8/3/98 to 13/4/99	175 m	½ hourly	86%	8/3/98 to 3/5/98 and 27/6/98 to 13/4/99

The 30 minute offshore wave data have been decimated to 3 hourly values by averaging. After careful analysis of the offshore wave data statistics, gaps in these wave data were filled by repeating limited periods of the measured data having exceedance statistics as similar as possible to those expected for the periods of missing data (e.g. data gaps in winter were replaced by measured data from a similarly active winter period). The resulting continuous time series is considered to be representative of the seasonal wave climate. The DSS requires an accurate representation of the covariance statistics of waves, currents and winds, thus it is necessary to treat the wind data in a similar manner to preserve the covariance in the measured and simulated data.

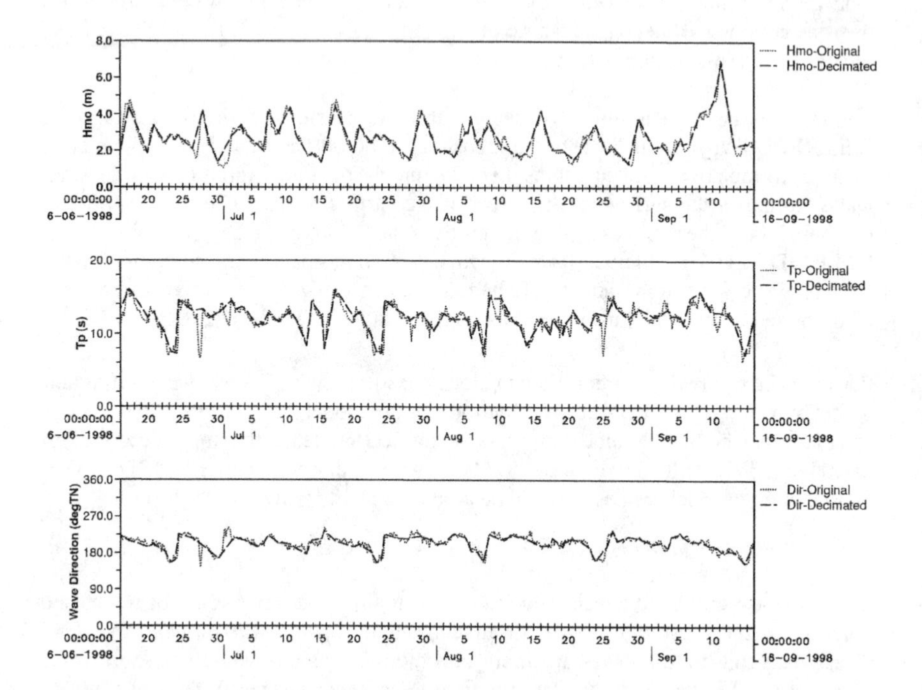

Figure 3: Original versus decimated wave time series for the period 16 June 1998 to 16 September 1998

Limitations in computational power available to the project necessitated the decimation of the offshore directional wave time series. It was therefore decided to reduce the number of wave conditions to be simulated in a manner that, whilst reducing the computational load, retained all relevant wave information contained in the offshore directional wave time series. Automated software was developed to reduce the possible 2920 wave conditions described in the original wave time series to a more manageable time series comprising 315 unique wave conditions to be simulated. Using Fast Fourier Transform (FFT) filtering techniques in MATLAB©, the offshore wave time series was decimated by identifying significant points of inflection in each of the time series of wave height, wave period and wave direction and then recording the time and full set of wave parameters for each significant point of inflection in any of the wave parameter time series. The priority was to retain sufficient significant wave height (H_{mo}) information to almost exactly reproduce the original undecimated wave height times series, as well as retain as much of the peak wave period (T_p) average wave directions as was feasible in the decimated data time series. Wave direction was found to be highly correlated with H_{mo} and T_p. This helped to limit the number of wave conditions that needed to be included in the decimated offshore wave time series to be simulated.

Decimation of the time series using this automated procedure was generally successful (see Figure 3), however intervention was required to make minor adjustments to the output series to register all recognised peaks, troughs and significant inflection points in the data.

The only *wind* data considered sufficiently representative of wind conditions at the sites of relevance are the hourly wind data measured at Alexander Bay by the South African Weather Services and the National Centre for Environmental Prediction (NCEP) hindcast data for two offshore locations in this region (Figure 2). These hindcast wind data are NCEP re-analysis wind data obtained from a global atmospheric numerical model operated by the National Center for Environmental Prediction in the United States of America. The data are available on a 6 hourly basis, representing the wind speed and direction at a height of 10 m above sea level. The speed and direction nominally represent six hourly averaged values. The details of the measured and modelled data are contained in Table 2 below.

The two NCEP wind data sets have been combined by taking the vector averages of the wind speed components to obtain a wind times series considered to be sufficiently representative of the offshore wind regime in the study area. The winds comprising this combined NCEP wind data time series, as expected, are more steady and overall stronger than the coastal winds measured at Alexander Bay. The Alexander Bay winds clearly display the strong diurnal wind variability typical of coastal winds in this region, i.e. low wind speeds early in the day that rapidly increase to a maximum in the mid- to late afternoon. While the NCEP winds display a similar diurnal variability, it is much less pronounced. It should be noted that the NCEP winds are from a model of limited spatial resolution and thus also tend not to reflect the high N to NW wind speeds associated with the propagation of coastal lows southwards along the West Coast of southern Africa. This is a severe limitation when determining extreme wind speeds for design (i.e. for survival conditions) but less so when determining operational windows, as is done in this study.

As the offshore (NCEP) winds in the evenings are more persistent and do not decrease to the same extent as the winds measured at the coastal station at Alexander Bay, the NCEP winds are used in the model simulations of the combined wind- wave and tidally driven currents as this is likely to constitute a conservative approach in that these more persistent winds are more likely to result in stronger and more persistent wind-driven currents. Given the strong diurnal variability of the winds and peak in wind speeds recorded at the coastal wind site at Alexander Bay, it was decided to use these wind data in the DSS to ensure conservative estimates of operational time based on wind speed exceedances.

Table 2: Table of available wind measurements in the vicinity of Oranjemund.

Station	Coverage period	Position	Height / Water depth	Sampling interval	Number of samples	% Coverage
Alex. Bay	Jul 1998 - Dec 2001	Lat: 28° 35.82' N Lon: 16° 31.98' E	29 m	hourly	67 200	>98
NCEP 1	Jan 1990 - Dec 2001	Lat: 27° 37.42' N Lon: 14° 59.13' E	10 m	6 hourly	17 533	100
NCEP 2	Jan 1990 - Dec 2001	Lat: 29° 31.92' N Lon: 16° 51.92' E	10 m	6 hourly	17 533	100

The tidal levels used in the model are predicted tidal levels based on the tidal constituents reported by Rosenthal and Grant (1989). Given the large extent of the domain being modelled, different tidal amplitudes and phases have been imposed at the north and southern boundaries of the model.

Computational grid and model bathymetry

A curvi-linear grid of 110 x 61 grid nodes was set up to cover the total area to be modelled. This model area extends to 60 km SE and 35 km NW of the regions of interest to the study and approximately 65 km offshore. This large model extent was required to ensure that boundary effects in the models did not interfere which the results at the sites of interest. A curvilinear model computational grid has been set up such that a coarse grid is used further offshore where less detail is required, while a much higher resolution grid is used in nearshore areas where the greatest detail is required. The curvilinear grid features grid cell dimensions ranging from 5500 m in the offshore to less than 30 m (cross-shore) in the nearshore.

The bathymetry used in the wave and current modelling was extracted from the following sources:

- SA Navy hydrographic charts
- CSIR coastline surveys from Orange River extending northwards
- CSIR surveys in the regions of interest comprising extensive surveys of the nearshore, midshore and offshore area
- Airborne laser beach surveys
- High resolution offshore surveys

The bathymetry data from these surveys were deficient in the surfzone and were thus supplemented by extrapolations of bathymetry comprising typical bar and trough formations suggested by the existing database of cross-shore surveys in the regions of interest. While is possible to enhance the coverage of bathymetric data in the critical surfzone area in this manner, residual uncertainties in the bathymetry required that the model sensitivity to changes (real and/or postulated) in these assumed nearshore bathymetries be tested (see “Limitations of the DSS”). The bathymetry data were all referenced to mean sea level (MSL). For computational reasons, the minimum depths in the computational grid are limited to 1.25 m.

Wave model Set-up

The decimated time series of deep sea wave conditions were applied at both the offshore and cross-shore boundaries of the model, although this does not strictly hold for the cross-shore boundaries, particularly in shallow waters.

In order to simulate the natural decrease of wave height as the wave approaches the shore, the wave heights at the cross-shore open boundaries were appropriately reduced. This was done by using SWAN in a one-dimensional mode to simulating shoreward wave propagation using the bathymetry along the boundary and assuming no longshore variability in bathymetry. It was found in the study area that, on average, waves between the shoreline and adistance of 3000 m offshore have a significant wave height of 75% of the offshore deep water wave height, and that the waves between 3000 m and 5000 m offshore have a significant wave height of 95% of the offshore wave height. This was implemented in the model by setting up “sheets” at the boundary that effected these reductions in wave height on the boundaries. These reduction factors were biased towards the smaller, shorter-period southerly waves to ensure that these waves would enter the model domain at the appropriate amplitude and be represented accurately in the areas of interest in the model. However, this resulted in the overestimation of wave heights along the cross-shore boundaries for the larger, longer period SW to SSW waves that typically are refracted and attenuated to a greater extent on approaching shallower waters than are the generally smaller, shorter period southerly waves. This approach resulted only in local errors near the boundary because of the near parallel approach of the larger, longer period SW to SSW waves to the cross-shore boundaries.

Wave simulation runs were conducted using a standard set of parameters describing each offshore wave condition in the input data series. These parameters are: significant wave height (H_{m0}), peak wave period (T_p), mean wave direction ($?_m$) and the directional spreading of waves ($?_{spread}$). In addition, an idealised parameterization of the frequency spectrum (spectral shape) was used. The parameters describing the spectral shape of the offshore wave data input waves are derived from the offshore directional wave data (denoted DSS in Figure 2). A JONSWAP spectral shape (Hasselmann *et al.*, 1973) with a peak enhancement factor of $\gamma = 2.0$ was found to best represent the offshore wave spectra. The physical processes included in the wave model runs were: depth-induced refraction using linear wave theory, bottom friction and depth-induced breaking. The form of bottom friction used was that of Madsen *et al.* (1998), and the formulation for depth-induced breaking is that of Battjes and Janssen (1978). The values of the coefficients for these formulations were determined during the calibration of the wave model. Considering the short fetch distance between the offshore directional wave buoy (DSS) and the project sites with respect to the dominant S to SE wind, generation of sea by wind was not included in the wave model. As a result, the processes of whitecapping, and nonlinear wave-wave interactions were also neglected.

To limit computational effort, the range of wave periods modelled were limited to a bandwidth of 25 s to 1 s (using 24 discrete periods), and the range of wave directions to a sector of waves arriving from 135 to 315 deg TN (using 10-degree intervals). These computational limits did not negatively affect the nearshore model solution at the sites of interest. All simulations were conducted at a tide level of +0,01 m MSL, which is the mean sea level between Lüderitz and Port Nolloth (S A Hydrographical Office, 1998). The assumption of a steady water level in the wave modelling has the potential to induce significant errors in the simulation of waves in very shallow waters, thus this expediency will not be adopted in future model implementations where water level variations will be explicitly included in the wave simulations.

Flow model set-up

The hydrodynamic modelling is limited to two-dimensional simulations. The reason for this is both pragmatic (i.e. 3D simulations would place too great a burden on the computational and storage resources) and due to the limited information with which to calibrate the simulations (i.e. there is some theoretical uncertainty as to the validity of three-dimensional simulations in the surfzone that would not have been possible to eliminate without additional calibration and validation data).

Both local and remote wind forcing are included in the model by specifying sea levels at the open boundary that are consistent with wind-driven flow under the scaling assumption of an infinitely long coastline with negligible alongshore changes in bottom topography. Added to these synoptic changes in water level at the open boundaries are the predicted tides. The influence of larger scale offshore flows are not included in the model as the west coast of southern Africa is an upwelling zone where the nearshore currents (water depths < 15 m) in the region of interest are primarily wind-and wave-driven and little affected by large-scale offshore flows. The model domain was made deliberately large to minimise the impact of any boundary effects/errors.

Calibration and verification of the models

The *wave model* has been calibrated by using the offshore directional wave data as model input and comparing the nearshore wave model solution to directional wave data measured in the study area using a SEAPAC instrument deployed in a water depth of 22 m. For the purpose of calibration, concurrent data are required from the offshore and nearshore measurement sites, thus a record length of only five months (July to November 1998) is available for calibration of the wave model. During this period, peak wave directions measured at the offshore wave buoy range from SSE to W, with significant wave heights of up to 7.6 m. For the final calibration simulations, the wave model settings are as follows: Processes included in the wave model are depth-induced refraction using linear wave theory, bottom friction and depth-induced breaking. The form of bottom friction used is that proposed by Madsen *et al.* (1998)

with a relatively large roughness length scale of 0.08 m. The formulation for depth-induced breaking is that of Battjes and Janssen (1978), with the coefficient $\alpha = 1$ and a breaking parameter of $\gamma_{BJ} = 0.73$.

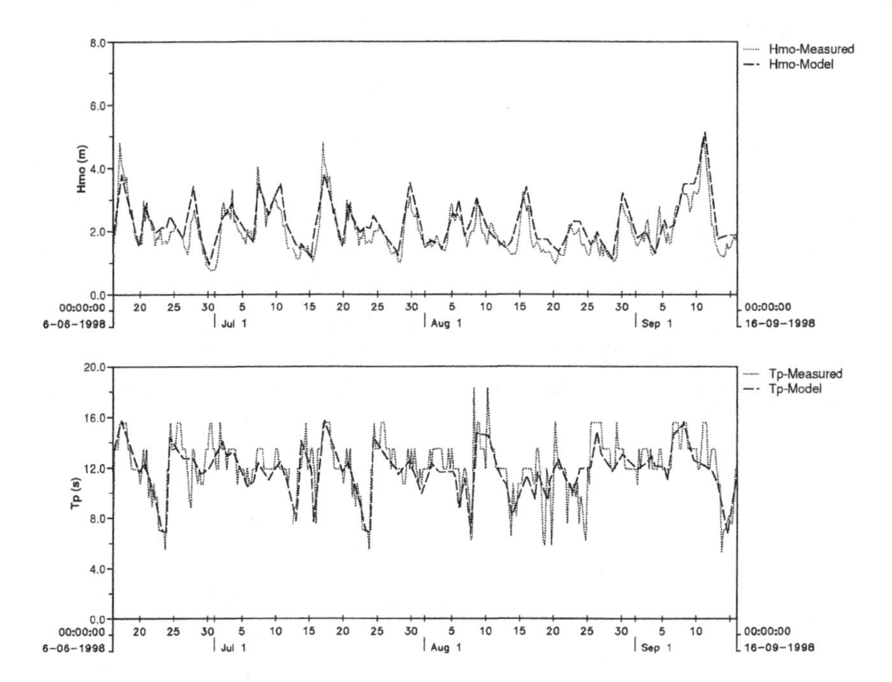

Figure 4: Modelled versus measured nearshore wave time series for the period 16 June 1998 to 16 September 1998

In general the simulated wave heights accurately represented the measured wave heights at a water depth of 22 m. Figure 4 contains a typical calibration plot. On occasion there are discrepancies between the model simulations of wave height and the nearshore calibration data, however these are often related more to the decimation of the input wave time series (and at times the parameterisation of the input wave spectra) than to deficiencies in the wave model. The calibration process also shows that the generation of waves by wind is negligibly small over the small fetch between the offshore model boundary (where the input offshore wave data were measured and applied) and the nearshore region of interest.

No data are available for the calibration of the hydrodynamic model simulations of nearshore currents. The simulated currents are however compared with current measurements made in an approximate water depth of 20 m some 7 km south of the study area. These measured data indicate mean current speeds of approximately 0.1 m/s with a standard deviation of > 0.1 m/s. The peak current speeds measured are of the order of 0.4 m/s which compares well with those simulated by the hydrodynamic model at a similar depth. The simulated current speeds in the surfzone also are comparable with other modelled and anecdotal evidence of current speeds in the surfzone, that indicate typical current speeds of between 1.0 m/s to 1.5 m/s (with extremes exceeding 2m/s in the surfzone for high wave conditions in this region). It should be noted that the lack of measured currents (and water levels) for the calibration of the hydrodynamic model poses a potentially significant limitation in providing an accurate one year time series of simulated currents.

Model Sensitivity

Model sensitivity tests have been conducted to investigate the sensitivity of the DSS outputs to uncertainties in the various input parameters and consequently the required accuracy for input parameters to the DSS. Specifically, this has provided information on:

- The impact of limited availability of bathymetry measurements across the surf zone on the accuracy of the model results and consequently the DSS outputs.
- The relevance of wind data to the accuracy of predictions both in terms of the location and nature of the wind data chosen (e.g. hindcast or measured coastal data) for use in the model simulations and the DSS.

To assess the model sensitivity to changes or uncertainty in the bathymetry, the hydrodynamics were simulated for both the assumed and a significantly modified alternate bathymetry. As expected, the differences between the two simulations is largest in the very shallow waters inshore of the surfzone. The differences in current speeds for the two simulations ranged from insignificant to moderately large at times. While these differences may seem significant in terms of absolute velocities, the impact on the outcomes from the DSS is limited due to the wide intervals used for current thresholds in the DSS.

Given the fairly substantial differences between the NCEP hindcast winds and those measured at Alexander Bay, a sensitivity analysis was undertaken whereby model simulations using the two sets of wind data were compared for the period 16 December 1998 to 16 March 1999. Comparison of the simulated currents at near beach, surfzone and offshore locations in the study area indicates that the nearshore and surfzone currents are predominantly wave-driven and that the changes in the wind regime have little impact at these cross-shore locations. The effect of the changes in wind regime are observable at offshore locations due to the larger wind-driven component of the currents in these deeper waters, however the differences remain small. In general, in simulations using NCEP winds the wind-driven currents in the offshore zone are larger than those in simulations using Alexander Bay winds, except for in the late afternoon when the peak in the wind speed recorded at Alexander Bay is reflected as a short duration peak in the current speeds.

Input time series to the DSS

The time series inputs into the DSS comprise:

- ***directly simulated times series*** (time series of peak wave period, T_p, from DELFT3D-WAVE simulations). These time series, used to determine the operational windows, are linearly interpolated where necessary to provide time series with a nominal temporal resolution of one hour.
- ***measured time series*** (measured coastal wind speeds at Alexander Bay that include both the strong late afternoon diurnal peaks in wind speed and strong NW winds associated with coastal lows that are typically absent from the NCEP data).
- ***derived times series***, namely
  - *maximum wave height* - the model simulations produce significant wave heights (H_{mo}) rather than the maximium wave heights (H_{max}) required in the DSS. The methodology proposed by Battjes and Groenendijk (1999) was used to convert the simulated root mean-squared wave height (H_{rms}) to the maximum wave height (H_{max}) or a wave height ($H_{0,1\%}$) with a very low exceedance (< 0.1%). The algorithm requires H_{rms}, wave direction and the local bottom slope as inputs.
  - *total water level* – the total water levels used in the DSS are a combination of the maximum wave heights and changes in water level (above MSL) due to wind and wave set-up and tidal forcing. The maximum water level was calculated to be sum of the water level (due to wind and wave set-up and tidal forcing - infragravity wave effects excluded) and 67% of the estimated maximum wave height (H_{max}).
  - *surface and bottom currents* - there is a requirement to compute both surface and bottom currents, however limitations in the theoretical framework for computing such three dimensional currents in the nearshore zone and the increased computational resources (processor speed and storage) required for full three-dimensional modelling, resulted a simpler approach being taken whereby two-dimensional model simulations were undertaken and analytical formulae used to determine vertical current profiles. A logarithmic profile (WL|hydraulics, 1999) was assumed and the surface and bottom velocities determined from the vertically averaged (two-dimensional) simulated currents.

The Decision Support System

The DSS comprises an easy to use and readily accessible database of quantitative measures of available operational time for user-specified environmental design thresholds in the region of interest (e.g. mining concession), accessed via a pull-down menu system running on standard web browser software in a PC-based Windows/NT environment. The quantitative measures of operational time include:

- contours and tabulations of the percentage operational time (in a year or season) in the area of interest, for a given set of environmental data thresholds of maximum wave height (H_{max}), peak wave period (T_p), current speed (near surface and near bottom), total water level (above MSL)[1] due to waves, tides and wind and wave set-up, and wind speed;

detailed information (both as histograms and tables) on the distribution of operation windows durations at selected cross-shore locations for specified environmental thresholds.

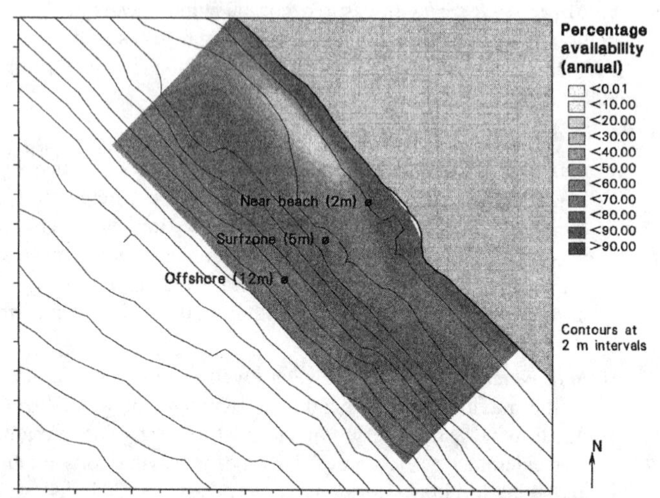

Figure 5: Total amount of operational time available in an assumed typical year for various user-specified thresholds (H_{max} < 5.0 m, Tp < 14 s, Total water level < 4 m, near bottom current < 0.7 m.s-1, near surface currents < 0.7 m.s-1 and wind speed < 10 m.s-1), expressed as a percentage of the total possible operational time in a year (i.e. 365 days).

The plots of *available operational time* (Figure 5) comprise spatial contour plots of the total amount of operational time available annually. An allowance has been made for initial mobilisation time (e.g. the time taken to set up mining equipment) as well as the demobilization time of the specific operation being considered. For example, mobilisation and demobilisation times of three hours each would result in operational windows of six hours or less being unusable and thus are ignored by the DSS. Also provided are tables containing the totals of the area accessible for mining operations in the study area (e.g. the mining concession area) for specific operational durations (e.g. area accessible for between 10% and 20% of the time during a one year period). Both annual and seasonal totals are contained in these tables.

[1] or total water depth if more relevant to the operation (e.g. total water depth is a more relevant parameter for jack-up rigs)

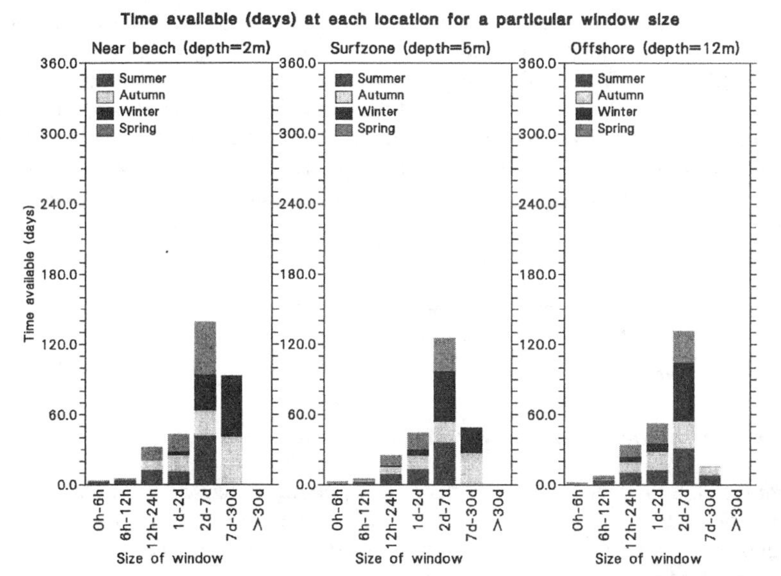

Figure 6: Total amount of operational time available in various window lengths a an assumed typical year for various user-specified thresholds ($H_{max} < 4.0$ m, $T_p < 14$ s, Total water level < 4 m, near bottom current < 0.7 $m.s^{-1}$, near surface currents < 0.7 $m.s^{-1}$ and wind speed < 10 $m.s^{-1}$).

Histograms of the *distribution of operational window durations* (Figure 6) comprising the total operational time available for a specified combination of environmental thresholds, are provided for a number of relevant cross-shore locations (i.e. the near-beach region, zone of wave breaking or mid surf-zone and the zone seaward of wave breaking or offshore zone). Thus for a specific set of input thresholds, the DSS provides a histogram of the distribution of the total operational time available annually for a number of discrete operational window durations in a specific region of interest. For example, Figure 6 shows that for the user-specified environmental thresholds of $H_{max} < 4.0$ m, $T_p < 14$ s, total water level < 4 m (relative to MSL), near bottom current < 0.7 $m.s^{-1}$, near surface currents < 0.7 $m.s^{-1}$ and wind speed < 10 $m.s^{-1}$, approximately 140 hours of operational time is available in windows of a duration of between 2 and 7 days in the near beach zone and approximately 90 hours of operational time is available in windows of a duration of between 7 and 30 days in the near beach zone. Accompanying these graphical outputs are tabulations of the processed data used to generate the histograms of operational windows for the various cross-shore locations of interest. The user of the DSS is thus able to select a mining area and for any discrete combination of the 5 parameters, obtain the corresponding operational time plot, histograms of the durations of individual operational windows and the summary tables.

While sediment transports (longshore transport, cross-shore transports and scour) are an important consideration in many nearshore operations (e.g. scour around structures and infill of trenches or mined areas), sediment transport parameters have not been included in the DSS. It was thought that calculated longshore and cross-shore sediment transport rates under natural conditions (i.e. without interference of mining operations) would not be sufficiently accurate and meaningful, compared to local effects induced through mining operations such as slumping into trenches, *etc.*

Potential Limitations of the DSS

Although considerable effort has been expended in ensuring that the DSS is as accurate as possible, the DSS has some limitations associated mainly with a number of simplifying assumptions made in the modelling as well as the development of the DSS.

Representivity of the input data to the DSS

For the DSS to produce useful results for engineering design and the design of mining operations, the input data to the DSS needs to be representative of the long-term environmental conditions in the region of interest. Specifically the one year duration of simulated wave time series and wind time series used in the DSS need to provide operational window durations and total operational durations that are statistically representative of the long-term environmental conditions in the region of interest.

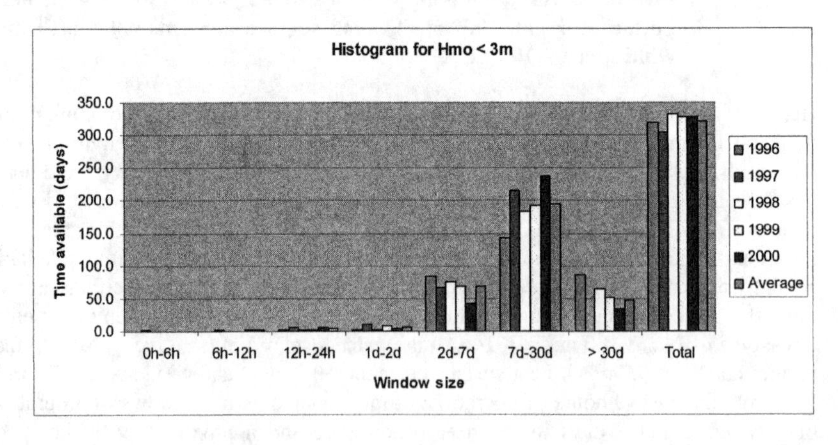

Figure 7: Histogram of available time below a wave height threshold of 3 m, for various window sizes (assuming a mobilisation/demobilisation time of 6 hours).

A five year hindcast of directional wave data was obtained to compare the annual distributions of operational windows over the five year period (1996 to 2000). Analysis of these data indicate that, although the annual distributions of the operational window durations for a number of wave height thresholds indicate some variability, the total available operational time per annum for each of the operational wave height thresholds remained remarkably steady over the 5 years of hindcast wave data (Figure 7). A similar analysis was not undertaken for *wave periods* or for *wave directions* as they are considered to be variables of secondary importance, however it should be noted that changes in these parameters may have a significant influence on the simulated currents in the nearshore zone.

In the absence of a long term measured.*coastal wind data* set at the measurement site of the of the wind times series used in the DSS, a similar analysis to that described above was undertaken for NCEP (averaged) 10 m wind data at an offshore location close to the coastal wind measurement site. For both high and low wind thresholds the distribution of both the operational window durations and the total operational time available per annum were similar for all years, indicating that the wind data used in the DSS (mainly 1998 wind data) were sufficiently representative of the average wind conditions in the region.

Accuracy of the input data to the DSS.

In addition to the requirement that the input data are representative of the long term environmental conditions in the region, the simulated time series need to be sufficiently accurate for the DSS to produce robust estimates of operational window durations and the total available operational time available per annum.

The simulated *wave data* have been calibrated using measured wave data in the region of interest at a depth of 20 m, consequently there is considerable confidence in the simulated wave heights and periods up to this depth. The only concern is the uncertainty in the nearshore bathymetry used in the model as the simulated wave height are somewhat sensitive to changes in the assumed bathymetry.

The *water level* used in the DSS is a combination of the maximum wave heights and changes in water level due to wind and wave set-ups and tides, all of which are adequately simulated in the hydrodynamic and wave models. The only potential source of significant error is the effect that infragravity waves (if present) may have on the simulated water levels as infragravity wave physics have not been included in the numerical model simulations.

Estimates of the *near surface and near bottom currents* have been derived from the simulated vertically averaged currents due to wind and wave forcing by assuming a logarithmic profile for current velocities. A wind-driven current profile may display more vertical shear near the surface than indicated by an such assumed logarithmic velocity profile, consequently the surface currents estimated assuming a logarithmic velocity profile may be underestimates of the true surface currents. However, given that i) the DSS was focussed on the nearshore waters (e.g. surf zone and further inshore) where the waters well-mixed and flows are predominantly wave-driven and, ii) the relatively wide increments in the surface current speed thresholds used in the DSS (i.e. < 0.4 m/s, < 0.7 m/s < 1.0 m/s, etc), the DSS outcomes are largely insensitive to the potential inaccuracies in the estimated near surface currents introduced by assuming a logarithmic profile for vertical current shear.

Of potential concern is the inaccuracies in the available *bathymetry* as changes in bathymetry can lead to relatively large local changes in the simulated currents, however global changes in current speeds are conservatively estimated to be < 0.3 to 0.5 $m.s^{-1}$. Due to these potential uncertainties we would recommend that current measurements be obtained both to calibrate the simulated currents and to provide the information necessary to more accurately determine the consequences of uncertainties in the specified bathymetry.

While the *wind* time series used are deemed to be adequate, coastal winds measured within the region of interest would provide greater confidence in the DSS results.

In summary, the greatest limitation in the accuracy of the time series used in the DSS is the poorly specified bathymetry, particularly the limited bathymetric information in the zone between the shoreline and the shallowest depth of the offshore bathymetric surveys used in this study.

Summary and Conclusions

It is concluded that the DSS as presently constituted is adequate to provide a preliminary assessment of the potential operational time available for the inshore mining operations under consideration. As a typical project progresses and a better knowledge is obtained of potential systems/equipment to be used, it is anticipated that aspects of the DSS may need to be refined. Such refinements are expected to include:

- Refinements in the various environmental parameters thresholds used in the DSS.
- Improvements in the databases used by the DSS. These could include the simulation of longer times series to be used as input data to the DSS (to ensure that these data are more representative of the longer term environmental conditions in the region of interest).
- Additional bathymetry data due to the sensitivity of the simulated waves and particularly the currents to the bathymetry used in the model.
- The measurement of currents to calibrate the simulated current time series given the sensitivity of the simulated currents to the detail of the potentially poorly defined nearshore bathymetry and the sensitivity of sediment transports (of major importance to coastal mining) to near bottom currents.
- The simulation of three dimensional currents if deemed necessary (i.e. due to the impact of surface currents on equipment such as umbilicals, *etc*).
- The inclusion of the effects of infragravity waves on total water level (or total water depth) in the DSS as seemingly the magnitude of the potential changes in total water depth in the nearshore due to infragravity waves may on occasion be significant. It is recommended that the contribution of infragravity waves to total water level in the region(s) of interest be better quantified to ensure robust decisions using the DSS.

***Acknowledgements**: The funding of the project by Namdeb Diamond Corporation (Pty) Ltd and their permission to publish this study, is gratefully acknowledged. We also thank Shell International Exploration and Production for their permission to use their offshore directional wave data in this project.*

References

Battjes, J.A. and J.P.F.M. Janssen (1978) Energy loss and set-up due to breaking of random waves, *Proceedings of the 16th International Conference on Coastal Engineering. ASCE*, pp569-587.

Battjes, J.A. and H.W. Groenendijk (1999) Shallow foreshore wave height statistics. Proceedings of Coastal Structures '99, Losada (ed.) Balkema, Rotterdam, 29-35. (ISBN 90 5809 092 2)

Booij, N, Ris, R C and Holthuijsen, L H (1999). A third generation wave model for coastal regions, Part I, Model description and validation, Journal of Geophysical Research, 104 (C4), 7649-7666.

Fredsøe, J. (1984) Turbulent boundary layer in wave-current interaction. Journal of Hydraulic Engineering, ASCE, 1000, 1103-1120.

Rosenthal, G. and S. Grant (1989) Simplified tidal prediction for the South African coastline, South African Journal of Science, 85, 104-107.

Hasselman, K., T.P. Barnet, E. Bouws, H. Carlson, D.E. Cartwright, K. Enke, J.A. Ewing, H. Gienapp, D.E. Hasselman, P. Kruseman, A. Meerburg, O. Müller, D.J. Olbers, K. Richter, W. Sell and H. Walden (1973) Measurement of wind-wave growth and swell decay during the Joint North Sea Wave Project (JONSWAP), Deutsche. Hydrogr. Zeit. Suppl., 12(A8)

Madsen, O.S., Y-K. Poon and H.C. Graber (1998) Spectral wave attenuation by bottom friction: Theory, Proceedings of the 21st International Conference on Coastal Engineering, ASCE, p492-504.

S A Hydrographical Office (1998) South African Tide Tables 1998, The Hydrographer, South African Navy, Private Bag X1, Tokai 7966.

Simpson, J. H., P. Hyder, T.P. Rippeth and I.M. Lucas. (2002). Forced Oscillations near the critical latitude for diurnal-inertial resonance. Journal of Physical Oceanography,32(1), 177-187.

WL|Delft Hydraulics (1999) DELFT3D-FLOW User Manual Version 3.05. WL|Delft Hydraulics, Delft, The Netherlands.

WL|Delft Hydraulics (2000) DELFT3D-WAVE User Manual Version 2.00. WL|Delft Hydraulics, Delft, The Netherlands.

Numerical Modeling Study of Cooling Water Recirculation

Jianhua Jiang[1] and David B. Fissel[1]

Abstract

A three-dimensional, finite difference coastal circulation numerical model, ASL-COCIRM, was adapted to examine the possible recirculation of cooling water in the Burrard Generating Station, Canada. The model resolved the near-field zone of the cooling water buoyant jet and intake using a very fine grid size (2.5 m by 2.5 m in the horizontal), which allowed the cooling water outlet pipe and intake to be represented in a realistic manner. The near-field model was then nested within the large model domain of the far-field zone of cooling discharge receiving water. Model calibration and verification results demonstrate that the model has the capability of adequately capturing the overall behavior of the buoyant jet and quantitatively investigating the waste heat recirculation into the intake and the consequent effect on the efficiency of power plants.

Introduction

The possibility of recirculation of waste heat into cooling water intakes is of particularly concern to coastal engineers in building thermal or nuclear power stations because of its adverse impacts on the plant efficiency and operation. However, quantitative identification of such potential recirculation remains difficult due to the complicated buoyancy processes and hydrothermal features within the near-field of cooling water discharges, especially with a submerged buoyant jet. In the presence of this submerged heated buoyant jet, the receiving water exhibits large gradients in the hydrothermal field and much different spatial scales between the near-field and far-field zones, on order of 10 – 100 m in the near-field and 1 – 10 km in the far-field. In the recent study of the Burrard Generating Station cooling water recirculation by the authors of this paper (Jiang, et al., 2001), the potential recirculation of waste heat into the shallow coastal waters was quantitatively examined using a three-dimensional coastal circulation model, ASL-COCIRM. The model incorporated a newly developed nested grid scheme to resolve the large hydrothermal gradient in the near-field and to represent the cooling water intake and outlet in a realistic manner. The near-field model was embedded within the spatially larger and coarser grid of the far-field model. The complete field was then solved with a single modeling procedure at

[1]ASL Environmental Sciences Inc., 1986 Mills Rd., Sidney, BC, V8L 5Y3, Canada, jjiang@aslenv.com and dfissel@aslenv.com

every time step. In this paper, we use the nested model formulation to investigate the degree of direct recirculation of discharged cooling water into the intake of the plant cooling water system. We present the methodology and key model results concerning cooling water recirculation.

Cooling Water Structure and Receiving Waters

The Burrard Generating Station (BGS) is located on the north shore of the Port Moody Arm (hereafter, simply called the Arm), at the eastern end of the Burrard Inlet, Canada (Figure 1). The Arm has a length of 6.5 km, a mean width of 0.9 km, and the mean water depth varies from 5 to 30 m in mid-channel. A mixed tide occurs with a mean tidal range of 3 m. Tidal currents are typically 10 – 20 cm/s or less. The tidal prism is 12.2×10^6 m^3, which represents approximately one-third of the total volume of the Arm (Fissel, et al., 1998). The BGS releases a large volume of heated cooling water into the Arm. Under existing operating permits, the maximum allowable discharge is 1.7×10^6 m^3/d with a maximum temperature of 27 °C. The cooling water temperature exceeds the temperature of the ambient water by about 4 – 10 °C in summer and up to 20 °C in winter. The maximum allowable daily discharge is about 14% of the tidal prism, and approximately 4% of the volume of water in the Arm.

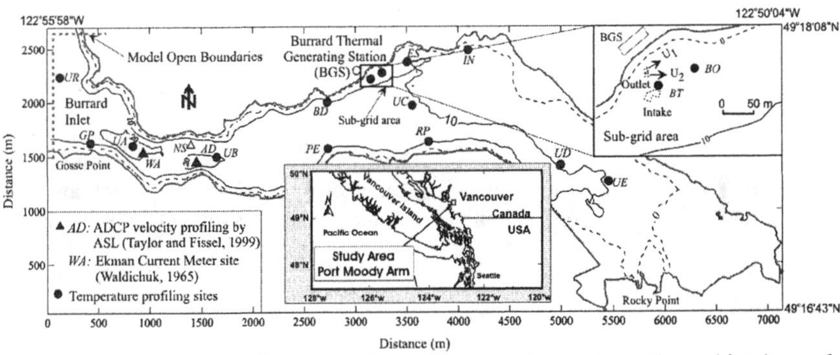

Figure 1. Location map of the Port Moody Arm, cooling water outlet and intake, and survey sites of temperature and currents. Depths are in meters below lowest astronomical tide.

The cooling water discharge is released into the Arm through two independent outlet systems, each consisting of a pair of adjacent horizontal pipes of diameter 2.45 m, respectively denoted the shoreward and seaward pipes, and takes the form of a submerged horizontal heated buoyant jet in a relatively shallow water depth of approximately 10 m (Figure 2). The two side pipes are set in an identical horizontal plane with a spacing of 3 – 4 m. The heights of the outlet pipes above the seabed are 6 m for the seaward pipes and about 3 m for the shoreward pipes. The exit flow from the seaward pipes is due east. The shoreward pipe flow is directed along the shoreline, resulting in an exit flow at an angle of 20° to the east (Figure 2). The cooling water intake is located about 30 m SSE of the outlet pipe (Figure 2) in a depth of about 11 m, and has an opening at the height of 1.54 – 5.88 m above the seabed with the sides facing the outlet being closed. For safety reason, a floating boom is installed at about 70 m eastern of the outlet pipe. The boom is always floating at the water surface and has no considerable effect on flows. Extensive data sets, including temperature profiling at near-shore locations, and an ADCP and other current meter time series data, have been collected in the Arm (Fissel, et al., 1998; Taylor, et al., 1999; Birtwell, et al., 2001). These data sets have been used for model calibration and verification.

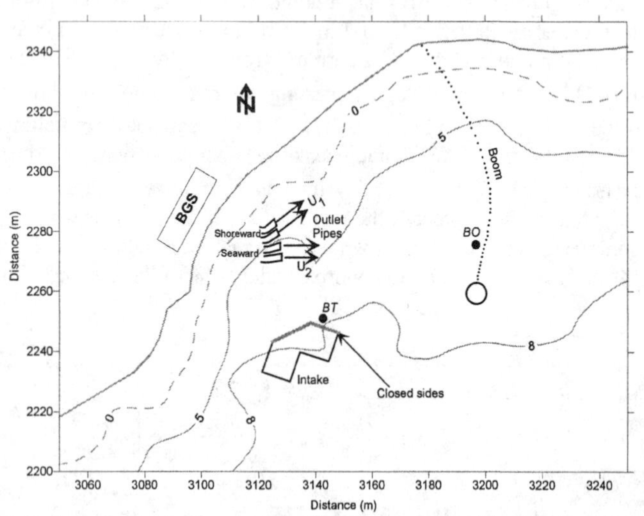

Figure 2. Enlarged map in the near-field of the outlet and intake. Symbols are the same as Figure 1. The boom is limited to the surface and has little or no effect on the flows below ~1 m depth.

Methodology

Governing Equations

The model solves the full three-dimensional, shallow water hydrodynamic and thermodynamic equations in a terrain following sigma-coordinate system (Blumberg and Mellor, 1987). The continuity, momentum, thermal energy and salinity equations, respectively are

Continuity equation:

$$\frac{\partial \zeta}{\partial t}+\int_{-1}^{0}\left(\frac{\partial Hu}{\partial x}+\frac{\partial Hv}{\partial y}\right)d\sigma=\int_{-1}^{0} q\,d\sigma \tag{1}$$

Reynolds momentum equation in the x-direction:

$$\frac{\partial u}{\partial t}+u\frac{\partial u}{\partial x}+v\frac{\partial u}{\partial y}+\omega\frac{\partial u}{\partial \sigma}-fv=-g\frac{\partial \zeta}{\partial x}-\frac{gH}{\rho}\int_{\sigma}^{0}\frac{\partial \rho}{\partial x}d\sigma$$
$$-\frac{g}{\rho}\frac{\partial H}{\partial x}\left(\sigma\rho+\int_{\sigma}^{0}\rho\, d\sigma\right)+\left[\frac{\partial}{\partial x}\left(A_x\frac{\partial u}{\partial x}\right)+\frac{\partial}{\partial y}\left(A_y\frac{\partial u}{\partial y}\right)\right]+\frac{1}{H}\frac{\partial}{\partial \sigma}\left(\frac{A_v}{H}\frac{\partial u}{\partial \sigma}\right) \quad (2)$$

Reynolds momentum equation in the y-direction:

$$\frac{\partial v}{\partial t}+u\frac{\partial v}{\partial x}+v\frac{\partial v}{\partial y}+\omega\frac{\partial v}{\partial \sigma}+fu=-g\frac{\partial \zeta}{\partial y}-\frac{gH}{\rho}\int_{\sigma}^{0}\frac{\partial \rho}{\partial y}d\sigma$$
$$-\frac{g}{\rho}\frac{\partial H}{\partial y}\left(\sigma\rho+\int_{\sigma}^{0}\rho\, d\sigma\right)+\left[\frac{\partial}{\partial x}\left(A_x\frac{\partial v}{\partial x}\right)+\frac{\partial}{\partial y}\left(A_y\frac{\partial v}{\partial y}\right)\right]+\frac{1}{H}\frac{\partial}{\partial \sigma}\left(\frac{A_v}{H}\frac{\partial v}{\partial \sigma}\right) \quad (3)$$

Conservation equation for temperature:

$$\frac{\partial T}{\partial t}+u\frac{\partial T}{\partial x}+v\frac{\partial T}{\partial y}+\omega\frac{\partial T}{\partial \sigma}=$$
$$\left[\frac{\partial}{\partial x}\left(D_x\frac{\partial T}{\partial x}\right)+\frac{\partial}{\partial y}\left(D_y\frac{\partial T}{\partial y}\right)\right]+\frac{1}{H}\frac{\partial}{\partial \sigma}\left(\frac{D_v}{H}\frac{\partial T}{\partial \sigma}\right)+\frac{Q_t}{\rho c_p H} \quad (4)$$

Conservation equation for salinity:

$$\frac{\partial s}{\partial t}+u\frac{\partial s}{\partial x}+v\frac{\partial s}{\partial y}+\omega\frac{\partial s}{\partial \sigma}=\frac{\partial}{\partial x}\left(D_x\frac{\partial s}{\partial x}\right)+\frac{\partial}{\partial y}\left(D_y\frac{\partial s}{\partial y}\right)+\frac{1}{H}\frac{\partial}{\partial \sigma}\left(\frac{D_v}{H}\frac{\partial s}{\partial \sigma}\right) \quad (5)$$

where u, v and ω are the flow velocity components in the x, y and σ directions, respectively, t is the time, q is the source or sink of mass, g is the gravitational acceleration, ζ is the instantaneous water surface elevation, H is the total water depth, $H=\zeta+h$, with h denoting the undisturbed water depth, f is the Coriolis parameter, ρ is the fluid density, A_x and A_y are the horizontal turbulent momentum diffusion coefficients in the x and y directions, respectively, A_v is the vertical turbulent momentum diffusion coefficient, T is the temperature, s is the salinity, D_x and D_y are the horizontal mass diffusion coefficients in the x and y directions, respectively, D_v is the vertical mass diffusion coefficient, Q_t is the thermal source or sink, and c_p is the specific heat of sea water.

A second order turbulence closure model, as described in Mello and Yamada (1982), is applied to calculate the vertical momentum and mass diffusion coefficients, A_v and D_v. The horizontal diffusion coefficients, A_x and A_y, are evaluated using

Smagorinsky's formula (Smagorinsky, 1963). The vertical velocity in the σ-coordinate, ω, is determined according to

$$\omega = \frac{D\sigma}{Dt} = \frac{1}{H}\left[-\int_{-1}^{\sigma}\left(\frac{\partial Hu}{\partial x} + \frac{\partial Hv}{\partial y}\right) d\sigma + (1+\sigma)\int_{-1}^{0}\left(\frac{\partial Hu}{\partial x} + \frac{\partial Hv}{\partial y}\right) d\sigma + \int_{-1}^{\sigma} q\, d\sigma - (1+\sigma)\int_{-1}^{0} q\, d\sigma \right] \tag{6}$$

Boundary Conditions

The boundary conditions of zero surface momentum flux (no wind stress) and bottom shear stress, as expressed in terms of a quadratic law, are employed. At the open boundaries, the water surface elevations are specified from measurements. Under inflows, open boundary conditions of temperature and salinity are specified using observed data, while for outflows, the conventional Sommerfeld radiation condition is applied (Sommerfeld, 1949).

Solution techniques

The governing equations (1) – (5) are solved by a semi-implicit finite difference method in a staggered C-grid, which discretizes the convective and horizontal diffusive terms by an Eulerian-Lagrangian scheme (Casulli and Cattani, 1994), and the barotropic and vertical diffusive terms by an implicit method. Combining the differential continuity and momentum equations, a linear, five-diagonal system of equations for the water surface elevation, ζ, is obtained as the following generalized form

$$a_{i,j}\,\zeta_{i,j}^{n+1} - a_{i+1/2,j}\,\zeta_{i+1,j}^{n+1} - a_{i-1/2,j}\,\zeta_{i-1,j}^{n+1} - a_{i,j+1/2}\,\zeta_{i,j+1}^{n+1} - a_{i,j-1/2}\,\zeta_{i,j-1}^{n+1} = b_{i,j} \tag{7}$$

where a and b are the coefficients dependent on hydrodynamic properties at time step n, the subscripts i and j denote the horizontal location indices and the superscript n represents the time step. This system is solved effectively by the pre-conditioned conjugate gradient method (Casulli and Cheng, 1992).

Nested Grid Scheme

In the presence of a submerged buoyant jet, a much finer mesh in the near-field zone is required in order to resolve the sharp gradients in the hydrothermal field. Such mesh configuration can be accomplished using either unstructured grid or structured multi-grid methods. The unstructured-grid methods have shown impressive flexibility in grid refinement (Mavriplis, 1997), while existing structured multi-grid methods have limitations on high grid refinement as required by a buoyant jet. To obtain the necessary resolution for a buoyant jet using the latter method, one has to apply an adaptive multi-step grid scheme in that the hydrothermal field at each grid has to be calculated in sequence (Berger and Colella, 1989), or an independent multi-block method in that the hydrothermal fields in the coarse and fine meshes are calculated separately and interpolated at the interfaces (Mavripli, 1992; Ramsey, *et al.*, 1996). To overcome the above limitations, a new nested grid scheme is applied in ASL-COCIRM, where the near-field zone of the cooling water buoyant jet, as well as the intake, is embedded within the far-field zone using an extremely high resolution model grid, and these two zones are coupled together at their interfaces. The coupled system of hydrothermal equations is then solved at the same time in the complete field. The nested domain has the spatial grid sizes of $dx' = dx/L$ and $dy' = dy/L$, with dx and dy denoting the spatial grid sizes of the main-domain and L representing the sub-divided step. This approach removes the constraint in a structured mesh on grid refinement and can apply extremely high refinement at a single step. It therefore reduces the coupled interfaces, and at the same time, allows realistic computation of the large gradients in the hydrothermal properties.

Inside the nested-domain, the differential continuity, momentum and thermal conservation equations are the same as in the main-domain. At the interfaces, a coupling scheme is applied in terms of mass, momentum and thermal conservation. To ensure momentum, heat and salinity conservation, the flux forms of the momentum and thermal conservation equations are applied at the interface. By considering mass conservation at the interfaces, the resulting continuity equations are related to both the nested-domain and the main-domain grid points. The generalized continuity equations at nested-domain interior grid points and interface grid points, as shown in Figure 3, are given in Eqs. (8) – (10), respectively. The differential continuity equations at other interfaces will have similar forms.

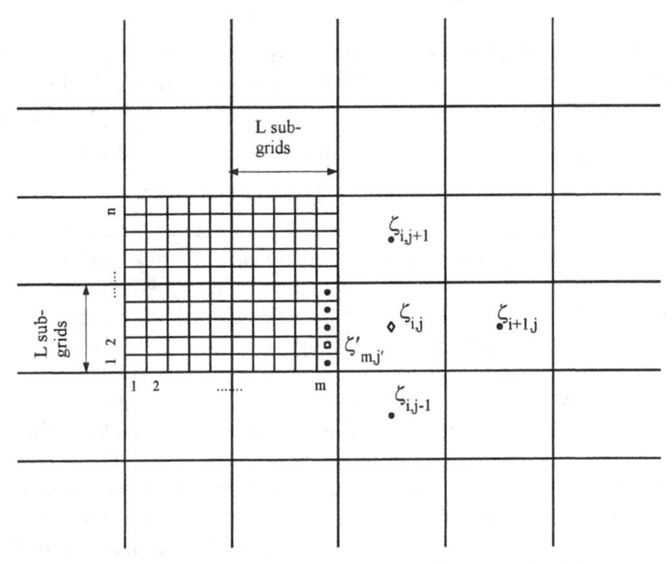

Figure 3. Schematic diagram showing nested-grid.

The interior grid point in the nested-domain:

$$a'_{i',j'}\zeta'^{n+1}_{i',j'} - a'_{i'+1/2,j'}\zeta'^{n+1}_{i'+1,j'} - a'_{i'-1/2,j'}\zeta'^{n+1}_{i'-1,j'} - a'_{i',j'+1/2}\zeta'^{n+1}_{i',j'+1} - a'_{i',j'-1/2}\zeta'^{n+1}_{i',j'-1} = b'_{i',j'} \quad (8)$$

The interface grid point at the nested-domain side (square point in Figure 3):

$$\left(a'^{it}_{m,2} + \frac{2a'_{m+1/2,2}}{L+1}\right)\zeta'^{n+1}_{m,2} - a'_{m-1/2,2}\zeta'^{n+1}_{m-1,2} - a'_{m,2+1/2}\zeta'^{n+1}_{m,3} - a'_{m,2-1/2}\zeta'^{n+1}_{m,1} - \left(\frac{2a'_{m+1/2,2}}{L+1}\right)\left(\beta'_{m+1/2,2}\zeta^{n+1}_{i,j-1} + \beta''_{m+1/2,2}\zeta^{n+1}_{i,j} + \beta'''_{m+1/2,2}\zeta^{n+1}_{i,j+1}\right) = b''_{m,2} \quad (9)$$

The interface grid point at the main-domain side (diamond point in Figure 3):

$$\left[a^{it}_{i,j} + \sum_{j'=1}^{L}\left(\beta''_{m+1/2,j'}\frac{2a'_{m+1/2,j'}}{L+1}\right)\right]\zeta^{n+1}_{i,j} - a_{i+1/2,j}\zeta^{n+1}_{i+1,j} - \sum_{j'=1}^{L}\left(\frac{2a'_{m+1/2,j'}}{L+1}\zeta'^{n+1}_{m,j'}\right) - \left[a_{i,j+1/2} - \sum_{j'=1}^{L}\left(\beta'''_{m+1/2,j'}\frac{2a'_{m+1/2,j'}}{L+1}\right)\right]\zeta^{n+1}_{i,j+1} - \left[a_{i,j-1/2} - \sum_{j'=1}^{L}\left(\beta'_{m+1/2,j'}\frac{2a'_{m+1/2,j'}}{L+1}\right)\right]\zeta^{n+1}_{i,j-1} = b''_{i,j} \quad (10)$$

where a' and b' are the coefficients derived from embedded grid, ζ' is the water surface elevation at the cell centers of the nested-domain, b'' is the combined coefficient of both the main and embedded grids at the interface, β is the weighting coefficient, with $\beta'+\beta''+\beta'''=1$, and a^{it} is a coefficient excluding the interface side, e.g., $a'^{it}_{m,2} = 1 + a'_{m-1/2,2} + a'_{m,2+1/2} + a'_{m,2-1/2}$ in Eq. (9) and $a^{it}_{i,j} = 1 + a_{i+1/2,j} + a_{i,j+1/2} + a_{i,j-1/2}$ in Eq. (10). Although the resulting differential continuity equations at the interface involve more than 5 grid points, with 7 points at the nested-domain interface grid [Eq. (9)] and (4+L) points at the main-domain interface grid [Eq. (10)], the differential Equations (7) – (10) are combined into a positive-definite linear system, and therefore, have an unique solution. Thus, the combined linear system is solved by the pre-conditioned conjugate gradient method in all grid cells and every time step with a single modeling procedure.

The buoyant jet entrainment and diffusion are represented in the model by Smagorinsky's formulation and the second order turbulence closure model. The Smagorinsky formula represents the horizontal diffusion as a function of velocity shear (Smagorinsky, 1963). To ensure that the buoyant jet entrainment will be represented appropriately in the 3D model, the Smagorinsky coefficient was adjusted in order that the jet entrainment agreed with the empirical formulation of Bemporad (1994). Further details on this optimization can be found in Jiang, et al. (2003). In the turbulence closure model, all empirical constants were assigned the values reported in Mellor and Yamada (1982).

In the present application, the main far-field domain included the whole area of Port Moody Arm from the mouth at Burrard Inlet (Figure 1), and was resolved using a grid-size measuring 50 m × 50 m with 10 equally spaced vertical sigma-layers. The nested near-field domain covers an area of 300×200 m^2 (Figure 1) with L=20, which results in a horizontal resolution of 2.5 m × 2.5 m and represents the cooling water outlet and intake in a realistic manner (Figures 4 and 5). It has the same vertical layers as the main-domain. Within the nested domain, selected cells of the equivalent area are used to represent the outlet pipes, i.e., 4.67 m^2 for each individual pipe (Jiang, *et al.*, 2002). By specifying the jet discharge, exit temperature and salinity in these cells, the tidal mean thermal flux and momentum are approximately equal to the actual thermal flux and momentum in the outlet pipes.

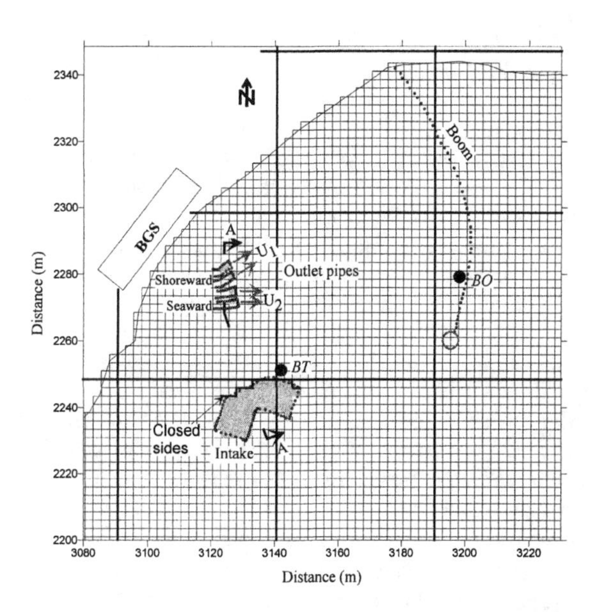

Figure 4. Numerical plane mesh (thin lines for nested-grid and thick lines for main-grid) at the near-field.

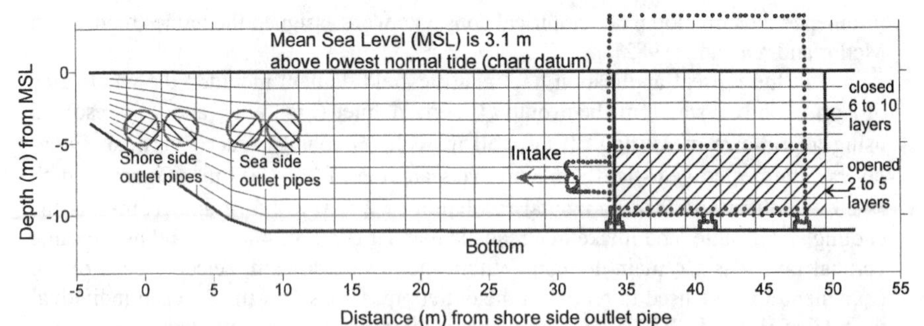

Figure 5. Vertical location of outlet and intake in numerical mesh grid along A–A cross-section (Figure 4).

Model Results

Model Calibration and Verification

Extensive model calibrations and verifications were carried out before examining the potential recirculation of the waste heat into the intake. At first, the simulated buoyant jet was calibrated and verified with an empirical relationship of jet entrainment (Bemporda, 1994) and an integral 1D model (Johnston, et al., 1994) to compute buoyant jet trajectory, dilution and impinging location at the surface. The integral model solves the radially integrated conservation equations following the jet axis. The jet entrainment was calibrated for the simple case of a neutrally buoyant jet discharging into a stagnant homogenous environment. The neutral jet was set at 4 m above bottom with an outlet velocity of 2 m/s. The empirical Smagorinsky coefficient was adjusted until the simulated entrainment rate realized a best fit with the empirical relationship. The correlation coefficient between modeled and empirical entrainment rates is found to be 0.89 (Jiang, et al., 2003). The model was then run with the optimized Smagorinsky coefficient for a buoyant jet discharging into a stagnant environment of typical summer stratification, with a cooling water discharge of 19.6 m^3/s, outlet velocity of U_2 = 2.0 m/s and outlet temperature of 26 °C. Figure 6 compares the numerical model outputs in the nested grid area with the corresponding integral model results and field measurement. It is seen that the model reproduces the buoyant jet trajectory, dilution and impinging location in a realistic fashion.

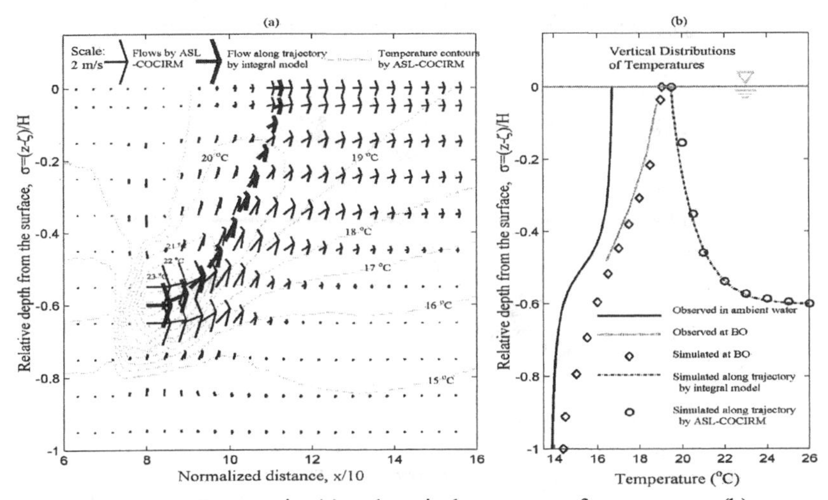

FIGURE 6. Buoyant jet (a) and vertical structures of temperatures (b).

The flow pattern around the buoyant jet was verified using historical dye tracing data (Hodgins and Webb, 1991; Jiang, et al., 2003). Modeled temperature profiles in the near-field, at the sites BT and BO as shown in Figure 2, were then compared with *in situ* observations to demonstrate the model's ability to reproduce the thermal field (Figures 7 and 8). The overall agreement between simulations and observations is very good (Figure 8), with the correlation coefficient of modeled versus observed results up to 0.95 or better (Jiang, et al., 2003), except at surface levels where the effect of wind mixing as a shallow surface layer of 2 – 4 m is not reflected in the model (which did not include wind inputs).

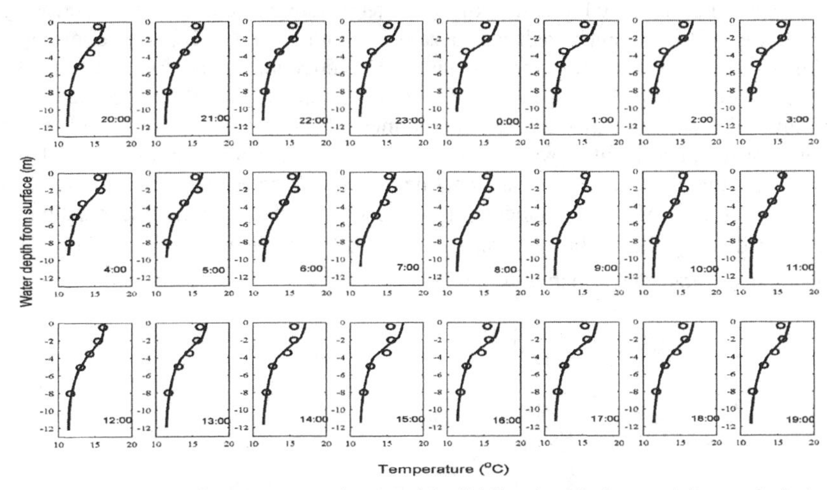

Figure 7. Comparison between simulated (solid lines) and observed (open circles) temperature profiles at BT (see Figure 2 for location) for October 24 – 25, 1998.

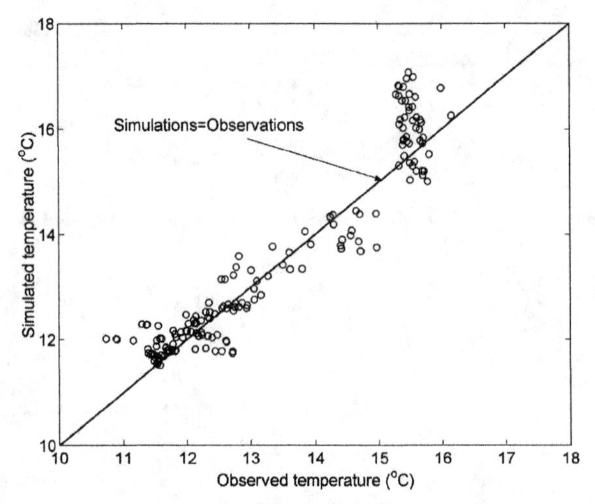

Figure 8. Simulated temperatures as a function of observations at survey elevations for the same site and period in Figure 7.

Cooling Water Recirculation

The simulated flows in the near field show no evident direct recirculation of the cooling water (Jiang, et al., 2001). Water flows at the intake structure are from the SW during flood tide, and originate from the east during ebb tide. Because of the higher water temperatures at the eastern side (Figure 9), which faces the buoyant jet, it is expected that cooling water intake temperature increases during ebb tide. However, the horizontal gradient of water temperature decreases with increasing water depth and is almost reduced to zero at depths below the stratification layer (Figure 7). Therefore, the effect of intake flow direction on cooling water intake temperature could be significant only as the water depth at the intake is reduced (i.e. at low tide) or the deeper part of the thermocline layer extends downward to the levels of the plant input flows. This condition could happen during the ebb of large tide when the intake takes warmer water not only from the eastern side but also from the deeper part of the thermocline layer. Such possible indirect recirculation is examined using the averaged cooling water intake temperature (T_{in}), a key plant operational parameter, which is defined as

$$T_{in} = \frac{\sum_{k=1}^{K}\left[\int_{1.54}^{5.88} T_k(z) q_k(z) dz\right]}{\sum_{k=1}^{K}\left[\int_{1.54}^{5.88} q_k(z) dz\right]} \tag{11}$$

where K is all the grid sides surrounding around the intake structure, the limits of integration mean the intake opening heights (m) above the bottom. $T_k(z)$ and $q_k(z)$ are respectively the simulated temperature and discharge profiles at each side, which are obtained through transformation of simulated temperature and discharge in σ-coordinate into z-coordinate.

The modeled results reveal considerable variations of cooling water intake temperature during large tides (Jiang, et al., 2001), and are in reasonable agreement with the data measured in the intake water mixing box (not shown). Typically, the cooling water intake temperature increases during ebb tide and reaches peak values at around low water (Figure 9). As the water level rises, temperatures decrease again. The variations of 0.3 – 1.0 °C are found in fall and winter seasons, and 1.0 – 1.5 °C variation for summer season (Jiang, et al., 2001). It is also observed that the variation of cooling water intake temperature appears to be very minor during low tide (Figure 9).

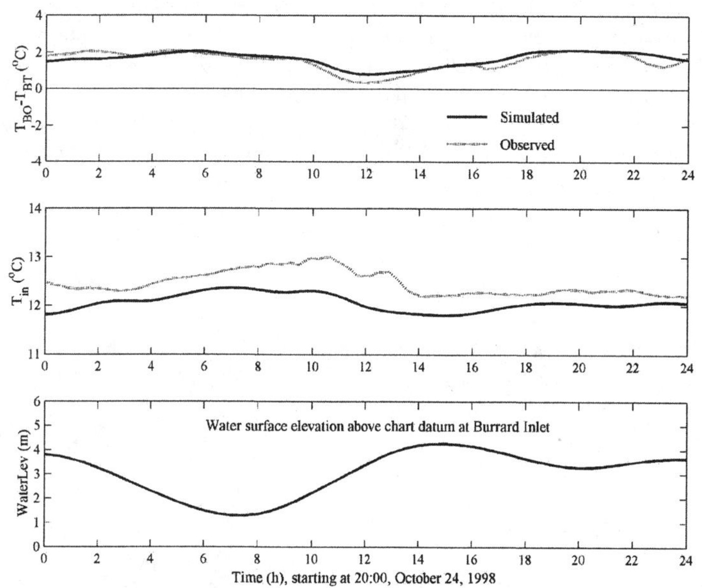

Figure 9. Comparisons between simulated and observed temperature gradients at 5 m depth and the cooling water intake temperature.

As stated above, this feature is mainly caused by stratification (or thermocline) of the vertical profile in water temperature, as shown in Figure 7. During summer (May – September), the thermocline penetrates to greater water depths (to 5 m but less than 8 m depth) and warmer water is found at the top layers and in the eastern side of the intake as well. As water depth decreases dramatically during the ebb from high tide, the thermocline is drawn down to the intake openings with the result that warmer water is taken into the cooling water system. As a result, the cooling water intake temperature gradually increases and reaches peak values around the time of low water, while it decreases over the flood tide. After summer, this effect becomes weaker because the thermocline is closer to the sea surface and further from the intake depths. Thus, there is less variability in cooling water intake temperature with tidal periods, as seen in the October simulation (Figure 9). During winter, this feature could be reversed, dependent on plant output and atmospheric conditions. For extreme weather conditions with air temperature well below zero °C, colder water occurs in the top layers most of the time, which is associated with salinity stratification at the top layers. As a consequence, the intake will take some colder water from the upper layers into the cooling water system when water depth decreases during ebb tide and has lower temperature. As the water depth increases during flood tide, it mainly takes warmer water from the near-bottom layers and has slightly higher temperature.

To better understand the simulated results, the relationship between the measured cooling water intake temperature, T_{in}, and water surface elevation were examined (Figure 10). It is observed that "low" T_{in}, which is defined here as the value at the "trough" of T_{in} tidal variation, occurs during high water, and lower values are usually found during higher high water (HHW) than during lower high water (LHW). While, the majority of peak values of T_{in}, defined in an analogous way as the value at the crest of T_{in} tidal variation, occurs during low water, usually with a higher value during lower low water (LLW) than during higher low water (HLW). Some maximum T_{in} occurs during low water with a lag of 0.5 – 1.5 hour (Figure 10). Therefore, the observations of cooling water intake temperature replicate the simulated results as stated above.

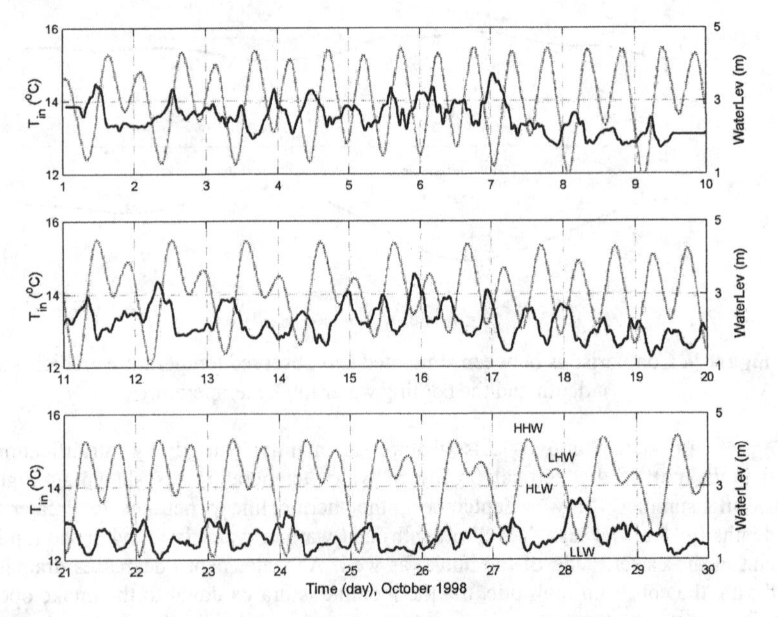

Figure 10. Measured cooling water intake temperature (solid line) and water elevation (dotted line) in October 1998.

Conclusion

The newly-developed nested grid scheme in ASL-COCIRM allows extremely high grid refinement at a single step. Such an approach is useful in representing the hydrothermal properties for a cooling water structure in a realistic manner and resolving the large gradients in the hydrothermal field in the presence of a submerged buoyant jet. By coupling the far-field and near-field and solving the complete zone in a single modeling procedure, the model is capable of adequately capturing the overall behavior of the buoyant jet at every time step and quantitatively examining the possible cooling water recirculation into the intake.

In the current application to model Burrard Generating Station cooling water recirculation, there was no direct evidence of any recirculation occurring. However, depending on the vertical structure of temperature, upper warmer stratified layers could be mixed down to the intake opening and be taken in as cooling water during low water. It is shown that cooling water intake temperatures increase during ebb tide and reach peak values around low tide, and decrease during flood tide. The lowest values of cooling water intake temperatures occur at high water. This effect is most significant during summer and becomes weaker in the fall and winter seasons. Generally, cooling water intake temperatures in summer are about 1.0 – 1.5 °C colder at higher high water than that at lower low water and about 0.3 – 1.0 °C colder in fall and winter seasons, for similar tidal conditions.

Acknowledgements

This study was funded by the Burrard Thermal Generating Station, BC Hydro, Canada and supervised by Mr. Al Brotherston. Measurements used in this study were provided by BC Hydro (Mr. Jeff Greenbank and Mr. Janmohamed Azim) and ASL Environmental Sciences. We wish to thank our colleagues at ASL including David Topham, Al Taylor, David Lemon, Rick Birch, Dave Billenness and Keath Borg for their contributions to the various phases of the measurement analysis and modeling portions of this overall study.

References

Bemporad, G. A. (1994). "Simulation of round buoyant jet in stratified flowing environment." *Journal of Hydraulic Engineering*, ASCE, 120, 529-543.

Berger, M. J., and Colella, P. (1989). "Local adaptive mesh refinement for shock hydrodynamics." *Journal of Computational Physics*, 82, 64-84.

Birtwell, K. L., Brotherston, A. E., Fink, R. P., Fissel, D. B., Greenbank, J. D., Heithaus, L. I., Korstrom, J. S. and Taylor, A. E. (2001). *Thermal Inputs into Port Moody Arm, Burrard Inlet, BC, and Effects on Salmon: a Summary Report*. Canadian Technical Report of Fisheries and Aquatic Sciences 2340, p35.

Blumberg, A. F., and Mellor, G. L. (1987). "A description of a three-dimensional coastal ocean circulation model." In: *Three-Dimensional Coastal Ocean Models*, N. S. Heaps ed., American Geophysical Union, Washington, DC, 1-16.

Casulli, V., and Cattani, E. (1994). "Stability, accuracy and efficiency of a semi-implicit method for three-dimensional shallow water flow." *Computers & Mathematics with Applications*, 27, 99-112.

Casulli, V., and Cheng, R. T. (1992). "Semi-implicit finite-difference method for three-dimensional shallow water flow." *International Journal for Numerical Methods in Fluids*, 15, 629-648.

Fissel, D. B., MacNeil, M., and Taylor, A. E. (1998). *A Study of the Thermal Regime of Port Moody Arm in Relation to the Burrard Thermal Generating Station*. ASL Technical Report, ASL Environmental Sciences Inc., Sidney, BC, Canada, 64p.

Hodgins, D. O., and Webb, A. J. (1991). *Determination of Residual Chlorine in Burrard Inlet Originating from the Burrard Thermal Generating Plant*. Technical Report, Seaconsult Marine Research Ltd., Vancouver, BC, Canada, 55p

Jiang, J., Fissel, D. B., Lemon, D. D., and Topham, D. (2002). "Modeling cooling water discharges from the Burrard Generating Station, BC Canada." *Proceedings of Oceans 2002 MTS/IEEE*, Biloxi, Mississippi, USA, 1515-1521.

Jiang, J., Fissel, D. B., and Taylor, A. E. (2001). *Burrard Generating Station Cooling Water Recirculation Study*. ASL Technical Report, ASL Environmental Sciences Inc., Sidney, BC, Canada, 45p.

Jiang, J., Fissel, D. B., and Topham, D. (2003). "3D numerical modeling of circulations associated with a submerged buoyant jet in a shallow coastal environment." *Estuarine, Coastal and Shelf Science*, 58, 475-486.

Johnston, A. J., Phillips, C. R., and Volker, R. E. (1994). "Modeling horizontal round buoyant jet in a shallow water." *Journal of Hydraulic Engineering*, ASCE, 120, 41-59.

Mavriplis, D. J. (1992). "Three-dimensional multi-grid for the Euler equations." *AIAA Journal*, 30, 1753-1761.

Mavriplis, D. J. (1997). "Unstructured grid techniques." *Annual Review of Fluid Mechanics*, 29, 473-514.

Mellor, G. L., and Yamada, T. (1982). "Development of a turbulence closure model for geographical fluid problems." *Review of Geophysics*, 20, 851-875.

Ramsey, J. S., Hamilton, R. P., and Aubrey, D. G. (1996). "Nested three-dimensional hydrodynamic modeling of the Delaware Estuary." In: *Proceeding of the 4th International Conference on the Estuarine and Coastal Modeling*, ASCE Waterway, Port Coastal and Ocean Division, 53-65.

Smagorinsky, J. (1963). "General circulation experiments with the primitive equations: I. The basic experiment." *Monthly Weather Review*, 91, 99-164.

Sommerfeld, A. (1949). "Partial differential Equations." *Lectures in Theoretical Physics*, Vol. 6, Academic Press.

Taylor, A. E. and Fissel, D. B. (1999). *A study of the thermal regime of Port Moody Arm, Volume I*. ASL Technical Report, ASL Environmental Sciences, Sidney, BC, Canada, 50p + figures and appendices.

Three-dimensional modeling of hydro-thermal structures in Lake Washington

Sung-Chan Kim[1], Billy H. Johnson[2], and Carl F. Cerco[1]

Abstract

A three-dimensional model—CH3D-Z—was applied to simulate the hydrodynamics of the Lake Washington—a deep, stratified temperate lake—over three-year period. The model was forced by winds and precipitation/evaporation as well as inflows from freshwater streams and the outflow through the lock connected to the Puget Sound. Surface heat exchange was also modeled. Realistic annual thermal-stratification cycle was simulated. The thermal structures were sensitive to the stability function of turbulence model related to Richardson number. We compared two approaches—one using non-equilibrium stability parameter and the other using empirical formulae. The former approach results in the better representation of the surface mixed layer depths. For the period of thermal stratification, mixing processes were studied by analyzing isothermal surfaces, buoyancy, and turbulence energy distribution. Basin scale motions represented by seiches and internal waves propagated through metalimnion showed typical response to wind forcing.

Introduction

Lake Washington is located in metropolitan Seattle in the Pacific northwest of the United States (Figure 1). The lake has a length of about 36 km with a width ranging from 1.5 to 6.5 km. The maximum depth of the lake is about 65 m, with an average depth of about 30 m. The total volume of the lake is approximately 29 km^3, and the surrounding drainage area is about 900 km^2. The only outlet of the lake is through Lake Washington Ship Canal and lock system, which connects the lake with Puget Sound.

Lake Washington is a temperate lake with seasonal stratification. The vertical exchange between the surface and bottom layer is a key conduit of nutrient and heat fluxes (e.g., Saggio and Imberger, 2001). Imberger (1998) postulated a conceptual model of vertical fluxes of mass in a stratified lake: Winds induce the surface layer mixing and energize basin-scale internal waves. Hodges et al. (2000) demonstrated the importance of the correct modeling of mixing and basin scale internal waves for a stratified lake by using a three-dimensional mixed-layer model.

---

[1] US Army Corps of Engineers, Engineer Research an Development Center, CEERD-EP-W, 3909 Halls Ferry Rd, Vicksburg, MS 39180
[2] Computational Hydraulics and Transport LLC, PO Box 569, Edwards, Ms 39066

Model setup

We set up a hydrodynamic model linked to a eutrophication model for the lake. We implemented a three-dimensional hydrodynamic model, the CH3D-Z, which has been applied on many water quality studies, most notably in the Chesapeake Bay (Johnson, et. al 1991, 1993 and Cerco et al., 2002). CH3D-Z is a general-purpose 3D hydrodynamic model for simulating flows in rivers, lakes, and coastal areas. The numerical grid is boundary-fitted in the horizontal with the vertical dimension being Cartesian (Z-plane) (Johnson, et al., 1991).

The model has been calibrated over a three year period between 1995 and 1997. In this study, we present the modeling of hydro-thermal structure. First, a brief description of model setup will be given, followed by the representation of varying scales of motion—from basin-scale to turbulence scale. The three-year calibration results will then be shown.

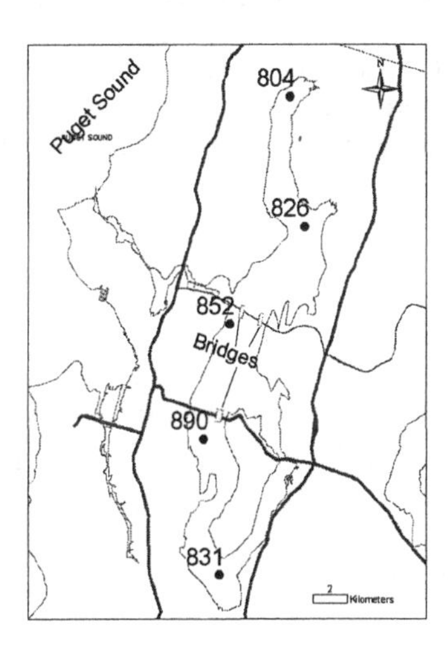

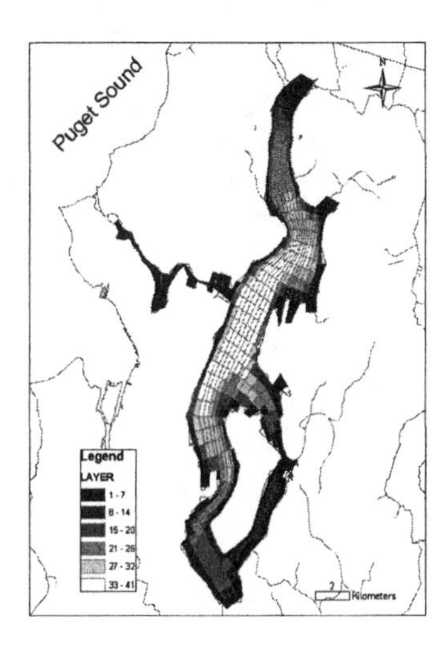

Figure 1. Location map. Three floating bridges were shown in the left panel. Also shown are the monitoring stations used for calibration in this study. The right panel shows surface cells of model grid. The number of layers was shown in different shades. Each layer is about 1.5 m thick.

The numerical grid contains 655 active surface cells with a maximum of 43 vertical layers, resulting in a total of 12,177 active computational cells (Figure 1). The lengths of the horizontal cells vary between 100 m (typically across the Lake axis) and 600 m (along the Lake axis). Each layer is about 1.5 m thick except for the top layer, which varies with water surface elevation. The computational time step was 60 s. At the locations of the two floating bridges in the lake, surface flow was blocked.

Hourly wind data measured at the nearby Seattle-Tacoma International Airport was taken as the major forcing. From the meteorological data at the same airport, we obtained precipitation and computed the surface heat exchange coefficient and the equilibrium temperature at 3-hr intervals following Edinger et al. (1974). The U.S. Army Corps of Engineers (1992) provided the monthly average evaporation from the lake. There are 13 inflow locations including two major contributors—the Sammamish River in the north and the Cedar River in the south. At the Hiram M. Chittenden Lock, an outflow discharge of water was specified based on the mass balance between inflows and lake level observation at the Lock. This was determined from a water balance in the lake. The total change in the volume of the lake over a time step was set to the sum of all inflows plus the precipitation minus the dam outflow and minus surface evaporation.

Basin-scale motion

Kalarius and Morgan (1967) observed that Lake Washington has surface seiches with periods of 54, 32.5, 20.8, 15.4, and 13.0 minutes oscillating about 1, 2, 3, 4, and 5 nodes, respectively. We analyzed spectra of computed water surface elevations at five locations along the major axis (NS-axis) of the lake every time step (60 sec) for three months during summer in 1995. Computed seiching periods were 52, 31, 21, 16 and 14 minutes, respectively (Figure 2). This substantiates the model's ability to resolve seiching at the surface.

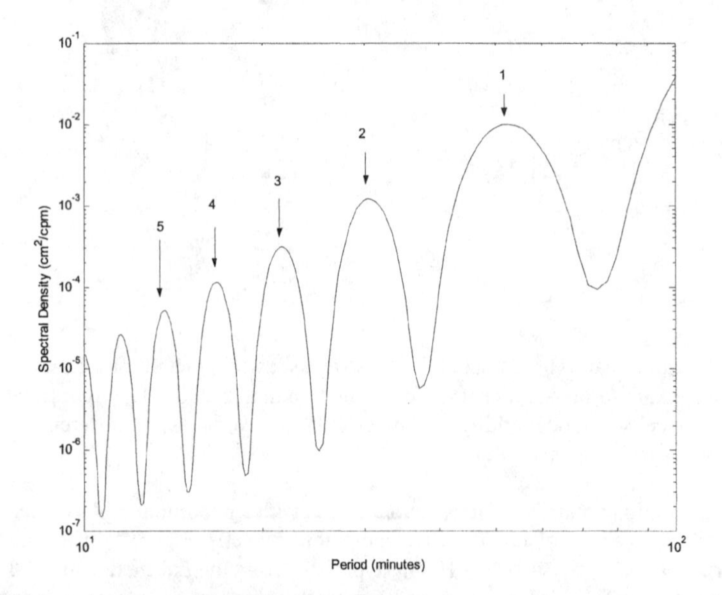

Figure 2. Spectral density of surface elevation averaged over 5 stations along the lake axis during June through August of 1995. The number denotes node.

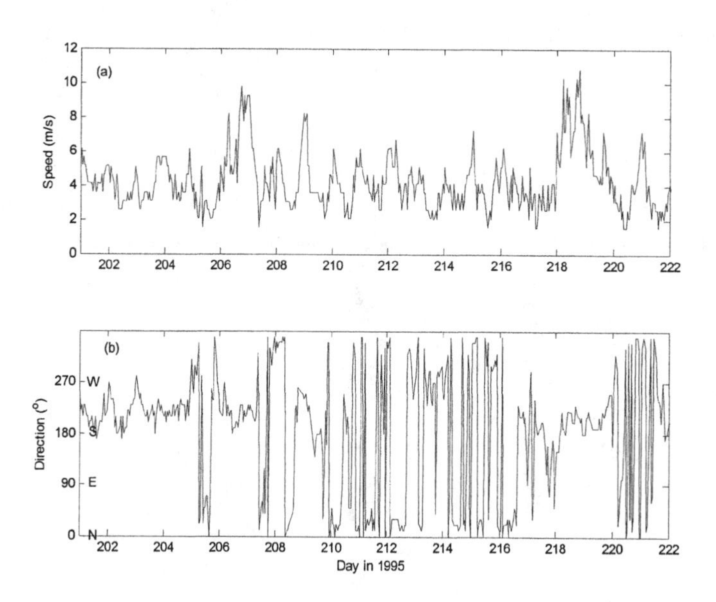

Figure 3. Wind speed (a) and direction (b) during 21 day period between day 201 and day 222 of 1995. Two southwest wind events were marked around day 207 and day 218.

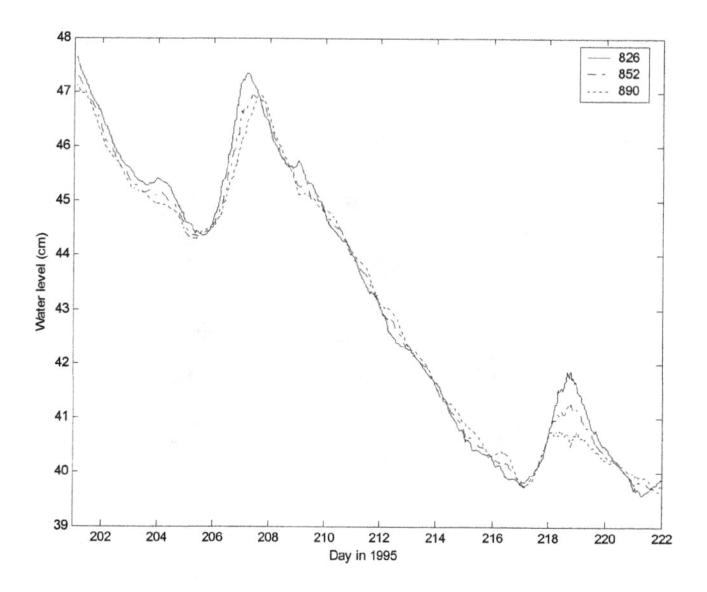

Figure 4. Surface elevations at 3 stations—826, 852, and 890 from north to south, respectively.

(a)

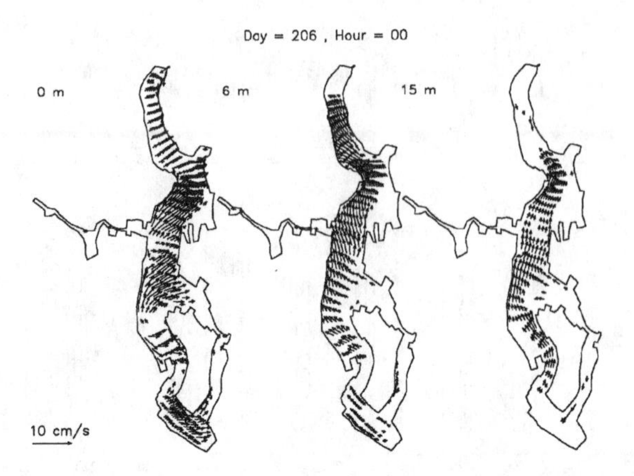

(b)

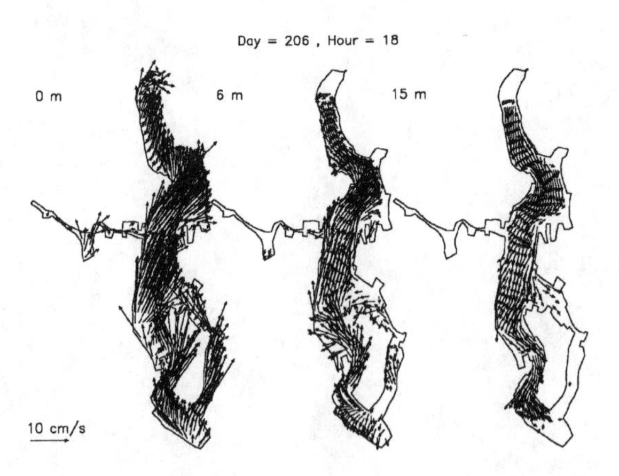

Figure 5. Currents at different layers during a wind event (continued)

(c)

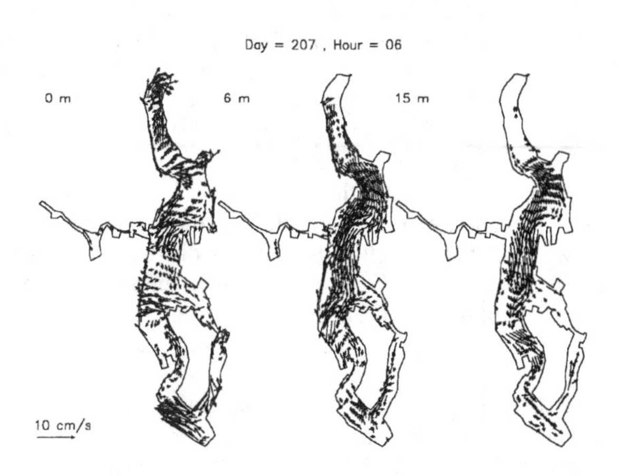

Figure 5. Currents at different layers during a wind event.

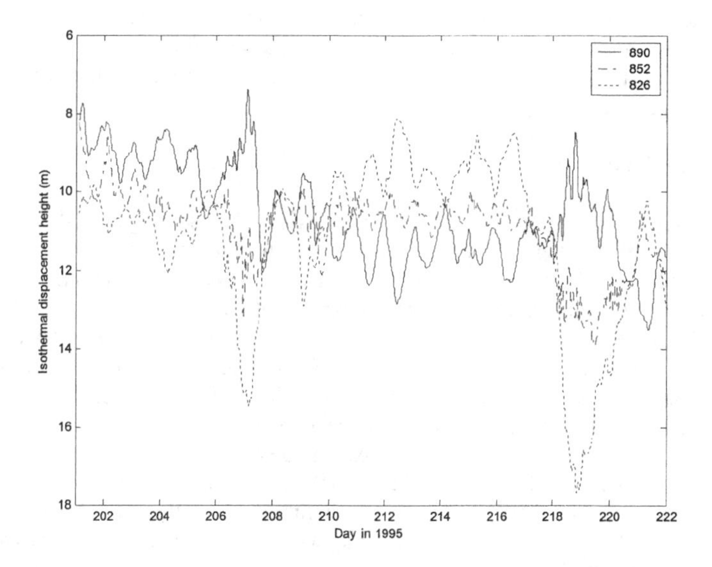

Figure 6. Isothermal depths of 20°C at Station 826 in the northern part of the lake, Station 890 in the southern part of the lake, and Station 852 near the center of the lake.

In order to further investigate the response of the lake to wind forcing, we focused on the period of wind events during summer when stratification is the maximum around day 207 and day 218 in 1995 (Figure 3). Both events were marked by strong southwest wind, which caused setup at the northeast corner of the lake (Station 826) and setdown at the southwest corner (Station 890) (Figure 4). After the event, the tilted surface was relaxed and seiching was shown as barotropic response of the water body to surface forcing. Figure 5 shows current velocities at three levels in different stages of an event. At initial stage (Figure 5a), whole water body responds to the surface wind as one layer. Direction of the flow is downwind everywhere in the horizontal and vertical planes. As the wind matured, downwelling was developed. Figure 5b shows two-layer flow structure. Surface flow is in the downwind direction and the flow 15 m below the surface is in the upwind direction. Figure 5c shows the relaxing stage at the end of the event. The two-layer structure is still maintained but the surface layer becomes thinner and the surface flow is reduced.

Isotherm depths of 20 °C (Figure 6) were out of phase with surface elevations (Figure 4). This is consistent with internal wave activities of stratified lakes (e.g., Antenucci et al., 2000). At the peak of the southwest wind events (e.g., day 207 and day 219), the isotherm depths were deeper in downwind direction (north) (Station 826) than in upwind direction (south) (Station 890). Downwelling at downwind station made the warmer surface layer thicker while upwelling at upwind station made the surface layer thinner. At the peak of the event, high frequency responses were shown. After the relaxation, the isotherm depths returned to the diurnal responses to the sea breezes, especially at boundary stations (Station 826 and Station 890). These two north-south stations showed out of phase relationships in the isotherm depths. The basin scale internal waves existed throughout the summer when the lake was stratified.

Turbulence mixing

Thermal structures vary spatially along the major axis of the lake. Figure 7 shows the time series of isotherm depths along the major axis of the lake. Near the lake boundaries (Station 826 and Station 890), all the isotherm depths were in phase to each other following upwelling and downwelling. But at the deep center of the lake (Station 852), the warmer isotherm depths and the colder isotherm depths are not in phase near the peak of events. The 20 °C and 17 °C isotherm depths were briefly out of phase at day 207 and day 218. This indicates the wind mixing at the surface layer. The more visible higher frequency motion at deeper parts of the lake (Station 852) also points to the importance of turbulence scale mixing. Figure 8 shows the temperature time series at the center of the lake (Station 852). At the peak of events, the mixing was evident not only at the top of metalimnion (less than 10 m) but also in the hypolimnion (below 30 m). In the following, we describe how turbulence mixing was handled in this model.

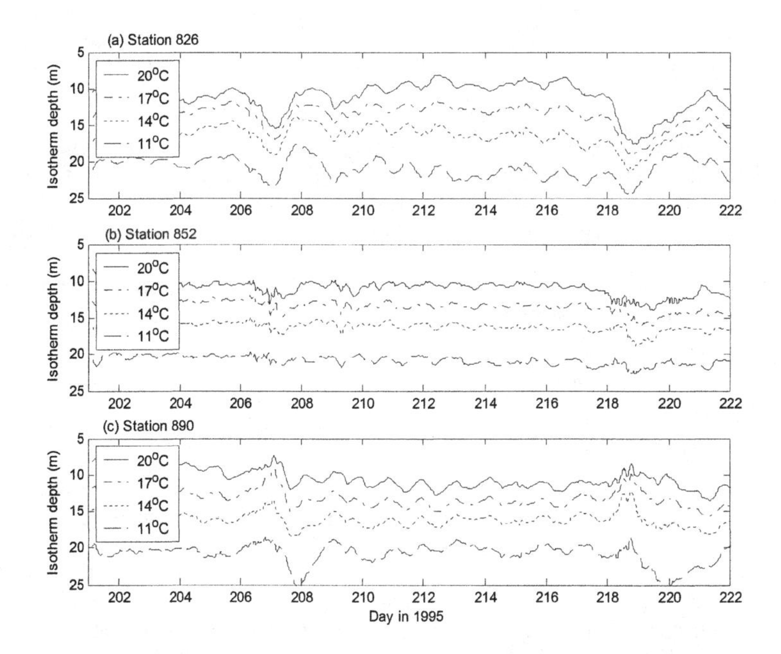

Figure 7. Isotherm depths of 20, 17, 14, and 11 °C at (a) Station 826, (b) Station 890,and (c) Station 852, from north to south, respectively.

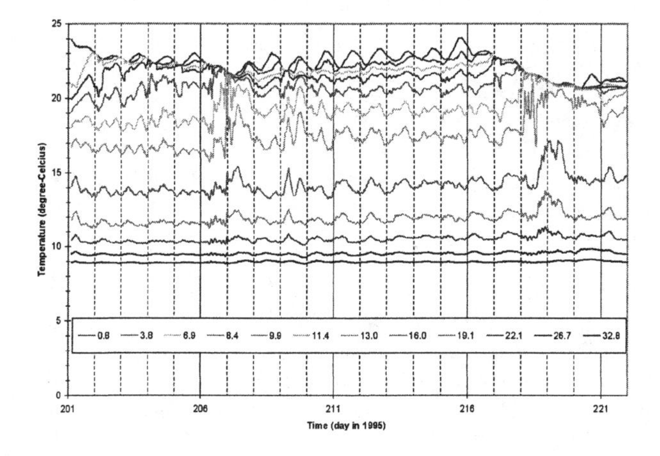

Figure 8. Temperature variations at different depths at Station 852.

We define eddy viscosity and diffusivity (ν_t) as functions of a velocity scale, $\sqrt{k}$, and a length scale, L:

$$\nu_t = C_\mu \sqrt{k} \cdot L$$

where ν_t is eddy viscosity/diffusivity, C_μ is a stability function, k is the turbulent kinetic energy, and L is the turbulence macro length scale. The turbulence energy dissipation rate, ε, is also defined as a function of the velocity and length scales:

$$\varepsilon = \left(C_\mu^0\right)^3 k^{3/2}/L$$

The energy, k, and dissipation, ε, vary in time and space and are related to the source terms and sink term (dissipation):

$$\frac{\partial k}{\partial t} - \frac{\partial}{\partial z}\left(\nu_t \frac{\partial k}{\partial z}\right) = P + B - \varepsilon$$

$$\frac{\partial \varepsilon}{\partial t} - \frac{\partial}{\partial z}\left(\nu_t \frac{\partial \varepsilon}{\partial z}\right) = \frac{\varepsilon}{k}\left(c_{\varepsilon 1}P + c_{\varepsilon 2}B - c_{\varepsilon 3}\varepsilon\right)$$

The major source terms are shear production, P, and buoyancy production, B:

$$P = \nu_{ts} M^2$$

$$B = \nu_{tm} N^2$$

where ν_{ts} is eddy viscosity and ν_{tm} is eddy diffusivity. The relative portion of buoyancy production to shear production is expressed through the Richardson number, Ri ($=N^2/M^2$) where

$$M = \frac{\partial U}{\partial z} \quad \text{: mean velocity shear}$$

$$N^2 = \frac{\partial b}{\partial z} \quad \text{: density gradient (Bruunt - Vaisala frequency)}$$

$$b = -g\left(\frac{\rho}{\rho_0} - 1\right) \quad \text{: bouyancy}$$

Ri represents the stability of the water column.

The solution requires boundary conditions at surface and bottom of the water column, related to friction:

$$k = u_*^2 / C_\mu$$

$$\varepsilon = u_*^3 / (\kappa z_0)$$

Here, u_* is the friction velocity. The governing equations can be solved throughout a water column. But, the distribution of energy is also controlled by the stability of the water column. A stable water column suppresses mass diffusion more than momentum diffusion. The Prandtl number, P_r, determines the ratio between momentum and mass diffusion

$$P_r = \frac{\nu_{ts}}{\nu_{tm}} = \frac{C_\mu}{C_\mu'}$$

Here ν_{ts} is eddy viscosity for momentum and ν_{tm} is eddy diffusivity for mass. C_μ and C_μ' are the stability functions for the momentum and mass, respectively. In this study, we adopted the non-equilibrium stability functions from Canuto et al. (2001):

$$C_\mu = \frac{0.1270 + 0.01526\alpha_N - 0.00016\alpha_M}{A}$$

$$C_\mu' = \frac{0.1190 + 0.00429\alpha_N - 0.00066\alpha_M}{A}$$

$$A = 1 + 0.1977\alpha_N + 0.03154\alpha_M + 0.005832\alpha_N^2 + 0.004127\alpha_N\alpha_M - 0.000042\alpha_M^2$$

The coefficients are defined as

$$\alpha_M = \frac{k^2}{\varepsilon^2}M^2$$

$$\alpha_N = \frac{k^2}{\varepsilon^2}N^2$$

$$Ri = \alpha_N / \alpha_M$$

In previous applications to estuaries (e.g., Johnson et. al., 1991), we have used empirical approaches of relating Pr to Ri. Munk and Anderson (1948) suggested

$$P_r = \begin{cases} P_r^{\,0} & \text{if } Ri < 0 \\ P_r^{\,0}\sqrt{\dfrac{(1+3.33Ri)^3}{1+10\,Ri}} & \text{if } Ri \geq 0 \end{cases}$$

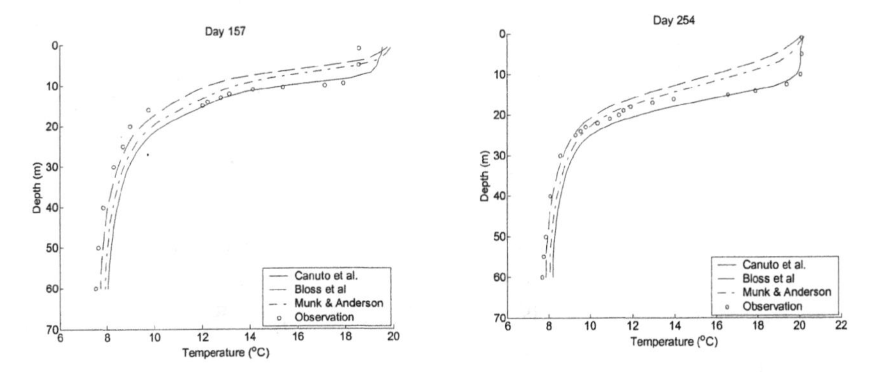

Figure 9. Examples of vertical profiles of temperature distribution at a deep Station 852. A cross represents observed data. A solid line represents the scheme from Canuto et al. (2001) while dotted and dashed lines represent the schemes from Bloss et al. (1988) and Munk and Anderson (1948), respectively.

In an estuary, the vertical momentum/mass transfer is more sensitive to the stability, giving a different formula (Bloss et al., 1988)

$$P_r = \begin{cases} P_r^0 & \text{if } Ri < 0 \\ P_r^0 (1 + 3Ri)^2 & \text{if } Ri \geq 0 \end{cases}$$

As can be seen in Figure 9, both empirical approaches didn't provide enough surface layer mixing, which is critical to the material fluxes in a stratified lake. One may argue that an estuarine model focuses on salt while a lake model focuses on heat, which may suggest the diffusion processes may be different for salinity and temperature.

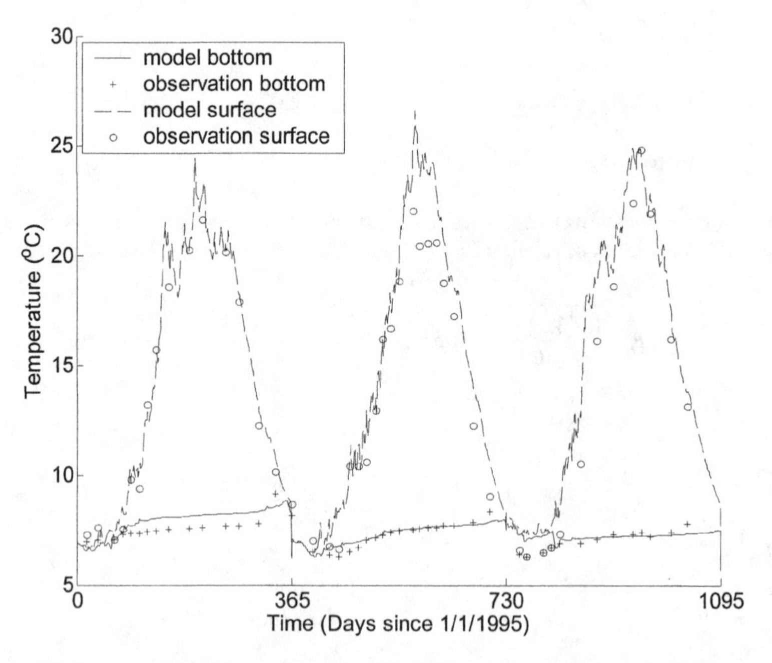

Figure 10. Temperature time series at Station 852 over 3 year time period between 1995 and 1997. Lines represent model results (dashed – surface, solid – bottom) and symbols represent observations (diamond – surface, plus – bottom).

Discussion and Conclusion

The two most important quantities for material fluxes in a lake—the flux path and the distribution of the rate of dissipation of turbulent kinetic energy—were investigated for Lake Washington three-dimensional hydrodynamic model. High frequency seiching with periods ranging from 15 to 60 minutes were resolved by the model, comparable to the observed values. The internal waves over a wide frequency range were investigated. Saggio and Imberger (2001) concluded there exist two classes of turbulence—one is energized by wave-wave interaction and the other is shear-driven—from the observations of turbulent fluxes in the metalimnion of a stratified lake. Our modeling approaches do not address the former mechanism. But with reasonable turbulence model such as the k-ε model adopted in our study, dynamics with varying scales can be represented.

This model is consistent with other short-term modeling (e.g., Hodges et al, 2000) to resolve basin-scale internal waves in a stratified lake as shown in the previous sections. But without comparable frequency of observation, it is only qualitative statement. We need more observation to improve the quantitative prediction of basin-scale motion. One of the important features of lake dynamics is the seasonality. There are not many long-term simulation studies of hydro-thermal structures in a deep stratified lake. Ahsan and Blumberg (1999) modeled Onondaga Lake in New York based on a z-grid version of ECOM. In their model, thermal deepening of the model was slower than observations while this study more closely matched observed results (figure 10). A further study is needed to find a reason for this better timing of thermal deepening by overturning of surface water in the fall.

References

Antenucci, J.P., J. Imberger, and A. Saggio, 2000. Seasonal evolution of the basin-scale internal wave field in a large stratified lake. Limnology and Oceanography, 45, 1621 – 1638

Ahsan, A.K.M.Q. and A.F. Blumberg, 1999. Three-dimensional hydrothermal model of Onondaga Lake, New York. Journal of Hydraulic Engineering. 125, 912 – 923

Bloss, S., R. Lehfeldt, and J. Patterson, 1988. Modeling Turbulent Transport in A Stratified Estuary. *Journal of Hydraulic Engineering,* 114, 1113-33.

Canuto, V.M., A. Howard, Y. Cheng, and M.S. Dubovikov, 2001. Ocean turbulence. Part I: One-point closure model. Momentum and heat vertical diffusivities. Journal of Physical Oceanography, 31, 1413 – 1426

Cerco, C.F., B.H. Johnson, and V.H. Wang, 2002. Tributary Refinements to the Chesapeake Bay Model. ERDC TR-0204, U.S. Army Engineer Research and Development Center, Vicksburg, MS.

Edinger, J., D. Brady, and J. Geyer, 1974. Heat Exchange and Transport in the environment. Report 14, Dept of Geography and Environmental Engineering, Johns Hopkins University, Baltimore, MD.

Hodges, B.R., J. Imberger, A. Saggio, and K.B. Winters, 2000. Modeling basin-scale internal waves in a stratified lake. Limnology and Oceanography, 45, 1603 – 1620

Imberger, J., 1998. Flux paths in a stratified lake: a review. (ed. By J. Imberger "Physical processes in lakes and oceans"). Coastal and Esturaine Studies, 54, 1 – 18

Johnson, B. H., K.W. Kim, R.E. Heath, B.B. Hsieh, and H.L. Butler, 1991. Development and Verification of A Three-Dimensional Numerical Hydrodynamic, Salinity, and Temperature Model of Chesapeake Bay. TR HL-91-7, U.S. Army engineer Waterways Experiment Station, Vicksburg, MS.

Johnson, B. H., K.W. Kim, R.E. Heath, B.B. Hsieh, and H.L. Butler, 1993. Validation of A three-Dimensional Hydrodynamic Model of Chesapeake Bay. Journal of Hydraulic Engineering, 119, 2-20.

Karalius, B.K., and C.W. Morgan, 1967. A Study of Surface and Thermocline Slopes and Seiches in Lake Washington during September and October of 1966. Project Report for Department of Oceanography, University of Washington.

Munk, W.H. and E.R. Anserson, 1948. Notes on the theory of thermocline. Journal of Marine Research, 3, 276 – 295

Saggio, A. and J. Imberger, 2001. Mixing and turbulent fluxes in the metalimnion of a stratified lake. Limnology and Oceanography, 46, 392 – 409

U.S. Army Corps of Engineers, 1992. Water Control Manual for Lake Washington Ship Canal. Seattle District, Seattle, WA.

The Influence of Centrifugal and Coriolis Forces on the Circulation in a Curving Estuary

Nickitas Georgas[1] and Alan F. Blumberg[2]

Abstract

A three-dimensional hydrodynamic model (ECOMSED – of the POM family of models) was used to investigate the physics of estuarine circulation and salt intrusion around a bending region of an estuary. The simulated estuarine channel modeled was long and had a flat bottom. The focus was the effect of transverse circulation created by channel curvature on the dynamics of salt transport, with and without Coriolis forcing. Model results show how, in the absence of Coriolis force, the centrifugal acceleration, directed toward the bend's outer bank during both flood and ebb, tends to give rise to a balancing cross-estuary, barotropic sea level gradient. The vertical shear stress then sets up a vertical plane transverse circulation cell directed toward the outer bank at the channel's surface and toward the inner bank at the channel's bottom. Shear in the along-channel velocity accompanies secondary circulation. In the presence of stratification, the secondary circulation tends to initially produce upwelling of salt at the inner part of the bend, giving rise to a baroclinic pressure gradient directed opposite to the barotropic one at the surface. During slack, when the centrifugal acceleration diminishes, the baroclinic pressure gradient slowly forces the salt at the inner bank towards the bend's outer bank. Thus, secondary circulation creates upwelling of salt, at both banks of the bend, at different times, decreasing the along-shore vertical steady shear dispersion of salt. It may also lead to overturning and intense vertical mixing. Both processes, in turn, lead to a decrease in salt intrusion. Results also show that the Coriolis force can significantly influence secondary circulation around a bend. The Coriolis force, by itself, can be a significant cause of secondary circulation and lateral mixing of salt, even in the absence of a bend. The creation of secondary circulation by Coriolis force alone can also reduce salt intrusion. If the channel bend is present with a curvature that creates centrifugal acceleration in the direction of the Coriolis acceleration, secondary circulation tends to increase, leading, in turn, to even less salt intrusion within the bend. The opposite is also true.

Introduction

The upstream extent of salt intrusion is an important factor influencing the physical characteristics of an estuary. Recently, it has also been recognized as a

[1]Project Scientist, HydroQual, Inc., 1200 McArthur Blvd, Mahwah, NJ 07430; phone 201-529-5151; ngeorgas@hydroqual.com

[2]George Meade Bond Professor of Ocean Engineering, Dept. of Civil, Environmental and Ocean Engineering, Stevens Institute of Technology, Castle Point on Hudson, Hoboken, NJ 07030; phone 201-215-5289; ablumber@stevens.edu

critical habitat indicator of estuarine populations (Jassby et. al., 1995), marking the boundary between salt and freshwater ecosystems. It is, thus, necessary for estuarine scientists to be able to make predictions on anthropogenic or climactic changes that could influence its location. To do so, the complex dynamics that control salt intrusion in an estuary need to be studied and modeled in sufficient detail. Assumptions and simplifications made and used in the past by leading scientists in the field of estuarine dynamics (Pritchard, 1956, Hansen and Rattray, 1965) have recently come under scrutiny by a number of researchers and may need to be reexamined.

The classic longitudinal balance between the pressure gradient and vertical mixing terms in the estuarine governing equations presented in Prichard's historic work in the 1950's has been questioned since the seminal work of Fischer (1972). Fischer suggested that salt intrusion was more a factor of lateral estuarine processes than vertical ones. Lateral (secondary) circulation can be created by a) Channel curvature, b) Coriolis force (for estuarine channels wider that ~5km, the internal Rossby radius of deformation), c) Varying lateral estuarine bathymetry, and d) Transient wind effects. The existence of any of the above will tend to create differential transport of salt along the major estuarine channel, which, in turn, will tend to create cross-channel density gradients. The appearance of the latter will result in baroclinic (gravitational) transverse circulation. The emergent secondary circulation will tend to interact with the longitudinal currents and associated transport of salt, affecting salt intrusion and the location of the turbidity maximum.

In this paper we are concerned with the character and mechanisms of salinity intrusion at the region of an estuarine channel bend. In particular, we are interested in the effect of transverse circulation around a bend of an estuarine channel on the upstream transport of salt. It is known that secondary circulation around river bends increases the rate of transverse mixing (Chang, 1971, Boxall et. al., 2003), and thereby to some extent reduces the alongshore dispersion coefficient (Fischer et al., 1979).

Methods

We approach the problem by means of three-dimensional, hydrodynamic simulations. The model used is ECOMSED, a well-established POM-variant, suitable for estuarine circulation problems (Blumberg and Mellor, 1987). Horizontal (Smagorinksy) and vertical (Mellor-Yamada) closure mixing schemes are included in the model. We neglect wind forcing since its effects on transverse circulation and salinity intrusion are transient (Wong, 1994). To simplify our analysis, we also neglect bathymetric changes. The effect of the Earth's rotation on secondary circulation around a bend is observed through comparison of a run that neglects the Coriolis force (Case 1), and two runs that include it: one for a Northern Hemisphere mid-latitude (Case 2: NH) and one for a Southern Hemisphere mid-latitude (Case 3: SH). By considering both hemispheres, the role of Coriolis acceleration acting with or against centrifugal acceleration at the region of an estuarine bend is clearly elucidated.

The channel modeled has an axial length L=180km, a uniform width B=5km, and a uniform depth H=15m. A sharp semicircular bend is present 45km upstream of the open boundary. The non-dimensional curvature of the bend is R/B=1.5, where R is the radius of curvature. The channel dimensions were chosen so that the work domain is typical of tidal rivers around the world, a local example being the Hudson River. Initially, the channel is filled with fresh water (0psu). A lock-exchange-type density adjustment ("LETDA") is created at the estuary's downstream boundary, by assigning a temporally constant boundary salinity of 30psu there. Salt is, thus, forced into the estuary by the resulting along-channel density gradient, and by a simple semidiurnal M_2 tide with amplitude of 1m.

The computational dimensions of the model domain are 10 segments wide by 120 segments long, by 10 uniformly distributed vertical sigma layers (Figure 1). The semicircular bend is represented with 25 segments in the axial direction to effectively capture the local dynamics. The bend starts at segment I=34, ~45km upstream of the open boundary, and ends at I=58, ~68km upstream of the open boundary. The model spin up is 1 day and the runs continue for another 17 days.

Results

Rozovskii (1957) conducted one of the first hydraulic experiments of secondary circulation created by a steady, open channel flow around a bend. In his important research he showed that as the river flow enters the bend, it veers due to centrifugal acceleration toward the outside bank. This, in turn, piles up water at the outer wall creating a balancing sea level slope in the cross-channel direction, pointing toward the inner bank, a phenomenon known in hydraulic engineering as superelevation. A secondary (lateral) circulation cell is thus set up by the 3D balance of centrifugal acceleration, barotropic pressure gradient, and vertical shear stress: water flows toward the outer bank at the surface, downwells, flows back toward the inner bank at the bottom, and upwells there. Superimposed on this secondary circulation is the primary movement of water downstream, completing the phenomenon known as helical motion. Hydraulic experiments with open channel flumes by Shukry (1950) have shown that this type of motion is observed in sharply curving channels with R/B<3.

The results of a modeling experiment – similar to Rozovskii's cases – with 50,000m^3/s of river flow entering from the upstream end of the simulated domain and no tidal forcing or LETDA at the downstream boundary are shown in Figure 2. The results, taken at the end of the 18-day simulation, display the typical characteristics of Rozovskii-type helical flow. It is important to note, however, that the set up of the cross-channel sea level slope balancing the centrifugal acceleration is not immediate, but rather takes several hours until the helical flow is fully developed. This is verified by the spin down time of secondary flows, $H/(C_DU)$ (Chant, 2002), where U is the river velocity. This is important to the discussion of the primary focus of this paper, which is of tidal flow around a bend. A well-developed, uniform, cross-channel sea surface slope does not occur during the strongly time-dependent forcing of a tidal regime in the absence of a river.

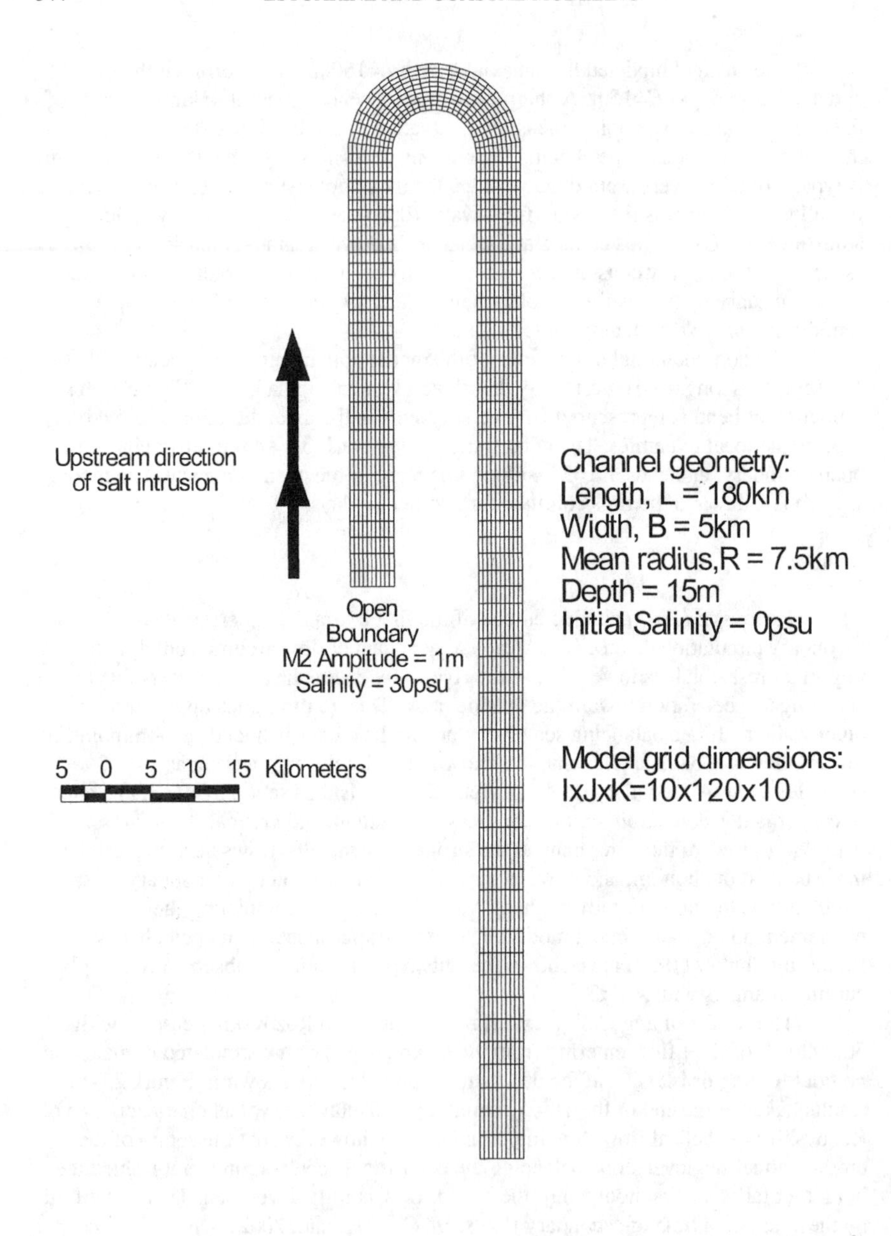

Figure 1. Model Grid

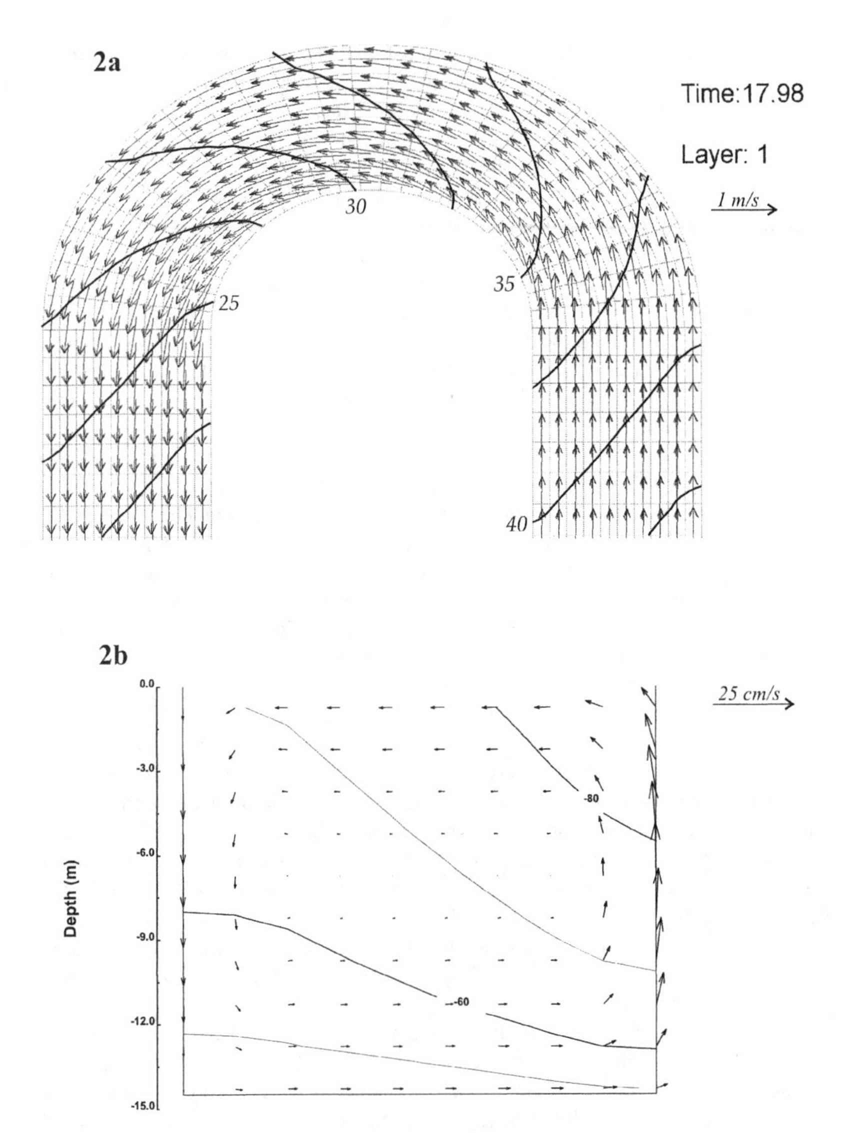

Figure 2. Secondary circulation induced by an 180° river bend. 2a) Surface velocities and sea level contours (cm). 2b) Cross-channel secondary circulation across the bend's middle transect (90°), with along-channel velocity contours in cm/s. Looking upstream. Vertical scale exaggerated by 450 times to match figure size.

Evidence of secondary circulation in the bend was, however, present in all three cases investigated in this work. Note that these runs did not include river flow at the upstream boundary to simplify the discussion that follows. Sensitivity runs with river flow show that the inclusion of such freshwater input decreases upstream salt transport, increasing the vertical shear, as expected. Thus, the model under consideration can be thought either as a tidal channel or as a tidal river at the beginning of a dry period following a storm.

Before the time the intruding salt reaches the bend, baroclinic flows are absent from the curving region of the domain, which, till then, experiences a tidally driven, barotropic circulation. For the first two simulation days, the salt did not reach the bend in any of the three model runs. Thus, we were given the opportunity to test whether a secondary circulation cell may exist at the region of a tidal channel bend, with, homogeneous, well-mixed waters lacking baroclinic gradients. Results during these first couple of days are discussed below.

The first run had tidal forcing, as well as LETDA on the open boundary, yet neglected the Coriolis force. Figure 3 shows that Rozovskii-type secondary circulation developed during both maximum flood and maximum ebb. A cross-channel sea level slope is evident, with higher water at the bend's outer bank. Water at the surface flows against that sea level slope, while water at the bottom flows with it, toward the bend's inner bank. As discussed above, the cause of that circulation is centrifugal acceleration. Since centrifugal acceleration does not reverse direction between ebb and flood and is always pointing toward the bend's outer bank, the transverse circulation cell does not change direction and is always present, becoming weaker and stronger with the fluctuating, alongshore, tidal current. Maximum alongshore velocities, u_s, occur near the inner bank, during both flood and ebb, as in the river-only case of Figure 2, and as in flat-bottom open flume experiments around 180° bends by Shukry (1950), and Rozovskii (1957). In the transition period around slack, the current first changes direction at the inner bank as well, and the waters along the two banks of the bend momentarily move in the opposite direction. Note also that a transition region exists immediately downstream and upstream of the bend (during ebb and flood, respectively). There, lateral circulation also forms, as inertia prohibits the current from being immediately streamlined with the straight axis of the channel upon leaving the bend, creating pressure differentials and flow separation with visible eddies.

Results taken before salt entered the bend from the simulations that included the Coriolis force (Case 2: NH, and Case 3: SH) show the impact of the Earth's rotation on the secondary circulation of homogeneous water around a channel bend subject to tidal forcing. While centrifugal acceleration is always directed toward the outside of the bend ($-u_s^2/R$), Coriolis acceleration ($+|f|u_s$ for the Northern Hemisphere) shifts direction with the alongshore current. In a straight channel, that shift in direction would lead to a reversal of the Coriolis-induced secondary circulation and cross-channel slope from ebb to flood and vice versa. In a curving channel, when the current is such that the Coriolis acceleration is pointing at the direction of the centrifugal acceleration, the secondary circulation cell is magnified. Conversely, secondary circulation weakens when the Coriolis and centrifugal accelerations are pointing in the opposite direction. The lateral circulation cell can

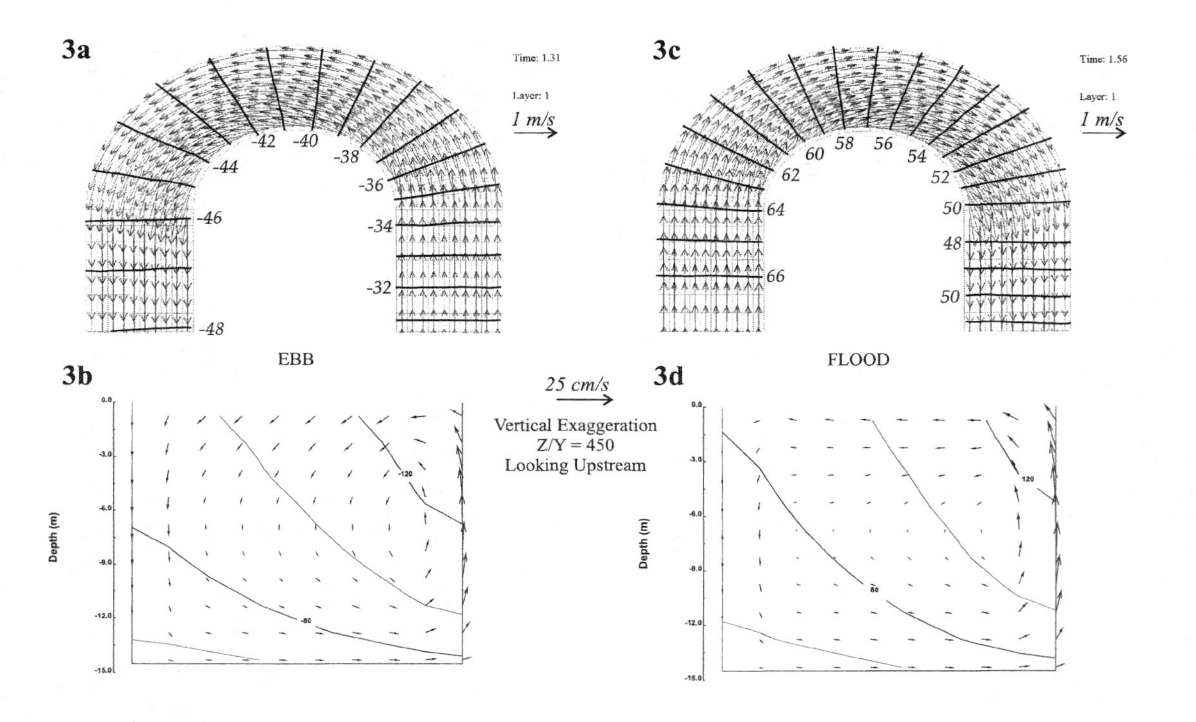

Figure 3. Secondary circulation induced by an 180° river bend in a well-mixed, homogeneous estuary. Case 1: Without Coriolis forcing. 3a) Surface velocities and sea level contours (cm) during ebb tide. 3b) Cross-channel secondary circulation across the bend's middle transect (90°) with along channel velocity contours in cm/s. Ebb tide. 3c) Surface velocities and sea level contours (cm) during flood tide. 3d) Cross-channel secondary circulation across the bend's middle transect (90°) with along channel velocity contours in cm/s. Flood tide.

even reverse in the beginning of each tidal cycle, when the Rossby number u_s/fR becomes small, and Coriolis acceleration dominates. Results are shown in Figure 4 for Case 2 only, with the Coriolis force taken to be for the Northern Hemisphere. In Case 3, with Coriolis forcing simulated for the Southern Hemisphere, the opposite effect is observed: The secondary circulation cell is weaker during ebb, and stronger during flood for that case. In Figure 4, note that the barotropic cross-bend pressure gradient during flood has reversed, and is directed opposite to the Coriolis force, even though the secondary circulation has not reversed. This may be due to the transient character of tidal forcing. The Coriolis force seems not to be balanced by the sum of pressure gradient and centrifugal force in the lateral direction at that time.

When salt coming from the open boundary reaches the bend introducing stratification, the baroclinic pressure gradient becomes important in the balance of forces in the cross-channel direction. The salt reaches the bend between days 3 and 4. It intrudes first along the bottom of the inner bank due to both the direction of secondary circulation in the lower layers, and due to higher alongshore velocities there. The strong secondary circulation then creates upwelling in the inner bank. Thus, salt is introduced in the middle layers of the estuarine bend not through longitudinal advection or vertical diffusion but through upwelling. A baroclinic pressure gradient is hence created directed both toward the outer bank, as well as downstream, opposite to the direction of salt intrusion. During the slack that follows the water is mostly subject to the baroclinic forcing, and starts moving cross-stream, parallel to isohalines. During this time, horizontal transport and dispersion of salt is not dominated by the streamwise dynamics alone, as energy is dissipated not only in the alongshore direction but, also, in the cross-stream one. Maximum cross-stream velocities reach 65 cm/s for a flooding streamwise current of 150 cm/s. As the lateral transport of upwelled salt will interfere with the longitudinal balance of the baroclinic pressure gradient and vertical mixing, the resulting estuarine circulation cannot be explained by the classical estuarine dynamics described in Pritchard (1956). The contribution of the lateral shear terms will, in times, be comparable, or exceed the contribution of the vertical shear terms to the alongshore transport of salt.

It should be noted here that similar dynamics would develop in a straight channel with Coriolis forcing. Coriolis-induced secondary circulation would tend to create upwelling along the sides of the straight estuary. The difference with the curved channel case with and without Coriolis is, again, that the direction of the cross-channel centrifugal acceleration is governed by the curvature and does not reverse, while Coriolis does. Thus, in a straight channel situated at a Northern Hemisphere mid-latitude, upwelling would happen at the right bank during flood, followed by upwelling at the left bank during ebb, looking upstream, caused by the tidal reversal of the Coriolis acceleration. In the curved channel case without Coriolis forcing a shift in the bank where upwelling happens from the inner to the outer will also occur, but not due to the reversal of centrifugal acceleration. It is rather driven by the time-variant nature of the centrifugal force, which first creates a cross-channel baroclinic pressure gradient during tidal peaks, and then allows it to adjust against friction when velocities diminish during slack. That adjustment may be abrupt, and may cause overturn of the water column.

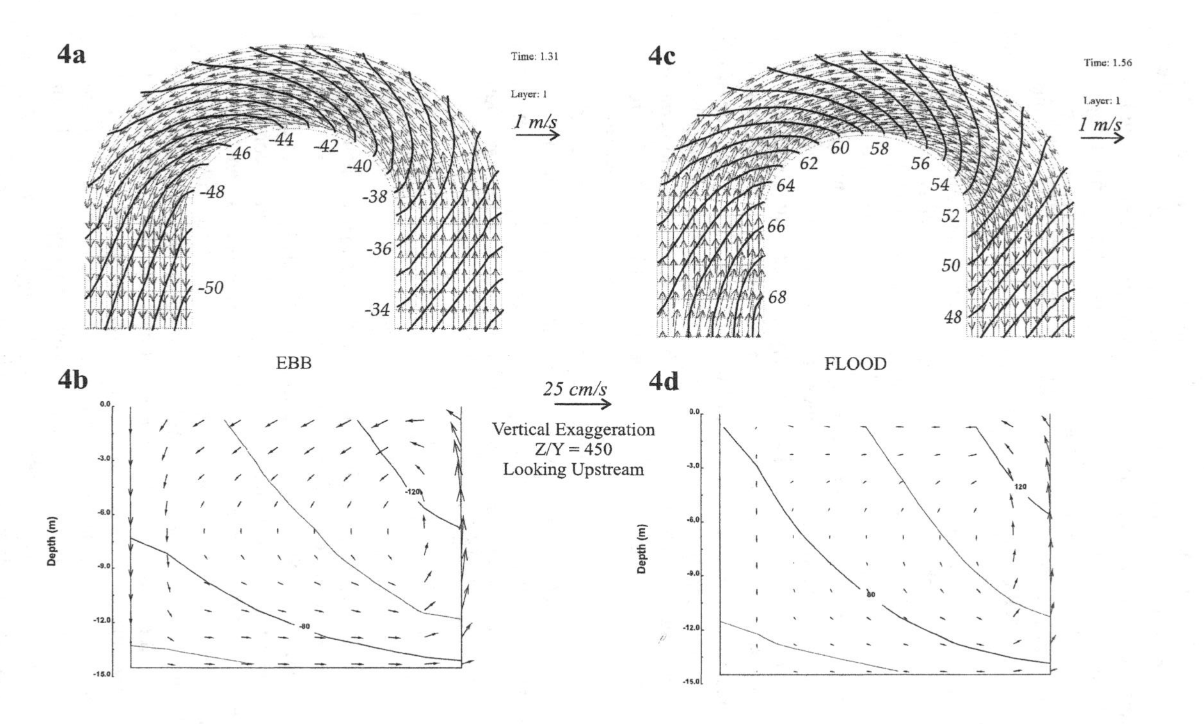

Figure 4. Secondary circulation induced by an 180° river bend in a well-mixed, homogeneous estuary. Case 2: Northern Hemisphere. 4a) Surface velocities and sea level contours (cm) during ebb tide. 4b) Cross-channel secondary circulation across the bend's middle transect (90°) with along channel velocity contours in cm/s. Ebb tide. 4c) Surface velocities and sea level contours (cm) during flood tide. 4d) Cross-channel secondary circulation across the bend's middle transect (90°) with along channel velocity contours in cm/s. Flood tide.

At day 3.85, during flood, the vertically- and cross-stream-averaged vertical eddy diffusion coefficient, K_M, computed by the Mellor-Yamada closure scheme across the middle of the bend increases sharply. During that flood, K_M peaks at a value equal to $0.4m^2/s$, two orders of magnitude higher than the value computed by the model during the rest of the simulation. This increase in K_M is mostly evident in the upper part of the water column. It is observed that this is due to an overturn of the water column causing intense vertical mixing, as the salt upwelled at the outer bank due to density adjustment during the previous ebb, does not have time to readjust, while the secondary circulation created by flooding starts upwelling in the inner bank. A void of salt is thus created in the middle of the channel, and the resulting converging cross-channel baroclinic pressure gradients cause an overturn of the water column in the center of the channel. A similar phenomenon has been observed in a triple junction of Puget Sound where R/B~1.67 (Seim and Gregg, 1997).

The richness of the dynamics is illustrated in Figures 5-7. In Figure 5, during flood, and in the absence of the Coriolis force, significant upwelling happens in the outer bank, coupled by downwelling in the inner bank, both caused by the centrifugally-induced secondary circulation. Five days after the salt reached the bend, circulation has reversed sign compared to the one in the absence of stratification, clearly as an effect of a change in the balance of forces due to the introduction of salt. The centrifugal acceleration appears to be now balanced by the baroclinic rather than barotropic pressure gradient, directed toward the inner bank. When Coriolis forcing is included, again, it tends to amplify or reduce secondary circulation, and the cross-channel baroclinic gradients it introduces (Figures 6-7), depending on whether it acts in the direction of the centrifugal force. For example, during flood, an increase in the baroclinic pressure gradient is observed when the Coriolis acceleration taken for the Southern Hemisphere acts in concert with the centrifugal one (Figure 7). Conversely, a decrease in the baroclinic pressure gradient is observed during flood for a Northern Hemisphere mid-latitude (Figure 6).

Discussion

Boxall et al. (2003), conducted hydraulic experiments to evaluate the effect of channel curvature on the transverse mixing coefficient of a meandering open channel. The researchers concluded that the transverse mixing coefficient varies in direct relation to channel curvature, the variation being cyclic with meander geometry: it increases at the bend's apex, and decreases in straighter regions. This analysis, of course, neglected the effects of the Coriolis force, which is not normally a factor on fluvial hydrodynamics. In the present investigation however, the effect of Coriolis force on estuarine salt transport is found to be significant. The effect of Coriolis force on a sinus estuarine channel tends to be asymmetric, reinforcing secondary circulation when centrifugal acceleration is at the direction of the Coriolis acceleration, and weakening, and reversing, secondary circulation on the next, opposite bend. This, in turn, will lead to a greater increase in the transverse mixing coefficient at one bend, and a lesser increase at the following, oppositely directed bend, compared to a case where the Coriolis force is not important.

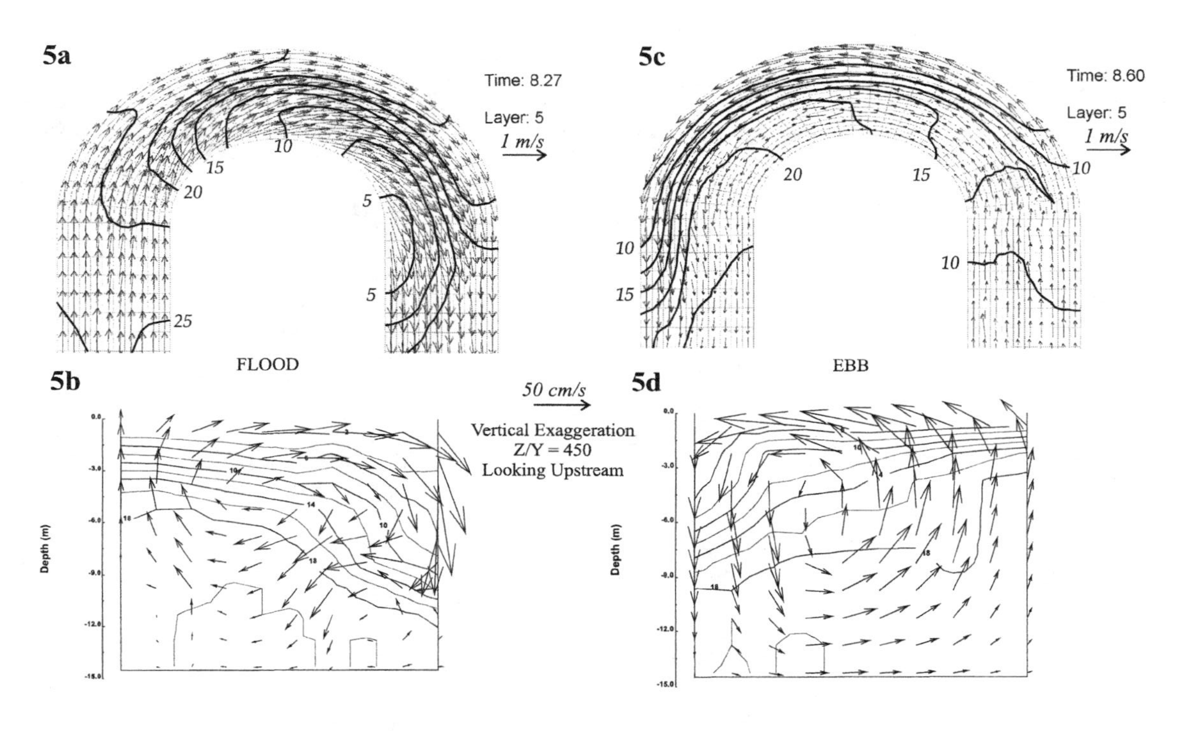

Figure 5. Secondary circulation induced by an 180° river bend in a stratified estuary. Case 1: Without Coriolis forcing. 5a) Mid-layer velocities and salinity contours (psu) during flood tide. 5b) Cross-channel secondary circulation across the bend's middle transect (90°) with along channel salinity contours in psu; Flood tide. 5c) Mid-layer velocities and salinity contours (psu) during ebb tide. 5d) Cross-channel secondary circulation across the bend's middle transect (90°) with along channel salinity contours in psu; Ebb tide.

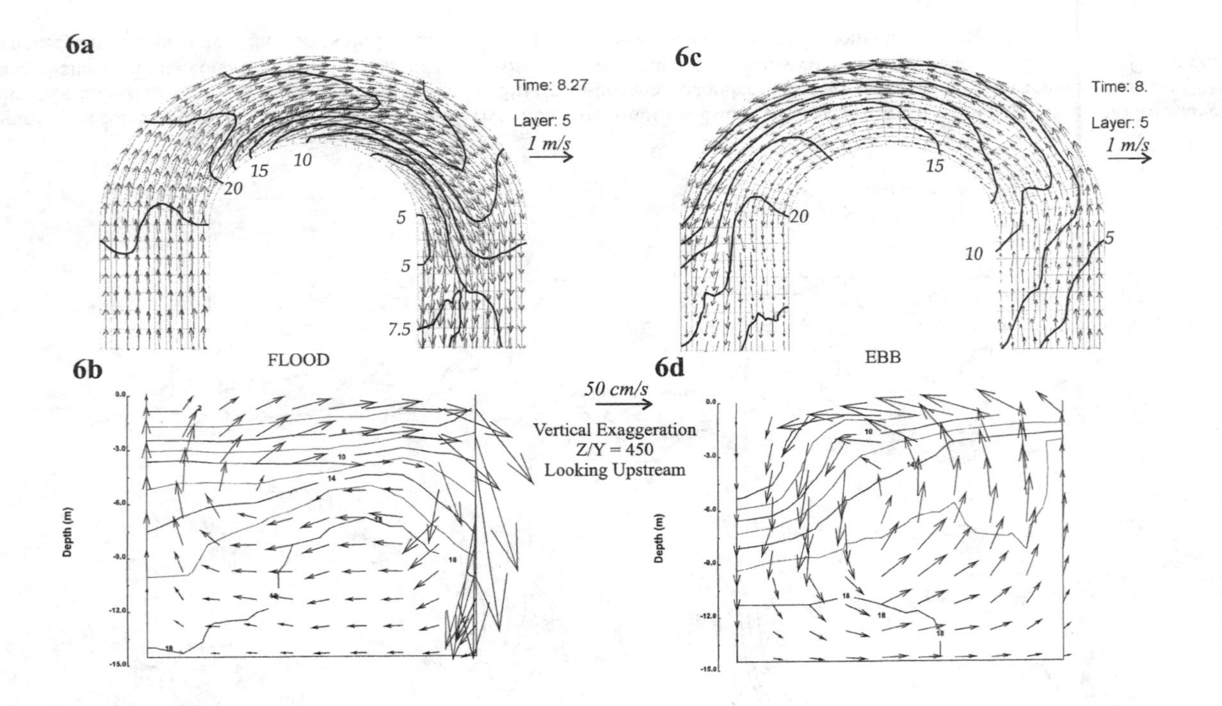

Figure 6. Secondary circulation induced by an 180° river bend in a stratified estuary. Case 2: Northern Hemisphere. 6a) Mid-layer velocities and salinity contours (psu) during flood tide. 6b) Cross-channel secondary circulation across the bend's middle transect (90°) with along channel salinity contours in psu; Flood tide. 6c) Mid-layer velocities and salinity contours (psu) during ebb tide. 6d) Cross-channel secondary circulation across the bend's middle transect (90°) with along channel salinity contours in psu; Ebb tide.

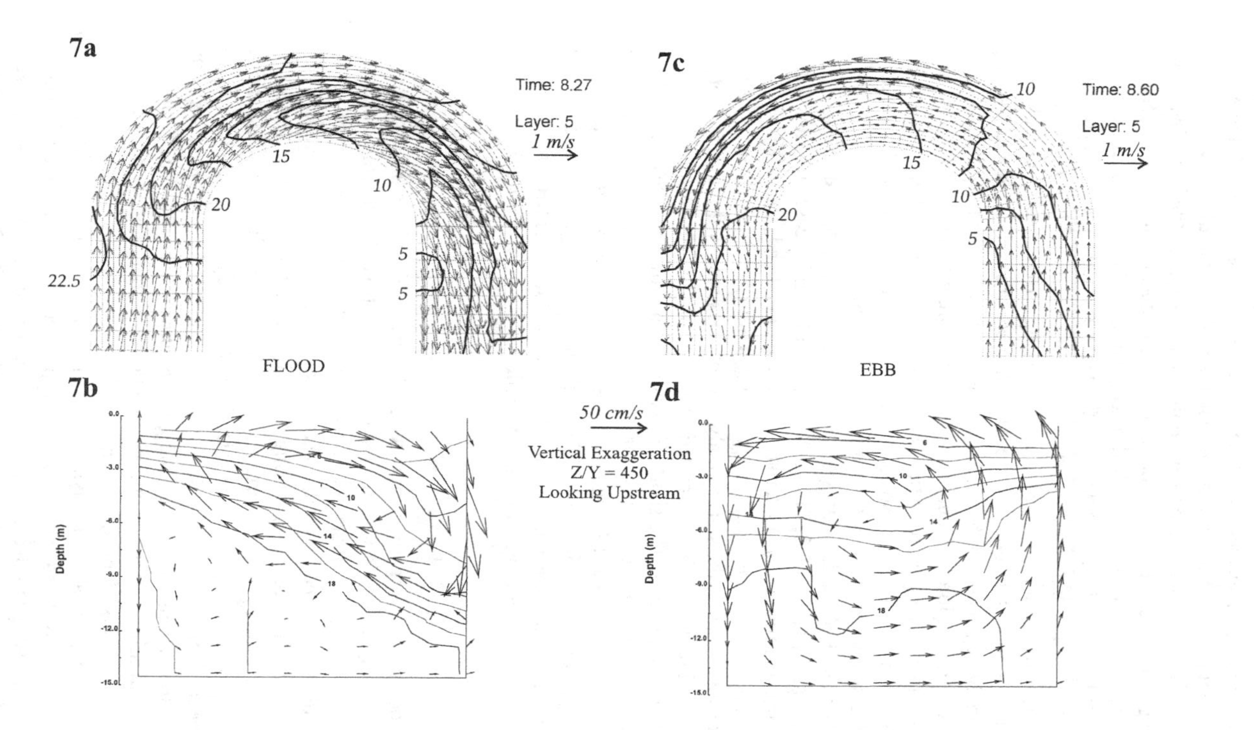

Figure 7. Secondary circulation induced by an 180° river bend in a stratified estuary. Case 3: Southern Hemisphere. 7a) Mid-layer velocities and salinity contours (psu) during flood tide. 7b) Cross-channel secondary circulation across the bend's middle transect (90°) with along channel salinity contours in psu; Flood tide. 7c) Mid-layer velocities and salinity contours (psu) during ebb tide. 7d) Cross-channel secondary circulation across the bend's middle transect (90°) with along channel salinity contours in psu; Ebb tide.

The emergence of secondary circulation in the region of an estuarine bend caused by local centrifugal acceleration is seen to create two processes of major importance to the upstream transport of salt. First, secondary circulation will create across channel density gradients due to upwelling at the banks of the bend that may be severe enough to cause overturn of the water column due to buoyancy effects. That overturn will dramatically increase vertical mixing simulated by the vertical turbulent mixing coefficient. This will tend to decrease the vertical steady shear considered by Pritchard as the major mechanism for upstream transport of salt. Through classic estuarine dynamics then, this increase in vertical mixing will decrease vertical stratification and, in turn, salt intrusion. Second, the vertically advected and laterally dispersed salt in the region of the bend will create a new along-channel density gradient. This emergent along-channel baroclinic pressure gradient will tend to disperse salt downstream, opposed to the LETDA forcing imposed at the open boundary. Finally, when the two opposing baroclinic salt fronts will merge, local upwelling will also happen, further decreasing stratification.

Figures 8 and 9 show salt intrusion as a function of time for the cases presented in this study (Cases 1-3; Figure 9), along with similar cases for runs with similar geometric and forcing characteristics, but without a bend (Cases 1S-3S; Figure 8). Results shown were time-averaged using a 34hr low-pass-filter to remove tidal oscillations. Of course, the salt front always intrudes mostly through the bottom layer. Thus these figures may be interpreted as a time series of the location of isohalines in the bottom layer. The short discussion of the straight cases serves the purpose of clarifying the effects of curvature. Compared to a straight channel case without Coriolis forcing (Case 1S; Figure 8), note the decrease in bottom salt intrusion through Coriolis forcing alone (Cases 2S-3S; Figure 8) acting on a straight estuary, as well, as the decrease in bottom salt intrusion due to centrifugal acceleration with and without geostrophic dynamics (Cases 1-3; Figure 9).

Conclusions

The modeling investigation presented in this paper elucidates the considerable influence of centrifugal and Coriolis forces on the circulation of a curving estuary. Salt transport is shown to be hindered by a sharp bend in an estuarine channel, as secondary circulation created by inertial and/or Coriolis accelerations increases the vertical eddy mixing and creates upwelling advection of salt, leading to a less vertically stratified estuary. It is important to note that the curvilinear grid the model used in this study made it possible to efficiently model the dynamics along and across an estuarine bend. This could also be done effectively with a finite element grid, but certainly not with a rectangular orthogonal grid. In addition, the use of a 3D model made it possible to fully explore secondary circulation, something that would be impossible with a 2D model, even in the homogeneous case.

In this modeling analysis bathymetric changes and the effect of different bend curvature were not considered. This will be incorporated in subsequent analysis. Preliminary results using similar methodology as the one used in this paper, as well as field investigations (Nunes Vaz and Simpson, 1985, Chant, 2002) and analytical modeling (Li and Valle-Levinson, 1999, and Li, 2001) have shown that varying

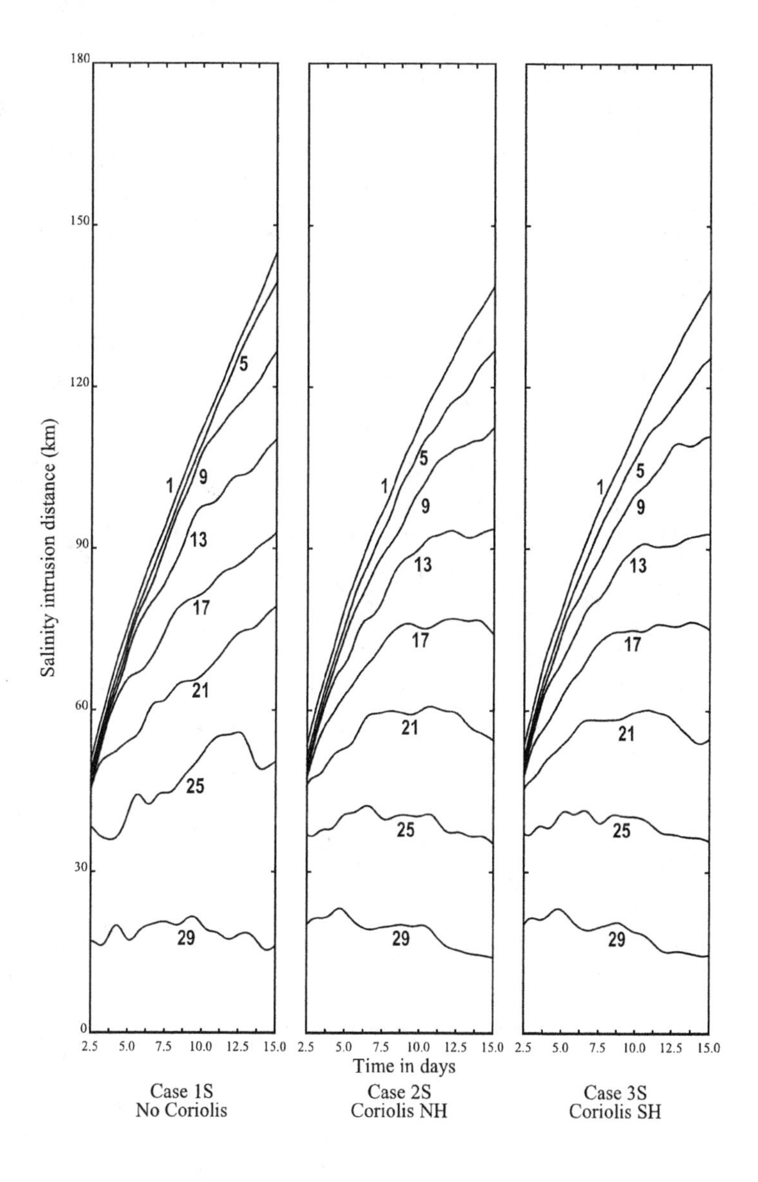

Figure 8. 34hr-low-pass-filtered salinity intrusion as a function of time. Values in psu. Straight channel cases for comparison with Figure 9.

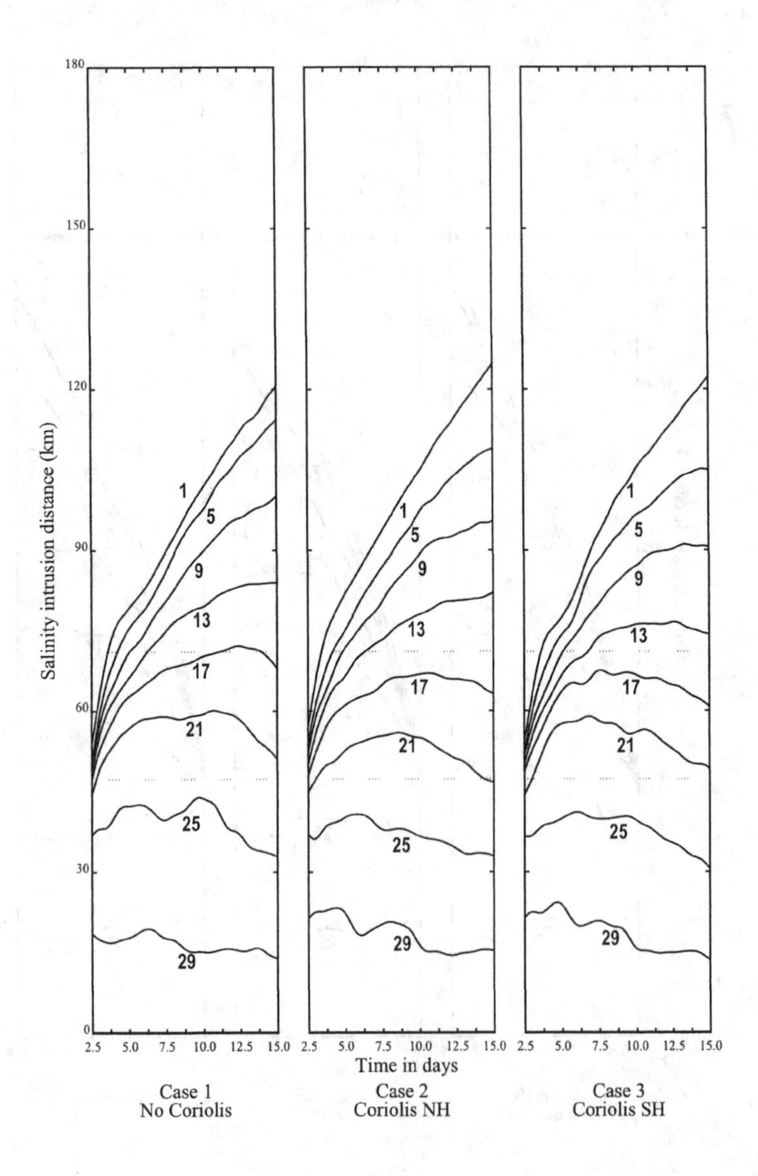

Figure 9. 34hr-low-pass-filtered salinity intrusion as a function of time. Values in psu. Curved channel cases. Dotted lines show average location of the bend.

bottom topography can significantly influence the upstream transport of salt. Future work will also focus on other relevant non-dimensional parameters, for example, the ratio of the tidal excursion distance to the bend radius, which may affect, among other things, the position of the flow separation regions.

Acknowledgements

The authors thank John Sondey and Christos Kavvadas at HydroQual, Inc. for their help with illustrations. Nickitas Georgas gratefully acknowledges his employer, HydroQual, Inc., for partial funding of this research. Alan Blumberg was funded under ONR grant, N00014-03-1-0633.

References

Blumberg, A. F. and Mellor, G. L. 1987. *A description of a Three Dimensional Coastal Ocean Circulation Model*. Three Dimensional Coastal Ocean Models: Volume 4, N. Heaps, Ed., American Geophysical Union, Washington, D.C., 1-16.

Boxall, J. B., Guymer, I., Marion, A. 2003. *Transverse mixing in sinuous natural open channel flows*. Journal of Hydraulic Research, Vol. 41, No. 2. pp. 153-165.

Chang, Y. C. 1971. *Lateral mixing in meandering channels*. PhD Thesis, University of Iowa, Iowa City, Iowa.

Chant, R. J. 2002. *Secondary circulation in a region of flow curvature: Relationship with tidal forcing and river discharge*. Journal of Geophysical Research, Vol. 107, No. C9, 3131. pp 14/1-14/11.

Fischer, H. B. 1972. *Mass Transport Mechanisms in Partially Stratified Estuaries*. Journal of Fluid Mechanics. Vol. 53. Pg.671-687.

Fischer, H. B., Imberger, J., List, E. J., Koh, R. C. Y., Brooks, N. H. 1979. *Mixing in Inland and Coastal Waters*. Academic Press, Inc. 483p.

Hansen, D. V. and Rattray, M. 1965. *Gravitational circulation in straits and estuaries*. Journal of Marine Research, 23. pp. 104-122.

Jassby, A. D., Kimmerer, W. J., Monismith, S. G., Armor, C., Cloern, J. E., Powell, T. M., Schubel, J. R., Vendlinski, T. J. 1995. *Isohaline Position as a Habitat Indicator for Estuarine Populations*. Ecological Applications, 5(1). pp. 272-289.

Li, C. and Valle-Levinson A. 1999. *A two-dimensional analytical tidal model for a narrow estuary of arbitrary lateral depth variation: The intratidal motion*. Journal of Geophysical Research, Vol. 104, No. C10, pp. 23,525-23,543.

Li, C. 2001. *3D analytical model for testing numerical tidal models.* Journal of Hydraulic Engineering. September 2001. pp. 709-717.

Lien, H. C., Hsieh, T. Y., Yang, J. C., Yeh, K. C. 1999. *Bend-flow simulation using 2D depth-averaged model.* Journal of Hydraulic Engineering. October 1999. pp. 1097-1108.

Nunes Vaz, R. A. and Simpson, J. H. 1985. *Axial convergences in a well-mixed estuary.* Estuarine Coastal Shelf Science, 20. pp. 637-649.

Pritchard, D. W. 1952. *Salinity distribution and circulation in the Chesapeake Bay estuarine system.* Journal of Marine Research, 11. pp.106-123.

Pritchard, D. W. 1956. *A study of the salt balance in a coastal plain estuary.* Journal of Marine Research, 13. pp. 133-144.

Rozovskii, I. L. 1957. *Flow of water in bends of open channels.* Academy of Sciences of the Ukrainian S.S.R. Institute of Hydrology and Hydraulic Engineering, Kiev. Translated from Russian by Y. Prushansky, Israel Program for Science Translation, S. Monson, Jerusalem. PST Catalog No. 363. 233 pp.

Seim, H. E. and Gregg, M. C. 1997. *The important of aspiration and channel curvature in producing strong vertical mixing over a sill.* Journal of Geophysical Research, Vol. 102, No. C2, DOI: 10.1029/96JC03415.

Shukry, A. 1950. *Flow Around Bends in an Open Flume*. Transactions, American Society of Civil Engineers, Vol 115, pp 751-788.

Steffler, P. M. 1984. *Turbulent flow in curved rectangular channel.* PhD thesis, University of Alberta, Edmonton Alta., Canada.

Wong, K-C. 1994. *On the nature of transverse variability in a coastal plain estuary.* Journal of Geophysical Research, Vol. 99, No. C7. pp 14209-14222.

Wu, W., Rodi, W., Wenka, T. 1997. *Three-dimensional calculation of river flow.* Environmental and Coastal Hydraulics. American Society of Civil Engineers. pp. 779-784.

High Performance Programming of the 3-D Finite Element Hydrodynamic Model - QUODDY

Chunfang Chen[1], Scott A. Yost[2], and Zhiyu Shao[3]

Abstract

This study revolves around enhancing the predicative ability of the 3-D finite element hydrodynamic ocean circulation model QUODDY by integrating various high performance computing tools. The strategies included converting the existing Fortran 77 code to Fortran 90 code to take advantage of the advanced features in Fortran 90, and developing an efficient parallel code based on the Fortran 90 code using Domain Decomposition techniques. Based on the characteristics of the numerical model, the parallel strategy was applied to the most computationally expensive subroutine VERTICALQ5 using the standard library—MPI (Message Passing Interface).

Numerical tests for the original Fortran 77 code, the converted Fortran 90 code, and the parallelized code were conducted on the University of Kentucky HP Superdome cluster. Comparisons were made by testing two realistic simulation cases: Georges Bank and the Chesapeake Bay. Improved performances were obtained from both the Fortran 90 version and the parallel version.

1. Introduction

Numerical modeling of ocean circulation plays an important role in the prediction of climate evolution and operational oceanography. Real world ocean circulation modeling normally involves large temporal and spatial scales. Therefore, the major practical problem with the ocean circulation simulation is the large amount of computing resources required both of CPU time and memory.

The application of most serial codes for large-scale problems is not realistic because they are unable to simulate the necessary physics in a reasonable amount of time. In recent years, the extraordinary development in computer hardware has made it feasible to conduct practical simulations. Therefore, it is important to develop new computational tools that can enhance the computational performance of the ocean circulation model.

---

[1] Ph.D. candidate, Civil Engineering Department, 161 O. H. Raymond Building, University of Kentucky, Lexington, KY 40506-0281, USA, cchen0@engr.uky.edu

[2] Associate Professor, Civil Engineering Department, 354C O. H. Raymond Building, University of Kentucky, Lexington, KY 40506-0281, USA, yostsa@engr.uky.edu

3. Ph.D. candidate, Civil Engineering Department, 161 O. H. Raymond Building, University of Kentucky, Lexington, KY 40506-0281, USA, zshao0@engr.uky.edu

Fortran is a popular programming language which has been widely used in science and engineering. Most existing ocean circulation codes are written in Fortran 77. The modernization of programming languages has introduced some new features, and it is important to upgrade existing Fortran 77 code to take advantage of those new features. Meanwhile, due to the development of high-performance computer architectures, researchers' attention has been attracted on the exploitation of parallelism. Nowadays, parallel programming has been increasingly used in many fields because of its capability to perform more sophisticated computations. However, the applications to existing ocean circulation models are limited.

QUODDY, a 3-D finite element hydrodynamic ocean circulation model developed by Lynch et. al. [6, 7, 8] in 1996 at Dartmouth College, has been successfully applied in ocean circulation modeling. But neither its upgrading nor its parallelization has been utilized. Campbell and Blain [3] developed a parallel model of QUODDY using OpenMP. Although the directive-based OpenMP is easy to implement, it requires a shared memory system, which puts it in a disadvantage in terms of portability. The message passing model, on the other hand, scales well and can be run on all platforms because the explicit message passing can provide more efficient communication. Pershing [11] developed a parallel version based on QUODDY 4.0 using C. His work mainly focused on the implementation of the collective and point-to-point communications. In 2002, QUODDY 5.1 was released. In this research, we developed a high performance QUODDY model based on QUODDY 5.1. First we converted the code from Fortran 77 to Fortran 90 to take advantage of some new features provided by Fortran 90. Then we developed a parallel version of QUODDY. This paper will discuss the design, implementation, and performance of the Fortran 90 and parallel version of QUODDY. Our goal was to achieve sustained high performance while reducing the computational cost.

2. Description of the QUODDY model

QUODDY is a free surface, tidal-resolving hydrodynamic model that uses the Galerkin finite element method (FEM) discretized on linear elements with nodal quadrature to solve the 3-D shallow water equation under both the Boussinesq and hydrostatic assumptions [3]. The 2.5 turbulence closure scheme [9] is incorporated to represent the vertical mixing of momentum.

The QUODDY algorithm is based on the shallow-water equation. The depth-averaged continuity equation is expressed in "wave equation" form. The formulation of descretization is based on the Galerkin FEM method on linear elements: triangular in the horizontal and linear in the vertical. A general terrain-following vertical coordinate is used with a flexible FEM approach for vertical resolution. This provides continuous tracking of the free surface and proper resolution of surface and bottom boundary layers. A semi-implicit time-stepping procedure is used to solve the implicit wave equation for elevation. The velocity, turbulent kinetic energy, temperature, and salinity require only a tridiagonal matrix solution in each time step.

The basic strength of this model lies in the freedom to make use of unstructured, non-uniform computational grids that allows for greater flexibility to represent complex geometries and strong horizontal gradients [1].
The original code was written in Fortran 77 with about 63 subroutines and a large number of physical parameterization options. The core part of the code is the computations which are enclosed in a time step loop. Within this loop, there are basically four parts [2]:

1. Set up the current time level
   a. load atmosphere forces
   b. load point source information
   c. compute linearized bottom stress coefficient
   d. evaluate baroclinic pressure gradient
   e. evaluate horizontal eddy viscosity/diffusivity
   f. evaluate nonlinear advection and horizontal diffusion of momentum
2. solve wave equation for free surface elevation
3. solve for vertical structure of 3-D variables
4. update arrays and increment timing parameters
   a. store present information at time level K-1
   b. compute equation of state
   c. compute vertical velocities

Among these four parts, Part 1 fulfills the pre-process function, in which a series of subroutines are called with their output being used as the input of Parts 2 and 3. Parts 2 and 3 are the core part of the computation: the water surface elevation and velocity vectors are solved. In Part 4, the equation of state is solved and the vertical velocity is computed. These four steps form a large time step loop which is the kernel of the code. Before this time loop, the main program reads input data and performs initialization. The final results are collected and displayed after the completion of the time loop.

3. Methodology

To achieve the goals, the study was divided into two steps. Our first objective was to upgrade the Fortran 77 version to the Fortran 90 version making use of the powerful new features provided by Fortran 90. The second goal was to develop a parallel version based on the upgraded version.

3.1 High performance using Fortran 90/95

The modernization process of programming language has introduced some new features into Fortran in the last decade. Fortran 90 was standard in 1991 and work continued with Fortran 95, which was published in 1996. The new features brought the old Fortran 77 up to date and increased the range of problems that can be solved easily. The language provides a very good framework for modern software development. In this study, the new features that were used to optimize the Fortran 77 code include:

1). Parallel Array Operation

Parallelism can be expressed using whole array operations which involve extensive slicing and sectioning facilities. Arithmetic may now be performed on whole arrays or array sections. Conceptually, operations in an array valued expression can be performed in parallel. In Fortran 90 many parallel intrinsic functions have been designed to support this feature [10]. Those that were used in this research include:

a. SUM(array) - reduction operation which returns the sum of elements in an integer, real, or complex array.

b. MAXVAL(array) - scan all elements in an array and return the biggest value.

c. FORALL - parallel assignment of all array elements or array sections.

2). Facility package – Module

The MODULE is used for global definition of types, objects, and definitions. In this optimized version, the system parameters and procedure definitions are placed in a MODULE. Since the detailed contents are hidden from the subroutine, this provides the flexibility for modification as subroutines just need to address a MODULE's name when using it.

3). Derived-type enhancements

Parameterized derived-type allows the kind, length, or shape of a derived type's components to be chosen when the derived-type is used. This new feature is applied for sending and receiving operations in the parallel program, in which we pack a section of array into a new type. Thus it can be transferred in a simple one-to-one communication mode because it is treated as a single variable.

4). Dynamic storage

At a few places in the code, temporary arrays were constructed just to be copied for following use. Elimination of these temporary arrays yields additional optimization. By using the dynamic storage, allocatable arrays can be created as necessary and will be instantly freed when they are not needed.

5). Loop-unrolling

In some subroutines, it is often seen that the same simple arithmetic will be repeated in a single loop. Such repetition is tedious and can be avoided by storing the calculated values into variables and then moving them out of the loop if those calculations are index-independent.

6). Elimination of logical statements

Logical IF statements are extensively used in the original code, especially in the subroutine of VERTICALQ5. In this subroutine, nodes at the top and bottom layers are treated differently from those inside due to the boundary conditions. Therefore, the index within each DO loop over the vertical direction will be differentiated with an IF construction to decide which level the node belongs to. Many loops in VERTICALQ5 have similar operations (i.e. loops over the vertical direction), whereby the repeated IF statement leads to substantial time consumption due to the large overhead. Since the IF statement is expensive, it is obvious that eliminating the IF statement can reduce the total cost. However, the compiler is not 'smart' enough to analyze and simplify such a complicated structure; therefore, the code transformation has to be done manually. In this study, we modify the IF construction by treating the top, bottom, and internal nodes in each vertical line separately. From the top to the bottom, the nodes are processed sequentially with only the internal nodes being treated in loops.

7). Compiler level optimization

Using optimization flags in compiling is a necessary and efficient way to optimize the code. The compiling option used in this study is +O2. It is an option of full optimization within each subprogram in a file and provides marked improvement in runtime performance.

3.2 Parallel strategy: domain decomposition technique

Using the optimized Fortran 90 version we wanted to proceed to the next step and develop the parallel version of QUODDY. Study of the physics and numerical algorithm of the model tells us that the computation in this model consists of two parts. In the first part, the sea-surface height is computed over the 2-D horizontal domain, and in the second part, the remaining components are solved only in the vertical direction. This formulation allows for convenience of parallelization via domain decomposition of the horizontal mesh. So if the horizontal domain can be divided into N sub-domains, the same operation can be applied on each sub-domain as was done on the entire domain, the runtime will be reduced. The basic idea behind this technique is to permit the simultaneous computation in each sub-domain. We utilize this property in parallelization by performing the same algorithm in each sub-domain on a different processor.

Generally, it is possible to improve the overall performance of the code if the section(s) that take most of the total runtime can be parallelized. Thus, it is usually necessary to recognize the computationally intensive areas of a model. In this study, a program performance analysis tool, Cxperf [8], was used to measure the load distribution on the basis of the upgraded Fortran 90 version. Cxperf requires special profiling options when compiling a program. The options instruct the compiler to create a special executable file containing information that the profiler uses to collect performance data. The amount of CPU time needed for each functional subroutine can be measured. Such data is informative as it suggests which subroutines should be exploited for possible parallelization. To illustrate, a simulation is conducted on an application of the Chesapeake Bay model. It is carried out with a simulation period of 16 hours using a time step of 30 seconds. Therefore a total of 1920 time iterations are conducted. The results are presented in Table 1. In this table, the 12 largest subroutines which took up more than 90% of the total CPU time are shown. The other 54 subroutines, which took up less than 10% of the total CPU time, are comparably negligible and are not presented. The table contains the CPU time for each individual subroutine and its children (the subroutines called within that subroutine).

All the subroutines in this table, except e_start, are associated with the computation of the physical state variables and they are executed within the time iteration. E_start is the main frame of this program and it includes the pre and post processing.

Table 1: Routine Performance of Fortran 90 Version

Without children		With children		Routine name	Description of routine/function
CPU (sec)	%	CPU (sec)	%		
805.185	42.98	1273.554	67.98	verticalq5	Determine the vertical structure and solve the vertical diffusion equations
122.437	6.53	122.437	6.53	sprsmltin3	Matrix multiplication
107.666	5.74	107.666	5.74	cthomas	Solve tridiagonal matrix
101.616	5.42	101.616	5.42	elemcoefs	Store element-based values to 3-D arrays
97.397	5.19	103.996	5.55	rhoxyq4	Compute baroclinic force terms
93.596	4.99	93.596	4.99	galperinq5	Apply the level 2.5 turbulence closure model to compute the vertical mixing coefficients, the source and, the decay terms
81.648	4.35	81.648	4.35	sprsconv	Compute the horizontal derivatives
81.082	4.32	1873.294	100.00	e_start*	Main program
72.859	3.88	72.859	3.88	bansoltr	Symmetric band matrix equation solver
59.711	3.18	59.711	3.18	thomas	Solve tridiagonal matrices using THOMAS algorithm
51.269	2.73	72.060	3.84	smagor1	Solve triangularized matrix equation
42.434	2.26	62.283	3.32	vertvel3_2	Compute vertical velocity

*: Main program includes the CPU time of all the subroutines

From the table, it can be seen that the execution of subroutine VERTICALQ5 dominated the overall CPU time. A closer investigation showed that in this subroutine the nodes in each vertical column were solved on the entire horizontal domain. Because the number of time steps in a production run is relatively large, it is obvious that the increase of the domain size will significantly increase the runtime of this subroutine and therefore the overall computation time. This implies that if the domain decomposition technique is applied to this subroutine, one can expect a reasonable savings in the total computation time. Since the computation of variables in the vertical direction is independent of one another, the idea for parallelization is intuitive and straightforward: the grid can be statically decomposed in the horizontal direction into N pieces and assigned to N processors. Such domain decomposition manner will lead to an effective data flow. However, the computation on the processors is not always independent because at the end of each time level, the information on the boundaries has to be collected by the master processor. So communication between the working processors and the master processor will take place. Every processor will send its own specific boundary information to the master processor after the variables of the new level are obtained.

4. Parallel Implementation

Some of the basic issues that are involved in the implementation of parallel programming include efficiency, portability, and programmability. These issues are studied and addressed in this section.

4.1 Load balancing

Load balancing is an important issue in developing a parallel program. Load balancing aims to balance processor workloads while minimizing the communication between processors. In this study, in order for the splitting of the domain to equally distribute the computational work among the processors, we divide the total number of nodes by the number of processors and let each processor handle the same number of nodes. If the number can not be divided evenly, the load distribution would be:

```
NNSIZE=NN/NUMPROCS
NNMOD=mod(NN, NUMPROCS)
NNSIZELAST=NNSIZE+NNMOD
```

Where
NN - number of node
NNPROCS – number of processors
NNMOD – the remainder from NN/NUMPROCS
NNSIZELAST – number of node on the last processor

4.2 Scalability

The choice of a programming paradigm is another fundamental issue in parallel implementation. There are several paradigms that have evolved in the past twenty years including the shared memory paradigm, the message passing paradigm, data parallel, multithreading, etc. In this study, we selected the message passing paradigm and use MPI as the data transfer interface. MPI is a popular paradigm and can be transferred to a wide range of computer architectures because when using MPI, data is shared explicitly by sending and receiving messages. Such architecture is inherently scalable for not requiring system coordination to maintain address space.

4.3 Compatibility

Our goal was to achieve better performance and minimize the deviation from the existing sequential QUODDY code. In the parallel code, we retained the basic structure of the sequential code. Particularly, the calling procedure was kept unchanged in the time step loop as well as the pre and post process. For each individual subroutine, the number, type, and name of each variable were also maintained. Furthermore, since the script in each time level is done during the pre-process and post-process, the parallelization is hidden from the user.

A major advantage of the domain decomposition technique employed here lies in the convinience of parallelizing the code without modifying the original model. The parallel code performs the same operations and thus fulfills the same functions as the serial code does. For each processor, it simply needs to change the starting and ending indexes for each loop operation.

5. Results and comparisons

The original Fortran 77 version was first compiled and run to establish baseline results for later verification. A series of performance experiments were then conducted with the Fortran 90 version and the parallel version on the Superdome at the Center for Computational Science at the University of Kentucky. The Superdome has a 224-processor Hewlett Packard cluster, with each processor being rated at about 3 gigaflops (peak) and having 2 gigabytes of main memory. There are three 64-processors and one 32-processor symmetric multi-processor (SMP) server with five terabytes of attached disk storage.

First, the correctness of the converted versions was verified. For the basis of comparison, the Chesapeake Bay simulation was used. We selected this simulation because the output of this test included variables of velocities in the x-direction (U) and y-direction (V), water surface level (H), temperature (T) and salinity (S), which were more comprehensive than those from the Georges Bank's simulation.

In this simulation, the nodal points used in the horizontal direction were 7258 with 11 in the vertical direction. The simulation period is 16 hours with a time step of 30 seconds. The verification results are given in Table 2. The data demonstrates that the results of the Fortran 90 version agreed well with the original Fortran 77 version. Also, we found the relative errors of parallel version were larger than those of Fortran 90. That might come from the truncation error resulted from the communication between processors.

Table 2: Validation of the Fortran 90 and Parallel Program

	Average Relative error with Fortran 77	
	Fortran 90	Parallel
U(bottom)	7.56E-05	1.64E-02
V(bottom)	3.54E-05	8.75E-03
U(top)	3.87E-05	3.18E-02
V(top)	1.61E-04	4.43E-02
H	0.0	8.65E-04
T	0.0	0.0
S	0.0	0.0

Next, numerical experiments were conducted on the two test cases mentioned above. First, we wanted to see whether improved performance can be achieved by Fortran 90 over Fortran 77. So in the first test, both versions were applied to the simulation of Georges Bank. In the second test, we compared the parallel version with the upgraded serial Fortran 90 version by performing the circulation simulation of the Chesapeake Bay.

5.1 Comparison of Fortran 90 Version vs. Fortran 77 Version

The Georges Bank is one of the heavily studied coastal regions. A quantitative understanding of the seasonal difference of advection, tidal mixing, and stratification of this region is necessary to know how ocean physics affect biological processes and population dynamics in this area as well as the long term trend of structure and stability of the ecosystems. So the study of the tidal circulations of Georges Bank is economically and ecologically important. In this numerical experiment, the Fortran 77 version and the Fortran 90 version were applied to the simulation of Georges Bank. The domain is given in Figure 1.

The results are presented in the chart in Figure 2. This chart shows the comparison of exclusive CPU time (without children) required by those subroutines which need 5 seconds or more for each run. Since the CPU time is coincident with the number of operations that are performed, the subroutines included in this chart can be roughly taken as the most time-consuming subroutines of this program. Such information suggests that it is worth more effort to focus on the improvement of these subroutines because the overall performance will benefit greatly from enhancing the performance of these major subroutines. Further, VERTICALQ5 is the dominant subroutine among all the others. It takes about 44% of the total CPU time. Therefore special care must be taken for it. From the chart, substantial decrease in runtime is observed in VERTICALQ5 with the other subroutines having little or no improvements. But an overall better performance was obtained because of the dramatic decrease in VERTICALQ5.

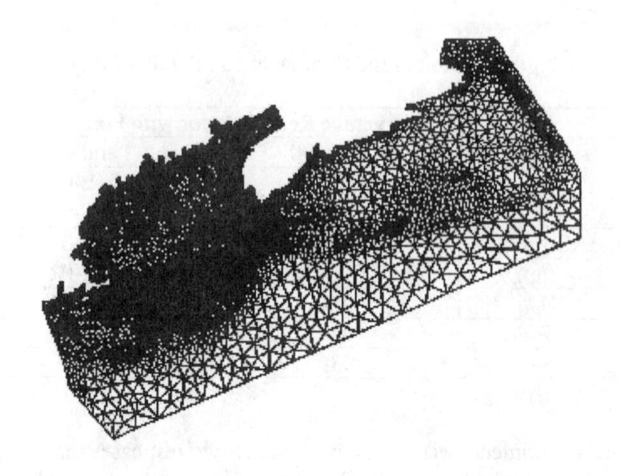

Fig.1 Domain of Georges Bank Simulation

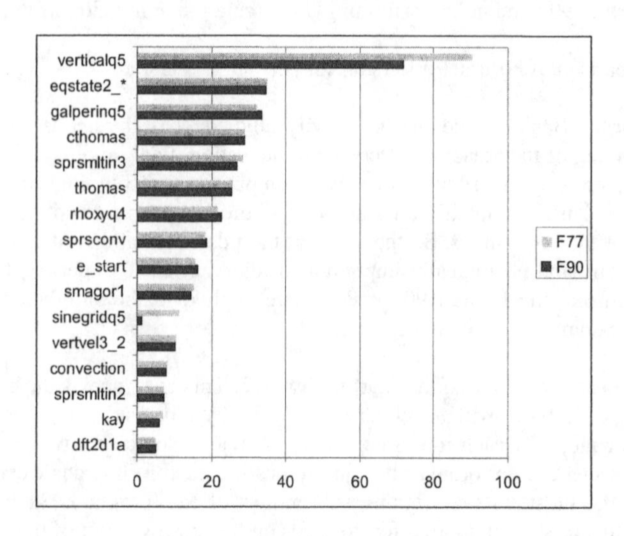

Fig.2 Comparison of CPU time (exclusive)

5.2. Comparison of Parallel Code vs. Serial Code

The Chesapeake Bay is located along the east coast of the United States and bordered Maryland Pennsylvania, Virginia, and the District of Columbia. It is 350 km long and has an average depth of 8.42m extending from the mouth of the Susquehanna River to its entrance into the Atlantic Ocean [5]. The simulation of water depths, salinity, temperature, and other related environmental variables is important for ship navigation and the study of biological activities in the bay. The domain under simulation is shown in Figure 3.

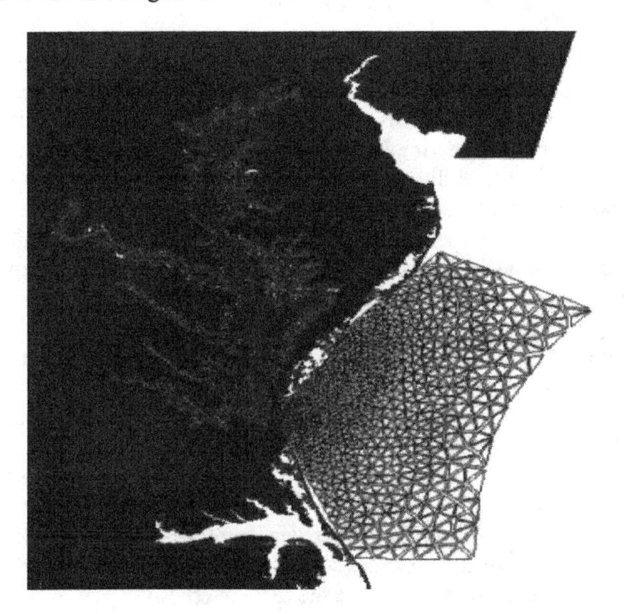

Fig.3 Domain of Chesapeake Bay Simulation

The QUODDY model has been applied to the nowcast and forecast system of the Chesapeake Bay study. In order to compare the performance of the Fortran 90 version and the parallel version of QUODDY, we applied them to the test case of the Chesapeake Bay. The parameters used in this test are given in Table 3.

Table 3: Parameters for Chesapeake Bay Simulation

Parameter	Value
Total number of nodes	7258
Total number of elements	13044
Total number of boundary elements	1476
Number of vertical nodes	22

In the experiment, we tested the program with 1, 2, 3, and 4 processors. The wall clock time of each processor was recorded. Then we calculated the total CPU time which is the summation of the wall clock time of all processors. In order to see how much performance gain is achieved by parallelizing the code, we further calculated the speedup (S) and efficiency (E). Speedup is a measure that captures the benefit of solving a problem in parallel and is defined as:

S = time for single processor run / time using N processors

Efficiency is closely related to speedup and is defined as:

E = speedup / number of processors

The results are shown in Table 3. In general, people are more concerned with the wall clock time. Better performances were observed by the multi-processor cases rather than the single processor case. For each test on multi-processor, the runtime of each processor is close, implying a balanced load among them, which is desirable. Also, the scalability in wall clock time is observed comparing two processors with four processors. It proves that the load is inversely proportional to the number of processors, which implies that the domain decomposition scheme developed in this model is feasible in this case. Thereby, we can expect to get further improvement with larger numbers of processors. Further, running on three processors provides the best results in total CPU time. The speedup and efficiency, however, turned out to be too good to truly represent the effectiveness of multi-processor. A main reason for this is the slow runtime of single processor (even slower than the serial code). Because the code is changed to accommodate the parallel implementation, the single processor not only needs to carry out all the computations but also has to initiate the environment for parallelism. Similar initialization applies for multi-processor but can be balanced off by the reduced time used by each processor. Moreover, single processor has to execute some commands that are not actually necessary in serial case, but in parallel they have to be performed which are known to be the extra overhead.

Table 4: Results of Chesapeake Bay Simulation

Processor Numbers	1	2	3	4
Total CPU time (sec)	2057.54	1611.37	1442.79	1743.56
Wall time (sec)	2057.54	813.165	488.957	441.631
P1	2057.54	813.165	488.957	441.631
P2		798.204	477.656	433.983
P3			475.31	432.723
P4				435.741
Speedup	1	2.53	4.20	4.65
Efficiency	1	1.26	1.4	1.16

6. Conclusion

In this paper, the high performance programming of the ocean circulation model QUODDY was studied. The FORTRAN 77 version was upgraded to FORTRAN 90 version. Better overall performance was observed, which demonstrated the applicability of the optimization techniques used in this study to enhance the performance of serial code. The parallel changes based on the upgrade Fortran 90 code provided ideal speedup and scalability using the Chesapeake Bay simulation. It can be concluded that the domain decomposition technique employed for parallelization was feasible.

7. Acknowledgments

The research described in this paper was performed at the Civil Engineering Department, University of Kentucky under contract to the DOE. The tests were conducted on the HP Superdome at the Center for Computational Science, University of Kentucky. The authors would like to acknowledge Tom Gross for providing the code and helpful discussion, and Prasada Rao for doing the initial parallel research.

References:

1. C. A. Blain. (2000). *modeling three-dimensional, thermohaline-driven circulation in the Arabian Gulf*, Estuarine and Coastal Modeling, Proceedings of the Sixth International Conference, M. L. Spaulding and H. L. Butler, eds., American Society of Civil Engineers, pp. 74-93.

2. T. J. Campbell, C. A. Blain. (2002). *Parallel Implementation of the QUODDY-3D Finite-Element Circulation Model*, Naval Research Laboratory Report.

3. T. F. Gross, Z. Li, S. A. Yost. (2002). *Simulation of Tides, Salinity and Temperature with Chesapeake Bay*, Estuarine and Coastal Modeling, Proceedings of the Seventh International Conference, M. L. Spaulding, American Society of Civil Engineers, pp. 152-164.

4. Hewlett-Packard Co., Cxperf Users' Manual. (1998). first edition.

5. Z. Li, T. F. Gross, C. W. Brown. (2002). *A Near Time Simulation of Salinity, Temperature and Sea Nettles (Chusaora quinquecirrha) in Chesapeake Bay*, Estuarine and Coastal Modeling, Proceedings of the Seventh International Conference, M. L. Spaulding, American Society of Civil Engineers, pp. 467-478.

6. D. R. Lynch, F. E. Werner. (1987). Three-Dimensional Hydrodynamics on Finite Elements, Part I: Linearized Harmonic Model, D.R. Lynch and F.E. Werner, *Int. J. Num. Meths. in Fluids*, **7**, pp. 871-909.

7. D. R. Lynch, F. E. Werner. (1991). Three-Dimensional Hydrodynamics on Finite Elements, Part II: Nonlinear Time-Stepping Model,. *Int. J. Numerical Methods in Fluids 12*, pp. 507-533.

8. D. R. Lynch, J. T. C Ip, F. E. Werner, E. M. Wolff. (1995). *Environmental Hydrodynamics: Comprehensive Model for the Gulf of Maine*, Finite Element Analysis of Environmental Problems: Surface and Subsurface Flow and Transport, John Wiley and Sons, G.F. Carey.

9. G.L. Mellor, T. Yamada. (1982). *Development of a Turbulence Closure Model for Geophysical Fluid Problems*, Rev. Geophys. Space Phys., 20, pp. 851-875.

10. M. Metcalf, J. K. Reid. (2002). *FORTRAN 90/95 explained*, Oxford.

11. A. J. Pershing. (2001). *MPQUODDY: A Portable, Parallel, Finite-Element Circulation Model.*
(http://www.eas.cornell.edu/geology/bioacoustics/AJP/MPQUODDY/MPqposter/mpqposter.html)

12. A. Sawdey, M. O'Keefe, R. Bleck, R.W. Numrich. (1994). *The Design, Implementation, and Performance of a Parallel Ocean Circulation Model.*
(http://www-mount.ee.umn.edu/~okeefe/papers/sawdey/ecmwf.94.pdf)

13. S. A. Yost, P. Rao. (1999). *Integrating High Performance Computing Strategies with Existing Finite Element Hydrodynamic Model*, Estuarine and Coastal Modeling, Proceedings of the Sixth International Conference, M. L. Spaulding and H. L. Butler, eds., American Society of Civil Engineers, pp. 738-754.

A parallel strategy applied in Projection method to simulate Free Surface Flow

Zhiyu Shao[1], Scott A. Yost[2]

Abstract

This paper describes a parallel strategy combing the Projection method and marker movement scheme to simulate free surface flow. The 2-D Navier- stokes equations were solved by using the two step Projection method while free surface location was tracked by putting massless markers on the surface. The main parallel approach is known as the Domain Decomposition. Stripe domain decomposition was applied for the parallel computation of momentum equation. A Schur domain decomposition scheme was used to parallelize the SOR algorithm for Pressure Poisson Equation. By analyzing the parallel computing procedure and the interprocessor communication cost, it can be found that the combination of the Projection method and Marker movement scheme has the potential to be an ideal structure for parallelism. The overhead cost for interprocessor communication is low leading to high efficiencies in the parallel computation.

1. Introduction

The free surface problem poses a difficulty in numerical simulation. In 1965, Harlow and Welch [1] first tackled this problem by using Marker and Cell method (MAC). The original MAC was introduced to solve time-dependent incompressible flow with a free surface. To track the movement of free surface, massless markers were introduced into flow field. By tracking the movement of the markers, the location and shape of the free surface can be determined. Marker movement scheme can be used in complex problems such as the overturning free surface case. However the Marker and Cell method is not efficient in solving the time-dependent Navier-Stokes equation. The velocity field computed from a single execution of solution procedure of the momentum equations and Pressure Poisson Equation (PPE) in MAC does not guarantee satisfaction of the divergence free constraint (continuity equation).The solution procedure for the momentum equations and the PPE in each time step must be repeated until the solution converges, which means that the divergence free constraint is satisfied. In contrast, in the Projection method, velocity field satisfies the divergence free condition immediately after the intermediate velocity field is corrected by pressures from the PPE. In other words, no iterations are required to satisfy the divergence free constraint. The Projection method gives a more direct and efficient way to obtain solutions for Navier Stokes equations.

---

[1] Grad. Student, Dept. of Civ. Engrg., Univ. of Kentucky, Lexington, KY 40508, (859) 257-4093, zshao0@engr.uky.edu

[2] Assoc. Prof., Member ASCE, Dept. of Civ. Engrg. , Univ. of Kentucky, Lexington, KY 40508, (859)257-4816, yostsa@engr.uky.edu

In this work, the free surface is tracked by the markers movement scheme described in Marker and Cell method [1]. But the Projection method is applied to solve the 2-D incompressible time dependent Navier-Stokes equations. Still procedures for solving the Navier-Stokes equations are still time consuming. With the development of supercomputer and high performance computational methods in recent years, in a parallel environment this problem is becoming less significant.

Some work on parallel computing for free surface flow has been reported. In a work by Cuminato et al. [2], a parallel technique based on a Simplify Marker and Cell method was introduced. They solved Navier-stokes equation explicitly by using a simplified Marker and Cell method which is similar to original MAC. The free surface locations were tracked by markers. The work of Sabau et al. [3] introduced a parallel computing approach to solve free-surface problems by a method analogous to a Solution Algorithm-Volume of Fluid (SOLA-VOF) like method. The Navier-stokes equation was solved by a Projection method while the free surface was tracked by Volume of Fluid (VOF).

Most of the parallelism strategies were based on an explicit time integration scheme. In the paper written by Winkelmann et al. [4], a parallel strategy was proposed to speed up the solution of Navier-stokes equation on highly complex computational domain such as domains of a complete aircraft, a spacecraft, or turbo machinery equipment. They applied a finite volume method to solve the Navier-stokes equation explicitly. In the work of Lappa et al. [5], the third-order Runge-Kutta integration in time was applied. Khan et al. [6] proposed a parallel algorithm using unstructured grid system based on a four-stage Runge-Kutta time integration scheme. In the work by Patel et al. [7], the Navier-stokes equations were solved in parallel using a MacCormack's explicit finite-difference method in more complex geometries. A semi-explicit scheme was introduced by Badcock et al. [8] in 1994. The method is based on an explicit treatment of streamwise fluxes and an implicit treatment of normal fluxes.

Explicit schemes perform better on a parallel algorithm, but the implicit scheme is more robust and stable, which is a desirable quality in computational fluid dynamics. Unfortunately, an implicit solution scheme is not easy to parallelize. In 1997, Grismer et al. [9] developed a parallel implicit Euler/Navier-stokes equations solver for unstructured grid. The solver is based on a Colbalt algorithm which solves the Navier-stokes equations in an integral format. A Goduno's first-order accurate Riemann method is the foundation of the Colbalt algorithm. The Riemann method is not widely used and not straightforward to parallelize. In the work by Van der Weide et al. [10], a parallel implicit Navier-stokes equations solver was introduced in which a Petrov-Galerkin Finite element method was applied. Since the implicit scheme was used, a full Jacobian matrix had to be computed in the whole computational domain. Then the resulting linear system was solved by using a Block Modified Incomplete LU decompostion (BMILU) preconditioned Krylov subspace method in a parallel fashion. The parallel algorithm for solving a large linear system is complicated because it requires a lot of interprocessor communication. Kalro et al. [11] [12] presented their parallel computation results based on an implicit finite element method. They solved a large linear system by using Krylov subspace method in a parallel environment. Similarly Habashi et al. [13] reported two parallel strategies to solve 3-D Navier-stokes equation implicitly based on a parallel algorithm for a large linear system. One strategy applied sparse matrix parallel algorithms and another strategy implemented a Gauss elimination algorithm optimized for vector-parallel processing on a Cray supercomputers. Averbuch et al. [14] developed a parallel solver based on a spectral method for Navier-stokes equation, called Multi-domain Local Fourier Method. In this method, special discretization was implemented by using Fourier method while the implicit time discretization equation is solved by using ADI scheme. All implicit parallel strategies mentioned above deal with parallelization of solving a large linear system directly instead of exploring the parallel potential of a sequential algorithm itself. Therefore, the advantages of an efficient sequential algorithm may not be explored or may be ignored in these parallel strategies.

Parallelizing the Pressure Poisson Equation (PPE) is a challenging task. The solution procedure of Poisson equation is fully coupled, which makes parallelism less straightforward to implement. SOR, preconditioned Conjugate Gradient method and Multigrid method are three most common ways to solve the PPE. Considering parallelism potential and feasibility, SOR seems to be the simplest and can potentially be a highly efficient parallel algorithm when the domain decomposition approach is implemented. Another way to solve the discretized Poisson equation is the direct solution procedure, which involves reversing and storing large sparse matrix. That is the approach used by Hoshino et al. [15]. To solve the linear discretized Poisson equation, Gauss Elimination was carried on in each processor and cyclic reduction in all processor. Their results showed that computation speed can be improved by a factor of 10 comparing to that consumed by a sequential optimal Black/Red SOR method. The parallel computer (PAX) they utilized has NNM (nearest-neighbor-mesh) connection architecture, which has the specific advantage for the 2-D Poisson equation. The discretized 2-D Poisson equation in five point stencil satisfies the proximity condition, which means that the value at each grid can be determined by the local adjacent grids. Portability and scalability of this parallel strategy to other parallel architectures still requires further studies.

SOR method is essentially a sequential method. But it can be parallelized in several ways. The most popular point-SOR can be directly parallelized by using a parallel technique called Pipelining. This approach speeds up computation by overlapping the computation processing. Several iterations are executed simultaneously in different processors. Each iteration is carried out by only one processor. Thus while one processor is busy executing current iteration, updated information can be sent to another processor to use for next iteration. Pipelining separates the parallel execution part from sequential execution part in SOR method. Since Pipelining procedure does not break the data ordering in SOR, Pipelining SOR has the same convergence rate as the corresponding sequential SOR. But it can be easily shown that large amount of data must be transferred between two processors in Pipelining strategy. In fact, if we have a computational domain with $N_x * N_y$ grids, the overall communication frequency will be $N_x * N_y$ per iteration.

Another widely used parallel procedure for Poisson equation is multi-color SOR. The simplest multicolor type of SOR is Red-Black SOR(R-B SOR). R-B SOR offers the advantage that the coefficient matrix is automatically consistently ordered. Numerical experiments [18] had previously shown somewhat higher convergence rates were achieved with natural ordering. R-B SOR offers a natural way to parallelize solution procedure of Poisson equation since it uncouples the computation of each grid in each iteration. Multicolor SOR is based on the similar idea used in B-R SOR. In multicolor type SOR, the solution vector is decomposed into two or more subvectors. Each color represents one subvector. Computation of each subvector is independent on all other subvectors. All the grids with the same color are calculated in one processor and the updated subvectors information will be sent to other processors to continue the execution of the other subvectors. Some work has been reported on the parallel multicolor SOR strategy [16]. The disadvantage of the parallel multicolor SOR strategy is similar to the pipelining approach. It indicates high communication cost. For Message-passing architecture type of parallel system, the overhead cost will be high.

A parallel Jacobi type SOR (JSOR) is also studied by some authors [23]. In JSOR, the computational domain is decomposed into several subdomains. In each subdomain, point-SOR was used while Jacobi iteration was implemented for all boundary grids. It was reported that JSOR has a higher convergence rate than Jacobi iteration but a lower convergence rate than SOR. In the work by Xie [17], a PSOR scheme is proposed. They showed that PSOR has the same convergence rate as the four-color SOR with a strip partition. PSOR can be seen as the combination of Pipelining and JSOR.

In this work, we introduced a parallel strategy based on Schur Domain Decomposition approach for SOR. Since domain decomposition scheme decreases the spectral radius of SOR type methods, asymptotic convergence rates on subdomains are greater compared to using on the original computational domain [18] which suggests less arithmetic, and thus more cost efficient.

2. Mathematical Formulation

2.1 Governing Equation

The governing equation for the studied free surface flow is 2-D incompressible Navier-stokes equation.

$$\frac{\partial u}{\partial x}+\frac{\partial v}{\partial y}=0 \tag{1}$$

$$\frac{\partial u}{\partial t}+\frac{\partial u^2}{\partial x}+\frac{\partial uv}{\partial y}=-\frac{\partial p}{\partial x}+\nu(\frac{\partial^2 u}{\partial x^2}+\frac{\partial^2 u}{\partial y^2})+B_x \tag{2}$$

$$\frac{\partial v}{\partial t}+\frac{\partial uv}{\partial x}+\frac{\partial v^2}{\partial y}=-\frac{\partial p}{\partial y}+\nu(\frac{\partial^2 v}{\partial x^2}+\frac{\partial^2 v}{\partial y^2})+B_y$$

where u and v are horizontal and vertical velocity respectively; p is pressure; B_x, and B_y are body force at x and y direction respectively.

2.2 Free surface boundary conditions

1. Kinetic condition: $$(v-v_B)\cdot n\,\big|_{fs}=0 \tag{3}$$
2. Dynamic condition: If curvature of the free surface is small enough, effect of surface tension can be ignored. Normal stress as well as tangential stress is also ignored in most cases. Thus, dynamic boundary of the free surface can be simplified as

$$p=p_0 \tag{4}$$

3. Sequential Solution Procedure

The sequential Algorithm for free surface flow can be summarized as follows:
Suppose *n* time steps have been completed. To advance to solution to time level *n+1*, perform the following steps:

1. Solve the "Burgers' equation" in form of momentum equation without the pressure term to get empirical velocities $\hat{u}$ and $\hat{v}$. Momentum equations are discretized in staggered grids. Quasilinearization and δ -form Douglas-Gunn time splitting scheme was implemented for the implicit scheme. On the free surface boundary, implement kinetic boundary condition; while on solid boundaries apply free slip boundary conditions.
2. Solve pseudo Pressure Poison Equation to get the pseudo pressure field p.

$$\Delta p = \frac{1}{k}(\frac{\partial \hat{u}}{\partial x} + \frac{\partial \hat{v}}{\partial y}) \qquad (5)$$

with boundary condition $\frac{\partial p}{\partial n} = 0$ at solid boundary and $p = p_0$ at free surface boundaries. PPE is discretized using the five grid stencil with central difference approximation. The discretized equation can be expressed as:

$$(D_{0,x}^2 + D_{0,y}^2)p_{ij} = \frac{1}{k}(D_{-,x}\hat{u}_{ij} + D_{-,y}\hat{v}_{ij}) \qquad (6)$$

where D is finite difference approximation defined the same way as in equation (10) and (11).

For grids near the free surface, an irregular computational stencil is applied to discretize PPE the same way as in Chan's work [23].

3. Project to get the mass conserved velocity field u and v:

$$u^{n+1} = \hat{u} - \delta t \nabla p \qquad (7)$$

4. Calculate velocity of markers and move the markers to their new locations. Velocity of markers can be calculated by using bilinear interpolation demonstrated in Figure 1.

$$u_k = \frac{A_1 u_1 + A_2 u_2 + A_3 u_3 + A_4 u_4}{hxhy} \qquad (8)$$

where, A is the area; hx, hy are the grid sizes.

If velocities are not available above the free surface, extrapolations were done to obtain image velocities in the same as in Chan's work [23].

New location of markers is determined by direct integration of motion equation.

$$x_k^{n+1} = x_k^n + u_k^{n+1}\, dt, \quad y_k^{n+1} = y_k^n + v_k^{n+1}\, dt \qquad (9)$$

(x_k, y_k): location of markers.

(u_k, v_k): velocities of markers.

5. Construct new free surface and prepare for the next time step.
6. Advance to the next time step.

4. Parallel Solution Procedure

4.1 Code Profiling

Figure 2a and 2b are code profiling of the sequential computation procedure. Both figures recorded CPU time consumed by all subroutines as well as the main program in sequential processing. CPU time in inclusive profiling includes all subroutines called in the current subroutine while CPU in exclusive profiling excludes all subroutines called in the current subroutine. Profiling of the sequential code showed that computation of the momentum equation (46%) and the pressure Poisson equation (50%) consumes most of the computation time. Therefore, focus of parallelism procedure will be put on solving the Momentum equations and PPE.

Performance of a parallel algorithm is evaluated by speedup and scalability. High efficient parallel strategy has low interprocessor communication cost. Three possible ways can be used to reduce the interprocessor communication cost, which include: 1. transfer short message each time in interprocessor communication; 2. reduce the interprocessors communication frequency; 3. use non-blocking interprocessor communication modes.

4.2 Message passing architecture

This parallel code algorithm was analyzed based on a distributed memory parallel system. A Message Passing Interface (MPI) is implemented whenever an interprocessor communication was required.

4.3 Parallel strategy for explicit/implicit time integration of momentum equation

Discretized momentum equations without pressure terms in x-direction and y-direction can be expressed in the following formula [18].

$$\begin{aligned}&\{\mathrm{I}-\theta\cdot\delta t[\nu(D_{0,x}^2+D_{0,y}^2)-2D_{\pm,x}(u^{(m)}\cdot)-D_{\pm,y}(\tilde{\bar{v}}^{(m)}\cdot)]\}\delta u=\\&u^n-u^{(m)}+\theta\cdot\delta t[\nu(D_{0,x}^2+D_{0,y}^2)u^{(m)}-D_{+,x}\bar{u}^{(m)^2}-D_{-,y}(\tilde{u}\cdot\bar{v})^{(m)}]\\&+(1-\theta)\delta t[\nu(D_{0,x}^2+D_{0,y}^2)u^n-D_{+,x}\bar{u}^{n^2}-D_{-,y}(\tilde{u}\bar{v})^n]\end{aligned}\tag{10}$$

$$\begin{aligned}&\{\mathrm{I}-\theta k[\nu(D_{0,x}^2+D_{0,y}^2)-D_{\pm,x}(\tilde{u}^{(m)}\cdot)-2D_{\pm,y}(v^{(m)}\cdot)]\}\delta v=\\&v^n-v^{(m)}+\theta k[\nu(D_{0,x}^2+D_{0,y}^2)v^{(m)}-D_{-,x}\tilde{u}\bar{v}^{(m)}-D_{+,y}\tilde{v}^{(m)^2}]\\&+(1-\theta)k[\nu(D_{0,x}^2+D_{0,y}^2)v^n-D_{-,x}\tilde{u}^n\bar{v}^n-D_{+,y}v^{n^2}]\end{aligned}\tag{11}$$

where, (u, v) are velocity vectors,

— is average value in horizontal direction,

~ is average value in vertical direction,

θ is the time integration factor,

θ = 0 is explicit scheme

θ = 1 is implicit scheme

θ = ½ is Crank-Nicholson scheme

$(u^{(m)}, v^{(m)})$ are velocity vectors from mth quasilinearization cycle,

(u^n, v^n) are velocity vectors from time step n,

D is the finite difference approximation symbol,

for super-index, 2 is 2nd order finite difference approximation,

for the 1st sub-index, 0 is central difference approximation,

\+ is forward difference approximation,

\- is backward difference approximation,

for the 2nd sub-index, x is in x- direction,

y is in y-direction.

It should be pointed out that in the above formula:

1. The discretization is under staggered grids in a rectangular domain with u being defined in the center of vertical cell wall while v is defined in the center of a horizontal cell wall. All scalars (divergence, pressure) are defined in the center of a cell.
2. The discretized formula was written for implicit scheme but can be easily adjusted for explicit scheme. Convective terms in the Left Hand Side (LHS) have been linearized by using Quasilinearization.
3. A δ-form Douglas-Gunn time splitting is used to solve the above equation. In quasilinearization cycle m, two-level time-splitting Douglas-Gunn scheme was implemented. To solve x-direction momentum equation without pressure, in the first time level, $\delta u^{(1)}$ is calculated from the following formula[18].

$$\begin{aligned}\{I-\theta\cdot\delta t[vD_{0,x}^2-2D_{\pm,x}(u^{(m)}\cdot)]\}\delta u^{(1)} &= \\ u^n-u^{(m)}+\theta\cdot\delta t[v(D_{0,x}^2+D_{0,y}^2)u^{(m)}-D_{+,x}\overline{u}^{(m)2}-D_{-,y}(\widetilde{u}\cdot\overline{v})^{(m)}] \\ +(1-\theta)\delta t[v(D_{0,x}^2+D_{0,y}^2)u^n-D_{+,x}\overline{u}^{n^2}-D_{-,y}(\widetilde{u}\overline{v})^n]\end{aligned} \quad (12)$$

In the second time level, $\delta u^{(2)}$ is calculated from

$$\{I-\theta\cdot\delta t[vD_{0,y}^2-D_{\pm,y}(\widetilde{\overline{v}}^{(m)}\cdot]\}\delta u^{(2)}=\delta u^{(1)} \quad (13)$$

Then $\delta u^{(2)}$ was used to update the velocity $u^{(m)}$ .The similar solution procedure can be applied to y-direction momentum equation without pressure. Since both δu and δv go to zero at the end of each time step, δ-form generally gives a stable and robust solution procedure. Later, we can see that application of δ- form gives the possibility of simplifying subdomain boundary conditions for implicit procedure.

The basic strategy used in the parallel processing for momentum equation is Domain Decomposition (DD), in which the whole computational domain is decomposed to P subdomains. This is demonstrated in Figure 3 and Figure 4. Each subdomain is assigned to one processor. Boundaries among subdomains do not overlap. If processor I deals with velocity variables in the x-direction with sub-index varying from *1* to *i*, then processor II deals with those in the x-direction with sub-index starting from *i+1*.

From discretized equation (10) and (11), one can see that the computation of velocity at each grid requires information from four adjacent grids. For an explicit scheme, this information is only a function of results from the previous time step. Computation of velocities in one time step at each grid is fully independent with the others. For an explicit scheme, all the information required for calculation of current time step is already available for every processor through interprocessor communication at the end of the previous time step. This process will be repeated for all time level. A fully parallel computation can be carried on without any impairment on convergent rate or results accuracy. Based on above analysis, we can give an estimate of the speed-up for explicit parallel solution procedure of the momentum equation. The speed-up S will be less than but close to the total number of processors P, or $S \approx P$.

At the end of each time step, updated information near subdomain boundaries must be sent to the processor in which the computation of the adjacent subdomain will be executed. This communication is necessary to update the information in that processor and receive the reciprocal updated information from another processor. Figure 4 demonstrates interprocessor communication stream for data transferring at the end of each time step. In figure 4, it can be observed that the parallel procedure for an explicit scheme has the following features:

1. At most 2 groups of data, which are two strips of velocity array, *U* or *V*, are required to be sent or received between two processors for each momentum equation at one time step. The size of one message is N_y, where N_y is the total grid numbers in Y direction. Total arithmetic of each momentum equation for one time step is $N_x \times N_y$. If N_x =100, N_y =100, only 1% of the whole vector size is required to be transferred during each data exchange between two processors.
2. Each processor is only required to communicate with at most two other processors at each time step. The maximum interprocessor communication frequency is two times per time step.
3. The communication is a one-to-one communication mode. The communication only involves either "sending" or "receiving" message. It is easy to implement the non-blocking communication to reduce the idle time. After receiving the updated information, current processor can go to the next iteration without having to wait to receive feedback information from another processor. No synchronizing is required.

For implicit schemes, computation of each grid is coupled with four other adjacent grids. Figure 4 demonstrates this coupled relationship. For implicit schemes, the same domain decomposition approach can be employed. Since the computation is coupled at inner boundaries between two subdomains, ghost cells are used to calculate the finite difference approximation of velocities. The basic requirement is to assume reasonable boundary conditions for all inner subdomains. δ-form is advantageous and a good assumption in the form of Dirichlet type of boundary conditions for inner subdomain boundaries. Since at the end of each time step, δu and δv will be asymptotic to zero, we can approximate the unavailable new values of the ghost grids with previous iteration results from adjacent subdomain. Thus at the end of each iteration, the value of δu and δv will be sent to adjacent subdomain as Dirichlet boundary condition for the next iteration. Therefore parallel procedure for implicit scheme is similar to explicit scheme except that the value of velocities increment δu and δv will be sent to another processor instead of the value of velocity u and v. This interprocessor communication is carried out at the end of each iteration. Data transfer streams between two processors for δu and δv are showed in Figure 3. Similar communication cost and frequency can be observed for implicit scheme, and is analogous to the explicit scheme.

4.4 Parallel strategy for Pressure Poisson Equation

The discretized Pressure Poisson Equation in 5-point stencil can be expressed as:

$$(D_{0,x}^2 + D_{0,y}^2)p_{ij} = \frac{1}{\delta t}(D_{-,x}u_{ij} + D_{-,y}v_{ij}) \tag{14}$$

The Domain Decomposition approach for Poisson equation is a little different from what was used for momentum equation. Here a Schur Domain Decomposition scheme is applied for pressure Poisson equation [18]. The computational domain is decomposed in a similar fashion as in solving momentum equation. But this time, one column of grids is overlapping two subdomains as showed in figure 5. Therefore, grids on the right boundary of subdomain I are the same grids on the left boundary of subdomain II. Pressure on the overlap boundaries will be calculated in both subdomains.

The computation stencil is demonstrated in figure 5. The discretized Poisson equation can be written in the matrix format [19] for two subdomains.

$$\begin{pmatrix} A_{11} & 0 & A_{13} \\ 0 & A_{22} & A_{23} \\ A_{31} & A_{32} & A_{33} \end{pmatrix} \begin{pmatrix} P_1 \\ P_2 \\ P_3 \end{pmatrix} = \begin{pmatrix} f_1 \\ f_2 \\ f_3 \end{pmatrix}, \qquad (15)$$

where, P_1: Pressure Vector in subdomain Ω_1

P_2: Pressure Vector in subdomain Ω_2

P_3: Pressure Vector on overlapping boundary Γ_{12}

Rewriting equation (15) into parallel format:

$$P_1 = A_{11}^{-1}[f_1 - A_{13}P_3] \qquad (16)$$

$$P_2 = A_{22}^{-1}[f_2 - A_{23}P_3] \qquad (17)$$

From equations (16) and (17), we can observe that if the value of P_3 is known, computation of P_1 and P_2 are fully independent of each other. From equation (15), we have

$$P_3 = A_{33}^{-1}[f_3 - A_{31}P_1 - A_{32}P_2] \qquad (18)$$

Substitution of the P_1 and P_2 into equation (18) yields

$$[A_{33} - (A_{31}A_{11}^{-1}A_{13} + A_{32}A_{22}^{-1}A_{23})]P_3 = f_3 - A_{31}A_{11}^{-1}f_1 - A_{32}A_{22}^{-1}f_2 \qquad (19)$$

We can solve P_3 accurately by inverting the left hand side of the above equation, which is called the Schur complement of the original matrix. The Schur complement is not sparse and preconditioning is usually used to solve for P_3. Many preconditioers for Schur domain decomposition are developed, but none of which is completely satisfactory [20]. Furthermore, a preconditioner of equation (18) can be really complicate when the domain is divided into more than two subdomains. Therefore, the approximate value of P_3 calculated from equation (18) is used in parallel computation.

In this approach, P_3 is calculated based on equation (18) ahead of time in all subdomains. This calculation requires results from the previous iteration. New value of P_3 on both subdomains will be the same since the computations are based on the same information as a result of interprocessor communication at the end of the previous iteration. Then, execution SOR on each subdomain can be carried out independently. At the end of iteration, updated information will be sent to adjacent domains. The message size of one communication in an iteration for PPE is the same as the x or y direction momentum equation. The interprocessor communication mode is also a one-to-one communication. The estimation of speed-up of parallel procedure for Poisson equation is not so straightforward. The decouple procedure used in Schur Domain Decomposition will have adverse effects on the overall convergence rate since inaccurate P_3 is used. On the other hand since the subdomain has a smaller domain size than the original domain, the spectral radius of subdomain is smaller comparing to that of the original domain. A smaller spectral radius indicates a larger asymptotic convergence rate for SOR. From this prospective, Domain Decomposition will speed up the convergence processing. Therefore, overall effects of Schur Domain Decomposition on the parallel processing of SOR can not be directly estimated. If the effects of "decreased spectral radius" overcome the effects of "decoupling", the overall speedup may exceed the number of processors, which is a phenomenon called "superlinear speedup".

4.5 Projection and Marker movement

In the Projection and Marker movement procedure, all updated information is already stored in every processor, thus no extra data transfer were required. Each processor executes the computation independently.

4.6 Parallel Algorithm:

The parallel algorithm can be summarized as follows:
Suppose calculation of n time steps have been completed. To advance to solution to time level *n+1*, performs the following steps in each processor:

1. Solve Burgers' equation explicitly or implicitly in all subdomains.

   For explicit scheme, velocities (u^n, v^n) from time level n are used as Dirichlet boundary conditions for all inner subdomain boundaries.

   For implicit scheme, velocity increments ($\delta u^{(m)}, \delta v^{(m)}$) from previous cycle m are used as Dirichlet boundary conditions for all inner subdomain boundaries. Repeated till $\delta u^{(m+1)}, \delta v^{(m+1)}$ are small enough. At the end of each cycle m+1, one-to-one communications of velocity vector increment ($\delta u^{(m+1)}, \delta v^{(m+1)}$) are required between two adjacent subdomains.
2. Perform one-to-one communications of velocities ($\hat{u}^{n+1}, \hat{v}^{n+1}$) are executed between two adjacent subdomains.
3. Prepare to solve Pressure Poisson equation iteratively. Computing pressures of all grids at inner overlapping subdomain boundaries P_3^{m+1} first by using equation (18). Solve Pressure Poison Equation (14) parallelly in all subdomains. Pressures (P_3^{m+1}) calculated at cycle m+1 initially are used as Dirichlet boundary conditions for all inner subdomain boundaries. One-to-one communications are required to transfer P^{m+1} between two adjacent subdomains at the end of iteration. Repeat this procedure till converge.
4. Project to obtain mass conserved velocity field.
5. Perform one-to-one communication of new velocities (u^{n+1}, v^{n+1}) are executed between two adjacent subdomains.
6. Calculate velocity of markers and move markers to new location.
7. Construct new free surface and smooth the surface if necessary.
8. Goto next time step.

5. Study Case

A numerical tank was used to verify the results of the sequential code and will be used check the performance of the parallel code. Figure 7 shows initial condition for the study case used by Armenio [21]. In this case, the rectangular tank (*1m* breadth, *0.5m* high) with fluid inside is at rest initially. Then it is suddenly accelerated along the horizontal direction with an acceleration *1 m/s²*.

The liquid motion is recorded until a steady state condition is reached. The body forces in the horizontal direction and the vertical direction are $B_x = -1$ *m/s²* and $B_y = -g$. Navier-stokes equation is dimensionless by defining Reynolds number Re= $(b^3 g)^{1/2} / \nu$. In this study, Re=80 was used. Results from sequential code shows that at steady state the free surface is a flat inclined plane making an angle of $\tan^{-1}(B_x / B_y) = 5.82°$ with the horizontal plan. From figure 8 and 9, we can observe that the results from the sequential code agree with theoretical prediction. Since the parallel code is still under development, information regarding speedup and the scalability of the parallel code is still unavailable at this stage.

6. Conclusion

In this work, a new parallel strategy for numerical simulation of free surface flow was described. From the analysis of this parallel strategy, the combination of Projection method and marker movement for free surface flow is a scheme with an ideal structure for parallelism and it requires little interprocessor communication cost in parallel processing. The analysis shows promise that it will be an efficient parallel approach to solve the 2-D free surface flow problem.

Reference:

1. Harlow F. H. and Welch J. E.(1965), Numerical Calculation of Time-dependent Viscous Incompressible Technique for solving Viscous, Incompressible Flow of Fluid with Free Surface, *Phys. Fluids*, vol. 8, 2182-2189
2. Cuminato J.A., Filho A.C.and Boaventura, M. (1999), Simulation of free surface flows in a distributed memory environment, *J. Comput. and Appl. Math.*, v 103, n 1, Mar. 15, 77-92
3. Sabau A and Tao, YX. (1997), Parallel implementation of the projection method for solving free-surface flows, *ASME, Fluids Engrg. Division FED*, v 20, FEDSM97-3507, *Numerical Developments in CFD*
4. Winkelmann R, Jochem H. and Williams, R.(1999), Strategies for parallel and numerical scalability of CFD codes, *Comput. Meth. Appl. Mech. and Engrg*, v 174, n 3-4, 433-456
5. Lappa. M.and Savino R.(1999), Parallel solution of three-dimensional Marangoni flow in liquid bridges, *Int. J. Num. Methods Fluids,* v 31, n 6, Nov, 911-935
6. Khan, M.M.S. and Atta, E.H.(1990), Prediction of laminar flows using tetrahedral meshes and massively-parallel computers, *ASME, Fluids Engrg. Division FED*, v 103, *Recent Advances and Applications in Computational Fluid Dynamics*, 57-66
7. Patel N. R., Sturek W.B.and Hiromoto, R.E.(1987). Parallel Compressible Flow Algorithm for Multiprocessors, *ASME, Fluids Engineering Division FED*, v 47, 49-64
8. Badcock K.J.(1994), Partially implicit method for simulating viscous aerofoil flows, *Int. J. Num. Methods Fluids*, v 19, n 3, Aug 15, 259-268

9. Grismer Matthew J., Strang W. Z., Tomaro R. F. and Witzeman, F. C.(1998), A parallel, implicit, unstructured Euler/Navier-Stokes solver, *Advances in Engineering Software*, v 29, n 3-6, Apr-Jul, 365-373
10. Van der Weide, E., Deconinck H., Issman E. and Degrez G.(1999), Parallel, implicit, multi-dimensional upwind, residual distribution method for the Navier-Stokes equations on unstructured grids, *Comput. Mech.*, v 23, n 2, 199-208
11. Kalro V. and Tezduyar T.(1997), Parallel 3D computation of unsteady flows around circular cylinders, *Parallel Computing*, v 23, n 9, 235-1248
12. Kalro V.and Tezduyar T.(1998), 3D computation of unsteady flow past a sphere with a parallel finite element method, *Comput. Meth. Appl. Mech. and Engrg*, v 151, n 1-2, 267-276
13. Habashi W.G., Nguyen V.N. and Bhat M.V.(1991), Efficient direct solvers for large-scale computational fluid dynamics problems *Comput. Meth. Appl. Mech. and Engrg*, v 87, n 2-3, Jun, 253-265
14. Averbuch A. , Ioffe L., Israeli M. and Vozovoi L.(1998), Two-dimensional parallel solver for the solution of Navier-Stokes equations with constant and variable coefficients using ADI on cells, *Parallel Computing*, v 24, n 5-6, Jun, 673-699
15. Hoshino T., Sato Y. and Asamoto Y.(1988), Parallel poisson solver FAGECR-implementation and performance evaluation on PAX computer, *J. Information Processing*, v 12, n 1, 20-26
16. Kuo C. J. and Levy B. C.(1987), Parallel 2-Level 4-Colour SOR method, *Proceedings of the IEEE Conference on Decision and Control Including The Symposium on Adaptive Pro*, 1445-1449
17. Xie D. and Adams, L.(1999), New parallel SOR method by domain partitioning, *SIAM J. Sci. Comput.*, v 20, n 6, 2261-2281
18. McDonough J. M., lectures in *Computational Fluid Dynamics of Incompressible Flow*, available at the following URL: http://www.engr.uky.edu/~me691
19. McDonough J. M., lectures in *Computational Numerical Analysis of Partial Differential Equations*, available at the following URL: http://www.engr.uky.edu/~me690
20. Barry F. Smith, Petter E. B., and William D. G. (1996), *Domain decomposition: parallel multilevel methods for elliptic partial differential equations*. Cambridge University Press, 101-144.
21. Armenio V. (1997), an Improved MAC Method (SIMAC) for Unsteady High-Reynolds Free Surface Flows, *Int. J. Numer. Methods fluids,* vol. 24,185-214
22. Chan R K. and Street R. L., A Computer Study of Finite-Amplitude Water Waves, *J. Comput. Phys.,* vol. 6, pp68-94,1970

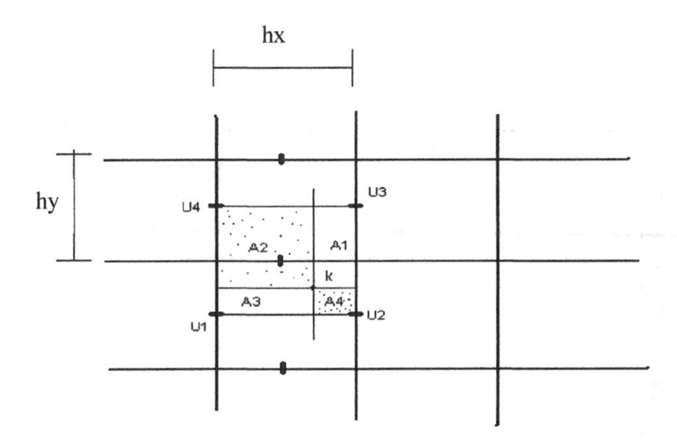

Figure 1. Area weighting scheme for marker[1]

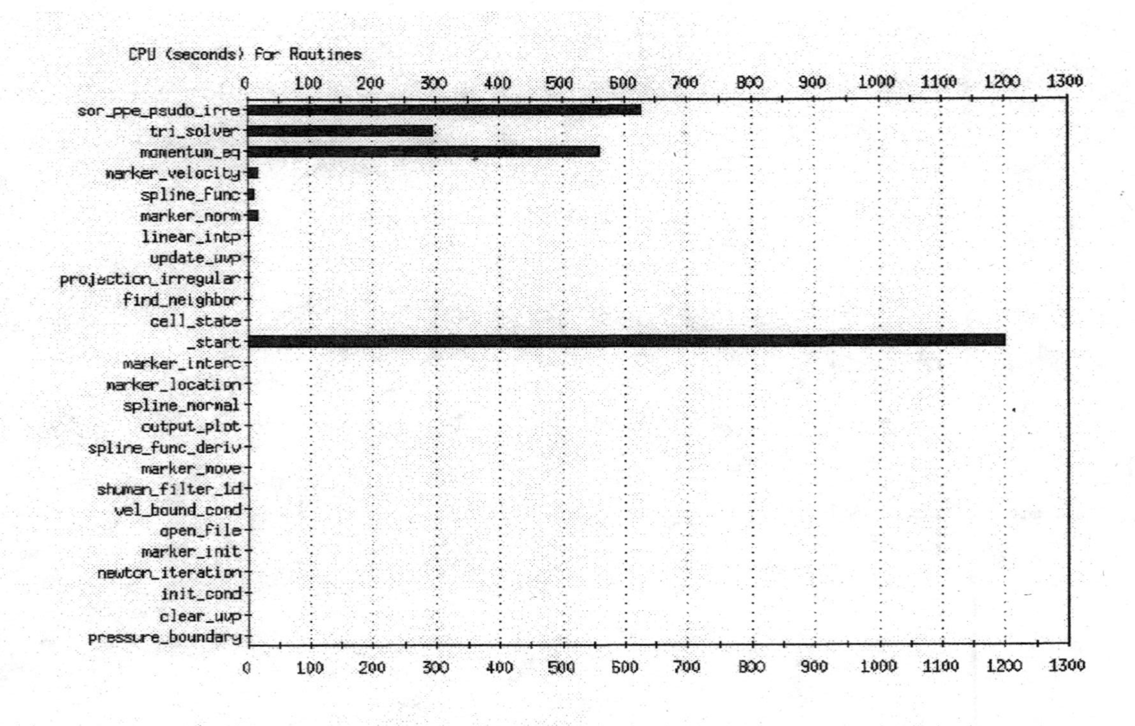

Figure 2a. Inclusive Profiling

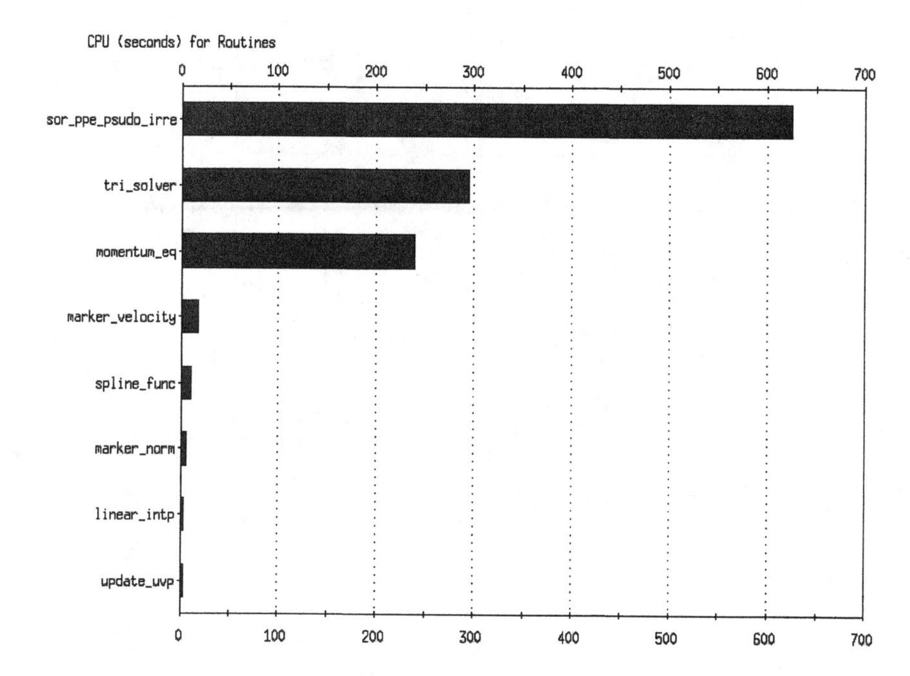

Figure 2b. Exclusive Profiling

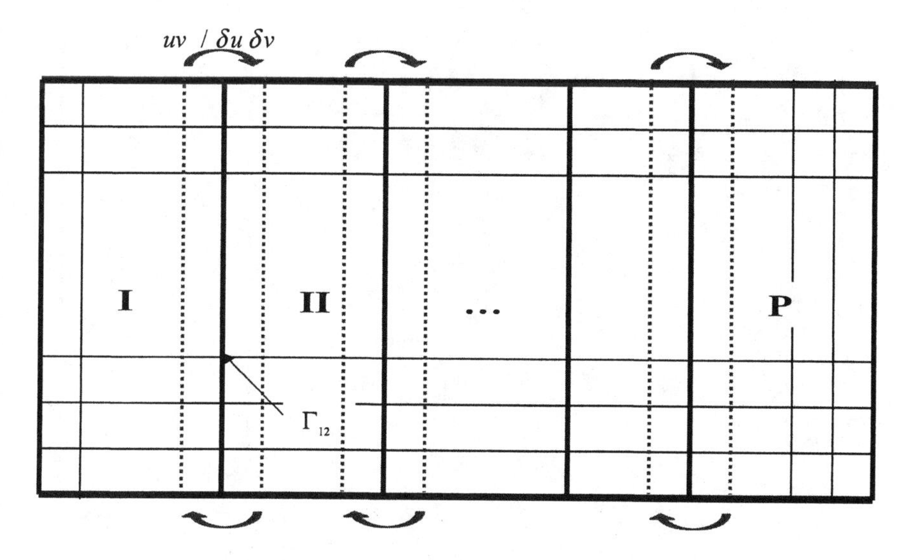

Figure 3. Data Transfer in Parallel Solution Procedure of Momentum Equation

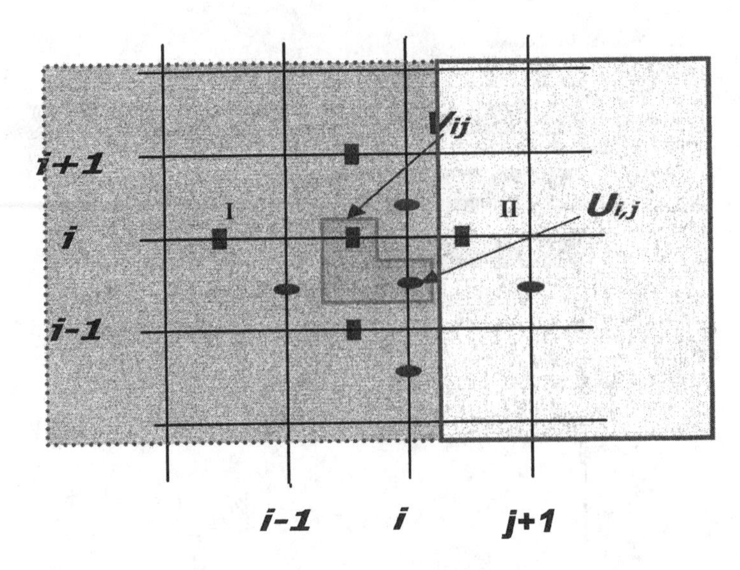

Figure 4. Velocity computation couple for implicit time integration

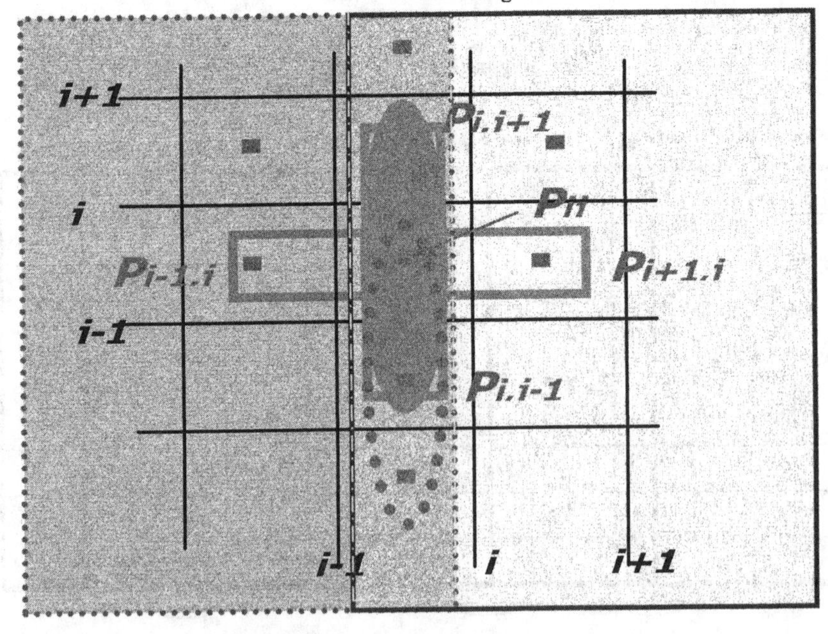

Figure 5. Computation grid stencil for SOR and LSOR

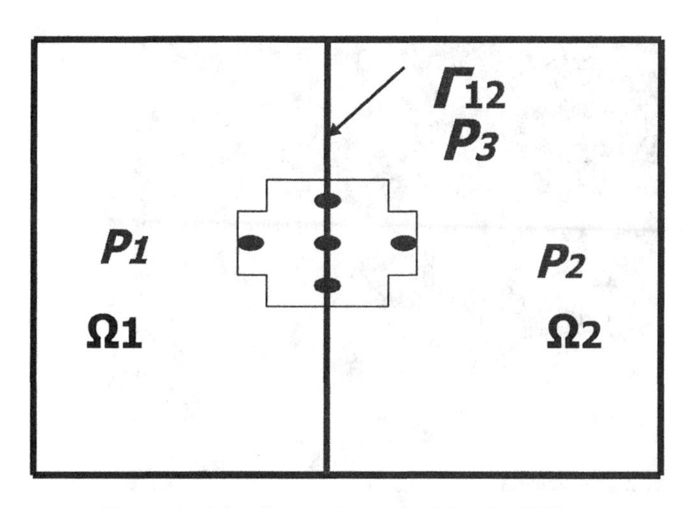

Figure 6. Schur Domain Decomposition For PPE

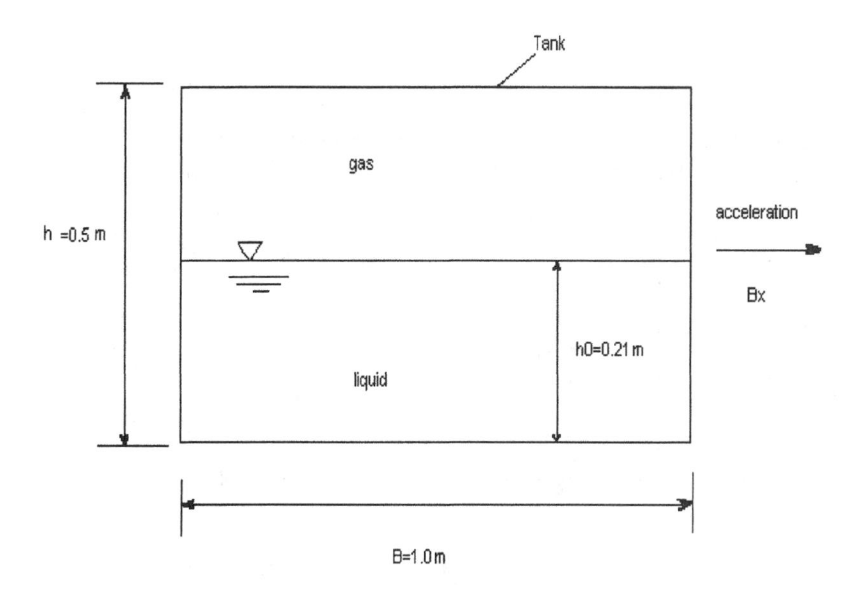

Figure 7. Initial State of numerical tank

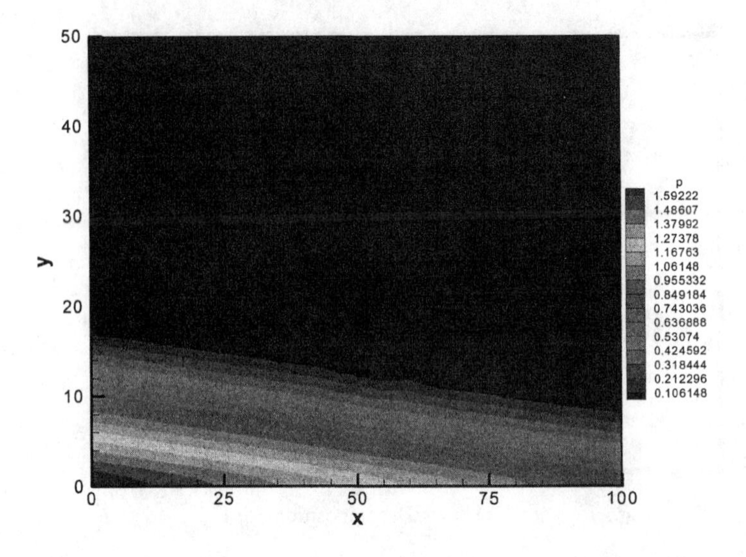

Figure 8. Pressure Contour at Steady State

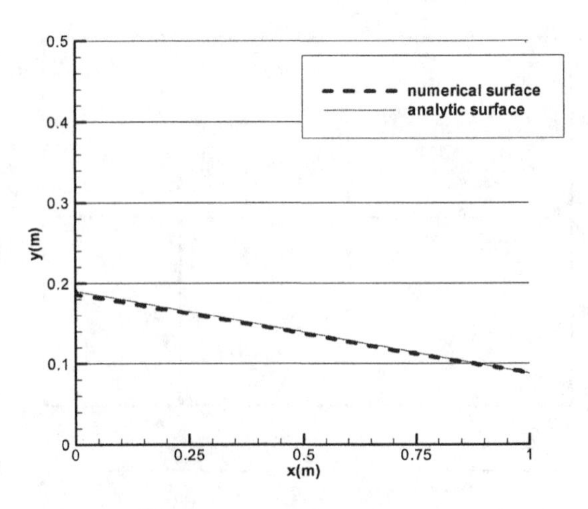

Figure 9. Free Surface at Steady State

A Fully Non-hydrostatic Three-Dimensional Model with an Implicit Algorithm for Free-Surface Flows

Chin H. Wu* and Hengliang Yuan
Department of Civil and Environmental Engineering, University of Wisconsin, Madison

ABSTRACT

A three-dimensional (3D) implicit numerical model based upon the Navier-Stokes equations (NSE) for free-surface flows is developed. To avoid solving the fully 3D Pressure Poisson Equation (PPE), an algorithm is developed by decoupling the original matrix system into a series of sub-matrix systems corresponding to two-dimensional (2D) vertical planes. The features of the model include (1) a novel treatment for the non-hydrostatic pressure at the top-layer cell in a staggered grid, consistent with the overall non-hydrostatic pressure discretization; (2) the option of partial-cell treatment on bottom topographies, enabling the model to simulate free-surface flows interacting with uneven bottoms with a reasonable vertical resolution; (3) simultaneously solving all flow-field components (e.g., the velocities, pressure, and free-surface elevation) within each time step; and (4) capability of simulating both 3D short waves and long waves with a second-order accuracy in both time and space. The developed model is validated by two free-surface flow problems, in which the vertical accelerations are considerable and thus the pressure fields are in non-hydrostatic distributions. The numerical results are in excellent agreements with the analytical solutions and experimental data, illustrating the capability of the algorithm for simulating 3D free surface flows.

* Correspondence to Chin H. Wu, Department of Civil and Environmental Engineering, 1269D Engineering Hall, University of Wisconsin, Madison, U.S.A.
Email: chinwu@engr.wisc.edu

1. INTRODUCTION

Numerical modeling of 3D free-surface flows becomes more computationally affordable and widely used nowadays. Most 3D models are based upon the hydrostatic pressure distribution assumption that neglects the vertical acceleration and viscosity terms in the NSE. It is known that hydrostatic models are valid for the flows where the horizontal scale of the motion is much larger than the vertical one (Blumberg and Mellor 1987, Cheng *et al.* 1993). However, in the case of short-period waves, abruptly changing bed topographies, or stratification due to strong density gradients, hydrostatic models are not capable of accurately resolving vertical flow motions. Therefore the complete NSE non-hydrostatic models for simulating large-scale 3D geophysical free-surface flows are desirable (Mahadevan *et al.* 1996).

In recent years, computationally feasible 3D non-hydrostatic models for single-valued free-surface flows have been developed. Different from the Marker-and-Cell (Kim *et al.* 2001), volume-of-Fluid (Hur and Mizutani 2003), and arbitrary Lagrangian-Eulerian methods (Hodges and Street 1999), this type of model utilizes an efficient top-layer moving grid technique for tracking free-surface motion. For example, Casulli (1999) developed a semi-implicit fractional step method to solve the NSE. Mahadevan *et al.* (1996) and Kocyigit *et al.* (2002) further developed 3D non-hydrostatic models in the sigma coordinate system. Li and Fleming (2001) used the explicit projection (or fractional-step) method to solve for free surface wave propagation. In general, the models using semi-implicit fractional step method or the explicit projection method provide efficient algorithms to solve for PPE and track 3D free-surface motions. However, splitting intermediate velocity and pressure correction steps usually degrades the solution to first-order accuracy in time (Armfield and Street 2002). In addition, the cost of solving the PPE remains a critical issue.

To avoid solving the PPE, Namin *et al.* (2001) developed a different implicit method that simultaneously solves the two-dimensional (2D) NSE with free-surface boundary conditions. The method is based upon the implicit Crank-Nicolson scheme by expressing the vertical velocity and pressure as functions of unknown horizontal velocities, which are then substituted into the horizontal momentum equation to yield a block tri-diagonal matrix system. An efficient matrix solver therefore can be used without any iteration procedure. The implicit PPT-free 2D model was further extended to the sigma coordinate system (Yuan and Wu 2003). To date, the implicit 3D model has yet been developed and challenges of extending this method to free-surface flow problems remain.

In this paper, an implicit 3D fully non-hydrostatic model for free-surface flows was developed. The Crank-Nicolson scheme is used to discretize the complete 3D NSE. At the free-surface cell, a new treatment of the non-hydrostatic pressure condition is applied, instead of employing the hydrostatic representation. For flows over uneven bottom topographies, partial-cell treatment (Pacanowski and

Gnanadesikan, 1998) is used. The unique characteristic of the method enables all the flow-field components to be solved simultaneously within each time step. An efficient algorithm is developed to decouple the 3D problem into a series of 2D problems, minimizing the computational cost. The developed 3D non-hydrostatic model is validated to analytical solutions and found to accurately simulate free-surface flows. In following sections, mathematical formulation is given first. Numerical method is presented next. Finally, the model was validated by two examples, including 3D standing wave oscillation and progressive wave shoaling on a submerged bar.

2. MATHEMATICAL FOMULATION

The governing equations for 3D free-surface flows are the incompressible NSE, which are based upon conservation of mass and momentum. In the Cartesian coordinate (x, y, z) system, these equations are

$$\frac{\partial u}{\partial x}+\frac{\partial v}{\partial y}+\frac{\partial w}{\partial z}=0, \tag{1}$$

$$\frac{\partial u}{\partial t}+u\frac{\partial u}{\partial x}+v\frac{\partial u}{\partial y}+w\frac{\partial u}{\partial z}=-\frac{\partial P}{\partial x}+\nu_x\frac{\partial^2 u}{\partial x^2}+\nu_y\frac{\partial^2 u}{\partial y^2}+\nu_z\frac{\partial^2 u}{\partial z^2}, \tag{2}$$

$$\frac{\partial v}{\partial t}+u\frac{\partial v}{\partial x}+v\frac{\partial v}{\partial y}+w\frac{\partial v}{\partial z}=-\frac{\partial P}{\partial y}+\nu_x\frac{\partial^2 v}{\partial x^2}+\nu_y\frac{\partial^2 v}{\partial y^2}+\nu_z\frac{\partial^2 v}{\partial z^2}, \tag{3}$$

$$\frac{\partial w}{\partial t}+u\frac{\partial w}{\partial x}+v\frac{\partial w}{\partial y}+w\frac{\partial w}{\partial z}=-\frac{\partial P}{\partial z}+\nu_x\frac{\partial^2 w}{\partial x^2}+\nu_y\frac{\partial^2 w}{\partial y^2}+\nu_z\frac{\partial^2 w}{\partial z^2}-g, \tag{4}$$

where $u(x,y,z,t)$, $v(x,y,z,t)$, and $w(x,y,z,t)$ are the velocity components in the horizontal x, y, and vertical z directions, respectively; t is time; P is a the normalized pressure, defined as the pressure divided by the constant water density; g is the gravitational acceleration; and ν_x, ν_y, and ν_z are constant eddy viscosities in the x, y, and z directions, respectively.

Various surface boundary conditions are needed for solving Equations (1)-(4). The kinematic boundary condition at the impermeable bottom is

$$u\frac{\partial h}{\partial x}+v\frac{\partial h}{\partial y}+w=0, \tag{5}$$

where $h(x,y)$ is the water depth measured from the undisturbed mean water level (Figure 1). Similarly, the kinematic boundary condition at the free surface is

$$\frac{\partial \eta}{\partial t}+u\frac{\partial \eta}{\partial x}+v\frac{\partial \eta}{\partial y}=w, \tag{6}$$

where $\eta(x,y,t)$ is the free-surface elevation measured from the undisturbed mean water level. Integrating continuity equation (1) over the water depth and applying the kinematic boundary conditions (5) and (6) give a conservative form of the free-surface equation

$$\frac{\partial \eta}{\partial t} + \frac{\partial}{\partial x}\int_{-h}^{\eta} udz + \frac{\partial}{\partial y}\int_{-h}^{\eta} vdz = 0. \tag{7}$$

At the free surface the pressure is equal to the atmospheric pressure P_a, taken as zero here, *i.e.*, $P = P_a = 0$. The continuity of stress at the free surface gives $(\tau_s)_{water} = (\tau_s)_{air}$, and $(\tau_n)_{water} = (\tau_n)_{air}$, where the subscript s and n denote the shear and normal stresses, respectively. In the case of no wind effect, both $(\tau_s)_{air}$ and $(\tau_n)_{air}$ are zero. For simulating free-surface wave propagation, inflow and outflow boundary conditions are also needed. At the inflow boundary, the velocity components and free-surface elevation can be specified using either analytical solutions or laboratory inflow conditions. At the outflow boundary, an open boundary condition is critical so to minimize wave reflection. In this paper, the sponge layer concepts (Kim *et al.* 2001) using the artificial damping terms are used for damping outgoing waves. At the end of the damping zone, the Sommerfeld radiation condition is used to further enhance wave absorption.

3. NUMERICAL METHOD

A finite difference approximation is adopted to solve the governing equations. The computational domain is discretized by N_1, N_2, and N_3 cells in the x, y, and z directions, respectively. A staggered grid system is adopted so that pressure and free-surface elevation are arranged in the cell center and the velocity components were defined at the side of each cell (Figure 1a, b). A uniform grid interval is applied in the horizontal x and y directions, *i.e.*, $\Delta x = \Delta y =$ constant. Similarly, a uniform grid interval, Δz, is used for vertical layers with exception of the bottom and top ones. For the bottom layer, Δz_{bt} is the distance between the top of the bottom layer and the real bottom, being a function of horizontal locations, *i.e.*, $\Delta z_{bt} = \Delta z_{bt}(x,y)$. While the vertical interval for the top-layer cell, Δz_{top}, is allowed to vary with space and time and updated by

$$\Delta z_{top}(x,y,t) = \Delta z + \eta(x,y,t). \tag{8}$$

Equation (9) results in a relatively thick top layer for the free-surface elevation. For cases involving bottom topographies, a partial cell approach is used to discretize bathymetries. The basic idea of this approach lies in relocating flow-field components for bottom cells by taking into account of topography effects. The values of relocated components are obtained from linear interpolations, and the discretized governing equations can then be applied to bottom cells (Figure 1b). Details of the partial cell approach can be found in Pacanowski and Gnanadesikan (1998). For simplicity, only the numerical schemes for general cells are given in this paper.

The overall procedure of the numerical method can be summarized in three steps. First, using the Crank-Nicolson scheme, the vertical velocity and pressure were expressed as functions of horizontal velocities. Second, eliminating the vertical velocity and pressure terms in the horizontal momentum equations yields a matrix system with the unknowns of the horizontal velocities. Finally, an efficient algorithm was developed to solve the horizontal velocities, and other flow field components were then updated. The developed model has second-order accuracy in both time and space.

3.1 Implicit discretization

The governing equations (1)-(4) and (7) are discretized as follows:

$$\left(\frac{\partial u}{\partial x}\right)_{i,j,k}^{n+1}+\left(\frac{\partial v}{\partial y}\right)_{i,j,k}^{n+1}+\left(\frac{\partial w}{\partial z}\right)_{i,j,k}^{n+1}=0, \tag{9}$$

$$\begin{aligned}&\frac{u_{i+1/2,j,k}^{n+1}-u_{i+1/2,j,k}^{n}}{\Delta t}+\theta\left(u\frac{\partial u}{\partial x}\right)_{i+1/2,j,k}^{n+1}+(1-\theta)\left(u\frac{\partial u}{\partial x}\right)_{i+1/2,j,k}^{n}+\theta\left(v\frac{\partial u}{\partial y}\right)_{i+1/2,j,k}^{n+1}+(1-\theta)\left(v\frac{\partial u}{\partial y}\right)_{i+1/2,j,k}^{n}\\&+\theta\left(w\frac{\partial u}{\partial z}\right)_{i+1/2,j,k}^{n+1}+(1-\theta)\left(w\frac{\partial u}{\partial z}\right)_{i+1/2,j,k}^{n}=-\theta\left(\frac{\partial P}{\partial x}\right)_{i+1/2,j,k}^{n+1}-(1-\theta)\left(\frac{\partial P}{\partial x}\right)_{i+1/2,j,k}^{n}\\&+\theta\left(\nu_x\frac{\partial^2 u}{\partial x^2}+\nu_y\frac{\partial^2 u}{\partial y^2}+\nu_z\frac{\partial^2 u}{\partial z^2}\right)_{i+1/2,j,k}^{n+1}+(1-\theta)\left(\nu_x\frac{\partial^2 u}{\partial x^2}+\nu_y\frac{\partial^2 u}{\partial y^2}+\nu_z\frac{\partial^2 u}{\partial z^2}\right)_{i+1/2,j,k}^{n}\end{aligned}, \tag{10}$$

$$\begin{aligned}&\frac{v_{i,j+1/2,k}^{n+1}-v_{i,j+1/2,k}^{n}}{\Delta t}+\theta\left(u\frac{\partial v}{\partial x}\right)_{i,j+1/2,k}^{n+1}+(1-\theta)\left(u\frac{\partial v}{\partial x}\right)_{i,j+1/2,k}^{n}+\theta\left(v\frac{\partial v}{\partial y}\right)_{i,j+1/2,k}^{n+1}+(1-\theta)\left(v\frac{\partial v}{\partial y}\right)_{i,j+1/2,k}^{n}\\&+\theta\left(w\frac{\partial v}{\partial z}\right)_{i,j+1/2,k}^{n+1}+(1-\theta)\left(w\frac{\partial v}{\partial z}\right)_{i,j+1/2,k}^{n}=-\theta\left(\frac{\partial P}{\partial y}\right)_{i,j+1/2,k}^{n+1}-(1-\theta)\left(\frac{\partial P}{\partial y}\right)_{i,j+1/2,k}^{n}\\&+\theta\left(\nu_x\frac{\partial^2 v}{\partial x^2}+\nu_y\frac{\partial^2 v}{\partial y^2}+\nu_z\frac{\partial^2 v}{\partial z^2}\right)_{i,j+1/2,k}^{n+1}+(1-\theta)\left(\nu_x\frac{\partial^2 v}{\partial x^2}+\nu_y\frac{\partial^2 v}{\partial y^2}+\nu_z\frac{\partial^2 v}{\partial z^2}\right)_{i,j+1/2,k}^{n}\end{aligned}, \tag{11}$$

$$\begin{aligned}&\frac{w_{i,j,k+1/2}^{n+1}-w_{i,j,k+1/2}^{n}}{\Delta t}+\theta\left(u\frac{\partial w}{\partial x}\right)_{i,j,k+1/2}^{n+1}+(1-\theta)\left(u\frac{\partial w}{\partial x}\right)_{i,j,k+1/2}^{n}+\theta\left(v\frac{\partial w}{\partial y}\right)_{i,j,k+1/2}^{n+1}+(1-\theta)\left(v\frac{\partial w}{\partial y}\right)_{i,j,k+1/2}^{n}\\&+\theta\left(w\frac{\partial w}{\partial z}\right)_{i,j,k+1/2}^{n+1}+(1-\theta)\left(w\frac{\partial w}{\partial z}\right)_{i,j,k+1/2}^{n}=-\theta\left(\frac{\partial P}{\partial z}\right)_{i,j,k+1/2}^{n+1}-(1-\theta)\left(\frac{\partial P}{\partial z}\right)_{i,j,k+1/2}^{n}\\&+\theta\left(\nu_x\frac{\partial^2 w}{\partial x^2}+\nu_y\frac{\partial^2 w}{\partial y^2}+\nu_z\frac{\partial^2 w}{\partial z^2}\right)_{i,j,k+1/2}^{n+1}+(1-\theta)\left(\nu_x\frac{\partial^2 w}{\partial x^2}+\nu_y\frac{\partial^2 w}{\partial y^2}+\nu_z\frac{\partial^2 w}{\partial z^2}\right)_{i,j,k+1/2}^{n}\end{aligned}, \tag{12}$$

$$\frac{\eta_{i,j}^{n+1}-\eta_{i,j}^{n}}{\Delta t}+\theta\left[\frac{\partial}{\partial x}\left(\int_{-h}^{\eta}udz\right)\right]_{i,j,k}^{n+1}+(1-\theta)\left[\frac{\partial}{\partial x}\left(\int_{-h}^{\eta}udz\right)\right]_{i,j,k}^{n}$$

$$+\theta\left[\frac{\partial}{\partial x}\left(\int_{-h}^{\eta}vdz\right)\right]_{i,j,k}^{n+1}+(1-\theta)\left[\frac{\partial}{\partial x}\left(\int_{-h}^{\eta}vdz\right)\right]_{i,j,k}^{n}=0, \tag{13}$$

where Δt is the time step; the superscript n represents the time step; the subscript i, j, and k are grid indexes in the x, y, and z directions, respectively; and the implicit weighting factor θ is set to be 0.5 for the Crank-Nicolson scheme.

The nonlinear advection terms in Equation (10), (11), and (12) are linearized by employing the procedure of (Beam and Warming, 1978) with second-order temporal accuracy. For example,

$$\left(v\frac{\partial u}{\partial y}\right)_{i+1/2,j,k}^{n+1}=\left[v^{n+1}\left(\frac{\partial u}{\partial y}\right)^{n}+v^{n}\left(\frac{\partial u}{\partial y}\right)^{n+1}-v^{n}\left(\frac{\partial u}{\partial y}\right)^{n}\right]_{i+1/2,j,k}. \tag{14}$$

Using Equations (9) and (12), the vertical velocity w^{n+1} and pressure P^{n+1} are expressed as functions of horizontal velocities u^{n+1} and v^{n+1}. Substituting these two expressions into the horizontal momentum equations (10) and (11) gives a matrix system

$$\begin{bmatrix}\mathbf{A}_{\mathbf{uu}} & \mathbf{A}_{\mathbf{uv}}\\ \mathbf{A}_{\mathbf{vu}} & \mathbf{A}_{\mathbf{vv}}\end{bmatrix}^{n}\times\begin{bmatrix}\mathbf{u}\\ \mathbf{v}\end{bmatrix}^{n+1}=\begin{bmatrix}\mathbf{b}_{\mathbf{u}}\\ \mathbf{b}_{\mathbf{v}}\end{bmatrix}^{n}, \tag{15}$$

where $\mathbf{A}_{\mathbf{uu}}^{n}$, $\mathbf{A}_{\mathbf{uv}}^{n}$, $\mathbf{A}_{\mathbf{vu}}^{n}$, and $\mathbf{A}_{\mathbf{vv}}^{n}$ are coefficient block matrixes with the dimension of $(N_3\times N_1\times N_2)$ by $(N_3\times N_1\times N_2)$; $[\mathbf{u}^{n+1},\mathbf{v}^{n+1}]^{\mathrm{T}}$ represents the unknown vector of horizontal velocities; and $[\mathbf{b}_{\mathbf{u}}^{n},\mathbf{b}_{\mathbf{v}}^{n}]^{\mathrm{T}}$ is a known vector. Since no diagonal-dominant features occur in $\mathbf{A}_{\mathbf{uu}}^{n}$, $\mathbf{A}_{\mathbf{uv}}^{n}$, $\mathbf{A}_{\mathbf{vu}}^{n}$, and $\mathbf{A}_{\mathbf{vv}}^{n}$, directly solving Equation (15) is computationally expensive. In this paper an efficient algorithm is thus developed to solve the above matrix system.

3.2 Description of the efficient algorithm

Equation (15) is rearranged to decouple the 3D matrix system into a series of 2D sub-matrix systems, which can be efficiently solved by a block tri-diagonal matrix solver. To achieve this purpose, some nonlinear advection terms and diffusion terms in Equations (9)-(13) are temporally treated explicitly in the sense of using an intermediate value approach. In other words, some unknown terms, denoted by the superscript $n+1$ in Equation (15), are moved to right-hand side as being the intermediate values, denoted by superscript $n+*$. As a result, $\mathbf{A}_{\mathbf{uu}}$ and $\mathbf{A}_{\mathbf{vv}}$ become block diagonal matrixes, and $\mathbf{A}_{\mathbf{uv}}$ and $\mathbf{A}_{\mathbf{vu}}$ become null. Equation (15) thus can be rewritten as

$$\left[\begin{array}{ccc|ccc} \mathbf{A}_{\mathbf{u}_1} & & & & & \\ & \ddots & & & \mathbf{0} & \\ & & \mathbf{A}_{\mathbf{u}_{N2}} & & & \\ \hline & & & \mathbf{A}_{\mathbf{v}_1} & & \\ & \mathbf{0} & & & \ddots & \\ & & & & & \mathbf{A}_{\mathbf{v}_{N1}} \end{array}\right]^{n+*} \times \left[\begin{array}{c} \mathbf{u}_1 \\ \vdots \\ \mathbf{u}_{N_2} \\ \hline \mathbf{v}_1 \\ \vdots \\ \mathbf{v}_{N_1} \end{array}\right]^{n+1} = \left[\begin{array}{c} \mathbf{b}_{\mathbf{u}_1} \\ \vdots \\ \mathbf{b}_{\mathbf{u}_{N2}} \\ \hline \mathbf{b}_{\mathbf{v}_1} \\ \vdots \\ \mathbf{b}_{\mathbf{v}_{N1}} \end{array}\right]^{n+*}, \tag{16}$$

where each $\mathbf{A}_{\mathbf{u}_j}^{n+*}$ is a $(N_3 \times N_1)$ by $(N_3 \times N_1)$ sub-block matrix corresponding to the unknown $(N_3 \times N_1)$ by 1 sub-block vector, $\mathbf{u}_j^{n+1}$, in the $x-z$ plane. Similarly, each $\mathbf{A}_{\mathbf{v}_i}^{n+*}$ is a $(N_3 \times N_2)$ by $(N_3 \times N_2)$ sub-block matrix corresponding to the unknown $(N_3 \times N_2)$ by 1 sub-block vector, $\mathbf{v}_i^{n+1}$, in the $y-z$ plane. The sub-block vectors, $\mathbf{b}_{\mathbf{u}_j}^{n+*}$ and $\mathbf{b}_{\mathbf{v}_j}^{n+*}$, are the known values including previous time, n, and intermediate time, $n+*$, information. Further decoupling the matrix system (16) gives

$$\mathbf{A}_{\mathbf{u}_j}^{n+*} \times \mathbf{u}_j^{n+1} = \mathbf{b}_{\mathbf{u}_j}^{n+*}, \tag{17a}$$

for the horizontal velocity u in each $x-z$ plane ($\mathbf{j} = \mathbf{1}, \cdots, \mathbf{N}_2$), and

$$\mathbf{A}_{\mathbf{v}_i}^{n+*} \times \mathbf{v}_i^{n+1} = \mathbf{b}_{\mathbf{v}_i}^{n+*}, \tag{17b}$$

for the horizontal velocity v in each $y-z$ plane ($\mathbf{i} = \mathbf{1}, \cdots, \mathbf{N}_1$). An efficient block tri-diagonal solver (Tannehill *et al.* 1997) is used to directly solve (17a) and (17b) because $\mathbf{A}_{\mathbf{u}_j}^{n+*}$ and $\mathbf{A}_{\mathbf{v}_i}^{n+*}$ are both block tri-diagonal matrixes. An iteration procedure is applied to have the advanced intermediate unknown ones (denoted by $n+1$) converge to the known intermediate values (denoted by $n+*$). This procedure also ensures the result obtained from the efficient solver for equations (17a) or (17b) converges to the one from directly solving Equation (15).

3.3. Pressure at the top-layer cell

In the staggered grid system, a hydrostatic representation is generally used to express the top-layer pressure as

$$P_{top} = g[\eta + \Delta z / 2]. \tag{18}$$

Difficulties of treating the non-hydrostatic component for the top-layer pressure in the staggered grid system remain (Casulli 1999, 2002). In this paper, a new method of including non-hydrostatic component in the top-layer pressure was applied. Integrating the vertical momentum equation (4) from $z = -\Delta z / 2$ to $z = \eta$, and applying the free-surface boundary condition (5) give

$$P_{top} = g[\eta + \Delta z/2] + \frac{\partial}{\partial t}\int_{-\Delta z/2}^{\eta} w dz + \frac{\partial}{\partial x}\int_{-\Delta z/2}^{\eta} uw dz + \frac{\partial}{\partial y}\int_{-\Delta z/2}^{\eta} vw dz - w^2\Big|_{z=-\Delta z/2}, \quad (19)$$

which consists of the summation of a hydrostatic surface elevation component (the first term of the right-hand side) and a non-hydrostatic vertical acceleration component (the rest of the terms in the right-hand side). The advantage of using Equation (19) is to algebraically represent the top-layer pressure by the free-surface elevation and the neighboring velocities. The finite difference discretization of Equation (19) gives

$$P_{i,j,N_3}^{n+1} \approx -P_{i,j,N_3}^{n} + g\left(\eta_{i,j}^{n+1} + \eta_{i,j}^{n} + \Delta z/2\right) + w_{i,j,N_3}^{n}\frac{\eta_{i,j}^{n+1} - \eta_{i,j}^{n}}{0.5\Delta t} + \left(\eta_{i,j}^{n} + \Delta z/2\right)\frac{w_{i,j,N_3}^{n+1} - w_{i,j,N_3}^{n}}{0.5\Delta t}$$
$$+ \left\{\frac{\partial[uw(\eta + \Delta z/2)]}{\partial x}\right\}_{i,j,N_3}^{n} + \left\{\frac{\partial[uw(\eta + \Delta z/2)]}{\partial x}\right\}_{i,j,N_3}^{n+*}$$
$$+ \left\{\frac{\partial[vw(\eta + \Delta z/2)]}{\partial y}\right\}_{i,j,N_3}^{n} + \left\{\frac{\partial[vw(\eta + \Delta z/2)]}{\partial y}\right\}_{i,j,N_3}^{n+*} - \left(w^2\right)_{i,j,N_3}^{n} - \left(w^2\right)_{i,j,N_3}^{n+*}, \quad (20)$$

where $\eta_{i,j}^{n+1}$ can be evaluated from Equation (13)

$$\eta_{i,j}^{n+1} = \eta_{i,j}^{n} - \frac{\Delta t}{2}\left[\frac{\Delta z_{top_{i+1/2,j}}^{n} u_{i+1/2,j,k}^{n+1} - \Delta z_{top_{i-1/2,j}}^{n} u_{i-1/2,j,k}^{n+1}}{\Delta x} + \frac{\Delta z_{top_{i,j+1/2}}^{n} v_{i,j+1/2,k}^{n+*} - \Delta z_{top_{i,j-1/2}}^{n} v_{i,j-1/2,k}^{n+*}}{\Delta y}\right.$$
$$\left. - w_{i,j,N_3-1/2}^{n+1}\right] - \frac{\Delta t}{2}\left[\frac{\Delta z_{top_{i+1/2,j}}^{n} u_{i+1/2,j,k}^{n} - \Delta z_{top_{i-1/2,j}}^{n} u_{i-1/2,j,k}^{n}}{\Delta x} + \frac{\Delta z_{top_{i,j+1/2}}^{n} v_{i,j+1/2,k}^{n} - \Delta z_{top_{i,j-1/2}}^{n} v_{i,j-1/2,k}^{n}}{\Delta y}\right.$$
$$\left. - w_{i,j,N_3-1/2}^{n}\right]. \quad (21)$$

3.4 Block tri-diagonal matrix for horizontal velocities

For each fixed j ($j = 1,\cdots,N_2$), Equation (17a) is a block tri-diagonal matrix system for each $x - z$ plane for unknown horizontal velocity, *ie.*

$$\mathbf{CL}_{i,j}^{n+*} \times U_{i-1/2,j}^{n+1} + \mathbf{CM}_{i,j}^{n+*} \times U_{i+1/2,j}^{n+1} + \mathbf{CR}_{i,j}^{n+*} \times U_{i+3/2,j}^{n+1} = \mathbf{d}_{i,j}^{n+*}, \quad (22)$$

where $\mathbf{CL}_{i,j}^{n+*}$, $\mathbf{CM}_{i,j}^{n+*}$, and $\mathbf{CR}_{i,j}^{n+*}$, with the dimension of $(N_3 \times N_3)$, are components of the sub-block tri-diagonal matrix, $\mathbf{A}_{\mathbf{u}_j}^{n+*}$, in Equation (17a); $U_{i-1/2,j}^{n+1}$, $U_{i+1/2,j}^{n+1}$, and $U_{i+3/2,j}^{n+1}$, with the dimension of $(N_3 \times 1)$, are the unknown column vectors corresponding to the $\mathbf{u}_j^{n+1}$ in Equation (17a). Detail description of the matrix arrangement in the 2D vertical plane can be found in Yuan and Wu. (2003) and is not given here. The matrix system (22) has the only unknown of u^{n+1}, which can be solved by the direct solver, *i.e.*, double-sweep method with the appropriate boundary conditions. Similar procedures can also be applied to solve the horizontal velocity component, $\mathbf{v}_{\mathrm{i}}^{n+1}$, for each fixed index $\mathbf{i}=\mathbf{1},\ \mathbf{2}$, or $\mathbf{N}_1$ in the $y-z$ plane. In cases involving bottom topographies, a flag is introduced to identify the corresponding locations of land cells in matrix systems. A unity value is imposed to the locations on the left-hand side coefficient matrixes (i.e., $\mathbf{CL}_{\mathbf{i,j}}^{\mathbf{n+*}}$, $\mathbf{CM}_{\mathbf{i,j}}^{\mathbf{n+*}}$, and $\mathbf{CR}_{\mathbf{i,j}}^{\mathbf{n+*}}$ in Equation (22)), while a zero value is imposed to the locations on the right-hand side coefficient vectors (i.e., $\mathbf{d}_{\mathbf{i,j}}^{\mathbf{n+*}}$ in Equation (22)). As a result, the calculated velocities are all zero for the land cells. The reconstruction of matrix system is also consistent with overall solution procedure.

3.5 Overall procedure

The overall procedures of the algorithm for each time step are summarized as below:

1. Initialize the computation domain and update the boundary conditions.
2. Set $u^{n+*}=u^n$, and $v^{n+*}=v^n$.
3. Arrange the block tri-diagonal matrix system in each vertical plane, *e.g.*, Equation (22), and solve u^{n+1} and v^{n+1}, respectively.
4. If $\left|u^{n+1}-u^{n+*}\right|<\varepsilon$ and $\left|v^{n+1}-v^{n+*}\right|<\varepsilon$, where ε is an acceptable convergence criterion, go to step 5. Otherwise, update $u^{n+*}=u^{n+1}$ and $v^{n+*}=v^{n+1}$, and go to step 3.
5. Calculate the vertical velocity w^{n+1}, pressure P^{n+1}, and free-surface elevation η^{n+1} from equations (9), (12), and (13), respectively.
6. Advance to next time step by replacing n value by $n+1$ values, and go back to step 1.

4. MODEL VALIDATIONS

To test the developed 3D non-hydrostatic model, two examples: (1) standing wave oscillation, and (2) progressive wave shoaling on a submerged bar are chosen. The model is verified using available analytical solutions for example (1) and experimental data for example (2).

4.1. Standing waves in a 3D closed basin

Small amplitude standing waves in the 3D closed basin can oscillate without damping if the viscosity is neglected. In this study, the horizontal dimension of the basin is chosen to be $L=W=5$ m in the *x* and *y* directions, respectively. The still water depth is $h=5$ m. The wave amplitude is set to $A=0.05$ m, 1% of the water depth. Based upon the linear wave theory, consider the free-surface elevation

$$\eta = A\cos(k_x x)\cos(k_y y)\cos(\omega t), \tag{24}$$

where $k_x = m\pi / L$, $k_y = n\pi / W$, and $\omega = \sqrt{gk\tanh(kh)}$ (frequency of the standing wave) with $k = \sqrt{k_x^2 + k_y^2}$ (wave number of the standing wave). The analytical solutions can be found in (Mei, 1983). The computational domain for the model is discretized by a set of uniform cells with $\Delta x = \Delta y = 0.25\,\text{m}$ and $\Delta z = 0.5\,\text{m}$ in the vertical interval. The top-layer cell, Δz_{top}, varies from 0.495 m to 0.505 m. A time interval, $\Delta t = 0.05$ s, is chosen.

The fully non-hydrostatic model is tested for the 3D standing wave oscillation. For $m = 1$ and $n = 1$, the wave number is $k = \sqrt{k_x^2 + k_y^2} = \sqrt{2}/5\,\pi$, and the corresponding wave period and wave celerity are $T = 2.13\,\text{s}$ and $c = 3.32$ m/s, respectively. The wave steepness is $Ak \approx 0.044$, to which the linear wave theory is valid. Figure 2 shows the simulated free-surface elevation at (*x, y*) = (0.125 m, 0.125 m) predicted by three different models. The hydrostatic model produces totally unrealistic results with a period about 0.6*T*, indicating its incapability of resolving the non-hydrostatic pressure for standing wave oscillation. Obvious phase errors are observed using the non-hydrostatic model with the hydrostatic representation at the top-layer cell. In contrast, the simulated free-surface elevation using the fully non-hydrostatic model are in excellent agreement with the analytical solutions over a long time (*e.g.*, over a 10 wave periods without any observable phase shift and amplitude damping), suggesting the significance of applying full representation of non-hydrostatic pressure in the model. In addition, a second-order time marching scheme of the developed non-hydrostatic model provides the minimum wave attenuation results. Figure 3 shows the excellent agreement between the analytical solution and the simulated velocity using the fully non-hydrostatic model at two different locations. The result indicates the capability of the model for simulating 3D free-surface flow field.

4.2 Progressive wave propagation over a submerged bar

The second example is periodic wave propagation over a submerged bar. The objective is to test the capability of the model to simulate the relative strong interaction between nonlinear wave and uneven bottom. It has been found that the shoaling would occur on the upward slope and the nonlinearity would generate significant higher harmonics, which travel phase-locked to the primary wave (Casulli 1999).

Figure 4 shows the experiments setup by Beji and Battijes (1994) for the numerical simulation. The water depth is $h_0 = 0.4$ m. At the inflow boundary a progressive wave with a wave height of $H_0 = 2.0$ cm and a period of $T_0 = 2.0$ s is specified. At outflow boundary the absorbing beach in the physical experiment setup is replaced by a radiation boundary in the model. The computational domain is discretized by a set of uniform $M \times N = 600 \times 10$ cells, and a time step of $\Delta t = 0.02$ s is chosen. Figure 5 shows the comparisons between the numerical results and experimental data of free-surface elevation at six wave gauge locations. At the position 1, the wave remains sinusoidal and the numerical results are in excellent agreement with experimental data. From $x = 6$ m to $x = 12$ m, the wave steepness is increased due to shoaling effects. At the position 2, the model predicted a value of 1.2 for shoaling coefficient (equivalent to the relative wave height η/H_0), which agrees well with the experimental data. The model also well simulates the wave riding over the bar at position 3 and the secondary wave mode at the position 5, 6, and 7. The excellent agreements between the numerical results and experiments indicate that the present model is capable to simulate the interaction between nonlinear short wave and uneven bottom.

4. CONCLUSIONS

A fully non-hydrostatic 3D model using an implicit algorithm is developed. The model solves the unsteady, incompressible NSE with the free-surface boundary condition. The model is based upon the implicit Crank-Nicolson scheme with second-order accuracy in both time and space. To avoid solving the 3D PPT, an efficient algorithm is developed which decouples the original matrix system into a series of sub-matrix systems so that the direct solver can be used. The model has an advantage of solving all flow-field components simultaneously within each time step. In addition, by employing the Crank-Nicolson scheme for overall discretization, the developed model is free of adjusting numerical parameters. The model uses a new treatment for the non-hydrostatic top-layer pressure in the staggered grid system without applying any hydrostatic pressure assumption. Compared to the hydrostatic representation, this treatment provides consistency with the vertical momentum equation. In addition, the model also uses partial-cell treatment on bottom topographies, which can simulate free-surface flows interacting with uneven bottoms.

The developed model is validated by two examples. For the standing wave simulation, different treatments to the top-layer pressure are compared, indicating the importance of using a fully non-hydrostatic pressure distribution in the model. For progressive wave propagating over a submerged bar, agreement between numerical results and experimental data is excellent, indicating that the model is capable of simulating interactions of nonlinear waves and bottoms with complicated geometry. In the near future, turbulence models will be added. The proposed implicit algorithm will be extended to the parallel computation for further increase the efficiency of solving the block tri-diagonal matrix.

REFERENCES

Apsley, D. and Hu, W. (2003). CFD simulation of two- and three-dimensional free-surface flow. *Int. J. Numer. Meth. Fluids,* 42, 465-491.

Armfield, S.and Street, R. L. (2002). An analysis and comparison of the time accuracy of fractional-step methods for the Navier-Stokes equations on staggered grids. *Int. J. Numer. Meth. Fluids* 38 (2), 255-282.

Beam, R.M. and Warming, R.F. (1978) An implicit factored scheme for the compressible Navier-Stokes equations, *AIAA Journal* 16, 393-402.

Beji, S. and Battijes, J.A. (1994). Experimental investigation of wave propagation over a bar. *Coastal Engineering*. 1993; **19:** 151-162.

Blumberg, A.F. and Mellor, G.L. (1987). A description of a three-dimensional coastal ocean circulation model, in: N.S. Heaps, ed., Three-Dimensional Coastal Ocean Circulation Models, *Coastal and Estuarine Sciences,* Vol. 4 AGU, Washington DC, 1-16.

Casulli, V. (1999). A semi-implicit finite difference method for non-hydrostatic, free-surface flows. *Int. J. Numer. Meth. Fluids* 30, 425-440.

Casulli, V. and Zanolli, P. (2002). Semi-implicit numerical modeling of nonhydrostatic free-surface flows for environmental problems. *Mathematical and Computer Modeling* 36, 1131-1149.

Cheng, R.T. Casulli, V. and Gartner, J. W. (1993). Tidal, residual, intertidal mudflat (TRIM) model and its applications to San Francisco Bay, California. *Estuar. Coast. Shelf Sci.* 36, 235-280.

Hodges, B. R. and Street, R. L. (1999). On simulation of turbulent nonlinear free-surface flows. *Journal of Computational Physics* 151(2), 425-457.

Hur, D-S and Mizutani. (2003). Numerical estimation of the wave forces acting on a three-dimensional body on submerged breakwater. *Coastal Engineering* 47, 329-345.

Kim, M.H., Niedzwecki, J.M., Roesset, J.M., Park, J.C., Hong, S.Y., and Tavassoli, A. (2001). Fully nonlinear multidirectional waves by a 3-D viscous numerical wave tank. ASME *Journal of Offshore Mechanics and Arctic Engineering* 23 124-133.

Kocyigit, M. B., Falconer, R. A. and Lin, B. (2002). Three-dimensional numerical modeling of free surface flows with non-hydrostatic pressure. *Int. J. Numer. Meth. Fluids* 40, 1145-1162.

Li, B. and Fleming, C. (2000). Three-dimensional model of Navier-Stokes equations for water waves. ASCE *Journal of Waterway, Port, Coastal, and Ocean Engineering*. January/February, 16-25.

Mahadevan, A. and Oliger, J. and Street, R. (1996). A non--hydrostatic mesoscale ocean model. Part 1: well posedness and acaling. *J. Phys. Oceanogr.* 26, 1868-1880.

Mei, C.C (1983). *The Applied Dynamics of Ocean Surface Waves*, Wiley Inter-science.

Namin, M., Lin, B. and Falconer, R. (2001). An implicit numerical algorithm for solving non-hydrostatic free-surface flow problems. *Int. J. Numer. Meth. Fluids,* 35, 341-356.

Pacanowski, R.C. and Gnanadesikan, A. (1998). Transient response in a z-level ocean model that resolves topography with partial cells, *Mon. Weather Rev.* 126 (12): 3248-3270.

Tannehill, J.C., Anderson, D.A. and Pletcher, R. H. (1997). *Computational Fluid Mechanics and Heat Transfer* (2nd ed.), in: W.J. Minkowycz, E.M. Sparrow eds., Series in Computational and Physical Processes in Mechanics and Thermal Sciences (Taylor & Francis: Hemisphere Publishing Corporation, Bristol, PA, 1997) 717-724.

Yuan, H.L. and Wu, C.H. (2003). A non-hydrostatic σ model with an implicit method for free-surface flows. To appear in *Int. J. Numer. Meth. Fluids* in 2003.

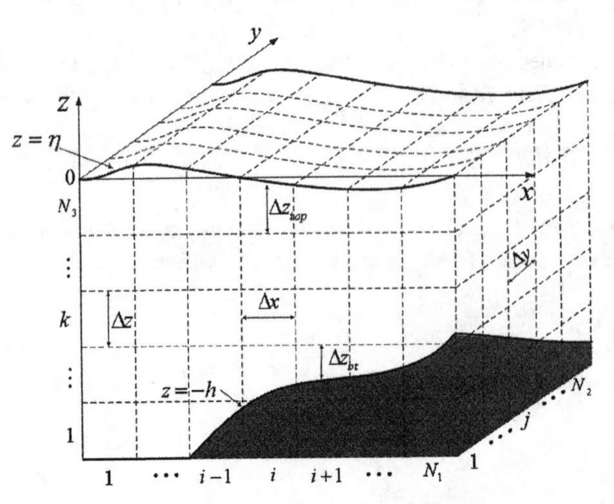

Figure 1a. A Cartesian coordinate system

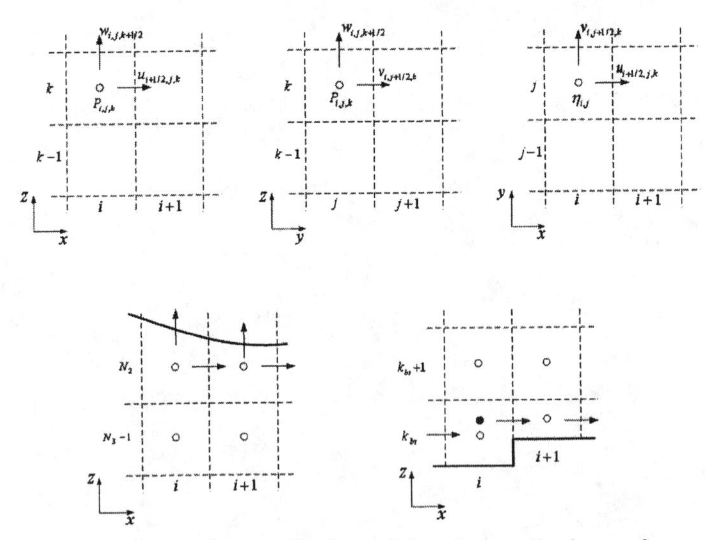

Figure 1b. A staggered grid system. The lower left represents the free surface cell and the lower right is the bottom cell.

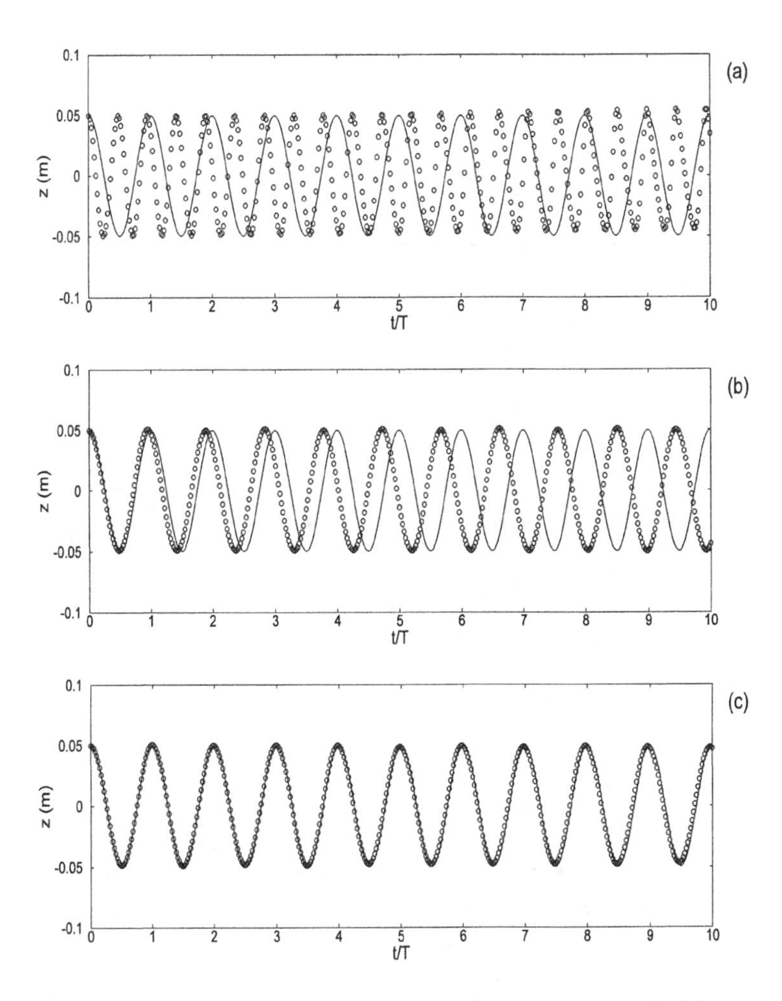

Figure 2. Comparison of 3D surface elevation at (x,y)=(0.125 m, 0.125m) between analytical solutions (solid lines) and numerical results (circles) of (a) hydrostatic model, (b) non-hydrostatic model with hydrostatic pressure distribution at the top-layer, and (c) fully hydrostatic model.

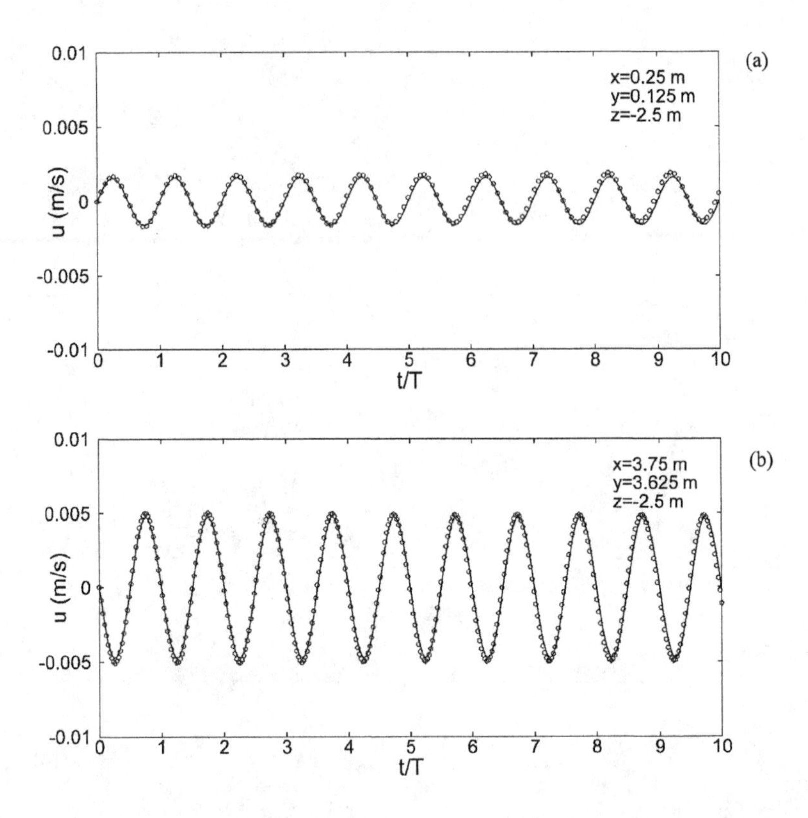

Figure 3. Comparison of 3D horizontal velocity at (a) (x,y,z)=(0.25m, 0.125m, 2.5m) and (b) (x,y,z)=(3.75m, 3.625m, 2.5m) between analytical solutions (solid lines) and numerical results (circles) using the fully non-hydrostatic model.

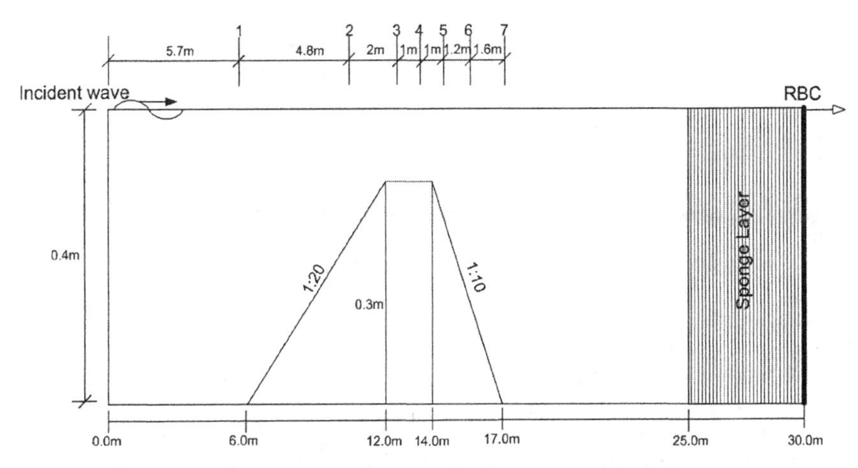

Figure 4. Sketch of periodic wave propagation over a submerged bar

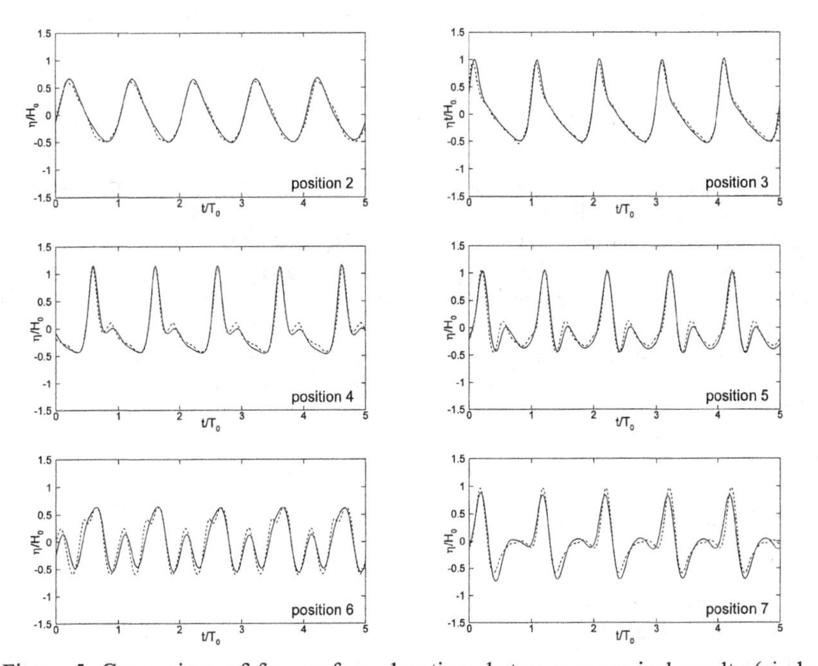

Figure 5. Comparison of free surface elevations between numerical results (circles) and experimental data (solid lines) at six different positions.

Modeling Hydrodynamic and Water Quality Processes in a Reservoir

Zhen-Gang Ji[1], Michael R. Morton[2], and John H. Hamrick[2]

Abstract

Despite the progress in three-dimensional (3D) hydrodynamic, water quality, and sediment diagenesis models and their successful applications in estuaries and bays, few similar 3D modeling studies on lakes and reservoirs have been published. In this study, a 3D hydrodynamic and water quality model has been developed and applied to Lake Tenkiller, Oklahoma. The model includes coupled hydrodynamic, eutrophication, and sediment diagenesis processes. Lake Tenkiller is a manmade reservoir up to 45 meters deep. The lake measures 48 kilometers (km) long, up to 3 m wide, and 70 km^2 in area. Its major water quality issues include nutrient enrichment, eutrophication, and hypolimnetic oxygen depletion. With large lateral variations, the lake needs a 3D model to simulate the hydrodynamic and water quality processes in detail. The model has 198 horizontal grid cells and 10 vertical layers. Measured data at 14 stations from February 1986 to September 1986 are available for hydrodynamic and water quality model calibration, including water elevation, water temperature, dissolved oxygen, chlorophyll *a*, 5-day biochemical oxygen demand, orthophosphorus, and nitrate-nitrogen. Comparisons between the modeled results and the measured data for all parameters were satisfactory. Seasonal variations of water quality variables in the lake were well replicated. A series of test cases was conducted to illustrate the importance of 3D modeling of lake hydrodynamic and eutrophication processes. It is shown that lake stratification and wind forcing are two major hydrodynamic processes controlling the hypolimnetic oxygen depletion. A seiche signal with a period of 2.36 hours is found in the lake. Theoretical estimation, time series plots, and the Fourier analysis all consistently support this finding. The model is used as a tool for water quality management in the lake.

---

[1]Tetra Tech. Inc., 10306 Eaton Pl., Suite 340, Fairfax, VA 22030. Current affiliation: Minerals Management Service, 381 Elden Street, Herndon, VA 20170.

[2]Tetra Tech. Inc., 10306 Eaton Pl., Suite 340, Fairfax, VA 22030.

Introduction

Water resource management requires scientifically credible numerical models for evaluations of management options. The dynamic simulation of eutrophication in a surface water system can be a very complicated and computationally intensive endeavor, due to the time variations of a large number of chemical, biological, and biochemical processes; reaction rates; and external inputs.

Despite the progress in three dimensional (3D) hydrodynamic, water quality, and sediment diagenesis models and their successful applications in estuaries and bays (Cerco 1999), few similar 3D modeling studies on lakes and reservoirs have been published. In modeling studies (such as Chung and Gu 1997; Tufford and McKellar 1999), the importance of 3D modeling of lakes and reservoirs is increasingly realized. In modeling hydrodynamics and sediment transport in rivers and reservoirs, Ziegler and Nisbet (1994, 1995) also illustrated the limitations of laterally averaged models and emphasized the needs for 3D modeling.

Lake Tenkiller (Figure 1) is a reservoir located in northeastern Oklahoma. The U.S. Army Corps of Engineers (USACE) constructed the lake in 1947 to provide flood control, water supply, fisheries, recreational activities, and electricity. The drainage area into the lake is 4,170 square kilometers (km^2), with the Illinois River serving as the major tributary inflow. Another major inflow to the lake is Caney Creek. The long-term, mean regulated outflow of the reservoir measures 44 cubic meters per second. The lake is 48 kilometers (km) long, up to 3 km wide, and 70 km^2 in area. Its depth varies from more than 45 meters (m) near the dam to less than 10 m in the northern section. The lake has a retention time of 1.76 years. With a width up to 3 km and large lateral bathymetry gradient, it is expected (and will be demonstrated in modeling results later) that this reservoir should have large 3D variability.

Lake Tenkiller is identified as a high-priority target for water quality improvement by the Oklahoma Department of Environmental Quality (ODEQ 2000). Major water quality issues include nutrient enrichment, eutrophication, and hypolimnetic dissolved oxygen (DO) depletion. Because of the 3D variability of Lake Tenkiller, it is critical to simulate the hydrodynamic and water quality processes using a 3D model, so that water quality parameters in the lake can be described in detail and cost-effective water management approaches can be proposed and evaluated.

The objectives of this study are: (1) to develop a 3D hydrodynamic and water quality model of Lake Tenkiller based on the Environmental Fluid Dynamics Code (EFDC) (Hamrick 1992 and Park et al. 1995); (2) to calibrate the Lake Tenkiller Model using the measured data graphically and statistically; and (3) to apply the model to the investigation of the hydrodynamic and water quality processes in the lake.

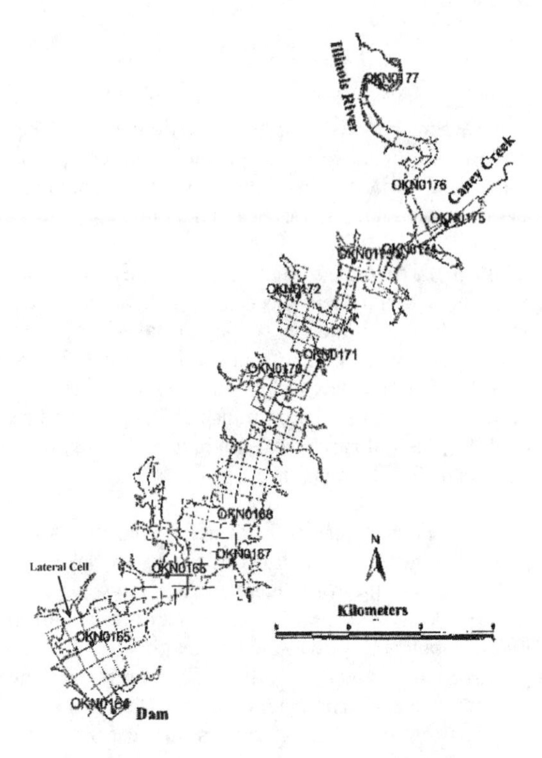

Figure 1. Lake Tenkiller study area and model grid.

Hydrodynamic and Water Quality Models

The Lake Tenkiller Model is developed within the framework of EFDC. The EFDC model is a public-domain modeling package for simulating 3D flow, transport, and biogeochemical processes in surface water systems (such as rivers, lakes, estuaries, reservoirs, wetlands, and coastal regions). The EFDC model has been extensively tested and documented in more than 80 modeling studies. The model is presently being used by a number of organizations, including universities, governmental agencies, and environmental consulting firms. Applications of the EFDC model include modeling of sediment and metals transport in Blackstone River (Ji et al. 2002); wetting and drying simulation of Morro Bay (Ji et al. 2000; Ji et al. 2001); and simulations of Lake Okeechobee hydrodynamic, thermal, and sediment processes (Jin et al. 2000, Jin et al. 2002; Jin and Ji 2003a; Jin and Ji 2003b). The EFDC model has also been applied to hydrodynamic modeling of Lake Billy Chinook Reservoir (Yang et al. 2000); study of tidal intrusion and its impact on larval dispersion in the James River estuary (Shen et al. 1999); and modeling estuarine front and its associated eddy (Shen and Kuo 1999). In addition, the EFDC model applications include investigation of bottom shear stress in estuaries (Kuo et al. 1996); hydrodynamic and sediment transport modeling in the Middle Atlantic Bight (Kim et al. 1997); and hydrodynamic and water quality modeling in Peconic Bay (Tetra Tech 1999).

The EFDC model includes four major modules: (1) a hydrodynamic model, (2) a water quality model, (3) a sediment transport model, and (4) a toxics model. The EFDC hydrodynamic model is similar to the Princeton ocean model (Blumberg and Mellor 1987) in many aspects. Hydrodynamic variables (such as water depth, velocity, temperature, salinity, and turbulent mixing) are directly coupled to the water quality, sediment transport, and toxics models. Details of the EFDC model are documented by Hamrick (1992) and Park et al. (1995). The Lake Tenkiller Model is based on the EFDC hydrodynamic, water quality, and sediment models.

The kinetic processes in the EFDC water quality model are mostly similar to the Chesapeake Bay water quality model (Cerco and Cole 1994). Park et al. (1995) and Tetra Tech (1999) described the kinetics of EFDC water quality model in detail. The EFDC water quality model has a total of 22 state variables; 16 are used in this study:

- cyanobacteria (B_c)
- green algae (B_g)
- refractory particulate organic carbon
- labile particulate organic carbon (LPOC)
- dissolved organic carbon (DOC)
- refractory particulate organic phosphorus
- labile particulate organic phosphorus
- dissolved organic phosphorus
- total phosphate
- refractory particulate organic nitrogen
- labile particulate organic nitrogen
- dissolved organic nitrogen
- ammonia nitrogen (NH_4)
- nitrate nitrogen (NO_3)
- chemical oxygen demand (COD)
- dissolved oxygen (DO)

The 5-day biochemical oxygen demand (BOD_5) is not directly simulated in the model and is calculated based on the following formula (HydroQual 1991, Tetra Tech 1999):

$$BOD_5 = 2.67 \cdot [LPOC \cdot (1 - e^{-5 \cdot K_{LPOC}}) + DOC \cdot (1 - e^{-5 \cdot K_{HR}}) + COD \cdot (1 - e^{-5 \cdot K_{COD}}) + \sum_{x = c \text{ and } g} B_x \cdot (1 - e^{-5 \cdot BM_x})] + 4.57 \cdot NH4 \cdot (1 - e^{-5 \cdot NIT}) \quad (1)$$

where K_{LPOC} = dissolution rate of LPOC (day^{-1})
K_{HR} = heterotrophic respiration rate of DOC (day^{-1})
K_{COD} = oxidation rate of COD (day^{-1})
BM_x = basal metabolism rate of algal group x (=c and g) (day^{-1})
NIT = nitrification rate (day^{-1})

A sediment diagenesis model with 27 state variables, developed by DiToro and Fitzpatrick (1993), is incorporated into the water quality model. The sediment diagenesis model, upon receiving the particulate organic matter deposited from the overlying water column, simulates the diagenesis and the resulting fluxes of inorganic substances (ammonium, nitrate, phosphate, and silica) and sediment oxygen demand back to the water column. The coupling of the sediment diagenesis model with the water quality model not only enhances the model's predictive capability of water quality parameters, but also enables it to simulate the long-term variations in water quality conditions in response to changes in nutrient loadings.

Model parameters used in this study are similar to the ones used in the Peconic Bay study (Tetra Tech 1999), Christina River study (Tetra Tech 2000a), the Long Island Sound study (HydroQual 1991), and the Massachusetts Bay study (HydroQual 1995). Di Toro (2001) concluded that the parameters used in the sediment diagenesis models are very similar to those used in the studies of Cerco and Cole (1994), HydroQual (1991), and HydroQual (1995). Di Toro (2001) further reported that the modeling of Lake Champlain (a large freshwater lake) also used very similar parameters.

Data Sources and Model Setup

Development of the Lake Tenkiller Model includes: (1) collecting and analyzing measured data for input and model-data comparison, (2) generating a model grid, (3) specifying model parameters, (4) setting up graphic and statistical packages, (5) calibrating the model, and (6) analyzing hydrodynamic and water quality processes in the lake.

The hydrodynamic inputs include meteorological forcings and lake inflows/outflows. The hourly meteorological data at Fayetteville, Arkansas, include wind speed and direction, air temperature, solar radiation, precipitation, relative humidity, and cloud cover. Hourly inflow and outflow data of the lake were provided by USACE (Miller 1999). The atmospheric deposition rates of nutrients are based on data from the National Atmospheric Deposition Program (http://nadp.sws.uiuc.edu/nadpdata/).

Approximately twice a month, the USACE (1988) monitored 14 stations in the lake area from February 1986 to September 1986. The collected data included water temperature (T), DO, chlorophyll *a* (CHL), BOD_5, orthophosphorus (PO_4), and nitrate-nitrogen ($NO_2 + NO_3$). These 14 water quality stations are shown in Figure 1, with the exception of OKN0169, which is not located on the lake. The data at OKN0177 and OKN0175 are used to represent tributary loadings from the Illinois River and Caney Creek, respectively. The remaining 11 stations in Figure 1 are used for model-data comparison and model calibration.

The study area was divided into a grid of discrete cells. To ensure that the grid conformed closely to the lake geometry, an orthogonal mapping procedure was used to represent the horizontal surface of the lake as a two-dimensional grid domain.

Figure 1 shows the model grid overlaying an outline of Lake Tenkiller. U.S. Geological Survey bathymetric data of Lake Tenkiller were interpolated to further define the lake in three dimensions. As shown in Figure 1, the inflows (Caney Creek and Illinois River) were each represented by one cell across the stream. To obtain adequate resolution in the lake, multiple cells were used in the lateral direction. The numerical grid consisted of 198 cells in the horizontal plane and 10 sigma layers in the vertical. It will be shown later that the 10 vertical layers are necessary and important to resolve the vertical temperature and DO profiles in the lake.

Hydrodynamic and Water Quality Simulation

Solutions to the hydrodynamic and water quality model were obtained using a 90-second time step. The time period used for calibration of the EFDC Lake Tenkiller Model was 262 days (January 5-September 24, 1986), which covers the USACE (1988) monitoring period. On a 2.4-GHz Pentium IV PC, approximately 5 CPU hours are required for a 262-day simulation.

Hydrodynamic Simulation

The hydrodynamic variables of the Lake Tenkiller Model include water surface elevation, water temperature, water velocity, and turbulent mixing. Hydrodynamic calibration is needed to properly characterize the nature, behavior, and patterns of water flow within the lake. The strong vertical temperature stratification and its effect on vertical mixing are also essential to eutrophication processes and water quality modeling.

Modeled versus observed water depths at OKN0164 are shown in Figure 2, in which the horizontal axis represents days from January 1, 1986, and the vertical axis represents water depth in meters. The dashed line represents the modeled water depth, and the solid line represents the measured water depth. It is evident that the modeled

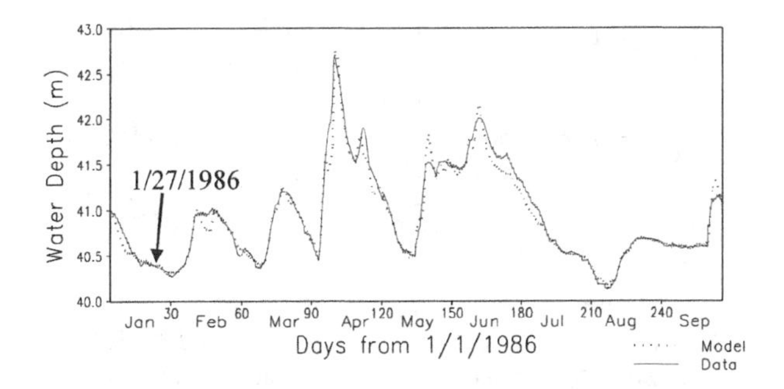

Figure 2. Model-data comparison of water surface elevation for 262 days at OKN0164. The solid line is the measured data, and the dashed line is the model results.

water depth closely matches the measured data. Figure 2 gives the daily averaged results. Modeled water depths at 15-minute time intervals (on 1/27/1986) will be presented in Figure 4.

Although numerous methods exist for analyzing and summarizing model performance, there is not a consensus in the modeling community on a standard analytical suite. A set of basic statistical methods is used in this study to compare model simulations and sampling observations. Statistical analysis on the modeled results and the measured data indicates that at OKN0164, the mean water depth is 40.9 m, the root mean square (RMS) error is 0.09 m, and the relative RMS error is 3.5 percent. The relative RMS error is defined as the RMS error divided by the observational water depth change, which is 2.57 m at OKN0164.

Temperature is also an important indicator of hydrodynamic behaviors. The 12 plots in Figure 3 show the vertical profiles of the modeled and measured temperatures at OKN0166. Also shown in Figure 3 are the DO profiles, which will be discussed later in the paper.

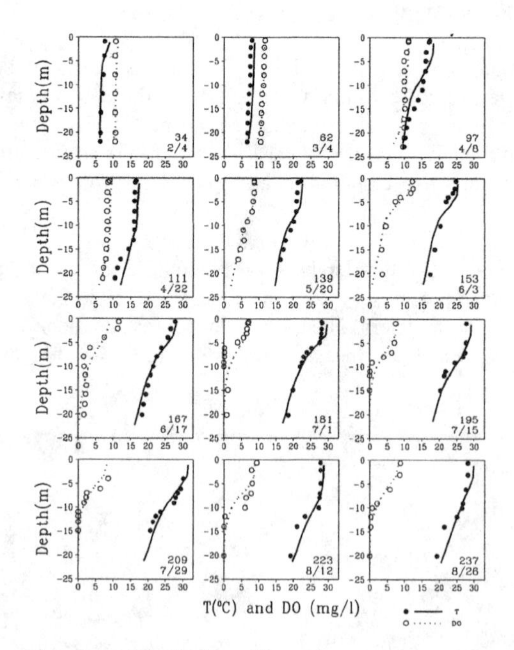

Figure 3. Vertical profiles of water temperature (T) and dissolved oxygen (DO) at OKN0166. The corresponding date and Julian Day are shown in the lower-right corner of each plot. Solid line = modeled T, closed circle = measured T, dashed line = modeled DO, and open circle = measured DO.

In Figure 3, both the model and the data indicate that the lake was well mixed in the winter (2/4 and 3/4), began to be stratified in the spring (4/8 and 4/22), was very stratified in the summer (7/1, 7/15, and 7/29), and became less stratified again later in late summer (8/26). The strongest vertical stratification occurred in the summer, with a surface-bottom temperature difference of more than 10 °C. Overall, the model simulated the temperature stratification and seasonal variation very well.

To conduct statistical analysis, modeled temperatures at the 11 stations were saved at the same times and water depths that the measured data were collected. Table 1 summarizes the statistical analysis of the observed and modeled temperature at the 11 stations. The number of observed temperatures varies from 42 at OKN0173 to 122 at OKN0164 and OKN0168. The modeled temperature has the smallest relative RMS error of 4.58 percent at OKN0172, and the largest relative RMS error of 7.97 percent at OKN0164. The RMS error varies from 1.11 °C at OKN0174 to 1.81 °C at OKN0164. Since there is no dam operation information available, the model is not able to describe the discharges from the dam in detail. Instead, the model describes daily discharge from the bottom layer near the dam. Overall, the water temperature profiles are simulated very well, with a mean relative RMS error of 5.85 percent among the 11 stations. Since DO modeling and eutrophication processes are closely linked to water temperature, accurate simulation of water temperature is also a vital step toward successful modeling of water quality processes.

Table 1. Statistical analysis of observed and modeled temperature at the 11 stations shown in Figure 1

Station Name	Obs. Data Number	Obs. Mean (°C)	Modeled Mean (°C)	Mean Abs. Err. (°C)	RMS Err. (°C)	Obs. Change (°C)	Relative RMS Err. (%)
OKN0164	122	19.09	20.47	1.47	1.81	22.7	7.97
OKN0165	113	18.32	19.09	1.18	1.41	22.5	6.25
OKN0166	120	19.13	19.99	1.17	1.40	23.9	5.86
OKN0167	97	20.63	21.68	1.33	1.58	23.7	6.69
OKN0168	122	19.73	20.62	1.14	1.38	23.8	5.79
OKN0170	112	20.47	21.07	0.91	1.20	24.7	4.85
OKN0171	94	21.26	22.15	1.23	1.65	24.0	6.88
OKN0172	94	21.97	22.74	0.94	1.13	24.7	4.58
OKN0173	42	22.14	23.28	1.14	1.26	23.4	5.37
OKN0174	77	22.28	22.71	0.92	1.11	23.0	4.83
OKN0176	63	24.23	25.16	1.00	1.15	21.7	5.30

No measured velocity data are available for model-data comparison. Due to its long retention time of 1.76 years, the inflows and outflows of the lake have a relatively small influence on circulation patterns in the lake. Water currents in the lake are primarily driven by the wind. Typically, the modeled surface velocities are about 2-4 percent of the wind speed. More discussions on velocity and on inflows and outflows will be presented later.

Seiches in the Lake

Lake Tenkiller has a typical length of 48 km and an average water depth of 14 m. The fundamental mode of seiches in the lake should be (Martin and McCutcheon 1999):

$$T_1 = \frac{2L}{\sqrt{gH}} = \frac{2 \times 48 \times 10^3}{\sqrt{9.8 \times 14}} = 2.36 \text{ hours} \qquad (2)$$

where T_1 = period of the fundamental mode of seiches
L = lake length in meters
H = mean water depth in meters
g = gravitational acceleration (= 9.8 m/s^2)

Therefore, it is expected that the lake should have a seiche period around 2.36 hours. The measured water depths shown in Figure 2 are at daily time intervals and are inadequate to resolve events within periods of a few hours. The modeled water depths, on the other hand, should be able to resolve seiche signals. By saving modeled results at 15-minute intervals, the modeled water depths on January 27, 1986, are shown in Figure 4, which is the same section of the time series on January 27, 1986 in Figure 2. The water depths in Figure 4 exhibit clear periodical behaviors. There are about 10 cycles during the 24 hours, which are consistent with the estimation in Eq. (2). Similar periodical signals are also found in other sections of the time series in Figure 2.

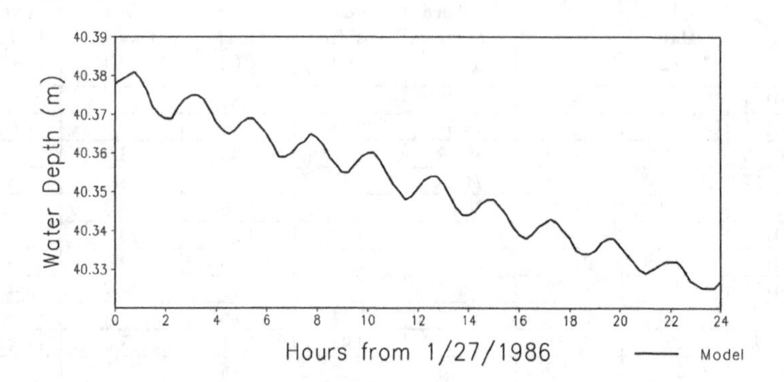

Figure 4. Modeled water surface elevation for 24 hours on January 27, 1986, at OKN0164.

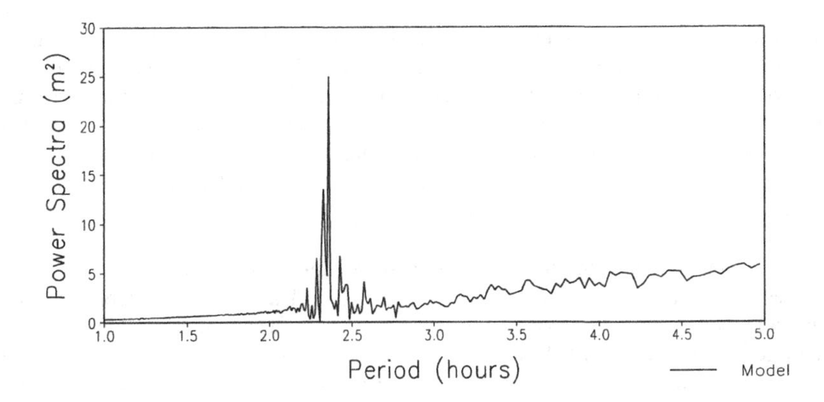

Figure 5. Power spectra of the modeled water surface elevation at OKN0164.

The Fourier analysis technique is used to identify periodical signals in water elevations. Figure 5 gives the power spectra of the modeled water elevation at OKN0164, the same station whose time series are shown in Figures 2 and 4. Figure 5 indicates that the strongest signal within the period range of 1 to 5 hours is at 2.36 hours, the same as the theoretical estimation given in Eq. (2). Fourier analysis is also applied to the modeled water elevations at other stations, and similar seiche periods are also found.

The Lake Tenkiller Model was originally developed for the seasonal and annual simulation of hydrodynamic and water quality processes in the lake. A primary application of the model is to support decisionmaking on water quality management, with a time scale of years and longer. It is surprising (and gratifying) to see that the model is capable of realistically simulating processes with time scale of hours.

Water Quality Simulation

Dissolved oxygen (DO) is controlled by complicated hydrodynamics, water quality, and eutrophication processes, such as reaeration, sediment oxygen demand, nitrification, denitrification, decay of organic substances, photosynthesis of algae, and respiration of algae. The DO is an important indicator of model performance in terms of water quality modeling. The 12 plots in Figure 3 show the vertical profiles of modeled and the measured DO at OKN0166. The dashed line represents the modeled DO, and the open circle represents the measured DO. Figure 3 shows that both the temperature and the DO exhibit similar vertical profile patterns. The DO is vertically mixed in the winter and is very stratified in the summer. For instance, on July 29, 1986 (Day 209), both the model and the data indicate that DO reduced to almost 0 mg/l at only 11 m below the water surface.

As shown in Figure 3, DO dynamics in Lake Tenkiller are typical of a lake system: shortly after the onset of thermal stratification, hypolimnetic oxygen content decreases. This condition begins at the water-sediment interface because of a high oxygen demand from sediment diagenesis. This process transfers nutrients between the sediment bed and the stratified lower water column, and progresses upward through the hypolimnion. Hypolimnetic anoxia occurred after stratification formed in the water columns, and continued through September.

While the model simulated the DO profiles reasonably in most times, the model did not resolve the DO stratification well on July 1, 1986 (Day 181). It appears that more vertical layers might be needed in this extremely stratified situation, in which DO was reduced to zero only 6 m below the surface. Errors in wind forcing might also cause too much mixing on this day. It will be demonstrated later that lake stratifications are sensitive to wind forcing. More accurate meteorological data might be able to improve the wind forcing and the model's vertical mixing.

Table 2 presents the statistical analysis of observed and modeled DO at the 11 stations. The relative RMS errors have a mean value of 16.34 percent, varying from 10.86 percent at OKN0170 to 28.26 percent at OKN0176. The modeled DO profiles at OKN0166 (shown in Figure 3) have a RMS error of 1.41 mg/l and a relative RMS error of 11.25 percent. Overall, as shown in Figure 3 and Table 2, the model simulated DO satisfactorily.

The USACE (1988) data also include CHL, NO_2+NO_3, PO_4, and BOD_5. But these data lack the quantity and quality to make model-data comparisons in vertical profiles (such as the ones in Figure 3). To deal with this kind of data lacking problem, one approach is to aggregate the measured data and modeled results in a certain area over a certain period of time, and then to compare the aggregated results (such as Cerco and Cole 1994; Tetra Tech 1999). However, this kind of spatial and time aggregation can filter out useful information, and in the present study, there are not many data for aggregation anyway.

Table 2. Statistical analysis of observed and modeled dissolved oxygen at the 11 stations shown in Figure 1

Station Name	Obs. Data Number	Obs. Mean (mg/l)	Modeled Mean (mg/l)	Mean Abs. Err. (mg/l)	RMS Err. (mg/l)	Obs. Change (mg/l)	Relative RMS Err. (%)
OKN0164	122	6.73	8.28	1.79	2.65	12.6	21.07
OKN0165	113	6.92	7.52	0.96	1.41	12.3	11.50
OKN0166	120	6.46	6.89	1.00	1.41	12.5	11.25
OKN0167	97	6.50	7.26	1.31	1.85	12.4	14.93
OKN0168	122	6.26	6.34	1.08	1.46	13.0	11.26
OKN0170	112	6.48	6.20	1.20	1.56	14.4	10.86
OKN0171	94	6.69	6.08	1.28	1.89	15.7	12.05
OKN0172	94	6.92	6.26	1.47	2.02	16.3	12.37
OKN0173	42	10.05	8.09	2.18	2.63	10.2	25.79
OKN0174	77	8.19	5.74	2.74	3.14	15.4	20.36
OKN0176	63	8.60	6.20	2.83	3.39	12.0	28.26

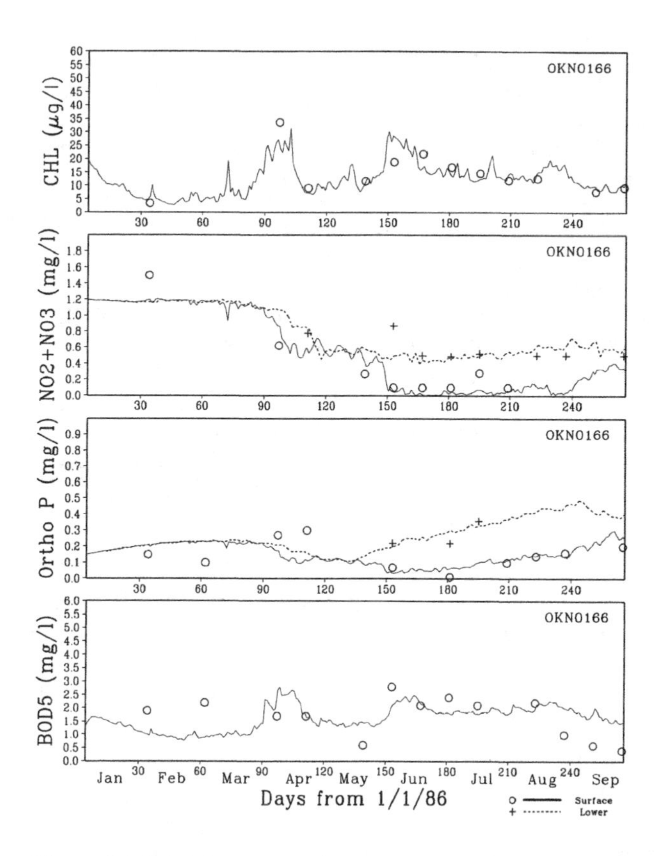

Figure 6. Model-data comparison of water quality variables for 262 days at OKN0166. First panel from the top: algae, second panel: nitrite-nitrate, third panel: orthophosphorous, and fourth panel: 5-day biochemical oxygen demand. Solid line = model results on the surface layer, open circle = measured data on the surface layer, dashed line = model results in the lower layer, and cross = measured data in the lower layer.

The four panels of Figure 6 show the modeled and the measured CHL, NO_2+NO_3, PO_4, and BOD_5 at OKN0166, the same station whose temperature and DO are shown in Figure 3. In Figure 6, the open circle represents the measured data on the surface, and the cross represents the measured data in the lower layer, which is 10 m below the water surface. The corresponding model results are represented by the solid line and the dashed line, respectively. In the first panel, the model captured the two algal blooms realistically, one in the spring and one in the summer, with algae concentration varying from less than 5 μg/l in the winter to more than 30 μg/l in the spring. In the second panel, both the model and the data indicate that the NO_2+NO_3

concentration is very stratified in the summer, as the result of thermal stratification (shown in Figure 3) and algae uptake. In the third panel, the PO_4 concentrations are also very stratified in the summer. In the fourth panel, the model and the data show that BOD_5 varies in the range of 0.5 mg/l to 2.5 mg/l. Overall, the model results in Figure 6 are generally consistent with the measured data and represent the seasonal changes reasonably.

There are no measured sediment fluxes data for model-data comparison in this study. Data from other waterbodies given by Di Toro (2001) and Lung (2001) are referenced. The modeled results are within the range reported in the literature. Di Toro (2001) detailed sediment flux data and modeling in Lake Champlain, a large freshwater system in North America. The average depth of Lake Champlain is 22.8 m, which is comparable with the depths of Lake Tenkiller. The sediment fluxes from the Lake Tenkiller Model are similar to the measured data and modeled results of ammonia nitrogen, NO_3, and PO_4 in the deep waters of Lake Champlain (Di Toro 2001), both in magnitude and in seasonal variation.

Sensitivity Tests

For understanding the sensitivity of model results to model configuration and external forcings, Figure 7 gives vertical profiles of water temperature (T) and DO in 6 different cases at OKN0165 on August 12, 1986. Plot 1 gives T and DO profiles from the 10-layer model, whose results at OKN0166 have been shown in Figure 3.

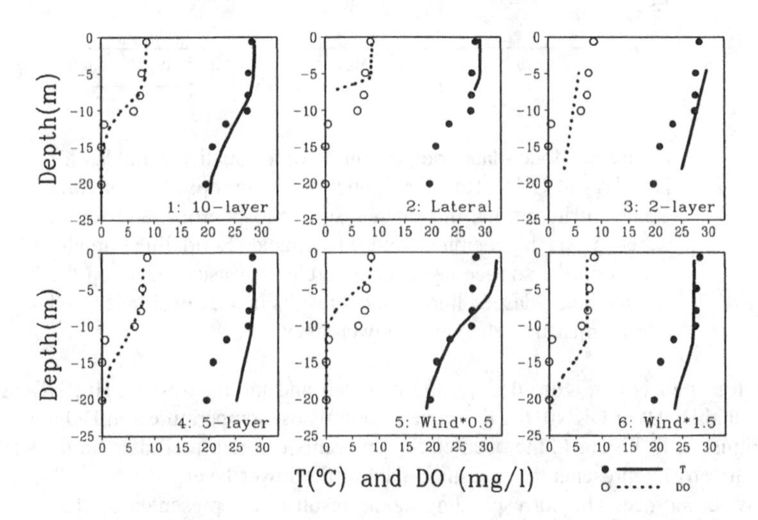

Figure 7. Vertical profiles of water temperature (T) and dissolved oxygen (DO) at OKN0165 on August 12, 1986, in 6 different cases. Case 1: 10-layer model, Case 2: lateral cell, Case 3: 2-layer model, Case 4: 5-layer model, Case 5: wind speed reduced by 50 percent, and Case 6: wind speed increased by 50 percent.

Plot 2 shows T and DO profiles at the lateral cell marked in Figure 1, which is on the west bank of the reservoir, 1.2 km away from OKN0165, and about 8 m in depth. The T and DO profiles in Plot 2 are very different from the ones in Plot 1. Hence, if the Lake Tenkiller Model is a laterally average model, instead of being a 3D model with multiple lateral cells, the model would have not been able to simulate the T and DO as well as the ones shown in Plot 1 (the 10-layer model).

Even though it is well known that high vertical resolution is needed to represent stratifications, there are actually limited publications on the influence of vertical layers on 3D simulations in lakes. To illustrate the importance of vertical resolution, the Lake Tenkiller Model is also constructed with 2 vertical layers and with 5 vertical layers. Plot 3 and Plot 4 show T and DO profiles from the 2-layer model and the 5-layer model, respectively. It is evident that neither 2 layers nor 5 layers are sufficient to represent vertical stratifications in the lake. For instance, both the data and the 10-layer model indicate very low DO at a water depth 12 m from the surface, while the 2-layer model and the 5-layer model fail to capture it. The modeled T from the 5-layer model is about 5 ^{o}C higher than the data near the lake bottom. A 20-layer run was also performed, and the results are very similar to the ones from the 10-layer model (shown in Plot 1).

Wind forcing can significantly influence lake stratification via vertical mixing processes. Plot 5 shows the modeled T and DO profiles when the wind speed is reduced by 50 percent. It is evident that reducing wind speed leads to less vertical mixing and strong T and DO stratifications. Plot 6 presents the T and DO profiles when wind speed is increased by 50 percent. The stronger wind made the lake much less stratified.

A test run without inflow and outflow was also conducted, and the results are similar to the ones in Plot 1. Since the lake has a retention time of 1.76 years, the inflows and outflows should have limited influence on the lake circulation and eutrophication processes.

Anoxic Volume in the Lake

The Oklahoma water quality standards (ODEQ 2000) state that a lake is not meeting designated beneficial uses if greater than 50 percent of the water column is anoxic (DO < 2.0 mg/l). It is critical to estimate the anoxic volume accurately to support decisionmaking on water quality management. Anoxic volume is often used as an indicator of a lake system's overall water quality condition. Figure 8 presents the anoxic volume (in percent of each water column) and surface current velocity on Day 253.5 (September 11, 1986).

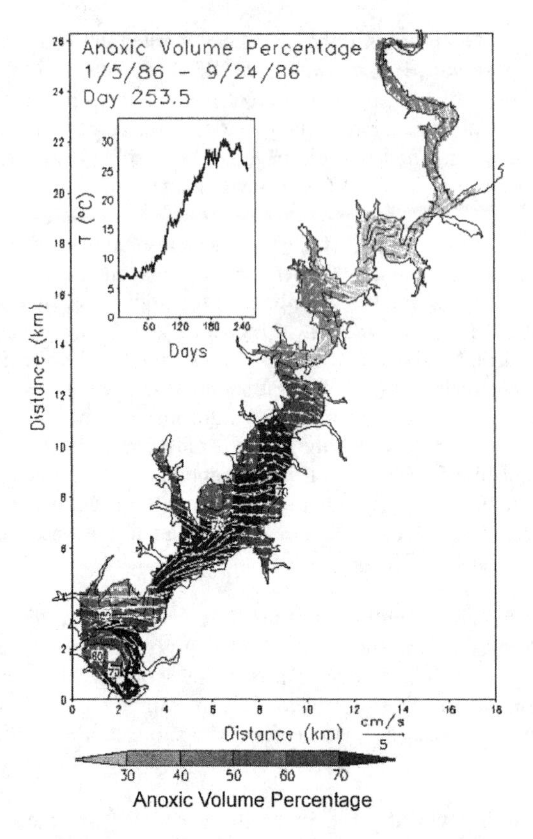

Figure 8. Anoxic volume (%) and surface current velocity (cm/s) at Day 253.5 (9/11/1986).

The small plot in the figure gives the air temperature from Day 4 to Day 253, which is a good indicator of thermal forcing from the atmosphere. Figure 8 shows that near the dam, more than 70 percent of the water column is anoxic. As the result of wind forcing, the surface currents are directing upstream in some areas on Day 253. This is clear evidence that the circulation patterns in Lake Tenkiller are primarily driven by the wind and not by inflows and outflows (which would have made the water velocity directing downstream). Because of the long retention time (more than a year), the influence of inflow/outflow forcings to the lake circulation is secondary during most times.

Figure 8 also shows strong lateral variation in the lower portion of the lake and indicates the need for 3D modeling. For example, in areas near the dam (where OKN0164 and OKN0165 are located), the anoxic volumes vary from less than 25 percent to more than 70 percent in the lateral direction, which is consistent with the results shown in Plot 2 of Figure 7. The primary cause of this phenomenon is that with a width up to 3 m, the lake in this section has water depths varying from less than 8 m to more than 40 m in the lateral direction. The lake circulations in this area also show large lateral variations and require 3D modeling.

Discussions and Conclusions

The primary purpose for developing the Lake Tenkiller Model was to use the model as a tool for proposing and testing load-management strategies aimed at limiting eutrophication processes in the lake (Tetra Tech 2000b). The model was calibrated against measured data in 1986 and represented existing hydrodynamic and water quality processes in the lake satisfactorily. In addition to being compared with the measured data graphically, the model results were also analyzed statistically. This provided a different perspective on model-data comparison that numerically quantified the state of model calibration.

This modeling study presents the following aspects:

1) Despite the progress in 3D hydrodynamic, water quality, and sediment diagenesis models and their successful applications in estuaries and bays, few similar 3D modeling studies on lakes and reservoirs have been published. In this study, a 3D hydrodynamic and water quality model has been developed and applied to Lake Tenkiller, Oklahoma. The importance of 3D modeling of lake hydrodynamic and eutrophication processes is discussed in detail and demonstrated through a variety of test cases.

2) A 262-day calibration period was simulated using the Lake Tenkiller Model. Particular attention was given to reproducing the seasonal cycles of temperature, algae, and nutrients in the lake. Comparisons of the modeled results and the measured data for all parameters were satisfactory. The seasonal variations of T, DO, CHL, NO_2+NO_3, PO_4, and BOD_5 were replicated reasonably well.

3) In addition to nutrient loadings and eutrophication processes, lake stratification and wind stress significantly affect hypolimnetic oxygen depletion. It was demonstrated that 10 vertical layers are needed to resolve the lake's stratifications, and that wind speed has a significant impact on the vertical mixing in the lake.

4) A seiche signal with a period of 2.36 hours was found in the modeled water elevation. The theoretical estimation from Eq. (2), the time series in Figure 4, and the power spectra analysis shown in Figure 5 all consistently support this finding. Further studies are needed to clarify the effects of seiches on the hydrodynamic and water quality processes in the lake.

The Lake Tenkiller Model is used as a tool for proposing and evaluating cost-effective approaches for water resources management (Tetra Tech 2000b).

Acknowledgments

The authors wish to thank Steve Webb and Paul Yu of the Oklahoma Department of Environmental Quality, and Philip Crocker of the U.S. Environmental Protection Agency for their helpful comments. Andrew Miller of the U.S. Army Corps of Engineers provided data of inflow, outflow, and water surface elevation. Technical support from Jian Shen, Terry Nelson, and John Riverson of Tetra Tech are greatly appreciated. The authors would also like to thank Eileen Lear of the Minerals Management Service for editing the manuscript.

References

Blumberg, A. F., and Mellor, G. L. (1987). "A description of a three-dimensional coastal ocean circulation model." In *Three-Dimensional Coastal Ocean Models, Coastal and Estuarine Science*, Vol. 4. (Heaps, N. S., ed.) American Geophysical Union, 1-19.

Cerco, C. F. (1999). "Eutrophication models of the future." *Journal of Environmental Engineering*, ASCE, 125(3), 209-210.

Cerco, C. F., and Cole, T. (1994). "Three-dimensional eutrophication model of Chesapeake Bay." Volume 1: Main Report. Technical Report EL-94-4. US Army Corps of Engineers Waterways Experiment Station.

Chung, S.-W., and Gu, R. (1997). "Two-dimensional simulations of contaminant currents in stratified reservoir." *Journal of Hydraulic Engineering*, ASCE, 124 (7). 704-711.

Di Toro, D. M. (2001). Sediment flux modeling. John Wiley & Sons, Inc., New York.

DiToro, D. M, and Fitzpatrick, J. (1993). "Chesapeake Bay sediment flux model." Contract Report EL-93-2, U.S. Army Engineer Waterways Experiment Station, Vicksburg, MS.

Hamrick, J. M., (1992). "A three-dimensional Environmental Fluid Dynamics Computer Code: theoretical and computational aspects." The College of William and Mary, Virginia Institute of Marine Science. Special Paper 317, 63 pp.

HydroQual. (1991). "Water quality modeling analysis of hypoxia in Long Island Sound." HydroQual Rep., HydroQual, Inc., Mahwah, N.J.

HydroQual. (1995). "A water quality model for Massachusetts and Cape Cod Bays: calibration of the bays eutrophication model (BEM)." HydroQual Rep., HydroQual, Inc., Mahwah, N.J.

Ji, Z.-G., Hamrick, J. H., and Pagenkopf. J. (2002). " Sediment and metals modeling in shallow river." *Journal of Environmental Engineering*, 128, 105-119.

Ji, Z.-G., Morton, M. R., and Hamrick, J. H. (2000). "Modeling hydrodynamic and sediment processes in Morro Bay." In *Estuarine and Coastal Modeling: Proceedings of the 6th International Conference* (Spaulding, M. L., and Butler, H. L., eds.), ASCE, 1035-1054.

Ji, Z.-G., Morton, M.R., and Hamrick, J. H. (2001). "Wetting and drying simulation of estuarine processes." *Estuarine, Coastal and Shelf Science*, 53, 683-700.

Jin, K.-R., Hamrick, J. H., and Tisdale, T. (2000). "Application of a three-dimensional hydrodynamic model for Lake Okeechobee." *Journal of Hydraulic Engineering*, ASCE, 126, 758-771.

Jin, K. R., Ji, Z. G., and Hamrick, J. M. (2002). "Modeling winter circulation in Lake Okeechobee, Florida." *Journal of Waterway, Port, Coastal, and Ocean Engineering*, 128, 114-125.

Jin, K. R., and Ji, Z. G. (2003a). "Modeling of sediment transport processes and wind-wave impact in a large shallow lake." *Journal of Hydraulic Engineering*, ASCE (tentatively accepted)

Jin, K. R., and Ji, Z. G. (2003b). "Application and validation of a 3-D model in a shallow lake." *Journal of Waterway, Port, Coastal, and Ocean Engineering* (tentatively accepted)

Kim, S., Wright, D. L., Maa, J. and Shen, J. (1997). "Morphodynamic responses to extratropical meteorological forcing on the inner shelf of the middle Atlantic Bight: wind wave, currents and suspended sediment transport." *Estuarine and Coastal Modeling*, Proceedings of the 5th International Conference, ASCE, 456-466.

Kuo, A. Y., Shen, J. and Hamrick, J. M. (1996). "The effect of acceleration on bottom shear stress in tidal estuaries." *Journal of Waterway, Port, Coastal, and Ocean Engineering*, ASCE, 122 (2), 75-83.

Lung, W.-S. (2001). Water quality modeling for wasteload allocations and TMDLs, John Willey & Sons, Inc, New York.

Martin, J. L., and McCutcheon, S. C. (1999). Hydrodynamics and transport for water quality modeling. Lewis Publishers, Boca Raton, Florida.

Miller, A. (1999). Personal communication, U.S. Army Corps of Engineers, Tulsa District.

ODEQ. (2000). "Continuing planning process." 1999-2000 edition. Oklahoma Department of Environmental Quality, the State of Oklahoma.

Park, K., Kuo, A. Y., Shen, J., and Hamrick, J. M. (1995). "A three-dimensional hydrodynamic-eutrophication model (HEM-3D): description of water quality and sediment processes submodels." Special report in Applied Marine Science and Ocean Engineering No. 327. Virginia Institute of Marine Science, Gloucester Point, Virginia.

Shen, J., Boon, J., and Kuo, A. Y. (1999). "A numerical study of a tidal intrusion front and its impact on larval dispersion in the James River estuary, Virginia." *Estuary*, 22(3A), 681-692.

Shen, J., and Kuo, A. Y. (1999). "Numerical investigation of an estuarine front and its associated eddy." *J. Waterways, Ports, Coastal and Ocean Engineering*, ASCE, 125 (3), 127-135.

Tetra Tech. (1999). "Three-dimensional hydrodynamic and water quality model of Peconic estuary." Technical Report for Peconic Estuary Program, Suffolk County, NY. Tetra Tech, Inc., Fairfax, Virginia.

Tetra Tech. (2000a). "Hydrodynamic and water quality model of Christina River basin." Technical Report for U.S. Environmental Protection Agency Region 3. Tetra Tech, Inc., Fairfax, Virginia.

Tetra Tech. (2000b). "Water quality modeling analysis in support of TMDL development for Tenkiller Ferry Lake and the Illinois River Watershed in Oklahoma." Technical Report for U.S. Environmental Protection Agency Region 6 and Department of Environmental Quality, State of Oklahoma. Tetra Tech, Inc., Fairfax, Virginia.

Tufford, D. L., and McKellar, H. N. (1999). "Spatial and temporal hydrodynamic and water quality modeling analysis of a large reservoir on the South Carolina (USA) coastal plain." *Ecological Modeling*, 114, 137-173.

USACE. (1988). "Water Quality Report on Tenkiller Ferry Lake, 1985-1986." U.S. Army Corps of Engineers, Tulsa, OK.

Yang, Z., Khangaonkar, T., DeGasperi, C., and Marshall, K. (2000) "Three-dimensional modeling of temperature stratification and density-driven circulation in Lake Billy Chinook, Oregon." In *Estuarine and Coastal Modeling: Proceedings of the 6th International Conference* (Spaulding, M. L., and Butler, H. L., eds.), ASCE, 411-425.

Ziegler, C. K., and Nesbitt, B. (1994). "Fine-grained sediment transport in Pawtuxet River, Rhode Island." *J. Hyd. Engrg.*, 120, 561-576.

Ziegler, C. K., and Nesbitt, B. (1995). "Long-term simulation of fine-grained sediment transport in large reservoir." *J. Hyd. Engrg.*, 121, 773-781.

Hydrodynamic and Water Quality Modeling of Ward Cove, Alaska

Hugo N. Rodriguez[1], Ben Cope[2] and Steven J. Peene[3]

Abstract

Ward Cove is a small estuary located on the north side of Tongass Narrows, approximately five miles north of Ketchikan, Alaska. Three activities–pulp mill effluent discharges, log rafting, and seafood processing facility discharges–have introduced contaminants to Ward Cove. A pulp mill began operating in the cove in 1952 and ceased operations in 1997. The bottom of the cove has been covered with fiber mats, sunken logs, and enriched sediments. In addition, the seafood processing and log rafting operations continue discharging.

Numerical modeling techniques were developed to simulate the loading of organic material and nutrients, and the resulting response of dissolved oxygen within Ward Cove. A system of models was developed to allow the determination of the circulation and transport within Ward Cove, and the response of critical water quality parameters. The system of models includes the Environmental Fluid Dynamics Code (EFDC) to simulate the circulation and transport of material and the Water Quality Analysis and Simulation Program (WASP) to simulate the response of critical water quality parameters to the loads.

The EFDC and WASP models provided a good framework for predicting responses of hydrodynamic and water quality parameters. The results show that the models capture the primary temporal and spatial distribution of temperature, salinity and dissolved oxygen in the system, especially the vertical stratification conditions.

Introduction

Ward Cove is an estuary located on the north side of Tongass Narrows, approximately 8 km north of Ketchikan, in southeastern Alaska (Figure 1). The area surrounding the cove is mountainous and largely forested. Ward Creek drops quickly from the nearby mountains to enter the cove at its east end. Tongass Narrows connects to Ward Cove at its west end. The predominant orientation of the cove is northeast to southwest.

----

[1]Tetra Tech, Inc., 2110 Powers Ferry Road, Atlanta, GA 30339. USA. Hugo.Rodriguez@tetratech-ffx.com

[2]Environmental Protection Agency, Region 10, 1200 6th Avenue, Seattle, WA 98101, USA. Cope.Ben@epamail.epa.gov

[3]Formerly Tetra Tech, Inc., 2110 Powers Ferry Road, Atlanta, GA 30339. USA. speene@earthlink.net

Ward Creek is the cove's primary source of freshwater and it is tidally influenced by currents in Ward Cove. Walsh Creek, a much smaller stream, enters Ward Cove along its southeast shoreline. Runoff from precipitation also enters Ward Cove from its adjacent watershed.

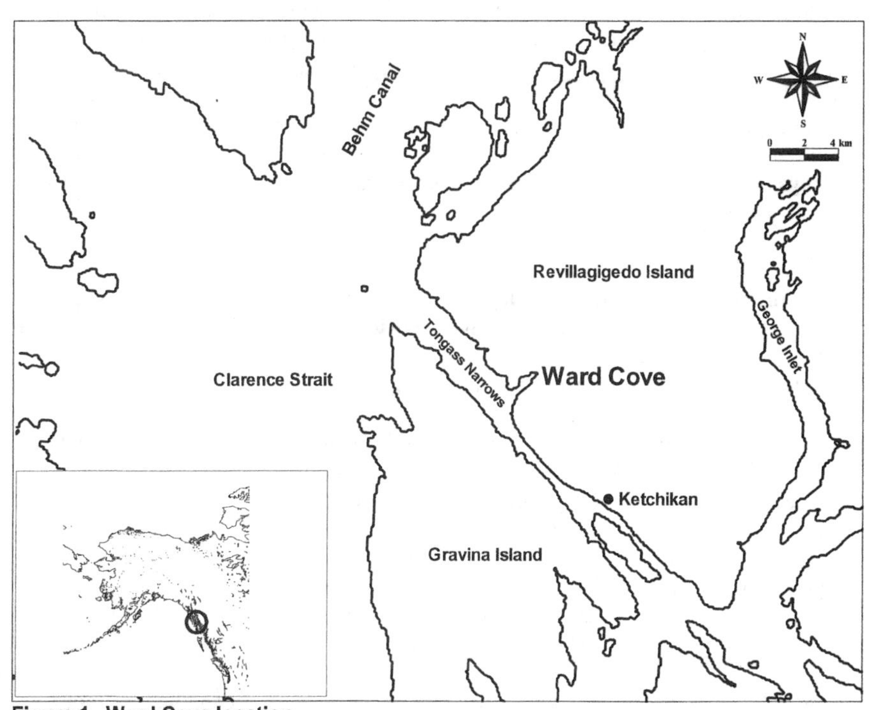

Figure 1. Ward Cove location.

The cove itself is approximately 1.6 km long and 0.8 km wide at its widest point. Water depth ranges from 3 m below MLLW at the northeast end of the cove to 61 m below MLLW at the southwest end of the cove, which opens to Tongass Narrows. The shoreline is mostly rocky (i.e., basalt) and relatively steep. Portions of the north shoreline are especially steep; the slope of some areas exceeds 25 percent.

The tides at Ward Cove are semidiurnal (a tidal period of 12.4 hours). The mean tidal range is 4.05 m with spring tides reaching values of 4.78 m. Velocities inside the cove are low compared to the velocities in Tongass Narrows. Average current velocities within the cove are on the order of 3 cm/s compared to 30 cm/s in Tongass Narrows. Field measurements of currents within Ward Cove identified a counterclockwise residual circulation pattern with flows into the cove along the southeastern shoreline and flows out of the cove along the northwestern shore. Superimposed on this horizontal residual circulation pattern is a standard estuarine flow condition caused by the mixing of the saline waters from the narrows with freshwater coming out of Ward Creek. The pattern superimposed shows residual flows into the cove along the bottom and flow out at the surface (Exponent, 1999).

As an embayment of Tongass Narrows, Ward Cove exhibits estuarine circulation with saline waters at depth and less saline waters near the surface. A salinity density gradient, or pycnocline, separates surface water from deeper water in summer months. Fresh water remains on top when there is no strong mixing impetus and flows outward, while deeper salt water migrates inward to maintain a mass balance of water. Near the entrance to Ward Cove, there is no bilayer flow or other dominant pattern. This is likely caused by influences of the swifter currents and eddies in Tongass Narrows. Because Ward Cove's tidal range is large, its currents are largely tidally driven.

Average flow velocities in Ward Cove vary with depth. In its main section, average velocities decrease slightly with depth (2.4 cm/s near the surface to 1.2 cm/s near the bottom), while near its mouth, average velocities increase slightly with depth (3.0 cm/s near the surface to 3.4 cm/s near the bottom). The rapid currents in Tongass Narrows may cause the higher velocities at depth.

There appears to be no area of stagnation within Ward Cove. In combination with the bilayer flow, this indicates that dissolved or suspended material introduced to the surface water layer is likely to be transported directly to the mouth of the cove and then mixed into the flow of Tongass Narrows. Dissolved or suspended material introduced to deeper waters likely would be subject to circulation below the pycnocline. However, material that has settled on the floor of the cove is not expected to travel. Residual velocities below the pycnocline are less than 5cm/s with negligible resuspension and transport capacity.

The only surficial features of the cove, other than sunken logs and bark residues, are exposed bedrock and a mound of seafood waste offshore of the seafood processing facility on the southern shore (Exponent, 1999). Sunken logs and bark residues are present over roughly 75 percent of the bottom of the cove. The concentration of logs on the sea floor varies from more than 500 per 10,000 m^2 in the center of the cove to fewer than 100 per 10,000 m^2 near the mouth of the cove (Exponent, 1999). Underwater video and side-scan sonar data have indicated that there are multiple layers of logs in the area of highest concentration, as well as numerous partially buried logs. The pile of fish processing waste is located directly offshore from the cannery, and, under the NPDES general permit for seafood processors, is limited to one acre in area. The debris pile forms hills, valleys and stair steps that progress rapidly down at approximately a 45-degree grade from the discharge point (Wards Cove Packing, 1999).

An extensive area of non-native sediments that are primarily the result of pulping effluent discharges from the paper mill occurs on the seafloor along the northwest shore from the former mill site out toward the entrance of the cove. These organic sediments may contain bark wastes, wood chips and other wood wastes. These pulp residues range in size from undetectable to greater than 3 m, with a typical thickness of about 1.2 m. Throughout the cove, the upper material is generally a watery, black, flocculent material with a strong sulfide odor, containing varying amounts of wood debris (e.g., wood chips, bark). Both the 1989 *Ward Cove Water Quality Assessment* (Jones & Stokes and Kinnetic, 1989) and the Sediment Remediation Project (Exponent, 1999) indicate that surface sediments throughout the cove contained bark or other wood debris, with a higher percentage of wood present in cores collected near the pulp mill docks. The underlying native sediments are olive-green to gray silty clays and clayey silts with imbedded roots, shells, and schist fragments. The combined thickness of the organic-rich sediments and native clay sediments ranges from 20 to 30 m.

The Ketchikan area's maritime climate is characterized as relatively mild and very wet. Average seasonal temperatures range from -2 to 4 degrees Celsius in winter, 7 to 13 degrees Celsius in the spring and fall, and 11 to 18 degrees Celsius in the summer. Winds are from the southeast and the northwest due to the funneling effect of the Tongass Narrows. The predominant winds are from the southeast as a result of the presence of low-pressure cells to the northwest, which draw air from the Gulf of Alaska to the Ketchikan area. The annual average wind speed is 0.45 m/s. The average annual rainfall at Ketchikan, Alaska is 3850 mm, making it one of the wettest locations in the United States. Normally the annual rainfall is spread over the year with a dryer period during July and August. Evapotranspiration is approximately 61 cm per year (Exponent, 1999).

Site-specific data from a local meteorological station at the Gateway facility indicate that, depending on the season, local winds come from the west, south, east, and northeast at an average wind speed of approximately 9.7 km per hour (Exponent, 1999).

Modeling Methodology

Numerical modeling techniques were developed to simulate the loading of organic material and nutrients, and the resulting response of dissolved oxygen within Ward Cove. A system of models was developed to allow the determination of the circulation and transport within Ward Cove, and the response of critical water quality parameters. The system of models includes the Environmental Fluid Dynamics Code (EFDC) to simulate the circulation and transport of material within Ward Cove, and the Water Quality Analysis and Simulation Program (WASP) to simulate the response of critical water quality parameters to the loads.

EFDC is a hydrodynamic modeling package for simulating three-dimensional flow and transport in surface water systems including: rivers, lakes, estuaries, reservoirs, wetlands and nearshore to shelf scale coastal regions. The EFDC model was originally developed at the Virginia Institute of Marine Science for estuarine and coastal applications and is public domain software. EFDC uses a stretched or sigma vertical coordinate and cartesian or curvilinear, orthogonal horizontal coordinates (Hamrick, 1992; Hamrick, 1996).

The Water Quality Analysis and Simulation Program (WASP) modeling system is a generalized modeling framework for contaminant fate and transport in surface waters. Based on flexible compartment modeling, WASP can be applied in one, two, or three dimensions and predicts dissolved oxygen, carbonaceous biochemical oxygen demand (CBOD), phytoplankton, chlorophyll a, ammonia, nitrate, organic nitrogen, and orthophosphate in bed and overlying waters. WASP is a compartment water quality model, and needs the transport and exchange between the model cells as an input; these come from the EFDC three-dimensional simulations. The EFDC model is externally linked to the WASP model. EFDC generates WASP input files describing cell geometries and connectivity as well as advective and diffusive transport fields.

Ward Cove Hydrodynamic Model Application

The EFDC hydrodynamic application for Ward Cove is a three-dimensional simulation with tidal forcing at the ocean boundaries (North and South boundaries), and freshwater inflow from Ward Creek at the upstream boundary. Figure 2 presents the model grid used to simulate the hydrodynamics in Ward Cove. The model grid covers part of Tongass Narrows in order to correctly simulate the hydrodynamics at the cove mouth. In order to properly simulate the hydrodynamics of the Narrows in the area, only the main channel of Tongass Narrows was modeled. The area known as Refuge Cove consists of several islands and rock outcrops that is only partially covered with water during part of the tidal cycle and was thus not modeled. The tidal wave propagates along the Narrows from north to south generating strong currents, which in turn generate eddies at the cove mouth. Placing the model boundary at the mouth of the cove would prevent the correct simulation of the hydrodynamic interaction between the Narrows and the cove by weakening these eddies which generate the main flow pattern in the zone. The grid resolution is fine to medium with an average cell length of 100 m in the cove, and 200 m in Tongass Narrows. In order to properly capture stratification, 20 sigma layers were considered. Figure 2 also presents a graphical representation of the bathymetric conditions used in the model simulations. The model is able to capture the variations in bathymetric conditions both in the cove and Tongass Narrows.

The simulation period covers the summer of 1997 from May 1 to November 15. This period was chosen based upon the availability of data. The data collected were salinity, temperature, and dissolved oxygen vertical profiles at thirteen stations throughout the system (Figure 3).

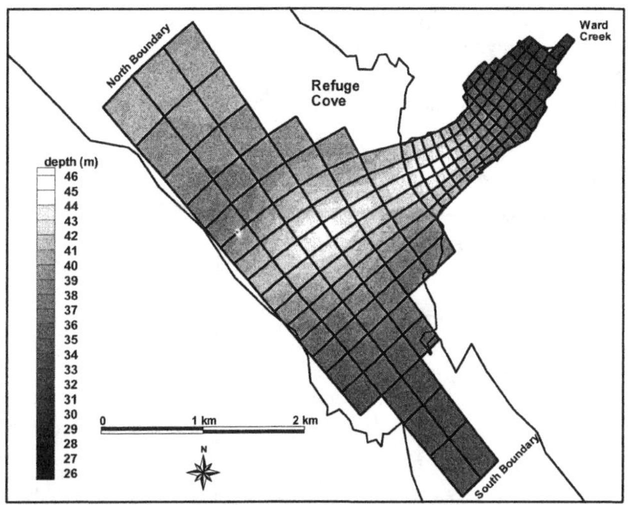

Figure 2. Ward Cove model grid and bathymetry.

The hydrodynamic model was forced based on astronomical tide projections for 1997 at Ward Cove. Wind and atmospheric pressure were not considered in the tidal forcing boundary conditions, but wind effects on local water currents were taken into account. Given the dominance of the astronomical tides on water surface elevation, with tide ranges on the order of 4 m, the components of the water surface elevation caused by meteorological forcing is assumed negligible.

To determine the boundary conditions on the north and south boundaries of the model domain, the propagation of the tidal wave along Tongass Narrows was analyzed. Projected astronomical tides for Ketchikan, south of Ward Cove on Tongass Narrows, and Vallenar Point, north of the Cove at the confluence of Tongass Narrows and Behm Canal, were used to this effect. The tidal signals from these two locations show that the tidal wave propagates from north to south along the Narrows. To correctly simulate the tidal wave propagation and minimize boundary reflections, the Ward Cove Model was forced with a non-radiating water surface elevation boundary condition at the north boundary. On the south boundary a free-surface non-radiating boundary condition was set up. This type of boundary condition lets the tidal wave freely propagate through it.

The freshwater inflow for Ward Creek was computed by the rational method ($Q = C\ A\ I + Q_{base}$), using a runoff coefficient of C = 0.3, a watershed area A = 55 km^2, and the rain intensity measured at the National Climatic Data Center (NCDC) Ketchikan station (WBAN No. 25325). The watershed area was graphically obtained from a topographic map of Revillagigedo Island. Historical flow data measured by the U.S. Geological Survey (USGS) at Perseverance was used to compute the base flow. The Perseverance gage was located at the outlet of Lake Perseverance and is within the Ward Creek watershed. The base flow obtained at Perseverance was multiplied by the

ratio of the Lake Perseverance watershed to the overall watershed to define the base flow used in the model. The freshwater inflow to the model then is the summation of the base flow and the calculated rainfall driven flows.
Temperature and Salinity data from Station TDP were used as boundary conditions for the hydrodynamic simulation.

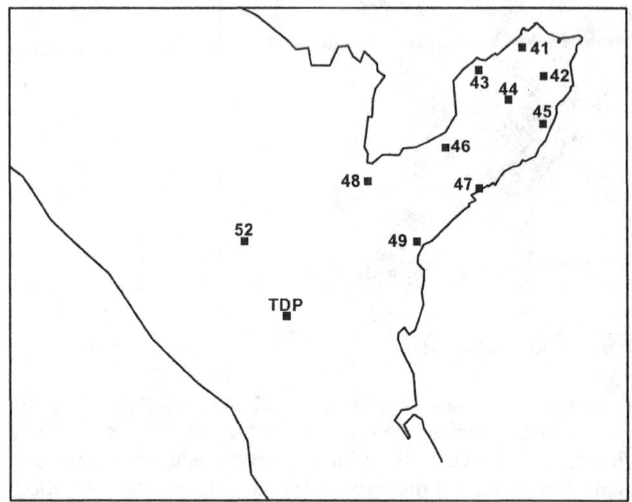

Figure 3. Ward Cove profile station locations.

The model parameterization was accomplished by adjustment of bottom friction and vertical and horizontal diffusion coefficients within acceptable literature ranges to achieve the best fit between the model results and observed data. The final values of the bottom roughness coefficient and the horizontal dispersion were 0.1 m and 90 m^2/s, respectively, and were uniform throughout the model. The vertical dispersion was determined internally within the model using a turbulence closure scheme, which calculated the vertical dispersion based on local velocity shear. A minimum vertical mixing coefficient of 0.01 m^2/s was utilized to limit the allowable degree of stratification.

Water surface elevation projected at Station 46 was used to calibrate the tidal amplitude and phase. Figure 4 presents a comparison between the astronomical projected tide and the model simulation for the period from June 9, 1997 (Julian Day 160) to June 24, 1997 (Julian Day 175). Due to the small area modeled, no appreciable differences in phase appear inside the domain.

The model accuracy in representing mass transport and stratification conditions within the system was determined through comparison of salinity and temperature vertical simulated profiles with measured profiles. The comparisons were done for Stations 41 through 47 inside the Cove and 48, 49 and 52 in the Narrows. Figures 5 and 6 present comparisons of the vertical profiles of salinity over the entire period of the model simulation at Stations 52 (in the Narrows) and 44 (at the center of the Cove), respectively. Figures 7 and 8 present comparisons of the vertical profiles of temperature for the same stations, respectively. Table 1 presents the root mean square (RMS) errors for all the stations. The errors for most of the stations are less than 1°C for temperature and less than 1ppt for salinity. Station 41 has the largest errors due to the innacuracies in the boundary data for Ward Creek. This also explain larger erros in station 42 and in a lesser degree station 46, cause the circulation inside the Cove indicate that most of the outflow is done on the northern coast of the Cove.

Station	Temperature RMS Error (°C)	Salinity RMS Error (ppt)
41	1.2	2.0
42	1.0	1.3
43	0.4	0.9
44	0.8	1.0
45	0.9	1.3
46	1.3	1.8
47	1.1	1.6
48	0.6	0.9
49	0.7	0.9
52	0.5	0.7

Table 1. RMS Errors for Temperature and Salinity.

The results show that the model does simulate the degree of stratification and the vertical distribution of salinity and temperature overall, although certain events are not captured well. Based upon the limited temporal data for the vertical distribution of salinity and temperature within the narrows, some of the errors may also be a function of the limited boundary forcing data.

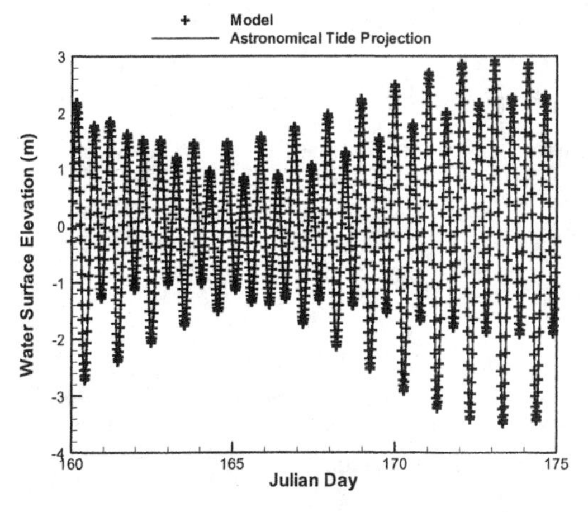

Figure 4. Projected and Simulated water surface elevation in Ward Cove.

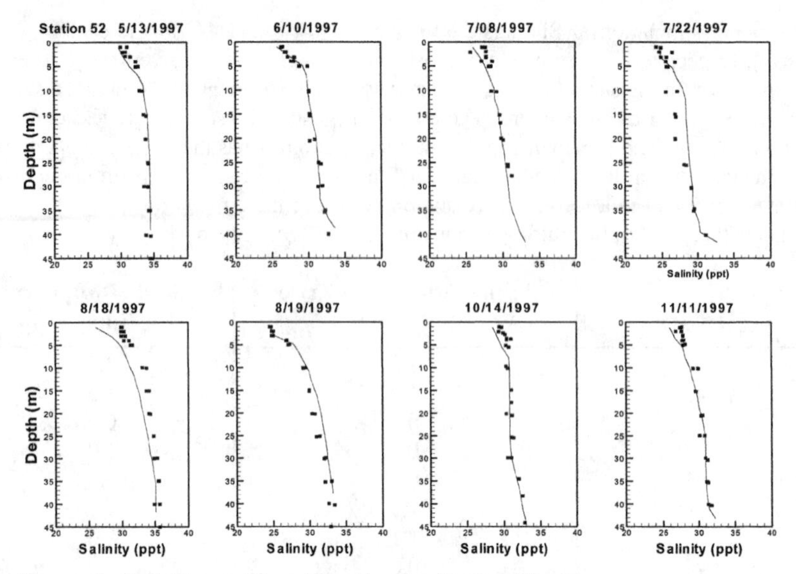

Figure 5. Salinity profile calibration for Station 52.

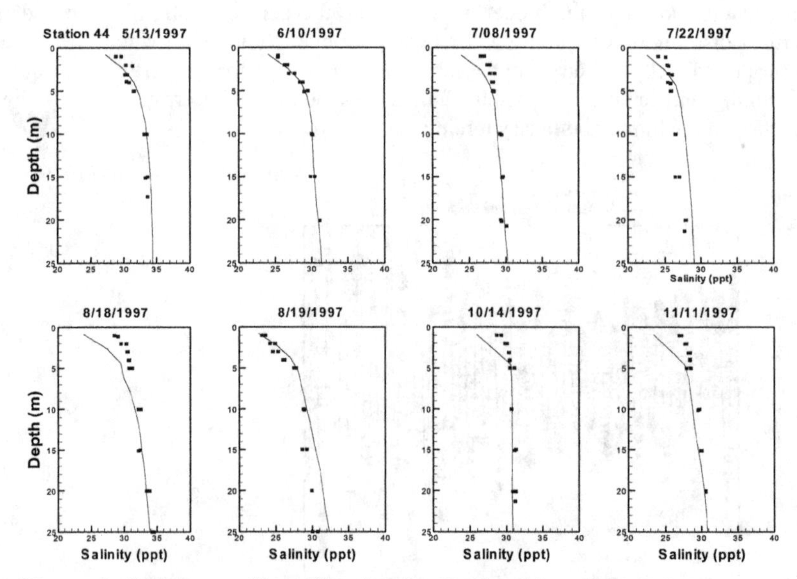

Figure 6. Salinity profile calibration for Station 44.

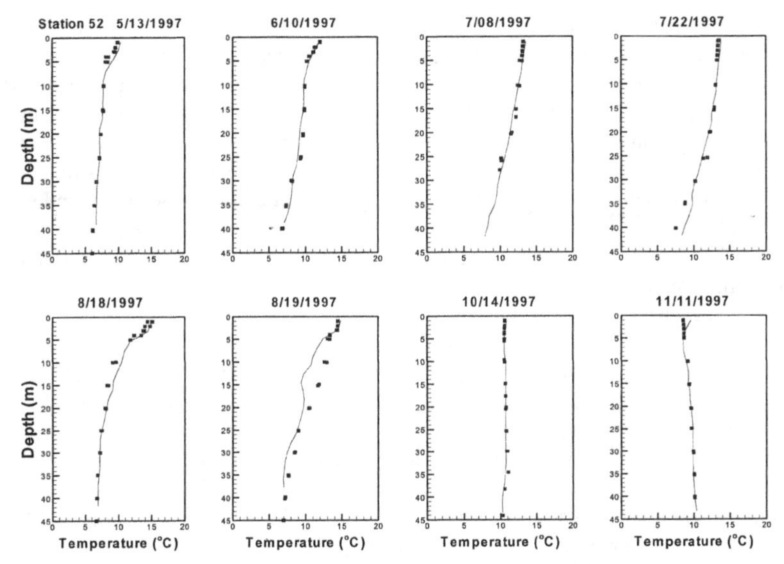

Figure 7. Temperature profile calibration for Station 52.

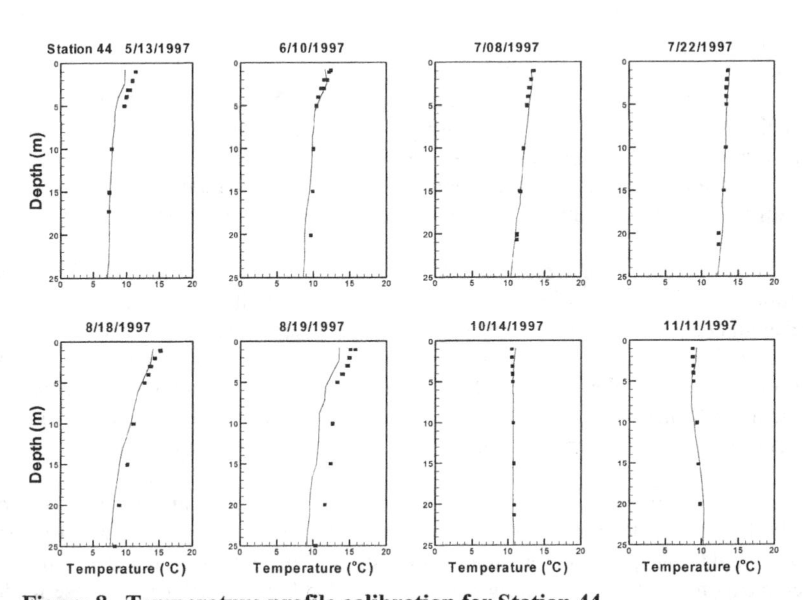

Figure 8. Temperature profile calibration for Station 44.

Ward Cove Water Quality Model Application

The water quality application for Ward Cove utilized the WASP component model in a dynamic three-dimensional simulation. The hydrodynamic conditions are imported as an external forcing file generated by the EFDC 3-D hydrodynamic model discussed previously. The advective and dispersive transport solutions within the WASP simulation were performed on a subgrid of the model grid used to simulate the hydrodynamics (Figure 9). The temporal variations in cell volumes and depths, and the flow and dispersion between cells were transferred. The flow and dispersion between boundary wasp cells and adjacent efdc cells were transferred as external for wasp and boundary conditions using water quality from stations 48 and 49 were associated to these external flows.

Though the primary state variable of concern was dissolved oxygen, a full eutrophication simulation was performed including the fate and transport of ammonia, nitrate/nitrite, organic nitrogen, total phosphorus, ortho-phosphorus, ultimate biochemical oxygen demand (BOD_u), and chlorophyll a.

Limited data were available for the development of concentration boundary conditions, determination of the model kinetic rates and constants, and comparisons of model projections of nutrients and other parameters. The data available included: vertical profiles of dissolved oxygen, historic measurements of nutrients, and BOD at nearby stations and sediment chemistry studies (Exponent, 1999). Based on the limited data sets, model parameterization was conducted by providing a baseline input set of boundary conditions, and model kinetic rates and constants taken from historic reports and literature values (Bowie et al., 1985; Wool et al., 2002). As no specific data for the time period of our study were available for water quality concentrations in the Tongass Narrows, typical values representative of similar conditions throughout Alaska were utilized. As with the boundary concentrations in the narrows, no data were available for the water quality concentrations entering from the creek for the period of simulation. Therefore, representative values for Ward Creek based on measurements in similar creeks were utilized as input to the model.

A primary aspect of the Ward Cove model is the oxygen demand associated with the bottom sediments. Historic discharges from the paper mill resulted in large areas of accumulated waste in the bottom of the cove. These wastes have been shown to produce fluxes of ammonia (NH_3) and total organic carbon (TOC), both of which can impact the oxygen demand in bottom waters of the cove. Additionally, there is a baseline sediment oxygen demand associated with the waste discharges due to organic decomposition and anaerobic conditions at the soil water interface.

Maps of the distribution of ammonia, ultimate carbonaceous oxygen demand ($CBOD_u$), and TOC concentrations within the sediments were generated (Exponent, 1999). Additionally, model simulations of the flux of ammonia and TOC from the sediments to the water column were conducted for Exponent (1999). These model simulations defined the rate of ammonia and TOC flux to the surface waters and the associated changes in sediment concentration over time.

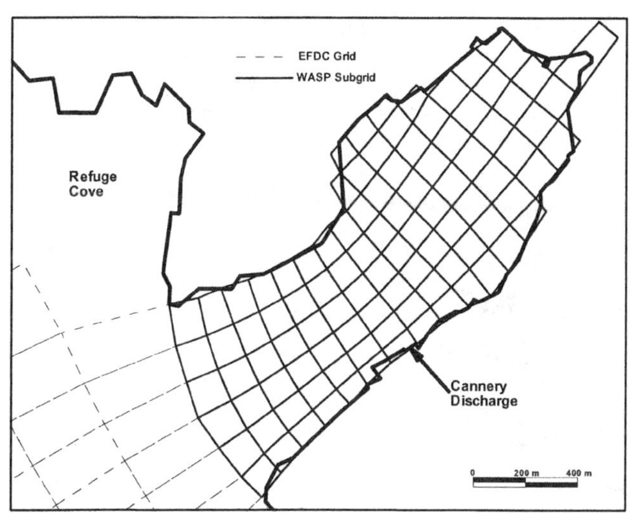

Figure 9. Water Quality model subgrid.

For the 1997 model simulations, the assumption was made that a spatial distribution of the sediment oxygen demand (SOD) existed that was proportional to the distribution of CBOD, ammonia, and TOC in the pore water. This provided the basis for the establishment of a spatial distribution of SOD multiplier in the system. The next step was to establish a baseline sediment oxygen demand, or assumed average SOD on which to apply the SOD multipliers. Historically, under the conditions of full discharge from the paper mill, SOD measurements were taken within Ward Cove (Jones & Stokes and Kinnetic, 1989). Utilizing measured SOD, along with the spatial multiplier determined from Exponent (1999), a spatial distribution of SOD within the system was determined (Figure 10). The average SOD was 3.3 $g/m^2/d$ (Jones & Stokes and Kinnetic, 1989). This condition represents the potentially worst case condition immediately following the removal of the paper mill discharge for SOD within Ward Cove.

A typical seafood processor 5-day biological oxygen demand (BOD_5) concentration of 155 mg/L was assumed to determine the ultimate biochemical oxygen demand (BOD_u) associated with the cannery located on the south side of Ward Cove (Tetra Tech, 1994). To determine a BOD_u concentration, an f-ratio (the ratio of the BOD_u to the BOD_5) of 3.0 was utilized to get the final concentration used in the calculations (465 mg/L). The BOD_u discharged was calculated based on the Round Production assuming a discharge flow of 1 MGD for every 23,980 kg of Round Fish processed with BOD_u of 465 mg/L. The flow rate by Round Fish processed was based on typical flow rates associated with salmon processing plants. The Round Production was obtained as reported by the cannery to EPA on the Annual Certification of Compliance. The seafood processing plant typically operates in June, July, August, and September. The discharge pipe from the plant is located near the bottom in approximately 30 m of water. Due to the circulation pattern within Ward Cove (residual flow into the cove at the bottom and counterclockwise within the cove) this material is carried into the cove and moves into the areas of low dissolved oxygen.

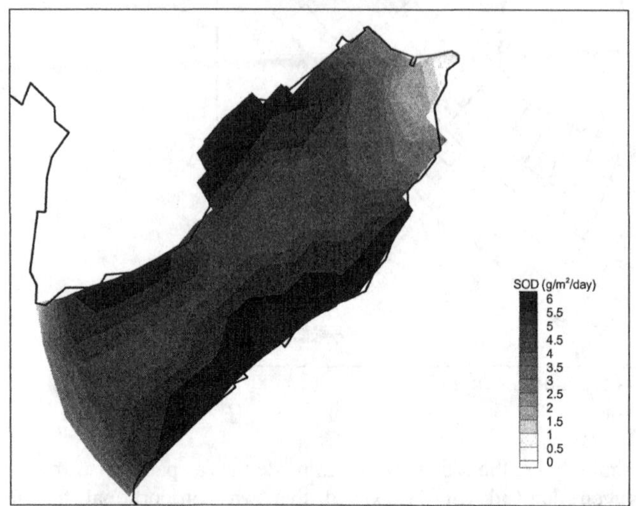

Figure 10. Spatial Distribution of Sediment Oxygen Demand (SOD).

To evaluate model performance the primary comparison of the water quality model results to observed data were based on observed vertical dissolved oxygen profiles. To accomplish the model parameterization the following kinetic rates were adjusted: CBOD decay rate, nitrification rate, and denitrification rate. The final values of the CBOD decay, nitrification, and denitrification rates were 0.3, 0.2 and 0.05 day^{-1}, respectively. These simulations represent the best overall results achieved under this model parameterization and the limited available boundary and interior data. The model parameterization goal was to capture the primary temporal and spatial distribution of dissolved oxygen in the system, with special focus on capturing the vertical stratification conditions.

Table 2 presents the RMS errors for dissolved oxygen for all the stations. Figures 11 and 12 present comparisons of the vertical profiles of dissolved oxygen for the central station 44 and for station 42, respectively. These results show that the model does simulate the degree of stratification and the vertical distribution of dissolved oxygen, although certain events are not captured well. Based on the limited temporal data for the vertical distribution of dissolved oxygen within the narrows, some of the errors may be a function of the limited boundary forcing data.

Station	DO Mean Square Root Error (mg/l)
41	0.7
42	0.9
43	0.8
44	0.8
45	1.0
46	1.2
47	0.7

Table 2. RMS errors for dissolved oxygen.

Relative Impact Evaluation

Evaluation of the historic data identified that a net deficit of dissolved oxygen exists within Ward Cove relative to the waters within the Tongass Narrows. Utilizing the water quality model described above, under a set of predetermined conditions, a relative impact evaluation was performed to identify the effects of the assumed SOD conditions presented above, and the cannery discharge during the critical months of July and August. The goal of this analysis was to quantify the relative impact of each of the two primary dissolved oxygen deficit sources within the cove.

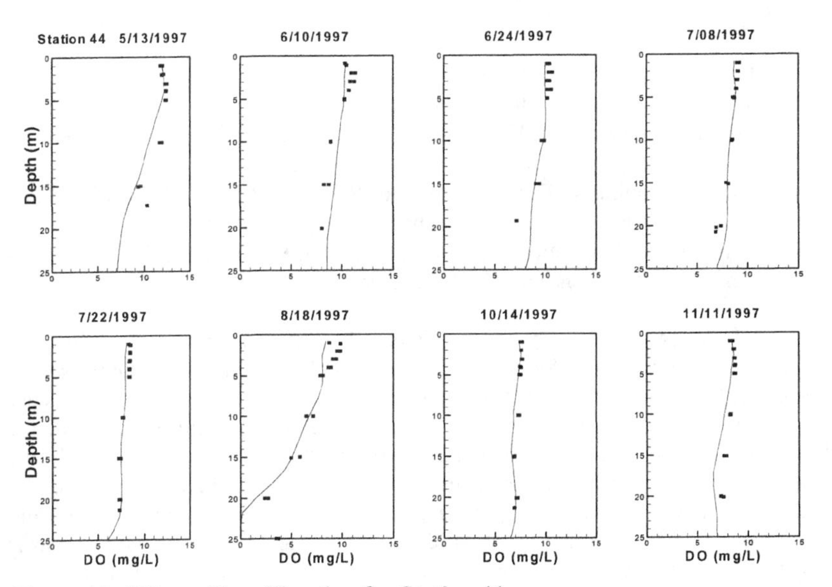

Figure 11. DO profile calibration for Station 44.

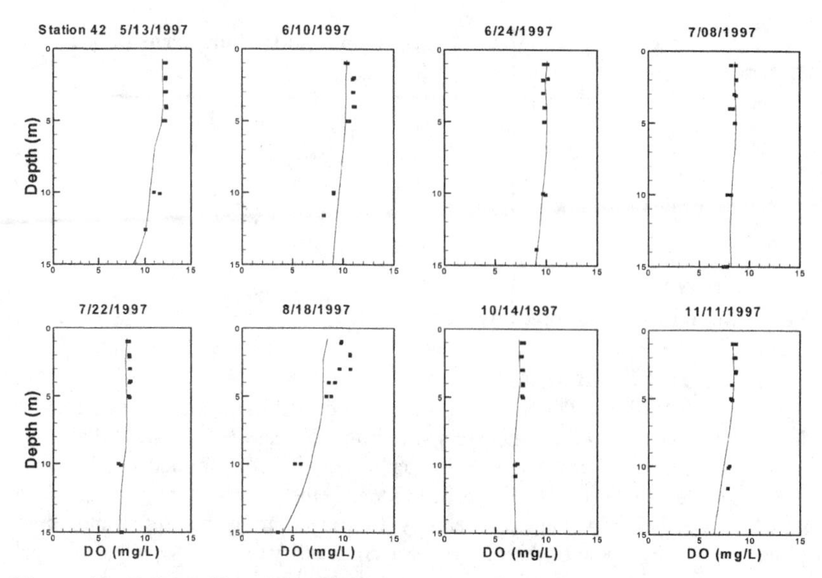

Figure 12. DO profile calibration for Station 42.

The model simulation was run under the 1997 conditions from April through September. Three separate simulations were run:

- Baseline Simulation: 1997 conditions with SOD set at a typical baseline level of 1.5 g/m^2/day and no discharge from the cannery.
- SOD Simulation: 1997 conditions with SOD set at the spatially distributed values described above and no discharge from the cannery.
- Point Source Simulation: 1997 conditions with SOD set at a level of 1.5 g/m^2/day and the measured discharge from the cannery.

Figures 13 and 14 present snapshots of the dissolved oxygen profile under the three conditions described above for July 19 and August 18. In addition to the dissolved oxygen profiles, the cannery discharge and surface to bottom density differences are shown. The bottom-surface density differences highlight the critical nature of the degree of stratification on the development of the dissolved oxygen deficit in the bottom waters. Greater impacts are present during higher stratification periods. The results also show that the most significant relative impact to the system come from the cannery discharge during high stratification periods and higher discharges. This is consistent with measurements showing the highest deficit existing at the time of the cannery discharge peaks when interior and exterior historical dissolved oxygen profiles are compared. SOD impacts are comparatevely small and isolated to the near bottom waters.

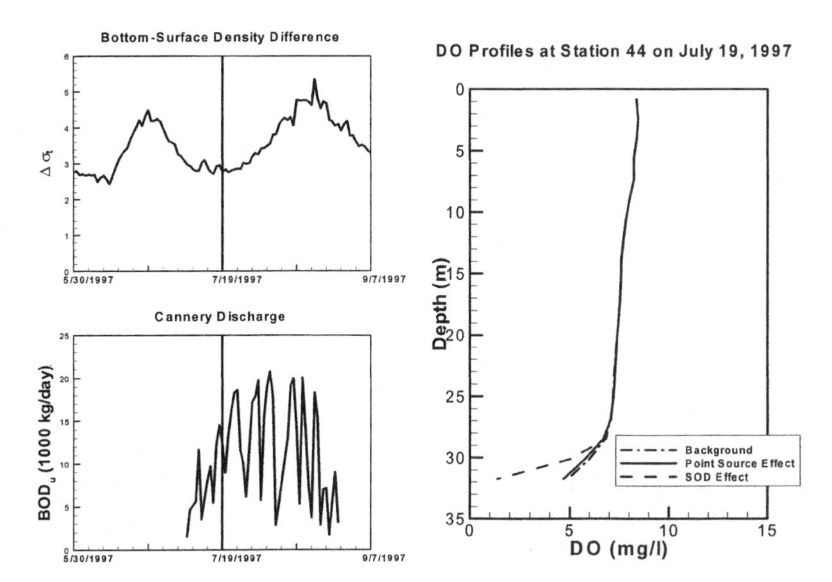

Figure 13. Simulated dissolved oxygen profile, vertical density difference, and cannery discharge on July 19, 1997 at Station 44.

The relative impact evaluation presents the net deficit in dissolved oxygen created by the bottom SOD and discharges within Ward Cove. This net deficit is superimposed on the background water quality conditions entering the cove from Tongass Narrows. Evaluation of historic data has shown that dissolved oxygen levels in the lower waters of Tongass Narrows are at or below the water quality standard of 5.0 mg/L at times.

Summary and Conclusions

EFDC and WASP models integrated together provided a good framework for predicting responses of hydrodynamic and water quality parameters in Ward Cove, Alaska. Results show that the models capture the primary temporal and spatial distribution of temperature, salinity, and dissolved oxygen in the system, especially the vertical stratification conditions.

The models were useful in evaluating the relative impact on dissolved oxygen created by the bottom SOD and point source discharges within Ward Cove. With the aid of the integrated models, it was possible to determine the relative causes of a net deficit in dissolved oxygen between the waters inside Ward Cove and the waters in Tongass Narrows.

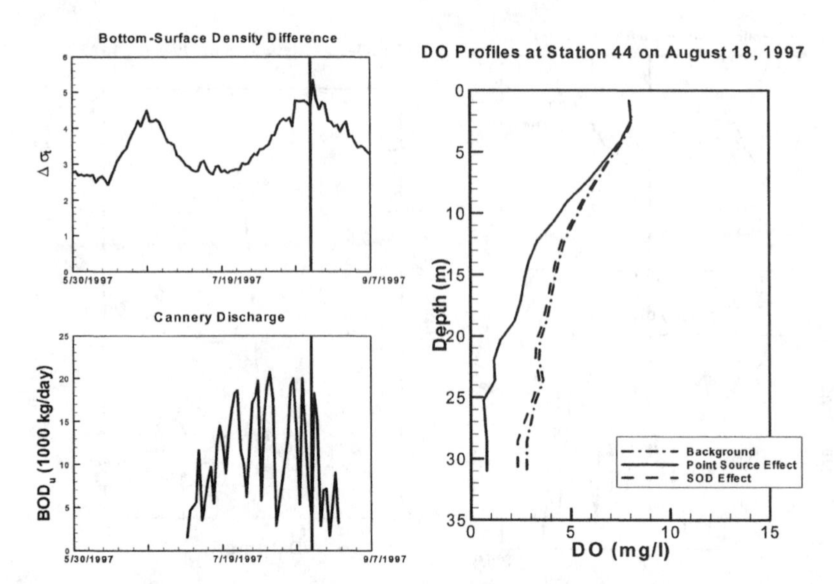

Figure 14. Simulated dissolved oxygen profile, vertical density difference, and cannery discharge on August 18, 1997 at Station 44.

References

Bowie, George L.; Mills, William B.; Porcella, Donald B.; Campbell, Carrie L.; Pagenkopft, James R.; Rupp, Gretchen L.; Johnson, Kay M.; Chan, Peter; Gherini, Steven A. and Charberlin, Charles E. 1985. *Rates, Constants and Kinetics Formulations in Surface Water Quality Modeling (Edition 2).* Environmental Research Laboratory. Office of Research and Development. U.S. Environmental Agency. Athens, Georgia.

Exponent. 1999. *Ward Cove Sediment Remediation Project: Detailed Technical Studies Report, Volume I, Remedial Investigation and Feasibility Study.* May 1999. Bellevue, Washington.

Hamrick, John M. 1992. *A three-dimensional environmental fluid dynamic computer code: theoretical and computational aspects.* Virginia Institute of Marine Science. The College of William and Mary, Gloucester Point, Virginia.

Hamrick, John M. 1996. *User's manual for the environmental fluid dynamics computer code.* Virginia Institute of Marine Science. The College of William and Mary, Gloucester Point, Virginia.

Jones & Stokes and Kinnetic. 1989. *Ward Cove water quality assessment.* Final. EPA Contract No. 68-02-4381. Jones & Stokes Associates, Inc., Bellevue, Washington and Kinnetic Laboratories, Inc., Santa Cruz, California.

Tetra Tech, Inc. 1994. *Ocean Discharge Criteria Evaluation for the NPDES General Permit for Alaskan Seafood Processors.* EPA Contract No. 68-C1-0008.

Wards Cove Packing Company. 1999. *Certification of Compliance for General NPDES Permit No. AD-G-52-0060.* To EPA Region 10, Water Compliance Section, Seattle, WA.

Wool, Tim A.; Ambrose, Robert B.; Martin, James L. and Comer, Edward A. 2002. *Water Analysis Simulation Program (WASP). Version 6.0 User's Manual.* US Environmental Protection Agency – Region 4. Atlanta, Georgia.

Three-Dimensional Hydrodynamic Modeling of Tomales Bay, California

Edward S. Gross[1] and Mark T. Stacey[2]

Abstract

The TRIM model is applied to simulate hydrodynamics and salinity in Tomales Bay during winter conditions. The model is calibrated to current meter data. Then it is applied, without modification of model parameters, to simulate the seasonal variability of salinity along the axis of Tomales Bay. The salinity is accurately predicted during a six month winter and spring period with multiple flow events that cause stratification. The gradual vertical mixing of a surface lens of freshwater is predicted accurately. The model is then applied with idealized forcing in order to study dispersion mechanisms in Tomales Bay. Appropriate dispersion coefficients for a 1D tidally-averaged model are estimated from the three-dimensional model results for different steady freshwater (creek) flow rates. The flux through several cross-sections is analyzed in detail by decomposition of velocity and salinity profiles. Tomales Bay is frequently divided into two regions, the "Inner Bay" and the "Outer Bay," distinguished primarily by differences in bathymetry. This study indicates that different transport mechanisms are important in the two regions. In the outer portion of the bay, which is characterized by a channel-shoal structure and strong tidal currents, tidal dispersion mechanisms dominate. In the "Inner Bay," which has a more uniform cross-section and reduced tidal currents, baroclinic flow is the dominant flux mechanism. In addition to this spatial variability, the strength of some transport mechanisms is tied strongly to freshwater flow entering Tomales Bay.

Introduction

The complexity of coastal estuarine environments, where tidal forcing interacts with both horizontal and vertical density gradients, creates a challenging environment for the analysis of transport and dispersion of contaminants or other scalars. The fact that many coastal waters are significantly affected by either anthropogenic or natural contamination makes this analysis a paramount concern for the management of our coastal resources. In Tomales Bay, a relatively pristine coastal estuary north of San Francisco, a variety of water quality concerns, including high levels of fecal coliforms and mercury input from an abandoned mine in the watershed, have led to the estuary being considered an 'impaired' waterbody, which motivates the analysis of transport and mixing in the Bay. The overall goal of the study is to understand tidal and long-term transport of dissolved and suspended material that enters Tomales Bay from the adjoining watersheds.

A variety of approaches have been used to develop simplified models of estuarine transport, each applying a different method of averaging (tidal, depth, lateral or cross-sectional). The appropriate approach to modeling transport of dissolved and

[1] Environmental Consultant, 1777 Spruce Street, Berkeley, CA,94709 ed.gross@baymodeling.com
[2] Department of Civil & Environmental Engineering, University of California, Berkeley 94720-1710 mstacey@socrates.berkeley.edu

suspended material in an estuary depends on the mechanisms that drive dispersion along the axis of the embayment. Baroclinic circulation creates steady (tidally-averaged) vertically sheared flow that combines with a stratified water column to create a net salt flux up-estuary. Subtle phase shifts between currents and salinity, created through tidal trapping as the oscillatory tides interact with a complex bathymetry (Fischer et al. 1979), can result in a net flux either up-estuary or down-estuary. Finally, horizontally sheared steady circulation can be created through the interaction of complex bathymetry with tidal flows, wind forcing, or even density gradients (Fischer et al. 1979). The ability to effectively model an estuary with a reduced dimensionality model – either through vertical averaging, lateral averaging or cross-sectional averaging – depends on the particular environment and on the model's ability to effectively parameterize the unresolved dispersion mechanisms.

In order to understand the complete dynamics of the system, and to facilitate analysis of the transport mechanisms in Tomales Bay, we have applied a three-dimensional modeling approach. The ultimate goal of this modeling work, however, is to develop an appropriate management model that effectively models the system but with reduced complexity and dimensionality. The work presented here examines the dynamics that dominate dispersion and to begin considering which modeling approaches will be most appropriate.

Physical Setting

Tomales Bay is a narrow estuary located along the San Andreas fault zone in Marin County, California. It extends 20 kilometers southeast from Tomales Point where Tomales Bay meets Bodega Bay. The bay contains two distinct regions (Johnson et al. 1961). The "Outer Bay" extends from Tomales Point to Pelican Point and has a system of channels with adjacent shoal regions. In contrast, the "Inner Bay" does not have a channel-shoal structure but, instead, relatively uniform depth over the cross-section. Approximately 6 km of the up-estuary portion of the Inner Bay is shallow with broad mudflats.

Tides are mixed diurnal and semi-diurnal in Tomales Bay. Observed tidal range in Bodega Bay is similar to the tidal range observed at Point Reyes and San Francisco. Tidal range decreases as tides propagate into Tomales Bay. Tidal velocities vary substantially in Tomales Bay, with large peak tidal currents (~1 m/s) in the channels of the Outer Bay and relatively small currents (~0.2 m/s) in the Inner Bay (Smith et al. 1971).

Weather conditions in Tomales Bay watershed follow a pattern of dry summer and fall conditions and wet winter and spring conditions with large inter-annual variations in rainfall. The largest creek draining to Tomales Bay is Lagunitas Creek, which enters at the head of the estuary. The other substantial creek is Walker Creek, which enters Tomales Bay on the eastern shore, southeast of Toms Point. The location and shape of Tomales Bay, Lagunitas Creek and Walker Creek are shown on Figure 1.

Salinity in outer Tomales Bay is similar to oceanic salinity during most of the year although it can be depressed following large flow events. Salinity is more variable in the Inner Bay, with hypersaline conditions during dry summer and fall conditions and stratified conditions in the Inner Bay following winter and spring flow events (e.g., Johnson et al. 1961).

Hydrodynamic Model

The three-dimensional TRIM (Tidal, Residual, Intertidal & Mudflat) model (Casulli 1990; Casulli & Cattani 1994) is a well-established tool for long-term simulation of hydrodynamics in estuaries that include intertidal regions.

Governing Equations. The TRIM model used in the simulations of Tomales Bay solves the Reynolds-averaged Navier-Stokes equations. The model can be applied to non-hydrostatic flows, but the flow in Tomales Bay was assumed to be hydrostatic. A scalar conservation equation is solved for each dissolved constituent in the water (e.g., salinity). This equation describes how the scalar mass is advected by tidal currents and diffused by turbulent mixing. The TRIM model uses the QETE turbulence closure model (Galperin et al. 1988), a modification of the Mellor-Yamada level 2.5 model (Mellor & Yamada 1982), to estimate vertical turbulent mixing.

Numerical Method. The three-dimensional TRIM model uses a semi-implicit finite-difference method to solve the three-dimensional shallow water equations on a staggered grid (Casulli & Cattani 1994). The central feature of the TRIM method (Casulli 1990) is a highly efficient semi-implicit solution that is proven to be stable even at a large computational time step. A preconditioned conjugate gradient method is used to solve the matrix of equations at each time step. For uniform density simulations this iterative matrix solver is guaranteed to converge without any restriction on the computational time step (Casulli & Cattani 1994). When density gradients are present, a mild restriction on the time step applies (Gross et al. 1998).

The TRIM model used in the simulations of Tomales Bay solves the governing equations of the QETE turbulence closure model by a method described in detail by Gross et al. (1999b). The horizontal advection scheme is a widely used flux-limiting method that guarantees monotonicity preservation (Gross et al. 1999a) and the vertical advection scheme is an unconditionally stable semi-implicit method (Gross et al. 1998).

Tidal Hydrodynamics Simulation

The TRIM model was calibrated against current meter data by adjustment of bottom friction coefficients. The bottom friction coefficients were specified as a function of water column depth.

Bathymetry. The Tomales Bay model grid includes all of Tomales Bay with the open boundary located at Tomales Point. The downstream portions of Walker Creek and Lagunitas Creek are also represented by the model grid. The Cartesian grid has 30 meter resolution in the horizontal dimensions. The bed elevation was specified using NOAA soundings from 1993 and 1994. The most recent available soundings were used in each region and questionable NOAA soundings were identified and discarded.

A weighted-average interpolation algorithm was used to specify bed elevation on the model grid and NOAA shoreline data was used to specify the location of the shoreline.

The grid was rotated counterclockwise by 53.4 degrees to roughly align the axis of the Tomales Bay with the x coordinate direction of the model grid, and the vertical datum was converted from MLLW to meters NGVD 1927 using NOAA tide data. There are 35,211 active water columns in the Tomales Bay grid and 885,293 potentially wet grid cells.

Open Boundary Conditions. Water surface elevation and salinity are specified on the open boundary of the model grid, located at Tomales Point. Water surface elevation was specified using observed tides at NOAA station 9414290 in San Francisco, shown in Figure 2. No corrections were applied because observed tidal amplitude in Bodega Bay is similar to the observed tidal amplitude at the San Francisco. Observations show that the tidal amplitude decreases as tides approach and pass Sand Point, the first constriction in Tomales Bay, Therefore it is not appropriate to use tidal corrections based on observed tides inside Tomales Bay to specify water surface elevation at the open boundary.

Model Setup. The hydrodynamic calibration period was summer of 1960 because calibration data were available during this period. The salinity field was considered to be uniform. Both freshwater inflows and wind forcing were neglected in the model calibration because Johnson et al. (1961) report that during summer currents are tidally dominated and not substantially affected by wind or freshwater inflows. The simulation began on June 23, 1960 with quiescent initial conditions. A spin-up period of 4 days was allowed before the predicted tidal currents are compared to model results.

The only calibration parameter adjusted in the model calibration is the bottom roughness coefficient. The bottom friction was specified as a function of water column depth as given in Table 1.

Table 1. Bottom roughness coefficient (z_0)

Depth (m)	**z_0 (cm)**
2	0.003
4	0.0002
8	0.00005
9	0.00001

The z_0 used in the shallow regions is realistic while the z_0 coefficient used in deep regions is unrealistically low. This is commonly the case in applications of TRIM (Cheng et al. 1993, Gross et al. 1999b) and may partially be to compensate for a limited amount of numerical diffusion inherent in the model. Some numerical diffusion of momentum from the channel regions to shoal regions occurs due to the Eulerian-Lagrangian method used by TRIM to handle the advection terms in the momentum equations.

Tidal Current Data. Few tidal elevation and velocity data have been collected in Tomales Bay. The field study with the best spatial coverage was performed in 1960 (Johnson et al. 1961). In this field study, peak ebb and flood current speeds were measured and reported at 22 locations in Tomales Bay. At each location the currents were measured at several depths and found to be quite uniform over the depth range of observations. Tidal range was also reported at each station based on the 1 to 2 days of data, but these data were not reliable.

Because the data collected in 1960 has very limited temporal coverage, the University of California, working with the San Francisco Regional Water Quality Control Board, is currently planning a considerable data collection effort involving continuous observations of currents, salinity and suspended sediment concentrations at several locations.

During the time from the collection of hydrodynamic data in 1960 and collection of NOAA soundings in 1993 and 1994, upon which the model grid is based, some changes in the bathymetry of Tomales Bay have occurred. However, bathymetric data indicates that these changes have been small (Philip Williams and Associates 1993) and, therefore, the 1993-94 bathymetry adequately represents 1960 conditions.

Model Results. At each station, the predicted currents were depth-averaged during the simulation and compared with observed currents. The locations of a subset of the stations are shown in Figure 3.

The observed and predicted peak tidal current speeds in the channel near Toms Point are shown on Figure 4. The predicted peak tidal current speed is somewhat smaller than the observed tidal current speed for both flood tide and ebb tide.

The observed tidal currents are substantially weaker in the Inner Bay. The peak tidal current speeds near Marshall are shown on Figure 5. Near Marshall the predicted peak tidal current speed is larger than the observed speed during both flood tide and ebb tide. This underestimate of tidal current strength was also noted at other Inner Bay stations.

On Figure 6, all current speed data available from the 1960 field study is compared with the corresponding predicted current speed. The larger observed current speeds are typically predicted accurately while the smaller observed current speeds are not all predicted accurately.

It is difficult to make extensive quantitative comparisons between the model output and observations for several reasons. First of all, the reported data included magnitude and direction at a location without notation of the depth of the observation or the method of averaging. In low energy flows, averaging of magnitude and direction could result in the calculated magnitude of observed current speed being biased high, which is consistent with the error seen in Figure 6. Perhaps more importantly, we are comparing depth-averaged velocity from the model to the reported velocities from the 1960 data set. While the observations reported that the flow was uniform over the depth, there will still be shear in the water column and the location of the observation could bias the observation above or below the depth-averaged value.

The predicted currents are generally similar to the observed currents. When more detailed tidal elevation and tidal current data are collected it may be possible to improve the calibration.

Salinity Simulation

The salinity field in Tomales Bay was simulated for a winter period during which surface salinity observations were available for comparison. The model parameters were not changed from the hydrodynamic calibration but a number of additional model inputs were required to simulate the salinity field.

Creek Flows. The total watershed area of the Tomales watershed is 561 km^2. The two major creeks that discharge into Tomales Bay are Lagunitas Creek and Walker Creek. Flow from these two creeks is estimated to be 87% of the total flow to Tomales Bay (Fischer et al. 1996) and their watersheds cover the majority of the Tomales Bay watershed. Both the Walker Creek watershed and the Lagunitas Creek watershed contain reservoirs. Data on reservoir operation, including removal from the reservoirs, was provided by North Marin Water District.

During the simulation period, Lagunitas Creek was gaged at both USGS Station 11460600 near Point Reyes Station and at USGS Station 11460400 at Samuel P. Taylor State Park. Additional flow data were collected on Olema Creek, a tributary to Lagunitas Creek, by Questa Engineering from 1986 to 1989. Walker Creek is gaged at USGS Station 1460750 near Marshall.

Although flow was gaged at these locations, a portion of the flow from Lagunitas Creek and Walker Creek to Tomales Bay was ungaged because it enters Lagunitas Creek or Walker Creek below the gaging stations. In order to estimate the ungaged flow in Lagunitas Creek and Walker Creek at Tomales Bay, gaged flows and measured releases from reservoirs were used to estimate the unimpaired flow per unit area. All of the ungaged flow was unimpaired flow (from regions without reservoirs). The unimpaired flow per unit area was frequently different for the ungaged portion of Walker Creek than the ungaged portion of Lagunitas Creek, which is not surprising because both the topography and the average rainfall varies significantly between the two watersheds. The ungaged flow in each creek was estimated and added to the gaged flow to form the total flow to Tomales Bay. The estimated flow from Lagunitas Creek into Tomales Bay is shown on Figure 7.

Flow from minor creeks, which typically contribute 13% of the total flow to Tomales Bay (Fischer et al. 1996) were estimated using the calculated unimpaired flow per unit area.

Precipitation and Evaporation Precipitation and evaporation over the area of Tomales Bay were specified in the salinity simulations. The specified evaporation and precipitation were time-varying but spatially uniform.

The daily precipitation used in the simulation was the average of the daily precipitation measured collected near Nicasio Reservoir and the Soulajule Reservoir by the North Marin Water District. Mean annual rainfall varies substantially over Tomales Bay watershed with higher rainfall in the Southern portion of the watershed. The average of the precipitation at the Nicasio Reservoir and the Soulajule Reservoir is believed to be representative of rainfall on Tomales Bay based on maps of isohyets in the Tomales watershed.

The monthly evaporation was computed from measured evaporation at the Soulajule Reservoir using a relationship between evaporation at Soulajule Reservoir and Tomasini Point (Tomales Bay) developed by Smith et al. (1989).

Wind. A wind stress was specified at the water surface of Tomales Bay. The wind was specified based on wind speed and direction data measured at Dillon Point as part of the LMER/BRIE program (Smith et al. 1989). The wind speed and direction was assumed to be uniform over Tomales Bay.

Boundary Conditions. Water surface elevation was specified on the open boundary based on observed tides at NOAA station 9414290 in San Francisco near the Golden Gate. NOAA observations indicate that tides in Bodega Bay are very similar to tides at the Golden Gate.

Salinity was specified on the open boundary from observed NOAA surface salinities measured in Bodega Bay. Observed salinity at this location was in a small range near oceanic salinity during the simulation period (Dever & Lentz 1994). Both salinity and water surface elevation were assumed to be uniform across the open boundary of the model.

Initial Conditions. Quiescent initial conditions were assumed with water surface elevations at 0 NGVD and no velocities in all grid cells in the model domain. The starting time of the simulation was 12:00am on November 10, 1987. The velocities were allowed to spin up for over 36 hours before the salinity field was specified.

The initial salinity field was specified based on salinity data from the LMER/BRIE surface salinity data on November 11, 1987 at 12:15pm (Smith et al. 1989). Because only surface salinity was available, the salinity field was assumed to be vertically and laterally uniform. November 11, 1987 was chosen as the date to initialize the salinity field because observed salinity was very uniform at this time. The observed surface salinity along the axis of Tomales Bay ranged from 33.2 ppt to 33.9 ppt and, therefore, it is likely that a good deal of lateral and vertical uniformity was present in Tomales Bay on this date.

Model Setup. The model simulation period was from November 10, 1987 to May 11, 1988. The salinity simulation used the same bottom friction coefficients as the model calibration. No additional calibration parameters were employed. Because a high-resolution three-dimensional model was employed in the simulation, a horizontal dispersion coefficient was not necessary.

Salinity Data. The salinity data used in the salinity simulation was collected as part of the LMER/BRIE program (Smith et al. 1989). The data was collected at stations spaced every 2 kilometers along the axis of Tomales Bay from station 0, located near

Sand Point to station 18, located near Millerton Point, 18 kilometers from Sand Point as shown on Figure 8. On 17 dates from November 11, 1987 to May 11, 1988 salinity samples were collected in Tomales Bay. On most days only surface salinity was measured but, on some days, mid-depth salinity was also measured. The surface water was sampled using a bucket and the salinity was measured using a laboratory salinometer.

Several storms occurred during the simulation period causing substantial variability in measured Inner Bay salinity during this period. Figure 9 shows the surface salinity observed at several LMER/BRIE stations during the simulation period.

Model Results. The predicted salinity was compared with observed salinity for each salinity sample collected as part of the LMER/BRIE program during the simulation period. During each day of data collection the salinity was typically measured at each station within a two-hour period. However the exact time of data collection on some days was not reported. In these cases the observations were assumed to have been made at noon.

The initial salinity field is shown on Figure 10. The predicted salinity is compared with observed salinity at a subset of the observation dates in Figures 11 to 13. For each of these figures, the predicted salinity at each vertical layer of the model was output at the LMER/BRIE stations and interpolated longitudinally to provide the salinity field in these figures. The observed surface salinity values at the LMER/BRIE stations are shown in the buckets on these figures to represent surface salinity sampled using buckets.

The predicted salinity field on December 29, 1987, following a flow event, is shown on Figure 11. The inflow causes a stratified surface layer to form in the model. This stratification is also noted in the observations. On December 29, mid-depth salinity observations were also collected at a subset of the salinity stations and match the predicted salinity at all stations to within 1 ppt.

The predicted salinity on January 2, 1988, shown on Figure 12, matches the observations precisely. Only four days after the strongly stratified conditions, shown on Figure 11, were present, the model indicates that the water column is well-mixed vertically in Tomales Bay. The strong increase in surface salinity suggests that the relatively low salinity water in the thin stratified layer has been mixed over the water column.

The predicted salinity on May 11, 1988, shown on Figure 13, is nearly identical to observed salinity. On May 11, the salinity has returned to the uniform conditions present at the beginning of the simulation.

The comparison of model results to salinity observations is good during this dynamic period of periodic stratification. Both the model results and observations show a rapid decrease in surface salinity following flow events (storms). The model results show that this decrease in surface salinity is due to the formation of a thin freshwater lens at the water surface that is generally mixed over the water column in a period of days after the end of a flow event. This conclusion is supported by a limited number of mid-depth salinity observations.

Using all salinity observations during the simulation period, the average error in the predicted surface salinity is –0.21 ppt while the average error in predicted mid-depth salinity is 0.04 ppt, indicating that there is little bias in the predictions. The average magnitude of error in predicted surface salinity is 0.77 ppt while the average magnitude of error in predicted mid-depth salinity is 0.29 ppt. The surface salinity is predicted accurately during most of the simulation period but some error is evident during the high flow periods where a thin surface water lens forms. During this period the predicted surface salinity is highly sensitive to the estimated freshwater inflow and the vertical mixing predicted. Mid-depth salinity varies less abruptly and is predicted very accurately at all times during the simulation.

The predicted salinity was found to be somewhat sensitive to the wind shear stress following flow events. It is believed that wind is a significant factor influencing the rate of vertical mixing of the freshwater lens that forms following flow events.

Because strong stratification occurs following flow events, it is possible that shear instabilities were present in Tomales Bay during part of the simulation period in local regions. The vertical mixing that may be caused by shear instabilities was not represented by the turbulence closure. Including a parameterization of this mixing may significantly decrease the time required for vertical mixing. If continuous observations of salinity were available in Tomales Bay, it would allow a turbulence closure to be tested with and without a parameterization of shear instabilities and other processes that result in vertical mixing.

Transport Simulations

The TRIM model was applied to study transport in Tomales Bay. The goal of the transport simulation is to provide insight into the relative importance of different transport mechanisms in Tomales Bay and how the importance of these mechanisms varies in space, particularly between the Inner Bay and Outer Bay, and with flow, ranging from a low winter flow (baseflow) to peak winter flow. These simulations provide much insight to the hydrodynamics of Tomales Bay.

The model inputs were simplified so that periodic steady state conditions would be reached in which the tides, tidal velocities and salinity repeat the same cycle each day. At periodic steady state conditions the "salt balance" equation (Fischer et al. 1979) applies

$$QS = KA\ dS/dx$$

where Q is the net (freshwater) flow rate , S is the cross-sectional averaged and tidally-averaged salinity, K is the dispersion coefficient, A is the cross-sectional area, and dS/dx is the longitudinal salinity gradient.

This equation shows a balance between the net flow, which acts to advect salt mass out of the estuary, and dispersion mechanisms, which act to transport salt into the estuary. At each cross-section these terms exactly balance when periodic steady state conditions are reached.

The dispersion coefficient in the salt balance equation represents many processes that act to transport salt mass up-estuary. In the following analysis the magnitude of the dispersion coefficient is estimated in different locations for different flow rates and the relative strength of different dispersion mechanisms is discussed.

Model Input. In order to simplify the interpretation of model results, idealized conditions were simulated. Wind, evaporation and precipitation were neglected. The tides were idealized mixed diurnal and semi-diurnal tides. The period of M_2 was changed to exactly 12 hours and the period of K_1 was exactly 24 hours so that the tide repeats the same two cycles each day. The only inflow considered is Lagunitas Creek and the flow was constant in each simulation. Four simulations were conducted at different flow rates in Lagunitas Creek. The flow rates range from flows representative of low winter flow to peak winter flow.

After several months of simulation time for each simulation a periodic steady state was reached in which the tides, tidal velocities, and salinity repeat the same cycle each day. After steady state conditions are reached the model results were analyzed to provide insight to the strength of transport mechanisms.

Dispersion Coefficients. Dispersion coefficients at each cross-section can be estimated from the salinity field using the salt balance equation. Typical dispersion coefficients estimated from the model results are shown in Table 2. The dispersion coefficients are generally larger in the Outer Bay than the Inner Bay, suggesting that more active mixing occurs in the Outer Bay, which is expected because tidal velocities are much larger in the Outer Bay and the bathymetry is more varied and, therefore, more likely to result in "tidal trapping" (Fischer et al. 1979) and other tidal dispersion mechanisms. The strength of dispersion also increases with increased freshwater flow suggesting that the up-estuary dispersion mechanisms become stronger during flow events, perhaps due to a strengthening of the baroclinic circulation.

Table 2. Estimated dispersion coefficients from salt balance equation

Simulation	Flow (cms)	Dispersion Coefficient (m^2/s)	
		Inner Bay	Outer Bay
Low Flow	3	100	200
Medium Flow	11	150	250
High Flow	46	250	600
Peak Flow	65	300	1000

Cross-Section Analysis. The dispersion mechanisms that lead to salt flux are analyzed at several cross-sections in Tomales Bay. The velocity and salinity was saved at each grid cell in the section for each time step of the simulations to allow detailed analysis once periodic steady state conditions have been reached.

Two sections will be discussed in detail. The first section, located near Marshall, is representative of the Inner Bay because it has a relatively uniform depth over the cross-section, as shown in Figure 14. The second section, located at Toms Point, is representative of the Outer Bay because it has a distinct channel-shoal structure, as shown in Figure 16.

Figures 14 and 15 show the daily-averaged (tidally-averaged over two M_2 periods and one K_1 period) salinity distribution and velocity distribution over the cross-section at Marshall for the medium flow simulation and the high flow simulation. In both simulations, the salinity field is uniform laterally and stratified vertically but stronger stratification is present in the high flow simulation. This stratification, coupled with vertical shear in the daily-averaged velocity distribution, results in a down-estuary salt flux near the surface and an up-estuary salt flux at depth.

Figure 16 shows the daily-averaged salinity distribution over the cross-section at Toms Point for the medium flow simulation and the high flow simulation. In the medium flow simulation the salinity is fairly uniform over the section, while, in the high flow simulation, significant vertical stratification is present. Table 2 shows that Outer Bay dispersion coefficients were typically 250 m^2/s for the medium flow simulation and 600 m^2/s for the high flow simulation. Therefore, the increased stratification apparently allows substantially more dispersion of salt up-estuary to compensate for the increased advection of salt down-estuary by the higher net (freshwater) flow.

Conclusions

Several authors have analyzed field data or model results (e.g., Oey et al. 1985) to provide insight to the physical mechanisms that maintain the "salt balance" in an estuary. While actual estuaries are dynamic with constantly changing tidal conditions and frequently changing inflows, the idealized model inputs used in these simulations lead to an exact periodic steady state in which the dispersion mechanisms which transport salt up-estuary are exactly in balance with the net flow which advects salt down-estuary.

It appears that the dynamics of Tomales Bay can be best understood by considering the Inner Bay and Outer Bay separately. In the Inner bay, the uniform bathymetry and reduced tidal energy creates an environment in which the baroclinic circulation dominates the net along-estuary salt flux. In the Outer Bay, on the other hand, tidal trapping, which establishes a phase shift between velocity and salinity, dominates the along-estuary transport of scalars. This is not entirely unexpected, as the bathymetry undergoes a pronounced transition at Pelican Point, switching from a braided channel-shoal structure in the Outer Bay, which creates extensive tidal dispersion, to a flat-bottom nearly prismatic environment in the Inner Bay.

From a modeling perspective, the three-dimensional model effectively simulated the evolution of the salt field in a coastal estuary through an entire winter season characterized by intermittent freshwater forcing. Not only was the freshening of the Bay correctly predicted during the freshwater flow events, but the recovery of the Bay's salinity, which is driven by both tidal trapping and baroclinic dispersion mechanisms, was also correctly simulated. Because both of these dispersion mechanisms were important to the recovery of the salt field, albeit in different sub-regions, we do not believe that a single two-dimensional model would have performed adequately. In the inner bay, the dynamics are dominated by vertical structure, while in the Outer Bay, horizontal structure dominates. As a result, we would expect that a laterally averaged model may perform reasonably well in the Inner Bay but poorly in the Outer Bay. In contrast, a vertically averaged model would be a more appropriate two-dimensional model for the Outer Bay but would not perform well in the Inner Bay. Unless a reliable method for linking laterally averaged and vertically averaged models is applied, therefore, a three-dimensional model has clear advantages over two-dimensional models for Tomales Bay and other coastal estuaries that are characterized by similar bathymetric variability.

References

Casulli, V. (1990). "Semi-implicit finite difference methods for the two-dimensional shallow water equations." *Journal of Computational Physics*. 86, 56-74.

Casulli, V., and Cattani, E. (1994). "Stability, accuracy and efficiency of a semi-implicit method for three-dimensional shallow water flow." *Computers Math. Applic.* 27(4), 99-112.

Cheng, R.T., Casulli, V., and Gartner, J.W., (1993). "Tidal residual intertidal mudflat (TRIM) model and its applications to San Francisco Bay, California." *Estuarine, Coastal and Shelf Science*, 36, 235-280.

Dever, E.P., and Lentz, S.J. (1994). "Heat and salt balances over the northern California shelf in winter and spring." *Journal of Geophysical Research*, 99(C8), 16,001-16,017.

Fischer, D.T., Smith, S.V., and Churchill, R.R., (1996). "Simulation of a century of runoff across the Tomales watershed, Marin County, California." *Journal of Hydrology*, 186, 253-273.

Fischer, H.B., List, E.J., Koh, R.C.Y., Imberger, J., and Brooks, N.H. (1979). *Mixing in Inland and Coastal Waters*. New York: Academic Press.

Gross, E.S., Casulli, V., Bonaventura, L. and Koseff, J.R. (1998). "A semi-implicit method for vertical transport in multidimensional models." *International Journal for Numerical Methods in Fluids*, 28, 157-186.

Gross, E.S., Koseff, J.R., and Monismith, S.G. (1999a). "Evaluation of advective schemes for estuarine salinity simulations." *Journal of Hydraulic Engineering*, 125(1): 32-46.

Gross, E.S., Koseff, J.R., and Monismith, S.G. (1999b). "Three-dimensional salinity simulations of South San Francisco Bay." *Journal of Hydraulic Engineering*, 125(11): 1199-1209.

Johnson, R.G., Bryant, W.R., and Hedgpeth, J.W., (1961). "Ecological survey of Tomales Bay." Pacific Marine Station Research Reports No. 1.

Mellor, G.L., and Yamada, T. (1982). "Development of a turbulence closure model for geophysical fluid problems." *Reviews of Geophysics and Space Physics*, 20, 851-875.

Oey, L.Y., Mellor, G.L., and Hires, R.I., (1985). "A three-dimensional simulation of the Hudson-Raritan Estuary. Part III: Salt flux analyses." *Journal of Physical Oceanography*, 15, 1711-1720.

Philip Williams and Associates (1993). "An evaluation of the feasibililty of wetland restoration on the Giacomini Ranch, Marin County." 80 pp.

Smith, E.H., Johnson, R.G., and Obrebski, S., (1971). "Final report, environmental study of Tomales Bay, Volume 2, 1966-1970." Pacific Marine Station Research Reports No. 9.

Smith, S.V., Weibe, W.J., Dollar, S.J., and Vink, S. (1989). "Tomales Bay, California: A case for carbon-controlled nitrogen cycling." *Limnology and Oceanography*, 34(1), 37-52.

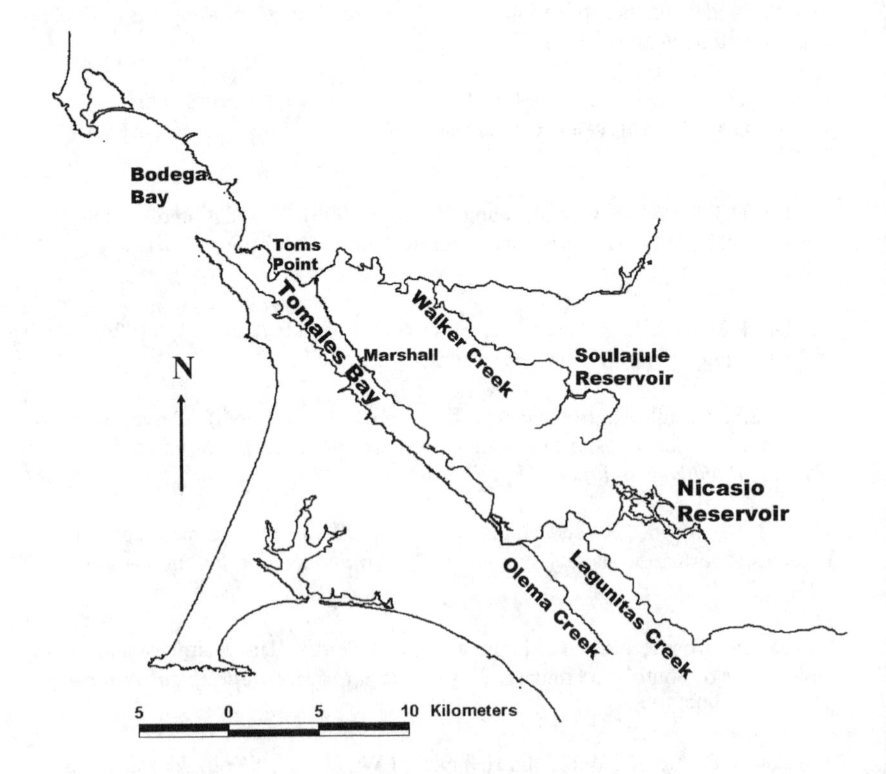

Figure 1. Location of Tomales Bay, associated tributaries and landmarks.

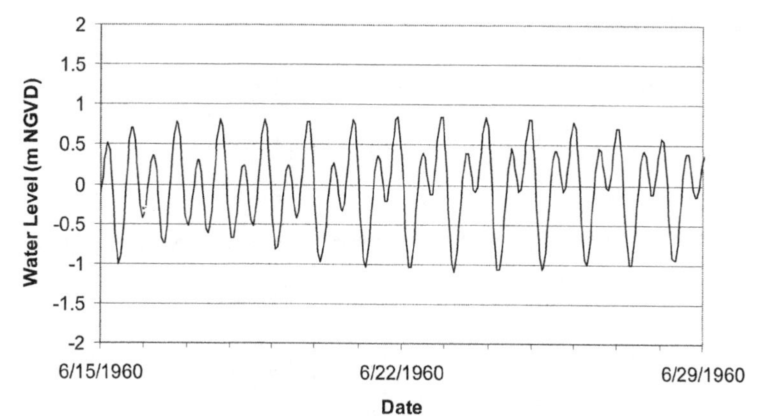

Figure 2. Observed tides at NOAA Station 9414290 in San Francisco

Figure 3. Locations of selected current meter stations.

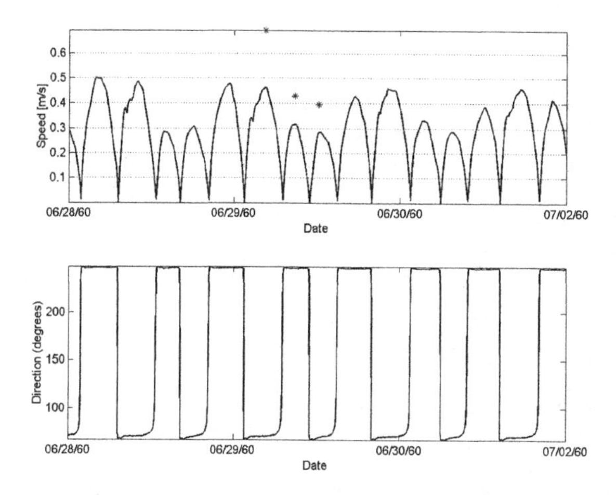

Figure 4. Predicted velocity and observed peak current speed at Station 3, near Toms Point.

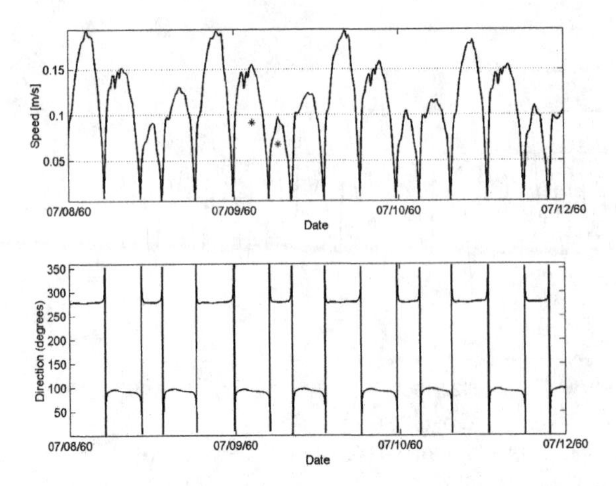

Figure 5. Predicted velocity and observed peak current speed at Station 8 near Marshall.

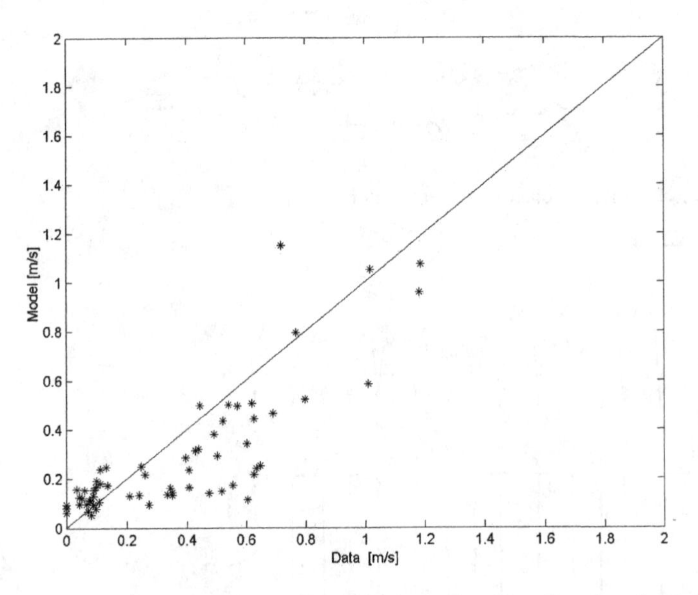

Figure 6. Comparison of observed peak tidal speed and predicted peak tidal speed at all Stations.

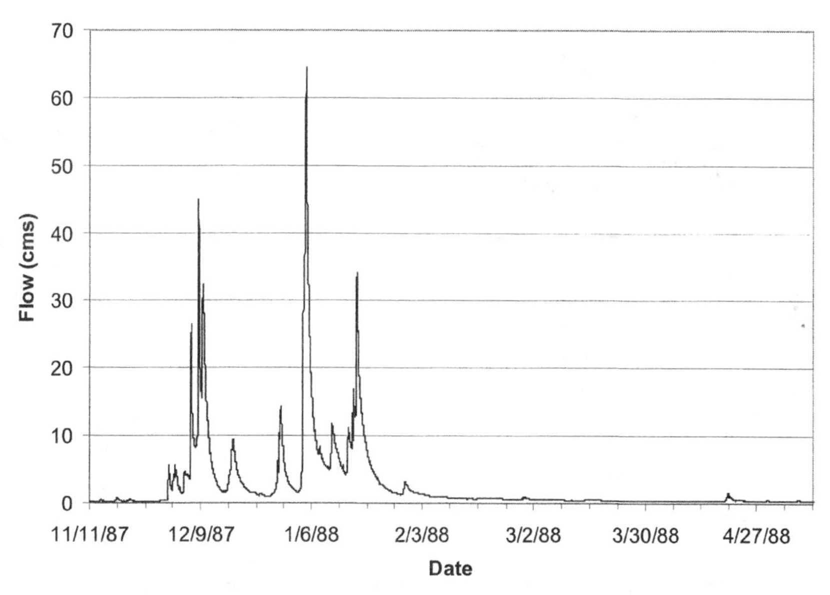

Figure 7. Lagunitas Creek flow into Tomales Bay during salinity simulation period.

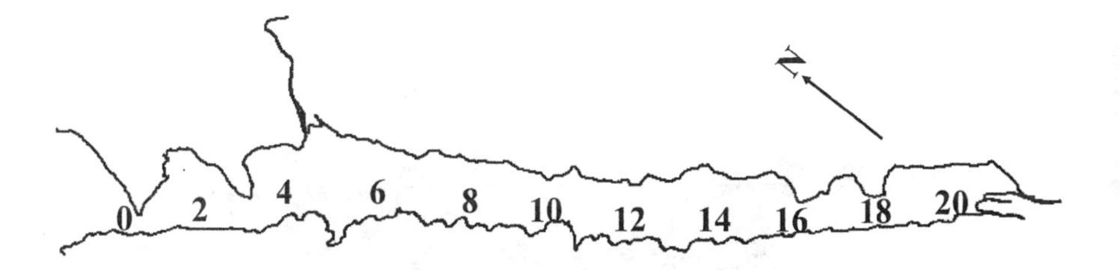

Figure 8. LMER/BRIE salinity monitoring stations.

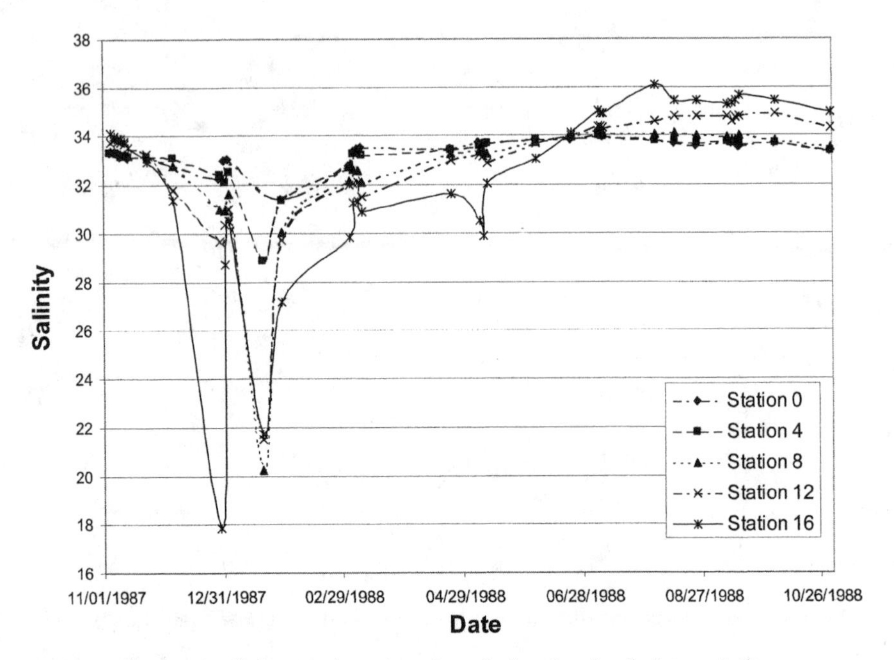

Figure 9. Observed salinity at selected stations during the simulation period.

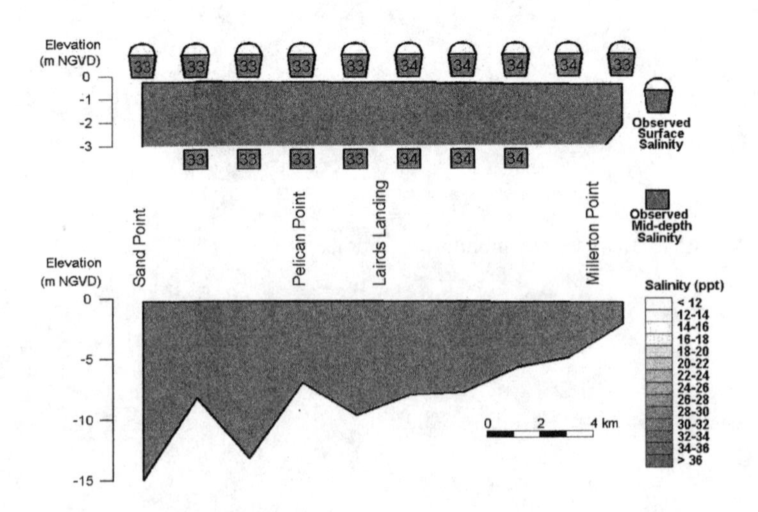

Figure 10. Observed surface salinity on 11/11/1987 and initial salinity field in simulation.

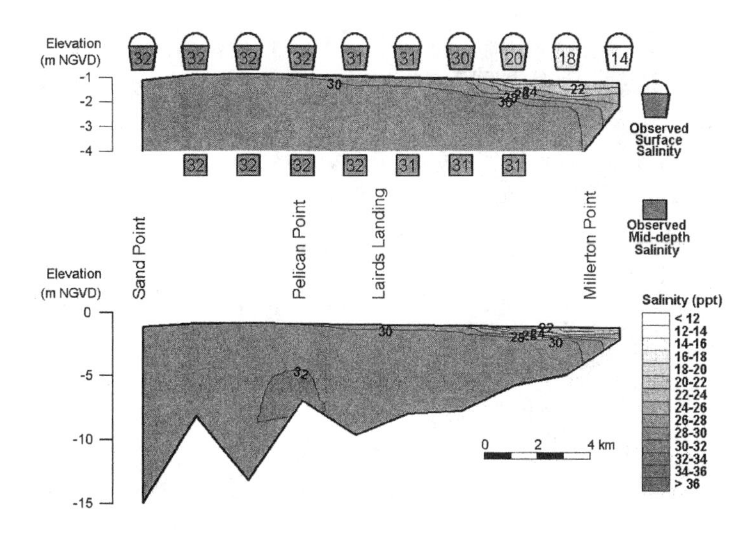

Figure 11. Comparison of Predicted Salinity Field on 12/29/1987 with Observed Surface Salinity.

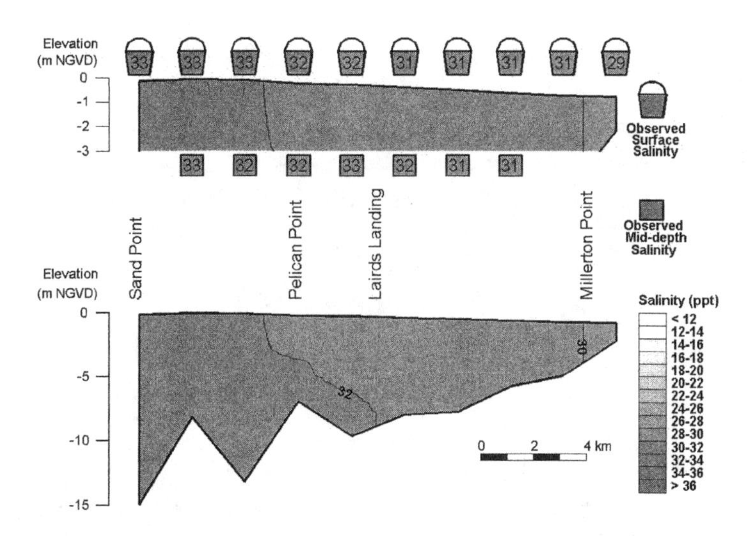

Figure 12. Comparison of predicted salinity field on 1/02/1988 with observed surface salinity.

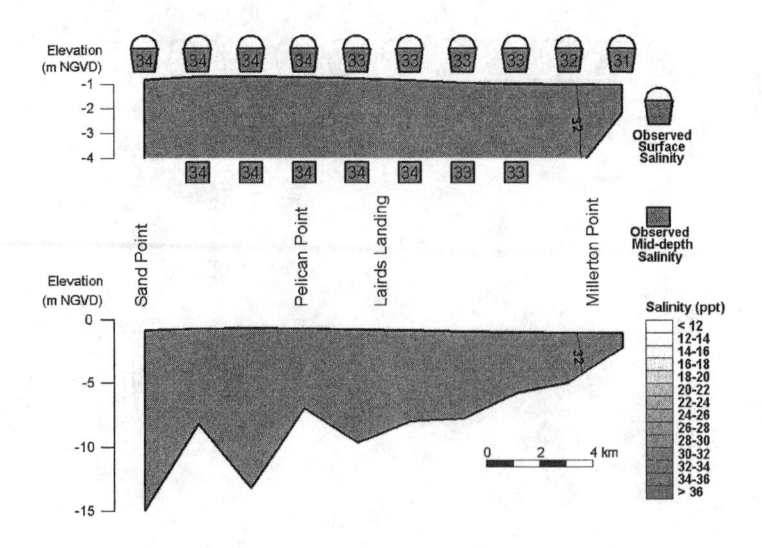

Figure 13. Comparison of predicted salinity field on 5/11/1988 with observed surface salinity.

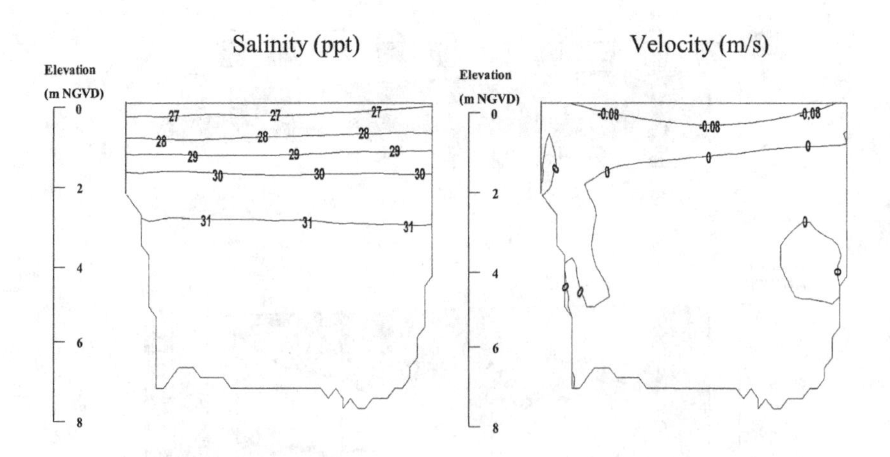

Figure 14. Daily-averaged salinity and velocity at the Marshall cross-section for the Medium Flow Scenario.

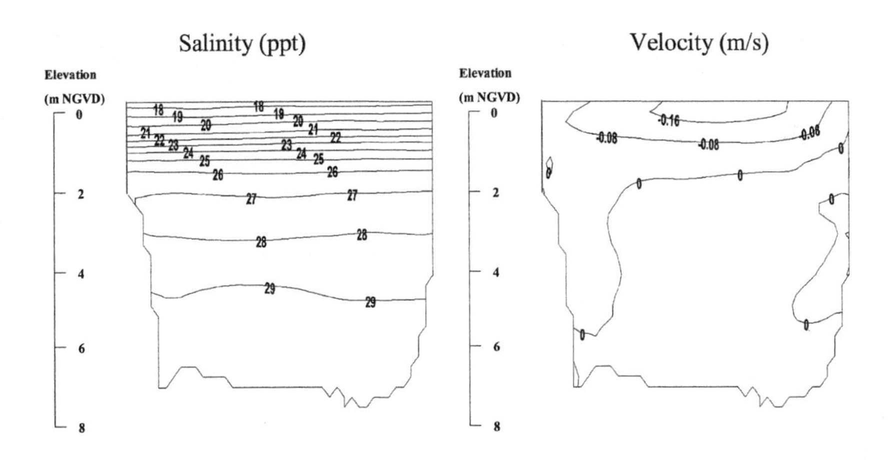

Figure 15. Daily-averaged salinity and velocity at the Marshall cross-section for the High Flow Scenario.

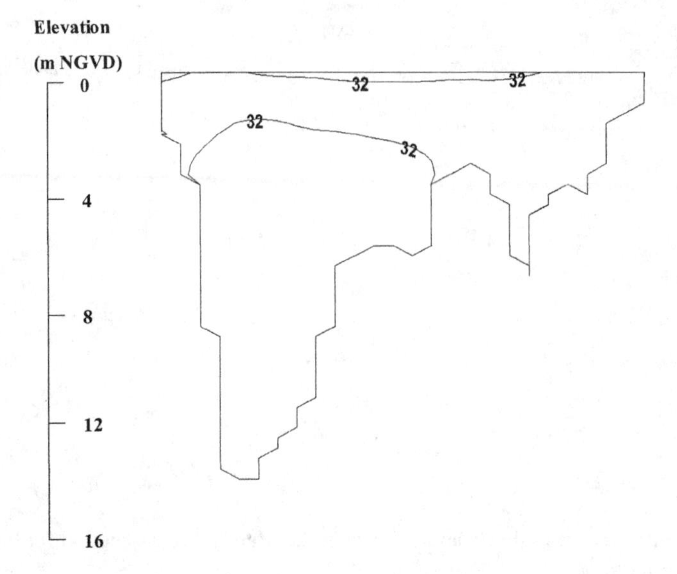

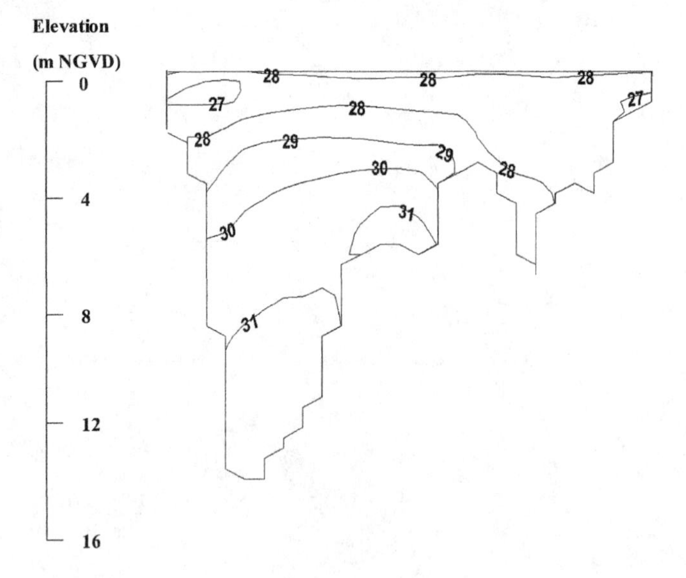

Figure 16. Daily-averaged salinity and at the Toms Point cross-section for the Medium Flow Scenario and the High Flow Scenario.

Modeling Study of Saltwater Intrusion in Loxahatchee River, Florida

Guangdou Gordon Hu, Ph.D., PE[1]

Abstract

The upstream advance of saltwater into the historic freshwater reaches of the Loxahatchee River has altered the floodplain cypress forest community along the river and some of its tributaries. A hydrodynamic/salinity model was developed to study the influence of freshwater input, tidal inlet deepening and sea level rise on the salinity regime in the estuary (Hu, 2002). The model was recently updated with new bathymetric data. The updated model was tested against the tide and salinity data that had been collected in 2003. The model output and field data were examined and compared to assess the possibility of establishing a relationship between freshwater inflow and daily average salinity.

Introduction

The Loxahatchee River estuary empties into the Atlantic Ocean at Jupiter Inlet in southeastern Florida. The estuarine system is comprised of three forks: the Southwest Fork, North Fork, and Northwest Fork (Figure 1). Estuarine conditions extend from the Jupiter Inlet to about 5 river miles up the Southwest Fork, 6 river miles up the North Fork, and 10 river miles up the Northwest Fork. Four major tributaries: the upper Northwest Fork of Loxahatchee River, Cypress Creek, Hobe Grove Ditch, and Kitching Creek discharge to the Northwest Fork. Canal 18 (C-18), built in 1957 – 1958, is the major tributary to the Southwest Fork. The North Fork has several small unnamed tributaries. Rainfall in the area is seasonal, 5 inches per month is common during the wet season from May through October. Amounts near 2.5 inches per month generally occur during the dry season from November to April (Russell and Goodwin, 1987).

The upstream migration of salt water into the historic freshwater reaches of the Loxahatchee River is the likely cause of the altered floodplain cypress forest community along the Northwest Fork and some of its tributaries (McPherson, Sabanskas, & Long, 1982). A hydrodynamic/salinity model was developed to study the influence of freshwater input on the salinity conditions in the river and estuary. The model was applied to scenarios with varying amounts of freshwater inflow. Both

[1] Lead Engineer, South Florida Water Management District, 3301 Gun Club Road, West Palm Beach, Florida 33406, USA; Phone: (561) 682-6720; Email: ghu@sfwmd.gov

the field data and model simulation indicated that there is a strong correlation between freshwater inflow and the salinity regime in the estuary. Based on model output and field data analysis, a relationship was developed to predict salinity at various points in the estuary with respect to freshwater inflow rates and tidal fluctuations. The model also provided a preliminary assessment of the impacts that inlet deepening and sea level rise have had on the salinity regime in the estuary (Hu, 2002).

In parallel with the preliminary model setup, a data collection program was implemented. A bathymetric survey was conducted in Northwest Fork and North Fork in early 2003. Two flow gauges were established on Cypress Creek and Hobe Grove Ditch in November 2002. Combined with flow gauges that were previously established at the Lainhart Dam on the upper Northwest Fork and Kitching Creek, the four gauges monitor a majority of freshwater input to the Northwest Fork which has been the focus of the salinity study. Four tide and salinity stations have been deployed in the estuary since November 2002. For current velocity measurements, two bottom mount ADCP units have been deployed at various locations through out the estuary since June 2003. The estuary model was updated recently using the new bathymetry and freshwater inflow data. The updated model was tested against tide, salinity and velocity data collected in the period from May to August 2003. This paper outlines the results of the three-month model simulation.

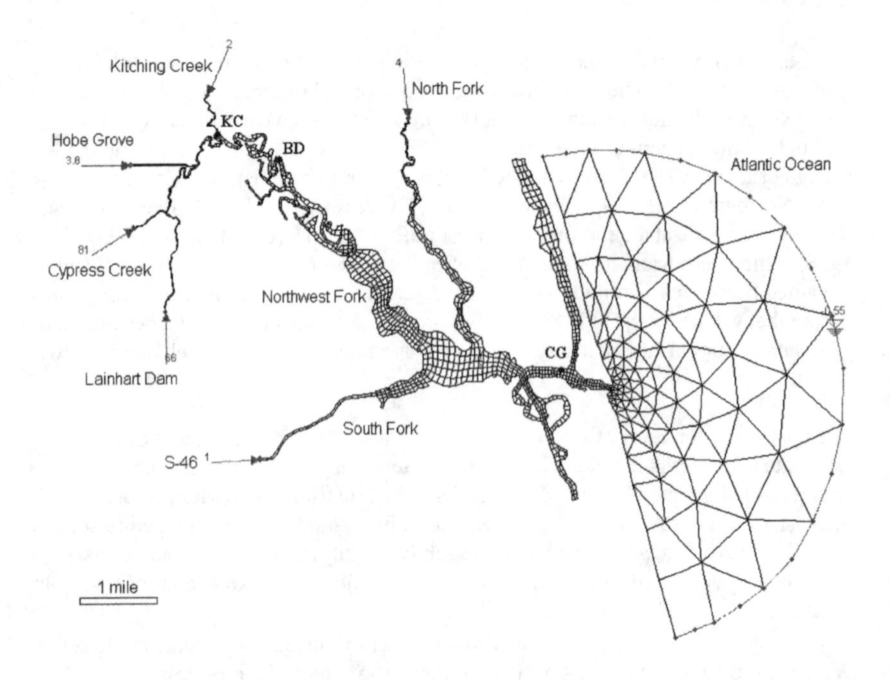

Figure 1. Model domain and finite element mesh

Model Description

The Loxahatchee River Hydrodynamics/Salinity Model was setup using two computer programs: RMA-2 and RMA-4 (USACE, 1996). RMA-2 is a two dimensional depth averaged finite element hydrodynamic numerical model. It computes water surface elevations and horizontal velocity components for subcritical, free-surface flow in two dimensional flow fields. RMA-2 computes a finite element solution of the Reynolds form of the Navier-Stokes equations for turbulent flows. Friction is calculated with the Manning's or Chezy equation, and eddy viscosity coefficients are used to define turbulence characteristics. The water quality model, RMA4, is designed to simulate the depth-average advection-diffusion process in an aquatic environment. In this application, RMA-4 was used for salinity simulation only. The finite element mesh was recently updated with new bathymetric data. The current model mesh includes a total of 4956 nodes with elevations derived from survey data. Figure 1 shows the model mesh with 1075 quadrilateral elements and 231 triangular elements. Arrows in the figure indicate the locations where flow boundary conditions are applied. The model mesh was extended three miles offshore into the Atlantic Ocean in order to obtain a relatively stable salinity boundary condition. Tide and salinity data collected from three stations were used in the initial model testing. The locations of the three stations are marked in the mesh map as CG – Coastguard Station, BD – Boy Scout Dock Station and KC – Kitching Creek Station.

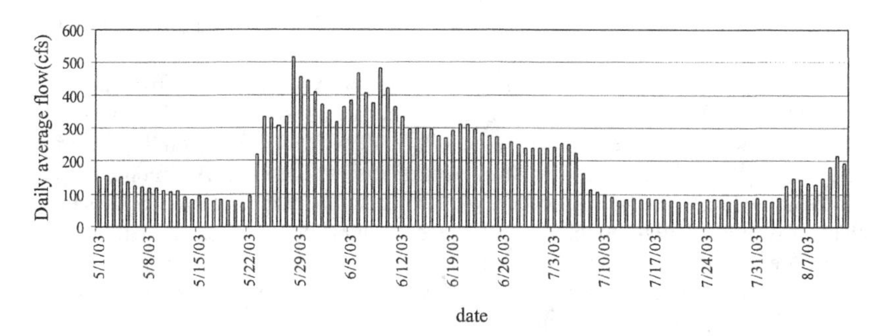

Figure 2. Freshwater inflow from major tributaries to the Northwest Fork

Figure 2 shows the combined freshwater inflow from four major tributaries to the Northwest Fork for the period from May 1 to August 12, 2003. Daily averaged flow rates in terms of cubic feet per second from flow gauges on upper Northwest Fork at Lainhart Dam, Cypress Creek, Hobe Grove and Kitching Creek were used for the calculation. Discharge from S-46 into South Fork was based on measurements at the discharge structure for the model simulation period.

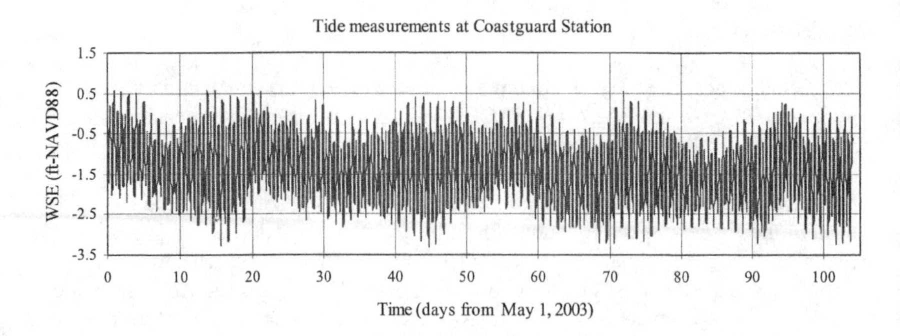

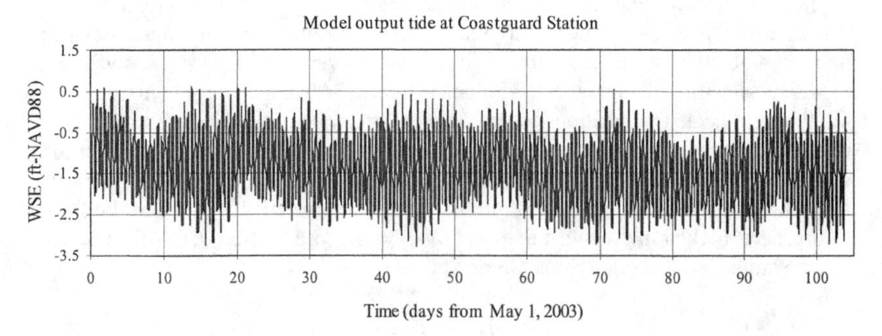

Figure 3. Tide at the Coastguard Dock station – field data and model output

The water surface elevation on the ocean side was based on tidal data from the Coastguard Station near the Jupiter Inlet. Tidal data was collected at 15-minute intervals. The model time step was set at 30 minutes in this application. Figure 3 is a comparison of tidal data from the Coastguard station with the RMA-2 model output for the same location. The two curves overlap each other when printed in the same chart. It is more difficult to tell the difference when printed in black and white. Therefore the model and field data were plotted in separate charts with same scale and grid lines for purpose of comparison. For RMA-4 applications, a constant salinity of 35.5 ppt was applied on the ocean boundary.

Results and Discussion

Field data and model output of tides at Boy Scout Dock and Kitching Creek are plotted in Figure 4 for comparison. The two stations are approximately 2 river miles apart and there is no major tributary in between. Both field data and model output indicate that the tidal regimes at these two sites are similar in terms of range. Flow velocity measurements were also taken at these two sites in late June and July of 2003. The instruments were bottom mount ADCP units with transducers at approximately two and a half feet over the bed surface. Since the water depth in the area is only about 6 to 10 feet deep, the records provided by these instruments are in fact the flow

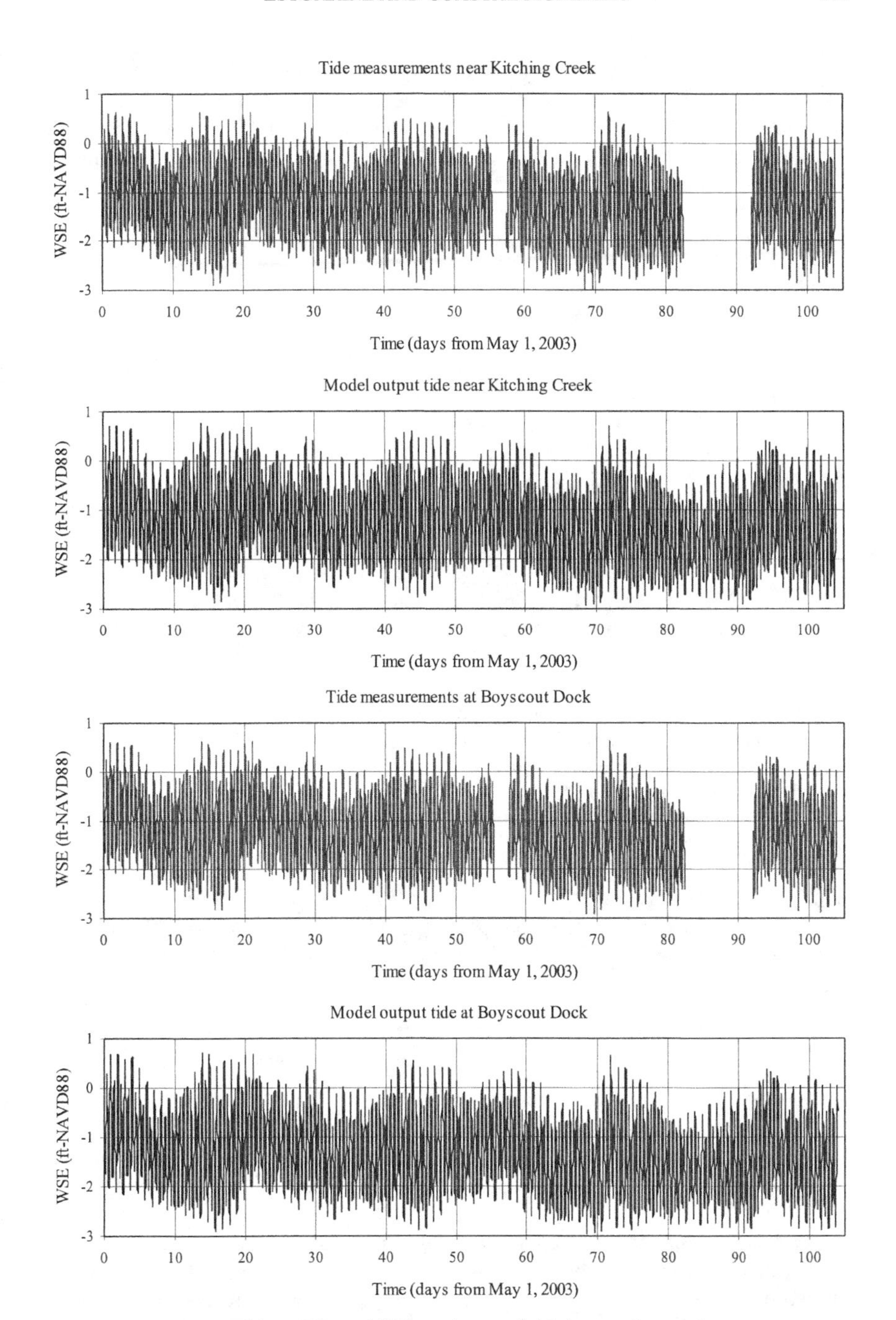

Figure 4. Tide at BD and KC stations – field data and model output

velocity of the faster moving upper layer in the water column. This is probably one of the reasons that the depth averaged velocity from RMA-2 is lower then the ADCP measurements. Figure 5 is a comparison of model results and ADCP measurements at Kitching Creek site from July 20 to July 30 of 2003. Flow direction here is dominated by North-South flow that is parallel with the river channel at this location. It is expected that the depth averaged model velocity is lower than ADCP measurements since the instrument does not read lower flow velocity in the near bed region.

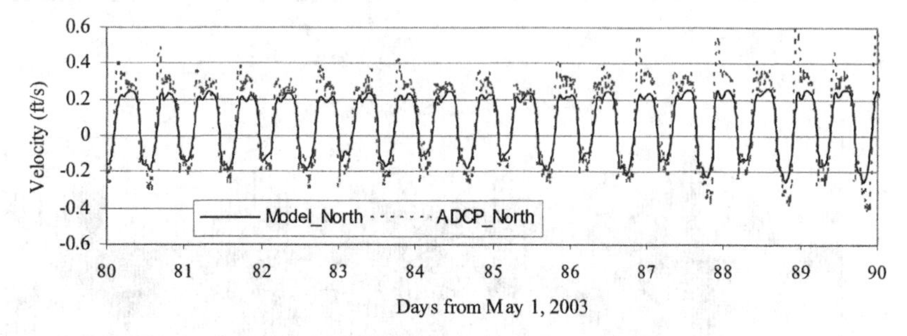

Figure 5. Flow velocity (North-South component) – ADCP measurements of upper water column flow velocity and model output of depth averaged velocity

The magnitude of the East-West component is much smaller than the North-South component (Figure 6). Both Figure 5 and Figure 6 show that the model simulated the phases of tide movement accurately.

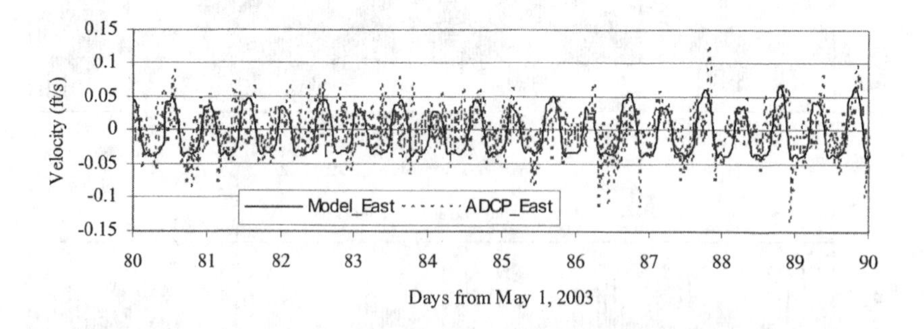

Figure 6. Flow velocity (East-West component) – ADCP measurements of upper water column flow velocity and model output of depth averaged velocity

In order to speed up model execution for long-term simulations, the Northwest Fork is represented in the current model mesh with a single row of elements. Therefore the elevation variation of bed surface in the lateral direction was not represented in detail by the current model mesh. The flow velocity produced by the model should be considered as mean velocity over a cross section. On the other hand, the up-looking

ADCP units can only cover a small portion of the river channel depending on the spread angle of acoustic signal and the water depth. The instrument was deployed near the center of the channel where flow velocity is relatively high. This is another reason that the model flow velocity is smaller than ADCP measurements. This difference between model output and ADCP measurement is more apparent in the river reach near the Boy Scout Dock where channel is wider than the Kitching Creek section.

Figure 7 compares model output of depth-averaged salinity with salinity measurements from instruments at a fixed elevation. While these two quantities are not exactly the same physically, the comparison reveals both the limitations and capabilities of a depth averaged model such as RMA-2 and RMA-4.

The salinity record at Boy Scout Dock was around 10 ppt between day 60 and 70. This sudden salinity increase does not seem to be related or supported by data from other field records. Salinity around 10 ppt at this site usually occurs when freshwater inflow is below 100 cfs. The flow gauges actually recorded over 200 cfs for that period. The salinity record from the adjacent Kitching Creek station is also inconsistent with this salinity increase in the Boy Scout Dock station record. Previous studies indicated that 10ppt at Boy Scout Dock station would have raised salinity at Kitching Creek station to 2 ppt or above (Russell & McPherson, 1983). The station at Kitching Creek did not record such an increase for that period (See the Kitching Creek chart in Figure 7).

Both RMA-2 and RMA-4 are two dimensional depth averaged models. The models do not simulate the formation of a saltwater wedge and salinity stratification. When the system is less stratified, such as the condition near the inlet at the coastguard station, the model output salinity tracks field data rather closely. On the other hand when the system is stratified in certain areas the model output for that area tends to give a smaller salinity variation between high tide and low tide that is the most evident in this simulation at the Boy Scout Dock station.

The RMA-4 output is depth averaged salinity, which is different from salinity read by a transducer at a fixed elevation. The conductivity transducers were installed at elevations that would remain below the water surface at low tide. Since the range between higher high and lower low water is close to 4 feet and the overall water depth is only about 6 feet to 10 feet, the conductivity transducers would be situated in the lower water column during high tide. Such a setup would make the instrument take measurements in the surface layer at low tide and in the bottom layer at high tide. If the system is well mixed (no stratification), there should be no difference between depth averaged salinity and salinity at a fixed elevation. On the other hand, when the system is stratified, the daily variation range recorded by the instruments would be wider than the daily variation range of the depth averaged salinity.

The intended application of the model is to predict daily mean salinity for a number of locations in the estuary. It is interesting to observe how a depth averaged model would perform for such applications. Figure 8 compares field data with daily mean salinity from the model simulation.

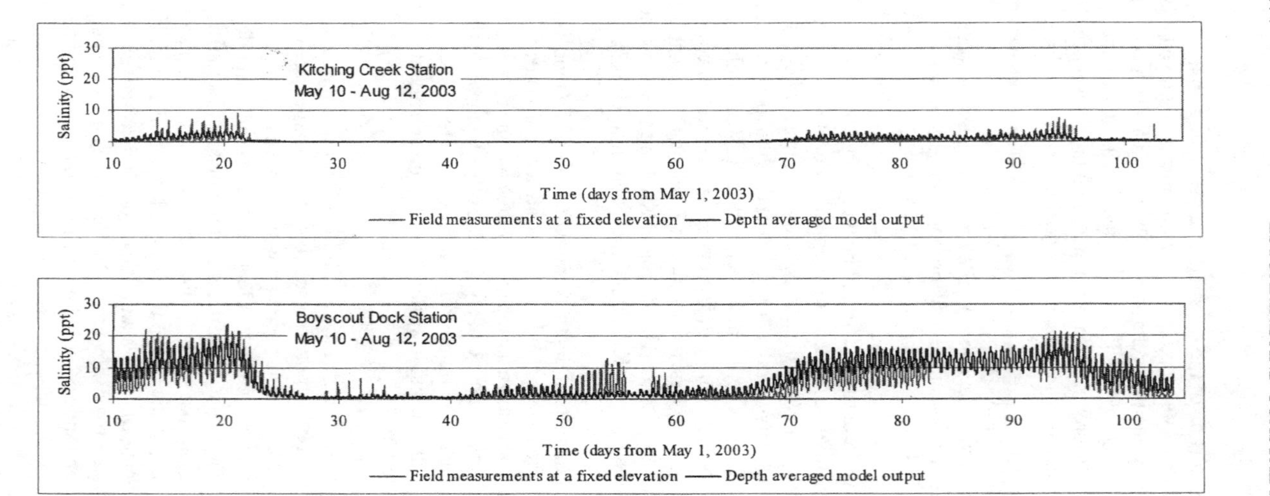

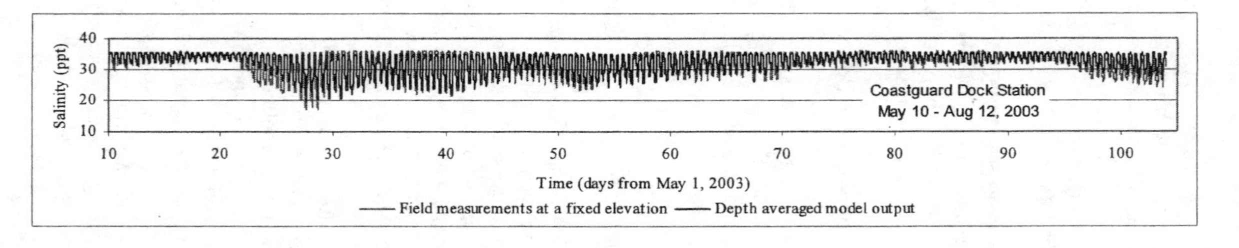

Figure 7. Model output of depth averaged salinity and field measurements at fixed elevations in the water column

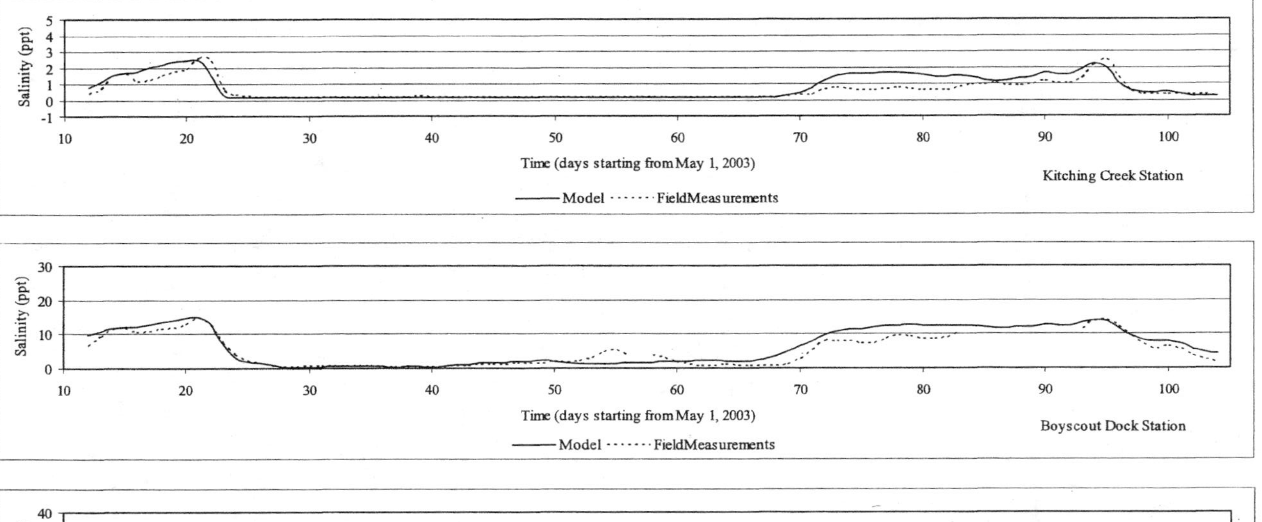

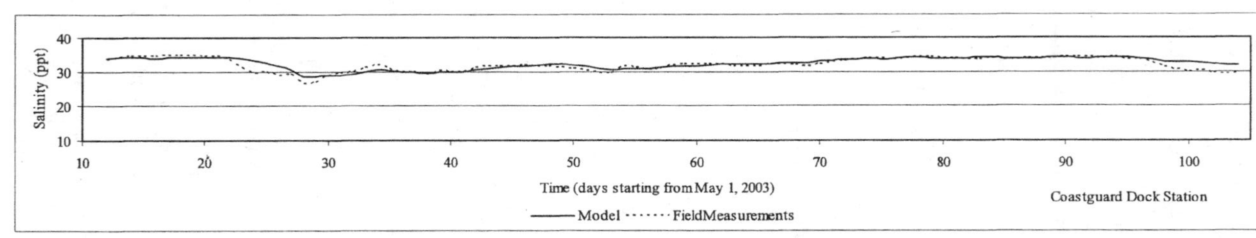

Figure 8. Daily average salinity based on model output and field measurements

The watershed model for the study runs on daily time steps. The estuary model is required to produce a relationship between freshwater flow and daily averaged salinity. The model output in Figure 8 was averaged over calendar days although tide regime does not evolve exactly over calendar days. Model output followed the overall trend of salinity changes over the three month period. At the Kitching Creek station, the model prediction is higher by 0.5 to 1 ppt. The difference is 2 to 3 ppt at the Boy Scout Dock station. The most likely reason is that there was additional freshwater inflow to the system that bypasses the four stations on the major tributaries. Such additional source of freshwater may includes overland flow and groundwater seepage into the system. A groundwater monitoring network that was established in 2003 indicates active exchanges between the river and the groundwater. In the next phase of the study, model input will include groundwater input to the system. The amount of groundwater input will be based on field data and a groundwater model.

The objective of this preliminary modeling study is to assess the possibility of establishing a relationship between the amount of freshwater inflow from major tributaries and daily average salinity at various locations in the estuary. Figure 9 shows plots of freshwater inflow versus daily average salinity at the Boy Scout Dock station. The first chart in the figure was based on model output salinity. The second chart was based on field data. There are certainly driving forces other than freshwater inflow that affect salinity such as tides. Therefore the salinity record shows a wide range of variation for the same freshwater input. On the other hand, the chart does show a clear trend, which indicates that freshwater inflow affects salinity at this site significantly. The range of salinity variation in the model output was narrower since the model in its current form is driven by tide and freshwater inflow only. The scattering of model data points is mostly due to the variation of tidal conditions. Another factor causing deviation from a single freshwater ~ salinity relationship was the transition of the system from one salinity regime to another in response to the changes in the freshwater input.

Data points based on field measurements are more scattered, which is an indication that there are other factors at work. For example, additional freshwater inflow to the system from overland flow and discharge from small tributaries bypasses the four flow gauges. Given the intensive rainfall in summer, the significance of direct rainfall and evaporation on the estuary also needs to be assessed. Wind is another factor that needs to be included in the next phase of the model improvements.

Similar freshwater flow versus salinity plots are shown in Figure 10 for the Kitching Creek station. The relationship based on model output resembles more closely the relationship from the field data. The data points are also less scattered comparing to the Boy Scout Dock station plots. Kitching Creek station is two miles upstream from the Boy Scout Dock station. Apparently freshwater inflow is a more dominating factor near the head of the estuary. Therefore the freshwater inflow and salinity appears more closely correlated at the upstream station.

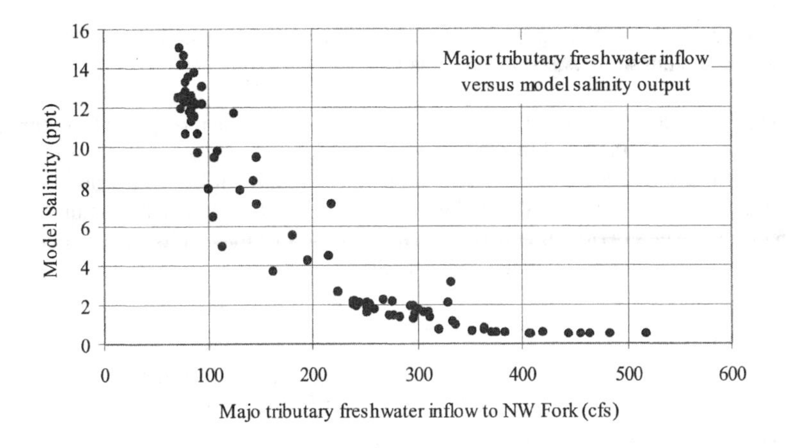

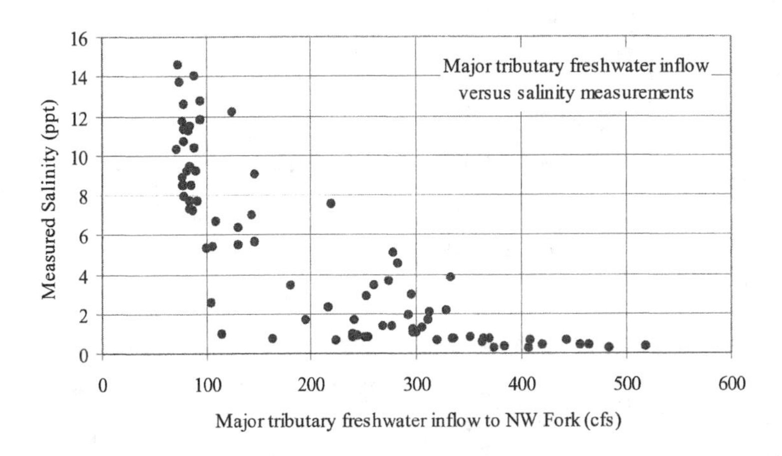

Figure 9. Freshwater inflow versus daily average salinity at Boy Scout Dock Station

Due to insufficient freshwater inflow data, the previous modeling study and statistical analysis estimated total freshwater inflow to the Northwest Fork using flow records at the Lainhart Dam. The estimation was based on a percentage established by USGS and South Florida Water Management District field measurements in 1980s and 1990s (SFWMD, 2002). Over the three month model simulation period in 2003, freshwater from Lainhart Dam was about 30 – 70% of the total from all four freshwater flow gauges. The average percentage for the three month period was 52%. The freshwater inflow versus salinity relationships shown in Figure 9 and 10 are consistent with the previous modeling study and statistical analysis.

It is apparent that salinity is influenced by multiple factors. The model results outlined above seem to indicate that tide and freshwater inflow are two most dominating factors. While it is possible to produce a freshwater ~ salinity relationship based on multiple model simulations under various freshwater input scenarios, the simulations will have to cover a period of at least a lunar month so that the results can present salinity regimes under an "average" tidal condition.

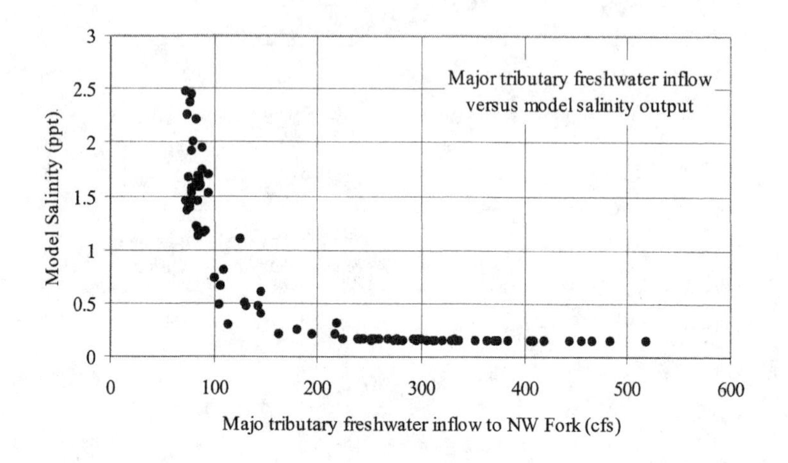

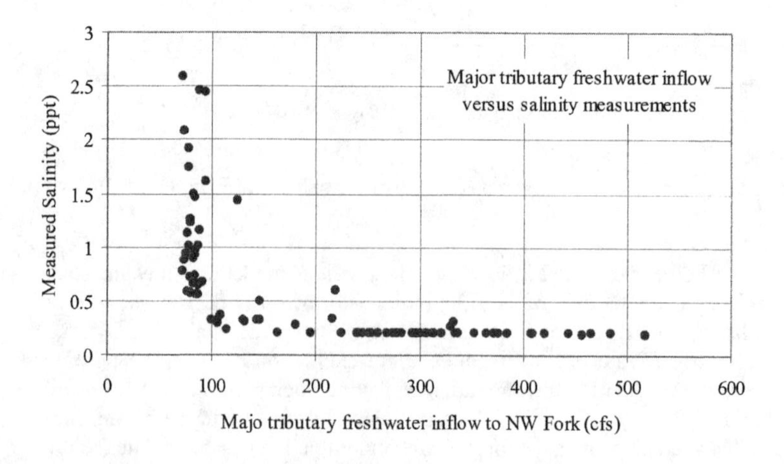

Figure 10. Freshwater inflow versus daily average salinity at Kitching Creek Station

Conclusions

The current model does not include driving forces such as wind, precipitation/evaporation and the exchange between the river and the groundwater which can be significant in the upper river reaches during dry season. While sensitivity analysis will be conducted to assess the significance of these factors the current model, which was only driven by major tributary freshwater input and ocean tide, was able to predict the tide regime rather accurately and predict the trend of salinity changes over the three month simulation that include both low and high freshwater input to the estuary. This seems to indicate that the amount of freshwater inflow to the estuary and tide are the two most dominant factors that affect the salinity regime in the estuary.

The depth averaged model does not simulate salinity stratification in the system. It appears that this will probably hinder the capability for the model to predict the full range of daily salinity variation between high and low water. On the other hand, the daily mean salinity from the model output follows the field record rather closely. At the Kitching Creek station, the difference during the three month simulation was 1 part per thousand or less.

The model was able to predict the overall tide and salinity regime over the three month period that includes both low and high freshwater input to the system. For applications where an accurate description of flow field (spatial distribution of flow velocity) is required in the upper estuary such as Northwest Fork, the model mesh needs to increase its resolution to describe the river channel geometry in greater detail. For simulations of high flow scenarios, floodplain should also be included in the model mesh.

Both model output and field data were examined to assess the possibility of establishing a relationship between salinity and freshwater inflow. While both data plots for the two stations on the Northwest Fork show a clear pattern of freshwater influence, it appears that fresh water is a more dominating factor at the upstream station. The relationships drawn from both the field data and model simulation over the three month period in 2003 (shown in Figure 9 and 10) are consistent with the results of previous modeling study and statistical analysis.

References

Hu, G. (2002). "The Effects of Freshwater Inflow, Inlet Conveyance and Sea Level Rise on the Salinity Regime in the Loxahatchee Estuary." Proceedings of *2002 Environmental Engineering Conference, Environmental & Water Resources Institute, American Society of Civil Engineers-Canadian Society of Civil Engineers*, Niagara Falls, Ontario, Canada, July 21-24, 2002.

McPherson, B.F., Sabanskas, M., & Long, W. (1982). "Physical, Hydrological, and Biological Characteristics of the Loxahatchee River Estuary, Florida." USGS Water-Resources Investigations, Open file report, 82-350.

Russell, G.M., McPherson, B.F. (1983). Freshwater Runoff and Salinity Distribution in the Loxahatchee River Estuary, Southeastern Florida. USGS Water-Resources Investigations Report, 83-4244.

Russell, G.M., Goodwin, C.R. (1987). Simulation of Tidal Flow and Circulation Patterns in the Loxahatchee River Estuary, Southeastern Florida. USGS, Water-Resources Investigations Report 87-4201. Tallahassee, Florida, 1987.

SFWMD (2002). Technical Documentation to Support Development of Minimum Flows and Levels for the Northwest Fork of the Loxahatchee River. South Florida Water Management District, Water Supply Department, November, 2002.

USACE, (1996). Users' Guide to RMA2 Version 4.3. US Army Corps of Engineers, Waterways Experiment Station, Hydraulic Laboratory.

A progress report on the control of the free surface in a 3D Navier Stokes solver for coastal applications

Igor Gejadze and Graham Copeland*, Department of Civil Engineering, University of Strathclyde, 107 Rottenrow, Glasgow, G4 0NG, UK tel: +44 141 548 3252, email: igor.gejadze@strath.ac.uk, g.m.copeland@strath.ac.uk
and
Simon Neill, Centre for Applied Oceanography University of Wales, Bangor Marine Science Laboratories Menai Bridge Anglesey LL59 5EY. s.p.neill@bangor.ac.uk
and

Fangxin Fang and Chris Pain, Department of Earth Science and Engineering,South Kensington Campus, Imperial College London, SW7 2AZ, UK. Tel: +44 20 7594 9322, email: f.fang@imperial.ac.uk, c.pain@imperial.ac.uk

* communicating author

Abstract
This paper reports work in progress on the development of a data assimilation capability for an existing 3D Navier Stokes solver with application as an ocean model. The project is a collaboration between University of Strathclyde and Imperial College, London with the assistance of Dr Neill from the University of Wales, Bangor. The paper reports work carried out at Strathclyde University. The 3D NS solver developed by the Imperial team uses finite element methods with adaptive meshing, is non-hydrostatic and includes the free surface. However, proper treatment of the free surface in the adjoint problem is not straightforward. This paper describes a method of controlling the forward solution using either surface elevation or current data by solving an appropriate adjoint problem This is formulated in a new way using a co-ordinate transformation such that the control problem appears in a fixed domain even though the forward problem has a variable domain due to the presence of the free surface.

Introduction
Surface elevation is an essential variable in coastal circulation modelling especially in regions where tidal phenomenon dominate the flow behaviour. Additionally, elevation measurements obtained by remote sensing represent one of the most important new sources of data for ocean study. Surface elevation is naturally incorporated into the shallow water equations (SWE), which are commonly used for data assimilation into coastal flows, Copeland (1998) and into channel flows Sanders & Katapodes (2000). However, the application of the SWE is limited to well mixed coastal waters. A more general approach to modelling requires a 3D model which can describe both coastal and deep water phenomena. Such a model may be formulated from the Navier-Stokes equations with free surface (FSNS). Inclusion of the free surface into the solver is achieved either by using the free surface boundary conditions or by using a depth integrated continuity condition that involves the free surface

elevation directly. Both methods have been demonstrated by the project team from Imperial, Piggott (2003) .

In order to solve the data assimilation problem for FSNS we need the gradient of the objective function to be calculated. The method reported here is based on the adjoint to the FSNS. The derivation of the adjoint in this case is non-trivial due to the fact that the domain itself depends on the solution through changes in the free surface height. However, a method of formulating the adjoint control problem, developed by the principal author, is described which makes use of a co-ordinate transformation. The adjoint system together with boundary conditions and gradient formulas are presented.

The overall objective of the project is to build a data assimilation capability into an existing finite element ocean model that uses a 3D unstructured mesh that can adapt to topography and to flow features, see http://amcg.th.ic.ac.uk:80/research/. However, this paper is concerned with the correct development of the adjoint problem of the FSNS. The approach is, for now, 2D in section including the free surface and variable bed topography. The trial numerical implementation presented here uses a finite difference solution of the FSNS equations with a fixed regular mesh similar to the well known SOLA algorithm, Neill (2002). The numerical trials test the stability of the method and produce values of the adjoint sensitivities at the control boundary.

Problem Statement

Let us consider a 2D free surface, inviscid flow in a channel, where x-axis is directed along the channel (in the case considered here from deep water towards the shore), and y-axis is directed vertically upwards. Velocities $u(x,y,t)$ and $v(x,y,t)$ are oriented along the x and y axes respectively as it is shown on Fig.1.

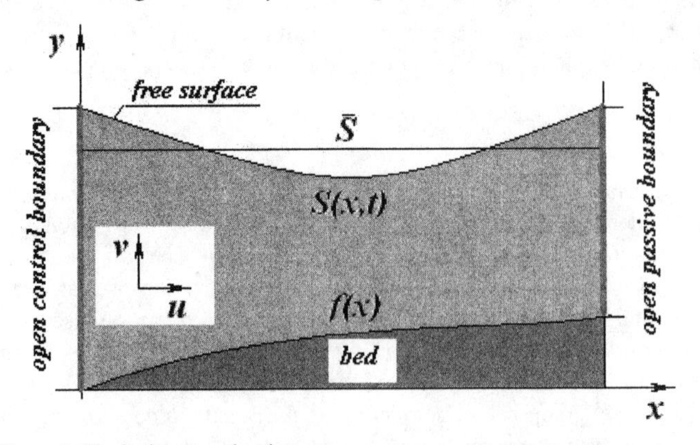

Figure 1. The bathymetry sketch

The governing equations are as follows:

$$E_1 =: \frac{\partial S}{\partial t} + u(x, S(x,t), t)\frac{\partial S}{\partial x} - v(x, S(x,t), t) = 0 \quad (1)$$

$$E_2 =: \frac{\partial u}{\partial x} + \frac{\partial v}{\partial y} = 0 \quad (2)$$

$$E_3 =: \frac{\partial u}{\partial t} + u\frac{\partial u}{\partial x} + v\frac{\partial u}{\partial y} + \frac{\partial p}{\partial x} = 0 \quad (3)$$

$$E_4 =: \frac{\partial v}{\partial t} + u\frac{\partial v}{\partial x} + v\frac{\partial v}{\partial y} + \frac{\partial p}{\partial y} + g = 0 \quad (4)$$

$$x \in (0, X), \quad y \in (f(x), S(x,t)), \quad t \in (0, T),$$

where S(x,t) is the elevation function describing the free surface, f(x) is the channel bed shape function, p is the relative pressure and ρ is the density.

Let us denote as $\hat{S}_k =: \hat{S}(x_k, t)$ elevation measurements given at some points $x_k \in (0, L)$; and as $\hat{u}_{l,m} =: \hat{u}(x_l, Y_m, t)$ - u-velocity measurements at some points $x_l \in (0, L)$ along trajectories $Y_m =: Y_m(x_l, t) \in (f(x), S(x,t))$. We will use boundary control to assimilate the data given by considering the objective functional:

$$J = \frac{1}{2}\sum_k \int_0^T \left(S(x_k, t) - \hat{S}_k\right)^2 dt + \frac{1}{2}\sum_l \sum_m \int_0^T \left(u(x_l, Y_m(x_l, t), t) - \hat{u}_{l,m}\right)^2 dt \quad (5)$$

The control boundary is assumed at x = 0, and the passive boundary at x = L, although this could be changed. Then, the initial conditions are simply:

$$S(x,0) = \bar{S}, u(x,y,0) = 0, v(x,y,0) = 0 \quad (6)$$

Boundary conditions for the free surface and for the bed both follow from certain physical considerations. These conditions at the bed and surface are

$$u(x, f(x), t) = 0, \quad v(x, f(x), t) = 0 \quad (7)$$

$$p(x, S(x,t), t) = 0 \quad (8)$$

The boundary conditions for elevation and for velocities at the control boundary should be of the Dirichlet type i.e.

$$S_0 =: S(0,t), \quad u_0(y,t) =: u(0,y,t), \quad v_0 =: v(0,y,t), \quad y \in (0, S(t)) \quad (9)$$

and for the passive boundary they could be different depending on the problem being considered. For a progressive wave model a radiation conditions could be used:

$$\frac{\partial u(L,y,t)}{\partial t} + c\frac{\partial u(L,y,t)}{\partial x} = 0, \quad y \in (0, S(L,t), t) \quad (10)$$

where c is the wave speed, otherwise a Dirichlet or Neumann type can be used.

It is intended to consider $S_0(t)$ and $u_0(y,t)$, $v_0(y,t)$, $y \in (0, S_0(t))$ as boundary control functions. The main difficulty here is that an independent variation in $S(x,t)$ cannot be made. Indeed the surface is part of the fluid and cannot exist without the body of fluid. This is a distinguishing point between the problem (1)-(4) and shape design problems (see Jameson, 1995). This can be expressed as $\partial\tilde{u}/\partial\tilde{S} \neq 0$ and $\partial\tilde{v}/\partial\tilde{S} \neq 0$, here $\widetilde{S}$, $\widetilde{u}$ and $\widetilde{v}$ variations in S, u and v respectively. In general the simplest form of boundary control follows from the assumption that $S_0(t)$ is a know function, i.e. $\widetilde{S}(0,t)=0$. In such a case the flow can be controlled by velocities u_0(y,t) and v_0(y,t), assuming that we know the appropriate sensitivities. The problem becomes less clear when S_0(t) is unknown; the sensitivity $\partial J/\partial\widetilde{S}$ may not be applied, even if it could be found. The remedy is to re-formulate the problem in a new co-ordinate system in which the domain boundary is fixed. That is the original domain with a variable and initially unknown surface boundary S(x,t) is replaced by a transformed domain which has a fixed boundary.

Co-ordinate Transformation

Let the original co-ordinate system be denoted by $\mathbf{R}(x,y,t)$ and then introduce a new co-ordinate system $\mathbf{R}'(x',y',t')$ related to $\mathbf{R}$ by the transformation Q: $\mathbf{R}\Rightarrow\mathbf{R}'$ as follows:

$$t'=t,\quad x'=x\quad y'=\frac{y-f(x)}{S(x,t)-f(x)} \tag{11}$$

assuming $S(x,t)>f(x)$. The vertical datum (origin of y) is arbitrary but is below f(x); $(S-f)=H$ which is the total depth of water. The transformation (11) is unique and the inverse transformation $Q^{-1}:\mathbf{R}'\Rightarrow\mathbf{R}$ exists. Using (11) the transformed derivatives can be written:

$$\frac{\partial\bullet}{\partial t}=\frac{\partial\bullet}{\partial t'}+\frac{h_1'}{S-f}\frac{\partial\bullet}{\partial y'},\quad \frac{\partial\bullet}{\partial x}=\frac{\partial\bullet}{\partial x'}+\frac{h_2'}{S-f}\frac{\partial\bullet}{\partial y'},\quad \frac{\partial\bullet}{\partial y}=\frac{1}{S-f}\frac{\partial\bullet}{\partial y'} \tag{12}$$

where

$$h_1'=y'\frac{\partial S}{\partial t'},\quad h_2'=\frac{\partial f}{\partial x'}+y'\frac{\partial(S-f)}{\partial x'},$$

and the inverse transformation of the derivatives is:

$$\frac{\partial\bullet}{\partial t'}=\frac{\partial\bullet}{\partial t}+h_1\frac{\partial\bullet}{\partial y},\quad \frac{\partial\bullet}{\partial x'}=\frac{\partial\bullet}{\partial x}+h_2\frac{\partial\bullet}{\partial y},\quad \frac{\partial\bullet}{\partial y'}=(S-f)\frac{\partial\bullet}{\partial y} \tag{13}$$

where

$$h_1=\frac{y-f}{S-f}\frac{\partial S}{\partial t},\quad h_2'=\frac{\partial f}{\partial x}+\frac{y-f}{S-f}\frac{\partial(S-f)}{\partial x}$$

The transformed domain $\mathbf{R}'$ is a rectangular spatial domain $[0,L]\times[0,1]$, i.e. the varying S(x,t) free-surface is replaced by a fixed boundary.

The Transformed Problem and Controlability

As mentioned above in the problem statement, an independent variation in S does not exist in $\mathbf{R}$. So in order to control the solution using data $\hat{S}_k$, the problem must be reformulated in $\mathbf{R}'$ using the transformation Q.

In order to proceed, the free surface must be described by a generalised function

$S(x,y,t)$. This is required so that E_1 can be rewritten in integral form using δ - function

$$E_1 =: 2\int_f^S \left(\frac{\partial S}{\partial t} + u\frac{\partial S}{\partial x} - v \right)\delta(y-S)dy = 0 \tag{14}$$

The objective functional, eq. 5, can also be represented with δ -function as follows:

$$\begin{aligned} J &= \frac{1}{2}\sum_k \int_0^T\int_0^L \left(S-\hat{S}_k\right)^2 \delta(x-x_k)dxdt \\ &+ \frac{1}{2}\sum_l\sum_m \int_0^T\int_0^L\int_f^S \left(u-\hat{u}_{l,m}\right)^2 \delta(x-x_l)\delta(y-Y_m)dydxdt \end{aligned} \tag{15}$$

Applying the transformation (12) to (15) and (2)-(4) gives

$$E'_1 =: 2\int_0^1 \left(\frac{\partial S}{\partial t'} + u'\frac{\partial S}{\partial x'} - v' \right)\delta(y'-1)dy' = 0 \tag{16}$$

$$E'_2 =: \frac{\partial u'}{\partial x'} - \frac{h'_2}{S-f}\frac{\partial u'}{\partial y'} + \frac{1}{S-f}\frac{\partial v'}{\partial y'} = 0 \tag{17}$$

$$E'_3 =: \frac{\partial u'}{\partial t'} + u'\frac{\partial u'}{\partial x'} + \frac{v'-h'_1-u'h'_2}{S-f}\frac{\partial u'}{\partial y'} + \frac{\partial p'}{\partial x'} - \frac{h'_2}{S-f}\frac{\partial p'}{\partial y'} = 0 \tag{18}$$

$$E'_4 =: \frac{\partial v'}{\partial t'} + u'\frac{\partial v'}{\partial x'} + \frac{v'-h'_1-u'h'_1}{S-f}\frac{\partial v'}{\partial y'} + \frac{1}{S-f}\frac{\partial p'}{\partial y'} + g' = 0 \tag{19}$$

$$x' \in (0,L),\quad y' \in (0,1),\quad t' \in (0.T),$$

where p', u' and v' are new variables related to p, u and v by mapping:

$$u'(x',y',t') = u(x, f+y'(S-f), t),\ \text{etc.} \tag{20}$$

The objective functional (15) becomes:

$$\begin{aligned} J' &= \frac{1}{2}\sum_k \int_0^T\int_0^L \left(S-\hat{S}_k\right)^2 \delta(x'-x'_k)dxdt \\ &+ \frac{1}{2}\sum_l\sum_m \int_0^T\int_0^L\int_0^1 \left(u'-\hat{u}_{l,m}\right)^2 \delta(x'-x'_l)\delta(y'-Y'_m)dy'dx'dt' \end{aligned} \tag{21}$$

It can be seen now that the velocities $u'(x',y',t')$ and $v'(x',y',t')$ are defined in the space region $(0,L)\times(0,1)$. That is, the free surface height $S(x',t')$ is no longer part of the velocity definition. Therefore an independent variation in $S(x',t')$ now exists in the transformed domain. It may be concluded that the control problem may be stated and solved for the transformed problem which is to minimise (21) subjected to the contraints of (16)-(19).

The Tangent Linear Model

The forward model, TLM and the adjoint model could all be formulated and solved in a consistent way in domain $\mathbf{R}'$. However, in practice, some existing solvers for free surface flows work in the original domain $\mathbf{R}$. Therefore, for practical reasons and to simplify the numerical implementation, the TLM and related adjoint problem derived in $\mathbf{R}'$ will be re-formulated in $\mathbf{R}$ using the inverse transformation Q^{-1}.

First express the complete local variation in the transformed equations in $\mathbf{R}'$ in a general form as follows. Let us assume that $\mathrm{A}\times w$ denotes an operator A acting at a variable w, and E_1' is one of eqs. (17)-(19). Then

$$\partial E_i' = \frac{\partial E_i'}{\partial u'} \times \widetilde{u}' + \frac{\partial E_i'}{\partial v'} \times \widetilde{v}' + \frac{\partial E_i'}{\partial p'} \times \widetilde{p}' + \frac{\partial E_i'}{\partial S} \times \widetilde{S}' \tag{22}$$

where $\widetilde{S}', \widetilde{p}', \widetilde{u}', \widetilde{v}'$ are variations in variables S', p', u', v' respectively and formally $w = \{S', p', u', v'\}$.

The complete local variation in **R** is found by the inverse transformation as follows:

$$Q^{-1} \partial E_i' = \frac{\partial E_i}{\partial u} \times \widetilde{u} + \frac{\partial E_i}{\partial v} \times \widetilde{v} + \frac{\partial E_i}{\partial p} \times \widetilde{p} + Q^{-1} \frac{\partial E_i'}{\partial S} \times \widetilde{S}' \tag{23}$$

It is emphasised that ∂E_i does not exist so $Q^{-1} E_i'$ must be considered instead. So from this point of view the TLM is really a pseudo-TLM and the adjoint to be derived is a pseudo-adjoint.

Looking at (22) and (23) shows that the form of the local variations with respect to *u*, *v* and *p* in **R′** and **R** are equivalent. It is only the variations with respect to *S*; that is, the terms relating to the free surface that are different. Therefore, it is possible to derive the TLM directly in **R** for the variations in *u*, *v* and *p*. So this part will be the well known TLM of the Navier-Stokes equations without free surface. It is only the part dealing with variations in *S* that needs to be first found in **R′** and then transformed back to **R**.

Noting, for i=1 in (23) and for E'_1 given in (16), that the operator created by differentiation of E'_1 comprises:

$$\frac{\partial E_1'}{\partial u'} = 2\int_0^1 \left(\frac{\partial S}{\partial x'} \bullet \right) \delta(y'-1) dy', \quad \frac{\partial E_1'}{\partial v'} = -2\int_0^1 \bullet\, \delta(y'-1) dy'$$

$$\frac{\partial E_1'}{\partial S} = 2\int_0^1 \left(\frac{\partial \bullet}{\partial t} + u' \frac{\partial \bullet}{\partial x'} \right) \delta(y'-1) dy'$$

Then the local variation in E_1 is

$$\partial E_1' = 2\int_0^1 \left(\frac{\partial \widetilde{S}'}{\partial t'} + u' \frac{\partial \widetilde{S}'}{\partial x'} + \frac{\partial S}{\partial x'} \widetilde{u}' - v' \right) \delta(y'-1) dy' = 0 \tag{24}$$

and in **R** it is

$$Q^{-1} \partial E_1' = 2\int_f^S \left(\frac{\partial \widetilde{S}'}{\partial t'} + u \frac{\partial \widetilde{S}'}{\partial x'} + \frac{\partial S}{\partial x'} \widetilde{u} - \widetilde{v} \right) \delta(y-S) dy = 0 \tag{25}$$

Following (23) first calculate

$$\frac{\partial \bullet}{\partial p} \times \widetilde{p} + \frac{\partial \bullet}{\partial u} \times \widetilde{u} + \frac{\partial \bullet}{\partial v} \times \widetilde{v}$$

which is the TLM for the NS equations without free-surface in **R**. This gives

$$Q^{-1} \partial E_2' = \frac{\partial \widetilde{u}}{\partial x} \times \widetilde{u} + \frac{\partial \widetilde{v}}{\partial y} \times \widetilde{v} + Q^{-1} \frac{\partial E_2'}{\partial S} \times \widetilde{S}' \tag{26}$$

$$Q^{-1}\partial E_3' = \frac{\partial \tilde{u}}{\partial t} + u\frac{\partial \tilde{u}}{\partial x} + \tilde{u}\frac{\partial u}{\partial x} + v\frac{\partial \tilde{u}}{\partial y} + \tilde{v}\frac{\partial u}{\partial y} + \frac{\partial \tilde{p}}{\partial x} + Q^{-1}\frac{\partial E_3'}{\partial S}\times \tilde{S}' = 0 \qquad (27)$$

$$Q^{-1}\partial E_4' = \frac{\partial \tilde{v}}{\partial t} + u\frac{\partial \tilde{v}}{\partial x} + \tilde{u}\frac{\partial v}{\partial x} + v\frac{\partial \tilde{v}}{\partial y} + \tilde{v}\frac{\partial v}{\partial y} + \frac{\partial \tilde{p}}{\partial y} + Q^{-1}\frac{\partial E_4'}{\partial S}\times \tilde{S}' = 0 \qquad (28)$$

$$x \in (0,L), y \in (f,S), t \in (0,T)$$

The extra terms in the free surface S $Q^{-1}\frac{\partial E_2'}{\partial S}\times \tilde{S}'$, $Q^{-1}\frac{\partial E_3'}{\partial S}\times \tilde{S}'$ and $Q^{-1}\frac{\partial E_4'}{\partial S}\times \tilde{S}'$ require special treatment which reveals that the operators are:

$$Q^{-1}\frac{\partial E_2'}{\partial S} = (S-f)\left(\frac{\partial u}{\partial y}Q^{-1}\frac{d}{dS}\left(\frac{h_2'}{(S-f)}\right) + \frac{\partial v}{\partial y}Q^{-1}\frac{d}{dS}\left(\frac{1}{(S-f)}\right)\right)$$

$$Q^{-1}\frac{\partial E_3'}{\partial S} = (S-f)\left(\frac{\partial u}{\partial y}Q^{-1}\frac{d}{dS}\left(\frac{v'-h_1'-u'h_2'}{(S-f)}\right) + \frac{\partial p}{\partial y}Q^{-1}\frac{d}{dS}\left(\frac{h_2'}{(S-f)}\right)\right)$$

$$Q^{-1}\frac{\partial E_4'}{\partial S} = (S-f)\left(\frac{\partial v}{\partial y}Q^{-1}\frac{d}{dS}\left(\frac{v'-h_1'-u'h_2'}{(S-f)}\right) + \frac{\partial p}{\partial y}Q^{-1}\frac{d}{dS}\left(\frac{1}{(S-f)}\right)\right)$$

where

$$\frac{d}{dS}\left(\frac{1}{(S-f)}\right) = -\frac{1}{(S-f)^2},\quad \frac{d}{dS}\left(\frac{h_2'}{(S-f)}\right) = \frac{y'}{(S-f)}\frac{\partial \bullet}{\partial x'} - \frac{h_2'}{(S-f)^2},$$

$$\frac{d}{dS}\left(\frac{v'-h_1'-u'h_2'}{(S-f)}\right) = \frac{y'}{(S-f)}\left(\frac{\partial \bullet}{\partial x'} + u'\frac{\partial \bullet}{\partial x'}\right) - \frac{v'-h_1'-u'h_2'}{(S-f)^2}$$

Using the inverse transform fomulae for Q^{-1} and collecting terms, the free surface terms in the TLM for E_2, E_3 and E_4 in domain **R** can be expressed neatly as:

$$Q^{-1}\frac{\partial E_i'}{\partial S}\times \tilde{S}' = a_{i,1}\frac{\partial \tilde{S}'}{\partial t} + a_{i,2}\frac{\partial \tilde{S}'}{\partial x} + a_{i,3}\frac{\partial \tilde{S}'}{\partial y} + a_{i,4}\tilde{S}' \quad i \in (2,4) \qquad (29)$$

where the coefficients $a_{i,j}$ can be evaluated from the expressions given above.

The (pseudo) Adjoint model

Define a set of adjoint variables $w'^{*} = \{ S^{*}, p'^{*}, u'^{*}, v'^{*} \}$ in **R′** and like S it is the case that S^{*} is independent of y', i.e. $\partial S^{*}/\partial y' = 0, \forall y'$. Then construct the total variation in **R′** in the usual way:

$$\Lambda' = \int_0^T\int_0^L S^{*}\partial E_1' dx'dt' + \int_0^T\int_0^L\int_0^1 \left(p'^{*}\partial E_2' + u'^{*}\partial E_3' + v'^{*}\partial E_4'\right)(S-f)dy'dx'dt' + \frac{\partial J'}{\partial S}\times \tilde{S}' + \frac{\partial J'}{\partial u'}\times \tilde{u}' \qquad (30)$$

The corresponding variation in **R** is:

$$\Lambda = Q^{-1}\int_0^T\int_0^L S^{*}\partial E_1' dx'dt' + \int_0^T\int_0^L\int_0^1 \left(p^{*}Q^{-1}\partial E_2' + u^{*}Q^{-1}\partial E_3' + v^{*}Q^{-1}\partial E_4'\right)dydxdt + Q^{-1}\frac{\partial J'}{\partial S}\times \tilde{S}' + Q^{-1}\frac{\partial J'}{\partial u'}\times \tilde{u}' \qquad (31)$$

The first term in (31) after transforming back into **R** produces the adjoint terms:

$$-2\left(\frac{\partial S^*}{\partial t}+\frac{\partial uS^*}{\partial x}+h_2\frac{\partial u}{\partial y}S^*\right)\delta(y-S)\widetilde{S}' \quad (32)$$

$$+2\frac{\partial S}{\partial x}S^*\delta(y-S)\widetilde{u} \quad (33)$$

$$-2S^*\delta(y-S)\widetilde{v} \quad (34)$$

Applying transformations $Q^{-1}E_2'$, $Q^{-1}E_3'$, $Q^{-1}E_4'$ given by (26)-(28) to all the terms in those equations except the last ones in $\widetilde{S}'$, performing integration by parts and collecting results around $\widetilde{p},\widetilde{u},\widetilde{v}$, while remembering the terms in (33) and (34), gives the following adjoint equations:

$$\frac{\partial u^*}{\partial x}+\frac{\partial v^*}{\partial y}=0 \quad (35)$$

$$-\frac{\partial u^*}{\partial t}-u\frac{\partial u^*}{\partial x}-v\frac{\partial u^*}{\partial y}-u^*\frac{\partial v}{\partial y}+v^*\frac{\partial v}{\partial x}-\frac{\partial p^*}{\partial x}+r_u+2\frac{\partial S}{\partial x}S^*\delta(y-S)=0 \quad (36)$$

$$-\frac{\partial v^*}{\partial t}-u\frac{\partial v^*}{\partial x}-v\frac{\partial v^*}{\partial y}-v^*\frac{\partial u}{\partial x}+u^*\frac{\partial u}{\partial y}-\frac{\partial p^*}{\partial y}-2S^*\delta(y-S)=0 \quad (37)$$

$$x\in(0,L), y\in(f,S), t\in(0,T)$$

where $r_u=\sum_l\sum_m(u-\hat{u}_{l,m})\delta(x-x_l)\delta(y-y_l)$

Note that without the terms from (33) and (34) these are the familiar adjoint equations to the NS equations, i.e. adjoint to the set (2)-(4). The process of integration by parts also produces the boundary set that is used for deriving adjoint boundary conditions. Now the adjoint equation for S^* must be found by collecting variations about $\widetilde{S}$. These are collected from terms in (32) and (29). The adjoint equation for S^* is

$$\frac{\partial S^*}{\partial t}-\frac{\partial uS^*}{\partial x}-\frac{\partial S}{\partial x}\frac{\partial u}{\partial y}S^*-\sum_2^4 F_i+\sum_k(S-\hat{S}_k)\delta(x-x_k)=0 \quad (38)$$

where from (29)

$$F_i=\int_f^S\left(\frac{\partial a_{i,1}w_i^*}{\partial t}+\frac{\partial a_{i,2}w_i^*}{\partial x}+\frac{\partial a_{i,3}w_i^*}{\partial y}-a_{i,4}w_i^*\right)dy \quad i\in 2,4 \quad (39)$$

where vector $w^*=\{S^*,p^*,u^*,v^*\}$

The adjoint initial (terminal) and boundary conditions are

$$S^*(x,T)=0,\quad u^*(x,y,T)=0,\quad v^*(x,y,T)=0 \quad (40)$$

$$u^*(x,f,t)=0,\quad v^*(x,f,t)=0 \quad (41)$$

$$p^*(x,S,t)=0 \quad (42)$$

The radiative adjoint boundary condition similar to (10) has to be applied to u^*,v^* and S^* at $x=L$.

Taking all of this into account it is now possible to write down what is left of the total variation in **R** originally stated in (30), and then finally in $\mathbf{R}'$. As a result we have the formula for sensitivities at the control boundary as follows

$$\frac{\partial \Lambda'}{\partial \widetilde{u}_0'} = -[(S-f)(p'^* + u'u'^*)]_{x'=0} \quad (43)$$

$$\frac{\partial \Lambda'}{\partial \widetilde{v}_0'} = -[(S-f)u'v'^*]_{x'=0} \quad (44)$$

$$\frac{\partial \Lambda'}{\partial \widetilde{S}'} = -u'(0,1,t')S^*(0,t') - (S(0,t') - f(0))\int_0^1 \left[a_{2,2}p'^* + a_{3,2}u'^* + a_{4,2}v'^*\right]_{x'=0} dy' \quad (45)$$

Trial Implementation

In summary, the above discussion describes the following situation and methodology to be used to assimilate data into a forward FSNS model:

- The physical problem exists in the original domain **R**. Existing solver are available in **R**
- However, the control problem with free-surface cannot be solved in **R** because there are no independent variations in u and S.
- Therefore, the domain is transformed into a fixed domain **R′** where the control problem can be solved.
- The TLM and adjoint are formulated in **R′** and transformed back into **R**.
- Both the TLM and adjoint have equivalent forms in **R′** and **R** except for the terms in the free surface S which are treated specially as described above.
- The adjoint equations are solved in **R** using adaptations of existing solvers for the forward FSNS problem.
- This produces values of adjoint variables $w^* = \{\ S^*, p^*, u^*, v^*\ \}$
- The values $w^*_{x=0} = \{\ S^*, p^*, u^*, v^*\ \}_{x=0}$ at the control boundary are transformed back into **R′** to produce $w'^*_{x=0} = \{\ S'^*, p'^*, u'^*, v'^*\ \}_{x=0}$
- The sensitivites are found from $w'^*_{x=0}$ and used to adjust u_0', v_0', S_0'.
- New values of u_0', v_0', S_0' are transformed back to give new values of the control variables u_0, v_0, S_0 in the forward model which is run again.
- Objective functional J is evaluated.
- Iterations progress until J reaches a minimum.

A trial implementation was made using a finite difference solution of the FSNS in 2D vertical section by Neill at al (2004). The test problem was a simple rectangular domain of length 2km (L =2,000m) and uniform depth 40m ($(S - f)$ = 40 m). The driving boundary condition is the inlet u-velocity uniform through the depth that varies in time as follows

$$u_0(t) = U_0(1 + \sin(\pi(3/2 + t/T_0)))/2.$$

where T_0 is the wave period, and U_0 is the amplitude. The component $v(0,y,t)$ is related to $u(0,y,t)$ assuming the incoming perturbation is a wave as follows

$$v(0,y,t) = \frac{y}{c}\frac{\partial u_0(t)}{\partial t}$$

where $c=\sqrt{g(S(0,t)-f(0))}$ is a wave speed, and the boundary condition on the elevation function is $\partial S(0,t)/\partial x=0$ (long wave condition). For the given U_0, T_0 the forward solution $u(x,y,t)$, $S(x,t)$ taken at some points x_k, y_m forms pseudo-measurements $\hat{u}(x_k, y_m, t)$ and $\hat{S}(x_k, t)$ that are used as driving sources for the adjoint problem. This approach is often referred to as 'twin numerical experiment'. Although many cases were considered, we have shown here a simple case when $U_0 = 1.0 m/\sec$, $T_0 = 120.0 \sec$, and both the elevation and velocity sensors are located at $x_1 = L/2 = 1000.0m$, and the velocity sensor is located at $y_1 = (S-f)/2 = 20.0m$. The adjoint problem is solved for a trivial initial guess that is $u, v \equiv 0$, $S = \bar{S}$ and $p = (S-y)g$. The results are presented at Fig.2.

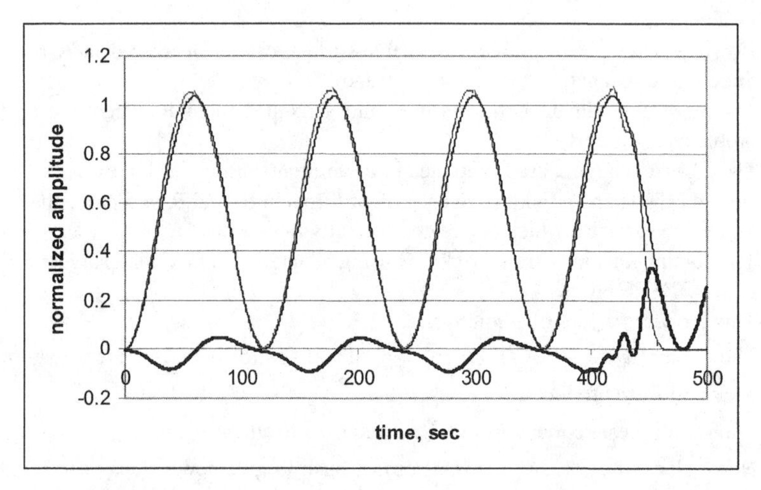

Figure 2. The driving boundary condition u_0' (single solid line), related sensitivity $\partial J'/\partial u_0'$ (dotted line) and the mismatch (double solid line), calculated for a trivial initial guess. Both velocity and elevation sensors are used.

The normalized value $u_0/k_n(u_0)$, where

$$k_n(u)=\frac{1}{T}\int_0^T |u|dt$$

is given by a solid single line, the corresponding normalized sensitivity (see eq.43) - by dotted line, and the mismatch between them – by solid double line. One can apparently see a "blind" spot located at the end of the time interval between 470-500 sec., where there is no 'signal'. This is because the perturbation from the sensors has not reached the control boundary. Then oscillations (~420-470 sec) are related to the shock input. After these oscillations pass beyond the domain of interest, the mismatch becomes a regular, smooth function about 10% of u_0. This mismatch is due to a non-linearity related to a wrong initial guess, mostly in S. This is proven by the fact that if

the pseudo-measurements are generated by the TLM solver based on the same initial guess (which of course removes the non-linearity), the regular mismatch is less that 0.1%. The residual value can be explained by small numerical effects, such as numerical viscosity, etc.
The results presented correspond to the case when both the elevation and velocity sensors are simultaneously used. If they are used separately, the results are essentially the same and could not be distinguished from results given in Fig.2. A structure of the adjoint pressure field delivering the sensitivity information in the two cases is given in Fig.3 and Fig.4. The adjoint pressure is taken at $y = 20.0m$, but it is almost uniform through the depth. This will be not the case for short waves nor for more complicated situations including sudden changes in the bed shape, for example. In Fig.3 we can see p^* - field, when the adjoint solution is driven by the source in u^* - momentum equation (36). One can clearly see the wave and the anti-wave running in opposite directions from the source located at x=1000.0m . For more details the cross-section of the field at $t \approx 130$ sec. is also presented, which shows the strong shear in p^* at x=1000.0m . In contrast, the adjoint solution driven by the elevation source in (38), see Fig.4, is remarkably smooth, and therefore requires less computational effort to be obtained properly. Other experiments performed later show an essential superiority of elevation measurements for producing stable and accurate gradients.
Further work is being carried out to determine the response of the adjoint system and its effectiveness in adjusting the control variables in the forward model. Its effectiveness in assimilating water surface data and in controlling the free-surface will be of particular interest.

Acknowledgements

The authors are grateful for the help and support of the following:
Prof. Tony Goddard, Dr. Matthew Piggott and Gerard Gorman, Department of Earth Science and Engineering, South Kensington Campus, Imperial College London.
Prof. Michael Navon, Department of Mathematics and Supercomputer Computations Research Institute, Florida State University, Tallahassee, Florida, USA.
Prof. Victor Shutyaev, Institute of Numerical Mathematics, Russian Academy of Sciences, Russia.
The Engineering & Physical Sciences Research Council for funding through grants GR/R60881 and GR/R60898.

References

Copeland, G.J.M.,1998. *Coastal Flow modelling using an inverse method with direct minimisation*. Proc. Conf. Estuarine and Coastal Modeling, 1997. Ed. M.L.Spaulding & A.F. Blumberg, Pub. ASCE (ISBN 0-7844-0350-3) pp.279-292.

Ford, R., Pain, C.C., de Oliveira, C. R. E., Goddard, A.J.H. and Umpleby, A., *A 3-D finite element ocean model*, EGS XXV General Assembly, Nice, France, April 2002.

Jameson, A., 1995, *Optimum aerodynamic design using CFD and control theory*, AIAA Paper 95-1729-CP

Neill S.P., Copeland G.J.M., Ferrier G., Folkard A.M. (2004). *Observations and numerical modelling of a non-buoyant front in the Tay Estuary, Scotland*, Estuarine. Coastal and Shelf Science, 59, pp. 173-184.
Piggott, M., 2003. [http://amcg.th.ic.ac.uk/people/staff/MatthewPiggott/ matthew.piggott@imperial.ac.uk].
Sanders, B.F and Katopodes, N.D., 2000 *Adjoint Sensitivity analysis for shallow-water wave control.* J. Eng. Mech., ASCE, pp 909-919.

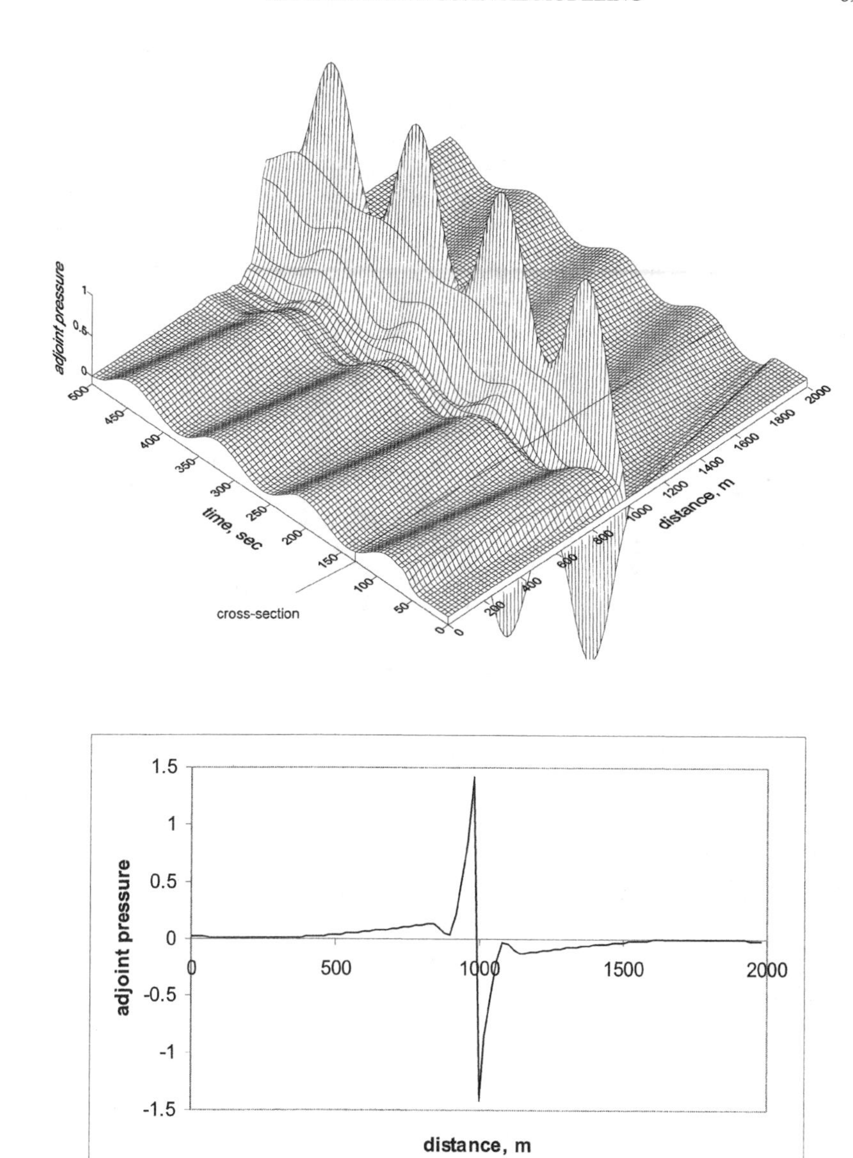

Figure 3. The adjoint pressure field p^*, in case u -velosity measurements are used. Below – the field cross-section at $t \approx 130$ sec.

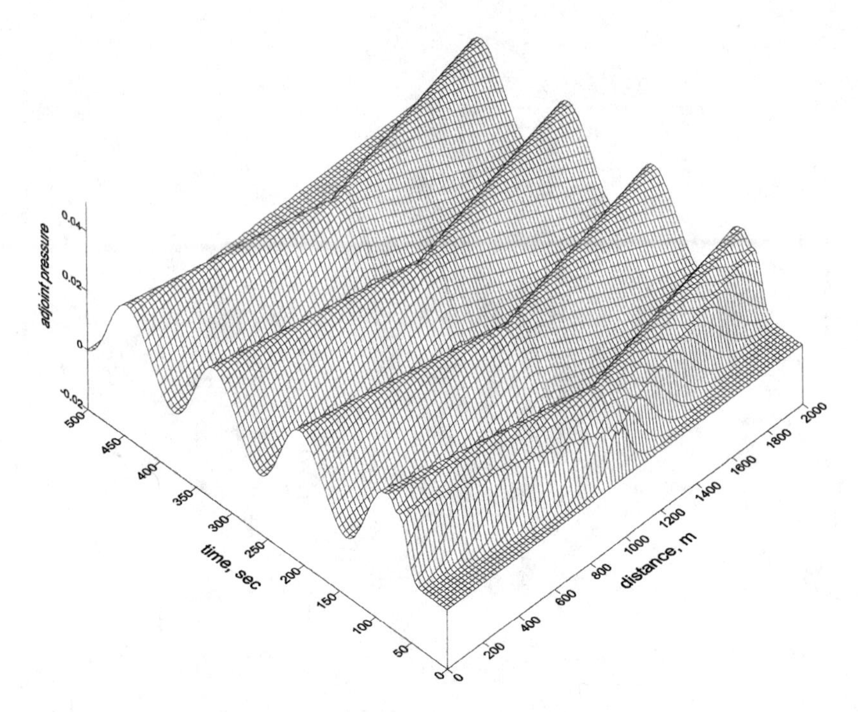

Figure 4. The adjoint pressure field p^*, in case the elevation measurements are used.

Volume Analysis for Attainability of Water Quality Criteria for Three Loading Constituents

Ping Wang[1], Lewis Linker[2], James Collier[3], Gary Shenk[2], Robert Koroncai[2], Richard Batiuk[2]

Abstract

This paper presents a method to analyze the outputs of an estuarine water quality model to determine the attainability of water quality criteria for designated-use-areas in the Chesapeake Bay estuary using different levels of reduction of nitrogen, phosphorus, and sediment loads. From a set of water quality model scenarios with different nitrogen (N), phosphorus (P) and sediment (S) loads, a relationship between a water quality index (W) and the loading-constituents N-P-S, i.e., W = f (N, P, S), in a designated-use-area, can be established through regression. In a designated-use-area, we may obtain a percentage of non-attainment (or violation, V) of a water quality criterion accumulated over time and water-volume versus the entire simulated time and volume for a model scenario. From a set of model scenarios, a relationship between W and V, i.e., W = g(V), can be established. The value of the water quality index (Wo) equivalent to near-zero violation (Vo) of the criteria can be projected, i.e., Wo = g(Vo). Using W = Wo as a constraint for the function W = f(N, P, S), we can plot a 3-dimensional surface of minimum water quality attainment, Wo, versus N, P, and S loads for the designated-use-area. This surface divides the 3-dimensional N-P-S "loading volume" into regions of attainment and non-attainment of the water quality criterion in the designated-use-area. This plot can be used to analyze tradeoffs among N-P-S load reductions that still ensure attainment of the water quality criteria, providing useful information for optimal nutrient and sediment controls in water quality management.

Introduction

Several methods have been developed to analyze the response of water quality to nutrient and sediment loads (Malanotte-Rizzoli, 1996; Thomann and Mueller, 1987; Emery and Thomson, 1998). Usually, a model scenario is used to estimate a water quality response to a specific loading condition. Thomann et al. (1994) used the surface analysis technique, which applied a statistical method on a set of model scenarios of the 8,000 cell Chesapeake Bay Estuarine Model (CBEM), to form a continuous function of response of anoxia volume to nitrogen and phosphorus loads. The surface plots provided a continuous function estimating how the dissolved

[1] University of Maryland Center for Environmental Science, Chesapeake Bay Program, 410 Severn Ave, Suite 109, Annapolis, MD 21403, USA; phone 410-267-5744; pwang@chesapeakebay.net
[2] US EPA, Chesapeake Bay Program Office, 410 Severn Ave., Annapolis, MD 21403, USA
[3] District of Columbia Department of Health, 51 N Street, NE 5th flow, Washington, DC 20002,USA

oxygen (DO) conditions of the Chesapeake Bay respond to total nitrogen (TN, or simplified as N) and total phosphorus (TP, or simplified as P) loads from the watershed. Wang et al. (2001) applied the surface analysis on the 10,000-cell CBEM to analyze the responses of DO, chlorophyll-a, water clarity, filter feeders, and submerged aquatic vegetation (SAV) to TN, TP, and total sediment (TS, or simplified as S) loads. Wang et al. (2001) also used the surface plots to analyze the interactions in the ecosystem and to explain the responses of water quality indices under various nutrient and sediment loading conditions. They also used the surface plots to estimate TN, TP, or TS load reductions corresponding to a specified set of water quality requirements, and to discuss possible tradeoffs between N and P reductions.

The Chesapeake Bay is the largest estuary in the United States, and is one of the most productive estuaries in the world. Degradation of water quality, particularly low dissolved oxygen conditions, was identified to be primarily due to excess nutrient inputs from the 166,000 square kilometer watershed (Fig. 1) (Gillelan et al., 1983). The 1987 Chesapeake Bay Agreement called for a 40% reduction in controllable nutrient loads by 2000 from a 1985 baseline (Chesapeake Executive Council, 1987). The Chesapeake 2000 Agreement set a goal of correcting all nutrient and sediment related problems sufficient to remove the Bay from the list of impaired water by year 2010 (Chesapeake Executive Council, 2000). The Chesapeake 2000 Agreement established a set of management objectives requiring sophisticated computer modeling of the response of Chesapeake water quality to loading in order to establish achievable, equitable, and protective caps on nutrient and sediment load reductions.

The latest version (13,000 cell) three-dimensional Chesapeake Bay Estuary Model (CBEM) has been applied to study the Bay's water quality. Excessive loads of nitrogen, phosphorus, and sediment are causing impairments in the Chesapeake. However, the surface plots of Thomann et al. (1994) and Wang et al. (2001) estimate the response of a water quality criterion to only pairs of loading constituents, e.g., TN and TP, or TN and TS, or TP and TS. This is because a visual 3-D plot is commonly based on a function of a dependent variable (i.e., a water quality criteria such as dissolved oxygen) versus two independent variables (e.g., TN and TP). For the water quality attainment analysis, it is desirable to have a plot showing the relationship of a water quality criterion with the three major loading constituents (TN, TP and TS), so that we can analyze tradeoffs in N, P and S controls. This paper will first present the plot of water quality versus two loading components, then describe how to obtain the plot with three loading components.

Method

The surface analysis in this paper is based on statistics to establish a relationship between a water quality index and loading constituents from water quality models, i.e., using the regression method to derive a function of water quality as a dependent variable and loading constituents as independent variables. A sufficient set of water quality model scenarios with different loading conditions is needed for the regression. In this paper, we use quadratic functions for the best-fit regression, although sometimes a higher order of regression function may be desirable (Wang et al., 2001).

The Chesapeake Bay Estuarine Model (CBEM) is coupled CH3D finite-difference hydrodynamic model (Johnson et al., 1993) and CE-QUAL-ICM finite-volume water quality model (Cerco and Cole, 1994). The 2002 version of CBEM is used, which consists of 12,920 model cells and 2961 surface cells. The model is run for 10 years using 1985-1994 hydrology, with 15-minute time-step and outputs of daily or monthly water quality.

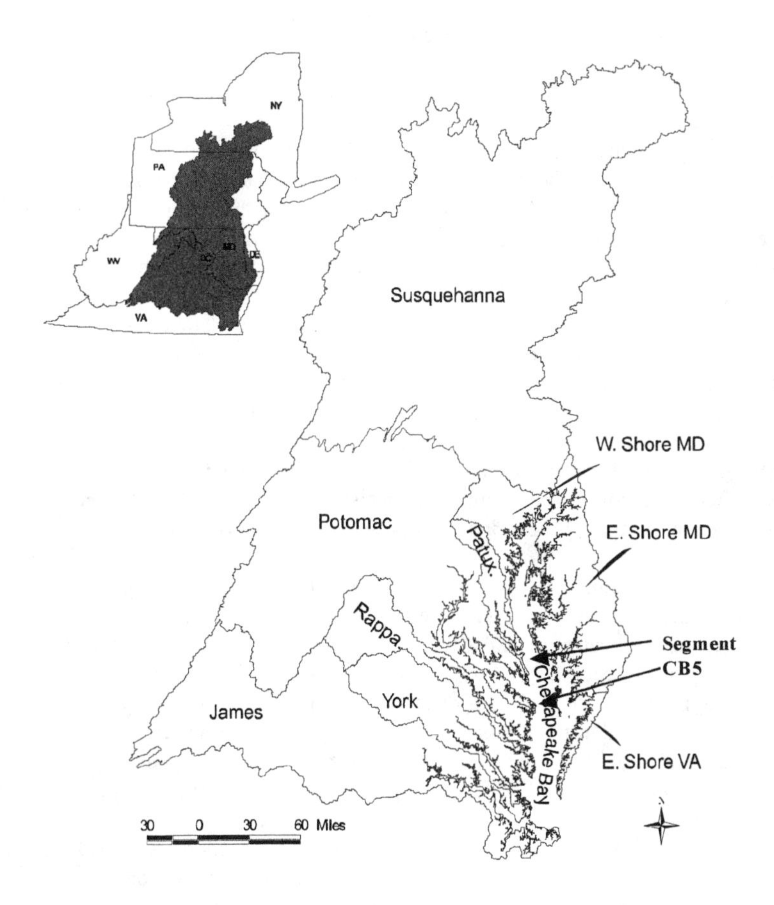

Figure 1. Nine major basins of the Chesapeake Bay watershed.

We used water quality criteria of DO, chlorophyll-a, and clarity to examine if the water quality standard is attained in each designated-use-area of the estuary. A designated-use-area (DU) is represented by a group of adjacent model cells, based on habitat, salinity, and depths (such as a migratory fish region, tidal fresh, mesohaline, or polyhaline regions, or open water, deep water, or deep trench regions).

Constructing Regression Functions for Two Loading Constituents

For a quadratic surface function of a water quality index (W) versus N and P, two loading constituents, we have:

$$W = a N^2 + b N P + c P^2 + d N + e P + f \quad (1).$$

At least 6 model scenarios are required to estimate the parameter values. We used 9 scenarios with different reductions in N and P loads from a reference condition (Figure 2). We chose the 2000 Progress Scenario as our reference condition, as that is the best estimate of the current loads to the Bay, taking into account land use, numbers of animals, point source loads, and best management practices estimated to be extent in the watershed in the year 2000. Though the reference condition is arbitrary, the 2000 Progress Scenario is a useful reference as that's estimated to the high end of the loads over the next two decades as management practices are put in place to remove all water quality impairments to the Bay. The other nine scenarios used to estimate the parameters of the quadratic equation are in some set proportion of the reference 2000 Progress Scenario (PR2000). For example, N60 means that the N load is 60% of the 2000 Progress Scenario; N60P60 is that N and P loads are 60% of the 2000 Progress Scenario; while all other constituents, including sediment, are 100% of the 2000 Progress Scenario. The reductions in the 9 scenarios are the same for all basins Bay-wide to eliminate inconsistency in the load reduction and water quality responses among scenarios (Thomann et al., 1994; Wang et al., 2001).

Figure 3 is a response of annual chlorophyll-a (Chl) concentration in the surface water (i.e., depth under 2 meters) of Segment CB5 (Figure 1) to N and P loads to the Bay. The surface water (or called the open water) of CB5 is denoted as CB5-OW. From the plot we can see how chlorophyll-a concentration changes with N and P loads. Figure 4 is a cross section of the chlorophyll response surface from Figure 3 cut by a plane of chlorophyll-a =5 ug/L. If, for example, the criterion for chlorophyll is <=5 ug/L, the surface below the plane will meet the water quality criterion. The Chl surface intersects with the Chl = 5 ug/l plane to yield a curve, called Chl=5 ug/l curve. The N-P loads at this curve would just meet the minimum water quality standard for chlorophyll. The curve also traces the tradeoffs of N-P loads for least-cost N-P control. Issues in tradeoffs in nitrogen and phosphorus controls are further discussed in Wang et al. (2001) and Wang and Linker (2002).

TN Loads Decrease →

TP Loads Decrease ↓

PR2000 (Prog 2000 N Prog 2000 P)	N70 (70%Pr2000 N Prog 2000 P)	N40 (40%Pr2000 N Prog 2000 P)
P70 (Prog 2000 N 70%Pr2000 P)	N70P70 (70%Pr2000 N 70%Pr2000 P)	N40P70 (40%Pr2000 N 70%Pr2000 P)
P40 (Prog 2000 N 40%Pr2000 P)	N70P40 (70%Pr2000 N 40%Pr2000 P)	N40P40 (40%Pr2000 N 40%Pr2000 P)

Fig. 2. Nine scenarios with different N-P loads used for a surface analysis of two loading constituents. Sediment load = 100% PR2000.

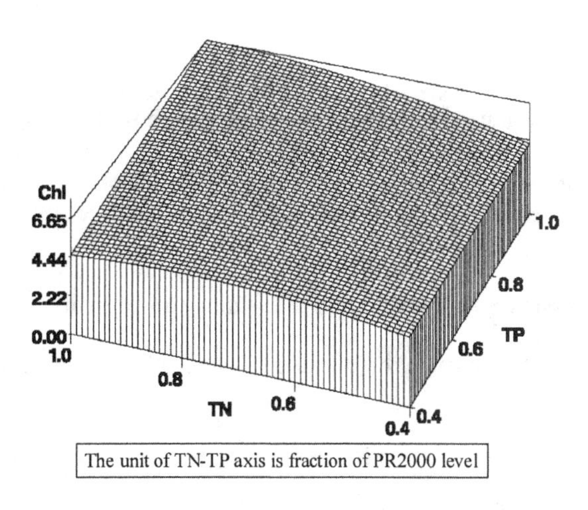

Fig. 3. Response surface of annual surface chlorophyll-a concentration (ug/l) in CB5-OW versus N-P loads to the Bay.
Regression equation: Chl = - 6.05506 N^2 – 0.7889 P^2 + 5.15625 N P + 7.41449 N =- 0.53854 P + 1.46443. $R^2 = 0.9913$.

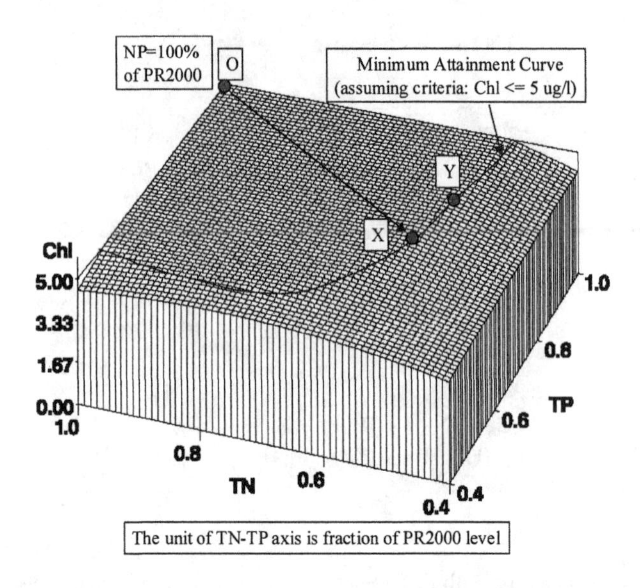

Fig. 4. Chl=5 ug/l curve versus N-P load, which is intersect of Chl=5 ug/l plane and the response surface of Figure 3.

To analyze the response of water quality to sediment load, we need a relationship of water quality, W, versus N-S or P-S loads. For W versus N-S loads, we need another set of 9 scenarios with variable N and S loads; for W versus P-S loads, we need another set of 9 scenarios with variable P and S loads. The following section explores how to plot a response surface versus the three loading variables of N-P-S.

Constructing Regression Functions for Three Loading Constituents

To construct a quadratic function of a water quality, W, versus three loading variables, N, P and S,

$$W = a\,N^2 + b\,P^2 + c\,S2^2 + d\,N\,P + e\,N\,S + f\,P\,S + g\,N + h\,P + i\,S + j \qquad (2)$$

at least 10 scenarios are required to estimate the values of all ten regression parameters a-j. Equation 2 will be symbolized as W = f(N,P,S) in the following discussions. In order to get a better regression function, we designed 27 scenarios with different reductions in N, P and S loads from the 2000 Progress Scenario reference condition, as shown on Figure 5. The notation for the scenarios is similar to that in Figure 2. For example, N70S40 means that N and S loads are at 70% and 40% of PR2000, respectively, while P are at 100% of PR2000. Among the 27 scenarios, the scenario of highest loading is PR2000, i.e., N100P100S100, which is also denoted as NPS in Figure 5. The scenario of lowest loading is N40P40S40, i.e., N, P and S are all at 40% of PR2000.

If the chlorophyll-a criteria is still set at Chl <=5 ug/L, we can use Chl=5 ug/L as a constraint for Equation 2, getting

$$5 = f(N,P.S) \qquad (3).$$

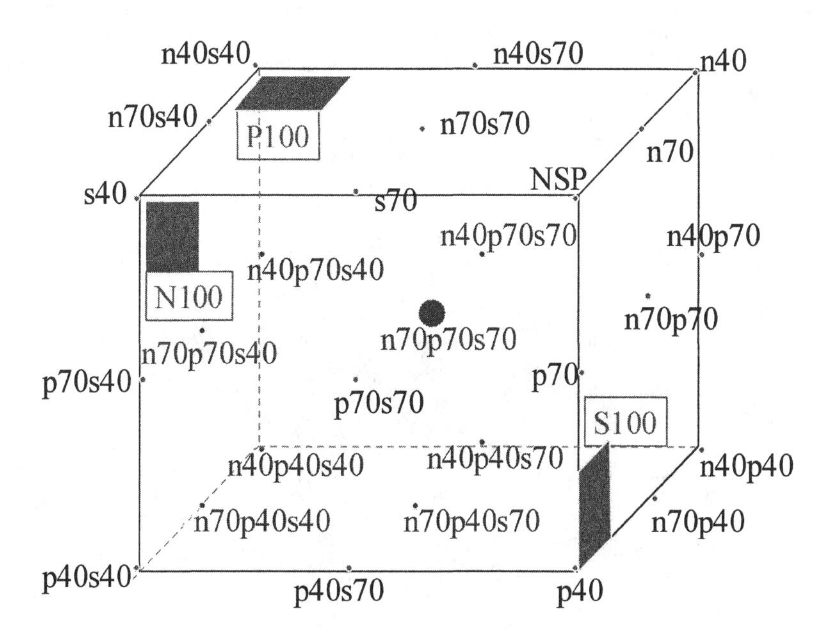

Fig. 5. Twenty-seven scenarios with different N-P-S loads for volume and surface analysis.

Equation 3 can be used to plot a surface of chlorophyll-a minimum attainment versus N, P, and S load (Figure 6) for a designated-use-area. On this surface, the chlorophyll-a concentration of 5 ug/l just meets the minimum required standard. This surface divides the N-P-S loading space into two regions. The loading region at lower left from this surface (i.e., with less N and/or P and/or S load) complies with the criterion (i.e., Chl =< 5 mg/L). Whereas, the loading region at the upper-right from this surface represents non-compliance of the chlorophyll-a criterion.

Let's compare Figures 6 and 4 to further understand their relationship. The minimum attainment surface in Figure 6 has Chl = 5 ug/l. The intersection of this surface and S=1.0 plane (i.e., S load at 100% of PR2000) yields a curve. At this curve, the water meets the minimum requirements of chlorophyll-a criteria. Each point at this curve defines a specific N-P load. This curve displays the tradeoff in N and P load reductions (under a fixed S load at 100% of PR2000) to meet the water quality criteria. The information from Figure 6 is similar to that from Figure 4. The Chl=5 ug/l curve on S=1.0 plane of Figure 6 is equivalent to the Chl=5 ug/l curve in Figure 4. Please note the orientations of the point of NP=100% of PR200 in Figures 4 and 6 are different, therefore, the two Chl=5 ug/l curves in the two figures are concave in opposite directions.

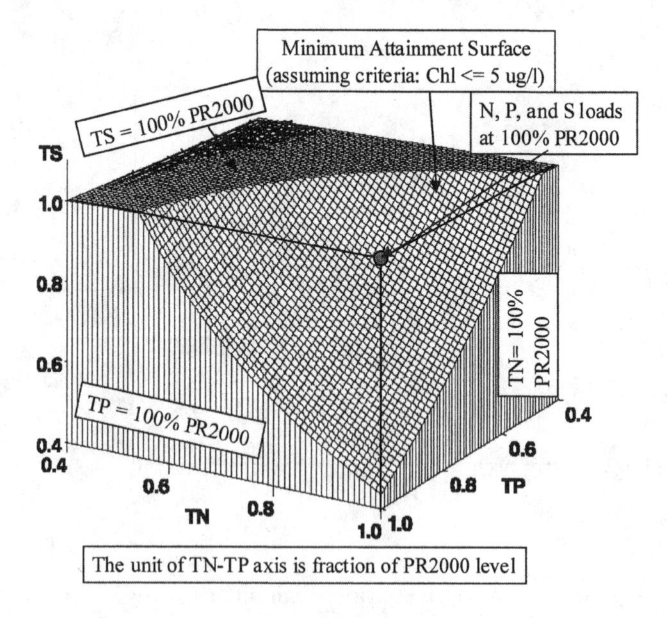

Fig. 6. Minimum attainment surface of chlorophyll-a for CB5-OW versus N-P-S loads to the Bay. Regression equation: $Chl = -4.7403\,N^2 - 0.60107\,P^2 - 0.76098\,S^2 + 4.38582\,N\,P + 2.32756\,N\,S + 0.28994\,P\,S + 3.78113\,N - 0.53286\,P + 1.60750\,S + 0.87620$; $R^2 = 0.9873$.

Both Figures 4 and 6 provide the same information about the dependence of chlorophyll-a on N-P load under sediment loads of 100% of PR2000 condition. Figure 6 can provide more information, such as the information of the dependence of chlorophyll-a on N-P load under various S conditions. The estimated water quality response of reduced algal chlorophyll with reduced sediment is due to increased nutrient uptake in shallow regions with the improving clarity caused by sediment load reductions. We can, for example, use a plane of S=0.8 to cut the surface, yield a curve of Chl=5 ug/l, showing N-P load tradeoff under 20% sediment reduction (i.e., S=80% PR2000) as shown in Figure 7. Figure 7 also plots the Chl=5 ug/l curve at S=100% PR2000, which is comparable to the Chl=5 mg/l curve in the horizontal plane in Figure 6.

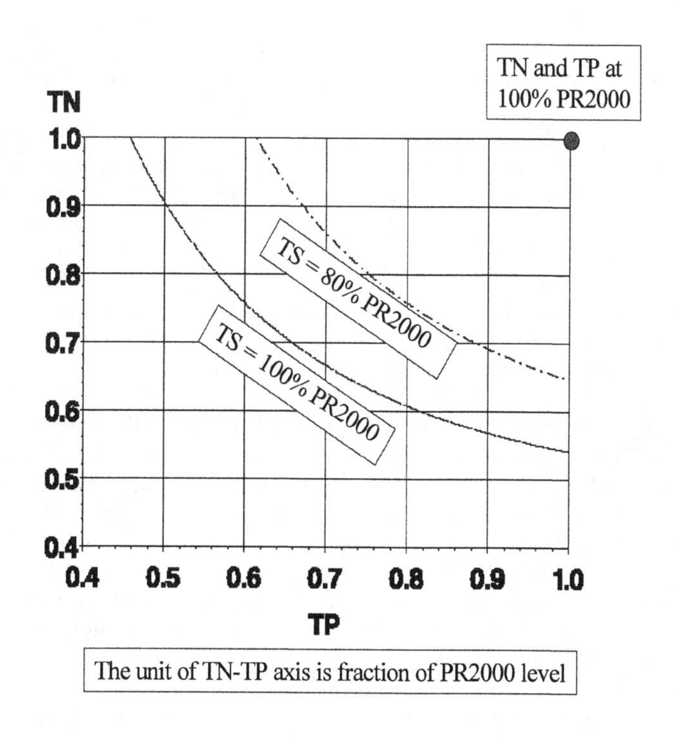

Fig. 7. Minimum attainment curves of chlorophyll-a for CB5-OW versus N-P loads for TS=100% PR2000 and TS=80% PR2000, respectively.

The curve of TS=80% PR2000 is closer to the point of TN-TP at 100% PR2000 than the curve of TS=100% PR2000 (Figure 7). This means that it requires less N and/or P load reduction for compliance of the chlorophyll-a criterion in CB5 for the condition of TS at 20% reduction, whereas, it requires more N and/or P reduction for the condition without TS reduction in order for CB5 to comply with the criteria (CBPO, 2003). Wang et al. (2003) used the surface of SAV versus N, P and S load to analyze the effect of N, P and S on SAV and suggestion of load reduction for SAV restoration.

We can also use a plane of N or P at certain percentage of PR2000 to cut the 3-D plot for other analyses. For example, the two vertical planes in Figure 6 show the dependence of chlorophyll-a on N-S load under P=100% of PR2000 condition, and the dependence of chlorophyll-a on P-S load under N=100% of PR2000 condition, respectively.

Constructing Attainment Surfaces for a Practical Management Case

The above approach uses a chlorophyll-a criterion (for example, less than 5 ug/L) to compare modeled average chlorophyll-a concentration in a designated-use-area (DU) to see whether the water quality standard is achieved. In practical management, a water quality criterion is used to exam individual cells in each time-unit defined by the criterion, instead of comparing the average in the whole DU, or the average in the entire time period. Chlorophyll-a concentration varies in different cells in different seasons. Some cells in some seasons may comply with the criterion, while other cells in other seasons may violate the criterion. An average of 5 ug/L in a DU may not guarantee full compliance of the chlorophyll-a criterion.

Similar situation exists for dissolved oxygen. Variations of DO concentration among model cells in a DU of "Deep Water" are evident. The current DO criterion in the summer season (June, July, August, September) for a Deep-Water DU set by the Chesapeake Bay Program is monthly concentration greater than 3 mg/L. In a month in one model cell if DO is below 3 mg/L, we call it an exceedence of the criterion. The cumulative volume for the model cells in the DU that have criteria exceedence in the 10-year simulation period versus the total cumulative volume over the entire simulation period for the DU is called the violation rate. Violation rate is a fraction, usually expressed as a percentage.

From each model scenario, we can get two pieces of information: 1) percent violation (V), and 2) average DO (W) in this DU. From the 27 scenarios we can establish a relationship between W and V:

$$W = g(V) \quad (5)$$

Table 1 lists W and V of DO in deep water of Segment CB5 (denoted as CB5-DW) for the 27 scenarios. Note: V=0.000%, i.e., non-violation, indicates that all cells in the DU compliance with the criteria in all time. The scenarios of zero violation are not used for the regression of Equation 5.

Table 1. Average DO and percent violation of criteria in CB5-DW for 27 scenarios

27 scenarios with various loads (at fraction of PR2000)			(W) DO	(V) violation
TN	TP	TS	mg/L	(%)
1.000	1.000	1.000	5.31	3.616
1.000	1.000	0.700	5.45	2.801
1.000	1.000	0.400	5.65	1.534
1.000	0.700	1.000	5.62	1.923
1.000	0.700	0.700	5.78	1.122
1.000	0.700	0.400	5.98	0.310
1.000	0.400	1.000	6.02	0.353
1.000	0.400	0.700	6.17	0.025
1.000	0.400	0.400	6.32	0.000
0.700	1.000	1.000	5.57	1.266
0.700	1.000	0.700	5.66	1.128
0.700	1.000	0.400	5.80	0.725
0.700	0.700	1.000	5.75	0.967
0.700	0.700	0.700	5.87	0.609
0.700	0.700	0.400	6.04	0.162
0.700	0.400	1.000	6.07	0.180
0.700	0.400	0.700	6.20	0.004
0.700	0.400	0.400	6.35	0.000
0.400	1.000	1.000	6.23	0.000
0.400	1.000	0.700	6.24	0.000
0.400	1.000	0.400	6.27	0.000
0.400	0.700	1.000	6.25	0.000
0.400	0.700	0.700	6.27	0.000
0.400	0.700	0.400	6.31	0.000
0.400	0.400	1.000	6.31	0.000
0.400	0.400	0.700	6.35	0.000
0.400	0.400	0.400	6.44	0.000

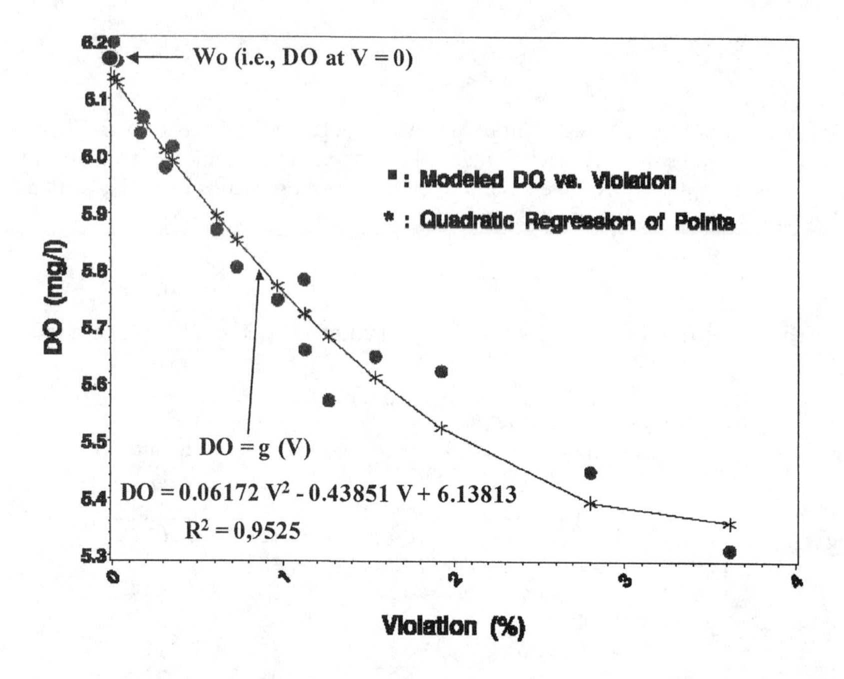

Figure 8. Relationship between average DO and percent violation in CB5-DW, and the regression curve.

Points in Figure 8 are the relationship of DO and violation based on the last two columns of Table 1. The stars connected with lines are based on the regressed quadratic fitting function.

Denoting Wo as the average DO in a DU when the violation, V, approaches to zero. From Equation 5 we have:

$$Wo = \lim_{V \to +0} g(V).$$

Wo is the minimum average DO in the DU that satisfies the attainment of the DO criterion. Wo can also be estimated visually from Figure 8 for DO at V=0.

Using the first 4 columns of Table 1 and using the same method for deriving Equation 2, we can establish the relation of average DO (i.e., W) of a DU and the N, P and S loads:

$$W = f(N, P, S)$$

Using W=Wo as a constraint for the above equation, we have

$$Wo = f\,(N, P, S) \tag{6}$$

Using Equation 6, we can plot a surface of DO = Wo as a function of the three loading constituents, N, P and S (Figure 9), in the similar way as we did for Figure 6.

On this surface, the DO criterion in CB5-DW is just in compliance. This surface divides the N-P-S loading space into two regions. The loading space at lower-left to this surface (i.e., with less N or P or S load) would comply with the DO criterion statistically on the modeled time span and the involved model cells. While, the loading space at upper-right of this surface could cause non-compliance of the criterion.

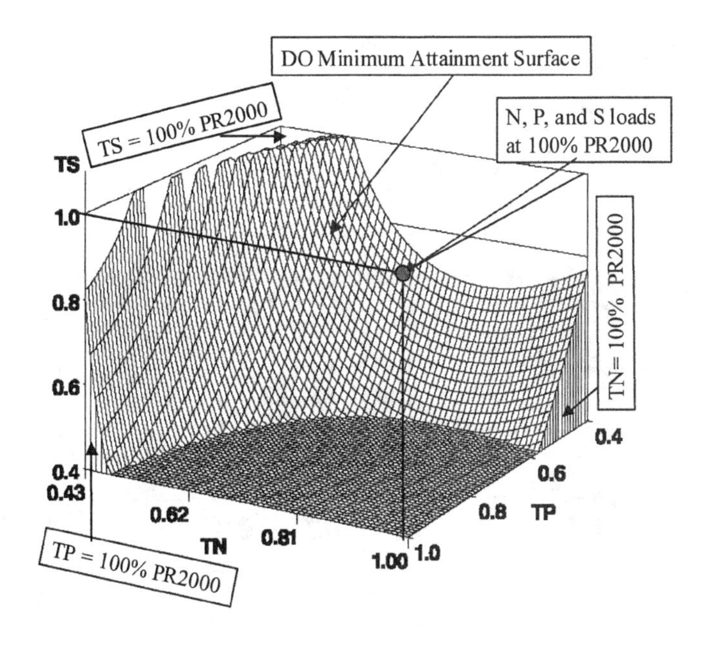

Figure 9. Minimum attainment surface of DO in CB5-DW versus N-P-S loads. Regression equation: $DO = 1.45224\ N^2 + 0.40672\ P^2 + 0.18667\ S^2 - 1.60065\ N\ P - 0.70543\ N\ S + 0.10917\ P\ S - 1.23053\ N - 0.27406\ P - 0.22275\ S + 7.17064$; $R^2 = 0.9914$.

Further Application of Surface/Volume Analysis in Water Quality Management

Sediment and nutrient reductions are usually implemented with different types of BMPs (Best Management Practice). Sediment control can be through the control of sediment from land covers, and also be through shoreline protections. 10% and 20% sediment reductions are assessed to see the impact of sediment control on SAV, as well as on dissolved oxygen, and are proposed for general management practice. We are interested to see the effect of DO attainment by various N-P loads under specified TS loading levels. We can obtain a curve of a water quality (e.g., DO) as a function of N-P loads under a constant TS load by using a plane of specific TS to cut the 3-D plot of Figure 8. Figure 9 shows two DO curves versus N-P loads under constant TS loads at 100%, and 80% of PR2000. The 80% TS curve is to the upper right from the 100% TS curve, i.e., closer to the 100% N and 100% P condition. It means that in order to meet the minimum DO criteria attainment, it requires less N or P reduction under 20% TS reduction than under 0% TS reduction.

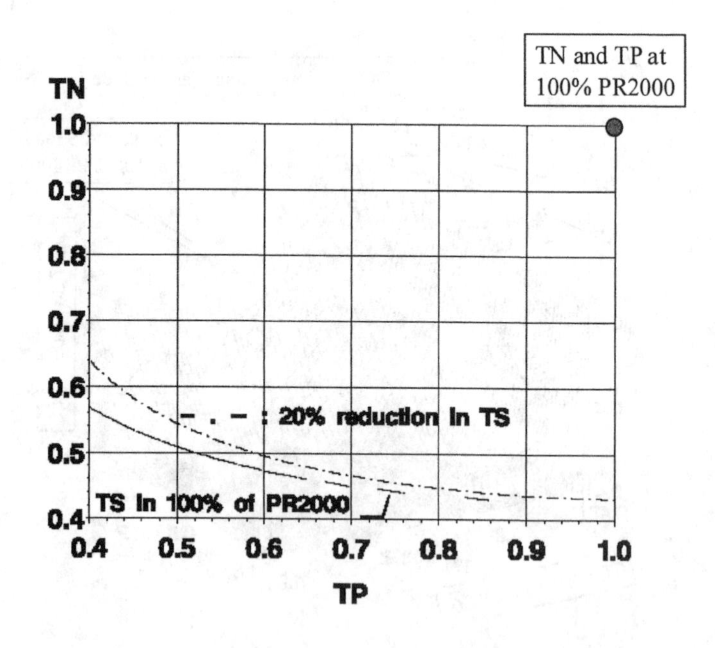

Figure 10. Contours of DO minimum attainment curve versus N-P load at specific S loads for CB5-DW.

Each of the contour lines shows the tradeoff between N-P reductions under a specific sediment reduction for water quality attainment. For example, let's look at the curve of 20% TS reduction. A reduction in 50% N and 40% P is estimated to comply with the DO standard. Alternatively, a reduction of 45% N and 50% P would also comply with the DO standard.

Discussion

The loading conditions among the scenarios for a surface analysis should have consistent relationships. For example, although load reductions from a reference state are different among scenarios, each scenario itself has "baywide same percent reduction" in all tributaries from the reference state. To construct the attainment surface o f a water quality index (W) versus N, P and S three loading constituents, e.g., Figures 6 and 9, it requires at least two scenarios to have certain degrees of criteria violation and at least two scenarios to have criteria compliance. There is no such a requirement to construct the response surface of a water quality index versus two loading constituents such as Figure 3, as long as there are sufficient scenarios.

The approaches in constructing Figure 6 and Figure 9 are similar. When a water quality criterion is used to compare the average concentration in a DU over times, the approach for Figure 6 is used. If water quality criteria violation is considered over space and time for individual cells in a DU, then the approach for constructing Figure 9 should be used. In the latter case, the average concentration in a DU is higher than the criterion since all cells in all times must comply with the minimum criteria. It involves an additional step, i.e., regression of W and V to project average W (e.g., DO) at minimum attainment. The scenarios with zero violation should not be used for the regression of W-V relation, because different loading scenarios with a same V=0 can have quite different average DO.

A Deep-Water DU consists of cells with different layers (depths). The cells in deeper layers usually have lower dissolved oxygen and are more likely to violate the criterion than the cells in the shallower layers. For a specific month, DO in some cells may be lower than 3 mg/L, while in some other cells may be higher than 3 mg/L. For a specific cell, its DO can be less than 3 mg/L in a month, but may be higher than 3 mg/L in some other months. The DO criterion may not be met despite the average DO in a DU over a long term can be higher than 3 mg/L. Therefore, the average DO of a DU at minimum attainment situation is usually higher than the criteria value.

For a specific value of average DO in a DU, there are many possible combinations of DO distribution among individual cells. Here we use a DU with two vertical cells as an example, and assume the average DO equal to 4 mg/L. Case 1: the lower cell is at 3 mg/L and the upper cell at 5 mg/l. In this case, the DU barely meets the criteria. Case 2: the lower cell is at 3.5 mg/L, and the upper cell at 4.5 mg/L. In this case the

DU is above the criteria. Case 3: the lower cell is at 2.5 mg/L, and the upper cell at 5.5 mg/L. In this case the DU violates the criteria. This example reveals the fact that for a same value of average DO in a DU, the extents of criteria violation can be different in different scenarios. Therefore, if loading conditions among scenarios do not follow a certain pattern systematically, the projected average Wo from function W=g(V) may not represent a minimum criteria attainment by the DU. In this project, each of the scenarios has same percentage, Bay-wide equal reduction from a reference. Therefore, the W-V relation obeys a certain trend among the scenarios, and the project Wo is likely to represent the average DO at minimum attainment for the designed scenario set. This can also be seen in Figure 8 that the individual points (as dots in the figure) are close to the regression curve. Since the scenarios having low V is more important to extrapolate W at Vo. The second order regression function provides a better extrapolation to Vo than a linear regression.

In a real ecosystem, if the exceedence of a water quality criterion is only in a small portion of a DU and in a short time period, the habitats may still survive and the ecosystem can still be considered healthy. USEPA (1997) considered a degree of criteria exceedence to be acceptable if it only involves less than 10% area and/or only at less than 10% time. Only when the exceedence is above the allowable exceedence, that portion of exceedence is then considered a real violation. The USEPA Chesapeake Bay Program Office applied statistics on benthic community indices (index of biological integrity – IBI) (Ransasinghe et al., 2002), and developed a biological-based-reference curve for DO criteria (USEPA, 2003) to provide an ecosystem basis to the principal of the 10% allowable time and space exceedence. A violation occurs only when the cumulative exceedence is above the IBI reference. The violations listed in Table 1 and used in the related plots (Figures 8 , 9 and 10) are calculated using the IBI reference. Detailed discussion of bio-reference curves is beyond the scope of this paper.

The attainment surface as a function of N-P-S three loading constituents provides a useful tool for management of nutrient and sediment loads in compliance with water quality standards.

References

Cerco, C.F., and Cole, T.M (1994). *Three-Dimensional Eutrophication Model of Chesapeake Bay*. Technical Report EL-94-4, U.S. Army Engineer Waterways Experiment Station, Vicksbutg, MS, USA.

Chesapeake Bay Program Office (2003) *Chesapeake 2000 agreement nutrient and sediment allocation reports*. Chesapeake Bay Program Office, Annapolis, MD, USA.

Chesapeake Executive Council (1987). *Chesapeake Bay Agreement*, Annapolis, MD, USA.

Chesapeake Executive Council (2000). *Chesapeake Bay Agreement, 2000 Amendments*, Annapolis, MD, USA.

Emery, W.J., and Thomson, R.E. (1998) *Data analysis methods in physical oceanography*, Elsevier, pp634.

Gillelan, M.E., Haberman, H., Mackieman, G, Macknis, B.J., and Wells, H.W. (1983). *Chesapeake Bay: A framework for action, USEPA Chesapeake Bay Program*, Program Report, Annapolis, MD, USA.

Johnson, B.,Kim, K., Heath, R., Hsieh, B., and Butler, L. (1993). "Validation of a three-dimensional hydrohynamic model of Chesapeake Bay." *J. Hydr. Engrg*., ASCE, 199(1), 2-20.

Malanotte-Rizzoli, R. (ed.) (1996) *Modern approaches to data assimilation in ocean modeling*, Elsevier, Amsterdam, pp435.

Ranasinghe, J.A., Frithsen, J.B., Kutz, F.W., Paul, J.F., Russell, D.E., Batiuk, R.A., Hyland, J,L., Scott, J., and Dauer. D.M. (2002). "Application of two indices of benthic community condition in Chesapeake Bay." *Environmetrics* 13, 499-511.

Thomann, R.V., Collier, J.R., Butt, A., Casman, E, and linker, L.C. (1994). *Response of the Chesapeake Bay Water Quality Model to Loading Scenarios.* Modeling Subcommittee, Chesapeake Bay Program Office, Annapolis, MD, USA. CBP/TRS 101/94.

Thomann, R.V., and Mueller, J.A.: (1987) *Principles of surface water modeling and contro.* Harper Collins Pub., New York, NY. (1987) pp644.

USEPA (1997). *Guidelines for preparation of the comprehensive state water quality assessments. (305 (b) Reports) and Electronic Updates*. Assessment and Watershed Protection Division, Office of Wetlands, Oceans and Watersheds, Office of Water, U.S. EPA, Washington, D. C.

USEPA (2003). *Ambient Water Quality Criteria for Dissolved Oxygen, Water Clarity and Chlorophyll a for the Chesapeake Bay and Its Tidal Tributaries.* USEPA Chesapeake Bay Program and Region 3 Water Protection Division, EPA 903-R-03-002, April 2003. pp170.

Wang, P., L.C. Linker, and R. Batiuk. (2002) "Surface analysis of water quality response to load." *Proc. 7th International Conference on Estuarine and Coastal Modeling*, St. Pete Beach, FL, USA. November 2001, p566-584.

Wamg, P., and Linker, L.C. (2002). "Application of surface and tracer analysis to assess the response of water quality of Chesapeake Bay to nutrient and sediment load, and allocation of load reduction for TMDL. " *Proc. Watershed 2002*, Ft. Lauderdale, FL, USA. February 2002, 27 pages.

Wang, P., Linker, L.C., and Batiuk, R.A. (2003) "The response of submerged aquatic vegetation to sediment and nutrient loads in the Chesapeake Bay and suggestions on load control." *Proc Coast Zoon 2003*. 5 pages.

Random-Walk Suspended Sediment Transport and Settling Model

Richard Argall[1], Brett F. Sanders[2] and Ying-Keung Poon[3]

Abstract

This paper describes a particle-based approach to simulate entrainment, transport, and settling of non-cohesive sediments in rivers and the coastal zone where the velocity distribution is characterized by a fully-developed turbulent boundary layer extending from the bed. Sediment distributions are modeled as a set of particles that are tracked on an individual basis by solving Lagrangian transport equations that account for advection by the mean flow, settling, and random turbulent motions. The vertical distribution of the mean flow is modeled by a power law, and turbulent motions are modeled as a random walk that is scaled by the turbulent diffusivity analogous to random molecular motions that constitute Fickian diffusion. Because it is not feasible to track individual sediment grains, each particle accounts for a large number of sediment grains of a single size and settling velocity. A suite of particles is used to account for a spectrum of grain sizes. The model resolves the vertical distribution of particles consistent with Rouse profiles for a wide range of Rouse numbers. Using an entrainment relationship from literature, model predictions of sediment loads compare favorably to established empirical models. The model is applied to predict the collection efficiency of a sedimentation basin in Ballona Creek, California, highlighting the effectiveness of the basin at collecting coarse sands but not finer grained material.

Introduction

The mobility of sediment within an open channel is responsible for local scouring/deposition around hydraulic structures, bank instability, burial of fish habitat and loss of clarity in the water column, which inhibits photosynthesis. Sediment mobility has also been associated with the transport of absorbed contaminants, and sediment re-suspension has a measurable affect on water column chemistry (Hu 1999). The recognized modes of sediment transport in rivers are bedload and suspended load. In the case of bedload, sediment grains roll, slide or saltate over each other, never deviating too far from the bed (Garcia 2001). In the case of

[1]Graduate Student Researcher, Department of Civil and Environmental Engineering, University of California, Irvine, CA 92697

[2]Associate Professor, Department of Civil and Environmental Engineering, University of California, Irvine, CA 92697

[3]Vice President/Principal Engineer, Everest International Consultants, Inc., 444 West Ocean Blvd., Suite 1104, Long Beach, CA 90802

suspended load, sediment particles are carried well up into the water column by turbulent fluctuations and further dispersed by mean-velocity gradients.

While theories of suspended sediment transport have largely been deterministic, a number of investigators have used probabilistic methods to derive time and space distributions of suspended particles. Indeed, the transport of suspended sediment has been regarded as an intrinsically random process. Alonso (1981) described two probabilistic approaches. One is based on the theories of turbulent diffusion by continuous movements, whereas the other arose from the random walk theory of stochastic processes. Li and Shen (1975) presented a random walk model for particle settlement in open channels. Pearson (1905) introduced the random walk concept although Rayleigh (1880) earlier considered a formally identical problem while studying the composition of vibrations of unit amplitude with phases distributed at random. The random walk theory has been applied to the solution of diffusion problems in two distinct manners. The classical approach consisted of obtaining the solution of complex differential equations as the limiting solution of an equivalent discrete probabilistic process. The classic example is the solution of the Fokker-Planck diffusion equation (Gnedenko 1968). The other approach is the direct simulation of the physical process by imitating the behavior of representative discrete particles through computer simulation (Alonso 1981).

Random walk particle tracking models have advantageously been applied to the simulation of transport of groundwater (Ahlstrom *et al.* 1977, Tompson and Gelhar 1990) where the random component represents unresolved spatial variation in velocity due to heterogeneity in medium properties such as hydraulic conductivity. Particle methods with or without a random component have also been used in coastal applications to characterize residual circulation (Signell and Geyer 1990). However, they are less common than concentration based models. Dimou and Adams (1993) describe a 2-D random walk particle-tracking model to simulate transport (advection-dispersion) in vertically well-mixed estuaries and coastal waters. Applied to a channel of uniform cross section and spatially variable dispersion coefficient, simulated average longitudinal concentration distributions and simulated residence times agreed well with analytical solutions. Li and Shen (1975) presented a random walk model for the simulation of solid particle settlement in open channels.

This paper describes a random-walk particle-based model for simulation of non-cohesive sediment transport and settling in rivers and the coastal zone where the vertical distribution of the mean velocity and turbulent diffusivity can be inferred under the assumption of a fully developed boundary layer. Boundary conditions and scaling relationships (number of grains per particle) are developed such that sediment loads predicted by the model compare well with established empirical formulas. This paper brings together traditional hydrodynamic concepts with a stochastic particle-based approach in order to simulate entrainment, transport, and settling of non-cohesive sediments. The inventive step necessary to facilate efficient sediment transport simulations is the use of a scaling rule for entrainment. The scaling is an important aspect of the model since it allows each particle to account for many grains of sediment and makes the overall method practical for application purposes.

Suspended Sediment Transport

Suspended sediment transport is described by the following equation,

$$\frac{\partial c}{\partial t}+\frac{\partial}{\partial x}(uc)+\frac{\partial}{\partial y}(vc)+\frac{\partial}{\partial z}(wc)-w_f\frac{\partial c}{\partial z} = \frac{\partial}{\partial x}\left(\varepsilon_x\frac{\partial c}{\partial x}\right)+\frac{\partial}{\partial y}\left(\varepsilon_y\frac{\partial c}{\partial y}\right)+\frac{\partial}{\partial z}\left(\varepsilon_z\frac{\partial c}{\partial z}\right) \tag{1}$$

where c = sediment concentration, u, v, and w = components of velocity, w_f = sediment fall velocity, and ε_x, ε_y and ε_z = sediment turbulent diffusivities and it is assumed that the coordinate directions are aligned with the principal axes of turbulent diffusion.

Velocity and Diffusivity Distribution The velocity of water flowing in channels is characterized by a turbulent boundary layer extending from the bed. A common model for the mean velocity (Reynolds averaged) is the following logarithmic formula,

$$u=\bar{u}\left[1+\frac{\sqrt{C_d}}{\kappa}\left(1+\log\frac{z}{h}\right)\right] \tag{2}$$

where κ is von-Karman's constant, C_d is the drag coefficient (one half the friction coefficient), and log is the natural logarithm. In wide channels, this model compares very well with observations (Rouse 1937, Fischer *et al.* 1979) though it does not account for secondary currents that are most significant at channel bends and near channel banks.

Near the bed Eq. 2 is not applicable and an alternative model with the property that $u \to 0$ as $z \to 0$ is the following power law,

$$u=\hat{u}(z/h)^{\alpha} \tag{3}$$

where $\hat{u}$ is the free surface velocity given by Eq. 2,

$$\hat{u}=\bar{u}\left(1+\frac{\sqrt{C_d}}{\kappa}\right) \tag{4}$$

and $\alpha=\sqrt{C_d}/\kappa$, which is close to the commonly used value of 1/7 (Crowe *et al.* 2001).

The sediment eddy diffusivity, ε_z, is derived from the power law velocity distributions by assuming a constant ratio, β, of sediment mass to momentum transfer,

$$\frac{\tau^x}{\rho}=\frac{\varepsilon_z}{\beta}\frac{\partial u}{\partial z} \tag{5}$$

where τ represents the shear stress in the $x - z$ plane which varies linearly from a value of zero at the surface to a value of τ_b at the bed. Rearranging for ε_z,

$$\varepsilon_z = \beta \frac{\tau/\rho}{\partial u/\partial z} = \beta \frac{(\tau_b/\rho)(1 - z/h)}{\bar{u}(1+\alpha)(\alpha/h)(z/h)^{\alpha-1}} \tag{6}$$

or in terms of the shear velocity $u_* = \sqrt{\tau_b/\rho} = \sqrt{C_d}\bar{u}$,

$$\varepsilon_z = \frac{\beta \kappa u_* h}{\alpha + 1}\left(1 - \frac{z}{h}\right)\left(\frac{z}{h}\right)^{1-\alpha} \tag{7}$$

The depth averaged value of the vertical eddy viscosity given by Eq. 7 is

$$\bar{\varepsilon}_z = \frac{\beta \kappa u_* h}{(\alpha+1)(6 - 5\alpha + \alpha^2)} \tag{8}$$

Interestingly, Eq. 7 and 8 are identical to those associated with the logarithmic velocity profile, $\varepsilon_z = \beta \kappa u_* z\,(1 - z/h)$ and $\bar{\varepsilon}_z = \beta \kappa u_* h/6$, respectively in the limit that $\alpha \to 0$. In this limit, however, the power law velocity distribution (Eq. 3) is not meaningful since it simply predicts a uniform velocity over the vertical.

Settling Velocity The settling velocity w_f is based upon the force balance between particle weight and particle drag. It is equivalent to the terminal speed of a grain of the given sediment in an infinite expanse of still water at a specified temperature. The fall velocity of a sediment is its most fundamental property governing particle motion and hence determines whether grains are carried as bed load or suspended load. It is a function of the shape, volume and density of the particle and the viscosity and density of the fluid. If the characteristics of the particle and fluid are known, the fall velocity of a sediment particle can be calculated.

Since the fall velocity is equivalent to the terminal velocity, the drag force must be equal and opposite to the submerged weight of the particle. By assuming the particle as a sphere the problem is identical to the classic fluid mechanics problem of estimating the drag around a submerged sphere, hence fall velocity can be estimated,

$$w_f = \sqrt{\frac{\frac{4}{3}(\frac{\gamma_s}{\gamma} - 1)gd}{C_D}} \tag{9}$$

A more convenient means of estimating w_f is given by Julien (1995),

$$w_f = \frac{8\nu}{d}\left[\sqrt{(1 + 0.0139 d*^3} - 1\right] \tag{10}$$

where $d*$ is a dimensionless grain size given by,

$$d* = \left[\frac{(\gamma_s/\gamma - 1)gd^3}{\nu^2}\right]^{1/3} \tag{11}$$

γ_s = unit weight of sediment, γ =unit weight of water, g =gravitational constant, and ν =kinematic viscosity.

Sediment Distribution In steady, uniform turbulent flow the vertical distribution of suspended sediment is described by the following equation,

$$\varepsilon_z \frac{dc}{dz} + w_f c = 0 \tag{12}$$

Substitution of Eq. 8 for ε_z and separating variables, Eq. 12 becomes,

$$\frac{dc}{c} = -\frac{w_f}{\beta \kappa u_*}(\alpha + 1)\frac{\zeta^\alpha}{\zeta(1-\zeta)}d\zeta \tag{13}$$

where $\zeta = z/h$. A solution is given by,

$$c(z) = c_a e^{-R_o(\alpha+1)[(z/h)^\alpha \Phi(z/h,1,\alpha) - (a/h)^\alpha \Phi(a/h,1,\alpha)]} \tag{14}$$

where c_a represents the sediment concentration a distance a above the bed, $R_o = w_f/(\beta\kappa u_*)$ is the Rouse number, and Φ is Lerch's function,

$$\Phi(\zeta, 1, \alpha) = \sum_{k=0}^{\infty} \frac{\zeta^k}{\alpha + k} \tag{15}$$

Rouse solved Eq. 12 using $\varepsilon_z = \beta\kappa u_* z(1 - z/h)$ to give the better known formula (the Rouse profile),

$$c(z) = c_a \left(\frac{a(h-z)}{z(h-a)}\right)^{R_o} \tag{16}$$

Predictions of sediment concentration using Eq 14 slightly over predict those using Eq. 16, as is shown in Figure 1. This is readily explained by the eddy diffusivity, which is maximum at mid-depth for the logarithmic velocity profile but maximum slightly below mid-depth for the power-law velocity profile (Figure 2). Larger diffusivities near the bed act to keep more material in suspension.

Particle Tracking Model

Particle tracking equations are solved to predict the position of many discrete particles, in contrast to concentration based equations which predict the number of particles inside a representative elemental control volume. Particle tracking equations account for a combination of deterministic and random motions. The former accounts for advection by the mean flow and settling, while the latter accounts for fluctuations in position caused by turbulence, or turbulent diffusion. Since turbulent diffusion is predicated on random eddy motions, just as Fickian diffusion is predicated on random molecular motions, there is a direct analogy between the physics of molecular diffusion and the numerics of particle tracking.

Following Dimou and Adams (1993), the particle tracking equations represent explicit update equations for particle position as follows,

$$\begin{aligned} x_p(t+\Delta t) &= x_p(t) + \left(u(\mathbf{x}_p, t) + \frac{\partial \varepsilon_x(\mathbf{x}_p, t)}{\partial x} \right) \Delta t \\ &\quad + R(0,1)\sqrt{2\varepsilon_x(\mathbf{x}_p, t)\Delta t} \end{aligned} \tag{17}$$

$$\begin{aligned} y_p(t+\Delta t) &= y_p(t) + \left(v(\mathbf{x}_p, t) + \frac{\partial \varepsilon_y(\mathbf{x}_p, t)}{\partial y} \right) \Delta t \\ &\quad + R(0,1)\sqrt{2\varepsilon_y(\mathbf{x}_p, t)\Delta t} \end{aligned} \tag{18}$$

$$\begin{aligned} z_p(t+\Delta t) &= z_p(t) + \left(w(\mathbf{x}_p, t) - w_f + \frac{\partial \varepsilon_z(\mathbf{x}_p, t)}{\partial z} \right) \Delta t \\ &\quad + R(0,1)\sqrt{2\varepsilon_z(\mathbf{x}_p, t)\Delta t} \end{aligned} \tag{19}$$

where $\mathbf{x_p} = (x_p,\ y_p,\ z_p)$ represents the position of the pth particle, Δt is the time step, and $R(0,1)$ represents a random number with a mean of zero and a standard deviation of one. In the limit of an infinite number of particles and an infinitesimal time step, the particle tracking equations give a solution to the Fokker-Plank equation, a generalizable stochastic partial differential equation describing advective-diffusive transport.

The particle tracking equations require both initial conditions, particle coordinates at time $t = 0$, and boundary conditions to enforce zero-flux conditions at the free surface and sediment entrainment and deposition at the bed. The zero-flux condition is enforced by a reflection procedure. Any particle predicted to move across the zero-flux boundary by the update equations is re-positioned equidistant from the boundary but on the interior side.

To model particle settling in the absence of entrainment, particles are reflected off the bed when the shear stress exceeds the critical value for settling defined by the Shields parameter and sticks to the bed otherwise. The Shields parameter is given by $\tau* = \tau_b/(\gamma_s - \gamma)d$ and its critical value is given by $\tau*_c = \tau_c/(\gamma_s - \gamma)d$.

To model sediment entrainment and settling, erodable layers of sediment on the channel bed are modeled with stacks of particles spaced a distance Δx apart in the streamwise direction. When, for a given stack of particles, the shear stress exceeds the critical value for motion defined by the Shields parameter, the top particle on the stack is repositioned a distance $w_f \Delta t$ above the bed at which point it becomes subject to the particle tracking equations. Particles predicted to settle are added to the top of the stack.

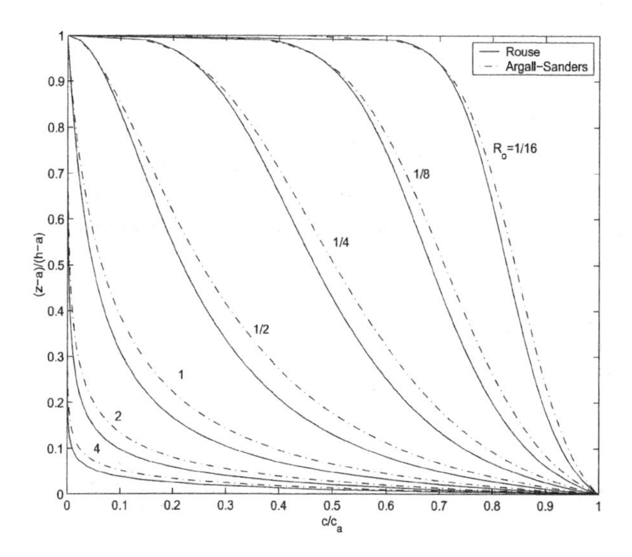

Figure 1: Comparison between Rouse profiles and profiles based on the power-law velocity profile. Here a is set arbitrarily to $0.05h$.

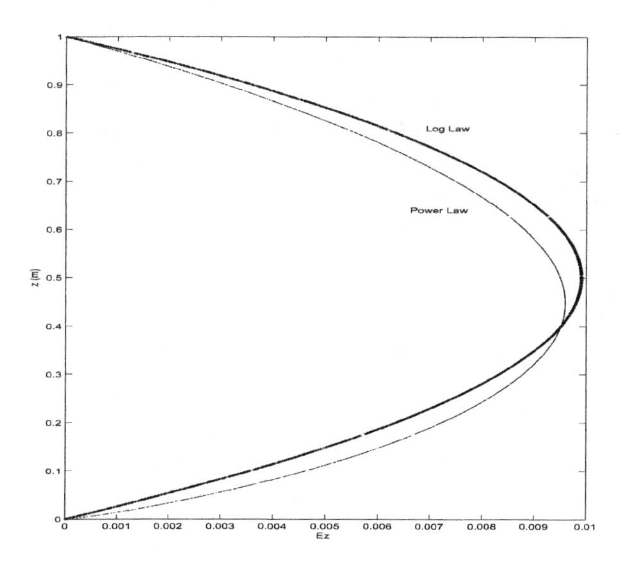

Figure 2: Comparison of eddy diffusivities predicted by the log-law and power-law velocity profiles.

In order to translate particle transport predictions into sediment transport predictions, a scaling relationship is needed. Physically, the scale represents the number of monodispersed sediment grains (all of the same diameter) per particle and is denoted n_g. This does not limit the method to monodisperse sediment distributions, but rather, requires that each particle account for a different grain size in order to predict transport of polydisperse sediments. The scale need not be the same for all particles, or even all particles representing a common sediment grain size. In fact, the power of a particle based transport method lay in the potential for each particle to take on its own unique scale. Here, erosion of sediment from an area Δx and over a time step Δt is modeled using one particle regardless of the entrainment rate. Therefore, the particle scale must be set to account for the corresponding mass of sediment and is given as follows,

$$n_g = \Delta t \, \Delta x \, E_g / m_g \tag{20}$$

where E_g is the grain-size specific entrainment rate (mass/time/surface area) and m_g is the grain-size specific mass per grain. Sediment load follows as,

$$L = \frac{1}{\Omega} \sum_{p \in \Omega} u_p (m_g)_p (n_g)_p \tag{21}$$

where Ω represents a channel reach and has dimensions of length. Usually only a bulk entrainment rate E is known based on the median grain diameter, and not a grain size specific rate, so the latter should be determined from the former based on the grain size distribution.

A number of expressions for entrainment are available from the literature and summarized by Garcia (2001). It is customary to describe these in terms of dimensionless entrainment $\hat{E} = E/\rho_s/w_f$, where w_f is based on the median diameter. The following is a convenient relation by Garcia and Parker (1991),

$$\hat{E} = \frac{A \, Z_u^5}{(1 + A Z_u^5/0.3)} \tag{22}$$

where

$$Z_u = \frac{u_*}{w_f} R_{ep}^{0.6} \qquad R_{ep} = \frac{(gRd^3)^{1/2}}{\nu}$$

$A = 1.3 \times 10^{-7}$, ν = kinematic viscosity, R_{ep} = particle Reynolds number, and R = submerged specific gravity. This relation was verified by Garcia and Parker (1991) using data covering the following ranges:

$$\hat{E} : 0.002 - 0.06, \quad u_*/w_f : 0.7 - 7.50$$
$$H/d : 240 - 2400, \quad R_{ep} : 3.50 - 37.00$$

These ranges correspond to a grain sizes ranging from 0.09mm to 0.44mm - typical of a sand bed stream.

Model Testing

Two tests and one application of the model are described. The first test is designed to characterize how well the particle tracking model resolves the vertical distribution of sediment concentration over a range of flow conditions. The second test is designed to characterize how well the model predicts sediment loads based on Eq. 21. The application of the model highlights how the model can be used to predict the transport and settling of a wide spectrum of grain sizes carried down a Southern California storm channel. Information about the settling of particles is important for harbor management, in terms of dredging operations, and to understand the fate of various grain sizes with adsorbed contaminants.

For the two tests, uniform flow at a depth of 1 m in an infinitely long channel with a 0.1 % slope and a Manning coefficient $n_m = 0.02$ is considered. The infinite channel length is modeled using a 100 m long reach and periodic boundary conditions. That is, particles transported past the downstream boundary are repositioned the same distance downstream of the upstream boundary and at the same vertical position relative to the bed. Common parameters used in these tests include a fluid viscosity $\nu = 1 \times 10^{-6}\, m^2/s$, and fluid density $\rho = 1 \times 10^3\, kg/m^3$, submerged specific gravity $R = 1.65$.

Prediction of Vertical Sediment Profiles For the first test, particles are initially distributed uniformly over the length and depth of the channel reach, and the particle tracking equations are integrated until the particle distribution reaches equilibrium (a simulation time of 2 min is used). A reflecting boundary condition is used at the bed, as if transport and settling was being modeled but the shear stress exceeded the threshold for settling. Sediment concentrations are obtained by counting the number of particles located within a set of evenly sized bins that span the distance from the free surface to the bed, and these are compared to the distribution described by Eq. 14, which is essentially the Rouse profile based upon the power law for the velocity distribution.

Figures 3 and 4 present suspended sediment concentration predictions for a range of Rouse numbers (Ro=1/16, 1/8, 1/4, 1/2, 1 and 2), compared to the Rouse profile. Figure 3 shows results from simulations using 2,000 particles and 20 particle bins. Figure 4 shows results simulations using 20,000 particles and 200 particle bins. These indicate that use of roughly 100 particles per bin captures the basic trend in the vertical sediment concentration profile, though scatter in predictions appears to increase with decreasing Rouse number corresponding to the more dominant role of turbulence. When the time step is not made sufficiently small, however, the Rouse profiles and the trend in the particle predictions diverge. Figure 5 illustrates how predictions change using larger time steps. Therefore, for accuracy purposes the time step needs to be constrained. By making predictions with a range of time steps and Rouse numbers, it was found that the trend in the vertical profile was captured within 10% when the time step is roughly constrained by $w_f \Delta t/h < 0.01$.

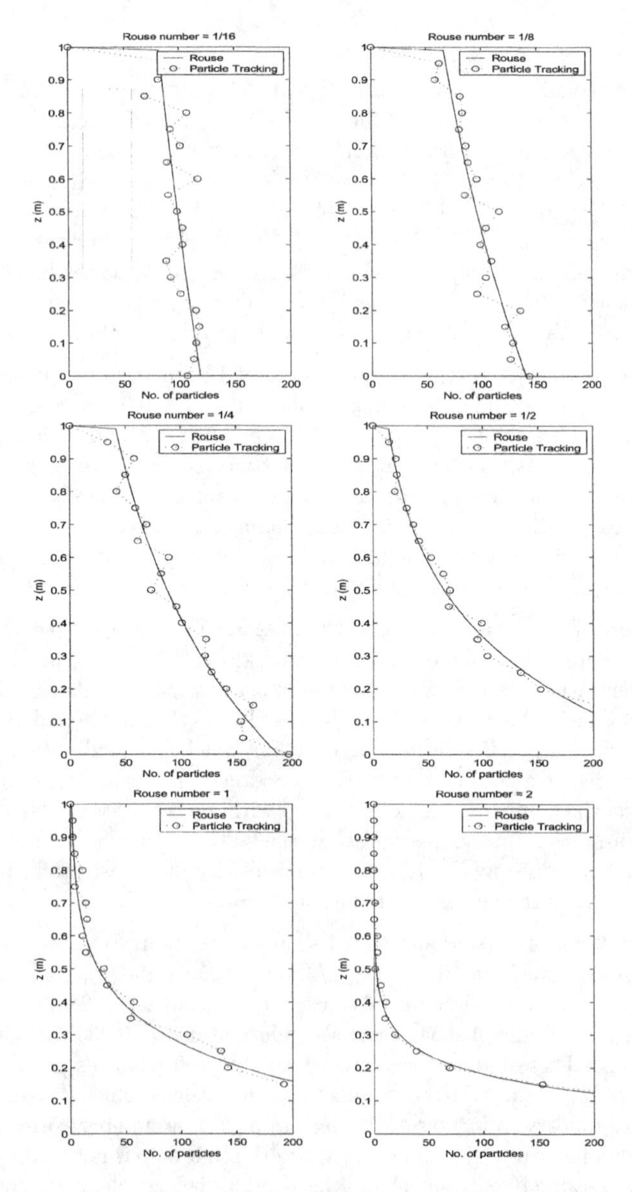

Figure 3: Comparison of suspended sediment concentration predictions - Rouse and Particle tracking (2000 particles, 20 bins).

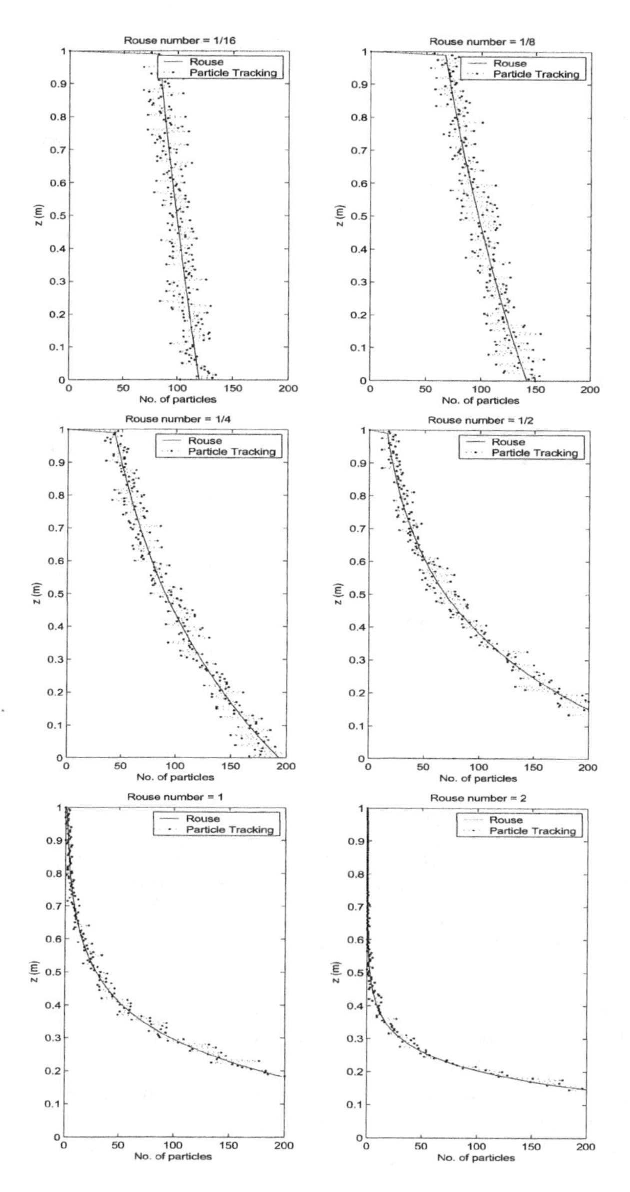

Figure 4: Comparison of suspended sediment concentration predictions - Rouse and Particle tracking (20,000 particles, 200 bins).

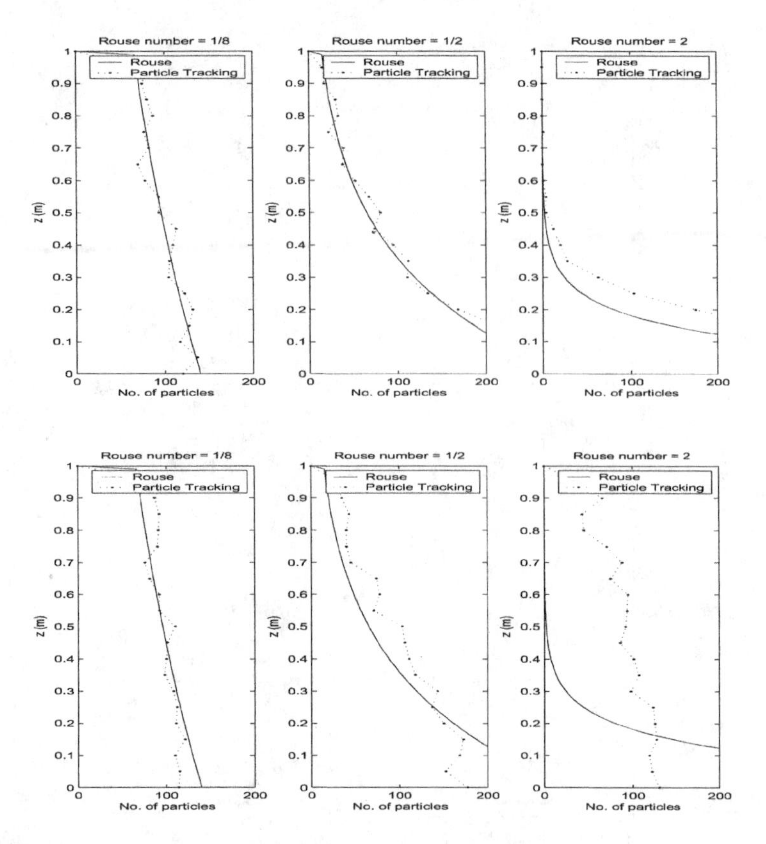

Figure 5: Comparison of suspended sediment concentration predictions - Rouse and Particle tracking , $\Delta t = 1s$ (top) and $\Delta t = 10s$ (bottom).

Prediction of Sediment Load For the second test, particles are initially at rest on the bed and the particle tracking equations are solved to evaluate whether, once the sediment distribution reaches equilibrium, the sediment load is comparable to that predicted by established empirical formulas. A total of 20 sediment stacks were used along the 100 m length of the channel, so each stack accounts for entrainment from and settling to a 5 m reach of the channel. Each stack consisted of 100 particles. A range of monodisperse sediment distributions were considered. Model predicted loads were computed using $\Omega = 100\,m$ (Eq. 21).

Figure 6 presents, for a range of grain sizes and roughness properties, model predicted loads as well as loads predicted by empirically-derived sediment functions (Yang, 1973 and Engelund-Hansen, 1967). For sediment in the range of 0.2-1 mm, loads predicted by the particle model fall between the predictions given by the Yang and Engelund-Hansen models. For smaller sized sediment, the particle model over-estimates the load. This is likely due to the fact that the entrainment relation used by the model is not applicable to smaller sized grains. For larger sized sediment, the particle model also over-estimates the load and this may be because bed load dominates transport, not suspended load.

Prediction of Settling in Ballona Creek Ballona Creek is a nine-mile long flood protection channel that drains the Los Angeles basin. The channel discharges into the ocean in the immediate vicinity of the harbor entrance to Marina-del-Rey. This coastal inlet experiences heavy vessel traffic and is protected by a robust detached breakwater. A large discharge of sediment from Ballona Creek during flood events coupled with the presence of the breakwater has necessitated costly dredging operations to remove deposited sediment. A sediment settling basin installed at some point along the Ballona creek flood protection channel has been postulated to reduce the need for dredging at the harbor entrance.

For the present study, we applied a depth-averaged hydrodynamic model by Bradford and Sanders (2002) to characterize flow along the final 6 km of the channel where the channel is rectangular and can be modeled as having a width of 50 m, a slope of 0.1%, and a Manning coefficient of 0.02. The channel was discretized with 300×10 computational cells. In the model, the settling basin is placed between 2100 m and 3000 m from the downstream boundary where the water level was specified to be mean sea level. Flow was forced at the upstream boundary by a 24 hr triangular hydrograph that peaks at 12 hr. The discharge Q specified by the hydrograph consisted of a baseflow of 10 m^3/s, and a peak flow 220, 500, 620, 780, 910, and 1,020 m^3/s which corresponds to the 1, 2, 5, 10, 25, 50, and 100 yr storm events. The total sediment load entering the upstream boundary of the model was assumed to follow the relation, $Q_s = 0.033Q$ where Q_s is the sediment concentration by weight in parts per thousand. The grain size distribution was assumed to be: 0.016 mm (20%), 0.088 mm (5%), 0.177 mm (15%), 0.354 mm (16%), 0.707 mm (16.5%), 1.414 mm (12.5%), 2.828 mm (10%), 5.657 mm (5%).

The particle tracking model was used to estimate the fate of sediment moving through the downstream reaches of Ballona Creek, across the settling basin, and into the coastal ocean. Particle fleets were released in the model once every 15 minutes over the course of the 24 hr event to simulate particle transport and settling over the duration of the flood. Each fleet consisted of 80 particles, 10 of each grain size, and a scaling factor n_g was assigned to each particle such that the fleet as a whole accounts for the sediment mass that enters the channel over the 15 minute period between fleet releases. A total of 96 fleets were released over the 24 hour duration of the storm. Hydrodynamic predictions of the depth averaged velocity and the bed shear were used to fit the power-law velocity profile (Eq. 3) to each cell and subsequently estimate the point-wise velocity and dispersivity. Since particles are in general not located precisely at cell centers where the model computes velocity, linear interpolation was performed.

As was described in the previous section for the case of particle transport and settling, particles were assumed to reflect off the bed when the shear stress

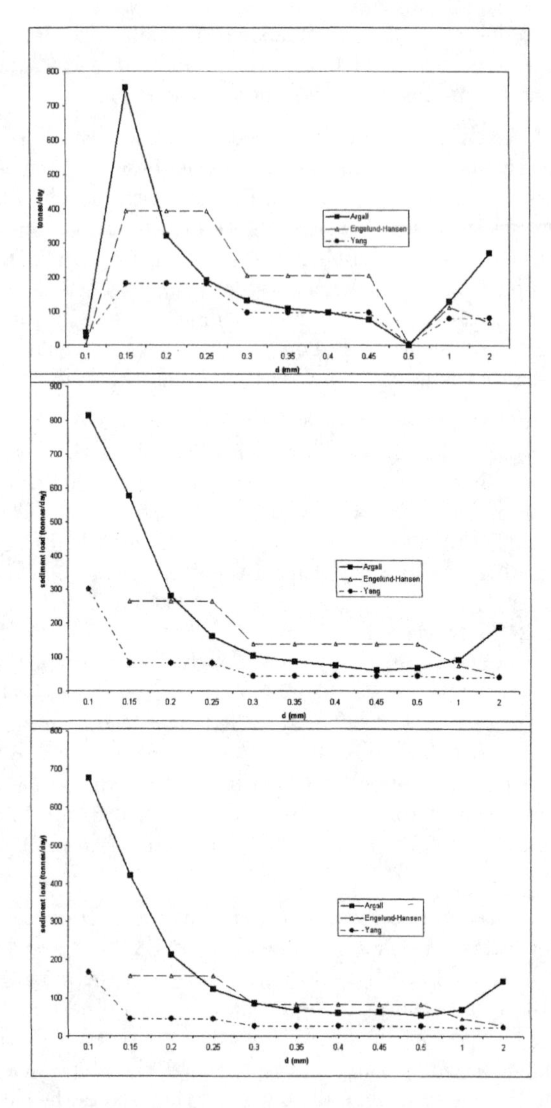

Figure 6: Comparison of sediment load predictions - Particle Tracking, Yang and Engelund-Hansen, n_m=0.02 (top), n_m=0.025 (middle), and n_m=0.03 (bottom).

exceeds the critical value for motion defined by the Shields parameter, and stick to the bed otherwise. Due to the unsteady nature of the flow however, particles initially predicted to settle under low shear conditions were once again suspended if, under high shear conditions later during the flood, the shear stress exceeded the threshold for motion. No attempt was made to model the erosion of the bed, which is likely to occur during the large flow events. That is, the model only tracks sediment from the upstream boundary. Consequently, for the largest flow events model predictions are less realistic due to signficant changes in the topography that aren't tracked by the model.

Figure 7 presents particle tracking model settling predictions for Ballona Creek in terms of the cumulative percent of the sediment mass trapped along the length of the channel. Note that this percentage increases substantially along the reach of the channel where the basin is located. The degree of trapping varies considerably with sediment size, and as would be expected, with return period of the event. The predictions suggest the basin is effective at catching larger sediment grain sizes but only during relatively frequent events, and poor at trapping finer sediments regardless of the event.

Summary and Conclusions

In this paper we present a particle based model for the prediction of non-cohesive suspended sediment transport. The model follows the position of a set of sediment grains that are representative of the sediment distribution as a whole, and scaling relationships are developed to relate particle transport predictions to sediment transport predictions. Particle advection by the mean flow is characterized under the condition of a turbulent boundary layer extending from the bed, each particle settles at a rate consistent with its weight and size, and random turbulent motions are resolved with a random-walk that scales with the turbulent diffusivity.

Model predictions of vertical sediment distributions map to Rouse profiles over a wide range of Rouse numbers using roughly 100 particles per unit volume of interest. In addition, the time step must be chosen based upon the Rouse number such that the near bed boundary layer is properly resolved. It was found that particle predictions follow the trend in the Rouse profile when the time step is roughly contrained by $w_f \Delta t / h < 0.01$.

Sediment loads predicted by the particle tracking scheme fall between predictions given by the Yang and Engluend-Hansen formulas for sediment sizes between 0.1 and 1 mm (silt and sand). For sediment sizes outside of these bounds, the particle model over-estimates loads. This error is not unexpected since the entrainment function used by the model is only suited to sediment sizes between 0.09 mm and 0.44 mm (Garcia 2001).

Particle tracking model settling predictions for Ballona Creek suggest the sedimentation basin is most effective at trapping larger grain sizes during the most frequent events. Model predictions pertaining to a flood event with a return pe-

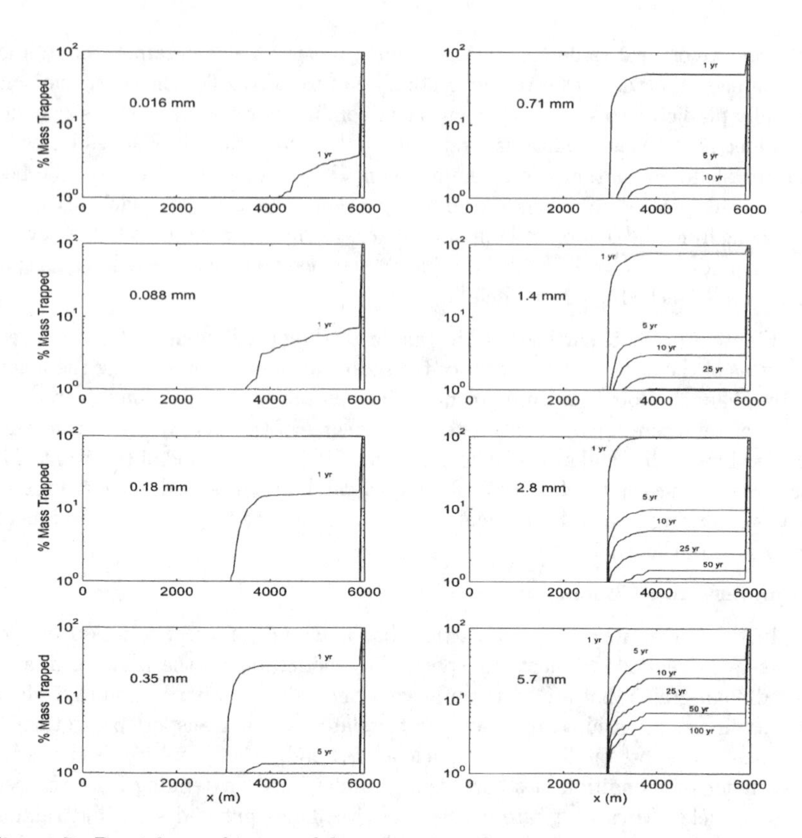

Figure 7: Particle tracking model predictions of cumulative percent of sediment mass trapped by Ballona Creek settling basin.

riod of 1 year show 100% mass trapped for the largest grain sizes (2.8 mm and 5.7 mm) and less than 5% mass trapped for the smallest grain sizes (0.016 mm and 0.088 mm). Settling predictions pertaining to the largest flood event (return period of 100 years) show no mass of sediment is trapped, with the exception of the largest class (5.7 mm) where less than 5% mass is trapped. In reality, for the larger events, the sediment accumulated in the basin is substantial relative to its storage capacity and therefore use of a more sophisticated model that tracks the movement of the bed and subsequent hydrodynamic changes is warranted.

Particle tracking models of sediment transport may offer an attractive alternative to concentration based models in a variety of settings. The chief advantage of the method is its ability to assign sediment grain attributes to individual particles, tracking each particle as if it was a single sediment grain, and then use the transport properties of a set of particles with different grain characteristics to infer information about the transport of polydisperse sediment distributions.

The settling basin analysis presented in this paper is a good example of this, highlighting the ability of the model to discriminate between grain sizes. Another advantage of the model is that the Lagrangian update equations are easy to solve and highly scalable in the numerical sense, facilitating computation in parallel computing environments. This is not necessarily the case should the model be modified to account for particle-particle interactions, which might be necessary to model cohesive sediment transport due to flocculation processes. Another attractive feature of the model is that it allows for quasi-three dimensional sediment transport modeling using hydrodynamic models of a reduced dimension, so long as the boundary layer is amenable to fitting by a distribution such as a power law. Again, the settling basin analysis is a good example of this. Using this approach, there is no need to model sediment dispersion as is the case in reduced dimension transport models, and there are no errors in the phase speed of particle transport predictions which might be the case in reduced dimension models because the depth-averaged sediment velocity is not necessarily the depth-averaged fluid velocity.

Acknowledgements

The research leading to this paper was made possible by a grant from the National Science Foundation (CMS-9984579), whose support is gratefully acknowleged. The authors also wish to thank C.V. Chrysikopoulos for his help in developing the particle tracking methodology.

References

Ahlstrom, S.W., Foote, H.P., Arnett, R.C., Cole, C.R. and Serne, R.J. (1977) "Multicomponent Mass Transport Model: Theory and Numerical Implementation," Report BNWL 2127, Battelle, Pacific Northwest Laboratories, Richland, Washington.

Alonso, C.V. (1981) "Stochastic Models of Suspended Sediment Dispersion." Journal of the Hydraulic Division ASCE 107(HY6), pp. 733-757.

Bradford, S.F. and Sanders, B.F. (2002), "Finite-Volume Model for Shallow-Water Flooding of Arbitrary Topography,", *Journal of Hydraulic Engineering*, 128(3), pp. 289-298.

Crowe, C.C. Roberson, J.A., and Elger D.F. (2001) *Engineering Fluid Mechanics*, 7th Edition, Wiley, New York, N.Y.

Dimou, K.N. and Adams, E.E. (1993) "A Random-Walk, Particle Tracking Model for Well-Mixed Estuaries and Coastal Waters," *Estuarine, Coastal and Shelf Science*, 37, pp. 99-110.

Engelund, F. and Hansen, E. (1967) "A Monograph on Sediment Transport to Alluvial Streams," Copenhagen, Teknik Vorlag.

Fischer, B.H., List, E.J., Koh, R. C.Y., Imberger, J., and Brooks, N. H. (1979) "Mixing in inland and coastal waters." Academic Press, Inc., New York, N. Y.

Garcia, M. (2001) "Modeling Sediment Entrainment into Suspension, Transport and Deposition in Rivers," Model Validation: Perspectives in Hydrological Science, Malcolm G. Anderson (Editor), Paul D. Bates (Editor), Wiley.

Garcia M, Parker G. (1991) "Entrainment of Bed Sediment into Suspension," Journal of Hydraulic Engineering-ASCE 117 (4): 414-435 APR 1991.

Gnedenko, B.V. (1968) ."Theory of Probability," Chelsea Publishing Co., New York, N.Y.

Hu, H. (1999) "Three-dimensional modeling of open channel flows and transport of suspended sediment.", Ph. D. Dissertation, Dept. of Civil and Environmental Engineering, University of Houston.

Julien, P.Y. (1995) "Erosion and Sedimentation." Cambridge University Press, Cambridge.

Li, R and Shen, H.W. (1975) "Solid Particle Settlement in Open-Channel Flow" Journal of the Hydraulics Division, ASCE, Vol. 101, No. HY7, pp 917-931.

Pearson, K. (1905) "The Problem of the Random Walk" *Nature*, Vol. 72, No. 1865.

Rayleigh, Lord (1880) "On the Resultant of a Large Number of Vibrations of the Same Pitch and Arbitrary Phase," *Philosophical Magazine*, Vol. X,pp. 73-78.

Rouse, H. (1937) "Modern conceptions of the mechanics of turbulence." Trans. ASCE, ASCE, 102, 463543.

Signell, R. P. and Geyer, W.R. (1990) "Numerical Simulation of Tidal Dispersion Around a Coastal Headland", in Proceedings of the International Conference on the Physics of Shallow Estuaries and Bays, Asilomar, CA, November, 1988, edited by R. Cheng, p. 210-222, Springer-Verlag, New York, N.Y.

Tompson A. and Gelhar L.W. (1990) "Numerical-Simulation of Solute Transport in 3-Dimensional, Randomly Heterogeneous Porous-Media," *Water Resources Research*, 26(10), pp. 2,541–2,562.

Yang, C. T. (1973) "Incipient Motion and Sediment Transport," Journal of the Hydraulics Division, ASCE, Vol. 99, No. HY10, pp 1679-1704.

Sediment Transport Associated with the Baroclinic Term in the 3D-Hydrodynamic Circulation Model; Study Case for Idealized Estuary

Wahyu W. Pandoe [1]and Billy L. Edge[2]

Abstract

This work provides a general hydrodynamic circulation model that can be used to understand density driven flows, which may arise in the case of suspension of fine-grained materials. The research is expected to provide a better understanding of the characteristics of spatial and temporal variability of current, which is to be associated with the period of ebb and flood tidal cycles with the inclusion of baroclinic term. The understanding of this process provides a basis for determining how the water circulation three-dimensionally controls the hydrodynamics of the system and ultimately the transport of suspended material and soluble materials.

The model development includes extending the existing three-dimensional (3D) ADCIRC Hydrodynamic Model with (1) baroclinic forcing term and (2) transport module of suspended and soluble materials. The transport module covers the erosion, material suspension and deposition processes for both cohesive and non-cohesive type sediments. In the study case of an idealized rectangular basin, the model provides less than 6% error of mass conservation for eroded, suspended and deposited material sediment. The influence of the baroclinic forcing on the distribution of suspended sediment, mainly for cohesive type sediment, under influence of the tidal current and normal flux fresh riverine water was demonstrated. The inclusion of the baroclinic term might increase the total net transport of suspended cohesive sediment.

1. Introduction

Shoaling processes in a bay is presumed to correlate with the flow pattern in the estuary driven by tide generating current and wind stress. Sediment conveyed by river runoff to the estuary might also be one of the sedimentation sources.

The equation of motion shows that water moves in response to differences in pressure, which are generated by two factors: water surface slope (barotropic) and horizontal water density differences (baroclinic). The determination of barotropic and baroclinic flow is important since it may cause a convergence zone, where the converging pressure gradients drive internal circulation patterns into a common point. The baroclinic term may also cause dissimilarities between ebb and flood tidal

[1] Graduate Research Assistant, Department of Civil Engineering, Texas A&M University, College Station, Texas 77843-3136; email: wpandoe@tamu.edu

[2] Professor, Department of Civil Engineering, Texas A&M University, College Station, Texas 77843-3136; email: b-edge@tamu.edu

currents. The presence of baroclinic-forced currents may strengthen the existence of the reverse estuarine flow.

The scope of this study is to implement and expand the hydrodynamic numerical modeling system to accommodate these density driven flows for an estuarine zone. The model used here is the three-dimensional (3D) version of Advanced Circulation (ADCIRC) Hydrodynamic Model with extended Transport module. ADCIRC is a finite element model used for hydrodynamic circulation problems. The model is based on the finite element codes that solve the shallow water equation on unstructured grids. Fine grid resolution can be specified locally to meet the accuracy requirements, and coarse resolution can be implemented in area distant from the region of interest. In the 3D-ADCIRC model, the baroclinic term might be appropriate to explain the variability of vertical current shear, which may alter sediment concentration vertically.

In the case of density stratification as a result of either little tidal action or with large fresh-water (river) flow, the flow profile can be separated into two portions in the seaward regions of the tidal intrusion: (a) the upper part is the fresh water flow with little mixing to the seaward, and (b) the lower part is the saline water practically similar to in the ocean water, that is called the 'salinity wedge'. With the inclusion of the baroclinic term in the model development for the idealized estuarine model, the existence of salinity wedge is distinctive penetrating to the river upstream. The objective of this research is to identify the circulation patterns of the water and sediment fluxes in an estuarine zone. Suspended materials in the water column are highly related to the hydrodynamic circulation. The interface of the salinity wedge or a convergence zone may be related to the presence of deposited sediments.

2. Model Description

Hydrodynamic. In this three-dimensional circulation model, the vertical finite element domain will be divided into a number of sigma (σ) layers ranging from -1 at the bottom to 1 at the surface. In some cases, high gradients may occur at the bottom and surface, thus normally the vertical grids have higher density around bottom and surface and lower in the mid layers. The governing equations for ADCIRC-3D are given as follows:

Continuity equation. Water surface elevation is solved in a vertically integrated continuity equation utilizing the GWCE (Generalized Wave Continuity Equation) formulation. The modified GWCE is derived as a summation of the time derivative of the continuity equation and the spatial gradient of the momentum equations. The GWCE is used to solve the sea level elevations. The complete formulation of GWCE is given in *Luettich et al.* (1992).

$$\frac{\partial H}{\partial t}+\frac{\partial (UH)}{\partial x}+\frac{\partial (VH)}{\partial y}=0 \tag{1}$$

where U and V are depth-averaged velocities, $H = h + \eta$, h is a bathymetric depth and η is a free surface elevation relative to the geoid.

Momentum equations. The three-dimensional version of ADCIRC applies the non-conservative form of the momentum equations to solve horizontal velocity u and v. The free surface elevation as described in Eq.1 is solved by substituting the vertically-integrated momentum equations into the continuity equation to form the GWCE. The momentum equations applied in ADCIRC-3D (Luettich and Westerink, 2003) are:

$$\frac{\partial u}{\partial t}+u\frac{\partial u}{\partial x}+v\frac{\partial u}{\partial y}+\omega\left(\frac{a-b}{H}\right)\frac{\partial u}{\partial \sigma}-fv=-g\frac{\partial \eta}{\partial x}+m_{x_\sigma}-b_{x_\sigma}+\left(\frac{a-b}{H}\right)\frac{\partial}{\partial \sigma}\left(\frac{\tau_{zx}}{\rho_0}\right) \tag{2a}$$

$$\frac{\partial v}{\partial t}+u\frac{\partial v}{\partial x}+v\frac{\partial v}{\partial y}+\omega\left(\frac{a-b}{H}\right)\frac{\partial v}{\partial \sigma}+fu=-g\frac{\partial \eta}{\partial x}+m_{y_\sigma}-b_{y_\sigma}+\left(\frac{a-b}{H}\right)\frac{\partial}{\partial \sigma}\left(\frac{\tau_{zy}}{\rho_0}\right) \tag{2b}$$

where u and v are velocities in the x and y direction; ω = vertical velocity in σ-coordinate; f = Coriolis force; g = gravity acceleration; $m_{x\sigma}$ = combined horizontal diffusion/dispersion momentum; $b_{x\sigma}$ and $b_{y\sigma}$= baroclinic pressure term in x- and y-directions; τ_{zx} and τ_{zy} = component of vertical shear stress; and ρ_o = reference density of water.

Velocities are determined from the non-conservative form of the momentum equation. The solution strategy for solving horizontal velocities u and v in Eqs. 2a,b includes finite element method for spatial and finite difference for temporal. The vertical grid nodes are defined vertically at each horizontal node, thus the horizontal and vertical integrations can be performed independently. The detail of the 3D-VS formulations is given in Luettich and Westerink (2003).

Vertical velocity is solved by the first derivative approach with the adjoint correction. Pandoe and Edge (2003) solved for ω in σ-coordinate, with essential boundary condition $\omega = 0$ at $\sigma = $ b, and natural boundary condition $\delta\omega=0$ at $\sigma=$a :

$$\omega_{k+1}-\omega_k=-\frac{1}{(a-b)}\int_k^{k+1}\left(\frac{\partial \eta}{\partial t}+\frac{\partial (u\,H)}{\partial x}+\frac{\partial (v\,H)}{\partial y}\right)\partial\sigma \tag{3}$$

where k is a node number of a vertical element. The solution ω_k will satisfy the bottom boundary condition only. In order to satisfy the free surface, the adjoint correction is applied to Eq. 3 based on Luettich and Muccino (2001) and Muccino et al. (1997).

Salinity//Tracer Transport model. The general governing equation for transport of salinity, temperature and tracer concentration (Blumberg and Mellor, 1987; Scheffner, 1999; Helfand et al., 1999; and HydroQual, 1998) is summarized as follows:

$$\frac{\partial R}{\partial t}+u\frac{\partial R}{\partial x}+v\frac{\partial R}{\partial y}+\frac{(a-b)}{H}(\omega-\omega_s)\frac{\partial R}{\partial \sigma}= \frac{\partial}{\partial x}\left(D_h\frac{\partial R}{\partial x}\right)+\frac{\partial}{\partial y}\left(D_h\frac{\partial R}{\partial y}\right)+\left(\frac{a-b}{H}\right)^2\frac{\partial}{\partial \sigma}\left[D_v\frac{\partial R}{\partial \sigma}\right] \tag{4}$$

where R is either salinity [psu] or a tracer concentration [g/l or kg/m^3]; D_h and D_v are horizontal and vertical dispersion coefficients [m^2/s]; ω is vertical velocities in σ-coordinate [m/s]; and ω_s is settling velocity of sediment in σ-coordinate [m/s]. The surface and bottom boundary conditions of Eq. 4 for vertical salinity gradient are zero.

In our development of the ADCIRC Transport, the horizontal diffusion coefficient, D_h, is assumed constant within the range of 0.1-10 m^2/s (Gross et al., 1999). The level 2.5 Turbulence Closure Model of Mellor-Yamada (1982) is used to calculate the vertical eddy diffusivity D_v.

Baroclinic Forcing Terms. Robertson et al. (2001) applied normalized density in order to reduce the truncation error in the computation of the baroclinic pressure gradient in Princeton Ocean Model (POM). Thus, the normalized baroclinic forcing terms in Eq. 2a,b on 3D-ADCIRC can be represented as:

$$b_x = \frac{g}{\rho_0} \int_z^{\eta} \frac{\partial}{\partial x}(\rho - \rho_0)\, dz = \text{baroclinic x - forcing}$$

$$b_y = \frac{g}{\rho_0} \int_z^{\eta} \frac{\partial}{\partial y}(\rho - \rho_0) dz = \text{baroclinic y - forcing}$$

where η is a free surface elevation, and ρ_o is a general mean density, which was set to 1000 (or 1025) kg/m^3. The baroclinic terms $b_{x\sigma}$ and $b_{y\sigma}$ are a function of density distribution. The variable density is determined from temperature T, salinity S and pressure p using the *International Equation of State of Sea Water, IES80.*

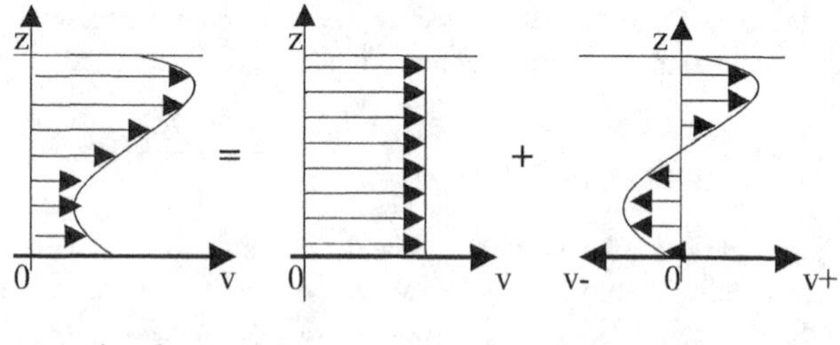

Velocity shear = Barotropic + Baroclinic

Figure 1. The influence of barotropic and baroclinic terms on the vertical velocity shear.

The barotropic term is constant with depth, in contrast to the baroclinic term which varies with depth (Figure 1). The interaction of these two terms can create tidal asymmetry between the ebb and the flood of tidal cycle. During the flood current, the baroclinic and barotropic term could be additive producing high acceleration near the bed, and vice versa during ebb current. In the region where the thermal gradient of the sea is most pronounced such as in estuarine and coastal regions, the condition of baroclinicity is the most extreme. Generally, the baroclinic situation can be found in the surface layer with the barotropic conditions being approached at greater depth.

Suspended Sediment Transport Model. The general governing equation for sediment concentration (HydroQual, 1998) is similar to Eq.(4), but with boundary conditions:

$$\left(\frac{a-b}{H}\right) D_h \frac{\partial C}{\partial \sigma} = -\omega_s C_k \qquad \text{at} \qquad z \rightarrow \eta$$

$$\left(\frac{a-b}{H}\right) D_h \frac{\partial C}{\partial \sigma} = E - D \qquad \text{at} \qquad z \rightarrow -h$$

where C is a sediment concentration [gr/lt or kg/m^3]; E is an erosion flux [kg/m^2/s]; and D is a deposition flux [kg/m^2/s]. The sediment transport covers both cohesive and non-cohesive types of sediment.

1) Non-Cohesive sediment (sand)

Above the threshold of motion, sand/sediment in the bed is lifted off into suspension, where it is carried by the current. The bottom friction governs the entrainment of sediment from the bed. In a sediment suspension, the settling of sediments towards the bed is counterbalanced by diffusion of sand upward near the bed. Soulsby (1997) describes the corresponding concentration profile, defined by van Rijn's profile (see Figure 2), as:

$$C_e(z) = C_a \left[\frac{z}{z_a} \cdot \frac{(H - z_a)}{(H - z)} \right]^{-b'} \qquad \text{for za} < z < H/2 \qquad (5a)$$

$$C_e(z) = C_a \left(\frac{z_a}{H - z_a} \right)^{b'} \exp\left[-4b' \left(\frac{z}{H} - \frac{1}{2} \right) \right] \qquad \text{for } H/2 < z < H \qquad (5b)$$

with b' is a modified form of Rouse number or suspension parameter; z is a height above sea bed [m]; z_a is a reference height near sea bed [m]; $C_e(z)$ is a sediment concentration at height z due to bed load [mass/volume]; and C_a is a sediment reference concentration at height z_a [mass/volume]

$$C_a = \frac{0.015 d T_s^{3/2}}{z_a D_*^{0.3}} \qquad \text{with: } T_s = \frac{\tau_b - \tau_{cr}}{\tau_{cr}} = \text{threshold shear stress} \qquad (6)$$

$H = h + \eta$; d is a mean grain size [mm]; b is the Rouse number or suspension parameter: $b = \frac{w_s}{\kappa u_*}$; κ is the von Karman's constant = 0.4; u_* is a friction velocity; τ_b is the bed shear stress; τ_{cr} is a critical shear stress; and D_* is dimensionless grain size.

In order to assure that the suspended sediment concentration in the water column C follows the distribution of C_e given in Eq. 5a,b, the erosion and diffusion terms, E and D, are taken as:

$$E = \frac{1}{\Delta t} \int (C_e(z) - C(z)) dz \qquad \text{for } C_e(z) > C(z)$$

$$D = -\frac{w_s}{H} \int (C_e(z) - C(z)) dz \qquad \text{for } C_e(z) \le C(z)$$

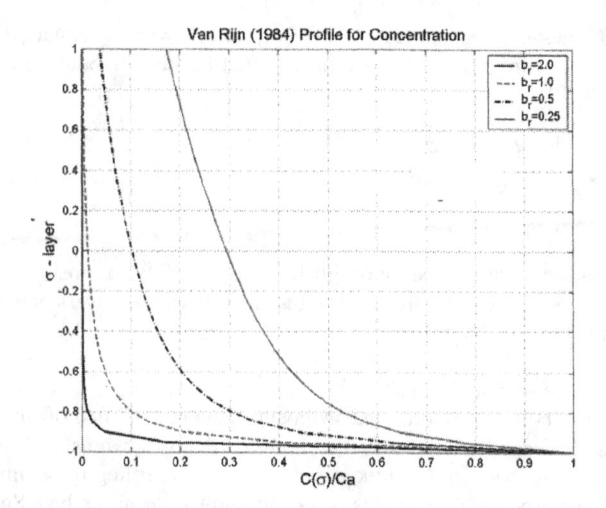

Figure 2. Van Rijn's vertical profile of suspended sediment concentration with variation in Rouse Number b_r, i.e. grain size and current speed. (After Soulsby, 1997)

Here we assume that the deposition removes an excess of suspended material in the water column with settling velocity w_S. In case $\tau_b < \tau_{cr}$ (ie. $C_e = 0$), the deposition flux term reduces to a general settling formulation:

$$D = w_s \, C_b$$

where D is a depositional flux of non-cohesive sediment [kg/m^{2}s]; C is a concentration of suspended non-cohesive sediment [kg/m^3]; C_b is the near bottom sediment concentration [kg/m^3]; and w_s is a sediment settling velocity [m/s]

The critical bed shear stress in Eq. (6) can be determined from the threshold Shields parameter θ_{cr} as given from Soulsby (1997) :

$$\tau_{cr} = \theta_{cr} g(\rho_s - \rho_w) D_{50},$$

where:

$$\theta_{cr} = \frac{0.30}{1+1.2D_*} + 0.055[1 - \exp(-0.02D_*)] = \text{ threshold Shields parameter}$$

$$D_* = \left(\frac{\Delta g}{\upsilon^2}\right)^{1/3} D_{50} \quad = \text{dimensionless grain size}$$

$$\Delta = \frac{\rho_s - \rho_w}{\rho_w},$$

Settling velocity of natural sand particle is evaluated using the formulation given by Cheng (1997) as follows:

$$w_s = \frac{\nu}{D_{50}}\left[\left(25 + 1.2 D_*^{\,2}\right)^{1/2} - 5\right]^{1.5} \tag{7}$$

where D_{50} = median grain diameter [m]; ν = kinematic viscosity [m^2/s]; ρ_s is a density of sediment; and ρ_w is a density of water.

2) Cohesive sediment (clay [1μm] – silts [50μm])

For cohesive type of sediment, Whitehouse et al. (2000) assumed that the flocs of cohesive sediment could be treated as low-density grains, where aggregation of flocs, break-up of flocs and water-flow within flocs are neglected. The corresponding formula of settling velocity w_s is given as:

$$w_s = \frac{\upsilon}{d_e}\left\{\left[10.36^2 + 1.049\left(1 - C_f\right)^{4.7} D_*^3\right]^{1/2} - 10.36\right\} \tag{8}$$

where: $D_* = d_e\left[\frac{g(\rho_e - \rho)}{\rho \upsilon^2}\right]^{1/3}$ is a dimensionless floc diameter; d_e is the effective diameter of a floc that increase with the volume concentration C of the suspension, ρ is water density; ρ_e is effective density of the floc; C_f is the volume concentration of flocs in water [non-dimension], and C_k is the mass concentration of the suspension [mass/volume].

The depositional flux rate, D, is computed using general settling formulation for cohesive sediment:

$$D = P\, w_s\, C_b$$

where:

$$\text{P = probability of deposition} = P = \begin{cases} -\left(1 - \dfrac{\tau_b}{\tau_{cd}}\right) & \text{for} \quad \tau_b < \tau_{cd} \\ 0 & \text{for} \quad \tau_b \geq \tau_{cd} \end{cases}$$

w_s = sediment settling velocity (m/s); τ_b is the bed shear stress ; τ_{cd} is a critical bed shear stress and is estimated from laboratory tests to be between 0.06 and 0.10 N/m^2; and C_b is the near bottom sediment concentration [kg/m^3].

The analysis to estimate the erosion flux rate (E) expressed as dry mass of mud eroded per unit area per unit time (kg m^{-2} s^{-1}) as a function of shear stress is given by Partheniades (1992), see Whitehouse et al (2000):

$$E = m_e\left(\tau_b - \tau_{ce}\right) \quad , \quad \tau_b > \tau_{ce}$$
$$E = 0 \quad , \quad \tau_b \leq \tau_{ce}$$

where m_e is an experimental/site specific erosion constant with m_e = 0.0002 - 0.002 kN^{-1}s^{-1}; τ_{ce} is the critical bed shear stress for erosion given in Whitehouse et al. (2000) as:

$$\tau_{ce} = 0.015\left(\rho_b - 1000\right)^{0.73}$$

Typical τ_{ce} is around 0.1-0.2 N/m^2 but it should not exceed 1.0 N/m^2, where ρ_b is the bulk density of the bed over the density range 1000 to 2000 kg/m^3.

3. Model Validation

The results of hydrodynamic velocity components *u*, *v* and *w* are compared to some analytical solutions. Three-dimensional analytical solution was originally developed by Lynch and Officer (1985), which is also available in Muccino et al. (1997). The Quarter Annular Test Problem (QATP) is applied for comparison between the model and analytical solutions with a good agreement for the hydrodynamic solution (Pandoe and Edge, 2003).

The benchmarking of the stratification due to the presence of baroclinic terms will be performed using the theoretical 'saline wedge' formulation provided by Partheniades (1992). The comparison against theoretical length of saline wedge will be discussed in the next section.

For model validation of non-cohesive type sediment transport, the model is compared with an analytical model provided by Van de Kreeke et al. (2002). The initial cross section of the trench in the Van de Kreeke formulation is approximated by a Gaussian distribution:

$$d = \frac{A}{\sqrt{2\pi}\sigma_o} e^{-x^2/2\sigma_o^2}$$

where *d* is a depth of the trench relative to the surrounding sea bottom, A is the cross sectional area of the trench, and σ_o is the half-width of the initial cross section.

The model domain is a long-flat rectangular channel with uniform depth 3m, and 5 m deep of a 'Gaussian shape' trench located across the channel width in the mid point of the channel. The model is driven by the 0.5 m amplitude of M2 tide on the left side and an influx normal flow 0.5m/s from the right downstream to the left. The M2 tidal range is required to generate the M2 tidal current. Investigating the model results, the maximum amplitude of M2 tidal current on the shelf around the trench is 0.12m/s. The comparison of the deposition and erosion at the bottom across the trench for median grain size 0.3mm is shown in Figure 3. The top figure shows the amount of erosion (-) and deposition at 5 days. The bottom figure represents the change of the bathymetric depth at t = 5-day relative to its initial depth profile at t = 0-day.

There is good agreement in the magnitude of both erosion and deposition; however there are also shifts of deposition and erosion, where the erosion occurs shifted to the left and the deposition shifted to the right. Van de Kreeke (2002) assumes that the water level is assumed uniform across the channel, so the solution will be symmetric between erosion and deposition.

The assessment on the model indicates that the water level and velocity in both edges of the channel are not uniform (figure is not shown). The water level in the edge of the upstream direction is slightly higher than the downstream site, thus, the velocity in the downstream site is slightly higher than the upstream site to the order of 0.07m/s. This velocity difference may drive different amounts of suspended sediment C_e between both sides. Thus, considering the transport balance, the erosion is more likely to occur more in the downstream site than in the upstream. Consequently, the velocity difference across the channel may explain the discrepancies between the model and the analytical models.

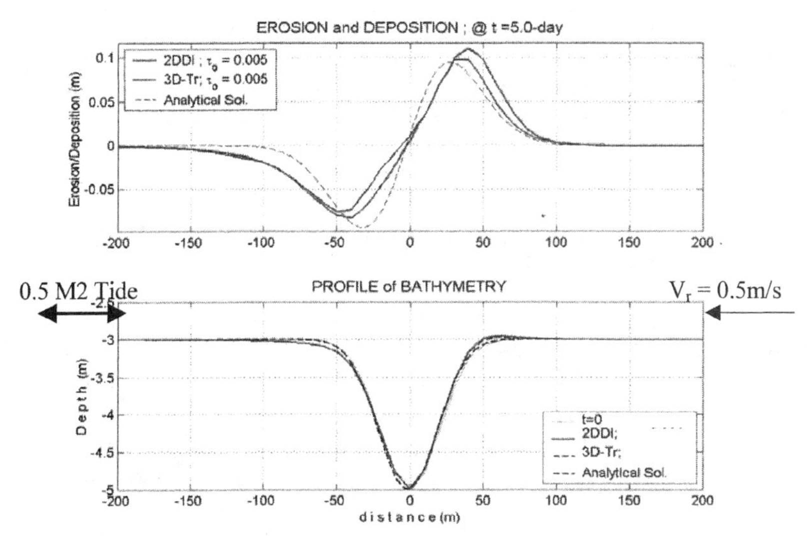

Figure 3. Comparison of erosion and deposition of non-cohesive sediment for d_{50}=0.3mm between ADCIRC 3D, ADTRANS-2DDI and Van de Kreeke's analytical models.

The comparison was also made against the ADCIRC-Transport (ADTRANS) 2DDI version. ADTRANS 2DDI is a two-dimensional ADCIRC and Transport model (Scheffner, 1999). The 3D result slightly improves the magnitude of erosion and deposition closer to the analytical model. Both models, however, over-estimate the magnitude of deposition in the channel upstream direction and erosion in the downstream direction.

4. Saline Wedge

Assuming that the distribution of constituent is symmetrical between upper-half and lower-half of the idealized estuary, and with neglecting the coriolis force, the simulation of developed saline wedge is performed only in the upper-half of an idealized estuary with the domain given in Figure 4.

In this case, the model is driven by the 0.8m amplitude of M2 tide in the open ocean boundary (left side), and by the influx normal riverine flow with a constant velocity 0.2 m/s. The initial condition of the domain has uniform salinity S_o = 35 psu and temperature T=19°C in the ocean part, while the river flow produces "fresh water" inflow with salinity S_r = 3 psu and temperature T_r = 19°C. The salinity gradient between ocean and river generate the pressure gradient due to salinity (i.e. density) difference. In this case, the prevailing stratification and existence of density gradient will drive a baroclinic flow towards the river that contributes to the development of a saline wedge. The balance between outward river flow to the ocean and inward flow into the river due to the baroclinic term provides a nearly steady point at the bottom of the penetrated saline wedge upstream. When the stability of hydrodynamic flow is achieved, the end tip of saline wedge consequently should shift up- and down-stream periodically in coherence with the period of the driving tidal current. It is found that the distance range of that periodic shifting is within the range of less than 500m.

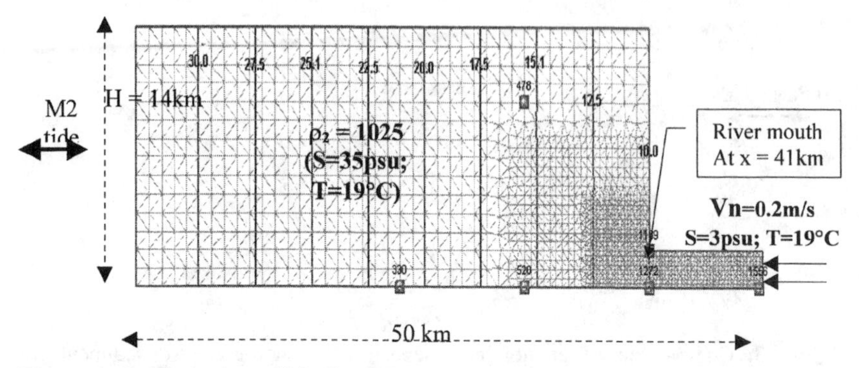

Figure 4. Configuration of Idealized Estuary (upper-half part only)

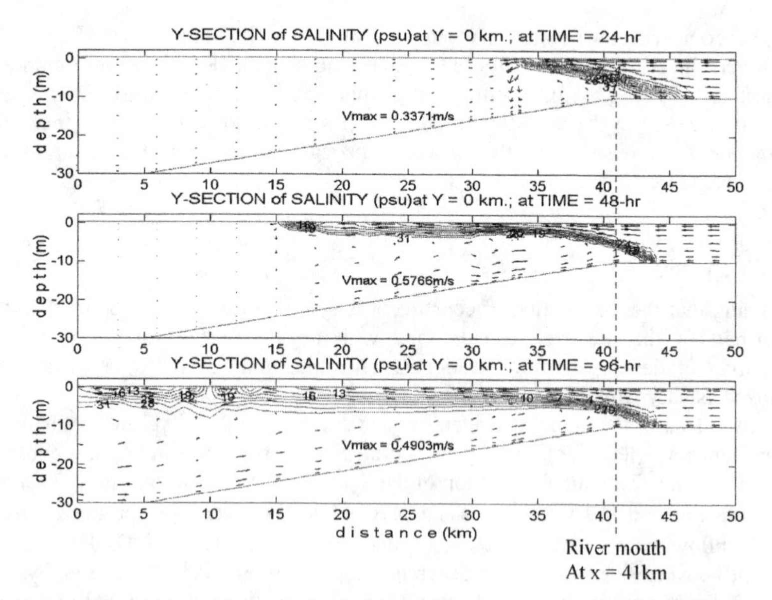

Figure 5. Side-view profiles of salinity along the transect indicated in Figure 4, during the developed saline wedge at t = 1st-day (top), 2nd-day (middle) and 4th-day (bottom).

Figure 5 (a,b and c) shows vertical profiles of developed saline wedge due to freshwater influx from the river out to the ocean at the end of first, second and fourth days, respectively. The profiles represent a transect along x-axis in Figure 4. The upper layers exhibit strong outward flow immediately outside the river mouth and weakened onshore. The maximum outward flow occurs at the river mouth with 0.5 m/s maximum velocity. Meanwhile the inward flow to the river develops the saline wedge with penetrating distance up to 3km upstream.

Empirical computation of saline wedge formulated by Partheniades (1990) giving various lengths of penetrating saline wedge based on the outward river velocity, is given in the Table 1 below and Figure 6. If we consider the velocity of normal flow 0.2m/s, then the length of saline wedge is about 20.7km, which is somewhat longer than the model result; however if we consider the maximum normal velocity around the tip of saline wedge, V_r=0.5 m/s, then the length of saline wedge is 2.4 km which is in good agreement with the numerical model applied here. The authors are currently investigating more detail to assess the problem.

Table 1. Length of Saline wedge

V_r (m/s)	L_W (km)
0.2	20.7
0.4	4.4
0.5	2.4

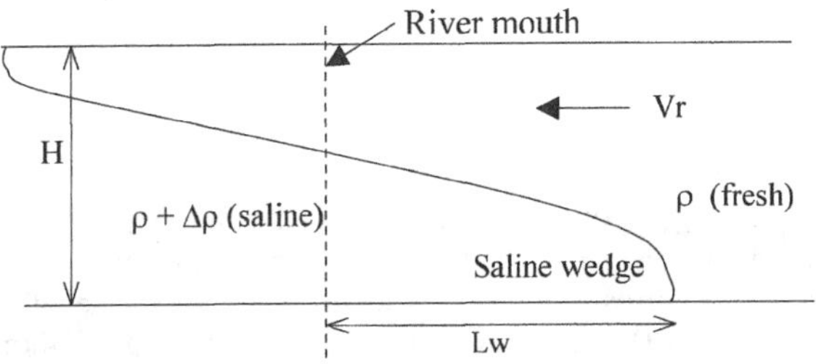

Figure 6. Saline wedge; L_W is the length of saline wedge computed from the river mouth.

5. Conservation of Mass

A simple sloped-bottom rectangular domain with a shallow bar located in the middle of the basin was developed as given in Figure 7. The open ocean boundary is located in the left (west) side of the domain; the normal flow boundary is located in the right (east) side, while the other two sides are defined as mainland (solid) boundaries. The dimensions are 12 km long and 4 km wide, with the 1 m depth of a sand bar located between 8 and 9 km from the ocean boundary. The bay dimension behind the bar (x = 9 - 12 km) has uniform depth of 2.5m, while from the bar to the open ocean boundary has a sloping bottom depth from 1m to 20m over 8 km. The triangular grid size was specified to be 125m along the x-axis and 250m along the y-axis, giving the number of nodes and elements of 1699 and 3079, respectively. Nine vertical layers are implemented in this simulation which are distributed uniformly in the vertical σ-layers from σ=-1 at the bottom to σ=1 at the surface.

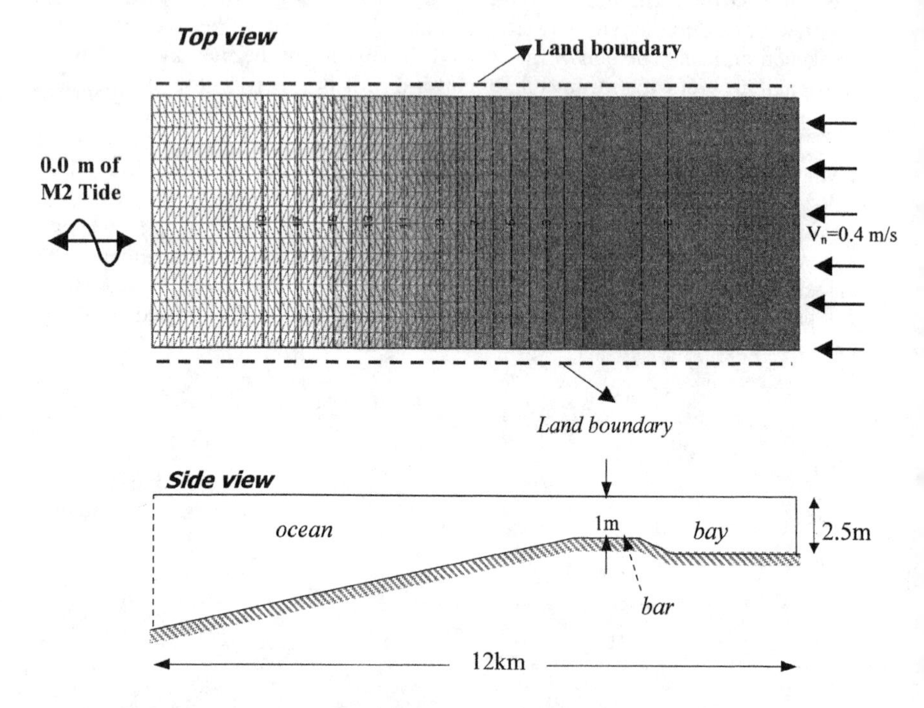

Figure 7. The configuration of barred rectangular basin. (Upper) The grid consists of 1699 nodes and 3079 elements and the bathymetric depth varies from 1.0 m to 19.0 m. Open ocean and normal flow boundaries are specified on the left and right side, respectively. (Lower) The side view of the basin, 12km long and 4km wide.

The grid was selected for the analysis of the conservation of mass for the sediment transport. To assess the conservation of mass for non-cohesive sediment, the simulation is driven with constant velocity 0.4 m/s of normal inflow, and zero tide at the open boundary. Starting at the bottom slope of the bar, a velocity will increase causing the increase of the bottom stress. Erosion occurs when the bottom shear stress τ_b has exceeded the critical shear stress for erosion τ_{ce}. For this case, erosion occurs starting near the top of bayside slope of the bar, continuously along the bar to the other tip of the bar. Figure 8 shows the along-basin vertical profile of the sediment transport results at 3-day simulation for a non-cohesive sediment type with uniform grain size (d_{50}) of 0.1 mm. The profiles represent the mid-line of the basin (i.e. y=2000m).

As shown in the top of Figure 8, the distribution of suspended sediment due to erosion is mostly contained near the bar. Arrows indicate the direction and magnitude of three dimensional flow components. Using Eq. 8, the settling velocity of 0.1mm grain size is ~0.6 mm/s or ~2.2m/hr. The existing very shallow 1m bar produces maximum horizontal velocity of about 1.1m/s near the left tip of the bar where the maximum erosion occurs. With these conditions, it is expected that the suspended sediment will not be advected far after it has been suspended into the water column. The deposition process occurs immediately; after the τ_b is less than τ_{cre}. The corresponding profile of erosion and deposition is given in the bottom Figure 8.

Figure 8 (middle) describes the depth integrated concentration (C_i) of suspended sediment in the water column at each node. The concentration units in g/l in Figure 8 (top) is integrated in vertical and lateral directions, and then converted to volume given in units of m^2, which represents the mass volume per width (m^3/m). Assuming the distribution of transport is uniform across the basin, the simple way to estimate the conservation of mass is to compute the area along the profile with the expected relationship:

$$\textit{Erosion } (m^2) = \textit{Depth integrated suspended sediment } (m^2) + \textit{Deposition } (m^2)$$

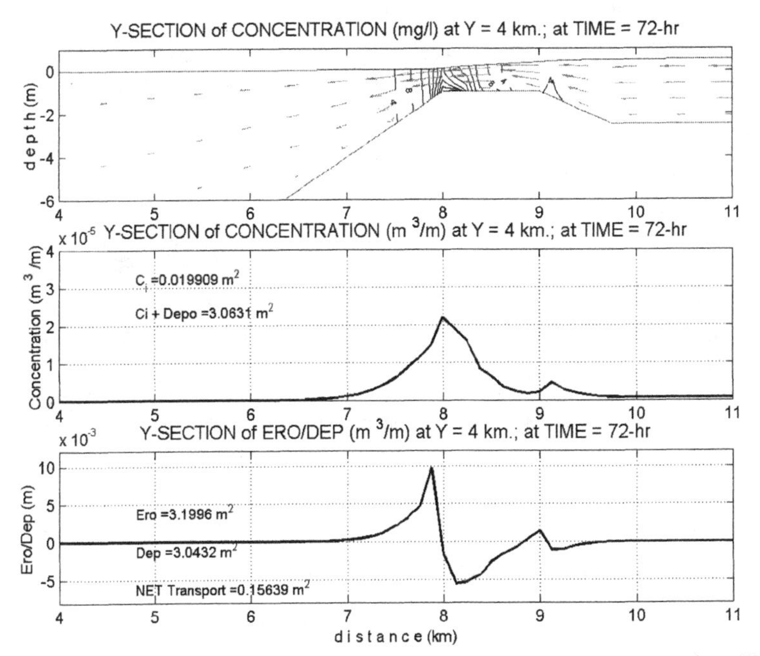

Figure 8. Profile of non cohesive sediment transport, with d_{50} = 0.1mm and t = 72h. (Top) Contour plot of suspended sediment concentration with arrows representing the relative magnitude and direction of current velocity. (Middle) Depth integrated concentration along the profile in term of m^2/m, and (bottom) the corresponding erosion (-) and deposition (+) along the profile line.

Thus, from Figure 8 bottom, we should expect that the sum of C_i + *Depo* should equal to *Ero*, but the results are not perfectly met. In this case, the error was $0.13m^2$ or about 4.3% of the total transported mass. The error of conservation of mass is usually less than 6%, especially in the sloping bottom. The sources of the error are commonly due to accumulative numerical accuracy either from hydrodynamic or transport modules. The finite element method for solving the governing equations is usually conserved locally, but not globally. Small numerical rounding errors might propagate from element to element and develop accountable errors. Conversion of a vertical reference from z- to σ-level can introduce another source of conservation errors.

For the cohesive sediment type, a similar way of assessment for conservation of mass is applied in the same grid and bathymetric configuration. The results are given in Figure 9 at time t = 28h. The erosion is arbitrarily stopped after 24h, then it is expected that the amount of depth-integrated suspended sediment plus deposited sediment must be constant. It is known that the properties of cohesive sediment in a bed is very easily eroded and suspended when τ_b exceeds τ_{cre}, transported in a relatively long distance, and deposited back to a bed at particular location whenever the τ_b is less thanτ_{crd}. The amount of suspended sediment is high and settles slowly due to slow settling velocity as given in Eq. 9.

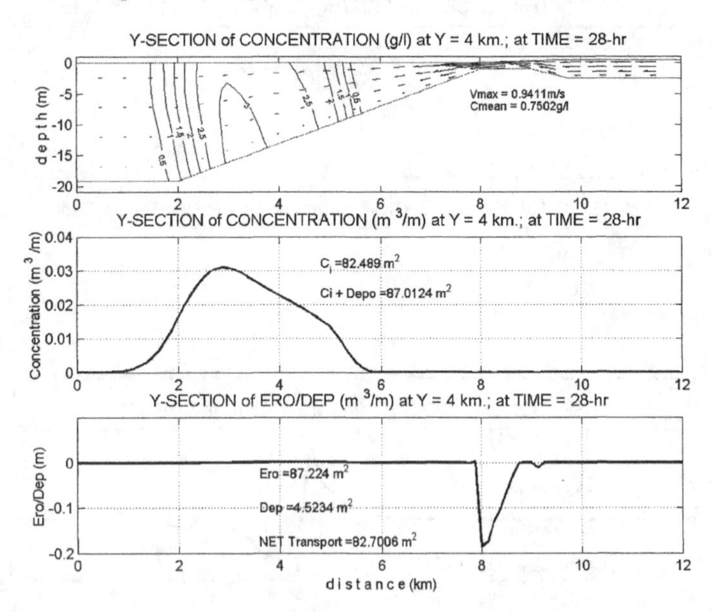

Figure 9. Profile of cohesive sediment transport taken at time t = 28-h. (Top) Contour plot of suspended sediment concentration with arrows representing the relative magnitude and direction of current velocity. (Middle) Depth integrated concentration along the profile in term of m^2/m, and (bottom) the corresponding erosion (-) and deposition (+) along the profile line.

At time t=28h, it is shown that the conservation mass of coincidently is better than the non-cohesive sediment. The error in this simulation is less than 1% of the transported mass volume. In the top panel of Figure 9, initial erosion occurs around the bar, and after the erosion is forced to stop at t=24h, then the cloud of cohesive suspended sediment will travel offshore while the particles tend to settling down. Thus the prevailing sediment concentration should be high at near bottom decreasing upward to the surface.

6. Barotropic versus Baroclinic Modes

Coupling the baroclinic hydrodynamic circulation and sediment transport in the stratified water is the main goal of this study. A comparison between barotropic and baroclinic sediment transport is performed to assess the importance of the baroclinic term.

6.1. Barotropic Mode

Figure 10a describes a snapshot taken at t=48h of the salinity contour of barotropic only transport for cohesive type sediment, hence so-called *barotropic mode*. The setup configuration of the model is similar to the previous cases with the barred rectangular basin, but the open boundary is driven with 0.5 m M2 tidal amplitude, normal flow V_n=0.2m/s, and the erosion occurs continuously. The freshwater discharge from the normal flow boundary on the right propagates towards onshore in nearly vertically uniform transport, indicated by the uniform density (i.e. salinity) distribution. Consequently, the eroded cohesive sediment along the top of the bar is also suspended and transported nearly uniformly, but there is a relatively small settling velocity taken place during their transport causing higher concentration near bottom immediately after it was suspended (Figure 10b). The vertical profile of horizontal velocity has higher velocity in the mid to top column than at bottom. Thus, the upper layers will travel faster than the bottom layers. Combining with the diffusion and vertical advection taking place during the horizontal transport, the sediment distribution is more likely become more uniform at the location far away from its origin. Here, the critical stresses for erosion (τ_{ce}) and deposition (τ_{cd}) are 0.2 N/m^2 and 0.1 N/m^2, respectively. The three-dimensional hydrodynamic model applies one-day ramp function to reach its designed hydrodynamic flows. The initial erosion occurred before the full ramp function achieved, and the site of deposition took place at some distance onshore from the erosion site due to the difference in the τ_{ce} and τ_{ce}. After the hydrodynamic ramp function was achieved, the erosion site occurred further onshore than previously deposition occurred. Since the erosion and deposition processes occur almost continuously, as shown in Figure 10c, there is likely continuous erosion followed by immediate deposition onshore.

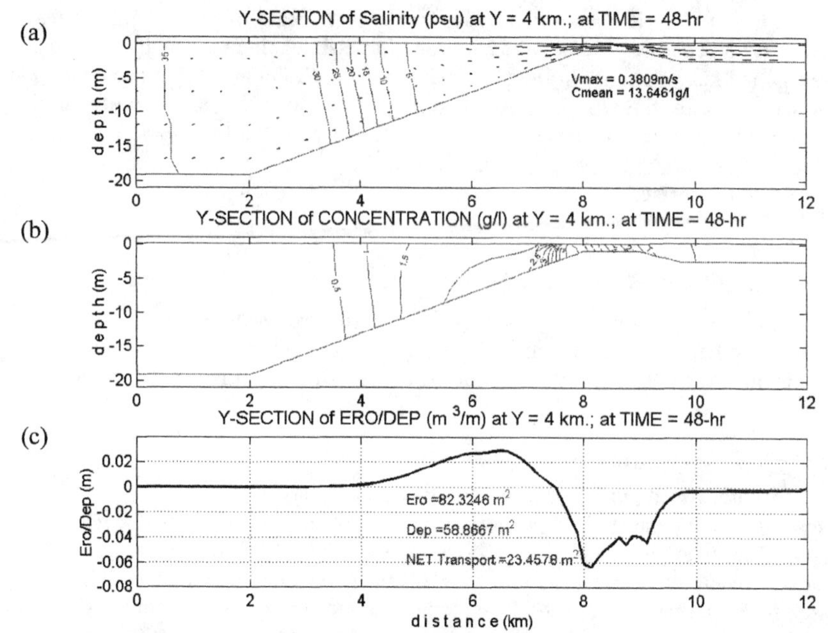

Figure 10. Profile of cohesive sediment transport in ***barotropic mode***, taken along the midline of the domain rectangular basin at time t = 48-h. (a) Contour plot of salinity in psu unit with arrows represent the relative magnitude and direction of current velocity. (b) Contour plot of suspended sediment concentration, and (c) the corresponding erosion (-) and deposition (+) along the profile line.

6.2 Baroclinic Modes

Another simulation was performed similar to the formerly discussed case with similar setup and configuration, but with activating the baroclinic and barotropic terms, hence so-called *baroclinic mode* (Figure 11). It will be a significant different of mechanism of sediment transport, mainly for the cohesive type sediment. Due to fresh water (S=3psu) discharge from the normal flow boundary on the right side of domain into the saline ocean water (S=35 psu), the presence of stratification is clearly seen when the baroclinic term is activated (Figure 11a).

The tip end of saline wedge is located in the x-axis between 6 and 6.5 km. At the end tip of saline wedge, the near bottom horizontal velocity is small, but upward vertical velocity increases causing the near bottom sediment tend to either settle down or be lifted up. In this case of cohesive sediments, the concentration of suspended sediment is small, thus the settling velocity is relatively smaller than its vertical lift velocity. Once the sediment is suspended and lifted up, the path of transport will be mainly driven by baroclinically hydrodynamic flow causing a cloud of suspended sediment concentrated in the top layers (Figure 11b). The suspended sediment cloud

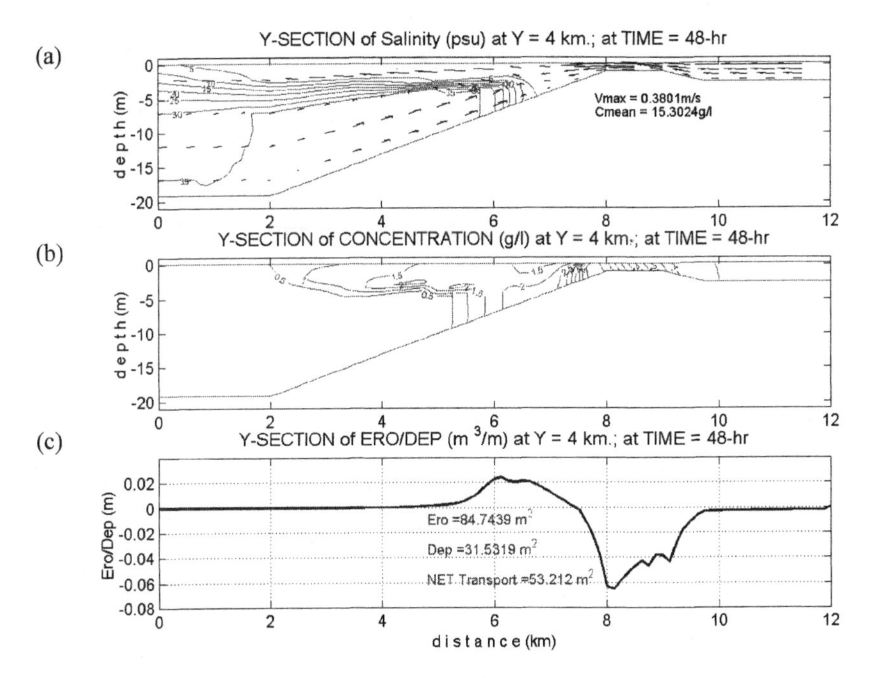

Figure 11. Profile of cohesive sediment transport in ***baroclinic mode***, taken at time t = 48-h. (a) Contour plot of salinity in psu unit with arrows represent the relative magnitude and direction of current velocity. (b) Contour plot of suspended sediment concentration, and (c) the corresponding erosion (-) and deposition (+) along the profile line.

will be either carried away outside the domain or deposited at some points further away offshore.

The processes of erosion and deposition around the top of the bar are almost similar to the barotropic case; however at point x = 5.8 – 6.2 km in Figures 9c and 10c, it is clearly seen that less deposition occurs in the baroclinic case around the tip of saline wedge, some amount of suspended sediment is raised up becoming a cloud of upper layer suspended sediment.

The amount of erosion, deposition and net sediment transport between them is summarized in Table 2. Therefore, in this case, the inclusion of the baroclinic term may transport the cohesive suspended sediment more than 60% of eroded materials, meanwhile the barotropic only model transports only about 28% of eroded materials.

Table 2. Comparison of sediment transport between barotropic and baroclinic transport at t = 48-h

	Erosion (m^3/m)	Deposition (m^3/m) (% of erosion)	Net Transport (m^3/m) (% of erosion)
Baroclinic	84.744	31.532 (37%)	53.212 (63%)
Barotropic only	82.325	58.867 (72%)	23.456 (28%)

7. Discussion and Conclusion

Application of a numerical model for simulation of hydrodynamic circulation and sediment transport in an idealized estuary with respect to baroclinic forces is described in this study.

Vertical distribution of suspended sediment of non-cohesive type is highly related to the Rouse Number (b_r). High b_r leads higher near bottom concentration than the layers above that produce a high vertical gradient. To reach a vertical stability solution of concentration, more vertical layers or smaller time step is required. On the other hand, depending on the grain size and depth of the water, the more layers will provide closer distance between the vertical layers, and may affect a large vertical Courant Number. Thus, to reach vertical stability, the optimal number of vertical layers should be estimated considering the grain size, minimum depth of the water, and time step. For example in the barred rectangular basin, with minimum depth of 1 m around the bar, six to nine vertical layers and a 6-second internal time step were appropriate. Also a threshold of $b_r<4.0$ was implemented to avoid instability due to large Rouse number b_r that may cause a high vertical gradient of concentration.

Based on the model in this study, it can be inferred that the three-dimensional sediment transport may explain the distribution of sediment considering the flow and transport pattern among the vertical layers may propagate non-uniformly. The upper layer could travel faster or slower relative to the bottom layer, due to both the typical vertical profile of velocity and the effect of the baroclinic pressure term. It is clearly seen that the deposition between barotropic and baroclinic modes provides different pattern and zone of near bottom high concentration. In the barotropic mode, the deposition is uniformly distributed along the channel, while in the baroclinic mode higher concentration occurred at one particular converging zone.

The inclusion of the baroclinic term in the stratified estuary contributes to a wider distribution pattern of suspended sediment, mainly for cohesive type sediment. However, to reach the stability condition for a large horizontal salinity gradient (i.e. large density gradient) between fresh and saline water, the baroclinic ramp function is required to smooth the baroclinic acceleration term; and a high horizontal diffusion coefficient, D_h; in most cases of stratified water is mostly less than $10 m^2/s$.

In case of cohesive sediment, it has relatively much smaller settling velocity than a non-cohesive sediment. Thus, the selected number of layers can be higher, or a larger internal time step can be implemented.

Even though there was a discrepancy between the numerical and analytical model, the transport model is able to describe a realistic sediment transport, both for cohesive and non-cohesive types with the influence of the baroclinic term. Near future development includes a benchmarking of the model against the laboratory and observational results.

Acknowledgement

The study described in this paper has been supported in part by the U.S. Environmental Protection Agency through the South and Southwest Hazardous Substance Research Center, the National Science Foundation, and the Texas Engineering Experiment Station at Texas A&M University.

References

Blumberg, A.F. and G.L. Mellor, G. L. (1987) ."A Description of a Three-Dimensional Coastal Ocean Circulation Model" *Three-Dimensional Coastal Ocean Models*, ed. N.S. Heaps, pp. 1-16, American Geophysical Union, Washington, D.C.

Cheng, N. S. (1997). "Simplified Settling Velocity Formula for Sediment Particle" *J. Hydraulic Engineering*, 123(2), 149-152.

Gross, E. S., Koseff, J. R. and Monismith, S. G. (1999) "Three-dimensional Salinity Simulations of South San Fransisco Bay" *J. Hydraulic Engineering*, 125(11), 1199–1209.

Helfand, J. S., Podber, D. P. and McCormick, M. J. (1999) "Effect of Heat Flux on Thermocline Formation" *Estuarine and Coastal Modeling*, Proc. 6th Int'l Conference, Spaulding and Butler (ed.), ASCE, New Orleans, LA.

Hydroqual Inc. (1998) "Development and Application of a Modeling Framework to Evaluate Hurricane Impacts on Surficial Mercury Concentrationns in Lavaca Bay" Project Report, Aluminium Company of America, Point Comfot, Texas, 1998.

Luettich, R. L., and Westerink, J. J. (2003) "Formulation and Numerical Implementation of the 2D/3D ADCIRC Finite Element Model Version 43.XX" Paper presented at *ADCIRC Workshop 27-28 Feb 2003*, NRL, Stennis, MS.

Luettich, R. A. Jr. and Muccino, J. (2001) "Summary of Vertical Velocity Calculation Methods as Considered for 3D ADCIRC Hydrodynamic Model" Paper presented at *ADCIRC Workshop 20-21 Feb 2001*, NRL, Stennis, MS.

Luettich, R. A. Jr., Westerink, J. J. and Scheffner, N. W. (1992) "ADCIRC: An advanced Three-Dimensional Circulation Model for Shelves, Coasts, and Estuaries. Report 1", *Technical Report DRP-92-6*, Dredging Res. Prog., USACE, Washington DC.

Lynch, D. R. and Officer, C. B. (1985) "Analytic Test Case for Three-dimensional Hydrodynamic Models." *Int. J. Num. Meth. Fluids*, 5, 529-543.

Mellor, G.L. and Yamada, T. (1982) "Development of a Turbulence Closure Model for Geophysical Fluid Problems" *Reviews of Geophysics and Space Physics*, 20, 851-875, 1982.

Muccino, J. C., Gray, W. G. and Foreman, G. G. (1997) "Calculation of Vertical Velocity in Three Dimensional, Shallow Water Equation, Finite Elements Model" *Int. J. Num. Meth. Fluids*, 25, 779-802.

Pandoe, W. W. and Edge, B. L. (2003) "Three-dimensional Hydrodynamic Model, Study Cases for Quarter Annular and Idealized Ship Channel Problems" *J. Ocean Engineering*, 30, 1117-1135.

Partheniades, E. (1992) "Estuarine Sediment Dynamics and Shoaling Processes" *Handbook of Coastal and Ocean Engineering*, Vol. 3, Chapter 12, Herbich, J. B. (ed.), Gulf Publishing Co.

Robertson, Padman, R., L. and Levine, M. D. (2001) "A Correction of the Baroclinic Pressure Gradient Term in the Princeton Ocean Model" *J. Atmos. Oceanic Technol.*, 18(6), 1068 – 1075.

Scheffner, N. (1999) "A Large Domain Convection Diffusion Based Finite Element Transport Model" *Estuarine and Coastal Modeling*. Proc. 6th Intern. Conf., Eds. M.L. Spaulding and H. L. Butler, ASCE, pp.194-208.

Soulsby, R.(1997) *Dynamics of Marine Sands*, Thomas Telford Services Ltd., UK.

Van de Kreeke, J, Hoogewoning, S. E. and Verlaan, M. (2002) "An Analytical Model for the Morphodynamics of a Trench in the Presence of Tidal Currents" *Continental Shelf Research*, 22, 1811-1920.

VanRijn, L.R. (2003) "Sand Transport by Currents and Waves; General Approximation Formulae" *Proc. Conference of Coastal Sediments 2003*, Florida.

Whitehouse, Soulsby, R., R., Roberts, W. and Mitchener, H. (2000) *Dynamics of Estuarine Muds*, H. R. Wallingford, UK, 2000.

Modeling Circulation and Mixing in Tidal Wetlands of the Santa Ana River

Feleke Arega[1] and Brett F. Sanders[2]

Abstract

The tidal wetlands of the Santa Ana River in Orange County, California are a combination of engineered flood control channels and partially restored salt marshes. The tidal channels extend several kilometers inland with depths comparable to the amplitude of the tides which force circulation, roughly one meter. As is typical of tidal wetlands, there is extensive wetting and drying of sand-bars and mudflats which is a challenge to resolve with numerical models. We report on the application of an unstructured-grid version of an explicit, high-resolution, monotonicity preserving Godunov-type finite volume scheme to solve depth-integrated equations and predict circulation and mixing. Finite volume methods have proven advantageous for applications involving surges and shocks in terms of conservation, monotonicity, stationarity properties, but little has been reported on tidal applications. We find that predictions of water level and velocity at a number of monitoring stations compare well with field measurements, and model predictions of a dispersing dye cloud compare well with measured values when physically meaningful mixing coefficients are used. The model accurately predicts the flooding and drying of mudflats without loss of fluid mass, but artificial dilution of dissolved scalars is predicted at the wet/dry interface. The model uses a small time step consistent with the CFL condition and this contributes towards its ability to resolve sharp fronts and impulse loads.

Introduction

Most of California's coastal wetlands are estuarine salt marshes that include tidal channels and mudflats. These systems are a critical component of coastal ecosystems, providing habitat for migrating waterfowl, exporting nutrients to the coastal ocean, and providing protection for juvenile fish and other species (California Coastal Commission 1987). Tidal wetlands in California are to a large extent physically and ecologically perturbed from their natural state as a result of factors such as dredging for harbor operations, filling for development, channelization for flood control, invasion of exotic species, larger scale climatic and

[1]Postdoctoral Researcher, Department of Civil and Environmental Engineering, University of California, Irvine, CA 92697

[2]Associate Professor, Department of Civil and Environmental Engineering, University of California, Irvine, CA 92697

ecological changes, and pollution. This situation has placed a high priority on restoring the condition of these systems and creating new ones wherever possible. Circulation and transport models play an important role in this process.

In this paper, we describe a depth-integrated modeling framework, the California Tidal Wetland Modeling System (CalTWiMS) for circulation and transport modeling in tidal wetlands and present an application of the model to the Santa Ana River (SAR) wetlands. CalTWiMS is based upon the hydrodynamic model described by Bradford and Sanders (2002), which utilizes a Godunov-type finite volume method. Finite volume models have been shown to perform well in applications involving surges and shocks, and Bradford and Sanders (2002) reported that this approach permits flooding and recession of uneven topography to be resolved with excellent conservation properties. However, advantages of the method for tidal applications are not clear, particularly since the method is constrained by the CFL condition for stability purposes. Using a grid resolution of roughly a few meters, this requires roughly 10^5 time steps to integrate over one tide cycle. With an application, we test the hypothesis that the low dissipation and good conservation properties of the numerical method allow circulation and mixing to be accurately resolved with physically meaningful flow resistance and mixing parameters. The application also offers insight into model performance in the context of a water quality study where simulations of circulation and mixing are of interest. The ability of the model to resolve tide levels, velocities, and solute transport is examined by comparison between model predictions and observations.

Site Description The SAR wetlands include the tide flooded portion of the Santa Ana River and two sloughs connected by tide gates 4.3 m wide. The site and bed elevation are illustrated in Figures 1. The Santa Ana River Watershed covers 153.2 square miles but significant runoff reaches the outlet only during storms. The tidal channels of the wetland extend roughly 10 kilometers inland, and are characterized by depths less than one meter. The bed consists of sands and silts in SAR and silts and muds in the neighboring sloughs. The width of the main channel varies between 100 to 130 m. At high tide, the surface area of the wetland is approximately 0.8 km^2. Circulation in SAR is driven externally by Pacific ocean tides (see figure 1) for most of the year and by runoff during infrequent storms.

Mathematical Model

Hydrodynamics Due to the shallow depths, depth-integrated continuity and hydrostatic momentum equations are the basis of the hydrodynamic model. The integral form of these equations can be written for a spatial domain Ω with a boundary $\partial\Omega$ as,

$$\frac{\partial}{\partial t}\int_{\Omega} \mathbf{U} d\Omega + \oint_{\partial\Omega} (\mathbf{F} dy - \mathbf{G} dx) = \int_{\Omega} \mathbf{S} d\Omega \tag{1}$$

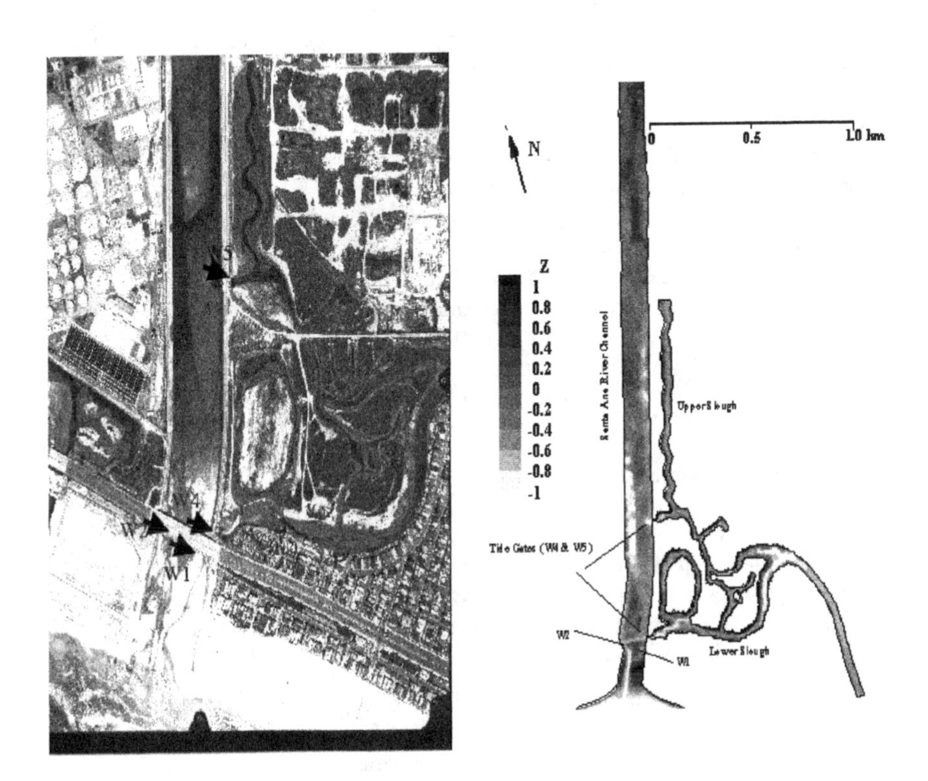

Figure 1: Site map and monitoring stations (left) and bed elevation contours (right).

where $\mathbf{U} = (h \quad h\bar{u} \quad h\bar{v})^T$ and

$$\mathbf{F} = \begin{pmatrix} h\bar{u} \\ h\bar{u}^2 + \frac{1}{2}g^*h^2 \\ h\bar{u}\bar{v} \end{pmatrix} \qquad \mathbf{G} = \begin{pmatrix} h\bar{v} \\ h\bar{u}\bar{v} \\ h\bar{v}^2 + \frac{1}{2}g^*h^2 \end{pmatrix} \qquad \mathbf{S} = \begin{pmatrix} 0 \\ -g^*h\frac{\partial z_b}{\partial x} - \frac{\tau_x^b}{\rho_o} \\ -g^*h\frac{\partial z_b}{\partial y} - \frac{\tau_y^b}{\rho_o} \end{pmatrix}$$

The terms $\bar{u}$ and $\bar{v}$ are the vertically-averaged velocities in the x and y directions, respectively. The term τ_b^x account for bed resistance in the x direction. Similar terms are used for stresses in the y direction. The elevation of the bed above an arbitrary datum is denoted by z_b and $g^* = g(1 + \Delta\rho/\rho_o)$ is a gravitational parameter that accounts for variations in fluid density, consistent with the Boussinesq approximation, where ρ_o is a reference density, $\Delta\rho = \rho - \rho_o$, ρ represents the depth-averaged fluid density, and g is the gravitational constant. For channelized wetlands with depths of a few meters or less, the Coriolis term is neglected. The bed stresses are computed in terms of a drag law where the bed resistance drag coefficient c_D^b is a function of the relative roughness of the bed k_s/h and the Reynolds number $Re = \sqrt{\bar{u}^2 + \bar{v}^2}h/\nu$, where ν is the kinematic viscosity of water and k_s is the roughness height of the bed. If the bed roughness is known in terms of a Manning coefficient, n, then the roughness can be computed in units of meters (e.g., Henderson 1966). The above equations do not include Reynolds stresses to account for lateral momentum transfer due to shear in the mean flow (Bedford *et al.* 1988). These terms are small in shallow water environments compared to the bed and vegetative resistance terms (Babarutsi *et al.* 1989).

Flow in SAR wetlands is externally forced by the ocean tide and storm water runoff during storms. Tidal forcing is modeled by specifying the depth at the ocean boundary of the model, while runoff forcing is modeled by specifying the volumetric flow rate at the head of SAR. Boundaries of the model that do not correspond to either ocean or inflow sections enforce no-normal flow boundary conditions. The present study focuses on a dry weather period when no significant runoff entered the SAR wetlands, so flow was forced exclusively by the ocean tide.

Solute Transport Solutes are modeled by solving depth-integrated transport equations that account for advection, turbulent diffusion, dispersion, and source and sink terms. These equations appear in integral form as,

$$\frac{\partial}{\partial t}\int_\Omega \mathbf{Q}d\Omega + \oint_{\partial\Omega}(\mathbf{F_Q}dy - \mathbf{G_Q}dx) = \int_\Omega \mathbf{S_Q}d\Omega \qquad (2)$$

where $\mathbf{Q} = (hc_1 \quad hc_2 \quad \ldots \quad hc_N)^T$ and

$$\mathbf{F_Q} = \begin{pmatrix} h\bar{u}c_1 - h(T_{uc_1} + D_{uc_1}) \\ h\bar{u}c_2 - h(T_{uc_2} + D_{uc_2}) \\ \vdots \\ h\bar{u}c_N - h(T_{uc_N} + D_{uc_N}) \end{pmatrix} \mathbf{G_Q} = \begin{pmatrix} h\bar{v}c_1 - h(T_{vc_1} + D_{vc_1}) \\ h\bar{v}c_2 - h(T_{vc_2} + D_{vc_2}) \\ \vdots \\ h\bar{v}c_N - h(T_{vc_N} + D_{vc_N}) \end{pmatrix} \mathbf{S_Q} = \begin{pmatrix} s_1 \\ s_2 \\ \vdots \\ s_N \end{pmatrix}$$

where c_i corresponds to the depth-averaged concentration of the ith dissolved solute for i=1,N (number of solute species to be modeled), T accounts for turbulent diffusion, D accounts for dispersion, and s_i is a generalized source/sink/reaction term. The sum of turbulent diffusion and dispersion in the longitudinal direction (Elder 1959) and transverse direction are given by,

$$T_{u_l c_i} + D_{u_l c_i} = E_L \frac{\partial c_i}{\partial s}; \qquad T_{u_t c_i} + D_{u_t c_i} = E_T \frac{\partial c_i}{\partial n}$$

where s,n,u_l and u_t represents the longitudinal and transverse direction and velocity components, respectively. $E_L = 5.93\,u^*h$, u^* represents shear velocity. E_T is correlated to the shear velocity and depth as, $E_T = \beta u^*h$. β varies in the range of 0.2-1.0 depending on the channel geometry (Ward 1974, Fischer *et al.* 1979). In CalTWiMS, the value of $\beta = 0.6$, typical of natural channels and observations is used (Ward 1974, Fischer *et al.* 1979). The turbulence and dispersion terms in Eq. 2 are resolved as follows,

$$T_{uc_i} + D_{uc_i} = E_{xx}\frac{\partial c_i}{\partial x} + E_{xy}\frac{\partial c_i}{\partial y}; \qquad T_{vc_i} + D_{vc_i} = E_{yx}\frac{\partial c_i}{\partial x} + E_{yy}\frac{\partial c_i}{\partial y}$$

where

$$E_{xx} = E_T + (E_L - E_T)\frac{\bar{u}^2}{\bar{u}^2 + \bar{v}^2}; \; E_{xy(=yx)} = (E_L - E_T)\frac{\bar{u}\bar{v}}{\bar{u}^2 + \bar{v}^2}$$

$$E_{yy} = E_T + (E_L - E_T)\frac{\bar{v}^2}{\bar{u}^2 + \bar{v}^2}$$

Boundary conditions for the solute transport model enforce a no-mass flux condition at all boundaries that are not inflow or outflow sections, specify solute concentration at inflow boundaries, and specify no dispersive flux conditions at outflow boundaries.

Particle Transport Particle transport is modeled using a quasi-three dimensional random-walk particle tracking method that is derived from the following transport equation,

$$\frac{\partial n_p}{\partial t} + \frac{\partial (un_p)}{\partial x} + \frac{\partial (vn_p)}{\partial y} + \frac{\partial (wn_p)}{\partial z} = \frac{\partial}{\partial x}\left(E_T\frac{\partial n_p}{\partial x}\right) + \frac{\partial}{\partial y}\left(E_T\frac{\partial n_p}{\partial y}\right) + \frac{\partial}{\partial z}\left(E_z\frac{\partial n_p}{\partial z}\right) \tag{3}$$

where n_p is the number of particles contained by a representative elemental volume. The coordinate system is aligned in with the principal axes of turbulent diffusion, so only the terms along the diagonal of the diffusion tensor are included in Eq. 3. Point wise velocities (u, v, and w) are recovered from depth-averaged velocities under the assumption that a turbulent boundary layer extends vertically from the bed. $E_z = \kappa u^*h/6$ represents the depth-average value of the vertical diffusivity where $\kappa = 0.41$. A vertically uniform diffusivity is used for simplicity. Boundary conditions enforce reflecting conditions on the bed, free surface, and lateral boundaries of the model. Particles that exit the outflow boundary are not subsequently tracked.

Numerical Methods

The model solves the hydrodynamic and scalar transport equations explicitly on an unstructured grid consisting of quadrilateral cells. Water levels, velocities, and scalars are stored at cell centers and fluxes are computed at cell faces. A second-order accurate predictor-corrector method is used for time stepping. The predictor solution is computed by solving the primitive form of the equations in generalized coordinates in cell j as

$$h_j^{n+1/2} = h_j^n - \frac{\Delta t}{2}\left(\bar{u}_\xi \overline{\Delta_\xi h} + \bar{u}_\eta \overline{\Delta_\eta h} + h(\xi_x \overline{\Delta_\xi \bar{u}} + \xi_y \overline{\Delta_\eta \bar{u}} + \eta_x \overline{\Delta_\xi \bar{v}} + \eta_y \overline{\Delta_\eta \bar{v}})\right)_j^n \quad (4)$$

$$\bar{u}_j^{n+1/2} = \bar{u}_{j,k}^n - \frac{\Delta t}{2}\left(\bar{u}_\xi \overline{\Delta_\xi \bar{u}} + \bar{u}_\eta \overline{\Delta_\eta \bar{u}} + g(\xi_x \overline{\Delta_\xi \zeta} + \eta_x \overline{\Delta_\eta \zeta}) - c_D \bar{u}\sqrt{\bar{u}^2 + \bar{v}^2}/h\right)_j^n \quad (5)$$

$$\bar{v}_j^{n+1/2} = \bar{v}_{j,k}^n - \frac{\Delta t}{2}\left(\bar{u}_\xi \overline{\Delta_\xi \bar{v}} + \bar{u}_\eta \overline{\Delta_\eta \bar{v}} + (\xi_y \overline{\Delta_\xi \zeta} + \eta_y \overline{\Delta_\eta \zeta}) - c_D \bar{v}\sqrt{\bar{u}^2 + \bar{v}^2}/h\right)_j^n \quad (6)$$

where ξ and η are the orthogonal components of each cell's coordinate system. The terms ξ_x, ξ_y, η_x, and η_y are the grid transformation metrics. They are computed by assuming a linear mapping of x and y onto ξ and η, which are assumed to vary from 0 to 1 in each computational cell. Also, Δt is the time step, $\zeta = h + z$ is the free surface elevation, $\bar{u}_\xi = \bar{u}\xi_x + \bar{v}\xi_y$, and $\bar{u}_\eta = \bar{u}\eta_x + \bar{v}\eta_y$.

The $\overline{\Delta}$ in Eqns. (4), (5), and (6) denotes cell-average gradients of the variables in the ξ and η directions.The Double Minmod limiter (Sweby 1984) is used when computing gradients in the velocity and free surface in order to preserve the monotonicity of the solution. Gradients in depth are computed by subtracting the bed elevation gradient from the limited free surface elevation gradient $\overline{\Delta h} = \overline{\Delta\zeta} - \Delta z$. Note that Δz is the non-limited change in bed elevation in the direction of $\overline{\Delta h}$.

Predicted values to the left and right of each cell face are linearly reconstructed using the Monotone Upstream Scheme for Conservation Laws (MUSCL) to obtain second order spatial accuracy (Van Leer, 1979). The fluxes at each cell face are then evaluated using Roe's (1981) Godunov-type upwind scheme in which a Riemann problem is solved across each cell face. The corrector solution follows as,

$$\begin{aligned}\mathbf{U}_j^{n+1} &= \mathbf{U}_j^n - \frac{\Delta t}{\Omega_i}[(\mathbf{F}_\perp^{n+1/2} \cdot \mathbf{n}\, s)_{f1} + (\mathbf{F}_\perp^{n+1/2} \cdot \mathbf{n}\, s)_{f2} \\ &\quad + (\mathbf{F}_\perp^{n+1/2} \cdot \mathbf{n}\, s)_{f3} + (\mathbf{F}_\perp^{n+1/2} \cdot \mathbf{n}\, s)_{f4}] + \Delta t \mathbf{S}_j^{n+1/2}\end{aligned} \quad (7)$$

where $\mathbf{U}$ and $\mathbf{S}$ are assumed to be cell-average values, $\mathbf{F}_\perp$ is the flux normal to the cell face, and the subscripts $f1$-$f4$ denote the indices of the four faces of each cell. Ω is the area of the cell, $\mathbf{n}$ is the unit outward normal vector of each face, and s is the face length. The flux $\mathbf{F}_\perp$ perpendicular to the boundary is given as

$$\mathbf{F}_\perp = \begin{pmatrix} h\bar{u}_\perp \\ h\bar{u}\bar{u}_\perp + \frac{1}{2}gh^2\cos\phi \\ h\bar{v}\bar{u}_\perp + \frac{1}{2}gh^2\sin\phi \end{pmatrix} + \begin{pmatrix} 0 \\ \frac{1}{12}g(\Delta_s h)^2\cos\phi \\ \frac{1}{12}g(\Delta_s h)^2\sin\phi \end{pmatrix} \tag{8}$$

where $\bar{v}_\perp$ is the velocity perpendicular to the cell face, $\Delta_s h$ represents the difference in h across the length of the face, and ϕ is the angle between the face normal vector and the x axis. The array on the left in Eq. 8 is the standard flux computed by Roe's method, while the array in the right is a correction term that is necessary to preserve stationary solutions (Bradford and Sanders 2002). The integral over the cell of the source terms is resolved by evaluating the source terms at the cell center and multiplying by the area of the cell, except in the case of the bed slope terms. These are resolved using finite element bilinear shape functions to represent h and z within the cell and applying 2×2 Gauss quadrature to evaluate the integral (Hughes, 1987).

The general strategy adopted tracking the movement of the wet/dry boundary without creating artificial disturbances is to solve the continuity equation in all cells, but only solve the momentum equation in completely wetted cells and for the other cells, set the velocity to zero. A wet cell is defined to be a cell with four wet nodes, and a wet node is defined as as having a depth greater than a tolerance of 10^{-6} m. Predictions were not found to be sensitive to this tolerance in the range of 10^{-6}-10^{-3} m.

Solute Transport The numerical scheme for solving the transport equation follows the hydrodynamic predictions and uses the same computational grid, time levels, and predictor-corrector scheme. The predictor for the ith scalar follows as,

$$\begin{aligned}(c_i)_j^{n+1/2} &= (c_i)_j^n - \frac{\Delta t}{2}\Big(\bar{u}_\xi\overline{\Delta_\xi c_i} + \bar{u}_\eta\overline{\Delta_\eta c_i} + \xi_x\Delta_\xi(T_{uc_i} + D_{uc_i}) + \eta_x\Delta_\eta(T_{uc_i} + D_{uc_i}) \\ &\quad + \xi_y\Delta_\xi(T_{vc_i} + D_{vc_i}) + \eta_y\Delta_\eta(T_{vc_i} + D_{vc_i}) - s_i\Big)_j^n \end{aligned} \tag{9}$$

where the $\overline{\Delta}$ again indicates the use of limited slopes to preserve monotonicity. In solute transport, the superbee slope limiter is used. Predicted values of c are linearly reconstructed using MUSCL to obtain second order spatial accuracy (Van Leer 1979). The corrector solution is then updated as,

$$\begin{aligned}\mathbf{Q}_j^{n+1} = \mathbf{Q}_j^n - \frac{\Delta t}{\Omega_j}[(\mathbf{F_Q}_\perp^{n+1/2}\cdot\mathbf{n}\,s)_{f1} + (\mathbf{F_Q}_\perp^{n+1/2}\cdot\mathbf{n}\,s)_{f2} + \\ (\mathbf{F_Q}_\perp^{n+1/2}\cdot\mathbf{n}\,s)_{f3} + (\mathbf{F_Q}_\perp^{n+1/2}\cdot\mathbf{n}\,s)_{f4}] + \Delta t\mathbf{S_Q}_j^{n+1/2}\end{aligned} \tag{10}$$

The flux $\mathbf{F}_{\mathbf{Q}\perp}$ is perpendicular to the cell boundary and defined as

$$\mathbf{F}_{\mathbf{Q}\perp} = \begin{pmatrix} h\bar{u}_\perp c_1 + h(T_{uc_1} + D_{uc_1})\cos\phi + h(T_{vc_1} + D_{vc_1})\sin\phi \\ h\bar{u}_\perp c_2 + h(T_{uc_2} + D_{uc_2})\cos\phi + h(T_{vc_2} + D_{vc_2})\sin\phi \\ \vdots \\ h\bar{u}_\perp c_N + h(T_{uc_N} + D_{uc_N})\cos\phi + h(T_{vc_N} + D_{vc_N})\sin\phi \end{pmatrix} \tag{11}$$

The advective fluxes are evaluated by multiplying the volumetric flux at the cell face ($F_\perp(1)$ from Eq. 8) by the reconstructed value of c_i on the upstream side of the face. The diffusive fluxes are evaluated by computing the turbulent diffusion tensor at the cell face based on the cell centered solution at the two neighboring cell centers, gradients of c normal to the cell face based on these two cells as well, and gradients parallel to the cell face using the four cells surrounding the cell face that only contact it with one node.

Resolving solute transport in the presence of drying and wetting is a challenge, which stems from the fact that fluid volume and dissolved mass do not simultaneously vanish when a cell drains. That is, it is possible for the fluid of a drying cell to vanish before the dissolved mass does which is not physically realistic. To our knowledge all contemporary methods based on fixed grids either predict unrealstically high dissolved concentrations at the wet/dry interface or suffer from dissolved mass conservation problems, i.e., create artificial dilution, as a result of setting the dissolved mass to zero when the fluid volume vanishes. Here to avoid unrealistic predictions, the latter of these options was adopted in the present model. Dissolved mass inside cells with an average depth less than 1 cm is set to zero. Hence, dissolved material is only lost in the model by cells at the wet/dry interface. Therefore, the percentage of dissolved mass that is artificially lost as a result of this approach will vary depending on the ratio of inter-tidal to sub-tidal surface area. That is, shallow tidal wetlands like SAR with substantial wetting and drying and relatively little storage are the most challenging to model. In a nearby wetland with similar hydrographic characteristics, this approach caused 0.7% of a unit impulse of dye to be artificially lost over one tide cycle (Arega and Sanders 2004). While further reduction of this percentage is of interest, for most water quality studies other factors affecting uncertainty are much more significant (e.g., non-conservative transport processes). In the absence of wetting and drying, the model conserves dissolved mass and fluid volume to numerical precision.

Particle Tracking A random-walk particle tracking model is also incorporated into CalTWiMS. This component is complementary to the solute transport capability, is useful for predictions of wetland residence times, and can also be used to assess potential limitations of the solute transport model. For example, the particle tracking model can be used to estimate the time scale associated with vertical mixing which is assumed to be short for the solute transport model. The quasi three dimensional update equations used in the model are given by,

$$x_p^{n+1} = x_p^n + (\Delta x_p)^n, \qquad y_p^{n+1} = y_p^n + (\Delta y_p)^n, \quad \text{and} \quad z_p^{n+1} = z_p^n + (\Delta z_p)^n \tag{12}$$

where

$$\Delta x_p = \left(u + \frac{\partial E_T}{\partial x}\right)\Delta t + Z_1\sqrt{2E_T\,\Delta t}$$
$$\Delta y_p = \left(v + \frac{\partial E_T}{\partial y}\right)\Delta t + Z_2\sqrt{2E_T\,\Delta t}$$
$$\Delta z_p = \left(w + \frac{\partial E_z}{\partial z}\right)\Delta t + Z_3\sqrt{2E_z\Delta t}$$

and Z_1, Z_2, and Z_3 are numbers randomly sampled from a normal distribution with mean of zero and a standard deviation of one (Tompson and Gelhar 1990). The vertical advection is resolved on the assumption that particle advection in the horizontal occurs along a constant dimensionless depth, which is essentially what would be predicted by potential flow theory. $\partial E_z/\partial z$ of Eq. 3 is necessarily zero and the terms $\partial E_T/\partial x$ and $\partial E_T/\partial y$ are ignored since these are just a small fraction of the point-wise velocities. Therefore, the three dimensional particle update equations used in the model are given by,

$$x_p^{n+1} = x_p^n + (\Delta x_p)^n, \qquad y_p^{n+1} = y_p^n + (\Delta y_p)^n, \quad \text{and} \quad \psi_p^{n+1} = \psi_p^n + (\Delta\psi_p)^n \tag{13}$$

where

$$\Delta x_p = u\Delta t + Z_1\sqrt{2E_T\,\Delta t}$$
$$\Delta y_p = v\Delta t + Z_2\sqrt{2E_T\,\Delta t}$$
$$\Delta\psi_p = \frac{Z_3}{h(x_p, y_p)}\sqrt{2E_z\Delta t}$$

where $\psi = (z - z_b)/h$. Vertical position is then computed as $z_p = z_b(x_p, y_p) + \psi_p h(x_p, y_p)$. The point-wise horizontal velocity components are computed from the depth-averaged values as follows,

$$u = f(\psi)\bar{u} \quad \text{and} \quad v = f(\psi)\bar{v}; \quad \text{where,} \;\; f(\psi) = 1 + \frac{u*}{\kappa\sqrt{\bar{u}^2 + \bar{v}^2}}(1 + \log\psi) \tag{14}$$

The hydrodynamic, solute transport, and particle transport equations are integrated with a common time step. Since the scheme is explicit, the time step is limited by the Courant-Friedrichs-Levy (CFL) condition for stability purposes and the rate limiting factor is the speed of gravity waves.

Model Application

Application of the model to the SAR wetlands is motivated by concern over coastal water quality. In particular, high levels of fecal indicator bacteria are common to this area, and it is generally thought that this is due to a combination of watershed runoff, feces from wildlife (notably birds), and interactions with sediments and vegetation. The monotone aspect of model predictions is particularly

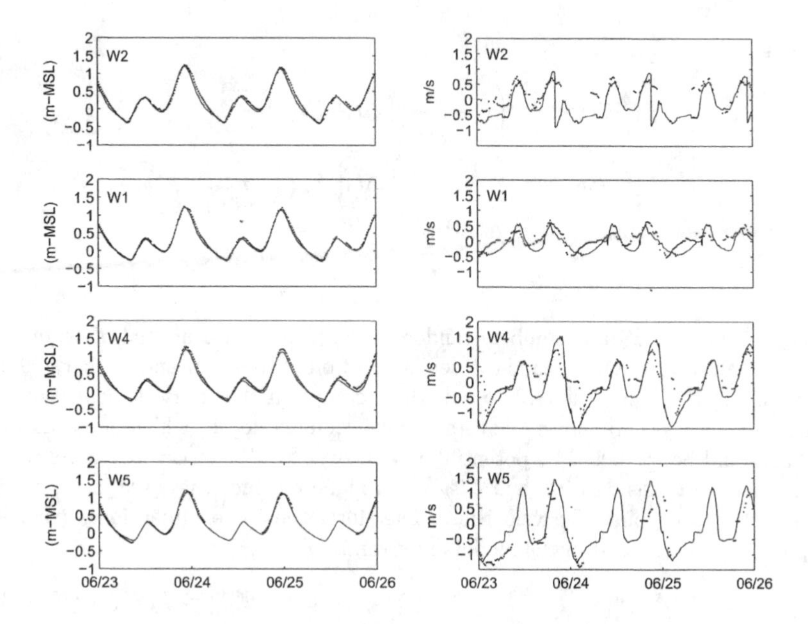

Figure 2: Calibrated model predictions at SAR monitoring stations.

attractive for biological pollutant modeling since predictions of concentration over several log cycles are needed. Here we report on the most fundamental aspects of the model prediction, its ability to resolve the flow field and predict the transport of a conservative tracer.

To apply the model, bed elevation data were surveyed and incorporated into an unstructured grid of 14818 quadrilateral cells. The grid was extended roughly 1 km offshore, and away from the outlet, to facilitate forcing of the wetlands by the ocean tide. Grid resolution was 3-4 m, though in some places the grid was further refined to resolve small (3 m wide) tidal channels with at least one fully wetted computational cell at low tide. Using tide measurements taken at NOAA gage 9410660 (off-shore of Los Angeles Harbor) to force the model, calibration of water level and velocity were made for the period of June 23- June 26, 2001. A time step of 0.15 s was used which corresponds to a maximum CFL of 0.7. A set of water level and velocity data were used to evaluate the performance of the model. The roughness coefficient (Manning coefficient) was the only calibration parameter. After a series of tests a uniform Manning coefficient of n=0.017 was chosen based on the agreement of model predictions and field observations. Referring to Figure 1, using acoustic Doppler velocimeters with integrated pressure transducers (*Model 4250, ISCO Inc., Lincoln, NE*) water level and velocity were measured during June and July of 2001 at four points in the SAR wetlands: SAR up coast (W2), SAR down coast (W1), tide gate 1 (W4) and tide gate 2 (W5).

Figure 2 presents comparisons between the measured and predicted water levels and velocities at all of the monitoring stations for the calibration stage. The agreement is good in all cases. It should be noted that the high velocity in at station W4 and W5 is due to the constriction caused by the tide gate. Relatively weak currents were predicted at W1, positioned in the smaller of two thalwags. Validation of the model was accomplished using data collected between June 27 -July 11 2001. Figure 3 presents model verification data for June 27 -July 11. Predictions of water level capture high and low tide levels at all stations within 5%, and also capture the distorted (from the roughly harmonic ocean tide) water level variations caused by the shallow topography. Note that many data points are missing, in particular velocity measurements. Many data points were removed as a result of instrumentation malfunction. Peak flood and ebb velocities are generally predicted within 25%, though in some cases the differences are closer to 50%. The basic trends in the velocity at each station are captured, except at W2 where the model predicts a change in flow direction at high tide (see Figure 2). This is due to a recirculation zone that is predicted by the model. The data do not indicate the presence of such a recirculation zone, though it should be noted that the instrument only measures the longtudinal component of velocity in the channel.

Dye measurements were used to evaluate the performance of the model with respect to mixing predictions. On September 20 2001, 100 mL and 50 mL of Rhodamine WT of dye were released at Tide gate W4 and W5 at 845 and 855 PDT, respectively, shortly after the beginning of a flood tide. Samples were collected hourly at each station for the next three days and taken to UC Irvine laboratories where the dye concentration was measured. Model predictions of dye concentrations were made and compared to these measurements. Circulation was forced in the model using tide measurements taken at NOAA gage 9410660, and the model was pre-run for several days to obtain an initial condition. The concentration of dye entering the wetlands during the flood was set to zero even though it is possible, but highly unlikely, that dye exiting the wetlands during the ebb later entered again during the flood. Figure 4 presents a comparison between the measured and predicted dye at W4 and W5 (top two panels). The predictions of dye concentration are reasonable, differing from measurements by at most 25%. The bottom panel of Figure 4 shows the total mass of dye in the system following the release. Dissolved mass is lost quickly during ebb tides and remains steady during flood tides, as one would expect. Upon closer inspection it can be seen that a small amount of dissolved mass is also lost during flood tides. This loss of mass occurs at the wet/dry interface and amounts to less than 1 % of the total mass per tide cycle. As a point of reference, regions of the wetland that experience wetting and drying account for roughly 55% of the wetted surface area at high tide. Considering that this is a relatively rigorous test of the wetting and drying feature of the model, we consider the 1% dissolved mass loss rate to be reasonable although a smaller loss rate remains desirable.

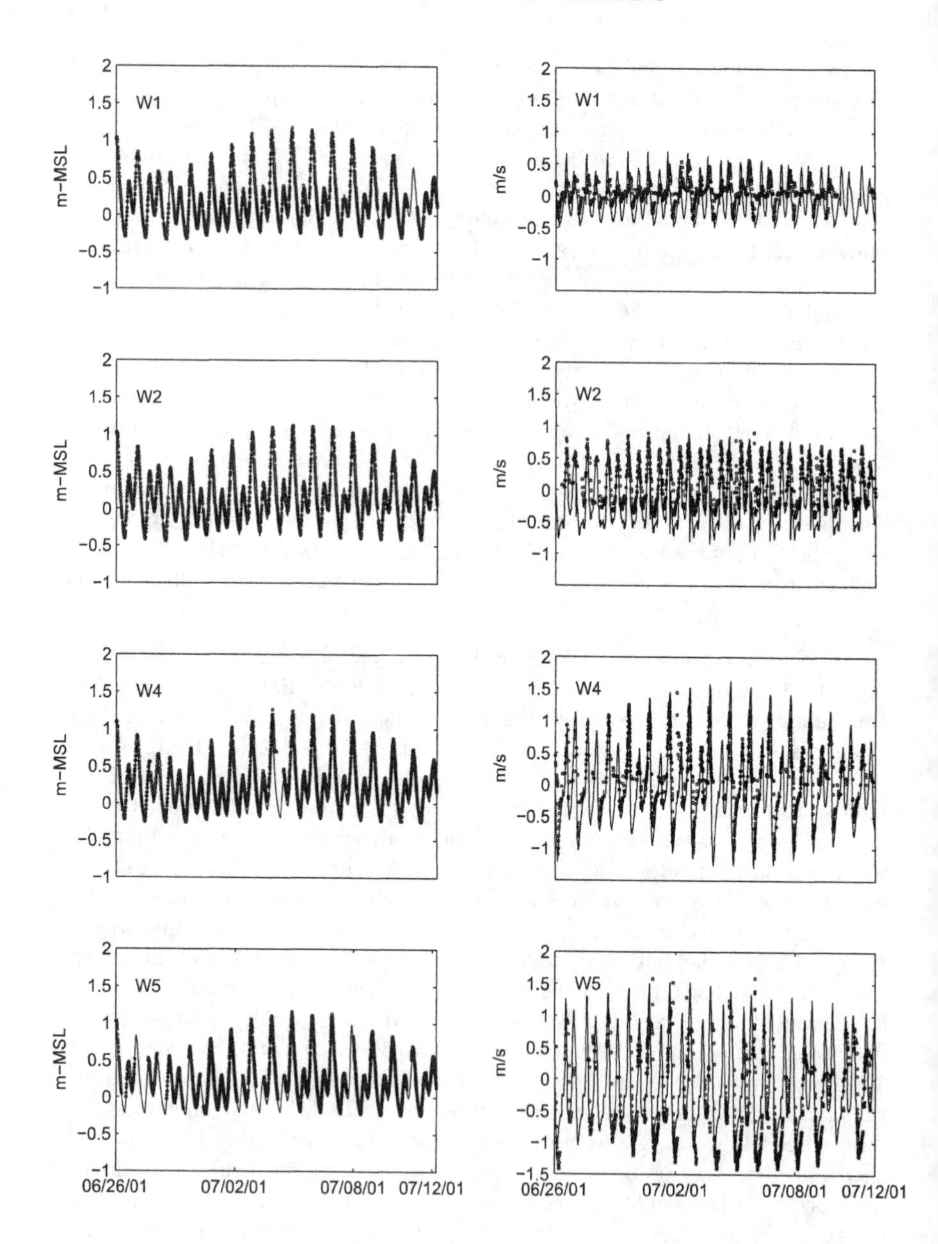

Figure 3: Validation of model predictions at SAR monitoring stations.

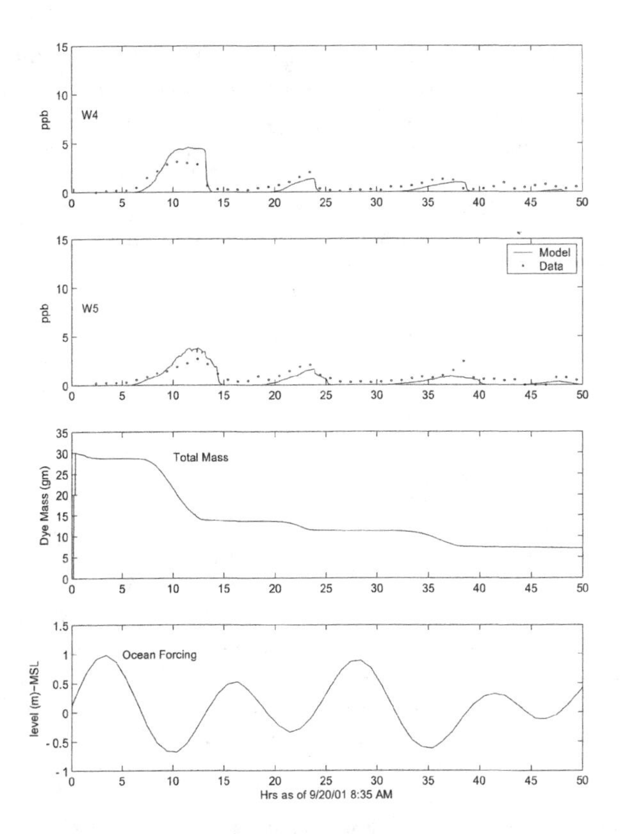

Figure 4: Model predicted and observed dye concentrations resulting from impulse load at tide gates.

This application highlights several interesting issues. First and most obvious is that the model resolves tidal circulation and mixing using physically meaningful mixing parameters. That is, it was not necessary to tune mixing coefficients in order to get reasonable predictions. Because advection is dominant in these systems, this is largely a reflection of the model's ability to resolve the flow field. Second, and more subtle is that the model is able to resolve transport resulting from impulse loads without the spurious oscillations that plague many models. Third, and more troubling is that the model does not conserve dissolved mass to numerical precision as it does for fluid volume. The problem is that it is possible for the volume of water in a computational cell to go to zero while the dissolved mass remains. How to synchronize the two is not obvious and remains a topic of research.

Particle Transport Simulations Using the validated model, residence times in the Santa Ana River wetlands were estimated. Residence time represents the time it takes a parcel of water to reach the outlet to the ocean (Monsen *et al.* 2002). Such information can be useful for water quality and ecological studies. To perform this calculation, the model was forced by a 1 m, 12 hr harmonic tide starting from a high tide initial condition and continuing for 12 tide cycles. Initially, 10 particles were placed in each of the computational cells and the position of each particle was subsequently tracked. The time each particle passes the outlet is recorded, after which the particle is no longer tracked. After the 12 cycle simulation ended, the residence time for each computational cell was computed based on the average of the 10 particles that originally departed from that cell. Particles that had yet to leave the system were assigned a residence time of 12 tide cycles. It should be noted that this calculation is somewhat biased by the release time of the particles, and can be made less so by repeating this simulation with releases at different tide stages.

Contours of residence times are shown in Figure 5. The main channel of the SAR, along with the lower reaches of the lower slough have very short residence times (1-2 cycles), which is expected since the tidal amplitude is comparable to the depth and therefore the volume of the tidal prism is comparable to the total storage at high tide. At the head of lagoon channels the residence time is longest, as the water is effectively trapped because it is shielded by any sort of residual circulation. In contrast, residence times in the network of small channels east of the SAR channel are generally smaller than a few days. (One exception is the northwestern corner of the oval loop just east of the SAR.) The channel network configuration permits residual circulation whereas dead end channels do not, particular when the network is connected at multiple points to a larger body of water (say the SAR) at points where the tide is slightly out of phase. To mitigate adverse water quality conditions that often accompany poor circulation, network features such as these can be incorporated into wetlands restoration designs.

Another use of the particle tracking component of the model is to assess the validity of the assumption that dissolved material mixes quickly over the depth. To illustrate this, fleets of 1000 particles were released at station W2, W4, and W5 at two different instances and tracked for one hour. Separate fleets were used to track transport from surface, mid-depth ($z/h = 0.4$), and bottom sources, and the mean vertical position of these fleets, as well as the standard deviation, was computed over time. Figure 6 presents the evolution of the the mean vertical position and indicates that the time scale for mixing over the depth appears to be roughly 10 minutes or less, although the time scale is longer (on the order of 30 minutes) during less energetic portions of the tide. Since predictions on the order of hours are of interest here, the assumption of rapid vertical mixing is good for energetic portions of the tide, but may be problematic for analysis of transport phenomena during neap tides when tidal currents are smaller.

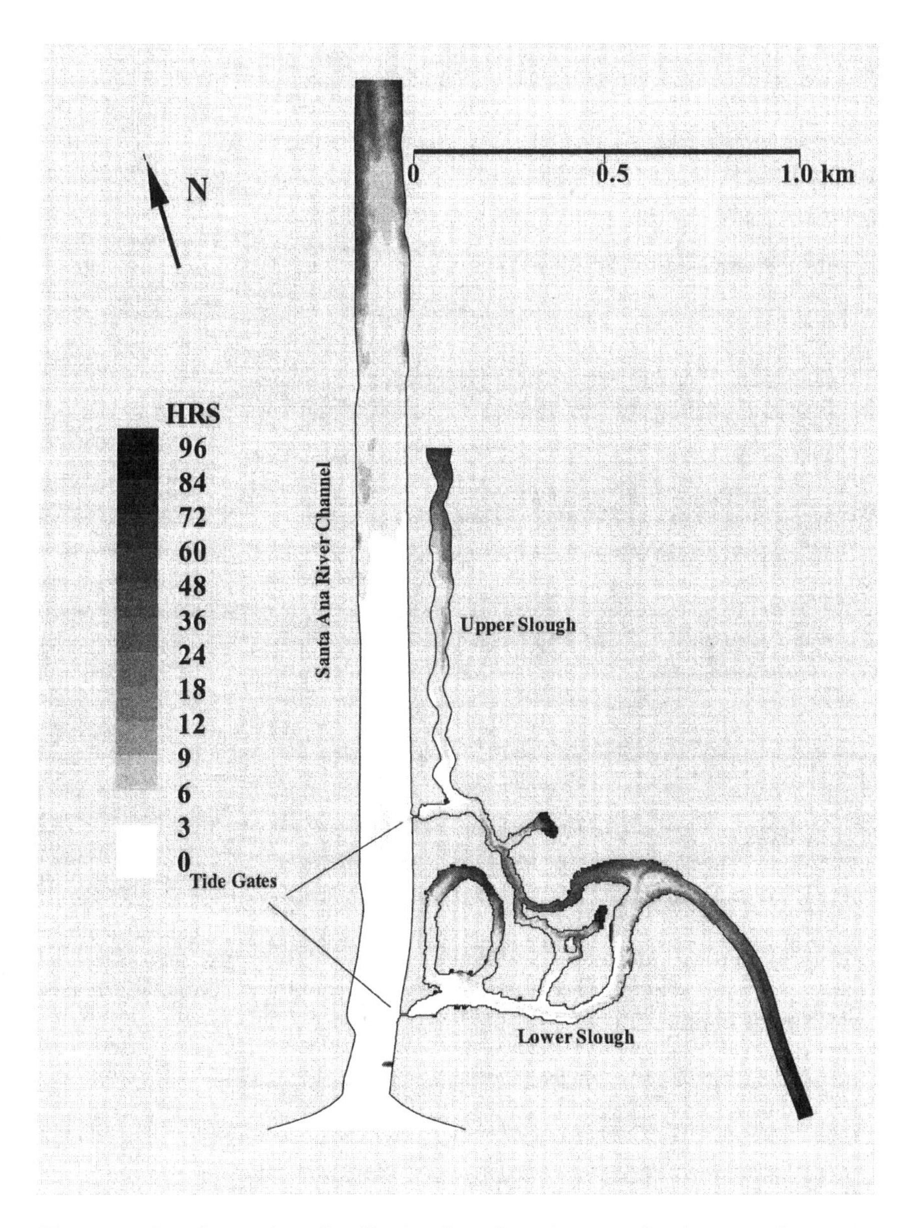

Figure 5: Residence time distribution based on 1 m amplitude monochromatic tides and particles released at high tide.

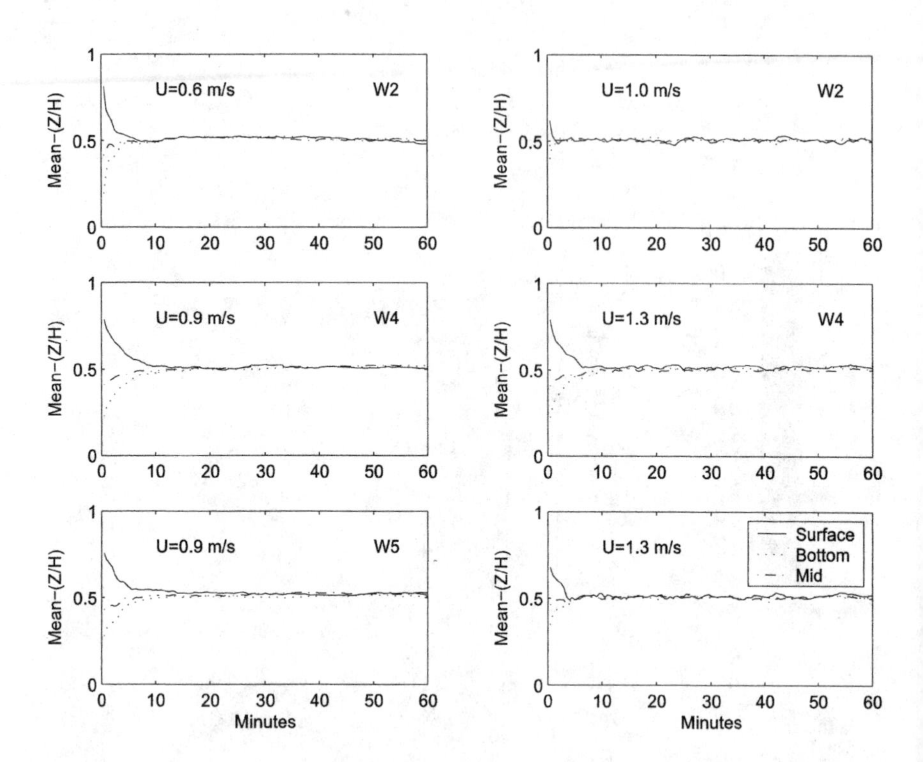

Figure 6: Vertical mixing of particle clouds released at surface, midwater, and bottom of water column. Shown is the mean vertical position of particle clouds, which tends towards $z/h = 0.5$ as the cloud becomes well mixed over the vertical.

Summary and Conclusions

A high-resolution finite volume method, initially developed for analysis of surges and shocks and later refined for flow simulations in tidal wetlands, has been implemented here in an unstructured grid format and applied to tidal wetlands of the Santa Ana River in Southern California. Circulation and mixing is resolved using a 3-4 m resolution grid and a time step that is less than a second so as to satisfy the explicit stability requirement of the method. The model accounts for the flooding and drainage of mudflats and sand bars using a fixed grid and does so without the artificial loss of fluid mass. The model also accounts for solute transport and mixing using physically meaningful mixing parameters, but is not able to do so in the presence of a moving wet/dry interface without artificial loss of dissolved mass. The monotonicity property of the model is advantageous for solute transport predictions because predictions are non-negative which is particularly important for water quality modeling. Use of a small time step is advantageous for resolving sharp fronts and impulse loads and avoids amplitude and phase errors that are common to implicit methods using a time greater than that associated with the CFL condition. However, such accuracy may not always be necessary and comes at the price of long simulation times.

References

Arega, F. and Sanders, B.F. (2004) "Dispersion Model for Tidal Wetlands," *Journal of Hydraulic Engineering,* (to appear).

Babarutsi, S. Ganoulis, J. and Chu V.H. (1989) "Experimental Investigation of Shallow Recirculating Flows" *Journal of the Hydraulics Division,* ASCE, 115(7) 906-924.

Bedford, K., Findikakis, A., Larock, B.E., Rodi, W. and Street, R.L. (1988) "Turbulence Modeling of Surface Water Flow and Transport, Parts I-V., *Journal of Hydraulic Engineering,* 114(9), 970-1073.

Bradford, S.F. and Sanders, Brett F. (2002), "Finite-Volume Model for Shallow-Water Flooding of Arbitrary Topography," *Journal of Hydraulic Engineering,* 128(3), pp. 289-298.

California Coastal Commission (1987) *The California Coastal Resource Guide,* University of California Press, Berkeley, California.

Dimou, K.N. and Adams, E.E. (1993) "A Random-Walk, Particle Tracking Model for Well-Mixed Estuaries and Coastal Waters," *Estuarine, Coastal and Shelf Science,* 37, pp. 99-110.

Elder, J. W.(1959) "The Dispersion of Marked Fluid in Turbulent Shear Flow," J. Fluid Mech., Cambridge, England, 5, pp. 544-560.

Fischer, B.H., List, E.j., Koh, R. C.Y., Imberger, J., and Brooks, N. H. (1979) "Mixing in inland and coastal waters." Academic Press, Inc., New York, N. Y.

Henderson, F.M., (1966), *Open Channel Flow*, Macmillan.

Hirsch, C., (1988) "Numerical Computation of Internal and External Flows, Vol. 1," John Wiley & Sons.

Hughes, T.J.R., (1987), *The Finite Element Method–Linear Static and Dynamic Finite Element Analysis*, Prentice Hall.

Monsen, N.E., Cloern, J.E., Lucas, L.V., and Monismith, S.G., (2002), "A comment on the use of flushing time, residence time, and age as transport time scales," *Limnol. Oceanogr.*, 47(5), pp. 1545-1553.

Roe, P. L. (1981), "Approximate Riemann Solvers, Parameter Vectors, and difference Schemes," Journal of Computational Physics, 43:357-372.

Sweby, P. K. (1984), "High resolution schemes using flux limiters for hyperbolic conservation laws," SIAM (Soc. Ind. Appl. Math.) J. Numer. Anal., 21, 995C1011.

Tompson A. and Gelhar L.W. (1990) "Numerical-Simulation of Solute Transport in 3-Dimensional, Randomly Heterogeneous Porous-Media," *Water Resources Research*, 26(10), pp. 2,541–2,562.

Van Leer, B., (1979) "Towards the Ultimate Conservation Difference Scheme," Journal of Computational Physics, 32:101-136.

Ward, P. (1974) "Transverse Dispersion in Oscillatory Channel Flow," *Journal of the Hydraulics Division*, 100(HY6), pp. 755–770.

Coupling a 3D model with a 2DV model using a free-surface correction method

XinJian Chen[1], Member, ASCE

Abstract

This paper presents an efficient method to couple a three-dimensional hydrodynamic model with a laterally averaged, two-dimensional hydrodynamic model. The methodology presented can be used for such a case where a narrow river, with a vertically two-dimensional flow pattern, flows to a relatively larger water body that has a three-dimensional flow pattern. The coupling is facilitated with a free-surface correction (FSC) method that is unconditionally stable with respect to gravity waves, wind and bottom shear stresses, and vertical eddy viscosity terms. The coupled model solves laterally averaged (2DV) RANS equations for the narrow river portion of the simulation domain and 3D RANS equations for the larger water body. The use of the FSC method allows the 3D equations and the 2DV equations to be solved simultaneously, instead of solving them in sequence. Model test runs show that the coupled model retains the efficiency of both the 3D and 2DV models, and can be run with a Courant number larger than 10.

Introduction

In many three-dimensional, estuarine modeling cases, it is sometimes necessary to extend the upstream boundary to a certain cross section in the upstream tributary where it becomes narrow and the downstream tidal effect is insignificant. In this case, a big portion of the simulation area is a narrow open channel flowing to a larger water body. Although the flow pattern in the downstream portion is three-dimensional, it is generally a vertically two-dimensional problem in the upstream tributary because of the narrowness. One may use a 3D model to simulate hydrodynamics for the entire simulation domain, including the narrow open channel in the upstream. However, it is often more desirable to use a laterally averaged 2D (2DV) model for the narrow upstream tributary, but a 3D model for the downstream estuary

[1] Southwest Florida Water Management District, 7601 Highway 301 North, Tampa, Fla. 33637, (813) 985-7481, xinjian.chen@swfwmd.state.fl.us

where it is wider.

One such case is the Peace River/Charlotte Harbor system in the southwest Florida Peninsula. The upstream part of the Peace River is quite narrow, while the downstream part of the river becomes wider. The river eventually drains to Charlotte Harbor, which is one of the largest estuaries in Florida. Figure 1 shows the lower portion of the Peace River from the City of Arcadia to Charlotte Harbor. As can be seen from the figure, the Lower Peace River is a very complex riverine system. Overall, the upstream portion of the Lower Peace River is very narrow, while the downstream portion is quite open. The Peace River is the major freshwater inflow to the Charlotte Harbor estuary. For the purpose of determining a regulatory minimum flow rate for the Peace River, it is necessary to simulate hydrodynamics in the Peace River/Charlotte Harbor system and the upstream boundary needs to be at a USGS (United States Geological Survey) measurement station in Arcadia where daily flow rate is measured.

Figure 2 shows a typical cross section of the narrow portion of the Peace River. As can be seen from the figure, the cross section is composed of a main channel and two floodplains at both sides of the river. The main channel is only in the order of 20 - 40m, while the floodplains can be as wide as over 1000m. When flow is low, water only exists in the main channel. However, during a major storm event, the floodplains will be submerged and used as conveyance for the flood. For a better understanding of the ecological system in the river, including both the main channel and the floodplains, it is critical to accurately simulate the emerging/submerging feature of the floodplain. In this circumstance, what one needs is information about the total flow rate and the water elevation, not the detailed velocity distribution in the narrow portion of the Peace River. As a result, a laterally averaged 2D (2DV) model is much more efficient than a 3D model for this part of the river, even if the 3D model has the ability to handle wetting/drying and to follow the change of the shoreline. As will be seen in the next section, a laterally averaged 2D model can automatically handle the emerging/submerging feature of the cross section without any special treatment often seen in a 3D model, simply because the river width is included in the governing equations for the 2DV model.

The objective of this study is to couple a three-dimensional hydrodynamic model LESS3D (Lake and Estuarine Simulation System in Three Dimensions) (Chen, 1999; Chen 2003b; Chen 2003c) with a laterally averaged, two-dimensional hydrodynamic model LAMFE (Chen and Flannery, 1998; Chen et al., 2000; Chen, 2003a). In the following, governing equations for the LAMFE and LESS3D models are first given, followed by a description of details on the coupling of the two models using the FSC method. An application of the coupled model to a test case is then presented, before conclusions of the study are drawn.

Governing Equations

Laterally Averaged RANS Equations

Because of the narrowness, hydrodynamics in the upstream riverine system often exhibit two-dimensional patterns. Flows and water quality parameters vary in the vertical and the longitudinal directions but are relatively homogeneous in the lateral direction. Using the hydrostatic pressure assumption and the Boussinesq approximation, laterally averaged continuity, momentum, and transport equations in narrow rivers and estuaries can be written as

$$\frac{\partial ub}{\partial x} + \frac{\partial wb}{\partial z} = v \tag{1}$$

$$\frac{\partial u}{\partial t} + u\frac{\partial u}{\partial x} + w\frac{\partial u}{\partial z} = -\frac{\tau_{wx}}{\rho_o b} - g\frac{\partial \eta}{\partial x} - \frac{g}{\rho_o}\int_z^{\eta}\frac{\partial \rho}{\partial x}d\zeta + \frac{1}{b}\frac{\partial}{\partial x}(bA_h\frac{\partial u}{\partial x}) + \frac{1}{b}\frac{\partial}{\partial z}(bA_v\frac{\partial}{\partial z}) \tag{2}$$

where t is time; x is the horizontal coordinate along the river/estuary; z is the vertical coordinate; u and w denote velocity components in x and z directions, respectively; v is the lateral velocity from lateral inputs (sheet flow of direct runoff, tributary, etc.); b, p, g, ρ, and η represent the width, pressure, gravitational acceleration, density, and the free surface elevation, respectively; ρ_o is the reference density; τ_{wx} represents the shear stress due to the friction acting on the side wall ($= \rho C_w u[u^2+w^2]^{1/2}$, where C_w is a non-dimensional frictional coefficient for side walls); A_h and A_v are kinetic eddy viscosities in the x and z directions, respectively; and ρ is the density which is a function of salinity and temperature.

Integrating Equation (1) over the water depth and considering the direct rainfall to the water surface, the equation for the free surface is obtained as

$$b_\eta \frac{\partial \eta}{\partial t} = -\frac{\partial}{\partial x}\int_{h_0}^{\eta} ubdz + \int_{h_0}^{\eta} vdz + r_n b_\eta \tag{3}$$

where h_o is the bottom elevation, r_n is the net rain intensity (rainfall minus evaporation) having the same units as the velocity, and b_η is the width of the river/estuary at the free surface. The above equation is obtained with the assumption that the flux through the bed is zero.

The LAMFE model solves the above equations using an efficient finite difference method that involves the use of a FSC method (Chen, 2003a). Because the river width is included in the above equations, the change of river width with the water level at each cross section is considered in the model. As a result, the conveyance of the floodplains is automatically included in the LAMFE model. No special treatment is required in the 2DV model.

Three-Dimensional RANS Equations

On the other hand, hydrodynamics in the downstream water body can be three-dimensional and require a 3D model for an accurate simulation of flow patterns. The LESS3D model solves the following set of RANS equations with the hydrostatic assumption and the Boussinesq approximation

$$\frac{\partial u}{\partial x} + \frac{\partial v}{\partial y} + \frac{\partial w}{\partial z} = 0 \tag{4}$$

$$\frac{\partial u}{\partial t} + \frac{\partial uu}{\partial x} + \frac{\partial vu}{\partial y} + \frac{\partial wu}{\partial z} = fv - \frac{1}{\rho_o}\frac{\partial p}{\partial x} + \frac{\partial}{\partial x}(A_h\frac{\partial u}{\partial x}) + \frac{\partial}{\partial y}(A_h\frac{\partial u}{\partial y}) + \frac{\partial}{\partial z}(A_v\frac{\partial u}{\partial z})$$
$$\frac{\partial v}{\partial t} + \frac{\partial uv}{\partial x} + \frac{\partial vv}{\partial y} + \frac{\partial wv}{\partial z} = -fu - \frac{1}{\rho_o}\frac{\partial p}{\partial y} + \frac{\partial}{\partial x}(A_h\frac{\partial v}{\partial x}) + \frac{\partial}{\partial y}(A_h\frac{\partial v}{\partial y}) + \frac{\partial}{\partial z}(A_v\frac{\partial v}{\partial z}) \tag{5}$$

$$p = g\int_z^{\eta} \rho d\zeta \tag{6}$$

where x, y, and z are Cartesian coordinates (x is from west to east, y is from south to north, and z is vertical pointing upward); u, v, and w are velocities in the x, y, and, z directions, respectively; f denotes Coriolis parameter; and A_h and A_v represent horizontal and vertical eddy viscosities, respectively.

Similar to Equation (3), the equation for the free surface in the 3D model is obtained by integrating Equation (4) over the water depth:

$$\frac{\partial \eta}{\partial t}+\frac{\partial}{\partial x}\left(\int_{h_o}^{\eta} u dz\right)+\frac{\partial}{\partial y}\left(\int_{h_o}^{\eta} v dz\right)=r_n \qquad (7)$$

Again, the above equation is obtained with the assumption that the flux through the bed is zero. The FSC method is also used in the LESS3D model to solve the above governing equations (Chen, 2003b).

Coupling of LESS3D and LAMFE

Using the FSC method, the 3D model LESS3D and the 2DV model LAMFE can be conveniently coupled in a straightforward manner. Similar to LAMFE and LESS3D, the coupled 3D-2DV model also involves a predictor step and a corrector step. In the predictor step, intermediate horizontal velocities are calculated with the explicit surface elevation gradient for both domains. Intermediate surface elevations in the 2DV and 3D domains are estimated using the calculated intermediate horizontal velocities from Equations (3) and (7), respectively. In the corrector step, the free-surface locations are corrected by simultaneously solving free-surface correction equations for both the 2DV and 3D domains. Finally, the final velocity fields are calculated after the final free-surface elevations are found. The FSC method is an efficient method that is unconditionally stable with respect to the external gravity wave, bottom stresses, and vertical eddy viscosity terms. Similar to the semi-implicit method of Casulli and Cheng (1992), the FSC method does not involve the mode-splitting technique used by many 3-D hydrodynamic models for free-surface flows (e.g., Sheng, 1986; Blumberg and Mellor, 1987; Hamrick, 1992).

The finite difference equations presented in this section are derived using Cartesian grid systems with staggered arrangements of model variables for both the 3D and 2DV domains. While pressure, density, and concentrations are defined at the center of a grid cell, horizontal velocities are defined at the centers of side faces of the grid cell. The vertical velocity is defined at the center of the top face of the cell. The water surface elevation is calculated at the center of the horizontal grid.

Predictor Step

In the predictor step, the intermediate horizontal velocities in the 2DV and 3D domains are calculated independently. In the 2DV domain, we have:

$$\frac{u_{i,k}^{n+*} - u_{i,k}^{n}}{\Delta t} = H_x^n - g\frac{\partial \eta^n}{\partial x} - \frac{C_w u_{i,k}^{n+*}}{b}\sqrt{(u_{i,k}^n)^2 + (w_{i+1/2,k-1/2}^n)^2} + \frac{1}{b}\frac{\partial}{\partial z}(bA_v\frac{\partial u^{n+*}}{\partial z}) \tag{8}$$

where Δt is the time step used in the computation; the superscript n represents the previous time step, while the superscript $n+*$ represents the intermediate solution at the new time step; η^n is the free surface location at the previous time step, and H_x^n is an explicit finite difference operator containing the convection terms, the baroclinic term and the horizontal eddy viscosity term:

$$H_x^n = -\left[u\frac{\partial u}{\partial x} + w\frac{\partial u}{\partial z} + g\int_z^\eta \frac{\partial \rho}{\partial x}d\zeta\right]^n + \frac{1}{b}\frac{\partial}{\partial x}(bA_h\frac{\partial u^n}{\partial x}) \tag{9}$$

For each i, Equation (8) is a tri-diagonal matrix system, which can be efficiently solved using the Thomas Algorithm (Ferziger and Peric, 1999). Once u^{n+*} is determined, the free surface can be computed by using Equation (3)

$$\Delta\eta_i^* = -\frac{\Delta t\theta}{b_\eta \Delta x_i}(U_i^{n+*} - U_{i-1}^{n+*}) - \frac{\Delta t(1-\theta)}{b_\eta \Delta x_i}(U_i^n - U_{i-1}^n) + \frac{\Delta t}{b_\eta}\sum_{k=k_n}^{k_m^n} v_{i,k}^{n+1/2}\Delta z_{i,k}^n + \Delta t\, r_n^{n+1/2} \tag{10}$$

where $\Delta\eta_i^* (= \eta_i^{n+*} - \eta_i^n)$ is the increment in free surface elevation that is estimated from the intermediate velocity field, or $\Delta\eta_i^*$ is the difference between the intermediate free surface and the free surface at the previous time step, θ is a model parameter varying between 0 and 1 (fully explicit for $\theta = 0$ and fully implicit for $\theta = 1$), and

$$U_i^{n+*} = \sum_{k_{un}}^{k_{um}^n} u_{i,k}^{n+*} b_{i+1/2,k}^n \Delta z_{i+1/2,k}^n \tag{11}$$

where k_n and $k^n{}_m$ are respectively the bottom and top k-indexes at the center of the horizontal grid, and k_{un} and $k^n{}_{um}$ are respectively the bottom and top k-indexes at the right face of the horizontal grid.

In the 3D domain, the intermediate velocity field is calculated from Equation (5) using the pressure field of the previous time step

$$\frac{u^{n+*} - u^n}{\Delta t} = fv^n - g\frac{\partial \eta^n}{\partial x} - \frac{g}{\rho_o}\left[\int_z^\eta \frac{\partial \rho}{\partial x}d\zeta\right]^n + H_x^n + \frac{\partial}{\partial z}(A_v\frac{\partial u^{n+*}}{\partial z})$$

$$\frac{v^{n+*} - v^n}{\Delta t} = -fu^{n+*} - g\frac{\partial \eta^n}{\partial y} - \frac{g}{\rho_o}\left[\int_z^\eta \frac{\partial \rho}{\partial y}d\zeta\right]^n + H_y^n + \frac{\partial}{\partial z}(A_v\frac{\partial v^{n+*}}{\partial z}) \tag{12}$$

For simplicity, the i-, j-, and k-indexes are omitted in the above finite difference equations.

In Equation (12), the vertical eddy viscosity terms are discretized implicitly, while the horizontal gradients of the free surface, H_x, and H_y, and the baroclinic terms are treated explicitly. Similar to Equation (8), Equation (12) represents two tri-diagonal matrix systems and can be efficiently solved using the Thomas Algorithm. Once u^{n+*} and v^{n+*} are solved, the intermediate free surface can be computed using Equation (7) as follows

$$\Delta\eta^{*}_{i,j} = \eta^{n+*}_{i,j} - \eta^{n}_{i,j} = -\frac{\Delta t(1-\theta)}{a^{\tau}_{i,j}}[U^{n}_{i+1/2,j} - U^{n}_{i-1/2,j} + V^{n}_{i,j+1/2} - V^{n}_{i,j-1/2}] - \frac{\Delta t\theta}{a^{\tau}_{i,j}}[U^{n+*}_{i+1/2,j} - U^{n+*}_{i-1/2,j} + V^{n+*}_{i,j+1/2} - V^{n+*}_{i,j-1/2}] + \Delta t\, r_{n}^{n+1/2} \quad (13)$$

where $a^{\tau}_{i,j}$ is the wet surface area of the horizontal grid with the indexes i and j at the previous time step; $\Delta\eta^{*}_{i,j}(=\eta^{n+*}_{i,j} - \eta^{n}_{i,j})$ is the increment of the free surface estimated from the intermediate velocity field; and U and V are vertically integrated fluxes:

$$U^{n}_{i+1/2,j} = \sum_{k=k_{un}}^{k_{um}} u^{n}_{i+1/2,j,k} a^{n}_{i+1/2,j,k}, \qquad V^{n}_{i,j+1/2} = \sum_{k=k_{vn}}^{k_{vm}} v^{n}_{i,j+1/2,k} a^{n}_{i,j+1/2,k}$$
$$U^{n+*}_{i+1/2,j} = \sum_{k=k_{un}}^{k_{um}} u^{n+*}_{i+1/2,j,k} a^{n}_{i+1/2,j,k}, \qquad V^{n+*}_{i,j+1/2} = \sum_{k=k_{vn}}^{k_{vm}} v^{n+*}_{i,j+1/2,k} a^{n}_{i,j+1/2,k} \quad (14)$$

where $a^{n}_{i+1/2,j,k}$ and $a^{n}_{i,j+1/2,k}$ are areas of the east and north faces, respectively, of the cell with the indexes i, j, and k at the previous time step, k_{un} and k_{um} are respectively the bottom and top k-indexes at the u-point, and k_{vn} and k_{vm} are the bottom and top k-indexes at the v-point.

If the superscript $n+*$ in Equations (8) and (12) is changed to $n+1$, the computation continues to the next time step (time step $n+2$) and the time step (Δt) will be restricted by the celerity of the gravity wave due to the explicit treatment of the free-surface gradient terms in the momentum equations. To eliminate this time step restriction, the second step of the FSC method is introduced.

Corrector Step

In the second step of the FSC method, the time step restriction is removed with the semi-implicit discretization of the barotropic pressure component in the horizontal momentum equations. In the 2DV domain, the semi-implicit scheme is

$$\frac{u^{n+1}_{i,k} - u^{n}_{i,k}}{\Delta t} = -g\theta\frac{\partial\eta^{n+1}}{\partial x} + \frac{1}{b}\frac{\partial}{\partial z}(bA_{v}\frac{\partial u^{n+1}}{\partial z}) + H^{n}_{x} - g(1-\theta)\frac{\partial\eta^{n}}{\partial x} - \frac{C_{w}u^{n+*}_{i,k}}{b}\sqrt{(u^{n}_{i,k})^{2} + (w^{n}_{i+1/2,k-1/2})^{2}}) \quad (15)$$

where u^{n+1} is the final velocity along the river at the new time instant. From Equations (8) and (15), one obtains

$$\frac{u^{n+1}_{i,k} - u^{n+*}_{i,k}}{\Delta t} = -g\theta\frac{\partial(\eta^{n+1} - \eta^{n})}{\partial x} + \frac{\partial}{\partial z}[A_{v}\frac{\partial(u^{n+1} - u^{n+*})}{\partial z}] \quad (16)$$

Integrating the above equation over the water column and inserting the resulting integral into Equation (3), we have

$$\eta_i^{n+1} - \eta_i^{n+*} = \frac{g\theta^2\Delta t^2}{b_\eta \Delta x_i}\left(A_{i+1/2}\frac{\Delta\eta_{i+1} - \Delta\eta_i}{\Delta x_{i+1/2}} - A_{i-1/2}\frac{\Delta\eta_i - \Delta\eta_{i-1}}{\Delta x_{i-1/2}}\right) \quad (17)$$

where $\Delta\eta_i (= \eta_i^{n+1} - \eta_i^n)$ is the final increment of the free surface over the time step Δt, and $A_{i+1/2}$ is the cross-section area at the right face of the horizontal grid (the sum of wetted areas of right faces for grid cells with the horizontal grid index *i*.) Re-arranging Equation (17), one obtains

$$-R_i^w \Delta\eta_{i-1} + (1 + R_i^w + R_i^e)\Delta\eta_i - R_i^e \Delta\eta_{i+1} = \Delta\eta_i^* \quad (18)$$

where

$$R_i^w = \frac{g\theta^2\Delta t^2}{\Delta x_i \Delta x_{i-1/2}}\frac{A_{i-1/2}}{b_{\eta i-1/2}}, \qquad R_i^e = \frac{g\theta^2\Delta t^2}{\Delta x_i \Delta x_{i+1/2}}\frac{A_{i+1/2}}{b_{\eta i+1/2}} \quad (19)$$

In the 3D domain, the final horizontal velocity components are calculated as follows

$$\begin{aligned} \frac{u^{n+1} - u^n}{\Delta t} &= fv^n - g(1-\theta)\frac{\partial\eta^n}{\partial x} - g\theta\frac{\partial\eta^{n+1}}{\partial x} - \frac{g}{\rho_o}\left[\int_z^\eta \frac{\partial\rho}{\partial x}d\zeta\right]^n + H_x^n + \frac{\partial}{\partial z}\left(A_v\frac{\partial u^{n+1}}{\partial z}\right) \\ \frac{v^{n+1} - v^n}{\Delta t} &= -fu^{n+*} - g(1-\theta)\frac{\partial\eta^n}{\partial y} - g\theta\frac{\partial\eta^{n+1}}{\partial y} - \frac{g}{\rho_o}\left[\int_z^\eta \frac{\partial\rho}{\partial y}d\zeta\right]^n + H_y^n + \frac{\partial}{\partial z}\left(A_v\frac{\partial v^{n+1}}{\partial z}\right) \end{aligned} \quad (20)$$

where u^{n+1} and v^{n+1} are final velocities at the new time step.

Subtracting Equation (12) from Equations (20) results in

$$\begin{aligned} \frac{\delta u}{\Delta t} &= -g\theta\frac{\partial(\Delta\eta)}{\partial x} + \frac{\partial}{\partial z}\left[A_v\frac{\partial(\delta u)}{\partial z}\right] \\ \frac{\delta v}{\Delta t} &= -g\theta\frac{\partial(\Delta\eta)}{\partial y} + \frac{\partial}{\partial z}\left[A_v\frac{(\delta v)}{\partial z}\right] \end{aligned} \quad (21)$$

where $\Delta\eta \; (= \eta^{n+1} - \eta^n)$ is the final increment of the free surface over the time step Δt, $\delta u = u^{n+1} - u^{n+*}$ and $\delta v = v^{n+1} - v^{n+*}$. Integrating both equations over the water column and inserting the integrals into Equation (8), one has

$$\begin{aligned} \eta_{i,j}^{n+1} - \eta_{i,j}^{n+*} = \frac{g\Delta t^2\theta^2}{a_{i,j}^\tau}\Big[& A_{i+1/2,j}^n\frac{\Delta\eta_{i+1,j} - \Delta\eta_{i,j}}{\Delta x_{i+1/2}} - A_{i-1/2,j}^n\frac{\Delta\eta_{i,j} - \Delta\eta_{i-1,j}}{\Delta x_{i-1/2}} \\ & + A_{i,j+1/2}^n\frac{\Delta\eta_{i,j+1} - \Delta\eta_{i,j}}{\Delta y_{j+1/2}} - A_{i,j-1/2}^n\frac{\Delta\eta_{i,j} - \Delta\eta_{i,j-1}}{\Delta y_{j-1/2}}\Big] \end{aligned} \quad (22)$$

where $A_{i+1,j}^n$ and $A_{i,j+1/2}^n$ are total areas of the east and north faces of the water column for a horizontal grid with the indexes *i* and *j* at the nth time step. Equation (22) can be rewritten in the following form

$$\begin{aligned} &-R_{i,j}^s\Delta\eta_{i,j-1} - R_{i,j}^w\Delta\eta_{i-1,j} + (1 + R_{i,j}^s + R_{i,j}^w + R_{i,j}^e + R_{i,j}^n)\Delta\eta_{i,j} - R_{i,j}^e\Delta\eta_{i+1,j} \\ &-R_{i,j}^n\Delta\eta_{i,j+1} = \Delta\eta_{i,j}^* \end{aligned} \quad (23)$$

where

$$R^w_{i,j} = \frac{g\Delta t^2 \theta^2}{a^\tau_{i,j}\Delta x_{i-1/2}} A^n_{i-1/2,j}\,, \qquad R^e_{i,j} = \frac{g\Delta t^2 \theta^2}{a^\tau_{i,j}\Delta x_{i+1/2}} A^n_{i+1/2,j}$$
$$R^s_{i,j} = \frac{g\Delta t^2 \theta^2}{a^\tau_{i,j}\Delta y_{j-1/2}} A^n_{i,j-1/2}\,, \qquad R^n_{i,j} = \frac{g\Delta t^2 \theta^2}{a^\tau_{i,j}\Delta y_{j+1/2}} A^n_{i,j+1/2} \tag{24}$$

<u>Dynamic Coupling</u>

While Equation (18) forms a tri-diagonal system, Equation (23) forms a five-diagonal system. If the 2DV domain is discretized with N grids along the river, Equation (18) represents N linear equations

$$\mathbf{r}\Delta\boldsymbol{\eta}_{2DV} = \Delta\boldsymbol{\eta}^*_{2DV} \tag{25}$$

where

$$\mathbf{r} = \begin{bmatrix} r_{11} & r_{12} & & & & \\ r_{21} & r_{22} & r_{23} & & & \\ & \cdot & \cdot & \cdot & & \\ & & \cdot & \cdot & \cdot & \\ & & & \cdot & \cdot & \cdot \\ & & & r_{(N-1)(N-2)} & r_{(N-1)(N-1)} & r_{(N-1)N} \\ & & & \cdot & r_{N(N-1)} & r_{NN} \end{bmatrix} \tag{26}$$

$$\Delta\boldsymbol{\eta}_{2DV} = [\Delta\eta_1 \quad \Delta\eta_2 \quad . \quad . \quad . \quad \Delta\eta_{N-1} \quad \Delta\eta_N\ ]^{\mathrm{T}}$$
$$\Delta\boldsymbol{\eta}^*_{2DV} = [\Delta\eta^*_1 \quad \Delta\eta^*_2 \quad . \quad . \quad . \quad \Delta\eta^*_{N-1} \quad \Delta\eta^*_N\ ]^{\mathrm{T}} \tag{27}$$

where $r_{i(i-1)} = -R^w_i$, $r_{i(i+1)} = -R^e_i$, and $r_{ii} = 1 - r_{i(i-1)} - r_{i(i+1)}$. On the other hand, if the 3D domain is discretized with N_x and N_y grids in the x- and y-directions, respectively, Equation (23) represents a maximum of $M = N_x N_y$ linear equations

$$\mathbf{q}\Delta\boldsymbol{\eta}_{3D} = \Delta\boldsymbol{\eta}^*_{3D} \tag{28}$$

where

$$\mathbf{q} = \begin{bmatrix} q_{11} & q_{12} & & & & q_{1(L+1)} & & & \\ q_{21} & q_{22} & q_{23} & & & & q_{2(L+2)} & & \\ & \cdot & \cdot & \cdot & & & & \cdot & \\ & & \cdot & \cdot & \cdot & & & & \cdot \\ & & & \cdot & \cdot & \cdot & & & \\ \cdot & & & & \cdot & \cdot & \cdot & & \\ & \cdot & & & & \cdot & \cdot & \cdot & \\ & & q_{(M-1)(M-L-1)} & & & & q_{11} & q_{11} & q_{11} \\ & & & q_{M(M-L)} & & & & q_{M(M-1)} & q_{MM} \end{bmatrix} \tag{29}$$

$$\begin{aligned} \mathbf{\Delta\eta}_{3D} &= [\Delta\eta_1 \quad \Delta\eta_2 \quad . \quad . \quad . \quad \Delta\eta_{M-1} \quad \Delta\eta_M \,]^{\mathrm{T}} \\ \mathbf{\Delta\eta}^*_{3D} &= [\Delta\eta^*_1 \quad \Delta\eta^*_2 \quad . \quad . \quad . \quad \Delta\eta^*_{M-1} \quad \Delta\eta^*_M \,]^{\mathrm{T}} \end{aligned} \tag{30}$$

where $q_{l(l-L)} = -R^s_{i,j}, q_{l(l-1)} = -R^w_{i,j}, q_{l(l+1)} = -R^e_i, q_{l(l+L)} = -R^n_{i,j}, q_{ll} = 1 - q_{l(l-L)} - q_{l(l-1)} - q_{l(l+1)} - q_{l(l+L)}$, and the index l is for the conversion between a 2D array and a 1D array. Let L be either N_x or N_y. If the i-loop is inside the j-loop, then $L = N_x$ and $l = i + (j-1)L$. Otherwise, $L = N_y$ and $l = j + (i-1)L$.

Equations (25) and (28) can be merged together to form a five-diagonal system:

$$\begin{bmatrix} \mathbf{q} & \mathbf{0} & \mathbf{q'} \\ \mathbf{0} & \mathbf{I} & \mathbf{0} \\ \mathbf{r'} & \mathbf{0} & \mathbf{r} \end{bmatrix} \begin{bmatrix} \mathbf{\Delta\eta}_{3D} \\ \mathbf{\Delta\eta}_{3D2DV} \\ \mathbf{\Delta\eta}_{2DV} \end{bmatrix} = \begin{bmatrix} \mathbf{\Delta\eta}^*_{3D} \\ \mathbf{0} \\ \mathbf{\Delta\eta}^*_{2DV} \end{bmatrix} \tag{31}$$

where **r'** and **q'** are rectangular matrices of order $N \times M$ and $M \times N$, respectively; **I** is a unit matrix of order $n \times n$ and $\mathbf{\Delta\eta}_{3D2DV}$ is a dummy matrix of order $n \times 1$. Matrices **I** and $\mathbf{\Delta\eta}_{3D2DV}$ are necessary in many cases to ensure that the above matrix system remains five-diagonal. Depending on the nature of the connection between the 3D and 2DV domains, n varies between 0 and L–1. Both **r'** and **q'** have just one non-zero element. The non-zero element in **q'** is $q'_{(M-L+n+1)1}$ and equals to one of the four coefficients in Equation (24), depending on how the 2DV and 3D domains are connected. The non-zero element in **r'** is $r'_{1(M-L+n+1)}$ and equals to one of the two coefficients in Equation (19), depending on how the 2DV and 3D domains are connected. For example, if the 2DV domain is connected to a northern boundary of the 3D domain, then $q'_{(M-L+n+1)1} = R^n_{i',j'}$ and $r'_{1(M-L+n+1)} = R^w_1$, where i' and j' are respectively i- and j-indexes of the 3DV grid where the 2DV domain is connected to. To ensure the mass conservation, the cross-sectional area of the internal boundary in $r'_{1(M-L+n+1)}$ should be consistent with the area A used in $q'_{(M-L+n+1)1}$.

The five-diagonal system of Equation (31) can be efficiently solved using a conjugate gradient method with incomplete Cholesky preconditioning (Golub and van Loan, 1990) or a bi-conjugate gradient method (Van der Vorst, 1992). After Equation (31) is solved, the final free surface location is found for the entire simulation area, including both the 3D domain and the 2DV domain. The final horizontal velocities in the 2DV and 3D domains are then calculated from Equations (16) and (21), respectively. As shown in Chen (2003a) and Chen (2003b), the following simpler equations can be used to calculate the final horizontal velocities, instead of using Equations (16) and (21):

2DV:
$$\frac{u^{n+1}_{i,k} - u^{n+*}_{i,k}}{\Delta t} = -g\theta \frac{\partial(\Delta\eta)}{\partial x} \tag{32}$$

3D:
$$\begin{aligned} \frac{\delta u}{\Delta t} &= -g\theta \frac{\partial(\Delta\eta)}{\partial x} \\ \frac{\delta v}{\Delta t} &= -g\theta \frac{\partial(\Delta\eta)}{\partial y} \end{aligned} \tag{33}$$

The vertical velocity is then calculated from the following flux-based finite difference equations that guarantee the mass conservation:

$$\text{2DV:}\quad w_{i,k}^{n+1} = \frac{1}{b_{i,k+1/2}^{n}}[\Delta z_{i,k}^{n} v_{i,k}^{n+1/2} + b_{i,k-1/2}^{n} w_{i,k-1}^{n+1} - \frac{u_{i,k}^{n+1} b_{i+1/2,k}^{n} \Delta z_{i+1/2,k}^{n} - u_{i-1,k}^{n+1} b_{i-1/2,k}^{n} \Delta z_{i-1/2,k}^{n}}{\Delta x_i}] \quad (34)$$

$$\text{3D:}\quad w_{i,j,k}^{n+1} = \frac{1}{a_{i,j,k+1/2}^{n}}[F_{i,j,k-1/2}^{n+1} - F_{i+1/2,j,k}^{n+1} + F_{i-1/2,j,k}^{n+1} - F_{i,j+1/2,k}^{n+1} + F_{i,j-1/2,k}^{n+1}] \quad (35)$$

where $a_{i,j,k+1/2}^{n}$ is the area of the top face of the cell with the indexes i, j, and k at the previous time step, $F_{i-1/2,j,k}^{n+1}$, $F_{i,j-1/2,k}^{n+1}$ and $F_{i,j,k-1/2}^{n+1}$ are fluxes of water flowing into cell (i, j, k) through the west, south, and bottom faces, respectively.

It should be pointed out that while the intermediate horizontal velocities and free-surface elevations in the 3D and 2DV domains are calculated independently in the predictor step, the final free-surface elevations in the two areas are found in a coupled manner by solving Equation (31). If the final free-surface elevations in the two domains were calculated in sequence from Equations (25) and (28), respectively, then it would be necessary to set the boundary conditions at the internal boundary between the 3D domain and the 2DV domain. For both domains, this internal boundary would be an open boundary. One might use either the surface elevation at the previous time step or the intermediate surface elevation at one side of the internal boundary as the boundary condition for the other side; however, this kind of treatment would not only introduce errors associated with the open boundary treatment, but also would weaken the integrity of the semi-implicit scheme. By solving Equation (31), however, this kind of internal boundary problem is avoided, because the final surface elevations at the two grids on both sides of the internal boundary are solved as unknowns in Equation (31).

A Test Case

To test the coupling method described in the last section, the coupled LESS3D-LAMFE model was tested with a hypothetical bay shown in Figure 3. The downstream portion is the 3D domain, while the upstream portion is the 2DV domain. The 3D domain has a length of 12km in the x-direction and 6km in the y-direction. The 2DV domain is a narrow river with a length of about 12km. In the model application, the 3D domain is discretized with $30 \times 15 = 450$ uniform grids in the x-y plane with the grid size of $\Delta x = \Delta y = 400$m. The 2DV domain is discretized with 30 segments that are 400m long. In the vertical direction, both the 3D and 2DV domains are discretized with the same vertical spacing that varies between 0.5m and 1.2m.

The 3D-2DV system is driven by tides at the downstream open boundary and the flow rate at the most upstream boundary. The downstream open boundary condition includes a K_1 tide (with a period of 23.93 hours) and an M_2 (with a period of 12.42 hours). Amplitudes for the K_1 and M_2 tides are assumed to be 0.4m and 0.2m, respectively. A time-independent vertical salinity distribution is assumed at the downstream open boundary. At the most upstream boundary, a constant freshwater flow rate of 100cfs is given.

The coupled model was used to run the test case for 60 days. Model results in the last 15 days at two locations (A and B in Figure 3) are shown in Figure 4. The top graph shows simulated surface elevations at the two locations. Because the entire simulation domain is not very large and the distance between A and B is only about 12km, the surface elevation at A is almost the same as that at B with only less than 30 minutes of time lag of the tidal signal. The middle graph of Figure 4 shows simulated velocities at A and B at a depth of z = -1.08m (NGVD). Simulated salinities at A and B at z = -1.08m (NGVD) are shown in the bottom graph of Figure 4. As can be seen from Figure 4, model results at the two locations are reasonable. Although no analytical solutions or measurement data are available to assess how good a job the coupled 3D-2DV model does for the test case, model results shown in Figure 4 at least indicate that the coupling of the 3D and 2DV models is successful.

The 3D domain is about 5.68m deep at its deepest area. In the 2DV domain, the maximum depth is 4.10m. The coupled model was run using a time step of 600 seconds, equivalent to a Courant number of 11.2. If salinity is not included in the simulation and is assumed to be zero everywhere, the model can be run with a time step of 1200 seconds without any problems.

Conclusions

The 3D model LESS3D has been coupled with the 2DV model LAMFE using a FSC method that is unconditionally stable with respect to gravity waves, wind and bottom shear stresses, and vertical eddy viscosity terms. The coupled 3D-2DV model is an efficient tool for simulating hydrodynamics in such an estuary where a narrow open channel that has a vertically two-dimensional flow pattern is connected with a larger water body that exhibits a three-dimensional flow pattern. This kind of a coupled model is especially desirable when the narrow open channel has large floodplains that can be submerged during a major storm event.

The coupled model solves laterally averaged (2DV) RANS equations for the narrow open channel. For the larger water body, it solves 3D RANS equations. The use of the FSC method allows a simultaneous solution of the free-surface elevation in both the 3D domain and the 2DV domain. As such, the coupled model does not have any problems associated with the internal boundary. The coupled 3D-2DV model was tested with a hypothetical case where a narrow riverine estuary flows to a bay. At the mouth of the bay, the open boundary is forced by a K_1 tide and a M_2 tide. At the most upstream of the river, a constant flow rate is given. The model test run shows that the method presented in this paper works very well for the coupling of the LESS3D model with the LAMFE model. The coupled model retains the efficiency of both LESS3D and LAMFE, and allows a Courant number of larger than 10 in the model simulation.

Although only one narrow tributary is assumed in this paper, the same coupling strategy can be used for a large water body with multiple narrow tributaries. In this circumstance, the 2DV model needs to be applied for each of the tributaries, resulting in multiple sets of linear equations for the free-surface change over the time step Δt (Equation (25)). All one has to do is to properly merge these equations with the linear equations for the free-surface change (Equation (28)) in the 3D area to ensure that the final matrix system (Equation (31)) is five-diagonal. Sometimes, the mesh layout for the 3D area and the numbering of the grids become tricky.

References

Blumberg, A. F. and Mellor, G.L. 1987. A description of a three-dimensional coastal ocean circulation model, *Three-Dimensional Coastal Ocean Models*, American Geophysical Union, pp. 1-16.

Casulli, V. and Cheng, R.T. 1992. Semi-implicit finite difference methods for three-dimensional shallow water flow. *Int. J. for Numer. Methods in Fluids*, 15(6): 629-648.

Chen, XinJian and Flannery, M.S. 1998. Use of a Hydrodynamic Model for Establishing a Minimum Freshwater Flow to the Lower Hillsborough River. In (M. L. Spaulding and A. F. Blumberg, eds.) *Estuarine and Coastal Modeling (V), Proceedings of the 5th International Conference,* ASCE, pp. 663-678.

Chen, XinJian. 1999. Three-Dimensional, Hydrostatic and Non-Hydrostatic Modeling of Seiching in a Rectangular Basin. In (M. L. Spaulding and H. L. Butler, eds.) *Estuarine and Coastal Modeling (VI), Proceedings of the 6th International Conference,* ASCE, pp. 148-161.

Chen, XinJian, M.S. Flannery, and D.L. Moore. 2000. Salinity Response to the Change of the Upstream Freshwater Flow in the Lower Hillsborough River, Florida. *Estuaries*, 23: 735-742.

Chen, XinJian. 2003a. An efficient finite difference scheme for simulating hydrodynamics in narrow rivers and estuaries. *International Journal for Numerical Methods in Fluids.* 42: 233-247.

Chen, XinJian. 2003b. A Free-Surface Correction Method for Simulating Shallow Water Flows. *Journal of Computational Physics*, 189:557-578.

Chen, XinJian. 2003c. Fitting Topography and Shorelines in a 3-D, Cartesian-Grid Model for Free-Surface Flows. *Computational Fluid and Solid Mechanics 2003, Second MIT Conference*, Vol. 2: 1892-1895, Elsevier Science.

Ferziger, J.H. and Peric. M. 1999. Computational Methods for Fluid Dynamics. 2nd rev. ed., Springer-Verlag, Berlin; Heidelberg; New York.

Golub, G.H. and C. van Loan. 1990. *Matrix computations*, John Hopkins University Press, Baltimore.

Hamrick, J.M. 1992. A Three-Dimensional Environmental Fluid Dynamic Computer Code: Theoretical and Computational Aspects. Special Report in Applied Marine Science and Ocean Engineering No 317. Virginia Institute of Marine Science, College of Willian & Mary, Gloucester Point, Virginia.

Sheng, Y.P. 1986. A Three-dimensional Mathematical Model of Coastal, Estuarine and Lake Currents Using Boundary Fitted Grid. Report No. 585, A.R.A.P. Group of Titan Systems, New Jersey, Princeton, NJ.

Van der Vorst, H.A. 1992. Bi-CGSTAB: A fast and smoothly converging variant of Bi-CG for solution of non-symmetric linear systems, *SIAM J. Sci. Statist. Comput.* 13: 631-644.

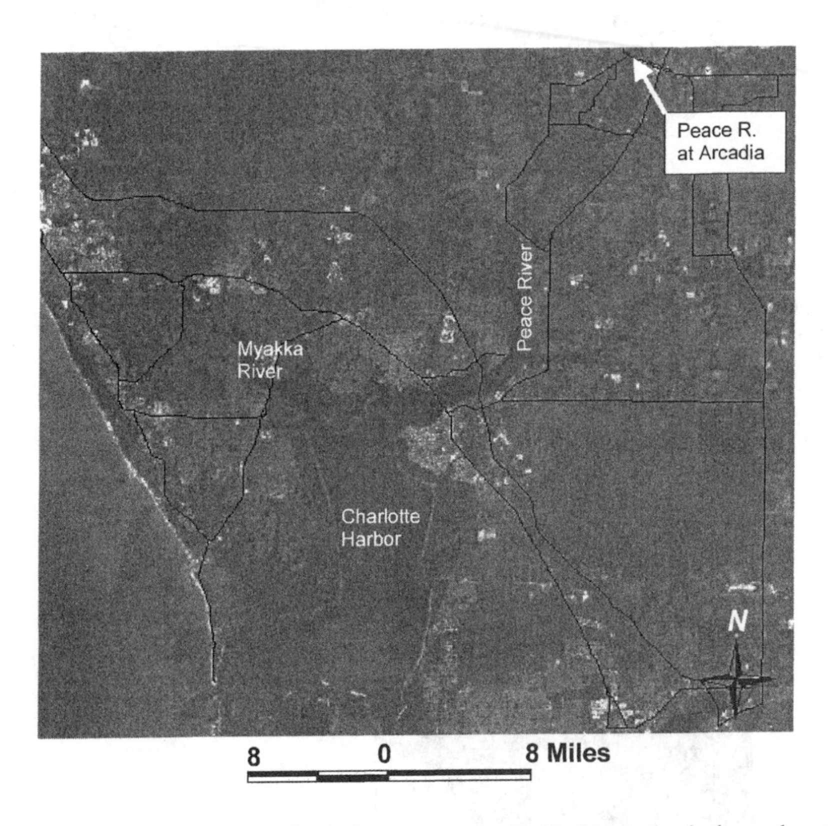

Figure 1 The Peace River is a major freshwater source to the Charlotte harbor in the southwest Florida Peninsula. The photo only shows the lower portion of the Peace River.

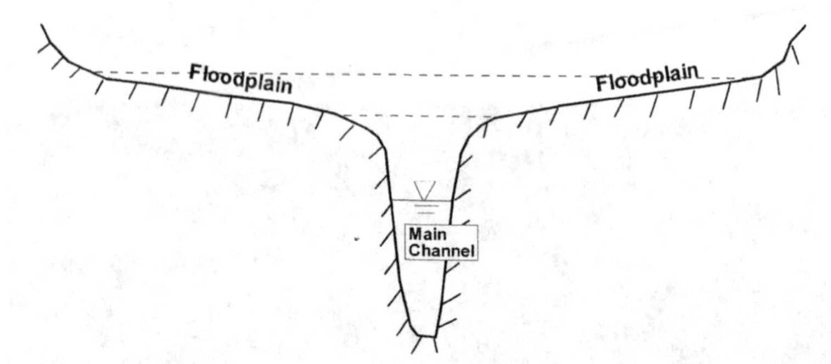

Figure 2 A typical cross section of the narrow part of the Peace River. It is comprised of a main channel and two floodplains at both sides. Most of the time, flow only exists in the main channel. During a major storm event, the floodplains can be submerged to convey the flood.

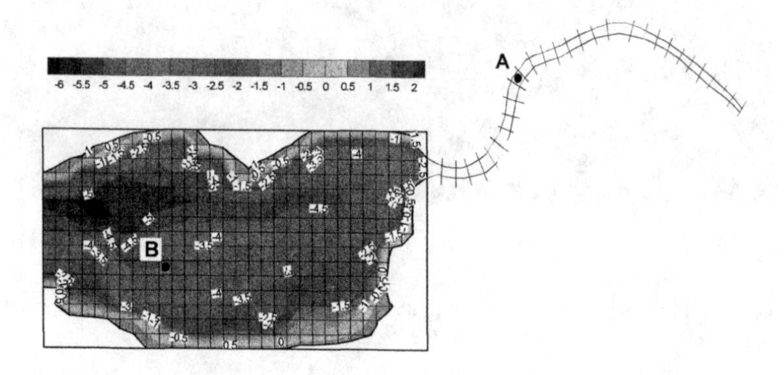

Figure 3 A hypothetical case where a narrow river is connected to a downstream water body that is large and exhibits a three-dimensional flow pattern.

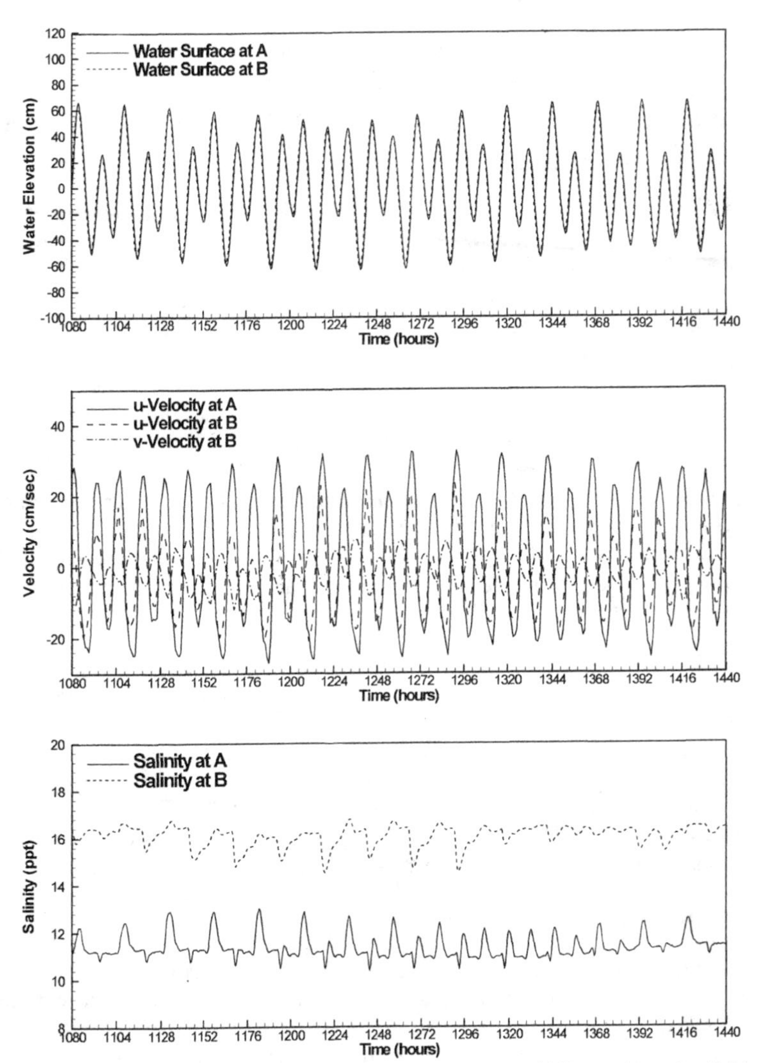

Figure 4 Simulated surface elevations (top graph), velocities (middle graph) and salinities (bottom graph) at Location A in the 2DV domain and Location B in the 3DV domain of a hypothetical case (Figure 3.)

Including a Near-Bed Turbid Layer in a Three Dimensional Sediment Transport Model With Application to the Eel River Shelf, Northern California

Courtney K. Harris[1], Peter Traykovski[2], and W. Rockwell Geyer[2]

ABSTRACT: Observations from the Eel River shelf, northern California show that during times of significant sediment supply and energetic waves, sediment flux is dominated by down-slope, gravitationally driven transport within a near-bed turbid layer. These fluid muds are supported by wave-induced turbulence. Their thickness scales with the height of the wave boundary layer, and therefore thins offshore. Flux within these layers is not resolved by conventional three-dimensional hydrodynamic and sediment transport models. To represent this transport mechanism we therefore modified an existing three-dimensional sigma-coordinate model to include a near-bed turbid layer by adding a wave-boundary layer grid cell beneath the conventional sigma grid. Sediment concentrations within the wave boundary layer represent a mass balance between erosion and deposition from the seafloor, entrainment into the overlying water column, sediment settling, and flux divergence in the across- and along-shelf directions. The transport velocity within the wave boundary layer is estimated by a balance between frictional drag and the negative buoyancy of the turbid layer.

The modified three-dimensional transport model replicates water-column and sea-bed observations of sediment dispersal and deposition on the Eel Shelf. Concentrations within the layer and depositional footprints of flood events proved to be sensitive to the settling and turbulence properties of the turbid layer. Development of a module for sigma coordinate models that includes this transport mechanism provides a means of comparing relative contributions of gravitationally-driven transport to conventional dilute resuspension, for studying other energetic, sediment-laden environments, and for comparing parameterizations of wave-supported fluid muds.

INTRODUCTION

The STRATA FORmation on Margins (STRATAFORM) program had as a goal to relate observable sediment delivery and reworking events to preservable stratigraphic deposits (see Nittrouer, 1999). This program focused on the continental shelf offshore of the Eel River in northern California (see Figure 1A), an area characterized by frequent energetic waves and fluvial sediment delivery during winter floods. While the mid-shelf mud bed provides a clear depositional footprint of flood events

---

[1] Corresponding Author: Department of Physical Sciences, Virginia Institute of Marine Science, P.O. Box 1346, Gloucester Point, VA 23062. Telephone: (804)684-7194. Email: ckharris@vims.edu. Contribution number 2583 of the Virginia Institute of Marine Science, College of William and Mary.

[2] Department of Applied Ocean Physics and Engineering, Woods Hole Oceanographic Institution, Woods Hole, MA.

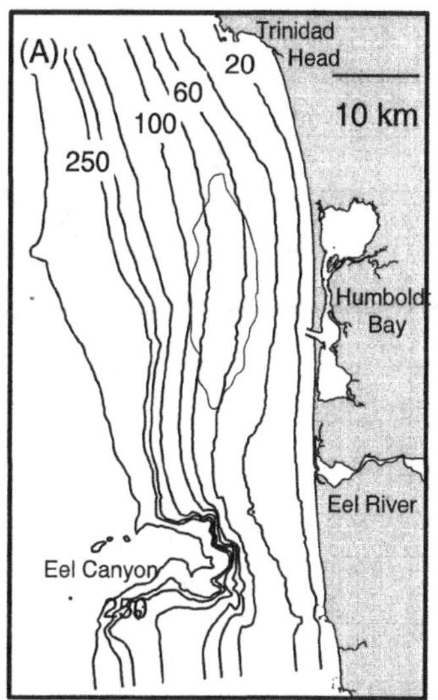

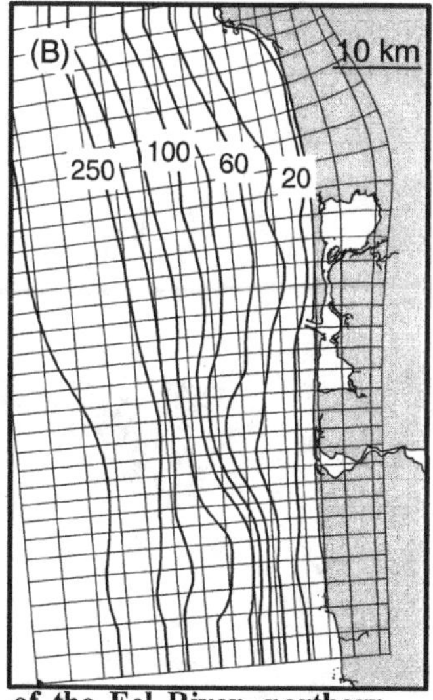

Figure 1: Continental shelf offshore of the Eel River, northern California showing (A) bathymetric contours and (B) smoothed bathymetry and model grid used for this study. Shaded area in (A) represents depocenter of 1997 and 1995 floods of the Eel River (see Wheatcroft and Borgeld, 2000). Each gray grid square in (B) represents 25 horizontal grid points.

(see Figure 1A), sediment within the mud bed accounts for only about 20% of the fine-grained material delivered by the Eel River over timescales ranging from a single flood event to centuries (Sommerfield and Nittrouer, 1999). Sediment delivery by the buoyant freshwater plume of the Eel River is confined to the inner shelf, and therefore does not explain depositional patterns (Geyer, et al. 2000). Conventional transport mechanisms, resuspension by energetic waves and currents, also does not account for the apparent mismatch between flood delivery and deposition (Harris, et al. 2002). Observations from the Eel River shelf by an Acoustic Backscatter Sensor (ABS) within a few centimeters of the bed indicated that sediment concentrations exceeded 10 g/L during energetic wave events (Traykovski, et al. 2000). During such times, velocities in the bottom boundary layer showed a down-slope component that increased near the bed (Traykovski, et al. 2000; Ogston, et al. 2000). These observations indicate that during times of significant sediment supply and energetic waves, transport is dominated by down-slope, gravitationally driven flux of near-bed turbid layers. The ABS measurements further indicated that the thickness of these layers scales with the height of the wave boundary layer (Traykovski, et al. 2000).

Subsequent theoretical work has shown that wave-induced turbulence and sediment-induced stratification at the top of the wave boundary layer are often sufficient to keep sediment from being dispersed upward into the overlying water column, under

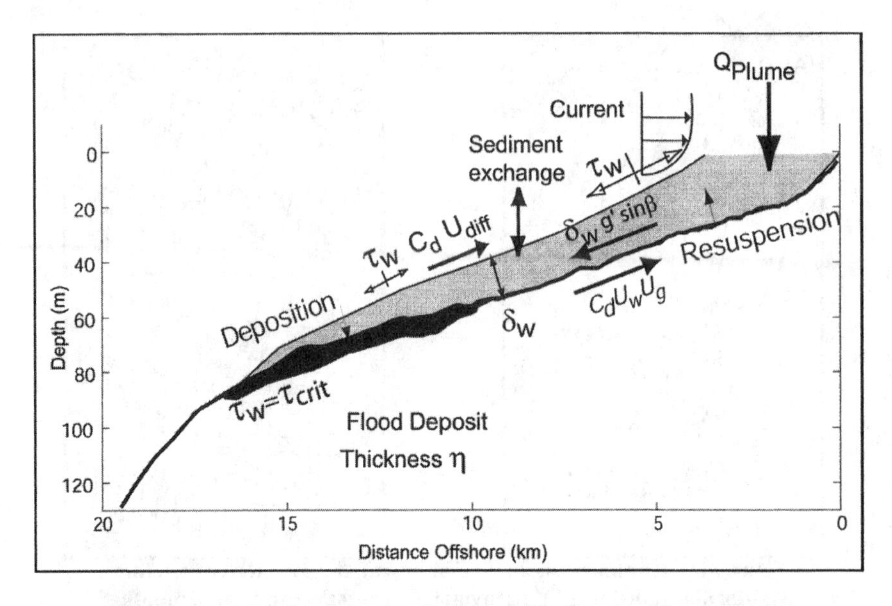

Figure 2: Schematic near-bed turbid flows and suspended sediment transport in the overlying water. Cross shelf view of transport and flux divergence, modified from Traykovski's submission to Parsons, et al. (in review) to illustrate exchange of sediment between turbid layer and overlying water column; frictional drag between the water column and fluid mud; and shear stress from oceanographic currents.

which conditions the added density of the layer drives downslope transport (see Figure 2; Traykovski, et al. 2000; Wright, et al. 2001; Scully, et al. 2002). Intense turbulence within the wave boundary layer allows it to hold large amounts of suspended sediment. Turbulence levels typically decrease above the wave boundary layer, where flows are dominated by quasi-steady currents. Both stable stratification caused by increased density of the turbid layer, and decreased turbulence above the wave boundary layer act to trap sediment within the thin, near-bed layer. At high concentrations (above ~10g/L), the added density of the turbid layer overcomes bed friction, and the layer flows downslope. Cross-shelf flux from this mechanism therefore relies on an abundance of erodible sediment, energetic waves, and sloping bathymetry. By accounting for both dense, gravitationally-driven flux and dilute suspension we compare transport and deposition from both of these mechanisms.

Fluid mud layers have been seen to play important roles in other continental shelf environments, but they have usually not been attributed to wave-induced turbulence. For many fluid muds, trapping mechanisms distinct from stratification at the top of the wave boundary layer have been recognized (see, e.g. Trowbridge and Kineke, 1994). Offshore of sediment-laden rivers, fluid mud transport is supposed to be driven by high fluvial inputs that create hyperpycnal river flows (see, e.g. Mulder and

Syvitski, 1996). Both large scale fluid muds, and hyperpycnal river flows might be resolved within conventional ocean models, run at sufficient vertical resolution, if sediment concentrations were included in density calculations.

Some two-dimensional models (cross-shelf and vertical) that include high resolution in the vertical (~cm) have succeeded in predicting downslope transport of these turbid layers (Reed et al., 1999; Fan, et al. 2004). The two-dimensional approach is unable to account for the inherent three-dimensionality of buoyant plume delivery, however. Wave-supported fluid muds can not be easily quantified by conventional sigma- or z-coordinate models for three reasons. First, typical implementations do not include sediment density in calculations of water column density. Furthermore, three-dimensional coastal ocean models usually do not have sufficient vertical resolution to resolve the wave boundary layer, which is less than 10 – 20 cm within continental shelf waters. Also, because the thickness of sigma coordinates decreases with water depth, while wave boundary layer thickness increases in shallow water, sigma grids will not effectively represent the vertical structure of the near-bed region across a continental shelf. Likewise, z-coordinate systems assume a constant thickness across a range of water depths, and can not resolve the wave boundary layer directly.

This paper describes an approach used to represent a wave-supported fluid mud within a sigma coordinate model that accounts for flux in the wave boundary layer, exchange between sediment suspended in the wave boundary layer and the overlying water column, and transport of suspended sediment above the wave boundary layer. ECOM-SED (the Estuarine Coastal Ocean Model -SEDiment; see Blumberg and Mellor, 1987; Hydroqual, 2002) was modified to account for gravitationally-driven transport of a thin, near-bed layer of fluidized mud. After summarizing the methods used to include the near-bed turbid layer within ECOM-SED, this paper discusses the sensitivity of the calculations to parameterizations within the model, and provides sample applications to the Eel River shelf, northern California.

METHODS

The model grid used to represent the Eel Shelf included 75 X 161 horizontal grid points arranged in a curvilinear orthogonal grid (see Figure 1), and 8 vertical sigma layers. Horizontal spacing of grid cells was about 200 m in the area of interest, and the thickness of sigma layers vary from 3—25 % of the flow depth, with the low-resolution cells in the mid-water column. Hydrodynamic calculations included turbulence following Mellor and Yamada (1982), at level 2.5; and wave-current interaction following Grant and Madsen (1979). For information on the conventional form of ECOM-SED, see Blumberg and Mellor (1987), and Hydroqual, 2002. Details of the sediment transport model follow, with particular emphasis on the representation of the near-bed turbid layer. Specifics of the forcing used for the simulations presented in this paper are provided in a later section.

Suspended Sediment Calculations: Calculations were limited to the dispersal of fine-grained silts and clays, because we focus on describing the formation of muddy flood deposits found on the mid-shelf (see Figure 1A). The simulations assume that the only source of mobile sediment is the Eel River outflow. Input sediment was transported as a tracer that travels along with the ambient oceanographic currents, modified by the addition of a prescribed settling velocity. Diffusion of sediment was predicted using the model's turbulence closure in the vertical (Mellor-Yamada, 1982, level 2.5), and a constant diffusion coefficient in the horizontal. Sediment that traveled out of the lateral grid to the north, south, or west was not replaced if current velocity reversed. At the bottom boundary of the sigma grid, sediment was exchanged between the bottom-most sigma coordinate and the wave-boundary layer at a rate determined by the difference between sediment settling and upward diffusion, as described below.

Multiple grain sizes were not dealt with explicitly in the model. Instead, it was assumed that different sediment grains would not interact with one another, and separate calculations were computed for each sediment type. Two types of sediment were considered, fine grained material that travels as single-grains and settles at 0.01 cm/s, and flocculated material that settles at 0.1 cm/s. Calculations indicated that the two sizes would interact very little because the faster settling material concentrates near the bed while the slower settling material is dispersed throughout the water column at low concentrations.

The wave boundary layer is modeled as a separate grid cell so that the model can account explicitly for stratification between the near-bed wave boundary layer and overlying water. To include the near-bed turbid layer within the model, we added a wave-boundary layer grid underneath the sigma coordinate system. The thickness of this layer at any location depends on the wave orbital velocity and frequency $(\delta_w = u_{*cw}/\omega)$. For the 3m, 12 second waves considered in the example calculations, the layer therefore ranged from 0 – 0.16 m thick.

Concentrations within wave boundary layer: Suspended sediment concentrations in the near-bed layer are determined by the balance between erosion and deposition from / to the underlying seabed, exchange with the overlying water, and advection from neighboring grid cells.

Exchange with the seabed is determined as the difference between an erosion term (E) and deposition term ($D = c_{wbl}\, w_s$) as follows. The erosion term (E) is estimated by the product of the equilibrium concentration (c_{eq}) and settling velocity (w_s). If the critical shear stress for motion and erosion are not both exceeded ($u_{*cw} < u_{*cr}$ or $u_{*cw} < w_s$), the equilibrium reference concentration is set equal to zero. Otherwise, it is estimated following Smith and McLean (1977):

$$c_{eq} = \frac{c_b \gamma_0 S}{1+\gamma_0 S}, \tag{1}$$

where bed concentration c_b=(1-porosity)ρ_s, excess shear stress is $S = (\tau_b - \tau_{cr})/\tau_{cr}$, and $\gamma_0 = 0.005$ is the resuspension coefficient (chosen following Cacchione, et al. 1999). The excess shear stress, S, depends on the relative sizes of the bed shear stress

(τ_b) and the critical shear stress for motion (τ_{cr}). An upper limit of c_{eq} = 200 g/L was applied, based on the assumption that turbulence would be suppressed at higher concentrations (Trowbridge and Kineke, 1994). Because of the difficulty in predicting critical shear stress for fine-grained silts and clays, τ_{cr} and u_{*cr} (the shear velocity) should be based on site-specific data, either from near-bed measurements of threshold erosion conditions, or from erosion chamber experiments. For this study, we assigned values of critical shear stress based on evidence of resuspension seen in observations obtained near the seafloor during floods of the Eel River. Net erosion during a timestep is estimated as

$$E - D = \left(c_{eq} - c_{wbl}\right) w_s \Delta t \tag{2}$$

where c_{wbl} is the calculated sediment concentration within the wave boundary layer. At equilibrium, the settling and entrainment terms balance, $c_{eq} = c_{wbl}$, and no net erosion occurs. The seabed will be net erosional when $c_{eq} > c_{wbl}$; and net depositional when $c_{wbl} > c_{eq}$.

Mixing between the wave boundary layer sediment and the overlying suspension is dampened by stable stratification at the top of the wave boundary layer (see Geyer, 1993). This is quantified using the stability parameter of Munk and Anderson (1948) based on the gradient Richardson number:

$$Ri = -\frac{(\rho_s - \rho) g}{\rho} \frac{\partial c_s / \partial z}{\left(\partial U / \partial z\right)^2} \tag{3}$$

where g is gravitational acceleration, ρ_s and ρ are sediment and fluid densities, $\partial c_s / \partial z$ is the vertical gradient of sediment concentration, and $\partial U / \partial z$ is the vertical gradient of velocity. Note that U represents the velocity vector whose alongshelf and across-shelf components are v, and u, respectively. The gradient Richardson number is estimated by the suspended sediment and velocity gradients over the interface between the wave boundary layer and overlying water. The interface thickness, δ, is assumed to be a small constant (1 cm) based on observations that the gradient between the turbid layer and overlying water is, in general, very sharp (Traykovski, et al. 2000). The suspended sediment and velocity gradients are therefore estimated as $\partial c_s / \partial_z \approx \left(c_{s,\sigma,K} - c_{s,wbl}\right) / \delta$, and $\left(\partial u / \partial z\right)^2 \approx \left[\left(u_{\sigma,K} - u_{wbl}\right)^2 + \left(v_{\sigma,K} - v_{wbl}\right)^2\right] / \delta^2$, where $c_{\sigma,s,K}$, $u_{\sigma,K}$ and $v_{\sigma,K}$ represent the suspended sediment concentration and velocity in the bottom-most sigma layer. The gradient Richardson number is calculated by substituting these into Equation (3):

$$Ri = -\frac{(\rho_s - \rho)}{\rho} \frac{g\left(c_{s,\sigma,K} - c_{s,wbl}\right)\delta}{\left[\left(u_{\sigma,K} - u_{wbl}\right)^2 + \left(v_{\sigma,K} - v_{wbl}\right)^2\right]} \tag{4}$$

Entrainment of sediment out of the wave boundary layer and into the overlying fluid (E_{wbl}, g/m^{2}s) is estimated by an eddy viscosity modified by a dependence on the Richardson number (Munk and Anderson, 1948):

$$E_{wbl} = f(Ri)\kappa u_* z_{wbl} \frac{\partial c_s}{\partial z}$$
$$f(Ri) = (1+\beta Ri)^{-\alpha} \quad (5)$$

whereVon Karman's constant, $\kappa = 0.408$, and $z_{wbl} = \delta_{wbl}/2$ is the height of the center of the wave boundary layer grid cell. Following Ross and Mehta (1989) parameters were chosen as $\beta = 3.6$ and $\alpha = 1.8$. The concentration gradient in Equation (5) is estimated by $\partial c_s/\partial z \approx (c_{s,\sigma,K} - c_{s,wbl})/(z_{\sigma,K} - z_{wbl})$ where $z_{\sigma,K}$ is the height of the mid-point of the bottom-most sigma layer.

Velocities within wave boundary layer: The velocity vector U_{wbl} is estimated using a linearized Chezy balance that considers the buoyancy anomaly of the turbid layer and frictional drag between the fluid mud, the bed, and the overlying water, following Wright, et al. (2001).

$$C_{d,b} u_{wbl} u_w + C_{d,wbl}(u_{wbl} - u_{\sigma,K})u_w = \delta_w g' \sin \mathrm{B}$$
$$C_{d,b} v_{wbl} u_w + C_{d,wbl}(v_{wbl} - v_{\sigma,K})u_w = \delta_w g' \sin \mathrm{B} \quad (6)$$

where u_{wbl} and v_{wbl} are the components of U_{wbl}, the velocity within the wave boundary layer, the buoyancy anomaly is $g' = g(\Delta\rho)/\rho = g c_{wbl}(\rho_{sed} - \rho)/\rho$, and $\sin \mathrm{B}$ is bathymetric slope. The orbital velocity of the waves, u_w, and the difference between the velocity of the wave boundary layer and overlying water column are the two velocity scales used.

Drag coefficients are specified between the turbid layer and the bed ($C_{d,b}$), and between the turbid layer and overlying water ($C_{d,wbl}$) using

$$C_{d,wbl} = f(Ri)\kappa^2 \left[\log(z_{\sigma,K}/z_{wbl})\right]^{-2}$$
$$C_{d,b} = \kappa^2 \left[\log(z_{wbl}/z_0)\right]^{-2} \quad (7)$$

where $f(Ri)$ is defined in Equation 5, and z_0 is hydraulic roughness. The hydraulic roughness, z_0=0.25 cm, was chosen to represent the outer roughness scale, which includes the wave boundary layer. Sensitivity tests using this value of roughness, and wave heights representative of the Eel Shelf yielded mean shear stresses that were similar to those observed by Strataform tripods (e.g. Cacchione, et al. 1999; Ogston, et al. 1999, 2000; Traykovski, et al. 2000). Using Equations 6 and 7, the velocity of the turbid layer can be estimated. The velocities are in turn used to estimate flux divergence within the wave-boundary layer in the along- and across-shelf directions where the horizontal fluxes per unit across- and alongshelf distance are $Q_{x,wbl} = c_{wbl} u_{wbl} \delta_{wbl}$, and $Q_{y,wbl} = c_{wbl} v_{wbl} \delta_{wbl}$.

Advection of sediment: The concentration within the near-bed turbid layer is estimated using conservation of mass and therefore responds to flux divergence in both the horizontal and vertical directions. Net sediment flux through the interface between the wave boundary layer and overlying fluid is estimated as the difference of entrainment from the wave boundary layer (E_{wbl}, Equation 5), and settling from the overlying fluid ($c_{s,\sigma,K} w_s$). Net exchange between the wave boundary layer and the seabed is provided by Equation 2, as $w_s\left(c_{eq} - c_{wbl}\right)$. Finally, horizontal and vertical flux divergence terms within the near-bed layer are summed to yield:

$$\frac{\partial}{\partial t}\left[\delta_w c_{wbl}\right] = w_s\left(c_{\sigma,K} + c_{eq} - c_{wbl}\right) - E_{wbl} + \frac{\partial Q_{x,wbl}}{\partial x} + \frac{\partial Q_{y,wbl}}{\partial y}. \quad (8)$$

Changes to sediment concentration within the wave boundary layer are estimated at each grid cell by conserving sediment mass through horizontal and vertical advection, vertical diffusion, and erosion / deposition (Equation 8). Within the framework of the numerical model, Equation 8 is solved using a semi-implicit upwind scheme.

Settling properties of turbid layer: Predicted and observed concentrations for the near-bed turbid layer often exceed 10 g/L, particularly during times of significant flux. At these concentrations, sediment within the turbid layer is not expected to settle as single grains. At concentrations in excess of 1—15 g/L we expect hindered settling effects to become important, under which conditions the rate of settling decreases. Little agreement exists for predicting the settling properties of such layers (see review in Mehta and McAnally, 2003, for example). To test the sensitivity of calculations to settling properties of the dense suspensions, predictions of depositional patterns were completed using a range of hindered settling relationships. For example, we included the function reported by Mehta and McAnally (2003) for Wolanski, et al. (1991), whereby the settling characteristics of a turbid layer are estimated from a power law that relates settling velocity to concentration. The parameterized settling velocity was applied only within the near-bed layer. Sediment throughout the rest of the flow was predicted to settle at the prescribed settling velocities of 0.01 – 0.1 cm/s.

Erosion and Deposition: Sediment erosion and deposition is predicted based on net exchange with the wave boundary layer overlying the sediment bed. The net difference between upward diffusion of sediment and downward settling is predicted to be $w_s\left(c_{eq} - c_{wbl}\right)$; where c_{eq} is the equilibrium reference concentration predicted by Smith and McLean (1977). Sediment availability is considered by limiting the amount to be eroded during a timestep $\left[\Delta t w_s\left(c_{eq} - c_{wbl}\right)\right]$ (see Equation 2) to the thickness of sediment that is available within a horizontal grid cell, as indicated by the bed height, η. The net erosion rate (E-D) is therefore constrained to be the minimum of $w_s\left(c_{eq} - c_{wbl}\right)$ and $\eta/\Delta t$, where η is bed elevation. Because the model assumes that there is initially no fine-grained sediment on the bed, any erosion reworks sediment deposited since the beginning of the flood event.

We neglected bed consolidation and multiple grain size effects, both of which may act to increase the capacity of the inner shelf to retain sediment. The timescale of the calculations is less than ~11 days, with peak sediment delivery occurring at day 3 in the calculation. It is possible that dewatering of fresh mud beds over these timescales (~1 week) could enable the flood beds to resist resuspension. The model also neglects interactions between the freshly delivered muds and sediment already present on the continental shelf. This prevents the model from accounting for the storage of fine grained sediment within inner shelf sands, for example, which has been observed on the Eel Shelf (Crockett and Nittrouer, 2003).

Treatment of Vertical Grid for Hydrodynamics: In building this prototype model for wave-supported fluid mud transport for continental shelves, we tried to avoid disruption of the hydrodynamic calculations. The grids used for sediment and water flux were therefore slightly different, in that the hydrodynamic grid calculations did not include the wave boundary layer grid cell. Wave – current interaction calculations are completed within the hydrodynamic calculations as they would if the fluid-mud layer were not present; following Grant and Madsen (1979). Velocity calculations for the sigma grid cells neglect the impact of frictional drag with the fluid mud layer. Attempts to include this failed to produce stable calculations. While observations from the Eel Shelf showed that occasionally the upper water column velocities were significantly impacted by the near bottom layer (Ogston, et al. 2000), these episodes were intermittent. They are unlikely to influence overall sediment budgets, because flux within the wave boundary layer accounts for 80% of the observed cross-shelf flux (Traykovski, et al. 2000).

This treatment will not capture feedbacks between the velocities of the near-bed layer and the overlying water, or on possible changes to bottom drag coefficient induced by the fluid muds. While this simplification is appropriate at continental shelf scales (10s of meters water depth), it might fail to reproduce hydrodynamics in shallower environments where wave-supported gravity flows might also be found. In such environments, the wave boundary layer accounts for a bigger proportion of the flow, and downslope transport might assert more influence on depth-averaged velocities.

SIMULATION OF EEL RIVER FLOODS

Calculations were completed for a flood event similar to a flood of the Eel River observed during the Strataform field program (Figure 3). Wave conditions are assumed to be steady and uniform during the event, with heights of 3m and periods of 12s. As is typical of the Eel River shelf, winds change from being from the south to the north during the flood event. Winds were modeled as being uniform over the domain, and during the event ranged from 15 m/s from the south, to northerly at 6 m/s. Sediment delivered to the shelf is estimated to be 60% flocculated (w_s = 0.1 cm/s), and 40% unflocculated (w_s = 0.01 cm/s), based on guidance from Paul Hill (Dalhousie University, pers. comm.). Sediments that deposit to the bed are assumed to form high-water content, easily eroded layers, and were therefore assigned a critical shear stress of 0.05 Pa, based on erosion threshold evidence from tripods deployed at the site.

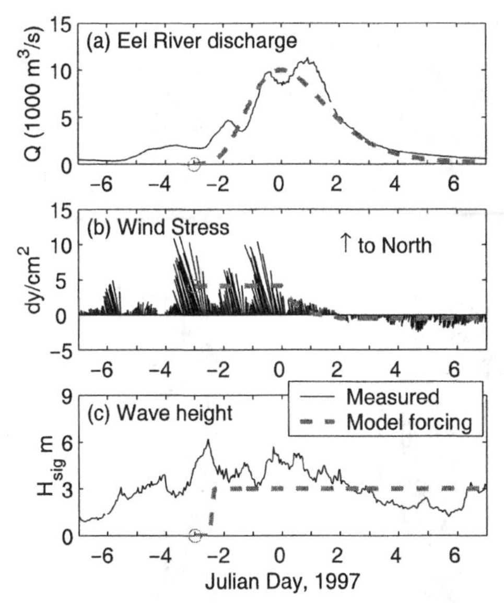

Figure 3: Forcing of (a) flood discharge, (b) winds, and (c) waves for the simulated event (dashed line) compared to conditions observed January, 1997 on the Eel River shelf (solid lines). The timeseries are provided for the week preceeding the flood and simulation; the simulation starts on Julian day -3 (December 29) and peaks on Julian day 0 (January 1).

Water column velocities are driven by a combination of wind- and buoyancy forcing (Figure 4). Tidal forcing was neglected for this simulation, based on sensitivity tests that included forcing by elevation fields derived from tidal models (i.e. Foreman, et al. 2000). Also neglected in these calculations are temperature and salinity gradients in shelf waters. Ocean water was assumed to be 32 ppt and 4°C. River water was given temperatures of 4° C, and 0 ppt salinity. This simplified water column structure captures the major source of stratification here, that of the gradient between river discharge and the coastal ocean. Open boundaries were specified using a radiation condition on the northern side, wind-forced currents on the southern side, and a partially clamped radiation condition on the offshore. The simulation proceeds for about 12 days before instabilities in the open boundary condition corrupt the solution.

These calculations, while not intended to simulate the exact conditions on the Eel River shelf during any particular event, do capture the essence of wave, wind forcing, and plume dynamics during a large flood event similar to the one observed in January, 1997. The plume hugged the coastline during the time of strong wind forcing, remaining confined to the inner shelf in a manner similar to that observed by Geyer, et al. (2000). Shear stresses and velocities predicted by the model are in qualitative agreement with tripod data taken during flood events, though a thorough validation is beyond the scope of this paper.

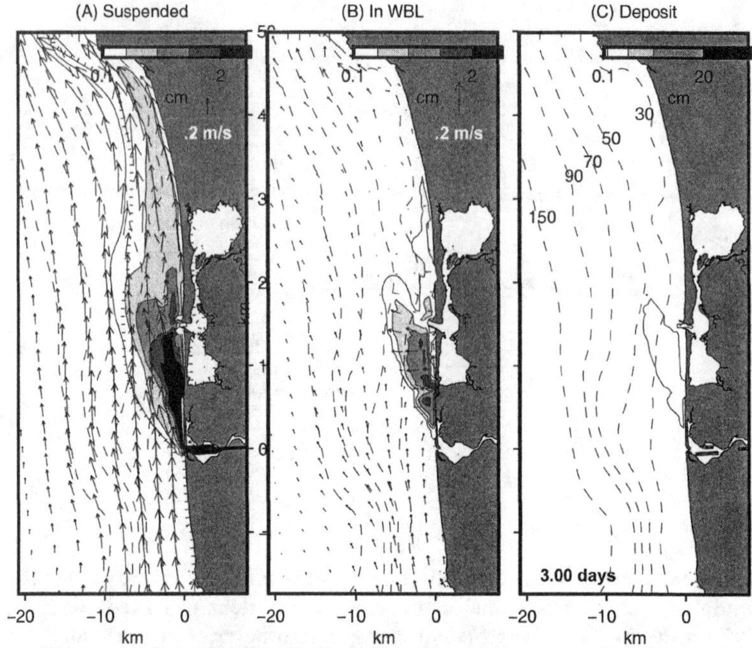

Figure 4: Calculated velocity and sediment concentration for peak of flood event (Julian day 0; model day 3). (A) Depth-averaged velocity (arrows) and depth-integrated sediment concentration (colorbar). Dotted line is extent of freshwater plume. (B) Near-bed turbid layer velocity (arrows) and depth-integrated sediment concentration (colorbar). (C) Thickness and location of sediment deposit. Dashed lines are model bathymetry.

Suspended sediment concentrations in the water column depend primarily on advection within the plume and settling from the plume (Figure 4). The along-shelf decrease in concentration occurs as sediment settles out of the plume, whereas the offshore decrease is mainly due to dilution with low-concentration seawater.

Within a week of peak flood waters, most of the sediment has settled out of the water column (Figure 5). Offshore transport due to upwelling conditions has advected the freshwater across the shelf, but there is little sediment flux in the water column (Figure 5A). More significant offshore flux of sediment continues to occur in the wave boundary layer where the velocity is directly offshore due to the action of gravity on the dense suspension (Figure 5B).

Predicted velocity and sediment concentrations within the near-bed turbid layer behave very differently from the overlying suspension (Figure 6). By model day 4 near-bed velocities are directed down slope and reach speeds ~ 1 m/s. Sediment concentrations within the wave boundary layer are on the order of 10—150 g/L. As

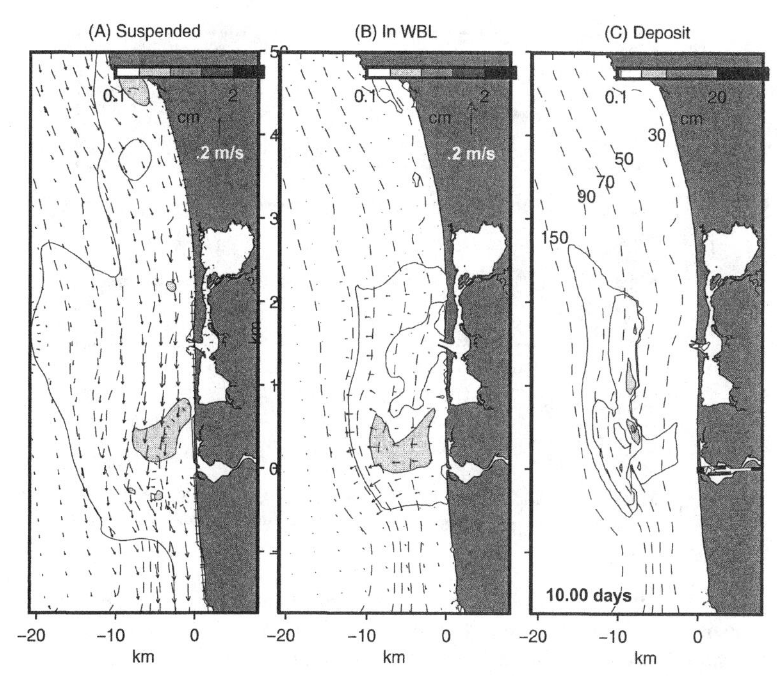

Figure 5: Calculated velocity and sediment concentration late in flood (Julian day 7; model day 10). Panels (A)—(C) as described for Figure 4.

shown by Figure 6, both sediment flux and concentration are dominated by material within the wave-boundary layer.

Sediment flux slows once sediment reaches deeper water. Here, wave energy and bed shear stresses are reduced, thereby decreasing c_{eq} (Equation 1). Also, the wave boundary layer thins, further decreasing flux within it. These lead to flux convergence, both from a reduction in the erosion term (Equation 2), and in horizontal convergence of flux (Equation 8). For the wave energy and settling properties assumed, deposition was centered on the mid-shelf, between 50 – 100 m water depths (Figure 5C).

The characteristics of the predicted flood deposit (Figure 5C) are consistent both with observations made shortly after the 1997 flood (Figure 1A), and results from predictions derived from models that included only transport within the wave boundary layer (Traykovski, et al. 2000; Scully, et al. 2003). The location of the flood deposit on the mid-shelf mirrors that of the observations, as does the overall thickness (~10 cm) and along-shelf extent. The calculations further predict significant sediment export due to dispersal of material that remains suspended and is transported in the water column. For example, during the first 4 days of the event

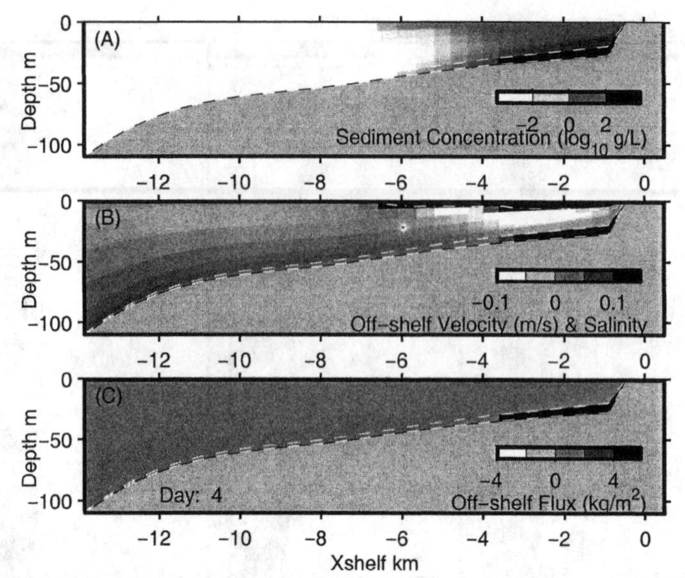

Figure 6: Cross-shelf view of (A) sediment concentration, (B) cross-shelf velocity, and (C) sediment flux shown at Julian day 1 (model day 4). Dashed line in (A)—(C) delineates wave boundary layer. Thickness of the wave boundary layer is exaggerated. Dash-Dot line in (B) shows plume location. Light colors in (B) and (C) indicate shoreward flux; dark indicates off-shore flux.

when winds are strong and from the south, significant sediment (10 – 20%) is lost from the proximal shelf by northward advection within the plume (see Figure 3C).

SENSITIVITY TO MODEL PARAMETERIZATIONS

Settling Velocity: The settling velocity of sediments within dense suspensions in the wave boundary layer is likely to be influenced by both aggregation /disaggregation processes and hindered settling. Predicted depositional patterns are sensitive to the manner in which these processes are modeled. For the calculations presented in the previous section (Figures 4, 5, and 6), we have chosen a simple approach, which is to neglect both floc breakup and hindered settling. These calculations use settling velocities that are independent of concentration and are based on field observations reported in Hill, et al. (2000). This results in a flood footprint that matches that seen on the Eel River shelf when steady, 3m waves are used (Figure 5C).

In a hindered settling regime, settling velocity in the turbid layer decreases, the wave boundary layer retains sediment and gains speed. This in turn shifts deposition offshore. For example, using the steady wave conditions shown in Figure 3C with a

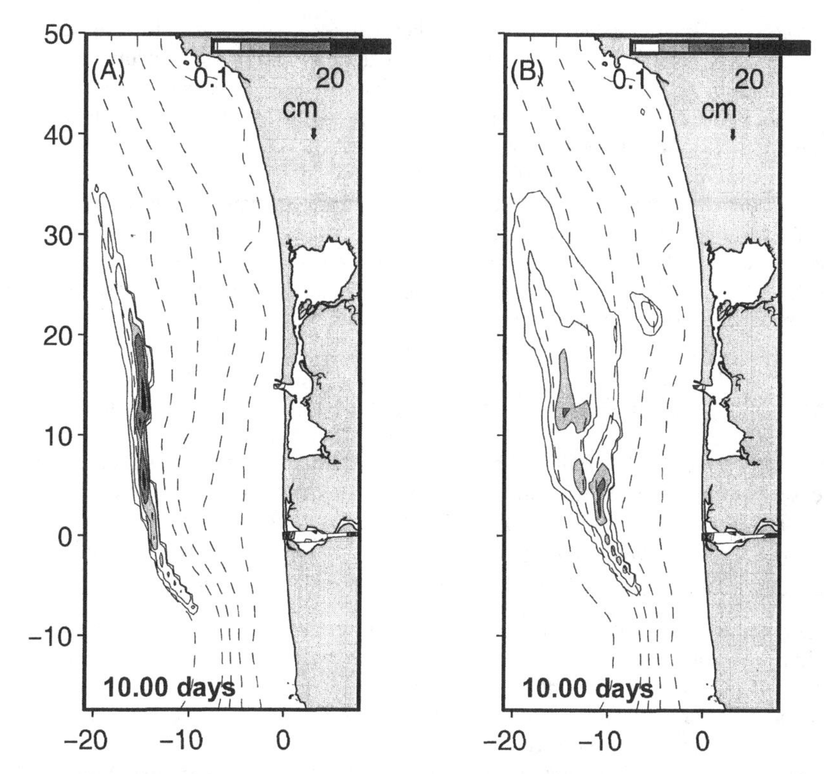

Figure 7: Depocenter plotted for (A) a model run that uses a hindered settling relationship, and steady wave energy; and (B) the hindered settling relationship with the realistic wave timeseries.

hindered settling relationship derived from the literature (Wolanski, et al. 1991) the depocenter shifts from the mid-shelf (Figure 5C) to the continental slope (Figure 7A).

Settling velocities of fine-grained sediments are also sensitive to the formation and breakup of flocs. This model neglects aggregation / disaggregation, and assumes that there is no exchange between the sediment class modeled as fast settling flocs (1 mm/s) and those modeled as slower settling grains (0.1 mm/s). This approach is based on findings that the settling characteristics of fine-grained sediment delivered by the Eel River are set at the river mouth (Hill, et al. 2000). More recent findings, however, indicate that aggregation processes are controlled by resuspension in the nearshore (Curran, et al. 2002), implying that floc fractions should be held constant in that region. To account for this would require the model to exchange sediment from the unflocculated to the flocculated state within energetic nearshore grid cells. This modification is beyond the scope of this paper, but would likely decrease the export of material from the shelf, and probably increase the volume deposited on the shelf.

Details of Wave Forcing: The wave shear stress calculations also impact the depositional footprint of the flood bed. For example, running the calculations with the actual time-series of wave heights (solid line in Figure 3C) results in a sediment bed that is inshore of the version shown in Figure 5C, predicted using steady waves. Using the realistic wave timeseries, sediment settles to the bed at a shallower depth because wave heights are reduced following day 3, a time at which substantial deposition is occurring (Figure 7C). This is consistent with Scully, et al. (2002, 2003), who report that deposition of gravitationally-driven fluid muds is sensitive to wave height; with less energetic waves shifting deposition to shallower water. When both the hindered settling relationship, and the actual wave timeseries were used, however, the results showed a mid-shelf mud bed (Figure 7B). This implies that hindered settling may be important during the transport and deposition of wave supported fluid mud beds. The difference between Figures 7A and 7B also illustrates the sensitivity of depositional patterns to temporal variations in the wave field.

Bottom Boundary Condition: The depositional structure of the mud bed is also sensitive to the parameterization of the bottom boundary condition for sediment, c_{eq}, which was set following formulations that have been successful for one-dimensional models on the Eel Shelf (Cacchione et al., 1999). The bottom boundary condition is set equal to the difference between entrainment and the settling term, $\left[w_s \left(c_{eq} - c_s \right) \right]$. Uncertainties in c_{eq} propagate to values predicted for concentrations and velocities within the wave boundary layer. Because of the dependence of c_{eq} on bed shear stress (Equation 1), reducing c_{eq} is equivalent to reducing the bed shear stress estimate in that it reduces concentrations of sediment in the wave boundary layer. This slows the across-shelf velocity and cross-shelf flux of sediment. Likewise, increasing c_{eq} acts to increase concentrations in the wave boundary layer and speeds deposition. Model runs that did not include gravitational transport of fluid muds showed sensitivity that the timescale of deposition was sensitive to c_{eq}, but that the spatial patterns were not because they were determined by flux divergence. With complex feedbacks between sediment concentration and velocity, implied in Equations 6 and 7, it is difficult to evaluate the impact of uncertainties in the equilibrium reference concentration, and to the estimate of bed shear stress. Future model runs should investigate this question.

Bathymetry: Calculations were somewhat sensitive to small-scale variability of seafloor bathymetry. Other model runs have been completed using the realistic bathymetry of the Eel River shelf was used (Figure 1A) instead of the smoothed bathymetry (Figure 1B). The results of these are preliminary, and therefore not presented here, but they showed a depositional pattern that was patchier than those shown using the smoothed bathymetry, but had similar volumetric budgets.

MODEL LIMITATIONS AND FUTURE DIRECTIONS

The model described herein attempts to account for the dominant sediment dispersal mechanisms that operate on the continental shelf during emplacement of a flood bed:

namely buoyant plume processes, sediment advection, settling, wave/current resuspension, and gravitationally forced flows. Other approaches have either neglected three-dimensional buoyant plume effects (e.g. Harris and Wiberg, 2001, 2002; Fan, et al. 2004; Reed, et al. 1999); neglected wave-current resuspension and water column transport (e.g. Scully, et al. 2002, 2003); or neglected the fluid mud layer (Harris and Wiberg, 2001, 2002). The model presented here quantifies a comprehensive suite of processes, potentially capable of simulating dispersal in a manner consistent with the observations.

Several mechanisms neglected in this model are also important for sediment dispersal. For example, the model assumes that the only source of sediment is that delivered by the Eel River during the flood, and thereby neglects resuspension of shelf sediment. Harris (1999) estimated that shelf sands are resuspended 85% of the time during energetic waves typical of floods on the Eel Shelf. Mixing of these mobilized shelf sands with recently delivered muds appears to account for 6—13 % of the fine-grained sediment budget on the Eel (Crockett and Nittrouer, 2004). Difficulties in including the erosion of shelf sediment, and mixing between fluvial and shelf sediment stem from two considerations. Accounting for these would require inclusion of sediment bed grading and mixing, such as that found in Harris and Wiberg (2001, 2002). Consideration of resuspension of sandy material would further require an advection scheme that is not limited by the Courant condition. Timesteps of ECOM-SED, as implemented in this project, are limited by the settling speed of the coarsest grains. Greater flexibility would come from employing and implicit advection scheme, as is now done within ECOM-SED, or by employing a piecewise-parabolic-method (see James, 2000) as is being done within the Regional Ocean Modeling System (ROMS) module for sediment transport (J.C. Warner, USGS, pers. comm.; see also Song and Haidvogel, 1994). Future model implementations that use more advanced advection schemes therefore promise to account for newly delivered sediment, resuspended sands, and mixing processes among them, and perhaps provide more accurate estimates of suspended concentrations and sediment budgets. The model published herein, however, is able to account for the majority of the transport of fine grained material during events that seem to dominate the recent depositional environment of the Eel shelf.

This model is further limited by the length of simulations that can be accomplished, ~2 weeks, which stems from difficulties in prescribing robust open boundary conditions for a shelf area impacted by a buoyant plume. Further efforts along these lines may attempt methods currently being used by Gan, et al. (2004a, 2004b) to represent wind-forced shelves.

A final limitation of the model is that it does not include the contribution of sediment to density of the water column outside of the wave boundary layer. This prevents it from dealing with high concentration input of sediment, such as described for hyperpycnal river plumes in Mulder and Syvitski (1996). Current efforts include adding this term, in the hope that this allows the model to resolve hyperpycnal plume events that occur when river input contains sediment concentrations in excess of ~40 g/L (Mulder and Syvitski, 1996). Resolution of these river-forced plumes may be amenable to sigma-coordinate models, with the wave-boundary layer grid cell approach only invoked for wave-dominated fluid mud transport events.

CONCLUSIONS

By accounting for dominant sediment transport mechanisms on the Eel Shelf, northern California, the model described provides a three-dimensional representation of flood deposition over timescales of weeks. Instrumentation deployed at the site observed a near-bed turbid layer (Traykovski, et al. 2000), but the mechanisms operating on the turbid layer had not been included within available models of sediment transport for the continental shelf (e.g. Harris and Wiberg 2000, 2002). The observations motivated and constrained a model of sediment flux within the wave boundary layer. This representation of the fluid mud layer was included within a sigma-coordinate model by modifying the vertical grid of the model. Using accepted parameterizations for settling speed, turbulence near the bed, and sediment entrainment, the model successfully predicted the location and volume of flood deposition for a flood of the Eel River.

Future work will improve our representation of the seafloor and sediment limits to suspension, use an advection algorithm more suited to a wide range of grain sizes, and include the suspended sediment concentration contribution to water column density. Because predictions of deposition are so sensitive to the settling behavior of the near-bed layer, it is hoped that further laboratory or observational work will guide representations of the settling properties of fine-grained material.

Acknowledgements
This project was funded by the Office of Naval Research's Geology and Geophysics Division as part of the StrataFORM project; award number N00014-01-1-0082. Some concepts were originally developed at the US Geological Survey in Woods Hole, MA, for whose support the first author is grateful. The early development of the model benefited from input by Rich Signell (then at the US Geological Survey). The manuscript has been improved by comments by T. Kniskern (VIMS) and an anonymous reviewer.

References

Cacchione, D.A., P.L. Wiberg, J. Lynch, J. Irish, and P. Traykovski. 1999. Estimates of suspended-sediment flux and bedform activity on the inner portion of the Eel continental shelf. *Marine Geology,* 154(1—4): 83—97.

Crockett, J. and C. Nittrouer. 2004. The sandy inner shelf as a repository for muddy sediment: an example from Northern California. *Continental Shelf Research, 24: 55—73.*

Blumberg, A.F. and G.T. Mellor. 1987. A Description of a Three-Dimensional Coastal Ocean Circulation Model, In: Three-Dimensional Coastal Ocean Models, N. Heaps, (ed), American Geophysical Union, pp 1—16.

Curran, K.J., P.S. Hill and T.G. Milligan. 2002. Fine-grained suspended sediment dynamics in the Eel River flood plume. *Continental Shelf Research*, 22: 2537-2550.

Fan, F., D. J. P. Swift , P. Traykovski, S. Bentley, A. W. Niedoroda , C.W. Reed, and J. Borgeld. 2004. River flooding, storm resuspension, and event stratigraphy on the northern California shelf: observations compared with simulations. *Marine Geology,* in press.

Foreman, M.G.G., W.R. Crawford, J.Y. Cherniawsky, R.F. Henry, and M.R. Tarbotton. 2000. A high-resolution assimilating tidal model for the northeast Pacific Ocean. *Journal of Geophysical Research,* 105: 28,629—28,652.

Gan, J. and J.S. Allen. 2004a. On open boundary conditions for a limited-area coastal model off Oregon. Part 1: Response to idealized wind forcing. *Ocean Modelling* (in press).

Gan, J., J.S. Allen and R. Samelson. 2004b. On open boundary conditions for a limited-area coastal model off Oregon. Part 2: Response to wind forcing from a regional mesoscale atmospheric model. *Ocean Modelling* (in press).

Geyer, W.R. 1993. The importance of suppression of turbulence by stratification on the estuarine turbidity maximum. *Estuaries,* 16(1): 113—125.

Geyer, W.R., P. Hill, T. Milligan, and P. Traykovski. 2000. The structure of the Eel River plume during floods. *Continental Shelf Research,* 20(16): 2095—2111.

Grant, W.D. and O. Madsen. 1979. Combined wave and current interaction with a rough bottom. *Journal of Geophysical Research,* 84: 1797—1808.

Harris, C.K. 1999. The importance of advection and flux divergence in the transport and redistribution of continental shelf sediment. Ph.D. Dissertation, University of Virginia, Charlottesville, VA. 155pp.

Harris, C.K. and P.L. Wiberg. 2001. A two-dimensional, time-dependent model of suspended sediment transport and bed reworking for continental shelves. *Computers and Geosciences,* 27 (6): 675-690.

Harris, C.K. and P.L. Wiberg. 2002. Across-shelf sediment transport: interactions between suspended sediment and bed sediment. *Journal of Geophysical Research,* 107 (C1): 10,1029/ 2000JC000634.

Harris, C.K., W.R. Geyer, and P. Traykovski. 2002. Flood layer formation on the northern California shelf by near-bed gravitational sediment flows and oceanographic transport. *EOS, Transactions, American Geophysical Union, 83(4): OS22K-02.*

Hill, P.S., T.G. Milligan, and W.R. Geyer. 2000. Controls on effective settling velocity in the Eel River flood plume. *Continental Shelf Research,* 20: 2095—2111.

Hydroqual, 2002. A primer for ECOM-SED, User's Manual Version 1.3. Hydroqual, Inc., Mahwah, N.J. 194pp.

James I.D. 2000. A high-performance explicit vertical advection scheme for ocean models: how PPM can beat the CFL condition. *Applied Mathematical Modelling,* 24: 1-9.

Mehta, A.J. and W.H. McAnally. 2003. Fine-grained sediment transport. Chapter 4 in M. Garcia (ed.) *Sedimentation Engineering.* ASCE.

Mellor, G.L., and T. Yamada. 1982. Development of a turbulence closure model for geophysical fluid problems. *Reviews of Geophysics and Space Physics*, 20, 851—875.

Mulder, T. and J.P.M. Syvitski. 1996. Climatic and morphologic relationships of rivers: Implications of sea level fluctuations on river loads. *Journal of Geology* 104: 509-523.

Munk, W. H., and E. R. Anderson. 1948. Notes on the theory of the thermocline, *Journal of Marine Research,* 3: 276—295.

Nittrouer, C.A. 1999. STRATAFORM: Overview of its design and synthesis of its results. *Marine Geology,* 154: 3—12.

Ogston, A.S. and R.W. Sternberg. 1999. Sediment-transport events on the northern California continental shelf. *Marine Geology,* 154(1—4): 69—82.

Ogston, A.S., D.A. Cacchione, R.W. Sternberg, and G.C. Kineke. 2000. Observations of storm and river flood-driven sediment transport on the northern California continental shelf. *Continental Shelf Research*, 20(16): 2141—2162.

Parsons, J.D., C.T. Friedrichs, P. Traykovski, D. Mohrig, J. Imran, J.P.M. Syvitski, G. Parker, P. Puig, and M.H. Garcia, in review. Sediment gravity flows: Initiation, transport and deposition. In: C.A. Nitrouer, J. Austin, M. Field, M. Steckler, J.P.M. Syvitski, and P. Wiberg (eds.) Continental Margin Sedimentation: Transport to Sequence.

Reed, C.W., A. W. Niedoroda and D.J.P. Swift. 1999. Sediment entrainment and transport limited by bed load armoring. *Marine Geology*, 154: 143—155.

Ross, M.R. and A.J. Mehta. 1989. On the mechanics of lutoclines and fluid mud. *Journal of Coastal Research,* 5: 51—61.

Scully, M. E. C.T. Friedrichs and L.D. Wright. 2002. Application of an analytical model of critically stratified gravity-driven sediment transport and deposition to observations from the Eel River continental shelf, northern California. *Continental Shelf Research*, 22: 1951—1974.

Scully, M.E., C.T. Friedrichs, and L.D. Wright. 2003. Numerical modeling of gravity-driven sediment transport and deposition on an energetic continental shelf: Eel River, northern California. *Journal of Geophysical Research.* 108(C4): 3120, doi:10.1029/2002JC001467,

Sommerfield, C. and C. Nittrouer. 1999. Modern accumulation rates and a sediment budget for the Eel River shelf, USA: a flood-dominated depositional environment. *Marine Geology*, 154(1—4): 227—241.

Song, Y. and D. Haidvogel. 1994. A semi-implicit ocean circulation model using a generalized topography-following coordinate system. *Journal of Computational Physics*, 115: 228—244.

Smith, J.D. and S. McLean. 1977. Spatially averaged flow over a wavy surface. *Journal of Geophysical Research,* 82(12): 1735—1746.

Traykovski, P., W.R. Geyer, J.D. Irish, and J.F. Lynch. 2000. The role of wave-induced density-driven fluid mud flows for cross-shelf transport on the Eel River continental shelf. *Continental Shelf Research*, 20(16): 2141—2162.

Trowbridge, J. H., and G. C. Kineke. 1994. Structure and dynamics of fluid muds on the Amazon continental shelf. *Journal of Geophysical Research*, 99: 865—874.

Wheatcroft, R.A. and J. Borgeld. 2000. Oceanic flood deposits on the northern California shelf: large scale distribution and small-scale properties. *Continental Shelf Research*, 20(16): 2059—2066.

Wolanski, E., R. Gibbs, P. Ridd, B. King, K.Y. Hwang, and A. Mehta. 1991. Fate of dredge spoil, Cleveland Bay, Townsville. *Proceedings of the 10th Australasian Conference on Coastal and Ocean Engineering*. Auckland, NZ, pp 63—66.

Wright, L.D., C.T. Friedrichs, S.C. Kim, and M.E. Scully. 2001. The effects of ambient currents and waves on gravity-driven sediment transport on continental shelves. *Marine Geology*, 175: 25—45.

Long-Term Sediment-Water Fluxes of Hydrophobic Organic Chemicals

Wilbert Lick[1], Craig Jones[2], and James Lick[3]

Abstract

Bottom sediments contaminated by hydrophobic organic chemicals (HOCs) are a major problem in rivers, lakes, and near-shore areas of the oceans. In order to determine the appropriate remedial action for these sediments, it is necessary to understand and quantify the flux of the HOCs between the bottom sediments and the overlying water not only for the near term, but also for the long term, as much as 20 to 50 years into the future. The emphasis here is on the quantitative prediction of the long-term fluxes of HOCs due to molecular diffusion and bioturbation as modified by time-dependent, non-equilibrium sorption. The effects of the partition coefficient on these fluxes are also illustrated. Numerical as well as approximate analytic solutions are presented and compared with each other and with experimental results. As a result of these calculations, it is demonstrated (a) that molecular diffusion is ubiquitous and more significant than ordinarily assumed and (b) effects of non-equilibrium sorption are significant and long-lasting.

Introduction

As a result of human activities, large amounts of hydrophobic organic chemicals (HOCs) have been disposed of in surface waters; they are then sorbed to suspended sedimentary particles and subsequently deposited in the bottom sediments. These HOCs are ubiquitous, often toxic, and may serve as a major source of contaminants to the overlying water for many years to come. In order to determine environmentally- and cost-effective remedial actions for these sediments and waters for the long term, the flux of HOCs between the bottom sediments and the overlying water must be understood and quantified. For this purpose, it is important to be able to quantitatively predict these fluxes as a function of time, not only for the near term, but especially for the long term, as much as 20 to 50 years into the future.

[1]Professor, Department of Mechanical and Environmental Engineering, University of California, Santa Barbara, CA 93106.
[2]Environmental/Ocean Engineer, Sea Engineering, Inc., Santa Cruz, CA 95060.
[3]Research Associate, Department of Mechanical and Environmental Engineering, University of California, Santa Barbara, CA 93106.

These fluxes are primarily due to sediment erosion and deposition, molecular diffusion, and bioturbation. In some cases, other processes such as pore water convection may also contribute. Each of these processes acts in a different way, and hence each must be described and modeled in a different way. For HOCs, these processes are continuously and often significantly modified by time-dependent, non-equilibrium sorption of the HOC between the solid particles in the sediment and the surrounding pore waters. This is especially true for HOCs with large partition coefficients, K_p (L/kg), where K_p is defined as

$$K_p = \frac{C_s}{C_w} \tag{1}$$

where C_s (mg/kg) is the concentration of the chemical sorbed on the solids and C_w (mg/L) is the concentration of the chemical dissolved in the water.

Preliminary work on the fluxes of HOCs due to sediment erosion and deposition as modified by non-equilibrium sorption has been done (Chroneer and Lick, 1997; Jones and Lick, 2001) and has shown large effects due to this sorption. In the present paper, the emphasis is on the long-term flux of HOCs between the sediments and overlying water due to molecular diffusion and bioturbation; because of its importance, time-dependent, non-equilibrium sorption is necessarily included in the analysis. It is assumed throughout that the equilibrium value of K_p is moderate to large, i.e., on the order of 10^2 to 10^6 L/kg; this range of values is typical of most PCBs and many other HOCs.

The analysis and calculations presented here are based on previous experiments and analyses concerned with non-equilibrium sorption; the relevant work is reviewed in the following section. Calculations of long-term fluxes of HOCs are presented in section 3. This is followed by a summary and concluding remarks.

Previous Work on HOC Fluxes Due to Molecular Diffusion and Bioturbation

Basic to the present model of the flux of HOCs between bottom sediments and the overlying water is the fact that the adsorption and desorption of HOCs (especially those with high partition coefficients) between sedimentary particles/aggregates and the surrounding water are often slow, with equilibration times as long as weeks to years. Both experiments and theory have demonstrated this repeatedly (e.g., see Karickhoff and Morris, 1985; Jepsen and Lick, 1996) and have also demonstrated that, as K_p increases, the sorption rate decreases and the sorption time increases.

For particles/aggregates, a quantitative model of the sorption process was developed by Wu and Gschwend (1986) and later extended by Lick and Rapaka (1996). In this model, the transport of the chemical within the particle/aggregate is described by a time-dependent diffusion equation in spherical coordinates with no reaction terms, but with an effective diffusion coefficient given by

$$D_e = \frac{D_m}{1 + \left(\frac{1-\phi}{\phi}\right)\rho K_p} \tag{2}$$

where D_m (cm^2/s) is the molecular diffusion coefficient of the chemical in the pore water of the particle, without consideration of any reactions but corrected for tortuosity; ϕ is the porosity of the particle; and ρ is the mass density of the solid particles (approximately 2.6 g/cm^3).

With the loss of some detail and accuracy, a simpler and computationally more efficient model was later developed (Lick et al., 1997). In this latter model, it is assumed that the time rate of change of the average contaminant concentration in the sediment particle or aggregate due to the transfer of the contaminant from the water to the solid is given by

$$\frac{dC_s}{dt} = -k(C_s - K_p C_w) \tag{3}$$

where k(s^{-1}) is a mass transfer coefficient that can be approximated by

$$k = \frac{D_e}{0.0165 d^2} \tag{4}$$

where d is the diameter of the particle/aggregate in centimeters.

The fluxes due to molecular diffusion of several HOCs into and out of consolidated bottom sediments have been investigated both experimentally and theoretically by Deane et al. (1999). Experimental results for the diffusion and sorption of dissolved hexachlorobenzene (HCB) from the overlying water into uncontaminated, consolidated Detroit River sediments are shown in Fig. 1. The HCB concentration on the solids, C_s, is normalized to the HCB concentration in the overlying water, C_{wo}, and is shown as a function of the depth at different time intervals from 4 to 512 d. Despite the length of the experiments, significant changes in C_s are limited to a few millimeters near the sediment-water interface; measured values of C_s/C_{wo} are less than 0.1 of their value at equilibrium (where $C_s/C_{wo} = K_p = 1.2 \times 10^4$ L/kg). A theoretical model was also used to analyze the results of the experiments. Results of the numerical simulation (which necessarily included non-equilibrium sorption) are shown as the solid lines in Fig. 1. Good agreement with the experimental results is apparent. In this calculation, the assumption of one size class of sediment aggregate was made. A calculation with three size classes gave even better agreement, especially at small and intermediate times.

This model has been simplified and has also been extended to include bioturbation (Lick, 2002a,b; Lick et al., 2003). The physical effects of bioturbation are approximated as a diffusion of solids and water with an effective diffusion coefficient due to bioturbation of D_b. The diffusion coefficient for contaminants sorbed to solids, D_s, is then given by D_b, while the diffusion coefficient for the contaminant dissolved in pore waters, D_w, is the sum of D_m and D_b, where D_m is the molecular diffusion coefficient for the contaminant in water. In general, D_b is dependent on depth in the sediments with its maximum value at the surface and decreasing with depth with a characteristic length scale, x_b, on the order of 5 cm for fresh water organisms and on the order of 10 cm or more for organisms in seawater (Boudreau, 1997). As a first approximation, D_b can be expressed as $D_{bo}\ e^{-x/x_b}$, where D_{bo} is the value of D_b at the surface.

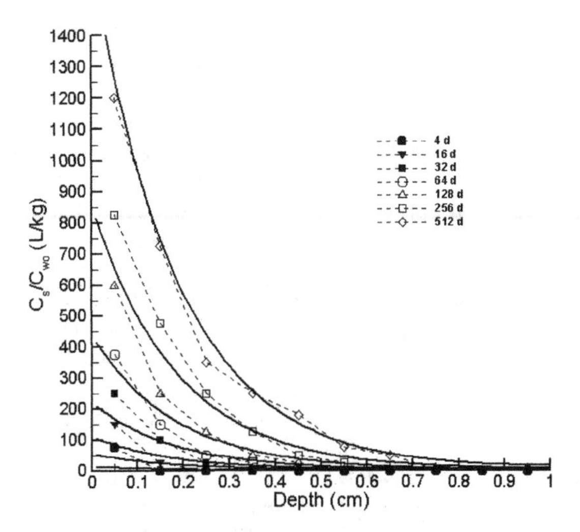

Fig. 1 Experimental and Calculated Results for Molecular Diffusion of HCB into Consolidated Detroit River Sediments. C_s/C_{WO} is shown as a function of depth at different times from 4 to 512 d. $D_w = 3 \times 10^{-6}$ cm^2/s. $K_p = 12{,}000$ L/kg.

With these approximations and including molecular diffusion, bioturbation, time-dependent sorption as described by Eq. (3), and one size sediment aggregate, the one-dimensional, time-dependent mass conservation equation for the contaminant dissolved in water (per unit volume of total sediment) is

$$\phi \frac{\partial C_W}{\partial t} - \phi \frac{\partial}{\partial x}\left(D_W \frac{\partial C_W}{\partial x} \right) = (1-\phi)\rho k (C_s - K_p C_W) \qquad (5)$$

while the conservation equation for the contaminant sorbed to the solids (again per unit volume of total sediment) is

$$(1-\phi)\rho \frac{\partial C_s}{\partial t} - (1-\phi)\rho \frac{\partial}{\partial x}\left(D_s \frac{\partial C_s}{\partial x} \right) = -(1-\phi)\rho k (C_s - K_p C_W) \qquad (6)$$

where ϕ is the porosity of the sediments. The flux of contaminant between the sediments and the overlying water due to diffusion of the dissolved contaminant is given by

$$q(t) = -\phi D_W \frac{\partial C_W}{\partial x}(0,t) \qquad (7)$$

It is assumed that there is no flux of contaminant from the solid particles directly into the overlying water.

In order to illustrate general characteristics of HOC fluxes between sediments and overlying water, solutions to a specific problem will be presented and discussed here in detail, i.e., the time-dependent flux of a dissolved HOC in the overlying water to clean bottom sediments. This and other related flux problems have been discussed previously (Lick et al., 2003) but only for short times, up to 512 d; the emphasis here is on longer times.

Model Results for Molecular Diffusion

For the flux due to molecular diffusion of a dissolved contaminant from the overlying water to clean bottom sediments, the boundary and initial conditions are that $C_w(0,t) = C_{wo}$ = constant, $C_w(x,t) \to 0$ as $x \to \infty$, $C_w(x,0) = 0$, and $C_s(x,0) = 0$. With these conditions, the present model, when used to describe the experimental results by Deane et al. (1999), gives essentially the same results as the more complex model used by Deane et al. In particular, the computational results shown as the solid lines in Fig. 1 are reproduced almost exactly by the present model. In the Deane et al. study, because of experimental limitations, C_w could not be measured as a function of depth and time. However, this quantity can be calculated by means of the numerical model. For times from 4 d to 512 d, calculated values of C_s and K_pC_w (both normalized with respect to C_{wo}) are shown in Fig. 2 as a function of depth in the sediments. Both C_s and K_pC_w monotonically increase with time and decrease with depth. It can be seen that (a) C_s and K_pC_w are quite different and are therefore not in local chemical equilibrium with each other, or even close to equilibrium, even after 512 d of HCB diffusion into the sediment and (b) C_w is remarkably almost independent of time and only a function of distance.

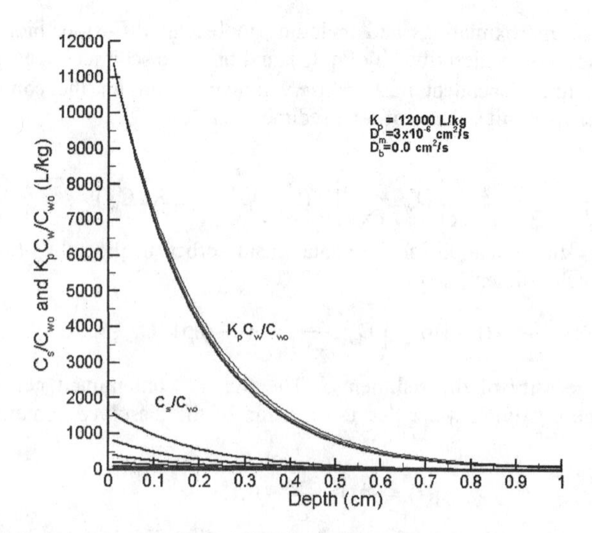

Fig. 2 Calculated Results for Molecular Diffusion of HCB into Consolidated Detroit River Sediments. C_s/C_{wo} and K_pC_w/C_{wo} are shown as functions of depth at times of 4, 16, 32, 64, 128, 256, and 512 d.

Approximate analytic solutions for this problem were also obtained (Lick, 2002a). From this analysis, it can be shown that characteristic time scales for this problem are t_s and t_d, where

$$t_s = \frac{1}{k^*} \tag{8}$$

$$t_d = \frac{1}{k} \tag{9}$$

$$k^* = k\left[1 + \left(\frac{1-\phi}{\phi}\right)\rho K_p\right] \tag{10}$$

From Eqs. (2) and (4), it can be seen that, for large K_p, k is inversely proportional to K_p and hence k* is independent of K_p. It follows that t_s is independent of K_p while t_d is proportional to K_p.

For HOCs with large K_p, $t_s << t_d$. Small time can therefore be defined as $t < t_s$, an intermediate time as $t_s < t < t_d$, and large time as $t > t_d$. For HCB and the sediments used by Deane et al., t_s is approximately 7.6×10^3 s or about two hours, while t_d is approximately 2.5×10^8 s or about 2900 d (8 years). Since the maximum duration of the experiments, t_{exp}, was 512 d, it follows that $t_s << t_{exp} << t_d$.

Times less than t_s are too small to be significant in most sediment-water flux problems. From asymptotic analyses, approximate analytic solutions for intermediate and large time were obtained. For intermediate time, the solutions for C_w and C_s are

$$C_w = C_{wo}e^{-\alpha x} \tag{11}$$

$$C_s = K_p C_{wo}(1 - e^{-kt})e^{-\alpha x} \tag{12}$$

$$\alpha^2 = k^*/D_w \tag{13}$$

Eq. (11) demonstrates that C_w is a function of x but is independent of time. The above results are essentially the same as those calculated by the numerical model for HCB and illustrated in Figs. 1 and 2; they are therefore in good agreement with the experimental results. From Eqs. (11) and (7), it follows that the contaminant flux is given by

$$q = \phi D_w C_{wo} \alpha \tag{14}$$

and is not a function of time. The experiments and more accurate numerical calculations with three size classes show that q is a slowly varying function of time. Since k* is independent of K_p, so are α and q.

For large time, $t > t_d$, it can be shown that the solution is described approximately by the diffusion equation

$$\frac{\partial C_w}{\partial t} - D^* \frac{\partial^2 C_w}{\partial x^2} = 0 \tag{15}$$

where D* is an effective diffusion coefficient given by

$$D^* = \frac{D_w}{1 + \left(\frac{1-\phi}{\phi}\right)\rho K_p} \tag{16}$$

For the present boundary and initial conditions, the solution to Eq. (15) is

$$C_w = C_{wo}\,\text{erfc}\left[\frac{x}{2\sqrt{D^* t}}\right] \tag{17}$$

where erfc is the complementary error function (Carslaw and Jaeger, 1959). In this limit of large time, C_s is in approximate local chemical equilibrium with C_w, and hence C_s/K_p is given by the same equation as above. From this equation and Eq. (7), the mass flux at the surface is given by

$$q = \frac{\phi D_w C_{wo}}{(\pi D^* t)^{1/2}} \tag{18}$$

Eqs. (17) and (18), although valid for large time, are not valid for intermediate time and do not agree with the experimental results shown in Figs. 1 and 2.

Model Results for Bioturbation

When bioturbation is present, the boundary and initial conditions on C_w are as above, while the boundary and initial conditions on C_s are that $\partial C_s(0,t)/\partial x = 0$, $C_s(x,t) \to 0$ as $x \to \infty$, and $C_s(x,0) = 0$. The solutions for intermediate time, Eqs. (10) to (13), are still valid with the substitution of $D_w = D_m + D_b$. For long time, Eqs. (15), (17), and (18) are valid with the substitution

$$D^* = \frac{D_m}{1 + \left(\frac{1-\phi}{\phi}\right)\rho K_p} + D_b \tag{19}$$

For molecular diffusion alone, it can be shown that the characteristic time t_d ($= 1/k$) is the time required for C_s and $K_p C_w$ to come to equilibrium in a surface layer with thickness $1/\alpha$ when the flux is given by Eq. (14). When bioturbation is included, the characteristic time t_d is now the time for C_s and $K_p C_w$ to come to equilibrium in the bioturbated layer with thickness x_b, again when the flux is given by Eq. (14). With this definition, t_d can then be determined from mass conservation, i.e.,

$$q t_d = \phi D_w C_{wo} \alpha t_d = (1-\phi)\rho C_{so} x_b \tag{20}$$

where $C_{so} = K_p C_{wo}$ at equilibrium. This then gives

$$t_d = \left(\frac{1-\phi}{\phi}\right)\frac{\rho K_p x_b}{\alpha D_w} \tag{21}$$

From this and the parameters as in the Deane et al. experiments for HCB, t_d is approximately 300 years.

Long-Term Fluxes

Calculations were made of the long-term flux (up to 100 years) of HOCs with a wide range of K_p to illustrate the long-term behavior of these fluxes and the influence of K_p. In these calculations, parameters associated with the sediment were essentially the same as those used in the study by Deane et al. (1999), i.e., $D_m = 3 \times 10^{-6}$ cm^2/s, $\phi = 0.5$ and $k = 3.3 \times 10^{-9}$ s^{-1}. From this, it follows that $\alpha = 5.4$ cm^{-1}, or $1/\alpha = 0.18$ cm. In these calculations, all results are normalized such that $C_{wo} = 1$ kg/L = 1 g/cm^3.

Molecular Diffusion

In order to accurately determine the long-term behavior and flux of HCB into bottom sediments due to molecular diffusion, calculations were made for longer times (from 512 d to 100 years) than those shown in Figs. 1 and 2. For these latter calculations, results for C_s/C_{wo} and K_pC_w/C_{wo} are shown as a function of depth with time as a parameter in Fig. 3. For 512 d and 3 years, K_pC_w/C_{wo} is approximately equal to that given by Eq. (11) and is almost independent of time; C_s/C_{wo} is approximately equal to that given by Eq. (12) and is increasing with time and approaching K_pC_w/C_{wo}. At and after 10 years, C_s/C_{wo} and K_pC_w/C_{wo} are approximately equal; their dependence on space and time is now equivalent to that given by Eq. (17), the quasi-equilibrium solution valid for long time.

For this case, q/C_o (cm/s) is shown as a function of time in Fig. 4. For intermediate time (up to about 8 years), Eq. (14) predicts a constant value for q/C_o of 8.7×10^{-6} $g/cm^2/s$. The numerical solution has this value initially, but slowly changes with time and decreases by about 30% in the intermediate time interval of about 8 years. The reason for this is that $C_s(0,t)$ is slowly increasing with time due to sorption from the surrounding pore water and is no longer negligible (as was assumed in the derivation of Eq. (14)); this decreases the forcing $(C_s - K_pC_w)$ that drives the flux. After this time, the flux can be approximated by Eq. (18) which is shown by the dashed line.

Calculations were also done for the molecular diffusion of an HOC with a K_p of 1.2×10^5 L/kg, ten times that of HCB. The solutions for C_s and C_w are shown in Fig. 5. It can be seen that the intermediate time solution (where C_w is independent of time) is now valid for more than 30 years, in agreement with Eq. (9), which indicates that t_d is directly proportional to K_p and hence should now be about 80 years. It can also be seen that the depth of penetration $(1/\alpha)$ is essentially the same as that for HCB (compare Fig. 3); this agrees with the asymptotic result for intermediate time that α is independent of K_p.

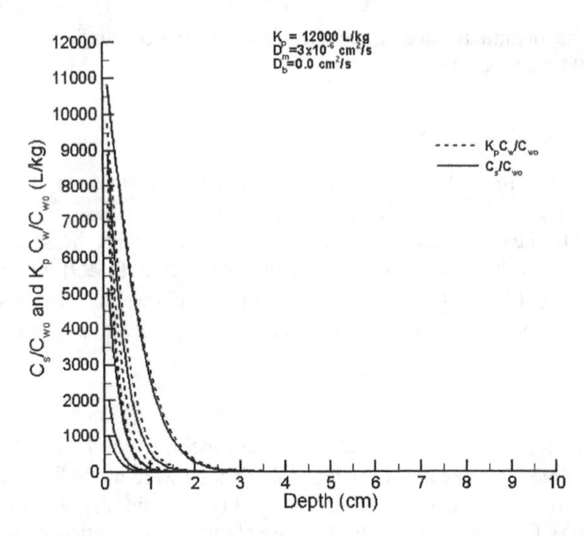

Fig. 3 Calculated Results for Molecular Diffusion of HCB into Consolidated Detroit River Sediments. C_s/C_{wo} and K_pC_w/C_{wo} are shown as functions of depth at times of 512 d and 3, 10, 30, and 100 years.

As K_p decreases, t_d should also decrease; for $t < t_d$, $1/\alpha$ should be constant. This is consistent with the solutions for C_w and C_s shown in Fig. 6 for a K_p of 1.2×10^3 L/kg, 0.1 of that for HCB. For this K_p, t_d should be about 0.8 years. After this time, C_w and C_s are in quasi-equilibrium, the solution is therefore given approximately by Eq. (17), and the intermediate time solution is no longer valid.

Mass fluxes for HOCs with $K_p = 1.2 \times 10^5$, 1.2×10^4, 1.2×10^3, and 1.2×10^2 L/kg are shown in Fig. 7 as a function of time; for these HOCs, the t_ds are approximately 80 years, 8 years, 0.8 years (290 d), and 0.08 years (29 d) respectively. The solutions for times greater than t_d are then given by Eq. (18), and there is a smooth transition between the intermediate and long-time solutions.

Molecular Diffusion and Bioturbation

The model has also been used to calculate numerical and analytical solutions when bioturbation is present. In this case, molecular diffusion is inherently present and interacts with the bioturbation process. For K_p = 12,000 L/kg, $D_m = 3 \times 10^{-6}$ cm^2/s, $D_b = 1 \times 10^{-7}$ cm^2/s, and x_b = 6 cm, the numerical results for C_s and K_pC_w for long times (512 d to100 years) are shown in Fig. 8. Due to bioturbation, the chemical on the solid has been diffused away from the surface so that, at and near the surface, C_s is much smaller than its value without bioturbation (compare Fig. 3) and of course is much smaller than K_pC_w. The solution for C_w near the surface is dominated by molecular diffusion and hence the quantities K_pC_w, α, and q are almost unchanged from their values without bioturbation. Because of the small values of C_s near the surface, these intermediate time solutions are now valid for a time much greater than 1/k, a time which is now given by Eq. (21).

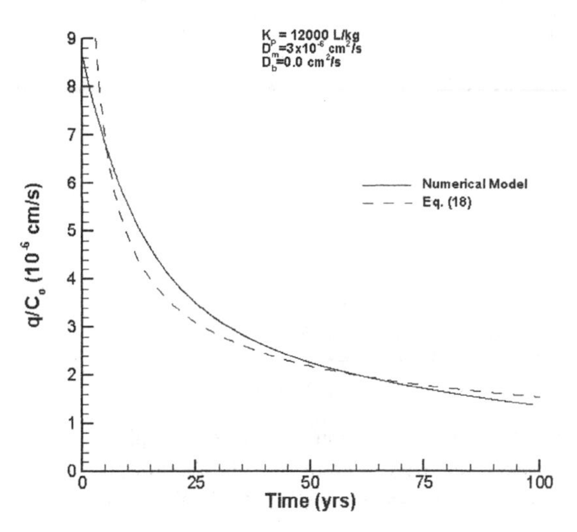

Fig. 4 HCB Flux Due to Molecular Diffusion as a Function of Time. Also shown is the flux as approximated by the quasi-equilibrium approximation, Eq. (18).

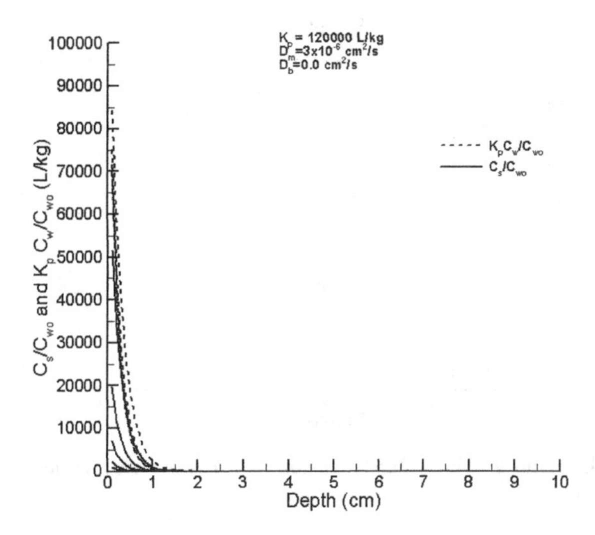

Fig. 5 Results for Molecular Diffusion of an HOC with K_p = 120,000 L/kg into Consolidated Detroit River Sediments. C_s/C_{WO} and K_pC_W/C_{WO} are shown as functions of depth at times of 512 d and 3, 10, 30, and 100 years.

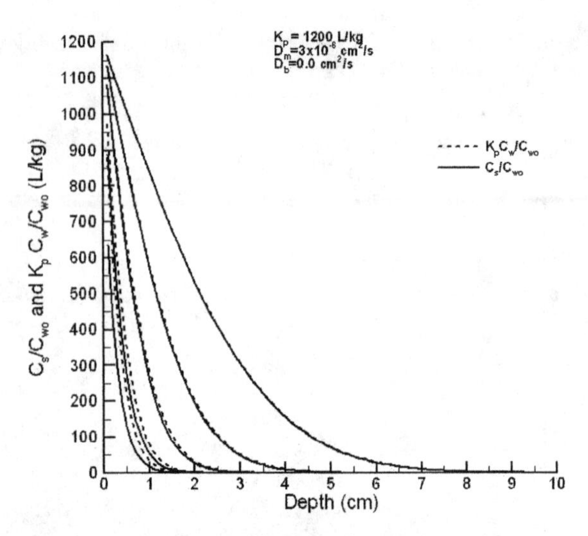

Fig. 6 Results for Molecular Diffusion of an HOC with K_p = 1200 L/kg into Consolidated Detroit River Sediments. C_s/C_{wo} and K_pC_w/C_{wo} are shown as functions of depth at times of 512 d and 3, 10, 30, and 100 years.

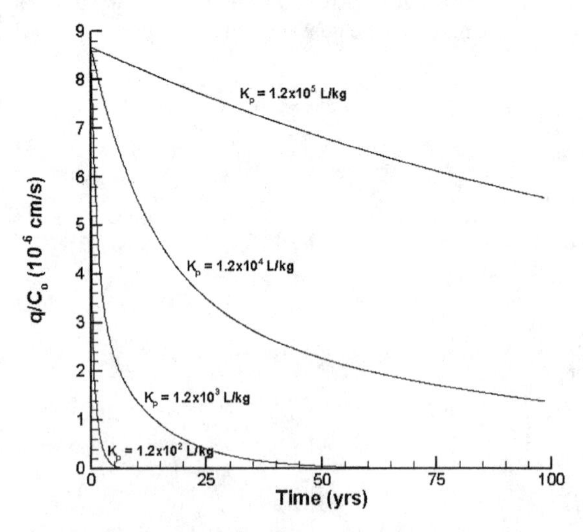

Fig. 7 HOC Flux Due to Molecular Diffusion as a Function of Time with K_p as a Parameter.

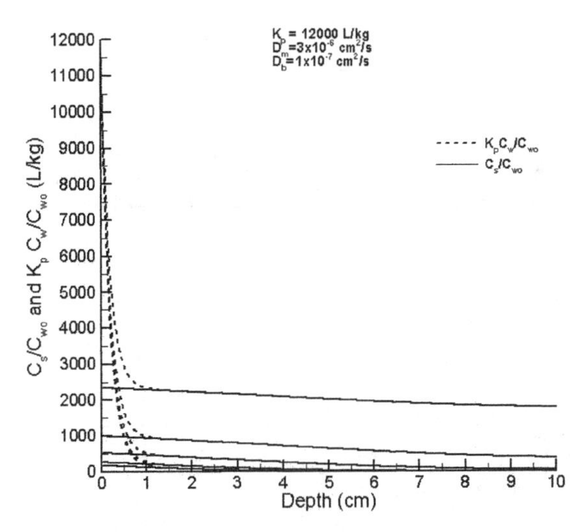

Fig. 8 Results for Molecular Diffusion and Bioturbation of HCB with $K_p = 1.2 \times 10^4$ L/kg and a relatively low value of D_b of 1×10^{-7} cm^2/s. C_s/C_{wo} and K_pC_w/C_{wo} are shown as functions of depth at times of 512 d and 3, 10, 30, and 100 years.

The value of 1×10^{-7} cm^2/s for D_b is low but in the range of what is normally observed. For a larger value of 1×10^{-5} cm^2/s (but still in the range of what is observed), the results for C_s and K_pC_w are shown in Fig. 9. Because of the greater mixing by organisms, C_s has been diffused more rapidly away from the surface. C_w near the surface and hence α and q are slowly changing functions of time.

For $D_b = 1 \times 10^{-5}$ cm^2/s, the fluxes due to molecular diffusion and bioturbation for HOCs with $K_p = 1.2 \times 10^5$, 1.2×10^4, 1.2×10^3, and 1.2×10^2 L/kg are shown in Fig. 10 as a function of time. Initially, the fluxes for all HOCs are given by Eq. (14). As time increases, $C_s(0,t)$ also increases and hence decreases the forcing $(C_s - K_pC_w)$ that drives the flux. Eventually, the flux can be described by Eq. (18), but only for $t > t_d$ (where t_d is proportional to K_p).

Summary and Concluding Remarks

For bottom sediments contaminated by HOCs, the flux of these chemicals between the sediments and the overlying water is a determining factor in predicting sediment and water quality. These fluxes must be known accurately, not only for the short-term, but also for the long term, 20 to 50 years into the future. In the present paper, a theoretical model has been used to investigate the general characteristics of HOC fluxes due to molecular diffusion and bioturbation as modified by time-dependent, non-equilibrium sorption. The emphases were on the effects of K_p and the long-term characteristics of the solutions.

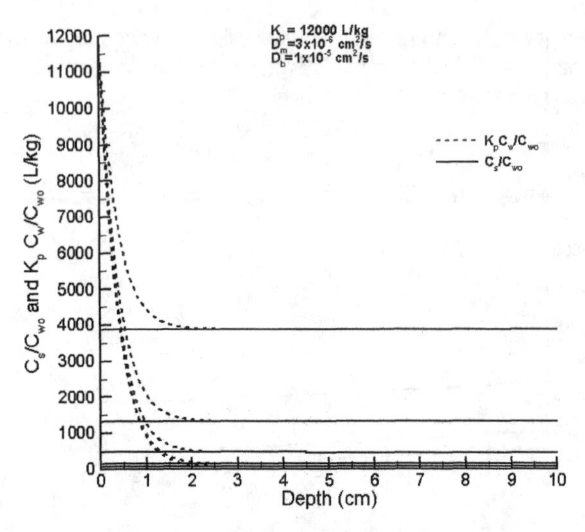

Fig. 9 Results for Molecular Diffusion and Bioturbation of HCB with $K_p = 1.2 \times 10^4$ L/kg and a relatively high value of D_b of 1×10^{-5} cm^2/s. C_s/C_{wo} and K_pC_w/C_{wo} are shown as functions of depth at times of 512 d and 3, 10, 30, and 100 years.

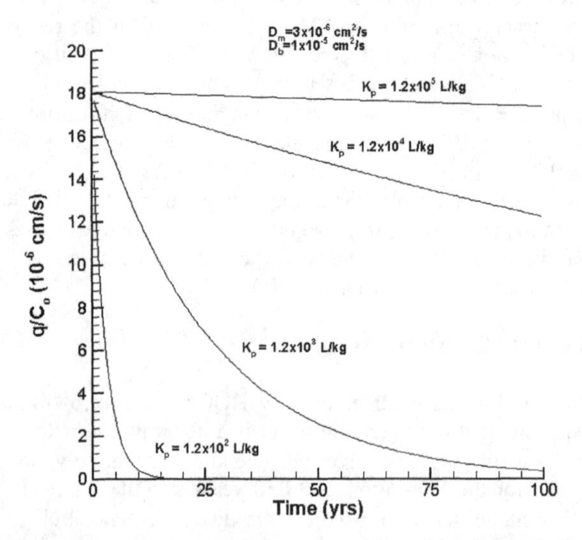

Fig. 10 HCB Flux Due to Molecular Diffusion and Bioturbation as a Function of Time with K_p as a Parameter.

In most water quality models, it is assumed that the contaminant flux due to molecular diffusion and bioturbation is determined by the gradient of the chemical acting over an effective distance of 2 to 10 cm (e.g., see Chapra (1997) and Schnoor (1996)). However, it is demonstrated here that, for intermediate time, the diffusion is balanced by time-dependent, non-equilibrium sorption such that the chemical gradient acts over an effective distance of $1/\alpha$, or about 2 mm. Because of this, the flux due to molecular diffusion is much larger than that ordinarily assumed and is now comparable to that due to bioturbation, even when organisms are plentiful. When bioturbation is present, molecular diffusion is inherently present and modifies the flux of the chemical.

In previous analyses by us, relatively simple analytic solutions for the flux were obtained for intermediate time ($t \ll t_d$) and large time ($t \gg t_d$); these solutions correspond to Eqs. (14) and (18) respectively. The numerical solutions are valid for all time and demonstrate that Eqs. (14) and (18) are quite accurate for their respective time intervals. The solutions also demonstrate effects of K_p. The time, t_d, varies directly with K_p so that, for example, when $K_p = 1.2 \times 10^5$ L/kg, t_d (for molecular diffusion alone) is approximately 80 years; it decreases to .08 years (29 d) for $K_p = 1.2 \times 10^2$ L/kg.

The bioturbation model as used here is relatively simple. More complex models have been suggested (e.g., see Boudreau, 1997). However, for the general description of the effects of organisms, the present model should be at least a good first approximation for HOCs, but may have to be modified when more detailed and long-term experiments with HOCs and a variety of organisms are performed.

The present calculations were meant to illustrate the general characteristics of the flux, especially as a function of K_p and time. For more general problems, the same general characteristics will be present but the solutions for C_s, C_w, and q will be more complex and will vary with time in a more complex manner depending on boundary conditions.

Acknowledgement

This research was supported by the U.S. Environmental Protection Agency.

References

Boudreau, B. (1997). *Diagenetic Models and Their Implementation*, Springer, New York, NY.

Carslaw, H.S., and Jaeger, J.C. (1959). *Conduction of Heat in Solids*, Oxford Press, London, Great Britain.

Chapra, S.C. (1997). *Surface Water Quality Modeling*, McGraw Hill Publishing, New York, N.Y.

Chroneer, Z. and Lick, W. (1997). "Parameters affecting the transport and fate of hydrophobic contaminants in surface waters," *Computing in Environmental Resource Management, Air and Water Management Association*, pp. 183-197.

Deane, G., Chroneer, Z. and Lick, W. (1999). "Diffusion and sorption of hexachlorobenzene in sediments and saturated soils," *J. Environmental Engineering*, 125, 8, pp. 689-696.

Jepsen, R. and Lick, W. (1996). "Parameters affecting the adsorption of PCBs to suspended sediments," *J. Great Lakes Res.*, 22, pp. 341-353.

Jones, C. and Lick, W. (2001). "Contaminant flux due to sediment erosion," *Estuarine and Coastal Modeling VII*, pp. 280-293.

Karickhoff, S.W., and Morris, K.R. (1985). "Sorption dynamics of hydrophobic pollutants in sediment suspension," *Environ. Toxicol. Chem.*, 4, pp. 469-479.

Lick, W. (2002a). "Modeling the sediment-water flux of hydrophobic organic chemicals due to diffusion and sorption," Report, Department of Mechanical and Environmental Engineering, University of California, Santa Barbara.

Lick, W. (2002b). "The sediment-water flux of hydrophobic organic chemicals due to diffusion, bioturbation, and sorption," Report, Department of Mechanical and Environmental Engineering, University of California, Santa Barbara, CA.

Lick, W., Lick, J., and Jones, C. (2003). "Effects of finite sorption rates on sediment-water fluxes of hydrophobic organic chemicals," Report, Department of Mechanical and Environmental Engineering, University of California, Santa Barbara.

Lick, W., Chroneer, Z., and Rapaka, V. (1997). "Modeling the dynamics of the sorption of hydrophobic organic chemicals to suspended sediments," *Water, Air and Soil Pollution*, Dordrecht, 99, pp. 225-235.

Lick, W. and Rapaka, V. (1996). "A quantitative analysis of the dynamics of sorption of hydrophobic organic chemicals to suspended sediments," *Envir. Toxicol. Chem.*, 15, pp. 1038-1048.

Schnoor, J.L. (1996). *Environmental Modeling*, Wiley Interscience, New York, N.Y.

Wu, S. and Gschwend, P.M. (1986). "Sorption kinetics of hydrophobic organic compounds to natural sediments and soils," *Envir. Sci. and Technol.* 20, pp. 717-725.

Coupled Circulation, Wave, and Morphology-Change Modeling, Shinnecock Inlet, New York

[1]Frank S. Buonaiuto, and [2]Adele Militello

Abstract

Analysis of five high-resolution bathymetric data sets collected at Shinnecock Inlet, NY indicates the evolution of ebb shoal morphology between 1994 and 2000 was primarily controlled by migration of the main navigation channel. Increased wave activity during the 1997 El Nino accelerated the rate at which the channel was deflected toward the west. These bathymetric data are applied in this study for assessment of morphology change calculation conducted within the Inlet Modeling System developed by the U.S. Army Corps of Engineers Coastal Inlets Research Program. Circulation, sediment transport, and morphology change were calculated by the two-dimensional finite-difference model M2D, which was coupled with STWave for computation of wave-driven currents. A simulation was conducted for August to November 1997 in which waves from NDBC Station 44025 were input as forcing for STWave. Tidal forcing for M2D was prescribed with water levels extracted from a regional ADCIRC model. Major observed changes in inlet morphology were reproduced by the modeling system. These changes are: scour and westward migration of the navigation channel, accretion along the eastern flank of the ebb shoal, and accretion of the seaward extent of the ebb shoal.

Introduction

Shinnecock Inlet, located on the south shore of Long Island, New York (Fig. 1), opened on September 21, 1938 during the passage of the Great New England Hurricane (Morang 1999). Since its opening, inlet evolution has been influenced by storms, dredging, beach nourishment projects, and construction and rehabilitation of two offset rubble-mound jetties originally built between 1953 and 1954. Morphology of the inlet and ebb shoal is controlled by tide and wave-driven transport, as well as by jetties and bathymetric features such as channels and shoals. Tide on the south shore of Long Island is semidiurnal with a mean range of 0.88 m at the inlet entrance (ocean side). Spring tide range is 1.1 m and the tidal prism is 3.29×10^7 m^3 (Militello and Kraus 2001). Incident waves have average height of approximately 1 m and

---

1) Coastal Oceanographer. Marine Sciences Research Center, SUNY at Stony Brook, Stony Brook NY 11794-5000. fsbuonaiuto@optonline.net.

2) Physical Oceanographer. Coastal Analysis LLC, 4886 Herron Road, Eureka, CA 95503. CoastalAnalysis@cox.net.

period of 7 sec, originating from the southeast. During large northeaster storms and hurricanes, wave height can exceed 4 m with periods in the range of 12-14 sec.

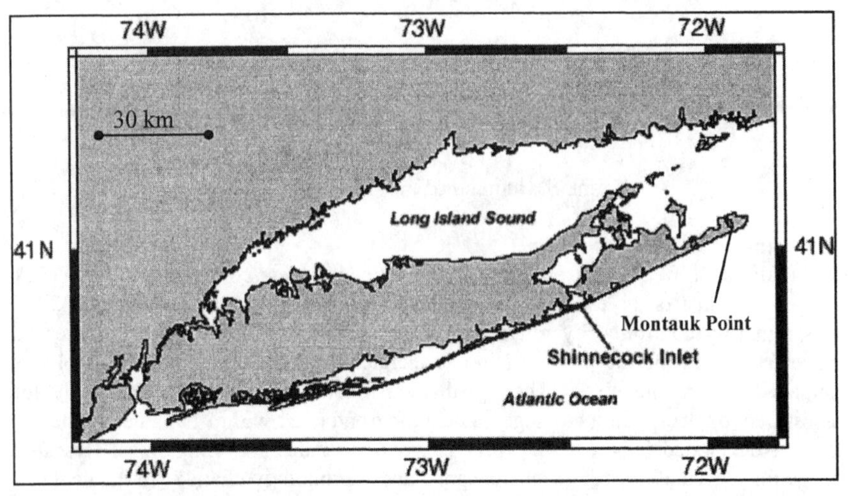

Figure 1. Location map for Shinnecock Inlet, NY.

During the 1997 to 1998 fall and winter season, increased transport and wave activity enhanced the natural westward migration of the main navigation channel (Buonaiuto 2003a). Channel reorientation potentially altered the transport conduits and delivery of sediment to the morphologic features that comprise the ebb shoal complex (Fig. 2). The movement of sediment was further complicated by an increase in annual wave energy from the east associated with larger climatic cycles.

The Inlet Modeling System (IMS) developed by the U.S. Army Corps of Engineers (USACE) Coastal Inlets Research Program (CIRP) was applied to Shinnecock Inlet with the main objective of simulating transport pathways and evolution of morphology arising from the increased supply of littoral sediments that took place from August 1997 to May 1998. The modeling system was used to gain insight into the complex hydrodynamics and transport processes that control the distribution of sediment around the inlet and ebb shoal. Acknowledging that the IMS was a relatively new tool available to coastal engineers and scientists, the numerical calculations were supplemented with more standard coastal engineering analyses of the inlet. The IMS replicated theoretical and observed trends in inlet behavior including deepening and deflection of the channel, increasing volume of the ebb shoal complex, and sediment bypassing across the inlet. Additionally the IMS furthered understanding of the natural rotation of the ebb jet and its interaction with wave-driven longshore currents, processes that could not be discerned from traditional engineering techniques.

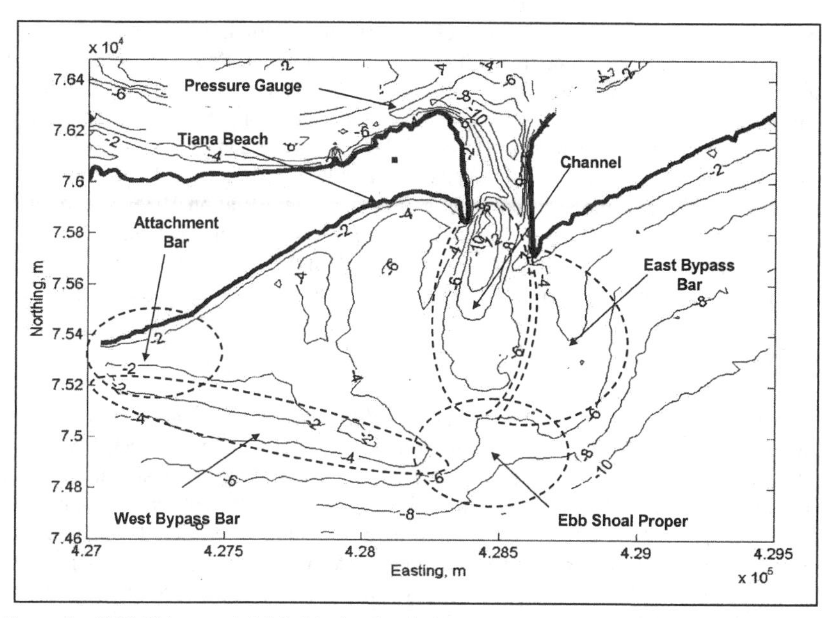

Figure 2. 1997 Shinnecock Inlet ebb shoal complex.

Morphology Change Analysis

Predictive relationships have been applied to Shinnecock Inlet to address the primary processes operating around the inlet and its general stage of development or evolution. Several investigators have identified a correlation between tidal prism of an equilibrium inlet and the cross sectional area of the throat (LeConte 1905; O'Brien 1931, 1969; Jarrett 1976). The empirical equation is,

$$A_c = CP^n \tag{1}$$

where A_c is the channel cross-sectional area, P is the tidal prism, and C and n are empirical coefficients. Values of C and n depend on the number of jetties present at the inlet and the general wave climate. For Shinnecock Inlet the coefficients C and n are 5.77 x 10^{-5} and 0.95 respectively. Application of the Jarrett (1976) equation to Shinnecock Inlet using the 1998 estimate of tidal prism indicates the minimum cross-sectional area in the throat of the inlet should approach 2,177 m^2 (Buonaiuto 2003b). A previous estimate gave a minimum inlet cross-sectional area of 2,694 m^2 (Militello and Kraus 2001). These empirically-estimated minimum cross-sectional areas are larger than those measured in 1994 (1,551 m^2) and 1998 (1,566 m^2) (Morang 1999). Between 1994 and 1998 the cross-sectional area of the channel increased, indicating that the inlet was scouring toward the predicted equilibrium flow area.

The dominant direction of longshore transport along the south shore of Long Island is from east to west (Taney 1961; Panuzio 1968). Seasonal winds from the west and southwest in summer can induce short-term transport reversals along the eastern half of Long Island. Littoral sediments on eastern Long Island have been assumed to be supplied either from bluff erosion at Montauk Point, or from offshore sources (Taney 1961; Kana 1995; Williams 1976; Schwab et al. 1997). Sediment budget analyses indicate that the inlet removes sediment from the littoral system at a rate of 70,000 to 115,000 m^3/year (USACE 1958, 1988; Kana 1995; Morang 1999; Rosati et al. 1999). Empirical relations are available to estimate the large volume of sediment contained in the ebb and flood shoals (Dean and Walton 1975; Walton and Adams 1976; Carr de Betts 1999). Properties of inlets along the Atlantic, Gulf, and Pacific coasts of the United States were examined, and the combined control by the tide and waves were related to the ebb shoal volume as (Walton and Adams 1976),

$$V_s = CP^n \tag{2}$$

where V_s is ebb shoal volume, and C and n are empirical coefficients that depend on wave energy. For Shinnecock Inlet, a moderately exposed inlet, the coefficients C and n are 10.5 x 10^{-5} and 1.23 respectively. Using the most recent estimate of tidal prism, application of Eq. 2 to the inlet gives an equilibrium ebb shoal volume of approximately 11,200,000 m^3. This empirically-estimated volume is larger than the amount of material contained in the ebb shoal in 1996 (5,803,000 m^3), 1997 (5,988,000 m^3), and 1998 (6,463,000 m^3) (Morang 1999). Increase in volume between 1996 and 1998 indicates the ebb shoal complex was trapping sediment prior to the dredging of the channel and deposition basin in 1998. By 1998, the ebb shoal had trapped 60 % of the theoretical equilibrium volume. Using the estimated rates of trapping, the ebb shoal is expected to attain equilibrium in approximately 42 to 70 years. A previous calculation of the theoretical ebb shoal volume determined equilibrium would be reached in approximately 75 years, although dredging of the channel and ebb shoal would prolong the evolution (Kraus 2001). As equilibrium is approached, it is expected that more persistent sediment transport pathways will be established, and the inlet will bypass sediment more effectively to the neighboring beaches.

Inlet morphology responds to variations in gross longshore sediment transport rates, which have been estimated to range from 230,000 m^3/year to 305,000 m^3/year in the vicinity of Shinnecock Inlet (Williams et al. 1998). Easterly transport predominates during the months of May through August, and during December. West-directed transport dominates the remainder of the year with the largest overall net transport rates (for either direction) occurring between the months of September and November. Inlet sediment bypassing controls the rate, location, and composition (Bruun and Gerritsen 1959, 1960; Liu and Hou 1997) of sand nourished to downdrift beaches. Sediment bypassing across inlets is achieved through the transport modes of 1) wave-induced transport along the periphery of the ebb shoal (terminal lobe), 2) transport through channels by tidal currents, and 3) migration of tidal channels and sand bars (Bruun and Gerritsen 1959, 1960). The dominant mode of sediment bypassing can be determined from the ratio,

$$r = P/M \tag{3}$$

which expresses the balance between the gross longshore sediment transport rate M brought to the inlet annually and the tidal prism. Inlets with high ratios ($r > 150$) bypass sand predominantly through tidal flushing, whereas inlets with low values ($r < 50$) bypass sand predominantly through channel migration and bar complex formation. Inlets with ratios ranging between 50 and 150 generally develop large ebb shoals and bypass sediment across the throat of the inlet (tidal flushing) and along the periphery of the ebb shoal (sand bridge transport). Application of Eq. 3 to Shinnecock Inlet using the most recent estimate of the gross longshore sediment transport rate (Williams et al. 1998) and tidal prism (Militello and Kraus 2001) resulted in an r parameter ranging between 107 and 143, which suggests natural sediment bypassing at Shinnecock Inlet should occur through a combination of wave-induced transport and tidal bypassing.

Principal Component Analysis (PCA) was applied to the bathymetry data to identify spatial patterns of morphology change (Buonaiuto 2003b). Bathymetric data used in the PCA were collected by the USACE's water-penetrating LIDAR system called SHOALS. The analysis indicated the locations of the bypassing bars, ebb shoal, and attachment point, were largely controlled by the position of the navigation channel. From 1994 through 2000, Shinnecock Inlet experienced a period of growth accreting almost 2,000,000 m^3 of sediment along the ebb shoal proper, bypass bar and attachment point (Fig. 3). Between 1994 and 1997 the inlet accreted sediment along the periphery of the ebb shoal, which included the eastern, up-drift flank of the main channel (eastern ebb shoal lobe), the ebb shoal saddle, the western, down-drift portion of the ebb shoal, the bypass bar and the attachment point (Fig. 4). Sediment was also deposited as a linear shoal extending from the tip of the western jetty along the west side of the channel. During this period the main channel appeared to scour between 1 and 2 m.

From 1997 to 1998, the Shinnecock Inlet system showed a net accretion of 530,000 m^3. Sediment was deposited along the periphery of the western lobe of the ebb shoal and along the bypass bar. As the channel was naturally redirected toward the west, the shoal that extended from the tip of the western jetty began to erode. Additionally the landward face of the bypass bar eroded as the deflected ebb jet drove sediment reserved in the bypass bar further offshore (Fig. 5). Dredging of the channel and deposition basin in 1998 re-oriented the entrance toward the south, promoting growth and seaward advancement of the ebb shoal proper.

As the channel naturally migrated, the westward movement of sediment deposited on the eastern flank was accelerated by increased wave driven-currents from the east. Analysis of ten years of directional wave data collected from National Data Buoy Center (NDBC) station 44025 located 33 nautical miles off the coast of Long Island (Buonaiuto 2003a) indicated an increase in wave activity from the east quadrant between the 1997 and 1998 surveys associated with the ENSO (El Nino southern Oscillation) cycle (Fig. 6).

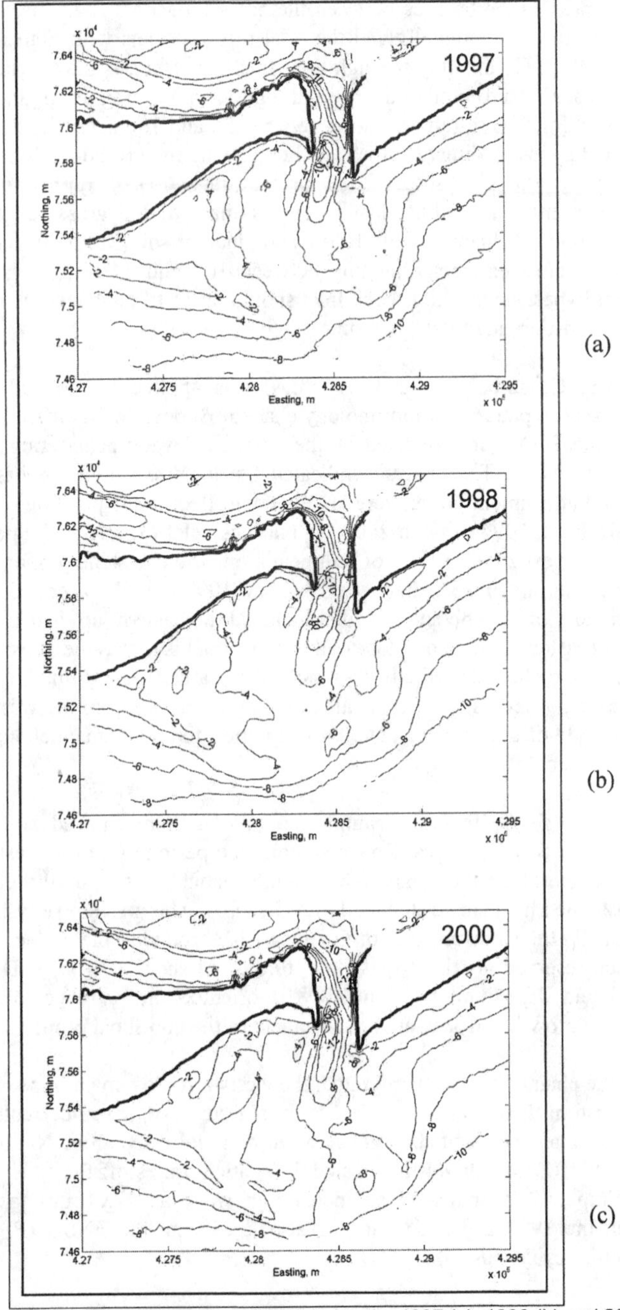

Figure 3. Shinnecock Inlet morphology captured during the 1997 (a), 1998 (b) and 2000 (c) SHOALS surveys.

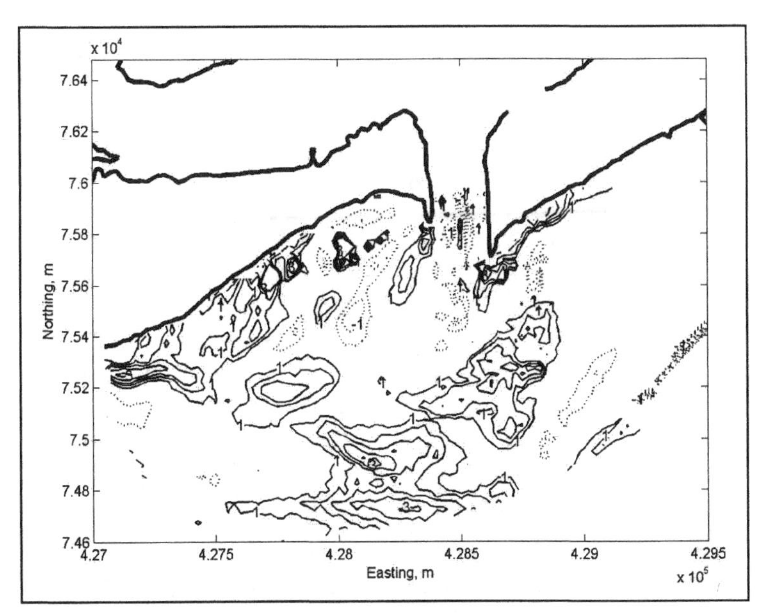

Figure 4. Depth changes between the 1994 and 1997 SHOALS surveys. Contour labels are in meters. Solid contours indicate areas of accretion and dotted contours mark zones of erosion.

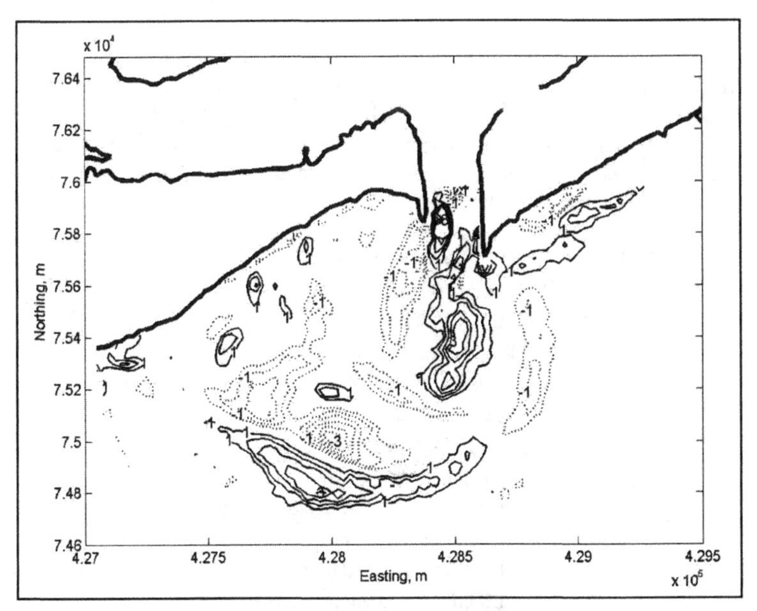

Figure 5. Depth changes between the 1997 and 1998 SHOALS surveys. Contour labels are in meters. See Fig. 4 for description.

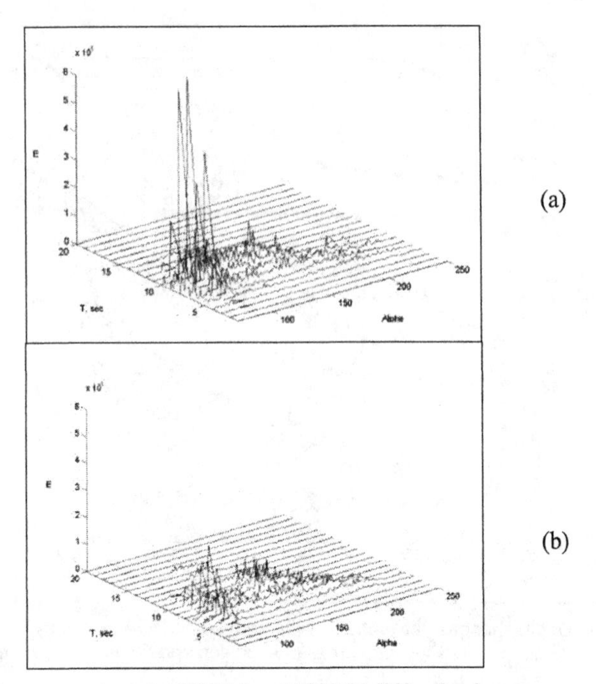

Figure 6. Incident wave spectrum for 1997 (a) and 1998 (b) JMA calendar years.

Prior to the 2000 survey the channel and deposition basin were dredged which effectively removed 336,404 m^3 of material (Table 1). During this period an additional 10,277 m^3 of sediment naturally eroded from the ebb shoal. Although there was deflation of the bypass bar and an overall net loss of material from the system, the periphery of the ebb shoal prograded seaward as a result of the re-orientation and deepening of the channel (Fig. 7).

Table 1: Recent Event and Activity Chronology

6/21/94	SHOALS LIDAR bathymetric survey
5/23/96	SHOALS LIDAR bathymetric survey
2/97 – 3/97	Shinnecock Flood Shoal Dredging 250,000 yd^3 removed from eastern flood shoal channel
8/13/97	SHOALS LIDAR bathymetric survey
5/28/98	SHOALS LIDAR bathymetric survey
6/27/98 – 7/11/98	Shinnecock Inlet Dredging 35,000 yd^3 removed from entrance channel and deposition basin
10/13/98 – 10/25/98	Shinnecock Inlet Dredging 405,000 yd^3 removed from entrance channel and deposition basin
7/3/00	SHOALS LIDAR bathymetric survey

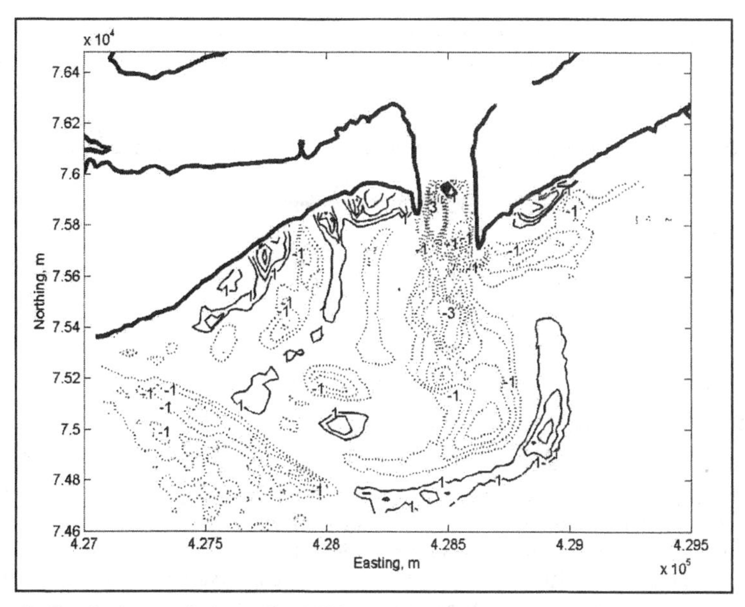

Figure 7. Depth changes between the 1998 and 2000 SHOALS surveys. Contour labels are in meters. See Fig. 4 for description.

Modeling Approach

The IMS was applied to simulate the natural migration of the inlet channel between the 1997 and 1998 surveys. The IMS consists of a suite of models implemented within the Surfacewater Modeling System (Zundel 2000) and designed to compute tidal hydrodynamics, wave transformation, sediment transport and morphology change. Models can be coupled for specific application requirements. For this investigation, water-surface elevations and current velocities were computed by M2D, a localized two-dimensional, depth-integrated, hydrodynamic model developed for shallow-water regions (Militello et al. 2004). M2D solves finite-difference approximations of the nonlinear equations of mass and momentum conservation on a variably-spaced rectilinear grid. Advection, mixing, and quadratic friction are represented in the model. Sediment transport and morphology change can be computed by M2D as a user-specified option.

Wave transformation was computed by STWave (Smith et al. 2000), a steady-state finite-difference spectral wave model based on the wave action balance equation. The model calculates depth- and current-induced wave refraction and shoaling, depth- and steepness-induced wave breaking, wave growth due to local wind stress, wave-wave interactions, and white capping. Calculations are conducted on a constant-spaced rectilinear grid.

M2D and STWave were coupled for calculation of wave-driven currents. Radiation stress gradients calculated by STWave were mapped onto the M2D grid to supply wave forcing. Total water depths calculated by M2D were mapped to the

STWave grid so that the temporal modulation of wave properties with changes in water level was represented. Coupling between M2D and STWave was set at 3-hr intervals.

Sediment transport and morphology change were calculated within M2D. Combined wave- and tide-driven velocities were entered into computations of local shear stress and sediment transport rates were calculated by the Watanabe (1987) total load formulation,

$$q_t = A\left[\frac{(\tau_b - \tau_{cr})V}{\rho_w g}\right] \tag{4}$$

where q_t is the total load (both suspended and bed), τ_b is the shear stress at the bed, τ_{cr} is the shear stress at incipient sediment motion, V is the depth averaged current velocity as provided by the coupled model, ρ_w is the density of water, g is the gravitational acceleration constant (9.81 m/s^2), and A is an adjustable coefficient (0.5 – 2 for regular to irregular waves). For this investigation the sediment was considered to be homogeneous medium sand. Transport rates for each M2D cell were computed every 100 seconds. Through mass conservation, the bathymetry was updated at 6-hr intervals giving morphology change of the inlet, shoals, and nearshore region. At each bathymetric update, new depths were provided to the hydrodynamic component of the model for calculation of velocity and water level response to changes in morphology.

Grid Development

Bathymetric data for the M2D and STWave grids were obtained from several sources including NOAA, GEODAS, Marine Science Research Center State University of New York at Stony Brook, and USACE. Bathymetry covering the region surrounding Shinnecock Inlet (ebb shoal, throat, and flood shoal) was obtained from the August 13, 1997 and the May 28, 1998 SHOALS surveys. Model grids cover all of Shinnecock Bay and extend up and downdrift of the inlet to regions unaffected by the ebb-tidal shoal (Fig. 8). Seaward boundaries extend offshore to 30 m water depth, beyond the zone where wave shoaling takes place and beyond the closure depth and farthest seaward extent of the ebb jet.

The STWave grid was specified to have 20-m spacing over its domain. This spacing provides several cells per wavelength and adequately resolves the radiation-stress gradients within the shoaling, breaking and surf zones, leading to more accurate calculation of water levels and currents there. The M2D grid was developed such that the cell spacing was 100 m in the deeper water offshore, and transitions to 10 m near the coastline, inside the wave shoal zone, around the flood and ebb shoals, and in the inlet throat (Fig. 9). Wave transformation in these regions takes place over relatively short distances and greater resolution was specified in areas so that details of the circulation and sediment transport would be reproduced. The two grids (STWave and M2D) covered identical geographical regions (Fig. 8).

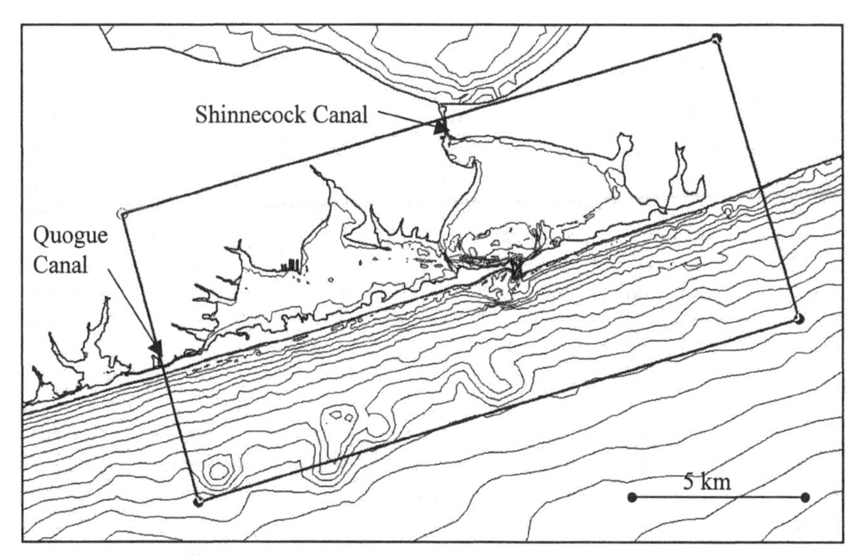

Figure 8. IMS Shinnecock Inlet domain.

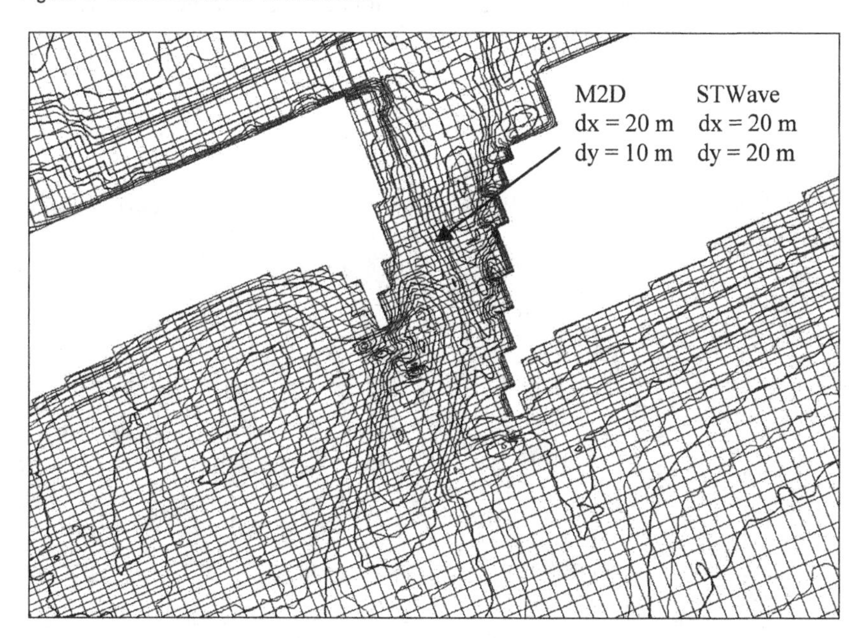

Figure 9. IMS-M2D grid of Shinnecock Inlet. Finer resolution was implemented throughout the surfzone and throat of the inlet.

Model Verification

In order to verify the M2D model, calculated water levels were compared to measurements collected during 1998 as part of a field monitoring system sponsored by the USACE New York District, the CIRP, and the New York State Department of Environmental Conservation. For the verification, M2D was forced with water-surface elevation values along the seaward boundary and at the Shinnecock and Quogue canals. Water levels were extracted from the regional ADCIRC model of Long Island (Militello and Kraus 2003) from November 4, 1998 through November 16, 1998. Wave forcing was not included in the verification, and calculation of sediment transport rates and morphology change was not invoked. Field data used for comparison were collected by a pressure gauge mounted along the bay side of Shinnecock Inlet near the western barrier. Results indicate the tidal elevation was well predicted (Fig. 10), and the calculated tidal range was within 12 percent of the observations. This error was comparable to that of the ADCIRC calculations for the same location (Militello and Kraus 2001a), indicating that the primary source of error was introduced by the boundary conditions obtained from the regional model. A meteorological event occurred during November 10 through November 14. The measured water level shows response to this atmospheric forcing. However, calculations for model verification were forced with tide alone so they do not contain deviations owing to wind events. Presently wind forcing can only be implemented as a constant stress and would not be applicable in a time varying simulation.

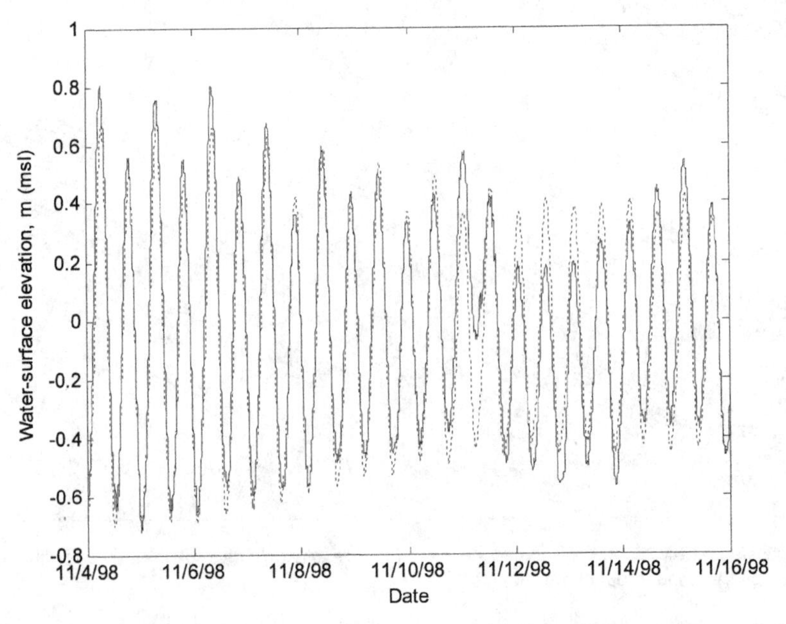

Figure 10. M2D model verification bay side of Shinnecock Inlet. Observed data indicated by solid black line, calculated water level shown as dotted line.

Model Forcing

Calculation of morphology change was conducted during a time in which controlling factors ranged from primarily tidal to storm dominated. The simulation time interval was specified to coincide with the natural deflection and migration of the main navigation channel through the ebb shoal. Data collected during two SHOALS surveys conducted on August 13, 1997 and May 28, 1998 provide bathymetric change information from which the calculations can be verified. M2D was forced by both water-surface elevation along the seaward boundary and wave-driven radiation stress gradients within the interior of the domain. Boundary conditions for the local model (M2D) were extracted from the regional ADCIRC model of Long Island and mapped to the seaward boundary, Quogue Canal boundary, and Shinnecock Canal boundary. This method of boundary specification preserves temporal and spatial variation of tidal properties at the boundaries. Thus, interior response to spatial amplitude and phase variation is included in the calculations. In the nearshore region, the coastal tidal current is represented which enhances the natural westward migration of the ebb jet.

Incident wave field characteristics (height, period, and direction) were obtained from NDBC Station 44025 and used to construct spectral energy input files for STWave. The spectrums were applied along the offshore boundary and propagated to the shoreline by STWave. During the simulation wave conditions were updated at 3-hr intervals and radiations stress gradients were calculated across the STWave model domain.

Simulation Results

Results presented in this section represent a 15-week-long time interval that started on August 13, 1997 and ended on November 30 1997. To reproduce morphology change the model simulation was condensed to a 4-week-long sequence during which incident wave fields with wave heights greater than 1 m and periods larger than 7 sec were modeled. These wave fields were chosen to represent events that forced peak transport of littoral material. By including August and September, when wave forcing was relatively weak, tidal control on morphology change could be evaluated. Sediment transport rates were computed each hour and the inlet morphology was updated four times each day based on 6-hr averages of the sediment transport rates. Calculated depth was output every 24 hr, and bathymetric c hange (Figs. 11-13) was plotted at the end of each month.

During the months of August and September, Shinnecock Inlet experienced relatively low wave energy. Therefore, during this time morphology change responded to tides and relatively weak waves. Wave heights rarely exceeded 1.5 m and periods ranged between 5 and 9 sec. The longest-period swell (11 to 14 sec) to reach the inlet between the 1997 and 1998 surveys occurred in mid-September. Although the swells approached from the south and lasted four days, wave heights did not exceed 1.75 m. The absence of long-period swells in the wave record during the simulation owes to a reduction in hurricane activity during El Nino years

(October 1, 1997 through September 30, 1998) (Bove et al., 1998). Morphology change calculations for the end of September indicate that active transport regions were primarily located in the throat of the inlet, and in the direct path of the ebb jet (Fig. 11). The channel extending from the jetty tips and across the ebb shoal eroded along its western flank, reproducing the measured trend of westward channel migration. Deposition took place updrift (east) of the channel. Small bar complexes offshore of Tiana Beach were translated along the western barrier within the wave-shadowed region of the ebb shoal.

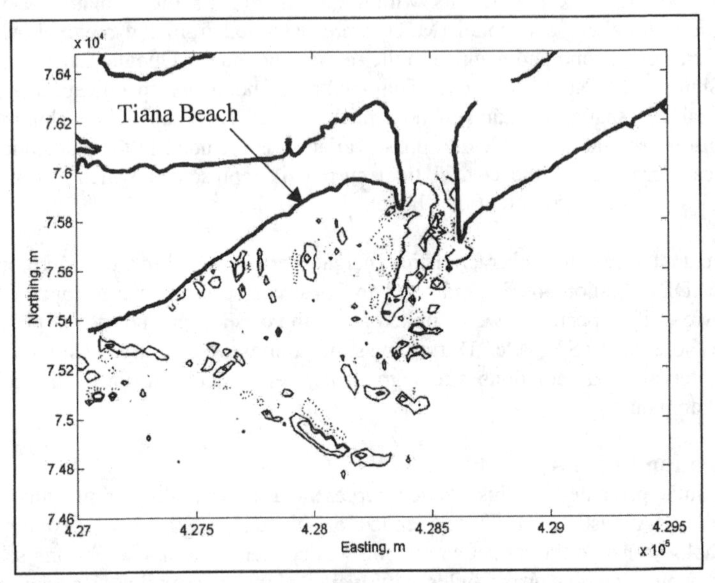

Figure 11. Calculated morphology change at end of September, 1997. Contours were constructed at the 0.05 and 0.1 m intervals. Solid contours indicate zones of deposition and dotted lines indicate regions of erosion.

Wave activity began to increase during the month of October from both the southwest and southeast quadrants. Four significant events occurred in which wave heights exceeded 2 and 3 m, and periods ranged from 9 to 12 sec. Two of the events were from the southwest and resulted in longshore transport reversals. Active regions of morphology change were located along the coast of the east and west barrier, and in the inlet throat and channel (Fig. 11). Sediment was deposited along the bypass bar, attachment point and at the seaward extent of the ebb jet. Patterns of morphology change were similar to those calculated for August and September, but with larger erosion and accretion signals.

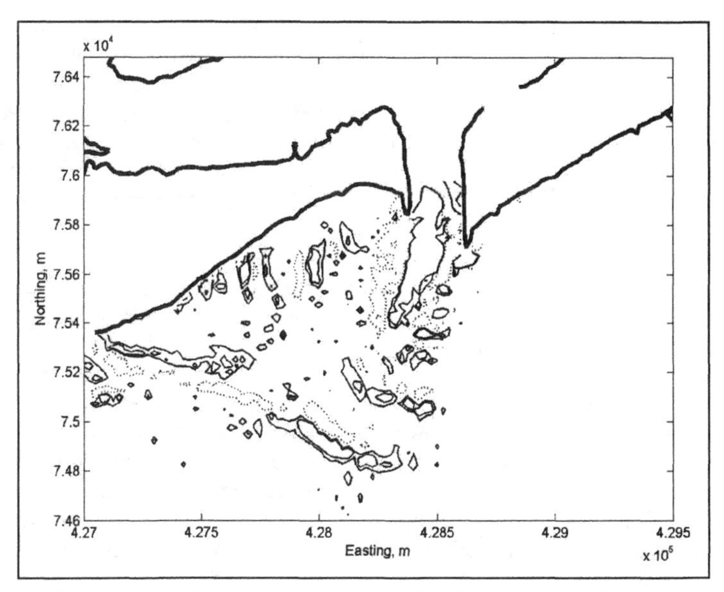

Figure 12. Calculated morphology change at end of October, 1997. See Fig.11 for description.

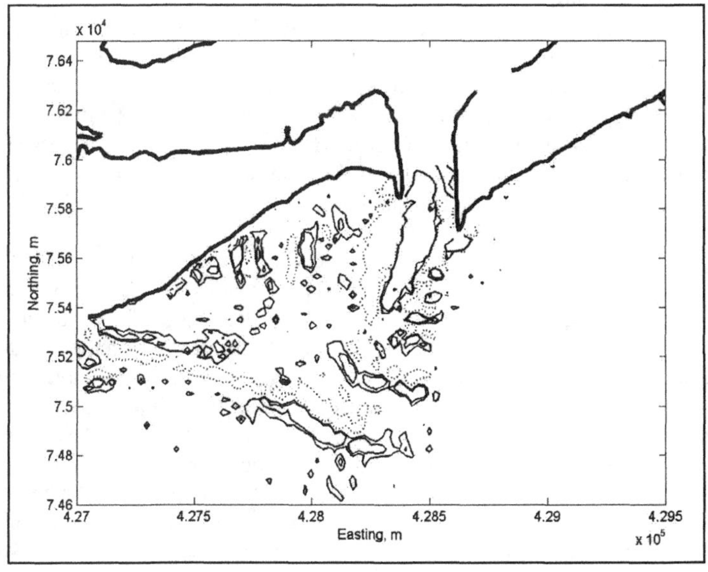

Figure 13. Calculated morphology change at end of November, 1997. See Fig.11 for description.

Shinnecock Inlet experienced five intense wave events during the month of November. For three of these events, wave heights exceeded 4 m and periods ranged between 10 and 11 sec from the southeast quadrant. The two smaller events (wave height of 2.5 to 3 m; period 7 to 9 sec) approached the inlet from the southwest. Bathymetry changes at the end of November indicated increased deposition along the bypass bar and seaward extent of the ebb jet, and continued westward deflection of the main channel (Fig. 12).

Discussion

Wave conditions for the simulation varied from low energy in the months of August and September to moderate and eventually high-energy environments for October, 1997 and November, 1997 respectively. Morphologic changes computed for the end of September indicated the regional influence of tidal transport was localized around the throat and mouth of the inlet and the pathway of the ebb jet. Westward migration of the channel thalweg was shown to occur during weak wave forcing. Two tidal processes are responsible for this migration. The periodic formation of the ebb jet scours the channel thalweg. In addition, the coastal tidal current advects the jet toward the west, which translates the erosional stresses of the jet westward. Together, these processes scour the western portion of the channel, which realigns of the thalweg.

Scouring of the navigation channel in the throat of the inlet throughout the simulation is consistent with the analytical relationship of Jarrett, 1976 (Eq. 1), and previous observations of the inlet (Militello and Kraus 2001b). Additionally the ebb shoal continued to evolve toward the equilibrium volume (Eq. 2) by accreting 14,370 m^3 of sediment over the fifteen-week period. Calculated modes of natural sediment bypassing were also in agreement with the theoretical stability ratio of Bruun and Geritsen (1959). Observed transport pathways were associated with channel flushing through the throat of the inlet and wave-driven transport along the perimeter of the ebb shoal and bypass bar.

Previous investigations have attributed the westward deflection of the main channel of Shinnecock Inlet attributes to: 1) continuous tidal deflection of the ebb jet (Militello and Kraus 2001a), and 2) increased deflection of the ebb jet by wave-driven longshore currents (Buonaiuto, 2003a). Other controls on the jet migration are the bay channel geometry, flood shoal, and offset jetties. Simulations conducted in this study indicate the continuous tidal deflection of the ebb jet is a dominant process controlling the migration of the channel, and it can be enhanced or accelerated by increased wave activity and littoral sediment supply. Further analysis is required to discern to what degree the observed changes in the morphology of Shinnecock Inlet between the 1997 and 1998 surveys were in response to an increase in sediment delivery. Sediment supplied to the inlet was most likely derived from wave driven erosion and transport of the beaches and surfzone along the eastern and western barriers.

Wind forcing was not included in the simulation, so wind-driven setup and setdown in the bay and near the coast were not represented. Although this forcing of water level would potentially alter tidal flushing through the inlet, breaking of waves and transport pathways of sediment on the ebb-shoal, its control on the hydrodynamics and sediment dynamics is expected to be secondary to waves and tides.

Additionally the sediment transport module was designed for a uniform grain size. In general the ebb and flood shoals, bypass bar and beaches are comprised of fine to medium sand (0.20 – 0.35 mm) and an assumed uniform grain size would produce reasonable results. However, the main navigation channel through the throat of the inlet could contain more coarse material including gravel and shell hash. This material would be more resistant to erosion and would not have scoured as rapidly as predicted by the present transport model.

There are certain weaknesses inherent in the condensed forcing approach of the modeling system, as well as limitations associated with the sediment transport module. Calculations of morphology and water levels are expected to improve by modeling the complete time period between surveys (8/13/97 through 5/28/98), and including time-varying wind stress in the momentum equations. Implementation of non-uniform grain sizes in the IMS is a planned enhancement, and once available simulations at Shinnecock Inlet will be conducted to take advantage of this capability.

Conclusions

Evolution of morphology at Shinnecock Inlet between August 13, 1997 and May 28, 1998 was found to be associated with the westward migration of the main channel. A coupled circulation, wave, and sediment transport modeling system was applied to calculate morphology change and to determine the primary controls on channel migration. General trends of natural sediment bypassing, growth of the ebb shoal, scour in the throat, and channel migration were reproduced by the IMS-M2D version forced by accurate tides and waves.

In summary, major measured morphologic change observed in at the inlet and ebb shoal was simulated successfully in the IMS. These changes were associated with the westward deflection of the channel seaward of the inlet, which include erosion of the western flank along the channel, shallowing of the eastern flank of the ebb shoal, and deposition along the seaward edge of the ebb shoal where the ebb jet terminates.

Channel deflection is controlled primarily by tidal processes consisting of periodic ebb jet formation and advection of the jet by the coastal tidal current. Waves originating from southeast enhance channel migration.

Acknowledgements

This research was sponsored by the Inlet Modeling System Work Unit of the Coastal Inlets Research Program, Coastal and Hydraulics Laboratory, U.S. Army

Engineer Research and Development Center. Permission was granted by Headquarters, U.S. Army Corps of Engineers, to publish this information.

References

Bove, M. C., Elsner, J. B., Landsea, C. W., Niu, X., O'Brien, J. J (1998). Effect of El Nino on U.S. landfalling hurricanes, *Bulletin of the American Meteorological Society*, v.79, 2477-2482.

Bruun, P., Gerritsen, F. (1959). Natural by-passing of sand at coastal inlets: *Journal of the Waterways and Harbors Division*, December, 75-107.

Bruun, P., Gerritsen, F. (1960). Stability of coastal inlets, Amsterdam, North Holland Publishing Co., (Elsevier), 123 p.

Buonaiuto, F.S. (2003a). Morphological evolution of Shinnecock Inlet, NY. Ph.d. dissertation, State University of New York at Stony Brook, Stony Brook, NY.

Buonaiuto, F.S. (2003b). Principal component analysis and morphology change at Shinnecock Inlet, NY, *Proceedings Coastal Sediments '03*.

Carr de Betts, E.E (1999). An examination of flood deltas at Florida's tidal inlets, University of Florida, Gainesville, FL thesis, Report No.UFL/COEL-99/015.

Dean, R.G., Walton T.L. (1975). Sediment transport processes in the vicinity of inlets with special reference to sand trapping, Estuarine Research, Vol. 2, L.E. Cronin ed., Academic Press, New York, 129-149.

Jarrett, J.T. (1976). Tidal prism – inlet area relationships, GITI Report No. 3, Coastal Engineering Research Center, U.S. Army Corps of Engineers, Fort Belvoir, Virginia, 32 p.

Kana, T.W. (1995). A mesoscale sediment budget for Long Island, New York. *Marine Geology*. v.126, 87-110.

Kraus, N.C. (2001). On equilibrium properties in predictive modeling of coastal morphology change, *Proceedings Coastal Dynamics '01*, ASCE, 1-15.

LeConte, L.J. (1905). Discussions of notes on the improvement of rivers and harbor outlets in the United States, Transactions of American Society of Civil Engineers, 55, 306-308.

Liu, J.T., and Hou, L. (1997). Sediment trapping and bypassing characteristics of a stable tidal inlet at Kaohsiung Harbor, Taiwan, *Marine Geology,* v.140, 367-390.

Militello , A., Kraus, N.C. (2001a). Shinnecock Inlet, New York, Site Investigation Report 4: Evaluation of Flood and Ebb Shoal Sediment source Alternatives for the West of Shinnecock Interim Project, New York, Coastal and Hydraulics Laboratory, Technical Report CHL-98-32.

Militello, A., and Kraus, N. C. (2001b). Re-Alignment of an Inlet Entrance Channel by Ebb-Tidal Eddies. *Proceedings Coastal Dynamics '01*, ASCE Press, New York, 423-432.

Militello, A., and Kraus, N. C. (2004). Regional Circulation Model for Coast of Long Island, New York. *Eighth International Estuarine and Coastal Modeling Conference* 2003. (This volume.)

Militello, A., Reed, C. W., and Zundel, A. K. (2004). *Two-Dimensional Circulation Model M2D: Version 2.0, Report 1, Technical Documentation and User's Guide.* Coastal Inlets Research Program Contractor's Report ERDC-CHL-CR-__-_, U.S. Army Engineer Research and Development Center, Vicksburg, MS (in press).

Morang, A. (1999). Shinnecock Inlet, New York, Site Investigation, Report 1, Morphology and Historical Behavior. Coastal Inlets Research Program, USACE Waterways Experiment Station, Coastal and Hydraulics Laboratory. TR-CHL-98-32.

O'Brien, M.P. (1931). Estuary tidal prisms related to entrance areas, Civil Engineering, v. 1, n. 8, 738-739.

O'Brien, M.P. (1969). Equilibrium flow areas of tidal inlets on sandy coasts, *Journal of the Waterways and Harbors Division*, ASCE, v. 95, n. WW1, 43-52.

Panuzio, F. L. (1968). The Atlantic Coast of Long Island, *Proceedings 11th Coastal Engineering Conference*, ASCE, 1,222-1,241.

Research Planning Institute (1983). Sediment Budget analysis, Fire Island Inlet to Montauk Point, Long Island, New York – refomulation study, Report prepared for U.S. Army Engineer District, New York, Research Planning Institute, Inc. , Columbia, SC.

Rosati, J.D., Gravens, M.B., Gray Smith, W. (1999). Regional sediment budget for Fire Island to Montauk Point, New York, USA, *Proceeding Coastal Sediments '99*, 802-817.

Schwab, W.C., Thieler, E.R., Foster, D.S., Denny, J.F., Swift, B.A., Danforth, W.W., Allen, J.R., Lotto, L.L., Allison, M.A., Gayes, P.T., Donovan-Ealy, P. (1997). Mapping the inner shelf off southern Long Island, N.Y.; Implications for coastal evolution and behavior. EOS Transactions, American Geophysical Union. v. 78.

Smith, J. M., Sherlock, A. R., Resio, D. T. (2000). STWAVE: Steady-State Spectral Wave Model User's Manual for STWAVE Version 3.0. USACE, ERDC/CHL SR-01-1.

Taney, N.E. (1961). Geomorphology of the South Shore of Long Island, New York. Beach Erosion Board, T.M. n. 128.

U.S. Army Corps of Engineers (1958). Moriches and Shinnecock Inlets, Long Island, New York, survey report, U.S. Army Engineer District, New York.

U.S. Army Corps of Engineers (1988). General Design Memorandum Shinnecock Inlet Project, Long Island, New York, Reformulaiton Study and Environmental Impact Statement, U.S. Army Engineer District, New York.

Walton Jr., T.L., Adams, W.D. (1976). Capacity of inlet outer bars to store sand, *Proceedings 15th Coastal Engineering Conference*, UACE, 1919-1937.

Watanabe, A. 1987. 3-dimensional Numerical Model of Beach Evolution. *Proceedings Coastal Sediments '87*, ASCE, 802-817.

Williams, S.J. (1976). Geomorphology, shallow bottom structure and sediments of the Atlantic inner continental shelf off Long Island. NY Tech Paper, 76-2. USACE, Ft. Belvoir, Virginia, 123 p.

Williams, G.L., Morang, A., Lillycrop, L. (1998). Shinnecock Inlet, New York, Site Investigation, Report 2: Evaluation of Sand Bypass Options, US Army Corps of Engineers, Waterways Experiment Station, Technical Report CHL-98-32.

Zundel, A.K. (2000). Surface-water modeling system reference manual, Brigham Young University, Environmental Modeling Research Laboratory, Provo, UT.

The Prediction of Dunes and Their Related Roughness in Estuarine Morphological Models

Andreas Malcherek [1] and Bert Putzar [2]

Abstract

The prediction of dune characteristics in coastal waterways is of great importance for at least two reasons. On the one hand the crests impose a lot of dredging problems and on the other hand dunes contribute to the hydraulic resistance of the system and to the damping of the tidal wave. Therefore it is a major research aim to simulate the development of dunes in morphological numerical models. In this paper an algorithm is developed to simulate the evolution of dune heights and lengths. In the first step the dune parameters are calculated using empirical relations (i.e. here van Rijn) for every hydrodynamical time step. The values obtained in that way are not the actual dune heights and lengths because they would grow during ebb and flood tide and they would disappear during slack water. Instead of this they are regarded as asymptotical values reached theoretically after a certain time. The actual dune height and length are calculated applying a relaxation method using the empirical values as asymptotical values. This algorithm gives good results for the location of dunes in the Weser estuary at the North sea coast of Germany. The predictions for the dune heights and lenghts are reasonable.

Introduction

The prediction of dunes in estuarine simulation systems is important for several reasons. A practical one is the maintenance of the navigation channels depth. Because dunes are generated more likely in deep waters they can be found especially in the navigation channels. Here their crests are the locations where dredging problems occur although the spatially averaged water depth fulfills the navigation channel standards.

Dunes are also important because of their effect on the flow itself. They increase the bed roughness significantly. This is usually be taken into account by dividing the effective bed roughness according to Nikuradse [13] into grain and form roughness whereby the latter one is connected to the presence of ripples and dunes. While the grain roughness lies in the range of a millimeter, the dune roughness scales with the dune

[1]Senior Research Engineer, Federal Waterways Engineering and Research Institute (BAW), Coastal Department, Wedeler Landstr. 157, 22559 Hamburg, Germany

[2]Diploma Student, Technical University Hamburg-Harburg, 21071 Hamburg, Germany

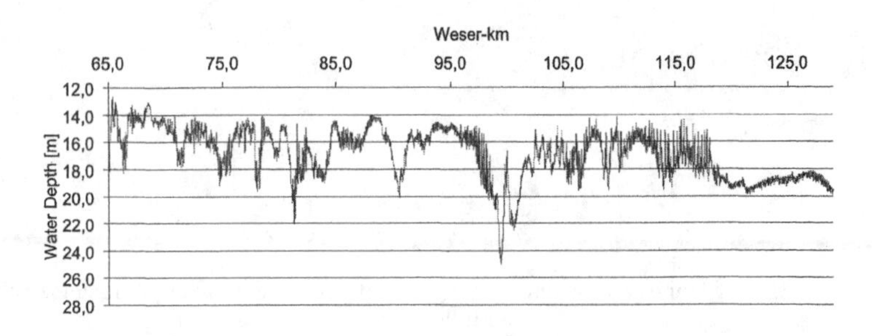

Figure 1: Bathymetry in the navigation channel.

height i.e. meters. Therefore in the presence of dunes the bed roughness is dominated by them.

The bed roughness affects the kinetic energy dissipation and the tidal wave height can decrease significantly over dune fields. Therefore estuarine hydrodynamic models can not be calibrated without knowing the locations of dune fields. Obviously direct measurements can be used for the calibration of the model. But in case of bathymetric changes due to construction measurements the question has to be answered whether the characteristics of the dune systems and their spatial extension will change, too. Therefore a method to predict the occurrence of dunes in numerical simulation models is needed.

Theories on the formation of dunes in steady flows i.e. flumes or rivers have a long tradition starting with EXNER [4] in 1925. He derived a formulation for the dune celerity but did not succeed to explain the generation of dunes by stability analysis. KENNEDY (1963 [8], 1969 [9]) used the linear Airy wave theory to describe the flow field over dunes. In his approach the free surface has a wave profile. Therefore good results are obtained for high Froude numbers as in the case of anti-dunes but bottom structures at very low Froude numbers can not be explained. This is also the case for the theories of ENGELUND and FREDSOE (1982) [3], who used the stream line function theory to describe the flow over dunes.

On the other hand a lot of empirical relationships have been developed to predict the dunes described by their height and length. Some of these dune height formulas are more or less only relations between water depth and dune height, e.g. (ALLEN, 1968 [1]), (YALIN, 1964 [16]), i.e. they do not contain information about the flow velocity or the actual bottom shear stress. Although such formulas correspond quite well to natural conditions they can not be used for the prediction of dunes in numerical simulation models because they would predict the same kind of dunes in deep lakes and rivers.

Therefore only formulas depending on the bed shear stress or the water depth and flow velocity can be used to predict the dune characteristics under natural water bodies.

Using such dune height and length formulas in an estuarine simulation model, they would perhaps do a good job at high flood or ebb velocities. But at slack water the dunes would disappear, because they depend simultaneously on the hydrodynamic conditions. Therefore in this paper a relaxation algorithm is presented which preserves the dune characteristics at slack waters.

Dune Systems in the Outer Weser Estuary

The Weser estuary disembogues into the North Sea at its southern coast. Its mouth is located at Bremerhaven (km 63 - 73). A satellite view of its outer parts indicating the shipping channel kilometers is shown in figure 2. In the center of the lower part of the picture the Jade bay can be seen. It is connected to the North Sea by a gorge, named the inner Jade followed by a wide natural inlet channel, the outer Jade. Figure 1 shows the bathymetry on a cross section through the navigation channel of the Weser estuary. Coming from the sea the largest dune system can be found between km 113 - 118. A very pronounced dune system is located at the trailing edge between km 97 - 99. Other dune systems can be seen between km 105 - 107, km 85 - 87 and km 71 - 75. In figure 3 an enlarged view on the bottom structures between km 116.4 and 118.0 is shown. Large dunes with lengths of about 200 m can be seen. Smaller structures of about 20 m length are superimposed on them.

The dune parameters are determined by the following algorithm. First a short scale spatial floating averaging $\overline{z}_B$ is calculated using an averaging interval of 43 m. As it can be seen in figure 3 the resulting curve does not represent the dune heights any more but the dunes length information is conserved. Therefore a second long scale floating average using an averaging interval of 200 m gives the gross trend of the bottom co-ordinate $\overline{\overline{z}}_B$. In summary the original bottom co-ordinate is split by the two averaging procedures as follows

$$z_B = \overline{\overline{z}}_B + z'_B + z''_B \tag{1}$$

The dune length is determined by the zero down crossing method. Wherever the dune signal z''_B passes from positive to negative values a new dune is defined to begin. The dune height on the other hand is determined as the difference between the maximum and the minimum value of the original bottom co-ordinate z_B over the dunes' length.

Comparison with existing Dune Formulas

The resulting dune heights are compared in figure 4 to several dune formulas. The water depth, bed shear stress, Froude number and sedimentological input data for such formulas are obtained from a calibrated numerical model at maximum ebb tide.

YALIN [16] proposed 1964 the formula:

$$\frac{\Delta_d}{h} = \frac{1}{6}\left(1 - \frac{\tau_c}{\tau'_B}\right) \tag{2}$$

Here Δ_d is the dune height, h is the water depth, τ_c the critical bed shear stress for the initiation of sediment motion according to Shields and τ'_B the effective bed shear stress acting on the sediment grains. The formula maximizes the dune height to one sixth of the water depth. As a matter of fact the Weser dunes are everywhere smaller than this limiting value.

GILL (1971) [6] presented the theoretical formula:

$$\frac{\Delta_d}{h} = \frac{1}{2n\alpha}\left(1 - \frac{\tau_c}{\tau'_B}\right)\left(1 - Fr^2\right) \tag{3}$$

He assumes $n = 3$ for the power to which the bed load transport rate increases with the bed shear stress. This is actually too high, when the transport formulas of e.g. Meyer-Peter and Müller [12] or van Rijn [15] are regarded. α is a shape factor having the value 0.5 for triangular and $2/\pi$ for sinusoidal dunes. In the Weser the dunes seem

Figure 2: The outer Weser estuary at low tide, Source: DLR, IRS-1C LISS III, 12.08.1997, 10:46 UTC.

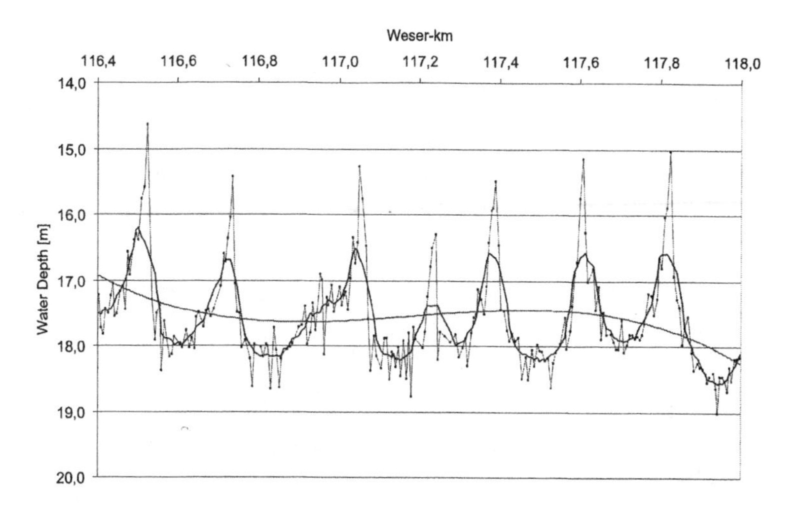

Figure 3: Determination of the dune parameters in the range of Weser-km 116,4 - 118 by double averaging. Blue line: original vertical co-ordinate of the bottom, red line: short scale averaging, green line: long scale averaging.

to have more a triangular than a sinusoidal shape. In contrary to Yalin, Gill takes into account the disappearance of dunes for critical flow conditions. As it can be seen in figure 4, Gills formula overestimates the dune height significantly.

The disappearance of dunes under higher shear stress conditions is much more pronounced in the formula proposed by YALIN in 1980 (see YALIN (1992) [17]):

$$\frac{\Delta_d}{h} = 0.023 \left(\frac{\tau_B'}{\tau_c} - 1 \right) \exp \left(1 - \frac{\tau_B'/\tau_c - 1}{12.84} \right) \tag{4}$$

The comparison with the Weser dunes is poor. The formula predicts increasing dune heights where the actual height decreases and vice versa.

ENGELUND and FREDSOE (1982) [3] derived the formula

$$\frac{\Delta_d}{h} = \frac{\left(1 - \frac{\tau_c}{\tau_B'}\right)}{3 + \frac{1}{2}\left(1 - \frac{\tau_c}{\tau_B'}\right)} \tag{5}$$

using a mathematical model. It overestimates the height of the Weser dunes about a factor of two or more.

VAN RIJN (1985) [14] developed the empirical formula

$$\frac{\Delta_d}{h} = 0.11 \left(\frac{d_{50}}{h} \right)^{0.3} \left\{ 1 - exp \left[-0.5 \left(\frac{\tau_B'}{\tau_c} - 1 \right) \right] \right\} \left(26 - \frac{\tau_B'}{\tau_c} \right) \tag{6}$$

taking into account a lot of data sets for natural dunes. Due to the last term his formula predicts the disappearance of dunes at quite low bottom shear stresses when compared to the other formulas. For the Weser estuary his formula fits the spatial average of the dune heights quite well whereas their local variations are not as pronounced as in nature.

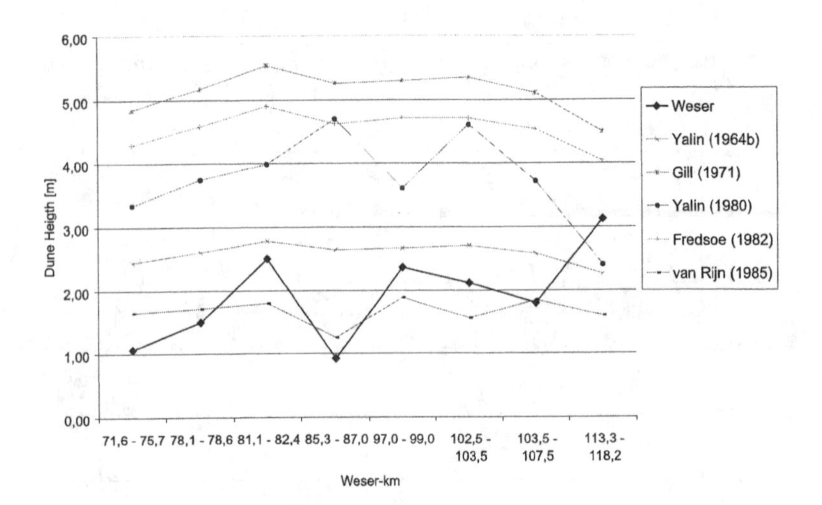

Figure 4: Comparison of the dune heights in the outer Weser with several formulas.

None of the formulas can predict the large dunes in the range of Weser km 113 - 118.

The second important dune parameter is the dune length λ_d. Here a lot of empirical formulations exist which all state that dune length and height are proportional to each other. In figure 5 the relations obtained from the Weser dunes are compared with the approach of FLEMMING (1988) [5]:

$$\lambda_d = \left(\frac{\Delta_d}{0.0677}\right)^{1/0.8098} = 27.8\,\Delta_d^{1/0.8098} \tag{7}$$

and that of YALIN (2001) [18]

$$\lambda_d = 6h = 36\Delta_d \tag{8}$$

Both formulas show a reasonable agreement, whereby the Weser dune heights seem to be too small with respect to their length.

The *SediMorph* Model

The *SediMorph* model is a three-dimensional morphological and fractionated sediment transport model. *SediMorph* is implemented a way that it can be coupled with different hydrodynamic models by well defined interfaces. The programming language is FORTRAN 90.

It works on a sediment classification defined by the user through a sediment classification file. This file contains the specification of the sediment classes by specifying their grain diameters and grain densities. In this way *SediMorph* can work on every kind of sediment classification. This could be the Udden-Wentworth scale but any other classification is possible too.

For this sediment classification the three dimensional morphological data set for the simulation area has to be constructed. It consists of a two dimensional unstructured mesh with several vertical layers. In this way a mesh of volume cells is defined. For every

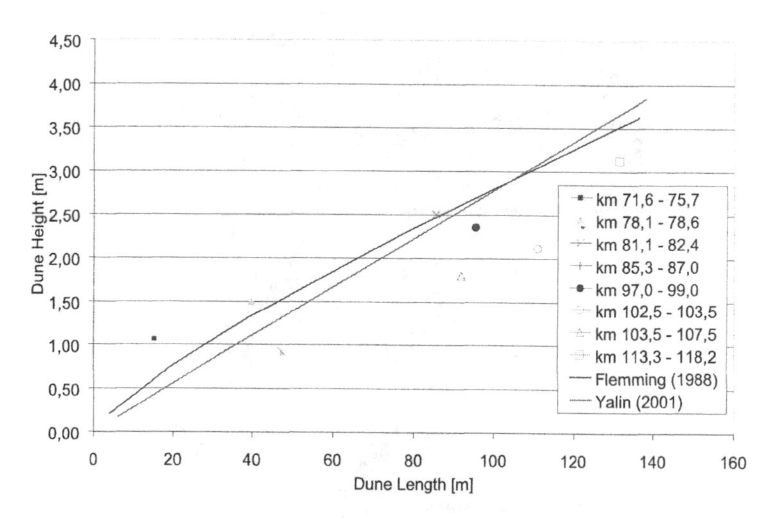

Figure 5: Comparison of the dune lengths in the outer Weser with two formulas.

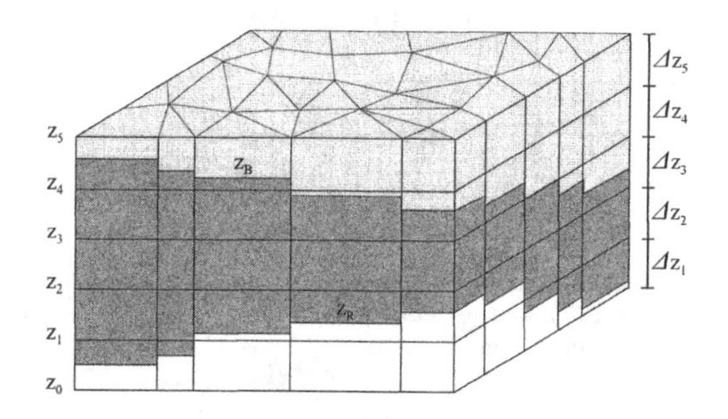

Figure 6: Structure of the morphological data set in *SediMorph*.

cell the volume fractions of every sediment class in the sediment classification file has to be specified.

During the simulation the hydrodynamic-numerical (HN) model passes a water height and the average velocity related to this water height to *SediMorph*. In this way *SediMorph* can be coupled to a 3D as well as a 2D HN model, because this water height can either be the total or the water depth of the first cell over the bed.

SediMorph is now able to calculate the critical shear stresses for the initiation of motion, the transport rates and the bed evolution from the sedimentological data.

The dune parameters are simulated by the following algorithm: For every new time t^{n+1} the equilibrium dune heights Δ_{eq}^{n+1} are calculated according the formula of van Rijn. This height would be achieved when the flow conditions would not change for a certain relaxation time T. Therefore the actual dune height at the time t^{n+1} is calculated using a simple relaxation strategy:

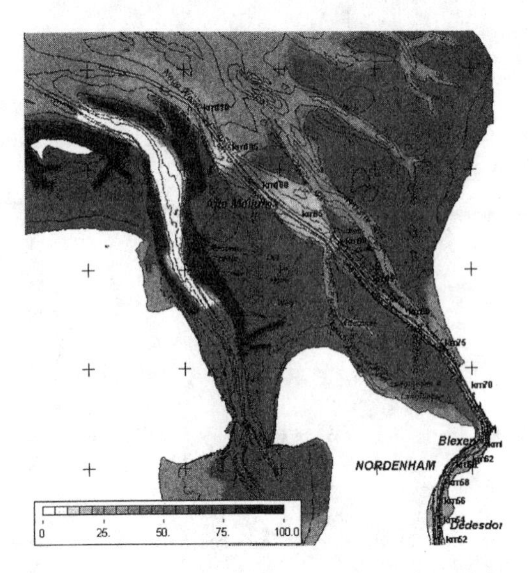

Figure 7: Assumed fine sand percentages in the morphological data set for the Jade-Weser estuary.

$$\Delta_d^{n+1} = \Delta_d^n + \left(\Delta_{eq}^{n+1} - \Delta_d^n\right)\frac{\Delta t}{T} \tag{9}$$

Starting from the dune heights Δ_d^n at the previous time the gap between the equilibrium and the actual dune heights is added. This gap is weighted by the ratio of the numerical time step Δt and the relaxation time T. Afterwards the dune length is calculated according to Flemmings formula.

Results for the Dune Simulation

For the numerical simulation of the Jade-Weser estuary *SediMorph* was coupled with the depth averaged two-dimensional hydrodynamic model *Telemac-2D* (HERVOUET and BATES, 2000 [7]). The horizontal mesh consists of 90375 nodes forming 174197 triangular elements. The sediment bottom is discretized by one vertical layer. The sediment itself is discretized by six classes: silt, fine sand, medium sand, coarse sand and gravel. The overall percentages are constructed combining data from different sources. No synoptic measurements are available for the whole area. The assumed resulting fine sand percentages are shown in figure 7.

A period of five days was simulated. Because this is too short for the full generation of dunes, two relaxation times were distinguished. The relaxation time for the generation ($\Delta_{eq}^{n+1} - \Delta_d^n > 0$) was set to one hour, and for the destruction ($\Delta_{eq}^{n+1} - \Delta_d^n < 0$) to one week.

In figure 8 a synoptic view on the simulated dune heights is shown. No dunes were generated in shallow areas as it is in reality. Otherwise everywhere in the deeper channels dunes are predicted. The dunes are strongly dependent on the bottom sediment composition as it can be seen in the outer Jade. On the medium sand bed of the outer Jade, large dunes are generated. This ends immediately over the fine sand bottom of the inner Jade which is the gorge to the Jade bay. Actually the dune heights in the outer Jade are as high as predicted by the numerical model.

A comparison of figures 7 and 8 shows that the results are very sensitive on the composition of the bottom sediments e.g. the median diameter d_{50} and the critical bed shear stress for the initiation of motion. As mentioned before this can be seen in the outer Jade, where the dunes immediately disappear when the sediment conditions change. The dunes are mainly located where the fine sand content is low and the medium and coarse sands dominate. Therefore the sediment composition of the bed has to be known quite good which is not the case for the whole estuarine system.

In figure 9 the simulated results are compared to the measurements. In principal the simulations indicate the trends of the measurements but the maximum values are not reached. A comparison between figures 9 and 4 shows that the numerical simulation including the tidal cycle and the relaxation strategy produces quite the same dune heights as the prior estimation using the conditions at maximum ebb flow. This means that the discrepancies lying between measurements and simulations originate from the sediment input data or the van Rijn formula but not from the numerical algorithm.

Finally in figure 10, the measured and simulated dune lengths are compared. There is a seaward trend to longer dunes in the measurements which is not predicted by the numerical model.

The Dune Roughness

After the dunes have been developed in the simulation area, the quality of the coupled hydrodynamic and morphological model has to be evaluated comparing measured to simulated free surface elevations. These are mainly controlled by the energy dissipation in the numerical model which stems from three different kind of sources (i.e. dispersion, numerical diffusion and bottom shear stress [11]). The first one is modeled by Elders law [2]. The numerical diffusion is dependent on the local resolution of the mesh and is unfortunately not negligible [10]. The bottom shear stress for the hydrodynamics is calculated using Nikuradse's law,

$$\frac{\vec{\tau_B}}{\varrho} = \frac{\kappa^2}{\left(\ln \frac{12h}{k_s}\right)^2} \|\vec{u}\|\vec{u} \tag{10}$$

whereby the effective bed roughness k_s is the sum of the grain and the dune roughness $k_s = k_s^g + k_s^d$. The grain roughness is assumed to be $k_s^g = 3d_m$ which is in agreement with a lot of measurements. The mean grain diameter is taken from the data stored in the *SediMorph* module. The dune roughness is calculated using van Rijns formula:

$$k_s^d = 0.77\Delta_d \left(1 - e^{-25\Delta_d/\lambda_d}\right)$$

Two kind of simulations are done. In the first, only the grain roughness and in the second the grain and the dune roughness are taken into account. The results for the tidal gauge Farge located at Weser-km 26 are shown in figure 11. Compared to the measurements the energy dissipation produced by the grain roughness is much too low (i.e. the tidal elevation is much higher than the measured one). Adding van Rijn's dune roughness the energy dissipation is too large because the tidal elevation is decreased below the measured one. Good results are obtained when the proportionality constant in van Rijn's formula is reduced from 0.77 to 0.5. As mentioned above this reduction must not be generally valid because the numerical model contains a certain amount of numerical diffusion.

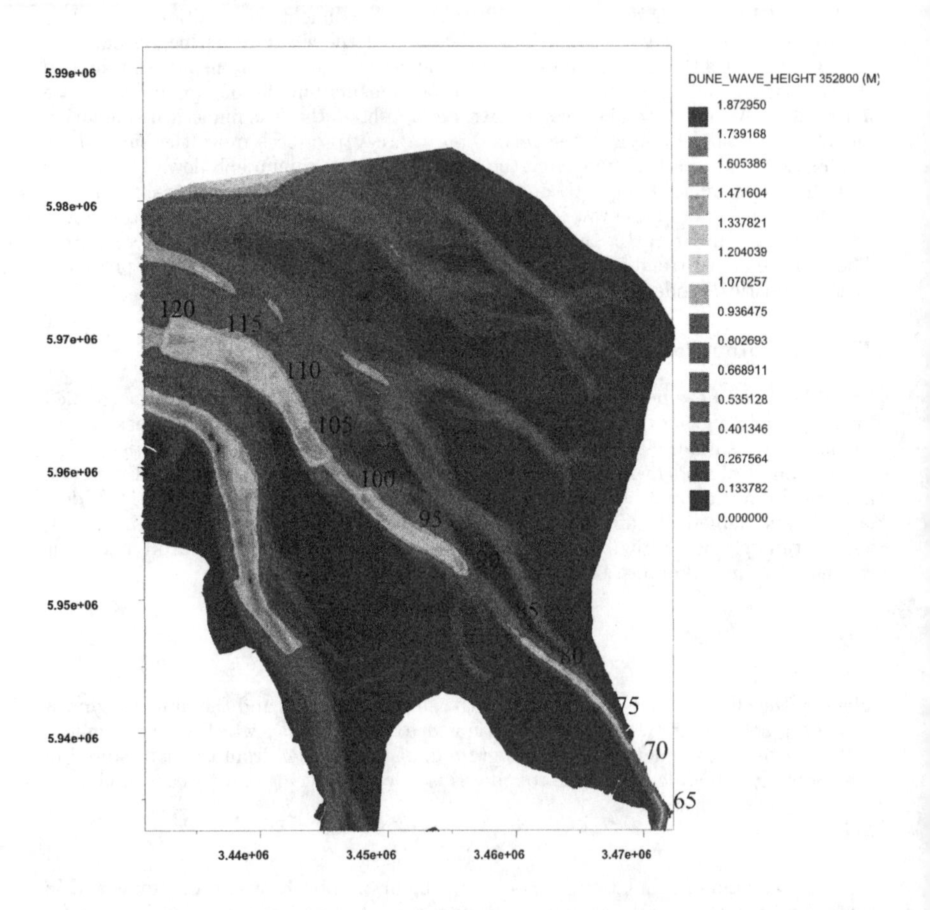

Figure 8: Synoptic View on the simulated dune heights in the outer Weser estuary.

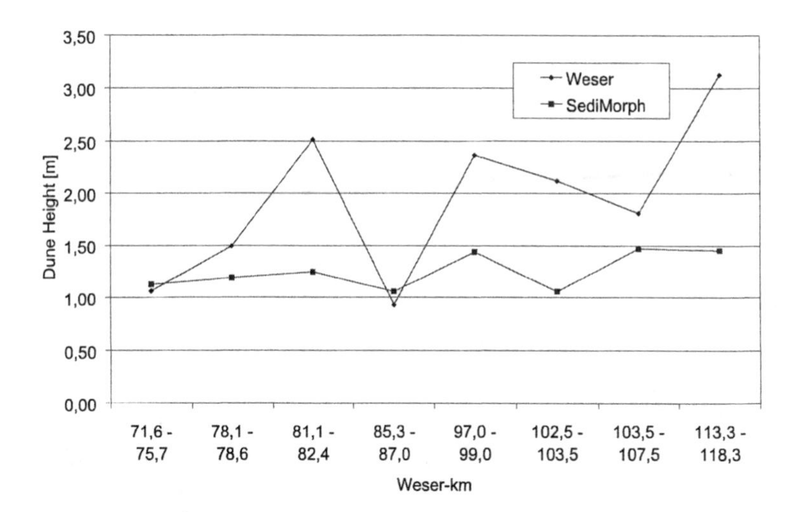

Figure 9: Comparison of measured and simulated dune heights in the shipping channel of the outer Weser estuary.

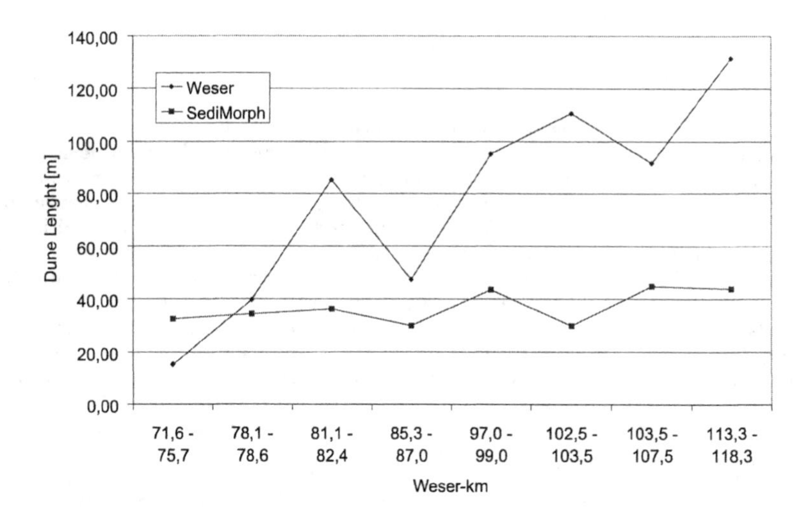

Figure 10: Comparison of the measured and simulated dune lengths in the shipping channel of the outer Weser estuary.

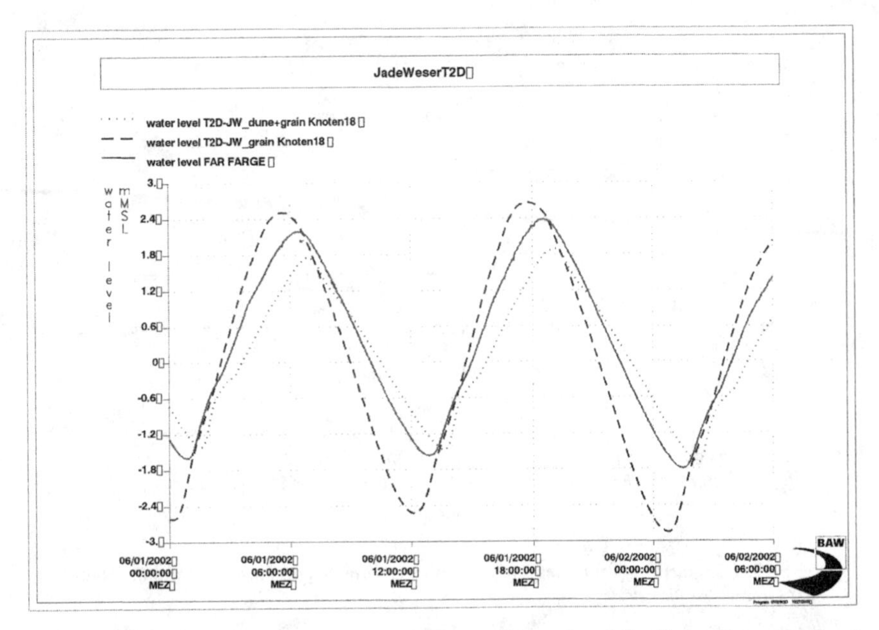

Figure 11: Tidal elevation at the gauge Farge (Weser-km 26). Green line: measured elevation, red line: with grain roughness, black line: with grain and dune roughness.

Conclusion

A simple dune predictor basing on empirical relations for the dune heights and lengths is presented. These relations are taken as equilibrium values which would be achieved in a stationary flow. The inertness of the dune parameters in tidal flows is represented by a linear relaxation method which accelerates the growth and retards the destruction of the dunes during slack water. As a matter of fact the relaxation times used in the simulation were chosen very pragmatically. As an improvement, the generation times found in some dredging experiments could be taken. These range from weeks to months. Unfortunately in this approach the simulation periods should then be much longer than the relaxation times which is related to very high computational costs. Theoretical approaches for the relaxation times can be obtained from stability theories. Here the relaxation time is directly related to the imaginary part of the complex wave frequency.

The results are reasonable with respect to the empirical dune prediction formula and the lack of sedimentological input data. Here more sedimentological data mining is needed before judging the dune formula.

Improvements of the modeling approach should be done in the calculation of the equilibrium dune parameters as well as in the determination of the relaxation times.

Acknowledgments

The author would like to thank the WSA Bremerhaven for the supply of the data, Dr. Reiner Schubert for the preparation of the data in MS Excel, and Dr. Andreas Kahlfeld, for preparing the morphological data set for the simulation.

References

[1] Allen, J.R.L. *Current Ripples; Their Relation to Patterns of Water and Sediment Motion.* North-Holland Publishing Company, Amsterdam, 1968.

[2] Elder, J.W. The Dispersion of Marked Fluid in Turbulent Shear Flow. *J. Fluid Mech.*, 5:544–560, 1959.

[3] Engelund, F. and Fredsøe, J. Sediment Ripples and Dunes. *Ann. Rev. Fluid Mech.*, 14:13–37, 1982.

[4] Exner, F.M. Über die Wechselwirkung zwischen Wasser und Geschiebe in Flüssen. *Sitzungsberichte der Akademie der Wissenschaften in Wien, Abt. II a*, 134:165, 1925.

[5] Flemming, B.W. Zur Klassifikation subaquatischer, strömungstransversaler Transportkörper. *Bochumer geologische und geotechnische Arbeiten*, 29:44–47, 1988.

[6] M.A. Gill. Height of sand dunes in open channel flow. *J.Hydr.Div.*, 97(HY12):2067–2074, 1971.

[7] Hervouet, J.M. and Bates, P. The TELEMAC Modelling System. Special issue of *Hydrological Processes* **14**, 2000.

[8] Kennedy, J.F. The Mechanics of Dunes and Antidunes in Erodible-Bed Channels. *J. Fluid Mech.*, 16:521–544, 1963.

[9] Kennedy, J.F. The Formation of Sediment Ripples, Dunes, and Antidunes. *Annu. Rev. Fluid Mech.*, 1:147–168, 1969.

[10] Malcherek, A. Application of TELEMAC-2D in a Narrow Estuarine Tributary. *Hydrological Processes*, 14:2293–2300, 2000.

[11] Malcherek, A. A Second Order Lagrangean Advection Scheme Without Numerical Viscosity for the Propagation of Tidal Energy. In Spaulding, M.L. and Cheng, R.T., editors, *Estuarine and Coastal Modeling - Proceedings of the 7th International Conference*, New York, 2002. ASCE.

[12] Meyer-Peter, E. and Müller, R. Eine Formel zur Berechnung des Geschiebetriebs. *Schweizerische Bauzeitung*, 67(3):29–32, 1949.

[13] Nikuradse, J. Strömungsgesetze in rauhen Rohren. Forschungsheft 361, VDI, 1933.

[14] L.C. van Rijn. Sediment transport part iii: Bed forms and alluvial roughness. *J. Hydr. Eng.*, 110(12):1733–1755, 1985.

[15] Rijn, L.C. van. *Principles of Sediment Transport in Rivers, Estuaries and Coastal Seas.* Aqua Publications, Amsterdam, 1993.

[16] M.S. Yalin. Geometrical properties of sand waves. *J. Hydr. Div, Proc. American Society of Civil Engineering*, 90(HY5):105–119, 1964.

[17] Yalin, M.S. *River Mechanics.* Pergamon Press, Oxford, New York, Seoul, Tokyo, 1992.

[18] Yalin, M.S. and Ferreira da Silva, A.M. *Fluvial Processes.* IAHR International Association of Hydraulic Engineering and Research, Delft, The Netherlands, 2001.

Seasonal changes of baroclinic circulation and water exchange in the Bohai Sea

Keiji Nakatsuji[1], Ryoichi Yamanaka[2], Shuxiu Liang[3] and Zhaochen Sun[4]

Abstract

In four principal sea areas in China, accelerated water pollution has recently become more widespread and severe with rapid economic growth. In enclosed areas such as the Bohai Sea, "red tides" are occurring with greater frequency, degrading ecosystems and causing eutrophication. As the first step in clarifying the mechanisms of ecosystem and coastal eutrophication and for considering concrete enforcement measures to reduce the damage, the present study focuses on physical phenomena such as hydrodynamics and transport processes. First, the flow structure and density field in the Bohai Sea are examined using a three-dimensional baroclinic flow model. Second, the seasonal change of wind and surface heat transfer on the residual current system and the water exchange through the Bohai Strait are examined using a Lagrangian particle tracking method. This computation makes clear that the residual currents in the majority of the Bohai Sea are very weak, less than 0.05m/s. The effect of seasonal winds on hydrodynamics such as residual currents in the Bohai Sea, water exchange between the Bohai Sea and the Yellow Sea, and stratification induced by heat transfer at the sea/water surface are studied.

----

[1]Professor, Department of Civil Engineering, Graduate School of Engineering, Osaka University, 2-1 Yamadaoka, Suita City, Osaka, 565-0871 Japan; Phone +81-6-6879-7603; Fax +81-6-6879-7607; nakatsuj@civil.eng.osaka-u.ac.jp

[2] Research Associate, Faculty of Environment and Information Sciences, Yokohama National University, Japan; yamanaka@ynu.ac.jp

[3]Lecturer, State Key Lab. of Coastal and Offshore Engineering, Dalian University of Technology, P.R.China; sxliang@dlut.edu.cn

[4]Professor, State Key Lab. of Coastal and Offshore Engineering, Dalian University of Technology, P.R.China; sunzc@dlut.edu.cn

Introduction

The Bohai Sea is the largest enclosed basin in China, encompassing the area on west of the line that extend from the south end of the Liaodong Peninsula to the north end of Shandong Peninsula, as shown in Fig.1. It contains three main bays, namely Liaodong Bay in the northeast, Bohai Bay in the west and Laizhou Bay in the south. The Bohai Sea connects with the Yellow Sea through the Bohai Strait, which is approximately 100 km wide and has a maximum depth of 70 m. The topography of the sea bed is rather flat and covered with mud. The water depth in most areas is less than 20 m except for the north area, where the water depth is more than 30 m. Laizhou Bay is the shallowest bay. The surface area of the Bohai Sea is about 77,000 km^2.

According to an official report of the Chinese Government, the number of red tide appearances in four principal Chinese sea areas rapidly increased from 1998 to 2002, that is, 22, 15, 28, 77 and 79 times per year in order. The problem of water pollution became more widespread and severe with rapid economic growth, especially in 2001 and. 2002. Eutrophication also seems to have become more serious in those areas because of increased influx of nutrients including phosphorus and nitrogen. Red tide is strongly related with the multiplication growth of phytoplankton owing to the excessive supply of nutrients discharged from land. Since all types of algae include Chlorophyll–a, Chlorophyll-a can be used as an index of eutrophication.

Figure 2 shows the horizontal distribution of surface Chlorophyll-a measured by SeaWiFS on 23rd October, 1998. In the case of Chlorophyll-a concentration larger than about 10 mg/m^3, eutrophication seems to become more serious. In the case of Chlorophyll-a concentrations larger than 50 mg/m^3, red tides may appear. It is clear from the figure that red tides appeared in the heads of three bays and that red tides spread throughout the whole basin except the central area and the Bohai Strait area. The scale of the red tide is surprisingly large and the event shows the seriousness of ocean water pollution. The occurrence of red tides damages the marine environment and affects suffering for fishes and shellfishes. In order to preclude damages from eutrophication in the Bohai Sea, it is necessary to control pollutants discharged from the land. There are, however, not so much ideas for promoting water quality conservation due to the lack of understanding of the mechanisms of tide and wind-induced flow and transport processes here.

Recently a number of researchers has interest in hydrodynamic features in the Bohai Sea. Most of them, such as Zhao and Maochong (1993) and Dou et al. (1994), have mainly concentrated on a tide-induced residual current system, because its system plays significant roles on long-term prediction of mass transport processes. Some researchers have studied that wind-induced current under the stratification caused by the cooling or warming at water/air heat balance from the view points of geological feature of very shallow and enclosed water body. (for example; Huang et. al.; 1999; Liang et al., 2003) Unfortunately there is a few reliable observation data, that is hard for researchers to develop refined modeling.

The present study focuses on the seasonal change of baroclinic circulations, especially the wind's effects on the hydrodynamics in the Bohai Sea and the water exchange through the Bohai Strait. A three-dimensional baroclinic flow model is applied to clarify the residual current system and density field, and a Lagrangian particle tracking method is applied to evaluate the wind effects on the water exchange between the Bohai Sea and the Yellow Sea. The novel feature of this computation is the simulation of the hydrodynamics of the Bohai Sea using the observed data from January and July of 2001.

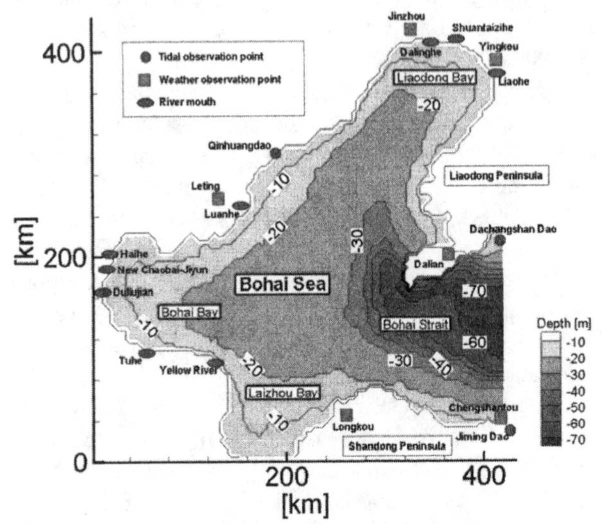

Figure 1. Bathymetric map of the Bohai Sea

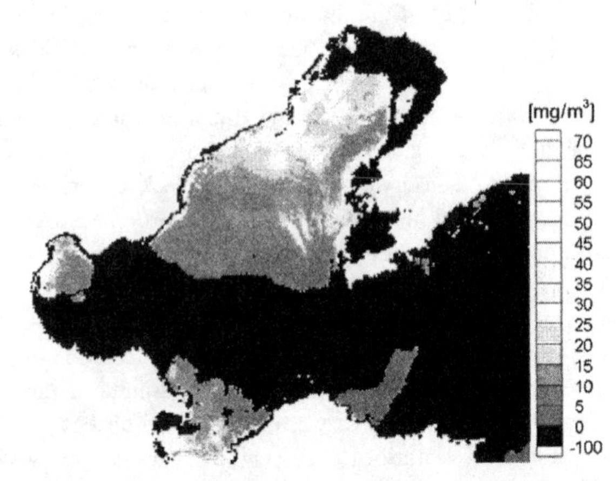

Figure 2. Horizontal distribution of surface Chlorophyll-a measured by SeaWiFS on 23rd October, 1998

Baroclinic Flow Model

The hydrodynamic model selected for this application was ODEM (Murota et al. 1988; Nakatsuji and Fujiwara, 1997). The basic hydrodynamic equations are based on the conservation of volume, momentum in three directions, and scalar quantities, namely temperature and salinity, under the Boussinesq assumption and hydrostatic approximation. The density is calculated as a function of temperature and salinity. The mathematical model is formulated by integrating the basic equations with respect to each control volume and by transforming them into finite-difference forms using a spatially-staggered grid system. Since the vertical momentum equation consists of only the gravity and vertical pressure gradient terms owing to the hydrostatic approximation, the vertical velocity component must be computed so that the continuity equation is satisfied for each control volume. The free surface variation, therefore, must be computed by applying the continuity equation to a vertical column of water from sea bottom to water surface.

Regarding the finite difference in time, an explicit leapfrog method is used; however, an implicit scheme is used for the computation of water level in order to prevent unexpected numerical instability. The authors have already applied ODEM to many coastal water bodies such as Osaka Bay, Tokyo Bay, Ise Bay and Mutsu Bay and clarified the flow mechanisms and density structure observed in the field (for example, Nakatsuji and Fujiwara, 1997; Nakatsuji et al., 1996; Sugiyama et al., 1994; Yamanaka et al., 2003).

The eddy viscosity and eddy diffusivity are used for representing the turbulent transport constituents. Since the vertical transport is reduced by stratification

Table 1. Harmonic constants

	M_2		S_2		K_1		O_1	
Observation point	H	θ	H	θ	H	θ	H	θ
Dachangshandao	1.32	265	0.42	321	0.33	342	0.25	304
Jimingdao	0.5	323	0.2	28	0.2	312	0.2	265

Unit: H(m),θ(degree)

Table 2. Main river discharge rate into the Bohai Sea

River name	**Diacharge rate(m^3/s)**
Liaohe	179
Shuantaizihe	141
Dalinghe	69
Lunhe	128
Haihe	164
New Chaobai-Jiyun	59
Duliujian	58
Tuhe	30
Yellow River	1550

caused by the density difference, the vertical eddy viscosity and diffusion coefficients must be a function of the Richardson number. Following the study of 3-D buoyant surface discharges by Murota et al. (1988), the present computation adopted Webb's formula (1970) for eddy viscosity and the Munk and Anderson formula (1948) for eddy diffusivity, respectively. Both empirical formulas are functions of the gradient Richardson number. Their values in unstratified conditions are set to be Av_0=0.001m^2/s. In addition, Henderson-Sellers' formula (1985) is applied to consider the wind-induced vertical mixing. On the other hand, for horizontal viscosity and diffusion coefficients, the sub grid scale (SGS) concept proposed by Smagorinsky (1963) is used. The MUSCL-TVD scheme (Van Leer, 1979) is used in differencing the advection term. These parameterization and their coefficients were determined based on the comparison with field surveys and hydraulic model experiments. (See Murota et al., 1988)

Numerical Arrangement for the Bohai Sea Model

Grid system

The computational domain is shown in Fig.1. This model's resolution is 4km x 4km in the horizontal plane with 16 layers in vertical, with their thickness varying from top to bottom as: 4m, 3m for 5 layers, 2m for 3 layers, 3m for 2 layers, 5m, 8m and 10m for 3 layers. This model adopts Arakawa-C grid and z-system in the vertical direction. The time step is 30 seconds.

Boundary and Initial Conditions

Tides: The sea surface elevation at the open boundary is determined according to two coastal tidal gauges; Dashangshandao in the north and Jimingdao in the south as shown in Fig.1. Four major tidal constituents M_2,S_2,K_1,O_1 are considered. The values of these harmonic constants are obtained from Admiralty tide tables from The United Kingdom Hydrographic Office. Table 1 is the common tidal amplitude and tidal phase of the major tidal constituents.

Salinity and Temperature: The initial salinity and temperature distributions are found by interpolating the multi-annual monthly mean data named GDEM-V, obtained from Naval Oceanographic Office (NAVOCEANO; see https://www.navo.navy.mil/). The values of temperature and salinity at the open boundary are determined by a zero-gradient condition.

Wind: A daily-averaged wind speed and direction over the sea surface are interpolated onto the grid points from measurements of the QuickScat back-scattered power (Wentz et al., 2001), using Kriging method (Deutsch and Journel, 1998). Two passes are used to calculate the daily values.

Meteorological effect: The heat balance at the air-sea interface is calculated according to the formula in Murakami et al. (1989). The heat balance is composed of four constituents: short-wave radiation, long-wave radiation, sensible heat and latent heat of evaporation. The distributions of other meteorological parameters (air pressure, air temperature and cloud cover) are given by interpolating the observation data of 2001, obtained from Japan Meteorological Business Support Center (JMBSC). This data was recorded every six hours at six points around the Bohai Sea: Chengshantou, Longkou, Leting, Jinzhou, Yingkou and Dalian as shown in Fig.1. Rainfall is not considered in this study.

River runoff: The river discharge from nine rivers around the Bohai Sea is considered. The position of all river mouths are shown in Fig.1 and the discharge rates of all rivers are shown in Table 2. They are yearly mean values.

Flow chart of Computation

There are four cases of numerical simulation as shown in Table 3. All cases have three steps of computations: The first computation uses the barotropic flow model for 2 days, aiming for the computation to be stable. The second computation uses the baroclinic flow model for 30 days to simulate the flow and density structure Averaged temperature and salinity distributions for December and January are used as the initial conditions in both Case 1 and Case 2. Likewise, an averaged temperature and salinity distributions for June and July are used as the initial conditions in both Case 3 and Case 4. The wind effect is taken into account in this step. All cases take the heat forcing into account and the examples of inputted meteorological values are shown in Fig. 3. The third computation evaluates the water exchange in the Bohai Sea using a Lagrangian particle tracking method.

Result and Discussion

Verification of ODEM Applied to the Bohai Sea

Figure 4 shows the comparison of computed and observed co-tidal and co-range charts of four dominant tidal constituents, namely, M_2, S_2, K_1 and O_1. A harmonic analysis was applied to time series of computed sea surface elevation during 15 days to compare with the chart presented by Nishida (1980). The upper figures show computed results, while the lower ones show observed results.

Table 3. Outline of Computation

Case	Season	Tide	Heat forcing	Wind+Air pressure
1	Winter (Jan/2001)	Yes	Yes	No
2	Winter (Jan./2001)	Yes	Yes	Yes
3	Summer (Jul./2001)	Yes	Yes	No
4	Summer (Jul./2001)	Yes	Yes	Yes

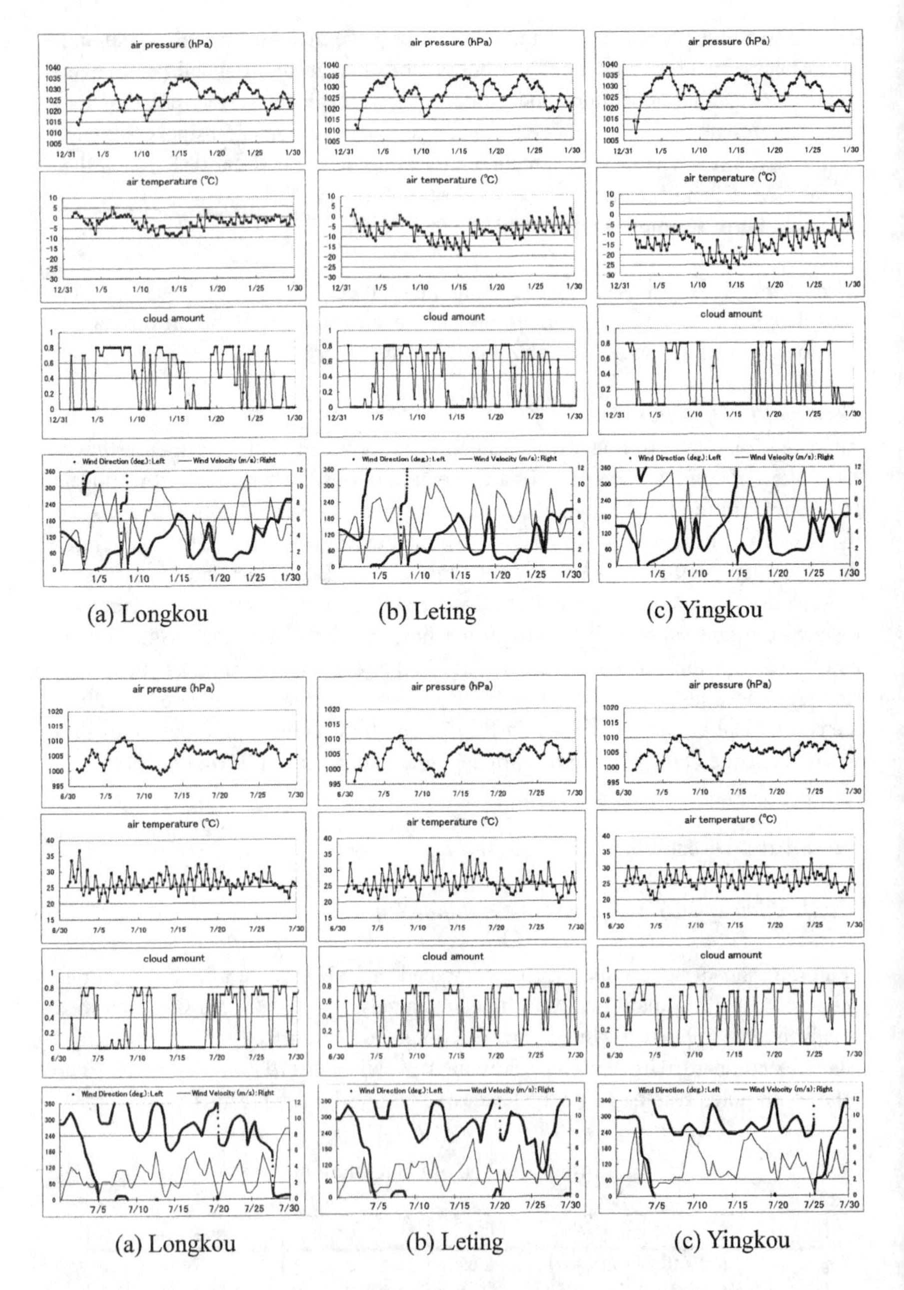

Figure 3. Time change of the meteorological boundary conditions; The upper and lower time series show data obtained in January and July, 2001, respectively.

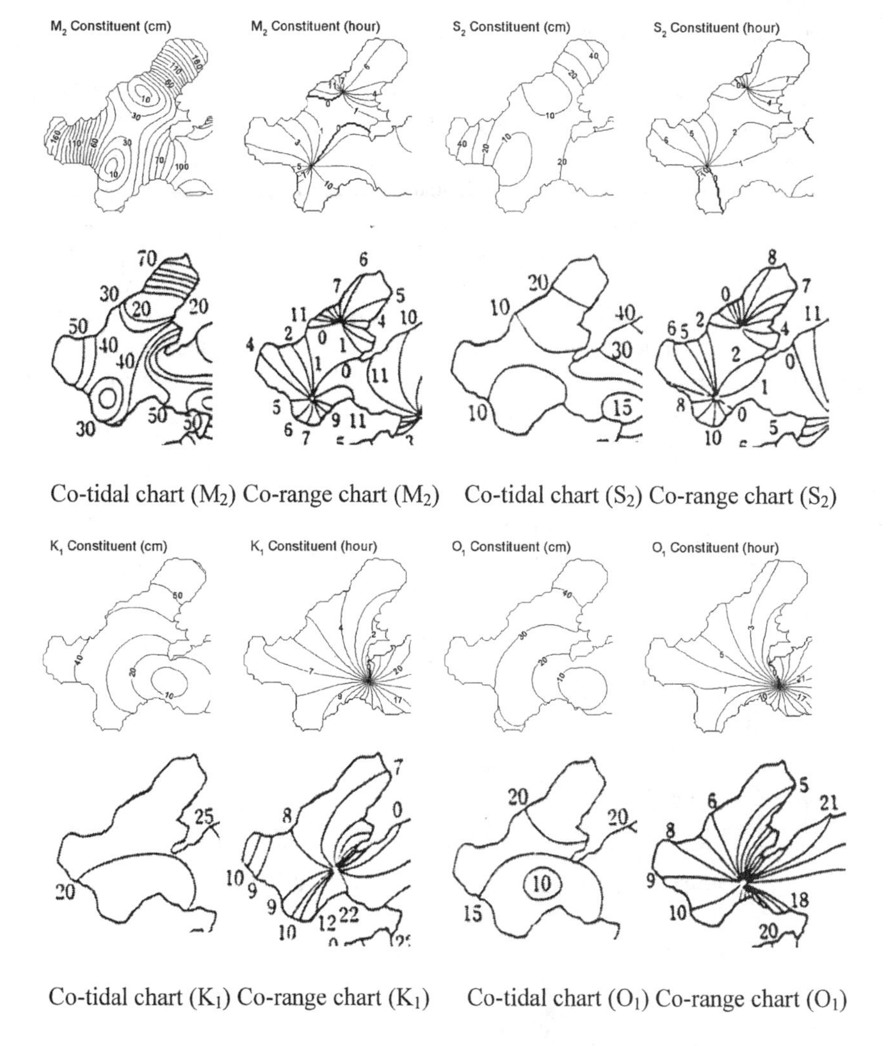

Figure 4. Comparison of simulated co-tidal and co-range charts compared with observations

The semidiurnal lunar tide, M_2, is most predominant among the four major tide constituents. Notable on the co-range charts of M_2 and S_2 tidal constituents is that there are two amphidromic points in the Bohai Sea. One appears in near the mouth of Yellow River, the other near Qinhuangdao. On the other hand, only one amphidromic point appears in co-range charts of the K_1 and O_1 tidal constituents.. The computed and observed values agree qualitatively.

Seasonal change of baroclinic circulations

In the present study, we define a residual current as a value integrated over 15 days of the second step of computation. Figures 5 and 6 show the horizontal distributions of tide-induced residual currents at the depths of 0.0m (sea surface) and 20.0m from the sea surface in January and July, 2001. The vector arrows denote flow direction and vector length denotes the velocity. Generally speaking, the residual current is very weak and average values are less than 0.05 m/s in the whole Bohai Sea except in the Bohai Strait, Liaodong Bay and Bohai Bay.

In the winter case (January) as shown in Fig.5, the flow direction of residual current at sea surface has several patterns, especially a weak anticlockwise flow that is induced in the northern area of Laizhou Bay and in the southern area of

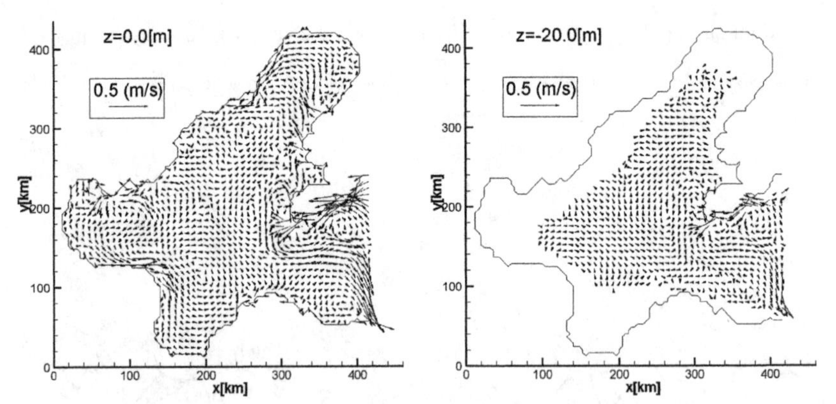

Figure 5. Computed residual current during 15 days: *Case 1*

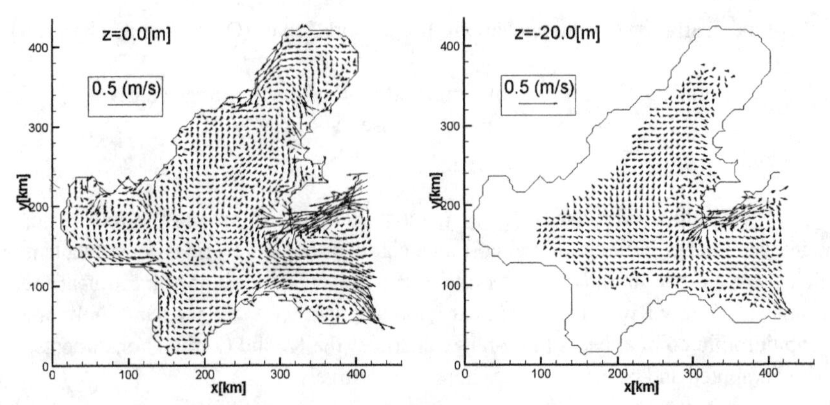

Figure 6. Computed residual current during 15 days: *Case 3*

Liaodong Bay. This flow pattern is kept at the depth of 20m because there is no thermohaline stratification. On the other hand, the flow structure at the Bohai Strait is simple; there is a relatively strong eastward flow, the velocity reached 0.15m/s, at the north area of the strait and a weak westward flow at the middle area of the strait. These flow patterns can be seen in the result at the depth of 20m, but the velocity is weak in comparison with the velocity at the sea surface

In the summer case (July) as shown in Fig.6, the residual current system is different from the winter case. The weak anticlockwise circulation in northern Laizhou Bay at the sea surface in winter is shifted to the north, and the structure is transformed to twin anticlockwise circulations. However, these twin circulations cannot be observed at the depth of 20m. It is for this reason that there is a thermal stratification as shown in Fig.14(a), which will be discussed in next section. On the

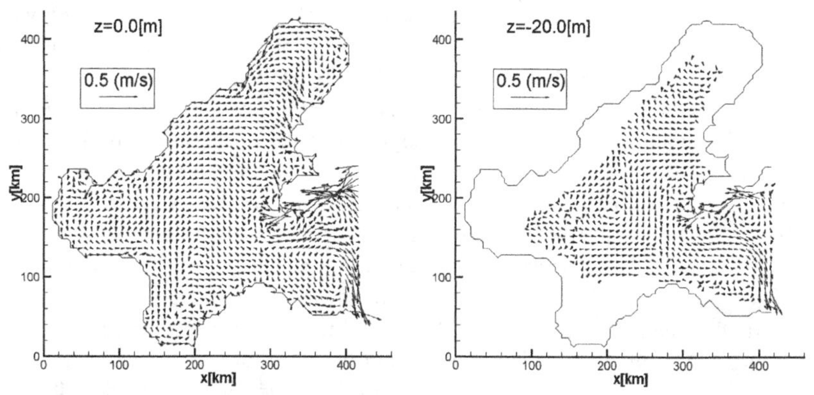

Figure 7. Computed residual current during 15 days: *Case 2*

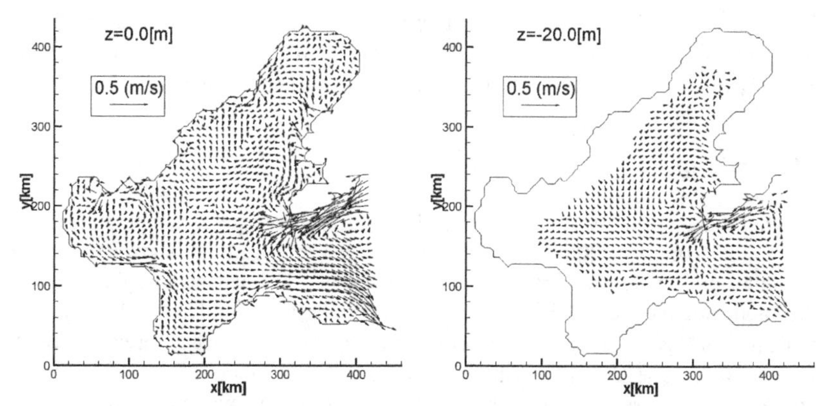

Figure 8. Computed residual current during 15 days: *Case 4*

other hand, the flow structure at the Bohai Strait is different from winter's flow; there is a strong westward flow in the northern end of the strait and eastward flow in the other areas of the Bohai Strait. The velocity of the both flows is faster than winter's flow and the velocity in the northern part reaches 0.45m/s at the sea surface. The flow structure of the Bohai Sea in summertime, especially in the central area of the Bohai Sea, is affected by the seasonal thermocline.

The wind's effects on hydrodynamics in the Bohai Sea

The distribution of residual currents under wind in winter and summer are shown in Figs.7 and 8. In the winter case (Case 2), the horizontal distribution of the residual current is drastically different from the result of Case 1 as shown in Fig.5. This is caused by strong wind as shown in Fig.3. The wind shear at the sea surface induced sea surface flow to accelerate toward to west, especially in the northern part of central area of the Bohai Sea and the eastern part of Laizhou Bay. The reason why the westward flow occurs is that an easterly wind is blowing on average during the period from January 15, 2001 to January 30, 2001. However, the flow direction of residual current at the depth of 20m is different from the surface one; an anticlockwise flow is induced in the central area of the Bohai Sea. This circulation is most likely caused by the change of flow structure at the west part of the central area of the Bohai Sea at the depth of 20m. Likewise, the anticlockwise circulation at the depth of 12.5m was induced by a north wind in winter (Yamanaka et. al. ; 2002). Since a north wind dominates in wintertime (Zhang, 1983)), the anticlockwise circulation dominates in the middle layer in winter.

In the summer case, the distribution of residual currents under the wind blowing condition (Case 4) is different from the result of Case 2, especially near the mouth of the Yellow River. The twin anticlockwise circulations described in Fig.6, disappear and change to a southward flow as shown in Fig.8. However, there is no characteristic change of the flow structure of the residual current at the depth of 20m as shown in Fig.8 in comparison with the no wind case as shown in Fig.6. The reason why the distribution of summertime residual currents does not change so much is that the wind in July, 2001 as shown in Fig.3 is weak: the maximum velocity of computed averaged winds is only 2.4m/s near Yingkou in July.

The wind's effects on distribution of temperature

Figures 9 and 10 show the horizontal distribution of computed temperature at the depths of 0.0m and 20.0m at the end of January. Figure 11 shows the cross-sectional profiles of temperature, salinity and density along the straight line from the mouth of the Yellow River to the mouth of the Liaohe River (A-A' line) as indicated in Fig. 9. The density is represented by σ_t. A distinct difference between Case 1 and Case 2 appears in the horizontal distribution of temperature; the temperature of Case 2 is obviously lower than Case 1. The maximum temperature

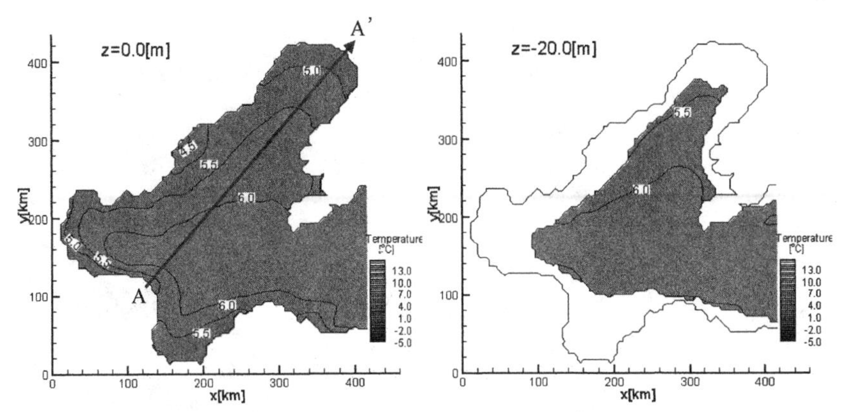

Figure 9. Horizontal distribution of computed temperature at the end of January: *Case 1*

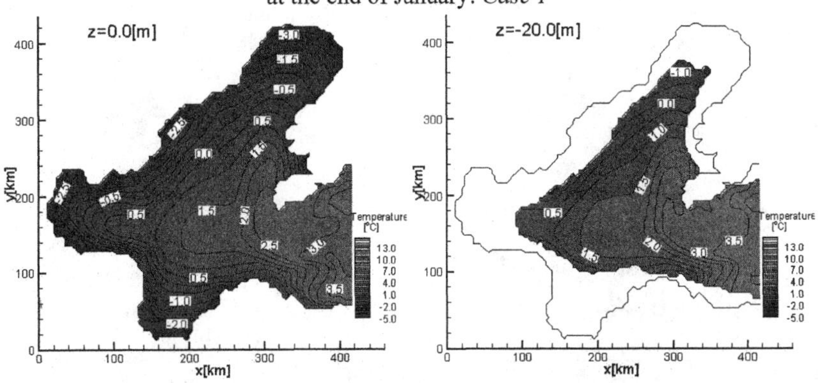

Figure 10. Horizontal distribution of computed temperature at the end of January: *Case 2*

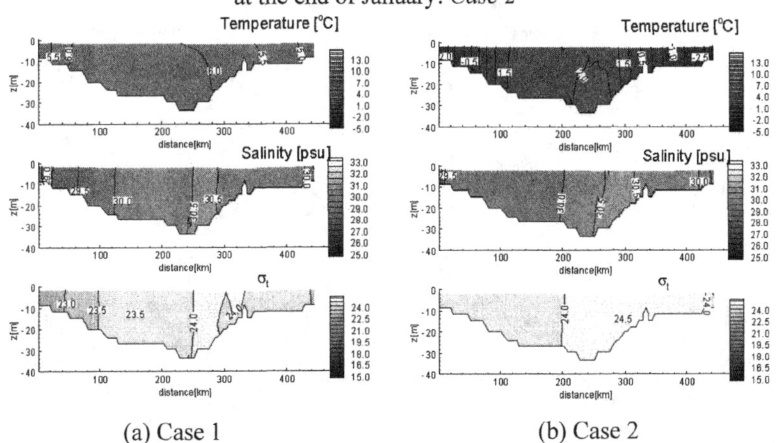

(a) Case 1 (b) Case 2

Figure 11. Vertical profiles of computed temperature, salinity and density at the end of January along the A-A' cross line as indicated in Fig.9

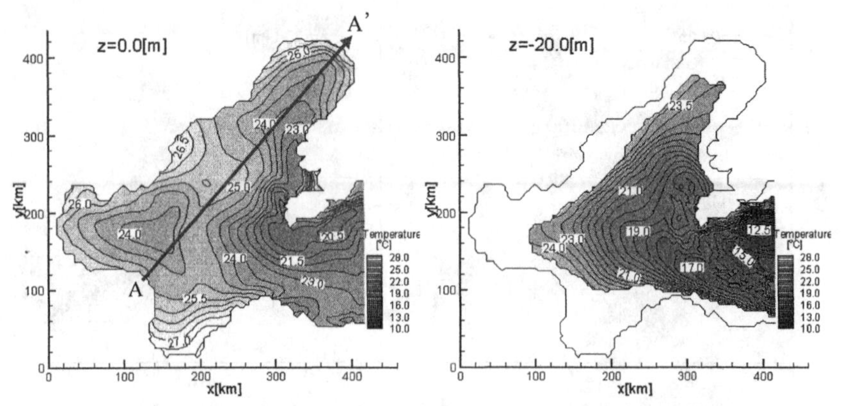

Figure 12. Horizontal distribution of computed temperature at the end of July: *Case 3*

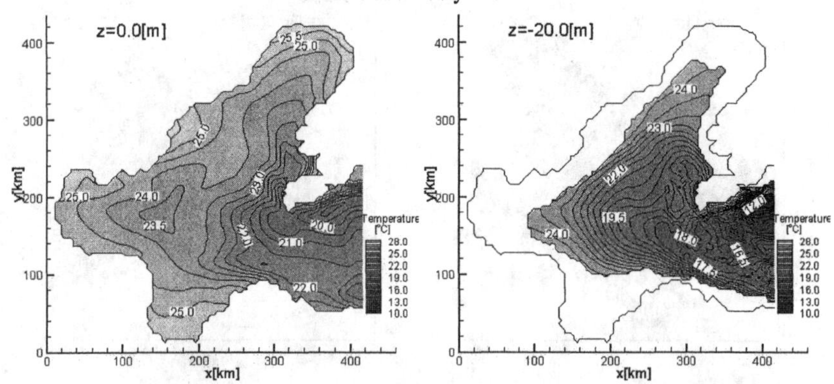

Figure 13. Horizontal distribution of computed temperature at the end of July: *Case 4*

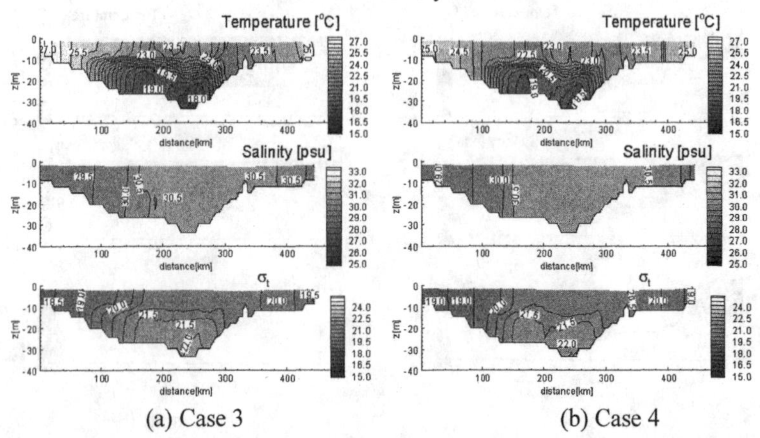

(a) Case 3 (b) Case 4

Figure 14. Vertical profiles of computed temperature, salinity and density at the end of July along the A-A' cross line as indicated in Fig.12

variation at the sea surface between Case 1 and Case 2 is 8 °C, in the northern area of Liaodong Bay. This difference in the temperature is caused by latent heat of evaporation and sensible heat transport, which depend on not only the difference in water vapor between atmosphere and sea, but also the wind speed. The observed sea surface temperature (SST) (from the MODAS (Modular Ocean Data Assimilation System) based on satellite data, and obtained by the Naval Research Laboratory at Stennis Space Center) on January 30, 2001 is quantitatively similar to Case 2's result. Hence, the wind blowing plays an important role in the determination of SST in winter in this region. On the other hand, the vertical profile of temperature is inverted in the vertical direction due to cooling caused by meteorological effects. The vertical distributions of salinity and density are well mixed and there is no density inversion.

In the summer case, Figs.12 and 13 show the horizontal distribution of computed temperature of Case 3 and Case 4. Since the wind velocity in summer-time is smaller than winter one as shown in Fig.3, SST differs only slightly between Case 3 and Case 4, dissimilar with results in winter. The vertical distributions of temperature, salinity and density along the A-A' line are shown in Fig.14. It becomes clear from these figures that thermal stratification developed in the central area of the Bohai Sea. Stratification was caused by temperature difference due to heat exchange at the sea surface, because salinity is well mixed and uniform in the vertical direction.

Wind's effect on water exchange

Lagrangian Particle Tracking Method

In order to quantitatively estimate the rate of water exchange through the Bohai Sea and the Yellow Sea, the particle tracer simulation method was applied. The outline of this method is as follows.

The movement of particles per time step Δt is given by

$$x_{i+1} = x_i + u(x_i, t) \cdot \Delta t + \nabla_i u(x_i, t) \cdot \Delta t^2 + u' \Delta t, \tag{1}$$

where the first term on the right side represents the position of a particle at the ith time step, the second and third terms represent convection and due to tidal flow and shear in tidal flow, and the fourth term represents movement induced by turbulence. The turbulence is assumed to follow the primary Markoff process (Kadoyu, 1984)

$$u_i' = \mu u_{i-1}' + N(0, \sigma_0), \tag{2}$$

where $\mu = \exp(-\Delta t / T_L)$, and $\sigma_o^2 = (1 - \mu^2) K / T_L$, N the normal random numbers, Kx, Ky and Kz eddy diffusivity coefficients given according to the SGS concept ; and T_L the integral time scale. T_L was set to 1 hour according to the tidal observation results obtained by Kadoyu (1984) for Osaka Bay, Japan. Because of the limitation that the computation time interval should be shorter than T_L and a one-step computation should remain within the grid interval $(\Delta x, \Delta y \; and \; \Delta z)$,

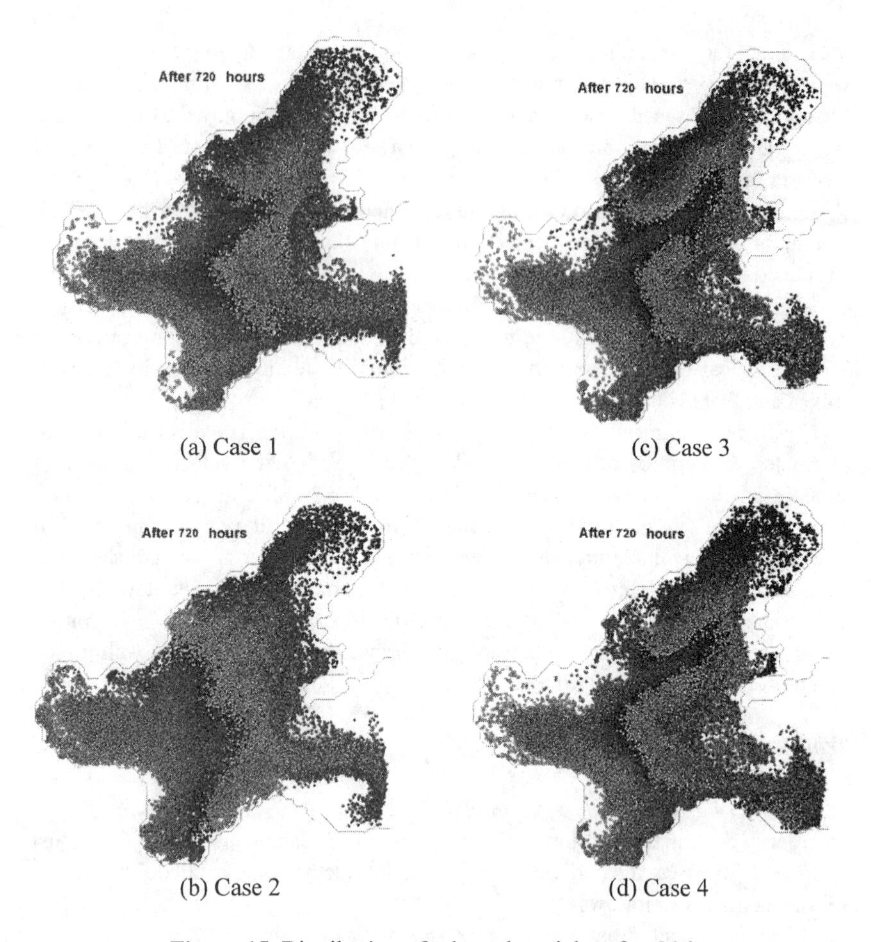

Figure 15. Distribution of released particles after 30days

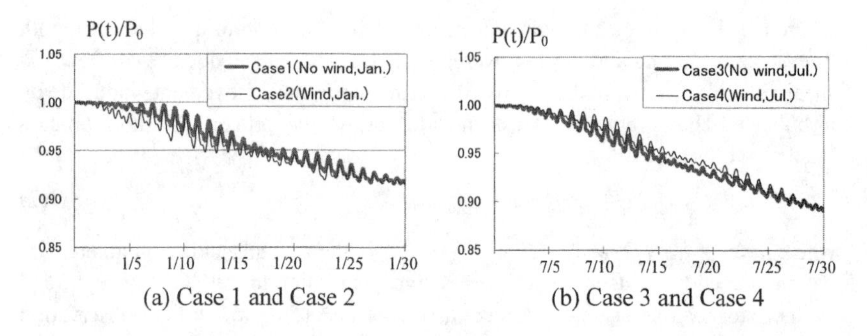

Figure 16. Time change of the particle retention ratio.

Initial Condition and Flow Chart of Computation

Particles were placed every 4km in the horizontal direction and every 1m in the vertical direction in the Bohai Sea as an initial condition. Then, the number of particles is 74739 in the whole Bohai Sea. In the present computation, a particle is treated as a water mass. The computation period was set to 30 days.

In this study, we examine four cases of particle tracking, according to the list as shown in Table 3. The velocities and eddy diffusivity coefficients at all computation grids are given at every hour from the hydrodynamic model. The velocity and turbulence coefficients of each particle are determined by the interpolation in time and space.

Computed Results

Figure 15 shows the computed particle distributions after 30 days in each case. In the case of winter, Case 1 and Case 2, there is a noticeable difference of the particle distribution in the central area of the Bohai Sea. It is because the residual current in the central area of the Bohai Sea is changed by east wind as shown in Fig.7. On the other hand, there is slight difference between the particle distribution of Case 3 and Case 4 in summer (when the residual currents also showed little difference).

Figure16 shows the time change of the particle retention ratio of the Bohai Sea defined as $P(t)/P_0$, where $P(t)$ is the number of particles at given time t and P_0 is the number of particles arranged initially in the Bohai Sea. In the case of winter as shown in Fig.16(a), the particle retention ratio of Case 2 is smaller than that of Case 1 during the computation period. In particular, the difference between the two cases is maximum on January 5, 2001 caused by the successive strong north wind as shown in Fig.3. This result agrees with the tendency that the north wind forces water exchange between the Bohai Sea and the Yellow Sea (See Yamanaka, 2002).

On the other hand, the ratio of Case 4 is larger than that of Case 3 in summertime as shown in Fig.16(b). It is clear that the west wind prevents water exchange between the Bohai Sea and the Yellow Sea because the west wind is dominant in July, 2001 as shown in Fig.3. The ratio of summer is larger than that of winter. The reason for the difference of the ratio between summer and winter is that the difference of density between sea surface and sea bottom in the Bohai Strait defines the circulation in summertime and this plays an important role for the water exchange between the Bohai Sea and the Yellow Sea.

Conclusion

The seasonal change of baroclinic circulations and the wind effects on the hydrodynamics in the Bohai Sea and water exchange through the Bohai Strait were investigated using a three-dimensional baroclinic flow model. It is confirmed that the present computation can predict well the co-tidal and co-range charts given by Nishida (1980), which are based on observation data. The tidal residual current is very weak (less than 0.05 m/s) in almost the whole Bohai Sea except in the Bohai

Strait in both winter and summer. Vertical profiles of temperature, salinity and density show that weak stratification occurs in the center of the Bohai Sea in summer. The strength of stratification depends on the temperature profiles. On the other hand, in winter, the vertical profile of temperature is almost uniform, that is there occurs no stratification in the Bohai Sea. The latent heat of evaporation and sensible heat transform cause water at surface layer cooling, which causes the inversion of temperature in the vertical direction. In particular, numerical result showed that the wind blowing plays an important role in the determination of sea surface temperature in winter.

The wind effect on the water exchange in the Bohai Sea is visually clarified by the Lagrangian Particle Tracking Method. There is noticeable difference of the particle distribution in the central area of the Bohai Sea in winter due to strong winter wind. According to the time change of the particle remaining ratio, the north wind forces water exchange between the Bohai Sea and Yellow Sea in winter. On the other hand, water exchange was prevented by the dominated west wind in summertime. The particle retention rate is therefore greater in summer than it is in winter.

The results of the present computation clearly demonstrate that the seasonal meteorological effects are important for modeling hydrodynamics in the Bohai Sea, and also the water exchange between the Bohai Sea and Yellow Sea.

Acknowledgement

This research is supported by the Grant-in-aid for Science Research (A) 14205073 of JSPS (Japan Society for the Promotion of Science).

References

Deutsch, C.V. and A.G. Journel (1998). "GSLIB Geostatistical Software Library and User's Guide Second Edition" Oxford University Press, p.369.

Dou Zhenxing, D., Y. Lianwu and J. Ozer (1994) "Numerical simulation of three-dimensional tidal current in the Bohai Sea", *Acta Oceanologica Sica*, Vol.3, No.3, pp.355-369. (in Japanese)

Huang, D., J. Su and J.O. Backaus (1999). "Modeling the seasonal thermal stratification and baroclinic circulation in the Bohai Sea", Continental Shelf Research, Vol.19, 1485-1505.

Henderson-Sellers.B. (1985). "New formulation of eddy diffusion thermocline models" *Appl. Math. Modeling*, Vol.9, pp.441-446.

Kadoyu, M. (1984). "Stochastic study on entrainment of floating particles with intake of cool water" *PhD thesis*, Osaka Univ., Osaka Japan. (in Japanese)

Liang, S.L., Z. Sun, R. Yamanaka and K. Nakatsuji (2003). "Typical seasonal circulation in the Bohai Sea" *Proc. 13th International Offshore and Polar Engineering Conference*, Honolulu Hawai, May 25-30, pp.282-289.

Munk, W. H., and Anderson E. R. (1948). "Notes on a theory of the thermocline" *J. Marine Research*, Vol. 7, pp.276-295.

Murakami, M., Y.Oonishi and K.Hayakawa (1989). "Heat and Salt Balances in the Seto Inland Sea" *Journal of the Oceanographical Society of Japan*, Vol.45, pp.204-216.

Murota, A., K.Nakatsuji and Huh J. Y. (1988). "A numerical study of three-dimensional buoyant surface jet" *Proceeding of 6th APD-IAHR Congress*, Vol. 3, pp.57-64.

Nakatsuji, K., Yoon, J.S., and Muraoka, K. (1996). "Three-dimensional Modeling of wind-driven upwelling of anoxic bottom water "A-oshio" in Tokyo Bay" *Estuarine and Coastal Modeling*, ASCE. pp.618-631.

Nakatsuji, K. and Fujiwara T. (1997). "Residual baroclinic circulations in semi-enclosed coastal seas" *J. Hydraulic Engineering*, ASCE,Vol.123, No.4, pp.362-373.

Nishida, H. (1980). "Improved tidal charts for the western part of the north pacific ocean" *Report of hydrographic researches 1980*, No.15, pp.55-70.

Smagorinsky, J. (1963). "General *ic researches 1980*, No.15, pp.55-70.
circulation experiments with primitive equations, *Monthly Weather Review*" Vol. 91, No.3, pp.99-164.

Sugiyama, Y., Takagi, F., Nakatsuji, K. and Fujiwara, T. (1998). "Modeling the baroclinic residual circulation in coastal seas under freshwater influence", *Estuarine and Coastal Modeling*, ASCE. pp.90-102.

Van Leer, B. (1977). "Toward the ultimate conservative difference scheme. 4, A New Approach to Numerical Convection" *Journal of Computational Physics*, 23, pp.276-299.

Webb, W. K.(1970). "Profile relationships, the log-linear range and extension to strong stability" *Quarterly J. Royal Meteorological Soc.,* 6, pp.67-90.

Wentz, F.J., Smith, D.K., Mears, C.A., and Gentemann, C.L. (2001). "Advanced

Algorithms for QuikScat and SeaWinds/AMSR" *In: GARSS'01 Proceedings 2001*, NASA, USA.

Yamanaka, R. and K.Nakatsuji.(2002). "Tide and Wind-induced currents in Bohai", *Proceeding 2nd International Work Shop on Coastal Eutrophication*, held in Nov. 21-24, 2002,in Tianjin, China, pp.142-150.

Zhao Jinping and S. Maochong (1993). "Study on short-range numerical forcastting of ocean current in the East China Sea – II Three-dimensional baroclinic diagnostic model and its application in the Bohai Sea", *Acta Oceanologica Sica*, Vol.13, No.2, pp.173-188.

Variability on flow and density structure near the mouth of Mutsu Bay forced by passage of migratory cyclone

Ryoichi Yamanaka[1], Shuzo Nishida[2] and Keiji Nakatsuji[3]

Abstract

The Mutsu Bay is a semi-enclosed bay located in northern Japan. In the present study, we clarify the influence of a passage of migratory cyclone on a flow and density structure around the bay and a water exchange through the bay mouth. A three-dimensional baroclinic flow model and wind calculation models are used. Five cases of simulations under various meteorological conditions are examined. Moreover a Lagrangian particle tracking model is also adopted in order to examine meteorological effects on the water exchange in the bay.

The numerical model was able to hindcast the observed sea surface setup and rapid change of density structure. The computed results show that the density structure is mainly affected by wind stress, while tidal anomaly is affected by time change of atmospheric pressure. The water exchange at the bay mouth is enhanced by the passage of migratory cyclone. These results suggest that meteorological disturbances may play an important role on conservation of water quality in Mutsu Bay.

Introduction

The Mutsu Bay is an enclosed bay approximately 50 km wide and 40 km long, located in northern Japan, as shown in Fig.1. It connects to the Tsugaru Strait through the Tairadate Strait of 10 km width. Field surveys have been carried out in every summer from 1996 to 2000 to clarify the water exchange in the mouth of

----

[1] Research Associate, Faculty of Environment and Information Sciences, Yokohama National University, 79-7, Tokiwadai, Hodogaya-ku, Yokohama, Kanagawa 240-8501, Japan; Phone/Fax +81-45-339-4097; yamanaka@ynu.ac.jp

[2] Associate Professor, Department of Civil Engineering, Graduate School of Engineering, Osaka University, Japan; nishida@civil.eng.osaka-u.ac.jp

[3] Professor, Department of Civil Engineering, Graduate School of Engineering, Osaka University, Japan; nakatsuj@civil.eng.osaka-u.ac.jp

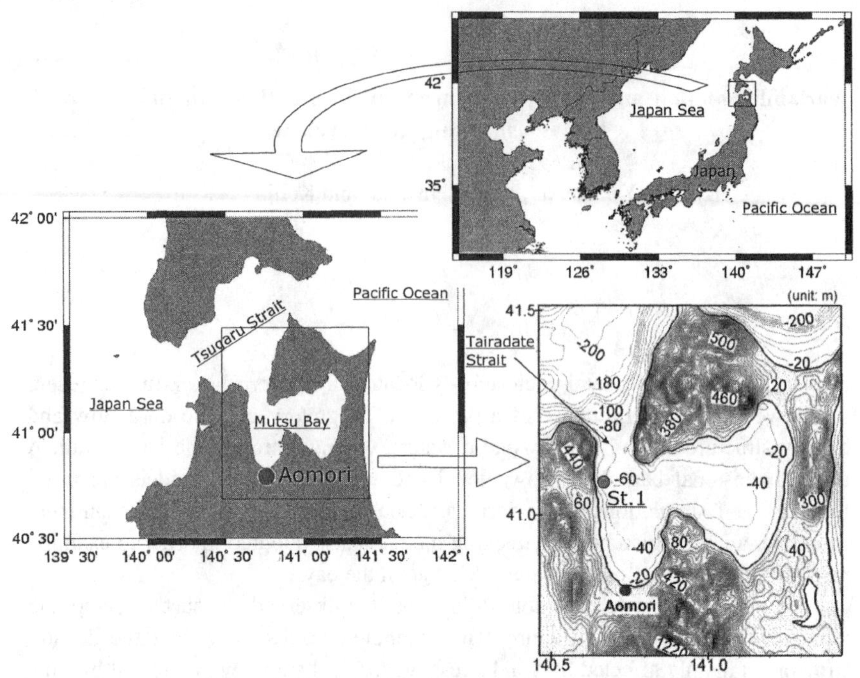

Figure 1. Topographic features around Mutsu Bay

Mutsu Bay. We have obtained some interesting results for a tide-induced residual current, its effects on water exchange and occurrence of internal waves (Nishida et al. (2000), Yamanaka et al. (2001; 2003)). Interestingly, the flow and density structures in the vicinity of the Tairadate Strait are affected by passage of migratory cyclones. Since the Tsugaru Strait (about 40 km wide and 120 km long) connects Japan Sea to the Pacific Ocean, migratory cyclones pass over the Tsugaru Strait and its vicinity changing the local climates and patterns of surface wind stress. However, there has been little investigation of such phenomena.

In the present study, we try to clarify how flow and density structure and water exchange are affected, not only in the vicinity of the Tairadate Strait, but also inside Mutsu Bay when a migratory cyclone passes.

Rapid changes in density profile at the mouth of Mutsu Bay

Figure 2(a) shows a vertical density profile observed on June 5-14, 1997 at St.1 shown in Fig.1. St.1 is one of the monitoring buoys arranged in the bay by Aomori Prefecture. Water temperatures and salinities are measured at four depths (1m, 15m, 30m and 45m) and flow velocities are measured at two depths (15m and 45m) in this station. It is found from Fig.2(a) that heavy water (σ_t>25.9) appears in the bottom layer, and the density structure appears well-mixed in the upper (15 m) layer from June 9-10. Concurrently with the rapid changes of density structure, the

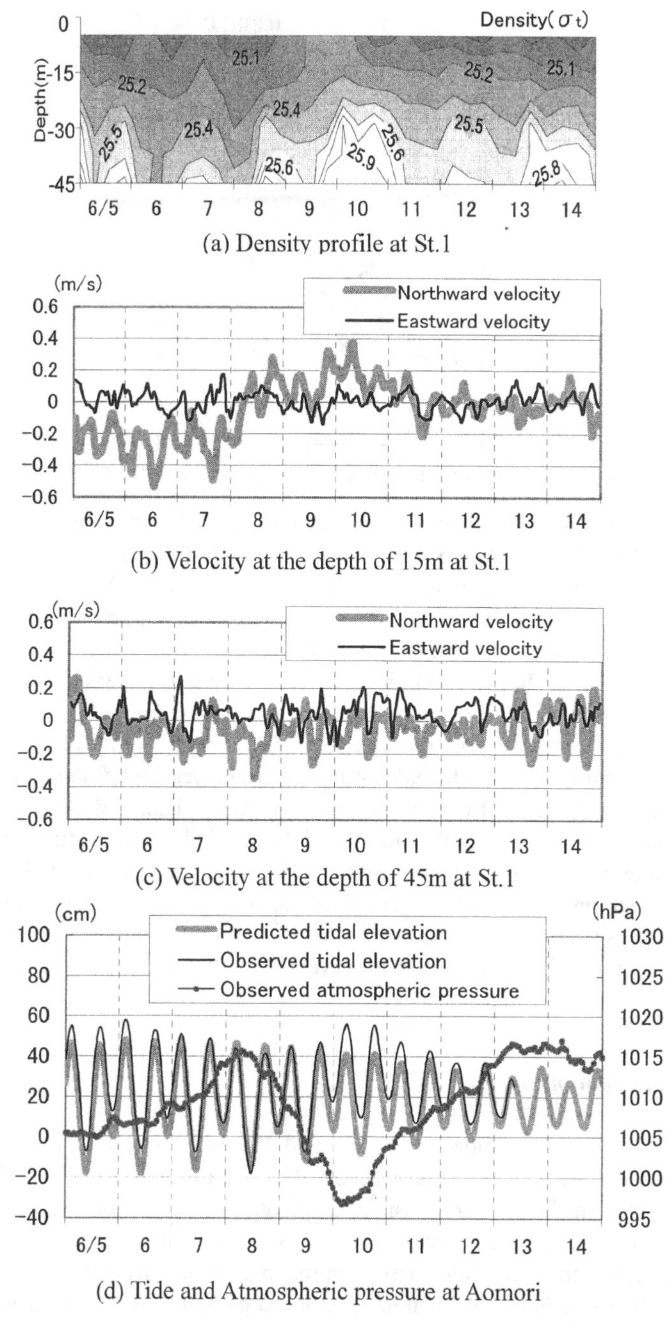

Figure 2. Density profile, velocity, sea level and atmospheric pressure from June 6-14,1997

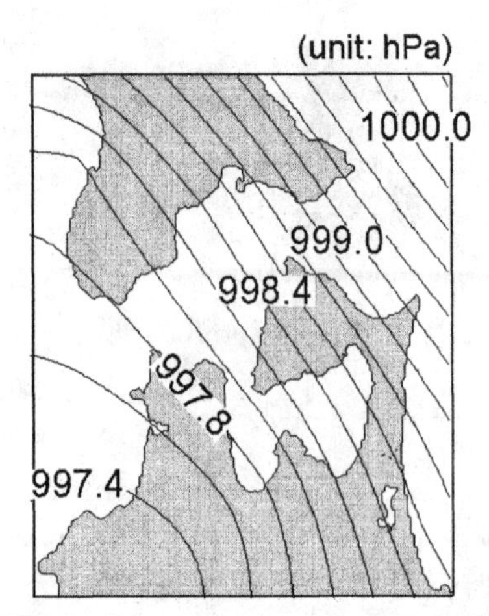

Figure 3. Distribution of atmospheric pressure at 3:00 June 10, 1997

flow structure varies, as shown in Figs 2(b) and 2(c): In particular, the northward velocity at the depth of 15m obviously shifts to positive value.

In order to examine a relationship between flow structure and meteorological external forces, we show observed sea level, predicted tide, and the observed atmospheric pressure at Aomori in Fig. 2(d). The predicted tidal elevation coincides to the observed sea level except during the period of decreasing atmospheric pressure due to the passage of a migratory cyclone. Figure 3 shows the distribution of atmospheric pressure when its lowest value was observed in Aomori. The migratory cyclone, located at the southwest end of this figure, moves to the northeast end of Fig.3 in one day. The rapid change of density profile, a northward shift in flow, and both the tidal elevation anomaly and atmospheric pressure drop, occurred at the same time. This suggests that the rapid changes in density distribution and flow structure are caused by a meteorological disturbance. In the following sections, we carry out numerical simulations to clarify the influence of the passage of a migratory cyclone on flow and density structures.

Methods

Baroclinic flow model

A quasi-3D baroclinic flow model, ODEM (Osaka Daigaku Estuary Model) is applied in this study, which was originally developed by Nakatsuji in 1987 (Murota et al., 1988). The basic hydrodynamic equations are based on the three-dimensional conservation law for mass, momentum and scalar quantities, namely, temperature and salinity under the Boussinesq approximation and hydrostatic assumption. Density is calculated as a function of temperature and salinity. The mathematical model is formulated by transforming them into finite-difference forms using the space-staggered grid system. A z-system is used for vertical discritization. Since the vertical momentum equation, subject to the hydrostatic assumption, consists of only the gravity and vertical pressure gradient terms, the vertical velocity component must be computed through the continuity equation. The free surface elevation may be computed by applying the continuity equation to a vertical column of water from the sea bottom to the water surface. Regarding the finite differencing in time, the explicit leapfrog method is used; however, the semi-implicit scheme is used for the computation of water level in order to prevent numerical instability. The authors have used the model on many marine areas such as Osaka Bay, Tokyo Bay and Ise Bay to clarify the structure of flow and density.

Eddy viscosity and eddy diffusivity are used for representing the turbulent transport constituents. Since the vertical transport is reduced by stratification caused by the density difference, the vertical eddy viscosity and diffusion coefficients must be a function of the Richardson number. According to a study of 3-D buoyant surface discharges by Murota et al.(1988), the present computation adopted the Webb's formula (1970) for eddy viscosity and the Munk and Anderson formula (1948) for eddy diffusivity, respectively. Both empirical formulae are functions of the gradient Richardson number. Their values in neutral condition are set to be Av_0=0.005m^2/s. In addition, the Henderson-Sellers formula (1985) is applied with regard to wind-induced vertical mixing. On the other hand, to account for horizontal viscosity and diffusion coefficients, the sub-grid scale (SGS) concept proposed by Smagorinsky (1963) is used.

Gradient-Wind model and MASCON

We calculate the wind on the sea surface using two kinds of models: the Gradient-Wind model and MASCON (Mass Consistent Atmospheric Flux Model). Gradient wind is computed through a distribution of atmospheric pressure. It is necessary to correct the computed gradient wind near the surface, because the wind is affected by surface friction. Some empirical methods for the correction of the wind near the surface have been suggested. In this study, the computed gradient wind is corrected, based on Takahashi (1947): the sea wind velocity is multiplied by 0.67, the sea wind direction is rotated 17 degrees to the low pressure side, the surface wind velocity is multiplied by 0.31 and the surface wind direction is rotated 33 degrees to low pressure side.

MASCON was developed by Dickerson (1978). This model can take account of the effects of land topography. A 3-D wind field is computed by correcting an initial wind field using the variational method based on constraint of mass conservation law. The initial wind field is computed under the assumption that the wind field is consisted of a surface boundary layer, an Ekman layer, and a free atmosphere layer. The initial wind field at the surface layer is set up using the computed gradient wind.

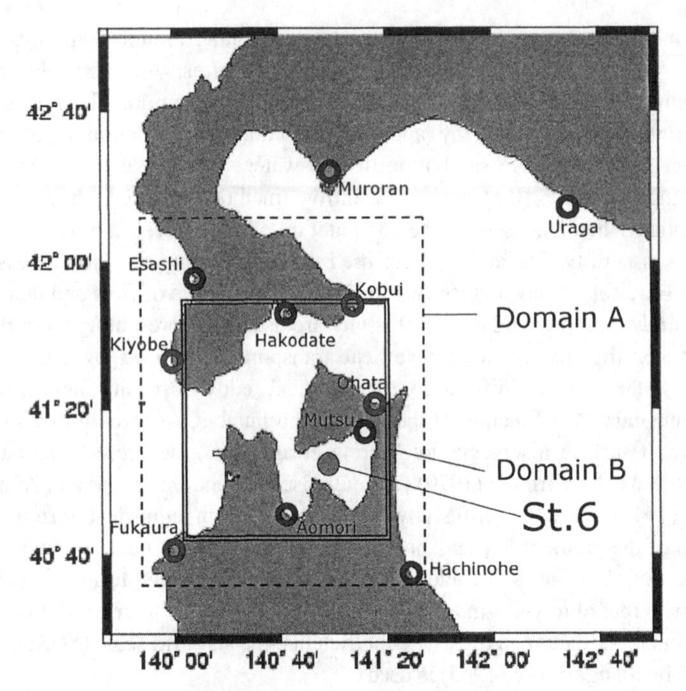

Figure 4. Computational domains, tidal gauges (gray circle) and weather observation points (black circle).

Lagrangian particle tracking model

A particle tracer simulation method (Nakatsuji, 1997) is applied in this study. The outline of this method is as follows:

The movement of particles per time step Δt is given by

$$x_{i+1} = x_i + u(x_i, t) \cdot \Delta t + [\nabla_i u(x_i, t) \cdot \{u(x_i, t) \cdot \Delta t\}] \cdot \Delta t + u' \Delta t .$$

where the first term on the right side represents the position of a particle at the ith step, the second and third terms represent advection due to tidal flow and shear in tidal flow, and the fourth term represents movement induced by turbulence. The turbulence is assumed to follow the primary Markoff process (Kadoyu, 1984).

$$u_i' = \mu u_{i-1}' + N(0, \sigma_0) .$$

where $\mu = \exp(-\Delta t / T_L)$, and $\sigma_o^2 = (1 - \mu^2) K / T_L$, N the normal random numbers and T_L is the integral time scale. T_L was set to 1 hour according to the tidal observation results obtained by Kadoyu (1984) for Osaka Bay, Japan. Because the computation time interval should be shorter than T_L and a one-step computed movement should remain within the grid interval $(\Delta x, \Delta y \; and \; \Delta z)$, Δt was set to 30 seconds in this study. The coefficient of eddy diffusivity, K, was given according to the SGS concept.

Numerical Arrangement for the Mutsu Bay

The computational domain is shown in Fig. 4 (Domain B). The grid size is 1km x 1km in the horizontal plane with 25 layers in vertical. The thickness of each layer from top to bottom is 2m x 3 layers, 4m x 1 layer, 5m x 15 layers, 10m x 2 layers, and 15m x 1 layer. The time step is set to 20 seconds.

The sea surface elevation at each open boundary is determined by tidal gauge data near the boundary; Kiyobe and Fukaura for the west boundary and Kobui and Ohata for the east boundary (Fig.4). The six major tidal constituents (i.e., M_2,S_2,K_2,K_1,O_1,P_1) and variation of atmospheric pressure are taken into account. Initial conditions for salinity and temperature are given by interpolating the data observed by Aomori Prefectural Fisheries Research Center and Naval

Table 1. Outline of Numarical Cases

Case	Tide	Atomospheric pressure	Gradient Wind	MASCON
1	Yes	No	No	No
2	Yes	Yes	No	No
3	Yes	No	Yes	No
4	Yes	Yes	Yes	No
5	Yes	Yes	No	Yes

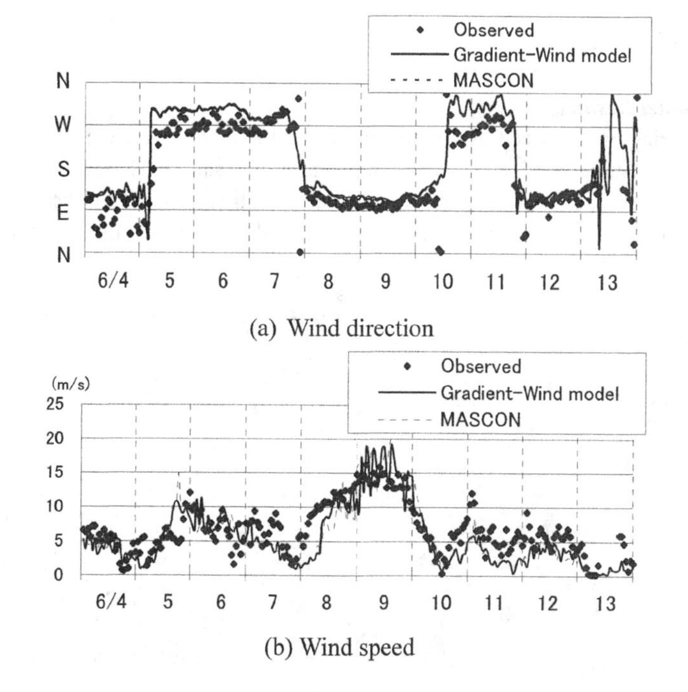

(a) Wind direction

(b) Wind speed

Figure 5. Comparison between observed and computed wind at St.6

Oceanographic Office.

The surface wind stress is calculated using results from the Gradient-Wind or MASCON models. The heat budget at the sea surface is calculated according to the formula shown in Murakami et al. (1989). The heat budget is composed of four components: short-wave radiation, long-wave radiation, sensible heat and latent heat of evaporation. These constituents are simulated with a bulk formula using meteorological data, obtained from Japan Meteorological Business Support Center. The distribution of sea surface pressure is obtained by interpolating data observed at weather stations around Mutsu Bay. The river discharges from 53 rivers around Mutsu Bay are considered.

Computational Approach

Five cases are simulated (Table 1). Four computational steps are used. The first is the computation of the barotropic flow for 4 days. The second is the computation of the baroclinic flow for 4 days to simulate the flow and density structure without meteorological forces. The third is the computation using the baroclinic flow model from June 4-13, 1997 considering the meteorological condition. All cases take the heat budget into account. The last is computation of a Lagrangian particle tracking to evaluate meteorological effect on the water exchange in the Mutsu Bay.

Results and Discussion

Wind model Results

A monitoring buoy, placed by Aomori Prefectural Fisheries Research Center at St.6 (Fig. 4) measures wind, sea water temperature and salinity. Figure 5 shows the comparison of computed and observed wind at St.6 (Fig. 4). The observed rapid changes in wind direction were simulated well, as shown in Fig. 5(a). The

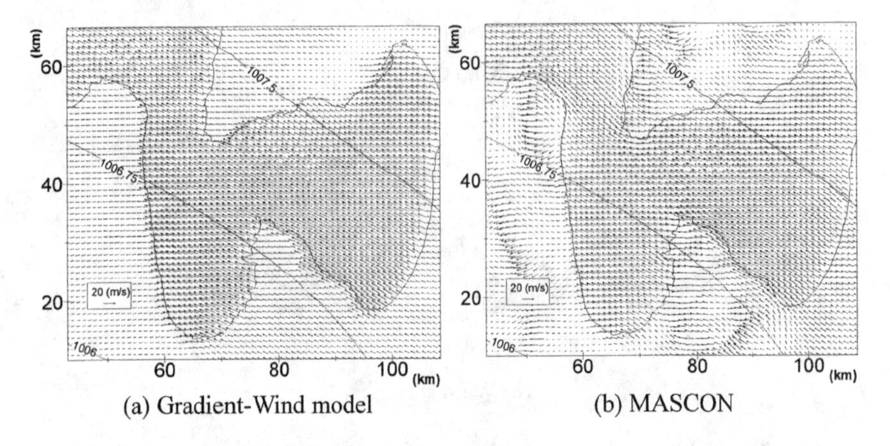

Figure 6. Distributions of computed wind vectors at 12:00 June 9, 1997

computed wind speed also agreed with observations as shown in Fig. 5(b). At first, it appears that there is no significant difference between the result of Gradient-Wind model and that of MASCON at St.6. For spatial distributions of wind vectors, however, the result of MASCON is different from that of Gradient-Wind model near the coastline, especially around the bay mouth (Fig. 6). The difference is caused by mountains reaching a height of about 400m above sea level on both east and west sides of the Tairadate Strait. MASCON can include effects of land topography, which are important, especially around the Tairadate Strait. The wind distribution of the MASCON may, thus, be considered more reliable than that of the Gradient-Wind model.

Simulation of the oceanographic phenomena and contributing factors

To examine a relationship between the unusual oceanographic phenomena (rapid change of density structure and sea level anomaly) and the passage of migratory cyclone, we compare results from Cases 1-5 with observations.

Figure 7 shows time series of sea level at Aomori and density profile at St.6 during June 8-13, 1997. The computed results for Case 1 in Fig.7(b) are different from the observed data in Fig.7(a): the observed sea surface rise and rapid change of density profile on June 10 do not appear in the computed result for Case 1. On the other hand, Case 2 in Fig.2(c), without wind effect, simulates well the tidal anomaly. However, Case 2 cannot reproduce the changes in the observed density structure. These results suggest that the observed sea surface rise is mainly induced by variation of atmospheric pressure. On the other hand, Case 3 in Fig.7(d), without the variation of atmospheric pressure, simulates well the density profile but fails to simulate the tidal anomaly. Consequently, it is reasonable to suppose that the rapid change of density profile is mainly caused by wind stress. Accordingly, Case 4 in Fig. 7(e) reproduces well both tide and density profile when the migratory cyclone passes. Therefore, it is concluded that these oceanographic phenomena (rapid change of density structure and sea level anomaly) are caused by the passage of migratory cyclone. Generally, a sea level anomaly is caused by both change in atmospheric pressure and wind set-up. However, there is no significant wind set-up in this case, because the fetch to Aomori observatory is very short for east wind on June 10 as shown in Fig.5(a)

Differences between using Gradient-Wind model and MASCON, are shown in Fig.7(e) and Fig.7(f). Since the density distribution of Case 5, especially near the sea bottom, is better than that of Case 4, MASCON is more effective in providing the boundary condition for wind.

Figure 8 shows time series of northward velocity at the depth of 15m at St.1 during June 8-13, 1997. The computed results for Case 5 also agree with observations except on June 13.

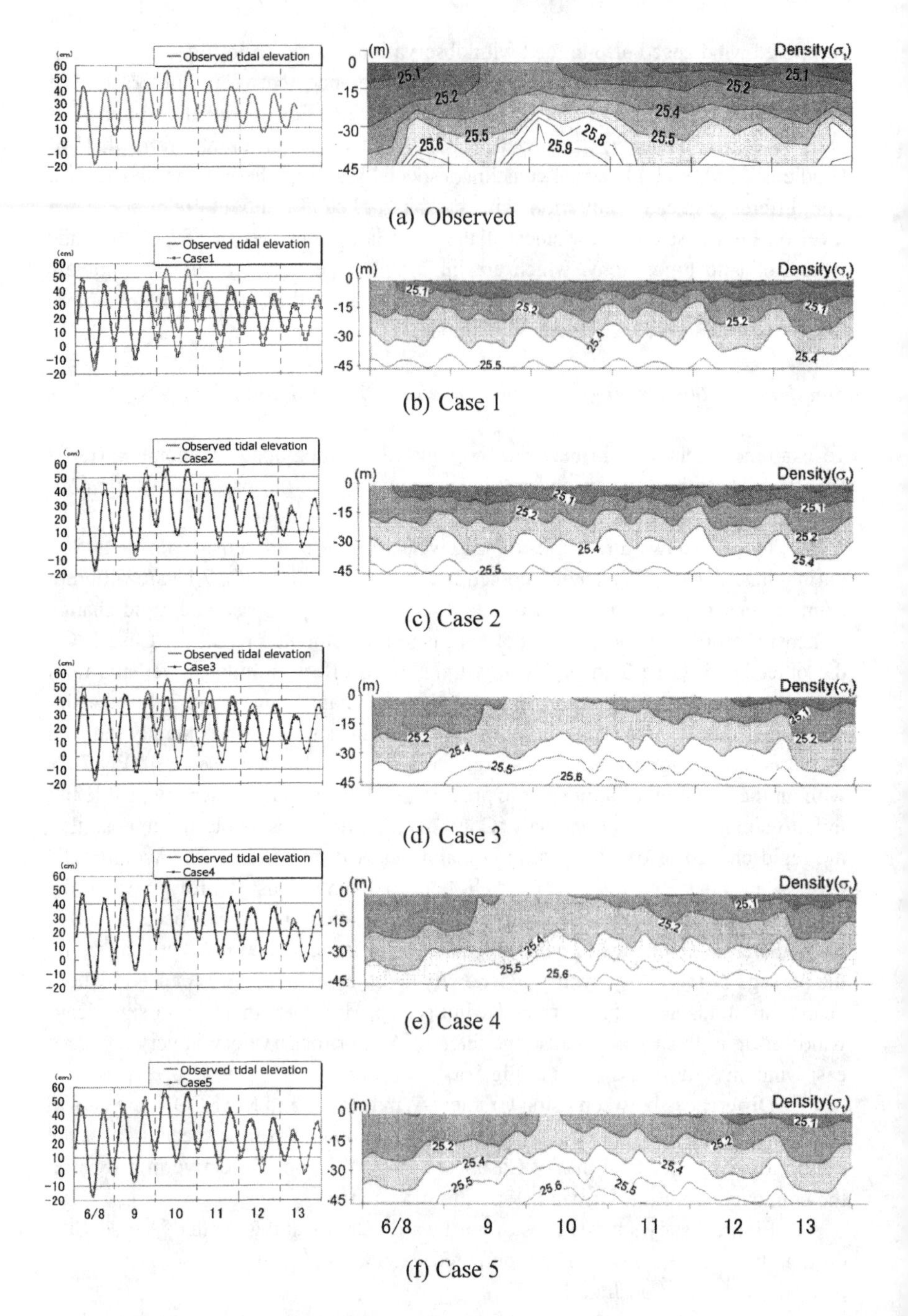

(a) Observed

(b) Case 1

(c) Case 2

(d) Case 3

(e) Case 4

(f) Case 5

Figure 7. Sea level time series at Aomori and density profile at St.6 from June 8-13, 1997

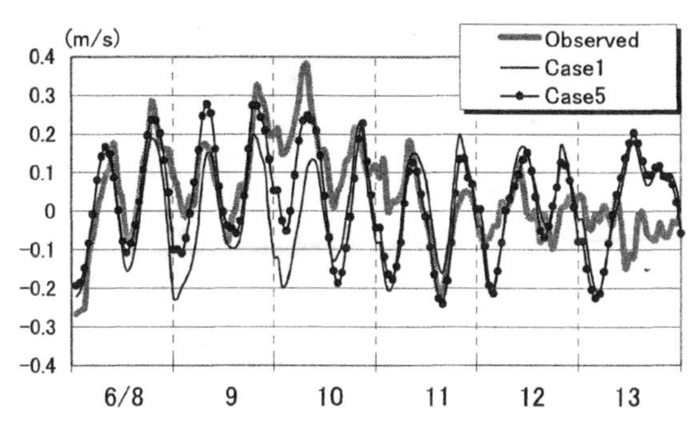

Figure 8. Northward velocity at the depth of 15m at St.1 from June 8-13, 1997

Meteorological effect on water exchange

The particle tracer simulation method is applied to Mutsu Bay to estimate the meteorological effects on the water exchange between Mutsu Bay and Tsugaru Strait.

Figure 9 shows an initial distribution of particles arranged at the sea surface. Particles are set every 2 km in horizontal direction and every 3 m in vertical direction inside Mutsu Bay. Then, the total number of particles is 3445 in the whole area of Mutsu Bay. In the present computation, a particle is treated as water mass. The computation period was set to 8 days (June 6-13). In this study, we examine the two cases of particle tracking, using computed results for Case 1 and Case 5.

Figure 10 shows the results of particle distributions after 8 days. This figure shows the locations of particles for all the levels. Some particles colored black, initially located in the west part of Mutsu Bay, move to the east along the south coast of the bay, while some other particles move out to Tsugaru Strait. Concurrently, almost all particles colored white, initially located in the east part of Mutsu Bay, remain in the same part. That means that the water exchange in Mutsu Bay caused by tidal currents alone is very small.

On the other hand, the distribution of particles in Case 5 is different from that of Case 1 due to the meteorological effect: a lot of black colored particles move out to Tsugaru Strait and some white colored particles move outside of the bay along the east coast of the bay mouth.

To estimate the effect of the passage of a migratory cyclone to the flushing of Mutsu Bay, Fig.11 shows the time series of the remaining particle ratio defined by $P(t)/P_0$, where $P(t)$ is the number of remaining particles at given time t and P_0 is the initial number of particles inside the bay. The remaining particle ratio for Case 1 decreases slightly from 1.00 to 0.97 after 8 days. On the other hand, the remaining particle ratio for Case 5 decreases to 0.90 after 8 days. Hence, it is concluded that the passage of migratory cyclone greatly affects the water exchange between Mutsu Bay and Tsugaru Strait, and that meteorological disturbances may play an important role on conservation of water quality in Mutsu Bay.

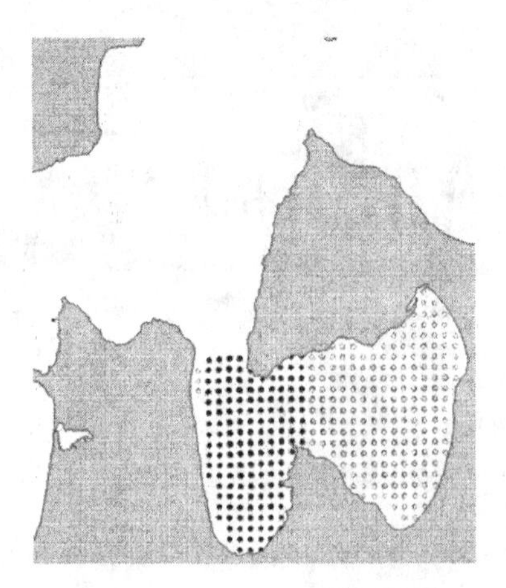

Figure 9. Initial distribution of particles

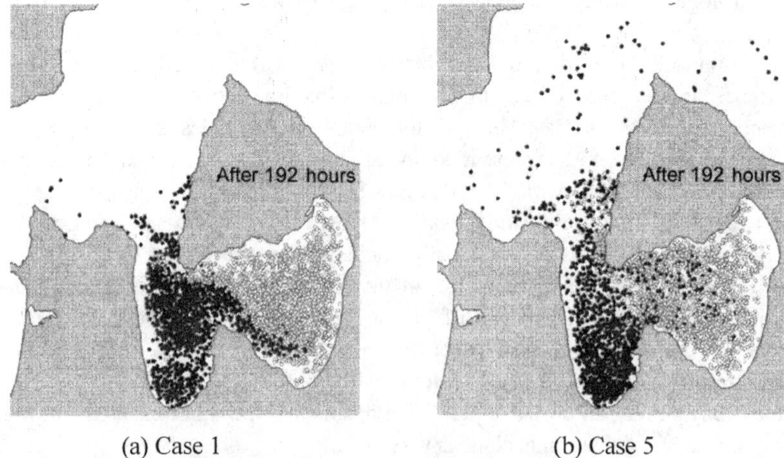

(a) Case 1 (b) Case 5

Figure 10. Distributions of particles at 24:00 June 13, 1997

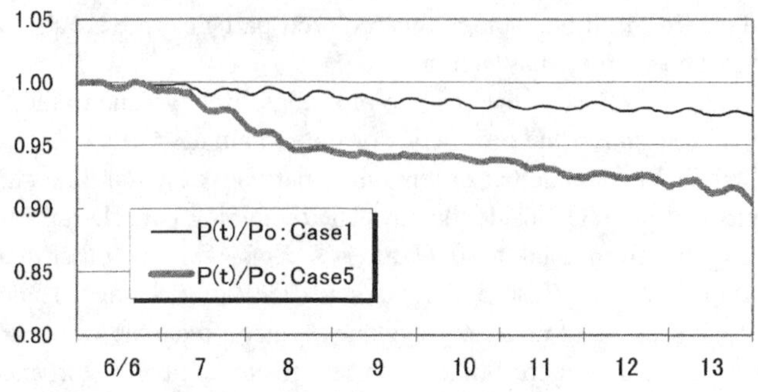

Figure 11. Remaining particle ratio from June 6-13, 1997

Conclusion

In this study, to clarify the influence of the passage of a migratory cyclone on the velocity and density structure around a semi-enclosed bay, numerical analyses were made using a three-dimensional baroclinic flow model under various oceanographic and meteorological conditions. In addition, a Lagrangian particle tracking model was used to examine the water exchange across the bay mouth.
The main conclusions obtained are as follows:

(1) The numerical model simulated well the observed sea surface setup induced by low pressure.
(2) With regard to wind and atmospheric pressure, the numerical model was able to hindcast the observed rapid change of density structure at the mouth of Mutsu Bay.
(3) The density structure and water exchange at the mouth of Mutsu Bay are affected by the passage of migratory cyclone over the Tsugaru Strait. The density structure is mainly affected by wind stress.
(4) Meteorological disturbances play an important role for the water exchange and flushing of Mutsu Bay.

Acknowledgment

This research was supported by the Grant-in-aid for Science Research (C) 11650528 of JSPS (Japan Society for the Promotion of Science).

References

Henderson-Sellers, B. (1985). "New formulation of eddy diffusion thermocline models" *Appl. Math. Modeling*, Vol. 9, pp. 441-446.

Kadoyu, M. (1984). "Stochastic study on entrainment of floating particles with intake of cool water" *Ph.D. thesis, Osaka Univ.*, Osaka, Japan (in Japanese).

Munk, W. H. and E. R. Anderson. (1948). "Notes on a theory of the thermocline" *J. Marine Research*, Vol. 7, pp.276-295.

Murakami, M., Y. Oonishi and K. Hayakawa (1989)."Heat and Salt Balances in the Seto Inland Sea" *Journal of the Oceanographical Society of Japan*, Vol. 45, pp. 204-216.

Murota, A., K. Nakatsuji, and Huh J. Y. (1988). "A numerical study of three-dimensional buoyant surface jet" *Proceeding of 6th APD-IAHR Congress*, , Vol. 3, pp. 57-64.

Nakatsuji, K. and T. Fujiwara (1997). "Residual baroclinic circulations in semi-enclosed coastal seas" *J. Hydraulic Engineering*, ASCE, Vol.123, No. 4, pp.362-373.

Nishida, S., K. Nakatsuji, H. Fukushima, K. Nishimura and T. Tashiro (2000)."Flow and Residual Flow in the Mouth of Enclosed Bay Opening to Strait "*The 4th International Conference on Hydro-Science and Engineering* (Published in CDROM).

Smagorinsky, J. (1963). "General circulation experiments with primitive equations" *Monthly Weather Review*, Vol. 91, No. 3, pp. 99-164.

Takahashi, K. (1947). "Quantitative Weather Forecast based on extrapolation method" *J. of Meteorological Research*, Vol. 13 (in Japanese).

Webb,W. K. (1970). "Profile relationships, the log-linear range and extension to strong stability" *Quarterly J. Royal Meteorological Soc.*,6, pp. 67-90.

Yamanaka, R., S. Nishida, S. Suzuki, K. Kawasaki, T. Tashiro and K. Nakatsuji (2001). "The effect of the passage of migratory cyclone on flow and density structure in Mutsu Bay", *Proceedings of Coastal Engineering*, Vol.48, pp 401-405(in Japanese).

Yamanaka, R., S.Nishida and K. Nakatsuji (2003). "Current Structure and Internal Waves Development at Stratified Strait", *Annual Journal of Hydraulic Engineering*, Vol. 47, pp. 1255-1260 (in Japanese).

Inverse Estimation of Estuary Flux

Jay Chung Chen[1] and Lai Ah Wong[1]

Abstract

The velocity field of a shallow bay is strongly influenced by two factors; the tides and the fresh water flux from the estuaries within the bay. The tidal motions can be accurately measured because of its periodical and consistent nature. While the volume of fresh water disposed from the estuary into the bay can be monitored in general, however, to do it continuously is not a trivial matter. In the present study, a method has been developed to indirectly estimate the estuary flux by using the continuous observations of surface velocity field of the nearby coastal waters. The approach is based on the concept of "system identification" in which a continuous relationship is assumed such that the surface velocities can be expressed as Taylor series functions of the estuary flux. Based on a rigorous mathematical procedure, the estuary flux can be estimated from the measured surface velocities. The method was applied to a continuous surface velocity data sets from a CODAR system located at the Pearl River Estuary for estimating the time history of fresh water flux from the river exits.

Introduction

The Pearl River is the third largest river in China. It discharges into the northern shore of the South China Sea through eight distributaries. The four eastern-most distributaries discharge their fresh water into the Lingdingyang which is a triangular water area also known as "The Pearl River Estuary" (PRE). The apex of this triangle area is at north approximately 50 km southeast of the city of Guangzhou. The other two corners of the baseline of the triangle are Hong Kong at the east and Macau at west. The fresh water discharges into the Lingdingyang triangular bay at the apex (Humen) and three other places at western shore (Jiaomen, Hongqili and Hengmen).

----

[1]Center for Coastal and Atmospheric Research, The Hong Kong University of Science and Technology, Clear Water Bay, Kowloon, Hong Kong; phone: (852) 2358-6901; fax: (852) 2358-1582; Email: jcchen@ust.hk

The circulation in the Pearl River Estuary (PRE) is complex owing to its complicated topography and multiple forcing functions such as river outflows, tidal motions, coastal currents and meteorological conditions (Figure 1). There are two distinct circulation patterns for PRE, one for summer and the other for winter because of the seasonal river discharge and seasonal climate.

There have been a few numerical studies on the circulations in the PRE and its nearby coastal waters (Ye et al, 1986; Ye and Preiffer, 1990; Wang et al. 1992; Huang et al, 1992). These studies have been limited to mostly tidal currents in the estuary which is insufficient for understanding the hydrodynamic characteristics and their variations in the PRE. Recently, 3D models have been established to simulate the estuary flow and the behavior of the plumes (Wong et al, 2003; Wong et al, 2003). However, the values for the estuary's fresh water fluxes within the Lingdingyang were approximate quantities based on the historical data of monthly averaged values. In principle, the fresh water discharge can be monitored, however, to do it on a continuous basis is not a trivial matter. The best available data are the annual or semi-annual average values plus estimated monthly variations.

In the present study, a method has been developed to indirectly estimate the estuary fresh water flux by using the continuous hourly observations of surface velocity field of the Lingdingyang. The results are the continuous daily fresh water discharges at the four estuarine inlets.

Surface Velocity Measurement

CODAR, an acronym for Coastal Ocean Dynamics Application Radar, measures the ocean currents remotely and continuously (Barrick, 1972; Puduan and Rosenfeld, 1996). During a project for the Pearl River Estuaries and adjacent coastal waters, a CODAR system was installed (Chen et al, 2000) and hourly surface velocity field measurement has been taken continuously for the past 3 years. Figure 2 shows the output of the CODAR system for the surface velocity field from 10:00am to 3:00pm on Feb 11, 2001. The 6 sets of images show clearly the high tide in the morning and the onset of low tide in the afternoon. Fig. 3 shows the 3 CODAR sites together with the fresh water discharge points of the PRE. Four sampling locations were selected for establishing the surface current time histories as also shown in Figure 3. A period of 3 months, from Sept 1 to Nov 30, 2001, was chosen for the hourly data sets. Figure 4 shows the time history of the selected sampling point 1 (upper figure) and the FFT frequency spectrum plot (lower figure). It is very important to note that the surface velocity is dominated by the periodic motions and the velocity amplitudes are changing rapidly. From the frequency spectrum, one may see that the major periodic motions are the once daily and twice daily harmonics. These can be readily attributed to the diurnal and semi-diurnal tides. Based on the spectral analysis, narrow band-pass filters with center frequencies at each of the diurnal and semi-diurnal tides, was constructed. The velocity data were then filtered by these band-pass filters. The results are the velocity without the harmonic components, i.e., the residual velocity. One may immediately notice that the residual velocity is very slowly varying in temporal coordinate. Figure 5 shows the residual velocity for the 4 selected points for the month of October, the end of the wet season and the beginning of the dry season. In general, the residual velocities show a decreasing trend.

The selected sampling points as shown in Figure 3 are the grids for the outputs of CODAR system. After applying the band-pass filtering process to the outputs of all the grids, a residual surface velocity map can be generated. Figure 6 shows the surface velocity maps at 12:00 and 19:00 of December 12, 2000. The maps at left show the velocity field of high tide at 12:00 (upper) and low tide at 19:00 (lower), respectively. After the narrow band-pass filtering process, the residual velocity maps are shown at the right for these two cases. The amplitudes of residual velocity are quite smaller and as expected the directions are going outward to open ocean for both high tide and low tide situations.

Using the technique, the entire time history of the residual velocity field can be obtained. It is these data which will be used for the estimation of fresh water discharges in the Lingdingyang bay.

System Identification

The approach used in the present study for the estimation of river fresh water discharges has been developed previously for aerospace applications (Chen and Garba, 1979; Chen et al, 1987). The method is based on the fact that the output parameters of any a system must be related to the system characteristics and its input parameters. Therefore, using the output parameters, in principle, the input parameters can be obtained, or the system characteristics can be calculated. The methods for these two procedures are referred to as "inverse solution" or "system identification", respectively. Mathematically, the exact solutions are not possible, thus only the approximate values can be estimated. Here it is assumed that the surface residual velocity field is the function of fresh water discharge only.

Consider now a residual surface velocity in Lingdingyang bay V, that is a function of variables, Q_1, Q_2, Q_3 and Q_4, representing the river discharges. Furthermore, it is assumed that V is an analytical function of the Q_i's such that it is continuous and all the derivatives exist. Then V can be written in the form of a Taylor series expansion:

$$V(Q_1.......Q_4) = V_0 + \frac{\partial V}{\partial Q_1} dQ_1 + \frac{\partial^2 V}{\partial Q_1^2} (dQ_1)^2 + ...$$

$$+\frac{\partial V}{\partial Q_2}dQ_2+\frac{\partial^{2V}}{\partial Q_2^2}(dQ_2)^2+............ \tag{1}$$

$$+\frac{\partial V}{\partial Q_4}dQ_4+\frac{\partial^2 V}{\partial Q_4}(dQ_4)^2+............$$

where V_0 is the averaged "monthly or seasonal" residual velocity and represents the contribution of other dynamic factors (Wind and density gradient field).

Based on the time history of observed residual surface velocities, it is reasonable to approximate that the changes of residual velocity are small with respect to the discharge quantities Q_i's, such that,

$$\frac{\partial^2 V}{\partial Q_i^2},\ \frac{\partial^3 V}{\partial Q_i^3}\ ....<<1 \tag{2}$$

and for small increment,

$$(dQ_i)^2,\ (dQ_i)^3\ ....\ <<1 \tag{3}$$

Then Eq. (1) can be approximated as

$$V=V_0+\sum_{j=1}^{j=4}\frac{\partial V}{\partial Q_j}dQ_j \tag{4}$$

a linearized Taylor series expansion. The term $\frac{\partial V}{\partial Q_j}$ is called "influence coefficient" whose physical meaning is the rate of change of the surface velocity due to changes of the discharges. These quantities will be estimated theoretically.

At this point, it should be mentioned that the influence coefficients and the averaged residual velocities are calculated by using a hydrodynamic model. The detailed descriptions of the model can be found in references (Wong et al, 2003; Wong et al, 2003). The averaged monthly discharges Q_0's are given from which the averaged residual velocity V_0 can be calculated. The influence coefficient can be defined mathematically as the ratio of the velocity increment ΔV and discharge increment ΔQ.

$$Lim\frac{\partial \mathrm{V}}{\partial \mathrm{Q}_\mathrm{j}} = \frac{\Delta V}{\Delta Q_j} \text{ and } dQ_j \rightarrow \Delta Q_j \tag{5}$$

ΔV is obtained from the model by the increment of the discharge quantity ΔQ_j at a specific discharge location keeping all other discharges at nominal values. If both the incremental values of ΔQ and the corresponding velocity variations ΔV are kept small and the corresponding velocity changes are also small, Eq. (5) is mathematically valid. Therefore, we can determine the influence coefficient matrix $\left[\frac{\partial V}{\partial Q}\right]$ by using the least square method and the simulated data from the numerical model for given a certain input variation of ΔQ .

For any an instantaneous $t_k, k = 1,2,\cdots,K$, Eq. (4) can be written as

$$\delta V_{qi}(t_k + \tau_i) = a_{ij}\delta Q_j(t_k)\text{, k=1,2, ...,K, i =1,2,...} \tag{6a}$$

The matrix form of the Eq. (6a) is

$$\{\delta V\}_i = [\delta Q]\{a\}_i\text{, i =1, 2,...} \tag{6b}$$

Where the matrix $\delta Q = \lfloor \delta Q_j(t_k) \rfloor_{KxJ}$ is specified in the model. Both $\{\delta V\}_i$ and $\{a\}_i$ are the column vectors and $\{\delta V\}_i = \lfloor \delta v_{ri}^j(t_1+\tau_i) \quad \delta v_{ri}^j(t_2+\tau_i) \quad \cdots \quad \delta v_{ri}^j(t_K+\tau_i) \rfloor^T$, can be obtained by simulated current data from numerical model sensitive analysis experiments. $\{a\}_i = \lfloor a_{i1} \quad a_{i2} \quad \cdots \quad a_{iJ} \rfloor^T$ is a unknown influence coefficients which can be determined by solving the Eq. (6b) using the least square method.

After the influence coefficient matrix between the river discharge variation and the variation of the residual current, then substituting the variation of the residual currents and the Eq. (4), the unknown variation of the river discharge can be identified. Eq. (4) can be expressed in a matrix form as:

$$\{V\}=\{V_0\}+\left[\frac{\partial V}{\partial Q}\right]\{\Delta Q\} \tag{7}$$

In this equation the term $\{V\}$ contains residual velocity measured at four selected sampling points. Each point has two velocity components, namely, u_i in east-west direction and υ_i in north-south direction. Therefore, the $\{V\}$ column matrix contains eight components. The $\{V_0\}$ is obtained from the averaged value of the residual current in our considering the time domain. $\left[\frac{\partial V}{\partial Q}\right]$ is the influence coefficient matrix as described previously and its dimension is 8 x 4, it can be determined by using the least square method to the given input variation of the river discharge and simulated residual current variation from the numerical model. $\{\Delta Q\}$ is the unknown column matrix containing four components representing the discharge variations at the 4 river exits. Eq. (7) contains more governing equations than the number of unknown to be solved. This is an over-constrained case and no exact solution exists. However, an approximate solution is possible. By using a minimization procedure (Chen and Garba, 1979; Chen et al, 1987), the approximate solution can be obtained as,

$$\{\Delta Q\}=\left(\left[\frac{\partial V}{\partial Q}\right]^T[W]\left[\frac{\partial V}{\partial Q}\right]\right)^{-1}\left[\frac{\partial V}{\partial Q}\right]^T[W]\{\Delta V\} \tag{8}$$

where $\{\Delta V\}=\{V\}-\{V_0\}$ and

$[W]$ is a positive definite compatible weighting matrix whose purpose is to emphasize the important measurements in the procedure.

Validation of Method – Simulation

Before applying the method to the real data, the proposed method is first validated by using simulated data. The simulation was from the same model outputs as described in references (Wong et al, 2003; Wong et al, 2003) but with the assumed freshwater discharge functions which is shown in Figure 7. The assumed discharge function consists of a constant level and then a 30 day period oscillations. Figure 8 shows the model output of residual velocities at 4 selected points after filtering out the much shorter period tidal motions. These results will be used as the simulated data. Applying these simulated data to the previously described method, the discharges at the four river exits were estimated. The results are compared with the assumed values as shown in Figure 9 which indicates very good agreement. It should be noted that the influence of discharges arriving at different points within the Lingdingyang bay at different times. This can be seen from Figure 8 in which the responding residual velocity to the input discharges shows phase difference for each of the four sampling points. During the computing of influence coefficients, these time delays must be considered. Similarly, attention is paid during the comparison of results.

Discharge Estimation

Finally, the residual velocity data obtained from CODAR measurement as shown in Figure 5 were applied to the procedure as described previously. The discharge quantities at each of the river exits, namely, Humen, Jiaomen, Hongqili and Hengmen were estimated as time functions (daily). Figure 10 shows this result in which the vertical axis is the time varying portion of the discharge rate i.e. the total discharge minus the monthly average value, $\{\Delta Q\}=\{Q\}-\{Q_0\}$.

The result does show that the discharges are varying in time and the net changes within the month could be significant, as much as 40% of the monthly average values. While no actual daily discharge measurements exits for comparison, the results appear to be reasonable. The two river exits at Humen and Jiaomen shows much higher variations than the other two at Hongqili and Hengmen. This is consistent with the monthly averaged values in which the first two discharges are 2 to 3 times bigger than the other two. Also, the characteristics of the discharge time functions at Hongqili and Hengmen are similar, this is again consistent with the fact that these two river exits are very close and their catchment area are similar.

Concluding Remarks

A preliminary attempt has been made to estimate the fresh water discharge time function for the Pearl River Estuary by using a continuous data set of surface current measurement. The estimation was made possible by several assumptions. These assumptions, while appeared reasonable, have not been verified vigorously. Ultimately, the results of estimated discharge values must be compared with the in-situ measurement continuously over a period. It is hoped that this will be done in the future.

In addition to the hydrodynamic models for the coastal waters, more and more requests for water quality and eco-system models are being formulated. These models require higher resolutions and higher accuracies for the boundary conditions and driving forces in order to simulate the reality. Detailed time function requirement for river discharges are certainly necessary for these modelling tasks.

Acknowledgement

The present study was supported by the Pearl River Estuary Pollution Project funded by the Hong Kong Special Administrative Region Government/Hong Kong Jockey Club, the 863/818-09-01 grant from the Ministry of Science and Technology of China, and the N_HKUST609/02 project funded by NSFC/RGC Joint Research Scheme.

References

Barrick, D.E. (1972). "First-order theory and analysis of MF/HF/VHF scatter from the sea." *IEEE Trans. Antennas Propag.*, AP-20, 2-10.

Chen, J.C., and Garba, J.A. (1979). "Analytical model improvement using modal test results," *AIAA Journal.*, Vol. 18, No. 6. pp. 684-690.

Chen, J.C., Peretti, L.F., and Garba, J.A. (1987). "Spacecraft structural model improvement by modal test results." *Journal of Spacecraft and Rockets.*, Vol. 24, No.1.

Chen, J.C., Wong, L.A., Dong, L.X., Xu, W., Guan, W.B., and Xu, D.F. (2000). "The Pearl River Estuary integrated observation system." *Tech. Rep.*, Cent. For Coastal and Atmos. Res., Kowloon, Hong Kong, 7, pp.1-7.

Huang, B.X., Guao, Z.S., Xian, K.K., and Huang, F. (1992). "Numerical simulation of tidal current of the Pearl River Estuary." *Chin. J. Oceanol. Limnol.*, 23, 475-484.

Paduan, J.D., and Rosenfeld, L.K. (1996). "Remotely sensed surface currents in Monterey Bay from shore-based HF radar (Coastal Ocean Dynamics Application Radar)." *J. Geophys. Res.*, 101, No. C9, pp 20,669-20,686.

Wang, J., Yu, G., and Chen, Z. (1992). "Numerical simulation of tidal current in Neilingding Yang of the Pearl River Estuary." *Acta Oceanol.*, 14, 26-34.

Wong, L.A., Chen, J.C., Xue, H., Dong, L.X., Su, J.L., and Heinke, G. (2003). "A model study of the circulation in the pearl River Estuary (PRE) and its adjacent coastal waters: 1. Simulations and comparison with observations." *J. Geophys. Res.*, 108, doi:10.1029/2002JC001451.

Wong, L.A., Chen, J.C., Xue, H., Dong, L.X., Guan, W.B., and Su, J.L. (2003). "A model study of the circulation in the Pearl River Estuary (PRE) and its adjacent coastal waters: 2. Sensitivity experiments." *J. Geophys. Res.*, 108, doi:10.1029/2002JC001451.

Ye, J., He, Z., and Zhou, Z. (1986). "Numerical model of salt transport under tidal current for the Pearl River Estuary." *Renmin Zhujiang*, 6, 7-15.

Ye, L., and Preiffer, K.D. (1990). "Studies of 2D & 3D numerical simulation of Kelvin tide wave in Neilingding Yang at the Pearl River Estuary." *Ocean Eng.*, 8, 33-44.

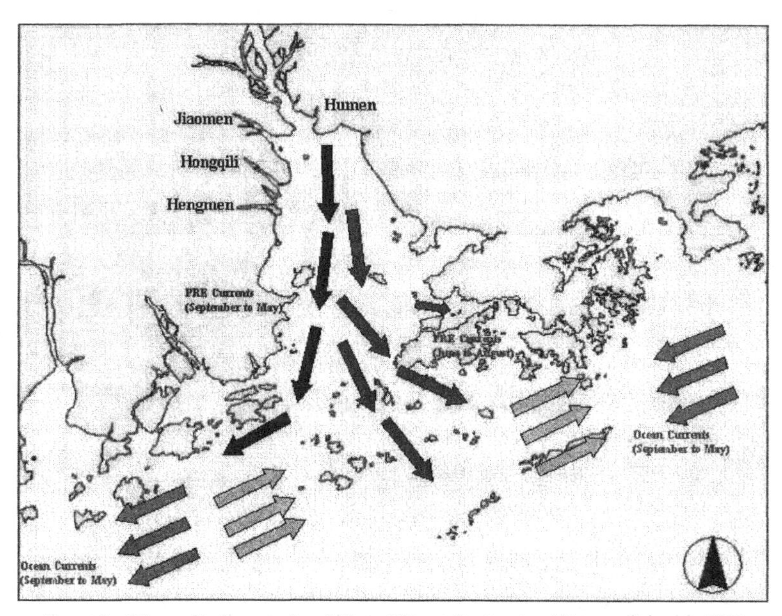

Figure 1. Schematic of seasonal variation of Ocean Currents and Estuary Currents, PRE

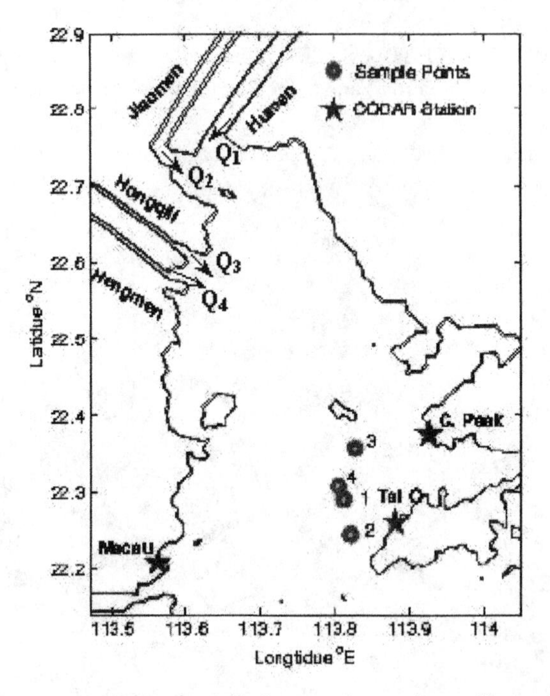

Figure 3. CODAR sites, discharge points and sampling points

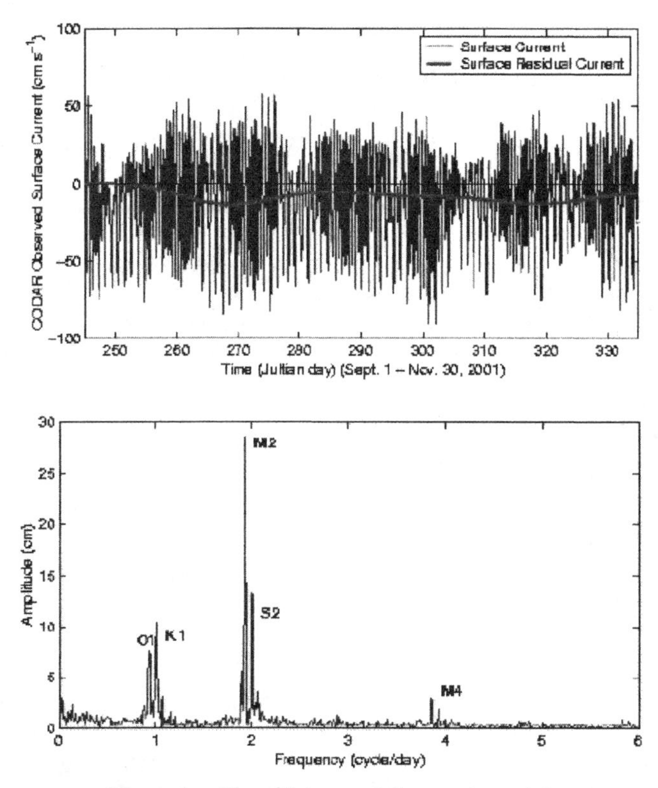

Figure 4. Time history and the spectrum data

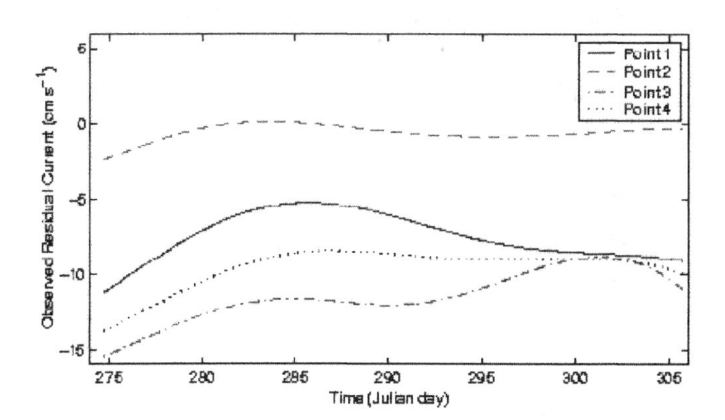

Figure 5. Observed Residual Current in Lingdingyang

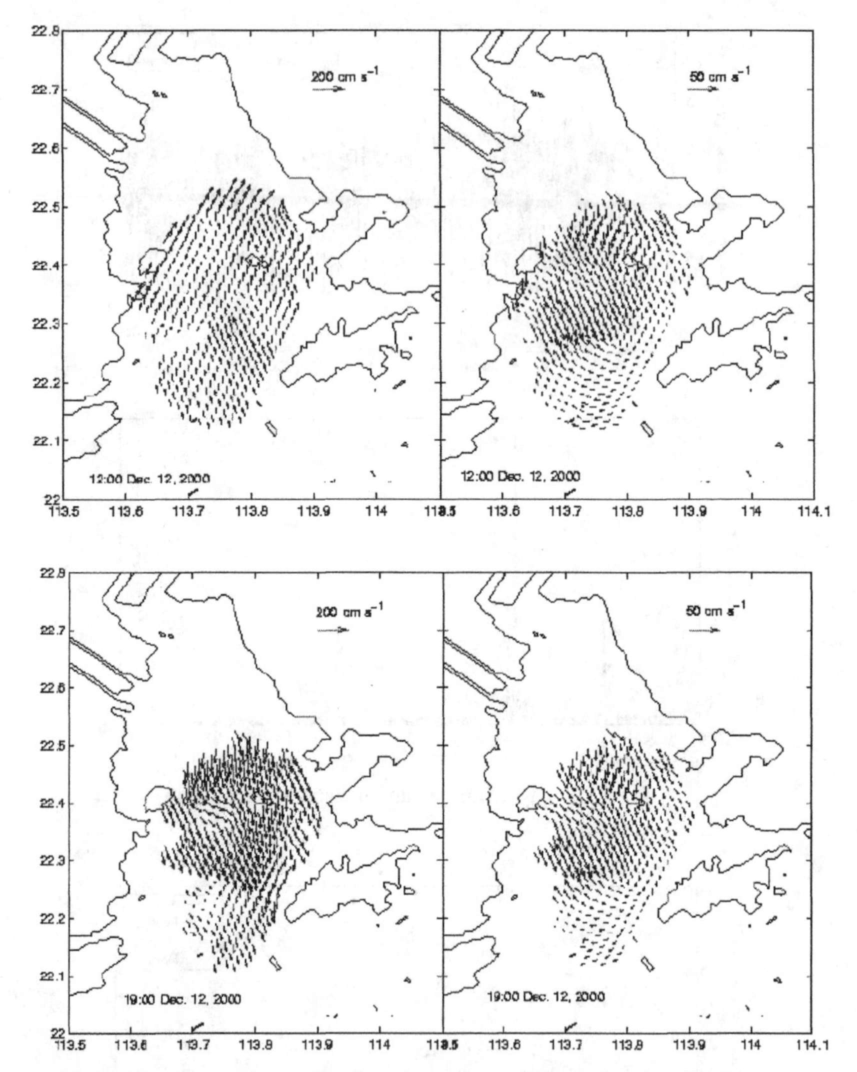

Figure 6. Comparison of surface current and the associated residual current during the tidal motion.

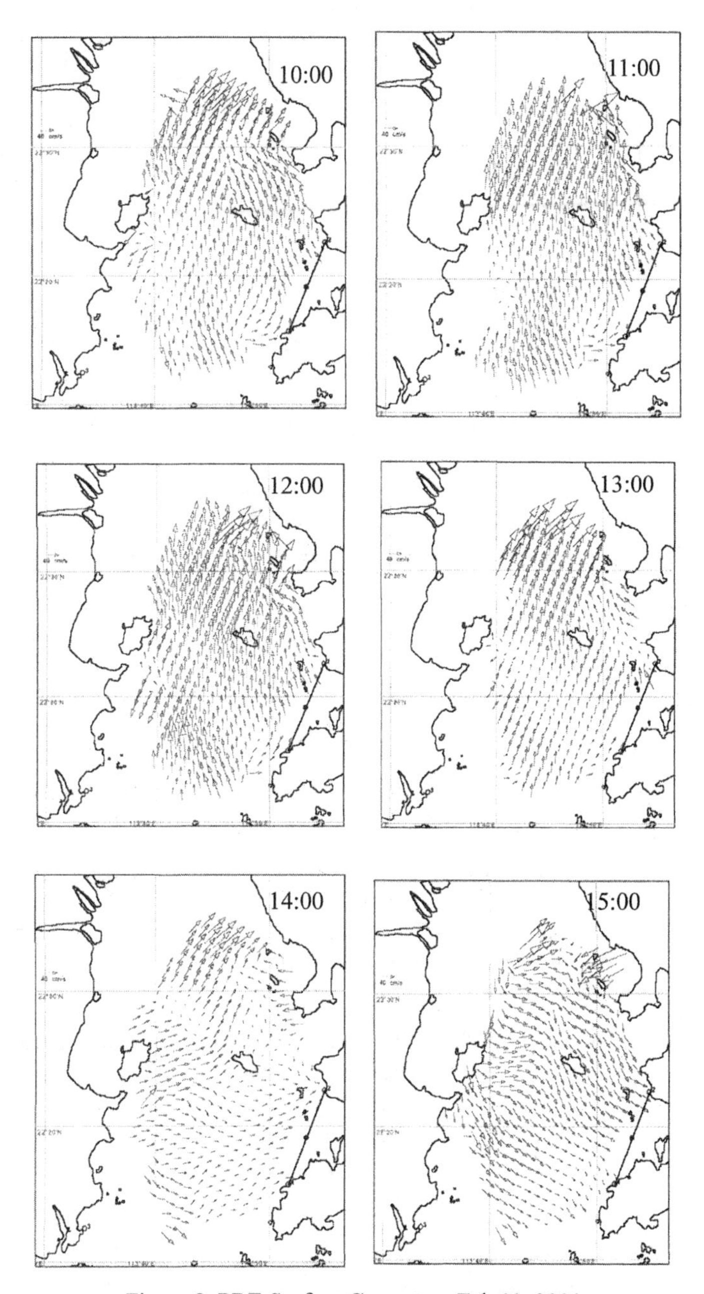

Figure 2. PRE Surface Current on Feb 11, 2001

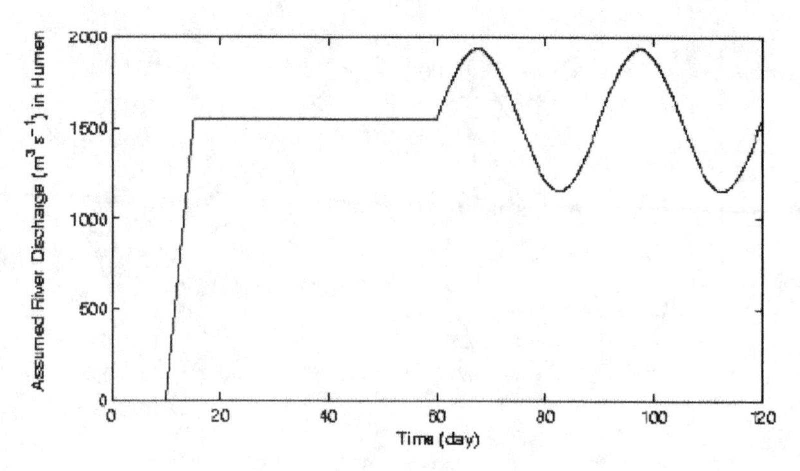

Figure 7. Assumed Discharge

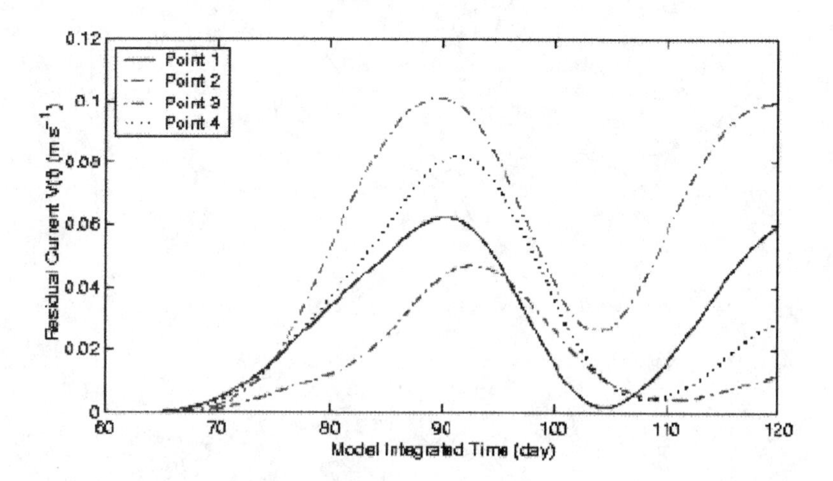

Figure 8. Variation of residual current in Lindingyang from assumed discharge variation

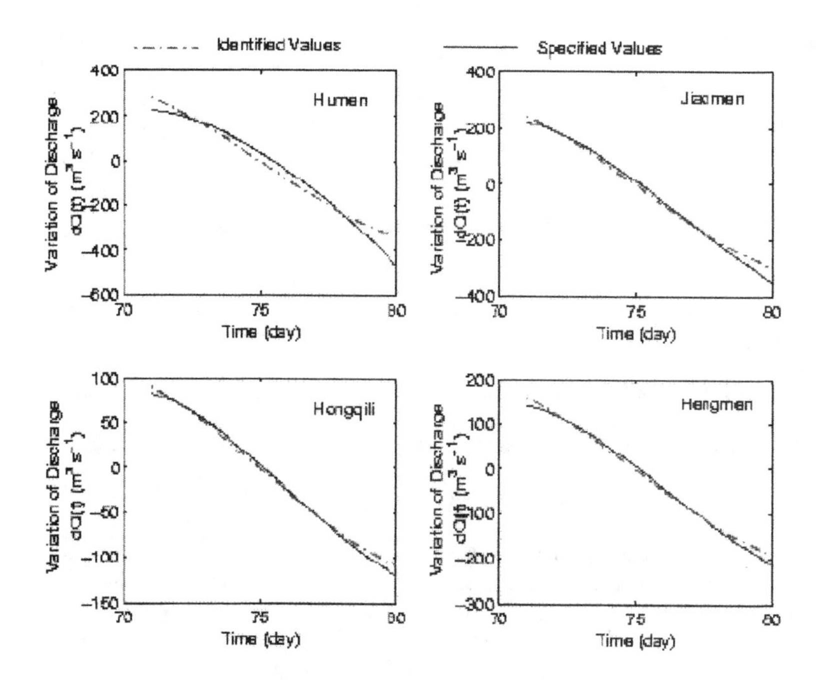

Figure 9. Comparison of Discharge Variation

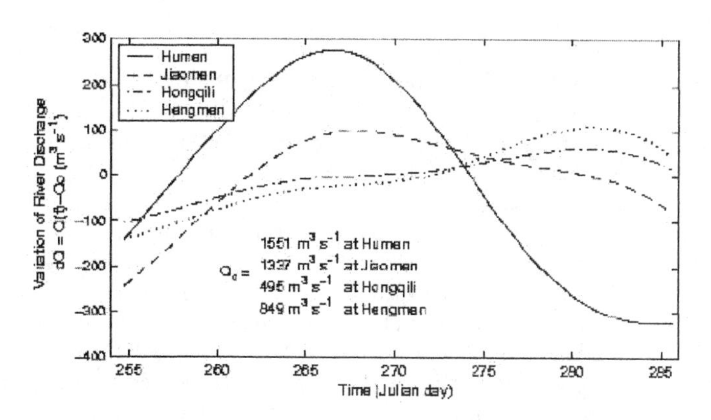

Figure 10. Estimated Discharge Variation at PRE

"Grid"-based Particle Tracking in Florida Bay

Justin R. Davis[1], Y. Peter Sheng[12] and Renato J. Figueiredo[3]

Abstract

In recent years, grid computing has emerged as an attractive approach to solve large problems through the coordinated sharing of resources. While successful in other fields, e.g. high energy physics, few studies focusing on coastal and estuarine problems have been performed; thus, a pilot study was performed to illustrate how coastal and estuary modeling could use grid computing technologies. To this end, several coastal and estuarine models were dynamically coupled and then executed in a local computational grid environment composed of Silicon Graphics computers connected by a fast ethernet network. The models: a robust hydrodynamic model (CH3D), a particle tracking model, and a visualization model communicated through the network using a grid-enabled version of MPI, MPICH-G2. The grid-enabled, coupled modeling system was then tested on a real estuary, Florida Bay, USA. Using simulated hydrodynamics (Node 1), the trajectories of 2000 particles were simulated by two independent instances of the particle tracking model (Nodes 2 and 3). Updated particle positions were then visualized (Node 4) and compared with those calculated with a statically coupled version of the modeling system running on a single node. The dynamically coupled system was shown to reproduce the particle trajectories exactly with some speedup due to the two particle tracking models running in parallel. A graphical interface to the modeling system was then created using the advanced web portal and grid middleware software In-VIGO. The robustness of the results shown along with relative ease of converting an existing modeling system into a dynamically coupled system running in a grid environment demonstrated the feasibility of using grid technologies on coastal and estuarine modeling.

---

[1]Department of Civil and Coastal Engineering, 365 Weil Hall, P. O. Box 116580, University of Florida, Gainesville, FL 32611-6580.

[2]Member, ASCE.

[3]Advanced Computing and Information Systems Laboratory, Department of Electrical and Computer Engineering, 339 Larsen Hall, P. O. Box 116200, University of Florida, Gainesville, FL 32611-6200

Introduction

In recent years, grid computing has emerged as an attractive approach to solve large problems through the coordinated sharing of resources. The sharing of resources in grid computing is analogous to the sharing of power in an electrical power grid and can be defined as the "secure, coordinated resource sharing among dynamic collections of individuals, institutions, and resources" with the caveat that the sharing "...is not primarily file exchange but rather direct access to computers, software, data and other resources ..." (Foster et al., 2001). For example, resources can include such things as: computational resources (CPU cycles), storage resources (disk space), network resources, code repositories and catalogs. Whereas traditional parallel message passing paradigms focus on local systems or clusters, grid computing is focused more on the multi-institutional, geographically disparate sharing of computational resources.

While grid technologies have proven successful in other fields, e.g. high energy physics, few studies focusing on coastal and estuarine problems have been performed; thus, a pilot study was performed to illustrate how coastal and estuary modeling (e.g. integrated scale and integrated process modeling (Sheng et al., 2004)) could use grid computing technologies through the creation of a coastal and estuarine application solved in a grid environment. In the remainder of this paper, the following issues will be discussed: (1) how the coupled grid-based modeling system was developed, (2) how the system was tested on a real estuary, Florida Bay, USA, and finally (3) how the system was interfaced through a grid-enabled Internet-based web portal. In addition, based on the experiences of this study, recommendations on the use of grid-technologies in coastal and estuarine modeling will be provided.

Development of the Coupled Modeling System

The hydrodynamic and particle tracking modeling system is composed of three component models: a hydrodynamic model, a particle tracking model and a visualization model, which are coupled together in a local computational grid. The computational grid is created using grid middleware software while the model coupling is achieved through minor modifications in the component models' source code.

The hydrodynamic model is composed of the circulation and salinity transport components of the CH3D-IMS. CH3D, a three-dimensional curvilinear-grid hydrodynamic model, was originally developed by Sheng (1987, 1990, 1994). Early versions included only a hydrodynamic model, but the present CH3D Integrated Modeling System (CH3D-IMS) (Sheng, 2000) includes sediment transport, flushing and water quality components, as well. Circulation and salinity transport modeling parameters (boundary and initial conditions, friction coefficients, *etc.*) related to the example application study area, Florida Bay, are based on a previous study of Florida Bay (Sheng et al., 1995).

The particle tracking model uses a simple Lagrangian methodology which includes a simple random walk assumption. The particle tracking model is designed to read a set of initial particle locations followed by CH3D output files to produce a set of particle trajectories. Each individual particle's velocity is determined from the three-dimensional velocity at the end of some time increment as determined by the CH3D model; hence, this time increment is user defined and can be any multiple of the time step used by the hydrodynamic model.

The visualization model is a simple interface to the particle tracking model designed to convert particle properties (e.g. location and velocity) to a format suitable for use by the commercial plotting software, Tecplot. The functionality of the visualization model is detached from the particle tracking model to maximize portability of the particle tracking model thus allowing particle trajectory information produced by the particle tracking model to remain generic.

Three steps were taken in the model development and application process: (1) development and configuration of the local computational grid environment, (2) modification of the component hydrodynamic, particle tracking and visualization model's source code to include communications to pass data between nodes, and (3) setup and execution of the simulation followed by assessment of the particle trajectories as produced by the visualization model.

The computing platform chosen for creation of the local grid environment consisted of a four Silicon Graphics computers (Nodes A, B, C and D) connected by a fast ethernet network. A homogeneous system was chosen to more easily facilitate installation of necessary software. The local computational grid environment which connects the nodes through the network was established through the installation of the core grid middleware software, the Globus Toolkit[4] (Foster and Kesselman, 1999). Model inter-communication was achieved using a grid-enabled version of MPI (Message Passing Interface Forum, 1994), MPICH-G2[5] (Karonis et al., 2003). MPICH-G2 enables communications between component models to occur within the grid using standard MPI constructs.

Based on the four available nodes, the following algorithm was developed illustrating the interaction of the component models within the local computational modeling grid:

- Node A simulates bay circulation using the CH3D hydrodynamic model.
- Node A broadcasts (*MPI_BCAST*) the simulated flow field to the particle tracking models located on Nodes B and C periodically.
- The particle tracking models on Nodes B and C each receive (*MPI_BCAST*) the broadcasted flow field and then simulate the movement of their respective particles during the time period.
- The particle tracking models on Nodes B and C send (*MPI_SEND*) the updated particle position and velocity data to the visualization model located on Node D.

[4]http://www.globustoolkit.org
[5]http://www.niu.edu/mpi

- The visualization model on Node D receives (*MPI_RECV*) the updated particle position and velocity data, assembles the data together, and then converts the data to a format usable by the plotting software Tecplot.

Implementing the algorithm using the hydrodynamic, particle tracking and visualization models described previously, required only minor modifications of input and output procedures within each models' source code. The hydrodynamic model required no changes to the input procedures while the flow field output procedures were modified from Fortran *write* statements to MPI broadcast (sending) statements. The input procedures in the particle tracking model were modified from Fortran *read* statements to MPI broadcast (receiving) statements. Output procedures in the particle tracking model were modified from Fortran *write* statements to MPI send statements. The input procedures in the visualization model were modified from Fortran *read* statements to MPI receive statements while the output procedures remained unchanged.

Setting up and starting the coupled system within the Globus grid environment, requires several steps. First, each of the component models and their associated data files need to be placed on the nodes of the computational model grid. The hydrodynamic model executable and its associated input files (boundary conditions, initial conditions, coefficients, *etc.*) need to be placed on Node A, the first particle tracking model executable and its associated initial particle locations need to be placed on Node B, and so on.

Next, on the node where the simulation will be started (this could have been any of the nodes), a file needs to be created which gives executable names and locations of the working directories for all the component models to be used in the coupled simulation. This file is written in the Globus Globus Resource Specification Language (RSL) and is commonly known as a "RSL script".

Finally, with the coupled system now setup, the simulation is started using the "globusrun" command which is similar to "mpirun" command commonly used when running MPI programs. Once the simulation has started, output of the visualization model can be monitored to assess the progress of the simulation.

Application to Florida Bay

As a test of the coupled modeling system, a study of particle tracking motion in Florida Bay, USA was performed. Florida Bay is the triangular, shallow-water estuary located directly south of the Florida peninsula (Figure 1a). The bay is bordered to the southeast by the Florida Keys and to the west by the Gulf of Mexico and is very shallow with average depths of less than 1 m in the interior of the bay (Figure 1b). The horizontal grid system used for simulation is a boundary fitted, curvilinear grid (96 $\times$ 73) (Figure 1c) (Sheng et al., 1995). In the vertical direction, a vertically stretched, sigma grid is used with 4 vertical layers. Bathymetry in the boundary-fitted grid system is compiled from high resolution (20 m $\times$ 20 m) National Park Service data.

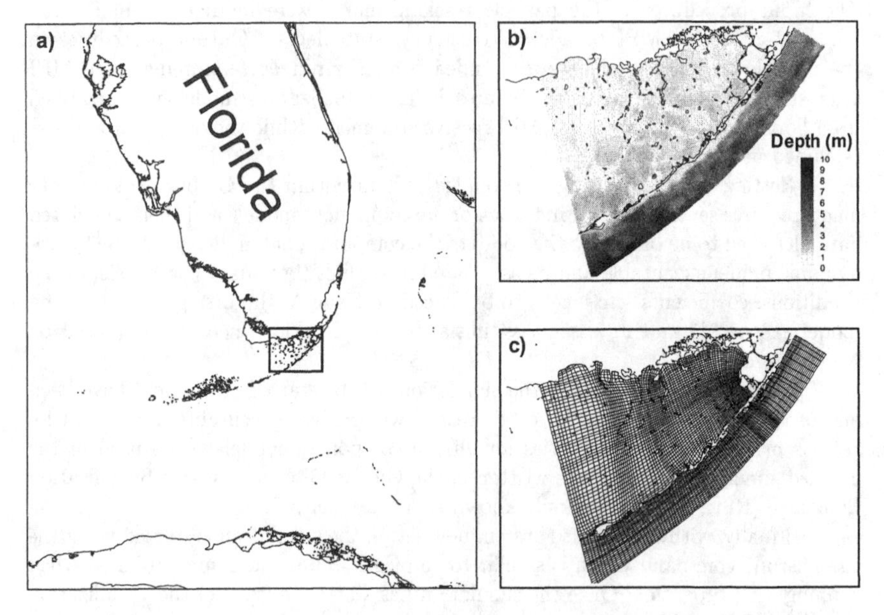

Figure 1: a) The Florida Bay study area. b) Bathymetry in the interior of Florida Bay. c) The boundary-fitted, curvilinear-grid system (96 × 73) used for model simulations.

A five day simulation was performed using the dynamically coupled system. The hydrodynamic model used a 30 s time step and broadcasted the flow field every 10 time iterations (5 min of simulated time, approximately 3 s of wall clock time). Tidal forcing was used as the primary driving force of hydrodynamic model and was imposed using harmonic constants derived from measured sources (Smith, 1995; Smith and Pitts, 1995). Specifically, three constituents were used: M_2, K_1, and O_1. These constituents comprise more than 75% of the total tidal signal. Along the western boundary, constituent data was obtained from co-amplitude and co-phase charts (Smith and Pitts, 1995). Along the southeastern boundary, constituents were calculated from measured tidal data collected by the Harbor Branch Oceanographic Institution (HBOI) and the National Ocean Service (NOS) at three stations along the Florida Reef Tract: Tennessee Reef (HBOI), Alligator Reef (NOS) and Carysfort Reef (NOS) (Smith, 1995).

The overall flow of data in the Florida Bay particle tracking example is shown in Figure 2. Based on the number of wet cells in the Florida Bay boundary-fitted (approximately 40% of the total number of cells) hydrodynamic grid system, 558 kb of field data (water levels, velocities, etc.) are broadcast from the hydrodynamic model to the particle tracking models. Based on the number of particles being simulated, the particle tracking models send 23.4 kb of updated particle properties (positions, velocities, etc.) to the visualization model. Since data is being passed through a local network and the total amounts are small (581 kb) and occur infrequently (every 3 s), bandwidth and latency issues are unimportant in this test. However, in general, such issues need to be considered and will be a topic of future study.

Initially, the 2000 particles (1000 particles on each Nodes A and B) are randomly distributed within the boundaries one cell of the hydrodynamic model (Figure 3). The simulation is then started allowing the particles to move with the simulated currents. Output of the visualization model was then monitored to determine the particle locations after 1, 3 and 5 days (Figures 4-6, respectively). The coupled modeling system running in the local computational grid was able to exactly reproduce particle movements as compared to running the particle tracking model in a post-processing manner after the completion of the hydrodynamic simulation. Although both the particles' initial position and simulated paths were based on randomness, exact reproduction was possible due to identical seed values provided to the "pseudo" random number generators used by the models. A slight decrease in wall clock time for the entire coupled simulation was experienced due to the parallel nature of the particle tracking models although further study is necessary to quantify exact speedups.

System Interface via an Internet Web Portal

Interactive access to the grid-based coupled modeling system was created through an innovate approach that relies on using virtualization of physical and information resources to hide infrastructure heterogeneity and deliver the desired functionality to end-users. The heart of this approach is the enabling middleware technology,

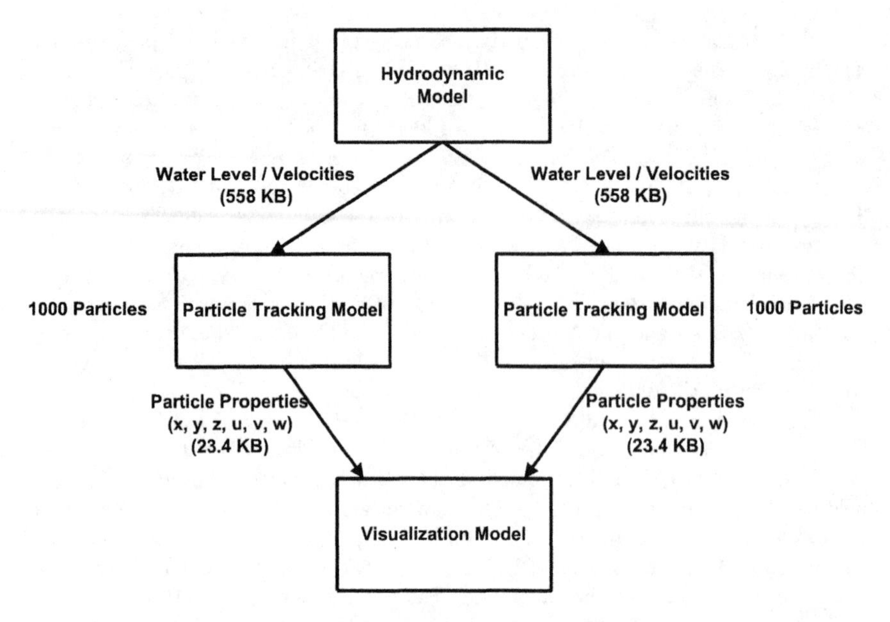

Figure 2: Data flow in the coupled hydrodynamic-particle tracking modeling system.

In-VIGO[6] (Adabala et al., 2003) developed at the Advanced Computing and Information Systems (ACIS) Laboratory at the University of Florida. In-VIGO is web portal and grid middleware software which provides software, simulation services and educational resources in an easy-to-use environment. In-VIGO controls all aspects of model setup and simulation as well as subsequent usage post-processing programs and visualization tools. In-VIGO is the next generation of PUNCH[7] (Kapadia and Fortes, 1999; Kapadia et al., 2000a,b). Whereas PUNCH was based on custom written PERL code, In-VIGO is based on Java and reuses software such as Apache, MySQL and Globus.

Some of the advantages of the In-VIGO system are its ability to directly use virtual machines and provide graphical access to access through virtual network computing software. Virtual machines are a software emulation of a physical computing environment. By using virtual machines, In-VIGO can interface with any application regardless of the operating system under which is operates, e.g. Windows, Linux, OS2, etc. Virtual network computing software (e.g. TightVNC) allows remote network access and control of graphical desktops. VNC is based on the concept of a remote frame buffer (RFB) with images transferred to a remote machine pixel-by-pixel. By using virtual network computing software, graphical interfacing to applications is possible.

Users of the system are first authenticated through a login page (Figure 7)

[6]http://invigo.acis.ufl.edu

[7]http://punch.purdue.edu

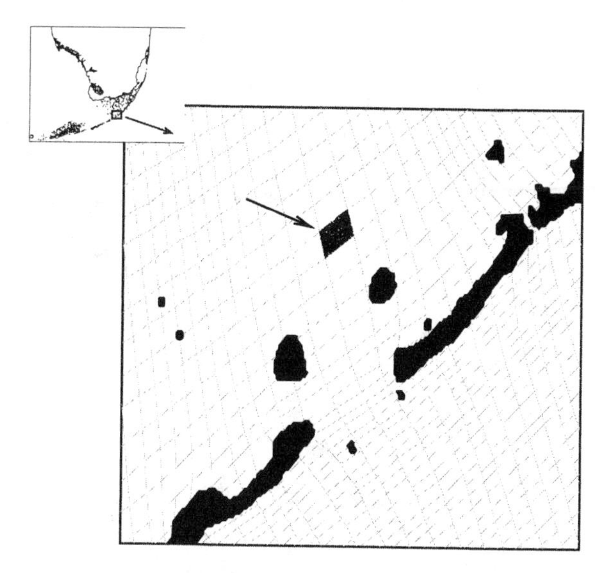

Figure 3: Initial particle distribution in cell (25×33) is shown with a arrow.

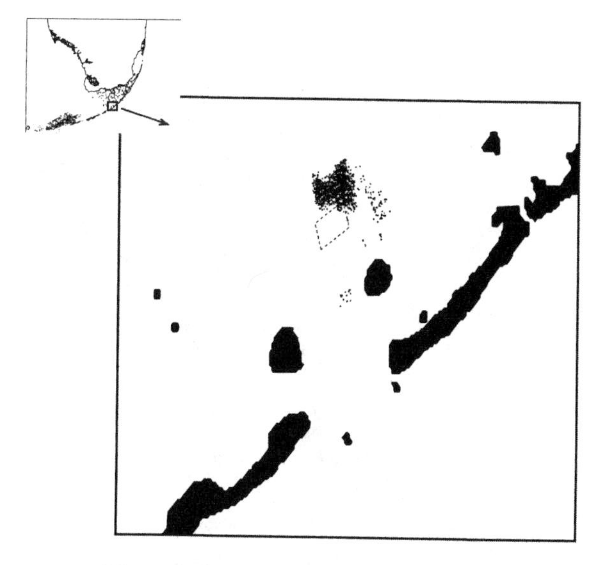

Figure 4: Particle locations after 24 hr.

Figure 5: Particle locations after 72 hr.

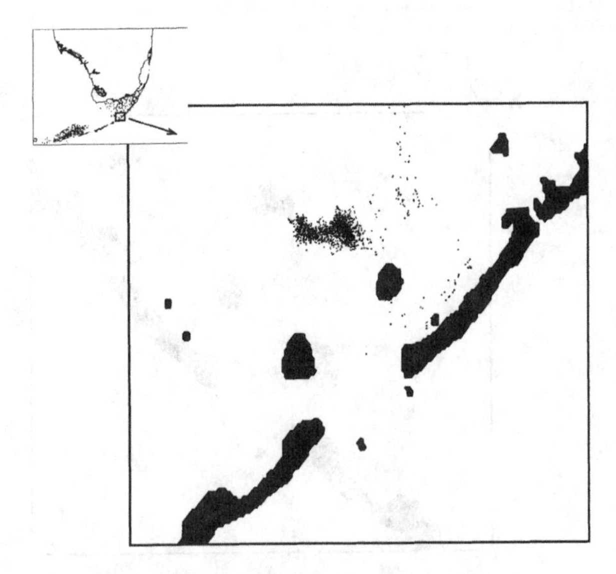

Figure 6: Particle locations after 120 hr.

after which the In-VIGO environment is created (Figure 8) and the user is provided with a list of applications (Figure 9). An interface to the coupled hydrodynamic-particle tracking modeling system is provided through the "Ocean Demo". After the user has chosen to start the Florida Bay particle tracking demo, In-VIGO creates a virtual filesystem workspace and then starts the coupled modeling system. The original coupled modeling system required the user to separately view output from the visualization model. For the purposes of demonstrating the capabilities of In-VIGO, another module was coupled to the system which allows for interactive, graphical access to the visualization model output. This access was achieved through the addition of an "Add-on" to the plotting software Tecplot. Add-ons extend the basic functionality of Tecplot through the use of executable modules. An Add-on, called the Dynamic Data Loader (DDL), was developed to dynamically read output from the visualization model and re-display new particle locations. In-VIGO then provides access to this visualization tool through the VNC software (Figure 10). The user can then directly observe the simulation as it progress. Using the Tecplot software, selected regions can be then highlighted and static images or animations created as needed.

Conclusions and Recommendations

The robustness of the results shown in this pilot study along with relative ease of converting an existing modeling system into a dynamically coupled system running in a grid environment demonstrated the feasability of using grid technologies on coastal and estuarine modeling. Thus, the authors are investigating the possibilities of expanding the local modeling grid used in this study into a multi-institutional modeling grid among the University of Florida, Louisiana State University, and the Virginia Institute of Marine Science. The development of this larger "prototype" modeling grid will facilitate more detailed research into the applications of modeling grids for coastal and estuarine studies. For example, the prototype modeling grid would provide an open and collaborative environment to allow these institutions to share models, computing and networking hardware, and resident expertise. Its is also anticipated that large scale coastal research initiatives, such as the Southeastern Coastal Ocean Observing and Prediction (SCOOP) Program sponsored by the Southeastern Universities Research Association (SURA), would greatly benefit from such modeling grid technologies (Sheng and Davis, 2003).

To summarize, a local computational grid was created using a collection of Silicon Graphics computers and a coupled modeling system developed linking three models: a hydrodynamic model (CH3D), a particle tracking model, and visualization model. Dynamic model coupling was obtained through minor model modifications which enabled grid-based communications via MPICH-G2. As a test of the coupled modeling system, 2000 particles were released in Florida Bay and tracked using the system. One node of the computational grid simulated bay hydrodynamics, two nodes each simulated the movement of 1000 particles while the fourth node handled visualization. Comparisons between particle trajectories calculated with the coupled

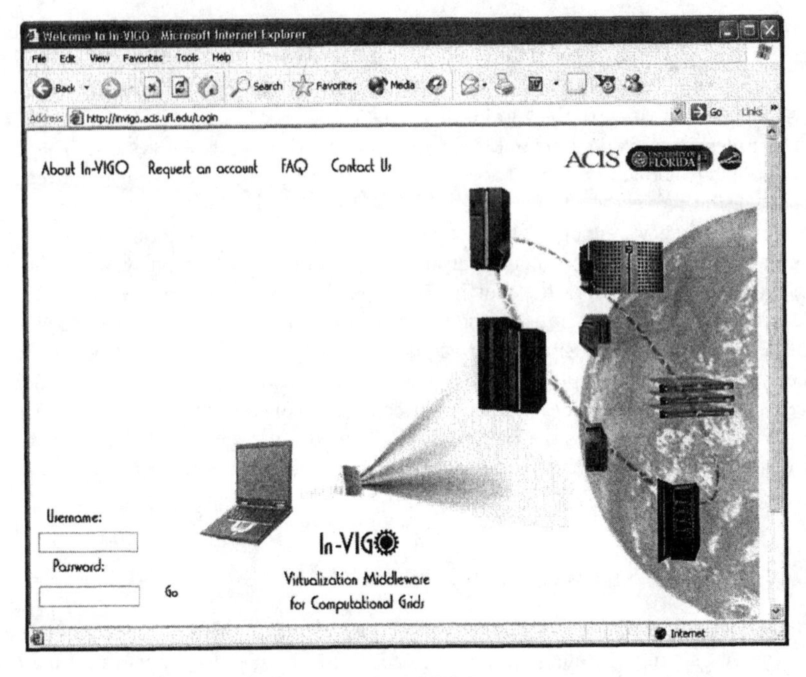

Figure 7: The In-VIGO login screen

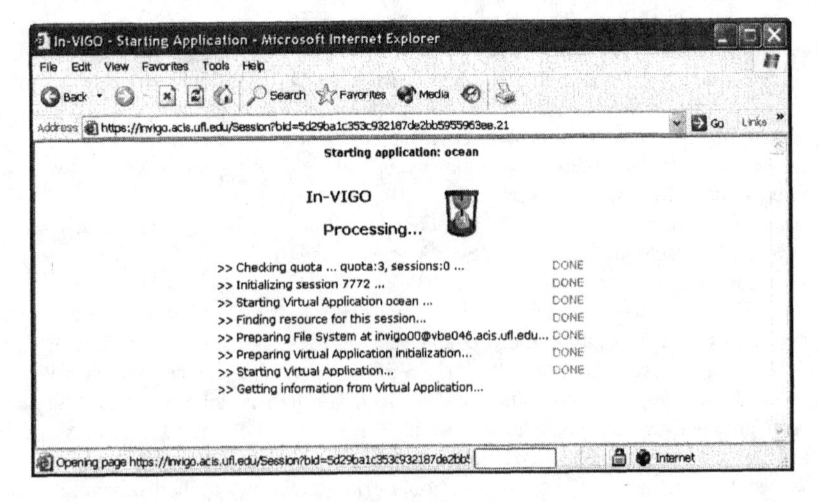

Figure 8: In-VIGO startup

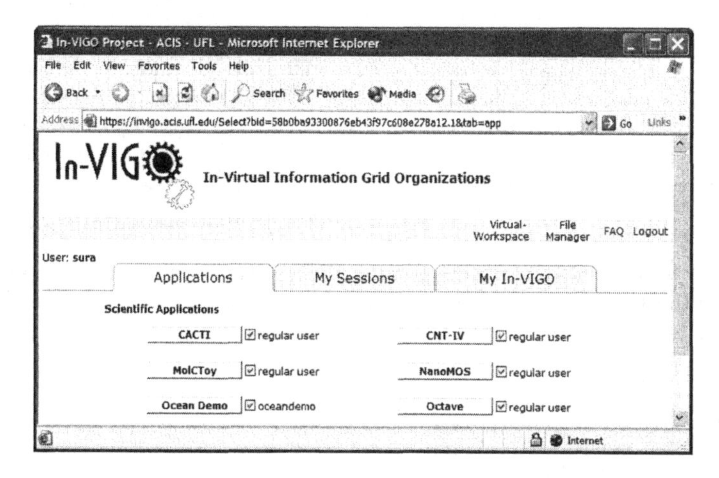

Figure 9: In-VIGO applications

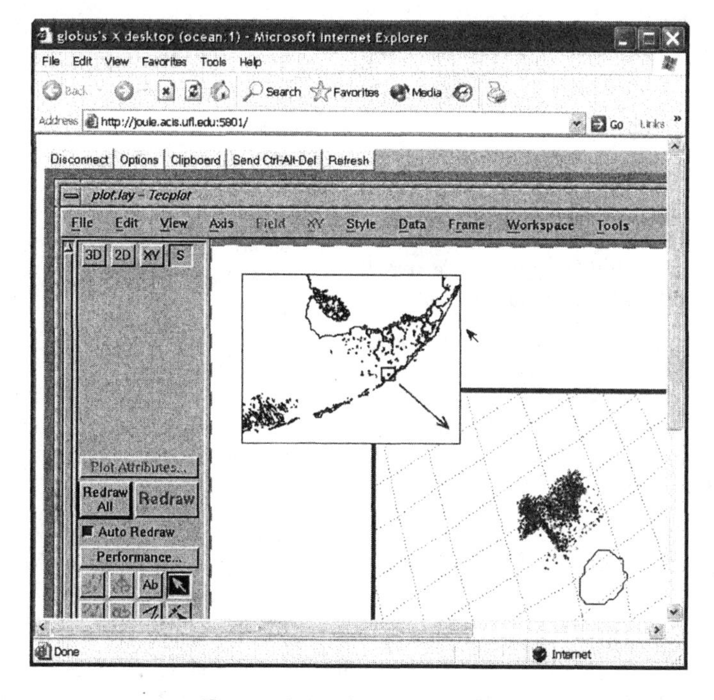

Figure 10: application interface

system and with a more traditional post-processing approach showed that the coupled system was able to reproduce trajectories exactly while providing a slight decrease in overall simulation time due to the particle tracking models running in parallel. High level access to the coupled system was achieved through linking of the coupled modeling system with the enabling middleware technology In-VIGO. In-VIGO hides infrastructure heterogeneity through the virtualization of physical and information resources and would make an ideal platform for the development of coastal and estuarine applications which traditional run on a variety of platforms and operating systems. Windows, Linux, OS2 applications could run in tandem in a virtual grid environment allowing model researchers and end users to focus on processes instead of underlying technologies.

Acknowledgments

The authors wish to gratefully acknowledge all the members of "The In-VIGO Team" within the Advanced Computing and Information Systems Laboratory at the University of Florida who aided in the incorporation of the grid-based particle tracking modeling system into the In-VIGO system. The members of the team are (in alphabetical order): Sumalatha Adabala, Renato J. Figueiredo, Ivan Krsul, Mauricio Tsugawa, Liping Zhu, and Xiaomin Zhu.

References

Adabala, S., Chadha, V., Chawla, P., Figueiredo, R., Fortes, J. A. B., Krsul, I., Matsunaga, A., Tsugawa, M., Zhang, J., Zhao, M., Zhu, L., and Zhu, X. (2003). From virtualized resources to virtual computing grids: The In-VIGO system. *Future Generation Computing Systems*. to appear in the Special Issue on Complex Problem-Solving Environments for Grid Computing.

Foster, I. and Kesselman, C., editors (1999). *The Grid: Blueprint for a New Computing Infrastructure*. Morgan Kaufmann.

Foster, I., Kesselman, C., and Tuecke, S. (2001). The anatomy of the Grid: Enabling scalable virtual organizations. *International Journal of Supercomputer Applications*, 15(3):200–222.

Kapadia, N. H., Figueiredo, R. J., and Fortes, J. A. B. (2000a). PUNCH: Web portal for running tools. *IEEE Micro Special issue on Computer Architecture Education*, 20(3):38–47.

Kapadia, N. H., Figueiredo, R. J., and Lundstrom, M. S. (2000b). The Purdue University network-computing hubs: Running unmodified simulation tools via the WWW. *ACM Transactions on Modeling and Computer Science (TOMACS)*, 10(1):39–57.

Kapadia, N. H. and Fortes, J. A. B. (1999). PUNCH: An architecture for Web-enabled wide-area network-computing. *Cluster Computing*, 2(2):153–164. Special Issue on High Performance Distributed Computing.

Karonis, N., Toonen, B., and Foster, I. (2003). MPICH-G2: A grid-enabled implementation of the Message Passing Interface. *Journal of Parallel and Distributed Computing*, 63(5):551–563.

Message Passing Interface Forum (1994). MPI: A message-passing interface standard. Technical Report MCS-P342-1193, Computer Science Department, University of Tennessee.

Sheng, Y. P. (1987). On modeling three-dimensional estuarine and marine hydrodynamics. In Nihoul and Jamart, editors, *Three-dimensional models of marine and estuarine dynamics*, pages 35–54. Elsevier, Amsterdam.

Sheng, Y. P. (1990). Evolution of a three-dimensional curvilinear-grid hydrodynamic model for estuaries, lakes and coastal waters: CH3D. In *Estuarine and Coastal Modeling*, pages 40–49. American Society of Civil Engineers.

Sheng, Y. P. (1994). Modeling hydrodynamics and water quality dynamics in shallow waters. In *Keynote Paper, Proceedings of the First International Symposium on Ecology and Engineering*, Taman Negara, Malaysia. University of Western Australia and Malaysian Technical University.

Sheng, Y. P. (2000). A framework for integrated modeling of coupled hydrodynamic-sedimentary-ecological processes. In *Estuarine and Coastal Modeling*, volume 6, pages 350–362. American Society of Civil Engineers.

Sheng, Y. P. and Davis, J. R. (2003). Towards a prototype SURA SCOOP grid. SURA Workshop on Coastal Ocean Grid Computing, University of Florida, Gainesville, Florida, July 31, 2003.

Sheng, Y. P., Davis, J. R., and Liu, Y. (1995). Development of a preliminary numerical model of circulation and transport in Florida Bay. Final Report to the National Park Service, Coastal and Oceanographic Engineering Department, University of Florida, Gainesville, Florida.

Sheng, Y. P., Davis, J. R., Paramygin, V., Park, K., Kim, T., and Alymov, V. (2004). Integrated-process and integrated-scale modeling of large coastal and estuarine areas. In *Estuarine and Coastal Modeling*, volume 8. American Society of Civil Engineers.

Smith, N. and Pitts, P. A. (1995). Semidiurnal and diurnal tidal constituents in Florida Bay. The First Report in Connection with Cooperative Agreement CA 5280-4-9022, Harbor Branch Oceanographic Institution, Fort Pierce, Florida.

Smith, N. P. (1995). Personal communication. Harmonic constants from the NOS and HBOI tide studies along the Florida Reef Tract.

Wind-driven Transport in the Labrador Current

Guoqi Han[1]

Abstract

Relative importance of density and wind in generating the Labrador Current and the roles of local vs. remote wind forcing have not been conclusively resolved (Loder et al., 1998). In this study the wind-driven transport of the Labrador Current is investigated using finite-element ocean circulation models and NCEP-NCAR wind data for the 1990s. Long-term monthly-mean and individual seasonal-mean wind-driven circulation is computed with the NCEP-NCAR wind stresses prescribed at the sea surface and sea levels specified at the open boundary determined from a North Atlantic model. The annual-mean model solutions clearly show wind-driven equatorward inshore, shelf-edge, and deep-ocean branches of the Labrador Current. There are strong seasonal variations in the branches of the Labrador Current with the largest in December-February and smallest in June-August. The interannual changes are also significant in the 1990s, with the winter transport in positive correlation with the wind stress curl associated with the Icelandic Low. The model shelf-edge and deep-ocean Labrador Current is dominated by the remote forcing, while the local wind forcing plays an important/dominant role in the inshore current. Role of wind-driven currents is evaluated, which indicates that the wind forcing contributes significantly to the inshore branch and substantially to the shelf-edge branch.

1. Introduction

The Labrador Current off Newfoundland and Labrador (Fig. 1), with significant contributions from both buoyancy and wind variations, is highly variable at various time and space scales. It flows equatorward along the coast and along the shelf edge and the upper continental slope and has a strong direct influence on the shelf circulation (Loder et al., 1998). So far, the relative importance of buoyancy and the wind stress curl in the Labrador Current's forcing has not been conclusively resolved. The roles of local vs. remote wind forcing have not been clearly identified.

----

[1]Fisheries and Oceans Canada, Northwest Atlantic Fisheries Centre, St. John's, NL, A1C 5X1, Canada; phone 709-772-4326; HanG@dfo-mpo.gc.ca

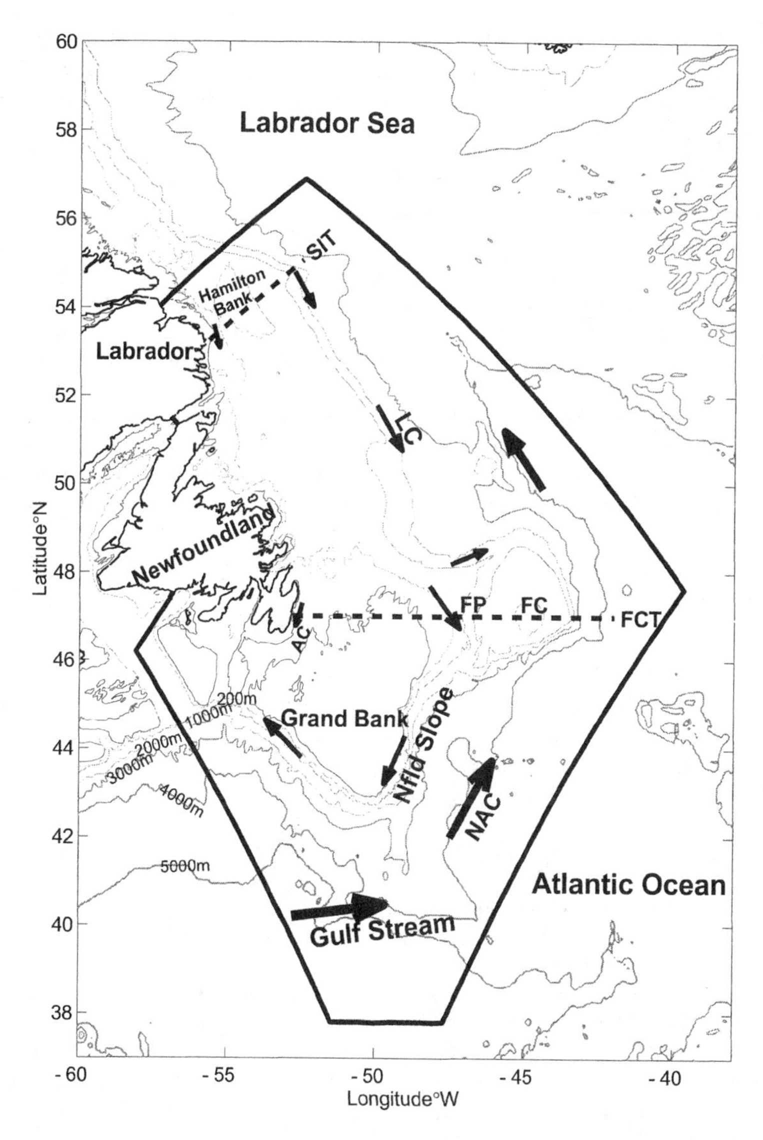

Figure 1. Map showing the Labrador and Newfoundland Shelf and adjacent NW Atlantic Ocean. The isobaths shown are 100, 200, 1000, 2000, 3000, 4000 and 5000 m. SIT: Seal Island Transect; FCT: Flemish Cap Transect; LC: the Labrador Current; FP: Flemish Pass; FC: Flemish Cap; NAC: the North Atlantic Current; AC: Avalon Channel. The model's open boundary is shown as thick line.

During the past decade, significant interannual variations in physical environments and biological productivity were observed off Atlantic Canada, which are presumably directly or indirectly related to the dominant mode of the atmospheric variability in the North Atlantic, the North Atlantic Oscillation (NAO). It was found that the Labrador Current transport variation in the 1990s was positively correlated with the wind stress curl fluctuation associated with the Icelandic Low and the NAO (Han and Tang, 2001).

There have been earlier studies of seasonal transport variability in the Labrador Current in response to wind forcing. Thompson et al. (1986), using the topographic Sverdrup relationship, obtained a seasonal range of 6 Sv in the Labrador Sea circulation and indicated the transport being mainly over the deep slope. Greatbatch and Goulding's (1989) 1° by 1° wind-driven barotropic model of the North Atlantic showed an increase of the Labrador Sea transport by 5 Sv from July to January.

In this paper, we have used a high-resolution 3-D finite element circulation model (Naimie and Lynch, 1993) to study seasonal, interannual, and along-shore variability of the Labrador Current. Long-term monthly-mean and individual seasonal-mean wind-driven circulation were computed with the wind stresses from the NCEP-NCAR reanalysis data prescribed at the sea surface and the large-scale remotely forced sea level specified at the open boundary. Relative importance of local vs. remote forcing in the annual-mean transport and seasonal and interannual variations is examined. Role of wind in generating the Labrador Current is also discussed.

In Section 2 we briefly describe the circulation model, its boundary conditions, and forcing data. Section 3 presents annual-mean circulation, seasonal, and interannual variations, and examines sensitivity of the model transport to friction. Section 4 summarises the major results.

2. Methodology

2.1 Finite Element Model and Mesh

The model uses a harmonic method to solve the linearized 3-D hydrodynamic equations with hydrostatic and Boussinesq assumptions and a vertical eddy viscosity closure (Naimie and Lynch, 1993). We have used a horizontally uniform bottom friction coefficient of 0.0005 m/s and a horizontally and vertically uniform vertical eddy viscosity of 0.005 m^2/s in the baseline cases, based on Han et al. (1999). Sensitivity cases with the bottom friction coefficient of 0.001 m/s and the vertical eddy viscosity of 0.02 m^2/s are also discussed.

The model's horizontal mesh (Fig. 2) has about 10000 variably spaced nodes. It covers the southern Labrador Shelf, the Newfoundland Shelf, and adjacent deep oceans, with high resolution in shallow areas and those with small topographic length scale ($h/|\nabla h|$ where h is the local water depth). The inclusion of deep oceans is to account for interactions among the shelf, slope and deep ocean currents in future prognostic studies with wind, buoyancy and tidal forcing. Typical horizontal node

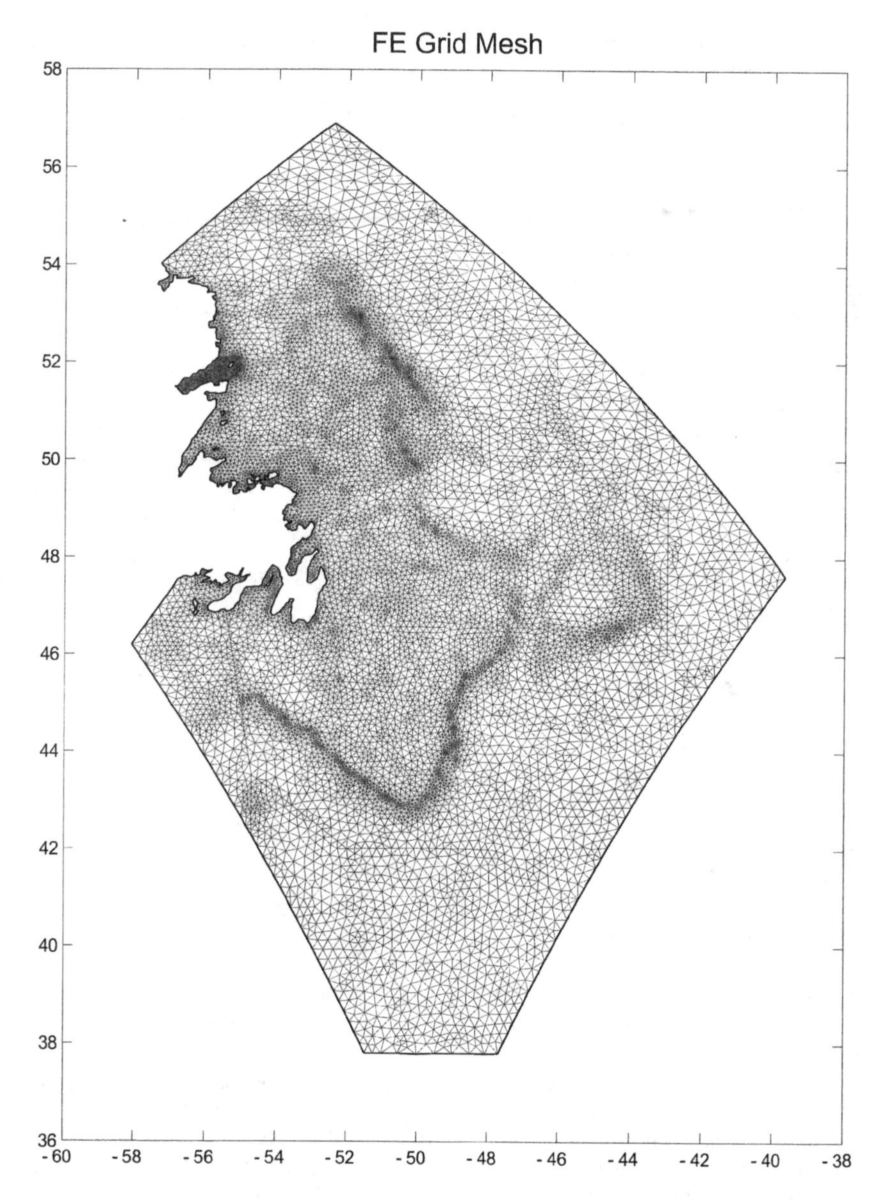

Figure 2. The horizontal finite-element grid (slns2) used in the numerical model.

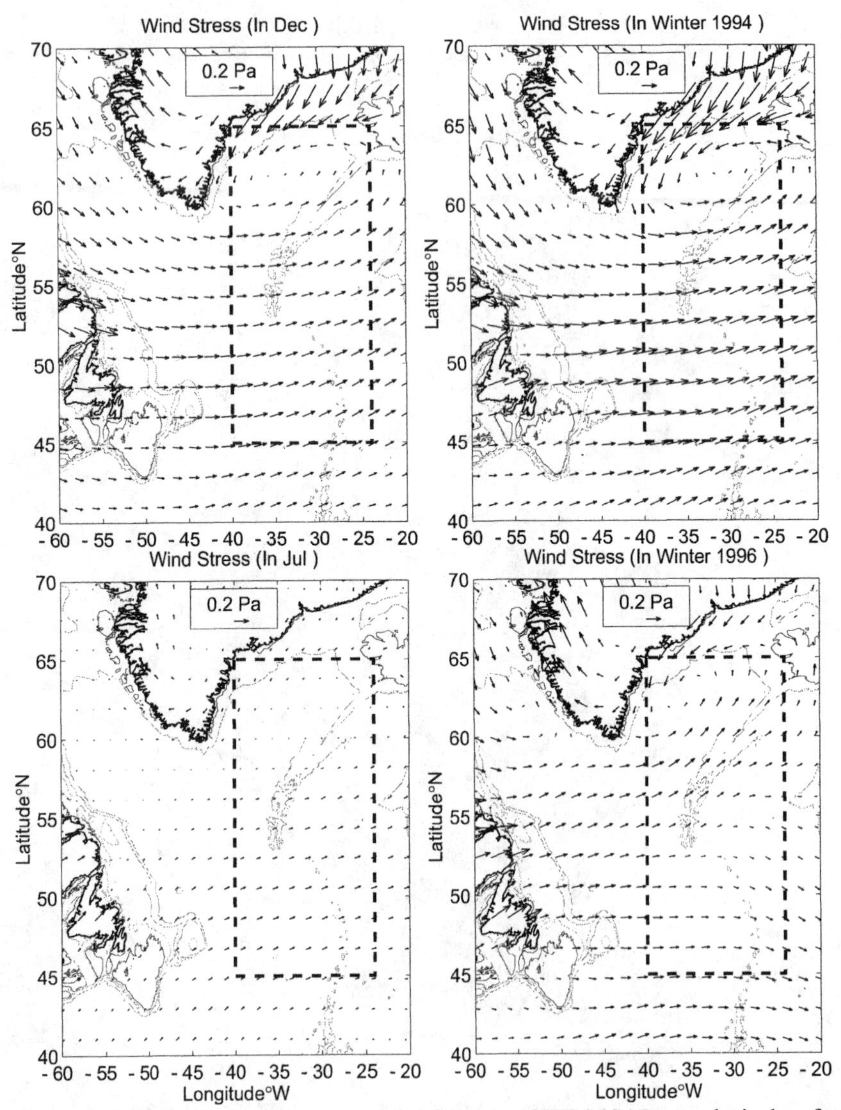

Figure 3. Surface wind stresses calculated from the NCEP/NCAR reanalysis data for (a) July and (b) December and for (c) winter 1994 and (d) winter 1996. The 100-, 200-, 1000-, and 2000-m isobaths (grey lines) are also displayed

spacing is 5 km over the shelf. The vertical mesh has 21 variably spaced nodes with higher resolution near the sea surface and the seabed. The model uses topography for the shelf from a database (the Canadian Hydrographic Service) with about 7-km resolution and that for deep oceans from the ETOPO5 bathymetry database.

2.2 Wind Data and Boundary Conditions

Wind stresses are computed from 6-hourly wind data of NCEP-NCAŘ reanalysis data for the period from 1990 to 1999. The climatological seasonal-mean stresses (Fig. 3) have seasonal variations in both magnitude and direction, with the stress in winter being stronger and directed more in the cross-shelf direction (offshore) than that during the other seasons. The typical magnitude of the wind stresses in the model domain is 0.1 Pa in winter and 0.02 Pa in summer. There was also significant interannual change in the wind stresses, for example, the winter (January to March) magnitude was much larger 1994 than in 1996.

For each month/season, a "local" wind solution is obtained first with zero-elevation conditions specified on all the open boundaries. Then "remote" forcing is specified in the form of elevations on all the open boundaries, and geostrophy is applied on the Strait of Belle Isle boundary. The condition of no normal depth-integrated flow is implemented on the land boundaries. The remote forcing is obtained from a North Atlantic model with the same model dynamics and the same frictional parameters driven by the corresponding wind stresses. The model grid and bathymetry are the same as Greenberg et al. (1998). We assume the model domain is completely enclosed by the land boundary in which no normal depth-integrated flow is allowed.

Generally, the present model response at month/season is in a steady equilibrium with the local wind forcing prescribed at the sea surface and/or the remote wind forcing specified as sea level along the open boundary. The bottom friction leads to diffusion across f/h (where f is the Coriolis parameter) contours in the form of the arrested topographic waves (Csanady, 1978).

3. Results

3.1 Annual-mean Transport

The annual-mean model results (Fig. 4) indicate dominant equatorward current along the shelf break and continental slope inshore of the 3000-m isobath and the inshore current along the coast. Vigorous cross shelf exchanges are apparent and anticyclonic circulation over the offshore Banks is evident, apparently associated with topographic steering. The wind-driven North Atlantic Current is located just offshore of the 3000-m isobath. Its features are not examined in this study.

The transport associated with the inshore current is about 0.7 Sv at the Seal Island transect (Table 1), accounting for most of the shelf transport. The estimated total transport is 0.8 Sv for the nearshore Labrador Current (Lazier and Wright, 1993). The transport associated with the shelf-edge current is about 1.5 Sv from the 200- to1700-m isobath. The total shelf-edge current transport was estimated to be 6.7

Sv by Lazier and Wright (1993), from the current meter observations and geostrophic methods. From the 1700- to 3000-m isobath, the model-driven transport is about 5 Sv, almost entirely by remote forcing.

There are strong cross-shelf exchanges after the flows enter the Newfoundland Shelf and Slope (Fig. 4). A portion of the inshore branch steers offshore to join the shelf-edge current. The bifurcation of the shelf-edge current north of Flemish Pass is evident, with a strong stream through the Pass and a significant current toward the east along the northern flank of the Flemish Cap. At the Flemish Cap transect the shelf transport from the coast to the 130-m isobath on the northeastern Grand Bank is 0.2 Sv, with 0.3 Sv from the local forcing and –0.1 Sv (northward) from the remote forcing (also Table 1). The total transport through the Avalon Channel associated with the inshore Labrador Current is estimated to be 0.39 Sv from current meter data (Greenberg and Petrie, 1988). The wind-driven transport through the Flemish Pass is 1.5 Sv, which is a substantial portion of the estimated total transport of 5.8 Sv (Loder et al., 1998). About 1 and 0.5 Sv of the model Flemish Pass transport are from remote and local forcing, respectively. The inshore branch after passing through the Flemish Cap transect moves offshore along a deep channel and merges with the shelf-edge current that rounds the Tail of Grand Bank. Over the Grand Bank the circulation is generally weak.

The results reveal that off South Labrador the shelf-edge transport is predominated by the remote forcing, and in generating the shelf transport the remote forcing plays a major role (2/3), but the local forcing is also substantial (1/3). Off Newfoundland, the shelf-edge current is mainly (2/3) forced by remote forcing with substantial (1/3) contribution from local forcing, while the shelf transport is mainly driven by local forcing.

Compared with literature estimates of the mean transports associated with the shelf-edge Labrador Current, the present barotropic wind-driven transport can account for 1/5~1/4 of the total transport at the Seal Island transect and 1/4 of that through the Flemish Pass. These values suggest the wind is a substantial forcing component of the Labrador Current along the edge of the Labrador and Newfoundland Shelf, which is consistent with Lazier and Wright's (1993) finding of substantial barotropic contribution in moored measurements. The missing majority is presumably buoyancy-driven, as suggested by Lazier and Wright (1993). The relative role of wind forcing seems more prominent in the inshore Labrador Current transport, accounting for more than half of the total estimated transports at both the Seal Island and Flemish Cap transects.

3.2. Seasonal Variation

The model results clearly demonstrate significant seasonal variations in the wind-driven circulation. The model Labrador Currents are strong in winter and fall and weak in summer and spring. The topographic-scale (bank) features are also intensified in winter and fall. The seasonal circulation changes result from variations in the westerlies over the model domain and of the wind stress curl in the northern North Atlantic, especially associated with the Icelandic Low (Fig. 3).

The transport at the Seal Island transect has a seasonal cycle of 1.2 Sv associated with the inshore current, with 0.9 and 0.3 Sv from remote and local forcing, respectively (Fig. 5). The transport is larger in December-February (1.4 Sv) and smaller in June-August (0.2 Sv). The shelf-edge current has a seasonal change of 2.3 Sv from the 200- to 1700-m isobath (the upper slope) and of 8.5 Sv from the 200- to 3000-m isobath (Fig.5, Table 2), predominantly driven by remote forcing. Lazier and Wright (1993) from the current meter data found the estimated annual range to be 4 Sv for the upper slope. They showed the annual cycle with the maximum in October and the minimum in March-April was dominated by the buoyancy-driven flow. The model transport range of 8.5 Sv is close to Han and Tang's (1999) estimation of 10 Sv from altimetry and hydrography, but significantly larger than Greatbatch et al.'s (1989) North Atlantic model result of 5 Sv and Thompson et al.'s (1986) topographic Sverdrup transport of 6 Sv. The present wind stress fields are from NECP and NCAR reanalysis data, while Greatbatch et al. (1989) used Hellerman and Rosenstein wind fields, and Thompson et al. (1986) used Thompson and Hazen's wind climatology. Furthermore, the seasonal change of 2.3 Sv at the shelf edge from our study is from two separate streams inshore and offshore of the1000-m isobath (not shown), which are not resolved by Thompson et al. (1986) or by Greatbatch and Goulding (1989). The present model indicates wind-driven current and its seasonal cycle are weak near the 1000-m isobath, consistent with Lazier and Wright (1993)'s current meter data at a depth of 985 m over the 1000-m isobath showing little seasonal variation.

Through the Flemish Pass, the present model indicates a seasonal transport variation of 2.5 Sv with 2/3 of the total accounted for by remote forcing (Fig. 5). The shelf transport from the coast to the 130-m isobath on the northeastern Grand Bank side has a seasonal range of 0.6 Sv, dominated by the local forcing. The northward transports by remote forcing occur from spring to early fall as do the transports dominated by local forcing in July and August.

Due to the absence of the wind-driven current near the 1000-m isobath at the Seal Island transect, the annual transport range of 4 Sv based on moored measurements over this water depth (Lazier and Wright, 1993) does not include the wind-driven barotropic component. The present high-resolution model reveals the existence of the wind-driven currents over the shelf edge and offshore and inshore of the 1000-m isobath, with a seasonal transport range of 2.3 Sv large in early winter and smaller in summer. Therefore, a reasonable seasonal cycle for the shelf-edge Labrador Current at the Seal Island transect may be constructed by superimposing Lazier and Wright's (1993) annual cycle and the seasonal cycle of the present study. As a result, the seasonal transport variability is dominated by buoyancy-driven component, with the important contribution from the wind-driven component as well.

3.3. Interannual Changes

In the 1990s the NAO experienced profound interannual variations. The NAO was intensified in the early 1990s with deepened Icelandic Low. Overall, the fall/winter wind stress curl associated with the Icelandic Low was large (except for 1993) and westerlies over the model domain were strong in these years (Fig. 6), especially in 1994. It was opposite in 1996, when the NAO was weakest for the decade.

We obtained the seasonal-mean solutions under both local and remote forcing by season and year. The wind-driven transports of the Labrador Current exhibit significant interannual variations in its three branches (Fig. 7).

At the Seal Island transect, the transport of the shelf-edge current (from the 200- to 1700-m isobath) was 3.2 Sv and 1.8 Sv in the winters of 1994 and 1996, respectively. The winter transport of the inshore current also decreased from 2.1 Sv in 1994 to 1.2 Sv in 1996. The deep-ocean branch (from 1700- to 3000-m isobaths) carried 11.5 and 6 Sv in the two periods.

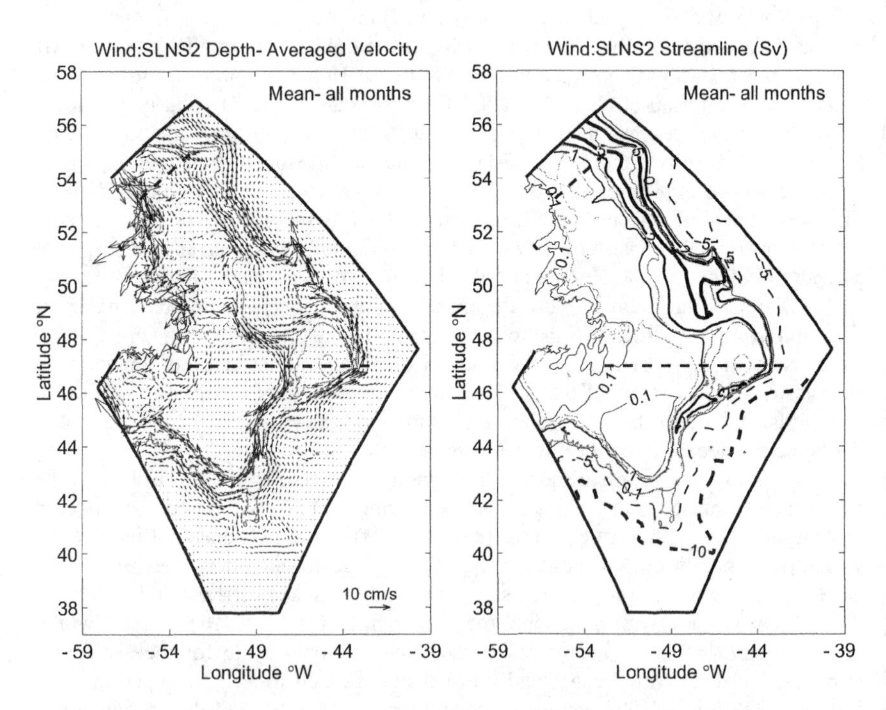

Figure 4. (a) Annual-mean depth-averaged currents. The model fields have been subsampled for clarity of presentation. (b) Transport streamfunction in units of Sv (10^6 m^3/s). The streamfunction value is zero along the Labrador and Newfoundland coast, with low values on the right looking downstream; i.e., the flow is cyclonic around local streamfunction maxima and anticyclonic around local minima. The 200-, 1000-, and 3000-m isobaths (grey lines) are also displayed.

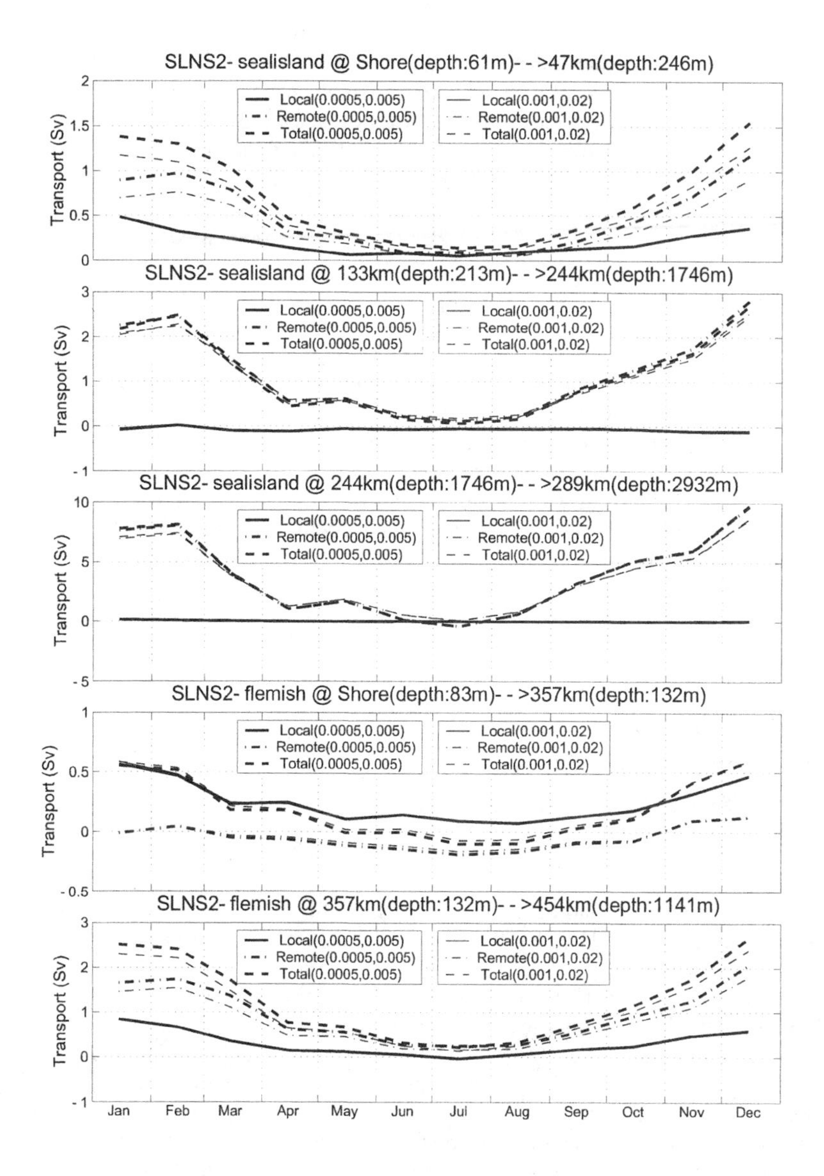

Figure 5. Monthly variations of the Labrador Current transport (positive southward) at the Seal Island transect and the Flemish Cap transect. See Fig. 1 for locations. Note that the two local forcing cases appear as one solid line.

Table 1. Annual mean transports (Sv, positive southward) of the wind-driven Labrador Current from the present model. Also shown in parentheses are observational estimates of the total transport from Lazier and Wright (1993), Greenberg and Petrie (1988) and Loder at al. (1998).

	Inshore	Shelf-edge
Seal Island transect	0.7 (0.8)	1.5 (6.7)
Flemish Cap transect	0.2 (0.39)	1.5 (5.8)

Table 2. Seasonal transport ranges (Sv) of the Labrador Current at the Seal Island transect. Thompson et al. (1986), Greatbatch and Goulding (1989) and this study provide wind-driven transports only. The wind driven transport is virtually excluded in Lazier and Wright's (1993) estimate (see text for more detail). Han and Tang (1999) provide a geostrophic estimate of the total transport.

	Shelf-edge	Shelf-edge to deep-ocean
This study	2.3	8.5
Han and Tang (1999)	-	10
Lazier and Wright (1993)	4	-
Greatbatch and Goulding (1989)	-	5
Thompson et al. (1986)	-	6

At the Flemish Cap transect, the transport of inshore current was 0.65 and 0.3 Sv in the winters of 1994 and 1996 and that through the Flemish Pass was 3.2 and 2 Sv, respectively.

The interannual transport variability is primarily associated with the remote forcing for the inshore, shelf-edge and deep ocean branches at the Seal Island and Flemish Cap transects (Fig. 7), presumably representing the change in the large-scale subpolar gyre of the North Atlantic in response to the wind stress variation. The wind stress curl associated with the Icelandic Low, calculated from the NCEP/NCAR wind data (Fig. 6), shows prominent interannual variations. We can see high positive correlation between the wind stress curl and the model transport. The high correlation is presumably exaggerated due to the lack of baroclinicity in the model. Han and Tang (2001) also demonstrated positive correlation between the NAO and the wind stress curl. It follows that the interannual wind-driven transport variability of the Labrador Current may be positively correlated with the NAO, somewhat similar to the barotropic and total transports of Han and Tang (2001).

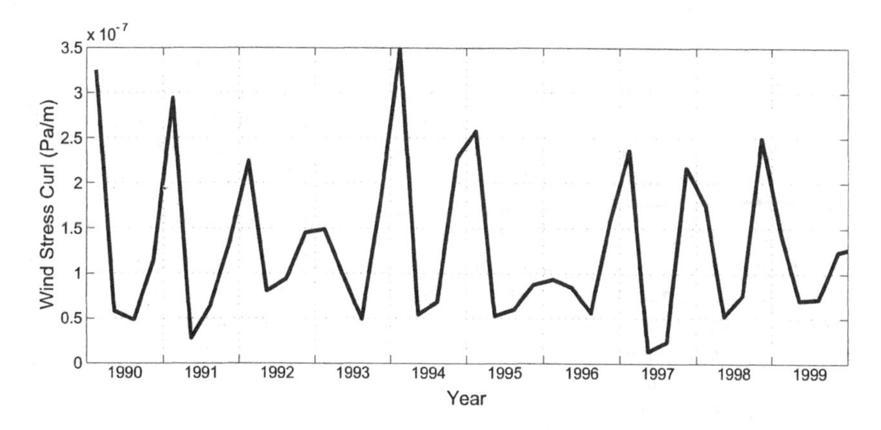

Figure 6. Nominal seasonal-mean wind stress curls associated with the Icelandic Low, calculated from the NCEP/NCAR reanalysis data for the rectangle area enclosed by the dashed lines in Fig. 3.

3.4 Sensitivity to Friction

In the baseline model runs we have used the bottom friction coefficient of 0.0005 m/s and the vertical eddy viscosity of 0.005 m^2/s^2. To examine the sensitivity of model results to the frictional parameters, we also computed monthly-mean circulation fields with the bottom friction of 0.001 m/s and the vertical eddy viscosity of 0.02 m^2/s^2.

Through the Seal Island transect the shelf transport in response to the remote forcing indicates limited changes in magnitude with friction (Fig. 5). The increased friction reduces the magnitudes of the annual-mean and seasonal cycle by less than 10%. In contrast, the shelf transport seems insensitive to friction under the local wind forcing. The transport associated with the shelf-edge current under the remote forcing also shows limited friction influence.

The shelf transport through the Flemish Cap transect, for both local and remote forcing, is insensitive to the change of the model friction parameters (Fig. 5). The transport through the Flemish Pass exhibits little sensitivity under the local forcing and limited sensitivity under the remote forcing.

Overall, the model transport of the wind-driven Labrador Current shows little to limited sensitivity to the frictional parameters. Major findings from the baseline solutions also hold in the sensitivity cases. However, the magnitude of the near-surface current driven by local forcing can be sensitive to the spatially uniform eddy viscosity, as governed by Ekman dynamics.

4. Summary

The model solutions indicate the wind-driven transport contributes significantly to the inshore branch and substantially to the shelf-edge branch of the Labrador Current. The study for the first time reveals prominent seasonal and interannual variations and variable importance of the local vs. remote wind forcing in the wind-driven Labrador Current.

The shelf-edge Labrador Current has a mean wind-driven transport of 1.5 Sv at the Hamilton Bank and 1.5 Sv through the Flemish Pass. The inshore Labrador Current has a mean transport of 0.7 Sv at the Seal Island transect and 0.2 Sv at the Flemish Cap transect. The offshore branch is dominated by remote forcing and the local forcing plays an important/dominant role in the inshore branch. The model transport for the shelf-edge current is about 1/4 of the total transport previously estimated from *in situ* observations (Lazier and Wright, 1993; Loder et al., 1998).

There are strong seasonal variations in the wind-driven transport of the Labrador Current. The model transport is stronger in December-February and weaker in June-August. The shelf-edge branch has a seasonal range of 2.3 Sv (between 200- and 1700-m isobaths) at Hamilton Bank and 2.5 Sv through the Flemish Pass. The inshore branch has a seasonal range of 1.2 Sv at the Seal Island transect and 0.6 Sv at the Flemish Cap transect. The seasonal variability of the shelf-edge current at the Seal Island transect is associated with remote forcing, while that of shelf-edge current through the Flemish Pass and the inshore current is with both remote and local forcing. The seasonal cycle in the model shelf-edge current is significant compared with that estimated from moored measurements (Lazier and Wright, 1993).

Interannual variations of wind-driven transport are also evident. The shelf-edge winter transport at the Seal Island transect was 3.2 Sv in 1994 and 1.8 Sv in 1996. Flow through the Flemish Pass carried 3.2 and 2 Sv in the winters of 1994 and 1996, respectively. The inshore and deep-ocean branches all showed decreases in transport between the two winters, to a similar relative extent. The transport variability results mainly from the remote forcing and is generally correlated with the wind-stress curl associated with the Icelandic Low.

The present model study reveals the role of local and remote wind forcing and in comparison with earlier observational transport estimates the significance of wind forcing in generating the Labrador Current. The sensitivity of its transport to model friction parameterization is limited and does not change the major findings drawn from the baseline solutions. Nevertheless, an improved friction formulation (such as flow dependent) may be useful for detailed 3-D structure of the wind-driven circulation in the region. Besides, baroclinic transport driven by density variations is surely another important component and has to be included in a more sophisticated model for a complete picture of the Labrador Current, which is to be addressed in separate papers.

Acknowledgments

I thank J. Li for assisting in running the model, D. Greenberg and Z. Xu for generating the model grids, and B. DeTracey for computing the NCEP/NCAR wind stress data. This research project is funded through the Offshore Environmental Factors program of the (Canadian) Program for Energy Research and Development and through Fisheries and Oceans Canada.

References

Csanady, G.T. (1978). "The arrested topographic wave." *J. Phys. Oceanogr.*, **8**, 47-62.

Greatbatch , R.J., and Goulding A. (1989). "Seasonal variations in a linear barotropic model of the North Atlantic driven by the Hellerman and Rosenstein wind stress field." *J. Phys. Oceanogr.*, **19**, 572-595.

Greenberg, D.A., and Petrie B.D. (1988). "The mean barotropic circulation on the Newfoundland shelf and slope." *J. Geophys. Res.*, 93, 15541-15550.

Greenberg, D.A., Werner F.E., and Lynch D.R. (1998). "A diagnostic finite element ocean circulation model in spherical-polar coordinates." J. Atmos. Oceanic Technol. 15, 942-958.

Han G., and Tang C.L. (1999). "Velocity and transport of the Labrador Current determined from altimetric, hydrographic, and wind data." *J. Geophys. Res.*, 104, 18047-18057.

Han G., and Tang C.L. (2001). "Interannual variation of volume transport in the western Labrador Sea based on TOPEX/Poseidon and WOCE data." *J. Phys. Oceanogr.,* 31, 199-211, 2001.

Han, G., Loder J.W., and Smith P.C. (1999). "Seasonal-mean hydrography and circulation in the Gulf of St. Lawrence and eastern Scotian and southern Newfoundland Shelves." *J. Phys. Oceanogr.*, 29(6), 1279-1301.

Lazier, J.R.N., and Wright D.G. (1993). "Annual velocity variations in the Labrador Current." *J. Phys. Oceanogr.*, **23**, 659-678.

Loder, J.W., Petrie B.D., and Gawarkiewicz G. (1998). "The coastal ocean off northeastern North America: a large-scale view." Ch. 5 In: The Global Coastal Ocean: Regional Studies and Synthesis. *The Sea*, Vol. 11, K.H. Brink and A.R. Robinson (eds.), John Wiley & Sons, Inc., 105-133.

Naimie, C.E., and Lynch D.R. (1993). *FUNDY5 Users' Manual*. Numerical Methods Laboratory, Dartmouth College, N.H., 40 pp.

Thompson, K.R., Lazier, J.R.N. and Taylor B. (1986). "Wind-forced changes in Labrador Current transport." *J. Geophys. Res.*, **91**, 14,261-14,268.

Summer Ekman pumping and its implications to the deep water renewal in Massachusetts and Cape Cod Bays

Mingshun Jiang[1], Meng Zhou[2]

Abstract

The area of Massachusetts and Cape Cod Bays (MB) is a semi-enclosed embayment surrounded by the Greater Boston Metropolitan Area, Cape Cod and the coastal communities from Boston to Cape Ann. An understanding of physical and biological processes in MB is critical to the local fishing industry, recreation and nature conservation. Our studies focus on wind driven currents, Ekman pumping induced coastal jets and water renewal in the deep basin of MB using both field observations and modeling. Results indicate that the characteristics of the deep water in Stellwagen Basin, the only deep basin in MB, are similar to the Gulf of Maine (GOM) water during early summer; and are significantly different from the GOM water during late summer. This suggests that the deep water exchange between MB and the GOM is limited during summer. The modeled results reveal that the Ekman pumping during a wind event induces coastal upwelling or downwelling, and produces fronts and coastal jets in response to baroclinic pressure gradients and surface slopes. The surface Ekman transport enhances the deep water renewal by pumping deep water from the GOM to Stellwagen Basin through the North Passage off Cape Ann, and by crossing over the shallow Stellwagen Bank. The time scales to spin up geostrophic currents during wind events and to renew the deep basin water in MB are also estimated.

Introduction

Massachusetts and Cape Cod Bays (MB) are important economical and recreational areas, which support maritime transportation, commercial fishing of lobster and shellfish, and a habitat for endangered North Atlantic Right whales. The public concerns about the potential impacts of the sewage outfall relocation on the MB environment led to a long term study of the water quality, contaminant levels and ecosystem dynamics over more than a decade using both field surveys and numerical modeling (Geyer et al., 1992; Libby et al., 2001; HydroQual and Signell, 2001). Many concerns focus on the potential eutrophication in MB due to direct nutrient loads, the fate of which is greatly affected by physical processes, e.g. advection and vertical mixing. The studies of these physical processes have been conducted by two-dimensional and three-dimensional models (Signell et al., 1996; Signell et al., 2000), and field surveys (Geyer et al., 1992). Results from these studies showed that the outfall relocation would dramatically reduce effluent loads in Boston Harbor. The effluent at the offshore Bay Outfall site would be greatly diluted by mixing and advection, while the nutrient influx from the Gulf of Maine (GOM) is the

---

[1] University of Massachusetts Boston, 100 Morrissey Blvd, Boston, MA 02125; phone 617-287-7416; mingshun.jiang@umb.edu

[2] University of Massachusetts Boston, 100 Morrissey Blvd, Boston, MA 02125; phone 617-287-7419; meng.zhou@umb.edu

predominant source for MB and varies seasonally (HydroQual, 2003). Thus, an understanding of water exchange processes between MB and the GOM is critical for predicting the water quality and ecosystem dynamics of MB.

Our study area is a semi-enclosed embayment with an eastern open boundary connecting to the GOM (Figure 1). It is approximately 100 km long and 50 km wide. The southern part of MB is referred to as Cape Cod Bay (CCB). The average depth of MB is approximately 35m. The western and southern parts of MB are shallower than the average. The deepest water (greater than 80 m) is found within Stellwagen Basin west of Stellwagen Bank. It is generally thought that the nutrient-rich deep water from the GOM can be transported into MB through the 70 m deep North Passage between Cape Ann and Stellwagen Bank, and the 50 m deep South Passage between Race Point and Stellwagen Bank.

Previous studies have indicated that the circulation in MB varies in response to short term and seasonal meteorological and boundary forcing (Geyer et al., 1992). The local and remote forces include wind stresses and heat fluxes at the sea surface, tides, fresh water runoff and mean surface slopes at the open boundary. The yearly mean circulation in MB is southward along the western side of the bay, and intrudes into CCB. This is primarily driven by both the intruding current through the North Passage associated with mean sea surface slopes, and baroclinic pressure gradients associated with the horizontal density gradients produced by both salinity and temperature differences (Bigelow, 1927; Geyer et al., 1992; Lynch et al., 1996). Tides are dominated by the semi-diurnal M_2 constituent, with tidal currents varying from 10 cm s^{-1} in the interior, to 50 cm s^{-1} in the South Passage off Race Point. The water column stratification varies seasonally. Stratification occurs in spring due to both freshwater runoff and surface heating. In summer, the stratification is strongest. The water column is destratified during late fall, and is well-mixed throughout the winter.

It is important to understand both the variability of the general circulation pattern, which affects biological and chemical processes in MB, as well as the fundamental mechanisms that determine the variability. Few analyses have been done on the low frequency variability of the circulation pattern in response to short-term, seasonal and climate variability. It is not understood how the deep circulation in Stellwagen Basin is forced by surface processes, and how important the deep circulation is to the nutrient influx, especially in the summer season. The renewal of the deep waters in Stellwagen Basin through the North and South Passages in late spring and summer may be limited (Geyer et al., 1992). Based on limited data, it is speculated that little exchange between MB and the GOM would occur through April to late summer. In this paper, we report the results from our study on the circulation under variable wind conditions in MB during the summer of 2000 using the Massachusetts Bay hydrodynamic model by Signell et al. (2000). We also use CTD data in summer, 1999 to understand the water property differences between MB and the GOM.

Survey and model descriptions

Hydrographic data have been collected in MB during monthly cruises by the Massachusetts Water Resources Authority (MWRA). Each cruise covers the entire bay in a period of 4~5 days during which hydrographic measurements, including temperature and salinity, and biochemical measurements, such as nutrients, chlorophyll and zooplankton abundances, are gathered. Discrete net tows and water samples are collected at five standard depths for each station. Detailed information about samples and data can be found in the MWRA report (Libby et al., 2001). Currents are measured by a bottom-mounted Acoustic Doppler Profiler (ADCP) associated with the United States Geological Survey (USGS) buoy maintained offshore of Scituate, Massachusetts (42° 9.8' N, 70° 38.4' W), where the water depth is approximately 22 m (Butman et al., 2002). In this study, hourly binned current measurements are smoothed by a low-pass filter to examine the low frequency variability. Conductivity, temperature and depth (CTD) data collected by the National Marine Fisheries Service at two locations (G1 and G2) in the western GOM are used to compare with data in MB collected by the MWRA (Figure 1). All CTD data used for comparisons are collected in 1999 because we do not have appropriate GOM data for late summer, 2000.

The hydrodynamic model is based on the Estuarine, Coastal, Ocean Model (ECOM-si) with Mellor and Yamada level 2.5 turbulence closure for the vertical mixing (Blumberg, 1991; Blumberg and Mellor, 1987; Signell et al., 1996). The incorporation of a semi-implicit scheme on the surface elevation eliminates the need to solve the external and internal modes separately in the Princeton Ocean Model (POM) and improves the computation efficiency. The model is forced with hourly meteorological data, including the observed wind stresses and heat fluxes, daily fresh water discharges, M_2 tides, and low frequency surface slope at the open boundary. The surface heat fluxes are estimated using the observed solar radiation, air temperature, air pressure and humidity at a NOAA buoy (42°24'N, 70°40'W) near the new outfall based on the bulk formulation (Large and Pond, 1981). The surface slope was estimated from the dynamic height related to a non-flow surface at 120m based on objectively interpolated temperature and salinity (TS) data collected throughout year 2000. For areas shallower than 120m, the zero-flow surface is set at the bottom. A detailed description and calibration of the model for MB can be found in the HydroQual report (HydroQual and Signell, 2001).

Results and discussions

Water properties

Though the upper layer in MB is influenced by the GOM water, local forcing such as surface heating and freshwater runoff play an important role in determining the temperature and salinity (Geyer et al., 1992). In spring, the freshwater runoff is greatest and produces strong horizontal salinity gradients. In summer, after the peak runoff season, the freshwater runoff creates weak horizontal salinity gradients in the upper layer. The water in shallow coastal areas is warmed up faster than in deep areas in spring and summer, which reduces or reverses the horizontal temperature gradients caused by winter cooling. During summer, weak horizontal gradients of temperature fields are typical in the bay except in the very shallow CCB area.

The water properties in Stellwagen Basin closely resemble the GOM water during early summer, 1999. The TS diagram at two stations in Stellwagen Basin (F12 and

F17) and a station in the east of Stellwagen Bank (G1) are essentially the same, with the temperatures between 4 and 14°C and salinities between 31.5 and 32‰ (Figure 2a). All three stations are located downstream of the West Maine Coastal Current (WMCC). Thus, the close agreement of these TS curves indicates a strong intrusion of the WMCC into MB during early summer.

The water in Stellwagen Basin has distinct features that are different from the GOM water during late summer, 1999, particularly at depth (Figure 2b). The temperature and salinity data at station G2, which is located within the main axis of the WMCC, are characteristic of the GOM surface water within the upper 40 m and the GOM intermediate water below 40 m (Brown and Irish, 1993). The deep water at Stations F12 and F17 in Stellwagen Basin has similar temperature with G2 but is fresher. In particular, the near bottom salinity in the basin was about 0.3 ‰ lower than the mid-depth (40-70m) salinity at G2. This indicates that the water exchange between MB and the GOM is limited during the summer period. The significant differences in bottom salinity between the three stations (F12, F17 and SB) in the basin further suggests limited horizontal mixing or advection in the basin as well, though temporal changes due to a half month delay in sampling time at SB may contribute to these differences. Interestingly, the water at station F12, which is further offshore, is fresher than F17. The reason for this is not clear. The water below 100m at G2 has salinities higher than 32.6%, which is characteristic of GOM bottom water formed from the previous winter (Brown and Irish, 1993).

Summer circulation

The general circulation pattern in MB has been described as a counterclockwise circulation (Geyer et al., 1992; Signell et al., 1996). The WMCC is considered to be the major driving force at the open boundary. It enters MB from the northeast along the shelf break, and splits into two branches off Cape Ann. The major branch flows along the eastern flank of Stellwagen Bank, and the minor one intrudes into MB through the North Passage with a speed of approximately 20 cm s^{-1}. This intruding current joins coastal currents along the western side of MB, and flows southward, often penetrating deeply into CCB (Figure 3a). In year 2000, the circulation follows this general pattern except during the summer season (Figure 3).

In summer 2000, the WMCC flows southward along the eastern flank of Stellwagen Bank with little intrusion into MB (Figure 3b). The magnitude of the flow on the eastern flank of Stellwagen Bank is approximately 20 cm s^{-1}. The southward coastal current produced by the freshwater input into Boston Harbor, which typically flows along the west coast of MB, is pushed offshore during this summer. A weak northward coastal current of cold water is found flowing next to the coast, and joining the southward current of the fresh water plume off Scituate. The water temperature in CCB is approximately 2 to 3°C warmer than the rest of MB, which indicates limited water exchange between MB and CCB. The warm water in CCB is associated with a weak anticyclonic eddy. In the northern part of CCB, this circulation joins the current from MB, and exits MB through the South Passage. This circulation pattern in summer 2000 is considerably different from spring (Figure 3a) and the general circulation pattern described above.

Temporal variability of temperature, salinity and currents

The atmospheric forcing at mid-latitude is featured by synoptic scale processes, e.g., wind changes its direction and magnitude every several days. The spin-up of a geostrophic current associated with Ekman pumping takes approximately 2 to 6 days in MB following a wind event (Figs. 4 and 5). For example, the north wind persists for about 8 days at the end of July, and shifts to southwest winds for 9 days at the beginning of August. Here we use the meteorological convention for wind directions, i.e., a north wind indicates wind blowing to the south. The currents observed at 5 m below the sea surface shows a delayed response to the surface wind. After 6 days of the southwest wind, the current starts to change its direction and magnitude, evolving to the opposite direction while the wind maintains the same direction. We interpret the change of the current as being caused by the setting up of sea surface slopes produced by Ekman pumping, i.e., the southwest wind produces an offshore Ekman transport, which leads to offshore elevation of the sea surface and the formation of a coastal front. The resulting front leads to northeastward geostrophic currents. As can be seen from the temporal evolution of the temperature profile at Scituate buoy (see below), the upwelling occurs almost immediately after the wind switches to a southwest wind. However, the bottom water takes several days to reach the surface. A similar Ekman pumping process occurs in the next wind event (north winds) and the spin-up of currents also takes about 6 days. Comparatively, the delayed time is about 3 days for the last wind event (south winds) in this month. The modeling results catch most of the features in the observations (Figure 4). Modeled and observed surface currents show very good agreement except for the first 7 days, when the modeled currents respond quicker to the south wind forcing. Under the north wind condition, the modeled currents show a similar phase and speed to the observed, and under the south wind condition, the modeled currents show a similar phase, but overestimate speed relative to the observed. It appears that the modeled currents respond to south wind more quickly and with stronger magnitudes. However, it is not clear why the model shows such a quick response to the south wind in the first week, while the observations do not.

Temperature and salinity fields show dramatic changes in response to the variable wind direction and speed. The time scale of temperature and salinity adjustments is determined by the Ekman pumping with a time-lag of between 2 and 6 days (Figures. 4, 5 and 6). The modeled temperature and salinity at the Scituate buoy station show a consistent response to the surface Ekman pumping (Figure 6). The water is weakly stratified at the beginning of August, which is due to earlier wind mixing. The temperature and salinity fields evolve in response to the southwest wind induced Ekman pumping, and the thermocline is uplifted. At day 10, a surface front is produced while the stratification becomes intensified at 10 m. The uplifting of the thermocline and halocline remains until day 12 after the wind switches its direction on day 9. During the upwelling period, the thermocline and halocline are lifted approximately 5 m. On day 9, the wind becomes downwelling favorable and lasts approximately a week with a peak at day 15. The thermocline and halocline are suppressed by the downwelling. The mixed layer depth reaches approximately 12 m. During the relaxation period, the upper water column is restratified again. On day 21, the wind changes to south winds, and the bottom water immediately shows signs of upwelling. About 3 days later, the bottom water reaches the surface and a surface front is formed.

The modeled currents in the offshore area respond differently to the wind forcing. As an example, we show the modeled currents at station F17, which is located at the western slope of Stellwagen Basin (Figure 5). The surface current is much weaker than at the Scituate buoy station (Figs 4 and 5). Both surface and near bottom currents at F17 are persistently southeastward during most of the August. The surface currents have a mean speed of approximately 5 cm s^{-1}, which is consistent with the seasonal mean current at this location (Figure 3). The near bottom currents have a mean speed of approximately 2 cm s^{-1}. These mean currents are associated with the mean monthly or seasonal barotropic and baroclinic forcing. By removing the mean currents, anomalies of the surface and near bottom currents should represent the current responses to wind events. The surface and near bottom currents at Station F17 respond differently to winds. For example, the south winds between day 5 and day 9 produce a northeastward current at the surface and a southeastward current at 50 m. During the strong north wind event in week 2, the surface current is southward while the near bottom current is northward. The difference between the surface and near bottom currents in Stellwagen Basin implies the different physical mechanisms responsible for these currents. The mixed layer depth at Stellwagen Basin is approximately between 10 and 20 m. Thus, the near bottom current at 50 m is not directly generated by wind stress. The mechanism of Ekman pumping for generating the near bottom current is discussed below.

Spatial variability of currents and temperature

Two horizontal snapshots of the surface and near bottom currents in MB on day 15 and day 25, respectively, elucidate the spatial and temporal evolution of the circulation field during strong north and south wind events (Figures 7 to 10). On day 15, the surface currents in MB are mostly southward or southwestward with intrusion into MB from the GOM in response to a north wind forcing prior to day 15. The western coastal areas have the strongest currents in the bay, which penetrate deeply into CCB without any surface return currents. Very weak horizontal temperature gradients exist in MB except in the eastern part of CCB where strong upwelling can be identified from the low temperature along with meandering features at scales from 10 to 20 km. At the southern end of Stellwagen Basin, weak upwelling is visible together with an intensified intruding current through the South Passage off Race Point, which is consistent with the observed upwelling event during a short northeast wind event in late September, 1998 (Yu et al., 2002). The wind changes to south winds on day 21. On day 25, a strong coastal jet has developed with northward currents and extending from CCB to the Merrimack River along the upwelling front indicated by the temperature gradient from 16°C in near shore areas to 17.5°C in offshore areas. The coastal current exits MB through the Northern Passage, and bifurcates into two branches. The northern branch goes around Cape Ann and continues northward, while the southern branch turns southward, flowing along the eastern flank of Stellwagen Bank, and joins the WMCC. The surface water in MB also exits through the South Passage. As a consequence, the GOM water is restricted to the east of Stellwagen Bank. An anticyclonic eddy of relatively warm water is formed in CCB.

The sea surface elevations show patterns consistent with surface currents (Figure 8). On day 15, water accumulates in the western part of MB and southern part of CCB with the maximum surface elevations between 6 and 7 cm along the coasts. On day 25, the surface elevation along the western edge is depressed to 2 cm which leads to strong gradients in the offshore direction. The maximum surface elevations are found in the central basin, corresponding to an anti-cyclonic circulation pattern for the entire bay. Such rapid sea surface elevation change is simply caused by the offshore transport produced by Ekman pumping.

The near bottom circulation at 50m shows very different patterns from the surface circulation patterns on both day 15 and day 25 in Stellwagen Basin, though currents east of Stellwagen Bank also show a persistent southward flow of approximately 10 cm s^{-1}. On day 15, little horizontal exchange between Stellwagen Basin and the GOM are found. A weak inflow through the North Passage is evident, but the amount is insignificant relative to the water volume in Stellwagen Basin. In the South Passage, a weak inflow can also be identified. On day 25, a strong intrusion into Stellwagen Basin is found in the North Passage with a current speed between 5 and 6 cm s^{-1}. A narrow jet starts from the deep basin north of Stellwagen Bank, flows northward along the shallow flank, turns westward into Stellwagen Basin, and forms a southward jet in a confined 8 km band crossing the basin and penetrating into CCB. This jet appears to be separated from the WMCC, and is originated from the deep basin next to the northeastern flank of Stellwagen Bank. The mechanism for generating such an intruding jet into MB can be simply explained as the compensation flow needed for volume conservation of the surface offshore Ekman transport. The compensation mechanism is further depicted in temperature and current data (Figure 10) from the west-east transect shown in Figure 1. The current field on Day 15 clearly indicates that Stellwagen Bank operates as a barrier during a north wind event. Though vertical currents are enhanced on the bank, limited cross-bank currents are found. During a south wind event, upwelling occurs on the eastern flank, and downwelling occurs on the western flank of Stellwagen Bank while there are strong cross-bank currents intruding into Stellwagen Basin (Figure 10b).

The interactions between the Ekman transport, stratification and bottom topography lead to dramatic vertical circulation patterns in MB. On day 15, the southwestward Ekman transport produces the downwelling along both the west coast of MB and the eastern flank of Stellwagen Bank (Figure 10). An upwelling event occurs in the middle of Stellwagen Basin and forms a circulation cell over the western slope of the basin together with the downwelling at the mid-slope. A similar circulation cell is also found on the eastern flank of Stellwagen Bank. In western MB, warm surface water accumulates with weak stratification in the area shallower than the 20 m isobath. On day 25, the south wind drives an offshore Ekman transport and strong upwelling near the western coast, which leads to the tilting of the isothermals and a coastal upwelling front. The compensation flow is found approximately between 10 and 40 m. As the compensation flow reaches the coast, it bifurcates. The upwelled water joins the surface offshore transport flowing eastward into the GOM.

The downwelled compensation flow at the western coast forms a deep circulation cell with upwelling at the western flank of Stellwagen Bank.

Deep water renewal

The advective water exchange below 30 m between Stellwagen Basin and the GOM can be estimated using the water flux through the North Passage. Assuming a mean current of about 7 cm s^{-1} and a cross-sectional area of 5 km wide, and the vertical depth interval between 30 and 50 m, the influx is approximately 7×10^3 m^3 s^{-1}. To renew the entire 3×10^{10} m^3 of water between 30 and 50 m depth in Stellwagen Basin requires approximately 50 days. To examine the circulation pattern at this time scale, we take the monthly averaged circulation pattern at 50 m (not shown). Similar to the pattern seen on day 25, a narrow jet originates from the deep basin northeast of Stellwagen Bank, flows through the Stellwagen Basin, and penetrates into CCB. Thus, this jet is persistent in August, and should play a critical role in the deep water renewal in MB.

The offshore Ekman transport is the driving force that leads to the deep intruding current. Assuming a south wind of 1 dyne cm^{-2} and a coastline of 40 km long, the Ekman transport is 4×10^4 m^3 s^{-1}, which is one order of magnitude larger than the deep intruding flux through the North Passage. Taking this Ekman transport, the deep water will be renewed in approximately 8 days, which is similar to or less than the duration of a wind event. However, it appears that the shallow Stellwagen Bank significantly blocks the Ekman transport, as suggested by the downwelling on the western flank of Stellwagen Bank (Figure 10). The upper circulation cell during a south wind event can reach only down to 40 m with most compensation flow being found between 10 and 40 m. Thus, the deep water influxes through the North Passage and South Passage should be a portion of the total compensation flow. These results also indicate that the deep water crossing over Stellwagen Bank into MB may be an important process for deep water renewal (Figure 10), which can significantly shorten the deep water residence time.

Baroclinic response processes

The analysis of modeled hydrography and current fields shows significant baroclinic responses of water column and induced baroclinic currents to Ekman pumping. The horizontal scale of baroclinic jets and eddies can be analyzed by the internal Rossby deformation radius (R_i) (Pedlosky, 1987). Assuming a mixed layer depth (D_E) of 20 m at the coast, a cross-front density difference (? ?) of 2.5×10^{-3} g cm^{-3}, and the Coriolis constant (f) of 1×10^{-4} s^{-1}, we have $R_i = \sqrt{(\Delta\rho/\rho)gD_E}/f$ of approximately 7 km, which is consistent with the width of coastal jets seen in Figure 7.

The spin-up time scale of a baroclinic geostrophic current can be estimated from the uplifting of the thermocline and halocline by Ekman pumping. Within the internal Rossby deformation radius, the coastal upwelling volume flux can be estimated following Bowden (1983), i.e.,

$$R_i w_e = \frac{\tau}{f\rho} \qquad (1)$$

where w_e is the upwelling velocity driven by the alongshore wind stress τ, f is the Coriolis coefficient and ρ is the water density. If we assume the wind stress is 1 dyne cm^{-2} corresponding to a wind of 7 m s^{-1}, f is 10^{-4} s^{-1}, and R_i is 7 km, w_e is equal to 1×10^{-2} cm s^{-1}. Taking the mixed layer depth (D_E) of 20 m, the spin-up time scale can be estimated by (D_E/w_e) and is approximately 2.5 days. We note that this is a lower limit because the observed wind in summer is overall weaker than 7 m s^{-1} as seen in Figure 4. Moreover, as mentioned above the responding time between surface and bottom waters are significantly different. A correlation analysis by Geyer et al.(1992) found a 4 day time-lag between winds in Cape Cod Bay and surface currents measured at Broad Sound, which is close to new outfall site. By contrast, only 1 day lag was found for bottom currents. Such spin-up processes are also studied in other areas. Bowden (1983) estimated that a time scale of upwelling spin-up is on the order of 3.5 days in mid latitude. Halpern (1976) also reported a 2-day response time of upwelling along the Oregon coast during a strong north wind event in July, 1973.

4. Summary

During early summer, 1999, the observed TS characteristics of the water in MB are similar to the GOM water. This indicates strong horizontal advection and vertical mixing in the spring, which results in a deep mixed layer and strong barotropic currents such that the deep water in MB is efficiently renewed. During late summer, 1999, the observed TS characteristics of the deep water in Stellwagen Basin are significantly different from the GOM, implying reduced water exchange between MB and the GOM. Field measurements in MB indicate that both summers of 1999 and 2000 are normal years in terms of hydrographic conditions, therefore we believe year 2000 shows similar features as seen in 1999. The physical mechanisms responsible for limited deep water renewal are explored and interpreted using year 2000 simulation in terms of stratification, baroclinic responses and interaction with the shallow Stellwagen Bank. The shallow mixed layer during the summer separates direct effects of wind stresses on the deep water. The Ekman pumping due to surface winds leads to fronts, surface slopes, coastal surface jets, and vertical circulation cells with significant upwelling and downwelling along shallow coastal areas and the bank. The deep intruding current through the North Passage is enhanced during south wind events as part of a compensation flow. The deep current forms a jet up to 7 cm s^{-1} and intrudes southward into CCB. Modeled circulation indicates the deep water in the GOM must upwell and cross over Stellwagen Bank into Stellwagen Basin as a result of Ekman pumping. The time scale for the deep water renewal during the summer is estimated to be 8 to 50 days.

Acknowledgements

This project is supported by the cooperative agreement between the University of Massachusetts and MWRA. MJ and MZ would like to acknowledge R. Isleib, R. Signell, and J. Fitzpatrick for their assistance in transferring MB models to UMass Boston, R. Payne, B. Butman and others for providing observed atmospheric and hydrographic data, and G.B. Gardner and G. Wallace for their helps to the project. Two reviewers' constructive comments greatly improve the manuscript.

References

Bigelow, H.B. (1927). "Physical oceanography of the Gulf of Maine (Part II)." *Bull. of the Bureau of Fisheries*, 40: 511-1027.

Blumberg, A.F. (1991). *A Primer for ECOM-si.*, HydroQual, Inc. 73p.

Blumberg, A.F. and Mellor, G.L. (1987). "A description of a three-dimnesional coastal ocean circulation model." In: N.S. Heaps (Editor), *Three Dimensional Coastal Ocean Models.* American Geophysical Union, Washington, D.C. p1-16.

Bowden, K.F. (1983). *Physical oceanography of coastal waters.* Halsted Press, Chichester, New York.

Brown, W.S. and Irish, J.D. (1993). "The annual variation of water mass structure in the Gulf of Maine: 1986-1987." *Journal of Marine Research*, 51: 53-107.

Butman, B. et al. (2002). *Long-term oceanographic observations in western Massachusetts Bay offshore of Boston*, Massachusetts: Data Report for 1989-2000. DDS-74, U.S. Geological Survey Digital Data Series 74.

Geyer, W.G. et al. (1992). *Physical oceanographic investigation of Massachusetts and Cape Cod Bays*, U.S. Massachusetts Coastal Zone Management Office, Boston. MBP-92-03.

Halpern, D. (1976). "Structure of a coastal upwelling event observed off Oregon during July 1973." *Deep-Sea Res.*, 23: 495-508.

HydroQual, Inc., and Signell, R. (2001). *Calibration of the Massachusetts and Cape Cod Bays hydrodynamic model: 1998-1999*, MWRA ENQUAD report 2001-12, Massachusetts Water Resources Authority, Boston.

HydroQual. Inc. (2003). *Bays eutrophication model (BEM): model Verification for the period 1998-1999.* Boston: Massachusetts Water Resources Authority. Report 2003-03.

Large, W.G. and Pond, S. (1981). "Open ocean momentum flux measurements in moderate to strong winds." *J. Phys. Oceanogr.*, 11: 324-336.

Libby, P.S. et al. (2001). *1999 annual water column monitoring report.* Boston: Massachusetts Water Resources Authority. Report 2000-09.

Lynch, D.R., et al. (1996). "Comprehensive coastal circulation model with application to the Gulf of Maine." *Cont. Shelf Res.*, 12: 37-64.

Pedlosky, J. (1987). *Geophysical fluid dynamics.* Springer-Verlag, New York.

Signell, R.P., et al. (1996). *Circulation and effluent dilution modeling in Massachusetts Bay: model implementation, verification and results.* USGS Open File Report 96-015, U.S. Geological Survey, Woods Hole.

Signell, R.P., et al. (2000). "Predicting the physical Effects of relocating Boston's sewage outfall." *Estuarine, Coastal and Shelf Science*, 50, 59–72.

Yu, X.-R., et al. (2002). "The application of autonomous underwater vehicles for interdisciplinary measurements in Massachusetts and Cape Cod Bays." *Cont. Shelf Res.*, 22: 2225–2245.

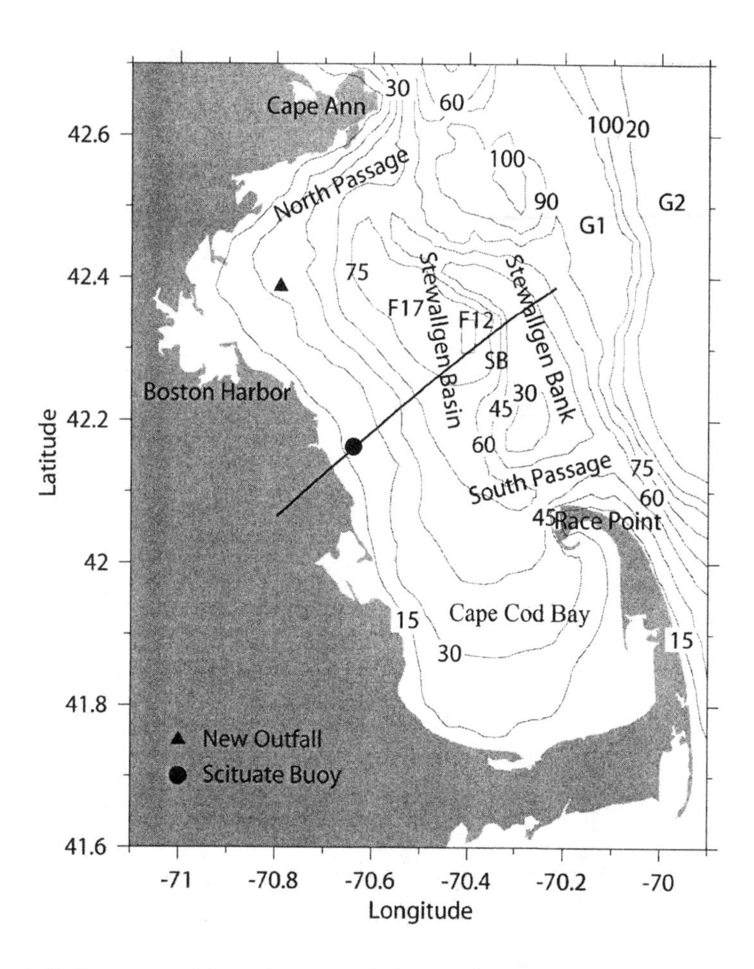

Figure 1. Bathymetry in Massachusetts and Cape Cod Bays. The bold line indicates the west-east cross section . The stations G1, G2 and SB are sampling sites of the National Marine Fisheries Services with the G1 and G2 located in western GOM and SB in the Stellwagen Basin. The F12 and F17 are MWRA sampling stations. The New Outfall site of sewage effluent is indicated by the black triangle and the USGS buoy station at Scituate is indicated by the black dot. The NOAA Buoy station is deployed next to the new outfall site (not shown).

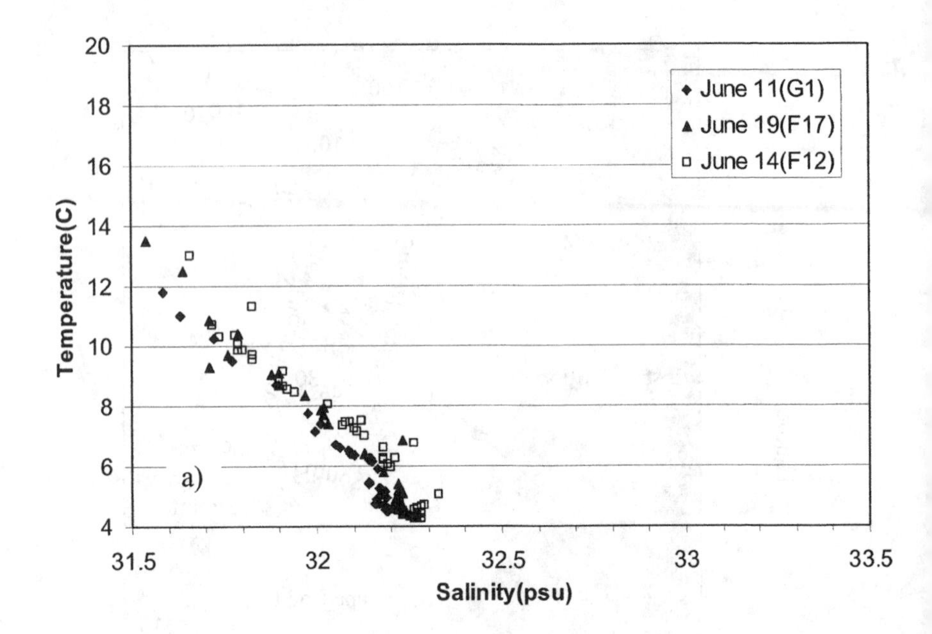

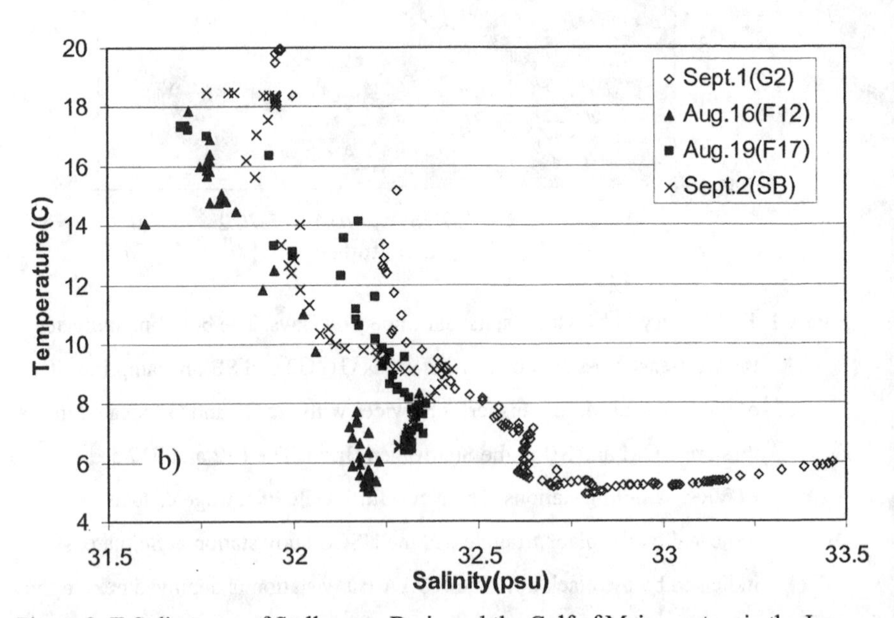

Figure 2. T-S diagrams of Stellwagen Basin and the Gulf of Maine waters in the June (a) and late summer (b) of 1999. The station locations are indicated in Figure 1.

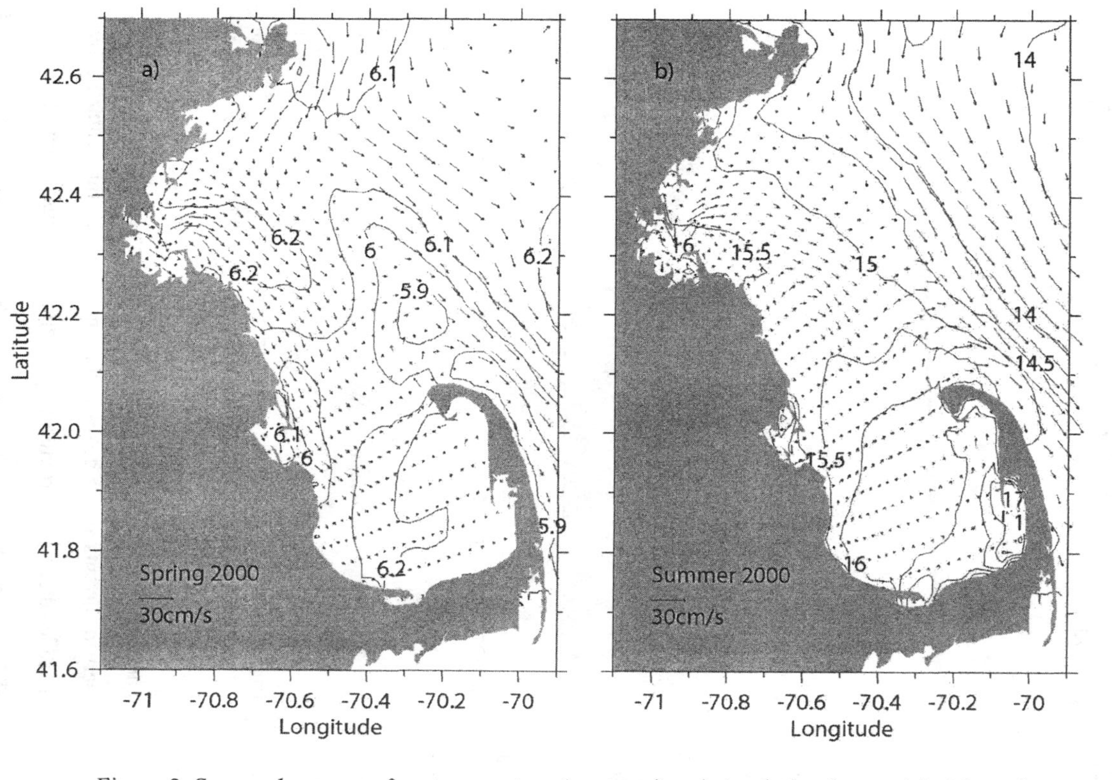

Figure 3. Seasonal mean surface temperature (contours) and circulation (arrows) in Massachusetts and Cape Cod Bays from numerical simulation: (a) the spring (a) and (b) summer of 2000.

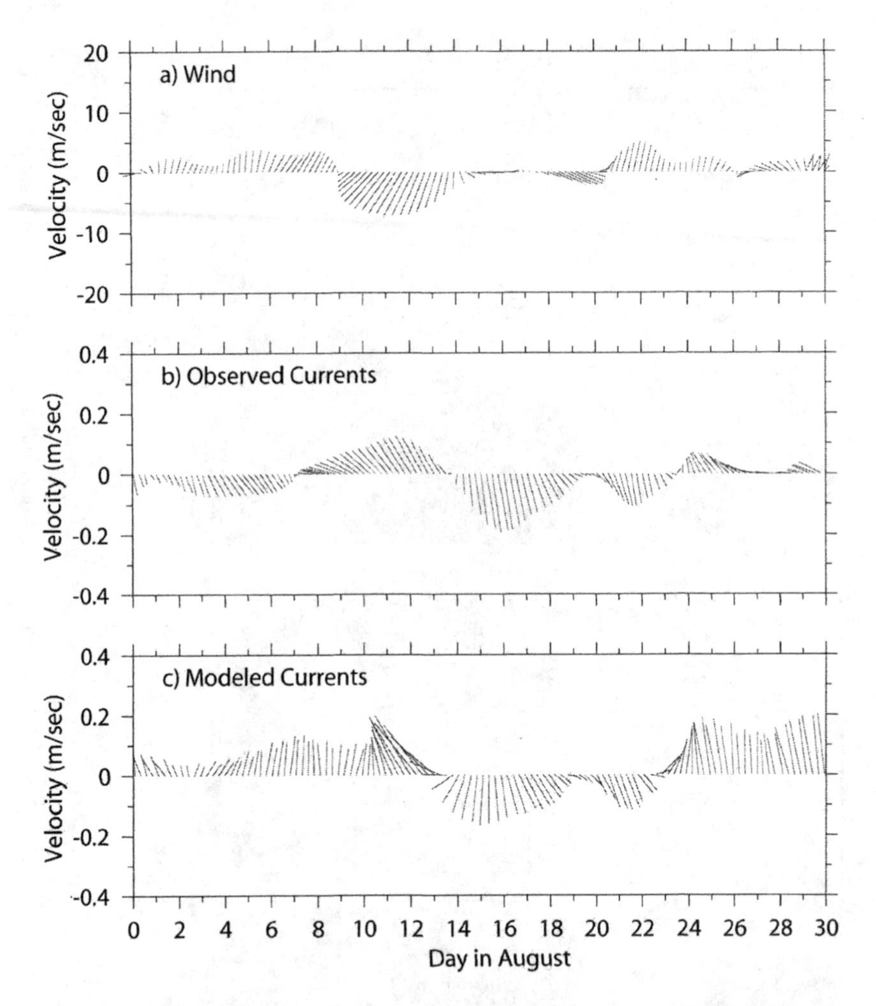

Figure 4. Wind and currents in August, 2000: (a) Observed wind at the NOAA buoy; b) Observed currents at 5 m below surface measured by the Scituate buoy; and c) Modeled surface currents at the location of Scituate buoy. The wind vectors are plotted using oceanographic convention, i.e., a vector stemming from the x-axis and pointing upward indicates wind blowing to the north.

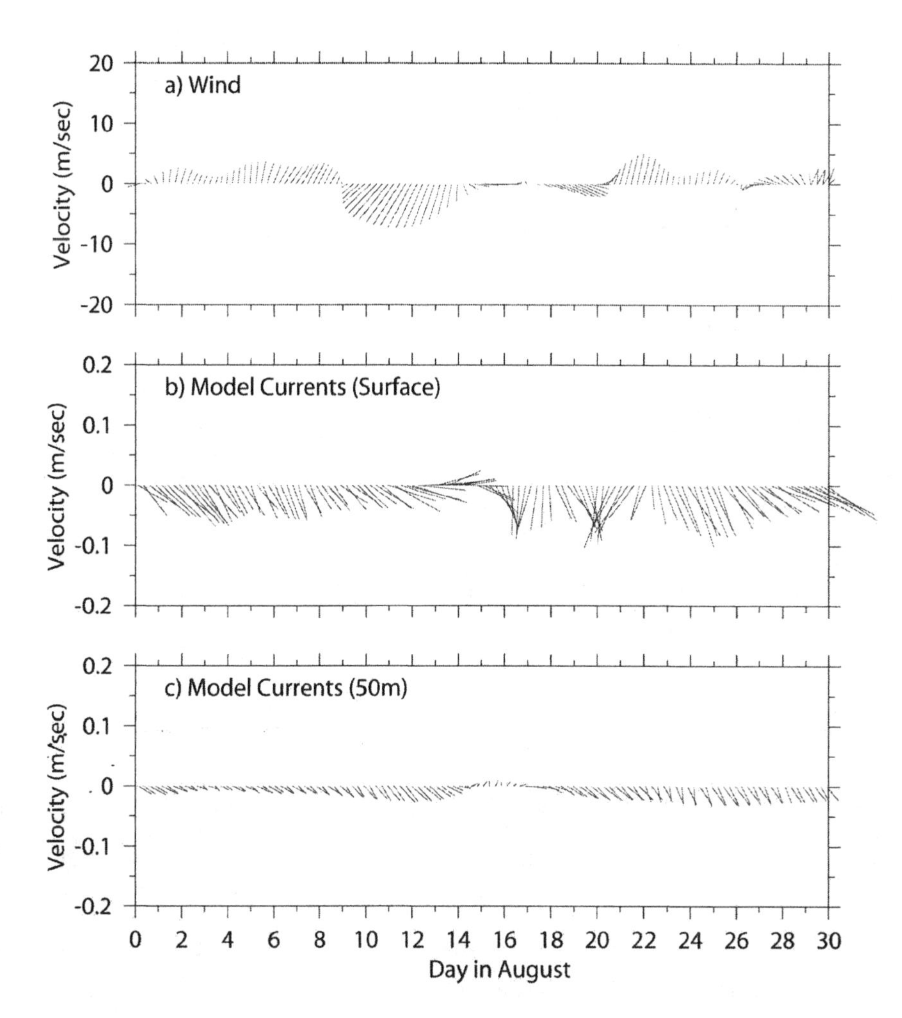

Figure 5. Wind and currents in August, 2000: (a) Observed wind at the NOAA buoy; b) Modeled surface currents at station F17; c) Modeled 50m currents at station F17.

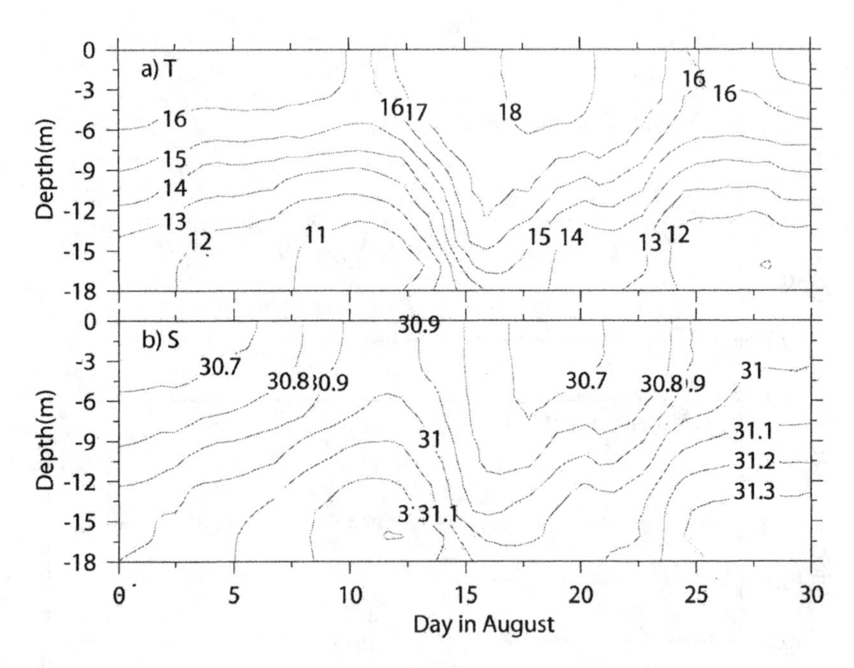

Figure 6. (a) Modeled temperature (°C) and (b) salinity (‰) at the Scituate buoy in August 2000. The model output is averaged every 2 hours and then filtered with a low-pass Lanczos filter.

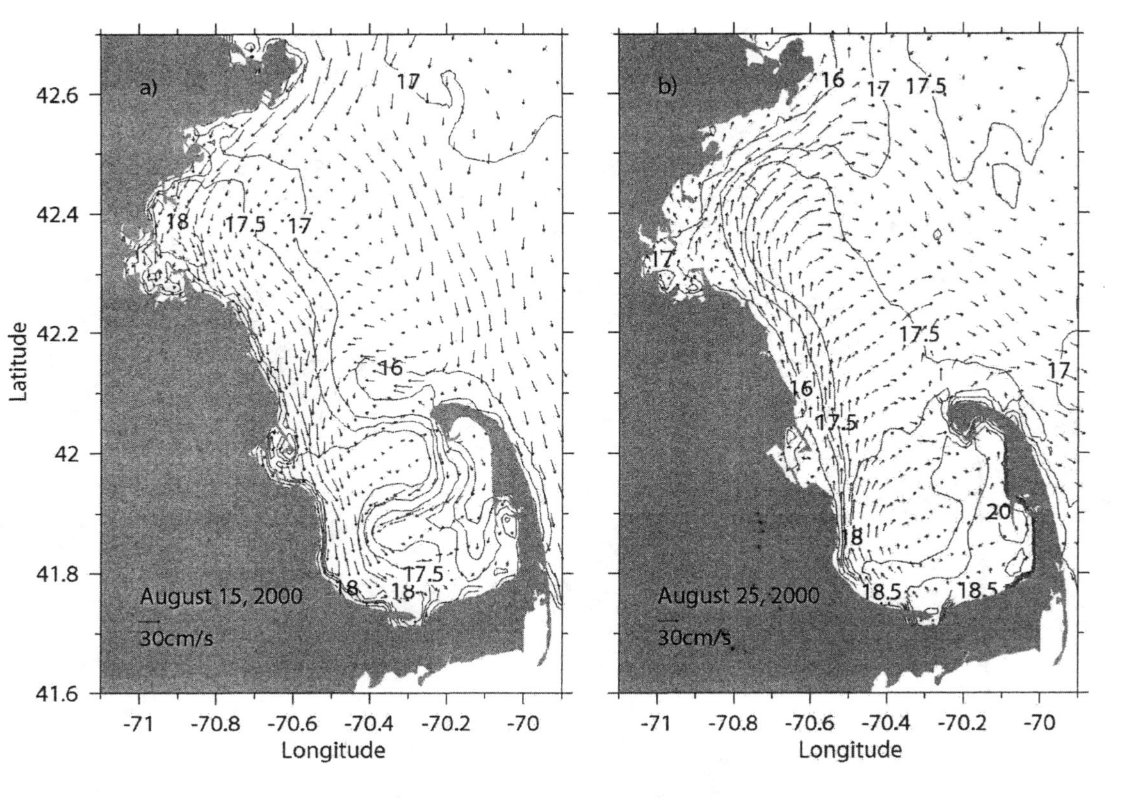

Figure 7. Modeled surface temperature(contonurs) and currents(arrows) on the 15th of August (a) and the 25th of August, 2000 (b). The model output is averaged over a M_2 tidal cycle.

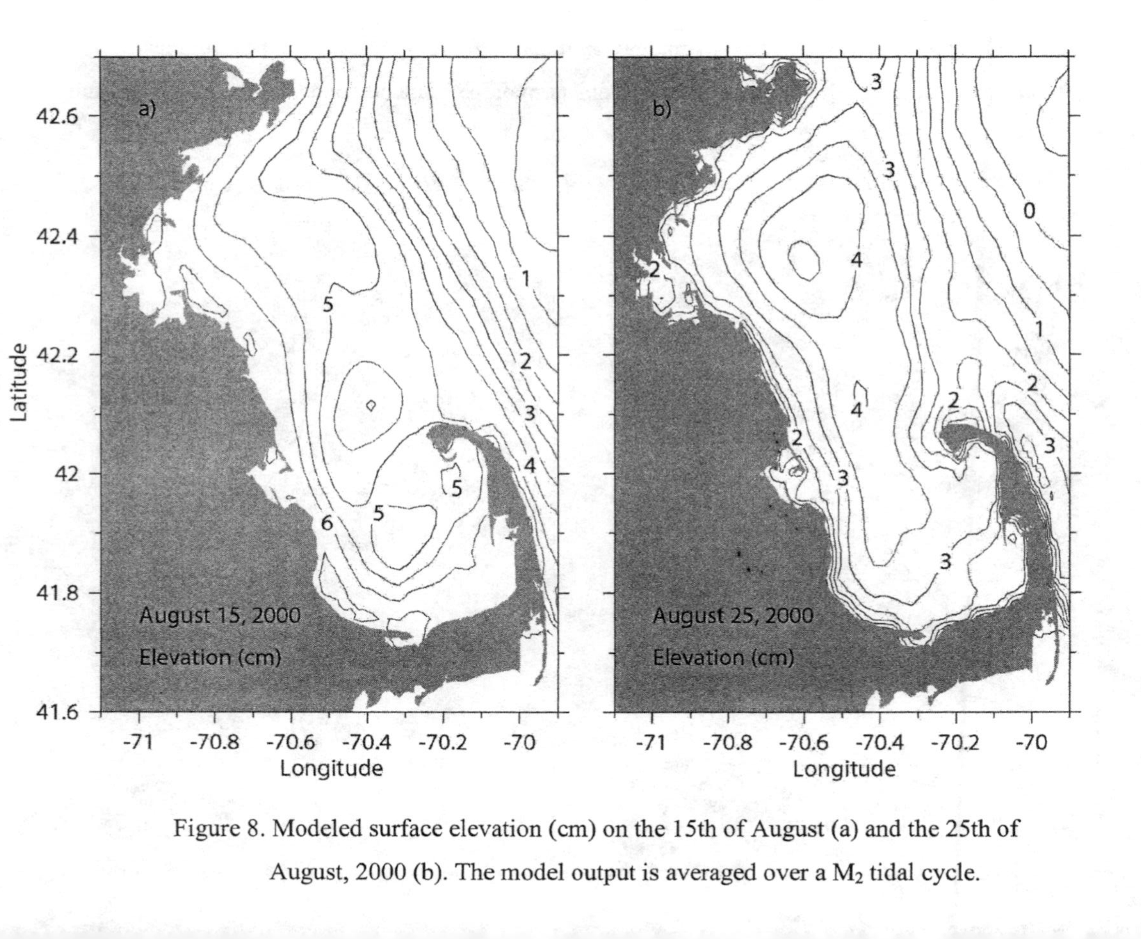

Figure 8. Modeled surface elevation (cm) on the 15th of August (a) and the 25th of August, 2000 (b). The model output is averaged over a M_2 tidal cycle.

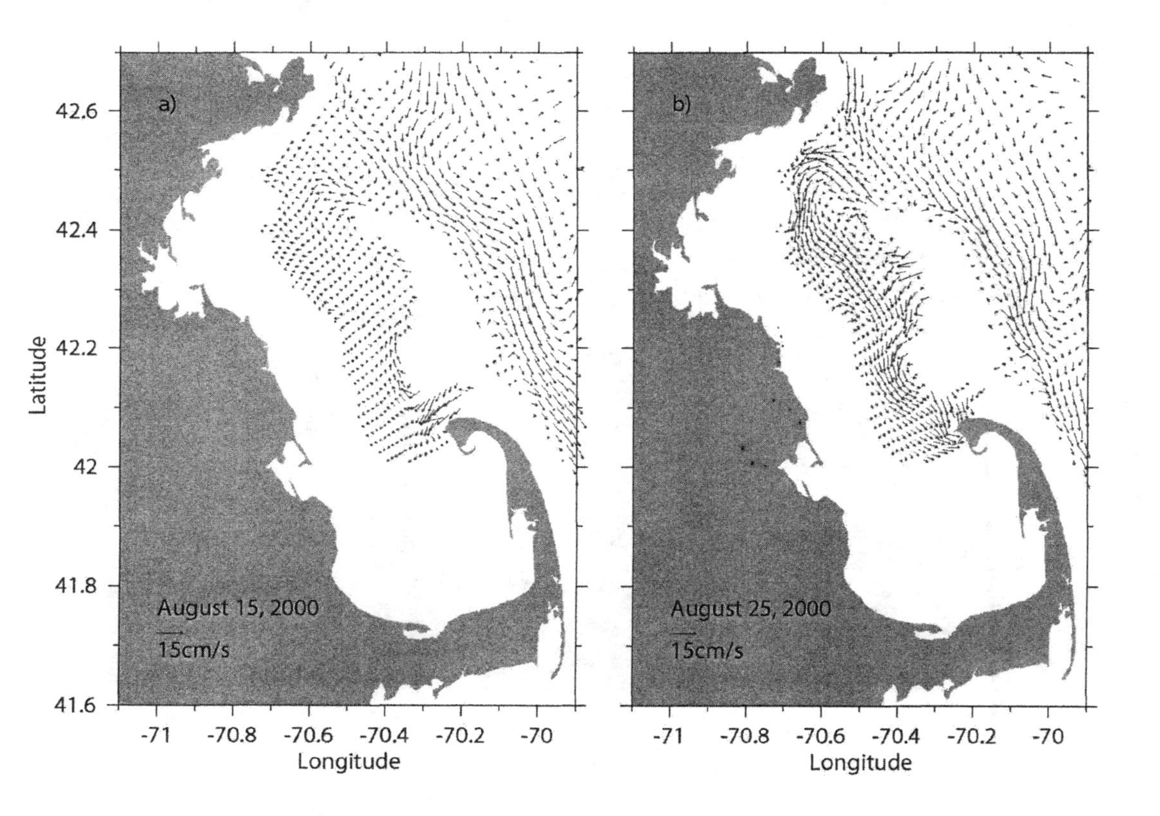

Figure 9. Modeled currents at 50m on the 15th of August (a) and the 25th of August, 2000 (b). The model output is averaged over a M_2 tidal cycle.

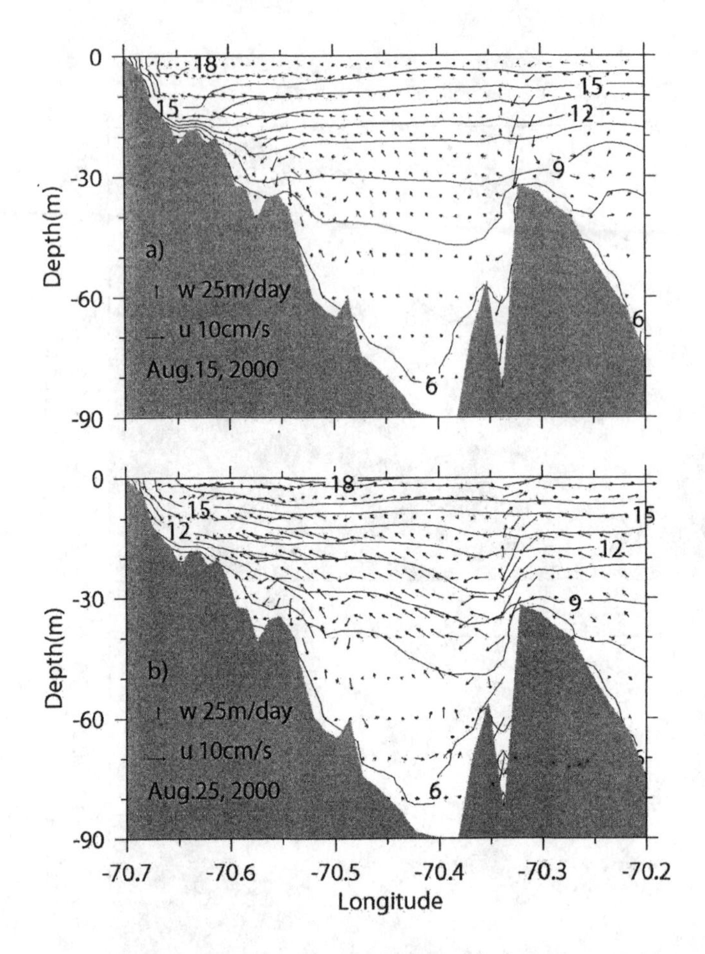

Figure 10. Temperature (contours) and currents (arrows) along the west-east transect from Scituate to Stellwagen Bank as indicated in Figure 1 on the 15th of August (a) and the 25th of August, 2000 (b). The model output is averaged over a M_2 tidal cycle.

Nested Hydrodynamic Modeling Using the MIKE 3 Model

E. B. Rasmussen[1], A. Driscoll[2], T. S. Wu[3], and R. Copp[4]

The numerical technique adopted in a hydrodynamic model using multiple grids of different spatial resolution is described. The nested grid configuration can reduce the CPU time compared to the standard approach of using only one fixed grid size. Typically applications of hydrodynamic models have a limited physical area of main interest. A standard hydrodynamic model can be set up using one grid size, but this will often result in a very large number of computational grid points beyond the area of main interest. This approach will require a large amount of computer time and memory.

When applying a multiple grid resolution set-up, it is necessary to ensure conservation of mass and momentum during the dynamic exchange between the coarser grid region and the finer grid regions. This nested model design included a simple test channel with dimensions of 12,000m x 300m x 10m. The initial condition was set up as still water. The driving force in the model is a simple 5 cm set-up and 5 cm set-down at the closed ends of the channel.

Fifteen 300m x 300m x 10m coarse grids were used in both channel closed ends, making the channel cross-section only one cell wide. Thirty 3-cell fine grid cross-sections were nested in the middle of the channel. The fine grid size was 100m x 100m x 10m, creating 3 cells in the channel cross-section.

The MIKE 3 model was set up with 10 vertical layers. The model was simulated until a steady state was achieved conservation of mass and momentum was then verified between a fine grid cross-section and two coarse grid cross-sections.

A real application in which the nested MIKE 3 model has been used is also described. This application illustrates that the nesting technique successfully can be adopted without compromising the model accuracy and performance. The applied nesting technique is thus an efficient tool to optimize simulations with otherwise excessive and potentially prohibitive run times.

[1] DHI Water & Environment, Agern Allé 5, DK-2970 Horsholm, Denmark
[2] DHI Inc, 301 S State Street, Newtown, PA 18940
[3] Florida Department of Environmental Protection, 2600 Blair Stone Road, Tallahassee, FL 32399
[4] DHI Inc. 311 S. Brevard Ave. Tampa, FL 33606

Introduction

Typical applications have a limited physical area of main interest, which covers only a smaller part of the total modeling area. To obtain a satisfactory spatial resolution within this area of interest, a model using the same horizontal resolution can be used. But this will often result in a very large number of computational grid points, many of which are often wasted in areas of only limited interest for the application, and accordingly this approach will require much computer time and memory. In response to this DHI Water & Environment has for its two-dimensional and three-dimensional models developed a special technique for embedding multiple and successive finer resolution models within the overall model. The main difference to other techniques focussing on a gradual refinement of the model resolution (nesting) is that this method is implicit of nature while most other techniques are explicit. The main differences compared to other dynamic nesting techniques are that the present scheme does not apply an relaxation zone and that the same time integration scheme is used for all grids. The advantage of applying the nested grid facility compared to the standard approach of using only one grid is mainly the reduced CPU requirements. Applying the nested module, the spatial resolution can be optimized in order to save computer time.

An alternative to the nesting facility is to apply a model based on a finite element technique where greater flexibility in the model set-up is inherent. Potentially a finite element technique can be superior to classical finite difference techniques despite the lower efficiency in terms of computational speed per node. Unfortunately there's no rule-of-thumb determining when a finite difference technique is to be preferred and vice versa. This is very much dependent on the actual application.

Curvilinear models may also be an attractive alternative but generally speaking they lack the full flexibility of finite element models, and often have some constraints in terms of applicability and numerical properties compared to standard finite difference models.

In the present paper the nesting technique used in DHI's two- and three-dimensional models MIKE 21 and MIKE 3 is described in detail and illustrated through an application from Denmark.

Nested Hydrodynamic Model

The MIKE 3 hydrodynamic module (DHI Software 2003, Rasmussen et al 1999) is a general numerical modeling system for the simulation of water level variations and flows, it simulates unsteady three-dimensional flows in fluids when presented with bathymetry and reverent conditions (e.g. bed resistance, wind field, baroclinic forcing, hydrographic boundary conditions).

This nested version of MIKE 3 HD (MIKE 3 NHD) has a built-in refinement-of-scale facility, which gives a possibility of making an increase in resolution in areas of special interest. MIKE 3 NHD solves the hydrodynamic equations simultaneously in a user-defined number of dynamically nested grids. The advantage of applying the nested grid facility compared to the standard approach of using only one grid is mainly the reduced CPU requirement. Typical applications of hydrodynamic module have a limited physical area of main interest, such as our residential canal system, which covers only a smaller part of the total modeling area. To obtain a satisfactory spatial resolution of the model within this area of interest, the standard hydrodynamic module can be used. But this will often result in a very large number of computational grid points, many of which are often wasted in the areas of only limited interest for the application, and accordingly this approach will require much computer time and memory. Applying the nested module, the spatial resolution can be optimised in order to save computer time. The nesting is in the horizontal direction only. The vertical resolution is the same throughout the entire model. The reason for this is that in any stratified environment a potential interface inevitably will be resolved poorly with a coarser vertical resolution compared with a finer vertical resolution. A distinct interface can thus only rarely be retained across two area borders whereby the advantage of the nesting is lost unless special sub-grid-scale parameterizations are applied. This is of course also true for the horizontal direction but from a physical point of view much less significant.

MIKE 3 is a finite difference model, which utilizes a rectangular grid in the horizontal directions and a z-level formulation in vertical direction. The vertical discretization allows for a bottom-fitted grid, in which the bottom grid cell can be stretched in thickness if the bed is deeper than the bottom of the lowest grid cell (see Figure 1). Similarly, the uppermost cell may stretch or compress as the water level varies in the model.

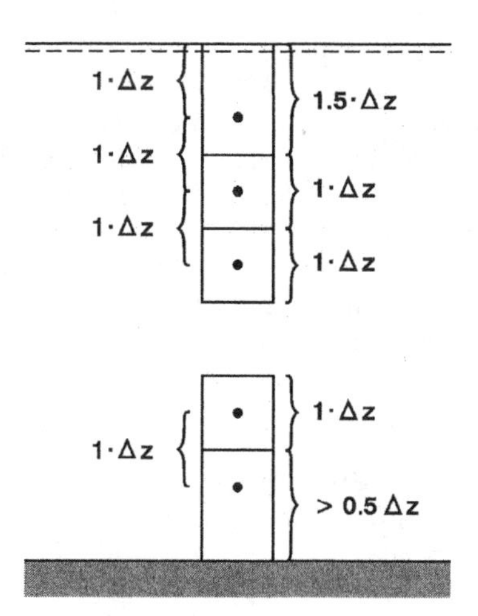

Figure 1 *Schematic diagram showing the vertical grid structure in MIKE 3. Both the topmost and bottom cells are variable in thickness. The topmost grid may vary from as little as the specified drying depth (typically 0.2m) to 1.5Δz or more as the water levels in the model change. Cells of less than the specified drying depth are dried and removed from the calculation.*

In the MIKE 3 NHD the grids are dynamically coupled and interact between the modeling grids of different resolution. The MIKE 3 NHD can work with up to nine bathymetries (model areas) of different resolutions. The bathymetries can be nested into each other with more than one model area at each spatial resolution shown in Figure 2. There are a number of rules to obey when preparing nested bathymetries to obtain compatibility between a fine grid and its coarse grid:

- The ratio between the horizontal spatial resolution at one level to the next level must be 3, i.e. $\Delta x_{coarse} = 3\ \Delta x_{fine}$
- Model area at the same level must not overlap. The distance between model areas at the same level should be at least three times Δx_{coarse}.
- Corners of sub-areas must be placed in grid points (integer values) of the respective enclosing grid. This means that, in each horizontal direction, every third grid point of the fine grid is common to a grid point in the coarse grid, a so-called common grid point
- Open boundaries are only allowed in the coarsest grid, referred to as the main area.
- Model sub-areas must be placed at least three grid points within the boundary of the enclosing area if the closest boundary point is a water point. If the boundary point is a land point, a distance of one grid point is sufficient.
- The water depths in common grid points along borders between areas must be equal in the coarse grid and in the fine grid. The water depths in the two border grid points of the fine grid next to the common grid point must have values which equal the water depth of the common grid point.
- To simplify the numerical implementation, it has been chosen to demand that the water depths across borders should be equal within a band on each side of the border. This means that the water depth in the coarse grid must be equal in three points orthogonal to the border (one point at on the border and one point on each side). In the fine grid, the water depth in the first four grid points orthogonal to the border should be equal. If the border is land, this rule does not need to be satisfied.
- Finally, all interior common points should have equal water depth in the coarse grid and in the fine grid. As the nested model does not perform any coarse grid calculations in the area covered by the fine grid, this rule is only included in order to ease pre- and post-processing. Fine grid solutions of any model quantity are always copied to the coarse grid.

A pre-processing tool for checking and possibly adjusting nested bathymetries may conveniently be applied prior to running the MIKE 3 model.

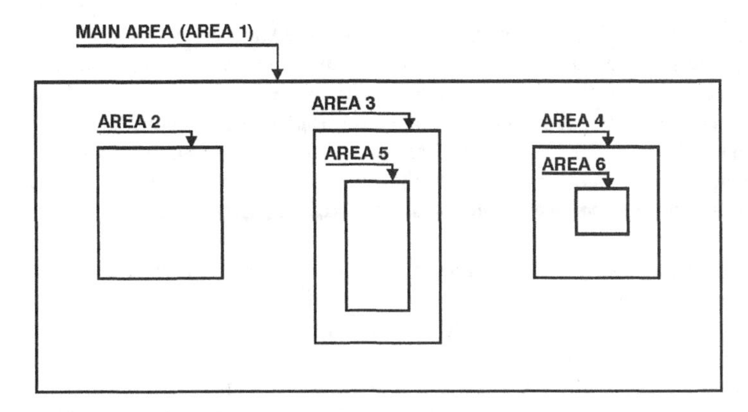

Figure 2 Sketch showing possible nesting of bathymetries.

Most model specifications in MIKE 3 NHD are identical to those in MIKE 3 HD. The major difference is that the nested model has to be specified by the most of the model parameters (coefficients, initial fields, maps, etc.) separately for each area. A few extra efforts for MIKE 3 NHD are listed below:

- For a "cold start", you need to specify how many areas you want to include in the simulation. The main area and all sub-areas supplied with origin in enclosing grid coordinates (integers) should be selected. The vertical grid spacing and the number grid points in the vertical direction should be specified. The vertical resolution is common to all areas. The model orientation and origin in geographical coordinates should be supplied with the data file for the main area bathymetry.
- For a "hot start", it is necessary to select a hot data file for each area. The other information mentioned above is contained in the hot data files and cannot be changed.
- Specifications related to the simulation period are common to all areas.
- The same turbulence formulation should be used in all areas. Separate parameters for eddy viscosity (e.g. an Smagorinsky formulation or a k-ε model) should be specified for each area. Bed roughness parameters (constants or maps) should also be specified for each area.
- The choice of density variation description (i.e. varying salinity and/or temperature and advection-dispersion scheme) is common to all areas. Initial fields, dispersion factors and dispersion limits for each area should be specified. The background value is common to all areas.
- Open boundary conditions in the nested MIKE 3 model are specified exactly as for the standard MIKE 3 model. The open boundaries must be located in the main area.
- Initial surface elevation should be specified for each area. Flooding and drying depths are common to all areas.

- Source/sink specifications are the same as for the standard MIKE 3 model, except the specified area in which the source/sink is located. This area number must correspond to the finest grid covering the source location.
- The wind, precipitation and heat exchange specifications are identical to those in the standard MIKE 3. Data files (wind field, precipitation map, air temperature map) are specified for the main area only and the respective quantities are interpolated automatically by MIKE 3 NHD to match the finer grids.
- Having selected particle tracking as in the standard MIKE 3, it is necessary to specify an area number corresponding to the location of each particle source.
- The discharge calculations in MIKE 3 NHD need additional information of the number of the area in which the discharge cross-section is placed. Discharge lines are allowed to cross area borders, and in this case you specify the coarsest of the involved areas. It is, however, recommended to keep the discharge lines inside as fine a grid as possible and, if necessary, to sub-divide the cross-section into minor sections which do not cross borders.
- In the output specifications, each output area - also hot files - must be related to an associated model area.

The mathematical foundation in MIKE 3 NHD is the mass conservation equation, the Reynolds-averaged Navier-Stokes equations, including the effects of turbulence and variable density, together with the conservation equations for salinity and temperature.

The equations are spatially discretised on a rectangular, staggered grid, the so-called Arakawa C-grid, as depicted in Figure 3. Scalar quantities, such as pressure, salinity and temperature, are defined in the grid nodes whereas velocity components are defined halfway between adjacent grid nodes in the respective directions.

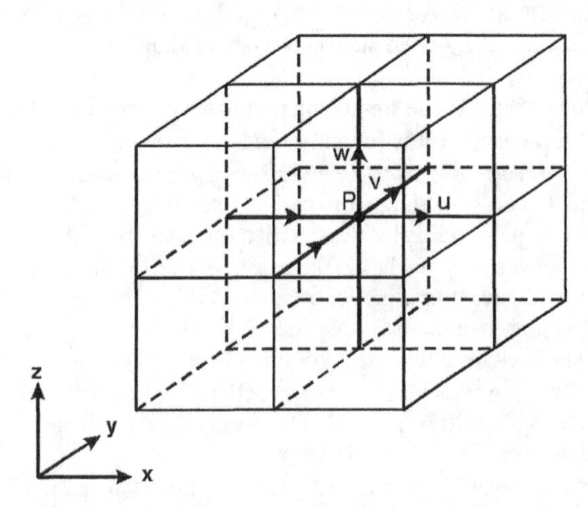

Figure 3 Arakawa C-grid.

Time centering of the hydrodynamic equations is achieved by defining pressure P at one-third time step intervals (i.e. n+1/6, n+3/6, n+5/6, etc.), the velocity component in the x-direction u at integer time levels (n, n+1, n+2, etc.), the velocity component in the y-direction v at time levels n+1/3, n+4/3, n+7/3, etc., and the velocity component in the z-direction w at time levels n+2/3, n+5/3, n+8/3, etc.

Using these discretisations, the governing partial differential equations are formulated as a system of implicit expressions for the unknown values at the grid points, each expression involving known but also unknown values at other grid points and time levels.

The applied algorithm is a non-iterative Alternating Direction Implicit (ADI) algorithm, using a "fractional step" technique (basically the time staggering described above) and "side-feeding" (semi-linearization of non-linear terms). The well-known double sweep algorithm solves the resulting tri-diagonal system of equations.

The method applied to ensure the dynamical nesting in MIKE 3 NHD, i.e. the two-way dynamically exchange of mass and momentum between the modeling grids of different resolution, is a relatively simple extension of the solution method used in the standard hydrodynamic module and it may basically be described as:

Along common grid lines shown in Figure 4, the mass and momentum equations (in the staggered grids) are dynamically connected across the borders by first setting up all coefficients for the double sweep algorithm in both coarse and fine grids. Hereafter the tri-diagonal system of equations is established.

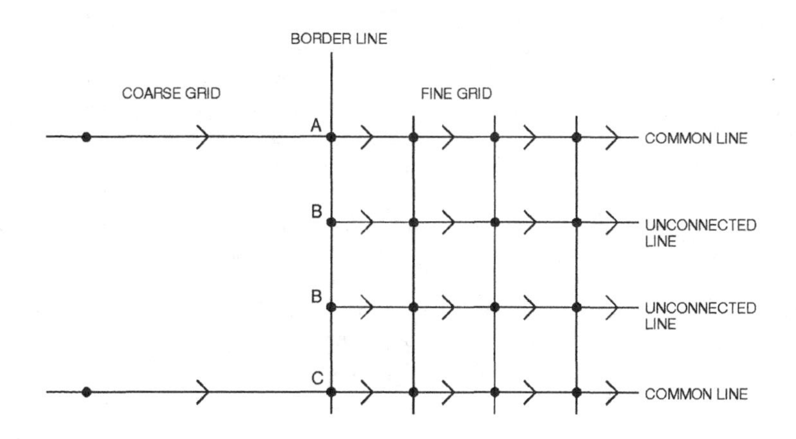

Figure 4 Coupling across a borderline between coarse and fine staggered grid during a x-sweep.

At the end points (points B in Figure 4) of the two unconnected grid lines in the fine grid, between the connected grid lines, a level (i.e. pressure) boundary condition is applied. The pressure boundary values are found by assuming that $\partial P/\partial t$ in the common mass point (point A in a down-sweep and point C in an up-sweep due to the applied ADI technique) on the border of the connected grid line is identical to $\partial P/\partial t$ at the end point (mass point B) of the unconnected grid line.

From a conceptual point of view the nesting technique in MIKE 3 NHD implies that the matrix systems for the coarse and fine grids are linked through the common nodes at the interface of the two grids such that the coarse grid equations in the area covered by a finer grid is replaced by the set of equations representing the fine grid before inverting the matrix systems. Other similar two-way nesting models (see for instance Ginis et al 1998) often apply a spatial smoothing technique to control potential numerical instabilities.

At the borders, explicit parameters and values of the prognostic variables at old time steps are found by interpolation between the fine grid and the coarse grid.

In MIKE 3 NHD, the equations for all the different spatial scales in the actual model are solved at the same time. As for the standard MIKE 3 HD, the computations are divided into two parts:

- the coefficient sweeps (traditionally referred to as EF sweeps), and
- the SP- or SQ-sweeps (a traditional notation: S for surface elevation, P and Q for the two horizontal fluxes)

According to the double sweep algorithm:

The horizontal sweep structure for the hydrodynamic module in the nested version of MIKE 3 is described in the following. It is emphasized that the nesting is in the horizontal direction only. For the vertical direction, the sweeps are conducted as for the standard MIKE 3 HD.

For odd time steps, the computations start in the upper-right corner with a x-sweep moving down in the model, followed by a y-sweep moving left in the model. These sweeps are referred to as down-sweeps. For even time steps so-called up-sweeps are performed, i.e. the computations start in the lower-right comer with a x-sweep moving up in the model, followed by a y-sweep moving right in the model. The superior computation structure is shown in Figure 5.

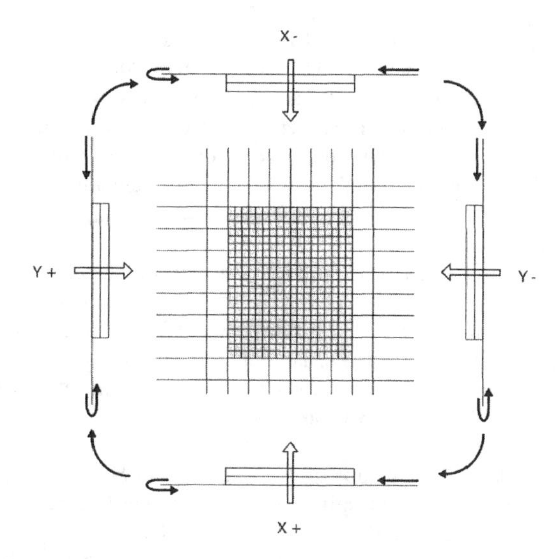

Figure 5 Superior structure of the computations for the nested MIKE 3 HD. X and Y indicate the sweep-direction, - indicates down-sweeps, + indicates up-sweeps.

The detailed computation structure for a down-sweep (odd time step) in the x-direction is shown in Figure 6. The equations at the different scales are solved at the same time. Thus, information can travel in two directions: Effects from the coarse grid influence the solution in the fine grids but also the effects in the fine grids can influence the solution in the coarse grid.

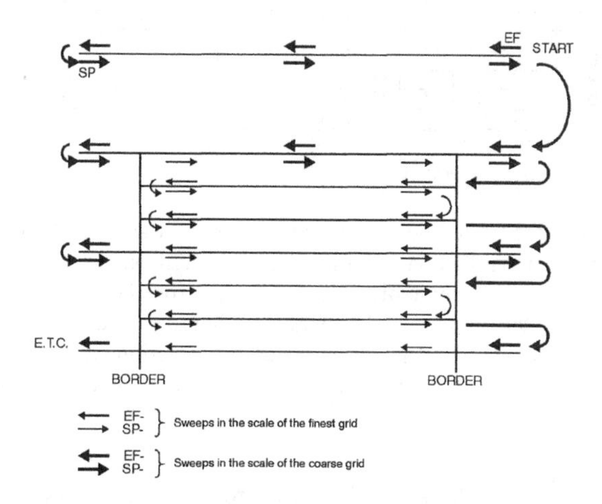

Figure 6 Computation structure for a down-sweep in the x-direction. The corresponding up- or y-sweeps are symmetrical.

The governing equations are discretized aiming at a second-order accuracy in both time and space. It can be shown, however, that the side-feeding technique at a specific time level leads to a first-order accuracy on the convective terms and eddy terms due to the time centering. However, the successive down-up-down-up order changes the sign of the first-order truncation error such that within two time steps (down sweep and up sweep) a second order accuracy is maintained.

MIKE 3 NHD Test Case

Much effort has been put into conservation of water mass, temperature, and salinity at the interfaces between the different model grid sizes. A simple test channel with dimensions of 12,000m x 300m x 10m illustrates MIKE 3 NHD's ability to ensure conservation of the different quantities. The initial condition was set up as still water. The driving force in the model is a simple 5cm set-up and 5cm set-down at the closed ends of the channel as the boundary condition.

Fifteen 300m x 300m x 10m coarse grids were used in both channel closed ends, making the channel cross-section only one cell wide. Thirty 3-cell fine grid cross-sections were nested in the middle of the channel. The fine grid size was 100m x 100m x 10m, creating 3 cells in the channel cross-section. The MIKE 3 model was set up with 10 vertical layers. The partial model domain is show in Figure 7.

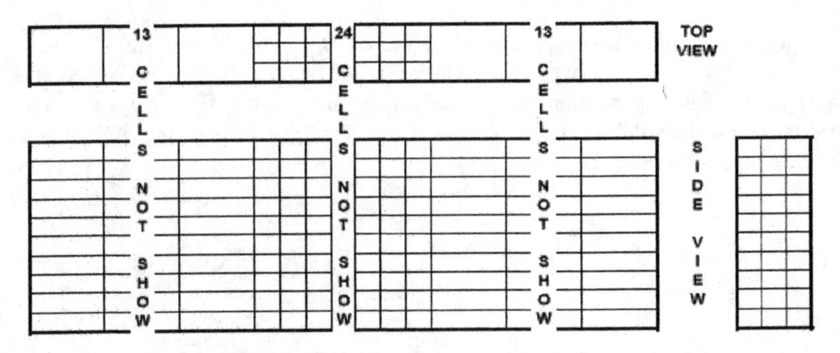

Figure 7 Partial Model Domain.

For time-stepping schemes, the 30-second time step was used to ensure a stable solution. It can be shown that MIKE 3 is numerically unconditional stable. However, too large time steps will inevitably lead to phase error. This is especially true for cross flow due to the applied side-feeding technique. The time step is usually determined from Courant-Friedrichs-Lewy stability criterion which must be less that unity for explicit models. The maximum CFL time step could be calculated by the following equation:

$$\Delta t_{\max} = \frac{\Delta x}{\sqrt{gH}} \approx \frac{100m}{\sqrt{9.82\dfrac{m}{\sec^2} \times 10m}} \approx 10\sec$$

Thus, the applied time step in the present test case corresponds to a Courant number of approx. 3.

The adopted turbulence closure is a mixed k-ε / Smagorinsky model where a standard k-ε model is used in the vertical while a Smagorinsky formulation is used in the horizontal directions. The relevant coefficients are shown in Table 1 are used for this test case study.

Table 1 Turbulence specification terms

Name	Value
Horizontal Smagorinsky coefficient	0.40
Empirical constant, c_μ	0.09
Empirical constant, $c_{1\varepsilon}$	1.44
Empirical constant, $c_{2\varepsilon}$	1.92
Empirical constant, $c_{3\varepsilon}$	0.00
Empirical k-diffusion constant, σ_k	1.00
Empirical e-diffusion constant, σ_ε	1.30
Empirical buoyancy constant, σ_t	0.90

The model was run for 12 hours and output the model results every 15 minutes (30 time steps). Figure 8 shows three model flows in 0.25, 0.5, and 0.75 channel lengths representing the upstream coarse grid, center finer grid, and downstream coarse grid respectively. These three flows show no difference and reach the steady state condition after 9 hours. It indicates no mass loss between the fine and coarse grids of the nested hydrodynamic module. Also, the partial horizontal velocity fields from the last time step (12 hours) are shown in Figure 9. Furthermore, the averaged horizontal cross-sectional velocity and water elevation are shown in Figure 10 and plot the channel water energy relations in Figure 11. Figure 11 is an extra energy conservation checking by using the following formula:

$$E = \eta + \frac{U^2}{2g}$$

where E is the water energy, η is water elevation, U is the averaged cross-sectional velocity, g is the gravity.

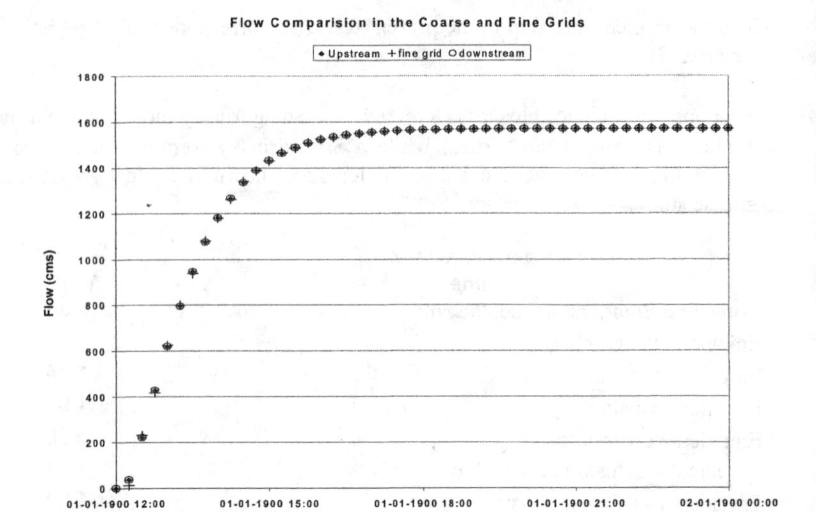

Figure 8 *Flow comparison between the fine grid and the coarse grid from cold start to steady state.*

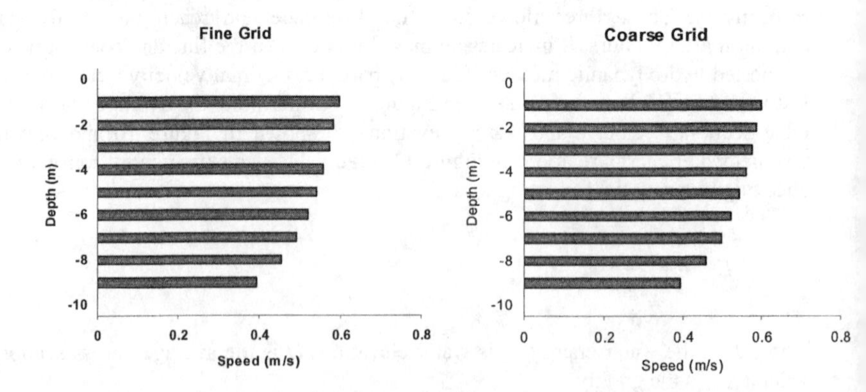

Figure 9 *Partial cross-sectional modeling vertical velocity profile at steady state in mid-channel fine grid model area (left) and mid-channel coarse grid model area (left).*

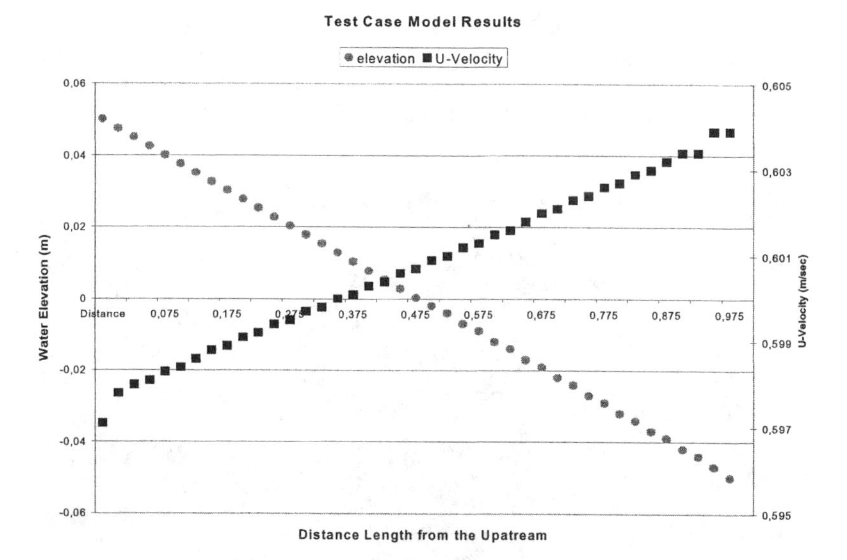

Figure 10 *Test model results at steady state situation.*

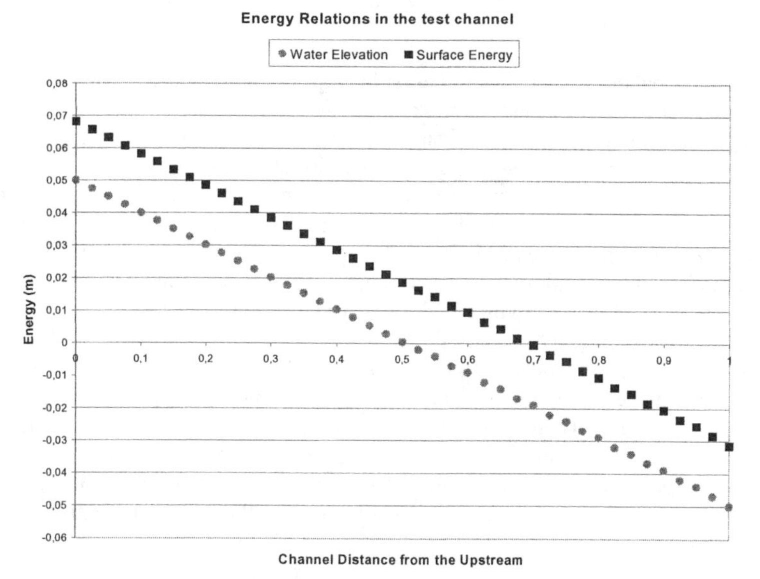

Figure 11 *Energy relations in the test channel at the last time step.*

Application of the Nested Version of MIKE 3 to the Sound, Denmark

The Sound (Øresund) is one of the three straits connecting the Baltic Sea with Kattegat and the North Sea (Figure 12). The flow through Øresund influences the salinity and the oxygen budgets of the Baltic Sea, and hereby the entire Baltic ecosystem.

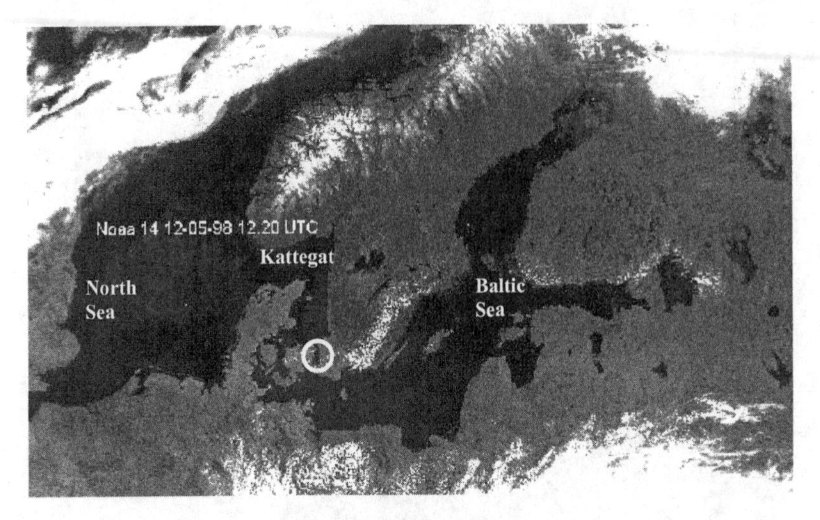

Figure 12 Satellite view of the North Sea and the Baltic Sea. The circle indicates the Sound.

The water balance of the Baltic Sea is determined by (i) inflow from the catchment with its various streams and rivers, and groundwater seepage; (ii) direct rainfall and evaporation; and (iii) water exchange (inflow and outflow) through the connecting straits. The water exchange is mainly forced by air pressure variations and by wind set-up in Kattegat and the western Baltic. During extended calm periods, there is a slight outflow from the Baltic. During normal weather conditions, periods with outflow shift with periods of inflow. The water exchange in turn controls the inflow and outflow of salt and oxygen. Occasionally, during particular storms, large inflows occur of North Sea water with a high salinity and a high oxygen content.

The Baltic Sea is brackish. Its overall salinity budget is entirely determined by a balance between salt lost by outflow and salt supplied by inflow through the connecting straits. There are several deep basins with dense bottom layers with a higher salinity and a diminutive vertical mixing. These water volumes can remain stagnant for several years until they are renewed during the occasional large inflows of saline water. The average salinity distribution and the general salt transport pattern are shown in Figure 13.

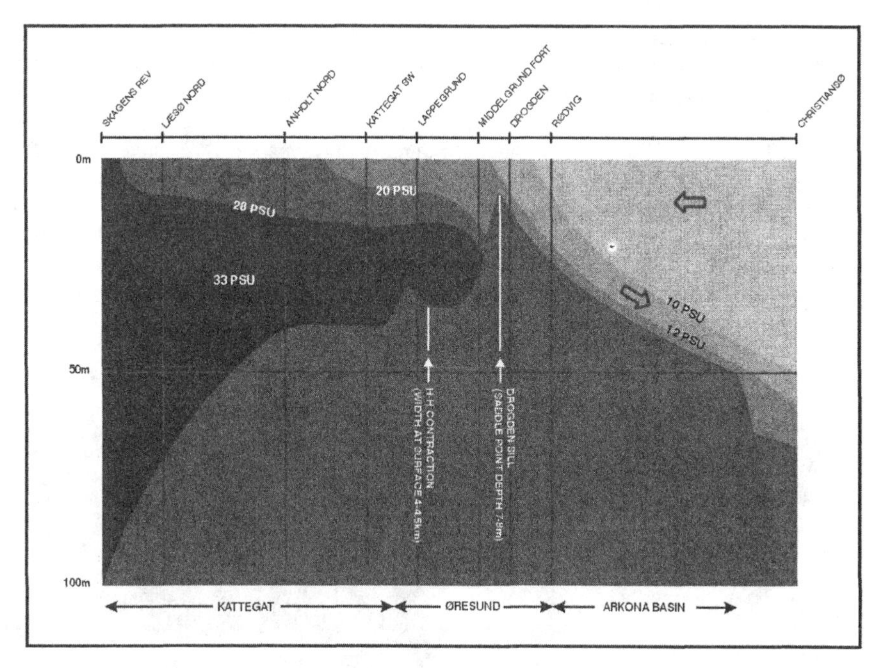

Figure 13 Regional salinity distribution and salt transports (PSU: Practical Salinity Units ~ kg salt per 1,000 kg seawater)

The Baltic oxygen budget is determined by a balance between oxygen production in the water column by photosynthesis; supplies from the atmosphere through the surface; supplies by inflows from rivers and from the connecting straits; and losses due to local oxygen consumption caused by mineralisation of organic matter and other oxidation processes. The vertical oxygen distribution is highly influenced by density stratification reflecting the salinity distribution, which can literally block any vertical exchange of solutes. Oxygen to the deep bottom layers is mainly supplied by the large inflows.

Following bilateral discussions between the Danish and Swedish governments an agreement on the construction of a fixed link between the two countries was signed in 1991. The significance of the Sound in relation to the ecological system of the Baltic Sea implied a number of constraints be put on the construction of the link. Essentially these constraints were to ensure that the link would have a negligible effect on the ecosystem of the Baltic Sea. Three parameters were used as measures for the impact of the link being the water, salinity and oxygen budgets. The modeling strategy that was adopted for the impact assessments were to run the calibrated MIKE 3 NHD model for a so-called design period reflecting the natural variability of the Sound for two scenarios being the preconstruction and final construction situations. The differences between the two scenarios in terms of the water, salinity and oxygen budgets would then be used to assess the impact of the link. In the present paper, however, focus is put on the hydrodynamic calibration and performance to illustrate the principles of the nesting technique to simulate a complex real-life application.

The MIKE 3 model was set-up in the Sound using dynamically nested grids with 900 m - 300 m - 100 m horizontal resolutions and 1 m vertical resolution. The model was forced by water level variations and salinity/temperature distributions at the open lateral boundaries and by wind stress and heat flux at the free surface. These forcings were prescribed based on measurements. A special field program was designed to support the model calibration and validation. In Figure 14 the model set-up and associated field program are shown.

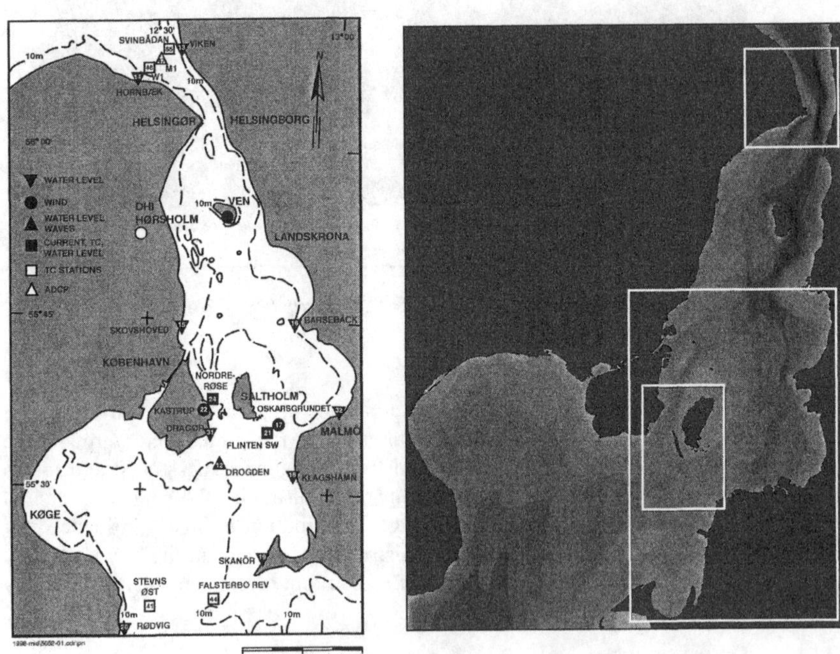

Figure 14 *Field program (left) and nested MIKE 3 (right) in a 900 m – 300 m – 100 m model set-up.*

MIKE 3 was calibrated and validated against measurements of water levels, currents and salinities. The model simulations have been assessed by means of simple statistical methods, visual inspection and analysis of both energy losses and correlation between velocity profiles. The model was calibrated until both a satisfactory visual agreement between the measured and modelled data was obtained and until the statistical methods exceeded predefined criteria. These criteria were set such that the correlation coefficients for water levels, currents and salinities at the available stations should yield 0.90, 0.90, and 0.85 for the three parameters, respectively.

Two ADCP and CTD stations (Nordre Røse and Flinten SW) located in the vicinity of the link alignment were used as the main calibration and validation stations. The main calibration parameters were the bed roughnesses and the parameters entering the Smagorinsky eddy viscosity formulation. While the wind stress often plays a significant role in the flow pattern this is not the case in the Sound where it primarily influences the secondary flow patterns. As part of the study a comprehensive sensitivity analysis was performed for all input parameters of the model to identify the significance of each of the parameters and furthermore, to determine the uncertainly of the model results. In Figure 15 a time series of the measured current from the uppermost bin level of the ADCP instrument and corresponding modeled current from a 72-day design period in 1993 is shown. The current has for simplicity been projected onto its main flow direction with positive values indicating northward flow. In Figure 16 a similar salinity comparison is shown.

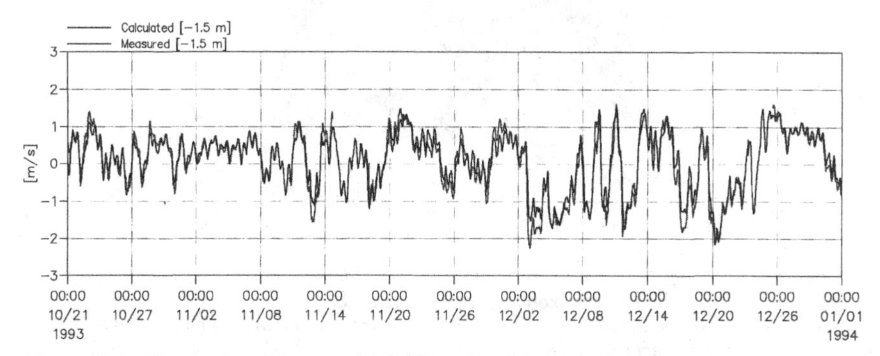

Figure 15 Time series of measured (ADCP) and modeled current at St. Flinten SW

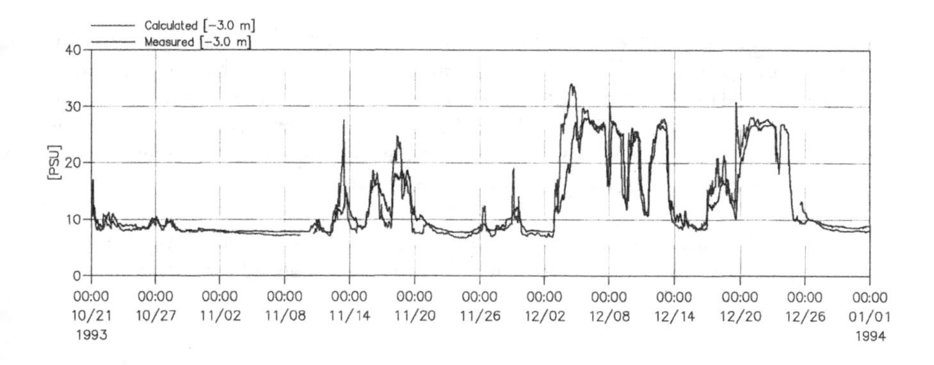

Figure 16 Time series of measured and modeled salinity at St. Flinten SW

In addition to the fixed station numerous ADCP transects were conducted using a ship-borne ADCP instrument. The data sets were subsequently compared to the corresponding modeled currents. An example of such a comparison of the top bin level of the ADCP instrument is shown in Figure 17.

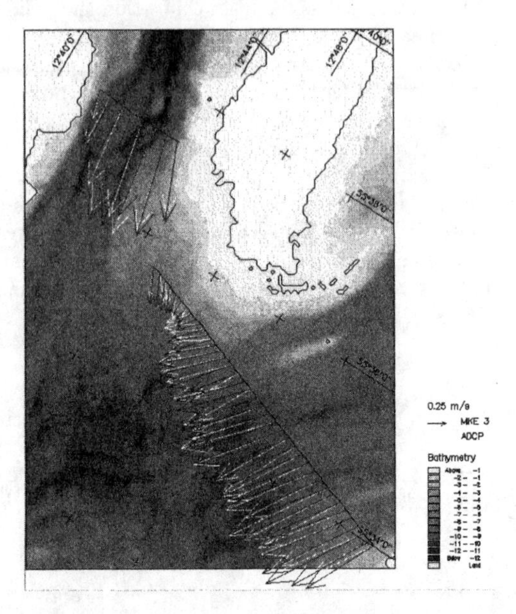

Figure 17 *Comparison of top bin level of two ADCP transects in the vicinity of the alignment of the fixed link and the corresponding modeled surface current. The modeled current (dark) has been interpolated in time and space to match the exact position and time of the measured (white) data. Grey shading indicates the depth.*

The statistical analysis was performed on all stations and for all levels. The overall performance of the MIKE 3 model is reflected in Table 2 showing the mean deviation, standard deviation and correlation coefficient for the water level, current and salinity. No calibration was performed on the temperature, as the density in this region is entirely dominated by the salinity and thus not important. For completeness, however, the analysis was also performed for temperature.

Numerous other analyses were carried out focussing on the transport of water and salt through the Sound. An example of such analysis is the so-called Hovmoeller diagram shown in Figure 18.

Table 2 Overall statistical properties of the MIKE 3 calibration.

	Mean deviation	Standard deviation	Correlation coefficient
Water level	0.02 m	0.06 m	0.97
Current	-0.04 m/s	0.19 m/s	0.95
Salinity	1.35 PSU	3.30 PSU	0.85
(Temperature)	-0.69 C	0.84 C	0.89

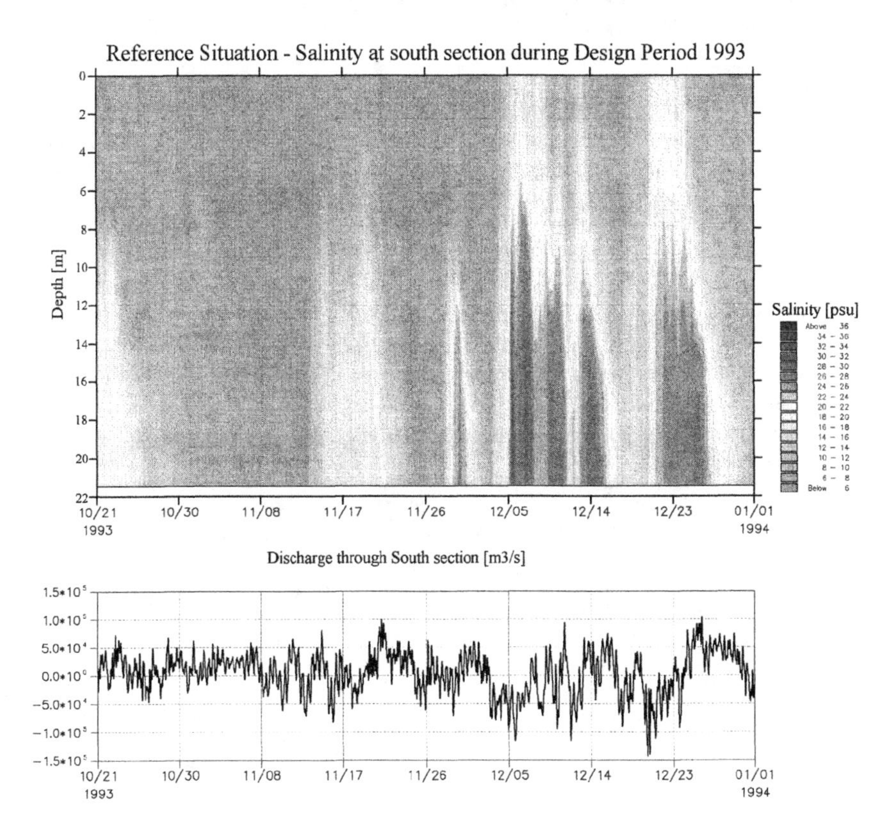

Figure 18 Hovmoeller diagram for a cross-section south of the link alignment (top) and the cross-sectional integrated discharge of salt (bottom) for the design period.

Conclusion

The dynamic (two-way) nesting technique applied in DHI's two- and three-dimensional models has proven to be an efficient mean to reduce the overall computational time in many practical applications. It is important though that the adopted nesting technique does not compromise the general accuracy and performance of the model set-up and that important conservation properties are retained across the borders of a coarse and fine grid area. The model characteristics with respect to conservation of mass, momentum and scalar quantities like salinity and temperature have been illustrated through a simple test case.

Likewise a real case application has illustrated that the dynamic nesting feature successfully can be applied and thereby reduce run times significantly compared to standard finite difference models.

References

DHI Software 2003: "MIKE 3 Estuarine and Coastal Hydraulics and Oceanography, Hydrodynamic Module, Scientific Documentation", DHI Water & Environment, Agern Allé 5, DK-2970 Horsholm, Denmark. (also available from www.dhisoftware.com)

Ginis, I, R. A. Richardson, and L. M. Rothstein, 1998: "Design of a multiply-nested primitive equation ocean model", Monthly Weather Review, 126, 1054-1079

Rasmussen, E.B., J. Pietrzak and R. Brandt (1999): "A coupled ice-ocean model for the Greenland, Icelandic and Norwegian Seas", Deep Sea Research II 46, 1169-1198

Modeling of Tidal Flow and Sediment Transport at Mitchell's Cut, Texas

Detong Sun,[1] Michael P. Walther[2]

Abstract: A two-dimensional hydrodynamic model and a sediment transport model were developed for Mitchell's Cut inlet management study. The models were used to assess the existing condition to address sedimentation and navigation problems at Mitchell's Cut. The models were also applied to evaluate inlet management alternatives. Model results show that sedimentation in the interior channels of the inlet system is likely caused mainly by sand transported from the Gulf by flood tide and partly by mud transported from East Matagorda Bay by ebb tide. The inlet is naturally stable. A jetty alternative shows promising performance as it can significantly reduce sedimentation with little change in the hydraulic regime of the study area.

1. Study Background

This model study is part of the Mitchell's Cut Inlet Management Study performed for Matagorda County, Texas. The project aims to investigate the existing condition of Mitchell's Cut and ultimately form a reliable and cost-effective inlet management plan for Mitchell's Cut.

Mitchell's Cut (Figures 1 and 2) is a man-made cut to the Gulf of Mexico near the southeastern limits of East Matagorda Bay. East Matagorda Bay is a shallow lagoon located between Colorado River and Caney Creek. Along its northern edge is the Gulf Intra-coastal Waterway (GIWW). The GIWW is separated from the Bay by narrow dredged-material islands and the Bay is separated from the Gulf of Mexico by mostly uninhabited barrier islands along the coast. Mitchell's Cut currently provides the sole direct connection and path for tidal exchange between the Gulf of Mexico and East Matagorda Bay. Caney Creek discharges into the Bay and GIWW to the north of Mitchell's Cut.

Historically, a natural cut also named Mitchell's Cut existed near the same location as it is today and was observed as early as 1875, but it had meandered along the Gulf shoreline. The modern Mitchell's Cut was excavated in 1989 in order to provide drainage of floodwaters from the areas surrounding Sargent Beach and the Caney Creek drainage basin. East Matagorda Bay had been used as a dredging disposal area by the US Army Corps of Engineers in association with initial construction of the

1) Coastal Technology Corporation, 1625 20th Street, Vero Beach, FL 32962, USA. dsun@coastaltechcorp.com..

2) Coastal Technology Corporation, 1625 20th Street, Vero Beach, FL 32962, USA. mwalther@coastaltechcorp.com.

GIWW during the 1940's and subsequent maintenance during the 60's, 70's and into the 80's. This Bay disposal appears to have affected sedimentation and turbidity within the Bay (Tipps, 2001). The Bay depth appears to have become shallower with average depths at about 2 to 4 feet. Caney Creek has been the major source of fresh water inflow into East Matagorda Bay since the 1940's. Other freshwater sources include Lake Austin, Peyton Creek and Live Oak Creek, which yield much less freshwater inflow than Caney Creek. Caney Creek drains an area of 344 square miles. The creek begins near the City of Wharton and flows in a southerly direction to the Bay. The creek is experiencing increased flow rate and higher turbidity in recent years in association with development in the drainage basin. (RUST Lichiliter/Jameson, 1994)

Since its construction, Mitchell's Cut and its surroundings have been undergoing substantial changes. The inlet itself has widened. An ebb shoal has developed at the inlet mouth. Sediment from the Gulf beaches and possibly Caney Creek has contributed to the formation of a tidal delta in the northeast corner of the East Matagorda Bay. A sand bar exists across the closet cut to Caney Creek between the GIWW and the Bay. This sand bar likely impedes freshwater flow into the Bay from Caney Creek and also affects currents in the study domain. Sedimentation has also occurred at the west intersection between Caney Creek and the GIWW affecting boat access from the lower Caney Creek area to the Gulf.

Figure 1. Regional project location map, arrow pointing to Mitchell's Cut.

Figure 2. Study domain of the Mitchell's Cut area (right frame) and part of the East Matagorda Bay area (left frame).

Although the modern Mitchell's Cut was constructed for drainage, it is now used for access to the Gulf on a daily basis by recreational and commercial boaters. High current velocities in Mitchell's Cut pose a dangerous situation for boaters during rough weather. The shifting channels and bars constitute a significantly dangerous situation for boaters.

The local community has expressed strong concern with the navigability of the unmanaged Mitchell's Cut, deterioration of water quality in East Matagorda Bay possibly due to heavy sedimentation, high turbidity and less freshwater input, increased flooding in the lower Caney Creek area, and concern for potentially significant erosion of the dredged material islands on the northern perimeter of East Matagorda Bay.

2. Physical Conditions at Mitchell's Cut

Wind: As noted by Lohse (1952) and Watson and Behrens (1970), winds in the study domain are strongly bimodal, consisting of persistent southeasterly winds occurring from March to April through August or September and north-northeasterly winds occurring from September or October through February or March. According to Watson and Behrens (1970) and Watson (1971), the annual resultant vector based on velocity, duration, and direction is perpendicular to the central Texas coast. Wind roses (USACE, 1992) constructed for Galveston, Texas, which is approximately 60 miles northeast of the project area, based on data provided by the National Climatic

Data Center, show that mean monthly wind speeds for the Galveston area range between 9.4 mph in August and 12.1 mph in April. The mean yearly wind speed is 11.0 mph. Thirty percent (30%) of the wind observations are from the southeast, 21 % are from the east and 21% are from the south.

Tide and Currents: Tide along the Texas coast is mixed, meaning that both semi-diurnal and diurnal signals are present, but typically, the diurnal component is predominant. Figure 3a shows hourly measurement of water level recorded at Mitchell's Cut by USGS between October 2000 and October 2001. Notice that the tidal range defined as the difference between daily high and low waters does not reflect the much larger range in water elevation that occurs seasonally.

Other factors will be present in a time series of water surface elevation. For example, storms produce surge, and the winter weather fronts that pass through the south Texas coast with an average periodicity on the order of 5 days will also change the water level through the accompanying wind, atmospheric pressure differentials, and movement of water into and out a particular region.

Currents are generally very strong in Mitchell's Cut. Figure 3b shows the current measured at Mitchell's Cut by USGS between October 2000 and October 2001. The maximum current velocity exceeded 3 ft/s. Currents are less strong in the GIWW and weak in the East Matagorda Bay (Kraus and Militello, 1996). Because the prevailing wind from east, currents in Matagorda Bay have a stronger E-W component.

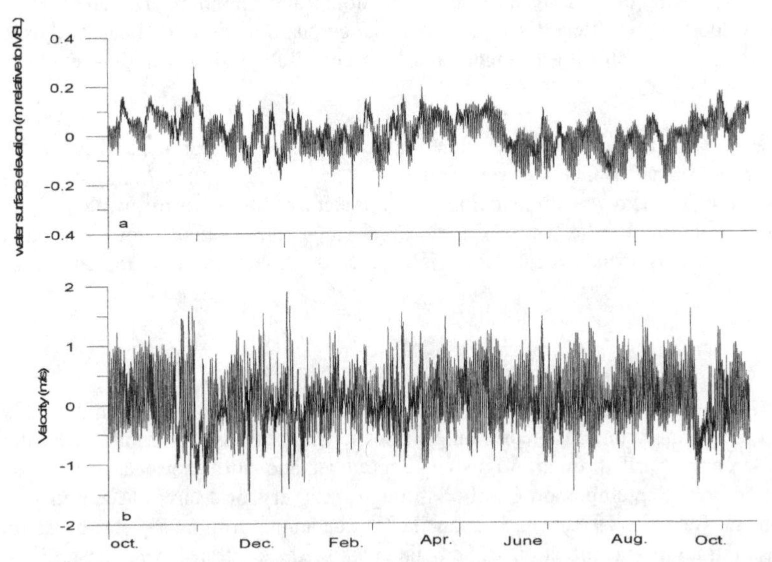

Figure 3: a: Tidal elevation and b: current velocity measured during Oct 2000 to Oct. 2001 by USGS at Mitchell's Cut

Waves and Longshore Transport: Measured wave data are not available for the Study Domain. One source of wave information is the Wave Information System (WIS) developed by Corps of Engineers Water Experiment Station (WES). The data was obtained by 20-year numerical hindcast for the Gulf of Mexico. WIS Station 9 is the closest station to Mitchell's Cut. It is approximately 17 miles offshore in about 84 ft. of water. Based on 20-year (1956 through 1975) statistics of WIS data, the mean significant wave height at Station 9 is 3.3 ft and 64% of the waves come from southeast. However, caution has to be used when applying the WIS wave data as pointed out by Kraus and Heilman (1996) that the WIS wave hindcast may over predict wave height on the west side of the Gulf.

Estimation of longshore sediment transport rates is best determined empirically because of the complex process of waves propagating towards the beach and breaking. Although seasonal conditions may cause the net direction of longshore transport in an area to change from year to year, the long-term predominant direction of transport along the coast of Matagorda Bay has been estimated to be to the south. Evidence supporting the south net longshore transport includes the observation by Price (1933) that Aransas Pass migrated southward prior to being stabilized by jetties. Through a study of geologically effective winds over Gulf waters, Lohse (1952) also concluded that a zone of net longshore transport convergence existed at approximately 27 deg north latitude. Based on wind observations which were used to develop wave hindcasts representative of the eastern and southern ends and the central Texas coast, Carothers and Innis (1971) supported the theory of net convergence near 27 deg north latitude after studying the concentration of shells which had been transported to the area from northern and southern beaches and the wind tidal-flat fill of the Laguna Madre.

There are few studies of the transport rates in the Matagorda area. Qualitatively, the net transport should be directed to the south (right-directed) as the area is located in the north of the convergence area discussed above. A study by Bureau of Economic Geology at the University of Texas at Austin (Gibeaut et al., 2000) estimated the net and gross longshore transport to be 64,000 and 249,000 m3/year respectively based on ten-year deep-water wave data at 17 locations between Sargent Beach and Sabine Pass. Historically, shoreline at the study area has been undergoing significant erosion with a long-term shoreline erosion rate as high as 8 to10 meters per year

2. Study Objectives

The objectives of this model study were to investigate the existing condition of the inlet-bay system, to understand the complicated processes in the study area, and to evaluate inlet management alternatives. A hydrodynamic model and a sediment transport model were developed to study the hydraulic condition and stability of the inlet and to study the mechanisms and pathways of sediment transport. The models were then used to evaluate the feasibility, effectiveness and the impact of the inlet management alternatives, which were proposed to address the navigation and sedimentation problems.

3. Study Methodology

For the study, historical data were gathered primarily from Texas Coast and Ocean Observation Network (TCOON) including wind, tide, and river discharge. A hydrographic survey was performed in March 2003 in the study area to collect bathymetric data, which was the first detailed survey for the area. Two field trips, one in Dec. 2002 and the other in May 2003 were taken to collect sediment samples and to measure current velocity and suspended sediment concentration in the study area. In addition, tide and current data collected at Mitchell's Cut by USGS from Oct. 2000 to Oct. 2001 were also used for the study. These data are not only important for modeling purpose, they are also important for the understanding of the inlet processes. Based on these data and survey results, a two-dimensional hydrodynamic model and sediment transport model were developed and calibrated. The hydrodynamic model and the sediment transport model were used to study inlet hydraulic condition and stability, sediment transport pathways and sedimentation pattern. And finally, the models were used to evaluate different inlet management alternatives.

4. The Hydrodynamic Model

In this study, the hydrodynamic model, TRIM2D developed by USGS (Cheng and Casulli, 1992) was used. TRIM2D is a depth-averaged tidal and residual circulation model. The model uses a semi-implicit, finite-difference numerical scheme to solve two-dimensional shallow water governing equations. Assumptions made in the model's formulation include: hydrostatic approximation, Boussinesq approximation and incompressibility. A Manning-Chezy approximation is used to estimate bottom shear stress. The model used in this study is slightly different from the original version. The current model allows varying horizontal grid spacing so that the model grid can be made such that it is very fine at Mitchell's Cut but coarser away from the study area of interest.

4.1 The model grid

Figure 4 shows the model grid. The model domain covers from the west of the Colorado River to east of the Brazos River, south from about 1 mile offshore of the Gulf of Mexico to the north of the GIWW, including the entire East Matagorda Bay, Mitchell's Cut, part of the Gulf near shore area and part of the GIWW. The grid spacing ranges from 25 m in the Mitchell's Cut area to 1000 m in the offshore area of the Gulf of Mexico. The grid in the nearshore area is 15 m while it is much coarser in the offshore area. Except for the Mitchell's Cut area, the bathymetry data was taken from the Mouth of Colorado River Project model developed by Coastal Hydraulic Laboratory, US Army Corps of Engineers. The bathymetry data in Mitchell's Cut area were updated with the most recent survey data acquired by Coastal Tech.

4.2 Boundary conditions

Boundary conditions for the hydrodynamic model are as follows: measured tidal elevation data by TCOON from Rawlings and Freeport were used at the south boundary (Gulf of Mexico). Wind data measured at East Matagorda Bay and Port O'Connor were used at the surface. River discharge data collected by USGS at Colorado River and Brazos River were applied at the rivers as boundary conditions. There's no river discharge flow available for Caney Creek. The discharge at Caney Creek was assumed to be 1/5 of Brazos River according roughly to the sizes of the watershed areas. At the GIWW, where water flows out of the model domain, a radiation boundary condition was applied.

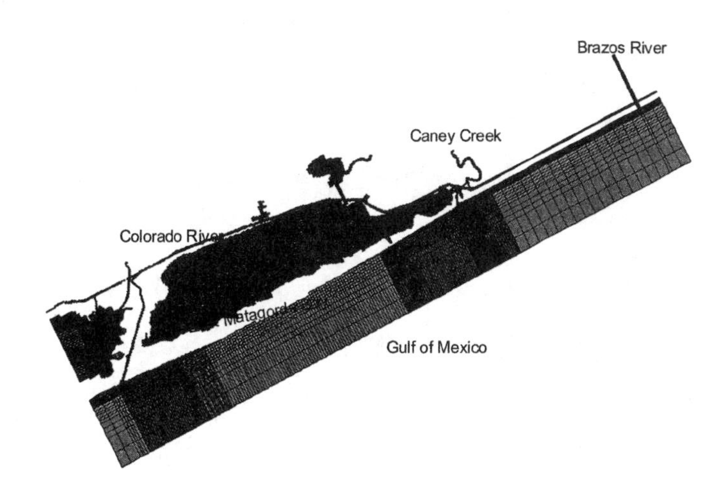

Figure 4. The model grid for this study

4.3 Calibration and verification of the hydrodynamic model

The hydrodynamic model was calibrated with data collected by USGS at Mitchell's Cut (Station 2 in Figure 6) in 2001. Figure 5 shows the comparison of model predicted surface elevation and current velocity from March 6 (Julian day 66) to March 24 (Julian day 83), 2001. For both the surface elevation and current velocity, the model prediction and USGS observation agree very well. Under normal wind and tide condition, the maximum current velocity at the inlet can reach as high as 1.5 cm/s or 4.5 ft/s but usually not exceeding 2.0 m/s or 6 ft/s.

The model was further verified with the current data collected by Coastal Tech from May 14 (Julian day 134) to May 16 (Julian day 136), 2003 during which current velocities were measured at twelve stations (Figure 6). Figures 7 and 8 show the model predicted currents vs. observed currents from May 14 (Julian day 134) to May

16 (Julian day 136) at Station 1 (inlet throat), 4 (Caney Creek at the GIWW), 6 and 7 (GIWW). Again, the agreement between the model results and measured data is very well. The currents were the strongest at the inlet throat, where, the maximum currents reached 0.8 m/s or 2.5 ft/s. Currents were weak in Caney Creek, where, currents were only a few centimeters per second. Currents were stronger at the GIWW, where, maximum currents can reach more than 0.4 m/s or 1.3 ft/s.

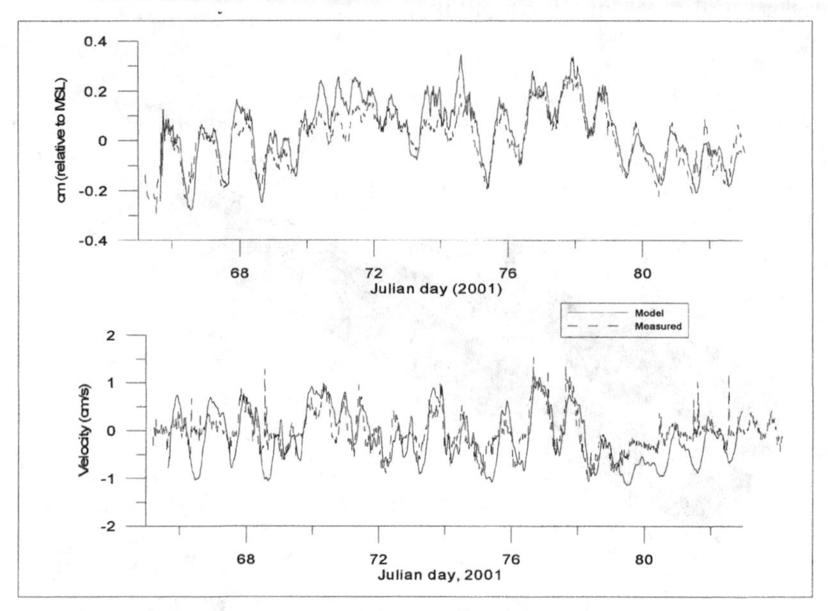

Figure 5. Modeled (solid) and measured (dashed) surface elevation (upper panel and current velocity (lower panel) at Mitchell's Cut.

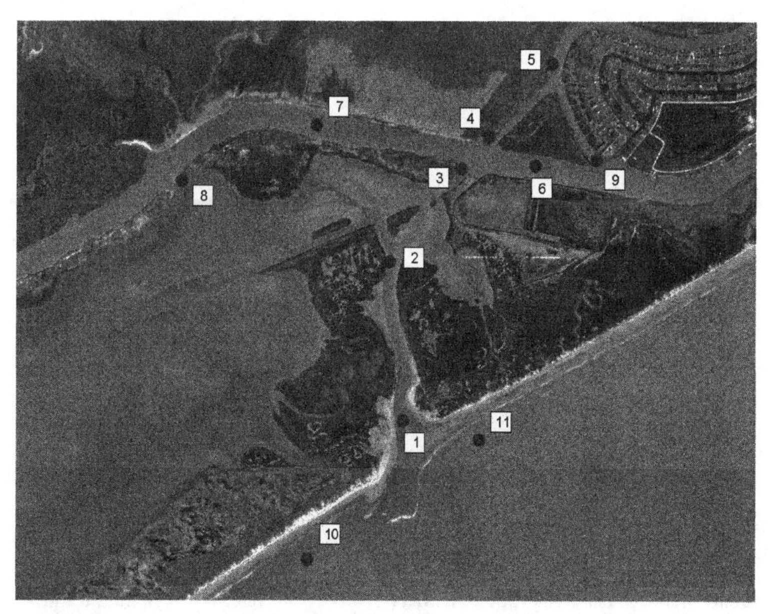

Figure 6. Observation stations for measurement of current velocities and suspended sediment concentrations in the study area during the field trip in May 2003.

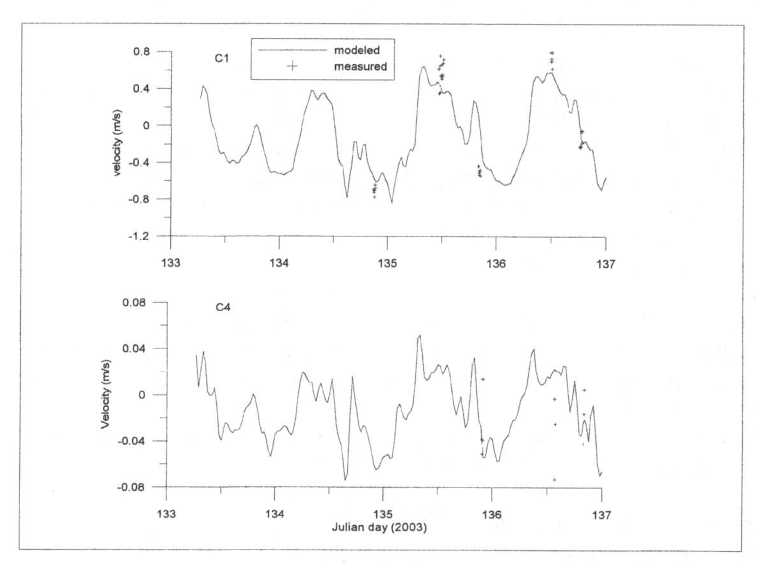

Figure 7. Model predicted (solid line) and observed (cross) current velocities at stations 1 (inlet throat) and 4 (Caney Creek) from May 13 (Julian day 133) to May 16 (Julian day 136), 2003.

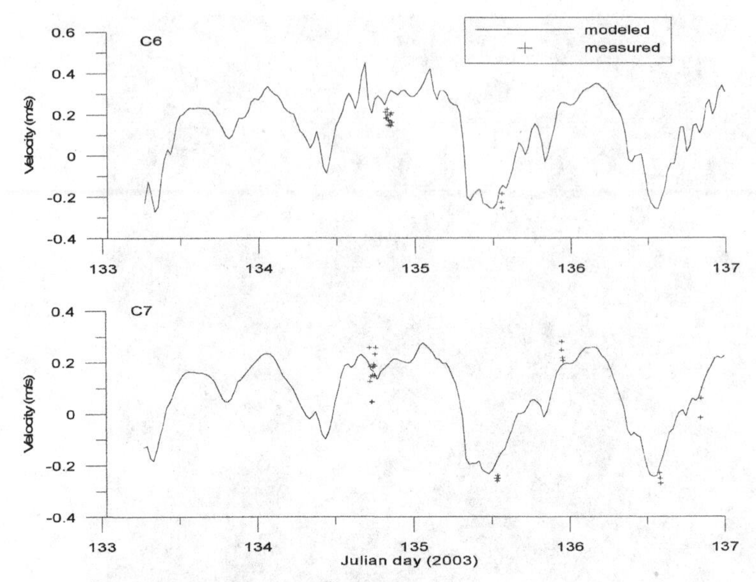

Figure 8. Model predicted (solid line) and observed (cross) current velocities at stations 6 and 7 (GIWW) from May 13 (Julian day 133) to May 16 (Julian day 136), 2003.

5. The Sediment Transport Model

The sediment transport model (Sun, 2001) was originally coded in CH3D. Based on the sediment samples recently collected sediment samples, sediments can be categorized into two groups for the study area, a fine group (mud) and a coarse group (sand). Most of the East Matagorda Bay and the GIWW are muddy. Fine sand can be found on the Gulf Coast. A mixture of sand and mud can be found in the inlet area and the flood shoal on the northeast corner of East Matagorda Bay. Therefore, in this study, a two-group approach was used for the sediment transport modeling. Basic assumptions for the sediment transport model are as follows:

1. Sediments in the study area can be categorized either as fine (mud) or coarse (sand).
2. There is no interaction in the transport of the two groups
3. The fine group follows the theory of cohesive sediment transport (e.g., Mehta, 2001) including the processes of settling, flocculation, erosion, deposition and bed consolidation
4. The coarse group follows the theory of sand transport (e.g., van Rijn, 1989) including the processes of settling, resuspension and deposition.
5. Except for the open beaches (including the ebb shoal), bed load transport is assumed small and negligible.
6. On the open beaches, sediment transport is driven by waves and wave-induced currents. The sediment transport model has a built-in SMB wave

model (Munk, 1950) to compute wind-generated waves on the open coast. While the sediment transport model can take into account wave-induced sediment resuspension, it does not explicitly take into account littoral transport. Longshore transport has to be considered separately to address beach morphology. While this is a limitation of simulating beach evolution, it is sufficient to address sediment interaction between the Gulf and East Matagorda Bay since the transport between the Gulf and the Bay through the inlet is predominantly suspended load transport.

7. Sediment sizes and composition were taken from the results of analysis of sediment samples taken in recent survey. Settling velocities were calculated using empirical formulations. Parameters such as critical shear stress and erosion rate were taken from similar studies, e.g., the study of Laguna Madre, Texas (Teeter, 2002) and the study of the Mouth of Colorado River (Brown et. al., 2003).

Calibration of the sediment transport model

The sediment transport model was calibrated with data collected by Coastal Tech in May 2003, during which measurement of suspended sediment concentrations (SSC) were taken. The SSC were measured at the same sites of current measurement, using an Optical Back Scatter (OBS) sensor. Figures 9 and 10 show the model predicted SSC vs. observation at four locations, Station 1 (inlet throat), 3 (inlet at GIWW), 4 (Caney Creek at GIWW) and 8 (East Matagorda Bay). The model prediction is roughly in agreement with observation. In general, under normal condition, the highest sediment concentrations occur in the East Matagorda (8), the lowest is at Caney Creek. Sediment concentrations are relatively high at the mouth of the inlet especially during flood tide and when strong wind occurs.

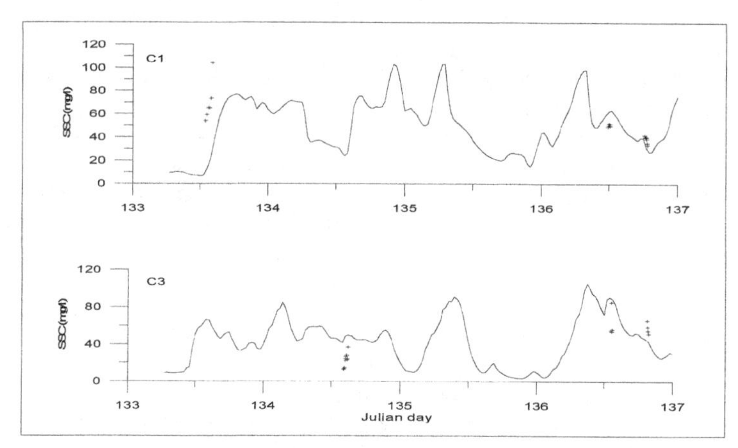

Figure 9. Model predicted (solid line) and observed (cross) suspended sediment concentrations at stations 1 (inlet throat) and 3 (inlet at GIWW) from May 13 (Julian day 133) to May 16 (Julian day 136), 2003.

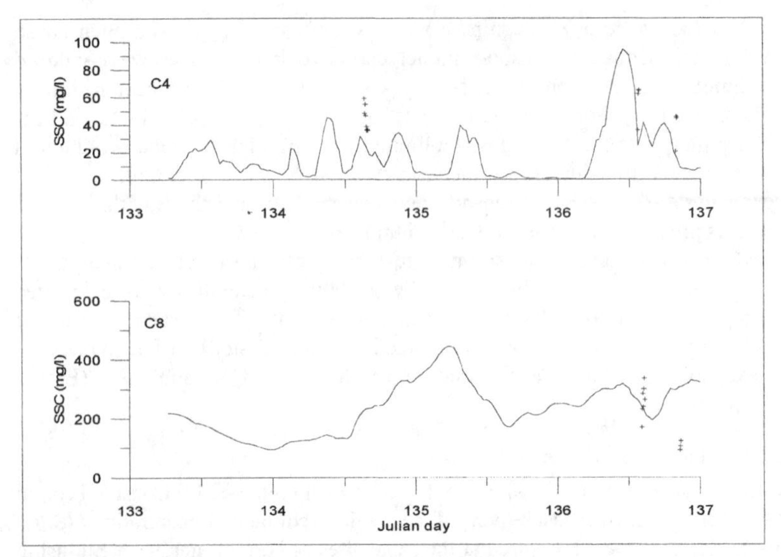

Figure 10. Model predicted (solid line) and observed (cross) suspended sediment concentrations at stations 4 (Caney Creek) and 8 (East Matagorda Bay) from May 13 (Julian day 133) to May 16 (Julian day 136), 2003.

6. Modeling the Existing Condition

The existing condition was modeled for a one-year period from Oct. 2000 to Oct. 2001 (during which period the USGS water level and current velocity data at Mitchell's Cut are available). Figures 11 and 12 show typical flow and sediment transport patterns during flood and ebb tide respectively. Currents are apparently strongest at the inlet throat, where peak current speed may exceed 1.5 m/s but seldom exceed 2.0 m/s. In fact, during most of the time (>90%), current speed remains below 1.3 m/s or 4 ft/s, which is considered as a safety criterion for small craft navigation. Peak current speed at the ebb shoal channel, where the water depth is less than 2 m, may exceed 1.2 m/s, which can be very dangerous considering strong waves are breaking at the ebb shoal during rough weather. Data and model results suggest that the inlet is healthy in its natural (unmanaged) condition, i.e., it is stable and hydraulically efficient.

Animation of the sediment model results demonstrates that during the flood tide, sand can be transported into the East Matagorda Bay through the inlet and part of the sand may deposit in the flood shoal area. Figure 11 shows the maximum sand penetration at the peak of flood tide. During the ebb tide (Figure 12), the muddy sediment can be transported through the channel system and part of the mud can reach the open coast in the Gulf. Both of the processes are supported by evidence of the expansion of the flood shoal system during the years since the inlet was open and the sediment plume often visible from aerial photos of the area.

Based on the model results, bathymetry survey and historical data, a sediment budget was developed for the study area (Figure 13). It has to be pointed out that the sediment budget includes the contribution from wave-induced longshore sediment transport along the beaches, which is not explicitly modeled but is an important part of this inlet management study. The longshore transport affects sediment budget mainly in the ebb shoal area. The ebb shoal receives approximately 47,000 yard^3/year from longshore sediment transport, the majority (approximately 32,000 yard^3/year) of which is retained in the ebb shoal. The rest of about 5,000 yard^3/year is transported to the inlet interior and the flood shoal area. The contribution from the northeast beach is higher than from the southwest beach due to the fact that the dominant longshore transport direction is from northeast to southwest. Among the 5,000 yard^3/year enters the channel, 4,000 yard^3/year goes to the flood shoal and 1,000 yard^3/year goes to the GIWW. The flood shoal receives a contribution of about 1,000 yard^3/year from the Bay as silty bay material is resuspended and transported seaward by ebb tidal currents. The contribution from Caney Creek is negligible. This is because, first, according to a study by the Matagorda County Drainage District (1994) that the suspended sediment load at Caney Creek is on average less than 2000 yard^3/year; second, the sediment from Caney Creek is very fine material (clay) and do not settle out immediately and would have an insignificant impact on the sedimentation in the inlet channel and the flood shoal area. However, during flooding events sediment from Caney Creek may form plume type suspension at the mouth of the inlet.

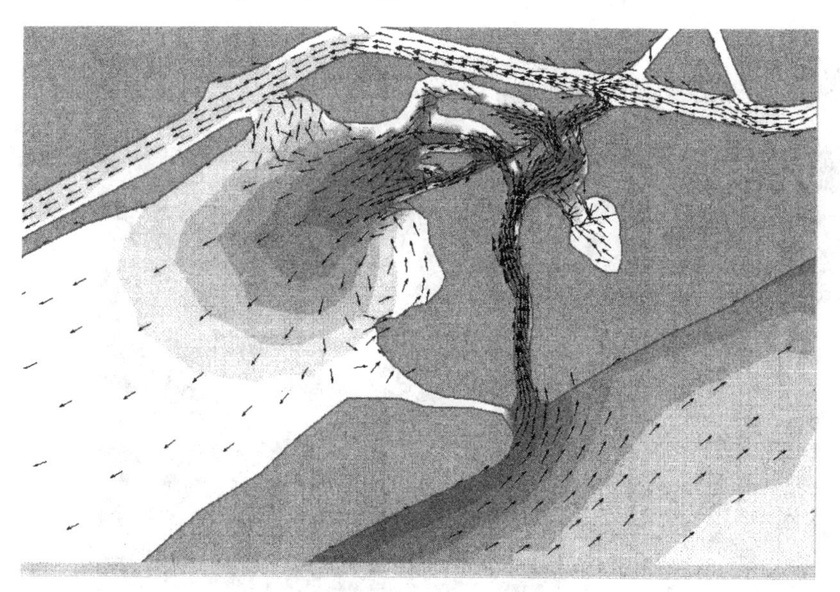

Figure 11. Modeled flow field (arrows) and sand concentration (contours) for the existing condition during the flood tide showing maximum penetration of beach material into the Bay area.

Figure 12. Modeled flow field (arrows) and mud concentration (contours) for the existing condition during the ebb tide showing bay material transported by ebb currents into the Gulf.

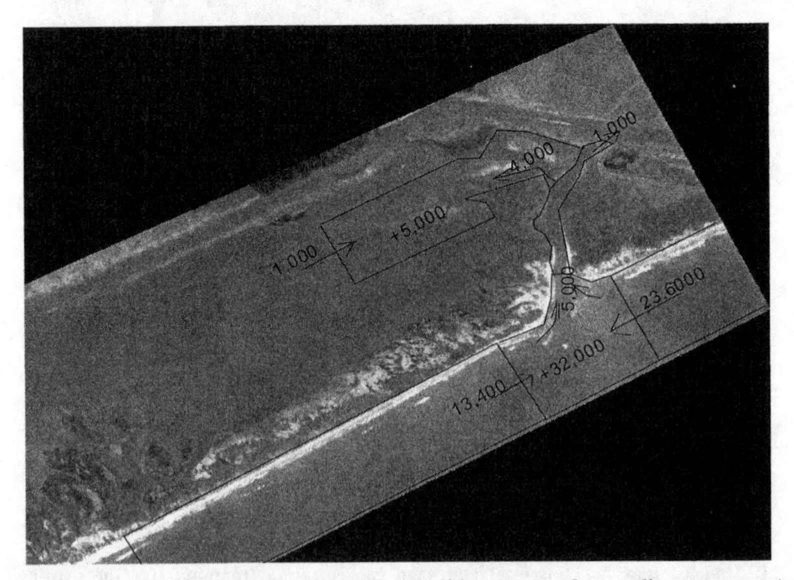

Figure 13. Sediment budget for the existing condition. Arrow shows sediment transport pathways. Quantities in cubic yards per year. + or – shows deposition or erosion rates for the cell.

7. Modeling Inlet Management Alternatives

To reduce the sedimentation and improve navigation, inlet management alternatives have been proposed based on study results of the existing condition and local interests. The alternatives were evaluated in terms of their technical feasibility and their effectiveness of achieving the project objectives. Specifically, each alternative was modeled and analyzed for possible physical impacts such as changes in hydraulic regime, changes in sediment transport pathways, and changes in erosion and accretion pattern. The modeling methodology was to model each alternative with exact the same condition with respect to a base condition except changes made according to the alternative designs. The base condition was chosen as the existing condition and the modeling results of the base condition has been discussed in the previous section. The design alternatives considered include:

- "No Action"
- Closure of the inlet
- Maintenance dredging with "beneficial use" - beach disposal
- Expansion of the existing channels
- Creation of a sedimentation basin
- Jetties and/or embankment stabilization
- Use of ebb shoal and flood shoal for beach nourishment

In this paper, only one design alternative is discussed here. That is the jetty alternative (Figure 17). For this alternative, two jetties will be constructed. The jetties will be curved and almost perpendicular to the shoreline. The lengths are 3115 ft and 3595 ft for the west and east jetty respectively. The tips of the jetties are to be built to the 12 ft depth contour line. In between the two jetties, a channel of 200 ft wide will be dredged to the 12 ft depth (NAVD88). In the flood shoal area, a channel of 100 ft wide and 10 ft deep will be dredged providing access to the East Matagorda Bay from the inlet. With the same forcing condition as for the existing condition, the hydrodynamic model and sediment transport model were run for this design alternative. Figure 14 shows flow field and sand penetration during flood tide. Compared with the base condition the currents around the jetty are substantially changed as the jetties force the currents go through the channel with eddies around the outside. Sand is effectively stopped by the jetties. The maximum penetration of the sand to the flood shoal area is effectively reduced comparing Figure 13 with Figure 10. Figure 14 shows that the SSC are reduced more than 60% at the inlet throat and the flood shoal area. However, very little changes can be seen in the current velocities at the throat and the old GIWW (Figure 16), which means the jetties would not affect the hydraulic efficiency given the design jetty configuration. Figure 17 shows the sediment budget developed for this alternative. Sedimentation rate in the flood shoal area can be reduced from 5,000 yard3/year to 2,600 yard3/year. The jetty alternative appears to be technically very promising.
Similar approach was applied to all the inlet management design alternatives. Eventually, the alternatives were evaluated for their technical performance, economic performance and environmental impact.

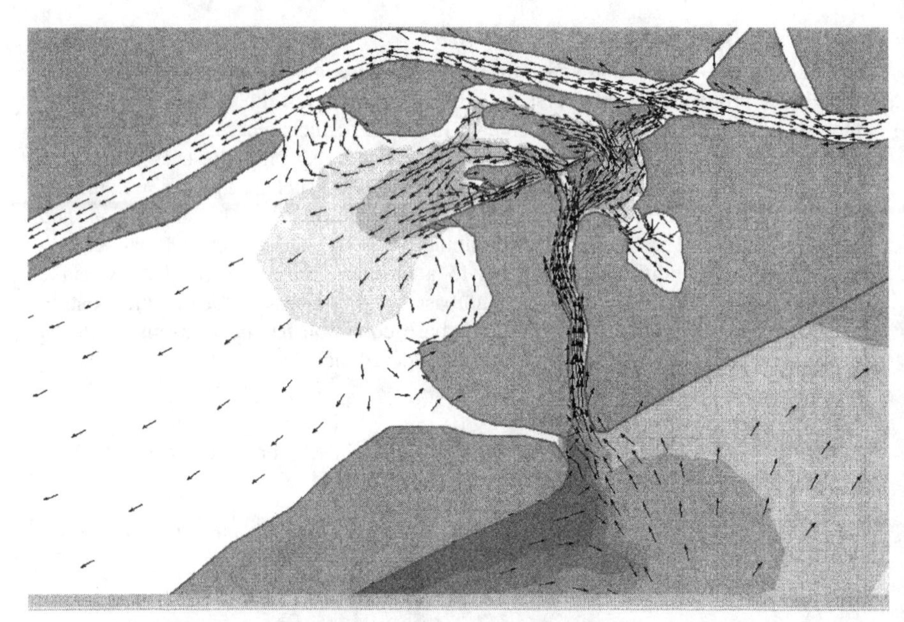

Figure 14. Modeled flow field (arrows) and sand concnetration (contours) for the jetty alternative during the flood tide showing maximum penetration of beach material into the Bay area.

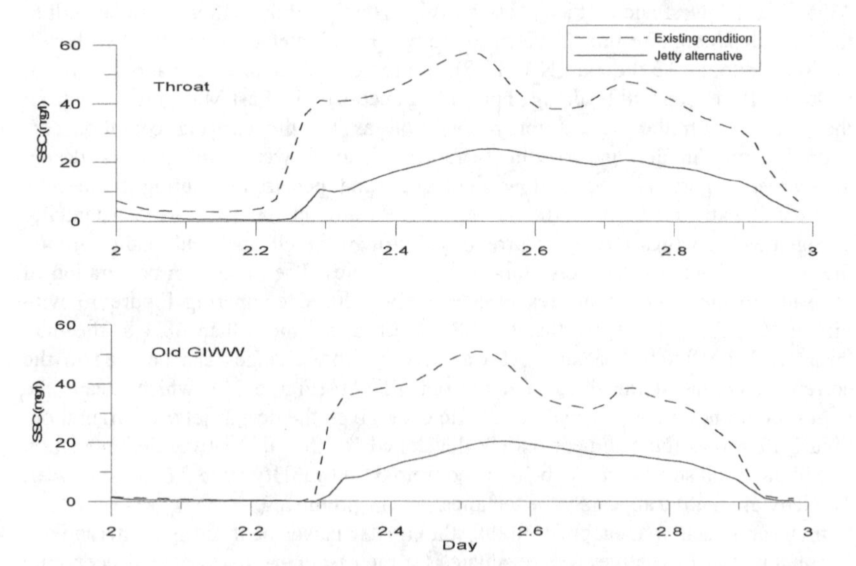

Figure 15. Comparison of modeled suspended sediment concentration (SSC) at the inlet throat and the old GIWW between the Jetty Alternative and the existing condition.

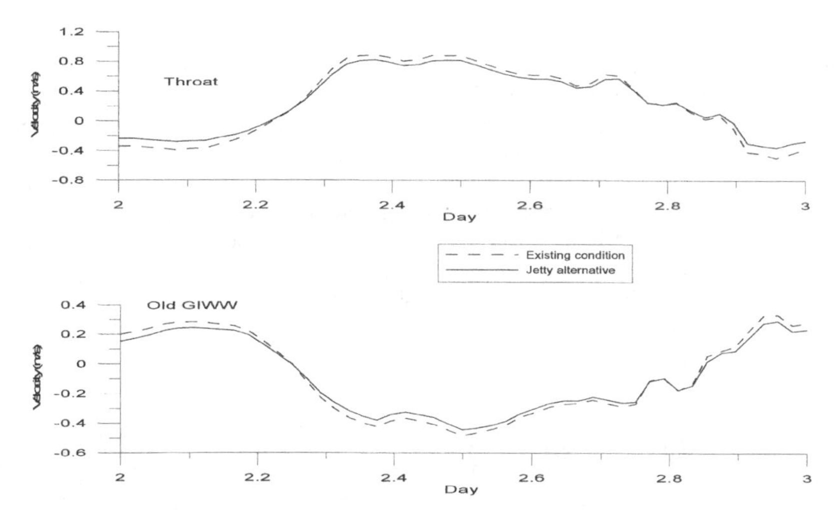

Figure 16 Comparison of modeled current velocities at the inlet throat and the old GIWW between the Jetty Alternative and the existing condition.

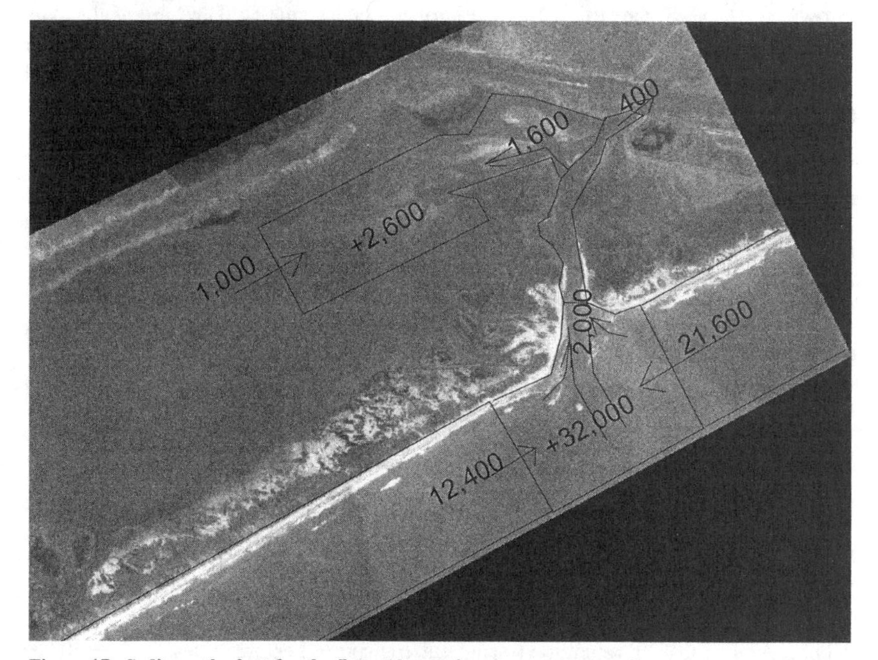

Figure 17. Sediment budget for the Jetty Alternative. Arrow shows sediment transport pathways. Quantities in cubic yards per year. + or – shows deposition or erosion rates for the cell.

8. Conclusions

In this study, a two-dimensional hydrodynamic model and sediment transport model were applied to inlet management study at Mitchell's Cut, Matagorda, Texas. The model results clearly demonstrate sediment transport pathways and sedimentation pattern in the study area. That is beach material can be transported into the flood shoal area and partly deposit there through the inlet by flood tide. And the bay material can be transported to the Gulf through the inlet by ebb tide. Caney Creek was found insignificant in this sedimentation processes for the inlet and flood shoal area. The inlet was found stable and hydraulically efficient. Assessment of the inlet management alternatives was performed using the numerical model and the jetty alternative was found technically very promising in reducing sedimentation and improving navigation.

Acknowledgement

The authors appreciated bathymetric data provided by Coastal Hydraulic Laboratory, US Army Corps of Engineers for this study.

Reference

Brown, G. L., M. S. Sarruff, T. L. Fagerburg and N. Raphelt, 2003: Numerical Model Study of Proposed Navigation Improvements at the Colorado River Intersection with the Gulf Intra-Coastal Waterway, TX. Coastal Hydraulic Laboratory, Vicksburg, Mississippi. Draft report.

Carothers, H. P. and Innis, H. C. 1960. Design of Inlets for Texas Coastal Fisheries. Journal of the Waterways and Harbours Division, Proceedings of the ASCE, 127 (IV): 231-259.

Cheng, R. T. and V. Casulli, 1992. Tidal, residual, inter-tidal, mud-flat (TRIM) model, using semi-implicit Eulerian-Lagrangian method. USGS open-file Report, 92-162

Gibeaut, J.C., W.A.White, T. Hepner, R.Gutierrez, T. A. Tremblay, R. Smyth, and John Andrews, 2000. Texas shoreline project, Gulf of Mexico shoreline change from Brazos River to Pass Cavallo.

Kraus, N. C. and Heilman, D. J. 1996. Packery channel feasibility study: inlet functional design and sand management. Technical Report TAMU-CC-CBI-96-06.

Kraus, N. C. and Militello, A., 1996. Hydraulic feasibility of proposed Southwest Cut, East Matagorda Bay, Texas. Final Report. Conrad Blucher Institute for Surveying and Science, Texas A&M University-Corpus Christi.

Lohse, E. A. 1952. Shallow-marine sediments of the Rio Grande Delta. Ph.D. Dissertation, University of Texas at Austin, Austin, Texas, 114p

Mehta, A. J., 2001. Fine-grained sediment transport. Lecture notes. University of Florida, Gainesville, Florida.

Price, W. A. 1933. Role of diastrophism in topography of Corpus Christi area, South Texas. American association of Petrolem Geologists Bulletin, 17(8): 907-962.

van Rijn, L.C., 1994. Handbook, Sediment Transport by Currents and Waves. *Delft Hydraulics Report H461*, Delft Hydraulics, Delft, the Netherlands.

RUST Lichiliter/Jameson, 1994. Hydrologic and Hydraulic Analysis of Caney Creek. Draft report prepared for Matagorda County Drainage District No. 1, Matagorda, Texas.

Sun, D., 2001. Modeling suspended sediment transport under combined wave-current actions in Indian River Lagoon. Dissertation, University of Florida, Gainesville, Florida.

Teeter, A.M., G. L. Brown, M. P. Alexander, C. J. Callegan,, M. S. Sarruff, and D. C. McVan,, 2002. Wind-Wave Resuspension and Circulation of Sediment and Dredged Material in Laguna Madre, Texas. U. S. Army Corps of Engineers, Engineer Research and Development Center, Draft Report.

Tipps, M. 2001. Background Summary of Mitchell's Cut, input from local community.

Watson, R.L. 1971. Roigin of shell beaches, Padre Island, Texas. Journal of Sedimentary Petrology, 41(4): 1105-1111.

Watson, R. L. and Behrens, E. W. 1970. Nearshore surface currents, Southeastern Gulf Coast. Contributions in Marine Science, 15: 133-143.

U.S. Army Corps of Engineers. 1992. Gulf Intracoastal Waterway Section 216 Study, Sargent Beach, Texas. Feasibility report and final environmental impact statement.

MODELING COASTAL SEDIMENT TRANSPORT AT THE MOUTH OF THE COLORADO RIVER, TEXAS

Lihwa Lin[1], Nicholas C. Kraus[2], and Ronnie G. Barcak[3]

Abstract: This paper describes diagnostic numerical simulations of waves, current, sediment transport, and morphology change at the mouth of the Colorado River, Texas. The simulations were performed with the Inlet Modeling System (IMS) developed by the Coastal Inlets Research Program at the Coastal and Hydraulics Laboratory, U.S. Army Engineer Research and Development Center. The IMS version applied consists of the circulation and sediment transport model M2D coupled with the wave spectral model STWAVE. The study site exhibits complex patterns of sediment transport over a weir jetty and deposition in an impoundment basin. Six alternatives are evaluated by comparison of calculated morphologic change and sediment impoundment coupled with the hydrodynamics over a 1-year simulation interval. Simulations are performed with and without a sediment training structure, different jetty lengths, and different weir lengths and elevations. The alternatives are examined for effectiveness in improving efficiency of the impoundment basin and in reducing sediment shoaling in the entrance channel to decrease dredging frequency.

INTRODUCTION

The mouth of the Colorado River (MCR) is located on the Texas Gulf of Mexico coast approximately midway between Galveston and Corpus Christi (Fig. 1). It is a federally maintained navigation channel that experiences relatively weak water exchange between the Gulf of Mexico and East Matagorda Bay, connected through the 10.5-km long Colorado River Navigation Channel (CRNC) and the eastward section of Gulf Intracoastal Waterway (GIWW). The flow system associated with the mouth of the Colorado River is separated from West Matagorda Bay. Dual rubble

---

1) Research Hydraulics Engineer. U.S. Army Engineer Research and Development Center (ERDC), Coastal and Hydraulics Laboratory (CHL), 3909 Halls Ferry Road, Vicksburg, MS 39180. Lihwa.Lin@erdc.usace.army.mil.

2) Senior Scientist. ERDC, CHL, 3909 Halls Ferry Road, Vicksburg, MS 39180. Nicholas.C.Kraus@erdc.usace.army.mil.

3) Civil Engineer. U.S. Army Corps of Engineers, Galveston District, 2000 Fort Point Road, Galveston, TX 77550. Ronnie.G.Barcak@swg02.usace.army.mil.

mound jetties constructed in 1985 protect the entrance to the CRNC. The west jetty extends 333 m from the shoreline. The east jetty has a 333-m long weir section on its landward side with a deposition basin located westward of the weir. The entrance channel, 5 m deep by 65 m wide, connects to the GIWW through a 4-m deep by 33-m wide channel.

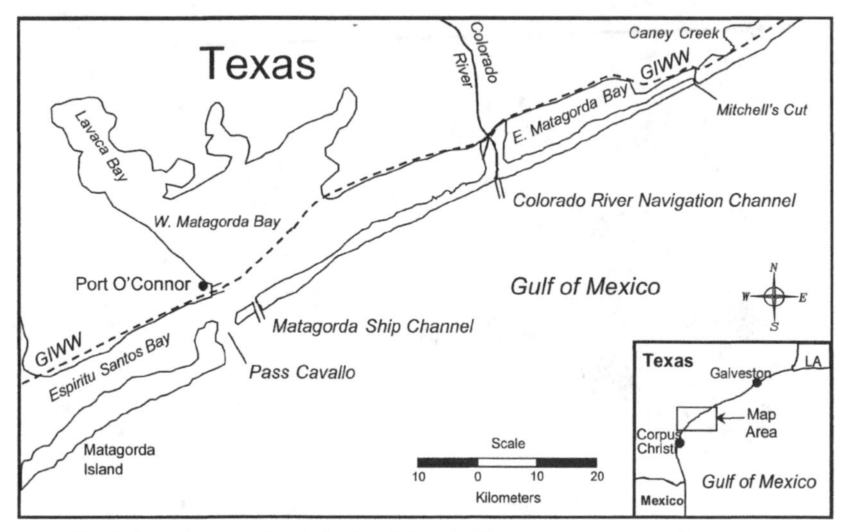

Fig. 1. Location map for study site, Colorado River entrance, central Texas coast

Before 1930, East Matagorda Bay and West Matagorda Bay comprised a single Matagorda Bay. The bay was divided in 1935, when a large river delta was formed and moved rapidly across the bay after a logjam was freed in the river between 1925 and 1929 (US Army Corps of Engineers (USACE) 1992). The Colorado River began discharging to the Gulf of Mexico in 1935. In 1992, the river was rerouted to discharge into West Matagorda Bay through a diversion channel as part of an environmental enhancement project. A pair of locks was added to the GIWW at the junction of the Colorado River to mitigate the crosscurrent at the intersection of the river and GIWW. The CRNC was separated from the main stream of the river, but it is connected to the eastward section of GIWW. As a result, the Colorado River was diverted into Matagorda Bay, eliminating the river discharge as a means of scouring sediment from the entrance channel to the Gulf. Sediment shoaling at the entrance has greatly increased since that time. Spits now tend to form on both sides of the inlet throat, and sediment accumulates on the entrance bar (Fig. 2).

The MCR includes a sediment impoundment basin located between the weir section and the entrance channel (Fig. 2). The deposition basin accommodates sand bypassing to the down-drift beaches by means of pipeline dredge in the impoundment basin, with discharge in the surf zone about 600 m on the west side of the MCR. The basin is designed to hold a 2-year supply of sediment or approximately 460,000 m^3 (600,000 cy) capacity (King and Prickett 1998).

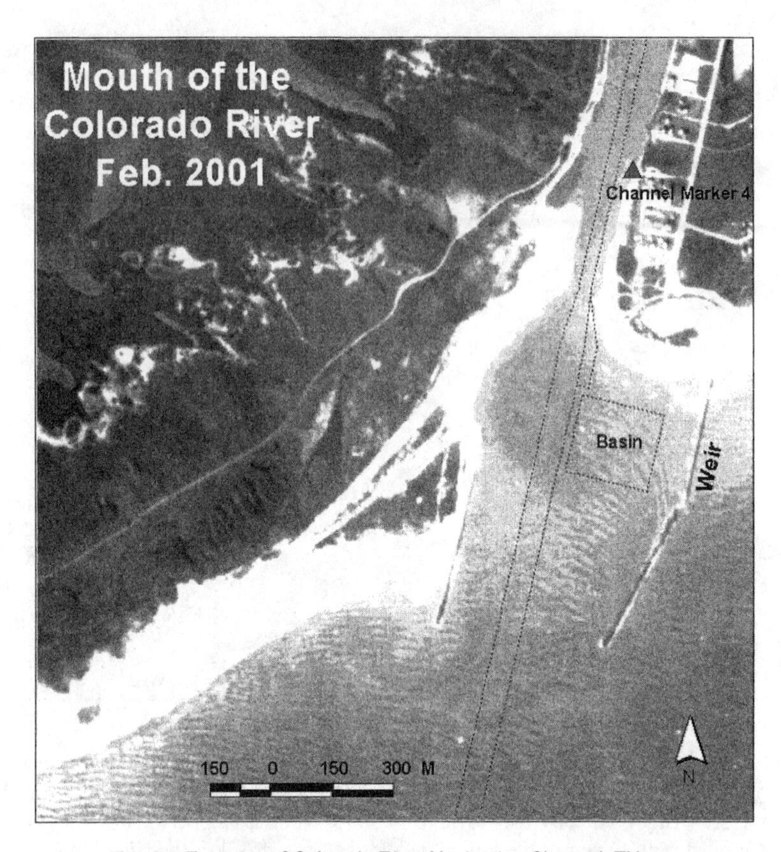

Fig. 2. Entrance of Colorado River Navigation Channel, TX

Initial dredging of the deposition basin was completed in 1990. Since then, the basin has tended to fill more rapidly than designed. The entrance mouth has experienced a large increase in sediment shoaling, and spits form on both the east and west sides of the landward ends of the jetties, encroaching in the navigation channel. Figure 3 shows the volume of sediment dredged in the basin and CRNC from 1990 to 2002 and indicates that the total volume of dredging annually is about 425,000 m^3 (556,000 cy) or double the design estimate.

To reduce dredging requirements, an impoundment basin training structure was constructed in September 2003. This structure and other incremental measures such as raising the weir jetty (Kraus and Lin 2004) are also being evaluated through application of the Inlet Modeling System (IMS) (Cialone and Kraus 2002; Militello and Zundel 2002; Brown *et al.* 2003) that consists of a circulation and sediment transport model M2D coupled with a wave spectral model, STWAVE. The IMS-M2D calculates total load transport under the combined waves and currents, and updates the bottom within the hydrodynamic simulation system according to gradients in the transport rate.

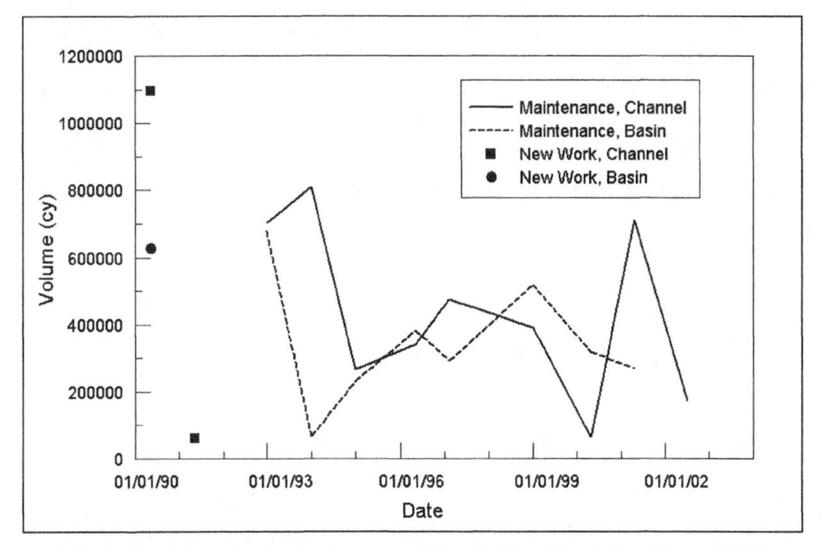

Fig. 3. Dredging volume at entrance of Colorado River Navigation Channel, TX

The tide on the central Texas coast is mixed, but mainly diurnal, having a period of approximately one tidal day (24.84 hr). The National Ocean Service (NOS) maintains a long-term tidal station at the Galveston Pleasure Pier, Texas. The diurnal tide range calculated from the record for the 5-year period 1990-1994 was 0.65 m. Strong wind along the Texas coast can often dominate the water level in the river, and river flows can be significant during times of heavy rain such as brought by tropical storms and hurricanes.

Nearshore wave data are available from gauges that were deployed at the 10-m depth contour 3.2 km offshore of the MCR, located 65 km south of the San Bernard River (King and Prickett 1998). For a 17-month period of data collected in 1991-1993 that somewhat under-represents the winter months, the mean significant wave height was 0.6 m and the mean peak period was 5.9 sec. These waves can generate a strong current near to shore to transport the littoral sediment into the deposition basin or along the shoreline to the entrance channel. Wind may also play a role in generating a longshore current, not accounted for in our diagnostic modeling.

Samples taken in the entrance channel and impoundment basin of MCR basin between February 1994 and September 1995 indicate that sediments are primarily sand in the entrance channel, with some silt and clay in the basin. The sediment in the entrance channel has a median grain size of 0.12 mm and silt and clay fractions of 35 percent. The sediment in the impoundment basin has a mean grain size of 0.03 mm and contains 87-percent silt and clay fractions. Sediment samples taken in the CRNC show mean grain size of 0.04 mm and contained 60 to 80 percent silt and clay. The fine materials in the basin and CRNC, silt- and clay-sized particles, are consistent with the sediment in the GIWW and from the river. The samples indicate that riverine sediments are reaching the MCR and likely settling into the impoundment basin.

Both the peak ebb and flood currents are weak, as shown by the data collected in January-February 2002 at Channel Maker 4 (Fig. 2). The ebb-tidal current dominates the flow in the CRNC. The mean velocity is equal to 0.013 m/sec and the averaged maximum current is 0.38 m/sec at the throat of the entrance channel in the primarily tidal condition. The mean velocity and the averaged maximum current become slightly greater, equal to 0.057 and 0.40 m/sec, respectively, in the combined tide and strong wind condition. Because of weak tidal flow and the active littoral process in the near shore, there is no appreciable ebb tidal shoal at the MCR.

This paper describes a diagnostic application of the IMS to examine alternatives for decreasing the frequency and, perhaps, volume of dredging at the mouth of the Colorado River. These simulations are part of a preliminary examination to examine relative merits of a large number of alternatives, done both to stimulate development of alternatives and to screen them to arrive at a subset that will be examined in more detailed modeling. The diagnostic modeling also serves to guide additional and effective data collection at the site for validation.

PREVIOUS STUDIES

Tidal hydrodynamics at the MCR were examined with a finite-element circulation model ADCIRC (Westerink *et al.* 2001) in a previous study (Lin *et al.* 2001). The model grid encompasses the MCR, East Matagorda Bay, West Matagorda Bay, and surrounding GIWW based on work of Kraus and Militello (1999), and published and new bathymetry surveys and aerial photographs. This grid was constructed as a regional model for the Texas regional coastal modeling study (Brown *et al.* 2004). The modeling effort includes representation of the surface wind field, river inflow, and ocean boundary water surface elevation. The model verification was based on water surface elevation measurements at two tidal stations, one at the lower section of Colorado River Navigation Channel and the other at the east end of East Matagorda Bay. The model was also verified with current velocity measurements made in October-November 1995 and in July 2001.

Heilman and Edge (1996) discussed sediment pathways and compiled estimates of longshore sediment transport at the mouth of the Colorado River obtained from several publications and by different means of measurement or inference. They concluded that an average annual amount of 230,000 m^3 is necessary for sediment bypassing at the entrance. King and Prickett (1998) estimated that the net rate of sediment transport at MCR to be 510,000 m^3/year to the west, with a gross transport rate of 670,000 m^3/year, based on the SPM (1984) CERC equation and wave measurements at the site. These CERC formula estimates (with K=0.35 for significant wave height) are higher than the average annual volume removal at 430,000 m^3, based on dredging volumes in the 5-year interval from 1990 to 1995. These records indicate that of the 430,000 m^3/year, about 280,000 m^3/year originates from longshore transport, in agreement with the conclusion of Heilman and Edge (1996), and the remainder, 150,000 m^3/year, is supplied by the river and GIWW. Therefore, the average west-directed long-shore transport rate was taken in the present study to be around 280,000 m^3/year, assuming material entering the channel originates primarily with the predominant direction of longshore transport.

MODELING APPROACH

The IMS allows selection of ADCIRC or M2D as the circulation and transport model. For this study with relatively small domain, the finite-difference model M2D was selected. Both ADCIRC and M2D were initially run at the site, and both were verified with water level and current velocity measurements made in January-February 2002 at Channel Maker 4 and at other locations.

The IMS-M2D version calculated total transport for combined waves and currents by the formula of Watanabe *et al.* (1986) with skin friction coefficients by Soulsby (1997). The wave radiation stresses computed from STWAVE are passed to M2D to calculate contributions to the current and water level. The wave height, period, and direction computed from STWAVE are also passed to M2D to calculate the total bed shear stress and wave skin-friction contribution to the sediment transport. The sediment model includes a threshold bed shear-stress formula of the sediment motion beneath waves and currents. The new threshold formula functions well for both fine and coarse sand grain sizes.

The grid domains and depth contours are shown in Fig. 4. The M2D grid consists of 250 (along-shore direction) by 210 (across-shore direction) square cells, with each cell 20 m by 20 m. The grid extends from 3.5 km updrift (east) of MCR to 1 km downdrift (west) of MCR, and from the 7-m contour offshore to about midway up the CRNC. The bathymetry information was based on a post-maintenance dredging survey conducted in December 2001. The STWAVE grid is a sub-domain of the M2D grid. The grid consists of 222 (along-shore) by 57 (across-shore) square cells. Each cell is 20 by 20 m. The grid covers from about 3 km updrift shoreline and 0.6-km downdrift shoreline, and from the 6-m contour offshore to inside the entrance channel throat where wave activity from the Gulf of Mexico is negligible. The water depth defined in the grid is based on the mean tide level (mtl). The local navigation tidal datum of the inland surveys, mean low tide was converted to mean tide level for the modeling by adding 0.3 m to the mean low tide data.

The IMS-M2D was applied to simulate the sediment transport and morphology change for year 2002. The incident wave condition was supplied by the transformation of directional buoy spectra, collected from Buoy 42019 (27° 54' 36" N and 95° 21' 36" W, maintained by the National Data Buoy Center) to the wave model offshore boundary. Water level records from the Galveston Pleasure Pier (29° 19' 36" N and 94° 41' 30" W) and Rawlings (28° 37' 24" N and 95° 58' 12" W) tide gauges provided input to M2D at the ocean and river boundaries, respectively. Measurements of the water level and current (using acoustic-Doppler profilers) at the entrance of the CRNC near Channel Marker 4 (Fig. 2) in January and February 2002, and at the deposition basin near the weir in October to December 2002, served for verification of the model hydrodynamics.

At the site, considerable longshore transport occurs in the inner surf zone, bypassing the weir and deposition basin to create a spit that eventually protrudes into the channel. Kraus *et al.* (1982) noted that relative or absolute peaks in the longshore sediment transport occur in the inner surf zone, where sediment concentrations are large. The Watanabe *et al.* (1986) formula was modified to give a persistent, but smaller relative peak in the inner surf zone, to be discussed in a future publication.

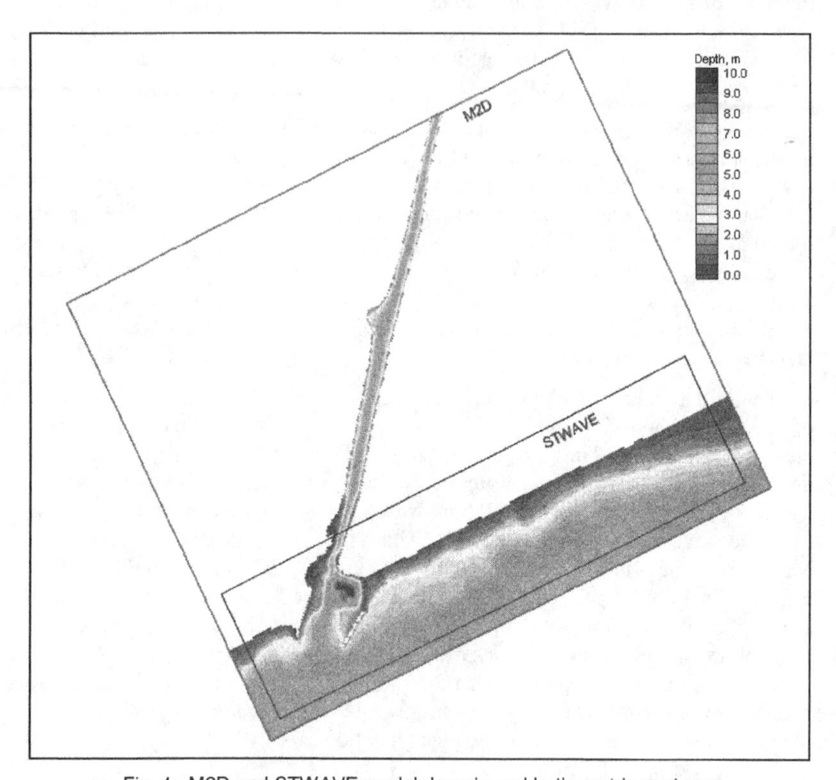

Fig. 4. M2D and STWAVE model domain and bathymetric contours

CALCULATION RESULTS

The 1-year model simulation was conducted for six structural alternatives, summarized in Table 1: (1) Alt 1 is the existing MCR before a training structure is constructed; (2) Alt 2 is the existing MCR with a training structure (Fig. 5), constructed in September 2003; (3) Same as Alt 2 with additional extension of the training structure by 56.5 m; (4) Same as Alt 2 with an extension of the west jetty by 56.5 m (Fig. 5); (5) Same as Alt 2 with raising of the landward half of the weir depth from 0.5 m to 0.25 m, mtl; (6) Same as Alt 2 with raising of the entire weir depth from 0.5 m to 0.25 m, mtl. All alternatives started with the same initial bathymetry, but with different structure configurations according to the alternative.

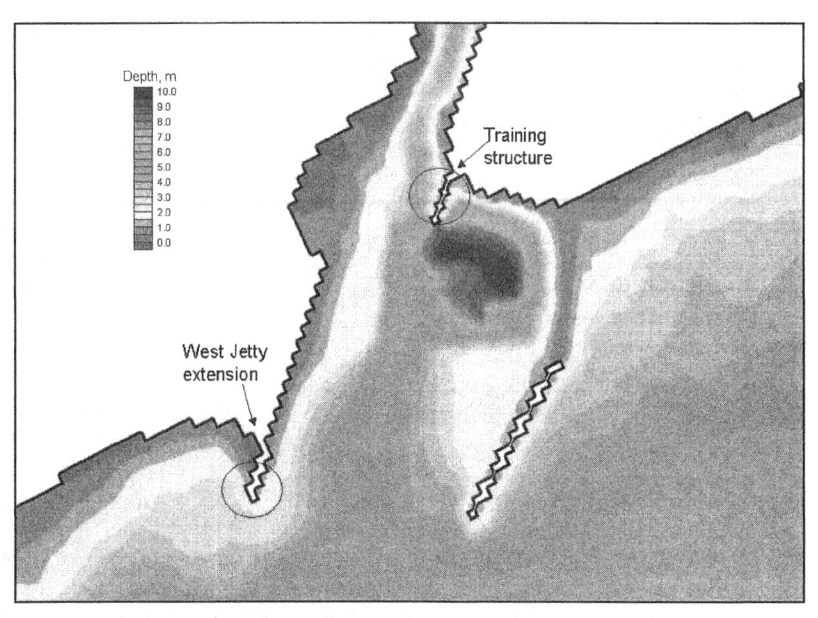

Fig. 5. MCR with training structure and west jetty extension (Alt 4)

Table 1. Summary Description of Structural Alternatives

Alt	Description
1	Existing MCR without impoundment basin training structure
2	Existing MCR with impoundment basin training structure
3	Same as Alt 2 with extension of training structure by 56.5 m
4	Same as Alt 2 with extension of West Jetty by 56.5 m
5	Same as Alt 2 and raising of landward half weir depth to 0.25 m MTL
6	Same as Alt 2 and raising of the entire weir depth to 0.25 m MTL

IMS-M2D was run using a 1-sec time step for the hydrodynamics, 90-sec time step for the sediment transport calculation, and 15-min time increment for the morphological change calculation based on the average of the calculated transport rates. STWAVE was run with 3-hr interval, with wave results including wave height, period, direction, and radiation stresses linearly interpolated in the 3-hr interval as the input to M2D. M2D updates the depth information in the 15-min morphological time interval and supplies the new depth information to the wave and hydrodynamic models. Sediment transport rates and morphology change were output to files at weekly intervals for the 1-year calculation.

Figures 6 to 11 illustrate sediment deposition and erosion patterns from the 1-year simulations for the six alternatives. Dark tones indicate areas of deposition, and light tones indicate deepening. For the existing condition (Fig. 6), the simulations show sediment shoaling in the entrance channel and deposition in the impoundment basin. The calculated great deposition in the impoundment basin, observed at the site as well, is the result of the westward longshore transport over the weir under the combined wave and tidal flow conditions, particularly during strong flood tide. Waves that cross the weir during high tide or propagate into the entrance channel can mobilize sediment in the basin and cause encroachment in the navigation channel, especially the transport occurring near the shoreline. This sediment transport pattern is qualitatively observed at the MCR.

It is possible to reduce the rate of shoaling in the navigation channel and to increase the efficiency of the impoundment basin function by modifying the existing jetty or weir or by adding an impoundment basin training structure as shown in the modeling alternatives. Tables A1 and A2 in the appendix compile the sediment accumulation in the entrance channel and in the deposition basin, respectively, on the monthly basis for the six simulations. The total volume accumulated in the basin and in the entrance channel over the 1-year period for Alt 1 (no training structure) is 255,175 m^3 (125,233 m^3 in the channel and 113,319 m^3 in the basin). The annual gross rate of longshore transport estimated by the CERC formula, using the STWAVE output of the same 1-year simulation at 3-hr interval, is about 240,000 m^3 (215,000 m^3 toward the west and 25,000 m^3 toward the east). Thus, the sediment transport calculation within the IMS gives a reasonable estimate of sediment volume arriving at the MCR that is in agreement with empirical engineering estimates.

The sediment supplied by the river, much finer material, is not included in the model. The fine sediment from the river can be carried into the impoundment basin by the ebb current. The fine sediment tends to be deposited in the deeper basin area where the current becomes weak. The volume of the sediment supplied by the river is not small, estimated to be approximately 150,000 m^3/year. Deposition of the river and estuarine sediment in the impoundment basin can reduce the efficiency of the basin. This fine river sediment does not settle in the entrance channel, where the current is strong, and so is only an indirect consequence to navigation reliability through deposition in the impoundment basin.

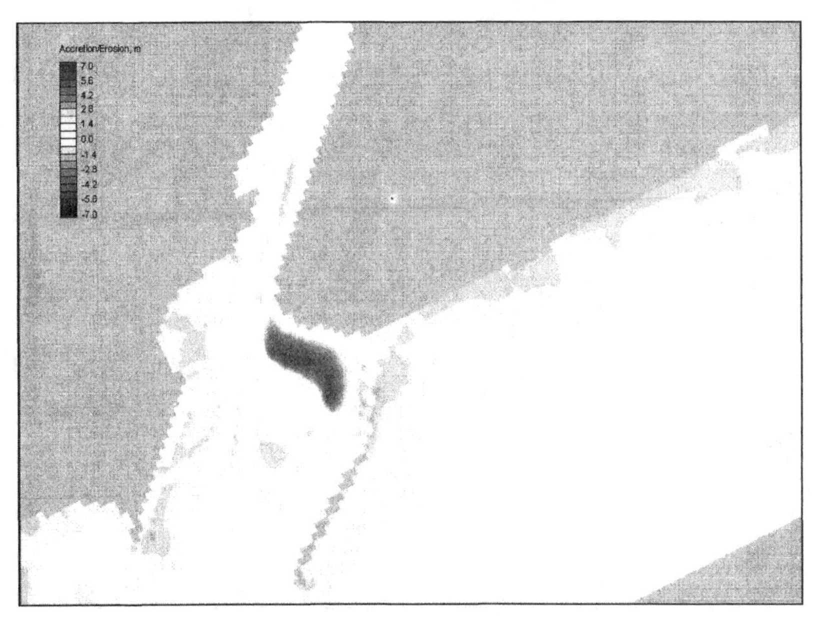

Fig. 6. Alt 1: sediment accretion and erosion map

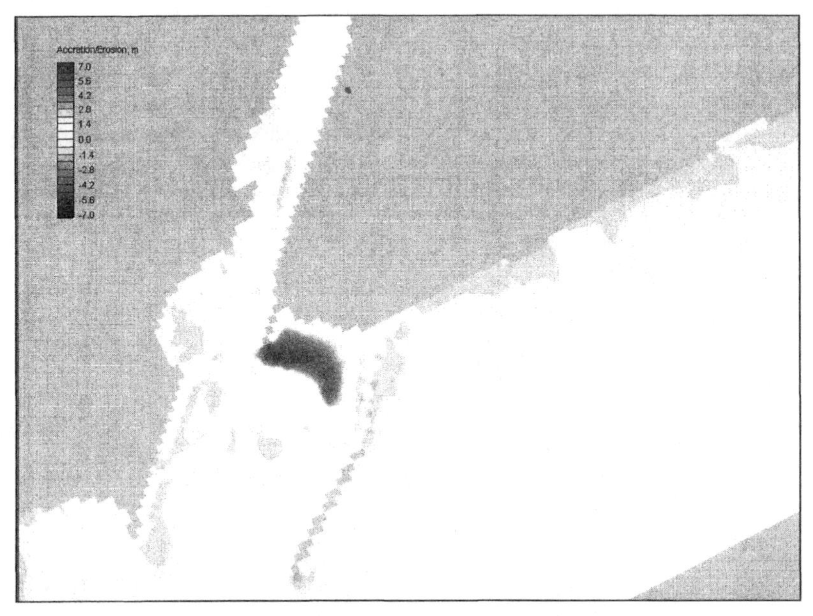

Fig. 7. Alt 2: sediment accretion/erosion map

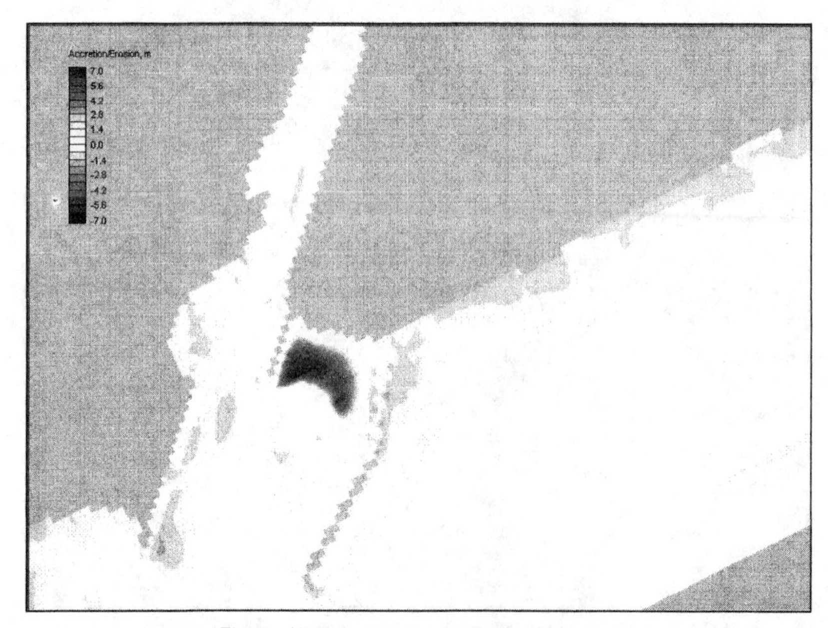

Fig. 8. Alt 3: sediment accretion/erosion map

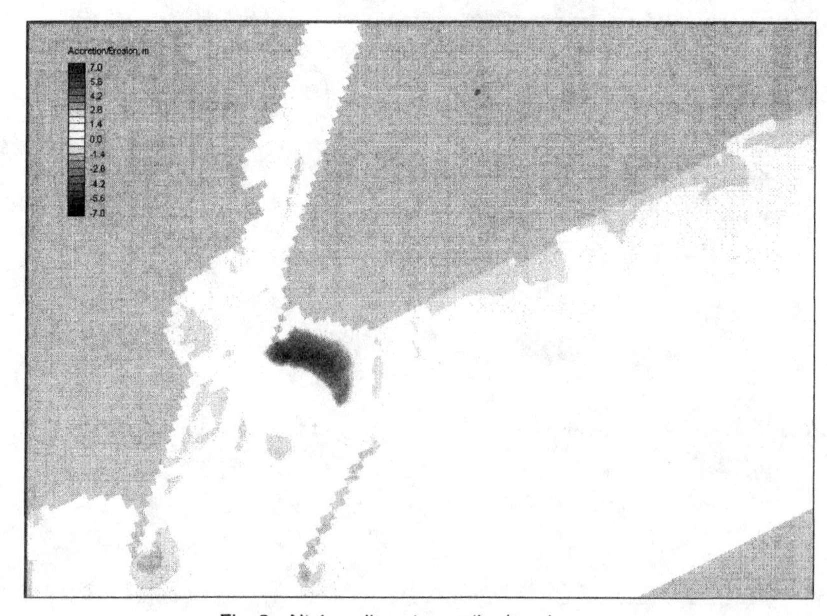

Fig. 9. Alt 4: sediment accretion/erosion map

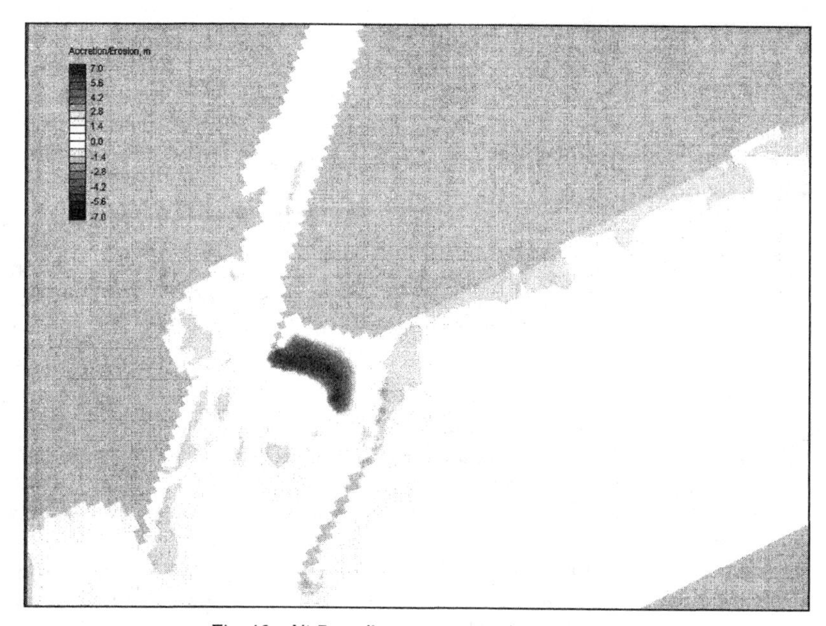

Fig. 10. Alt 5: sediment accretion/erosion map

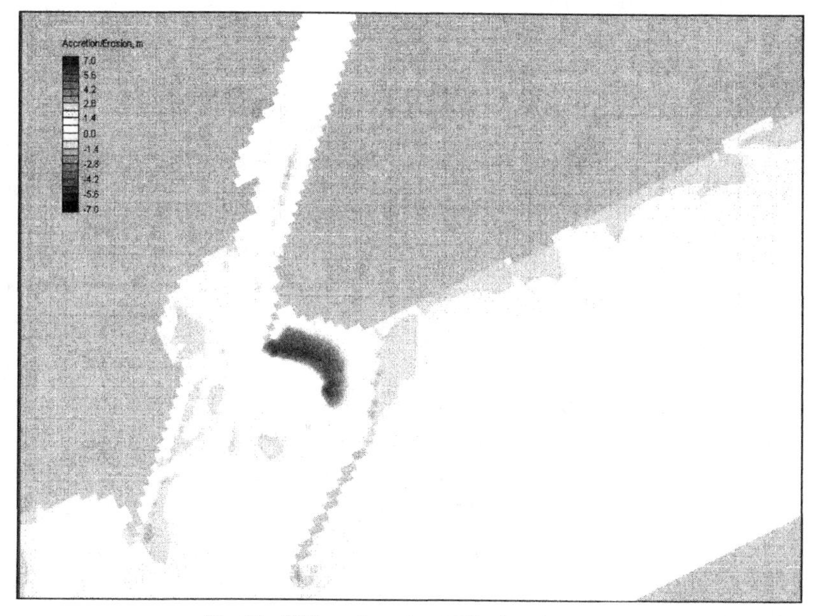

Fig. 11. Alt 6: sediment accretion/erosion map

EVALUATION OF STRUCTURAL ALTERNATIVES

Six structural alternatives are evaluated for their functioning to modify sediment transport pathways at the MCR. Alt 1 is the existing MCR condition before an impoundment basin training structure was constructed. Alt 2 is the same as Alt 1 with the newly constructed training structure. Alts 3 to 6 are expanded from Alt 2 with additional modification to the new training structure or existing jetty-weir system (Table 1). Alt 1 serves as the benchmark so that model results of Alts 2 to 6 are evaluated by comparing to Alt 1. Calculated results are compared for sediment volume accumulation in the entrance channel and in the impoundment basin. Volume accumulation in the entrance channel is of direct interest because it can cause excessive sediment shoaling in the channel and limit navigable depth and channel width, increasing dredging frequency.

Figures 12 and 13 compare sediment volume accumulation in the entrance channel and in the impoundment basin, respectively, from the 1-year simulation for the six alternatives. It appears that the rate of volume accumulation is large, in the channel and in the basin, for all alternatives in the first four months (January to April) of the simulation, when the sediment transport process is responding or adjusting to the initial bathymetry, representing a post-dredging condition, in the model. The rate of volume accumulation becomes smaller after the month of May and shows some volume reduction in the summer time (June to August) when the wave activity is relatively weak compared to other months in the year.

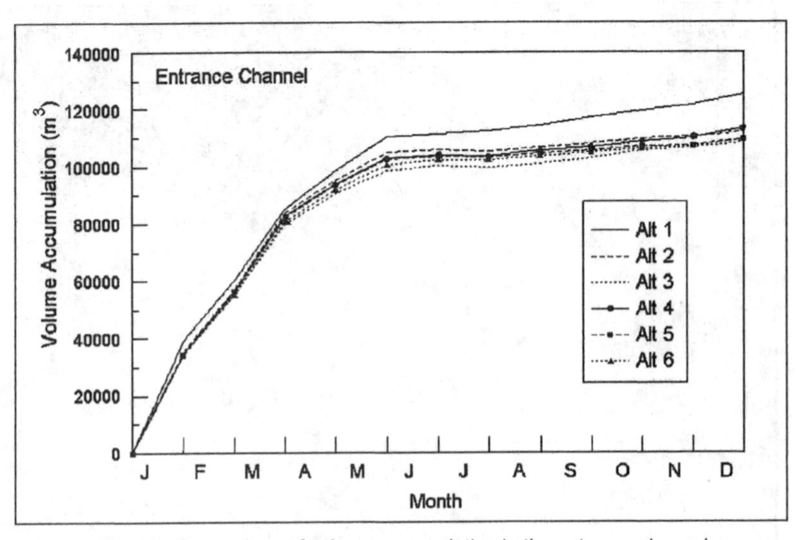

Fig. 12. Comparison of volume accumulation in the entrance channel

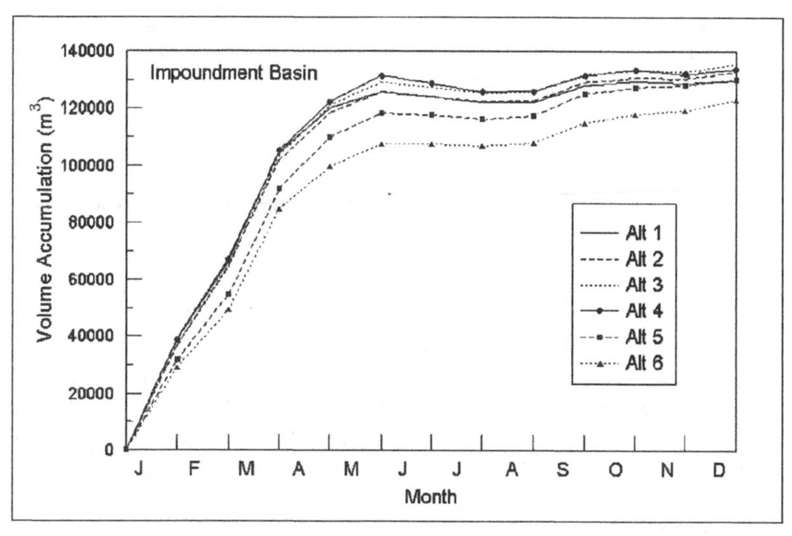

Fig. 13. Comparison of volume accumulation in the impoundment basin

For volume accumulation in the entrance channel, Alts 2 to 6 show a reduction as compared to Alt 1. At the end of the 1-year simulation, Alt 3 (long training structure) shows the most volume reduction at -13.7%, and Alt 4 (extended West Jetty) shows the least volume reduction at -9.5% (Table A1 in the appendix) in the entrance channel. The rate of volume reduction in the entrance channel is even greater if the rapid accumulation in the beginning month(s) is excluded in the calculation. Based on the March to December results, Alt 3 shows the largest rate of volume reduction at -18.6%, and Alt 4 shows the smallest rate at -12.2% in the entrance channel. The new training structure (appears in Alts 2 to 6) is predicted to function well in decreasing sediment encroachment from the basin and surf zone into the navigation channel. The new training structure is predicted to further reduce sediment accumulation in the channel, particularly if it is extended (Alt 3).

The amount of the sediment volume accumulated in the impoundment basin varies according to alternative. Infilling rates in the impoundment basin for Alts 2 to 4 are similar and somewhat higher than for Alt 1. Among these three alternatives, Alt 3 (long training structure) shows the highest infilling rate at 4.4% and Alt 2 (short training structure) shows the least rate at 2.2%, as compared to Alt 1. If the volume accumulation is calculated for March to December, to exclude the rapid and likely artificial adjustment in the initial stage, Alt 3 shows a rate at 13.8% and Alt 2 at 8.4%. Therefore, the training structure meets the desired design objective of retaining more of the sediment in the impoundment basin that enters over the jetty weir. In Alts 5 and 6, the amount of accumulation is less than other alternatives (Alts 1 to 4) as a result of raising the weir elevation to interrupt partially longshore transport from the east. The infilling rate is substantially smaller in the beginning of the simulation, particularly from January to March, when the local mean sea level is lowest during the year. The infilling rate is reduced to 0.2% for Alt 5 and is -5.1% for Alt 6 from the 1-year simulation. However, interrupting the longshore transport at the weir to reduce sediment volume directed toward the impoundment basin is a time-dependent process. After the sediment accumulates on the updrift beach over some time, the rate of sediment crossing the weir can resume and again increase the infilling rate. The calculated volume accumulation from March to December indicates that the infilling rate is 20.8% higher for Alt 5 and 18.1% higher for Alt 6.

The sediment modeling indicates that the training structure will reduce the channel shoaling and increase the efficiency of the impoundment basin. Extension of the training structure enhances the trapping efficiency of the structure. A longer training structure will increase the current flow in the entrance channel and, as a result, act to reduce sediment settling in the channel. It is also expected that the longer training structure will tend to block fine sediment from the river from entering the basin. The fine sediment from the river is likely flushed from the MCR under times of strongest current in the channel. Raising the landward half or entire segment of the weir (Alts 5 and 6) can reduce the infilling rate of the impoundment basin by longshore transport from the eastern beach. For the alternative involving raising the landward half of the weir, sediment tends to move gradually around the raised weir segment and enters the basin over the lower weir elevation. For extension of the west jetty (Alt 4), the response of the sediment transport in the model simulation is insignificant.

CONCLUDING DISCUSSION

The Inlet Modeling System (IMS) applied in this diagnostic study calculated sediment transport and morphology change under combined current and wave conditions. The sediment transport model estimated well the longshore sediment transport as compared to previous empirical estimates and the CERC formula based on a 1-year simulation driven by wave measurements in year 2002. Forced by wave buoy data and tidal currents, the model simulation indicates longshore sediment transport is directed to the west and into the impoundment basin over the existing weir. The sediment can then deposit in the basin or be transported across on the landward side of the basin and into the entrance channel. It has been demonstrated both qualitatively and quantitatively at this site that the IMS is capable of computing salient features of sediment transport and morphology change under the interaction of currents, waves, and structures.

A summary of the sediment accumulation from the 1-year model simulation at MCR associated with six structural alternatives discussed here is given in Table 2. Constructing of the impoundment basin training structure shows a good reduction of sediment accretion in the entrance channel as compared to the alternative without the training structure. Extension of the training structure 56.5 m more (Alt 3) is predicted to produce the least volume accumulation in the entrance channel by retaining more sediment in the impoundment basin. Raising the weir elevation shows a temporary significant blockage of the longshore sediment transport into the basin. However, sediment will eventually enter the basin over the weir at a later time in the normal rate when sufficient sediment is impounded on the updrift beach. For a short extension of the west jetty (Alt 4), the calculated response of the morphologic model is insignificant. It may be necessary to extend the west jetty beyond what was considered in the present investigation to obtain a beneficial result.

Table 2. Calculated Sediment Accumulation (m^3) at MCR for Year 2002						
Area	**Alt 1**	**Alt 2**	**Alt 3**	**Alt 4**	**Alt 5**	**Alt 6**
Channel	125,233	112,064	108,137	113,319	109,204	109,466
Basin	129,942	132,790	135,607	133,950	130,200	123,266
Total	255,175	244,854	243.744	247.269	239,404	232,732

Sediment supplied by the river was not included in the simulations. The fine sediment from the river tends to deposit in the impoundment basin and in the nearshore bars and shoals outside the MCR. The volume of the sediment supplied by the river accumulated at the MCR may exceed 150,000 m^3/year, according to interpretation of the dredging records. A large amount of this river sediment can deposit in the impoundment basin and reduce the efficiency of the basin. It is speculated in the present study that a longer training structure can block the river sediment transport into the deposition basin and, therefore, potentially reduce the channel and basin maintenance effort.

The modeling of six structural alternatives has shown to be useful to the study of sediment transport patterns in the MCR. These simulations were tested for the model sensitivity and served as the first step in evaluation of various structural modifications to the existing system. Because the impoundment basin was originally designed for the 2-year dredging cycle, it is necessary to extend the simulation to at least a 2-year interval. Extension of the west jetty and raising the weir fully or partially will be considered in future studies.

ACKNOWLEGEMENTS

This study was supported by the U.S. Army Engineer District, Galveston, in augmentation of technology under development in the Coastal Inlets Research Program. We appreciate assistance provided by Drs. Adele Militello, Ryuichiro Nishi, Zeki Demirbilek, and Frank Buonaiuto over the course of application of the Inlet Modeling System. Permission was granted by Headquarters, U.S. Army Corps of Engineers, to publish this information.

REFERENCES

Brown, M.E., Kieslich, J.M., Kraus, N.C. (2003). "Regional Circulation Modeling of the Texas Coast and Inland Coastal Waters," *Proc. 8th International Conf. on Estuarine and Coastal Modeling*, ASCE, submitted.

Cialone, M.A. and Kraus, N.C. 2002. "Coupling of Wave and Circulation Numerical Models at Grays Harbor Entrance, Washington, USA," *Proc. 28th Coastal Eng. Conf.*, World Scientific, 1279-1291.

Heilman, D.J., and Edge, B.L. (1996). "Interaction o f the Colorado River Project, Texas, with Longshore Sediment Transport," *Proc. 25th Coastal Eng. Conf.*, ASCE, 3,309-3,322.

King, D.B., and Prickett, T.L. (1998). "Mouth of the Colorado River, Texas, monitoring program," Monitoring Completed Navigation Projects Program. Technical Report CHL-98-2, U.S. Army Corps of Engineers, Waterways Experiment Station, Vicksburg, MS.

Kraus, N.C., and Lin, L. (2004). "Alternatives to Reduce Channel Shoaling, Mouth of Colorado River Navigation Channel, Texas," *Proc. Dredging 02*, ASCE, Reston, VA, CD Media.

Kraus, N.C., and Militello, A. (1999). "Hydraulic Study of Multiple Inlet System: East Matagorda Bay, Texas," *J. Hydraulic Research*, 125(3): 224-232.

Kraus, N.C., Isobe, M., Igarashi, H., Sasaki, T. and Horikawa, K. (1982). "Field Experiments on Longshore Sand Transport in the Surf Zone," *Proc. 18th Coastal Eng. Conf.*, ASCE, 969-988.

Lin, L., Kraus, N.C., and Barcak, R.G. (2001). "Predicting Hydrodynamics of a Proposed Multiple-Inlet System, Colorado River, TEXAS," *Proc. 7th International Conf. on Estuarine and Coastal Modeling*, ASCE, 837-851.

Militello, A., and Zundel, A. K. (2002). "Automated Coupling of Regional and Local Circulation Models through Boundary Condition Specification," *Proc. 5th International Conf. on Hydroinformatics*, IWA Publishing, London, 156-161.

Shore Protection Manual (SPM) (1984). 4th ed., 2 Vol, U.S. Army Engineer Waterways Experiment Station, U.S. Govt. Printing Office, Washington, DC.

Soulsby, R. (1997). "*Dynamics of Marine Sands*," DETR Environment Transport Regions, Thomas Telford Publications, London.

USACE, Galveston District. (1992). "Inlets along the Texas Gulf Coast," Section 22 Report, Planning Assistance to the States Program, U.S. Army Engineer District, Galveston, TX.

Watanabe, A. (1987). "3-Dimensional Numerical Model of Beach Evolution," *Proc. Coastal Sediments '87*, ASCE, 802-817.

Westerink, J.J., Luettich, R.A., and Militello, A. (2001). "Leaky internal-barrier normal-flow boundaries in the ADCIRC coastal hydrodynamics code," ERDC/CHL CHETN-IV-32, U.S. Army Engineer Research and Development Center, Vicksburg, MS. *http://chl.wes.army.mil/library/publications/chetn.*

APPENDIX A: Data Tables

Table A1. Sediment Accumulation (m^3) in the Entrance Channel, 2002						
Month	Alt 1	Alt 2	Alt 3	Alt 4	Alt 5	Alt 6
Jan	39,283	35,146 (-10.5)	34,363 (-12.5)	34,477 (-12.2)	34,431 (-12.4)	34,204 (-12.9)
Feb	60,472	56,967 (-5.8)	55,403 (-8.4)	56,450 (-6.7)	56,209 (-7.1)	55,487 (-8.2)
Mar	85,171	83,747 (-1.7)	79,623 (-6.5)	82,547 (-3.1)	82,061 (-3.7)	80,637 (-5.3)
Apr	98,504	95,005 (-3.6)	90,334 (-8.3)	93,626 (-5.0)	93,466 (-5.1)	91,884 (-6.7)
May	110,293	104,858 (-4.9)	98,570 (-10.6)	102,745 (-6.8)	102,429 (-7.1)	100,691 (-8.7)
Jun	111,229	105,993 (-4.7)	100,053 (-10.1)	103,938 (-6.5)	103,717 (-6.8)	102,333 (-8.0)
Jul	112,345	105,208 (-6.4)	99,612 (-11.3)	103,732 (-7.7)	103,136 (-8.2)	102,085 (-9.1)
Aug	114,406	106,696 (-6.7)	100,934 (-11.8)	105,596 (-7.7)	104,677 (-8.5)	103,592 (-9.5)
Sep	117,088	108,035 (-7.7)	103,036 (-12.0)	106,746 (-8.8)	105,671 (-9.8)	104,765 (-10.5)
Oct	119,533	109,637 (-8.3)	105,544 (-11.7)	108,748 (-9.0)	106,839 (-10.6)	106,405 (-11.0)
Nov	121,343	110,226 (-9.2)	106,547 (-12.2)	109,911 (-9.4)	107,350 (-11.5)	107,092 (-11.7)
Dec	125,233	112,064 (-10.5)	108,137 (-13.7)	113,319 (-9.5)	109,204 (-12.8)	109,466 (-12.6)
Feb-Dec	85,950	76,918 (-10.5)	73,774 (-14.2)	78,842 (-8.3)	74,787 (-13.0)	75,262 (-12.4)
Mar-Dec	64,761	55,097 (-14.9)	52,734 (-18.6)	56,869 (-12.2)	52,995 (-18.2)	53,979 (-16.7)
Remark: Values in parentheses are the percent change compared to volume in Alt 1. Number of significant digits is based on computation and does not represent physical accuracy.						

Table A2. Sediment Accumulation (m^3) in the Impoundment Basin, 2002

Month	Alt 1	Alt 2	Alt 3	Alt 4	Alt 5	Alt 6
Jan	39,545	36,716 (-6.4)	36,465 (-7.1)	38,919 (-0.8)	31,699 (-19.2)	28,928 (-26.3)
Feb	67,400	65,011 (-3.5)	64,445 (-4.4)	66,614 (-0.1)	54,640 (-18.9)	49,401 (-26.1)
Mar	104,690	101,714 (-2.8)	103,714 (-0.9)	105,172 (0.5)	91,917 (-12.2)	84,719 (-19.1)
Apr	120,136	118,352 (-1.5)	121,230 (0.9)	122,515 (2.0)	109,826 (-8.6)	99,376 (-17.3)
May	125,903	125,716 (-0.2)	129,072 (2.5)	131,517 (4.5)	118,548 (-5.8)	107,497 (-14.6)
Jun	123,985	124,171 (0.2)	127,457 (2.8)	128,741 (3.8)	117,861 (-4.9)	107,579 (-13.2)
Jul	122,111	122,379 (0.2)	125,626 (2.9)	125,841 (3.1)	116,342 (-4.7)	106,792 (-12.6)
Aug	122,067	122,842 (0.6)	125,684 (3.0)	126,214 (3.4)	117,276 (-3.9)	107,796 (-11.7)
Sep	128,290	129,294 (0.8)	131,957 (2.9)	131,542 (2.5)	125,012 (-2.6)	115,046 (-10.3)
Oct	129,501	130,864 (1.1)	133,385 (3.0)	133,438 (3.0)	127,587 (-1.5)	117940 (-8.9)
Nov	128,708	130,652 (1.5)	133,081 (3.4)	131,962 (2.5)	128,226 (-0.4)	119,577 (-7.1)
Dec	129,942	132,790 (2.2)	135,607 (4.4)	133,950 (3.1)	130,200 (0.2)	123,266 (-5.1)
Feb-Dec	90,700	96,074 (5.9)	99,142 (9.3)	95,031 (4.8)	98,501 (8.6)	94,338 (4.0)
Mar-Dec	62,542	67,780 (8.4)	71,162 (13.8)	67,336 (7.7)	75,560 (20.8)	73,865 (18.1)

Notes: Values in parentheses are percent change compared to the volume in Alt 1. Number of significant digits is based on computation and does not represent physical accuracy.

Calibration of A Sediment Transport Model for San Francisco Bay

Vivian Lee[1], Phillip Mineart[2], Louis Armstrong[3]

ABSTRACT

A sediment transport model of the San Francisco Bay estuary was developed to assess the potential impacts of projects in San Francisco Bay. The estuary comprises the North Bay, which receives freshwater inflow from the Delta, the Central Bay, which connects to the Pacific Ocean, and the South Bay, which is characterized as a semi-enclosed lagoon. The MIKE 21 cohesive sediment transport model was selected for the study. This paper describes the calibration of the model under baseline conditions. The focus of the modeling effort is the South Bay where the project is located. Previous research indicated that wind-wave resuspension is a key process in the estuary especially in the shallow regions. Therefore, a crucial element of the sediment model is the inclusion of wave-induced bottom shear stress. A calibrated nearshore wave model was coupled to the sediment model by creating a database which contained the wave parameters for every grid cell as a function of the input wind speed, wind direction, and water depth. Erosion and deposition are driven by the local sediment properties. In addition to horizontal spatial variability, the resistance of sediments to erosion increases with depth owing to consolidation. As such, the Bay bottom was prescribed by a five-layered bed model to represent the vertical profile of the sediment strength. In particular, the topmost layer was modeled as a fluid mud to reflect the water and seabed interface. The model was calibrated using time series of suspended sediment concentrations collected at seven different locations throughout the Bay. Both instantaneous (15-minute interval) and tidally averaged values were compared for simulations spanning a few months. Calibration and verification were performed over a range of hydrologic and meteorological conditions, reinforcing the ability of the model to capture the estuarine sediment processes. The results revealed that suspended sediment concentrations in the South Bay are mainly controlled by local erosion and deposition, with sediment influx playing a less significant role compared to the North Bay.

1 Vivian Lee, URS Corporation, 1333 Broadway, Suite 800, Oakland, CA 94612, (510) 874-1733, Vivian_Lee@urscorp.com

2 Phillip Mineart, URS Corporation, 1333 Broadway, Suite 800, Oakland, CA 94612, (510) 874-1785, Phillip_Mineart@urscorp.com

3 Louis Armstrong, URS Corporation, 1333 Broadway, Suite 800, Oakland, CA 94612, (510) 874-3034, Louis_Armstrong@urscorp.com

INTRODUCTION

The proposed San Francisco International Airport (SFO) expansion project involves construction of new runways by filling in 130 to 380 hectares of the San Francisco Bay estuary, in-bay excavation to obtain fill materials, and restoration of historical wetlands as mitigation measures. The sediment transport model is part of the integrated modeling study conducted to assess the potential physical, chemical and biological changes introduced by the project. The present paper focuses on the calibration of the sediment transport model under baseline conditions to define the existing system prior to construction.

The San Francisco estuary comprises the North Bay, which receives the freshwater inflow from the Delta, the Central Bay, which connects to the Pacific Ocean, and the South Bay, which is characterized as a semi-enclosed lagoon. Figure 1 shows the study area as represented by the numerical model. The shallow regions of the North Bay consist of mudflats composed of primarily poorly sorted (i.e. wide gradation of particle size) silt and clay, whereas bottom sediments in deeper waters are mostly sand and silty sand (Conomos and Peterson, 1977). Sector scanning sonar technology was employed in this study to examine the bedforms in the Central Bay. Results indicated that this part of the Bay is covered extensively by sand waves approximately 1 meter long and 0.5 to 1 meter high. Similar findings were reported in an earlier study (Rubin and McCulloch, 1979) regarding the presence of sand waves in the Central Bay. Bottom sediments of the South Bay are typically composed of silt and clay, while mixture of silt with sand and shell fragments were observed in the eastern shallow areas (Nichols and Thompson, 1985). Since the Bay, especially the South Bay where the proposed project is located, is composed of predominantly cohesive sediments, i.e. silt and clay, the MIKE 21 mud transport (MT) model was selected for the study.

Previous studies have shown that tidal currents, spring-neap tidal variation, wind-wave resuspension, Delta inflow, and advective transport are the most important elements driving sediment transport in San Francisco Bay (Schoellhamer, 1996; McDonald and Cheng, 1994). Because all the project elements are located south of the Bay Bridge, the emphasis is on calibration of the South Bay. However, Central and North Bay should be satisfactorily calibrated to ensure that sediment fluxes are properly transferred to the South Bay boundary.

MODEL DESCRIPTION

The MIKE 21 MT model chosen for this study is a two-dimensional, depth-averaged hydrodynamic and sediment transport model. Since the focus of the study is in the South Bay which is shallow and generally well-mixed, a depth-averaged model provides a reasonable representation of the system under most circumstances. The sediment model accounts for settling, deposition, erosion, flocculation, liquefaction, consolidation, bed property gradation, and wave-induced resuspension. A uniformly spaced grid with a 200-meter cell size was developed for the entire Bay extending into the Pacific Ocean at the western boundary with the eastern boundary located at the Delta as shown in Figure 1. A 200-meter grid was selected as a compromise between computer runtime and details required to estimate project impacts. The extent of the model domain allows for the specification of tidal forcing and freshwater inputs at the western and eastern boundaries, respectively. The model grid was rotated 35.4 degrees counterclockwise from north such that the primary flow direction is aligned with the grid axes in order to optimize the accuracy of the numerical model.

The bed is divided into layers with distinct densities and shear strengths. The multi-layer bed structure of the model accounts for depthwise variations in density and bed shear strength that control whether erosion or deposition takes place under different hydrodynamic conditions. This layering allows the inclusion of the fluid-like mud layer at the sediment-water interface, representing the sediments that are readily mobilized by wind-wave resuspension. The presence of the fluid mud is widely acknowledged to contribute significantly to sediment transport in San Francisco Bay (McDonald and Cheng, 1997).

The MIKE 21 MT model solves the advection-dispersion equation for suspended sediment concentration (SSC). The advective transport terms are calculated using depth-averaged velocities computed by a coupled hydrodynamic model. The hydrodynamic model was calibrated to observed water surface elevations and velocities at 19 and 31 locations throughout the bay under a range of hydrologic conditions. Sedimentation processes are handled by a sub-model with results incorporated into the advection-dispersion equation as source/sink terms. This approach was first developed by Krone (1962).

Cohesive sediments react to a complex array of physical, chemical and biological factors, which are understood to varying levels of completeness. Erosion, deposition, transport and consolidation are the four major processes controlling the dynamics of cohesive sediments. No deterministic description of the behavior of cohesive sediments has been developed to-date. Therefore, mathematical formulations developed for erosion and deposition are essentially empirical, although they are based on physical principles.

Sedimentation processes are driven by bottom shear stress generated by tidal currents as well as wave-induced orbital velocities. Therefore, a nearshore wave model that simulates the propagation, growth, and decay of short-period, spectrally averaged wind waves in the nearshore areas was used to estimate the bottom shear stress generated by wave-induced littoral currents. The formulation of wind-wave generation in the nearshore wave model is based on the fetch and water depth at each point, under the assumption that the wave field is fetch limited and fully developed regardless of the duration the wind blows. Measured time series of significant wave heights and periods were used in the calibration of the wave model, the details of which are outside the scope of this paper.

The calibrated nearshore wave model was then coupled to the sediment model by creating a database consisting of 72 distinct wave fields, which contained the wave parameters (i.e. significant wave height, period, and mean wave direction) for every grid cell as a function of the input wind speed (from 0 to 15 m/s at 5 m/s increment), wind direction (at 45 degrees increment), and water level (from –2 to 2 m NGVD at 2 meters increment). The bed shear stress thus consists of a steady component due to tidal currents and an oscillatory component due to waves. In laminar flows, the combined shear stress is a linear addition of the laminar current-alone and wave-alone shear stresses. However, under conditions of strong currents and wave motions as in San Francisco Bay, the flow will be turbulent leading to nonlinear interaction of the shear stresses (Whitehouse et al.,

1999). The MIKE 21 MT model uses Equation (1) to compute the bed shear stress under combined wave-current motion:

$$\tau_b = \frac{1}{2}\rho f_w (U_b^2 + U_d^2 + 2U_b U_d \cos\theta) \quad (1)$$

where U_b is the horizontal mean wave orbital velocity at the bed, U_d is the current velocity at the top of the wave boundary layer, θ is the angle between the tidal current direction and the direction of wave propagation, f_w is the wave friction factor.

Due to computational constraint, a spatially constant time series of wind was input into the sediment model. The effect of this simplified approach on model predictions was evaluated by performing a sensitivity analysis using different wind fields prescribed for the North Bay and South Bay.

CALIBRATION DATA

Two different periods were selected for calibration: Spring/summer of 1995 and Winter of 1995. The dataset for Spring/summer of 1995 was selected to separate the influence of Delta sediment inflow and vertical circulation resulting from density stratification. It is recognized that the two-dimensional model used in the study cannot replicate stratified flow conditions. Nevertheless, satisfactory results can still be achieved if the influence of vertical stratification on SSC is small compared to local erosion and deposition, and the sediment influx. Therefore, the second calibration period was chosen to be from January through March of 1995 covering the peak Delta inflow that occurred on January 12, 1995 to demonstrate the model performance during stratified conditions in winters. Finally, the model was verified by simulating the extreme El Nino event occurred during Winter 1997. Figure 2 shows the Delta inflows and winds inputs for the three calibration/verification periods.

Calibration dataset consists of continuous SSC data collected at 15-minute interval by the U.S. Geological Survey (Buchanan and Schoellhamer, 1995) using optical backscatterance sensors (OBS) at two sites in Suisun Bay (Mallard Island and Martinez), two sites in Central Bay (Point San Pablo and Pier 24), and three sites in South Bay (San Mateo Bridge, Dumbarton Bridge and Channel Marker 17). Most of the sites were equipped with 2 sensors: one placed near the surface or at the middle of the water column, and another one placed near the bottom. Location of the sediment stations is shown on Figure 1.

Although not described in this paper, the model's ability to predict long term changes in bed elevation was verified by comparing predicted to measured elevation changes over a 30-year period. A detailed error analysis was conducted on the comparison, and the model was determined to be adequate for the environmental study.

SEDIMENT PARAMETERS

The formulations used to describe erosion, deposition, and consolidation are based on several empirical parameters. Unlike variables that lump the entire phenomenon, these model parameters have physical meanings that govern the individual process. Since some of these parameters are interdependent, multiple combinations of parameters could possibly result in a good match between field data and predicted data. To facilitate the calibration procedure, most of the sediment properties were assigned a best estimate based on literature review and not varied throughout the calibration. Because SSC is most sensitive to the critical shear stresses for erosion and deposition as well as settling velocity, they were used as the primary calibration parameters. Adjustment of these parameters was made within reasonable bounds as reported in the literature, and was applied to the larger area assumed to have the same bed properties, not locally around the sediment calibration stations.

Bed Structure. The bottom of the Bay was represented using 5 sediment layers as a discretization of vertical variation in bed sediment properties. Initial layer thickness and layer properties specified for all the layers are shown in Table 1. Due to the lack of bed strata data in the uppermost one meter, definition of layer thickness and other properties is to some extent arbitrary and developed as part of the calibration process. The depletion or accretion in layer thickness at each grid cell is accounted for in the model as erosion or deposition takes place. However, there is no automatic feedback mechanism built in the MT model to update the local bathymetry resulting from sedimentation. The error in hydrodynamics is considered negligible since the predicted change in bed elevation is typically less than 0.1 to 0.2 m for simulations spanning a few months.

The top layer is the "fluffy" layer and provides most of the sediment source resuspended by tides and waves. This layer was estimated from the calibration to be 50 mm thick. It behaves almost like a fluid or fluid-mud mixture with solid concentration exceeding 10,000 mg/L (Whitehouse et al., 1999). The existence of this layer can be explained by the periodic weakening of the sediment matrix by wind-wave actions, as well as the bioturbation caused by organisms digging and feeding in the substrate. Freshly deposited sediment has been reported to have dry densities in the range of 100 to 300 kg/m^3. The layer thickness was in line with the data collected by Rivera-Duarte and Flegal (1997) that showed changes in chemical properties in the top 50 to 150 mm range for several sites in San Francisco Bay. At three sites they estimated the location of the suboxic/anoxic front to be between 40 and 80 mm, and one at 220 mm. Based on these limited data, a 50 mm thick top layer in the shallow areas seems reasonable. The layer was assumed to be 2 mm thick in the deeper waters greater than 6 meters deep where waves are insignificant and hence the mechanism for the formation of the fluid-mud layer does not persist.

The four underlying layers consist of more consolidated material, with each layer more resistant to erosion than the layer above it. Studies have shown that densities increase with depth from 500 kg/m^3 to as high as 1600 kg/m^3 for over-consolidated soils at a depth of 1 meter (HR Wallingford, 1991). In the absence of vertical sediment profile data, typical values ranging from 450 to 900 kg/m^3 were assigned to the four layers accounting for varying degrees of consolidation. Similar values of 450 kg/m^3 (McDonald and Cheng, 1997), and 400 to 500 kg/m^3 (Ogden Beeman and Krone, 1992) have also been used for the Bay.

Erosion Parameters. As discussed earlier, critical shear stress for erosion, τ_{ce}, is a key factor for determining the onset of erosion. However, in-situ measurement of critical shear stress is difficult, while values determined from laboratory tests that are indicative of an undisturbed soil matrix are hard to obtain. In addition, given the scope of the modeling (over 1,500 km^2) it was not economically feasible to collect enough data to characterize the model domain. Geotechnical explorations of soils in the bay were mainly focused on deeper, consolidated bay mud and no direct measurement of critical shear stress in the upper erodible layers have been carried out.

Some laboratory studies showed that τ_{ce} was related to the dry density of the soil with proportionality coefficients varied according to the degree of consolidation (Thorn and Parsons, 1980). However, field measurements demonstrated that in-situ values of τ_{ce} were substantially smaller than those predicted from laboratory-derived expressions (Amos et al., 1998; Mitchener et al., 1996) because the full environmental conditions found in natural estuarine muds could not be reproduced in laboratory tests. Nonetheless, critical shear stress in general increases with dry density for pure cohesive sediments.

Sternberg et al. (1986) estimated τ_{ce} to be 0.049 N/m^2 for fine sediments using empirical curve fitting. Krone (1962) had a similar estimate of 0.06 N/m^2, whereas values ranging from 0.20 N/m^2 (Hauck et al., 1990) to 0.40 N/m^2 (Teeter, 1987) had also been reported. Over-consolidated soil can have values exceeding 1 N/m^2 (Parchure and Mehta, 1985). Association of these values with gradation of the layered beds or with spatial heterogeneity of the bed material is not clear. The approach in this study is to use a depthwise increase in shear strengths with local adjustment as part of the calibration procedure.

Uncertainty in the determination of τ_{ce} is further complicated by spatial variability of sediment properties in the bed. For example, sediments in the South Bay are finer than in the North and Central Bays as a result of winnowing of fine particles from repetitive resuspension as sediments originated from Central Valley migrate through the system (Krone, 1991). In certain parts of North Bay, sediment type is distinctly different in shallow and deep waters (Conomos and Peterson, 1977). It was also found that sand increases binding between clay particles and forms a more compact and dense matrix, thereby making the mixture more resistant to erosion (Whitehouse et al., 1999). This spatial variability is accounted for in the model by specifying a spatially variable τ_{ce}.

For the soft, unconsolidated layer, a value of τ_{ce} of 0.06 N/m^2 was selected based on calibration for the south of Dumbarton region which is dominated by fine silt and clay, with higher values (0.15 to 0.3 N/m^2) specified towards the Central Bay where coarser sediments exist. San Pablo Bay has a considerable supply of fine sediments from its large area of broad shoals and 0.18 N/m^2 was used. The higher value of 0.3 N/m^2 was applied to Carquinez Strait and the deep channels in Suisun Bay. Spatially uniform critical shear stresses were used for each of the underlying layers (but gradual strengthening with depth) except for the Central Bay region because of its higher bottom irregularities.

The rate of erosion in the fluffy layer is controlled by the excess shear stress (difference between bottom shear stress and τ_{ce}, and normalized by τ_{ce} in the case of consolidated layers) and the rate constant, E, as described by Equation (2) (Parchure and Mehta, 1985).

$$S = E \exp\left[\alpha\left(\tau_b - \tau_{ce}\right)^{1/2}\right] \tag{2}$$

where E is an erosion rate constant (in $g/m^2/s$), which can be a constant or a function of bed shear stress, τ_b is the bed shear stress calculated at any one point, τ_{ce} is the critical shear stress for erosion for that layer, α is an empirical coefficient based on comparison with measurement.

Again, wide range of values of E has been published for the Bay. Flume studies by Teeter (1987) estimated E to be between 0.01 and 0.05 $g/m^2/s$, while limited field study by Hauck et al. (1990) came up with E in the order of 0.1 $g/m^2/s$. In this study, a value of 0.03 $g/m^2/s$ was used for the more energetic uppermost layer, and a slightly lower value of 0.02 $g/m^2/s$ for the other four layers. Note that the rate of erosion is a function of both E and τ_{ce}; therefore, if a different value of E was used, the model might have been calibrated to a different of τ_{ce} values.

Deposition Parameters. The rate of deposition is a function of excess shear stress (difference between bottom shear stress and critical shear stress for deposition, τ_{cd}, and normalized by τ_{cd}), near-bed sediment concentration, and settling velocity, as described by Equation (3) (Krone, 1962).

$$S = w_s C_b \left(1 - \frac{\tau_b}{\tau_{cd}}\right) \tag{3}$$

where w_s is the settling velocity, C_b is the near bed concentration related to the depth-averaged concentration, τ_b is the bed shear stress calculated at any one point, τ_{cd} is the critical shear stress for deposition.

As for the critical stress for erosion, limited field data are available on τ_{cd}, which is strongly dependent on sediment characteristics and is site-specific. Values of 0.06 N/m^2 (Krone, 1962) and 0.15 N/m^2 (Hanck et al., 1990) were used for different areas of the bay. A larger range from 0.005 to 1.0 N/m^2 have been proposed by van Rijn (1989) in other estuarine environments. Spatially variable τ_{cd} with similar zoning as τ_{ce} for the top layer was specified in the model.

Settling velocity, w_s, of cohesive sediments depends on grain size, temperature, salinity, flocculation, and mineralogy of the sediment. In the MIKE 21 MT model, settling velocity is not spatially variable. Therefore, the same settling velocity is used for sediments that are eroded from different parts of the Bay with different bed composition. Nonetheless, flocculation of sediment was accounted for by using Equation (4) (van Rijn, 1989):

$$w_s = k \bullet SSC^m \tag{4}$$

where k and m are empirical coefficients varying with the site. Whitehouse et al. (1999) compiled data from eight estuaries and showed that the coefficients k and m could be approximated by 1×10^{-6} and 1.0, respectively. These values were adopted in the model for SSC between 50 and 400 mg/L. Settling velocity was assumed constant outside of the range. Thus, the maximum settling velocity prescribed was 4×10^{-4} m/s, same as the constant settling velocity used in simulations performed by McDonald and Cheng (1997).

This upper bound on settling velocity used in the MT model is profoundly lower than that reported by Sternberg et al. (1986). In Sternberg's study the average settling velocity of flocs was approximated to be 2.8×10^{-3} m/s based on size distributions of suspended sediment samples collected at a tidal channel site in the Central Bay. Several factors may account for this difference. First, sediments in the Central Bay are in general coarser than in other parts of the bay. In the sediment transport modeling conducted by McDonald and Cheng (1997), an increase in settling velocity from 4×10^{-4} to 1×10^{-3} m/s was found to produce better fit for one of the simulations where sediment composition was believed to be coarser. Secondly, the settling velocity of 2.8×10^{-3} m/s was estimated for sediments collected at depths of 4 to 8 meters. The same value thus may not be applicable to the large mudflats with less than 3 meters of water. Lastly, a lower settling velocity may be warranted to account for the longer traveling time for particles in the upper part of the water column, whereas in a depth-averaged model sediments throughout the water column settle at the same speed.

Bed Roughness. The bed roughness z_0 used in the MIKE 21 MT model is decoupled from hydrodynamics calculations. Bed shear stress under combined current-wave motion was computed using mean current velocity obtained from the hydrodynamic module, wave parameters calculated from the nearshore wave model, together with the z_0 value specified. The separation of z_0 from Manning's n used in hydrodynamics calculations conveniently solves the issue of conflicting roughness required in the calibration of the hydrodynamic and sediment models (Inagaki, 2000). A constant value of 0.001 m was used for the bay despite spatial variation in sediment properties.

Consolidation. Consolidation is the gradual strengthening of soil that has settled onto the bed in a loose state. It leads to an increase in sediment density with time as the void ratio decreases when trapped porewater is expelled from the sediment, thereby leading to a density profile within the bed. It is a complex process that is difficult to address mathematically. To include this effect, a transition was specified between the first and second layers in the deeper waters. Consolidation takes time and therefore it was assumed to be negligible in the shallow mudflats that are subject to continual resuspension at subtidal time scale.

RESULTS

Summer 1995 Calibration. Figure 3 shows the time series comparing measured and predicted SSC at Channel Marker 17 (see Figure 1 for location) over a 30-day period, along with the local tide (denoted as Water Surface Elevation, WSEL, in the upper panel) to illustrate the phasing between SSC and tide. This is the southernmost calibration station and is located in a narrow channel surrounded by a large area of mudflats. The SSC was observed to vary with the spring-neap cycle and the peaks in SSC and spring tide were almost in phase with each other. The phenomenon of increasing minimum SSC during spring tide was observed during some but not all of the spring tides. Similar observations were reported in other studies (Schoellhamer, 1996). Concentration peaks occurred about 2 to 3 times a day, usually at slack water after ebb tides. The same trend was reflected in the modeled time series, suggesting that the important sub-tidal transport mechanisms were captured correctly in the model. Calibration results at sub-tidal scale are shown in Figure 3a, while Figure 3b presents calibration results for daily averaged SSC over a longer span of 3 months. During water year 1995, SSC reached its peak around mid-April which is believed to be a result of a concurrent spring tide and strong wind period. Daily averaged SSCs were about 400 mg/L and 800 mg/L at the upper and lower sensors, respectively, and were almost twice as large as other spring tides in this period. Therefore, the high SSC recorded during the mid-April spring tide must be accentuated by factors other than those occurred on a cyclic basis, for instance wind-wave induced resuspension. The model captured the monthly variability observed in the data and successfully reproduced the mid-April SSC peak, indicating that the model is reproducing the large-scale sediment transport processes.

Runs were conducted with and without the fluid mud layer. Predicted SSC was substantially lower than observed values and temporal variability much reduced when the fluid-mud was omitted. This supports the assumption that fluid-mud is important in the short time scale sediment dynamics in the Bay.

Figure 4 show the instantaneous and daily averaged SSC calibration plots for the Dumbarton Bridge calibration station. This station is located at a constriction in South Bay with a width of about 2 km or 10 grid cells. Similar to Channel Marker 17, a spring-neap pattern was noted in the data and the minimum concentration increased during some of the spring tides. No distinct lag occurred between the SSC peak and spring tide peak. Again, SSC was highest during the mid-April spring tide. Strong semi-diurnal periodicity was observed in the data and the daily concentration peak typically occurred at slack water after ebb tide. The hypothesis is that sediment suspended during flood tides in the northern part of South Bay enters the area south of Dumbarton Bridge through the shoals as water is advected from the channel to the shallows. During ebb tide sediment-laden water recedes from the mudflats and enters the channel. The calibration of this station revealed the significance of bathymetric features in simulating suspended sediment transport. In the initial simulations, the predicted SSC compared reasonably well to the measured data at Channel Marker 17, but the model substantially underestimated SSC at Dumbarton Bridge and predicted a SSC peak four times a day. Examination of the bathymetry south of the bridge from aerial photos revealed a network of small narrow channels not resolved by the 200-meter grid. Therefore, the model underestimated draining of the mudflats. To enhance drainage of the area, the bathymetry was smoothed south of Dumbarton Bridge which drastically increased the peak SSC at the Dumbarton Bridge. Moreover, wetting and drying depths (i.e. the depth at which a model grid cell is considered to have drained completely at ebb tide or rewetted at flood tide) were reduced which has the effect of increasing the drainage area. This adjustment further improved the predicted SSC at Dumbarton Bridge.

Concentrations are significantly lower at the San Mateo calibration station compared to the other two South Bay stations as shown on Figure 5. This could be due to the change in bed composition in the seaward direction in South Bay. Large mudflats located on the eastern side of the main channel are more resistant to erosion because the sediments contain mixture of sands and shell fragments. Trends exhibited at the other two sites such as spring-neap variations and increasing minimum SSC during spring tides were more pronounced at this site. The SSC peak lagged the spring tide peak by about 1 to 2 days, similar to the observation made by Schoellhamer on the 1993 data (Schoellhamer, 1996). Diurnal periodicity was observed during spring tides with one large concentration peak occurring at lower low water each day. This phenomenon was generally captured by the model. The smaller volume of water exchanged between the shallows and channel during the smaller ebb tide did not cause a notable increase in SSC.

During neap tides, a semi-diurnal signal was apparent with concentration peaks observed twice a day at low water after ebb tides. The lower sensor data were unavailable for this period.

Station Pier 24 is located on the western end of the Bay Bridge where strong tidal currents are found. Bottom sediments are predominantly sands and bedform movement is believed to be one of the key transport processes. Since the MIKE 21 MT model only handles cohesive sediments, it cannot accurately address the sediment transport mechanisms in this part of the Bay. Figure 6 shows the instantaneous and daily averaged SSC at Pier 24. Modeled SSC demonstrated regular spring-neap variations because the tide was the primary forcing function at this deep water site. Such a pattern was not observed in the measured data indicating that other factors not represented in the model were contributing to the variations in concentration.

Since the focus of model calibration is in the South Bay, results for the North Bay stations will only be briefly described. Figure 7 presents the calibration plots for the Point San Pablo Station located between Central Bay and San Pablo Bay. As opposed to the South Bay stations, SSC peaks occurred mostly at ebb tides when current velocity and thus bed shear stress are at maximum. This is consistent with the findings reported by McDonald and Cheng (1994) using data collected from February 1979 to June 1980. The model did not capture the sub-tidal phasing correctly at this station; nonetheless, the general spring-neap trend was reproduced. Figure 8 shows the SSC comparison at Martinez Station between model output and the upper sensor (1 meter below water surface) measurement. The tidal range is smaller at this site because it is located further away from the tidally dominated zone. Nonetheless, a spring-neap variation could be observed. Model predicted SSC peaks were similar in magnitude to the measurements, but the daily averaged values were overestimated. SSC showed narrower peaks in the data than in the modeled time series, indicating that sediments might be coarser at this site than found in South Bay as coarse sediments have higher settling velocity.

Table 2 provides the statistical comparison for the daily averaged SSC for both the measurements and modeled results. The model tends to overpredict SSC by less than 20 to about 30 percent. Overall, the values are in good agreement in view of the lack of precise description of the different sediment processes, spatial variation and uncertainty in sediment properties, and quality of the measured SSC data. Given the temporal variability in both the measured and predicted SSC's, a combination of visual and statistical comparison was considered adequate for assessment of the model's capability to predict project impacts on SSC in the Bay.

Winter 1995 Calibration. As aforementioned, the winter calibration was conducted to evaluate the response of SSC to freshwater inflows and the accompanying sediment loads. Although all the North Bay stations showed an increase in SSC due to increases in Delta inflow during the Winter of 1995, none of the South Bay stations showed a corresponding increase in SSC with Delta discharge. SSCs were comparatively low in typical winter months owing to the lack of strong persistent winds.

Verification using El Nino Event. The model was verified using data from the winter of 1997-98 to attest the validity of the model under extreme hydrologic conditions. During this period a series of storm events occurred that significantly raised the average water level in the Bay for several months. A large storm occurred around January 3, 1997, with a peak inflow of 16,000 m^3/s (average daily inflow is about 600 m^3/s). SSC measured at Mallard Island showed an increase from the pre-storm level of 20 mg/L to 250 and 750 mg/L at the upper and lower sensors, respectively. Plots of tidally averaged SSC for the South Bay and North Bay stations are given on Figures 9 and 10 for this storm event.

Overall, verification results are in good agreement with the data. The first pulse was accurately reproduced at Point San Pablo, but the predicted SSC was lower in the subsequent period. This could be due to resuspension of sediments deposited in the initial storm period. In spite of a much larger sediment inflow, no increase in SSC in association with the storm was observed at the South Bay stations nor predicted by the model. The model underestimated SSCs at Dumbarton Bridge and Channel Marker 17 during the neap tides in mid-January and late-January, the cause of which could be inaccurate prescription of sediment loads from local rivers.

Effect of Simplified Wind Input. Wind speed and direction vary over different parts of the Bay depending on the origin of the wind and the local terrain. Also, they will be modified as the wind traverses over water. However, wind gages are usually located on land. Considering the lack of information to interpolate wind properties over the Bay and the vast amount of data input required to specify a spatially varying wind field (e.g. for a 30-day simulation, 2 gigabytes of wind data would be needed to prescribe a spatially variable wind field), a single wind time series collected at SFO was applied to the entire Bay. To determine if this simplification in wind description would jeopardize the model predictions of wind-generated waves in the project area, a comparison was made between simulations conducted using only wind data collected at SFO versus a wind field comprising SFO and San Pablo Bay wind data for the South Bay and North Bay, respectively.

A one-month simulation period in November 1993 was selected for this test case because concurrent data were available in the two embayments. The wind speed and direction at SFO and San Pablo Bay during this simulation period are notably different. Results indicate that the sediment flux at the Bay Bridge is the same for the two simulations with different wind fields. Therefore, it is concluded that using the SFO wind data for the entire Bay would provide approximately the same result for the South Bay as using respective winds in the different embayments.

CONCLUSIONS

Calibration and verification of a two-dimensional sediment transport model developed for San Francisco Bay were performed over a range of hydrologic and meteorological conditions to demonstrate the ability of the model to capture the estuarine sediment processes. Results reveal that suspended sediment concentrations in the South Bay are mainly controlled by local erosion and deposition, with sediment influx playing a less significant role compared to the North Bay. The specification of a fluid-mud layer is essential to reproduce the instantaneous resuspension on a tidal basis. Inclusion of wind-wave resuspension is important because of the large areas of mudflats. Model calibration demonstrates that smoothing bathymetry and reducing wetting and drying depths can be useful techniques to circumvent impaired drainage resulting from low grid resolution.

ACKNOWLEDGMENT

The work summarized in this paper was funded by the City of San Francisco and performed as part of the environmental impact assessment for the SFO Runway Reconfiguration Project administered by the Federal Aviation Administration.

Table 1. Sediment Properties Used in MIKE 21 Mud Transport Model

Layer	Dry Density (kg/m^3)	Critical Shear Stress for Erosion, τce (N/m^2)	Erosion Rate Constant, E (g/m^2/s)	Critical Shear Stress for Deposition, τcd (N/m^2)	Settling Velocity (m/s)	Layer Thickness (mm)
1	180	0.06 - 0.30	0.03	0.04 - 0.25	ws = 1x10^{-6}C, 50<C<400 mg/L, i.e. 5x10^{-5} < ws < 4x10^{-4}	2 - 50
2	450	0.40 - 0.43	0.02			50
3	600	0.43 - 0.46	0.02			50
4	750	0.46 - 0.50	0.02			50
5	900	0.50 - 0.55	0.02			1000
References						
Cheng and McDonald (1996)	450	0.3 - 0.4	0.01 - 0.05	0.15	0.4 - 1x10^{-3}	one layer only
Hauck et al. (1990)		0.2	0.1	0.15	1x10^{-3}	
Krone (1962)				0.06		
Teeter (1987)		0.1 - 0.4	0.013 - 0.047			
Sternberg et al. (1986)		0.049			2.8x10^{-3}	
Kranck and Milligan (1992)					1 - 10x10^{-3}	
Ogden Beeman (1992)	400 - 800					

Table 2. Statistics on Daily Averaged SSC for Observed and Predicted Data – Summer 1995 Calibration

Sediment Station	SSC Mean (mg/L)		SSC Median (mg/L)		SSC Lower Quartile (mg/L)		SSC Upper Quartile (mg/L)	
	Observed	Predicted	Observed	Predicted	Observed	Predicted	Observed	Predicted
Channel Marker 17	178	237	152	187	115	138	215	263
Dumbarton Bridge	159	152	122	122	80	87	249	181
San Mateo	46	65	36	48	26	33	60	93
Pier 24	46	57	36	51	25	36	57	74
Point San Pablo	72	75	63	63	53	47	88	96
Martinez	52	78	49	70	40	64	58	90
Mallard Island	31	60	30	57	27	53	33	66

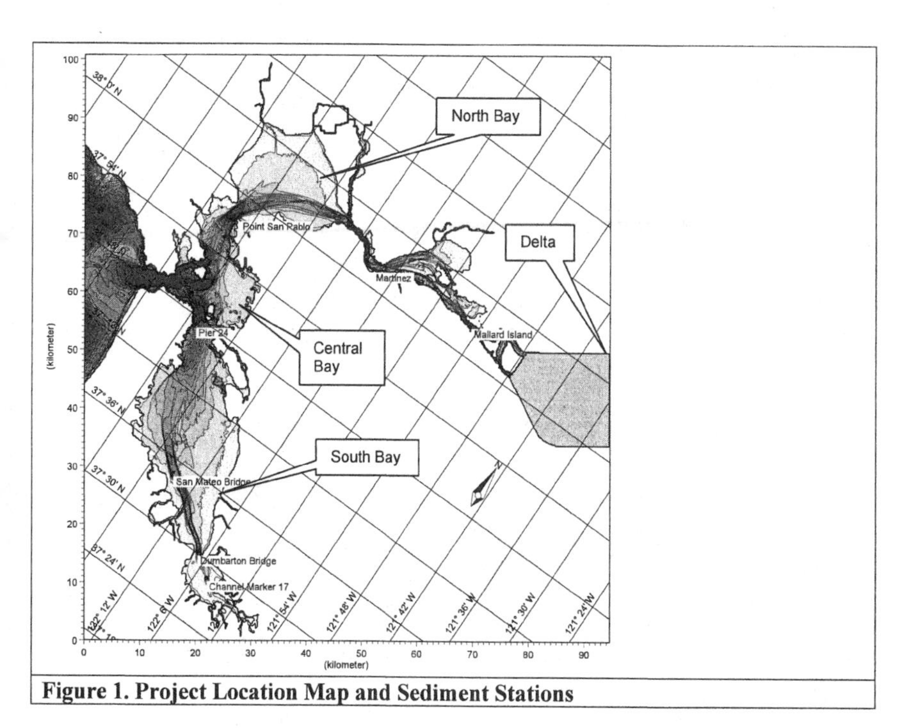

Figure 1. Project Location Map and Sediment Stations

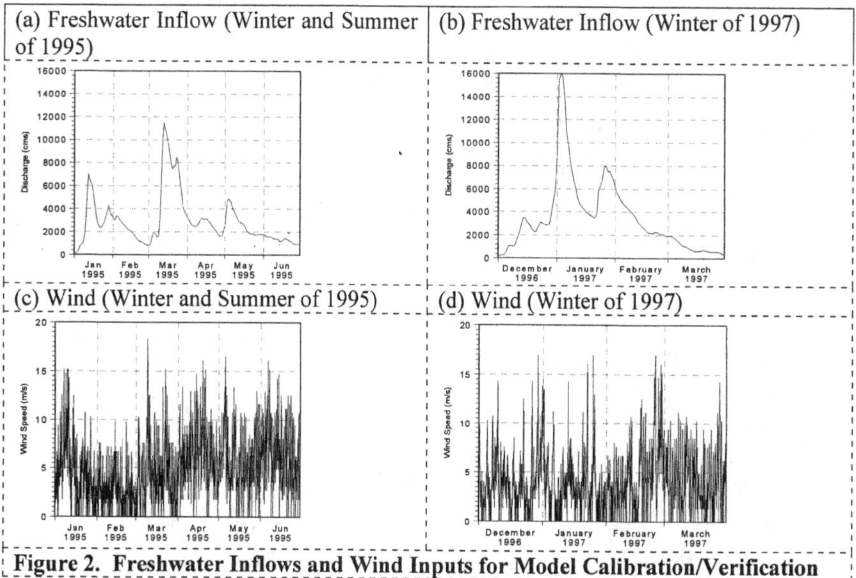

Figure 2. Freshwater Inflows and Wind Inputs for Model Calibration/Verification

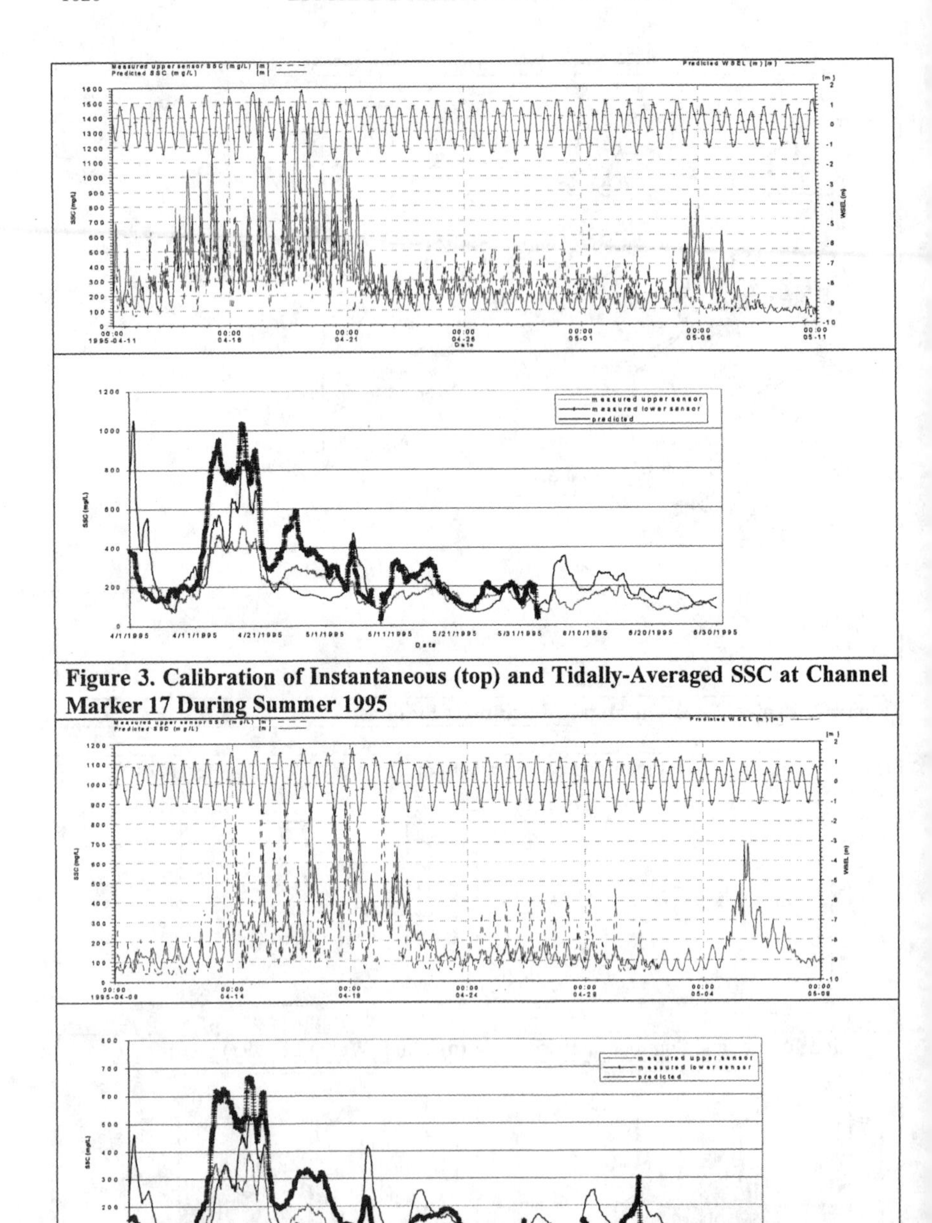

Figure 3. Calibration of Instantaneous (top) and Tidally-Averaged SSC at Channel Marker 17 During Summer 1995

Figure 4. Calibration of Instantaneous (top) and Tidally-Averaged SSC at Dumbarton Bridge During Summer 1995

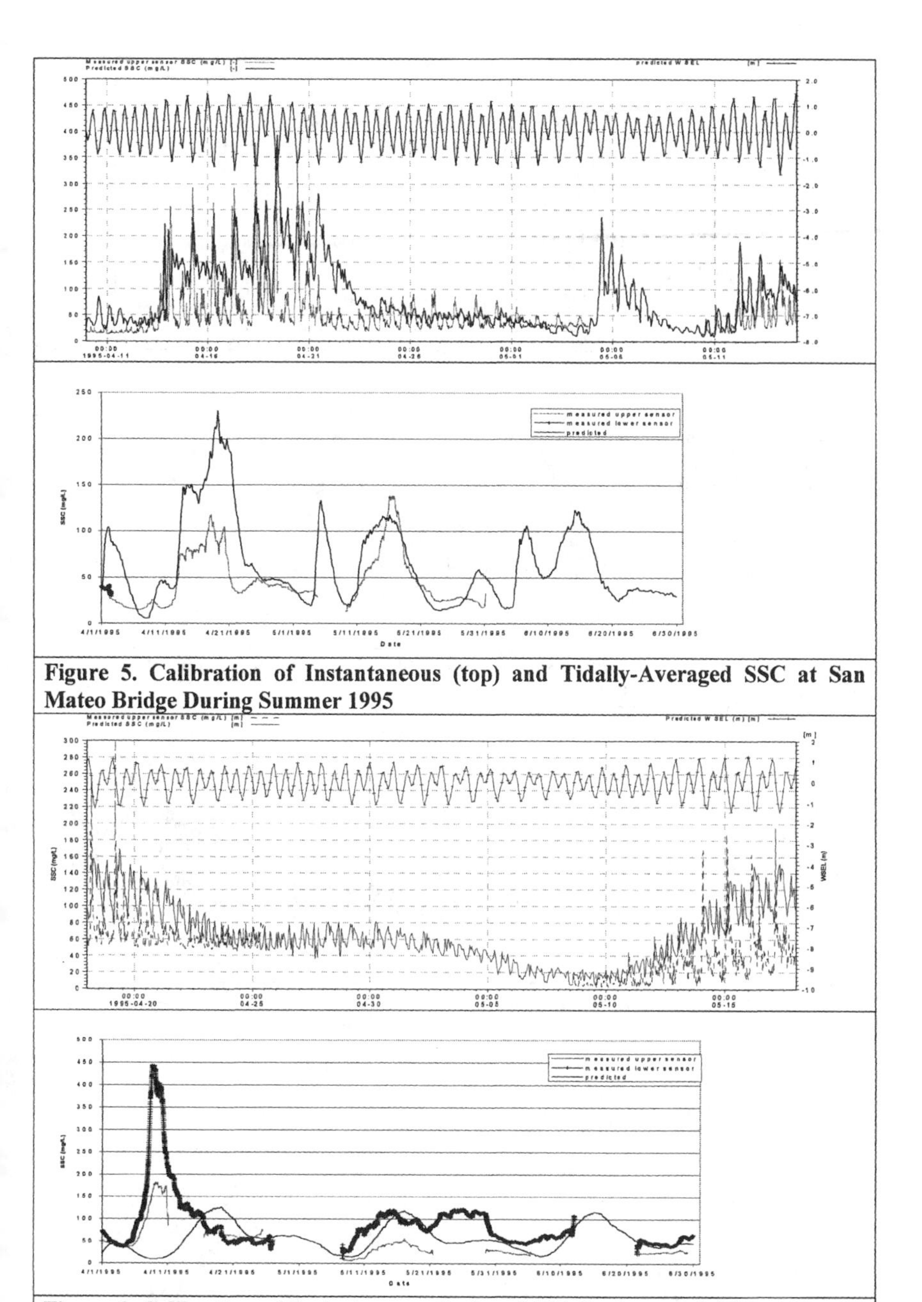

Figure 5. Calibration of Instantaneous (top) and Tidally-Averaged SSC at San Mateo Bridge During Summer 1995

Figure 6. Calibration of Instantaneous (top) and Tidally-Averaged SSC at Pier 24 During Summer 1995

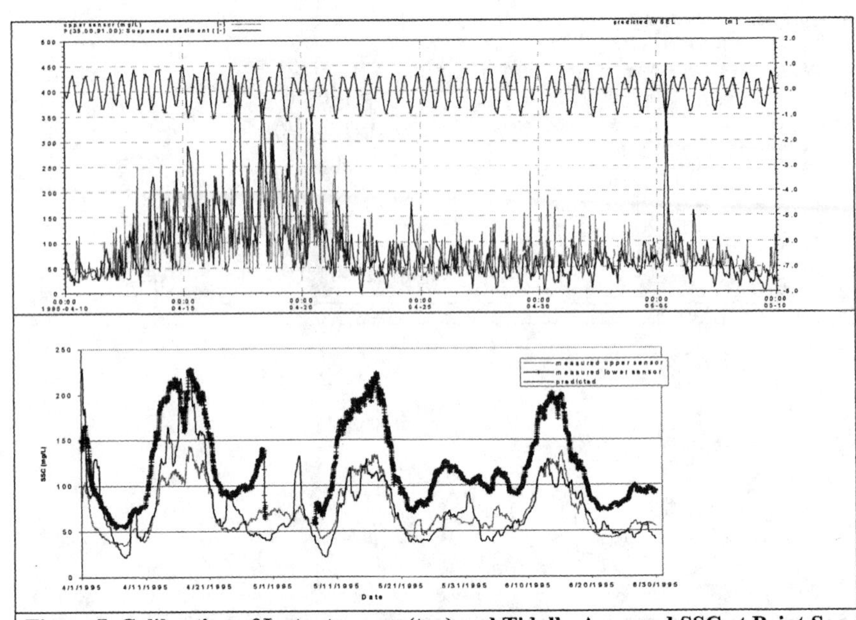

Figure 7. Calibration of Instantaneous (top) and Tidally-Averaged SSC at Point San Pablo During Summer 1995

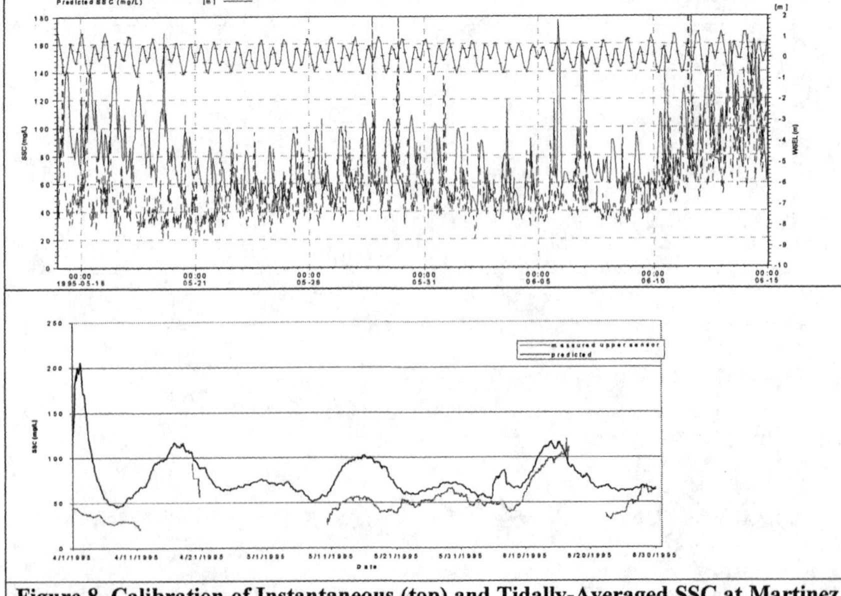

Figure 8. Calibration of Instantaneous (top) and Tidally-Averaged SSC at Martinez During Summer 1995

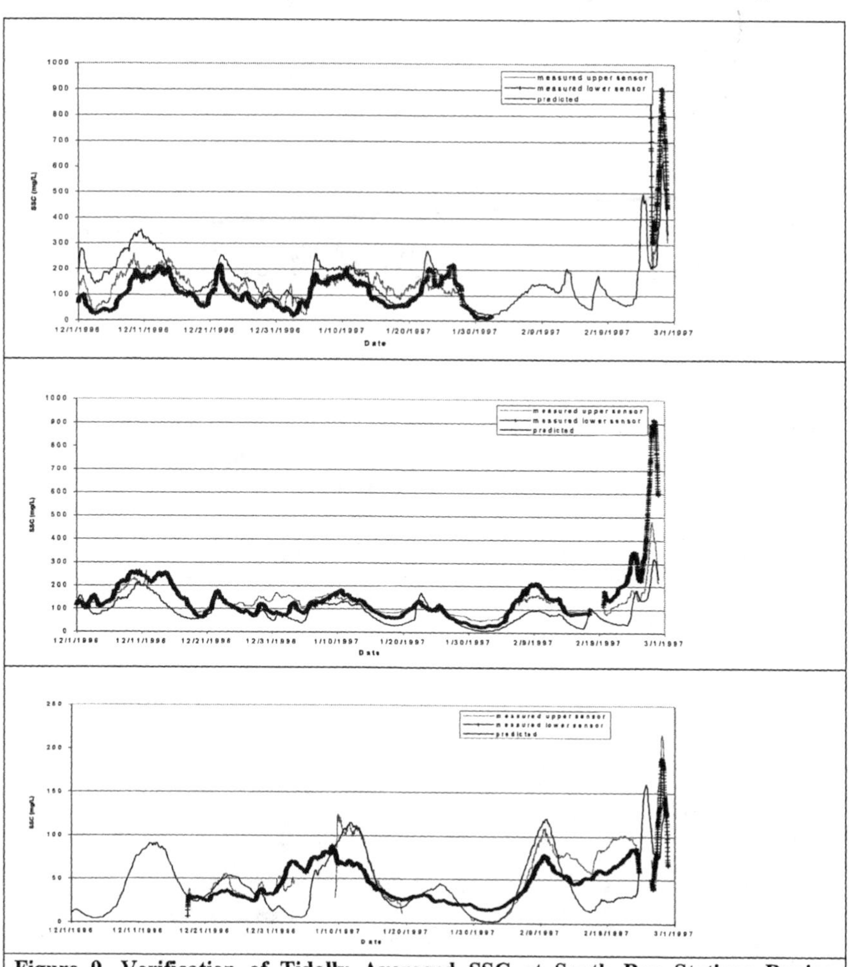

Figure 9. Verification of Tidally Averaged SSC at South Bay Stations During Winter 1997: Channel Marker 17, Dumbarton Bridge and San Mateo Bridge (Top to Bottom)

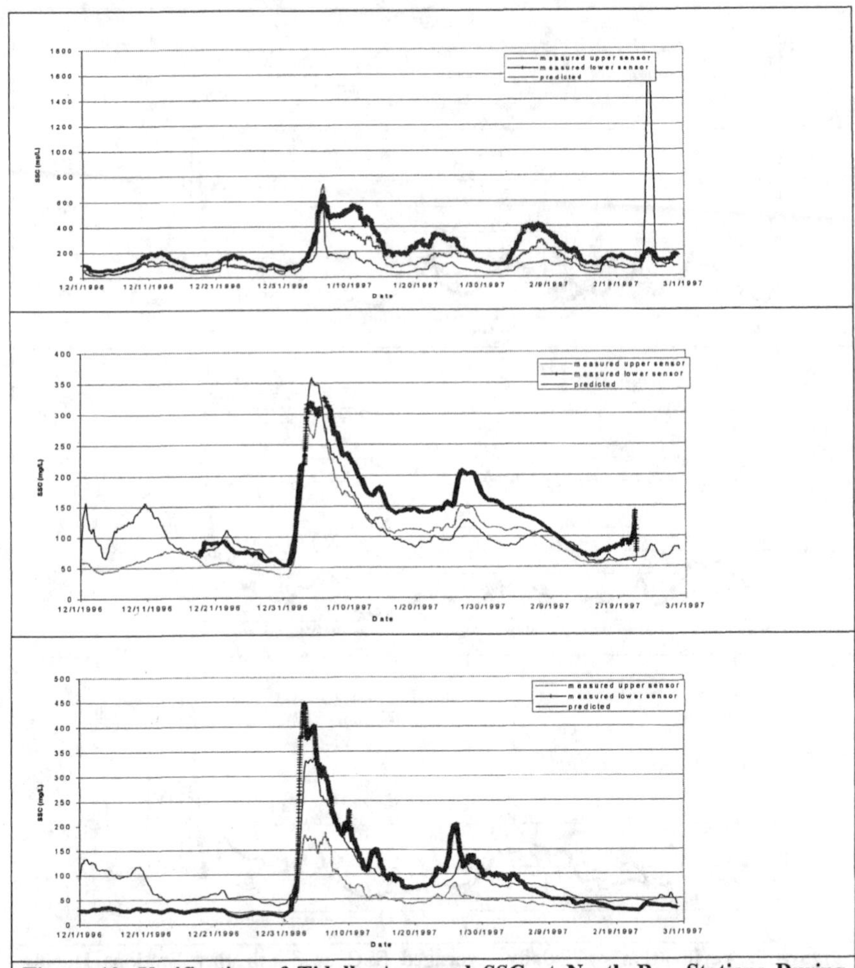

Figure 10. Verification of Tidally Averaged SSC at North Bay Stations During Winter 1997: Point San Pablo, Benicia and Mallard Island (Top to Bottom)

REFERENCES

Amos, C.L., Brylinsky, M., Sutherland, T.F., O'Brien, D., Lee, S and Cramp, A., The stability of a mudflat in the Humber Estuary, South Yorkshire, U.K., Geological Society, Special publication, 139, pp.25-43, 1998

Buchanan, P.A. and Schoellhamer, D.H., Summary of suspended-solids concentration data, San Francisco Bay, California, Water Year 1995, U.S.G.S. Water Resources Division, Sacramento, CA, 1995

Conomos, T.J. and Peterson, D.H., Suspended-particulate transport and circulation in San Francisco Bay, an overview, Estuarine Processes, Academic, San Francisco, Calif., v2, pp.82-97, 1977

Hauck, L.M., Teeter, A.M., Pankow, W. and Evans, R.A. Jr., San Francisco Central Bay Suspended Sediment Movement, Report 1: Summer condition data collection program and numerical model verification, U.S. Army Corps of Engineers, Tech. Rep. HL-90-6, 81 pp., 1990

Inagaki, S., Effects of a proposed San Francisco Airport Runway Extension on Hydrodynamics and Sediment Transport in South San Francisco Bay, thesis, Stanford University, 2000

Krone, R.B., Flume studies of transport of sediment in estuarial shoaling processes, Final Report, Hydraulics Engg. Res. Lab., U. of Cal., Berkeley, Calif., 1962

Krone, R.B., Sediment Inflows to the San Francisco Bay System, prepared for Ogden Beeman & Associates, Inc., 1991

McDonald, E.M. and Cheng, R.T., Issues related to modeling the transport of suspended sediments in northern San Francisco Bay, California, Proceedings of the third Inter. Conf. On Estuarine and Coastal Modeling, ASCE, Chicago, IL., pp.551-564, 1994

McDonald, E.M. and Cheng, R.T., A numerical model of sediment transport applied to San Francisco Bay, California, J. Marine Env. Engg., v4, pp.1-41, 1997

Mitchener, H.J., Properties of Dredged Material, A review of available measurement techniques for determining physical properties, HR Wallingford Report TR 21, 1996

Nichols, F.H. and Thompson, J.K., Time scales of change in the San Francisco Bay benthos, Hydrobiologia, 129, pp.121-138, 1985

Ogden Beeman & Associates, Inc. and Krone, R.B., Sediment Budget Study for San Francisco Bay, Final Report, 1992

Parchure, T.M. and Mehta, A.J., Erosion of soft cohesive sediment deposits, J. of Hydraul. Engg, v111, n10, pp.1308-1326, 1985

Rubin, D.M. and McCulloch, D.S., The movement and equilibrium of bedforms in Central San Francisco Bay, San Francisco Bay: The Urbanized Estuary, Pac. Div. Am. Assoc. for the Adv. of Sci., San Francisco, Calif., pp.97-113, 1979

Schoellhamer, D.H., Factors affecting suspended-solids concentrations in South San Francisco Bay, California, J. of Geophysical Research, v101, no. C5, pp.87-95, 1996

Sternberg, R.W., Cacchione, D.A., Drake, D.E. and Kranck, K., Suspended sediment transport in estuarine tidal channel within San Francisco Bay, California, Mar. Geol., 71, pp.237-258, 1986

Teeter, A.M., Alcatraz disposal site investigation, San Francisco Bay; Alcatraz disposal site erodibility, Report 3, Misc. Paper HL-86-1, U.S. Army Corps of Engineer Waterways Experiment Station, Vicksburg, MS., 1987

Thompson, J.K., Nichols, F.H. and Wienke, S.M., Distribution of benthic chlorophyll in San Francisco Bay, California, February 1980 – February 1981, U.S. Geol. Surv. Open File Rep. 81-1134, 1981

Thorn, M.F.C. and Parsons, J.G., Erosion of cohesive sediments in estuaries: An engineering guide, 3rd Int'l Symp. On Dredging Tech., Bordeaux, pp.349-358, BHRA, Cranfield, 1980

van Rijn, L.C., Handbook on Sediment Transport by Current and Waves, Delft Hydraulics, Report H461, pp.12.1-12.27, 1989

Whitehouse, R.J.S., Soulsby, R.L., Roberts, W. and Mitchener, H.J., Dynamics of Estuarine Muds: A Manual for Practical Applications, HR Wallingford, Report SR 527, 1999

Modeling a Three-dimensional River Plume over Continental Shelf Using a 3D Unstructured Grid Model

Ralph T. Cheng
U. S. Geological Survey
Menlo Park, California

Vincenzo Casulli
Department of Civil and Environmental Engineering
University of Trento
Trento, Italy

Abstract

River derived fresh water discharging into an adjacent continental shelf forms a trapped river plume that propagates in a narrow region along the coast. These river plumes are real and they have been observed in the field. Many previous investigations have reported some aspects of the river plume properties, which are sensitive to stratification, Coriolis acceleration, winds (upwelling or downwelling), coastal currents, and river discharge. Numerical modeling of the dynamics of river plumes is very challenging, because the complete problem involves a wide range of vertical and horizontal scales. Proper simulations of river plume dynamics cannot be achieved without a realistic representation of the flow and salinity structure near the river mouth that controls the initial formation and propagation of the plume in the coastal ocean. In this study, an unstructured grid model was used for simulations of river plume dynamics allowing fine grid resolution in the river and in regions near the coast with a coarse grid in the far field of the river plume in the coastal ocean. In the vertical, fine fixed levels were used near the free surface, and coarse vertical levels were used over the continental shelf. The simulations have demonstrated the uniquely important role played by Coriolis acceleration. Without Coriolis acceleration, no trapped river plume can be formed no matter how favorable the ambient conditions might be. The simulation results show properties of the river plume and the characteristics of flow and salinity within the estuary; they are completely consistent with the physics of estuaries and coastal oceans.

I. Introduction

River derived fresh water discharging into an adjacent continental shelf forms a trapped river plume that propagates in a narrow region along the coast. These river plumes are real and they have been observed in the field (Hicky et al., 1998; Berdeal et al, 2002; Fong and Geyer, 2002). The trapped river plumes transport fresh water, nutrients, and pollutants over a large coastal region that could affect the regional ecosystem. A trapped river plume can propagate for hundreds of kilometers. The importance of the river plume dynamics prompted numerous past and present investigations. Because the region of interest is very large, direct measurements of water properties over such a large domain of interest are quite difficult. Although satellite images can provide evidence of the presence of river plumes, their properties are somewhat qualitative. Therefore numerical modeling has become a powerful tool in the investigations of trapped river plume dynamics.

Using a relatively coarse grid, Chao and Boicourt (1986) modeled the coastal ocean assuming a constant depth basin of limited extent. Neither tidal motions nor ambient currents were used. At the river mouth, estuarine circulation was artificially imposed in the form of a boundary condition. Although the numerical model was crude by today's standard, their model reproduced essential features of a trapped river plume. In a follow up paper, Chao (1988) added the effects of a sloping bottom in the ocean. Hyatt and Signell (2000) used a Princeton Ocean Model (POM) (Blumberg and Mellor, 1987) over a variable grid mesh to represent the coastal ocean with a sloping bottom. The river was modeled as a simple fresh water source discharging into the coastal ocean. Neither tides nor wind forcing was used. In order to reproduce the essential features of the river plume, an ambient current of 5 cm/sec was imposed in the direction of Kelvin wave propagation; i.e. anticyclonically in the northern hemisphere. The convective transport of fresh water would mimic that of a trapped river plume. It is not clear whether a trapped river plume would be formed if an ambient current were not introduced. However, Hyatt and Signell (2000) emphasized that the objective of their study was on the investigation of a few numerical convective schemes used in the numerical model.

Garvine (1999, 2001) has studied trapped river plumes since the mid-1980; main results of Garvine's findings are summarized in these two papers. Through dimensional analysis, Garvine (1999) pointed out that the properties of a trapped river plume are functions of river width, depth, discharge, exiting velocity, degree of stratification (mixing in the vertical water column), intensity of Coriolis acceleration (latitude of the river), bottom slope of the coastal ocean, winds, and upwelling and downwelling of coastal currents. Although Garvine's studies were comprehensive and extensive, there are some weaknesses in the numerical model that produced these conclusions. In Garvine's model (1999), the river was represented as a fresh water source without tidal motion. The flow and salinity structures near the river mouth play a very important role in the eventual formation and propagation of the fresh water plume along the coast. Proper simulations of river plume dynamics cannot be achieved without a realistic representation of the flow and salinity structure near the river mouth. When the tides are neglected, the simulations are equivalent to reproducing residual flow and salinity transported by residual currents. While this approach might be justifiable, a proper turbulence closure should be used to represent the vertical mixing in a residual time scale. Therefore, if the dynamics in tidal time-scale is not considered, the adaptation of Melloer-Yamada closure in residual calculations is questionable.

Garvine (1999, 2000) also used POM (Blumberg and Mellor, 1987) in his investigations. The turbulence closure scheme was based on the Mellor-Yamada 2.5 level model (1982). In most of the simulations, tidal motions were neglected, and yet the kinetic energy based Mellor-Yamada 2.5 turbulence closure was used. Therefore the turbulent mixing in the vertical was not properly represented in these simulations. Since there was no direct measurement of river plume properties that can be used for comparison with model simulations, it is hard to judge the correctness of the model results. In real world situations, tidal motions are always present and important. It is important to note that the Mellor-Yamada turbulence closure is derived from the balance of turbulent kinetic energy (TKE), production and dissipation of TKE, and the transport of a local mixing length. When the tides are neglected, the simulations represent temporal variations of residual currents not the kinetic energy associated with tidal currents; thus the use of the Mellor-Yamada closure in the 'residual' model might be, at least, questionable. In the end, the results reported by Garvine (1999, 2001) are likely to be qualitatively correct; a quantitative capturing of the plume properties is not yet available.

Other investigations also used numerical models to examine the driving mechanisms and the balance of momentum of a river plume (Zhang et al, 1987; Garvine, 1987; Chao, 1988; O'Connor, 1990; Oey and Mellor, 1993; Kourafalou et al, 1996; Yankovsky and Chapman, 1997; Berdeal et al, 2002, and Fong and Geyer, 2002). Nearly all the models assumed the river inflow as a freshwater source and neglected the astronomical tides. Most models also assumed a constant depth ocean. They investigated the resulting transport of a low salinity plume driven by upwelling or downwelling due to the superimposing wind forcing. Quite often an ambient coastal current had to be imposed in order to reproduce the transport of freshwater in a river-derived plume. Numerical modeling of the dynamics of river plumes is very challenging because the complete problem involves a wide range of vertical and horizontal scales. A typical river is about 10 - 20 m deep, a few kilometers wide, and up to a hundred of kilometers long, while over the continental shelf, the water is a few hundred meters deep and the region of interest is a few hundred kilometers in length. Many authors represent the plume problem by a simple source function and justify such a representation by stating that "The establishment of a realistic estuarine flow which requires a much larger channel is beyond the scope of the present study." for example, Kourafalou et al (1996).

In this study, an advanced 3D numerical model known as UnTRIM (Casulli and Walters, 2000; Casulli and Zanolli, 2002; Cheng and Casulli, 2002) is used in simulations of a river plume. The numerical model uses an unstructured grid in the computations that allow the use of a very high-resolution model grid to study the mechanisms that are responsible for the formation of a river plume over the continental shelf, and the use of coarse grids for far field to allow extensive coverage of coastal-ocean. In the vertical, fine fixed levels are used in surface layers in the top 20 m, and coarse vertical levels are for the remaining domain covering the continental shelf. The main objective of this study is to show the correct physics with which the formation and propagation of a trapped freshwater plume in the coastal ocean. In addition, the properties of the river plume are related to the circulation structure in the river-estuary-ocean system, and the circulation near the mouth of estuary (river).

II. The UnTRIM Numerical Model

One of the common characteristics of the TRIM family of models is that the numerical solutions are computed in physical space without any coordinate transformation in the x-, y-, or z-directions. Semi-implicit finite-difference schemes are used to control the propagation of surface gravity waves that could cause numerical instability (Casulli, 1990; Casulli and Cheng, 1992; Casulli and Cattani, 1994). The latest addition to the TRIM family of models uses an orthogonal unstructured grid while a semi-implicit finite-difference method is still used for solving the 3D shallow water equations (Casulli and Walters, 2000; Casulli and Zanolli, 1998, 2002). The combined use of a semi-implicit algorithm for stability and an unstructured grid for flexibility in solving the shallow water equations constitutes the essence of the UnTRIM model. The governing equations include the 3D conservation equations of mass, momentum, a transport equation for each scalar variable, an equation of state, and a kinematic free-surface equation. The estuarine and coastal system is assumed to be sufficiently large so that a constant Coriolis acceleration term is included in the momentum equations. Furthermore, the water is assumed to be incompressible; the pressure is assumed to be hydrostatic; and the Boussinesq approximation applies. Cheng and Casulli (2002) has reported an evaluation of the UnTRIM model for applications to estuarine and coastal circulation; the details of the governing equations and solution algorithm are referred to Casulli and Walters(2000), Casulli and Zanolli, (2002) and Cheng and Casulli (2002).

III. Turbulence closure

For three-dimensional barotropic flows (constant density), the solute transport equation is uncoupled from the momentum equations. For baroclinic flows, the transport equations are coupled with the momentum equations through the density gradient terms and the choice of a turbulence closure sub-model. All transport variables including salinity, temperature, and turbulence properties among others are solved lagged one time-step. Thus, the baroclinic forcing terms (density gradients) in the momentum equations are effectively solved explicitly, resulting in a numerical scheme that is subjected to a weak Courant-Friedrich-Lewy (CFL) stability condition based on the propagation of internal waves. Turbulence transport equations are solved in a similar manner; the resulting turbulence closure properties do not affect the numerical stability of the solution scheme as long as the effective vertical eddy viscosity remains non-negative. A non-negative bottom friction coefficient is typically specified by the Manning-Chezy formula, or by fitting to a bottom turbulent boundary layer.

Vertical mixing mechanism and the resulting stratification near the mouth of an estuary is an important aspect of the river plume dynamics; the correct choice of a turbulence closure sub-model is necessary in order to properly reproduce the freshwater stratification near the mouth of the estuary and the gravitational circulation within the estuary/river. A stratified ambient condition just out side of the river mouth creates a favorable condition for the formation of a freshwater river plume that is forced by Coriolis acceleration, resulting in trapping of a thin freshwater layer along the coast in the direction of Kelvin wave propagation. Because the horizontal length scale in the mixing process is several orders of magnitude larger than the length scale of vertical mixing, the horizontal eddy viscosity and diffusivity play a less important role. Several different approaches have been reported for treating vertical eddy viscosity, N_v, and vertical diffusivity, K_v, with varying degrees of success. The k-ε type of closure is commonly used in the European modeling community (Rodi, 1984; Uittenbogaard et al., 1992). Direct field verifications of the k-ε closure in estuaries are rare.

Two types of turbulence closure sub-models have been implemented for the present model application. For a type I closure model, the vertical eddy and diffusivity coefficients are written as neutral (unstratified) mixing coefficients modified by a damping function due to stratification. Thus,

$$N_v = N_o\,\psi_N(Ri) \quad \text{and} \quad K_v = K_o\,\psi_K(Ri) \tag{1}$$

where N_o and K_o are the neutral values of eddy viscosity and diffusivity coefficients; Ri is a local gradient Richardson number,

$$Ri = -\frac{g}{\rho}\frac{\frac{\partial \rho}{\partial z}}{\left[(\frac{\partial u}{\partial z})^2 + (\frac{\partial v}{\partial z})^2\right]} \tag{2}$$

and ψ_N (Ri) and ψ_K(Ri) are damping functions. In Eq.(2), ρ is density, g is gravitational acceleration, $\frac{\partial \rho}{\partial z}$ is the vertical density gradient, and $\left[(\frac{\partial u}{\partial z})^2 + (\frac{\partial v}{\partial z})^2\right]^{1/2}$ is the velocity gradient in vertical direction. Based on theoretical arguments or empirical data, the damping functions are given in a variety of forms usually dependent on the local gradient Richardson number. Table I gives some commonly used type-I turbulence closure models including the constant value (CV) model, the classic Munk-Anderson (MA) formulation (1948) of the damping function, and the turbulence models proposed by Lehfeldt and Bloss (LB) (1988), and Pacanowski and Philander (PP) (1981). The turbulence models LB and PP have been used in a study of periodic stratifications in San Francisco Bay (Cheng and Casulli, 1996).

Table I. Some Turbulence Closure Models
($N_v = N_o \psi_N$ and $K_v = K_o \psi_K$)

Mixing Model	N_o (m^2/sec)	K_o (m^2/sec)	ψ_N	ψ_K	Remarks
Constant (CV)	0.0001 ~0.001	0.0001 ~0.001	1.0	1.0	No Ri dependence
Munk & Anderson (MA)	N_o	K_o	$(1+10Ri)^{-1/2}$	$(1+3.33Ri)^{-3/2}$	
Lehfeldt & Bloss (LB)	$l^2\left\|\frac{\partial u}{\partial z}\right\|$	$l^2\left\|\frac{\partial u}{\partial z}\right\|$	$(1+3Ri)^{-1}$	$(1+3Ri)^{-3}$	$l = \kappa z(1-z/H)^{1/2}$
Pacanowski &Philander (PP)	0.01	K_o	$(1+3Ri)^{-2}$	$(1+3Ri)^{-1}$	

The neutral values of eddy viscosity and diffusivity are taken from a mixing length turbulence model from Rodi (1984),

$$N_o = K_o = (C_\mu'^{\,3}/C_d)^{1/2}\, l^2\, |dU/dz| \quad \text{with } l = \kappa z(1-z/H)^{1/2} \qquad (3)$$

where $C_\mu' = 0.58$, $C_d = 0.1925$ and l is the mixing length, κ is von Karman constant, z and H are the distance from bed and the total water depth, respectively.

Type-II turbulence closure models consider the transport of mixing length, turbulent kinetic energy, and dissipation. The k-ε model and the model proposed by Mellor-Yamada (1982) fall in this category. When a higher level of physical processes is considered, the turbulent eddy viscosity and diffusivity must be solved along with the flow variables, thus incurring higher computational expense. As reported in the literature, all models have reproduced certain flow conditions with some degree of success. In this study, a Mellor-Yamada 2.5 level turbulence closure scheme has been implemented in the UnTRIM model. The turbulent kinetic energy ($q^2/2$) and the quantity ($q^2 l$) are solved from transport equations for q and l. The vertical eddy viscosity and diffusivity are then defined as functions of (ql), and stability functions are derived by Mellor-Yamada (1982). Details of the Mellor-Yamada 2.5 level turbulence closure sub-model and its stability functions are well documented, they are not repeated here.

IV. Model configuration of a river and coastal ocean system

Since the main objective of this study is to show the interactions of river/estuarine flows with the coastal ocean, both the coastal ocean and the connecting river/estuary must be well represented. An estuary (river) with a constant depth of 20 m, 80 km long and 4 km wide is assumed. The coastal ocean is 140 km x 440 km with the river located at 120 km north from the southern boundary of the ocean, Figure 1. The coastal ocean has a constant slope; the water depth varies from 20 m at shore to 200 m at the western boundary of the study region. An M_2 tide of 1 m amplitude is specified at the western ocean open boundary where the salinity is 30 ppt. At the northern and southern open boundaries of the ocean, a zero normal gradient for all variables are assumed. At shorelines, the normal velocity is zero, and at the head of the river an open boundary is used, where a constant stage is assigned.

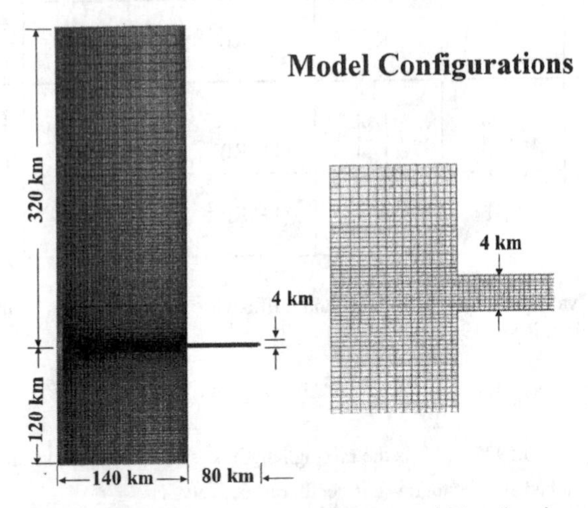

Figure 1. A river/estuary of 4 km wide, 80 km long and 20 m deep is connected to a coastal ocean of 140 km x 440 km. Depth in the ocean varies from 20 m at shore to 200 m at 140 km from shore.

Furthermore, at all open boundaries, a radiation boundary condition

$$\frac{\partial \eta}{\partial t} + \frac{\partial}{\partial x}\left[\int_{-h}^{\eta} u dz\right] + \frac{\partial}{\partial y}\left[\int_{-h}^{\eta} v dz\right] = (\eta^* - \eta)/\tau_{rex} \quad (4)$$

is applied, where η^* is the specified water level, and the actual water level η is determined by Eq.(4). In Eq.(4), (u,v) are velocity components in (x,y) directions; h is water depth, and τ_{rex} is the specified relaxation time. At the head of the river, salinity is assumed to be 0 ppt. For a given water level at the head of the river, the net river discharge is not known a priori. The tide propagating from the open ocean into the estuary is allowed to pass through the head of the river (radiation boundary condition); the entire river reach is influenced by tidal fluctuations. Due to tides, the volumetric flux of water in a river cross-section is a function of time. The net river discharge is the tidally averaged volumetric flux, which is determined during the simulation.

The UnTRIM model performs computations over an unstructured orthogonal grid-mesh. A combination of three-sided or four-sided polygons can be used. To assure orthogonality, a variable sized rectangular computational mesh shown in Figure 1 is used. The river-ocean interactions are expected to be most intense near the river mouth, therefore a very fine mesh is used in this region. A fine mesh is also used along the coast of the ocean. A coarse grid-mesh is used for regions away from the river mouth to optimize computing time because the river-ocean interactions are expected to decrease with distance from the river mouth. The computational grid has 125 node-points in the east-west direction and 200 points in the north-south direction. Excluding dry points, there is a total of 13,471 nodes on the horizontal plane. Fixed z-levels are used to represent the model domain in the vertical. In order to capture the thin fresh water lens, the vertical levels are defined in finer intervals near the free surface, and distributed more sparingly in deep water so that the z-levels are located at z = 0.5, 1.0, 2.0, 4.0. 6.0, 8.0, 10.0, 12.0, 14.0, 16.0, 18.0, 20.0, 23.0, 27.0 m,....etc. A total of 30 layers is used to cover the ocean up to 200 m. Within the top 20 m, 12 layers are used to resolve the vertical structure of the dependent variables. The overall grid-mesh includes 262,846 3D computational cells.

The velocities are assumed to be zero initially, and the salinity varies from 30 ppt at the western boundary of the ocean to 0 ppt at the head of the river. Initially the salinity distribution in space is obtained by linear interpolation between these two extreme values. During simulations computed results of all dependent variables are saved at selected intervals (every 2 hours). Time-series of the dependent variables are saved at pre-selected stations. At each station, time-series of the 3D velocity and salinity structures, depth-averaged velocity (speed and direction), salinity, and the water level are recorded for further analysis.

V. Results

A. Choice of turbulence closure sub-model and net river discharge

The entire coastal ocean and estuary system is driven by the tides at ocean boundaries and by river outflow. Figure 2 shows a typical example of the time-series of depth-averaged salinity, sea-level, depth-averaged speed and direction at two stations 30 km (dotted line) and 50 km (solid line) from the head of the river. The tidal current speed varies from ~75 cm/sec (flood) to ~ -100 cm/s (ebb), and the tidal current direction is nearly bi-directional. The salinity fluctuates with the tide, with the mean salinity value decreasing as the station is located closer to the head of the river. Most previous numerical studies of trapped river plumes neglected the tides; the river discharge was represented by a freshwater source. The circulation in the modeled estuary was the residual current with which the salinity was transported. This formulation may give rise to a qualitatively correct description of the plume if a correct turbulence sub-model representing the mixing processes at the residual time scales was used. Clearly the kinetic energy distribution and its temporal variations, along with the spatial distribution of velocity shear, dissipation and production of turbulence in a residual time-scale would differ by an order of magnitude when examined in a tidal time-scale. As previously noted, the Mellor-Yamada 2.5 level turbulence closure is based on the transport of turbulent kinetic energy ($q^2/2$) and a length scale quantity ($q^2 l$), this sub-model should <u>only</u> be used in simulations of tidal time-scale processes. When the Mellor-Yamada 2.5 level sub-model is used in simulations of residual processes, the correct physics are NOT represented in the computations, and the model results are suspect.

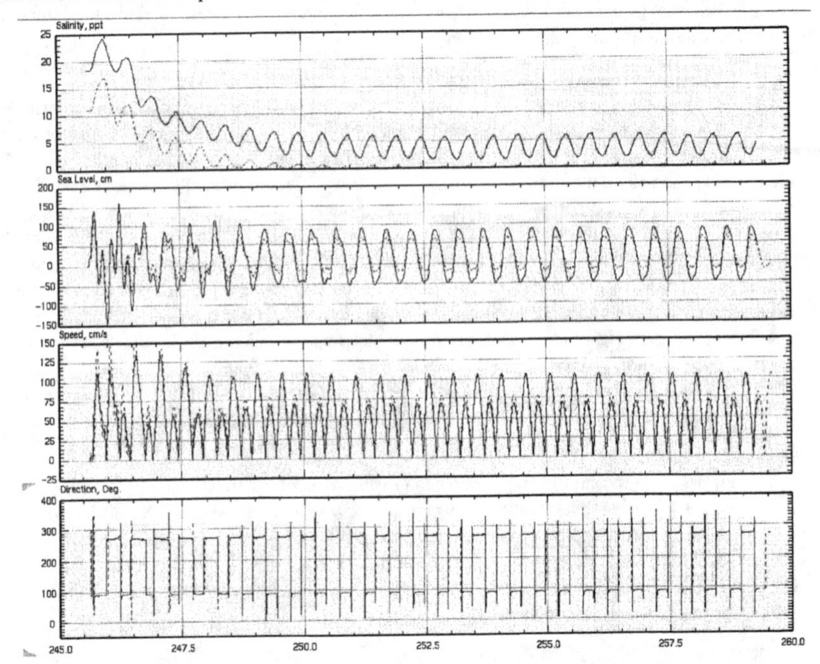

Figure 2. Time-series of depth averaged salinity, sea-level, speed and direction at two stations 30 km (dotted line) and 50 km (solid line) from the head of the river.

Figure 3 shows the instantaneous volumetric flux at a cross section 20 km from the head of the river. Positive values indicate upstream pointing flux. The tidal fluctuations give volumetric fluxes on the order of 5.0×10^4 m^3/s, while the low pass-filtered net flux converges to 7.2×10^3 m^3/s. It is reasonable to expect the tidal volumetric flux to be an order of magnitude larger than the net flux (residual). For each simulation, a stage level at the head of the river is specified; the net river discharge is determined in the course of simulation.

B. Coriolis forcing

Coriolis acceleration is the main driving force responsible for the formation of trapped river plumes; all other factors such as tides, river discharge, and turbulence mixing acting together in the ocean-estuary system are responsible for creating a condition conducive to the formation of a trapped plume. However, without Coriolis forcing, the mechanism for the formation of river plume does not exist. To demonstrate this assertion, several simulations were carried out using a range of river

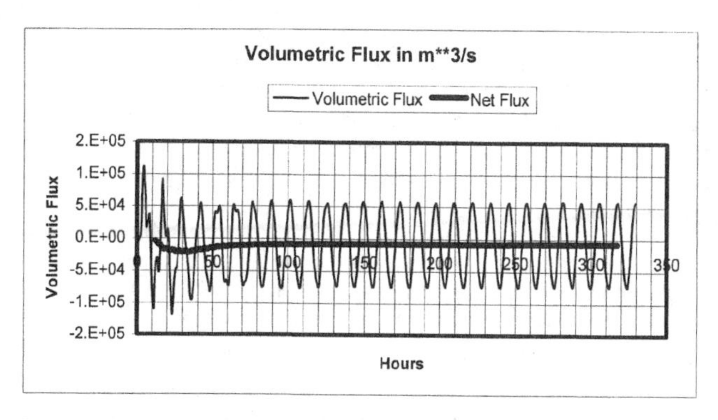

Figure 3 Instantaneous volumetric flux passing a cross-section of the river is on the order of 5.0×10^4 m^3/s, while the low pass-filtered net flux converges to 7.2×10^3 m^3/s, an order of magnitude smaller.

values and without Coriolis forcing. A symmetrical freshwater plume is seen in the region adjacent to the mouth of the river. In a few days, the mean-size of the plume appears to have reached a steady state although small fluctuations can be observed as the flooding and ebbing processes progress in time. The size of the freshwater zone is a function of river discharge, but under none of these scenarios was a trapped river plume along the coast found. Shown in Figure 4 is the symmetric freshwater zone along with the depiction of the velocity field for a medium river discharge value.

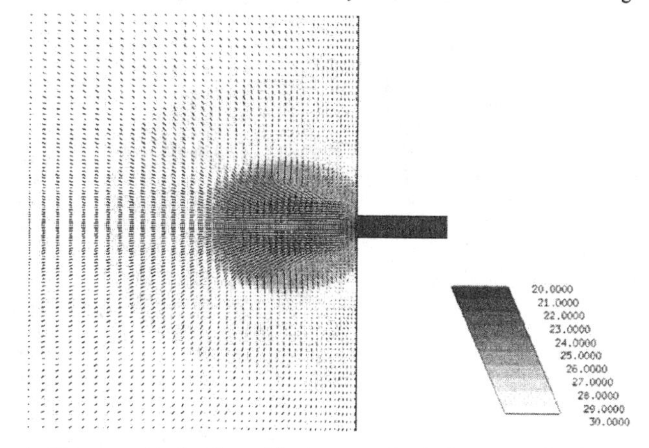

Figure 4. The symmetrical pattern of freshwater region is shown near the river mouth, where the black and white shade scheme displays only salinity between 20 and 30 ppt with lighter shade representing higher salinity. The velocity distribution is also symmetric with respect to the axis of the estuary.

It is instructive to examine the cycle of tidal circulation near the river mouth. The velocity field just outside of the river mouth assumes a "jet-like" flow distribution during an ebbing cycle, with high momentum flow exiting along the direction of the river axis resembling a jet flow. The intensity of momentum decreases away from the centerline of the main jet and away from the river mouth. During flood, the velocity field behaves as a "sink" flow; the momentum is roughly evenly distributed as flow entering into the estuary from all directions. Because of the net river discharge as well as vertical stratification, the exiting momentum in the surface layer (ebbing) is substantially higher than the momentum of the returning flow (flooding). Both the ebb dominance of flows and the distinctly different patterns of momentum distribution between flooding and ebbing cycles have direct bearings on the formation of trapped river plumes. In the northern hemisphere, Coriolis acceleration forces the flow to bend clockwise; the magnitude of the acceleration is proportional to the magnitude of the velocity itself $\left(\vec{\Omega}\times\vec{V}\right)$, Figure 5. During ebb, the high momentum "jet-like" flow is pushed clockwise with a maximum acceleration. During flood, the Coriolis acceleration acts more evenly because the momentum distribution is more even (sink flow). When the momentum is weak, the Coriolis acceleration still plays a role, however, the net effect of Coriolis acceleration is proportionally weak.

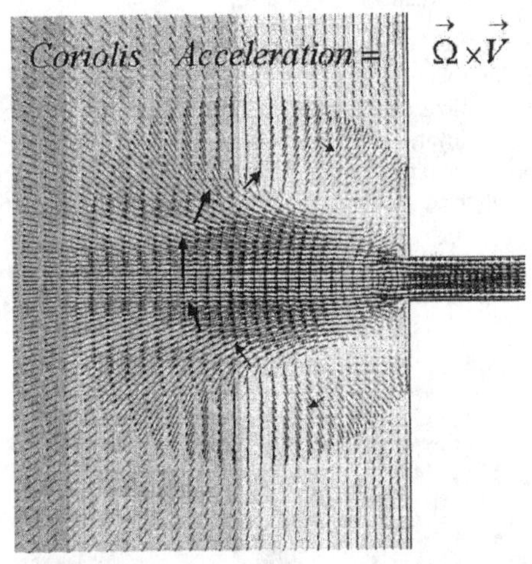

Figure 5. Schematic diagram illustrates the properties of Coriolis acceleration (arrows normal to velocities) in an ebbing cycle with flow exiting from the mouth of the estuary.

Figures 6a-6f show the surface velocity distribution at the mouth of the estuary for a tidal cycle (12 hours). Over a tidal cycle, the dominating exiting flow is pushed first to the north and then landward against the coast north of the river mouth, forming a trapped river plume. The residual circulation south of the river mouth spreads freshwater at a much slower rate, Figure 7. After 14 days into the simulation, the surface freshwater plume is shown propagating to the north trapped in a coastal zone by Coriolis acceleration. Some freshwater is transported south starting from a mixing zone near the south-side of the estuary. The southward transporting freshwater is NOT trapped to the coast by Coriolis acceleration, and is spreading over a broader region.

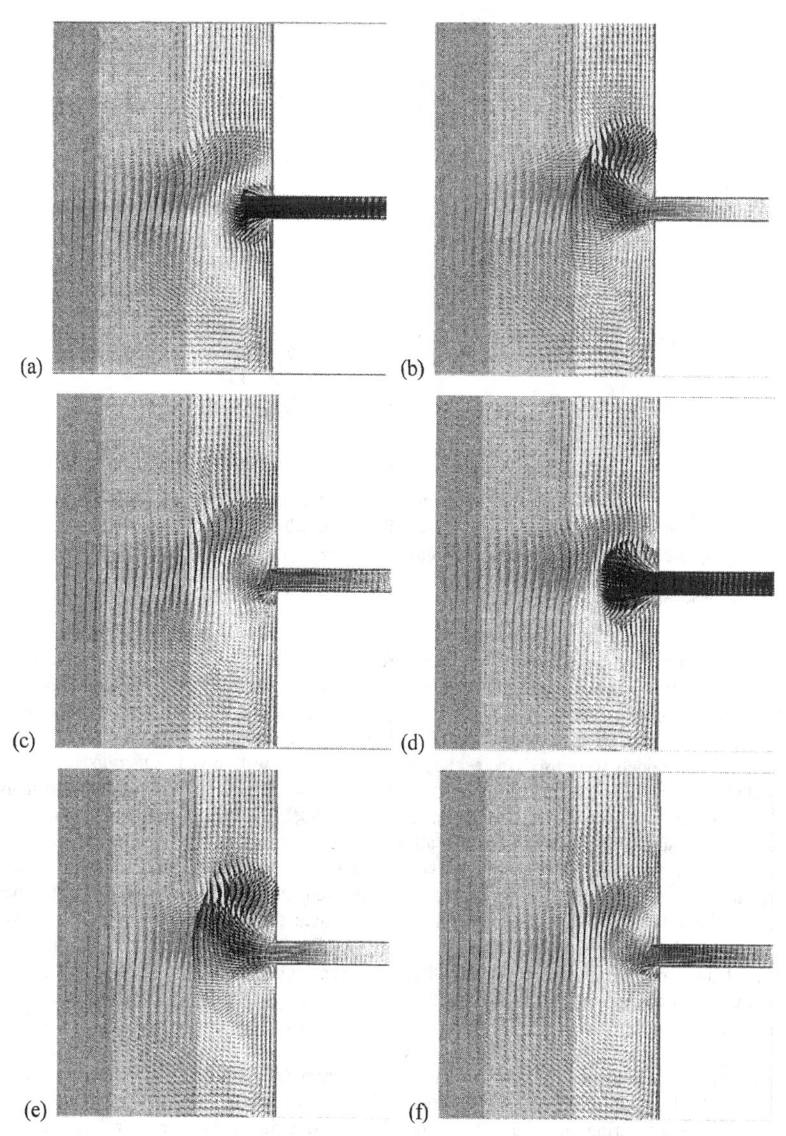

Figure 6. Surface velocity distribution near the mouth of an estuary over a 12-hour cycle. Each snapshot of the velocity distribution is taken at 2-hour interval.

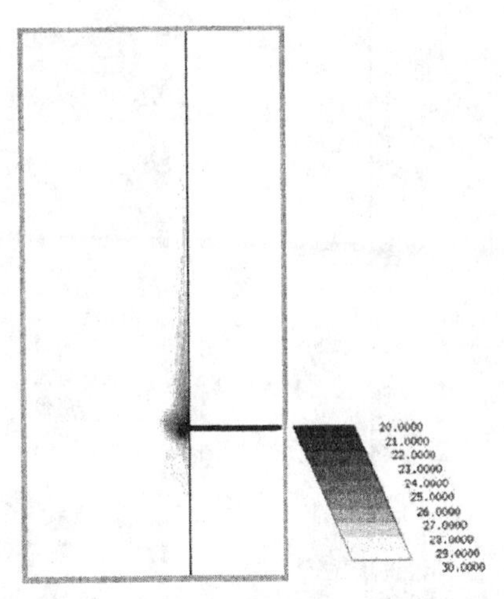

Figure 7. Trapped surface river plume at medium river discharge (7.2 x 10^3 m/s) 14 days into the simulation. The shaded black and white scheme for surface salinity ranges between 20 and 30 ppt with lighter shade representing higher salinity.

C. River Discharge

Although the principal mechanism for the formation of a trapped river plume is Coriolis forcing, other parameters determine the gross properties of the trapped plume. Turbulence mixing controls the development of vertical stratification. Highly stratified flow near the river mouth concentrates higher momentum in the surface layer; such a condition is more conducive to the development of trapped river plumes. Of the few turbulence sub-models discussed in Section III, if type-I closure models were used, the estuary would tend to be well mixed. Of course, this situation could also be adjusted by changing the coefficients in the closure models. While there is no analytical solution for this problem to guide calibration of the model, the type-II Mellor-Yamada 2.5 level turbulence closure sub-model was used for all remaining simulations. This closure was used because the sub-model includes more complete consideration of the mixing mechanisms. Numerical simulations were conducted for three river discharges ranging from low (1.6 x 10^3 m^3/s), medium (7.2 x 10^3 m^3/s) to high (10.6 x 10^3 m^3/s) flows. In all cases, the gross properties of the trapped river plume show the same pattern but differ quantitatively. The simulated properties of the river plumes after 14 days are summarized in Figures 8 and 9. The higher the river discharge, the further the trapped freshwater plume propagates northward. The "bulged head" of the river plume is also proportional to river discharge. The transport of fresh water to the south of the river mouth shows similar properties except the freshwater spreads over a broader region. This region grows with increasing river discharge. Within the estuary, the river discharge interacts with tides to determine the location of the salt-water front in the estuary. In a tidal cycle, the location of 2 ppt salinity moves toward the mouth during the ebbing cycle, and moves toward the head of the river during flood. The tidal cycle mean location of the salt-water front (represented by the location of 2 ppt) is determined to be a function of river discharge, Figure 9. The salinity distribution in the vertical cross-section of the estuary-coastal-ocean system is shown in Figure 9 in which the salinity values are depicted in black and white shades between 0.0 and 30 ppt. The locations of 2 ppt varies from 58.5km, 35 km, and 25 km for the respective low (1.6 x 10^3 m^3/s), medium (7.2 x 10^3 m^3/s), and high (10.6 x 10^3 m^3/s) river discharges. Near the mouth of the estuary, an upwelling advection cell develops and intensifies with higher river discharge to counter-balance the exiting fresh water momentum flux from the river.

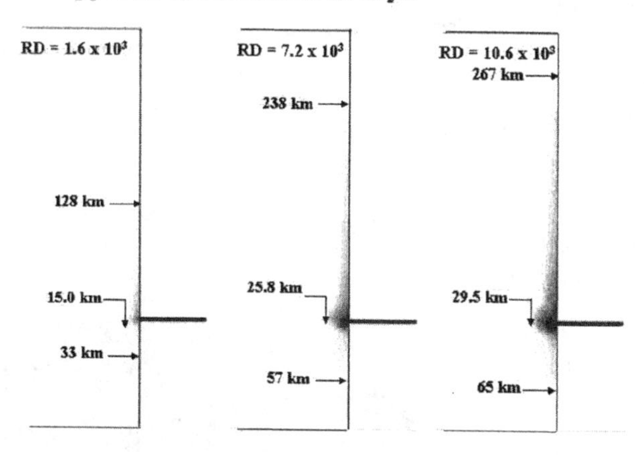

Figure 8. Simulated tidal current river discharge interactions in an estuary-coastal-ocean system. The surface river plume is trapped to the northern coast. The overall property of the river plume is a function of river discharge RD in m^3/s. The black and white shades show salinity values between 20.0 and 30.0 ppt with lighter shade representing higher salinity.

VI. Summary and Conclusion

The formation of trapped river plumes has been investigated by the UnTRIM numerical model. River derived fresh water discharging into an adjacent continental shelf forms a trapped river plume that propagates in a narrow region along the coast. Previous investigations of trapped river plume neglected tidal circulation or down played the importance of tidal circulation in the estuary. In spite of tidal currents being neglected, often the turbulent kinetic energy based closure model was used (such as the Mellor-Yamada 2.5 level turbulence closure). These weak assumptions prompted a complete investigation of the river plume dynamics by considering an estuary-coastal-ocean system driven by tides, river discharge, and stratification. The UnTRIM model allows the use of an orthogonal unstructured grid as the computational mesh, thus fine meshes are used in the entrance region of the estuary and in the region adjacent to the river mouth.

Model results from this study have clearly shown that the formation of the trapped river plume is due to Coriolis acceleration; all other factors would only create conditions conducive to the formation of a trapped river plume. It has shown further that no mater how favorable the ambient flow conditions might be, without the Coriolis acceleration, there will not be any river plume.

Without the presence of Coriolis acceleration the freshwater exits and partially returns back into the estuary in a tidal cycle forming a symmetric freshwater zone near the mouth of the estuary. In the presence of Coriolis acceleration, the high momentum jet exiting from the estuary is turned clockwise (northern hemisphere), and further confined by the coastal shore forming a long and narrow trapped river plume which propagates in the direction of Kelvin waves. The properties of trapped river plume are related to the magnitude of river discharge and the degree of stratification. Within the estuary, typical flooding and ebbing tidal currents evolve in response to incoming tides from the ocean. The net river discharge must be determined by applying a low-pass filter to the tidal volumetric flux fluctuations in a cross-section. The salinity front or the location of 2 ppt in the estuary is shown to be dependent upon river discharge. The complete properties of velocity and salinity distributions in the estuary and in the coastal ocean have been simulated with a fine resolution unstructured grid numerical model, UnTRIM. The model results are in agreement with general features of trapped river plumes reported in the literature. Direct comparisons of these model results with results reported in the literature are not made because the problem setup and formulation are different from those reported in the literature. In this study, the river plume dynamics is formulated on the basis of sound physical arguments. The numerical results are consistent with what would be expected of such a flow process from a physical point of view.

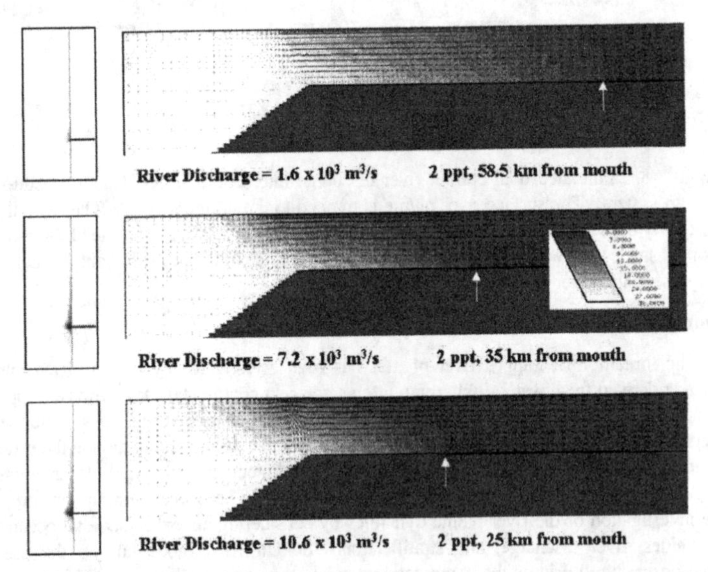

Figure 9. The salinity distributions in a vertical cross-section of the estuary-coastal ocean are shown with the horizontal to vertical scale ratio of 1: 500. The distance of 2 ppt salinity from the river mouth is shown to be inversely proportional to river discharge. The shaded black and white scheme shows salinity values between 0.0 and 30.0 ppt with lighter shade representing higher salinity value.

Acknowledgements: The authors appreciate the detailed editorial suggestions made by an anonymous reviewer.

VII. References

Berdeal, I. G., B. M. Hicky, and M. Kawase, 2002, Influence of wind shear and ambient flow on a high discharge river plume, J. of Geophysical Research, Vol. 107, No. C9, 3130, doi: 10.1029/2001JC000932, 2002.

Blumberg, A. F., and G. L. Mellor, 1987, A description of a three-dimensional coastal ocean circultion model, in Three-Dimensional Coastal Ocean Models, N. Heaps, Ed., Am. Geophys. Union, p. 1-16.

Casulli, V., 1990, Semi-implicit finite-difference methods for the two-dimensional shallow water equations, J. Comput. Phys., Vol. 86, p. 56-74.

Casulli, V. and R. T. Cheng, 1992, Semi-implicit finite difference methods for three-dimensional shallow water flow, Inter. J. for Numer. Methods in Fluids, Vol. 15, p. 629-648.

Casulli, V. and E. Cattani, 1994, Stability, accuracy and efficiency of a semi-implict method for three-dimensional shallow water flow, Comput. Math. Appl., Vol. 27, p. 99-112.

Casulli, V. and R.A. Walters, 2000, An unstructured grid, three-dimensional model based on the shallow water Equations. Inter. J. for Numer. Methods in Fluids, Vol. 32, p. 331-348.

Casulli, V. and P. Zanolli, 1998, A three-dimensional semi-implicit algorithm for environmental flows on unstructured grids, Proc. of Conf. On Num. Methods for Fluid Dynamics, University of Oxford.

Casulli, V. and P. Zanolli, 2002, Semi-implicit numerical modeling of non-hydrostatic free-surface flows for environmental problems, Mathematical and Computer Modeling, Vol. 36, p. 1131-1149.

Chao, S. Y. and W. C. Boicourt, 1986, Onset of estuarine plumes, J. of Physical Oceano., Vol. 16, p. 2137-2149.

Chao, S. Y., 1988, River-forced estuarine plume, J. of Physical Oceano., Vol. 18, p. 72-88.

Cheng, R. T. and V. Casulli, 1996, Modeling the periodic stratification and gravitational circulation in San Francisco Bay, in Proceedings of 4-th Inter. Conf.on Estuarine and Coastal Modeling, Spaulding and Cheng (Eds.), ASCE, San Diego, CA, October 1995, p.240-254.

Cheng, R. T. and V. Casulli, 2002, Evaluation of the UnTRIM model for 3-D tidal circulation, Proceedings of the 7-th International Conference on Estuarine and Coastal Modeling, St.Petersburg, FL, November 2001, p. 628-642.

Fong, D. A., and W. R. Geyer, 2002, The alongshore transport of freshwater in a surface-trapped river plume, J. of Physical Oceano., Vol. 32, p. 957-972.

Garvine, R. W., 1987, Estuary plumes and fronts in shelf waters: A layer model, J. of Physs. Oceanogr., Vol. 17, p. 1877-1896.

Garvine, R. W., 1999, Penetration of buoyant coastal discharge onto the continental shelf: A numerical model experiment, J. of Physical Oceano., Vol. 29, p. 1892-1909.

Garvine, R. W., 2001, The impact of model configuration in studies of buoyant coastal discharge, J. of Marine Research, Vol. 59, p. 193-225.

Hickey, B. M., L. J. Pietrafesa, D. A. Jay, and W. C. Boicourt, 1998, The Columbia River plume study: Substantial variability in the velocity and salinity field, J. Geophys. Res., Vol. 103, p. 10,339-10368.

Hyatt, J. and R. P. Signell, 2000, Modeling surface trapped river plumes: A sensitivity study, in Proceedings of the 6-th International Conference on Estuarine and Coastal Modeling, St.Petersburg, FL, November 3-5, 1999, p. 452-465.

Kourafalou, V. H., L. Y. Oey, J. D. Wang, and T. N. Lee, 1996, The fate of river discharge on the continental shelf: I. Modeling the river plume and the inner shelf coastal current, J. Geophys. Res., Vol. 101, No. C2, p. 3415-3434.

Lehfeldt, R. and S. Bloss, 1988, Algebraic turbulence model for stratified tidal flow, in Physical Processes in Estuaries, Edited by J. Dronkers and W. van Leussen, p. 278-291, Speringer-Velag.

Mellor, G. L., and T. Yamada, 1982, Formation of a turbulent closure model for geophysical fluid problems, Rev. Geophys. Space Phys., Vol. 20, p. 851-875.

Munk, W.H. and E. R. Anderson, 1948, Notes on a theory of the thermocline, J. Mar. Res., Vol. 7, 276-295.

O'Donnell, J., 1990, The formation and fate of a river plume: A numerical model, J. Phys. Oceanogr., Vol. 20, p. 551-569.

Oey, L. Y., and G. L. Mellor, 1993, Subtidal variability of estuarine outflow, plume and coastal current: A model study, J. Phys. Oceanogr., Vol. 23, p. 164-171.

Pacanowski, R. C., and S. G. H. Philander, 1981, Parameterization of vertical mixing in numerical models of tropical ocean, J. Phys. Oceanogr., Vol. 11, p. 16143-16160.

Rodi, W., 1984, Turbulence models and their applications in hydraulics - a state of the art review, 2nd edition, IAHR, The Netherlands, pp.104.

Uittenbogarrd, R. E., J. A. Th. M. van Kester, and G. S. Stelling, 1992, Implementation of the three turbulence models in TRISULA for rectangular horizontal grids, Delft Hydraulics Report Z162-22.

Yankovsky, A. E., and D. C. Chapman, 1997, A simple theory for the fate of buoyant coastal discharges, J. Phys. Oceanogr., Vol. 27, p. 1386-1401.

Zhang, Q. H., G. S. Janowitz, and L. J. Pietrafesa, 1987, The interaction of estuarine and shelf waters: A model and applications, J. of Physical Oceano., Vol. 17, p. 455-469.

Addressing Design Issues of an Orthogonal Unstructured Grid Using the Grid Generator JANET Demonstrated on the Delaware Estuary

T. Kutay Celebioglu[‡], Michael Piasecki[†]

Abstract

Concentrated efforts over the past 50 years have dramatically improved the water quality of the Delaware Bay Estuary. From conditions that saw part of the estuary being "dead", the system has improved to a level where a significant amount of fauna and flora has reinstated itself; in fact, many water quality parameters reach state and federal standards. Yet, the diversity and sheer volume of discharges that enter the estuary continue to pose a challenge for regulators and dischargers alike to achieve the stipulated water quality objectives. Typically, numerical models assist in determining current conditions and assessing what-if-scenarios. The Delaware Bay has a very complex geometry that combines fjord-like areas with very narrow stream segments, resulting in vastly different flow regimes. In addition, it is used heavily by the dense agglomeration of population and industry, all of which result in very different and sometimes contradicting and competing water quality objectives. Consequently, the numerical models used for the Bay need to be able to respond to the imposed complexity of tasks by being able to map the governing physics and associated processes appropriately on the scales and directions required. The UnTRIM module (Unstructured TRIM), an enhancement of the models comprising the TRIM family, is capable solving the equations of motion in all three dimensions on an unstructured grid. As such, it is ideally suited to map out the highly irregular geometry of the bay including all of its tributaries, capturing all processes in the horizontal and vertical directions. Because of its efficiency and speed in solving the governing equations, the model can afford to use a very fine meshing pattern further enhancing the accuracy of the mapping. This paper focuses on the development of a high resolution unstructured orthogonal grid for the Delaware Bay estuary that is mostly driven by the requirements of the pollutant fate and transport scenarios (those are quite different if compared to the needs resulting from the computation of the hydrodynamics only) using a program package called JANET.

[‡] Graduate Student, Drexel University Dept. of Civil & Architectural Engineering, 32nd & Chestnut Street Philadelphia, PA 19104, Student Member, ASCE e-mail: Kutay@drexel.edu

[†] Assistant Professor, Drexel University Dept. of Civil & Architectural Engineering, 32nd & Chestnut Street Philadelphia, PA 19104, Member, ASCE. e-mail: Michael.Piasecki@drexel.edu

Introduction

The Delaware Bay Estuary is located in the Mid-Atlantic region of the United States, and it includes portions of Pennsylvania, New Jersey and Delaware, through which the Delaware River flows. Delaware Bay Estuary is fed by 216 tributaries and drains at least half of Pennsylvania. Supplying water for more than 7 million people in Philadelphia Metropolitan area, the Delaware Bay Estuary plays a crucial role for drinking water supply and as a recipient of industrial and municipal wastewater effluent.

The Bay is comprised of three different regions: i) the outer Bay which is characterized by large and relatively shallow areas and adjacent marshes, ii) a middle section that provides the narrowing-down segments from a large bay to the narrow riverine section, and iii) the narrow channel that reaches upstream to the head of tide at a dam in Trenton. The latter is largely comprised of a deep navigation channel that needs to be dredged in order to permit 40ft draft vessels to reach the upstream port facilities. Because of the funneling effect of the midsection the velocities in the narrow channel are substantially higher than in the outer regions and the tidal signal is magnified due the preservation of wave energy (minus attenuation) leading to a fundamentally different flow regime than in the outer bay.

Because of the dual use as water supply and also a waste effluent recipient water quality issues remain a concern. For example, the contamination of the sediments with PCBs (metals as well) has received much attention in recent years and is currently being addressed in a Total maximum Daily Load (TMDL) development and implementation. Because of the importance of sediment dynamics, the grid development is largely based on the need to capture these dynamics. What this means is that regions of deposition and re-suspension must be represented in detail even though they may not necessarily play an important role in computing the hydrodynamics because mass and momentum exchange takes predominantly place in the deeper section of the channel. Hence, grid resolution is dependent on both the proper overall representation of the domain hydrodynamics but also on the need to map sub-region flow effects like counter flow circulations behind bars and islands, and local eddies at pier heads, confluences, and in the shadow of jutting land masses.

While the above mentioned aspects determine the necessary grid refinements in specific regions, numerical aspects determine shape, alignment, and the up- and down sizing of grid elements when transitioning from fine to coarser regions of the grid. These aspects are a direct result of the way the governing equations are being solved and cannot be viewed independently of the grid development. In fact, the spatial and temporal accuracy of the numerical integration technique of the governing equations directly depends on the element shape and how neighboring elements connect to each other. Therefore, much attention must be paid during the grid development to not only satisfy the requirements necessitated by the physics, but also the requirements imposed by the numerical solution technique to maintain a as high as possible degree of accuracy.

Computational Model

The hydro-system, being so complex, requires a stable and efficient way of solving the governing equations. The numerical model UnTRIM (Casulli and Zanolli, 1998; Casulli, 1999b; Casulli and Walters, 2000) is capable of solving Reynolds Averaged Navier-Stokes equations (RANS) together with salinity and temperature (scalar transport) on unstructured grids and shares the same philosophy with the family of TRIM models originally developed by Casulli (1990). The numerical approach uses a semi-implicit method which incorporates an Eulerian-Lagrangian approximation for the advective terms, making the method unconditionally stable for barotropic flows (Casulli and Walters, 2000). The numerical scheme is subject to a weak Courant-Friedrichs-Lewy (CFL) stability condition for baroclinic flows. This property allows the modeler to use high time step values in simulations and as a consequence, UnTRIM can be very efficiently used for long term and forecast simulations. The unique way of building the finite difference expressions on an orthogonal unstructured grid produces a system of equations which can be solved efficiently by a preconditioned conjugate gradient method. Moreover, the model allows wetting and drying of elements that is crucial to the different simulation scenarios. These properties make UnTRIM attractive for simulation in Delaware Bay Estuary.

The use of unstructured grids offers flexibility to the designer to concentrate on the areas where fine resolution is necessary and coarser resolutions for the remaining part of the domain. This not only increases the accuracy but also fortifies the model efficiency. The UnTRIM model uses a special version of unstructured grids: the orthogonal unstructured grid. An orthogonal unstructured grid is obtained by covering the domain with convex polygons. The element or polygon center coincides with the center of the circum-circle of the element, which is not necessarily the geometric center. Connecting each polygon center, the line joining the center of each adjacent polygon is orthogonal to the shared side of these two polygons (Figure 1), which is the main idea behind the orthogonal grid configuration.

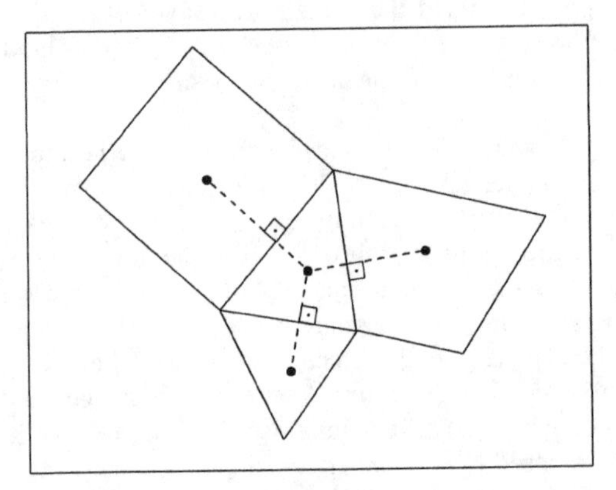

Figure 1. Orthogonal unstructured grid on a 2-D plane.

The orthogonal unstructured grids can be obtained efficiently and qualitatively by Delaunay triangulation (Spragle *et al.*, 1991). In this method the domain is decomposed into polygons where each polygon is linked with a single point (marked by black squares in Figure 2) which is called "generating point". A domain can be decomposed into polygons where any point inside the polygon is closer to its own generating point than the generating point of any other polygon. The polygons generated are referred as Dirichlet tessellation or Voronoi tessellation (Weatherill, 1988). The boundaries of polygons are perpendicular bisectors of the lines joining the neighboring generating points which form a perfect unstructured orthogonal grid (see Figure 2).

The accuracy of the numerical model depends on the perpendicularity of the segment joining the centers to the shared face. Also, the centers of the adjacent elements must be as equal in distance as possible from the shared face to increase accuracy. The special cases of unstructured orthogonal grids, obtained by squares and equilateral triangles where the polygon centers coincide with the geometric center, produce a perfectly second order accurate spatial discretization of the governing equations. On the other hand, the convergence rate depends on the diagonal dominancy of the coefficient matrix for the preconditioned conjugate gradient solver. An increase in diagonal dominance in the coefficient matrix can be obtained by either increasing the grid size or by decreasing the time step. While the former decreases spatial accuracy, the latter increases the computational execution time. The modeler must seriously consider these constraints and reach a trade off between these two parameters for the study.

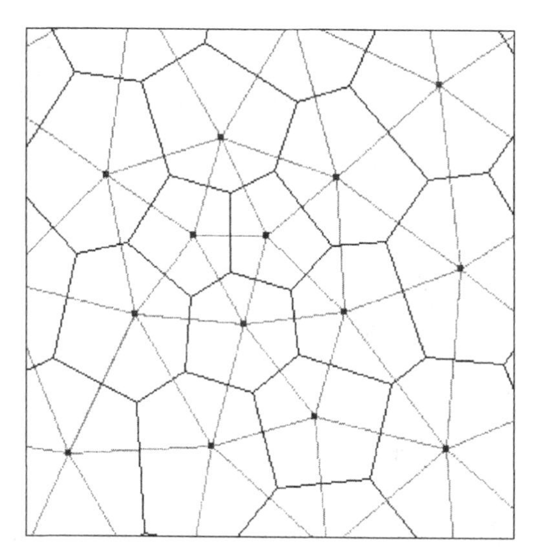

Figure 2. Delaunay Triangulation with Voronoi tessellation.

Generation of an unstructured orthogonal grid for Delaware Bay

The applicability of different grid generation techniques was also outlined by Cheng and Casulli (2001). As stated by Cheng and Casulli, the grids generated for finite element methods can be used for UnTRIM applications. However, there is no guarantee that the generated mesh is orthogonal because finite element type solution methods do not require orthogonality. The two common schemes for generating unstructured grids are the Advancing Front method and the Delaunay method. While the former lacks in supplying orthogonality, the mentioned properties of the second scheme can be used to produce an orthogonal mesh for UnTRIM.

The grid for Delaware Bay is generated by a commercial program called JANET which stands for Java net generator. Since the program is written in Java programming language, it is platform independent and runs on any type of operating system. The grid generator not only generates UnTRIM conforming grids by Delaunay triangulation but also reads and writes the grid file that meets the I/O requirements of UnTRIM.

As a first step, a boundary, which is 2.0 m above the MLLW, is selected with all the tributaries included. Then major tributaries and point sources are identified to produce the computational domain while minor tributaries were removed because their existence in the modeling domain will have no discernable effect on the hydrodynamic simulations. Next, 1.25 million sounding values are extracted from NOS (National Ocean Service) GEODAS-CD to extract the depth soundings. These points were then triangulated to generate a digital terrain model (DTM) (see Figure 3). The DTM is required by the grid generator to calculate the depth values for grid points that do not coincide with a bathymetry point. In addition, JANET needs the DTM to execute criteria based grid generation, like the automatic inclusion of finer grid resolution in regions of steep bathymetry gradients. The next step is to generate an initial grid. Janet provides a number of powerful tools that help in generating a "good" orthogonal grid. For example, it permits the restriction of maximum permissible inside angles, the definition of minimum and maximum areas an element can have, and also a number of smoothing functions available. In addition, working polygons can be set along with pre-specified nodes and their spacing permit the control of element sizes, which is important for the gradual increase or decrease edge lengths of elements. Finally, the layer structure of JANET allows isolating certain areas within the grid for which separate refining or coarsening routines can be applied. In fact, the grid generation itself can be carried out sequentially, i.e. the domain can be divided into several sub domains and then later be connected to form the final mesh. For this work, a grid is generated containing 47464 elements in the horizontal direction using a combination of triangular and quadrilateral elements.

The domain is split into four sub-domains starting from the upstream boundary Trenton and extending to the downstream boundary at Atlantic Ocean. A grid for each sub-domain is generated using different criteria and the generated grids for each sub-domain are then merged to form the grid for the whole domain (see Figure 4).

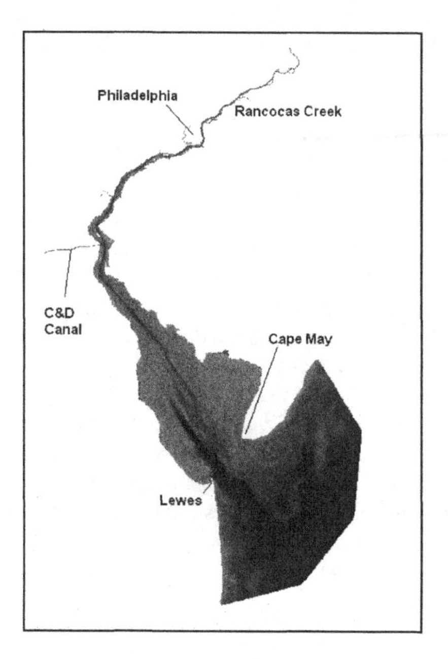

Figure 3. Bathymetry (DTM) of the domain.

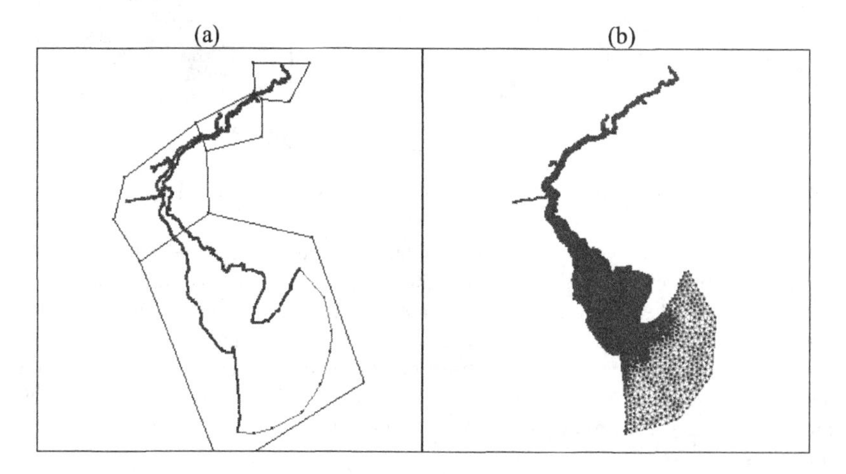

Figure 4. (a) Polygons used to split the domain into four sub-domains, (b) The merged grid of the four sub-domain grids.

In the top two sub-domains, the estuary, most of which is covered by navigational channel, is narrow. These parts also include treatment plants, fresh water intakes and man-made structures that complicate the flow field. Maximum area and maximum edge length criteria are used for these parts to achieve a high degree of resolution.

In the lower parts where the width is wide enough, it is better to use equilateral elements with refinements following bathymetry changes. Also, limiting the minimum angle in triangles produces high quality spatial discretization. The Gradient criterion forces refinement of elements at sharp gradients (user specified) of bathymetry while the minimum angle criterion forces the minimum angle of elements to be higher than a specified value. This produces triangles that are closer to equilateral triangles and increases accuracy. The first criterion is mostly applied to the bay entrance while the second is applied inside the bay where the width of the bay has a maximum value in the lowest sub-domain.

The Delaware River is heavily used by maritime transportation for which a navigation channel of 40 feet is maintained. As a consequence most of the flow passes through the deep navigational channels. Aligning the sides of elements to these channels not only prevents unrealistic fluxes out of the channel but also decreases the numerical diffusion. The alignment is made by building construction polygons which follow the navigational channel so that the element sides can be aligned to these polygons (see Figure 5).

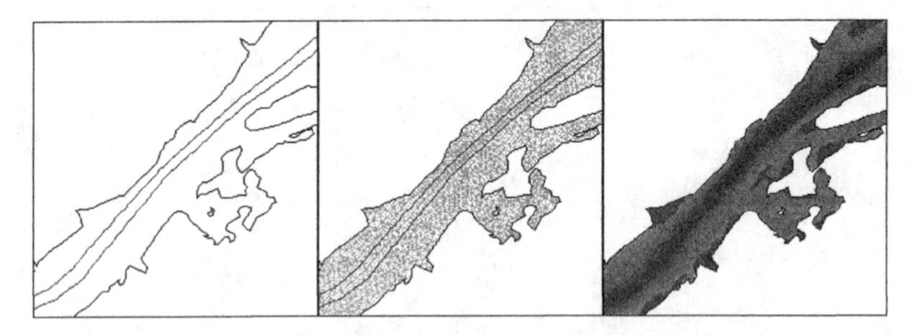

Figure 5. Alignment of elements to the navigational channel near Rancocas Creek.

The grid was generated such that the element size is smallest (in average) throughout the entire upper half of the estuary, i.e. the region replete with confluences, intakes and outfalls, side bays, shallow areas and sections with very rapid changes in the bathymetry due to the navigational channel. The average grid size in this area is approximately 40 meters, with some of the elements being larger and the smallest size being approximately 12 meters. The grid size then gradually increases to 1.5 – 2.0 km in size in the outer bay and approximately 3km at the open boundary.

Second order accuracy in spatial discretization can only be achieved if the polygon centers are equally distant from the shared side. Thus the gradual change in grid size keeps the center of polygons nearly the same distance from the shared side and minimizes the deviation from second order accuracy. This is achieved by forcing a gradual change of element size at the boundary by applying the mask options that permit the gradual transition from finer to coarse and coarse to finer gird sizes, for which an example is shown in Figure 6.

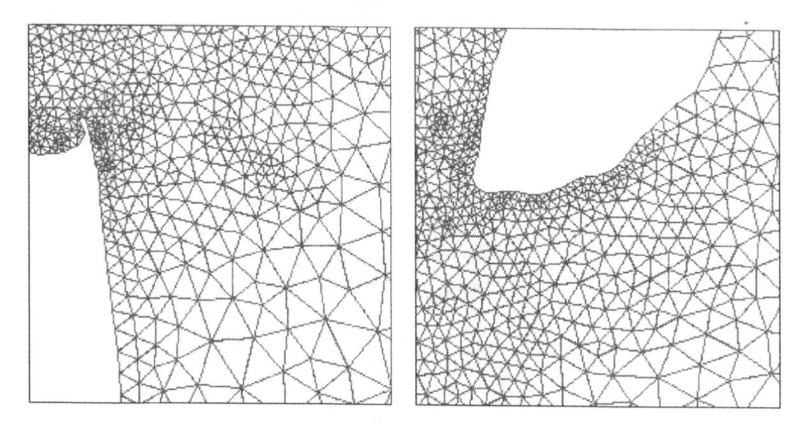

Figure 6. Gradual change in grid size near Lewes and Cape May.

Discussions

The Unstructured orthogonal grid required by UnTRIM is generated with the commercial program JANET. Using this tool, a grid for Delaware Bay Estuary shown in Figure 4(b) is generated. The program is installed on PENTIUM 4 processor machine having a 1GB of RAM. This configuration is sufficient to handle several million bathymetry soundings, to execute a triangulation to generate the DTM, and to generate the grid using the layer structure of JANET.

Although JANET generates high quality grids, some of the elements are of poor quality. The grid generator marks the non-orthogonal elements for visual inspection and subsequently allows the user fix those elements by moving and adjusting the location of element nodes. Typically, the poor quality is a result of the location of the polygon centers and not so much the result of a violation of orthogonality constraint. Close proximity of element centers to the shared edge results in an decrease of accuracy in the spatial discretization. To alleviate this problem JANET permits the conversion of two adjacent triangle elements into a single quadrilateral. The fourth corner of the generated quadrilateral may not coincide with the circumcircle, however, resulting in a deviation from perfect orthogonality. As a result, the demand for perfect orthogonality may need to be relaxed in those cases (see Figure 7). Also, the current version of JANET does not have an efficient algorithm that allows automatic generation of quadrilaterals. It can handle quadrilateral elements during I/O and also recognizes mixed element grids, but quadrilaterals need to be generated by hand. Some options are currently being tested to permit automatic generation of this type of elements in open and well defined sub-regions with regular boundaries that could then be nested with triangular grids.

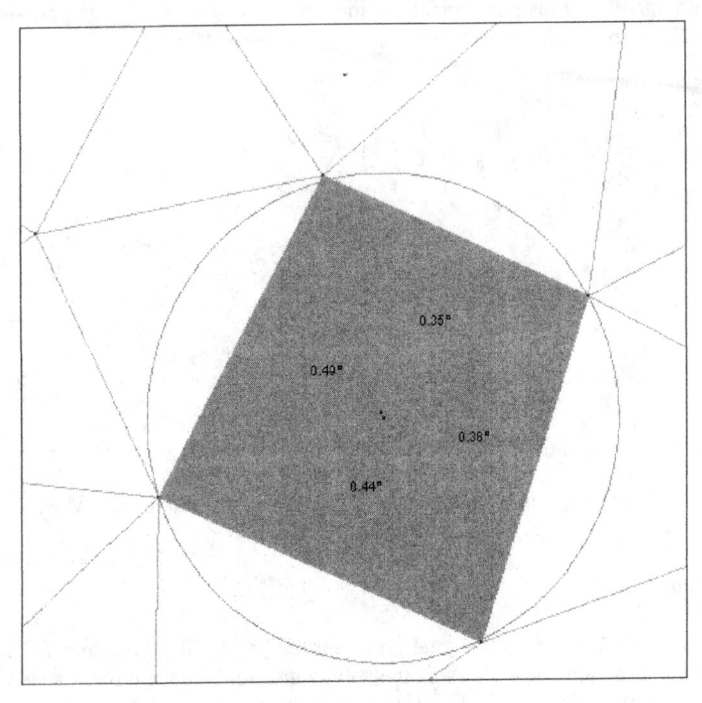

Figure 7.Level of deviation (in degrees) from orthogonality marked on a quadrilateral element.

Acknowledgements

The authors would like to acknowledge the generosity of SMILEconsult GmbH, Hannover (Germany) to provide us with a ß-version of JANET. The program is currently not commercially available but continues to be developed on a contractual basis with a German government institution. In view of these circumstances, the willingness of the company to share their product with us can not be lauded enough.

References

Casulli, V. (1990), "Semi-implicit Finite Difference Methods for the Two-dimensional Shallow Water Equations", J. Comput. Phys., Vol. 86, p. 56-74.

Casulli, V. and G. S. Stelling (1998), "Numerical Simulation of 3D Quasi-hydrostatic Free-surface Flows", ASCE, J. Hydra. Eng., Vol 124, p 678-686.

Casulli,V and Zanolli, P. (1998), "A Three-Dimensional Semi-Implicit algorithm for Environmental Flows on Unstructured Grids" Proceedings of conf. on Num. Methods for Fluid Dynamics. University of Oxford.

Casulli, V. (1999b), "A Semi-implicit Numerical Method for Non-hydrostatic Free-surface Flows on Unstructured Grid", Proceedings of Inter. Workshop on Numerical modeling of Hydrodynamic Systems, Zaragoza, Spain, June 1999, p. 175-193.

Casulli, V. and R. A. Walters (2000), "An Unstructured Grid, Three-dimensional Model based on the Shallow Water Equations", Inter. J. for Numerical Methods in Fluids, Vol. 32, p.331-348.

Cheng, Ralph T. and V. Casulli (2001), "Evaluation of the UnTRIM Model for 3-D Tidal Circulation", Proceedings of the 7th Inter. Conf. on Estuarine and Coastal Modeling, St. Petersburg, FL, November 2001, p. 628-642.

Spragle, G. S., McGrory, W. R., Fang, J. (1991), "Comparison of 2D Unstructured Grid Generation Techniques", AIAA-91-0726.

Weatherill, N. P. (1988), "A Method For Generating Irregular Computations Grids in Multiply Connected Planar Domains", Inter. Journal of Numerical Methods in Fluids, Vol. 8, pp. 181-197.

Storm Impacts on Sediment Deposition at a River-Coastal Confluence

Yong Guo,[1] Keith W. Bedford,[2] and David Welsh [2]

Abstract

To remain a viable harbor the Maumee River/Lake Erie confluence requires continuous dredging of two permanent bed mounds; the goal of this paper is to research the sediment transport characteristics under storms in the confluence region, with the objective of seeking the origin and physics of the two mobile sediment mounds in the navigation channel of the river.

An integrated three dimensional, hydrodynamic, sediment transport, wave current bottom boundary layer, and wave model is applied to simulate the hydrodynamics and sediment transport. A curvilinear grid system with 208 by 79 cells in the horizontal direction and 12 sigma layers in the vertical direction is used to cover the whole lake and the 12 km long dredged channel of the Maumee River. Four sediment transport application cases are constructed under strong and weak storms in spring and fall.

The numerical simulation results reveal the sediment transport and formation mechanisms of the two mobile sediments mounds in the Maumee River/Lake Erie confluence under storms. Under strong storms/seiches, Mound 1, the upstream one, is mainly formed by riverine sediment inflow and settlement while Mound 2, the downstream one, is mainly formed by sediment resuspension in the river bottom upstream of Mound 2. During weak storms both Mounds 1 and 2 are maintained by riverine sediment settlement with bottom sediment upstream of Mound 1 being the second largest contributor to the formation and maintenance of Mound 2. Strong vertical thermal stratification exists in spring, but not in fall. This stratification influences the sediment transport in the river/coastal confluence. Under both strong and weak storms/seiches, the riverine sediment deposition at Mound 1 is more in spring than in fall, but more in fall than in spring at Mound 2; the total sediment transport into the Maumee Bay is more in fall than in spring.

INTRODUCTION

Lake Erie is one of the Great Lakes in North America. It is 395 km long and 115 km wide, with 1400 km of shoreline (Figure 1). Its average depth is only about 19 m that makes it the shallowest of the Great Lakes. The Maumee River starts from Fort Wayne, Indiana and flows north-easterly towards the Western Basin of Lake Erie. The navigation channel in the Maumee River extends from Lake Erie upstream approximately 12 km to a ship turning basin (Figure 2).

---

[1] Senior CEA, Los Angeles County Dept. of Public Works, 900 S. Fremont Ave., Alhambra, CA 91803, USA.

[2] Department of Civil and Environmental Engineering and Geodetic Science, The Ohio State University, Columbus, Ohio 43210, USA.

A large amount of suspended sediment flows into the lower Maumee River, settles down to the bottom, and shallows the navigation channel each year. Further, the deposition is confined to two well formed mounds within the river channel (Figure 3).

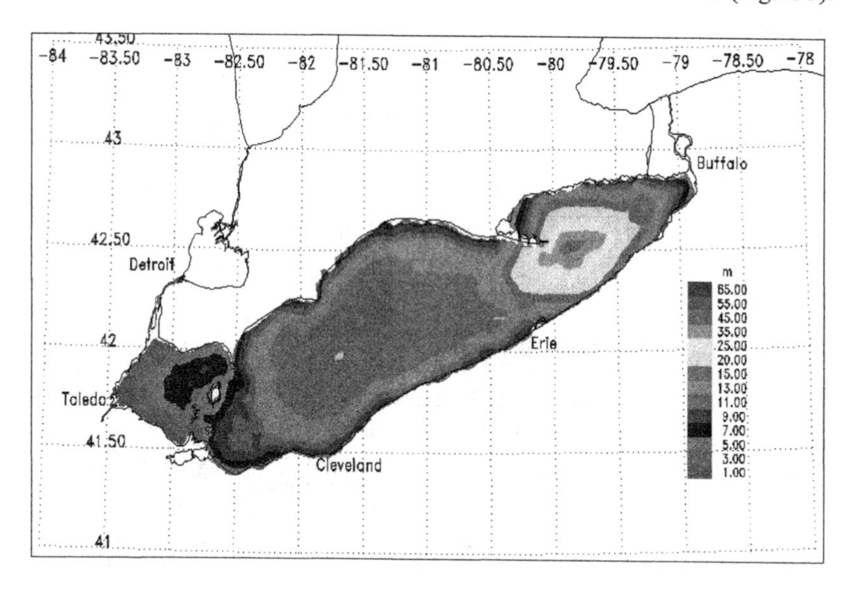

Figure 1: Lake Erie Bathymetry

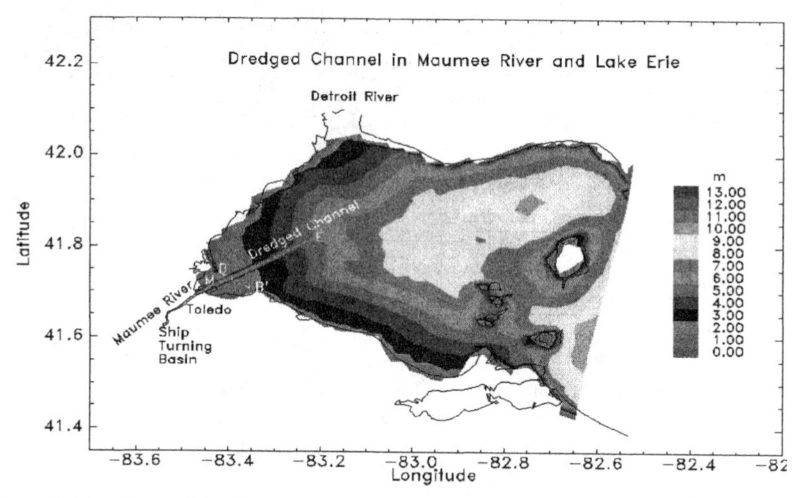

Figure 2: Positions of the Maumee River Mouth (Point M, End of the Maumee River), the Dredged Channel End (Point E), and the Maumee Bay Cross-Section (Dashed Line BB′)

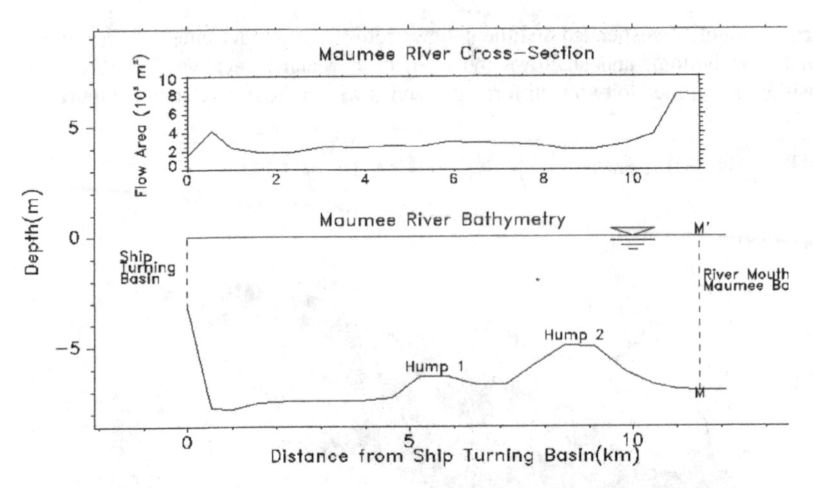

Figure 3:Bathymetry of the Dredged Maumee River Channel (Maumee River mouth cross-section MM′, the end of the Maumee River, is projected to Point M in Figure 2)

As a result, almost continuous dredging in this channel is necessary to ensure navigation in the Maumee River.

In a remote sensing and modeling study of neighboring Sandusky Bay, Bedford (1989, 1992) speculated that there were marine-like turbidity maxima dynamics in the rivers draining to Lake Erie especially during and after storms. Lake Erie is particularly responsive to storms, often giving rise to storm surges and seiches that cause flow reversals in the rivers tributary to the Lake. These turbidity maxima were hypothesized to exist as a result of the interplay of fresh water discharge during stratification, periodic (14 hr) flow reversals introduced by surge/seiche mechanisms, and selective sorting and deposition of sediment resulting from the asymmetric velocity patterns during flood and ebb portions of the flow reversals. In the marine turbidity maxima, larger grained sediment material is deposited at the furthest upriver extent of the flood portion of the tidal currents and Bedford proposed that similar sediment deposition dynamics in Great Lakes harbors could appear in seiche/surge affected confluences as well. This hypothesis would suggest that the origin of the sediments requiring dredging in the Great Lakes could result not only from the traditional downstream transport of watershed sediments but also sediment transported to and deposited in the River due to the flow reversing seiche/surge mechanism

Storm surges, the forced mode response of the Lake, are temporary rises in water level caused by sustained high winds from one direction blowing across open water and driving water towards the down-wind shore. Correspondingly lower water levels occur on the upwind shore. On Lake Erie, the set-up at Buffalo on the eastern end and the set-down at Toledo on the western end (recorded as water level difference) have historically been as large as 3.3 m (Schwab, 1978). Seiches, also called the free mode response, occur when winds causing set-up subside. There are five longitudinal seiche modes and one transverse mode in the Lake with the 14 hour mode being the strongest and having the most impact on the Maumee River. Because

Lake Erie is very shallow and is aligned with prevailing winds, it has the strongest seiches among all five Great Lakes (Schwab, 1978; Libicki and Bedford, 1991). The surge/seiche systems can be strong enough to cause flow reversals in the Maumee River that, for several seiche periods results in periodic (14 hour) flow reversals with exponentially diminishing amplitudes. Such motions result in tidal like behavior during their occurrence.

Stratification is a very important component of mound dynamics during the spring and fall seasons. Because the Maumee River is very shallow, it warms up more quickly than the dredged portion of the Channel and the Lake in the spring and thus the density of the river discharge will be lower than the water density in the Western Basin. It is therefore expected that the warmer river discharge into the lake will flow at the top of the lake and strengthen stratification (Bedford, 1992). In the fall, the river water cools more quickly than the deeper lake, therefore the density of the river discharge is lower than that of the upper part of the lake water. Consequently, the river water is expected to plunge and sink into the deeper part of the lake. This enhances the stratification in the areas reached by Maumee River outflow.

There are no data available with which to calibrate and deploy a traditional model for quantitative hypothesis assessment and decision support. What is therefore attempted here is a model based assessment as to whether the above hypotheses are credible. To do so a full 3-D modeling implementation of the Maumee River/Bay system is used which incorporates the complete thermal physics required to reproduce the hourly heating and cooling behaviors of day/ night cycle. A widely applied 3D hydrodynamic model is CH3D (Curvilinear Hydrodynamics in 3-Dimensions) developed by the U.S. Army Corps of Engineers (Sheng, 1983; Chapman et al., 1996), which has been used to simulate hydrodynamics in estuaries and lakes (Johnson et al., 1995; Yassuda and Sheng, 1997; Davis and Sheng, 1999; Velissariou et al., 1999; Chu et al., 1999). Based on CH3D, a new parallel code called COMAPS (COupled MArine Prediction System) (Welsh et al., 2001, 1999) has been developed at The Ohio State University. COMAPS is a parallel code that couples CH3D with a sediment model, a wave model and a wave current bottom boundary layer model. This model has been successfully used to simulate the current fields in the Adriatic Sea (Welsh et al., 2001) and the sediment resuspension and transport characteristics of spring plume events in Lake Michigan (P.Velissariou, in progress). In this Maumee River simulation, 10 different sediment classes are used in an Eulerian Particle Tracking formulation to represent the bottom and riverine sediment distributions originating from different locales in the confluence. Because strong storm derived currents and waves are to be simulated this model must include bottom morphology changes and their coupling with surface wave fields and bottom shear stress calculations.

Based on COMAPS, a sequentially coupled version of COMAPS, SCOMAPS, which basically keeps the features of COMAPS but does not use MPI or OpenMP techniques, is developed and applied in the sediment transport simulation in this article.

MODEL STRUCTURE AND IMPLEMENTATION

Hydrodynamic Model

The model governing equations are transformed and discretized on a curvilinear grid that makes the model easier to use for domains with irregular boundaries. A sigma stretched coordinate (Chapman et al., 1996) is used in the vertical direction for CH3D. This model mesh follows irregular bottom features. The solution algorithm for CH3D uses a mode-splitting method that consists of an external mode and internal mode; the external mode solves the vertically averaged momentum equations for the free surface fluctuation and vertically averaged velocities, and the internal mode solves for the vertical velocity, temperature and the 3D horizontal velocity perturbations. These perturbations are added to the vertically averaged horizontal velocities to get the full 3D horizontal velocities.

The governing equations for CH3D under a Cartesian coordinate system (Chapman et al., 1996) are not reproduced here, as they are standard equations for this class of problem. The variables are defined as follows: u, v, w are velocities in x, y and z directions respectively, t is time, f is the Coriolis parameter, ρ is density, p is pressure, A_h and K_h are horizontal turbulent eddy viscosity/diffusivity coefficients, A_v and K_v are vertical turbulent eddy viscosity /diffusivity coefficients, g is the gravitational acceleration, and T is temperature.

To solve the discretized equations describing a physical process on a physical domain, a grid system to cover the irregular domain along with high resolution on certain features must be constructed. In the Maumee River and Maumee Bay, the dredged river channel extending into Lake Erie is difficult for a Cartesian grid to model. In addition, because the Maumee River is of very small scale compared to Lake Erie, it is impossible to use a regular Cartesian Grid with reasonable grid points to cover the Maumee River with enough resolution to meet our simulation accuracy. Therefore, a curvilinear grid system is necessary. The grid generator used in this paper is SeaGrid (http://woodshole.er.usgs.gov/staffpages/cdenham/public_html/seagrid) that uses an elliptic method of grid generation.

The generated grid has 208 cells along the lake, 79 cells across the lake and 2 cells across the Maumee River. There are 12 layers in the vertical direction. The governing equations are transformed horizontally under the curvilinear coordinate system and vertically under the terrain following sigma-stretched grid system (Chapman et al., 1996). To close the governing equations, the standard $k-\varepsilon$ turbulence model is used to calculate the vertical turbulent eddy viscosity/diffusivity; the horizontal turbulent eddy viscosity coefficient A_h for the momentum equation is taken from the research results by the International Field Year for the Great Lakes (IFYGL), which is 10^{-4} m^2s^{-1}; the horizontal eddy diffusivity for heat transport, K_h, is obtained by $K_h = A_h / P_r$, where P_r is the Prandtl number which is taken as 7 (Fischer et al., 1979).

Sediment Transport Model

The sediment transport model used in this simulation, was initially developed by Spasojevic and Holly (1990, 1994). The model consists of two parts, bedload and suspended load transport, and can be coupled with CH3D to iteratively calculate the sediment size distribution change in the bed, the bed elevation change, and the sediment concentration change.

Bedload Simulation

CH3D-SED divides the bottom sediment into several layers. The uppermost layer that is in contact with water is called the active layer; the layer right under the active layer is called the active stratum layer; and other layers under the active stratum layer are called stratum layers. All sediment particles inside the active layer are assumed to be equally exposed to the flow.

The mass-conservation equation for a size class of sediment particles in an active-layer control volume is

$$\rho_s(1-p)\frac{\partial(\beta E_m)}{\partial t}+\nabla \bullet q_b+S_e-S_d-S_F=0 \tag{1}$$

Where; ρ_s is the sediment density, p is the porosity of the bed material, β is the mass fraction of a particular sediment class in a control volume of the active layer, E_m is the active layer depth, t is time, q_b is the bedload flux, S_e is the erosion source from the bedload for suspension, S_d is the deposition source from suspended load, and S_F is the floor source from the active stratum under the active layer. These source terms are evaluated by the empirical equations (Spasojevic and Holly, 1994).

For a control volume in the active stratum, the mass-conservation equation for a particular sediment class is

$$\rho_s(1-p)\frac{\partial}{\partial t}[\beta_s(z_b-E_m)]+S_F=0 \tag{2}$$

In equation (2), β_s is the active-stratum size fraction and z_b is the bed-surface level.

Summing all mass-conservation equations for all size classes in the active layer and active stratum (equations (1) and (2)) and making use of the fact that $\sum\beta=1$ and $\sum\beta_s=1$, the following global mass-conservation for the active layer and active stratum results in:

$$\rho_s(1-p)\frac{\partial E_m}{\partial t}+\sum(\Delta\bullet\vec{q}_b+S_e-S_d-S_F)=0 \tag{3}$$

$$\rho_s(1-p)\frac{\partial(z_b - E_m)}{\partial t} + \sum S_F = 0 \quad (4)$$

Adding equations (3) and (4) together gives the following global mass-conservation equation for bed sediment

$$\rho_s(1-p)\frac{\partial z_b}{\partial t} + \sum(\Delta \bullet \vec{q}_b + S_e - S_d) = 0 \quad (5)$$

In CH3D-SED, equations (1) and (5) are nondimensionalized, discretized, and transformed under the curvilinear coordinate system. Interested readers can refer to Spasojevic and Holly (1994) for the detailed discretized equations that form a system of nonlinear algebraic equations that are solved by the Newton-Raphson method.

Suspended Sediment Simulation

The mass-conservation equation for one particular size class of suspended sediment for a control volume is

$$\begin{aligned} &\frac{\partial(\rho C)}{\partial t} + \frac{\partial(\rho C u)}{\partial x} + \frac{\partial(\rho C v)}{\partial y} + \frac{\partial(\rho C w)}{\partial z} - \frac{\partial(\rho C w_f)}{\partial z} \\ &= \frac{\partial}{\partial x}(D_H \frac{\partial(\rho C)}{\partial x}) + \frac{\partial}{\partial y}(D_H \frac{\partial(\rho C)}{\partial y}) + \frac{\partial}{\partial z}(D_V \frac{\partial(\rho C)}{\partial z}) \end{aligned} \quad (6)$$

Where; ρ is the density of the mixture of water and suspended sediment (all size classes), C is the sediment concentration (the ratio of the mass of the particular size sediment to the total mixture mass in the control volume), w_f is the fall velocity of a particular size particles in suspension, D_H is the horizontal mass diffusivity, and D_V is the vertical mass diffusivity.

The governing equation (6) is nondimensionalized and transformed under the curvilinear coordinate system and then discretized into a set of implicit algebraic equations along the vertical sigma coordinate (Spasojevic and Holly, 1994).

The boundary conditions at the surface set advection, fall-velocity and diffusion fluxes equal to zero. At the bottom, the following conditions apply: the advection flux through the bed surface is equal to zero; the fall-velocity flux through the bottom face of the control volume is the deposition of suspended sediment into the control volume of the active layer; the diffusion flux through the bottom face of the suspended sediment control volume represents the erosion source of the active layer control volume which entrains into the control volume for suspended load.

The simulation of bedload and suspended load are coupled together by the deposition and erosion terms in the bed sediment and suspended sediment control equations.

The solution algorithm for the discretized sediment equations is

(1). Bed elevation and the active-layer size fractions at each node are solved from discretized equations of (1) and (5) in which sediment concentrations are assumed known from previous iterations.

(2). Sediment concentrations for each sediment size along a vertical sigma coordinate are solved from the discretized equation of (6) in which source terms in the control volumes next to the bed are evaluated from the size fractions values from step (1).

(3). Repeat steps (1) and (2) until the convergence criteria is met.

(4). Repeat steps (1) to (3) through the whole computational domain.

Wave and Current Bottom Boundary Layer Model

Short-period surface gravity waves and certain types of short-period internal wave activity are very energetic in Lake Erie. In the Western Basin and in the entire nearshore zone this activity can extend to the bottom, thereby creating a high frequency wave boundary layer. The undulating current field in Lake Erie also creates a current bottom boundary layer. It is found that a non-linear interaction exists between the two kinds of boundary layers and enhances the shear stresses at the bottom (Glenn, 1983; Grant and Madsen, 1986; Glenn and Grant, 1987; Keen and Glenn, 1994). The high shear stresses associated with the wave-current bottom boundary layer must be accurately represented to realistically model sediment transport

According to the bottom boundary layer theory by Glenn (1983), the wave boundary layer is embedded in the current boundary layer. The magnitude of the characteristic shear velocity within the wave boundary layer is

$$u_{*cw} = (\frac{1}{2} f_{cw} \alpha)^{1/2} u_b \tag{7}$$

Where; f_{cw} is a friction factor, u_b is the maximum wave orbital velocity at the top of the wave boundary layer, and α is given by

$$\alpha = 1 + 2\frac{u_a}{u_b}\cos\phi_c + \left(\frac{u_a}{u_b}\right)^2 \tag{8}$$

In which u_a and ϕ_c are the current reference speed and direction relative to the wave.

The total roughness in a combined wave and current flow is comprised of three components: sediment grains, bedforms, and near bed transport, and is given by (Glenn, 1983)

$$k_b = k_{bN} + k_{bB} + k_{bT} \tag{9}$$

Where; k_{bN} is the roughness caused by sediment grain size, k_{bB} is the roughness caused by bedforms and is related to wave-generated ripples, and k_{bT} is the roughness caused by transport layer. The last roughness component is a function of the maximum shear stress caused by the nonlinear wave-current interaction. Detailed equations for these roughness terms can be found in Glenn (1983).

The shear velocity in the current boundary layer above the wave layer is

$$u_{Xc} = \left(\frac{1}{2} f_{cw} V_2\right)^{1/2} u_b \tag{10}$$

In which V_2 is a nondimensionalized variable evaluated by the integral over a wave period and is dependent on ϕ_c and u_a / u_b (Glenn, 1983; Keen and Glenn, 1994).

The wave-current combined shear velocity u_{*cw} is the shear velocity actually applied to the sediment in the bed within the wave-current boundary layer and plays an important role in sediment resuspension. The bottom roughness k_b, which also reflects the nonlinear interaction between wave and currents, is felt by the current field and plays an important role in the velocity field simulation. Therefore, in the simulation of currents and sediment transport in Lake Erie and the Maumee River, k_b and u_{*cw} should be used to replace their counterparts in CH3D-SED.

Coupling of CH3D-SED with the Wave Bottom Boundary Layer Model

In SCOMAPS, three models (CH3D-SED, WAM wave model (Gunther et al., 1992), and a wave-current bottom boundary layer model (WCBL)) have been coupled together. The execution of the SCOMAPS is: first, CH3D predicts surface velocities and fluctuations; second, WAM uses these output data and wind fields to calculate wave amplitude, wave speed and direction; third, the wave information produced by WAM is fed to WCBL in CH3D-SED-WCBL to get the enhanced bottom shear stress and velocity which are then used for updating bottom shear stress and roughness in the hydrodynamic and sediment transport simulation.

The parameter exchanges among submodels and implementation of SCOMAPS are shown in Figure 4 (Note that SED represents the sediment transport submodel in CH3D-SED, and CH3D represents the hydrodynamics submodel in CH3D-SED. The dashed line processes are implemented first through the simulation duration, and output the wave information for WCBL, then the double solid line processes are implemented).

Simulation Conditions

There are three main rivers entering Lake Erie: Maumee River, Detroit River and Niagara River. Average flow rate values will be taken for these rivers: 146.3 m^3s^{-1} for Maumee River, 5380.2 m^3s^{-1} for Detroit River, and 5805.8 m^3s^{-1} for Niagara River. According to Podber (1991), flow reversals in the Maumee River would occur at a flow rate less than 200 m^3s^{-1} for moderate to high lake seiches. Therefore, it is expected that flow reversals would occur under an average flow rate of 146.3 m^3s^{-1}. Since the gravity wave speed for Lake Erie is about 13.4 ms^{-1}

(Bedford, 1989), a high surge and long duration seiche is generated if a 13.4 ms wind speed is applied over the lake. As a result, the wind speed for the flow reversal test is taken as 13.4 ms^{-1}, blowing along the main axis of Lake Erie (27.33 degrees east north), and progressing from Toledo to Buffalo. Since it takes the associated gravity wave about 9 hours to travel from one end of the lake surge will be created. For the weak seiche test, the wind is assumed to be5 ms^{-1} starting

Case No.	Season	River Inflow	River T (°C)	Lake T (°C)	Wind (ms^{-1})	Duration of Wind (Hrs)
1	Spring	Average	7	4	13.4 NE	9
2	Spring	Average	7	4	5.0 NE	23
3	Fall	Average	12	15	13.4 NE	9
4	Fall	Average	12	15	5.0 NE	23

Table 1: Forcing Conditions for Sediment Transport Simulation

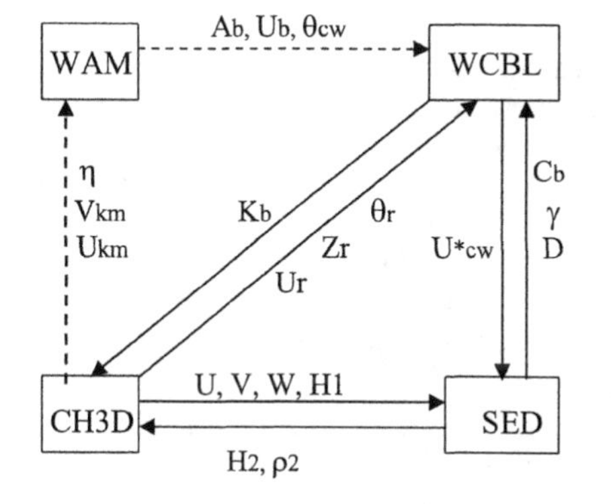

Ukm, Vkm, η: Water surface velocity and fluctuation;
Ab, Ub, θcw: Wave amplitude, critical velocity, and direction;
Kb: Bottom roughness;
Ur, Zr, θr : Reference current velocity, height, and direction;
U*cw: Wave current shear velocity;
D, γ, Cb: Sediment diameter, specific weight, and near bottom concentration;
U, V, W, H: Velocities, and water depth;
H2, ρ2: New water depth and mixture density.

Figure 4: SCOMAPS Implementation

From Buffalo and progressing with the speed of 5 ms^{-1} along the main lake axis towards Toledo, and stopping after 23 hours when the wind reaches Toledo.

The lake surface heat fluxes for spring and fall are generated using two sets of air temperature, dew point temperature, cloud cover and wind speeds representing typical weather conditions in the spring and fall. The flux data are then parameterized to reflect the seasonal heat budget characteristics and oscillatory changes between day and night. The typical temperatures in spring are 7 °C for the rivers and 4 °C for the lake; the typical temperatures in fall are 12 °C for the rivers and 15 °C for the lake. These values are used as river boundary conditions for the simulations in spring and fall. The four numerical test cases are listed in Table 1.

The bottom sediment is represented by two sediment diameters, 8 micron and 60 microns. Based on these two diameters, 10 sediment classes are constructed by varying the two diameters in the 3 rd significant figure of grain size and writing a bedload and suspended load conservation equation for each size. These sediment classes are used to represent the sediment inflow from the Maumee River and the bottom sediment in Lake Erie, and also the bottom sediment in the Maumee River shipping channel (Figure 5) for the purposes of tracking the sediment transport in the two mounds, between the two mounds and upstream of the two mounds (Figure 6). The initial proportion of the two sediment classes is 50% each of the total sediment in the bottom. The initial riverine suspended sediment is plotted in Figure 7 (Bedford et al., 1999).

RESULTS AND DISCUSSION

Storm Strength Impacts

The differences between Cases 1 and 2, or 3 and 4, are the storm strength: the storm causes flow reversals for Cases 1 and 3, but no flow reversal for 2 and 4. Therefore, these two pairs of cases are used to examine the storm impacts on sediment transport.

The 9-hour duration strong storm creates a lake seiche with a period of ~14.5 hours (Figure 8) which coincides with the first seiche mode described by prior researchers (Libicki and Bedford, 1991; Platzman and Rao, 1964). The strong storm also causes flow reversals in the Maumee River after 9 hours when the wind reaches Toledo (Figures 9A and 9B), and large amounts of sediment from the river bottom are entrained into the water column (Figure 10A and 10B). In contrast, no flow reversal exists under weak storm when the wind reaches Toledo after 24 hours (Figure 11A and 11B) and the maximum sediment concentration in the river is several times lower than that under strong storm (compare Figures 10A and 12A).

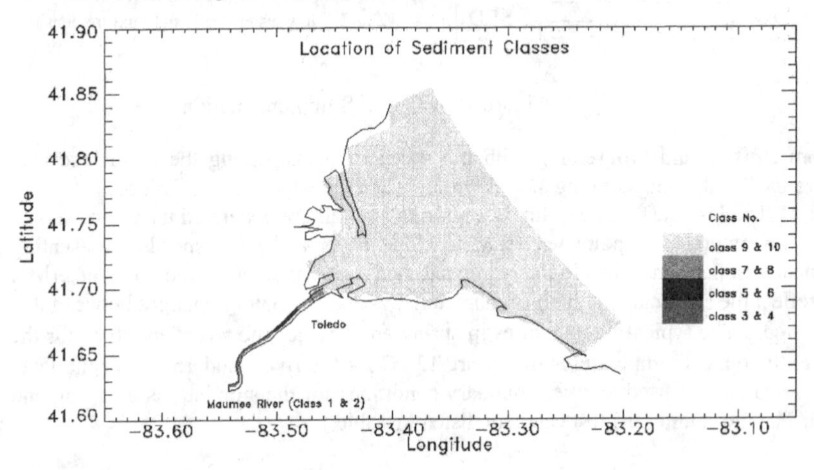

Figure 5: Location of Sediment Classes

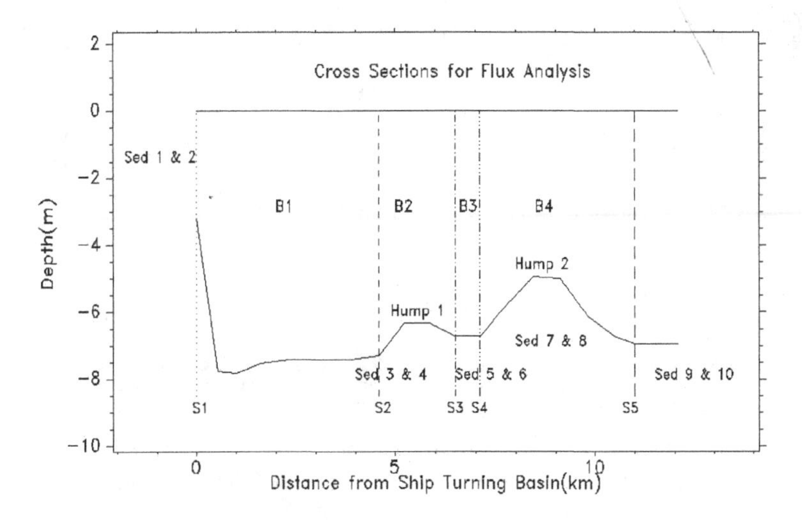

Figure 6: Cross Sections for Sediment Flux Analysis

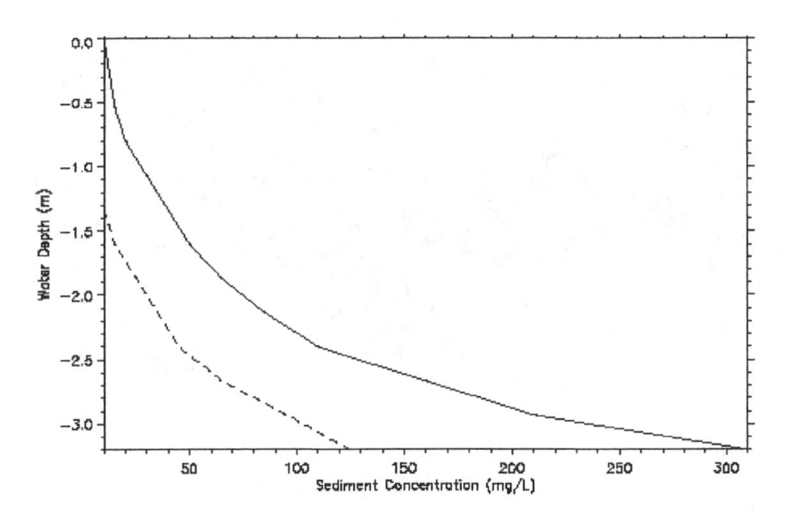

Figure 7: Riverine Suspended Sediment (Solid Line: Sediment Class 1, 8 micron; Dashed line: Sediment Class 2, 60 micron)

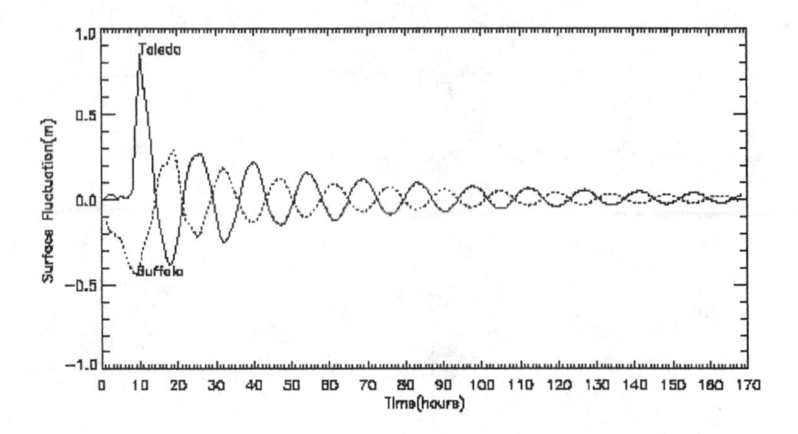

Figure 8: Case 1: Water Surface Fluctuation at Toledo and Buffalo

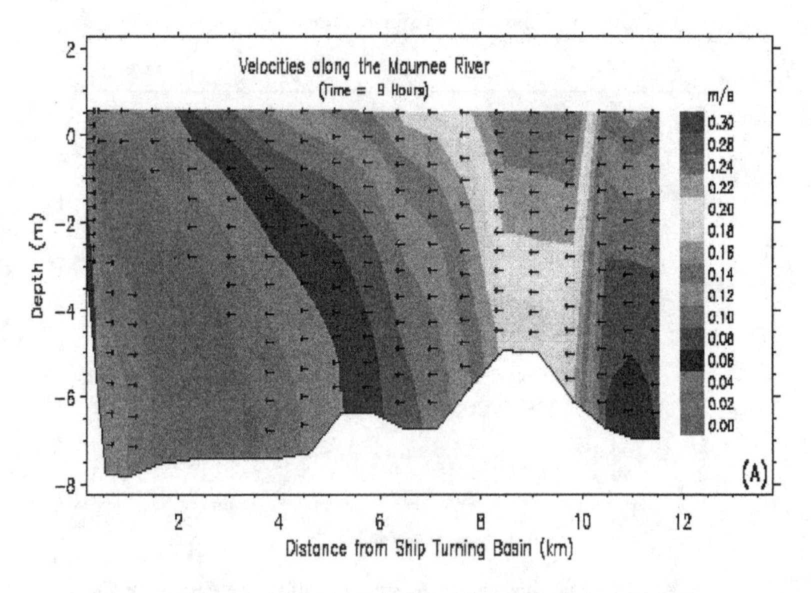

Figure 9A: Case 1: Velocities in the Maumee River

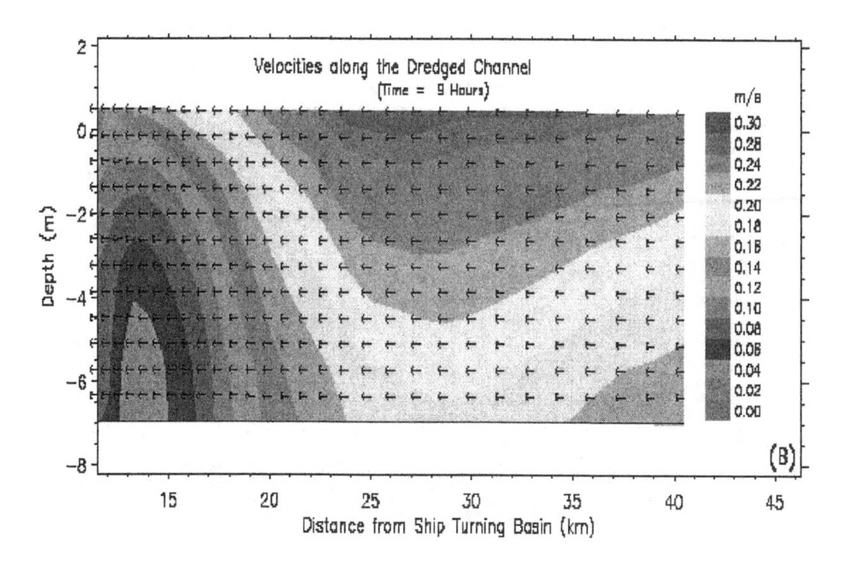

Figure 9B: Case 1: Velocities in the Dredged Channel

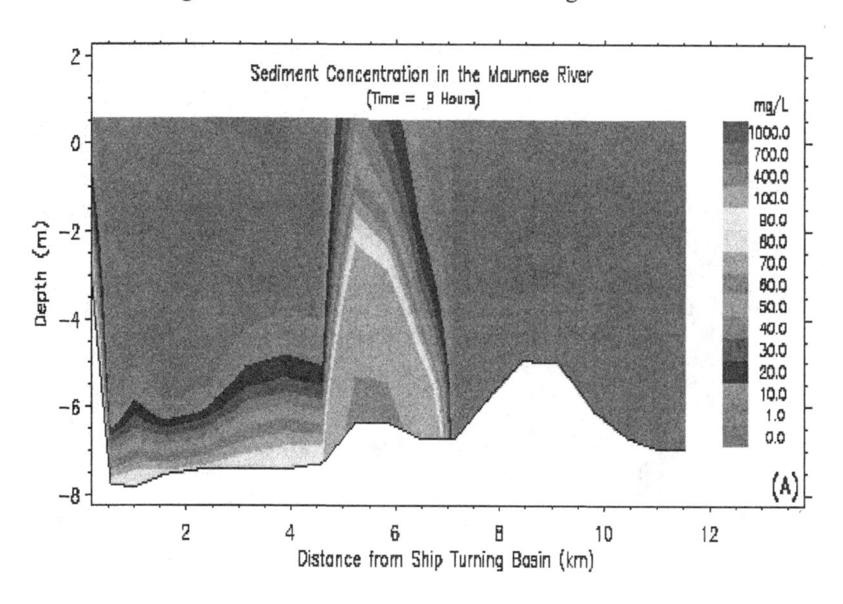

Figure 10A: Case 1: Sediment 3 Concentration in the Maumee River

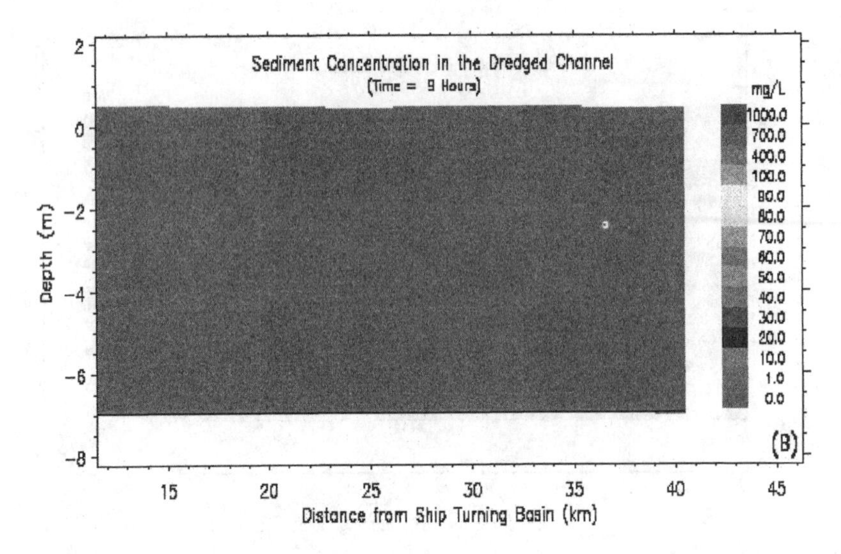

Figure 10B: Case 1: Sediment 3 Concentration in the Dredged Channel

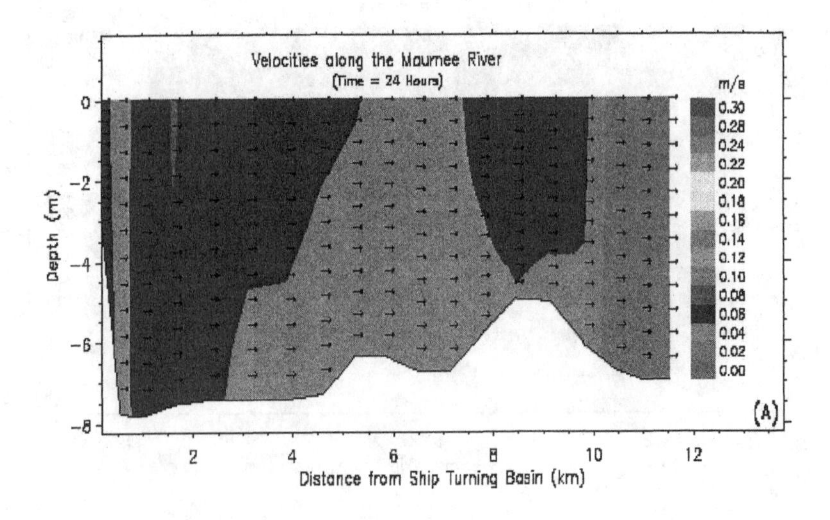

Figure 11A: Case 2: Velocities in the Maumee River

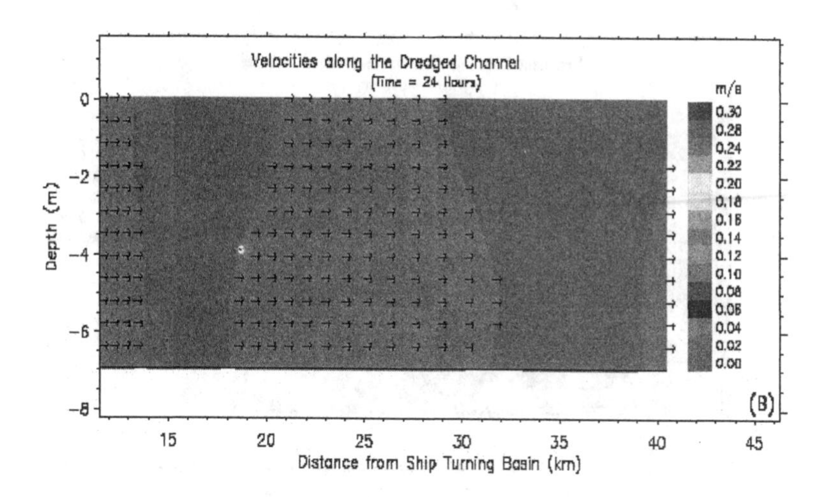

Figure 11B: Case 2: Velocities in the Dredged Channel

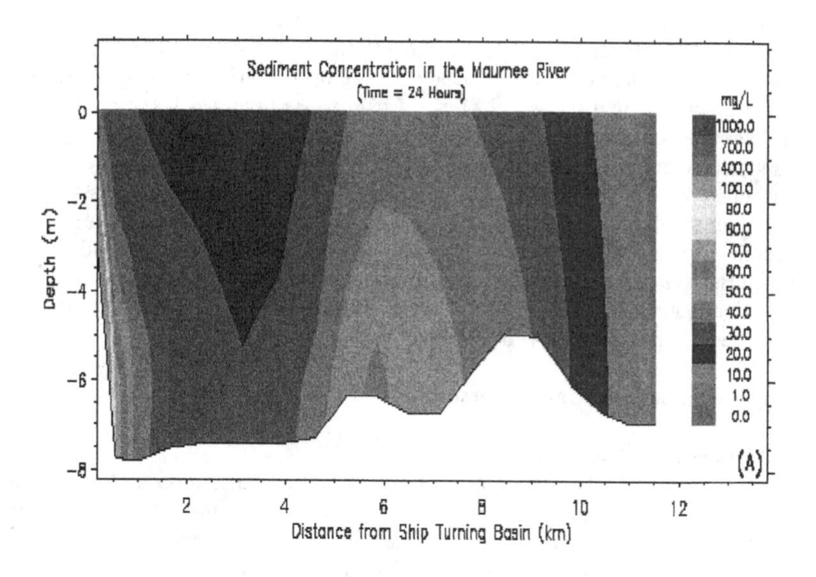

Figure 12A: Case 2: Sediment 3 Concentration in the Maumee River

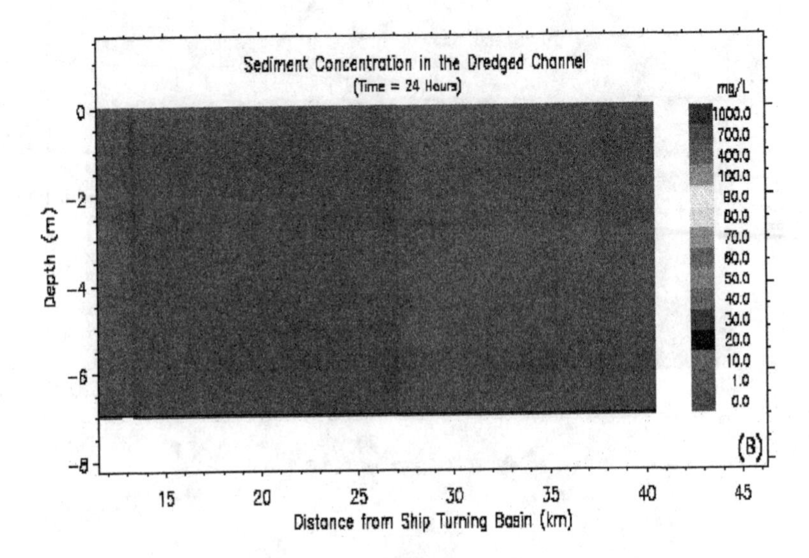

Figure 12B: Case 2: Sediment 3 Concentration in the Dredged Channel

In order to investigate the origination of the sediment in the two mounds, several control volumes are constructed as shown in Figure 6. The mass changes within the ten sediment classes in the two mounds are shown in Tables 2-4 These tables show the amount of sediment deposited or eroded from the first mound (classes 3/4), the second mound (classes 7/8), and Maumee Bay (classes 9/10) respectively. As an example depicting how the tables are to be read consider Table 3 that shows the fine and coarse grain erosion and deposition at Mound number 2 from all the zones in the tributary reach. During the spring 252,337 tons of Mound 2 (class 7) fine sediment are eroded from mound itself, as are 249,757 tons of coarser grained (class 8) sediment. Some 8,770 tons of fine sediment are deposited at Mound 2 from Maumee Bay (class 9) as are 653 tons of the coarse material (class10) Finally 55,132 tons of fine material are deposited at Mound 2 from Mound 1 (class 3) as are 4185 tons of coarse material (class 4) from Mound 1.

Strong Storm/Seiches: Cases 1 versus 3

Tables 2-4 show that, under strong storms

- Sediment from the river (classes 1 and 2) is the overriding source for mound 1. The second source is the sediment between the two mounds. Contributions from other sediment sources are negligible. This means that mound 1 is mainly formed by the deposition of riverine sediment.
- Sediment from the mound 1 (classes 3 and 4) upstream of mound 2 contributes the most to the formation of mound 2; the second source is sediment between the two mounds (classes 5 and 6); the third is lake sediment

(classes 9 and 10); the fourth is riverine sediment (classes 1 and 2) (Figure 14).

- It is noted that coarse sediment and fine sediment pairs from the same origin always display different seasonal changes. For example, classes 3 and 4 are both from the river bottom upstream of mound 1, but class 3 contributes more in fall than in spring, whereas class 4 contributes more in spring than in fall. The same is true for the pairs of classes 5 and 6, and 7 and 8.The differences are not dramatic.
- Table 4 shows that the sediment mass conveyed into the lake in fall is more than in spring.

Class No.	1	2	3	4	5	6	7	8	9	10	Total
Spring	1237	32	-89733	-81436	1	11	0	0	0	0	-169887
Fall	1238	24	-85690	-84805	1	18	0	0	0	0	-169213

Table 2: Bottom Sediment Mass Change (in tons) in Mound 1 in 7 Days under Strong Seiches

Class No.	1	2	3	4	5	6
Spring	1782	0	55132	4185	20532	25241
Fall	1793	1	57051	3273	19627	26267

Class No.	7	8	9	10	Total
Spring	-252337	-249757	8770	653	-385799
Fall	-245363	-242136	8635	763	-370090

Table 3: Bottom Sediment Mass Change (in tons) in Mound 2 in 7 Days under Strong Seiches

Class No.	1	2	3	4	5	6
Spring	945	0	31189	322	15042	233
Fall	993	0	33440	308	14456	199

Class No.	7	8	9	10	Total
Spring	27786	5497	24732	2648	108104
Fall	27416	5379	24646	2832	109393

Table 4: Sediment Mass (in tons) into Maumee Bay in 7 Days under Strong Seiches

Weak Storm/Seiches: Cases 2 versus 4

Tables 5-7 show that, under weak storms

- The main contributor for mound 1 is riverine sediment (class 1).
- The main contributors for mound 2 are riverine fine sediment (class 1) and the fine sediment upstream of mound 1 (class 3), more in fall than in spring.
- The riverine and fine sediment contributions from upstream of mound 1 are relatively equal, which is not the case under strong storm conditions.
- The sediment into Maumee Bay in fall is more than in spring. This is similar to the results under strong storms.

Class No.	1	2	3	4	5	6	7	8	9	10	Total
Spring	585	0	-1787	-460	0	0	0	0	0	0	-1661
Fall	509	0	-1898	-449	0	0	0	0	0	0	-1837

Table 5: Bottom Sediment Mass Change (in tons) in Mound 1 in 7 Days under Weak Seiches

Class No.	1	2	3	4	5	6	7	8	9	10	Total
Spring	1599	0	1490	0	38	30	-10660	-1712	76	0	-9140
Fall	1627	0	1791	0	42	36	-9659	-1193	54	0	-7302

Table 6: Bottom Sediment Mass Change (in tons) in Mound 2 in 7 Days under Weak Seiches

Class No.	1	2	3	4	5	6	7	8	9	10	Total
Spring	1614	0	1652	0	248	0	1855	0	1003	0	6371
Fall	1747	0	1883	0	263	0	1784	0	1050	0	6728

Table 7: Sediment Mass (in tons) into Maumee Bay in 7 Days under Weak Seiches

CONCLUSIONS

The numerical simulation results reveal the sediment transport and formation mechanisms of the two mobile sediment mounds in the Maumee River/Lake Erie confluence. Strong storms can cause flow reversals in the Maumee River and entrain several times higher sediment concentration in the river than under weak storms. The upstream mound (number 1) is mainly formed by riverine sediment settlement and the downstream mound (number 2) is mainly formed by the sediment resuspension in the river bottom upstream of Mound 2 under strong storms/seiches. During weak storms Mounds 1 and 2 are mainly formed by riverine sediment under weak storms/seiches; bottom sediment upstream of Mound 1 is the second largest contributor to the formation of Mound 2. Strong vertical thermal stratification exists in spring scenario, but not in fall. This stratification influences the sediment transport in the river/coastal confluence. Under both strong and weak storms/seiches, the riverine sediment (classes 1 and 2) deposition at Mound 1 is more in spring than in fall, but more in fall than in spring at Mound 2; the total sediment transport into the Maumee Bay is more in fall than in spring. In all the stratification tests the fall and spring differences are not dramatic. Finally, only during strong, flow reversing storms is the deposition at Mound 2 of Lake sediments of sufficient strength to suggest that the hypothesized marine turbidity maxima physics was present.

REFERENCES

Bedford, K. (1989). The effect of Lake Erie on its tributaries-resulting tributary transport mechanisms and contrasts with marine estuaries, Lake Erie Estuarine Systems: Issues, Resources, Status, and Management. Proceedings of the National Oceanic and Atmospheric Administration Estuary-of-the-Month Seminar Series, No. 14, May 4, 1988, United States Department of Commerce, Washington, D. C.

Bedford, K. W. (1992). The physical effects of the Great Lakes on tributaries and wetlands: A summary, Journal of Great Lakes Research 18(4): 571-589.

Bedford, K. W., Velissariou, P., Velissariou, V. and Guo, Y. (1999). Integrated Analysis of the Impact of Unconfined Placement Activities on Near-Shore Sensitive Areas, Contract DACW39-95-K-0018, The Ohio State University.

Chapman, Raymond S., R. C., Associates, Johnson, B. H. and Vemulakonda, S. R. (1996). User's Guide for the Sigma Stretched Version of CH3D-WES, *Technical report hl-96-21*, US Army Corps of Engineers Waterway Experiment Station.

Chu, Y.-F. P., Bedford, K. W. and Welsh, D. J. (1999). Lake Michigan Forecast System and the prospects for a sediment transport model, *in* M. L. Spaulding and H. Butler (eds), *Estuarine and Coastal Modeling, Proceedings of the 6th International Conference*, American Society of Civil Engineers, New Orleans, Louisiana, pp. 755–764.

Davis, J. R. and Sheng, Y. P. (1999). High performance estuarine and coastal environmental modeling: the CH3D example, *in* M. L. Spaulding and H. Butler (eds), *Estuarine and Coastal Modeling, Proceedings of the 6th International Conference*, American Society of Civil Engineers, New Orleans, Louisiana, pp. 470–484.

Fischer, H., List, E., Koh, R., Imberger, J. and Brooks, N. (1979). *Mixing in Inland and Coastal Waters*, Academic Press, Inc. Glenn, S. M. (1983). *A Continental Shelf Bottom Boundary Layer Model: The Effects of Waves, Currents, and a Moveable Bed*, PhD thesis, Woods Hole Oceanographic Institute of Technology, Massachusetts Institute of Technology, Woods Hole, Massachusetts 02543.

Glenn, S. M. and Grant, W. D. (1987). A suspended sediment stratification correction for combined wave and current flows, *Journal of Geophysical Research* **92**: 8244–8264.

Grant, W. D. and Madsen, O. S. (1986). The continental-shelf bottom boundary layer, *Ann. Rev. Fluid Mech.* **18**: 265–305.

Gunther, H., Hasselmann, S. and P.A.E.M.Janssen (1992). Wamodel Cycle 4, *Technical report no.4*. edited by Modell Betreuungsgruppe.

Herdendorf, C. E. (1990). Great Lakes estuaries, *Estuaries* **13**(4): 493–503.

Johnson, B., Wang, H. and Kim, K. (1995). Can numerical estuarine models be driven at the estuary mouth, *in* M. L. Spaulding and R. T. Cheng (eds), *Estuarine and Coastal Modeling, Proceedings of the 4th International Conference*, American Society of Civil Engineers, San Diego, California, pp. 255–267.

Keen, T. R. and Glenn, S. M. (1994). A coupled hydrodynamic-bottom boundary layer model of Ekman flow on stratified continental shelves, *Journal of Physical Oceanography* **24**: 1732–1749.

Libicki, C. and Bedford, K. W. (1991). Sudden, extreme Lake Erie storm surges and the interaction of wind stress, resonance, and geometry, *Journal of Great Lakes Research* **16**(3): 380–395.

O'Neil, S. (2002). *Three Dimensional Mobile Bed Dynamics for Sediment Transport Modeling*, PhD thesis, The Ohio State University, Columbus, Ohio.

Platzman, G. W. and Rao, D. B. (1964). Spectra of Lake Erie water levels, *Journal of Geophysical Research* **69**(12): 2525–2535.

Podber, D. (1991). *Modeling Great Lakes tributary flow and transport*, Master's thesis, The Ohio State University, Columbus, Ohio.

Schwab, D. J. (1978). Simulation and forecasting of Lake Erie storm surges, *Monthly Weather Review* **106**: 1476–1487.

Sheng, Y. (1983). Mathematical modeling of three dimensional coastal currents and sediment dispersion: Model development and application, *Technical Report CERC-83-2*, U. S. Army Engineer Waterways Experiment Station, Vicksburg, MS.

Spasojevic, M. and Holly, F. M. (1994). Three-dimensional numerical simulation of mobile-bed hydrodynamics, *Contract Report HL-94-2*, Iowa Institute of Hydraulic Research, The University of Iowa.

Spasojevic, M. and Holly, F. M. J. (1990). 2-D bed evolution in natural watercourses: New simulation approach, *Journal of Waterway, Port, Coastal, and Ocean Engineering* **116**(4): 425–443.

Velissariou, P., Velissariou, V., Guo, Y. and Bedford, K. (1999). A multi-size, multi-source for-mulation for determining impacts of sediments on near-shore sensitive sites, *in* M. L. Spaulding and H. Butler (eds), *Estuarine and Coastal Modeling, Proceedings of the 6th International Conference*, American Society of Civil Engineers, New Orleans, Louisiana, pp. 59–73.

Welsh, D., Bedford, K. W., Guo, Y. and Sadayappan, P. (2001). A Coupled Wave, Current, and Sediment Transport Modeling System, *Fourth International Symposium on Ocean Wave Measurement and Analysis*.

Welsh, D., R.Wang, P.Sadayappan and K.W.Bedford (1999). Coupling of Marine Circulation and Wind-Wave Models on Parallel Platforms, *Technical report*, The Ohio State University.

Yassuda, E. A. and Sheng, Y. P. (1997). Modeling dissolved oxygen dynamics of Tampa Bay during summer of 1991, *in* M. L. Spaulding and A. F. Blumberg (eds), *Estuarine and Coastal Modeling, Proceedings of the 5th International Conference*, American Society of Civil Engineers, Alexandria, Virginia, pp. 427–440.

Application of a Sediment Transport Model (BFSED) to Assess the Impact from Bridge Reconstruction in the Barrington River, RI

Christopher Galagan[1], C. Kirk Ziegler[2], Tatsusaburo Isaji[1], John King[3]

Abstract

A sediment transport model implemented within a hydrodynamic and water quality modeling system provides a tool that can be applied to sediment erosion and transport problems. The model, BFSED, simulates erosion, suspended sediment transport and deposition processes. BFSED is implemented within a map-based, graphical user interface and sophisticated, boundary-fitted hydrodynamic model (BFHYDRO). Tools for collecting and manipulating input data and analyzing and animating model results complete the modeling system. This paper shows the combined model's capabilities and the results of an application to a sediment transport problem associated with bridge reconstruction across a tide-dominated estuary.

The BFSED model predicted changes to sediment erosion and deposition resulting from the reconstruction of a highway bridge and causeway system across a narrow arm of a tidal estuary. The proposed removal of sections of the causeway adjacent to the bridge would increase the cross-sectional area of the estuary at the highway crossing, significantly changing tidal flow and sediment erosion, transport and deposition patterns in the area. The concern motivating the study was that these changes might be significant enough to affect small boat navigation and mooring. A field study supplied BFSED model input data requirements. The field study included collection of high-precision depth data, recording a month-long tide elevation time series, and collection of sediment cores and sediment surface samples.

The results of the coupled BFSED-hydrodynamic model predict a marked decrease in the speed of maximum flood currents across the bridge opening after construction of the replacement bridge. The hydrodynamic model also predicts an increase in tidal current speed in areas adjacent to the causeway sections to be removed, potentially changing the dominant process in these areas from net deposition to net erosion and causing a predicted redistribution of the eroded sediment to other parts of the estuary. The BFSED model was used to predict the maximum depth of erosion expected to occur in the areas of increased flow, the volume of sediment eroded from these areas, and the spatial distribution of the resulting sediment deposits.

[1]Applied Science Associates, Inc., 70 Dean Knauss Drive, Narragansett, RI 02882 www.appsci.com
[2]Quantitative Environmental Analysis, LLC, 305 West Grand Ave, Suite 300, Montvale, NJ 07645 www.qeallc.com
[3]Graduate School of Oceanography, University of Rhode Island, Narragansett, RI 02882

Introduction

WQMAP is a software system combining a number of analytical tools and numerical models used to determine pollutant transport and fate for pathogens, nutrients, thermal effluent, and toxic materials. Components of the WQMAP system include: 1) a state of the art boundary fitted grid generation capability; 2) hydrodynamic models (BFHYDRO) to solve water mass and momentum equations and predict time-varying surface elevations and velocity vectors; 3) pollutant transport models to solve 3D conservation of mass equations, and predict constituent concentration in time and space; 4) a eutrophication model (EPA WASP5) to predict the transport and transformation of multiple state variables in the phosphorous and nitrogen cycles, and other natural systems. The tools and models of WQMAP are implemented within a Geographical Information System (GIS) framework that supports the display and spatial analysis of model input data and model results. Some of the applications of the WQMAP system include dredging and dredge disposal impacts (Swanson and Mendelsohn, 1998; Swanson, et al, 2000), determining the extent of thermal plumes (Swanson, et al, 2003; Ward and Thomas, 2002; Swanson and Mendelsohn, 2000), predicting fecal coliform concentration in space and time (Swanson, et al, 1998), and simulating nitrogen loading from point and non-point sources (Rines, et al, 2001; Ward and Swanson, 2002).

A sediment transport model, BFSED, has been added simulate the erosion, suspended load transport and deposition of cohesive and non-cohesive sediment. The model predicts the spatial and temporal distribution and amount of sediment that accumulates on or erodes from the sediment bed, as well as simulating sediment concentrations in the overlying water column. BFSED is based on the work done by Ziegler (2002), Ziegler and Nisbet (1994, 1995), and Ziegler and Lick (1986). The addition of a sediment transport model (BFSED) to the WQMAP model suite makes available a complete modeling system for addressing sediment transport problems.

For this study, the BFHYDRO and BFSED models predicted changes in hydrodynamics, sediment erosion, transport and deposition processes associated with highway bridge reconstruction across the Barrington River at the head of Narragansett Bay, Rhode Island.

The replacement bridge design being considered would increase the river cross-sectional area, potentially changing the flow and sediment transport patterns in the river, and affecting nearby boat docks and moorings. Figure 1 is a map of the area showing the highway bridge (Route 114) and causeways.

The work included a field study to collect detailed, high-precision depth data; a sampling program to collect sediment samples from the river bottom; and a tidal height survey to record water elevations in the river over a complete lunar cycle. Data from the tide survey was used to assess the tidal dynamics in the area and to adjust the depth data collected to a mean sea level datum. The sediment sampling included the collection of cores up to 1.5 meters in length, from which sediment deposition rates were calculated using radiometric techniques. The BFSED-

BFHYDRO model system was applied to predict the changes in water flow in the river and subsequent changes in sediment transport occurring after proposed reconstruction of the Route 114 causeway. This paper presents a brief description of the model systems and the results of their application to the bridge reconstruction issues.

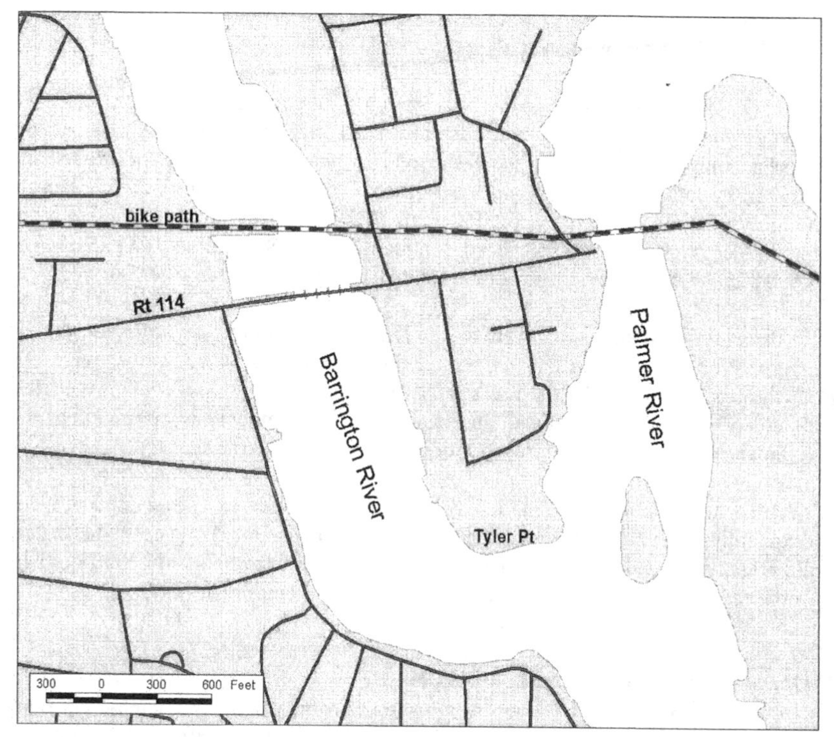

Figure 1. Map showing the location of the Route 114 bridge over the Barrington River and the causeway on the west bank of the River.

WQMAP BFHYDRO Hydrodynamic Model

ASA has developed and applied evolving versions of sophisticated model systems (Spaulding et al., 1999; Mendelsohn et al., 1995) for use in studies of coastal waters for more than a decade. WQMAP, as the model system is known, uses a three-dimensional boundary fitted finite difference model (BFHYDRO) developed by Muin and Spaulding (1997a and b). The hydrodynamic model predicts the four dimensional distribution (in space and time) of surface elevation, currents, temperature and salinity. It has been applied to many estuaries and coastal areas around the world. The calibrated and verified model has been used to simulate the currents in Boston Harbor, Massachusetts, for dredging studies (Swanson and

Mendelsohn 1996), for proposed port and dock developments at Quonset Point, Rhode Island, (Swanson et al., 1999), and for evaluation of potential dredged material disposal sites in the Providence River and upper, mid and lower Narragansett Bay, Rhode Island, (Swanson and Ward, 1999; Swanson and Mendelsohn, 1998). Most recently BFHYDRO was used in conjunction with SSFATE to estimate the extent of the suspended sediment concentrations from dredging operations in the Providence River, Rhode Island (Swanson et al., 2000).

The BFHYDRO boundary-fitted model matches the model coordinates with the shoreline boundaries and bathymetric features such as dredged channels to accurately represent study areas. This system also allows the user to adjust the model grid resolution to obtain more detail in areas of most concern. The approach is useful for the highly variable geometry of the Barrington River. Development of the boundary fitted model approach has proceeded for many years (Spaulding, 1984; Swanson, 1986; Swanson et al., 1989; Muin, 1993; and Huang and Spaulding, 1995a).

The three-dimensional conservation of mass and momentum equations, with approximations suitable for lakes, rivers, and estuaries (Swanson, 1986; Muin, 1993) form the basis of BFHYDRO. The model can be driven by tides, winds, river flow and density differences due to salt and temperature. The model can be run in either a two dimensional, vertically averaged mode, or in three dimensions.

A detailed description of the model with associated test cases is described in Muin and Spaulding (1997a) and (Muin, 1993). The model has undergone extensive testing against analytical solutions and has been found to perform accurately and quickly. Specific model comparisons are found in Huang and Spaulding (1995b).

WQMAP-BFSED Sediment Transport Model

The hydrodynamic model output can be used as input to any number of other specialized models. For this application a model jointly developed by ASA and Quantitative Environmental Analysis (QEA) was used to simulate the erosion, suspended sediment transport and deposition processes in the Barrington River. BFSED uses a two-dimensional, vertically averaged current and a Sigma coordinate system in the vertical dimension. The model has been applied to many river and estuarine environments (Ziegler, et al., 2000; Ziegler, 2002). The BFSED model uses the same boundary-fitted grid as the BFHYDRO hydrodynamic model for its computations. The primary output product of BFSED is a bed elevation change map showing the amount of erosion and deposition occurring on the riverbed over time, and a sediment concentration time series.

The BFSED model predicts the erosion, transport and deposition of sediment suspended by tidal and other currents. The sediment is divided into three size fractions: a cohesive fraction consisting of silt and clay size particles (less than 62 microns diameter); a non-cohesive fraction consisting of fine-to-medium sand particles (62 – 425 microns); and the coarse fraction containing coarse sand, gravel and cobbles (everything greater than 425 microns). Sediment is eroded in response to bottom shear stresses from a current passing over the sediment bed. Sediment erosion occurs if the bottom shear stress applied by the current exceeds the critical shear stress value for the bed. Below the critical shear no erosion is predicted. When bottom shear exceeds the critical shear, erosion takes place.

Once sediment particles are suspended, they are transported and allowed to deposit on the bottom if the current shear is below the critical shear value. BFSED treats suspended cohesive sediments as a single particle class in the deposition process. The deposition flux of cohesive sediment to the riverbed is a product of the probability of deposition, the settling speed of sediment flocs, and the sediment concentration in the water column. Settling speeds of cohesive sediment flocs, taken from measurements (Burban, et al, 1990), are a function of sediment concentration and bottom shear stress. The probability of deposition is a parameterization of the effects that near-bed turbulence and floc size have on the rate of deposition (Partheniades, 1992).

Non-cohesive sediments form a second class of particles in the deposition calculations of BFSED. The rate at which non-cohesive sediment particles settle through the water column is controlled largely by particle size. A correction factor is applied to non-cohesive sediment particles to account for the stratification that occurs as fine and medium sands descend through the water column. (For a more complete description of the model refer to Ziegler and Lick, 1986; Ziegler and Nisbet, 1994, 1995; Ziegler, et al, 2000).

The BFSED model computes suspended sediment concentration, net erosion and deposition in each cell of the boundary-fitted grid at each time step for the specified simulation period. The results of the model include a bed elevation map showing the change in the riverbed over time due to erosion or deposition. A positive change in bed elevation indicates deposition, a negative change erosion.

Field Data Collection and Application of the Models

Setting

The area investigated is a 914-meter (3000 ft) long section of the Barrington River just north of its confluence with the Palmer River near the head of Narragansett Bay, Rhode Island. The area is a low-lying plain characterized by dense urban and suburban development. Several fresh and saltwater wetlands occur in the area, particularly north of the bridge.

Sediment surface and core samples collected under this and previous studies show that fine grain deposits occur on either side of the Route 114 bridge causeways on both the east and the west banks of the river; in the area of floating docks on the east bank of the river south of the bridge causeway; and in small areas on the fringes of the river from the causeway to Tyler Point to the south. Areas of active sediment transport occur along the center of the river where the channels are deeper (4.5-6 m

(15-20 ft)) than the areas of deposition (1-3 m (3-10 ft)), and are characterized by coarse sands, gravels and cobbles.

Tides dominate the flow of water through this portion of the Barrington River. There are two high and two low tides per day. The mean tide range at Warren, Rhode Island (approximately 1 km downstream from the Route 114 bridge) is 1.4 meters (4.6 feet), with a mean spring tide range of 1.7 meters (5.7 feet). The highest current speeds occur down the center of the River with maximum currents across the narrowest river cross-sections corresponding to the Route 114 and bike path bridge causeways. While the causeways result in increased flow through the center of the river, they act as barriers to flow at the edges of the river immediately up- and down-stream from the causeway structure. Some of the lowest current speeds are seen in the nearshore areas adjacent to the bridge causeways. These areas of low current velocity experience long term sediment deposition.

The areas of active sediment transport where higher current velocities prevail carry suspended sediment up and down the river with the tidal flow. This sediment load increases and decreases throughout the tidal cycle with the speed of the tidal current. The high bottom shear stress forces in these areas of active sediment transport prevent deposition of all but the coarsest material. Moving away from the high flow areas in the center of the river towards the shallower edges, a decrease in current speed results in lower shear stress and deposition of finer grained sediment carried in suspension. This distribution of sediment grain size in the river (a center channel dominated by coarse sand, gravel and cobble material that becomes progressively finer towards the river margins) corresponds well with the distribution of current velocities.

Field Studies

The BFHYDRO and BFSED models require good field data. Data collection included a high-precision depth survey; a tidal height survey; and collection and analysis of sediment surface and core samples.

Depth Survey

Bottom profiles of the Barrington River within the Route 114 bridge span and the area north and south of the bridge were completed during June 2002. The depth data were acquired using a precision depth sounder tied to DGPS hardware that provided digital depth data with sub-meter accuracy horizontal positions. The depth sounder was calibrated at the start of each survey day and checked against manual depth measurements to allow recording depths at an estimated vertical accuracy of +- 15cm. Figure 2 shows the track lines followed during the depth survey and the resulting depth grid processed from the corrected soundings and used by the hydrodynamic and sediment transport models.

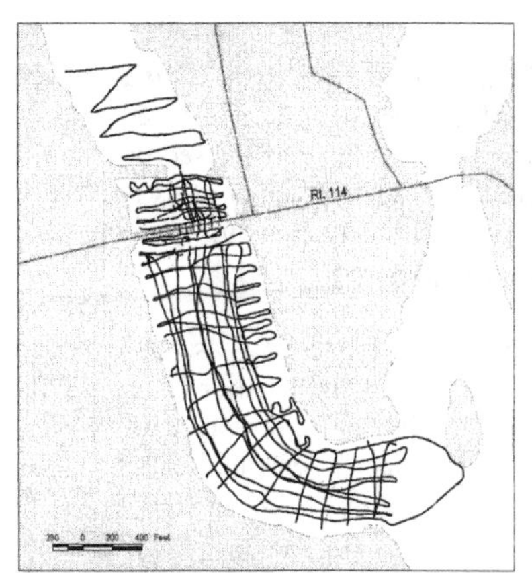

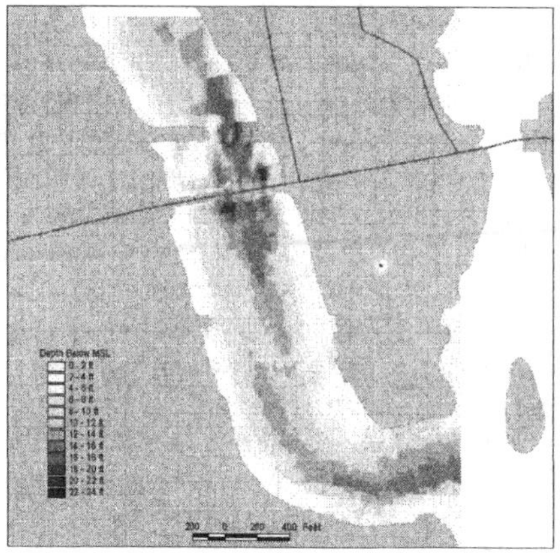

Figure 2. Track lines from depth survey performed during June, 2002, and resulting depth grid generated from the corrected soundings. Depths are in feet below Mean Sea Level.

Tide Height Survey

A tide gauge deployed during the bottom profiling accurately determined the water level in the river and enabled correcting the depth measurements for tidal stage. The gauge was installed at the east end of the Route 114 highway bridge across the Barrington River and recorded the water level at 10-minute intervals for the period June 4, 2002 to July 5, 2002. The elevation of the gauge was tied to the VGVD 1929 vertical datum during a survey on June 18, 2002, using survey instruments and a nearby reference benchmark. Figure 3 shows the water surface elevation data collected during the tidal survey at the Route 114 bridge.

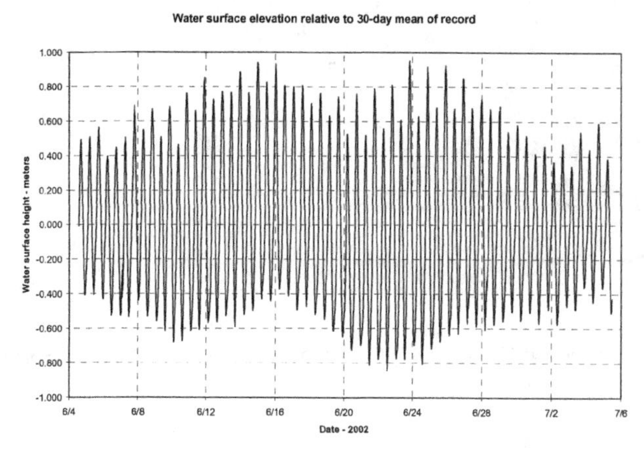

Figure 3. Water surface elevation data collected at the Barrington River bridge, June 4, 2002 through July 5, 2002.

Sediment Samples

Sediment cores and bottom grab samples in the area north and south of the proposed bridge were collected in July 2002. Thirteen surface grab samples and 11 cores were collected (Figure 4). Ten of the surface grab samples and three of the sediment cores collected were analyzed for grain size distribution. The sediment in this section of the river is primarily sand and silty-sand with very little clay content. (See the table in the appendix for a description of the sediment samples and cores.)

The sediment core at site "L" was dated using radiometric techniques. A sediment deposition rate of 3 mm/yr (0.12 in/yr) was determined for that location (King, 2002). This rate was measured using a Cesium 137 signal in the core and represents 40 years of sediment accumulation at the site of the core. The deposition rate in similar river environments around Narragansett Bay ranges from 0.2 – 2.0 cm/yr (0.079 – 0.79 in/yr) (King, 2002).

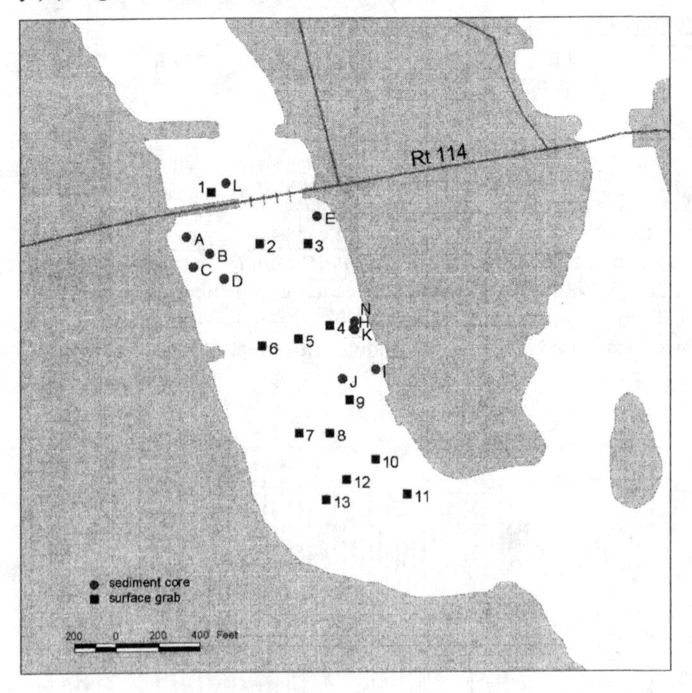

Figure 4. Locations of surface samples and sediment cores taken in the Barrington River.

Model Applications to the Barrington River

Hydrodynamic Modeling

The BFHYDRO model was used to simulate the tidal currents in the Barrington, Palmer and Warren Rivers in upper Narragansett Bay, Rhode Island. The model simulated three months of tidal currents in the river using tidal elevation forcing from the mouth of the Warren River and was calibrated using the tidal elevation data collected during the field program of this study. Currents were developed in a two-dimensional vertically averaged mode. One model simulation used the present causeway and bridge configuration to determine existing currents in the area. A second simulation used the proposed replacement bridge configuration to determine the tidal currents to be expected after the new bridge is in place. Hydrodynamic model results were used to drive the sediment transport model calculations.

Figures 5 and 6 show the simulated tidal velocities generated from the hydrodynamic model for spring flood tide conditions in the vicinity of the Route 114 bridge. Maximum flood conditions are shown for the existing bridge (Figure 5) and the proposed replacement bridge (Figure 6). Note the change in the current pattern immediately around the west causeway resulting from the new bridge configuration. There is a predicted decrease in current velocity across the bridge span that is seen under the new configuration due to the increase in the cross-sectional area of the river at this location.

In the area immediately up and downstream from the west bridge causeway there is a predicted increase in current velocity due to removal of this causeway section. Figure 7 shows the change in surface current velocity predicted by the hydrodynamic model after construction of the new bridge. Positive values show an increase in current speed while negative values depict a current speed decrease. The values displayed in Figure 7 were calculated by finding the difference between the surface current at maximum spring flood tide with the new and existing causeway configurations.

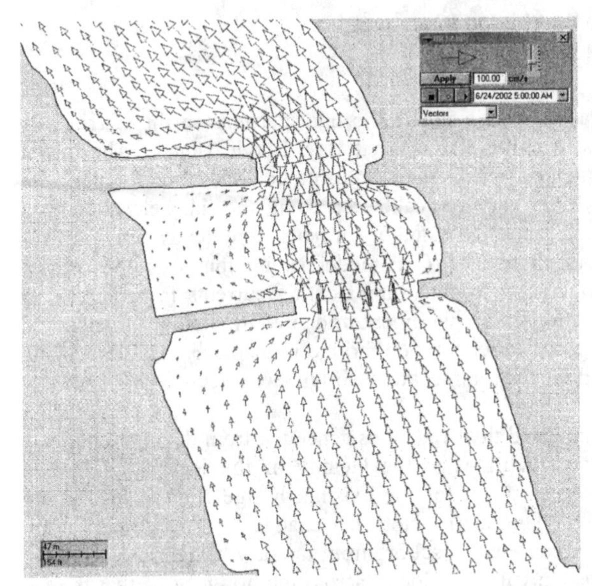

Figure 5. Hydrodynamic model simulated spring tide surface current velocities at maximum flood in the vicinity of the present Barrington River bridge. The box at the upper right shows a current vector representing a current speed of 100 cm/s (1.9 knots).

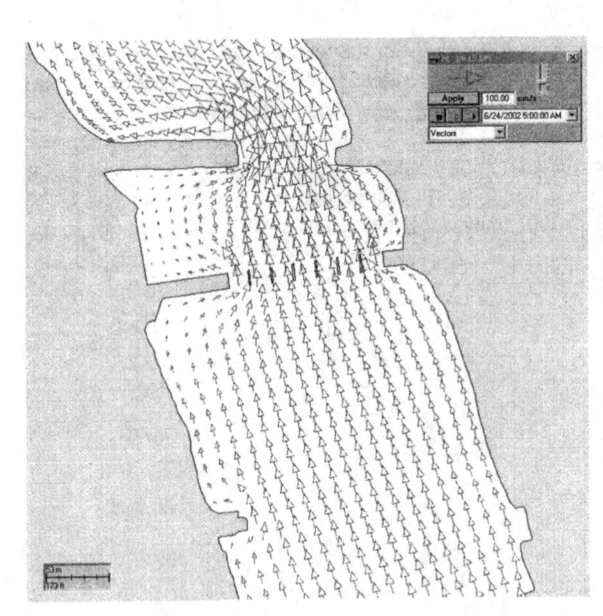

Figure 6. Hydrodynamic model simulated spring tide surface current velocities at maximum flood in the vicinity of the proposed Barrington River bridge. The box at the upper right shows a current vector representing a current speed of 100 cm/s (1.9 knots).

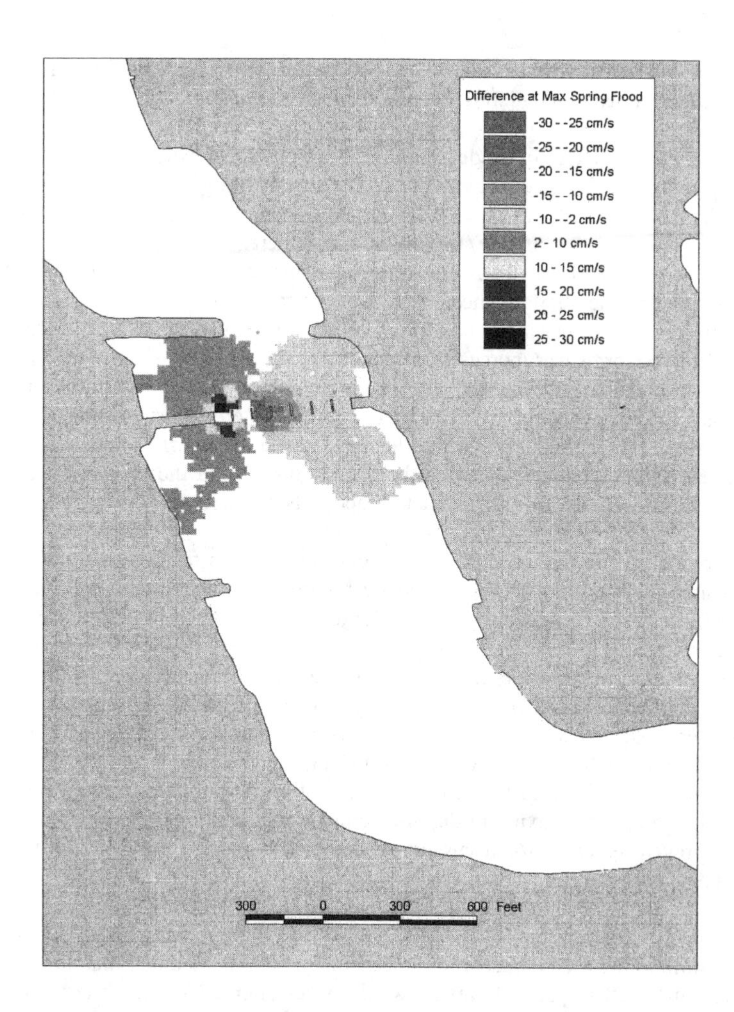

Figure 7. Difference between spring high tide flood surface currents in the Barrington River before and after construction of the proposed Route 114 bridge. Current predictions based on the BFHYDRO model. The green areas (negative values) indicate a decrease in predicted current speed. Positive values indicate areas where the current speed is predicted to increase.

Sediment Transport Modeling

The BFSED model was applied to the Barrington River to determine the change in sediment erosion and deposition patterns that would occur after construction of the proposed replacement bridge. Two-dimensional, vertically averaged currents generated in BFHYDRO were used to drive the sediment transport calculations. It is assumed that sediment bed load entering the bridge area from upstream is negligible.

The results of the hydrodynamic model show that bottom shear will increase in a small area adjacent to the west causeway of the Route 114 bridge (Figure 7) while other areas will see a decrease in current speed and associated decrease in bottom shear forces (green areas in Figure 7). The areas of increased bottom shear will experience either a decrease in deposition rate or an increase in the erosion rate. Within the area of increased bottom shear, areas where net deposition has occurred in the past may become erosional if the bottom shear becomes greater than the critical shear value. If the bottom shear increases but does not exceed critical shear, then the rate of deposition will likely decrease. Within the area of increased bottom shear, those areas that have been eroding in the past will continue to erode but potentially at a higher rate. The BFSED model was set up to simulate the erosion from the area of increased current speed (area of positive values in Figure 7) and the transport and deposition of sediment eroded from this area to other parts of the river.

The material eroded from this area is the silt, clay and fine-to-medium sand size particles that are small enough to be carried by bottom currents. The larger size particles remain in place because they are too large to be entrained by the current. This selective removal of the fine-grain particles causes the coarse fraction to increase as the bed erodes until all the fine-grain material is gone and the bed has armored itself with a layer of coarse sand and gravel. Once this armoring process is complete, the bed no longer erodes. The fraction of coarse material in the sediment bed and the erosion rate at the site control how quickly the armoring occurs and puts a limit on the depth of erosion. An analysis of the sediment adjacent to the bridge causeway was completed to determine the maximum depth of erosion that could occur before bed armoring will prevent erosion. A discussion of this analysis is presented here first, followed by a discussion of the BFSED results.

Analysis of a single sediment core was performed to determine the maximum depth of erosion possible in the area adjacent to the causeway. It was assumed that a bed-armoring layer sufficient to prevent erosion would occur after a 2 cm thick layer of coarse material formed on the sediment bed. Armored layers at the sediment surface were observed throughout the study area with thicknesses of 1-4 cm. The analysis procedure separated the coarse grain particles (greater than 420 microns) from the fine grain particles (less than 420 microns) in Core "L" for 5 cm sections of the core. The thickness of the coarse grain fraction from each 5 cm core section was measured to determine the armoring rate of the sediment deposit as it erodes. It was found that an armoring layer of coarse grain sediment of 2 cm would form after the erosion of 30 cm of the parent sediment bed. It is likely that armoring will occur before the depth of erosion reaches 30 cm, and that the 30 cm depth represents an upper limit on the depth of erosion, and thus a limit on the volume of sediment available for erosion from the area around the causeway. The maximum volume eroded was calculated by multiplying the armoring depth derived from the sediment core analysis (0.3 m) times the area of increased current speed depicted in Figure 7 (12,742 m^2) and subtracting the volume of the coarse fraction left behind armoring the bed (6.7%).

Max. Bulk Potential Erosion = erosion depth x erosion area
= 0.3 m x 12,742 m^2
= 3,822.6 m^3
Armoring Fraction = 0.067 x 3,822.6 m^3
= 256.1 m^3
Total Potential Erosion = 3,822.6 m^3 – 256.1 m^3
= 3,566.5 m^3

The total potential erosion volume (3,566 m^3) is the volume of sediment available to be transported to other parts of the river and represents the maximum that could be eroded because it assumes that the entire area of river bottom that experiences an increase in bottom shear will erode to a depth of 30 cm (11.8 in).

A simulation using the BFSED model was completed for the Barrington River using the proposed causeway configuration and hydrodynamics. The model was run and a bed elevation map produced as output. The bed elevation map shows the change in the sediment bed elevation as positive (deposition) or negative (erosion) for all parts of the modeled area. The bed elevation map in Figures 8a and 8b show that sediment is eroded from the area of increased current speed on either side of the west causeway of the Route 114 bridge, and deposited up and down the river. Table 1 summarizes the erosion and deposition areas, thicknesses and volumes predicted by the model.

Table 1. Area, thickness and volume of sediment eroded and deposited within the Barrington River as a result of modifications to the existing Route 114 bridge.

Area of Erosion (km^2)	Maximum Erosion Depth (cm)	Maximum Erosion Volume (m^3)	Area of Deposition (km^2)	Max Deposition Thickness (cm)	Maximum Deposition Volume (m^3)
0.013 [3.1acres]	30 [11.8 in]	3,566.5 [4665 yd^3]	2.3 @ 0-1 cm [587 acres] 0.057 @ 1-1.6cm [14 acres]	1.6 [0.6 in]	3,566.5 [4665 yd^3]

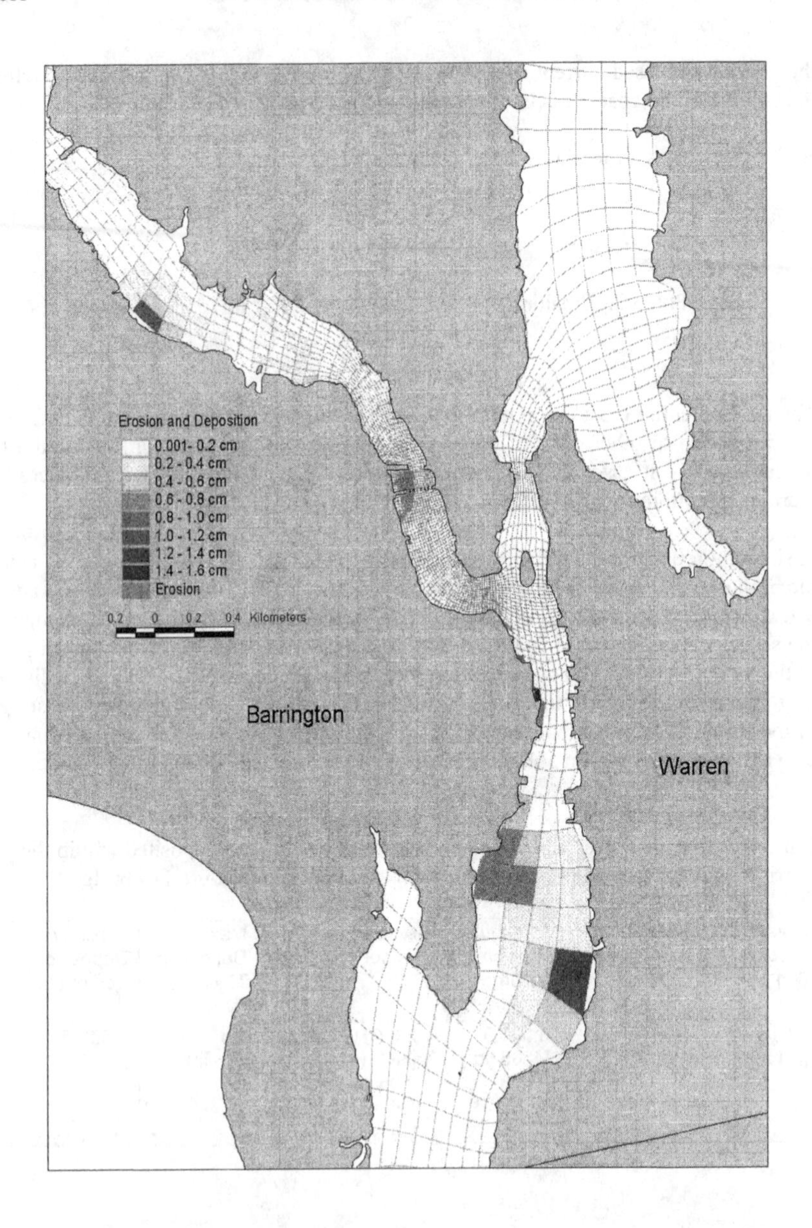

Figure 8a. BFSED results showing areas of predicted sediment deposition (green areas) and erosion (yellow areas) in the Barrington River after construction of the proposed replacement bridge and causeway.

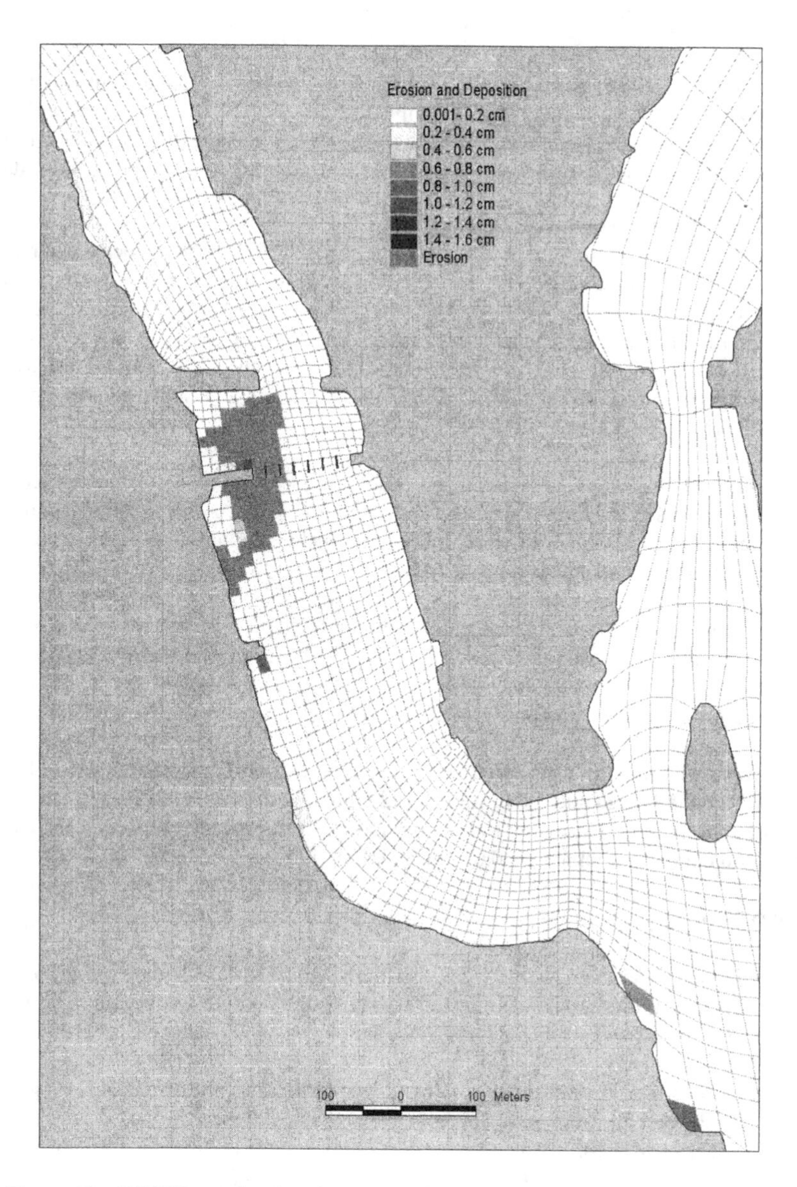

Figure 8b. BFSED results showing areas of predicted sediment deposition (green areas) and erosion (yellow areas) in the vicinity of the Route 114 bridge in the Barrington River

Conclusions

The study demonstrates a complete modeling system for sediment transport computations. The system contains an implementation of the BFSED model within a map-based, graphical user interface including a sophisticated, boundary-fitted hydrodynamic model (BFHYDRO), tools for collecting and manipulating input data and analyzing and animating model results.

The hydrodynamic model predicts that changes in the tidal current velocity as a result of the proposed bridge causeway configuration are limited to the area near the Route 114 bridge. Opening the existing west causeway results in a decrease in current speed across the bridge opening, but an increase in current speed over an area adjacent to the section removed from the west causeway. The area of increased current speed is a potential new source of sediment erosion and transport up and down the river.

An estimated maximum of 3,566.5 m^3 (4665 yd^3) of sediment will be subject to erosion as a result of the new bridge configuration. The eroded sediment will be transported to other parts of the river before the eroding bed armors itself with coarse grain material and erosion stops.

A maximum depth of 30 cm (11.8 in) of sediment erosion is predicted to occur in the area adjacent to the west causeway of the Route 114 bridge after the new bridge is in place. This depth of erosion is limited to an area of 12,742 m^2 (3.1 acres), representing a volume of 3,566.5 m^3 (4665 yd^3).

The eroded sediment is distributed in a thin deposit extending 3,460 meters (11,352 ft) north of the bridge and 3,840 meters (12,598 ft) south of the bridge, covering an area of 2,375,808 m^2 (587 acres). Maximum deposit thickness is 1.6 cm (0.6 in), and the average deposit thickness is 0.05 cm (0.02 in).

References

Applied Science Associates, Inc. 1996. Field investigations and hydrodynamic and sediment transport modeling study for temporary Route 114 bridges over the Barrington and Warren Rivers. ASA Project 93-097, December 1993.

Burban, P.Y, Xu, Y.J., McNeil, J., and Lick, W., 1990. Settling speeds of flocs in freshwater and seawater. Journal of Geophysical Research, 95(C10), pages 18,213 – 18,220.

Huang, W. and M. Spaulding, 1995a. Modeling of CSO-induced pollutant transport in Mt. Hope Bay. ASCE J. of Environmental Engineering, Vol. 121, No. 7, July, 1995, 492-498.

Huang, W. and M. L. Spaulding, 1995b. A three dimensional numerical model of estuarine circulation and water quality induced by surface discharges. ASCE Journal of Hydraulic Engineering, 121:(4) April 1995, p. 300-311.

King, John and C. Gibson, 2002. Barrington River Project Report. Report submitted to Applied Science Associates, Inc. in support of erosion and sediment redistribution calculations for proposed Route 114 bridge at the Barrington River.

Mendelsohn, D., E. Howlett, and J.C. Swanson, 1995. WQMAP in a Windows Environment. In: Proceedings of the 4th International Conference on Estuarine and Coastal Modeling, October 26-28, 1995, San Diego, CA.

Muin, M., 1993. A three dimensional boundary fitted circulation model in spherical coordinates. Ph.D. Thesis, Department of Ocean Engineering, University of Rhode Island, Kingston, RI.

Muin, M. and M.L. Spaulding, 1997a. Three-dimensional boundary fitted circulation model. In: *Journal of Hydraulic Engineering*, Vol. 123, No. 1, January 1997, p. 2-12.

Muin, M. and M.L. Spaulding, 1997b. Application of three dimensional boundary fitted circulation model to the Providence River. In: *Journal of Hydraulic Engineering*, Vol. 123, No. 1, January 1997, p. 13-20.

Partheniades, E., 1992. Estuarine sediment dynamics and shoaling processes. In: Handbook of Coastal and Ocean Engineering, Volume 3, Edited by J. Herbick, pp. 985-1071.

Rines, H.M., Isaji, T., Swanson, J.C., Mendelsohn, D., 2001. Total nitrogen modeling of coastal embayments in Chatham, Massachusetts. Submitted to the Town of Chatham Massachusetts by Applied Science Associates, Narragansett, RI, ASA Report 2000-011.

Spaulding, M. L., 1984. A vertically averaged circulation model using boundary fitted coordinates. Journal of Physical Oceanography, May, pp. 973-982.

Spaulding, M., D. Mendelsohn and J.C. Swanson, 1999. WQMAP: An integrated three-dimensional hydrodynamic and water quality model system for estuarine and coastal applications. Marine Technology Society Journal, Vol. 33, No. 3, pp. 38-54, Fall 1999.

Swanson, J.C., Kim, H-S, Subbayya, S., Hall, P., and Patel, J., 2003. Hydrothermal modeling of the cooling water discharge from the Vermont Yankee power plant to the Connecticut River. Submitted to Normandeau Associates, Bedford, NH by Applied Science Associates, ASA Report 2002-088.

Swanson, J.C., and D. Mendelsohn, 2000. Canal Station thermal plume modeling. Submitted to TRC Environmental Corporation, Lowell, MA by Applied Science Associates, ASA Report 1999-031.

Swanson, J.C., Isaji, T. and Ward, M., 2000. Dredged material plume modeling for the Providence River and Harbor maintenance dredging project. Submitted to SAIC, Newport, RI by Applied Science Associates, ASA Report 1999-063.

Swanson, J. C. and M. Ward, 1999. Extreme bottom velocity estimates for CRMC dredged material disposal sites in Narragansett Bay. Submitted to SAIC, Newport, RI. Submitted by Applied Science Associates, Narragansett, RI, ASA Project 98-044.

Swanson, J.C., Mendelsohn, D., and T. Isaji, 1998. Receiving water quality modeling of combined sewer overflow impacts to the Providence River. Submitted to Louis Berger & Associates, Inc., Providence, RI by Applied Science Associates, ASA Report 1996-105.

Swanson, J.C., and D. Mendelsohn, 1998. Velocity estimates for candidate dredged material disposal sites in Narragansett Bay. Submitted to SAIC, Newport, RI by Applied Science Associates, ASA Report 1997-059.

Swanson, J. C., and D. Mendelsohn, 1996. Water quality impacts of dredging and disposal operations in Boston Harbor. Presented at: ASCE North American Water and Environmental Congress '96 (NAWEC '96), Anaheim, CA, June 22-28,1996.

Swanson, J. C., 1986. A three dimensional numerical model system of coastal circulation and water quality. Ph. D. Thesis, Dept. of Ocean Engineering, University of Rhode Island, Kingston, RI.

Ward, M. and J.C. Swanson, 2002. Miacomet Pond nutrient loading model. Submitted to Massachusetts Executive Office of Environmental Affairs, Massachusetts Watershed Initiative, and Department of Environmental Protection, Bureau of Resource Protection by Applied Science Associates, Narragansett, RI, ASA Project 99-157.

Ward, M. and Thomas, R., 2003. Thermal dispersion modeling in support of the ACG Central Azeri project. Submitted to Azerbaijan International Operating Company by Applied Science Associates, ASA Project 2002-172.

Ziegler, C.K., P.H Israelsson, and J.P. Connolly, 2000. Modeling sediment transport dynamics in Thompson Island Pool, upper Hudson River. Water Quality and Ecosystem Modeling, Volume 1, pp. 193-222.

Ziegler, C. K., 2002. Evaluating sediment stability at sites with historic contamination. Environmental Management, Volume 29, No. 3, pp. 409-427.

Ziegler, C.K., and B.S. Nisbet, 1995. Long-term simulation of fine-grained sediment transport in a large reservoir. ASCE Journal of Hydraulic Engineering, 121(11), pp. 773-781.

Ziegler, C.K., and B.S. Nisbet, 1994. Fine-grained sediment transport in Pawtuxet River, Rhode Island. ASCE Journal of Hydraulic Engineering, 120(5), pp. 561-576.

Ziegler, C.K., and W. Lick, 1986. A numerical model of resuspension, deposition, and transport of fine-grained sediments in shallow water. Univ. of California, Santa Barbara, Report ME-86-3.

Appendix

Sediment grain size from sediment surface samples and cores. Non-organic, non-carbonate sediment grain size was determined by treating the sediment with hydrogen peroxide and acetic acid then wet sieving the sediment through a 63 micron sieve. A portion of the less-than 63 micron fraction was analyzed using an Elzone Model 180XY particle size counter and the percentages of sand, silt and clay were determined.

		DRY WEIGHT (g)		% Vol	% Vol				%SILT	
Sample Name	Interval (cm)	>63	<63	>3.9u	>15.6u	% SAND	% SILT	%CLAY	63-15.6u	<15.6u
Station 1	grab	17.975	1.242	99.61	54.18	93.5	6.4	0.0	3.5	3.0
Station 2	grab	6.469	0.127	98.57	29.22	98.1	1.9	0.0	0.6	1.4
Station 3	grab	4.717	1.332	98.85	25.72	78.0	21.8	0.3	5.7	16.4
Station 4	grab	10.020	1.826	98.93	35.03	84.6	15.2	0.2	5.4	10.0
Station 6	grab	13.076	1.005	99.71	67.44	92.9	7.1	0.0	4.8	2.3
Station 7	grab	6.349	0.099	98.50	45.56	98.5	1.5	0.0	0.7	0.8
Station 8	grab	9.701	0.285	97.54	54.52	97.1	2.8	0.1	1.6	1.3
Station 9	grab	9.213	1.111	98.39	44.01	89.2	10.6	0.2	4.7	6.0
Station 11	grab	11.301	0.253	98.41	46.90	97.8	2.2	0.0	1.0	1.2
Station 13	grab	15.872	0.289	98.32	49.35	98.2	1.8	0.0	0.9	0.9
Core A	0-2	0.666	1.092	98.15	45.10	37.9	61.0	1.1	28.0	34.1
Core L	0-2	3.996	0.601	99.12	39.10	86.9	13.0	0.1	5.1	8.0
Cor N	0-2	2.114	2.817	98.84	43.04	42.9	56.5	0.7	24.6	32.5

Analysis of the stratification in the Guadiana estuary

L. Pinto[1], A. Oliveira[2], A.B. Fortunato[3] and A.M. Baptista[4]

Abstract

The construction of new dams in the Guadiana River (Portugal/Spain) motivated the analysis of the hydrodynamics of its estuary under strong river flows. The main objectives were to determine under which combinations of river flow and tidal amplitudes stratification occurs, and to characterize the velocity and salinity structure under these conditions. High-resolution numerical simulations were performed using a three-dimensional baroclinic model (ELCIRC) with a level 2.5 Mellor and Yamada turbulence closure scheme. Field data collected in the scope of this study were used both to validate the model and to complement the analysis of the estuarine behavior.

The data were adequately reproduced and results were consistent with theory. Model results also provided additional details on the spatial distribution of salinity. For instance, the model shows that stratification decreases near the tip of the salt wedge at the end of flood. Some empirical criteria to predict stratification conditions were shown to be inadequate for the Guadiana estuary.

Both data and model results were used to verify the hypothesis that pelagic organisms able to migrate vertically during the tidal cycle can take advantage of the baroclinic circulation to remain in the estuary even at high river flows. Although residual velocities are directed seaward at all depths, it was shown that, under stratified conditions, the vertical phase lags in the velocity field are sufficient to allow an organism with vertical migration capabilities to move upstream.

1. Introduction

The construction of several dams in the Guadiana watershed has recently motivated an inter-disciplinary study to understand how changes in the freshwater flow affect the estuary. Salt-water intrusion and stratification, which are directly affected by the changes in river flow, are of primary concern. Salinity controls to a

[1] Research Fellow, Estuaries and Coastal Zones Division, National Laboratory of Civil Engineering, Av. do Brasil, 101, 1700-066 Lisbon, Portugal; phone 351-218443613; fax 351-218443016; llpinto@lnec.pt.

[2] Research Officer, Estuaries and Coastal Zones Division, National Laboratory of Civil Engineering, Av. do Brasil, 101, 1700-066 Lisbon, Portugal; aoliveira@lnec.pt.

[3] Principal Research Officer, Estuaries and Coastal Zones Division, National Laboratory of Civil Engineering, Av. do Brasil, 101, 1700-066 Lisbon, Portugal; afortunato@lnec.pt.

[4] ASCE member, Professor, Oregon Health & Science University, OGI School of Science & Engineering, Center for Coastal and Land-Margin Research, 20,000 NW Walker Road, Beaverton, Oregon 97006-8921, U.S.A.; baptista@ccalmr.ogi.edu.

large extent the spatial distribution of the biota, while stratification has implications on sediment dynamics, water quality and residence times.

The goal of this work is to improve the understanding of the hydrodynamic and saline stratification of the Guadiana estuary. This paper focuses on the analysis of the conditions and factors that influence the stratification in this estuary, as well as on the characterization of the salinity field under stratified conditions. In addition, the consequences of stratification on the retention of pelagic organisms with autonomous vertical mobility are also analyzed.

The study was conducted through a combination of data analysis and the application of a three-dimensional numerical model, to mitigate the limitations of each tool. Field data provided a detailed representation of reality, although with limitations in the temporal and spatial coverage and in the parameter range, while model results provided a better spatial and temporal coverage, as well as control over the parameter space.

The present study was preceded by efforts reported elsewhere. Historical data were first compiled and analyzed (Silva *et al.*, 2000), then used to setup and validate two-dimensional depth-averaged hydrodynamic and transport models (Fortunato *et al.*, 2002a; Oliveira and Fortunato, 2001). The limitations of the historical data, revealed by these analyses, led to the measurement of additional field data in 2001. Finally, satellite images were used to study the plume behavior and to define the required extent of the model grid for large river flows (Pinto and Ferreira, 2002; Pinto, 2003).

Six sections follow this introduction. First, the study site (Section 2), the field data (Section 3) and the model setup (Section 4) are described. The data and model results are then used to characterize the stratification (Section 5) and to analyze the influence of stratification on the retention of some active organisms in the estuary (Section 6). Section 7 summarizes the major conclusions.

2. The study system

The Guadiana forms the southern border between Portugal and Spain (Figure 1). At Moinho dos Canais, 76 km from the mouth, a weir determines the limit of tidal propagation. The width decreases exponentially from 800 m at the mouth to 70 m at Mértola. The average depth is 5 m, and reaches 10-15 m in the main channel.

The flow is forced mostly by tides and river flow. Tides are semi-diurnal, ranging from 0.8 to 3.5 m. The river flow enters the estuary mostly from the upstream boundary, as the tributaries that reach the estuary are small and controlled by dams. Monthly averaged river flows vary between 7 and 57 m^3/s on average years, and reach 1500 m^3/s on very wet years (Fortunato *et al.*, 2002a).

Stratification conditions occur frequently in the Guadiana estuary, depending on the magnitude of the river flow and the tides. Empirical criteria indicate that, for average tidal amplitudes, the estuary is well-mixed for river flows below 10 m^3/s and stratified for values above 100 m^3/s (Gonzaléz, 1995; Fortunato *et al.*, 2002a). Little more is known about the stratification variability because previous analyses were based on limited data or depth-averaged models.

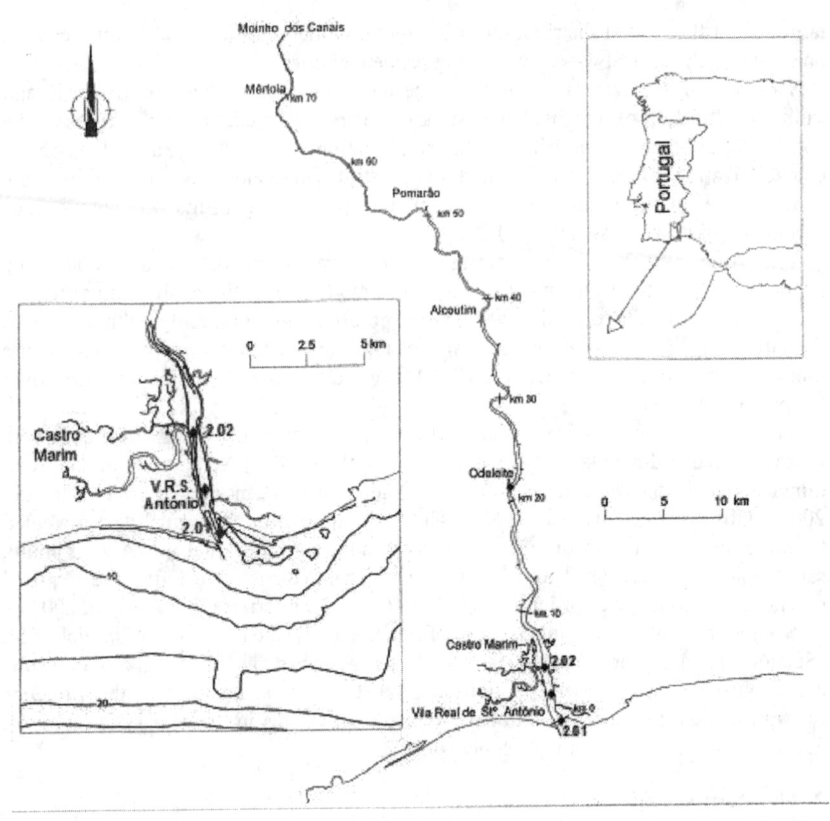

Figure 1 – The Guadiana estuary: geographical location, place names and location of the 2001 measurement stations.

3. Field data

Four field campaigns were performed in the Guadiana estuary in 2001, for various tidal and river flow conditions (Table 1). Hourly velocity and conductivity profiles were measured with an RCM9 current meter, for about 14 hours, with a vertical spacing of 1 m. The top measurements were taken 1 meter below the water surface. Two positions were occupied in each campaign (Figure 1): V.R.S. António and Odeleite (February and September) and stations 2.01 and 2.02 (May and October). In the February and September campaigns, 15-days velocity measurements were also made with an ADCP at a station near V.R.S. António. These observations were made vertically, from the bottom to the surface, with a bin length of 0.5 m and a sampling interval of 5 minutes. Further details on the data are given in Oliveira and Fortunato (2001) and Pinto (2003).

Table 1 – Summary of the field campaigns.

Leading institute	Date	Stations	River flow (m^3/s)	Tidal range (m)
Instituto Hidrográfico	2001/02/02	V.R.S. António Odeleite	384	1.4 (neap tide)
	2001/02/10		2005	3.1 (spring tide)
	2001/09/11		4	1.3 (neap tide)
	2001/09/18		4	3.3 (spring tide)
GEOSUB	2001/05/23	2.01 2.02	70	2.7 (spring tide)
	2001/05/29		35	1.6 (neap tide)
	2001/10/18		20	3.1(spring tide)
	2001/10/24		249	1.2 (neap tide)

4. Model establishment

4.1 Brief description of model ELCIRC

ELCIRC is being developed as an open source community model at the Center of Coastal and Land-Margin Research of the OGI School of Science & Engineering. The model and its numerical properties are described in detail in Zhang and Baptista, 2004, so only its main features are summarized here.

ELCIRC solves the fully non-linear, three-dimensional, baroclinic shallow water equations, coupled to transport equations for salt and heat. Forcings include tides, tidal potential, wind stress and solar radiation. Several turbulence closures are implemented, including a level 2.5 Mellor-Yamada scheme and the generic length scale (GLS) closure scheme recently proposed by Umlauf and Burchard (2003). Hence, from a physics viewpoint, ELCIRC is akin to well-established models such as POM (Blumberg and Mellor, 1987) or QUODDY (Lynch *et al.*, 1996).

Although with important implementation differences (e.g. treatment of baroclinicity and of tangential velocities), the numerical solution is inspired from that of UnTRIM (Casulli and Zanolli, 1998). The equations are solved with a finite volume technique for volume conservation and a natural treatment of wetting and drying. The horizontal domain is discretized with a triangular mesh for flexibility, and z-coordinates are used in the vertical. A semi-implicit time-stepping algorithm and the Lagrangian treatment of the advective terms ensure stability at large time steps.

4.2 Model set-up

The model domain extends from the tidal propagation limit to about 50 km south of the estuary mouth (Figure 2). The coastal area included in the grid was determined by the dimension of the river plume for high river flows observed in satellite images (Figure 2c). The model was forced by tidal elevations and river flows at the ocean and river boundaries, respectively. At the ocean boundary, the major tidal constituents (O_1, K_1, N_2, M_2, and S_2) were taken from the regional model of Fortunato *et al.* (2002b). The spatial variability of the mean water level (Z_0) was also taken from the regional model, while its amplitude was determined from data from the Vila Real de S. António station. Wind stress was neglected, since the conditions during the data collection periods were calm.

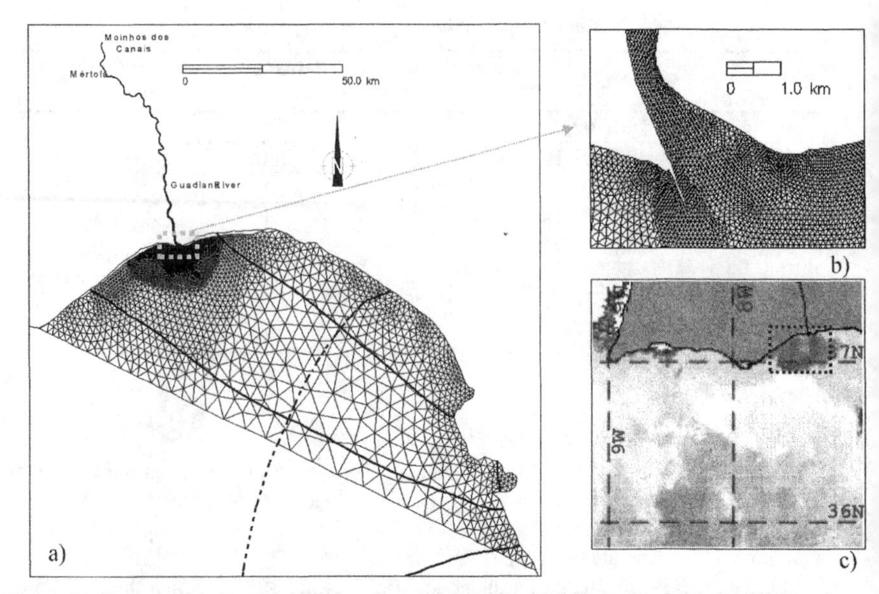

Figure 2 – Horizontal grid, with 12,000 nodes: a) full grid, with M2 amplitudes (solid lines) and phases (dashed lines) taken from the regional model; b) detail of the grid near the mouth; c) river plume from a Sea Surface Temperature Image (February 10, 2001) used to determine the coastal extent of the grid.

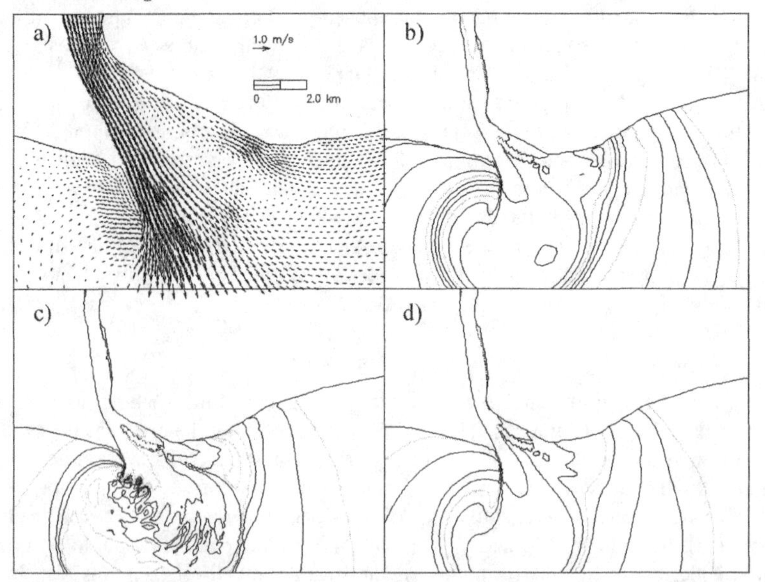

Figure 3 – Time step selection: snapshots of near-surface velocity and salinity fields, 4 days into the simulation. a) velocity, with $\Delta t = 1.5$ min; b) salinity, with $\Delta t = 1.5$ min; c) salinity, with $\Delta t = 5$ min; d) salinity, with $\Delta t = 2$ min. Salinity isolines are separated by 2.5 ppt, while gray isolines start at 5 ppt and are separated by 10 ppt.

The horizontal domain was discretized with an unstructured triangular grid with about 22,000 elements and 12,000 nodes. The spatial resolution varies between 9 and 3,400 m, being finer near the estuary mouth. The vertical direction was divided into 32 levels, spaced between 0.75 m in the upper 14 m (which include the whole estuary) and 200 m in the deepest layer.

All simulations started from rest, with boundary conditions spun-up during two days. Salinity initial conditions were specified as depth-independent, varying from 36.3 ppt in the ocean to 0 ppt at the upstream boundary. Temperature was held constant during the simulations, since both field data and satellite images indicate that the temperature contribution to density variations is negligible (Pinto, 2003).

Vertical diffusion was modeled with a level 2.5 Mellor and Yamada turbulence closure scheme. A sensitivity analysis showed that the minimum mixing length had little effect on the results. Values of 0.15 and 1 m were used for the estuary and the sea, respectively. The horizontal diffusion coefficient was set to zero.

Given the Eulerian-Lagrangian treatment of advection, large time steps can be used. A sensitivity analysis was done using $\Delta t = 5, 3, 2$ and 1.5 minutes. The larger time steps were computationally more efficient, but led to oscillations both in the velocity and salinity fields (Figure 3c, d). The time step was set to 1.5 minutes to avoid oscillatory results (Figure 3a, b).

4.3 Model validation

Calibration with a 2D model (Fortunato *et al.*, 2002a) and with a previous version of ELCIRC (Pinto, 2003) showed that elevations are very insensitive to friction. RMS errors were typically about 10 cm for all stations (V.R.S. António, Odeleite, Alcoutim and Pomarão), with variations with the friction coefficient below 1 cm. Hence, the validation focused on velocity and salinity.

Comparisons with salinity data from V.R.S. António and Odeleite indicate that stratification conditions are well reproduced qualitatively (Figure 4 and Figure 5). For well-mixed and partially mixed conditions, there is a quantitative agreement between the model and the data, with typical errors of about 2-6 ppt at the downstream station (V.R.S. António), and 6 ppt at the upstream station, Odeleite (Figure 4). Under stratified conditions, the model underestimates salinity by over 10 ppt (Figure 5). The analysis of the results and a comparison with 2D simulations suggest that these errors could stem from an incorrect representation of the plume behavior. Failure to reproduce the movement of the plume away from the estuary would lead to a reduction of salinity near the mouth, affecting the water that penetrates the estuary on flood. Indeed, while some of the mechanisms that affect the plume behavior (e.g., tidal residuals, Coriolis) are included in the model, others are not (e.g., shelf currents). Finally, the better results obtained for well-mixed conditions also suggest that the reproduction of the stratification could be improved. This would be consistent with recent experience with ELCIRC on the simulation of the Columbia River estuary and plume (Baptista and Zhang 2004), where the use of Umlauf and Burchard (2003) closures has shown advantages over the Mellor-Yamada closure.

The order of magnitude and the similarity between errors at the two stations suggest that the model provides an adequate representation of the maximum salinity intrusion.

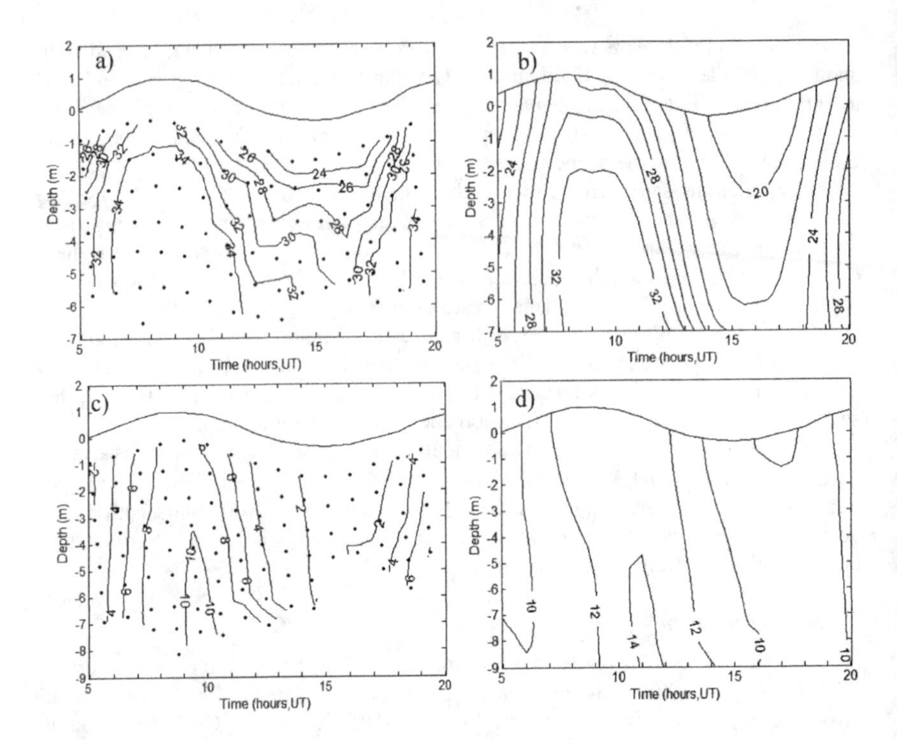

Figure 4 – Comparison between salinity data and model results for the September 11 campaign (ppt): a) V.R.S. António data; b) V.R.S. António model results; c) Odeleite data; d) Odeleite model results. Measurement points in a) and c) are indicated as black circles. Depths are relative to mean sea level.

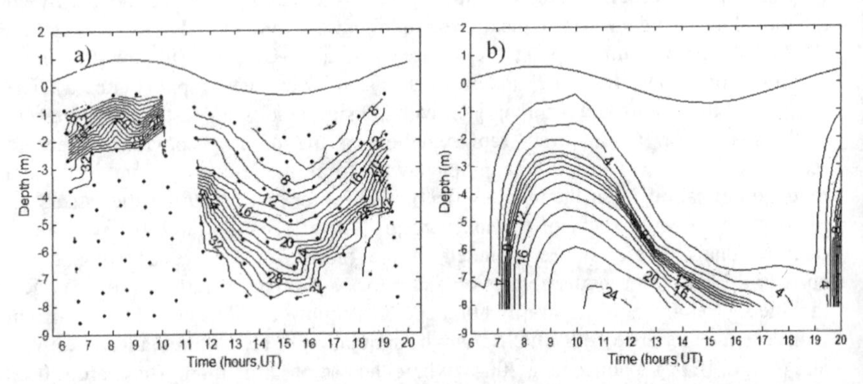

Figure 5 – Comparison between salinity data and model results for V.R.S. António for the February 2 cruise (ppt): a) data, with measurement points represented as black circles; b) model results. Depths are relative to mean sea level.

A comparison between the ADCP data and model results exhibits a reasonable representation of both magnitude and vertical structure of velocities (Figure 6). For instance, zero velocities occur at the correct time along the water column. Also, like the data, model results exhibit maxima near the surface on ebb and a few meters below the surface on flood. However, near-bottom velocities are under-predicted, possibly due to a poor vertical resolution in that area. Indeed, the proper representation of the bottom boundary layer requires a finer resolution (Fortunato and Baptista, 1996).

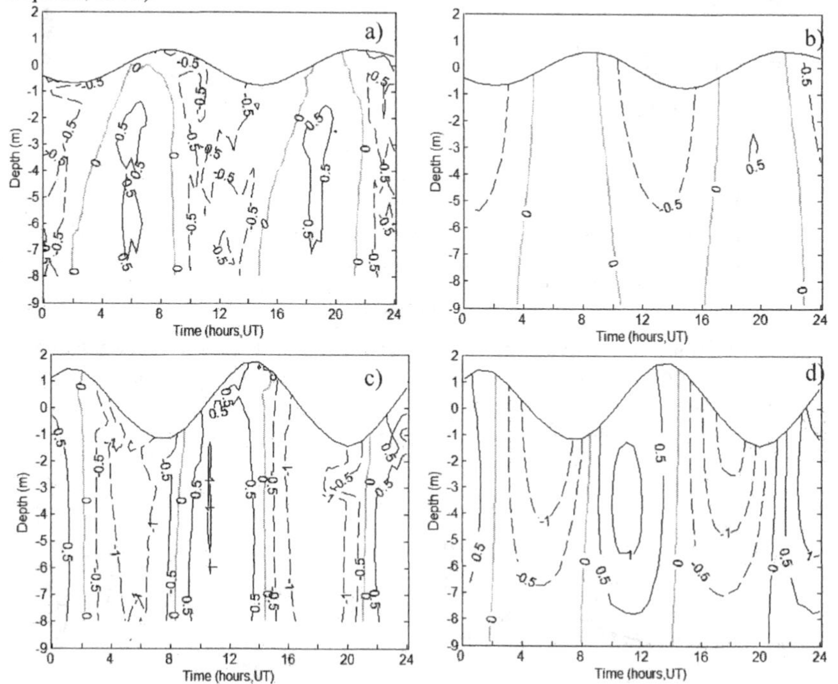

Figure 6 – Comparison between ADCP velocity data along the major axis and model results: a) February 2 data; b) February 2 model results; c) September 16 data; d) September 16 model results. Depths are relative to mean sea level.

5. Stratification analysis

5.1 Analysis based on field data

Stratification conditions were first characterized using the data described above. Prandle (1985) proposed the ratio $\delta S/\overline{S}$ to quantify the magnitude of stratification, where δS is the difference between salinity values at the bottom and at the surface, and $\overline{S}$ is depth-averaged salinity. The estuary is well-mixed if $\delta S/\overline{S} < 0.15$, stratified if $\delta S/\overline{S} > 0.32$ and partially mixed otherwise. Although Prandle (1985) computed this ratio with tidally-averaged values, instantaneous values are used here to highlight the variability of stratification during the tidal cycle.

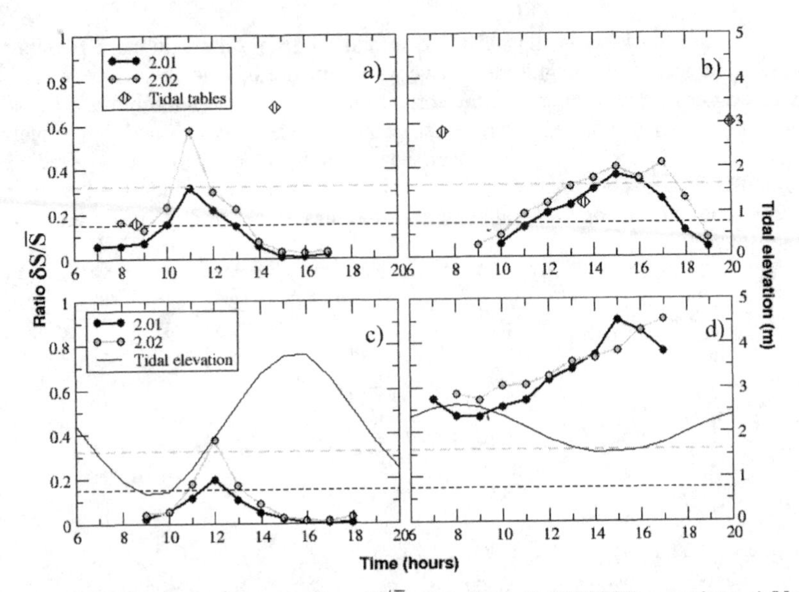

Figure 7 – Stratification analysis based on $\delta S/\overline{S}$ analysis for the GEOSUB campaigns: a) May 23; b) May 29; c) October 18; d) October 24. The estuary is well-mixed below the dashed black line and stratified above the long-dashed grey line.

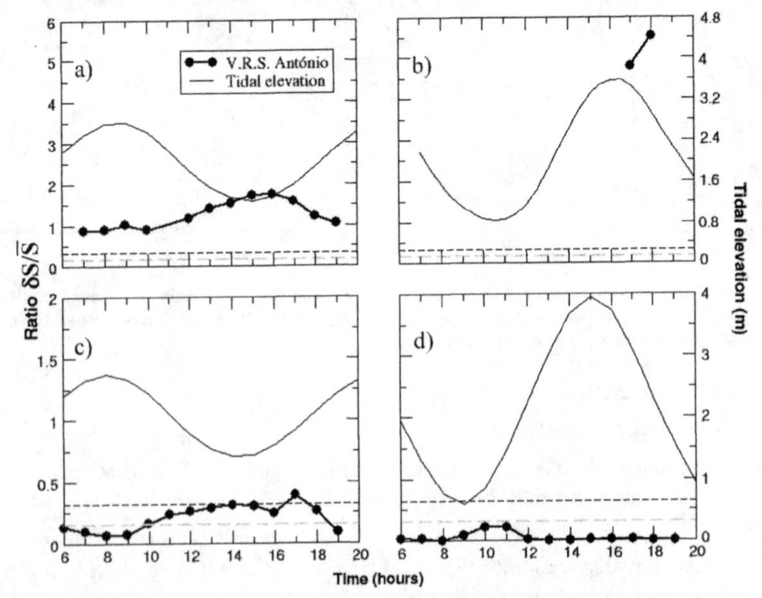

Figure 8 – Stratification analysis based on $\delta S/\overline{S}$ for the IH campaigns: a) February 2; b) February 10; c) September 11; d) September 18. The estuary is well-mixed below the long-dashed grey line and stratified above the dashed black line.

The ratio was calculated for all data sets using salinity values interpolated hourly (Figure 7 and Figure 8). Results were discarded when $\overline{S}$ is smaller than 3 due to the errors in the evaluation of the ratio. Indeed, an error propagation analysis shows that, when $\overline{S}$ goes to zero, a small error in $\overline{S}$ can generate a large error in $\delta S/\overline{S}$.

The ratio $\delta S/\overline{S}$ over the tidal cycle exhibits the expected behavior, highlighting the dependency of the stratification on the river flow and tidal conditions. Stratification increases during ebb and decreases on flood. Stratification also increases with the river flow and decreases with tidal amplitude. For instance, the river flow was similar in the two September cruises (4 m^3/s), but the estuary was well-mixed at spring tides (Figure 8d) and partially mixed during neap tides (Figure 8c).

The magnitude of stratification also varies in space (Figure 7). Data of the GEOSUB campaigns show that stratification is typically stronger for the upstream station, suggesting that a maximum of stratification will occur in different locations of the estuary depending on the magnitude of river flow and tidal amplitude.

5.2 Analysis based on model results

The effect of tidal amplitude and river flow on stratification was analyzed using five runs forced by constant river flows (2, 10, 50, 100 and 200 m^3/s) and three tidal constituents (Z_0, M_2 and S_2). Each simulation was run for 30 days, the first 15 being discarded. Results were analyzed along a 45 km longitudinal transect starting at the head of the breakwater (Figure 9). Selected results are shown in Figures 10 to 12.

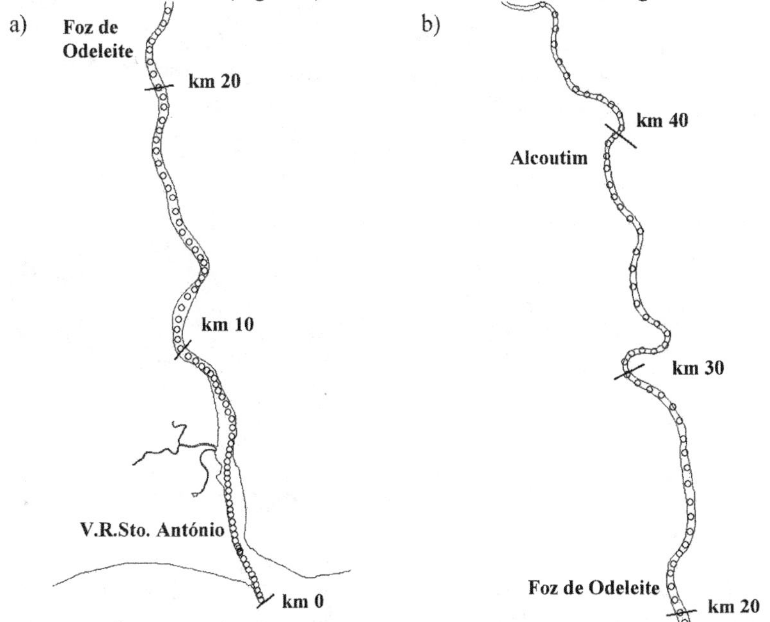

Figure 9 – Location of the longitudinal slice: a) from km 0 to km 20; b) from km 20 to km 40. Circles represent the location of the vertical profiles used in Figures 10-12.

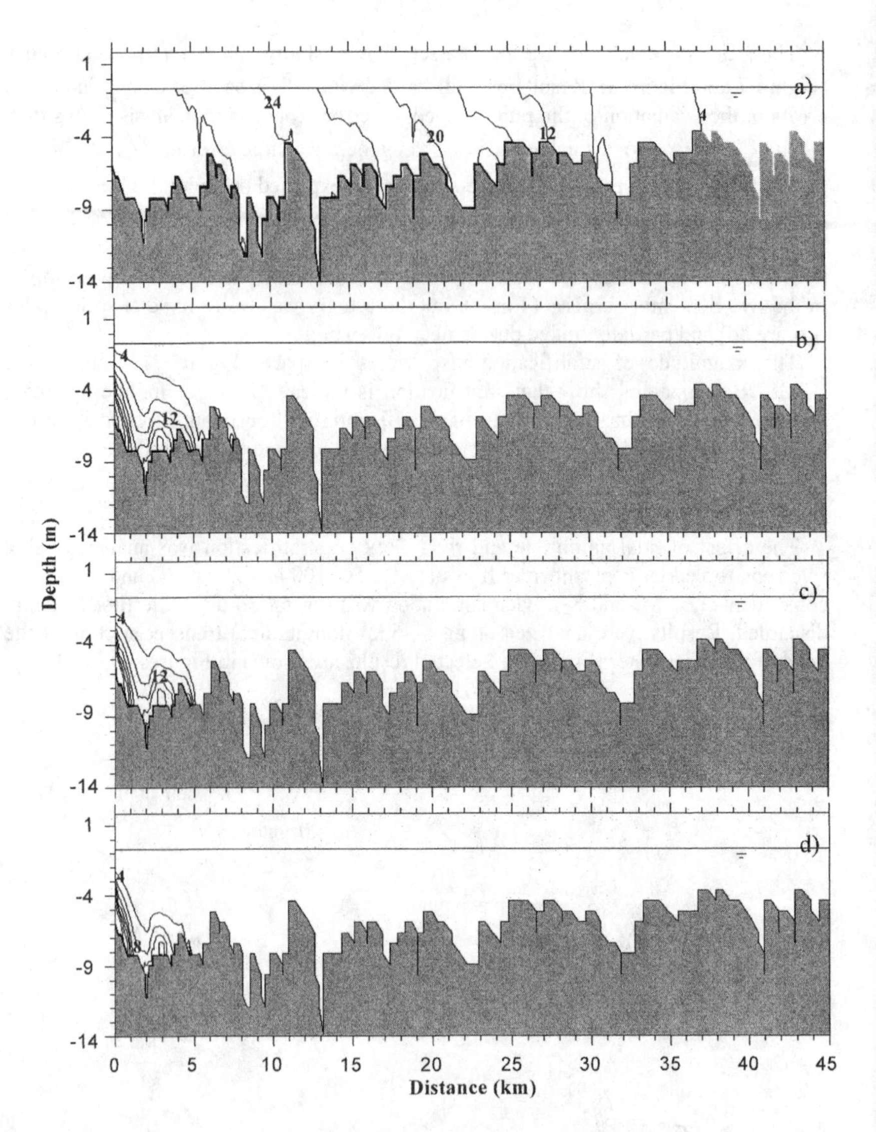

Figure 10 – Salinity field along the longitudinal slice of the estuary. Ebb slack for spring tides: a) Q = 10 m^3/s; b) Q = 50 m^3/s; c) Q = 100 m^3/s; d) Q = 200 m^3/s. Depth is relative to mean sea level.

Results agree with the expected behavior of the estuary. Stratification increases with the river flow and decreases with the tidal amplitude (e.g. Figures 11 and 12). Stratification is also larger in ebb than in flood (Figures 10 and 11), presumably due to the higher shear caused by the opposing baroclinic and barotropic pressure gradients (Jay and Musiak, 1996). Results also indicate that the saline front is less stratified than the downstream part of the estuary. At the end of flood, part of the estuary is vertically homogeneous while other parts exhibit significant salinity gradients (Figures 11 and 12).

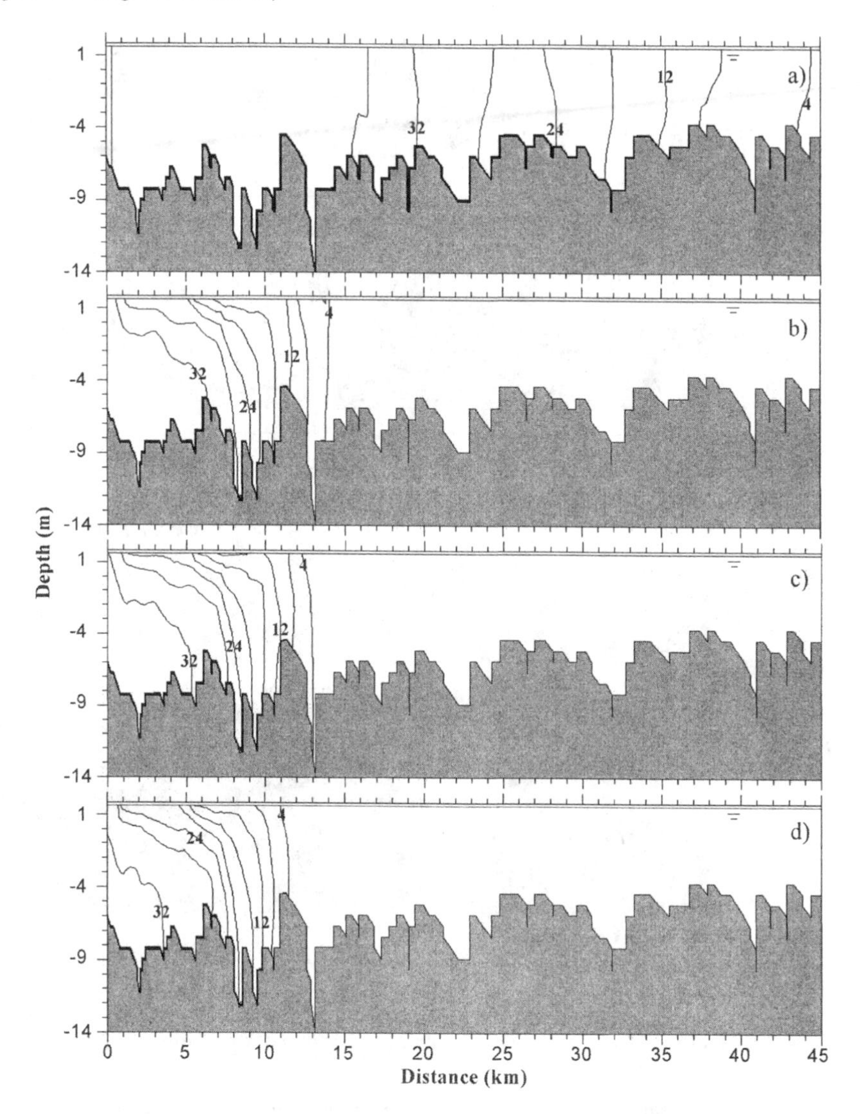

Figure 11 – Salinity field along the longitudinal slice of the estuary. Flood slack for spring tides: a) Q = 10 m³/s; b) Q = 50 m³/s; c) Q = 100 m³/s; d) Q = 200 m³/s. Depth is relative to mean sea level.

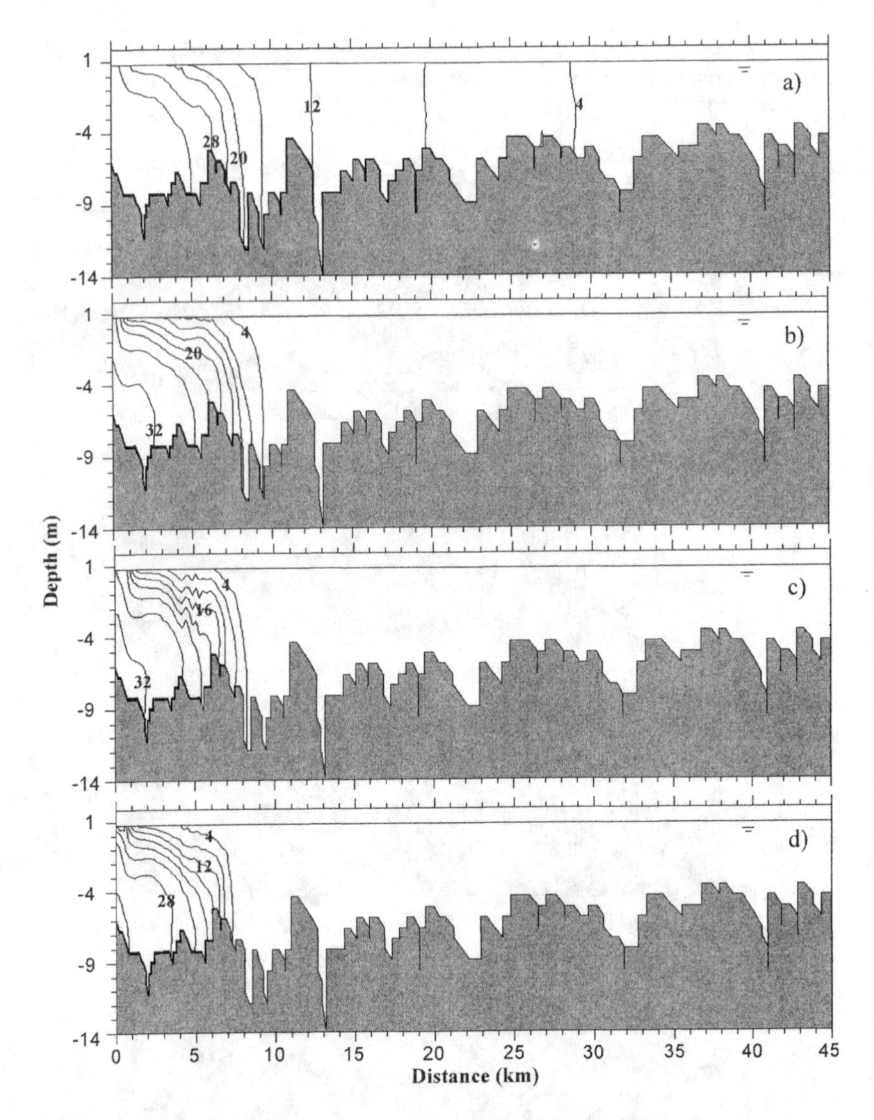

Figure 12 – Salinity field along the longitudinal slice of the estuary. Flood slack for neap tides: a) Q = 10 m³/s; b) Q = 50 m³/s; c) Q = 100 m³/s; d) Q = 200 m³/s. Depth is relative to mean sea level.

During ebb, salinity is trapped in the deep region near the mouth (2 km upstream), leading to a localized increase of stratification.

The ratio $\delta S/\overline{S}$ was again used to quantify stratification (Figure 13 and Figure 14). Again, results were discarded when $\overline{S}$ was smaller than 3.

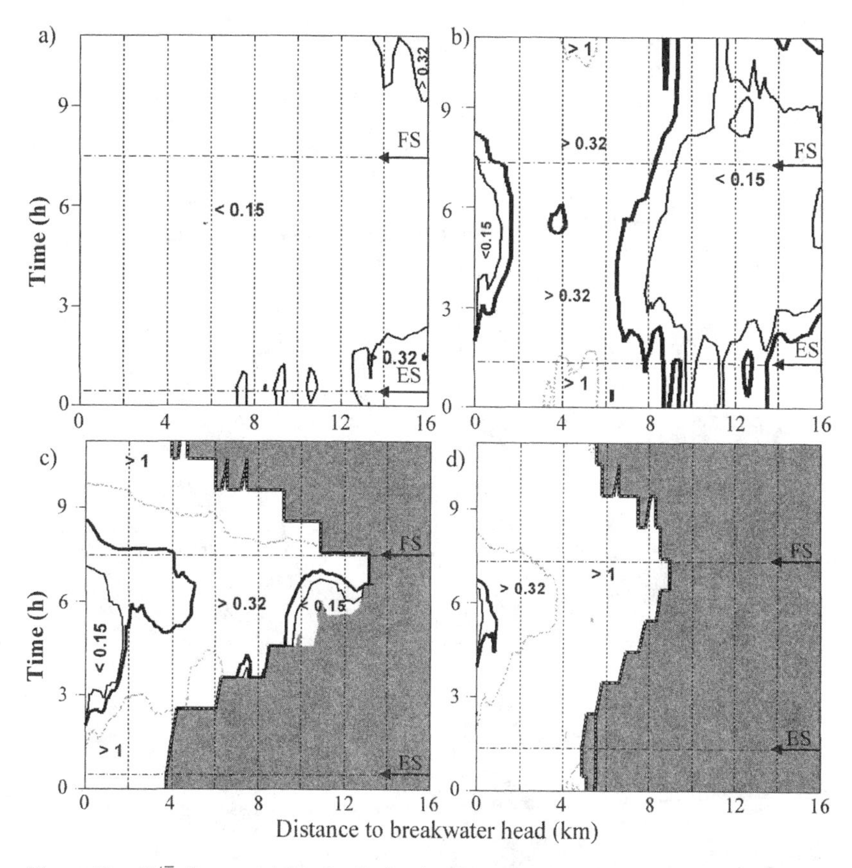

Figure 13 – $\delta S/\bar{S}$ time variability for the longitudinal slice of the estuary: a) Q = 10 m^3/s, spring tide; b) Q = 10 m^3/s, neap tide; c) Q = 50 m^3/s, spring tide; d) Q = 50 m^3/s, neap tide. The estuary is well mixed below the solid thin line and stratified above the solid thick line. Solid gray area: criterion is not valid. FS and ES indicate flood and ebb slack times, respectively.

This analysis confirms that the tip of the salt wedge becomes less stratified at the end of flood, in particular for spring tides. This behavior is only visible in Figure 13c and Figure 14a, due to the limited number of isolines shown, but occurs for river flows above 50 m^3/s. For Q = 200 m^3/s and neap tides, this effect is very small. The vertical phase lag of the velocities provides a likely explanation for this behavior. Due to the higher velocities near the surface, especially on spring tides, the inversion of tidal currents starts near the bottom. At the end of flood, as the near-bottom currents reverse, freshwater is brought underneath brackish water, generating baroclinic instabilities and promoting vertical mixing. In contrast, the reversal of tidal currents at the end of ebb brings denser water to the lower layer, further increasing stratification.

The analysis of $\delta S/\bar{S}$ indicates the conditions under which the estuary is stratified. For a river flow of 2 m^3/s, the estuary is always well-mixed (not shown), as predicted by empirical criteria. The river flow of 10 m^3/s marks the transition between well-mixed conditions and stratified flows. For this value of river flow, the estuary is well-mixed for spring tides, except for the tip of the salt wedge during ebb, where the flow is partially mixed. For neap tides, the estuary is always stratified between kilometers 2 and 6. Mixing at the tip of the salt wedge leads to well-mixed conditions during flood, and partially mixed to well-mixed conditions during ebb. For river flows above 50 m^3/s, the estuary is stratified even for spring tides. This result contradicts the empirical criteria used previously in the Guadiana (see Section 2), but agrees with available data (figures 7b and c), thus shedding doubt on the validity of these criteria.

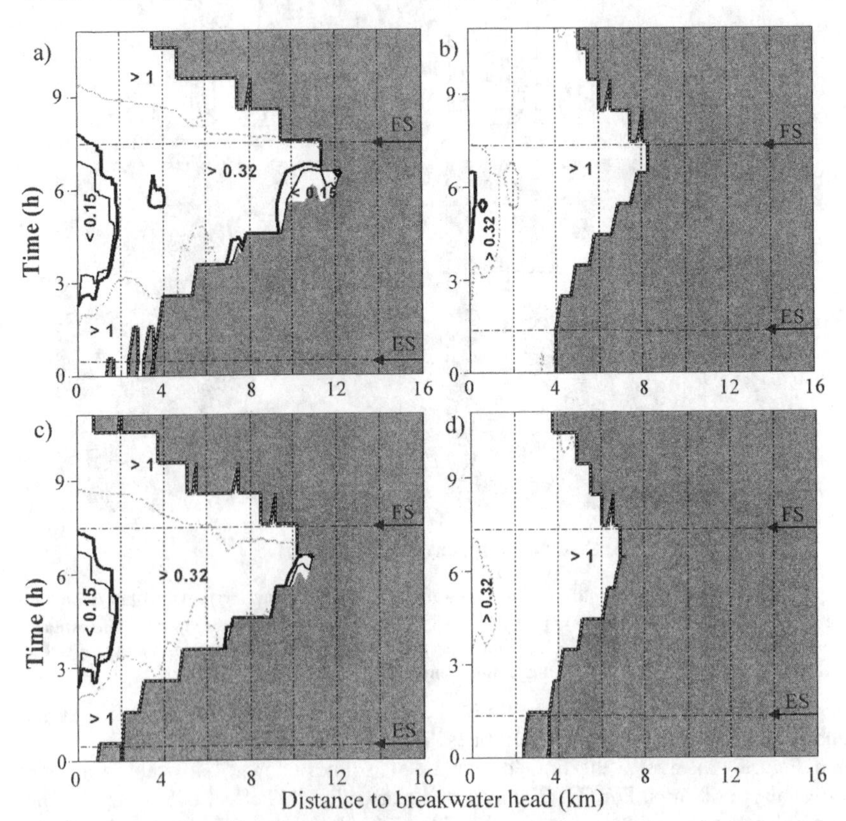

Figure 14 – $\delta S/\bar{S}$ time variability for the longitudinal slice of the estuary: a) Q = 100 m^3/s, spring tide; b) Q = 100 m^3/s, neap tide; c) Q = 200 m^3/s, spring tide; d) Q = 200 m^3/s, neap tide. The estuary is well mixed below the solid thin line and stratified above the solid thick line. Solid gray area: criterion is not valid. FS and ES indicate flood and ebb slack times, respectively.

Furthermore, the empirical criteria used in the Guadiana fail to represent the relative importance of river flow and tidal amplitude in the generation of stratification. Simmons (1955) proposed a criterion to predict estuarine stratification based on the ratio between river flow and tidal prism. This criterion, still widely adopted (e.g., Gonzalez, 1995; Bricker *et al.*, 1999), assumes that the river flow and the tidal amplitude have the same importance in the determination of the stratification strength. The criterion of Uncles *et al.* (1983), used in Fortunato *et al.* (2002b), is based on a similar hypothesis. The present results suggest that these criteria are inappropriate, at least for the Guadiana estuary. Indeed, a four-fold river flow increase leads to minor stratification differences (Figure 13c and Figure 14c), while a 50% reduction in tidal amplitude generates a significantly stronger stratification (Figure 13c and d). These results suggest that tidal amplitude is more important than river flow in the determination of the stratification strength, as recognized by other criteria (e.g., Harleman and Abraham, 1966, Prandle, 1985).

6. Analysis of the displacement of a fictitious organism

6.1 Analysis based on data

The two ADCP records were used to compute the residual velocities for the major velocity component (Figure 15). For both records, the residual velocity is always directed downstream throughout the water column.

The lack of residual velocities directed upstream during the stratified period seems to invalidate a hypothesis proposed by Chícharo *et al.* (2001). According to this hypothesis, some pelagic organisms would migrate vertically during the tidal cycle in order to take advantage of the baroclinic circulation. This migration would allow these organisms to remain in the estuary, even at high river flows.

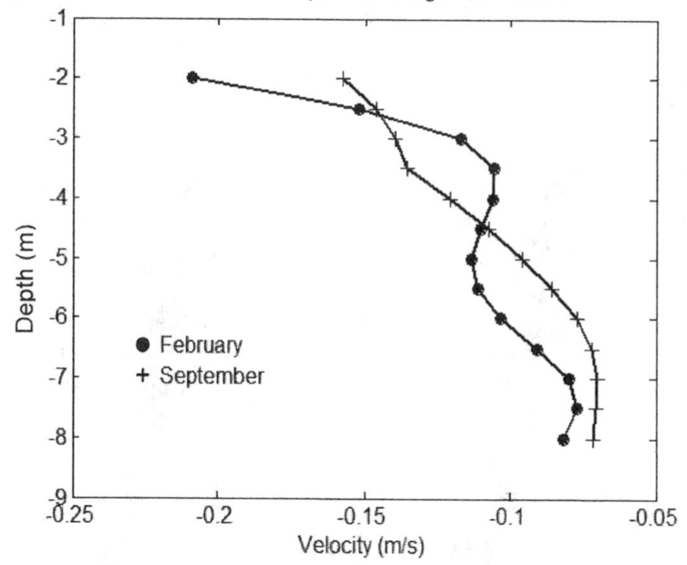

Figure 15 – Residual velocities computed with the ADCP data. Depth is relative to mean sea level.

In order to verify this hypothesis, we simulated the behavior of a fictitious organism that would be able to minimize its downstream velocity at each instant by migrating to the appropriate level. The residual velocity of this organism was evaluated as the time average of the maximum upstream velocity along the water column at each instant. Residual velocities of 0.06 m/s upstream and 0.00 m/s were obtained for the February and September data, respectively. These results show that the increase in the river flow can help the retention of organisms with vertical migration capabilities in the estuary. Indeed, in spite of the larger ebb velocities that occur for large river flows, stratification allows these organisms to remain in the estuary, due to the phase lags along the water column.

This analysis is incomplete by the limited coverage of the ADCP near the bottom. Indeed, the consideration of the bottom layer should further help these organisms to remain in the estuary.

6.2 Analysis based on model results

A similar analysis was performed using model results for organisms located at the nodes of the region shown in Figure 16. The residual velocity of these fictitious organisms was computed for the river flows of 10, 50, 100 and 200 m^3/s, for a 15-days period.

Maps of the residual velocity of the fictitious organism (Figure 16 and Figure 17) show that the Chícharo *et al.* (2001) hypothesis is valid in the deeper areas, where the retention of organisms occurs even at high river flows. In contrast, the residual velocities of the fictitious organism in the shallow banks at the mouth and near the margins are generally directed downstream. This behavior can be attributed to the non-existence of the lower layer in the shallower regions. A comparison of the results for the various runs suggests that, at least for the range of river flows analyzed, the magnitude of the river flow has little impact on the ability of the organisms to remain within the system.

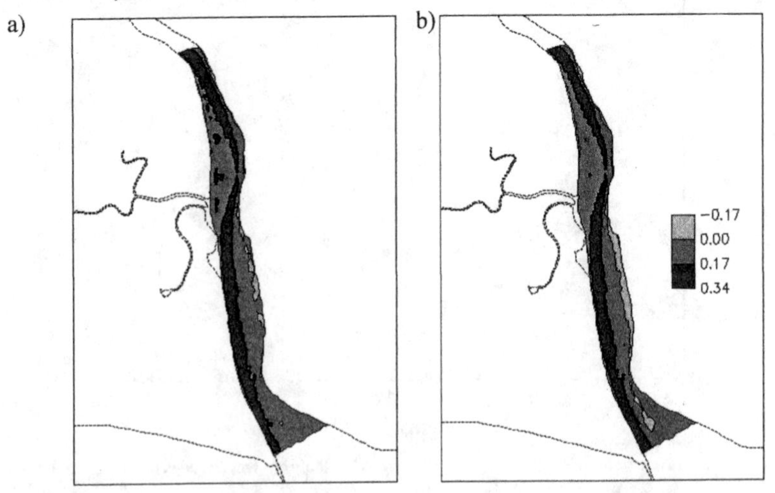

Figure 16 - Maps of residual velocity (m/s): a) Q = 10 m^3/s; b) Q = 50 m^3/s. Positive residual velocities are directed upstream.

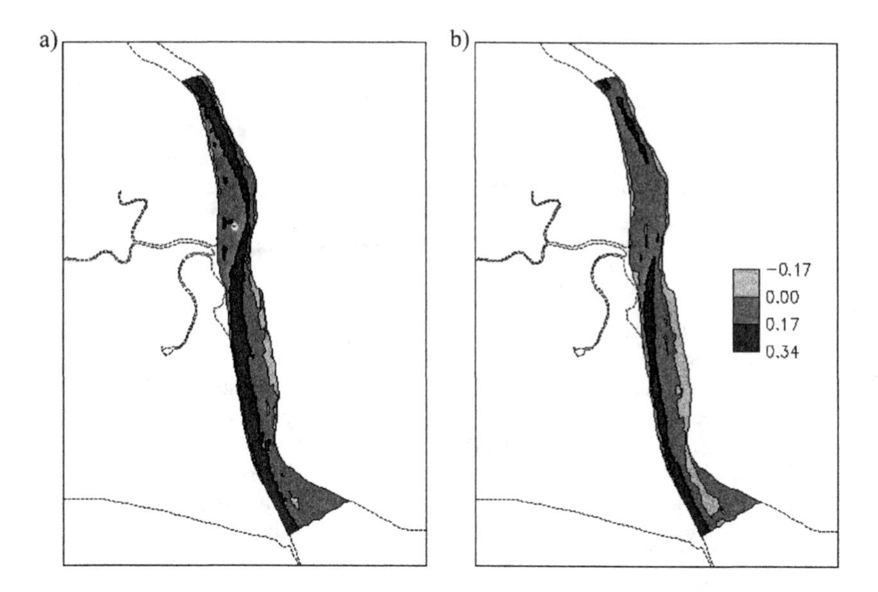

Figure 17 – Maps of residual velocity (m/s): a) Q = 100 m^3/s; b) Q = 200 m^3/s. Positive residual velocities are directed upstream.

An analysis of the vertical position of the organisms during the tidal cycle shows that, in order to take advantage of stratification, the organisms must be close to the surface during flood and remain close to the bottom during ebb (Figure 18).

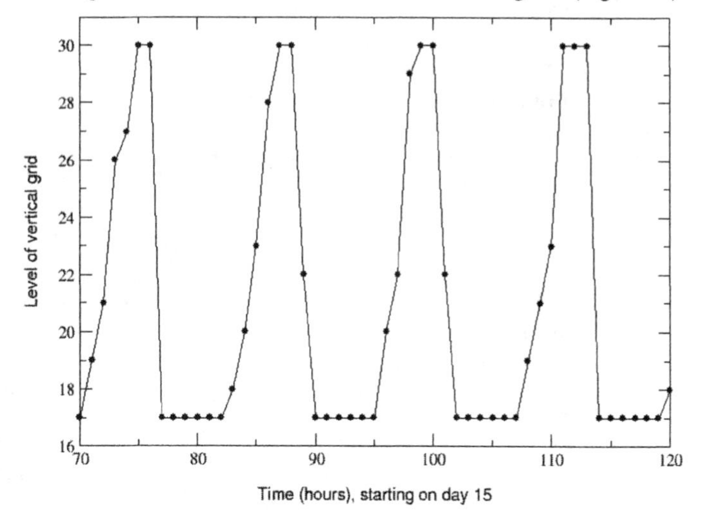

Figure 18 – Example of the vertical movement in time of a fictitious organism.

7. Summary and conclusions

The stratification conditions in the Guadiana estuary were analyzed using a combination of numerical modeling and observational data.

The analysis showed that the tidal amplitude is more important than the river flow in determining the stratification strength in the Guadiana estuary. This conclusion invalidates the use of some traditional empirical criteria (Simmons, 1955; Uncles *et al.*, 1983) for this system. Not surprisingly, the bounding values for stratification conditions determined from the present analysis disagree with those provided by the empirical criteria. Backed-up by the field data, stratification was shown to occur for river flows above 10 m^3/s, depending on tidal amplitude.

Model results were also used to investigate the hypothesis that some active pelagic organisms can remain in the estuary under high river discharges, by taking advantage of stratification. Our analysis showed that this behavior is possible from a physical viewpoint, depending on the location of the organism. Typically, organisms located in the deeper areas will be able to remain in the system, while those staying in the tidal flats near the margins and in the mouth banks will be flushed.

Finally, this application also served as a test for a new three-dimensional model (ELCIRC). The model proved to be sufficiently accurate, stable and efficient.

Acknowledgements

This work was sponsored by the *Fundação para a Ciência e a Tecnologia*, project *Protecção e Valorização da Zona Costeira Portuguesa* and by the *Laboratório Nacional de Engenharia Civil*, project *Modelação Matemática de Processos Estuarinos e Costeiros*. Field data were measured in the scope of the project *Estudo das Condições Ambientais no Estuário do Guadiana e Zonas Adjacentes*, funded by the *Instituto da Água*. The fieldwork was led by Instituto Hidrográfico (February and September), with additional funding from the *Fundação das Universidades Portuguesas* and the *Ministério da Defesa Nacional*, and by GEOSUB (May and October). Participation of one of the authors (Baptista) was funded by the National Science Foundation (ACI-0121475). We are grateful for the support on the use of ELCIRC given by Dr. Yinglong (Joseph) Zhang (Oregon Health & Science University). We thank the *Instituto de Oceanografia* for providing the satellite image.

References

Baptista A.M. and Y.-L. Zhang (2004). "Sensitivity to Turbulence Closure in a 3D Baroclinic Circulation Model", *Eos Trans. AGU, 84*(52), Ocean Sci. Meet. Suppl., Abstract OS41K-04, 2003

Blumberg, A.F. and Mellor, G.L. (1987). "A description of a three-dimensional costal ocean circulation model", in *Three-dimensional Coastal Ocean Models*, N. S. Heaps (editor), AGU, 1-16.

Bricker, S.B., Clement, D.E., Pirhalla, D.E., Orlando, S.P. and Farrow, D.R.G. (1999). "*National estuarine eutrophication assessment: effects of nutrient enrichment in the nation's estuaries*". NOAA, National Ocean Service, Special Projects Office and the National Centers for Coastal Ocean Service. Silver Spring, MD, 71 pp.

Casulli, V. and Zanolli, P. (1998). "A three-dimensional semi-implicit algorithm for environmental flows on unstructured grid", *Proceedings of the Conference on Numerical Methods for Fluid Dynamics,* University of Oxford.

Chícharo, M.A., Chícharo, L., Galvão, H., Barbosa, A., Marques, M.H., Andrade, J.P., Esteves, E., Miguel, C., Gouveia, C. and Rocha, C. (2001). "Status of the Guadiana estuary (South Portugal) during 1996-1998: an ecohydrological approach", *Aquatic Ecosystem Health and Management*, 4: 73-90.

Fortunato, A.B. and Baptista, A.M (1996). "Vertical discretization in tidal flow simulations", *Int. J. for Numerical Methods in Fluids*, 22/9: 815-834.

Fortunato, A.B., Oliveira, A. and Alves, E. (2002a). "Circulation and salinity intrusion in the Guadiana estuary", *Thalassas*, 18/2: 43-65.

Fortunato, A.B., Pinto, L.L., Oliveira, A. and Ferreira, J.S. (2002b). "Tidally-generated shelf waves off the western Iberian coast", *Cont. Shelf Res.*, 22/14: 1935-1950.

González, J.A.M. (1995). "*Sedimentología del Estuario del Río Guadiana*", Publicaciones Universidad de Huelva (in Spanish).

Harleman, D.R.F. and Abraham, G. (1966). "*One-dimensional analysis of salinity intrusion in the Rotterdam waterway*", Delft Hydraulics Lab, Pub. 44, 26 pp.

Jay, D.A. and Musiak, J.D. (1996). "Internal tidal asymmetry in channel flows: origins and consequences", C. Pattiaratchi (ed.), *Mixing Processes in Estuaries and Coastal Seas*, AGU, Coastal and Estuarine Sciences, 53: 219-258.

Lynch, D.R., Ip, J.T., Namie, C.E. and Werner, F.E. (1996). "Comprehensive coastal circulation model with application to the Gulf of Maine", *Cont. Shelf Res.*, 16/7:875-906.

Oliveira, A. and Fortunato, A.B. (2001). "Oscillation-removal schemes for salinity studies", M.L. Spaulding (ed.), *Estuarine and Coastal Modeling 7*, ASCE, 501-518.

Pinto, L. (2003). "*Haline stratification in the Guadiana estuary*", M.Sc. thesis in Geophysical Sciences - Oceanography, School of Sciences, University of Lisbon (in Portuguese).

Pinto, L. and Ferreira, T. (2002). "Study of the Guadiana plume by remote sensing". *6º Congresso da Água*, APRH, poster 48 (in Portuguese).

Prandle, D. (1985). "On salinity regimes and the vertical structure of residual flows in narrow tidal estuaries", *Estuarine, Coastal and Shelf Science*, 20: 615-635.

Silva, M.C., Fortunato, A.B., Oliveira, A. and Rocha, J.S. (2000). "Condições ambientais no estuário do Guadiana: o conhecimento actual". *5º Congresso da Água*, APRH, Lisbon, Portugal (CD-ROM – in Portuguese).

Simmons, H.B. (1955). "Some effects of upland discharge on estuarine hydraulics", *Proc. of the American Society of Civil Engineers*, 81/792.

Uncles R.J., Bale, A.J., Howland, R.J.M., Morris, A.W. and Elliott, R.C.A. (1983). "Salinity of surface water in a partially-mixed estuary, and its dispersion at low run-off", *Oceanologica Acta*, 6/3: 289-296.

Umlauf, L. and H. Burchard (2003). "A generic length-scale equation for geophysical turbulence models", *J. Mar. Res.*, 6(12): p. 235-265.

Zhang, Y.-L. and Baptista, A.M. (2004). "A cross-scale model for 3D baroclinic circulation in estuary-plume-shelf systems: I. Formulation and skill assessment", *Cont. Shelf Res.*, in press.

Subject Index

Page number refers to the first page of paper

Author Index

Page number refers to the first page of paper